D0099966

EVOLUTION

The Jones and Bartlett Series in Life Sciences

The Biology of AIDS
Hung Fan, Ross F. Conner, and Luis P. Villarreal, all of the University of California-Irvine

Basic Genetics
Daniel L. Hartl, Washington University School of Medicine; David Freifelder, University of California, San Diego; Daniel L. Hartl, Washington University School of Medicine

General Genetics
Leon A. Snyder, University of Minnesota, St. Paul; David Freifelder, University of California, San Diego; Daniel L. Hartl, Washington University School of Medicine

Genetics
John R.S. Fincham, University of Edinburgh

Genetics of Populations
Philip W. Hedrick, University of Kansas

Human Genetics: A New Synthesis
Gordon Edlin, University of California, Davis

Microbial Genetics
David Freifelder, University of California, San Diego

Experimental Techniques in Bacterial Genetics
Stanley R. Maloy, University of Illinois-Urbana

Cells: Principles of Molecular Structure and Function
David M. Prescott, University of Colorado, Boulder

Essentials of Molecular Biology
David Freifelder, University of California, San Diego

Introduction to Biology: A Human Perspective
Donald J. Farish, California State University at Sonoma

Introduction to Human Immunology
Teresa L. Huffer, Shady Grove Adventist Hospital, Gaithersburg, Maryland, and Frederick Community College, Frederick, Maryland; Dorothy J. Kanapa, National Cancer Institute, Frederick, Maryland; George W. Stevenson, Northwestern University Medical Center, Chicago, Illinois

Molecular Biology, Second Edition
David Freifelder, University of California, San Diego

The Molecular Biology of Bacterial Growth (a symposium volume)
M. Schaechter, Tufts University Medical School; F. Neidhardt, University of Michigan; J. Ingraham, University of California, Davis; N.O. Kjeldgaard, University of Aarhus, Denmark, editors

Evolution
Monroe W. Strickberger, University of Missouri, St. Louis

Molecular Evolution: An Annotated Reader
Eric Terzaghi, Adam S. Wilkins, and David Penny, all of Massey University, New Zealand

Population Biology
Philip W. Hedrick, University of Kansas

Virus Structure and Assembly
Sherwood Casjens, University of Utah College of Medicine

Cancer: A Biological and Clinical Introduction, Second Edition
Steven B. Oppenheimer, California State University, Northridge

Introduction to Human Disease, Second Edition
Leonard V. Crowley, M.D., St. Mary's Hospital, Minneapolis

Handbook of Protoctista
Lynn Margulis, John O. Corliss, Michael Melkonian, and David I. Chapman, editors

Living Images
Gene Shih and Richard Kessel

Early Life
Lynn Margulis, Boston University

Functional Diversity of Plants in the Sea and on Land
A.R.O. Chapman, Dalhousie University

Plant Nutrition: An Introduction to Current Concepts
A.D.M. Glass, University of British Columbia

Methods for Cloning and Analysis of Eukaryotic Genes
Alfred Bothwell, Yale University School of Medicine; Fred Alt and George Yancopoulous, both of Columbia University

Medical Biochemistry
N.V. Bhagavan, John A. Burns School of Medicine, University of Hawaii at Manoa

Vertebrates: A Laboratory Text, Second Edition
Norman K. Wessells, Stanford University and Elizabeth M. Center, College of Notre Dame, editors

The Environment, Third Edition
Penelope ReVelle, Essex Community College; Charles Revelle, The John Hopkins University

Medical Ethics
Robert M. Veatch, editor
The Kennedy Institute of Ethics--Georgetown University

Cross Cultural Perspectives in Medical Ethics: Readings
Robert M. Veatch, editor
The Kennedy Institute of Ethics--Georgetown University

100 Years Exploring Life, 1888-1988, The Marine Biological Laboratory at Woods Hole
Jane Maienschein, Arizona State University

Writing a Successful Grant Application, Second Edition
Liane Reif-Lehrer, Tech-Write Consultants/ERIMON Associates

EVOLUTION

MONROE W. STRICKBERGER

University of Missouri-St. Louis

Jones and Bartlett Publishers

BOSTON

Editorial, Sales, and Customer Service Offices
Jones and Bartlett Publishers
20 Park Plaza
Boston, MA 02116

Copyright © 1990 by Jones and Bartlett Publishers, Inc. All rights reserved. No part of the material protected by this copyright notice may be reproduced or utilized in any form, electronic or mechanical, including photocopying, recording, or by any information storage and retrieval system, without permission from the copyright owner.

Printed in the United States of America
10 9 8 7 6 5 4 3 2 1

Library of Congress Cataloging-in-Publication Data

Strickberger, Monroe W.
Evolution / Monroe W. Strickberger.
p. cm.
Includes bibliographical references.
ISBN 0-86720-117-7: $45.00
1. Evolution. I. Title.
QH366.2.S78 1989
575—dc20 89-19930
CIP

ISBN: 0-86720-117-7

Cover illustration: From *The Rise of Life: The First 3.5 Billion Years* by John Reader, p. 128, 1986. New York: Alfred A. Knopf. Photograph by John Reader, copyright 1990.

In memory of my parents, Samuel and Sarah

PREFACE

All biological phenomena derive from evolutionary relationships and interactions that have occurred in the past. As the great evolutionary geneticist Theodosius Dobzhansky stated, "Nothing in biology makes sense except in the light of evolution." Unfortunately, the unification of all biology under an evolutionary theme is still difficult to achieve, although the explosive increase in molecular, organismic, and populational information makes the realization of this goal more possible now than ever before.

The purpose in writing this book is to bring together some prevailing knowledge and ideas about evolution in order to help provide an informed evolutionary framework of thought for undergraduates. It is based on a course I have given for many years to biology majors who have had prior introductory biology courses. (Reviews of some basic biological and genetic concepts are, nevertheless, included in various places.) For these students, who ranged from sophomores to seniors, evolution has served to unite many previously specialized topics in science and has led, in my opinion, to conceptual insights and understanding that would have been impossible without such an approach.

I believe it is essential for students to understand that evolution takes place on many levels and encompasses many disciplines, from biochemistry to paleontology to population biology, each with its own mode of analyzing evolutionary change. A course in evolution has the responsibility of surveying as many of these disciplines as possible in order to provide prospective biologists with an overall evolutionary understanding and framework. It seems especially crucial now that evolution take a central place in the modern undergraduate science curriculum. At too many institutions of higher learning the topic of evolution is covered for the first and last time in the introductory biology course, and an evolution course may be taught every other year or not at all. My hope is that this text will help place evolution in its proper fundamental position as the discipline that binds biology together and provides the underlying structure for the understanding of so much of biology.

In general, my approach has been to consider evolution from a historical point of view both biologically and conceptually. That is, historical information passed on by transmitted genetic material connects the biology of organisms to events of the past, and evolutionary concepts derive from previous concepts transmitted culturally. In both these forms of transmitted information, "like" not only produces "like" but also produces "unlike" because of genetic changes in hereditary material and conceptual changes in the ideas of evolution. Almost every aspect of evolution indicates that knowledge of the past has become essential for fully understanding the present.

The realm of evolutionary science therefore includes both chronology and mechanisms—we seek concepts that explain both the sequence of events and their causes. For this purpose, evolutionary scientists have developed, and continue to develop, methods that provide reconstructions of evolutionary events and that enable us to understand not only biological chronology but also its genetic connections. That is, evolution follows logically understandable sequences of causes which provide us with rational explanatory powers and reliable knowledge of the past.

Evolution is an exciting subject, and I have found over the years that students often respond best when the textual material is generously illustrated. I have supplemented the teaching of the course with as much illustrative material as possible. The text contains close to 450 figures, tables, and diagrams. To further aid the student in mastery of the material, the text includes end-of-chapter summaries, lists of key terms, and a glossary. For research and reference complete bibliographies as well as separate author and subject indexes are provided.

Although the order of topics offered here has worked well for my own classes, I am aware that there are different ways to organize this material, and the chapters have therefore generally been written to allow considerable flexibility. I have avoided an overly theoretical treatment of the subject. Nothing beyond elementary algebra is needed to understand the mathematics used.

Since evolution is the broadest of biological fields, covering the greatest range of disciplines, even a brief survey of evolution is impossible without errors and ambiguities. To the extent that this book has been spared many such failings, I owe thanks to many reviewers, friends, and colleagues who commented on one or more sections:

Andrew Berry, Princeton University
Judy Bramble, University of Missouri-St. Louis
H. Jane Brockman, University of Florida
Robert Carroll, McGill University
Eloise Brown Carter, Emory University
Harriet Coret, St. Louis
Irving Coret, St. Louis University School of Medicine
The late Cedric Davern, University of Utah
David Deamer, University of California-Davis
Thomas Emmel, University of Florida
Ted Garland, Jr., University of Wisconsin-Madison
Laurie Godfrey, University of Massachusetts-Amherst
Michael Howgate, University College London
David Hull, Northwestern University
Thomas Jukes, University of California-Berkeley
Martin Kreitman, Princeton University
Michael LaBarbera, University of Chicago
M. Raymond Lee, University of Illinois at Urbana-Champaign
Ronald Munson, University of Missouri-St. Louis
Stephen Mulkey, University of Missouri-St. Louis
Frances Mussett, University College London
Ursula Rolfe, St. Louis University School of Medicine
Victoria Sork, University of Missouri-St. Louis
Jane Starling, University of Missouri-St. Louis
Robert Sussman, Washington University, St. Louis
Thomas Taylor, Ohio State University
Alan Templeton, Washington University, St. Louis
Jill Trainer, University of Missouri-St. Louis
Mark Wheelis, University of California-Davis

I am grateful to Joe Burns, Executive Editor of Jones and Bartlett, for the care and attention he devoted to gathering reviewers and for his continued interest and help. The CRACOM Corporation, in the persons of Joy Moore and Liz Rudder, saw the book through its various production phases. Sara Jenkins was especially helpful and perceptive in copyediting the manuscript. In various places, I made use of illustrative and textual material from my earlier work, *Genetics* third edition (1985) published by the Macmillan Publishing Company, New York.

M. W. S.

CONTENTS

PART 1

The Historical Framework

BEFORE DARWIN

Briefly described, biological evolution entails inherited changes in populations of organisms over a period of time which lead to differences among them. Essential to the present concept of evolution are the notions that a group of organisms is bound together by its common inheritance; that the past has been of sufficient duration for inherited changes to accumulate; and perhaps most essential of all, that the phenomena of evolution can be explained by discoverable natural laws. Each aspect has been studied and discussed at various times in human history, although it has only been during this last century, beginning with the work of Charles Darwin, that biological evolution as a whole gradually became socially accepted. This acceptance was based on many changes in how the world was viewed and how natural phenomena were explained. The purpose of the present chapter and the three that follow is to provide a review of some of the conceptual underpinnings which enabled the modern Darwinian concept of evolution to unfold.

IDEALISM AND THE SPECIES

Attempts to understand the world in a rational way, that is, by commonly acceptable methods of thought and logic, began about the fifth century B.C. in Greece. Plato (428-348 B.C.), the philosopher who, along with Aristotle (384-322 B.C.), had the greatest impact on Western thought, suggested that the observable world—our experience—is no more than a shadowy reflection of underlying "ideals" which are true and eternal for all time. Most things, according to Plato, were originally in the form of such eternal ideals, and any change represented disharmony. The Platonic goal for human society was to analyze experience in order to understand and strive for ideal perfection. The notions of "perfect circles" to explain the motions of the heavenly bodies (Fig. 1-1), "perfect numbers" such as 6 (1 + 2 + 3) and 10 (1 + 2 + 3 + 4), and the four "elements" (earth, water, fire, and air) to which all matter could be reduced were among the consequences of the search for perfection.[1] What are the sources of such **idealism**?

[1]According to Oken (1779-1851), one of the German Natural Philosophers, the highest mathematical idea is zero, and God, or the "primal idea," is therefore zero.

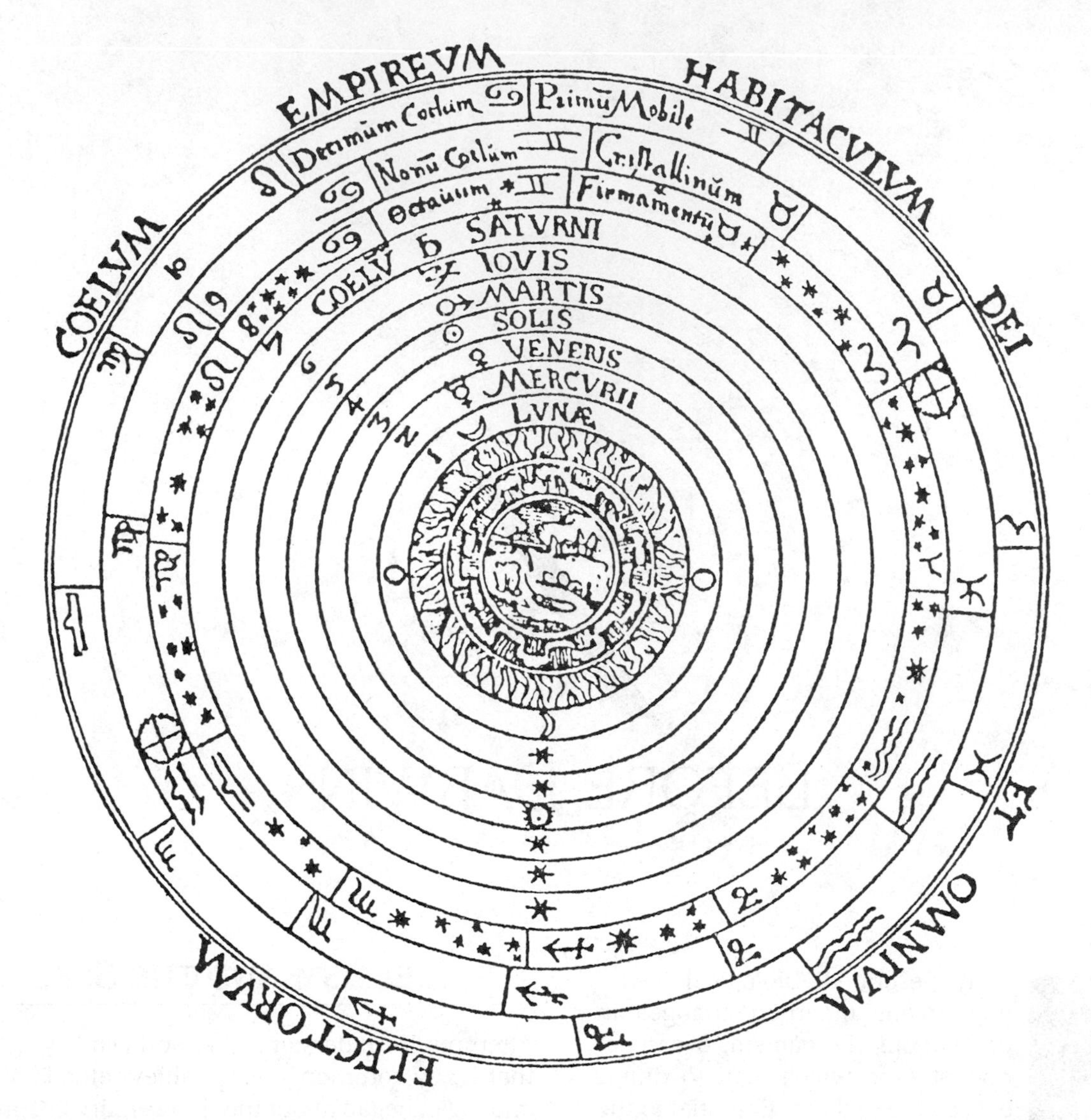

FIGURE 1-1 A medieval concept of the ten spheres of the universe with the Earth and its four elements (earth, air, fire, water) at the center, according to Apian's *Cosmographia* (published 1539, Antwerp). Surrounding the Earth are transparent crystal spheres containing in succession the Moon, Mercury, Venus, the Sun, Mars, Jupiter, Saturn, the fixed stars, and spheres involved in the motion of the stars and of the entire universe ("Primum Mobile"). Beyond these spheres lies Heaven ("The Empire and Habitation of God and All the Elect").

To a large extent, idealism originates from our often-used ability to abstract concepts from experience—to think, for example, of "cat" rather than one particular animal of specific size and head shape, with claws, tail, fur, and so on. Such abstraction enables us to generalize our experience, to differentiate between cat and tiger, to pet the cat and run away from the tiger, and to communicate these general concepts or universals to others through our symbolic language. In spite of these advantages, however, generalizations are not always reliable since the particulars of our experiences may modify the generalizations; not all cats or tigers are the same. In fact, the struggle between generalization and particularization is a continual one, since only by generalization can we conceive of regularity in nature and thereby consciously adapt ourselves to its needs, but only by particularization can we contact and observe reality. No sooner do we conceive of some new generality than we often discover further details and may thereby be forced to modify our original conception. Experience stresses continual change, and generalization stresses stability. The fact that few of the Greek thinkers, with the notable exception of Heraclitus (540-475 B.C.), attempted to incorporate change into their philosophies may indicate that the stability conferred by generalization is one of the prevailing comforts and prejudices of human thought.

Unfortunately, Plato and his successors assumed that only ideal generalizations are real while all else is illusion. From our present point of view, reality is not so narrowly defined but represents all the interactions of the universe. Thus it is true that different cats are imperfect reflections of our universal concept of "cat," but these pluralities are not imperfect reflections of reality: they are the realities which furnish and allow the generalization. Without cats there is no "cat"! Biologically, this viewpoint may extend to different groups of individuals in that they can interact as a group (e.g., as a population, race, or species) with other elements of reality. For example, cats have common features in the way they interact with prey and predators. The dilemma for biologists has therefore traditionally been to recognize the reality of differences among members of a group and yet to recognize the reality of the group itself. Idealism offered practically no means of reconciling these two aspects of reality.

Aside from its intellectual roots, an important source for Platonic idealism can be found in the underlying social structure from which it arose. The ethic of Pythagoras (c. 570-500 B.C.), which, in many ways, gave rise to Platonic idealism, considered the most exalted state of citizenship that of the philosopher-spectator who does not partake in activity but only contemplates it in order to understand it. This contemplative ideal must have derived, at least partially, from ancient social inequities in which superior gentlemen maintained their position by exploiting the activities of slaves or social classes deemed inferior. For example, the model for Plato's ideal Republic and its philosophers-statesmen-guardians is often suspected to have been the city-state of Sparta with its productively idle warrior-rulers supported by its serfs, or Helots. Even Athens, which pretended to democracy, restricted political participation to a relatively small portion of its population, since the disenfranchised majority was primarily composed of slaves and women. To be a Platonic idealist meant therefore to live in a world of cruel exploitation in which a large portion of reality may have been distasteful but was either ignored or accepted without serious question. In later periods, especially during the rigidly structured feudalism of medieval Europe, idealistic philosophies bolstered the concepts of idealized social classes and a perfectly ordered society (as had long been the case in the caste system of India) in order to help maintain the status quo.

Whatever its sources and sustenance and the guises under which it is hidden, idealism has been a persistent and pervasive philosophy and has had pronounced effects on biology and the study of evolution. To Plato, the form of a structure, biological or otherwise, could be understood from its function, since it was the function which dictated the form. Aristotle, who may be regarded as the founder of biology (among other sciences) extended this notion to the development of organisms, pointing out that the last stage of development, the adult form, explains the changes that occur in the immature forms. This type of explanation is called **teleological** because the adult represents the *telos*, or final attainment, of the embryo. Teleological explanations thus became commonly associated with mystical processes by which advanced stages, in some unknown manner, influenced and affected the earlier stages. Thus organs and organisms functioned for the sake of something which was yet to come, and each species was believed to have been created as an ideal in anticipation of its future use. Pliny the Elder (23-79 A.D.) carried this notion to the point of claiming that all species were created for the benefit of man. Some two hundred years later, Lactantius (c. 260-340 A.D.) wrote, "Why should anyone suppose that, in the contrivance of animals, God did not foresee what things were living, before giving life itself?" and helped cast the teleological origin of species more permanently into the religious form in which it was to remain in Christian Europe until the time of Darwin.

THE GREAT CHAIN OF BEING

Through idealism the concept of a species became strongly tied to its use in explaining the divine origin and design of nature. Plato had defined the species as representing the initial mold for all later replicates of that species: "The Deity wishing to make this world like the fairest and most perfect of intelligible beings, framed one visible living being containing within itself all other living beings of like nature." Aristotle expanded this view to indicate that there was a chainlike series of forms, each form representing a link in the progression from most imperfect to most perfect (Fig. 1-2). He called this the Scale of Nature, a concept which was to continue far into the history of European thought and merge with other ideas into the **Ladder of Nature** and the **Great Chain of Being**.

Philosophically satisfying as it was, the concept of the Great Chain of Being did not necessarily put humans on the highest, or even near the highest, rung of the Ladder of Nature. Many who contemplated the innumerable steps between humans and perfection (God) felt the despair of occupying a relatively lowly position and only consoled themselves with the thought that there were even more lowly organisms. However, even such consolations were unable to quell troubled feelings about a concept which suggested that the evils of nature are also part of the universal fabric and the special divine creation of everything may allow nothing to change. Nevertheless, despite its discomforts, the Great Chain of Being was generally accepted well into the eighteenth century.

In Germany this notion was fostered by Herder (1744-1803) and soon adopted by Goethe (1749-1832) and others of the Natural Philosophy (Naturphilosophie) school who tied it in strongly with an idealistic concept of biological forms. According to Goethe, the

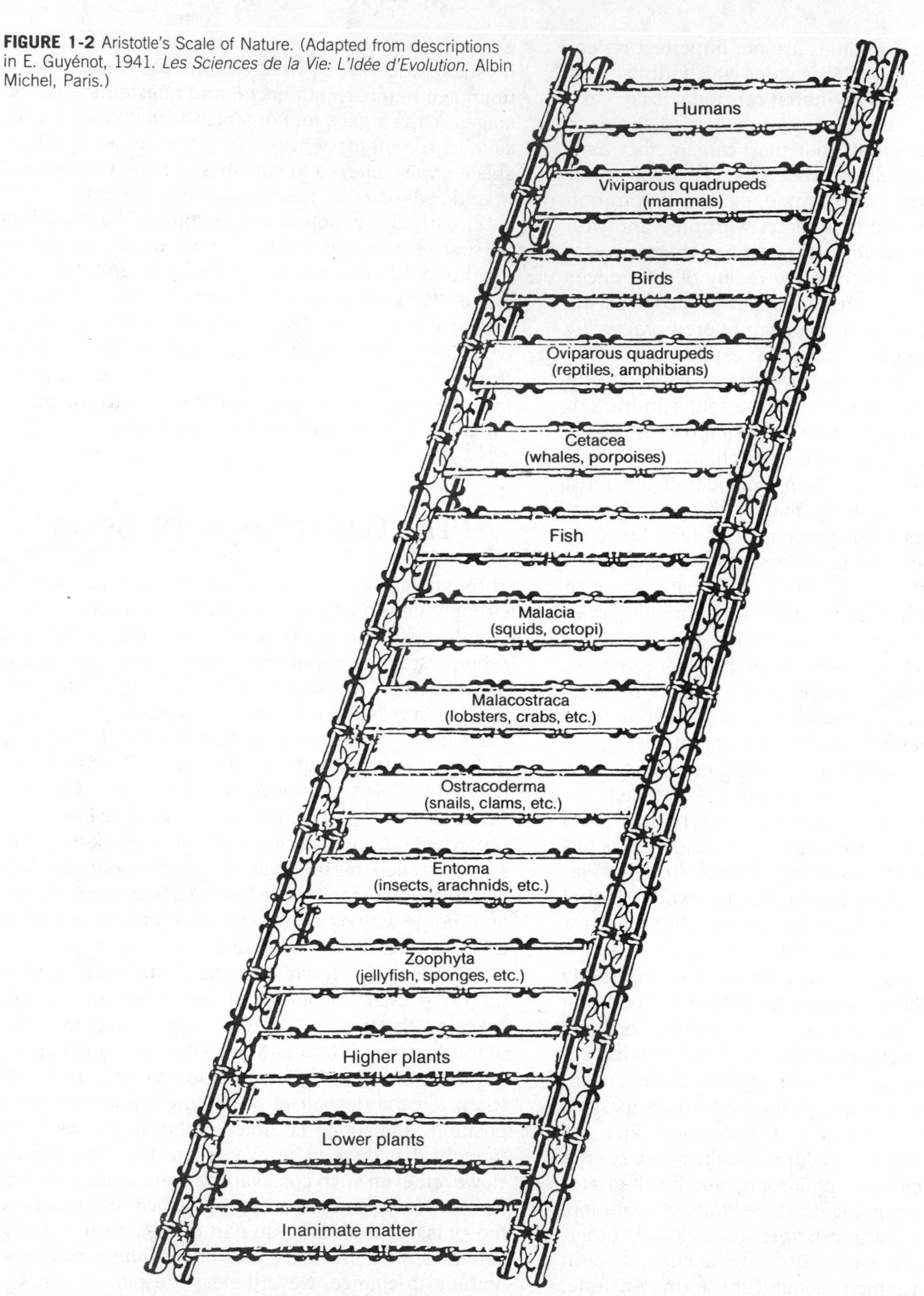

FIGURE 1-2 Aristotle's Scale of Nature. (Adapted from descriptions in E. Guyénot, 1941. *Les Sciences de la Vie: L'Idée d'Evolution.* Albin Michel, Paris.)

creation of each level of organisms was based on a fundamental primitive plan or **archetype**. The morphology of plants, for example, was conceived by Goethe to be founded on an "Urpflanze" which had only one main organ, the leaf, from which the stem, root, and flower parts derived as variations (Fig. 1-3(a)). Similarly, the bones of the skull were supposed to be merely modifications of the vertebrae of an animal archetype, or "Urskeleton," composed of only vertebrae and ribs (Fig. 1-3(b)).

To most of its exponents, the Ladder of Nature had the comforting qualities of stressing a precisely ordered regularity of relationships between organisms and could also be used to support and justify the pre-

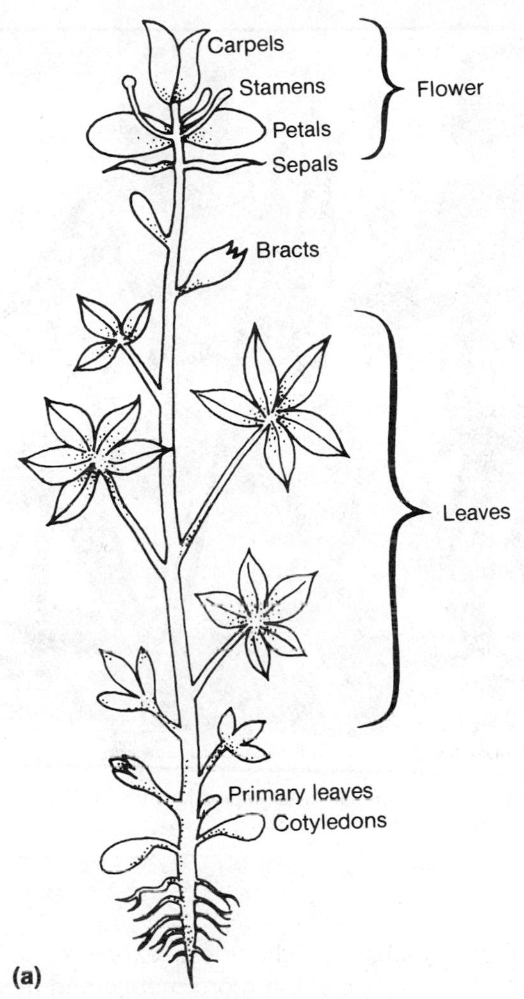

FIGURE 1-3 Archetypes of plants and vertebrate animals. The idealized plant (a) shows Goethe's concept of the derivation of all plant parts from the leaf. The segments in the vertebrate skeleton pictured by Owen (b) are alike from cranium to tail. (Adapted from C. W. Wardlaw, 1965. *Organization and Evolution in Plants.* Longmans Green, London; and R. Owen, 1848. *On the Archetype and Homologies of the Vertebrate Skeleton.* London.)

vailing social and political orders. As expressed by Soame Jenyns (1757),

> The universe resembles a large and well-regulated family, in which all the officers and servants, and even the domestic animals, are subservient to each other in a proper subordination; each enjoys the privileges and perquisites peculiar to his place, and at the same time contributes, by that just subordination, to the magnificence and happiness of the whole.

Among the relatively few who at first disputed this concept, Voltaire (1694-1778) incisively pointed to its earthly model:

> This hierarchy pleases those good folks who fancy they see in it the Pope and his cardinals followed by archbishops and bishops; after whom come the curates, the vicars, the simple priests, the deacons, the subdeacons; then the monks appear, and the line is ended by the Capuchins.

Voltaire also addressed the question of the many observed gaps between species, an observation which did not seem to be in accord with the expected innumerable steps in the continuous progression from imperfect to perfect. He proposed that although there were no living species to fill these gaps, such gaps were real, perhaps caused by the extinction of species. In this respect Voltaire essentially echoed the thoughts of the philosophers Descartes (1596-1650) and Leibniz (1646-1716). Leibniz had even proposed evolutionary changes to account for these gaps, suggesting that many species had become extinct, others had become transformed, and different species that presently share common features may at one time have been a single race.

To Leibniz, evolution of species was tied in with the perfection towards which the universe continually progressed, and his philosophy thus represented a major shift from a perfectly created universe to one in the

FIGURE 1-4 Presumed "missing links" between apes and humans in the Ladder of Nature. These individuals received binomial species designations, and Linnaeus made attempts to place them in his *Systema Naturae*. (From Taylor, reproduced from a 1760 work by Hoppius, a student of Linnaeus.)

process of becoming perfect. Progress towards the perfection of species was also expressed by biologists such as Bonnet (1720-1793) who maintained that the development of any organism from its "seed" was an unfolding of a preconceived plan inherent in the seeds of previous generations.[2]

Along with other changes in thought during the eighteenth century, these evolutionary forebodings were probably associated with some of the major changes then being undergone by society. That is, the progressive weakening of feudalism, which had begun in the fourteenth century with the rise of commerce and the new power of the merchant classes, was now accelerating because of rapid advances in technology and the Industrial Revolution. The old, rigid, land-based class structures were breaking up, and both social institutions and the ideas expressed by many thinkers were becoming more mobile and flexible.

The Great Chain of Being also had important effects on plant and animal classification since these derived partly from the search for that multitude of living organisms that many felt would be found to occupy all the various rungs of the Ladder of Nature. Even humans were believed to be linked to other species through the "wild man" (orangutan), which, according to some writers was of the human species (Fig. 1-4). Other authors saw the link between humans and animals in the South African Hottentots who were believed to be almost indistinguishable in reasoning power from apes and monkeys. In spite of the observed gaps between many species, they had all nevertheless been linked by the principle of continuity, expressed

[2]Bonnet predicted that in the future humans might reach the level of angels and animals reach the level of humans:

> Man—who will then have been transported to another dwelling place more suitable to the superiority of his faculties—will leave to the monkey or the elephant that primacy which he, at present, holds among the animals of our planet. In this universal restoration of animals, there may be found a Leibniz or a Newton among the monkeys or the elephants, a Perrault or a Vauban among the beavers.

Although this concept may seem quite advanced and evolutionary for the period, we should also keep in mind that Bonnet conceived of this process more as a perfectability of souls—that is, a series of progressive reincarnations.

by Leibniz as "Nature makes no leaps." Although not espousing the evolution of species as such, the philosopher Kant expressed this same idea as "the principle of affinity of all concepts, which requires continuous transition from every species to every other species by a gradual increase of diversity."

Thus, in spite of its idealistic nature, the Great Chain of Being led almost directly to the idea that the perfection of organisms may demand multiple intermediary stages. By the eighteenth century the basic concept of evolution, the actual transformation of one species into another, can therefore be said to have been merely awaiting the philosophical acceptance of actual change between the innumerable steps in the Great Chain of Being.

ORIGIN OF SYSTEMATICS

From the biological point of view, however, considerable difficulties still existed in respect to how species were to be defined and classified. As will be discussed in Chapter 11, without a rational system of classification, evolutionary relationships between most species would probably have been impossible to establish. But the recognition of the biological importance of species took considerable time. During the Middle Ages of Europe, species were generally collected and described on the basis of their culinary or medical properties. When the expansion of world-wide exploration and trade occurred in the sixteenth and seventeenth centuries, the discovery of many new species of plants and animals greatly increased the problems of classifying them. For example, Moufet (1553-1604), attempting to describe grasshoppers and locusts, writes:

> Some are green, some black, some blue. Some fly with one pair of wings, others with more; those that have no wings they leap, those that cannot either fly or leap, they walk; some have longer shanks, some shorter. Some there are that sing, others are silent. And as there are many kinds of them in nature, so their names were almost infinite, which through the neglect of naturalists are grown out of use.

Early attempts at classification were usually made in Aristotelian fashion by postulating a broad category (e.g., "substance") and then subdividing this into subsidiary categories (e.g., "body", "animal") until an individual species could be placed into a particular subdivision. Linnaeus (1707-1778), the founder of modern systematics, used a method of classification considerably more advanced by beginning with as precise a description of each species as possible. He then grouped species related by their morphology into **genera** (as had been foreshadowed more than a century earlier by Bauhin), grouped related genera into **orders** and these into **classes**. This helped to establish the system of **binomial nomenclature** in which each species name defines its membership in a genus and also provides it with its own unique identity, for example, *Homo* (genus) *sapiens* (species). This use of the species as the basic unit of classification enabled Linnaeus to arrive at groupings that were far more natural in their interrelationships than many of the previously proposed artificial groups. To use a somewhat simplified example, one early classification of animals was into those that can fly and those that cannot fly. Flying fish were therefore considered to be hybrids between birds and fish. However, by ignoring these "ideal" classes based on function and confining attention to a detailed description of the species itself, a flying fish shows first its fishlike relationships and then the change in its fins that enables it to glide. Therefore, except for those patterns shared by all vertebrate groups, there are obviously no special birdlike structures in such fish at all. Linnaeus' contribution to classification was thus an essential step leading to the discovery of natural evolutionary relationships among organisms.

For much of his career, however, Linnaeus conceived of the species as a fixed entity, deriving his concept essentially from John Ray (1627-1705) who defined a species on the basis of its common descent: "The specific identity of the bull and the cow, of the man and the woman, originate from the fact that they are born of the same parents." Ray had therefore attempted to separate different species on the basis of whether they could be traced to different ancestors: "A species is never born from the seed of another species." Thus, a species, with only rare exceptions, could never change, and its ultimate ancestor could only be divinely created. Linnaeus essentially adopted this view with the proviso that varieties within a species may show considerable nonheritable differences among themselves.

Under Linnaeus the art of systematics developed rapidly, and many species were described mainly on the basis of their reproductive parts and classified into groupings still valid today. Generally, however, classification was almost always based on appearance and not on observations of ancestry, since the classifiers usually described preserved specimens whose natural behavior and origins were often unknown. In accord with idealist concepts, "type" specimens of the species were deposited in museums or herbaria and used as the models for classification of further specimens.

Although Linnaeus placed special emphasis on the species as the practical unit of classification, it was Buffon (1707-1788) who codified the notion that species are the only biological units which have a natural existence ("Les espèces sont les seuls êtres de la nature"). Buffon introduced the idea that species distinctions should be made on the basis of whether there were reproductive barriers to crossbreeding between groups ("reproductive isolation") as indicated by whether fertile or sterile hybrids were produced.

> We should regard two animals as belonging to the same species if, by means of copulation, they can perpetuate themselves and preserve the likeness of the species; and we should regard them as belonging to different species if they are incapable of producing progeny by the same means.

To Buffon, considerable variation could occur between individuals of a species, perhaps eventually to produce even completely new varieties through time (e.g., different kinds of dogs). However, in spite of such variation, a species itself remained permanently distinguished from other species, although at times Buffon seemed to indicate the possibility that significant species changes could occur.[3]

Strangely enough, the eighteenth century barrier to the acceptance of evolution seemed to rest mostly on the reality of species. If species were real, then they seemed inevitably fixed. How then could new species arise? Buffon, who had proposed evolutionary events on both the cosmological and geological levels, had, at the same time, actually established three basic arguments against biological evolution which were used by anti-evolutionists well into the nineteenth century:

1. New species have not appeared during recorded history.
2. Although matings between different species lead to the inviability or sterility of hybrids, this mechanism could certainly not apply to matings between individuals of the same species. How then could individuals of a single species be separated from others of the same kind and become transformed into a new species?
3. Where are all the missing links between existing species if transformation from one to the other has taken place? (Numerous missing links had been imagined [see Fig. 1-4] but none had actually been found.)

It is therefore no surprise that one of the first serious proponents of biological evolution, Jean-Baptiste de Lamarck (1744-1829), felt that one must do away with the reality of species in order to establish the possibility of evolution. Lamarck proposed that all organisms are tied together by intermediate evolutionary forms and that species distinctions are man-made and arbitrary, although they may be helpful in classification. The observable gaps between species, genera, families, and so on, according to Lamarck, were only apparent, not real, since all intermediate forms existed someplace on earth although they were not necessarily easy to discover. Thus, although Lamarck shared the Great Chain of Being concept that species do not become extinct, he did not believe these forms were separately created but instead proposed that they had evolved from each other. In fact, his branching classification of animals (Fig. 1-5) introduced a direct challenge to the venerable doctrine of a Scale of Nature, which goes in only one direction, from imperfect to perfect: "In my opinion, the animal scale begins with at least two separate branches and ... along its course, several ramifications seem to bring it to an end in specific places."

As shall be discussed in Chapter 2, the mechanisms that Lamarck offered to account for these evolutionary changes were inadequate. However, even if Lamarck's explanations had seemed reasonable, a most serious impediment to evolutionary thought concerned the question of life itself: Is continuity between different generations of a species necessary at all?

Spontaneous Generation

Until perhaps the middle of the nineteenth century it had been common to believe that although most large organisms reproduced by sexual means, smaller organisms could arise spontaneously from mud or organic matter. Some folklore suggested that larger organisms decomposed into smaller ones when they died, and there were even common legends that magical transitions could change a living member of one species into another (e.g., werewolves). A classic expression of spontaneous generation was offered about 300 years ago by Van Helmont (1577-1644):

[3]In the fourth volume of his *Natural History* (1753) Buffon wrote:

> Not only the ass and the horse, but also man, the apes, the quadruped, and all the animals, might be regarded as constituting but a single family.... If it were admitted that the ass is of the family of the horse, and differs from the horse only because it has varied from the original form, one could equally well say that the ape is of the family of man, that he is degenerate man, that man and ape have a common origin; that, in fact, all the families, among plants as well as animals, have come from a single stock, and that all animals are descended from a single animal, from which have sprung in the course of time, as a result of progress or of degeneration, all the other races of animals.

However, in spite of this clear statement of an evolutionary view, Buffon felt forced to reject it because it is contrary to religion ("...all animals have participated equally in the grace of direct creation...") and because of the further arguments he offered.

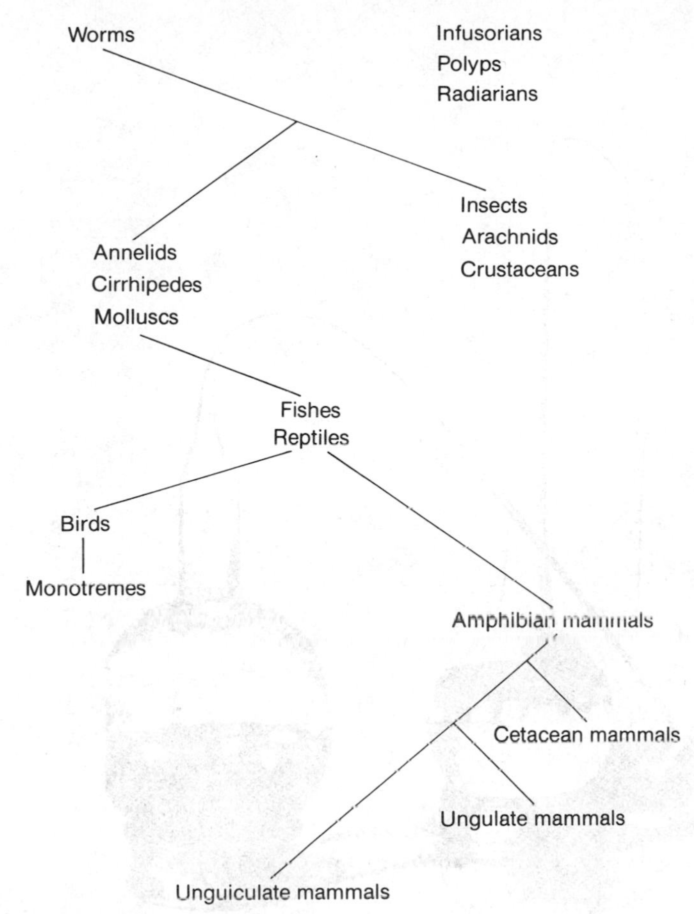

FIGURE 1-5 Evolutionary relationships among animals according to Lamarck.

> If you press a piece of underwear soiled with sweat together with some wheat in an open mouth jar, after about 21 days the odor changes and the ferment, coming out of the underwear and penetrating through the husks of wheat, changes the wheat into mice. But what is more remarkable is that mice of both sexes emerge, and these mice successfully reproduce with mice born naturally from parents.... But what is even more remarkable is that the mice which come out of the wheat and underwear are not small mice, not even miniature adults or aborted mice, but adult mice emerge!

Two serious and somewhat contradictory obstacles to the development of evolutionary concepts therefore prevailed almost simultaneously. The Linnaean contribution of constancy of species enabled the question of the origin of species to be asked but, by its insistence on species fixity, prevented consideration of any evolutionary transformations. Belief in spontaneous generation, on the other hand, seemed contrary to species fixity, but, at the same time, cast doubt on any permanent continuity between organisms. If species could arise *de novo* at any time or be capriciously changed into other species, could there ever be a rational mechanism to explain their origin or the sequence of their appearance?

Fortunately, in the late seventeenth century, use of the experimental method had begun in biology, and a number of the new scientists were able to show that, at least for insects, spontaneous generation was not taking place. In 1668, Redi (1621-1697) demonstrated that maggots (larvae) arise only from the eggs laid by flies, and flies arise only from maggots. If meat is protected so that adult flies cannot lay their eggs, then maggots and flies are not produced. A year later, Swammerdam (1637-1680) showed that the insect larvae found in plant galls arise from eggs also laid by adult insects. Within a century, further experiments demonstrated that even appearance of the microscopic "beasties" observed by Van Leeuwenhoek (1632-1723) in decaying or fermenting solutions and broth could be explained as deriving from previously existing particles. The Abbó Spallanzani (1729-1799) heated various types of broth in sealed containers and observed no growth of microscopic organisms. Only when the containers were open to the invasion of airborne particles did multiplication of organisms occur.

Although the theory of spontaneous generation was not generally abandoned until the crucial experiments of Pasteur (1822-1895) and Tyndall (1820-1893) in the nineteenth century (Fig. 1-6), serious attempts to replace it with a theory called **preformation** were begun much earlier. In the words of Swammerdam, preformation embodied the idea that "there is never generation in nature only an increase in parts." That is, when each embryonic organism is conceived, it is preformed as a perfect replicate of the adult structure, and then gradually enlarges through the nourishment provided by the egg and the environment. Some preformationists proposed that this miniature adult is contributed by the maternal egg (ovists) while others suggested that it was contained within the paternal seminal fluid (animalculists). In its most extreme form it led to the **emboitement** (encasement) theory espoused by Bonnet and others in which the initial member of a species encapsulates within it the preformed "germs" of all future generations. As Bonnet's critics (Dumas and Prevost, 1824) pointed out, "it seems easier [for preformationists] to imagine a time when nature, as it were, labored and gave birth all at once to the whole of creation, present and future, than to imagine continual activity." Thus, although preformation had the satisfying quality of explaining the many different plans of organismic growth and disputed the idea of spontaneous generation, it led once again to the fixity of species and brought the question of the origin of species back to a mystical, unknowable creation.

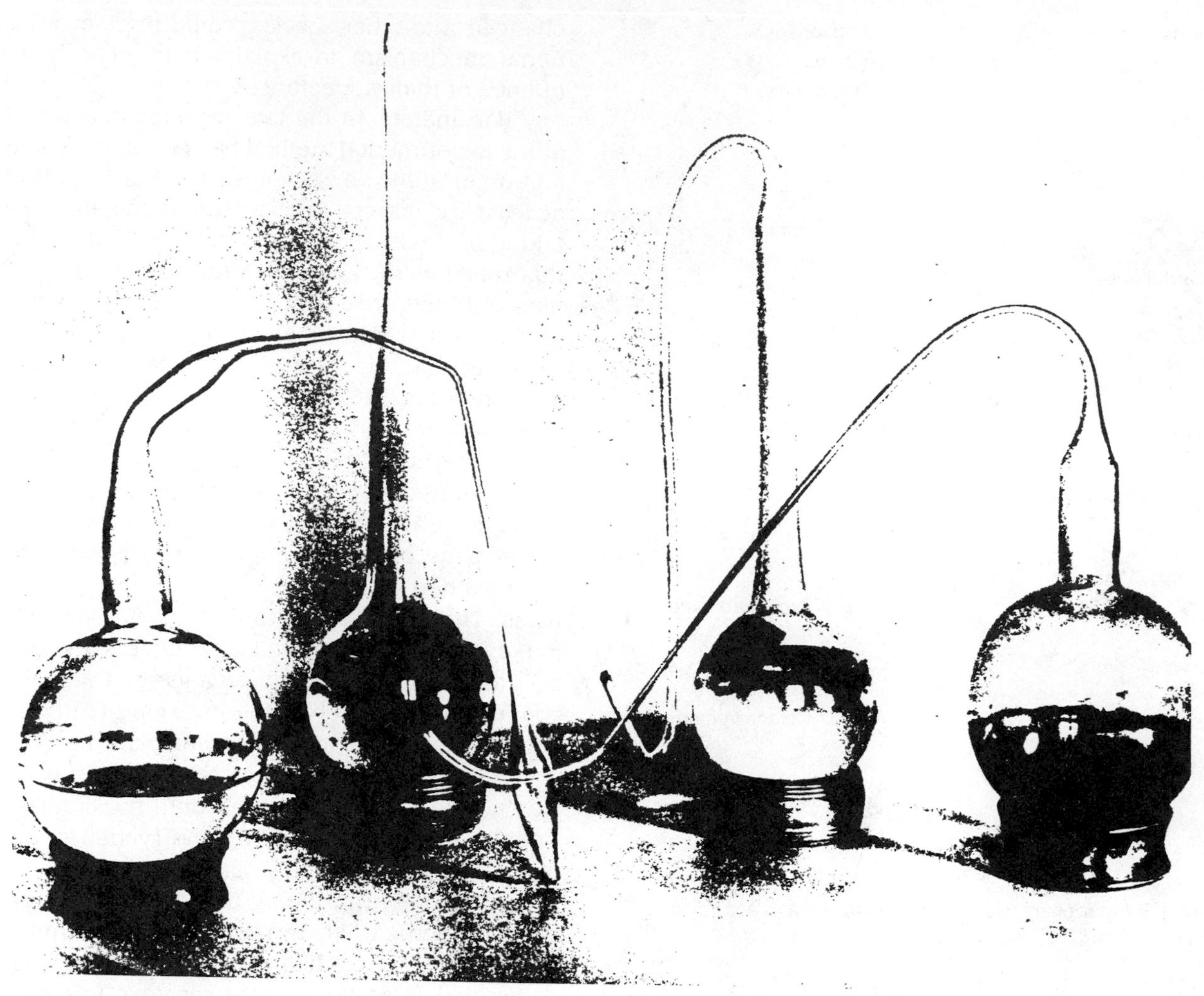

FIGURE 1-6 Pasteur's flasks used in demonstrating that spontaneous generation of bacteria does not occur even when nutrient broth is exposed to air. Because airborne bacteria are trapped in the lower bends of the curved necks of these flasks, the broth remains clear, and no fermentation occurs. Once the neck of a flask has been snipped off, bacteria enter directly into the broth, enabling them to multiply, thus causing fermentation. (From Taylor.)

By the nineteenth century, however, development of improved experimental techniques and microscopic observations led to replacement of preformation by the theory of **epigenesis**. According to epigenesis the development of an embryo proceeds by the gradual differentiation of uniform, undifferentiated tissues into organs which were not themselves present at conception. At first this differentiation of undifferentiated tissue was believed to occur because of mystical, nonphysical forces, such as Aristotle's suggestion of the contribution of "form" by the seminal fluid, or Harvey's (1578-1657) "aura seminalis," or Wolff's (1738-1794) "vis essentialis." These explanations were **vitalistic**, in that they ascribe to living beings the presence of a vital force which cannot be explained by any underlying physical or chemical principles. Fortunately, by the time of von Baer (1792-1876), the prevailing view of epigenesis had changed so that differentiation and growth was accepted as natural and as explainable a set of processes as any others. In addition, Wohler's (1800-1882) 1828 biochemical synthesis of an organic compound (urea), the first such extra-organismic synthesis, showed that there was no mystical essence in organic molecules which had to be understood outside the laws of chemistry. It was in this climate of rational biology that the evolutionary concepts of the nineteenth century were able to develop.

FOSSILS

An essential basis for understanding evolutionary relationships between organisms of the past and for gaining an appreciation of the long periods of time necessary to explain their history was a study of their fossil remains. The fossilized bones of animals which did not resemble existing species had long been noted, and even strange seashells had been found in the most unlikely places, such as mountaintops. The ancient Greeks were aware of such fossils, and a number of ancient writers, including Herodotus (484-425 B.C.), suggested that they could be explained by changes in the positions of sea and land. To Aristotle, there was no question that these changes occurred over considerable periods of time.

> The whole vital process of the earth takes place so gradually and in periods of time which are so immense compared with the length of our life, that these changes are not observed; and before their course can be recorded from the beginning to end, whole nations perish and are destroyed.

However, with the ascendancy of Christianity in Europe, the age of the world was defined by the number of generations since Adam in the biblical book of Genesis and was calculated to have begun no earlier than about 4,000 B.C. Limited to such a relatively short period, fossils could hardly be ascribed to a long historical process, and they were therefore commonly called *lusus naturae*, or "jokes of nature." Serious consideration of fossils as representing the remains of real organisms began only after the medieval "Dark Ages" had ended.

The breakup of feudalism and the expansion of trade and exploration led, as we have seen, to a number of important changes and challenges. Foremost among these was the challenge posed to the Great Chain of Being concept by the discovery of fossils in exposed riverbanks, mines, and eroded surfaces. Among the many who became engaged in fossil hunting was Thomas Jefferson (1743-1826), third president of the United States, who was a discoverer of the extinct clawed giant sloth *Megalonix jeffersoni* (Fig. 1-7), which he mistakenly thought to be a giant lion.

Did the fossils indicate possible errors in the plan of nature, causing some species to become extinct? Were there gaps in the Ladder of Nature caused by the loss of these extinct species? Jefferson, like many others who addressed themselves to these questions, proposed that these species were not truly extinct, only rare: "Such is the economy of nature, that no instance can be produced of her having permitted any one race of her animals to become extinct; of her having formed any link in her great works so weak to be broken." Other theories sought to explain fossils as caused by the biblical flood described in Genesis or having purposely been implanted into the earth at the time of creation in order to test man's faith in religion.

Contrary arguments, proposing the reality of fossil species, were offered by Robert Hooke (1635-1703) and Nicolaus Steno (1638-1686) and led to more naturalistic attempts to understand fossil origins. One of the commonly held theories among biologists during the late 1700's and early 1800's was called **catastrophism**, popularized largely by followers of Cuvier (1769-1832), the French comparative anatomist. According to catastrophism, the sharp discontinuities in the geological record—the stratifications of rocks, the layering of fossils, the transition from marine fossils to freshwater fossils—indicated sudden upheavals caused by catastrophes, glaciations, floods, and so on. Fossils were extinct species "whose place those which exist today have filled, perhaps to be themselves destroyed and replaced by others." To some of the upholders of the biblical account, catastrophism had the advantage of explaining each catastrophe as an obvious departure from "natural" laws which could be ascribed to divine intervention. Some writers, such as Agassiz (1807-1873), conjectured that there may have been as many as 50 to 100 successive special divine creations. This approach therefore justified both the prior existence of fossil species and the biblical flood and made it possible to conceive that all present organisms arose within the time span given by the Judeo-Christian Bible, although preceded by many geological ages.

In contrast to Cuvier's catastrophist position, Lamarck (p. 10) proposed that the geological discontinuities represented gradual changes in the environment and climate to which species were exposed, and it was through their effects on organisms that these changes led to species transformation. This **uniformitarian** concept, that the steady, uniform action of the forces of nature could account for the earth's features, had been foreshadowed by Buffon and others and was strongly developed in the work of the geologist Hutton (1726-1797). Later, Lyell (1797-1875), a geologist and contemporary of Charles Darwin, offered the uniformitarian reply to catastrophism through the following arguments:

1. Sharp, catastrophic discontinuities are absent if geological strata are examined over widespread geographical areas. In fact, any widely distributed stratum often shows considerable regularity in its structure and composition. It is only in specific localities that rapid shifts seem to appear and then because of local changes.
2. When changes occur in the geological record they arise from the action of erosive but natural forces such as rain and wind, as well as from volcanic upthrusts and flood deposits. The laws of motion and gravity which govern natural events are constant through time. Thus, phenomena which occurred in the past were es-

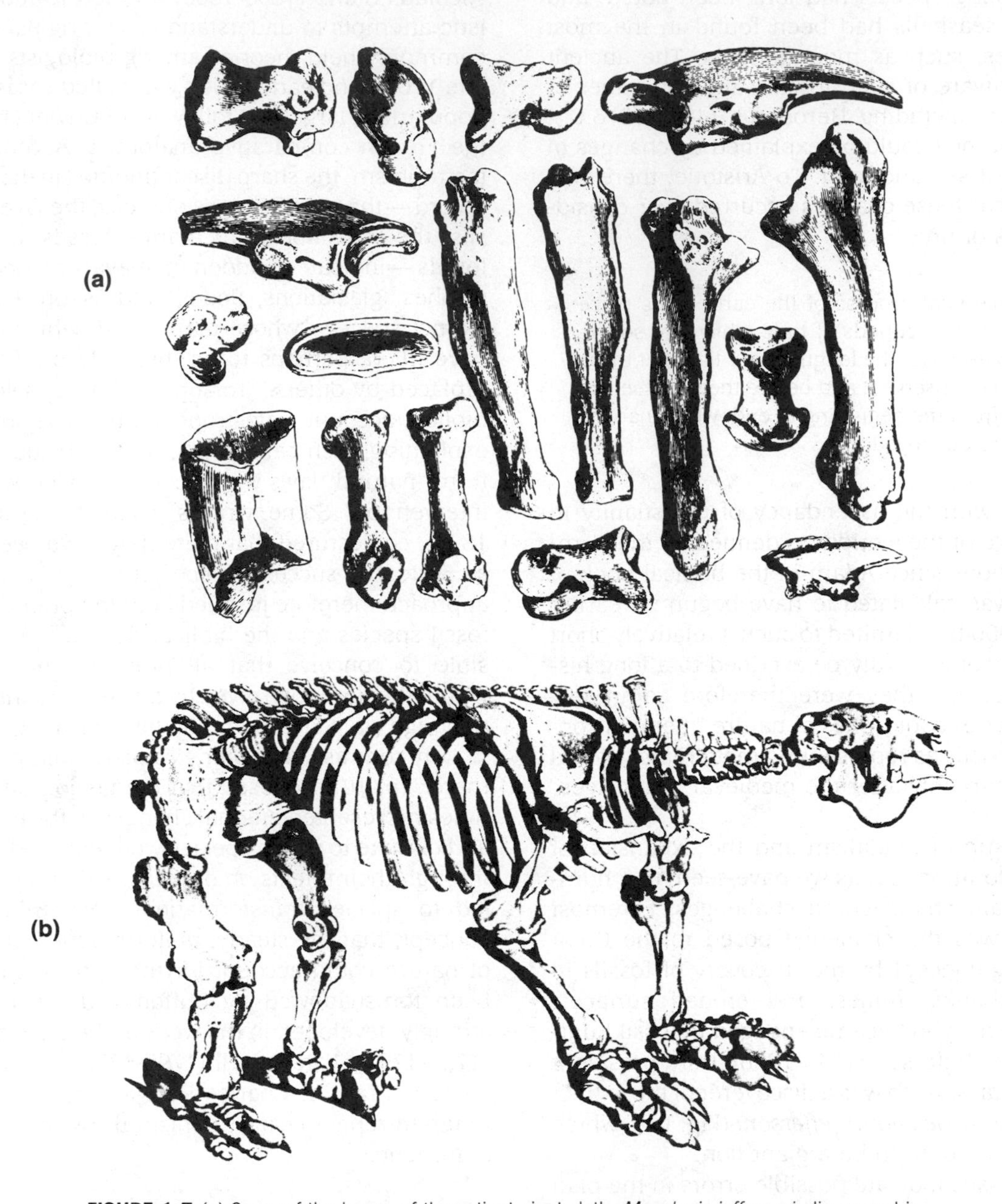

FIGURE 1-7 (a) Some of the bones of the extinct giant sloth, *Megalonix jeffersoni*, discovered in western Virginia in 1796. (b) Reconstruction by Cuvier of the skeleton of a similar extinct South American giant sloth, *Megatherium*. Both sloths were edentates, clawed mammals without cutting teeth. (From Greene.)

sentially caused by the same forces which produce phenomena at present, and the extent to which they occurred in the past may be no greater than their present activity. This means that all natural causes for phenomena should be investigated before supernatural causes are used to explain them.

3. The earth must be very old for its many geological changes to have taken place by such gradual processes.

The transition from catastrophism to uniformitarianism had profound effects, since it helped liberate scientific thinking from the concept of a static universe powered by capricious, unexplainable changes to one that is perpetually dynamic but more historically understandable. In biology, it was Charles Darwin who first offered an acceptable explanation for historical changes among organisms.

SUMMARY

Many intellectual threads led to the modern theory of evolution, which requires recognition that the earth is ancient, that there is a common inheritance within a biological group, and that natural events can be explained by discoverable natural laws. But it took a long time before these threads were brought together into an evolutionary concept.

Plato's idealistic concepts, according to which all natural phenomena are imperfect representations of the true essence of the natural world, was a long-prevailing philosophy in Western Europe that profoundly inhibited the development of evolutionary ideas. Since the world of essences is perfect, all change is illusory. Following Platonic ideas, Aristotle suggested that not only were species immutable but that there was a hierarchical order of species from most imperfect to most perfect, a concept refined over the centuries as the Great Chain of Being. This unchanging order remained unquestioned until inexplicable gaps in the chain of nature prompted philosophers such as Leibniz to propose that the universe was not perfect, only becoming so, and that it might go through successive intermediary stages on the pathway to perfection.

By the seventeenth and eighteenth centuries, new attention to living creatures and far-flung explorations led to an increasing interest in classifying organisms within the natural chain. Linnaeus revolutionized systematics by using the species as the basic unit and building his system from the species upwards to larger taxonomic categories. The naturalist Buffon went farther by implying that the species is not just a category in classification but the only natural grouping. But he remained wedded to the Platonic ideal of a species as a "real" unit, thereby precluding change or the formation of new species. Lamarck evaded this problem by proposing that species are arbitrary, not "real," and that there could and must be forms intermediate between species.

The idea that organisms could arise from nonliving materials by spontaneous generation or that organisms did not change during development but were already "preformed" in their ancestors further hindered the development of evolutionary thought. It was not until the nineteenth century that spontaneous generation was finally disproved and it was established that organisms develop epigenetically, by differentiation from undifferentiated tissues. At last biological phenomena became amenable to rational explanation.

The most severe blow to anti-evolutionary ideas was struck by the fossil record. The discovery of fossils of unknown types of organisms and the apparently inappropriate location of some fossils suggested that the surface of the earth and the organisms on it had existed for a long time. This, however, was in conflict with the Judeo-Christian view of a recent origin, and fossil data were interpreted to accord with biblical catastrophes such as the Noachian flood or as "jokes of nature." Geologists asserted that fossil evidence was only explicable if the earth were indeed old and its surface had been shaped by forces of nature. Changes on the earth's surface would then have led to alterations in the organisms which lived on it, and these changes would be reflected in their fossil remains. Charles Lyell, a contemporary of Darwin, was able to invalidate the idea that capricious catastrophic and miraculous events had influenced the geological structure of the earth and so helped establish the validity of a world which was comprehensible and rational.

KEY TERMS

archetype
binomial nomenclature
catastrophism
classes
emboitement
epigenesis
fossils
genera
Great Chain of Being
idealism
Ladder of Nature
orders
preformation
species
spontaneous generation
teleological
uniformitarianism
vitalistic

REFERENCES

Bowler, P. J., 1984. *Evolution: The History of an Idea.* Univ. of California Press, Berkeley.

Gasking, E., 1967. *Investigations into Generation: 1651-1828.* Hutchinson, London.

Glass, B., O. Temkin, and W. L. Straus, Jr. (eds.), 1959. *Forerunners of Darwin: 1745-1859.* Johns Hopkins Press, Baltimore.

Greene, J. C., 1959. *The Death of Adam.* Iowa State Univ. Press, Ames.

Hull, D. L., 1974. *Philosophy of Biological Science.* Prentice-Hall, Englewood Cliffs, N. J.

Lamarck, J. B., 1809. *Zoological Philosophy.* Translated into English by H. Elliot, 1914, Macmillan & Co., New York.

Lovejoy, A. D., 1936. *The Great Chain of Being.* Harvard Univ. Press, Cambridge, Mass.

Mayr, E., 1982. *The Growth of Biological Thought: Diversity, Evolution, and Inheritance.* Harvard Univ. Press, Cambridge, Mass.

Nordenskiold, E., 1928. *The History of Biology*. A. A. Knopf, New York.

Russell, B., 1945. *A History of Western Philosophy*. Simon & Schuster, New York.

Singer, C., 1959. *A History of Biology*, 3rd ed. Abelard Schuman, London.

Sirks, M. J., and C. Zirkle, 1964. *The Evolution of Biology*. Ronald Press, New York.

Smith, C. U. M., 1976. *The Problem of Life*. John Wiley & Sons, New York.

Taylor, G. R., 1963. *The Science of Life*. McGraw-Hill, New York.

Toulmin, S., and J. Goodfield, 1965. *The Discovery of Time*. Harper & Row, New York.

DARWIN

By the early nineteenth century, many of the basic concepts necessary to develop a belief in organic evolution were already present: (1) the age of the earth was conceived by geologists such as Hutton to be in the range of millions of years ("We find no vestige of a beginning—no prospect of an end"); (2) the reality of previously extinct fossil species had been accepted; (3) the close similarities between many different species had been noted through the efforts of systematists, comparative anatomists, and embryologists; and (4) most, if not all, organisms were believed to be descended through inheritance from previously existing organisms. The notion of a divine "common plan," which had been proposed to account for the relationships among species by supernatural acts of creation, was therefore only one step away from the materialist evolutionary notion that relationships are based on the common ancestry of different species: a change "from archetypes to ancestors." Nevertheless, for the materialist, at least two important questions remained to be answered: What natural cause or mechanism could explain why organisms change? What hereditary mechanism could enable organisms to change?

The first question was answered by Charles Darwin (1809-1882) in 1859 and thereby effectively helped transform biology into an evolutionary science. An acceptable answer to the second question had to await the twentieth century, although soon after Darwin, Gregor Mendel (1822-1884) provided the essential basis for understanding materialist biological inheritance (see Chapter 10).

CHARLES DARWIN

Charles Darwin (Fig. 2-1) was born to an English middle-class family whose fortunes derived largely from Darwin's father, Robert Darwin (1766-1848), and his paternal grandfather, Erasmus Darwin (1731-1802), both prosperous physicians. Erasmus Darwin was, in fact, an early popularizer of evolution but was considered wildly speculative in science by most of his contemporaries. Even Charles himself, when he was searching for evolutionary explanations in the 1830's, did not seriously take into account many of his grandfather's suppositions. In any case, it is believed that evolutionary ideas probably had little effect on Charles' early development.

FIGURE 2-1 Portrait of Charles Darwin in 1840 at the age of 31. (Painting by George Richmond, from the collection at Down House, Kent, Great Britain.)

At the age of sixteen Charles left grammar school in Shrewsbury and was sent to Edinburgh University to study medicine. Because of the brutality of surgical procedures at that time, Darwin found this experience distasteful and after two years transferred to Cambridge University with the intention of becoming a minister in the Church of England. However, his interests were not in academic or ministerial pursuits but in hunting, collecting, natural history, botany, and geology. He despised formal classical education and was usually no more than a mediocre student. His father apparently felt that Charles had betrayed the family trust of industrious professionalism and castigated him with the words, "You care for nothing but shooting, dogs, and rat-catching, and you will be a disgrace to yourself and all your family."

In 1831, through the recommendation of John Henslow (1796-1861), a botany professor at Cambridge, and the intercession of an uncle, Josiah Wedgewood (1769-1843), Darwin was able to put off further study for the ministry and instead undertook his now famous voyage around the world on the HMS *Beagle* (Fig. 2-2). His post on the *Beagle* was that of naturalist, a special unpaid position created by the British Admiralty on naval ships making broad geographical surveys. The *Beagle* voyage lasted approximately five years, and during this interval Darwin was transformed from a casual amateur to a dedicated geologist and biologist. His letters to Henslow on many of the observations made during the voyage, along with his collections of plants, animals, fossils, and minerals, excited considerable scientific interest even before his return to England. Darwin's account of the voyage, published later as his *Journal of Researches* and reprinted many times under the title *Voyage of the Beagle*, remains one of the most interesting and perceptive chronicles of exploration in the nineteenth century. On his return to England, a substantial income and inheritance enabled Darwin to forego financial pursuits and dedicate himself entirely to biology. He married his cousin, Emma Wedgewood (1808-1896), and in 1842 settled near the village of Down in the Kent countryside 16 miles from London. There, he and his wife, aided by servants, began to raise a large family.

For 40 years, to the time of his death in 1882, Darwin lived at home, mostly as a semi-invalid subject to heart palpitations, rashes, and gastric discomfort. The cause for his disability is not known, and conjectures have ranged from parasitic infection, heavy metal (arsenic) poisoning via some of the "cures" of his time, to psychosomatic illness (see Colp). Whatever the cause, his illness served to isolate him from most of the world about him except through letters and publications. However, in spite of his physical discomforts, Darwin appears to have lived a harmonious life, probably because of his own warm personality and behavior and the sympathetic concern of his wife. It has often been said that Darwin was the perfect patient and his wife the perfect nurse.

THE VOYAGE OF THE *BEAGLE*

The five-year voyage on the *Beagle* (Fig. 2-3) enabled Darwin to observe and think about a relatively wide range of organisms and geological formations. He collected birds, insects, spiders, and plants in the Brazilian tropical forests. At Punta Alta, on the coast of Argentina, he unearthed fossil bones of the 20-foot high giant sloth, *Megatherium*, the hippopotamuslike *Toxodon*, the giant armadillo *Glyptodon*, and other animals resembling present species yet recognizably different. The primitiveness and wildness of the Tierra del Fuego Indians at the southern tip of South America impressed him with the severity of their struggle for subsistence in a meager and unrelenting environment.

During the voyage, Darwin carried with him Lyell's *Principles of Geology* and assiduously noted the geo-

(a)

(b)

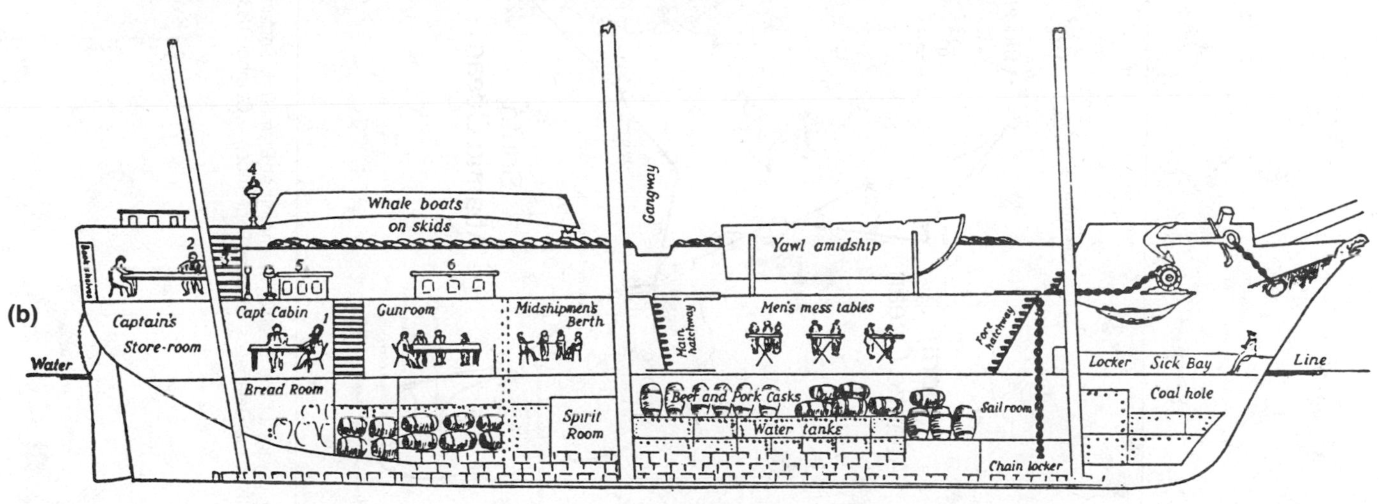

FIGURE 2-2 *Top*, HMS *Beagle* in the Strait of Magellan at the southern tip of South America. The ship was a 10-gun brig, 90 feet long, weighing 240 tons. In 1831, the year Darwin began his voyage, it had been refitted for circumnavigation in order to fix world longitudinal markings and chart the coast of South America. *Bottom*, Side elevation of the *Beagle* drawn by one of Darwin's shipmates showing the general plan of the ship and the cramped quarters which held a crew of 74. Darwin slept in the poop cabin at the stern of the ship which he shared with two officers. This cabin also held a 10- by 6-foot chart table and various chart lockers, as well as drawers for his own equipment and specimens. He wrote, "I have just room to turn around and that is all."

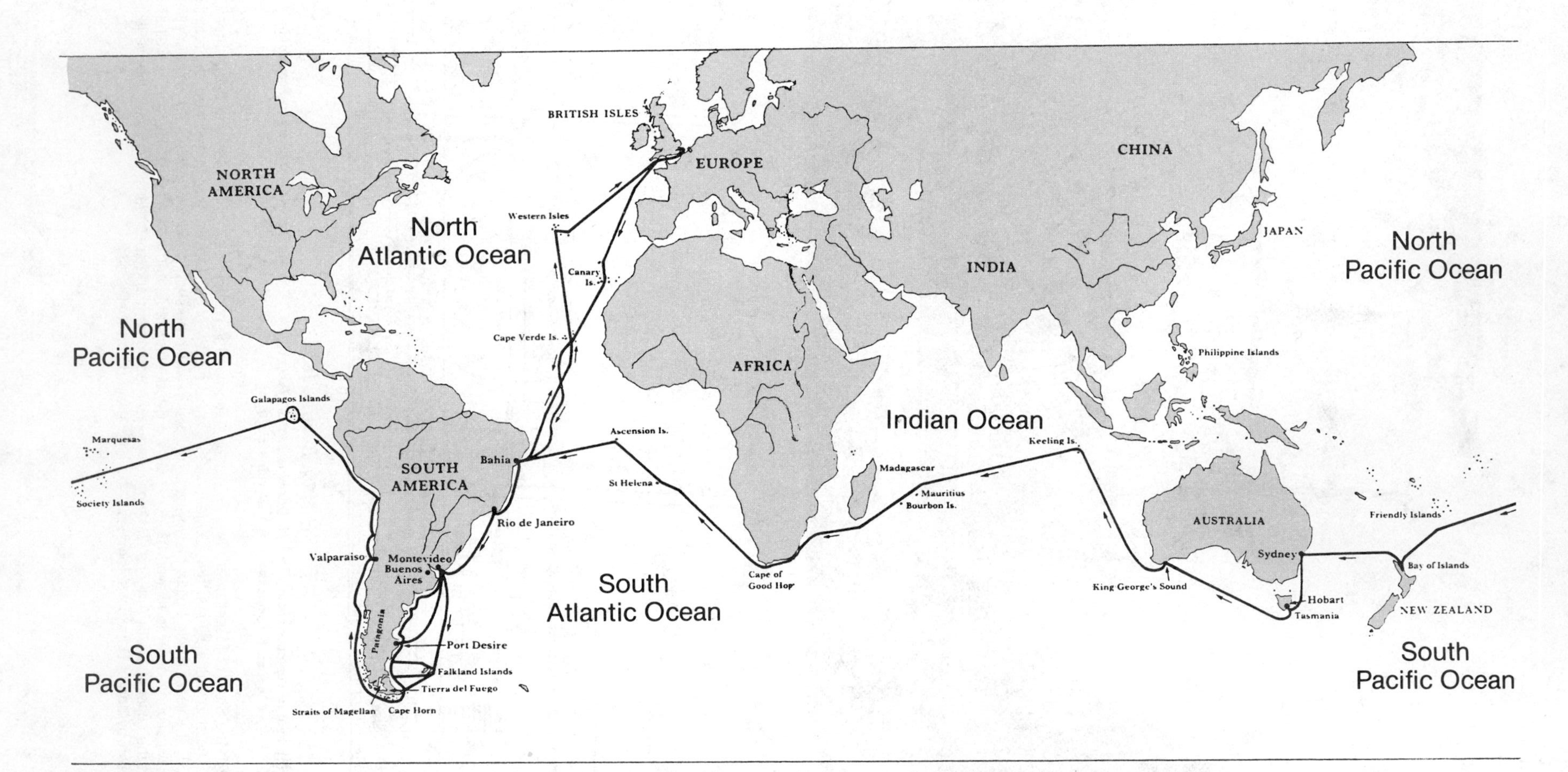

FIGURE 2-3 Track of the five-year voyage of the *Beagle*, beginning at Plymouth, England, in December 1831, and ending in Falmouth, October 1836. Almost four years were spent in South America, including one month in the Galapagos Islands (September – October 1835).

logical features of many terrains he covered. To explain some of the geological uplifting processes that affected the shape of the South American landscape, he gathered evidence showing the distribution of marine shells at various places above sea level, the loss of pigment of the older shells found at higher elevations, and the terracing of the land by erosion as it was lifted upwards. At the Bay of Concepcion, along the coast of Chile, he experienced a severe earthquake which raised the level of the land in some places from about 2 or 3 feet above sea level to as much as 10 feet. This had a deep effect on him.

> A bad earthquake at once destroys our associations: the earth, the very emblem of solidity, has moved beneath our feet like a thin crust over a fluid; one second of time has created in the mind a strange idea of insecurity, which hours of reflection would not have produced.

Perhaps his most significant experience was the month spent in the bleak, lava-ridden **Galapagos Islands** off the coast of Ecuador. Here, 500 miles from the mainland, was a most unusual collection of organisms: giant tortoises, yard-long marine and land iguanas, as well as many unusual plants, insects, lizards, and seashells. As he had noted previously on the mainland, different geographical localities, although possessing some environmentally similar habitats, were not always occupied by similar species. This was most striking in the Galapagos where insect-eating warblers and woodpeckers were absent but various species of finches, usually seed-eating, now assumed the insect-eating patterns of the missing species (Fig. 2-4). Also, the fact that each island appeared to have its own unique, closely related constellation of species raised the important question of what could account for this distribution of organisms. In Darwin's words,

> It is the circumstance that several of the islands possess their own species of the tortoise, mocking-thrush, finches, and numerous plants, these species having the same general habits, occupying analogous situations, and obviously filling the same place in the natural economy of this archipelago, that strikes me with wonder.[1]

Were there separate and different creations which made one species in one place slightly different from another species in another place? Why?

There is little doubt that the *Beagle* voyage stirred the seed of evolutionary thought in Darwin, leading him to begin his first notebook on the *Transmutation of Species* in 1837. He adopted the view that only changes between species could reasonably explain the facts that present species resemble past species and that different species share similar structures. "The only cause of similarity in individuals we know of is

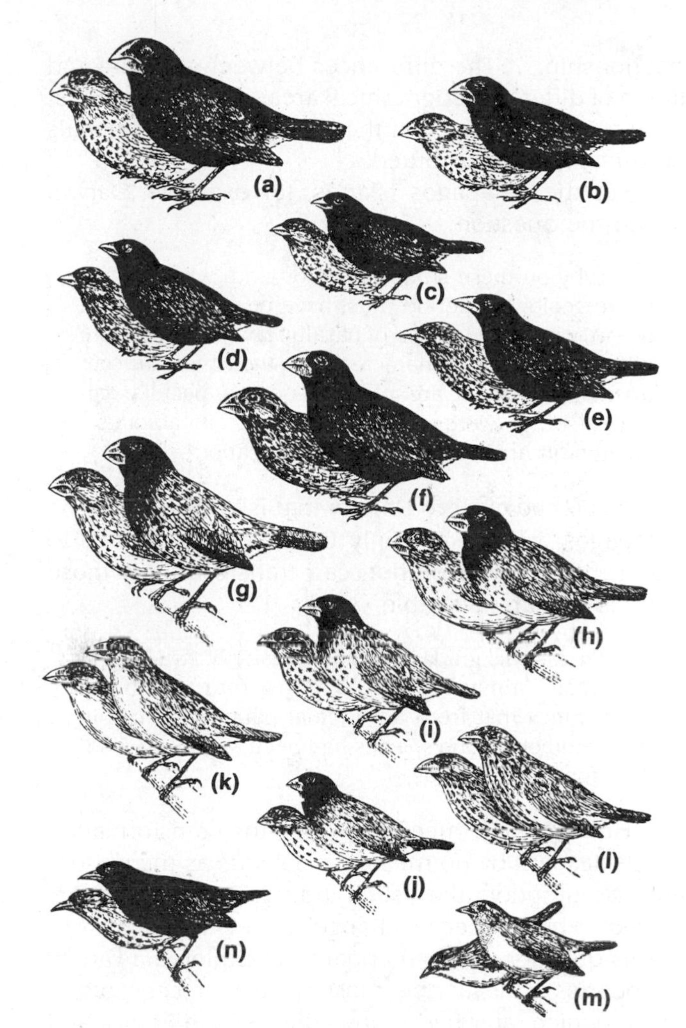

FIGURE 2-4 Species of finches (male on left, female on right; about twenty percent of actual size) that Darwin observed in the Galapagos Islands. (a) *Geospiza magnirostris* (large ground finch), (b) *Geospiza fortis* (medium ground finch), (c) *Geospiza fuliginosa* (small ground-finch), (d) *Geospiza difficilis* (sharp-beaked ground-finch), (e) *Geospiza scandens* (cactus ground-finch), (f) *Geospiza conirostris* (large cactus ground-finch), (g) *Camarhynchus crassirostris* (vegetarian tree-finch), (h) *Camarhynchus psittacula* (large insectivorous tree-finch), (i) *Camarhynchus pauper* (large insectivorous tree-finch on Charles Island), (j) *Camarhynchus parvulus* (small insectivorous tree-finch), (k) *Camarhynchus pallidus* (woodpecker-finch), (l) *Camarhynchus heliobates* (mangrove-finch), (m) *Certhidea olivacea* (warbler-finch), (n) *Pinaroloxias inornata* (cocos-finch). Evolutionary relationships among these finches are illustrated in Figure 3-3. (From Lack.)

[1]It should be noted that Darwin's account of his 1831-1836 voyage on the *Beagle* was first published in 1838 and revised some years later. The ornithologist David Lack and historians such as Sulloway have pointed out that although Darwin's *Journal of Researches* expresses these and other evolutionary forethoughts, the significance of his observations on the Galapagos Islands and elsewhere did not become apparent to Darwin until after his return to England. This was especially true for the various Galapagos finches, which were first classified in England by John Gould, a British ornithologist, whom Darwin met in 1837.

relationship...." The differences between the flora and fauna of different geographical areas, he thought, must have arisen from the fact that not all plants or animals are universally distributed.

For the Galapagos Islands, for example, Darwin raised the question,

> Why on these small points of land, which within a late geological period must have been covered with ocean, which are formed of basaltic lava, and therefore differ in geological character from the American continent, and which are placed under a peculiar climate,—why were their aboriginal inhabitants... created on American types of organization?

It seemed clear to Darwin that islands such as the Galapagos will contain only those organisms able to reach them, and evolution can transform only those species that are available.

> Seeing this gradation and diversity of structure in one small, intimately-related group of birds, one might really fancy that from an original paucity of birds in this archipelago one species had been taken and modified for different ends.

However, the mechanism for the transformation of species was by no means as obvious as the reasonable assumption that such transformation had occurred. Why do species change? In seeking an answer to this question, Darwin apparently explored a variety of theories. One of the most persistent concepts, a theory which later had many adherents in France and the United States, was that of Lamarck.[2]

THE LAMARCKIAN HERITAGE

Lamarck, the first biologist to become an active proponent of evolution, made the important leap from the simple extinction of species, as evidenced by fossils, to their gradual modification through time (p. 10). The means by which these evolutionary modifications occurred and the exquisite relationships through which organisms exploited their environments—**adaptations**—were areas which Lamarck began to explore at the beginning of the nineteenth century. He proposed that the variations among organisms originate through a response of the organism to the needs of the environment, and it is this ability to respond in a particular direction which accounts for the adaptability of a trait. For example, he suggested that the long legs of water birds, such as herons and egrets, have arisen through the following mechanism.

> We find ... that the bird of the waterside which does not like swimming and yet is in need of going to the water's edge to secure its prey, is continually liable to sink in the mud. Now this bird tries to act in such a way that its body should not be immersed in the liquid, and hence makes its best efforts to stretch and lengthen its legs. The long-established habit acquired by this bird and all its race of continually stretching and lengthening its legs, results in the individuals of this race becoming raised as though on stilts, and gradually obtaining long bare legs, denuded of feathers up to the thighs and often higher still.

Implicit in this process was that organisms possessed an unknown, inner "perfecting principle" which could sense the needs of the environment and respond directly by changing or developing traits in appropriate adaptive directions, mostly from simple to complex. The source for such directional orientation was not clear to Lamarck since he was not aware, as Darwin later was, that natural selection is the device which leads to continued improvement of adaptive mechanisms. At times Lamarck ascribed progressive evolution to some inner, mystical, vitalistic property of life (feu éthéré, an ethereal fire), whereas at other times he denied such supernatural causes. However, no matter whether their direction was caused by natural or supernatural events, the origin of organic changes and their transmission to further generations was believed by Lamarck to arise because of two universal mechanisms that he codified into two basic "Laws of Nature" (although both these concepts can be traced back to the folklore of antiquity and were also incorporated into *Zoonomia*, a popular work by Charles Darwin's grandfather, Erasmus Darwin).

1. **Principle of Use and Disuse:**

> In every animal which has not passed the limit of its development, a more frequent and continuous use of any organ gradually strengthens, develops and enlarges that organ, and gives it a power proportional to the length of time it has been so used; while the permanent disuse of any organ imperceptibly weakens and deteriorates it, and progressively diminishes its functional capacity, until it finally disappears.

---

[2]Although abandoned by practically all biologists in modern countries, a form of Lamarckism was adopted by the Soviet Union as official policy during the 1948-1963 period as a result of political demagoguery and experimental fabrications by the Russian agronomist T. D. Lysenko and his supporters. (For a review of this episode, see Joravsky.)

2. The Inheritance of Acquired Characters:

> All the acquisition or losses wrought by nature on individuals, through the influence of the environment in which their race has long been placed, and hence through the influence of the predominant use or permanent disuse of any organ; all these are preserved by reproduction to the new individuals which arise, provided that the acquired modifications are common to both sexes, or at least to the individuals which produce the young.

This remarkable hereditary plasticity by which organisms could adapt to their environments led Lamarck, as we have seen previously, to the notion that species exist in name only, since what is called a species must be merely a continuum between organisms that are at different points in the process of change. Thus, fossil species, according to Lamarck, were not truly extinct but had become modified in time and thereby evolved into later, more complex organisms.

It was Cuvier who marshalled what seemed at the time the most telling arguments against Lamarck's evolutionary proposals. Cuvier pointed out that no intermediate forms were found, either alive or as fossils, which bridged the gaps between different species. There was also the commonly observed fact that when a species hybrid was occasionally formed, such as the mule, it was inevitably doomed to sterility. The Lamarckian concept that organisms strive for perfection seemed ludicrous: What elements of consciousness could one ascribe to plants and lower organisms? Furthermore, Cuvier argued, in spite of four thousand years of recorded history, no new species had evolved. Why assume they can evolve? Like begets like!

Although Lamarck's theories fell into disfavor, it is important to note that the attitudes for which he was denounced during the nineteenth century were often attitudes which were eventually accepted. For example, Lyell, in his *Principles of Geology*, wrote of Lamarck:

> His speculations know no definite bounds: He gives the rein to conjecture, and fancies that the outward form, internal structure, instinctive faculties, nay, that reason itself, may have been gradually developed from some of the simple states of existence, —that all animals, that man himself, and the irrational beings, may have had one common origin; that all may be parts of one continuous and progressive scheme of development from the most imperfect to the more complex; in fine, he renounces his belief in the high genealogy of his species, and looks forward, as if in compensation, to the future perfectibility of man in his physical, intellectual, and moral attributes.

Although it was by no means obvious at the time, we can, with hindsight, summarize Lamarck's contribution as the concept that evolution depends upon natural processes. These processes, the inheritance of acquired characters and the effects of use and disuse, were later proved incorrect but nevertheless had the important advantage that they were uniformitarian in principle and did not immediately rely on supernatural or catastrophic events. Lamarck thus helped in developing the climate of opinion in which evolution could be understood in the same fashion as any other natural event. If we accept the basic idea that organisms can change through time and discard the Lamarckian explanation for this, the question then posed is, Why do they change?

NATURAL SELECTION

It remained essentially Darwin's task to elaborate a mechanism for evolution more acceptable to biologists than that of Lamarck. The mechanism he proposed, natural selection, was briefly defined by him as follows:

> As many more individuals of each species are born than can possibly survive; and as, consequently, there is a frequently recurring struggle for existence, it follows that any being, if it vary however slightly in any manner profitable to itself, under the complex and sometimes varying conditions of life, will have a better chance of surviving, and thus be *naturally selected*. From the strong principle of inheritance, any selected variety will tend to propagate its new and modified form.

Behind this simple explanation was a complex set of causative events which Darwin spent most of his life investigating, although some aspects of natural selection had been previously argued by others. In antiquity, Empedocles (c. 490-430 B.C.) had suggested that the initial appearance of life was in the form of parts and organs floating freely and combining together to form whole organisms. Those organisms that happened to be adapted to "some purpose" survived and those which did not "perish and still perish." From this original selective act, Empedocles proposed, all present organisms stem.

Aristotle disputed Empedocles' concept of the randomness upon which natural selection acts with an argument that has often been used since. He maintained that the attainment of the Scale of Nature, like any other teleological process, obviously arises through a fixed progression of steps from lowest to highest stages. There cannot, therefore, be anything arbitrary or random in this progression that would necessitate selection.

In the eighteenth century, Buffon saw natural selection (as well as selection by man) as the agent responsible for the extinction of species: "All the bodies imperfectly organized and all the defective species would vanish, and there would remain, as there remain today, only the most powerful and complete forms, whether plants or animals." However, Buffon did not see natural selection as responsible for the generation of new species, since he believed that new species could arise by spontaneous generation and that differences between species were probably caused by differences in the conditions under which their spontaneous generation occurred.

In the Lamarckian view, variations were initiated through the effect of the environment and occurred only in an adaptive direction. There could, therefore, be no extinction of "imperfect" or "defective" species, since organisms could always adapt themselves to changing environments via the inheritance of acquired characters. To Lamarck, variations were not separate from evolution, and therefore they could not be random. Thus, selection was not needed to choose adaptive traits.

In the early nineteenth century, a number of authors, including W. C. Wells (1757-1817) and P. Matthew (1790-1874), were able to separate the origin of variations from the forces responsible for their preservation and utilized the principle of natural selection in explaining changes within species. Unfortunately, their ideas seemed highly speculative, since they did not provide sufficient support, and their works were recorded in obscure publications that did not come to the general attention of biologists.

It is of interest that the source to which Darwin ascribes his notion for the tendency of a species to produce numbers in excess of the resources necessary to sustain them—that is, the primary populational pressure which leads to competition and selection—was not derived from biological literature but from the sociology of his time. During the Victorian period in England, a variety of social and economic problems had become apparent because of the rapid increase in poor people resulting from the Industrial Revolution. This tide of poverty had begun with the impoverishment of previous small handicrafts establishments and was continually fed by small farmers pushed from their lands by the Enclosure Acts. Among British economists, one attitude, expressed by Rev. Thomas **Malthus** (1766-1834), was that the fate of the poor is inescapable since their reproductive powers will always exceed their means of subsistence. Food supplies, Malthus pointed out, can at best increase arithmetically (1 → 2 → 3 → 4 → 5...) by the gradual accretion of land and improvement of agriculture, whereas numbers of poor will increase geometrically (1 → 2 → 4 → 8 → 16...), since the progeny of each family are always more numerous than the parents. Thus, famine, war, and disease inevitably become major factors among the controls which limit population growth. The only hope that Malthus held out for the poor was self-restraint: to delay marriage and refrain from sexual activity. All other solutions, such as the Poor Laws (welfare) or the redistribution of wealth and the improvement of living conditions, were held by Malthus to be inadequate, since such measures would stimulate the further increase in numbers of poor and begin again the cycle of famine, war, and disease.

Like many others of the time, Darwin was deeply impressed by the Malthusian argument, although Malthus was not an evolutionist. In fact, Malthus believed that limitations on population growth would prevent evolutionary change because individuals who departed from the population norm would be more susceptible to extinction. To Darwin, however, the importance of Malthus lay in bringing to light the conflict between a population's limited natural resources and its continued reproductive pressure. It was under such circumstances that selection could act by choosing for reproduction those individuals or types with increased chances of survival and therefore change the composition of the population. From his investigations on the breeding of domesticated species, Darwin had obtained clear evidence that selection (in this case, human or **artificial selection**) could have marked hereditary effects. As Darwin writes in his autobiography:

> I soon perceived that selection was the keystone of man's success in making useful races of animals and plants. But how selection could be applied to organisms living in a state of nature remained for some time a mystery to me.
>
> In October 1838, that is, fifteen months after I had begun my systematic enquiry, I happened to read for amusement 'Malthus on Population' and being well prepared to appreciate the struggle for existence which everywhere goes on from long-continued observation of the habits of animals and plants, it at once struck me that under these circumstances favourable variations would tend to be preserved, and unfavourable ones to be destroyed. The result of this would be the formation of new species. Here then I had at last got a theory by which to work; but I was so anxious to avoid prejudice, that I determined not for some time to write even the briefest sketch of it. In June 1842 I first allowed myself the satisfaction of writing a very brief abstract of my theory in pencil in 35 pages; and this was enlarged during the summer of 1844 into one of 230 pages, which I had fairly copied out and still possess.

Because of his need to gather supporting evidence and fears that his theory was not ready to be accepted,

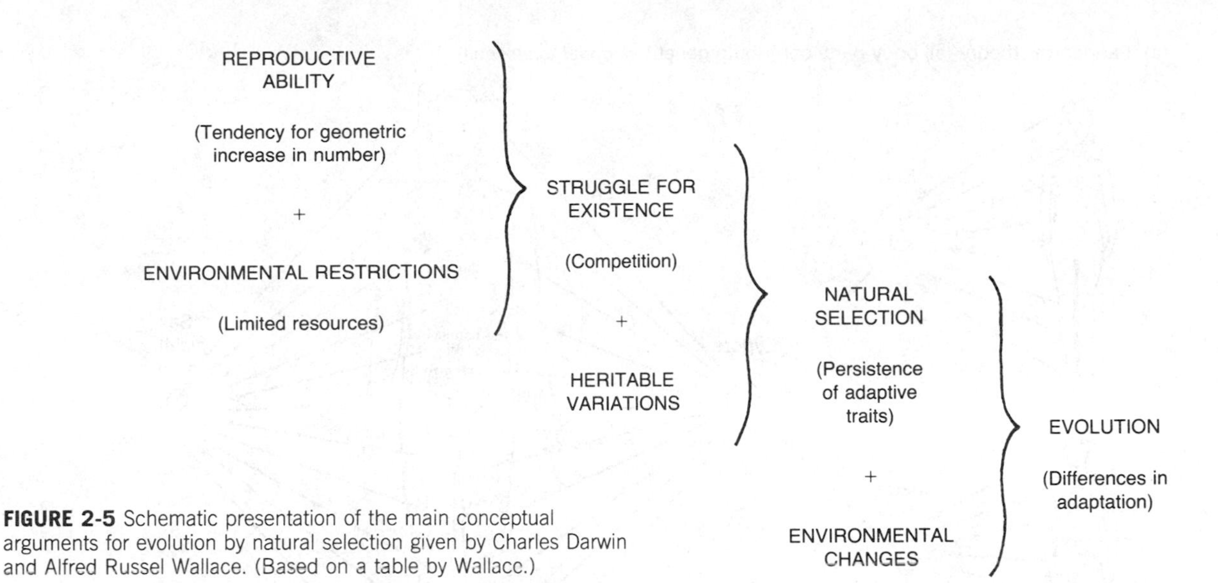

FIGURE 2-5 Schematic presentation of the main conceptual arguments for evolution by natural selection given by Charles Darwin and Alfred Russel Wallace. (Based on a table by Wallace.)

Darwin withheld publication for a considerable time.[3] But the idea was "in the air." In 1858 Alfred Russel Wallace (1823-1913), a naturalist then collecting mainly birds, insects, and mammals in the islands of Southeast Asia, sent Darwin a paper to be published in which he described the theory of natural selection in the essential form which Darwin had envisaged. Wallace too had puzzled about a mechanism for evolution and had also read Malthus. Remarkably, Malthus performed the same function for Wallace as he had for Darwin twenty years earlier. Wallace wrote:

> At that time [February 1858] I was suffering from a rather severe attack of intermittent fever at Ternate in the Moluccas...and something led me to think of the positive checks described by Malthus in his 'Essay on Population', a work I had read several years before, and which had made a deep and permanent impression on my mind. These checks—war, disease, famine, and the like—must, it occurred to me, act on animals as well as on man. Then I thought of the enormously rapid multiplication of animals, causing these checks to be much more effective in them than in the case of man; and, while pondering vaguely on this fact there suddenly flashed upon me the *idea* of the survival of the fittest—that the individuals removed by these checks must be on the whole inferior to those that survived. In the two hours that elapsed before my ague fit was over I had thought out almost the whole of the theory, and the same evening I sketched the draft of my paper, and in the two succeeding evenings wrote it out in full, and sent it by the next post to Mr. Darwin. (Introductory note to Chapter II of *Natural Selection and Tropical Nature*, revised edition, 1891.)

To prevent the loss of Darwin's priority in applying natural selection to evolution, his friends Lyell and Hooker arranged that short papers on the topic by both authors were published in 1858 in *The Journal of the Linnaean Society*.[4] Surprisingly, there was little response by either the scientific or nonscientific communities at that time, indicating perhaps that the theory itself, without supporting evidence, did not enlist serious interest and therefore did not seriously threaten established opinion. It was only with the publication of Darwin's heavily documented *Origin of Species* in November 1859 that the world took notice.

The evolutionary principle expressed by Darwin and Wallace is briefly outlined in Figure 2-5. Note that the evolutionary process is a continual one; the achievement of an adaptation by individuals leads to

---

[3]In 1844, a Scots writer, Robert Chambers, had published a book, *Vestiges of the Natural History of Creation*, that elaborated the idea that all matter, inorganic and organic, evolved out of inorganic dust. The mechanism of biological evolution that Chambers proposed was an accumulation of accidental mutations caused somehow by changes in nutritional or environmental conditions. Unfortunately, Chambers also espoused many questionable notions such as the spontaneous generation of organisms by electrical currents and phrenology (study of bumps on the head) in understanding how the mind works. Although popular for a time (there were twelve editions), Chambers' work engendered considerable religious and scientific denunciation, which made Darwin extremely fearful of exposing his own ideas to ridicule.

[4]Although some Wallace biographers (e.g., Brackman, Brooks) have suggested that Darwin either "borrowed" or "stole" Wallace's ideas, most historians agree that there is no evidence for this (see, for example, Beddall). As stated previously, natural selection and the evolutionary divergence of species were "in the air," and both Darwin and Wallace, although separated by many thousands of miles, were obviously original thinkers who could grasp these concepts independently.

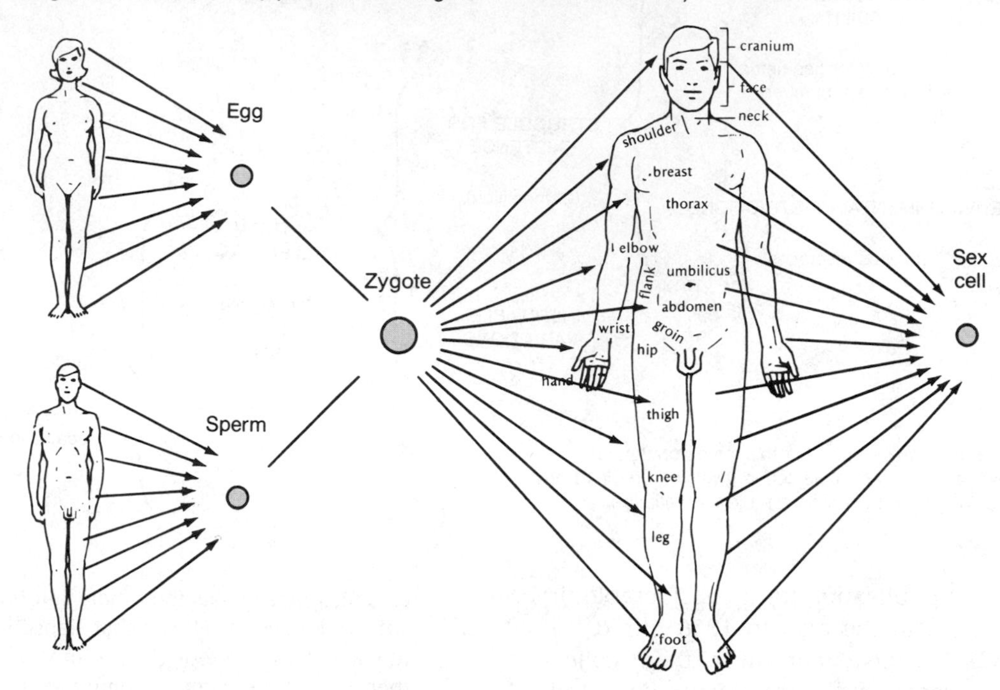

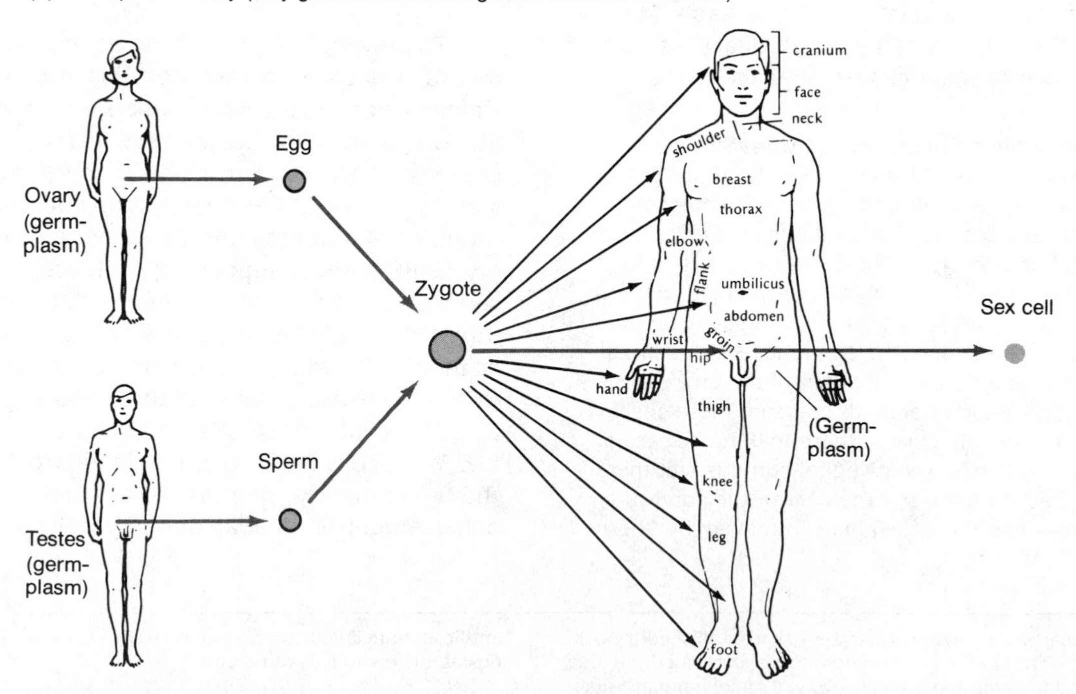

FIGURE 2-6 Comparison between (a) pangenesis and (b) germ plasm theories in the formation of a human. In pangenesis all structures and organs throughout the body contribute copies of themselves to a sex cell. In the germ plasm theory the plans for the entire body are contributed only by the sex organs. (From M. W. Strickberger, 1985. *Genetics*, 3rd ed. Macmillan, New York.)

TABLE 2-1 Comparison of Views on Variation and Heredity

	Creationist (Judeo-Christian)	Lamarck	Darwin	Present Biology
What accounts for the similarity among many species?	The divine plan of creation (purpose unknown) produced the basic "kinds" of organisms.	Descent from a common ancestor.	Descent from a common ancestor.	Descent from a common ancestor.
What accounts for the origin of variation among members of a species?	Although they can be environmentally caused,* they are part of the divine plan of creation.	Environmentally caused.	At times: unknown causation. At times: environmental changes cause new variations, although the variations may be in any direction.	Heritable differences are caused by random changes (mutation) in the genetic material. Noninheritable differences are caused by the environment.
What accounts for the presence of particular organs and structures through time?	They were initially designed so by the Creator. Many present creationists believe that organ defects, diseases, etc., are caused by the fall of man from divine grace and/or intervention by a devil.	Use enhances the development of adaptive variations, and disuse eliminates nonadaptive ones.	Natural selection perpetuates only adaptive traits and eliminates nonadaptive traits. At times: use and disuse.	Primarily natural selection but other forces may be involved, as discussed in Chapter 21.
What accounts for the variation among species?	The separate creation of each species. Many present creationists believe that the original "kinds" of organisms were perfect and variations leading to species differences have been degenerative.	Each species has responded to different environmental needs by developing new organs or discarding old ones.	At times: selective differences among species accounts for their changed inheritance. At times: differences in the use and disuse of particular organs has caused changed inheritance.	Changes occur in the genetic material of each species through the process of mutation and the various forces that change gene frequencies.
What accounts for the resemblance of organisms to their parents?	Mechanism unknown, but acquired characters are inherited as part of the divine plan.†	Those characters acquired through use and disuse are inherited through a pangenesis-like process.	At times: unknown. At times: pangenesis.	Transmission of genetic material through the germ plasm.

*See the story of Jacob and the sheep in Genesis 30: 37-39.
† "Visiting the iniquity of the fathers upon the children and the children's children unto the third and fourth generation." Exodus 34: 7.

enhanced reproductive ability relative to other individuals, followed by further competition for the limited resources and further natural selection. Since each evolutionary stage builds on the one before, the process spirals in the direction of improved adaptation for any particular environment.

At the base of the process is the mutational "fuel" of evolution, the continual introduction of new heritable variations upon which selection can act. Darwin did not know the biological basis for heredity or its variations, and his arguments were weakest in these areas. At times, he proposed that either environmental changes or a large increase in numbers may enhance the variability among a population of individuals, whereas at other times he adopted the Lamarckian view of the inheritance of use and disuse. At the bottom of his difficulties, as with other biologists of the time, was the commonly accepted theory that heredity is mostly a blend of the heredity of both parents, much as the dilution of red paint with white paint produces an intermediate pink product. As the following chapter will show, the idea of **blending inheritance** confronted evolutionary theory with the serious enigma of trying to explain how adaptive variations can be preserved by natural selection if they are blended out by the mating of their carriers with other members of the population.

To deal with the problem of blending inheritance, Darwin at one point reinstituted an old theory called **pangenesis**. He suggested that small atomic "gemmules" or "pangenes" derived from all of the tissues of a parent are incorporated into the parental gametes. When fertilization occurs and parental gametes unite, these gemmules would then spread out to form the tissues of the offspring (Fig. 2-6(a)). Pangenesis would therefore help account for the presumed effects of use and disuse by suggesting that changes can arise in the frequencies of particular gemmules and for the fact that not all traits become blended by postulating that the structure of gemmules can remain constant. However, there was no evidence for pangenesis, and Weismann (1832-1914) effectively disproved it some years later. Weismann cut off the tails of 22 generations of mice and showed that the length of tail was not affected by the presumed loss of tail gemmules in each generation. For pangenesis, Weismann substituted the modern **germ plasm** theory of inheritance in which only the reproductive tissues (testes and ovaries) transmit the heredity factors of the entire organism, and changes that occur in nonreproductive somatic tissues are not transmitted (Fig. 2-6(b)). Thus changes in heredity cannot be simply explained by inheritance of acquired characters or by use and disuse. With the development of the modern science of genetics, many of the difficulties faced by Darwin were resolved. For the present, a brief comparison of various views on variation and heredity is presented in Table 2-1.

SUMMARY

The basic ideas essential to the establishment of evolutionary theory were already present by the time Darwin made his famous voyage on the H.M.S. *Beagle*. The most important of these ideas were that the earth was ancient, that fossils represented the remains of extinct species, that many species showed close similarities, and that organisms descended from previously existing organisms. However, the mechanism for evolutionary change and the agents that allowed organisms to change remained undiscovered.

The information Darwin accumulated on his five-year voyage engendered his evolutionary ideas and later led him to a mechanism by which evolution might proceed. A keen observer of both geology and natural history, Darwin noted geological formations which gave evidence of historical transformation, as well as the peculiar geographical distribution of organisms and the close similarities of species. He recognized that the only rational explanation for these phenomena must be that species could be transformed. At first, however, he could find no mechanism by which transformation might occur.

Darwin rejected Lamarck's contention that structures survived or deteriorated through use or disuse and furthermore that traits so acquired could be inherited. It was left to Darwin to propose a more acceptable alternative, natural selection. From reading Malthus he derived the idea of a superabundance of progeny competing for limited resources, and this struggle provided Darwin with a scenario for changing the composition of a population. Those organisms which had traits better suiting them to their environment would tend to reproduce more prolifically than others, and their traits could then be passed on in higher proportion to future generations. Thus populations could continually improve their adaptations to environments to which they were subjected, and those populations with inadequate adaptations would become extinct. By coincidence, the naturalist Alfred Russel Wallace simultaneously proposed the same mechanism. The papers of both men were presented in 1858, and Darwin's *Origin of Species* was published in 1859.

However, the question of *how* organisms might change remained unresolved. The theory of natural selection depended upon the presence of inheritable variations on which selection could act. Neither Darwin nor Wallace knew how such variants might be produced. It was not until the development of the science of genetics that this difficulty was resolved.

KEY TERMS

adaptations
artificial selection
blending inheritance
Galapagos Islands
germ plasm theory
HMS *Beagle*
Inheritance of Acquired Characters
Lamarckism
Malthus
natural selection
pangenesis
Principle of Use and Disuse

REFERENCES

Beddall, B. G., 1968. Wallace, Darwin, and the theory of natural selection. *Jour. Hist. Biol.*, **1**, 261-323.

——, 1988. Darwin and divergence: the Wallace connection. *Jour. Hist. Biol.*, **21**, 1-68.

Brent, P., 1981. *Charles Darwin: A Man of Enlarged Curiosity*. Harper & Row, New York.

Brackman, A. C., 1980. *A Delicate Arrangement: The Strange Case of Charles Darwin and Alfred Russel Wallace*. Times Books, New York.

Brooks, J. L., 1984. *Just Before the Origin: Alfred Russel Wallace's Theory of Evolution*. Columbia Univ. Press, New York.

Burkhardt, R. W. Jr., 1977. *The Spirit of System: Lamarck and Evolutionary Biology*. Harvard Univ. Press, Cambridge, Mass.

Clark, R. W., 1984. *The Survival of Charles Darwin: A Biography of Man and Idea*. Weidenfeld & Nicolson, London.

Colp, R., 1977. *To Be an Invalid*. Univ. of Chicago Press, Chicago.

Darwin, C., 1845. *The Voyage of the Beagle*. (Originally published as *Journal of Researches*, it has now appeared in numerous editions.)

Darwin, F., 1887. *The Life and Letters of Charles Darwin*. Appleton, New York.

De Beer, G., 1963. *Charles Darwin*. Nelson & Sons, London.

Eiseley, L. C., 1958. *Darwin's Century: Evolution and the Men Who Discovered It*. Doubleday, New York.

Greene, J. C., 1959. *The Death of Adam*. Iowa State Univ. Press, Ames.

Irvine, W., 1955. *Apes, Angels, and Victorians*. McGraw-Hill, New York.

Joravsky, D., 1970. *The Lysenko Affair*. Harvard Univ. Press, Cambridge, Mass.

Keynes, R. D. (ed.), 1979. *The Beagle Record*. Cambridge Univ. Press, Cambridge.

Kohn, D. (ed.), 1985. *The Darwinian Heritage*. Princeton Univ. Press, Princeton, N. J.

Lack, D., 1947. *Darwin's Finches: An Essay on the General Biological Theory of Evolution*. Cambridge Univ. Press, Cambridge.

Lamarck, J. B., 1809. *Zoological Philosophy*. Translated into English by H. Elliott, 1914. Macmillan & Co., New York.

Lerner, I. M., 1959. The concept of natural selection: A centennial view. *Proc. Amer. Phil. Soc.*, **103**, 173-182.

McKinney, H. L. (ed.), 1971. *Lamarck to Darwin: Contributions to Evolutionary Biology, 1809-1859*. (Contains short excerpts from original writings of J. B. Lamarck, W. C. Wells, P. Matthew, C. Lyell, E. Blyth, R. Chambers, A. R. Wallace, and C. Darwin.) Coronado Press, Lawrence, Kan.

——, 1972. *Wallace and Natural Selection*. Yale Univ. Press, New Haven, Conn.

Millhauser, M., 1959. *Just Before Darwin: Robert Chambers and Vestiges*. Wesleyan Univ. Press, Middletown, Conn.

Moorehead, A., 1969. *Darwin and the Beagle*. Hamish Hamilton, London.

Ospovat, D., 1981. *The Development of Darwin's Theory: Natural History, Natural Theology, and Natural Selection, 1838-1859*. Cambridge Univ. Press, Cambridge.

Sulloway, F. J., 1982. Darwin and his finches: the evolution of a legend. *Jour. Hist. Biol.*, **15**, 1-53.

Zirkle, C., 1946. The early history of the idea of the inheritance of acquired characters and of pangenesis. *Trans. Amer. Phil. Soc.*, **35**, 91-151.

3
THE ARGUMENTS AND THE EVIDENCE

lthough the brief natural selection papers by Darwin and Wallace had been received quietly in 1858, in the following year Darwin's publication of his book, *The Origin of Species*, had profound effects. To many biologists, Darwin's detailed exposition of his theory supported by 20 years of thought and documentation proved impossible to overlook, and natural selection was rapidly recognized as an important if not primary mechanism for evolution. Thomas Huxley (1825-1895), who later became Darwin's main public defender, is reported to have exclaimed, "How extremely stupid not to have thought of that."

SCIENTIFIC OBJECTIONS

Objections, however, came rapidly in the nineteenth century and took various forms, including religious objections that will be discussed in Chapter 4. On the scientific level, a variety of major questions were raised about Darwin's theory on the following topics, which he and his supporters attempted to answer in various ways.

Blending Inheritance

Since the prevailing concept of inheritance was that of a blending of the maternal and paternal contributions, a number of critics raised the objection that the new adaptations would rapidly blend out with each generation of interbreeding. According to this argument natural selection would be incapable of maintaining a trait in the face of its continued dilution. Darwin had early been aware of this objection and attempted various replies, among which were the following:

1. A beneficial trait could maintain itself if the possessors of the trait became isolated from the remainder of the population. In support, Darwin pointed to the familiar practice among animal breeders of isolating newly appearing "sports" [called mutations by Hugo de Vries (1848-1935)] and their offspring as a mechanism commonly used to develop new stocks.
2. Some traits are "prepotent," or dominant, and appear undiluted in subsequent generations.
3. An adaptive trait is not confined to only a single incidence in a population, but rather such traits must arise fairly commonly, as witness the large amount of variability present in most pop-

ulations. Since variability is common, it cannot be diluted out as easily as if it were rare. Moreover, Darwin believed that some forms that carry a particular variation could pass on to future generations the tendency for the same variation to arise again.

4. Natural selection not only enhances the reproductive success of favorable variants but also diminishes the reproductive success of unfavorable ones. Thus the frequency of favorable variations is increased by removing unfavorable ones, and there is less chance of diluting out favorable variations.

5. As explained previously (p. 28), Darwin developed the concept of pangenesis by gemmules to help explain the inheritance of traits that he believed were affected by use and disuse and also to provide constancy for the determining agents of inheritance. There were presumably many gemmules for each particular trait, and their numbers could vary during passage from one generation to another; that is, gemmules could be lost but were not blended out.

Variability

In *The Origin of Species*, Darwin had explicitly confined evolution by natural selection to small, continuous variations and had (in the earlier editions of his book) excluded larger variations from being of any use. He had, in fact, literally adopted the Leibnizian dictum that "nature makes no leaps" (p. 9). A number of objections followed almost immediately.

The first objection, raised by various critics and emphasized by Fleeming Jenkin (1833-1885) in a review of 1867, concerned the limits of variability upon which selection could act. Except for monstrosities which were highly abnormal or sterile, most observed variations were only of small changes and did not depart from the species pattern. How then could new species arise? To this Darwin replied that there are really no limits to variability since each stage in evolution of a species is accompanied by further variability upon which selection then acts. Darwin maintained that it is the succession of changes through time, rather than a single simultaneous set of changes, which leads to species differences.

A second, more common objection was the difficulty in determining how each of the very small modifications proposed by Darwin would be recognized by selection. Certainly, in many instances, such as size, a very small modification might hardly be sufficient to confer a significant advantage upon an organism, whereas a large modification might well be selected. Although Darwin could not successfully reply to this argument, he doggedly held to his concept of gradual accretion of small modifications, and the findings of modern genetics (Chapter 10) later added considerable support to his position. Numerous traits have been discovered which are caused by small heritable changes ascribed to many different genes, each with small effect, called polygenes. For example, size differences are often distributed in populations so that some large individuals possess many genes which lead to an increase in size while others possess relatively few such genes and are therefore smaller. Thus, although differences in size may be based on genetic differences of small effect, selection may nevertheless act on accumulations of such differences in various individuals.

A further aspect of this same problem was the question of determining the initial adaptive level that a trait or organ would have to reach in order to be favored by selection. If the trait were already in existence before selection acted on it, perhaps some quantitative expression of the trait would suffice for further evolution; that is, a larger eye might function better than a smaller eye. But if the trait were not in existence, or only barely in existence, how could it be selected? For example, it seemed to many critics that if one looked at fully developed complex organs such as the eye, brain, liver, and so forth, the earliest incipient stages of such organs would appear to have no function at all and could hardly be selected. Can one conceive of an appropriate adaptive function in only one cell of an eye, brain, or leg?

An evolutionary answer to this question of the origin of new traits came from the concept of **preadaptation**, a principle of which Darwin was aware. According to preadaptation, a new organ need not arise *de novo* but may already be present in an organism where it is being used for a purpose other than that for which it is later selected. For example, in his monograph on barnacles Darwin suggested that the cementing mechanism by which present-day barnacles attach to their substrate is related to the cementing mechanism by which the barnacle oviduct coats its eggs in order to attach them to solid objects. That is, only after its earlier evolution in oviducts was this mechanism adapted for attachment of the barnacle itself. Similarly, Darwin pointed out that the evolution of lungs from swim bladders in fish illustrated "that an organ originally constructed for one purpose, namely flotation, may be converted into one for a wholly different purpose, respiration." Thus, a highly specialized organ like the vertebrate eye, for example, did not arise all at once but probably represents a succession of further evolutionary adaptations of a previous light-gathering organ that may have originally involved only a few cells. A turn-of-the-century illustration of one such progession is shown in Figure 3-1.

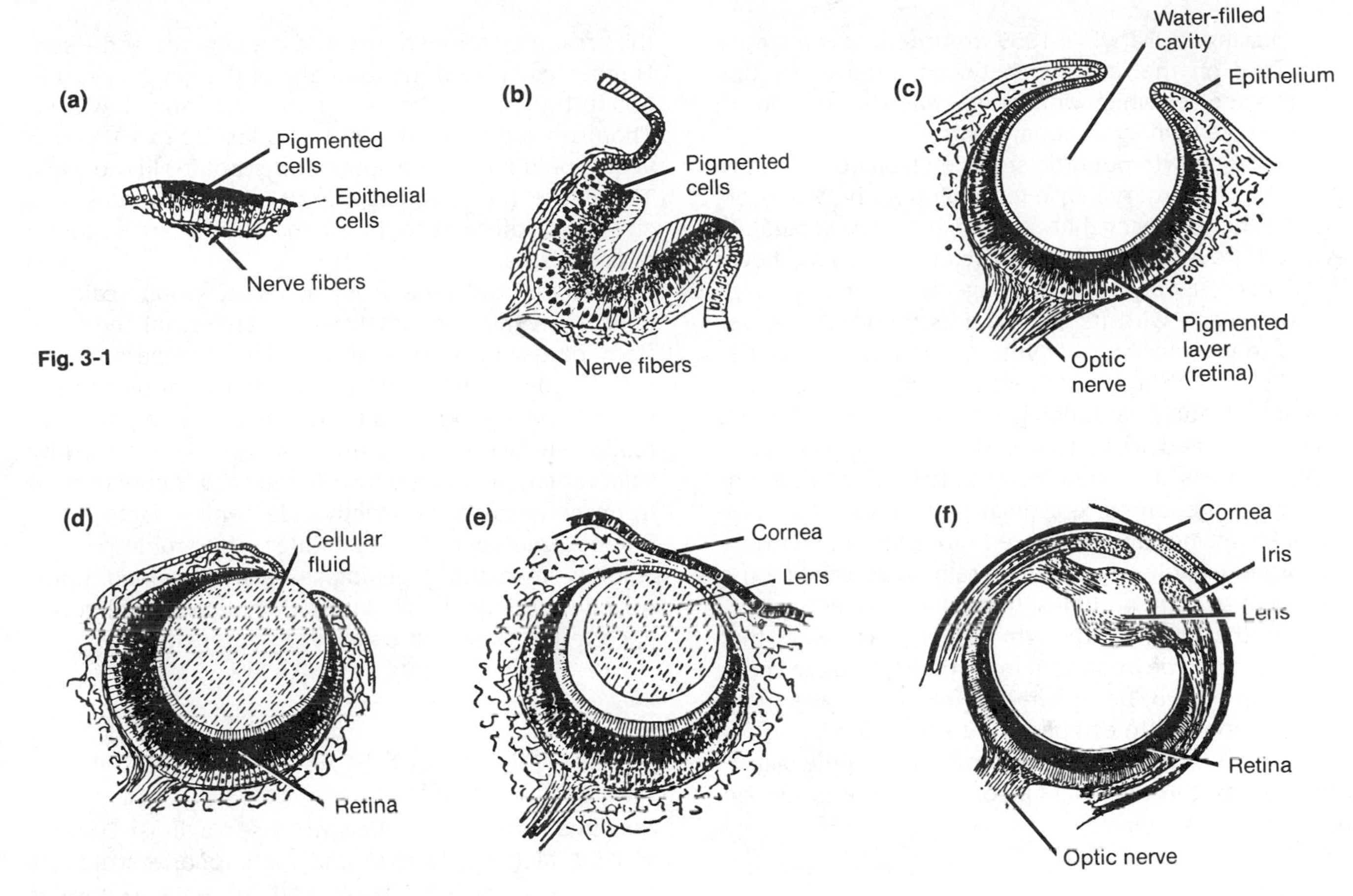

FIGURE 3-1 Some stages in the evolution of eyes as found in molluscs, a phylum whose various groups show different needs for vision and a wide range of light-gathering organs. (a) A pigment spot with neural connections that can be stimulated by light. (b) Folding of pigment cells concentrates their activity, thus providing improved light detection. (c) A partly closed, water-filled cavity of pigment cells which allows images to be formed on the pigmented layer as in a pinhole camera. (d) Secreted transparent cellular fluid is used instead of water, forming a barrier that protects the pigmented layer (retina) from external injury. (e) A thin film or transparent skin covers the entire eye apparatus, adding further protection. Also, some of the fluid within the eye hardens into a convex lens which improves the focusing of light on the retina. (f) A complex eye found in squids, which possesses an adjustable iris diaphragm and focusing lens. (Adapted from Conn.) For recent discussions of the evolution of eyes and photoreceptor pigments, see Ali and Wolken.

Unfortunately, Darwin's search for small modifications led him to place less emphasis on the fact that there are many traits which often show distinct steps and differences—such as differences in color, presence and absence of structures, differences in numbers of structures, and so on. These large variations may also be important for selection, as various biologists, including Huxley, suggested. Interestingly, it was traits such as these, showing large observable differences, which enabled Mendel to develop the basic laws that explain inheritance. In general, the problem of where, how, and to what extent variations originated was not explained by Darwin and remained the element in his theory that was most often attacked.

Isolation

Criticism was also leveled at Darwin because of his almost complete emphasis on the transformation of a single species into another single species (phyletic evolution; Chapter 11). It was pointed out that although Darwin's approach accounted for the evolution of a particular species in time, it did not easily account for the multiplication of species in geographical space. What explains the origin of many new species rather than the transformation of one old species?

Furthermore, argued Moritz Wagner (1813-1887), among others, even the evolution of a single species into a single new species was not fully explained by Darwin, since one could reasonably ask how a new species could possibly evolve in the same locality as its parents.

> Free crossing of a new variety with the old unaltered stock will always cause it to revert to the original type.... Free crossing, as the artificial selection of animals and plants uncontestably teaches, not only renders the formation of new races impossible, but invariably destroys newly formed individual varieties.

Missing in Darwin's 1859 argument was a strong emphasis on the isolation between different groups within a species which would allow each isolated group to follow its own evolutionary path.

Since Darwin put little stress on isolation between groups as a primary cause for evolution, he also overlooked the prospect that sterility between separately evolved groups might be beneficial. That is, as shown by Wallace, it would be advantageous for isolated populations, each with its special adaptations, to be selected to produce sterile hybrids when they meet, because such sterility would permit each group to maintain its unique adaptations without dilution. Darwin instead insisted that sterility was primarily an accidental process. This view thus made it difficult for him to explain the almost universal prevalence of sterility between species and prevented him from utilizing this important isolating mechanism to help account for the divergent evolution of closely related species.

On the whole, although Darwin was aware that isolation could be important in helping an isolated population to evolve, he apparently felt that it was more essential for him to establish that speciation could occur without isolation. It therefore remained the task of others to explore the role of isolation in the formation of species (Chapter 23).

Age of the Earth

Essential to Darwin's argument was a belief that the age of the earth extended beyond anything ever proposed before. As Darwin pointed out in his 1844 essay,

> The mind cannot grasp the full meaning of the term of a million or hundred million years, and cannot consequently add up and perceive the full effects of small successive variations accumulated during almost infinitely many generations.

This emphasis on a long duration of time for evolution ran counter to the timespans usually given. To many scientists, the heliocentric theory tied the earth's origin to the sun, and Newton had calculated that it would take about 50,000 years for a sphere the size of the earth to cool down to its present temperature. Since even such a short period was at variance with the 5,000 or so years of history given in the Judeo-Christian Bible, Newton piously rejected these calculations. Buffon, on the other hand, made calculations of approximately 75,000 years of age for the earth and reconciled this with the biblical time scale by interpreting each of the seven days of creation in Genesis as a separate geological epoch, varying in length from 3 to 35 thousand years.

In Darwin's time, William Thomson (Lord Kelvin, 1824-1907) reassessed the temperature gradients observed in mine shafts, the conductivity of rocks, and the presumed temperature and cooling rate of the sun. He then calculated the total age of the earth's crust at about 100 million years. Of this duration, however, Thomson suggested that only the last 20 to 40 million years could have been sufficiently cool for life to exist. This figure, although large by previous estimates, was still too small to account for the Darwinian evolution of organisms.

Darwin had no answer to these various calculations, since their underestimate of terrestrial age came from perceiving the earth's surface temperature as caused solely by radiation from hot interior sources which were the remains of the initial molten state. In reality, surface temperature is largely determined by solar radiation, and the heat of the earth's core derives from the decay of radioactive elements—facts which first became apparent much later. The problem of accurately measuring geological age remained unresolved until radioactive dating techniques were developed in the twentieth century (Chapter 6).

SUPPORT FOR DARWIN

Notwithstanding the criticisms raised against Darwin, and the fact that Darwin and his supporters did not always have the knowledge and the data to answer each objection satisfactorily, it was Darwin's works which primarily made the evolution of species an acceptable concept. One important reason for this was that although Darwin presented numerous hypotheses in his theory, such as the struggle for existence, natural selection, the divergence between species, and the improvement of adaptations, each of these mechanisms relied on natural processes and was capable of being supported by observations. This was in strong contrast to previous, more speculative, evolutionary theories such as Lamarck's, which often relied upon mystical concepts impossible to observe, such as an inner "perfecting principle" (p. 22).

Another attractive feature of Darwinism to many biologists was its expansion of the role of biology to include the study of relationships among all living creatures, including humans, who were formerly thought to be divine and separate. In every area of biology, from botany to paleontology to zoology, Darwinism opened new lines of thought and new areas of investigation. What are the relationships among different kinds of cells? Different parts of cells? Different flowers? How did orchids evolve? How did bone structures change? What are the steps in the evolution of circulatory systems? Nervous systems? Why are there species that mimic others? How did sterile insect castes such as worker bees evolve? What accounts for the geographical distribution of specific organisms? Al-

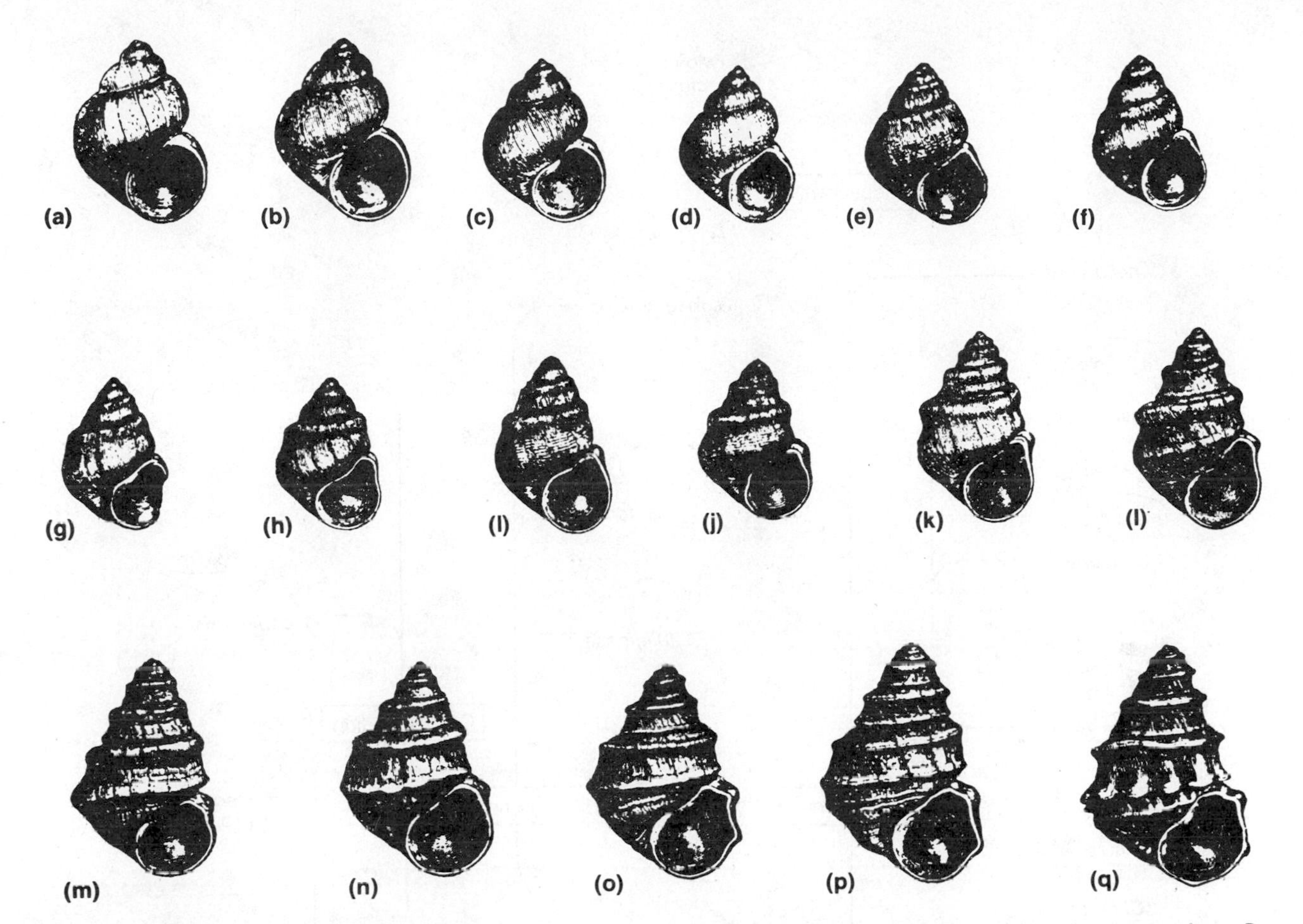

FIGURE 3-2 A nineteenth-century illustration of obvious evolutionary relationships among fossil species of the mollusc *Paludina*, ranging from the oldest form, *P. neumayri* (a), to the youngest form, *P. hoenesi* (q). (From Romanes.)

though each topic demanded separate techniques and study, they all sprang from an evolutionary source that could be explained by rational, understandable mechanisms. By offering an overall view of adaptation and evolution, Darwin helped begin the process by which all of biology could be tied together.

Of considerable importance to biologists of the time were also the many lines of both direct and indirect evidence that rapidly began to accumulate in support of an evolutionary view. Some of these may be briefly outlined as follows:

Systematics

Although the evidence was indirect, it seemed clear, after Darwin, that the gradation of different organisms observed in classification procedures, whether from simple to complex or from one type to another, could most easily be explained by evolutionary relationships (Fig. 3-2).

Geographical Distribution

Many biologists became aware that groups of organisms which are evolutionarily related are usually found, as expected, geographically connected. Large geographical barriers such as oceans and mountain ranges serve to isolate groups from each other and lead to considerable differences between the separated groups. Colonizers which then transcend such barriers often become the ancestors of entirely new groups. This could be shown most graphically in the wide evolutionary radiation of species which had their ancestry in the finches that reached the Galapagos Islands (p. 21). Beginning with what was probably an ordinary mainland finch, new kinds of finches evolved in the Galapagos that could function in habitats unoccupied by other bird species (Fig. 3-3). The name given to this process, **adaptive radiation,** signifies the rapid evolution of one or a few forms into many different species that occupy different habitats within a new geographical area. The marsupial radiation in Australia (Fig. 3-4) shows how this process can lead to an entire array of species with widely divergent functions, from herbivores to carnivores, when marsupials are protected from competition with placental mammals by the isolation of a continent.

Comparative Anatomy

This area of biology, the study of comparative relationships among anatomical structures in different

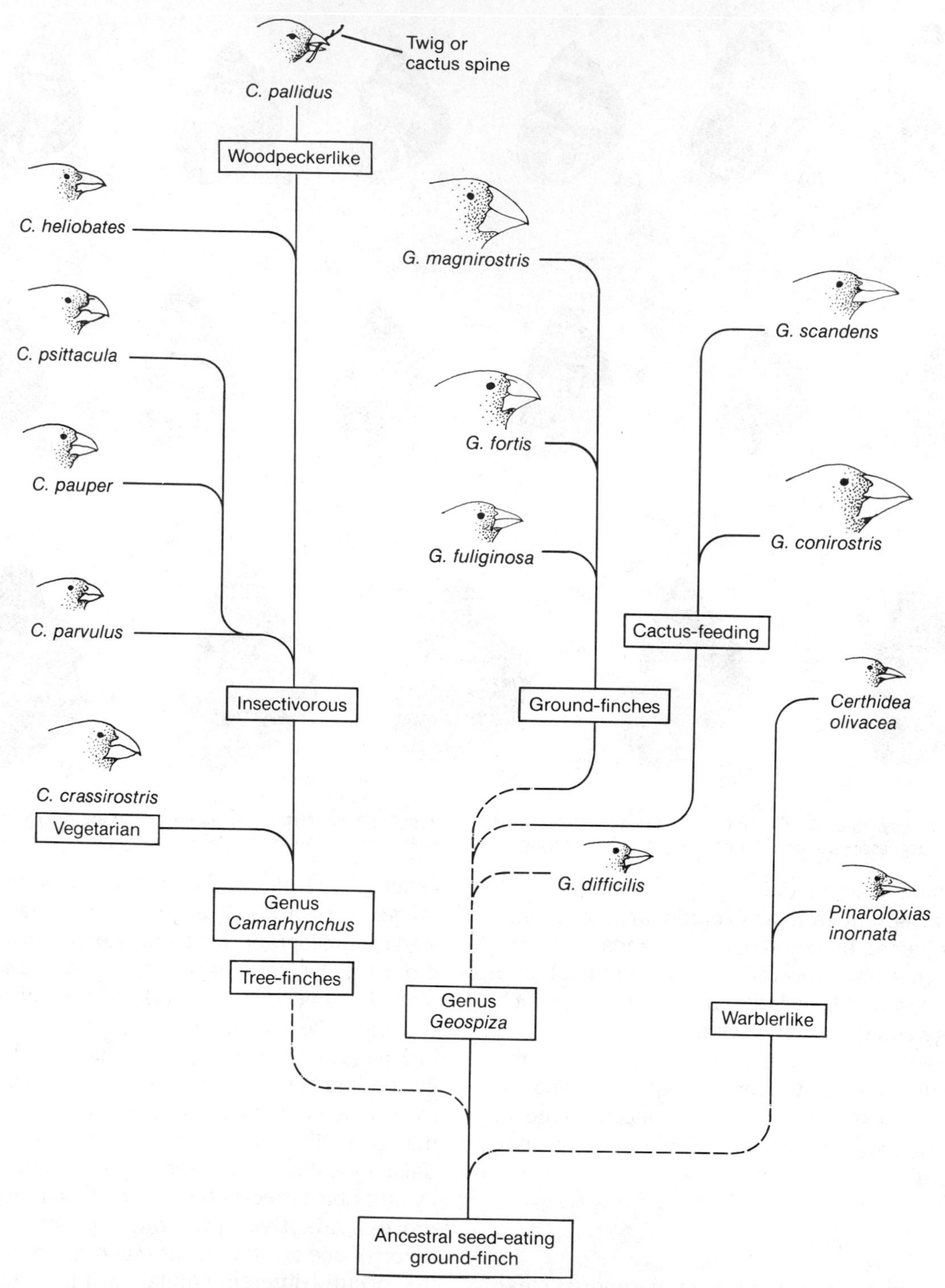

FIGURE 3-3 Evolutionary tree of Darwin's finches showing beak adaptations of the individual species. (Adapted from Lack. A somewhat different phylogeny of these finches has been presented by Schluter.)

species, became for a period of time after Darwin the most popular biological discipline. A search for evolutionary relationships made it possible to trace, especially in vertebrates, many stepwise changes in bones, muscles, nerves, organs, and blood vessels (Fig. 3-5). Such studies made clear that as each species and group of species evolved, previously inherited structures could become modified in entirely new ways.

By changing the meaning of the terminology introduced by Richard Owen (1804-1892) in the 1840's, organs which were related to each other through common descent, although now perhaps functioning differently, were called **homologous**. For example, a study of bones and muscles showed that the forelimbs

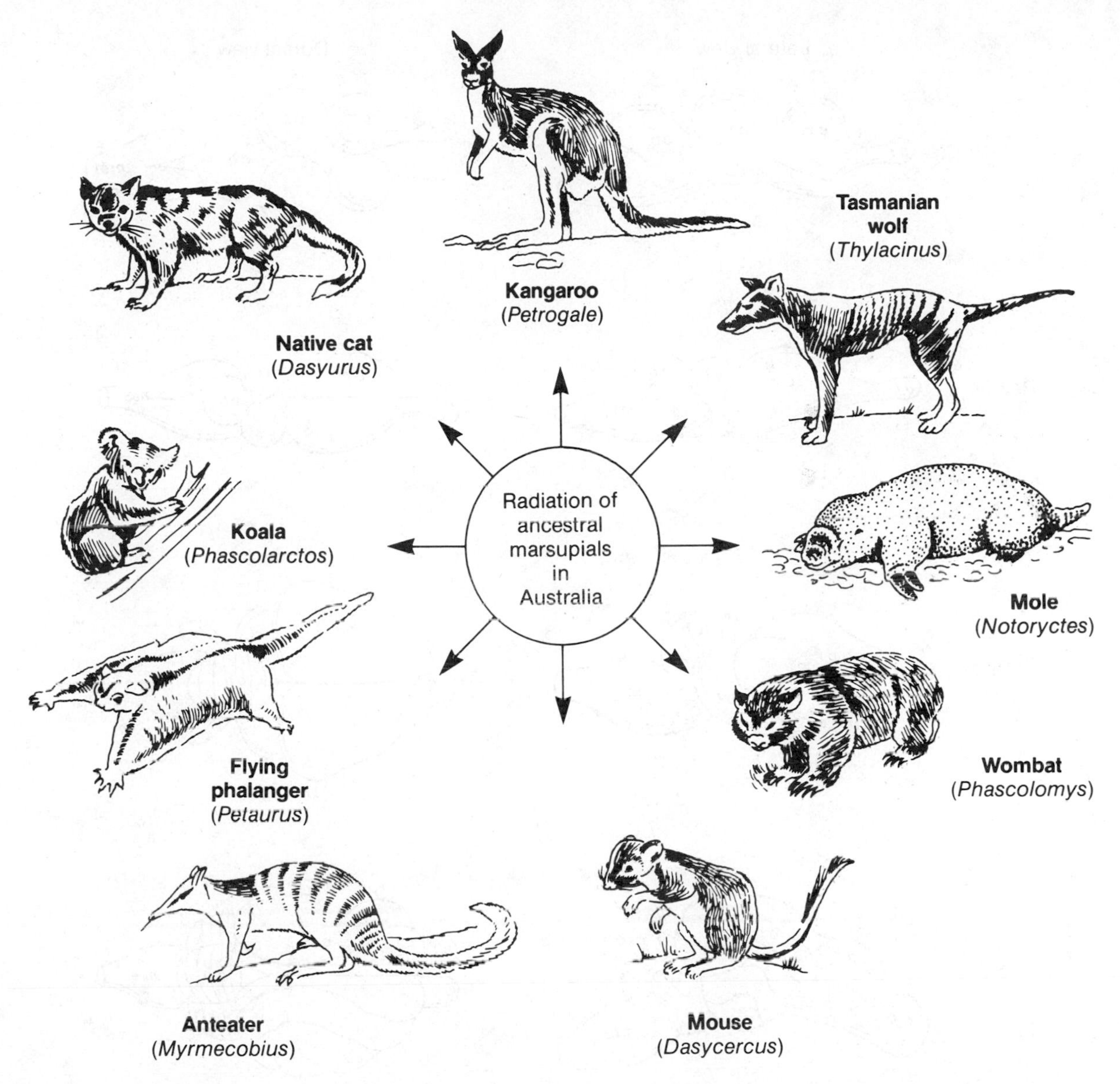

FIGURE 3-4 Adaptive radiation of Australian marsupial mammals showing the many divergent forms that evolved independently of changes occurring among placental mammals on other continents. The striking similarity between some of these marsupial mammals and placental mammals arises because selection for survival in similar habitats can lead to similar adaptations—that is, parallel or convergent evolution has taken place (see also Fig. 3-7 and p. 209). (Adapted from G.G. Simpson and W. Beck, 1965. *Life*, 2nd ed. Harcourt, Brace, and World, New York, with additions.)

of widely different vertebrates could all be explained on the basis of evolutionary homology (Fig. 3-6). On the other hand, **analogous** organs which performed the same function in different groups, such as the wings of bats and the wings of insects, do not show a common underlying plan of structure, since these organs were not evolutionarily derived from the same organ in a common ancestor. Even when analogous organs seemed strikingly similar, such as the eye of an octopus and the eye of a mammal, it could still be demonstrated that they had different evolutionary origins since these two structures arose embryonically from different tissues. The evolution of different organisms, or parts of organisms, in such similar directions was called **convergent evolution**, indicating that selection for similar habitats in different evolutionary lineages could occasionally lead to functionally similar, although not identical, anatomical structures (Fig. 3-7). However, with the general exception of such events, comparative anatomy followed the logic that organisms with shared structures derived from a common group of ancestors, whereas organisms with unlike structures represented divergent evolutionary pathways. Careful anatomical dissections and comparisons were therefore used to provide the criteria for constructing detailed evolutionary trees (Chapter 11).

Of considerable interest to comparative anatomists was the finding of structures that seemed to have

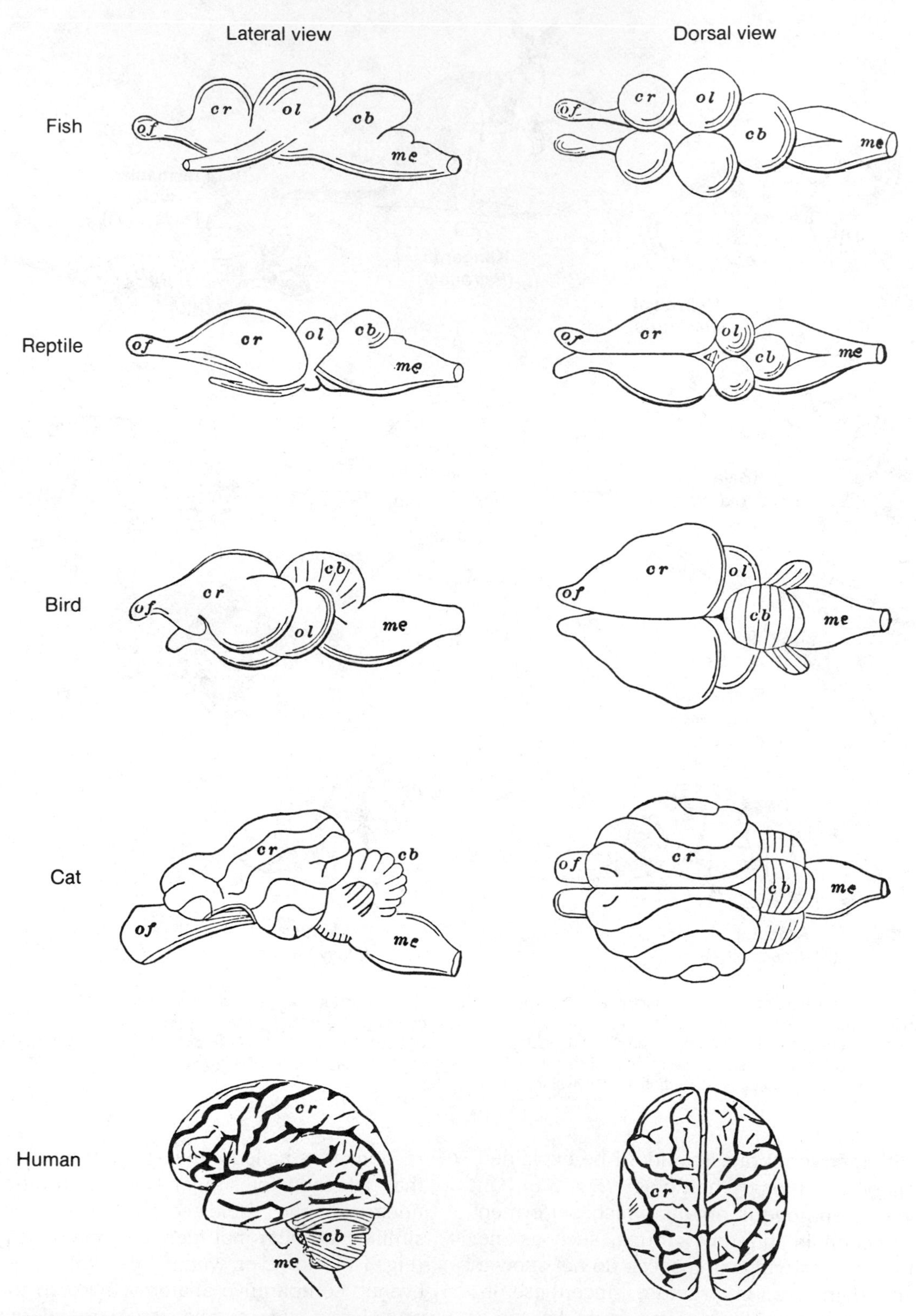

FIGURE 3-5 Nineteenth century diagrams of vertebrate brains, showing side and top views. Homologous structures shared by all these groups are *cb*, cerebellum; *cr*, cerebrum; *me*, medulla; *of*, olfactory bulbs; *ol*, optic lobes. In humans, overgrowth of the cerebrum covers the olfactory bulbs and optic lobes. (From LeConte.)

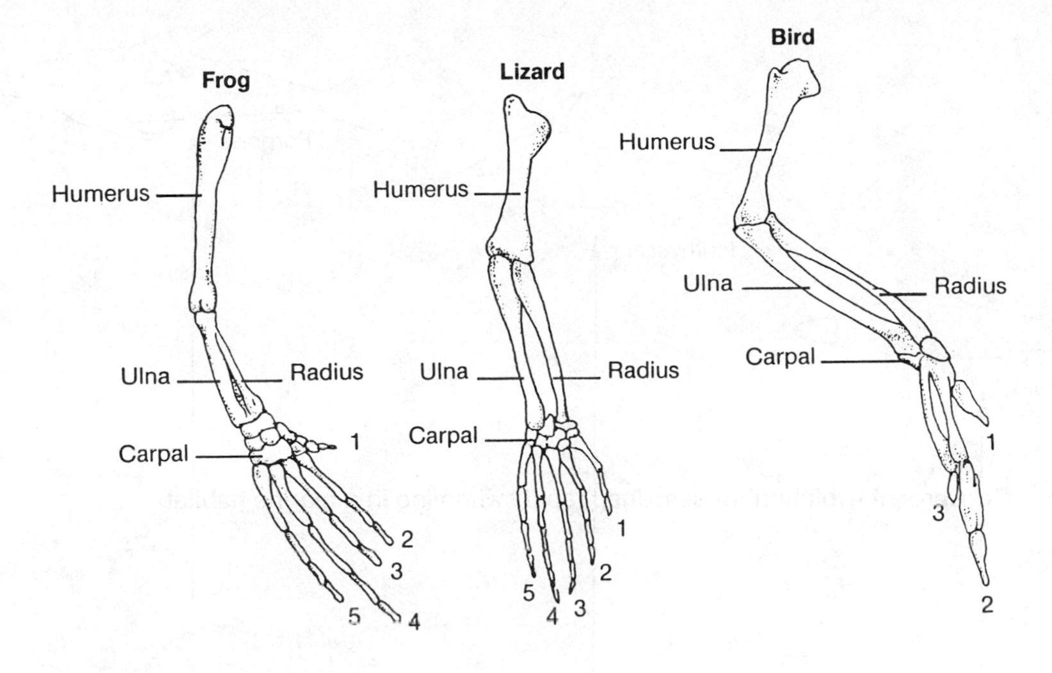

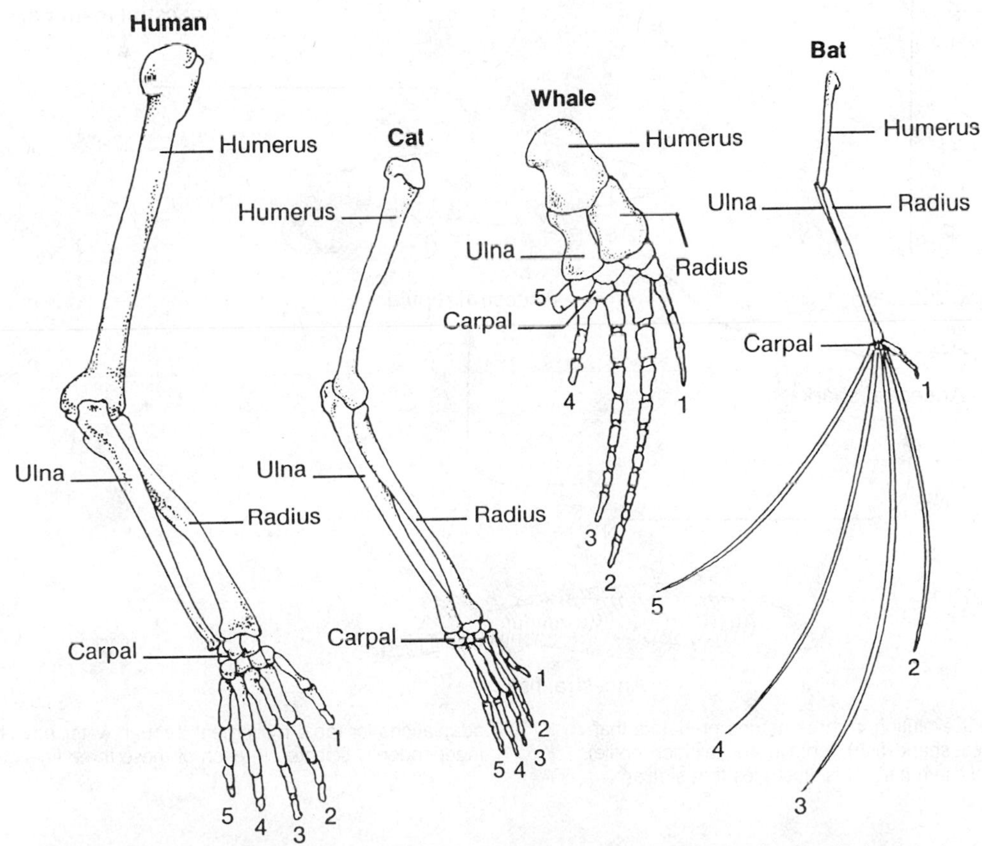

FIGURE 3-6 The skeletal structures of forelimbs in various vertebrate animals showing the homologies between bones. It became clear to comparative anatomists that although these vertebrates evolved in different directions, with changes in size, shape, and function, they all used common bone elements, as well as homologous nerve systems, blood circulatory systems, and other organ systems, thus indicating a common vertebrate ancestor. As Darwin noted, "What can be more curious than that the hand of man formed for grasping, that of a mole, for digging, the leg of a horse, the paddle of a porpoise and the wing of a bat, should all be constructed on the same pattern and should include similar bones and in the same relative positions?"

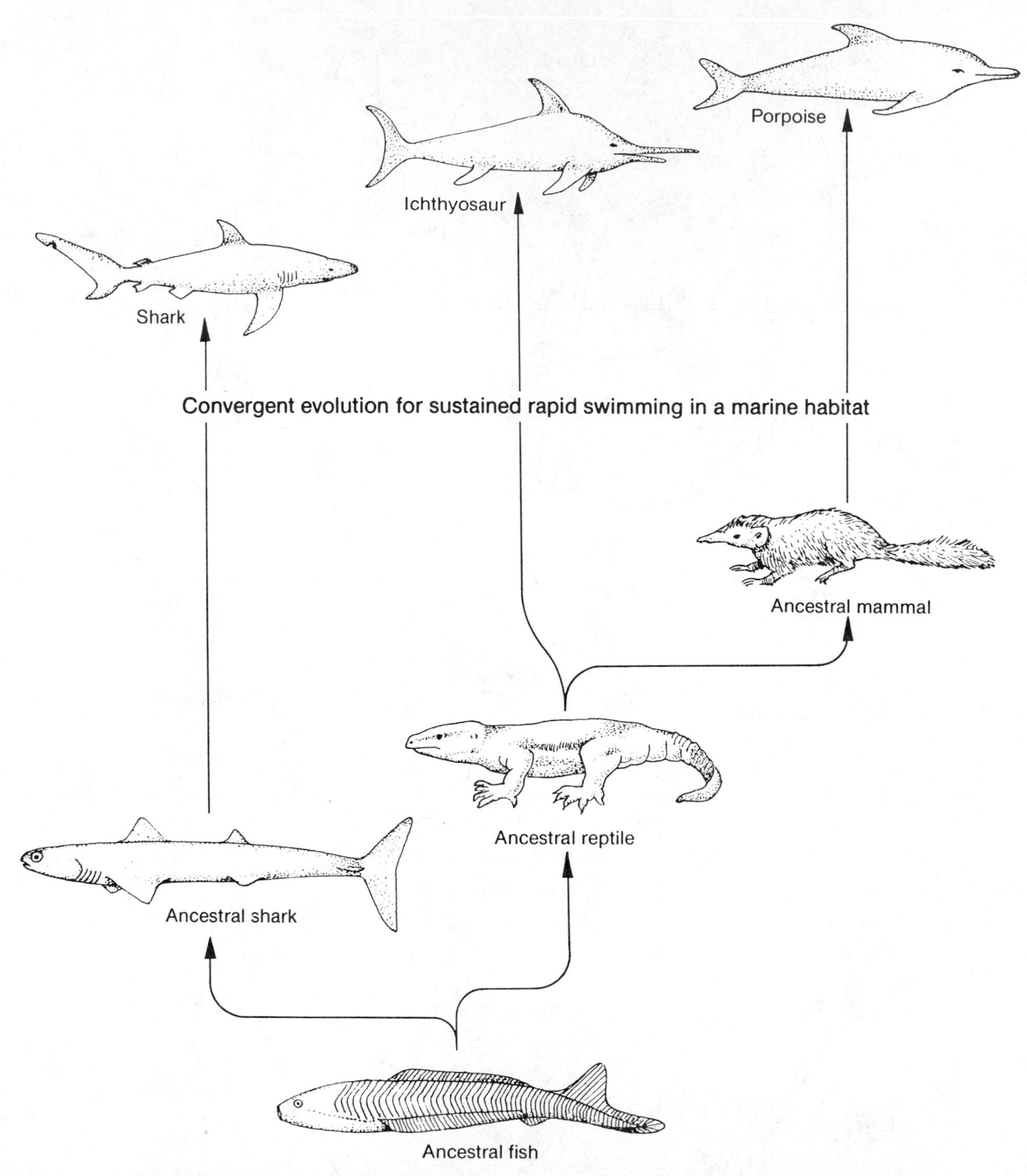

FIGURE 3-7 Convergent evolution in three marine predators that have different ancestries: shark (fish), ichthyosaur (extinct reptile), and porpoise (placental mammal). This illustrates that similar adaptations for rapid movement through water have been independently selected in each of these three lineages.

lost some or all of the function they possessed in earlier ancestors. From an evolutionary point of view the presence of such **vestigial** or rudimentary organs could be explained as arising from the fact that an organism adapting to a new environment usually carries with it some previously evolved structures which are now no longer necessary. According to the principle of natural selection, individuals that devote less energy to the specific elaboration and maintenance of such extraneous structures would be more reproductively successful than those that continue to maintain these structures. Moreover, some structures that were no longer necessary might well interfere with the function of new adaptations. Thus, as time went on, obsolete structures would tend to diminish, showing only traces of their former size and function. Examples of these

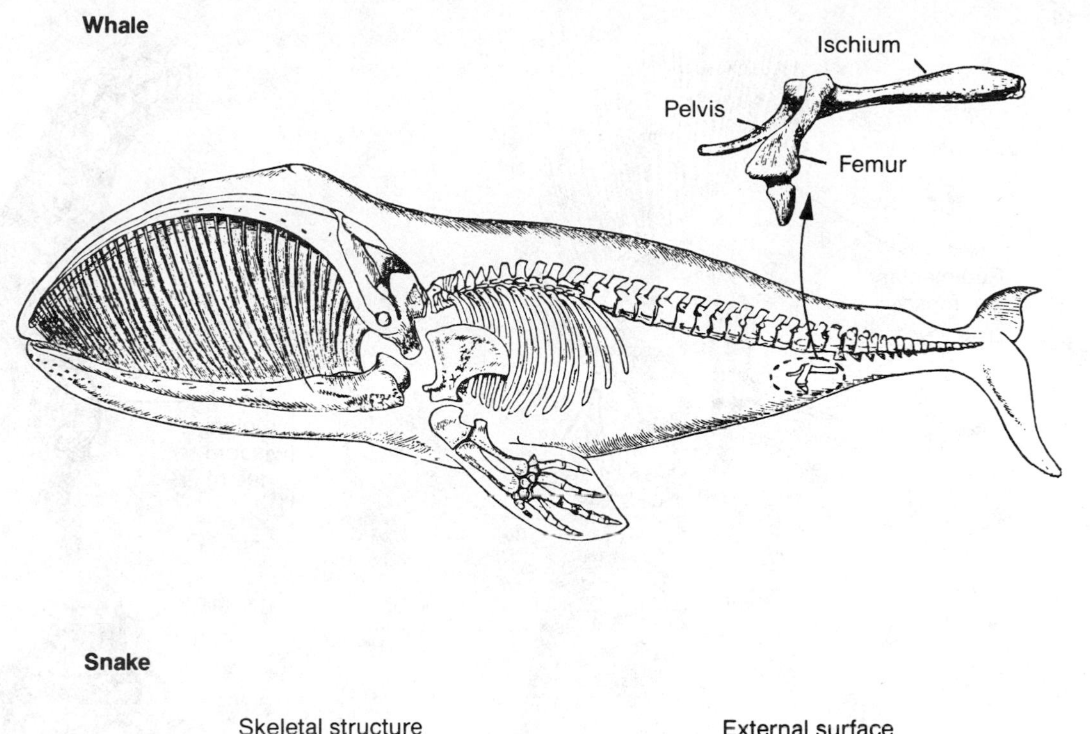

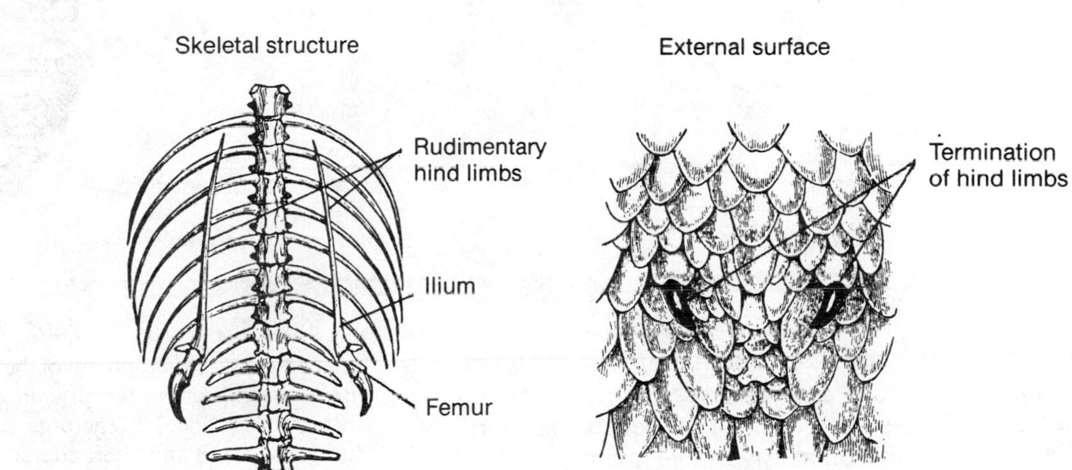

FIGURE 3-8 Rudimentary hind limbs in the Greenland whale and python snake. (From Romanes.)

were found in the rudimentary bones which were all that remained of the former hind limbs in the whale and snake species shown in Figure 3-8. The presence of reduced eye stalks in blind, cave-dwelling crustaceans were also indicative of the evolutionary process by which obsolete structures gradually became rudimentary.

In humans, a number of vestigial structures were found indicating not only the obsolescence of organs but also a relationship to other mammals and primates. For example, muscles of the external ear, as well as scalp muscles common to many mammals, are rudimentary in humans and often nonfunctional. The inflection of the feet for grasping, along with extension of the great toe, is a primate trait often expressed in human infants but which degenerates in human adults. The marvelous gripping power of the hands of human infants can also be considered a vestigial trait since in other primates it is even further developed for purposes of brachiation (grasping of tree branches) and to permit tight clinging of the infant to its mother. Obvious vestigial organs in humans and the other great apes are the reduced tail bones (*os coccyx*) and the remnants of a few tail muscles. To these one can probably add the nictitating membrane of the eye, the appendix of the cecum, rudimentary body hair, and wisdom teeth (Fig. 3-9). All of these are apparent vestiges of more developed structures present in earlier mammals.

Embryology

Early in the nineteenth century von Baer had noticed the remarkable similarity between the embryos of

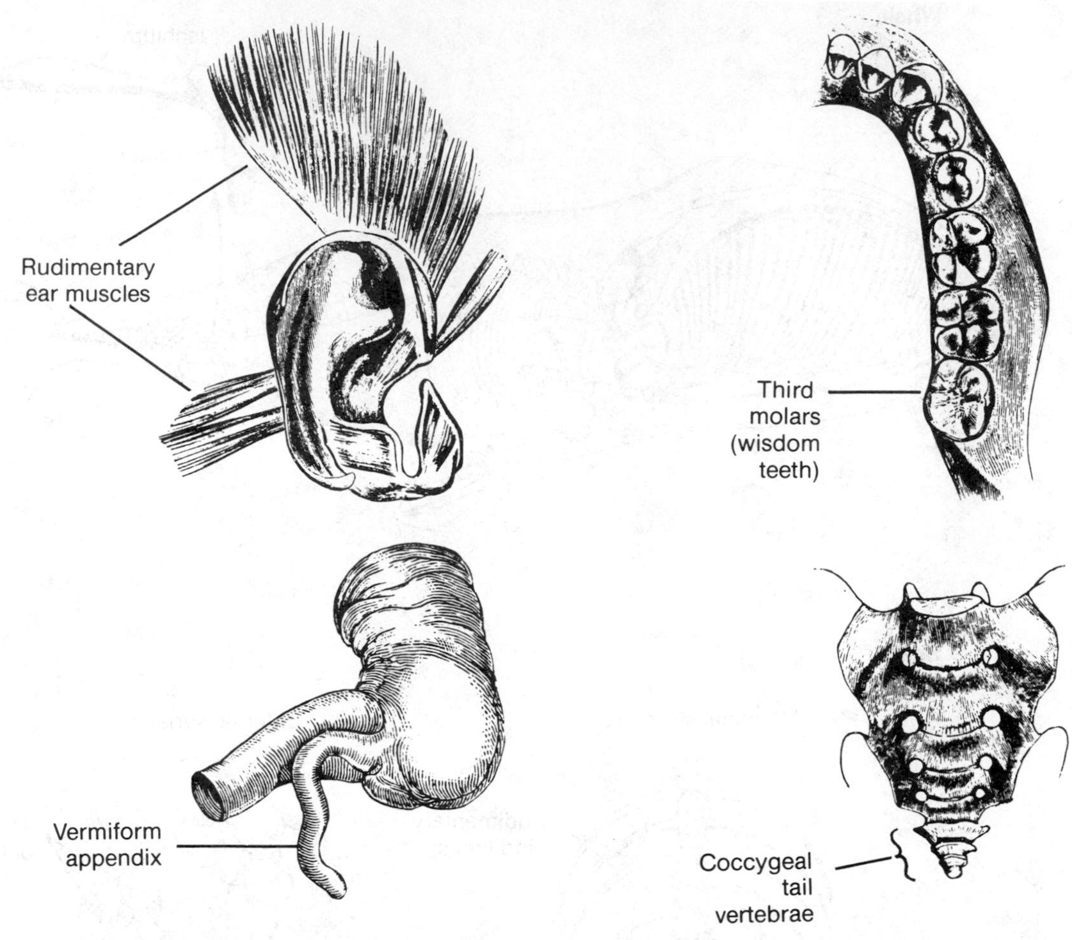

FIGURE 3-9 Some vestigial structures found in humans. (Adapted from Romanes.)

vertebrates which are quite different from each other as adults. To Darwin and others the early stages of development were therefore considered to be more conservative or evolutionarily stable than the adult stages. Remote evolutionary relationships could therefore be shown very strikingly by comparing embryos (Fig. 3-10). As Darwin stated,

> In two groups of animals, however much they may at present differ from each other in structure and habits, if they pass through the same or similar embryonic stages, we may feel assured that they have both descended from the same or nearly similar parents, and are therefore in that degree closely related. Thus community in embryonic structure reveals community of descent.

Ernst Haeckel (1834-1919), the main propagandist for evolution in Germany, further developed and popularized this concept into the **Biogenetic Law**:

> Ontogeny [development of the individual] is a concise and compressed recapitulation of phylogeny [the ancestral sequence].... The organic individual repeats during the rapid and short course of its individual development the most important of the form-changes which its ancestors traversed during the long and slow course of their paleontological evolution according the laws of heredity and adaptation (1866).

To Haeckel this apparently meant that early stages of development may also recapitulate the adult ancestral forms.

There has been considerable dispute on this point, and most biologists consider Haeckel's law an oversimplification.[1] According to present views, early stages of development probably recapitulate only early ancestral stages. It is as though organisms that share common descent make use of common underlying embryological patterns upon which to build later but different adult patterns.

---

[1]Difficulties in applying the concept that ontogeny recapitulates phylogeny are obvious for a number of developmental processes, especially those that show progressive but imperfect functional changes during the life cycle of an individual. Walking in humans, for example, is preceded by infantile stages of stumbling, falling, and sitting. It is difficult to imagine that these stages recapitulate pre-human stages in which populations fell or stumbled about. Admittedly, adaptations for walking took a long time to evolve, but those primate populations in which they were evolving must nevertheless have had functional mobility.

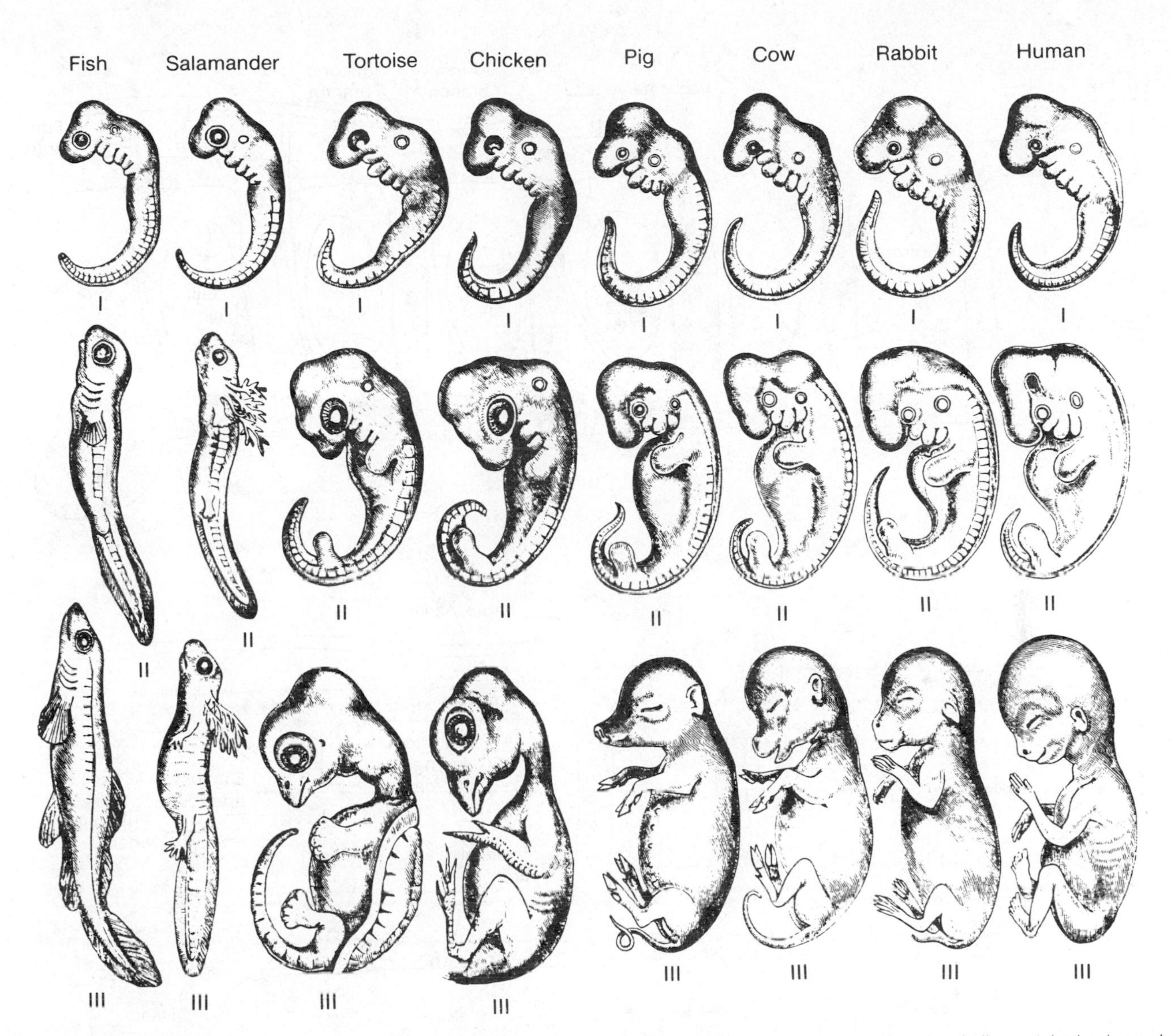

FIGURE 3-10 A series of embryos of different vertebrates at comparable stages of development. The earlier the stage of development, the more strikingly similar are the different groups. Note that each of the embryos begins with a similar number of gill arches (pouches below the head) and a similar vertebral column. In later stages of development these and other structures are modified to yield the various different forms. (From Romanes, adapted from Haeckel.)

Examples of the evolutionary persistence of underlying patterns have often been shown in human development where gill arches (see Fig. 3-10) serve as the basis for the further development of head and thoracic structures, yet no functional gills are ever formed. Similarly, the human embryo, like the embryo of the chick and other advanced vertebrates, goes through the stage of possessing a two-chambered heart similar to the fish, although the final functional human heart is four-chambered.

Even the anatomical positions of some adult nerves, blood vessels, and other structures are intelligible only by considering their evolutionary developmental patterns. For example, as shown in Figure 3-11(a), each branch of the vagus nerve in fish runs through an arterial arch pierced by a gill slit. From studies in many vertebrates it is clear that two of these vagal nerve branches eventually evolved in mammals for stimulating the larynx. The most anterior of these, called the anterior laryngeal nerve, loops around the third arterial arch, now the carotid artery, and a posterior nerve branch, called the recurrent laryngeal nerve, loops around the sixth arterial arch (Fig. 3-11(b)). However, in contrast to its function in fish, the left side of the sixth arterial arch in mammals is the *ductus arteriosis*, which is used embryonically to carry blood to the placenta until birth but then atrophies to become a pulmonary artery ligament. Since the ligamentous remnant of this old sixth arterial arch is close to the heart in adult mammals, the left side of the recurrent laryngeal nerve must travel from the cranium to the thoracic cavity and back to the larynx in order to complete its circuit. In mammals with long necks, such as giraffes, this extra loop is obviously

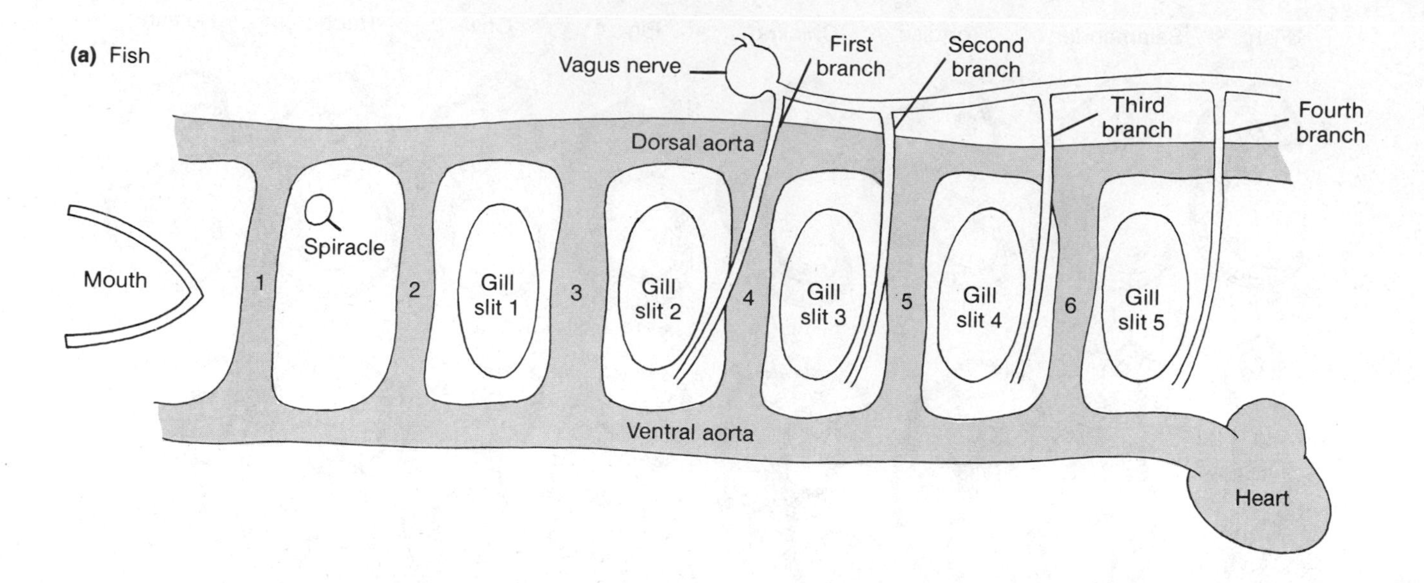

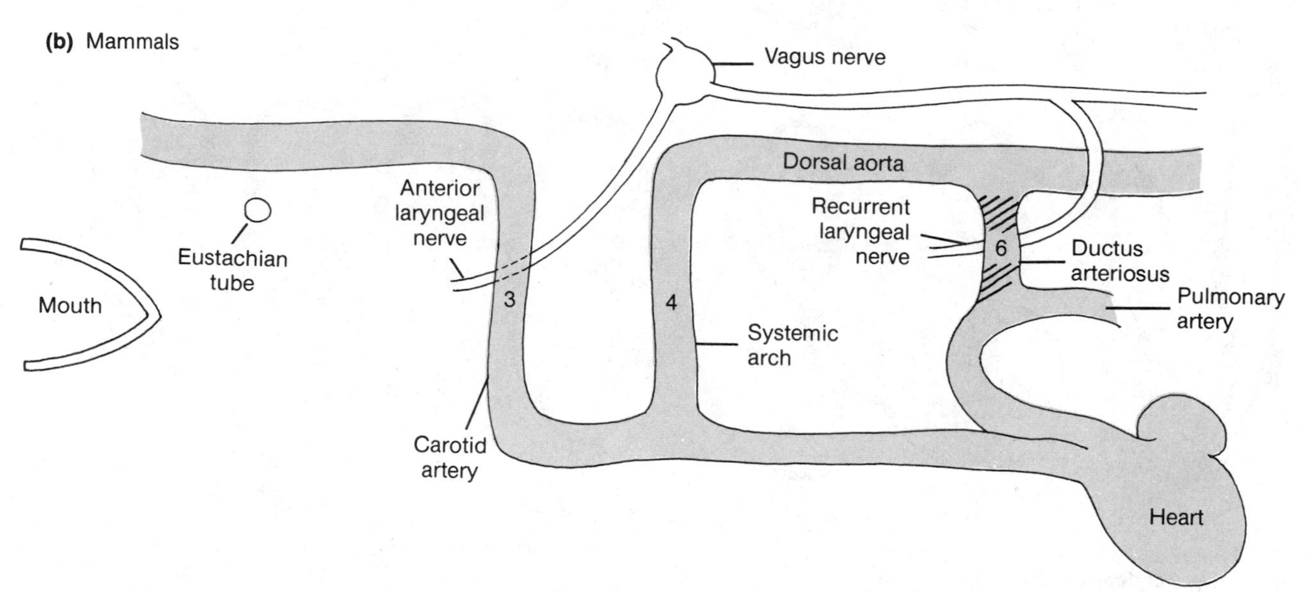

FIGURE 3-11 Schematic diagram showing the relationship between the vagus nerve and the arterial arches in fish (a) and mammals (b). Only the third, fourth, and part of the sixth arterial arches remain in placental mammals, the sixth acting only during fetal development to carry blood to the placenta. The fourth vagal nerve in mammals (the recurrent laryngeal nerve) loops around the sixth arterial arch just as it did in the original fishlike ancestor, but must now travel a greater distance since the remnant of the sixth arch is in the thorax. (Adapted from DeBeer.)

many feet longer than it would be were nerve development independent of evolutionary pattern.

All of these embryonic stages and anomalies therefore make sense to biologists only if we consider that humans and other terrestrial vertebrates had fishlike ancestors which provided them with some of their basic developmental patterns.

Fossils

In Darwin's day the fossil record, never too plentiful, was even more spotty than today. Fossil remains are predominantly found in sedimentary rocks which were originally laid down by a succession of deposits in seas, lakes, deserts, and so on and hence occur in some areas but not in others. Even in appropriate sedimentary environments, many dead organisms decompose before they can become fossilized or are later destroyed by the erosion of sedimentary rocks even when they have become fossilized. Also, since evolutionary changes between populations are primarily encouraged by isolation between them, it is rare to find transitional forms in the same place as the original forms. Thus, a complete evolutionary progression of fossils from most ancient to most modern has never been found in a single locality. In spite of these difficulties,

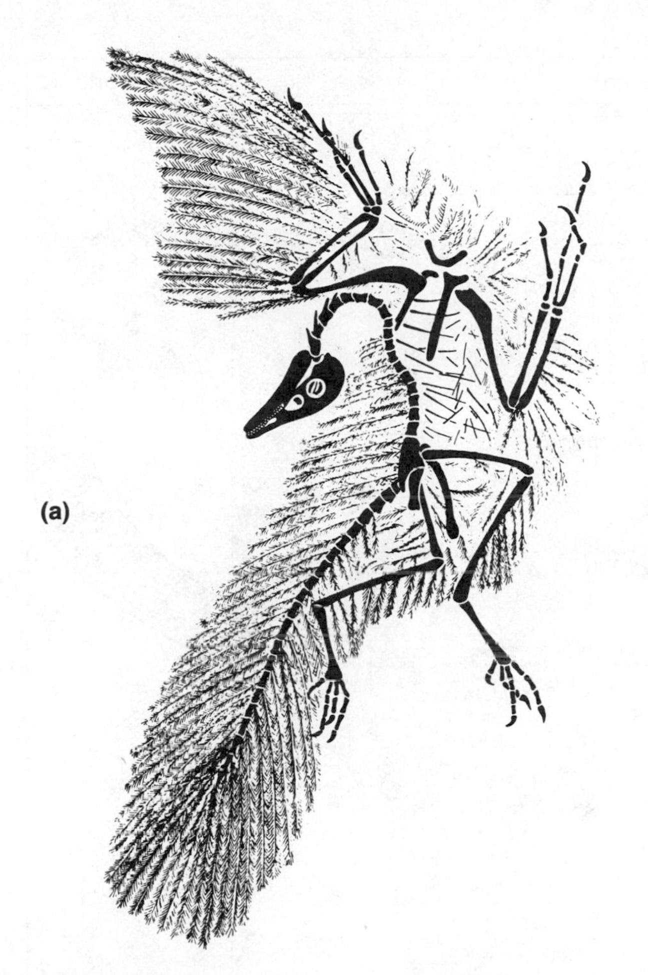

FIGURE 3-12 (a) The Berlin example of *Archaeopteryx* found in the Upper Jurassic limestones of Bavaria. (b) Nineteenth-century restoration of the bird's appearance during life. (Adapted from Romanes.)

the search for fossils engendered considerable interest since fossils could be used to provide the hard facts of evidence for evolution.

Fortunately, in Darwin's lifetime a few paleontological findings came to light which strongly supported the Darwinian position. One was the discovery in 1861 of a true "missing link": in this case, an animal that stood approximately midway between reptiles and birds. As shown in Figure 3-12, this fossil, *Archaeopteryx*, had a number of reptilian features, including teeth and a tail composed of twenty-one vertebrae, but also possessed a number of birdlike features such as wishbone and feathers. Huxley argued convincingly that it was evolution from reptiles to birds which could best explain the existence of *Archaeopteryx*—that such intermediates were predictable consequences of evolution that helped prove the theory.

By the 1870's, paleontologists such as Marsh (1831-1899) were able to use fossils of both North American and European horses to provide the first classic example of a stepwise evolutionary tree among vertebrates. The earliest known horselike mammal, first called *Hyracotherium*, had been described by Owen one year after the publication of *The Origin of Species*. It was about 20 inches high, weighing about 50 pounds, with four toes on its pad-footed front legs and three on its hind legs, and with simple teeth adapted for browsing on soft vegetation.

In the approximately 60 million year interval since *Hyracotherium*, horses have changed radically. They now run on hard ground, chew hard grasses, and show special adaptations for this particular environment. Their elongated legs are built for speed by bearing most of the limb muscles in the upper part of the legs, thus enabling a powerful, rapid swing. They now have the distinction of being the only vertebrates with a single toe on each foot, which, together with a special set of ligaments, provides them with a pogo-stick-like springing action while running on hard ground. Their teeth also show unique qualities adapted for chewing hard, abrasive grasses. The molars and premolars appear identical in shape and are very long, showing continued growth for the first eight years of life until the roots are formed. The high crowns of these grinding teeth have vertical layers of enamel and cement. As the tooth wears, the cement breaks down more rapidly

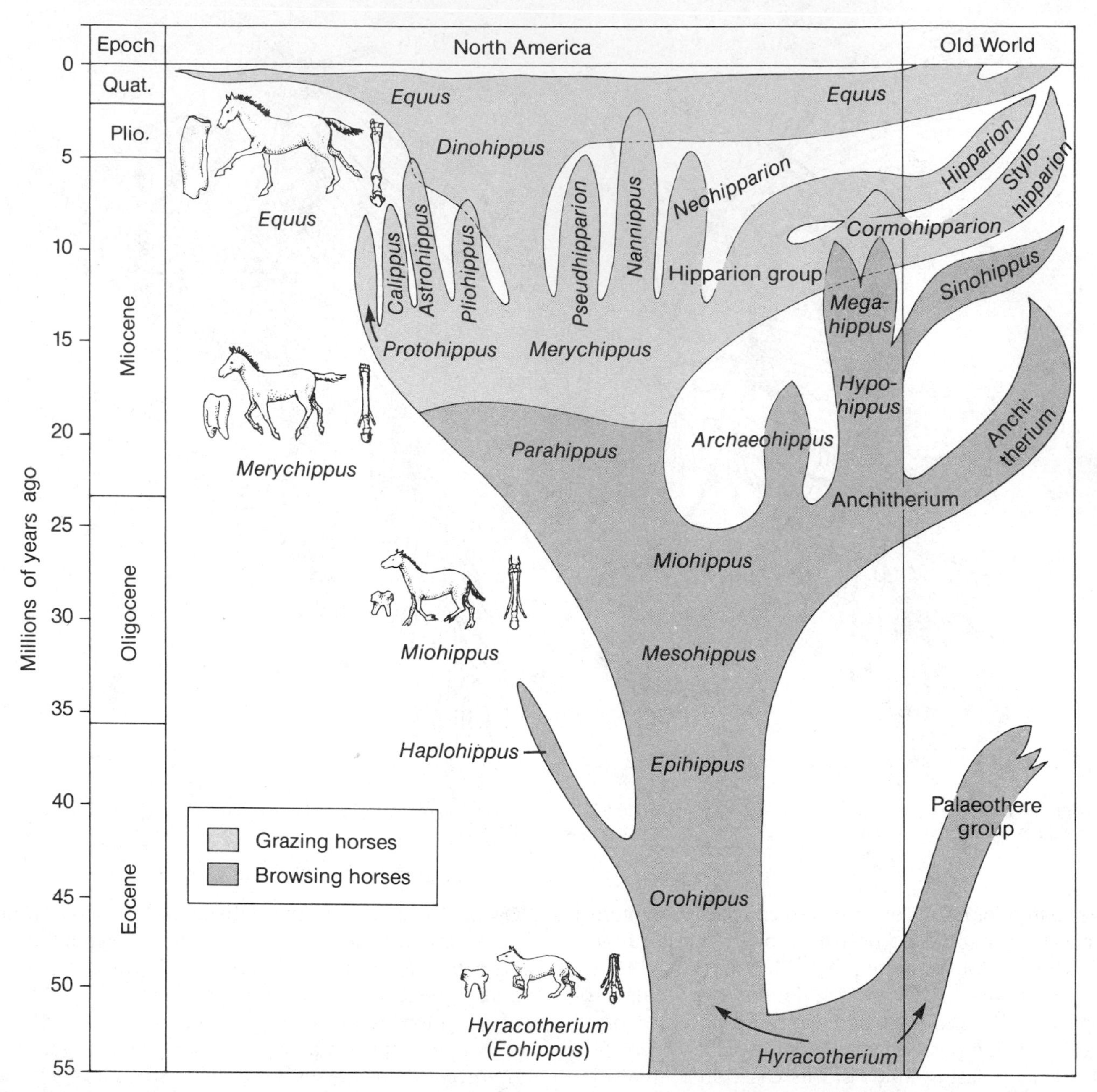

FIGURE 3-13 Evolutionary relationships between various lineages of horses, with emphasis on North American and Old World groups. Sample reconstruction of some fossil horses and modern *Equus* are shown, including diagrams of their molar teeth on the left side of the animal and bone structure of hind feet on the right side. (Adapted from MacFadden, with additions.)

than the enamel, enabling sharp grinding edges of enamel to remain above the cement.

Remarkably, almost all the intermediate stages between *Hyracotherium* and the modern horse, *Equus*, have become known: from low-crowned to high-crowned, from browsers to grazers, from pad-footed to spring-footed, and from small-brained to large-brained forms. However, as shown in Figure 3-13, the evolutionary progression among these forms is not a straight-line development. Horses evidently evolved adaptations for their habitats in different ways, with some individual branches maintaining fairly uniform structures until they become extinct. The rate of evolution for any particular trait among the various branches was also not constant. Size, for example, underwent relatively few changes for both the first 30 million years and for the last few million years (MacFadden). According to paleontologists, this finely detailed phylogeny encompasses hundreds of fossil species and is one of the best illustrations of some of the realities and complexities of evolution.

Interestingly, not all ancient organisms have been

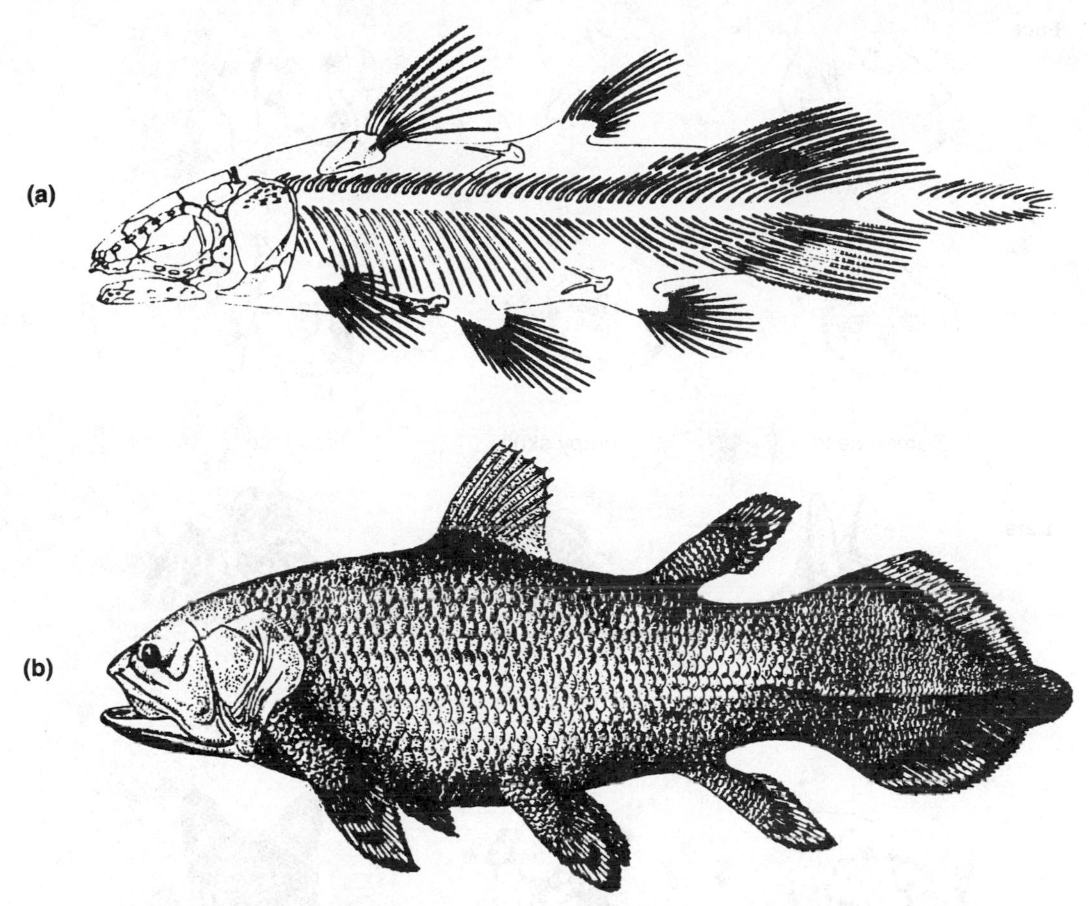

FIGURE 3-14 (a) Fossil lobe-finned coelacanth from the Triassic period, about 200 million years ago (from Carroll, adapted from Schaffer). (b) Modern coelacanth (*Latimeria chalumnae*) found off the eastern coast of South Africa.

replaced. **Living fossils** that have undergone long evolutionary histories with relatively little visible change include opossums, alligators, sturgeons, lungfish, horseshoe crabs, *Lingula* brachiopods, and ginkgo trees. Occasionally, species are discovered that are remarkably similar to organisms that were formerly believed to have become extinct many ages ago. For example, coelacanths are ancient lobe-finned fishes (Fig. 3-14) related to those which evolved into terrestrial vertebrates about 200 million years ago (Chapter 16). Although the fossil record of coelacanths seemed to have begun in the Devonian period about 380 million years ago and ended 80 to 100 million years ago, recent live coelacanths (*Latimeria chalumnae*) have been found in deep waters off the eastern coast of South Africa. Similarly, a very ancient form of segmental mollusc (*Neopilina*) believed extinct since the Devonian period, has been found in deep sea trenches off the coast of lower California, Costa Rica, and Peru. Although these findings support the validity of paleontological claims that fossils indicate the existence of real organisms, they also show the inadequacy of the paleontological record; that is, the disappearance of fossils of a particular type in the fossil record does not necessarily mean that this type immediately became extinct. However, aside from such rare "living relics," fossils are in almost every instance different from present forms, and the more recent geological strata generally show forms more like the present than do the older strata. Taken as a whole, fossil evidence therefore provides strong support for evolution.

Artificial Selection

To support his concept of evolutionary change, Darwin discussed a number of examples of evolution by selection, although the selection method involved human choice rather than natural events. In artificial selection as practiced before Darwin's time and later, the breeder, whether of dogs, cats, pigeons, cattle, horses, or whatever, selects the parents deemed desirable for each generation and culls, or destroys, the undesirable types. Since the selected parents may produce a variety of different offspring, the breeder can usually continue

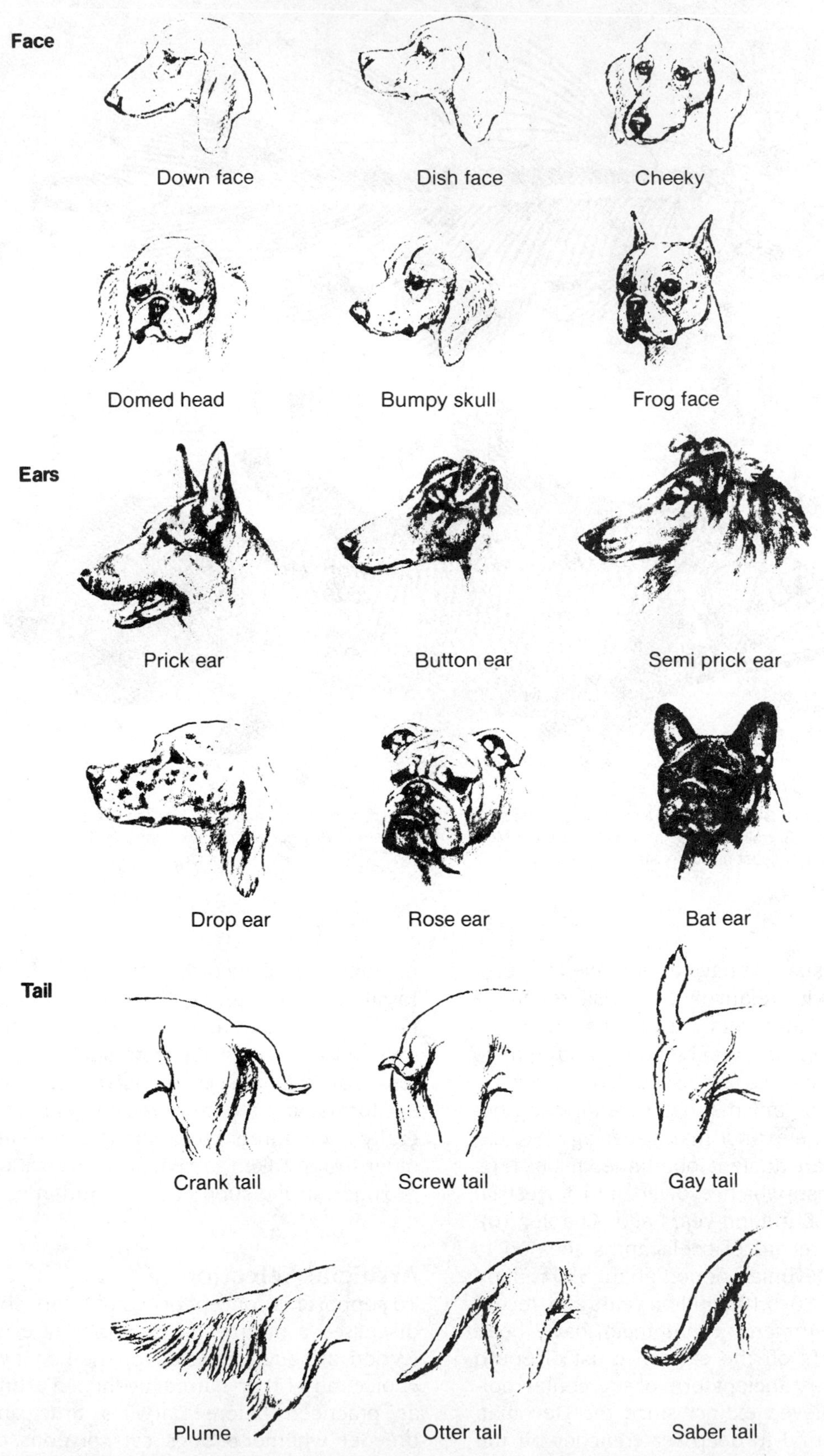

FIGURE 3-15 A few of the characters selected in various breeds by dog fanciers. (American Kennel Club, *The Complete Dog Book*, rev. ed. Garden City Books, Garden City, N.Y.)

to select in a particular direction until the traits in which he is interested appear considerably different than they did initially. Dogs, for example, have been selected by humans for thousands of years and now range in size from the St. Bernard to the chihuahua, and in features from the greyhound to the bulldog (Fig. 3-15). Pigeons, long bred by fanciers, now show a wide variety of beaks, shapes, and feathers. The same is true for sheep, cattle, and all the many different agricultural species of plants and animals. Artificial selection thus demonstrated to Darwin and his colleagues that continued selection was powerful enough to cause observable changes in almost any species. The Darwinian claim that natural selection for particular environments could do even more than artificial selection and lead to speciation seemed therefore a reasonable hypothesis, given the much longer periods of time in evolutionary history and the "unrelenting vigilance" of natural selection.

SUMMARY

The publication of *The Origin of Species* in 1859 aroused a variety of objections against Darwin's theory of evolution by natural selection. The prevalence of the idea that an individual's traits were a blend of those of the parents' made it difficult to see how traits could be selected for, since any trait would be diluted during successive generations by mating with non-adapted individuals. Darwin's insistence that the variations acted on by evolutionary forces are small and continuous led to criticism that such slight variations could not be recognized by selection or lead to the formation of new species. An additional question concerned the appearance of new traits, which Darwin apparently thought might form by preadaptation (the conversion of some structure already present to a new use). While selection might account for the evolution of a species into another (phyletic evolution), it was more difficult to understand how many species might arise from a single one. To account for this factor it was necessary to emphasize the isolation of groups from each other, but Darwin did not do so. Finally, there did not seem to be sufficient time for evolution to occur, since the great antiquity of the earth had not yet been established.

Nevertheless, Darwin's theory had the advantages of relying on understandable mechanisms, of being supported by much data, and of successfully explaining many features of natural systems. The transitional forms observed by taxonomists could be explained on this basis. Evolutionary theory could also explain the fact that organisms with similar characteristics are found geographically close to each other, whereas groups separated by geographical barriers have fewer characteristics in common. In comparative anatomy, structures found in different organisms but having an underlying similarity of plan could be explained by their relationship through a common ancestor (homology). Vestigal structures such as the rudiments of limb bones in snakes and whales would be the remnants of organs which had become obsolete. The similarities among vertebrate embryos during early developmental stages also suggested a common evolutionary past. Fossil evidence was particularly important. Some fossil forms intermediate between living groups (such as the bird-reptile *Archaeopteryx)* were found. But the horse provided perhaps the most complete and continuous fossil record of evolution of a group. It had evolved from a small, four-toed, browsing animal to a large, single-toed creature with hard teeth adapted to chewing tough grasses. Finally, Darwin recognized that man, while choosing to perpetuate favorable traits in domestic plants and animals, could produce over a relatively short period of geological time radical alterations in organisms by selection—albeit artificial rather than natural. The heavy accumulation of data made Darwin's claims for evolution appear reasonable to many scientists.

KEY TERMS

adaptive radiation
analogous
Biogenetic Law
comparative anatomy
convergent evolution
geographical distribution
homologous
isolation
living fossils
preadaptation
variability
vestigial organs

REFERENCES

Ali, M. A., 1984. *Photoreception and Vision in Invertebrates*. Plenum Press, New York.

Appleman, P. (ed.), 1970. *Darwin: A Norton Critical Edition*. W. W. Norton, New York.

Badash, L., 1989. The-age-of-the earth debate. *Sci. Amer.*, **261**(2), 90-96.

Barnett, S. A. (ed.), 1958. *A Century of Darwin*. Heinemann, London.

Carroll, R. L., 1988. *Vertebrate Paleontology and Evolution*. W. H. Freeman, New York.

Conn, H. W., 1900. *The Method of Evolution*. Putnam's, New York.

De Beer, G., 1966. *Vertebrate Zoology*. Sedgwick & Jackson, London.

Ghiselin, M. T., 1969. *The Triumph of the Darwinian Method*. Univ. of California Press, Berkeley.

Gould, S. J., 1977. *Ontogeny and Phylogeny*. Harvard Univ. Press, Cambridge, Mass.

Haber, F. C., 1959. *The Age of The World*. Johns Hopkins Press, Baltimore.

Haeckel, E., 1866. *Naturliche Schöpfungsgeschichte*. Reimer, Berlin.

——, 1905 (translated from the 5th German ed. by J. McCabe). *The Evolution of Man*. Watts & Co., London.

Hull, D. L., 1973. *Darwin and His Critics*. Harvard Univ. Press, Cambridge, Mass.

Lack, D., 1947. *Darwin's Finches*. Cambridge Univ. Press, Cambridge.

LeConte, A., 1888. *Evolution, Its Nature, Its Evidences, and Its Relation to Religious Thought*. Appleton, New York.

MacFadden, B. J., 1988. Horses, the fossil record, and evolution. *Evol. Biol.*, **22**, 131-158.

Romanes, G. J., 1910. *Darwin, and After Darwin*. Open Court Publ. Co., Chicago.

Rudwick, M. J. S., 1973. *The Meaning of Fossils*. Macdonald, London.

Ruse, M., 1979. *The Darwinian Revolution*. Univ. of Chicago Press, Chicago.

Schluter, D., 1984. Morphological and phylogenetic relations among the Darwin's finches. *Evolution*, **38**, 921-930.

Simpson, G. G., 1951. *Horses*. Oxford Univ. Press, New York.

Vorzimmer, P. J., 1970. *Charleś Darwin: The Years of Controversy*. Temple University Press, Philadelphia.

Wolken, J. T., 1986. *Light and Life Processes*. Van Nostrand Reinhold, New York.

THE DARWINIAN IMPACT: EVOLUTION AND RELIGION

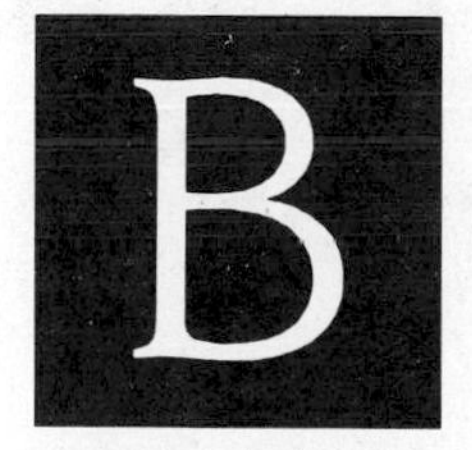

By making evolution an acceptable concept, Darwin's impact was profound. Darwinian evolution offered a vast historical framework in which biological change could be understood; it made clear that species fixity was not at all "natural"; and it proposed that the form and function of living organisms did not arise by creation but by selection. These were revolutionary ideas in the nineteenth century, and they not only revolutionized biology but also affected fields such as sociology (Herbert Spencer), anthropology (Lewis Henry Morgan), economics (Karl Marx, Thorstein Veblen), politics (Walter Bagehot), fiction (Joseph Conrad, Jack London, Jules Verne, Theodore Dreiser, H. G. Wells), poetry (Robert Browning, Alfred Tennyson, Walt Whitman), linguistics (William Dwight Whitney) philosophy (Charles S. Peirce, John Dewey, Henri Bergson), and psychology (William James, Sigmund Freud). The impact of Darwinism, however, was most dramatic in respect to religion. To many of Darwin's religious contemporaries and to others since, *The Origin of Species* as well as *The Descent of Man*, which Darwin published in 1871, raised controversial matters of vast proportion. It is, in fact, reasonable to say that one of the chief popular issues of the late nineteenth and early twentieth centuries was the struggle over the acceptance of evolution, with many scientists arrayed on one side and religionists on the other.

The RELIGIOUS ATTACK

A small model of the battles to come took place at Oxford soon after publication of *The Origin of Species*. In this debate Bishop Samuel Wilberforce (1805-1873) of the Anglican Church (Fig. 4-1) attacked Darwinian theory as incompatible with the Bible, and, coached by Richard Owen, a former student of Cuvier, attempted to destroy it through scientific arguments. Wilberforce's final point was made directly to the Darwinian defender, Thomas Huxley, when he asked whether it was through his grandfather or grandmother that Huxley claimed descent from a monkey. The wit

FIGURE 4-1 Caricatures of Bishop Samuel Wilberforce (left) and Thomas Huxley that appeared in the British magazine *Vanity Fair* some years after their debate at Oxford University.

of Huxley's response, recounted in a letter to a friend, has often been quoted:

> If, then, said I, the question is put to me would I rather have a miserable ape for a grandfather, or a man highly endowed by nature and possessed of great means and influence, and yet who employs these faculties and that influence for the mere purpose of introducing ridicule into a grave scientific discussion, I unhesitatingly affirm my preference for the ape.

Many theologians, however, were unimpressed with scientific arguments and continually hammered away at the heresy of evolution. Wilberforce accused Darwin of "a tendency to limit God's glory in creation." Cardinal Manning, a leader of English Catholicism, called Darwinism "a brutal philosophy—to wit, there is no God, and the ape is our Adam." The religious attacks were worldwide, frequent, harsh, and almost always focused on the same points. That is, Darwinists were accused of attempts "to do away with all idea of God," "to produce in their readers a disbelief of the Bible," and "to displace God by the unerring action of vagary."

If these claims were true, and there were good indications that there was more than reasonable cause for religious alarm, then one could well ask how such a heretical doctrine could have been developed in the midst of a religious European country and become acceptable to so many of Darwin's learned compatriots. Among the answers to this is one which helps to see the struggle between evolution and religion not as accidental but part of the historical framework of the time.

From a social point of view, we have already observed that the development of evolutionary theory was one aspect of that all-pervasive political and economic revolution in social behavior and thought which began with the overthrow of the rigidly ordered feudal class structures that had prevailed in Europe until the rise of capitalism. The economic challenges posed by capitalism and its new monied classes helped in many ways to allow ideological challenges to be posed to the prevailing religious and philosophical systems that had been the support of the old social order. The divine right of kings, for example, had to be overthrown socially, philosophically, and religiously. Without such changes, there is little question in the minds of many historians that European science could not have flourished as it eventually did.

However, not too surprisingly, capitalism in its triumph also sought ideological justifications for its power. As in feudalism, many of these justifications were based on religious concepts, and it was fairly common to find appeals to the inherent hierarchy of nature and the wisdom of the creator. In the words of Alexander Pope (1688-1744),

> Order is Heaven's first law, and this confest,
> Some are, and must be, greater than the rest,
> More rich, more wise.

Or, as stated in Darwin's century by C. F. Alexander in her hymn "All Things Bright and Beautiful" (1848):

> The rich man in his castle,
> The poor man at his gate,
> God made them, high or lowly,
> And order'd their estate.

Looked at historically, we can see that evolution, by threatening basic religious concepts of a fixed universal order, seemed to have exceeded the "game plan" for permissible ideological challenges. Nevertheless, it is interesting to note that there was no sustained political attempt to suppress evolutionary ideas, although they were attacked far and wide and even outlawed in some American states. One reason for the relative freedom which evolution enjoyed is probably its close ties with all other aspects of science. Science and technology were, after all, the mainstay of economic expansion. Many scientists were or became evolutionists, and it must have seemed to most social leaders of the time that evolution was a scientific foible that would have to be tolerated.

However, to many religionists, as we have seen, evolution was a deep, abiding threat. There was considerable cause for this concern, and in order to understand this it is valuable to review the development of those aspects of religion which were most directly threatened by evolution.

BASIS OF RELIGIOUS BELIEF

At present, different levels of religious development can be found in various cultures. Through the efforts of anthropologists and psychologists these cultures have provided considerable clues to the evolution of religion itself. In general, religion first develops in a culture as a form of behavior through which humans attempt to deal with those aspects of their experience which they cannot control or understand. Religions have apparently been sustained from their origins by the common feeling that what is outside one's control may nevertheless be humanlike and therefore subject to appeal and thus indirectly to control. In the earliest stages of religion, the forces of nature are directly endowed with the spirits of animals and humans. In the attempt to control these forces by sympathetic magic, particular events are enacted by imitation—hunting dances, rain dances, and so on—so that they will be guided towards desired ends. The development of ritual occurs when magical ceremonies are repeated to help ensure their efficacy. Ritualized behavior seems to have become especially important in the transition from hunting to agricultural societies in which crops had to be planted and harvested at appropriate seasons each year and where one's efforts could be either rewarded or condemned by forces which remained mysterious and arbitrary. In his advice to farmers, Hesiod (800 B.C.) wrote:

> Everyone praises a different day, but few fully understand their nature. One day may be like a stepmother, another like a mother. A man will be happy and lucky if, having an eye to all of these things, he completes his work without offending the gods, reads the omens of the birds, and avoids all transgressions.

Modern humans can ill afford to laugh at such beliefs. The uncertainty, hope, and wooing of good fortune expressed by Hesiod are, after all, basic feelings shared by many. They are just more baldly stated here than in the more sophisticated forms in which we usually acknowledge them.

Underlying many human anxieties about the world, and often at the base of the difficulty in treating nature as an independent object of study, was the feeling that it reflected, through punishment and reward—that is, through calamity and prosperity—a supernatural evaluation of human affairs. Since the forces of nature were seen as humanlike, the hope was sustained and encouraged by religion that nature's recrimination and punishment could be propitiated by those appeals which humans understood: submission, supplication, gifts, sacrifice, obedience, and loyalty.

In support of a supernatural view of events, religion has relied on two basic concepts that probably arose early in history, the **soul** and **God**. The soul, or the spirit, or what many would call the **personality**, is considered in most religions to be a humanlike, conscious entity without physical attachments or properties. According to various psychologists and anthropologists, it seems likely that the idea of the separation of the soul from the body has its origin in the separation of the mind from reality in dreams. The restrictions of space, time, and even death vanish in dreams, and one can then suppose that there is an aspect of human life, the soul, which is immune to such restrictions.[1] This attitude is reinforced by the fact that it is difficult to conceive that one's personality has only a limited existence. That is, it is a formidable task to realize that the mental organ of sensation and feelings which a person uses in perceiving his relationship to his environment, and through which he integrates, evaluates, and interacts with the world about him, has a beginning and an end. Although one can intellectually conceive of death, related to other creatures or phenomena or even to one's own body, it is impossible to "feel" death as one may feel and anticipate other events and sensations. To a human, and perhaps to other creatures as well, this organ of awareness and feelings, the personality, seems therefore to be eternally present since there is usually no self-knowledge of how it arrived or of how it ends.

Theologians have often attempted to provide a "scientific" argument for the non-material nature of the soul or the personality, claiming that intellectual processes cannot have a material origin. According to their

---

[1]One of the nineteenth-century founders of anthropology, E. B. Tylor, explained the soul concept among primitive tribes as follows:

> As it is well known by experience that men's bodies do not go on these excursions, the natural explanation is that every man's living self or soul is his phantom or image, which can go out of his body and see and be seen itself in dreams. Even waking men in broad daylight sometimes see these human phantoms, in what are called visions or hallucinations. They are further led to believe that the soul does not die with the body, but lives on after quitting it, for, although a man may be dead and buried, his phantom-figure continues to appear to the survivors in dreams and visions.

A description of these phenomena is given in greater detail by La Barre.

argument, freedom of choice, ethics, and all the many complex ideas which humans can think about must be considered the fruits of non-biological matter since they are so obviously separate from purely physical processes. However, the biological view is quite different: while it is true that we do not yet know the precise relationship between the matter of the brain (neurons, synapses, etc.) and the thoughts and feelings it produces, the fact that such a relationship exists is no mystery. Many creatures have thoughts, some even have personalities, and some even have dreams. While none of the thoughts in most creatures are probably as complex as those in primates—and among primates, none are apparently as complex as those of humans—there is every reason to believe that the complexity of thinking and feeling has evolved like any other trait.

Like the concept of the soul, the concept of God has many qualities that reflect human experience. Each of the various gods personify, often in human form, forces or tasks which seem beyond the capabilities of humans, in the fashion of a powerful parent seen through the eyes of a child. How better for primitive people to alleviate anxiety and fears about harvests, thunder, fertility, woods, rivers, and so on than to believe that these elements of nature embody extensions of human behavior? The progression from individual gods for each element of nature to gods that can be ranked in respect to their power and then to a God of gods, such as Jove, Jehovah, Allah, Brahma, and others, was only a succession of steps. At the heart of religion is therefore the feeling of reliance and dependence on what is held to be, and what is certainly desired to be, a wise and paternalistic nature. Different religions endow these feelings with structures derived from their own societies, often with considerable imagination. It is, for example, doubtful that reality could produce pleasures and torments as great as those provided by the imaginations of believers in heaven and hell.

CHALLENGES TO RELIGION: THE QUESTION OF DESIGN

It is easy to see that as long as active intervention by a divine power was believed necessary to explain most or all observed events and to allow the continued maintenance of the universe, it inhibited the attempt to discover natural laws that could account for changes in the present, past, or future. The first significant cracks in the theological armor of continued divine intervention in nature were made by Copernicus (1473-1543), Galileo (1564-1642), and Kepler (1571-1630) in their discoveries of natural laws regulating the motion of the solar system. These openings were considerably widened by the mechanistic explanations offered by Newton (1642-1727) on the motion of the solar system through the force of gravity and later by geologists such as Lyell who proposed how natural forces could be used to explain the appearance of the earth's surface. Although these scholars were not atheists, their findings about natural processes and other findings made through mechanics, optics, chemistry, and so on helped reduce God from a continually active, intervening agent to a prime force more like a master craftsman who has designed logically contrived, self-functioning machines.[2] This helped modify man's attitude towards the divine establishment from simple fear and subjection to the more comfortable attitudes of admiration and respect, but it also led to a closer examination of the nature of God, with many ensuing contradictions. If the universe is logical what was the logical purpose in creating it? An arbitrary God such as Jehovah obviously needs no understandable reason for creation, whereas the motivation of a logical and rational God is implicitly questionable. Was the motivation the pride of reaping adoration from his human subjects as they admired his works? Could God be vain? Was the motivation to create perfection? Did God need testimony for his perfect attributes? How could imperfection and evil arise from perfection? And, most pernicious of all paradoxes, if the world was not created perfect, what moral good could there have been in its creation?

However challenging these questions were to religion, they were not as damaging as the frontal attack made by Darwin and his evolutionists. Darwin's works made clear that it was no longer necessary to believe that biological relationships must be explained by the actions of a supernatural creator. Instead, he presented the notions that there are in nature continual change, unpredictable chance events, an unrelenting struggle for survival among living creatures, and no obvious guidance. He thus replaced what had seemed to many to be an understandable view of nature—that is, the creativity of a humanlike God—by the most heretical concepts of all, randomness and uncertainty, or the fear that now nothing can really be understood. An example of how essential it was to believe that each design had a creative purpose is reflected in the statement of Paley (1743-1805):

[2]In the words of one historian (Bernal), "God had, in fact, like his anointed ones on earth, become a constitutional monarch."

> There cannot be design without a designer; contrivance without a contriver; order without choice; arrangement without anything capable of arranging.... Arrangement, disposition of parts, subservience of means to an end, relation of instruments to a use, imply the presence of intelligence and mind.[3]

What this means in terms of the origin of humans is, of course, simply this: that the "silk purse," which is man, or his soul or personality, could not have been made out of a "sow's ear," that is, the rest of his animal kin, without the active intervention at some point, early or late, of an intelligent, sympathetic deity. This attitude gave man a reason for his creation—he was not born in vain. It also helped mollify the most unkind cut of all to an intelligent, sensitive creature—that he might die in vain—because the designer who gave him life was also responsible for his death and for the immortal preservation of his soul.

These arguments of supernatural design also seemed to be supported by what was conceived as simple common sense. Can there be a watch without a watchmaker, and by extension, can there be a man without a manmaker, laws without a lawmaker? To evolutionary theory, the essential challenge posed by religion has always been therefore, How from the disorder of random variability can nature achieve the beauty of adaptation without intelligent intervention? Darwin's contribution was to answer this question by means of a phenomenon which no one had thoroughly explored before—natural selection. We can illustrate the evolutionary view of selection in simple form as follows.

Although there are chance events in evolution it is primarily a historical process. That is, what evolves depends upon what has evolved before. Thus, no complex structure arises all at once by a lucky combination of events, but evolution builds and perfects new structures from old ones. For example, if one had a very large bowl full of ten different letters (A, C, E, I, L, N, O, T, U, V) with each letter present in equal frequency (Fig. 4-2), there are many millions of ways that nine letters randomly drawn from this bowl can be arranged. The chances of getting the exact word *EVOLUTION* from a random draw of nine letters is obviously small [$(1/10)^9$].

If we assume, however, that a selection mechanism exists which will perpetuate certain adaptive

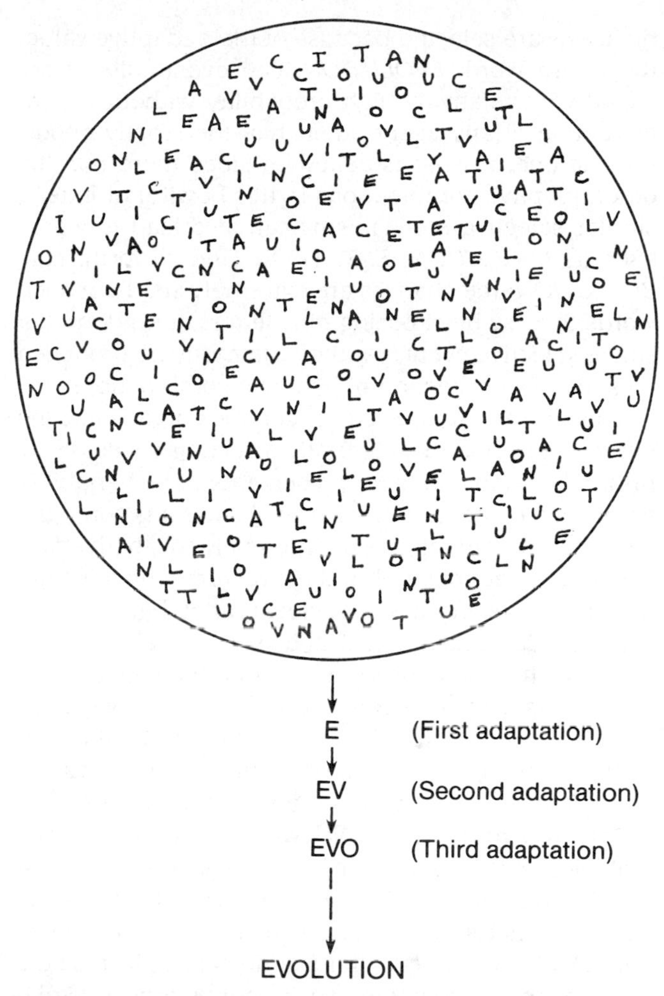

FIGURE 4-2 An illustration of how evolution can produce a "complex" structure (in this case, a word) by a succession of selective steps. If we assume a bowlful of ten letters (A, C, E, I, L, N, O, T, U, V) with each letter present (or generated by mutation) in equal frequency, there is a considerable chance of obtaining the word *EVOLUTION* within a short period if each letter is chosen one at a time and each new sequence of letters (e.g., *E, EV, EVO, EVOL,* etc.) is adaptive. That is, a succession of adaptive evolutionary events occurs, each stage having a reasonable probability (1/10).

combinations, then the "evolutionary" attainment of the word *EVOLUTION* is far greater than on first appearance. Thus, let us assume that *E* is the only letter which can survive by itself. The chance of drawing an *E* from the bowl is therefore 1/10. Let us assume next that a *V* in combination with the *E* is adaptive; then the chance of now drawing *V* is 1/10. Similarly, if we assume that the next adaptive combination consists of adding an *O* to *EV*, and that further successive com-

---

[3] Erasmus Darwin, Charles' grandfather, expressed this view as follows:

> Dull atheist, could a giddy dance
> Of atoms lawlessly hurl'd
> Construct so wonderful, so wise,
> So harmonised a world?

binations are selected because of their adaptive value, the entire word, *EVOLUTION*, can eventually be selected with relatively high probability without the intervention of any agent other than the strictly opportunistic one of what is adaptive. Since there may be other adaptive combinations in this bowlful of letters, similar selective mechanisms can also lead to words such as *EVOCATION*, *ELEVATION*, and so forth. Like *EVOLUTION*, the chances are extremely small for such words to arise by choosing nine letters at random in a single selective event, yet they can easily be produced by successive selection of adaptive combinations.

Evolution can thus create complex organs like the eye and the brain by naturally selecting successively improved adaptations for preservation. The variability upon which the choice is exercised is random (the letters, or mutations, are of different kinds, both adaptive and nonadaptive), but the structure which is built over many generations of selection has been historically molded and created and is not at all random. Thus, in the words of the population geneticist R. A. Fisher, "natural selection is a mechanism for generating an exceedingly high degree of improbability." Or to use a metaphor of Monod, mutation provides the noise from which selection draws out the music.

To repeat this in a somewhat different way: The biological changes produced by selection can occur whether humans are the selective agents (artificial selection) or nature is the selective agent (natural selection). Artificial selection, for example, has given us the many forms of dogs created by humans over a relatively short period of time from the variability present in that particular species. Natural selection, on the other hand, acting over much longer periods of time, has given us the various species we see, all created from the variability that accumulated in many different prior species. These prior forms, in turn, were selected from previous forms going back to those first primeval organisms of billions of years ago which could transmit their characteristics to their progeny. To the Darwinians, everything was accidental in the sense that hereditary variables arose at first randomly, and yet everything that survived was adaptive in the sense that only adaptive combinations of these random variables could perpetuate themselves. It was clear that purposeful function need not be purposefully designed.

The fear that Darwinism was an attempt to displace God in the sphere of creation was therefore quite justified. To the question, is there a special purpose for the creation of humans, evolution answered no. To the question, is there a special purpose for the creation of any living species, evolution answered no. According to evolution, the adaptations of species and the adaptations of humans come from natural selection and not from design. The properties that species possess do not exist as ideas or teleological causes prior to their evolution (except, of course, for those types of organisms for which man himself has now become the designer).

Surprisingly, the conflict between these seemingly irreconcilable points of view on the same subjects, the origin of species and the origin of humans, was not followed by either the defeat of religion or the defeat of evolution. Evolution suffered almost no loss at all. In fact, tied to the scientific method and supported by it, evolution became the main unifying force in biology and led to an expansion of research in almost every area. Although religious critics continued to argue that evolution could not be "seen," the evolutionary view became as strongly entrenched in biology as the atomic, molecular, and gravitational theories were in the physical sciences. None of these principles were seen, yet they were all working models indispensable to a consistent and scientific comprehension of events.

Sources for the Preservation of Religion

One might therefore think that, through the closely reasoned arguments offered by biologists, evolution had successfully undermined religion's prime justification for itself, the special creation of humans. The demise of religion should therefore have been just a matter of a short time. It turned out that this was not so, and some of the reasons for this are worth looking at.

First of all, the change from regarding nature as operating in terms of a kind of **anthropomorphic wisdom** to operating in terms of what the evolutionists viewed as a kind of **opportunism** expressed by the survival of the fittest was certainly most difficult for people raised in the bosom of paternalistic social structures and religious beliefs. For many centuries it was, after all, easy for many to follow a pattern of simple parallels between society and nature: the King rules, the Pope rules, God rules. It is hoped that each rules with wisdom, concern, and sympathy, but in all cases humans are dependent on the beneficences of their rulers and try to influence their judgements by special pleas, sacrifices, and rituals. Even for those who questioned the legitimacy of kingly rule over humans, it must have seemed far too audacious a leap to question the rule of nature by the wisdom of God. Even the philosopher Kant, who had an evolutionary approach to cosmology and to the origin of the solar system, found it at times abhorrent to admit that species could evolve and in one place described such notions as "ideas so monstrous that the reason shrinks before them." Darwin too is reported to have felt quite uncomfortable about his role in proposing the evolution of species and wrote in an 1844 letter to Joseph Hooker that "it is like confessing a murder."

We must remember that one source for these uncomfortably guilty attitudes was the staid, conservative milieu which held strongly to the notion that man was created in the image of God and endowed with the rule over other biological and social groups below him. Kinship to those below, whether ape or servant, was a repugnant idea. An oft-reported example of this repugnance is the response of the wife of the Bishop of Worcester when informed that Huxley had announced that man was descended from apes: "Descended from apes! My dear, let us hope that it is not true, but if it is, let us pray that it will not become generally known."[4]

Reinforcing these attitudes was the reaction of many of the European middle classes to the upheavals of the French Revolution and to the atheism of many of its leaders. (Darwinism, according to the critic St. George Mivart, led to "horrors worse than the Paris Commune.") These reactions probably helped to strengthen the religious climate among the middle classes in the nineteenth century and led, in various groups, at least for a time, to a more literal interpretation of the Bible.

Essential to the preservation of religion in the midst of the evolutionary bombardment was also the fact, as pointed out previously, that religion answers a series of strong emotional needs. We can identify these as the need to feel that one's life has a purpose; the need to feel that there is a powerful force ("someone up there") who is protective and upon whom one can depend, but who, in any case, can be appealed to; and the need, in some way, to fill in the important personal mysteries of birth and death (the question of identity, "Who am I?" and "Why me?"). Furthermore, religion has traditionally served as the main repository for the ethics and morals of society (what is right and wrong) and has also often served to maintain confidence in the social order and to unify nationalistic and provincially chauvinist sentiments ("God is on our side"). Evolution, on the other hand, deals with many basic questions of life that are of concern to religion, but as a science it did little to meet the emotional needs that were dealt with by religion until psychotherapeutic methods began to be developed by Freud and others. It is important to realize that as fears are diminished through knowledge in any particular area, the emotional needs that spring from such fears are usually also diminished. Unfortunately, the emotional feelings of humans arise from interaction among so many genetic, developmental, and social factors that fulfillment of human emotional needs is still quite difficult to achieve.

THE "TRUCE"

Society has therefore held on to both of these concepts to this day with a sort of armed truce between them. Religion, with the exception of some fundamentalists, essentially withdrew from the domain of biological evolution, leaving both the origin of species and the origin of humans in the hands of the evolutionists. The Judeo-Christian Bible was reinterpreted to permit evolution by either ignoring the creation story in Genesis or considering it allegorical. There was considerable sacrifice in such reinterpretations since, like any other religious document, the Bible is an attempt to explain the unknown in a religious framework. To concede that parts of the Bible can be known and understood outside the religious framework easily opens the door to further loss of religious credibility and was an important nineteenth-century cause for generating dissension among theologians (see, for example, Roberts). Among various compromises attempted was that by Asa Gray (1810-1888), the American evolutionist, who proposed that the variability upon which natural selection acts during evolution was itself specially created. To this Darwin replied that since not all variations are useful, it is inconceivable that evolution could be "designed" by such means.[5]

Some nineteenth-century writers, such as Owen, Campbell, Mivart, and others suggested that although some adaptations may have been caused by natural selection, many basic generalized patterns such as the vertebrate archetype and the parallel appearance of similar organs (e.g., eyes) in different groups, must have been caused by design. Essentially, these authors moved the hand of the "designer" from specifying minor adaptations for species to devising major plans for higher taxonomic groups. Most biologists, nevertheless, held to the concept that evolution operated on all levels and its mechanisms were sufficient to account

[4]It is ironic that many Western religionists were affronted by the concept of man's ascent from apes because it gave man a lowly ancestor, but were not offended by the concept of man's fall from the "state of blessedness" in the Garden of Eden, although that meant that humans live in a state of sin and relative degradation. Certainly, to arise from the beasts is a nobler attainment than to fall from the gods. What was at issue, of course, was not which of these positions was relatively higher or lower but which one allowed man the comfort of thinking he was a sort of angel, although a fallen one. Better, it seemed, to be a fallen angel than a risen beast!

[5]In a letter to Lyell, Darwin wrote:

> If you say that God ordained that at some time and place a dozen slight variations should arise, and that one of them alone should be preserved in the struggle for life and the other eleven should perish in the first or few generations, then the saying seems to me mere verbiage. It comes to merely saying that everything that is, is ordained.

for the similarities and differences among major groups. By the beginning of the twentieth century, arguments of supernatural design in biology were rarely used by biologists themselves and were considerably less popular in intellectual circles than they had been before.

However, although religion compromised or retreated in various places, it held fast to the domain of ethical and moral concepts by insisting that how humans should behave and the sanctity and the meaning of what has evolved (e.g., human personality, identity, and social relationships) can be considered apart from how human biology came into being. To varying extents, such religionists therefore share some of the reasonable claims made by cultural evolutionists that the means and goals of cultural evolution can differ from those involved in biological evolution (Chapter 24).

Challenges to religion have nevertheless continued; once the notion of fixity and design was removed from such a sacred concept as species and their creation, only one step further was needed to question and investigate the fixity and design of religion itself. Usually, the awe and mystery that surrounds an institution is diminished in direct proportion to "earthly" knowledge of its origin and development. Religion is no exception to this, and a considerable amount of its power, which depends upon awe and mystery, has been seriously undermined by the evolutionary approach expressed through various natural and social sciences. To evolution, nothing is necessarily permanent, even society and morality. In fact, it has been common from the time of Darwin for sociologists and anthropologists to point out that different standards of morality develop in different societies because of historical social reasons. Evolutionists such as Huxley and others extended these concepts to show that the types of morality that would be considered desirable in Western industrialized societies could be derived from human rational thinking and were not at all dependent on religious beliefs. This questioning evolutionary climate has undoubtedly been fed by the active social, scientific, and technological changes that have accelerated rapidly and radically in this century.

RELIGIOUS FUNDAMENTALISM AND "CREATION SCIENCE"

The uneasy truce between evolution and religious institutions in the Western world has generally not been accepted by fundamentalist religious groups. In the United States, many individuals who believe in the Judeo-Christian account of creation in the book of Genesis have banded together to form political pressure groups to impose their beliefs on public education. Their origins date back at least to the early 1900's when many fundamentalists were part of an anti-intellectual and anti-establishment populist movement with strong roots among economically threatened tenant farmers and small landholders, especially in the South and Southwest. It has been suggested by sociologists that a belief in the literal truth of the Bible, revivalism, and other aspects of fundamentalist religion popularized in a series of pamphlets called *The Fundamentals* served many of these rural groups as a means of defending their way of life against domination and control by the more intellectual but exploitative Northern and Eastern social and economic establishments.

Whatever their initial motivations, fundamentalist groups were fairly successful in pursuing their anti-evolution goals in the South and Southwest during the first few decades of this century. By the end of the 1920's anti-evolution bills had been introduced into a majority of American state legislatures, and some had been passed in various Southern states. Probably the most famous confrontation between evolution and biblical creationism during that period was the 1925 trial of a schoolteacher, John Scopes, who was convicted of ignoring the ban against teaching evolution in Tennessee schools. Although many evolutionists felt that the trial essentially defeated the intellectual validity of the creationist position,[6] creationists apparently lost little ground in these regions and managed to have an impact on public education far beyond the South and Southwest. As Nelkin points out, by influencing textbook adoption procedures in various local and state school boards, creationists successfully minimized evolutionary explanations in secondary school textbooks for a long time. One textbook publisher baldly stated, "Creation has no place in biology books, but after all we are in the business of selling textbooks"—a view reflected in the downgrading of evolution by other textbook publishers. By 1942, more than 50 percent of high school biology teachers throughout the country excluded any discussion of evolution from their courses.

The impetus for an increase of evolutionary teaching in secondary schools was the result of a movement to reform the science curriculum in the late 1950's and early 1960's, when it was realized that much of American science education lagged behind that of other

[6]The verbal interchanges between the creationist, William Jennings Bryan, and the defending evolutionist attorney, Clarence Darrow, were widely disseminated and were the subject of a popular play and movie film, *Inherit the Wind*. Although Scopes was fined by the court, the decision was later reversed by the Tennessee Supreme Court on the grounds that the fine had been wrongly imposed. The anti-evolution law in Tennessee, however, remained until 1967, when it was repealed by the state legislature.

countries (specifically Russia, which had launched the first space satellite, Sputnik, in 1957). Among these innovations were new high school textbooks in both the biological and social sciences ("Biological Science Curriculum Study," "Man as a Course of Study") that discussed evolution and analyzed changes in human social relationships. By the end of the 1960's anti-evolution laws were either repealed or declared unconstitutional. The fundamentalist response to these challenges was to intensify attacks on the teaching of evolution and even more importantly to adopt a new strategy claiming that creation was as much a science as evolution and therefore should be given equal time whenever evolution was taught. Within the last two decades, a number of societies and institutes established by fundamentalists for the propagation of "creation science" have entered the fray to further this approach.

However, in spite of the name, there is little, if any, recognizable science in creation science. One can ask: How or why did the divine creation of different species occur? What are the scientific mechanisms of creation science? How are proposed mechanisms for creation described, compared, and evaluated? How, for example, can a "creation scientist" determine which of the following creation stories are correct?[7]

1. God arose from the depths of the ocean, created dry land, and then created all creatures on the hill at Heliopolis at the center of the universe (Egypt).
2. God made sky and earth by splitting the powers of evil in half and then produced humans for purposes of worship (Mesopotamia).
3. God creates all that is good and struggles with an evil being that creates all that is bad. Each struggle lasts about 3,000 years and will continue until evil is vanquished, at which time creation will be complete and perfect (Iran).
4. God created himself from a golden egg, and from the various parts of his body everything was born. After a time, life is destroyed and the cycle begins again (India).
5. God created the universe in six days ending with the creation of humans, according to Genesis, Chapter 1; or, God first created Adam in the Garden of Eden and then created animals and birds and eventually Eve, according to Genesis, Chapter 2 (Israel).
6. God was a woman who produced twins—the sun and the moon. During various eclipses, the twins came together to create the various gods and spirits of earth and sky that rule over humans (Benin, Africa).
7. God created the world in four distinct periods, each separated by a flood (Yucatan).
8. God created the earth and its creatures from mud gathered in the webbed feet of ducks who swam on a primeval ocean (American Crow Indians).
9. The universe was originally in the shape of a hen's egg, out of which God emerged and chiseled its main physical features. After 18,000 years God died and the remainder of the world was derived from his body: the dome of the sky from his skull, rocks from his bones, soil from his flesh, rain from his sweat, plant life from his hair, and humans from his fleas (China).

If anything, creation science has been used only as a fundamentalist pseudonym for the religious creation myths in the Judeo-Christian bible. As stated by Henry Morris, a founder of the Institute for Creation Research:

> Since nothing in the world has been created since the end of the creation period [in Genesis], everything must then have been created by means of processes which are no longer in operation and which we therefore cannot study by any of the means of science. We are limited exclusively to divine revelation as to the date of creation, the duration of creation, the method of creation, and every other question concerning the creation.

By holding on to such extreme positions, it is easy to see that evolutionary discoveries and methodologies have little or no persuasive power with creationists. In fact, to many creationists, evolutionary concepts are the source of all evil (Fig. 4-3).[8] Thus, although con-

---

[7]This small sample of creation myths, in which "God" designates the creator, has been gathered from various religions, past and present. Additional myths can be found in collections by Farmer and Hamilton.

[8]From Morris (p.83):

> The origin of all the evil in the universe must have been coincident with the origin of the idea of evolution, both stemming from Satan's rejection of God's revelation of himself as Creator and Ruler of the universe. This primal act of unbelief and pride later led to the fall of man. Similarly, unbelief in God's Word and man's pride in his own ability to rule his own destiny have yielded the bitter fruits of these thousands of years of human sin and suffering on the earth. And today, this God-rejecting, man-exalting philosophy of evolution spills its evil progeny—materialism, modernism, humanism, socialism, Fascism, communism, and ultimately Satanism—in terrifying profusion all over the world.

siderable literature has been devoted to dealing with creationist attacks on evolution (see, for example, books by Futuyma and Kitcher and that edited by Godfrey), the implacable hostility of fundamentalist creationists toward evolution shows no promise of ever being resolved.

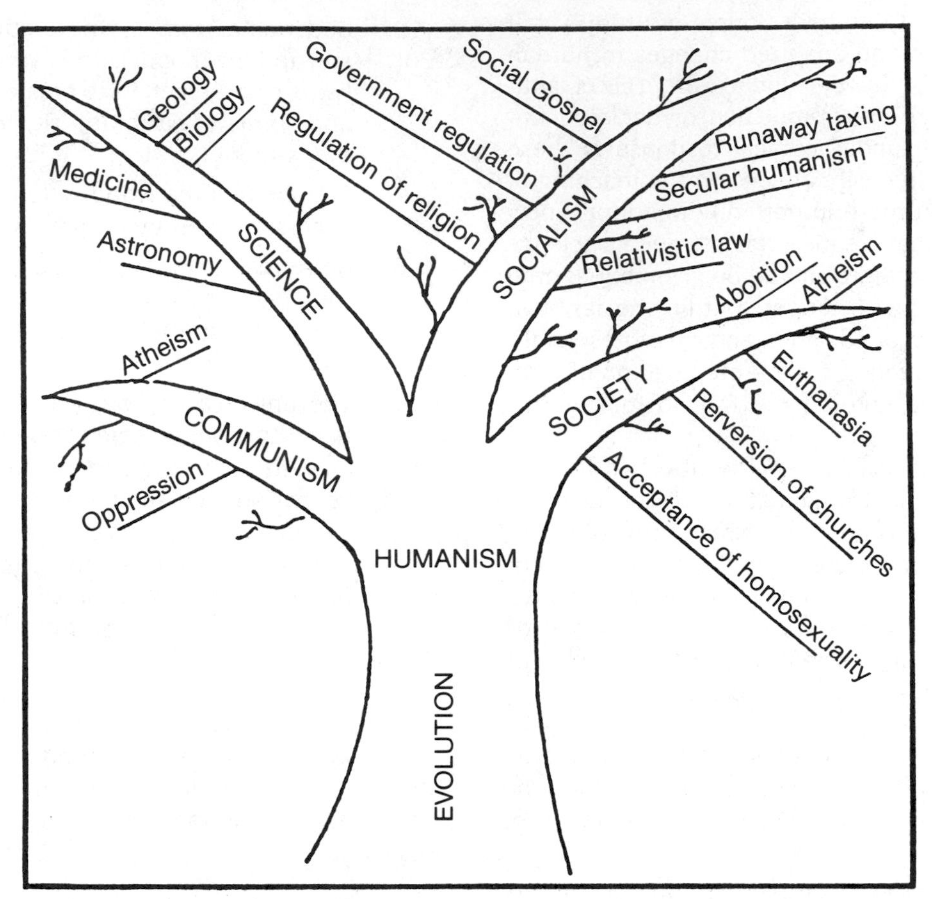

FIGURE 4-3 Creationist view of evolution as "the root of the tree of evil" that affects society. "Evolution is the taproot which is feeding the oppressive, murderous and infidel directions we see gaining acceptance around us" (P. A. Bartz, 1984. *Bible Science Newsletter*, **22**, 2). It should be obvious that such fanaticism is not open to rational scientific discussion.

SUMMARY

Darwinian evolution had a profound impact not only on biology but also on many other fields, religion in particular. Its extraordinary influence was due to social, economic, and technological developments which had helped overthrow the old social order of feudalism and monarchy and had brought about the rise of capitalism. The traditional religious rationale for social and biological systems was that the universe followed an exactly designed order established by an intelligent deity. Thus evolutionary theory, by denying that biological creatures were purposely designed, was perceived as a great threat to religious interests.

The roots of religious beliefs lie in human attempts to appeal to and control the forces of nature, which were long incomprehensible and thought to be supernatural and humanlike. From these roots arose the concept of God and soul, both of which were supposed to be eternal and immaterial. The idea of a powerful deity controlling the physical universe was not seriously challenged until the time of Copernicus and Galileo in the sixteenth century. In the new world view they brought about, God appeared as an initial creator rather than as an incessant manipulator of the solar system.

With the advent of Darwinism, it was no longer necessary to advocate a God even to account for manipulating biological relationships. It seemed to many that a deity with conscious, purposeful, human characteristics had been replaced by evolutionary randomness and uncertainty. The common religious view that there could be no design, biological or otherwise, without an intelligent designer was contradicted by the Darwinian view that evolution is a historical process and present-type organisms were not created spontaneously but were formed by a succession of selective events that occurred in the past. According to evolution, successful traits are selected through interaction between organisms and their environment, and these are then enhanced by further selective events. In this way adaptation to the environment can continuously modify organs and structures over long periods, and complexities which seem unlikely as singular spontaneous events become evolutionarily probable events. The variability that selection depends on may be random, but adaptations are not; they arise because selection chooses and perfects only that which is adaptive. In this scheme a God of design and purpose is not necessary.

Neither religion nor science has irrevocably conquered in their confrontation. Religion has been bolstered by paternalistic social systems in which individuals are dependent on the beneficences of those more powerful than they, as well as by the comforting idea that man was created in the image of a God to rule over the world and its creatures. Religion provided emotional solace, a set of ethical and moral values, and support for the established social system. Many Judeo-Christian denominations evaded interpretation of evolutionary biological events or attempted compromises between traditional religious explanations and scientific ones. Nevertheless, faith in religion has been eroded by natural explanations of its mysteries, by a deeper understanding of the sources of human emotional needs, and by the recognition that ethics and morality can change among different societies and that acceptance of such values need not be dependent on religion.

In the United States the attack on evolution was carried on with significant success by fundamentalist religious groups who have opposed the teaching of evolution in public schools and who have also influenced the reduction or elimination of evolution in many biology textbooks. However, in recent years, especially since the launching of the Russian satellite *Sputnik* in 1957, there has been a revulsion against this attitude and a wave of reform began in American science education. Rebuttal from fundamentalists has come in the form of the "creation science" movement which, despite its name, has nothing in common with scientific method. At present the positions of the creationists and the scientific world appear irreconcilable.

KEY TERMS

anthropomorphic wisdom
creation myths
creation science
evolution of religion
evolutionary opportunism
God
personality
religious fundamentalism
soul
supernatural design

REFERENCES

Bernal, J. D., 1969. *Science in History*. M.I.T. Press, Cambridge, Mass.

Bowler, P. J., 1977. Darwinism and the argument from design: Suggestions for a reevaluation. *Jour. Hist. Biol.*, **10**,29-43.

Farmer, P. (ed.), 1979. *Beginnings: Creation Myths of the World*. Atheneum, New York.

Futuyma, D. J., 1983. *Science on Trial: The Case for Evolution*. Pantheon Books, New York.

Gillespie, N., 1979. *Charles Darwin and the Problem of Creation*. Univ. of Chicago Press, Chicago.

Gillispie, C. C., 1951. *Genesis and Geology*. Harvard Univ. Press, Cambridge, Mass.

Glick, T. F. (ed.), 1974. *The Comparative Reception of Darwinism*. Univ. of Texas Press, Austin.

Godfrey, L. R. (ed.), 1983. *Scientists Confront Creationism*. W. W. Norton, New York.

Greene, J. C., 1961. *Darwin and the Modern World View*. Louisiana State Univ. Press, Baton Rouge.

Hamilton, V., 1988. *In the Beginning: Creation Stories from Around the World*. Harcourt Brace Jovanovich, San Diego.

Kitcher, P., 1982. *Abusing Science: The Case Against Creationism*. M.I.T. Press, Cambridge, Mass.

La Barre, W., 1970. *The Ghost Dance: Origins of Religion*. Doubleday, New york.

Lovejoy, A. O., 1959. Kant and evolution. In B. Glass, O. Temkin, and W. L. Straus, Jr. (eds.). *Forerunners of Darwin*. Johns Hopkins Press, Baltimore, pp. 173-206.

Monod, J., 1971. *Chance and Necessity*. A. A. Knopf, New York.

Moore, J. R., 1979. *The Post-Darwinian Controversies*. Cambridge Univ. Press, Cambridge.

Morris, H. M., 1963. *The Twilight of Evolution*. Baker, Grand Rapids.

Nelkin, D., 1982. *The Creation Controversy: Science or Scripture in the Schools*. W. W. Norton, New York.

Oldroyd, D. R., 1980. *Darwinian Impacts: An Introduction to the Darwinian Revolution*. Open Univ. Press, Milton Keynes, Oxford.

Paley, W., 1802. *Natural Theology: or Evidences of the Existence and Attributes of the Deity, Collected from the Appearances of Nature*. R. Faulder, London.

Roberts, J. H., 1988. *Darwinism and the Divine in America*. Univ. of Wisconsin Press, Madison.

Russell, C. A. (ed.), 1973. *Science and Christian Belief: A Selection of Recent Historical Studies*. Univ. of London Press, London.

Russett, C. E., 1976. *Darwin in America: The Intellectual Response 1865-1912*. W. H. Freeman, San Francisco.

Tawney, R. H., 1926. *Religion and the Rise of Capitalism*. Harcourt, Brace & Co., New York.

Tylor, E. B., 1881. *Anthropology: An Introduction to the Study of Man and Civilization*. Appleton, New York.

Wallace, A. C. F., 1966. *Religion: An Anthropological View*. Random House, New York.

Weber, M., 1963. *The Sociology of Religion*. Beacon Press, Boston.

White, A. D., 1896. *A History of the Warfare of Science with Theology in Christendom*. Appleton, New York.

PART 2

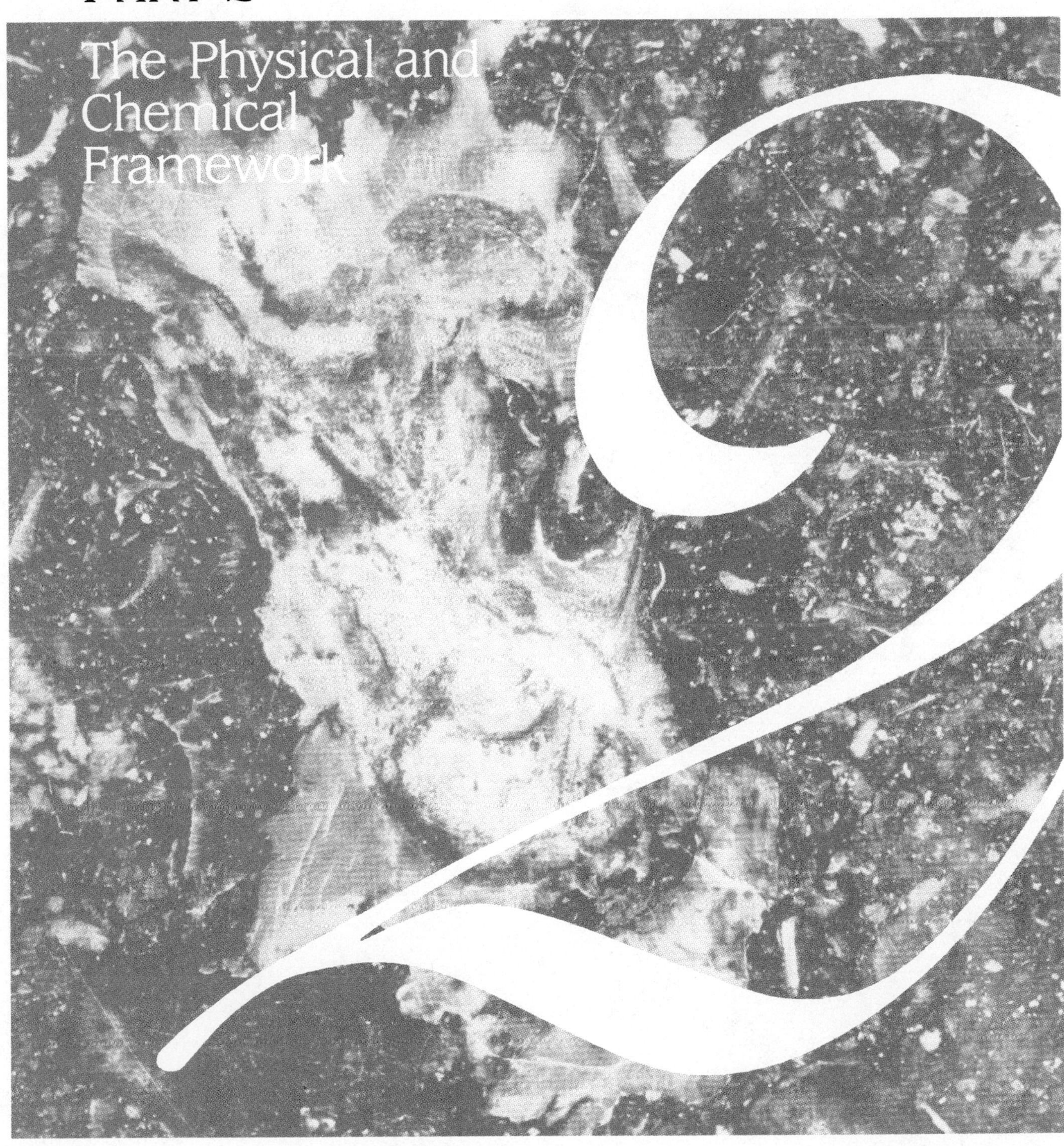

THE BEGINNING

he enormity of time that surrounds us is difficult to appreciate. We are almost always concerned with the here and now, and our direct experience of the past is limited to our own mortal histories. In the last five or six thousand years these limitations were somewhat lifted by written genealogies and historical accounts, but even then our notion of history usually extended no further than the dim legends of our culture. To explore even a small portion of the past beyond our culture, let us say 10 or 100 million years, long seemed an unnatural feat. What could have existed before our memories and traditions? As voiced in Ecclesiastes, "There is nothing new under the sun."

To conceive of a reality beyond our own records demands considerable evidence that such a reality existed. Fortunately, as we have seen, the evidence of fossils and of long-term geological changes ended the historical isolation of humans from the world about them. However, the question of the beginning remains: What, and how long ago, was the beginning?

THE ORIGIN OF THE UNIVERSE

Theories of the origin of the universe that have been developed by astronomers in recent years have attempted to deal with a few main observations that can be summarized as follows:

1. Hydrogen is the basic fuel used by stars when they begin radiating the large amounts of energy that makes them visible to us.
2. As hydrogen is burned in stars, other elements such as helium and carbon accumulate through various fusion reactions.
3. Stars then undergo an aging process which eventually produces the various elements.
4. The continual transformation of hydrogen indicates that this fundamental element will eventually diminish until no more stars can be created unless a new source of matter is present.

To many scientists these considerations all point to a time when the universe must have consisted mostly or entirely of hydrogen, and our present universe is only a stage in the evolution of this primordial mass. According to one group of astronomers which included Hoyle, Bondi, and Gold, as hydrogen diminishes, it is replaced by new matter from an unknown source. Matter is therefore never really lost, and the universe is preserved in a **steady state**. Another view proposed by Gamow and others suggests that at a distant time in the past all of the universe was confined to a small sphere of concentrated energy/matter. This substance then exploded in a **big bang** to form hydrogen at first and then eventually all of the galaxies and stars.

In the dispute between these two views, considerable evidence came to light which helped reconstruct various aspects of the origin of the universe. Much of this evidence supports the Big Bang theory, and includes the following:

- **Expansion of the Universe:** About 70 years ago, the American astronomer Slipher discovered that the wavelengths of light emitted from distant galaxies of stars were shifted in directions which indicate that these galaxies are rapidly receding from us. This effect, called the Doppler shift, arises because emitted wavelengths from any source appear to be longer (red) or shorter (blue) if an observer is moving, respectively, away from the source or towards it. As shown in Figure 5-1, absorption lines of a known element (calcium in this case) shift towards the red end of the spectrum as the distance from the earth increases. This indicates that some of the farthest galaxies may be receding from us at speeds which approach 25 to 50 percent of the speed of light. Also, the rate at which this recession occurs has probably been changing: the universe seems to have been expanding more rapidly in the past than it is at present.
- **Black Body Radiation:** Although energy seems to be radiating primarily from the galaxies and their stars, there is evidence that a background of fairly uniform low temperature radiation, 3 degrees Kelvin,[1] pervades the entire universe. This black body radiation is predictable if the original Big Bang occurred about 15 billion years ago and began with an initial temperature of about 10^{12} degrees Kelvin.
- **Radio Waves:** There is evidence that celestial radio wave sources are primarily associated with the presence of galaxies. These sources are greatly increased in number when they are observed at light-year distances which reflect the time period when galaxies were first created. That is, observations of galaxies that are about 7 or 8 billion light years from us indicate an increase in radio sources and point to a time when the universe was only about one third to one fifth its present age and considerably more compact than it is today. Before this period, few or no radio sources were present, indicating a time when galaxies had not yet formed.
- **Helium Proportions:** The proportion of helium in the universe (about 10 percent) is far greater than the amounts synthesized by galactic stars in which helium is produced by thermonuclear burning of hydrogen. Moreover, helium appears uniformly distributed throughout the universe in contrast to heavier elements that are localized to galactic centers in which most supernovae occurred (see p. 72). Most helium seems therefore to have been synthesized prior to the formation of galaxies, and its uniform universal distribution seems therefore to have been caused by the Big Bang.

Although these findings are considered strong evidence for the Big Bang theory, a number of astronomers have suggested that they do not exclude more than one Big Bang. That is, the universe may oscillate between an expanding state caused by a Big Bang to a contracting state (the "Big Crunch") followed by a subsequent Big Bang, perhaps *ad infinitum* (Fig. 5-2, bottom). Among other implications, this theory would mean that enough matter exists in the universe to enable its contraction through the force of gravity once expansion has ceased. Considerable attention has therefore been focused on measuring the amount of matter by measuring its average density. So far, calculations indicate that contraction of the universe can occur if the density is 10^{-29} gm/cm^3 (one proton per 10 cubic feet). Since the actual average density of all galactic matter is only 10^{-31} gm/cm^3, or about 100 times less than the density necessary for contraction, the oscillating universe seemed at first questionable. However, recent estimates of the amount of x-rays found in space suggest that they arose from a Big Bang radiation acting on 100 times the amount of matter observed in galaxies. Also, according to some claims, gravitational waves coming from outer space may eventually be shown to have an intensity derived from an amount of matter 100 times that in the galaxies. To date, the problem of deciding between the Big Bang and **Oscillating Big Bang** theories remains one of ac-

---

[1]Degrees Kelvin (°K) can be converted into degrees centigrade or Celsius (°C) by subtracting 273. Thus 3° K is -270°C or -454°F (Fahrenheit). The freezing and boiling points of water are respectively 273°K (0°C = 32°F) and 373°K (100°C = 212°F). Zero degrees Kelvin signifies absolute zero, at which all molecular motion ceases.

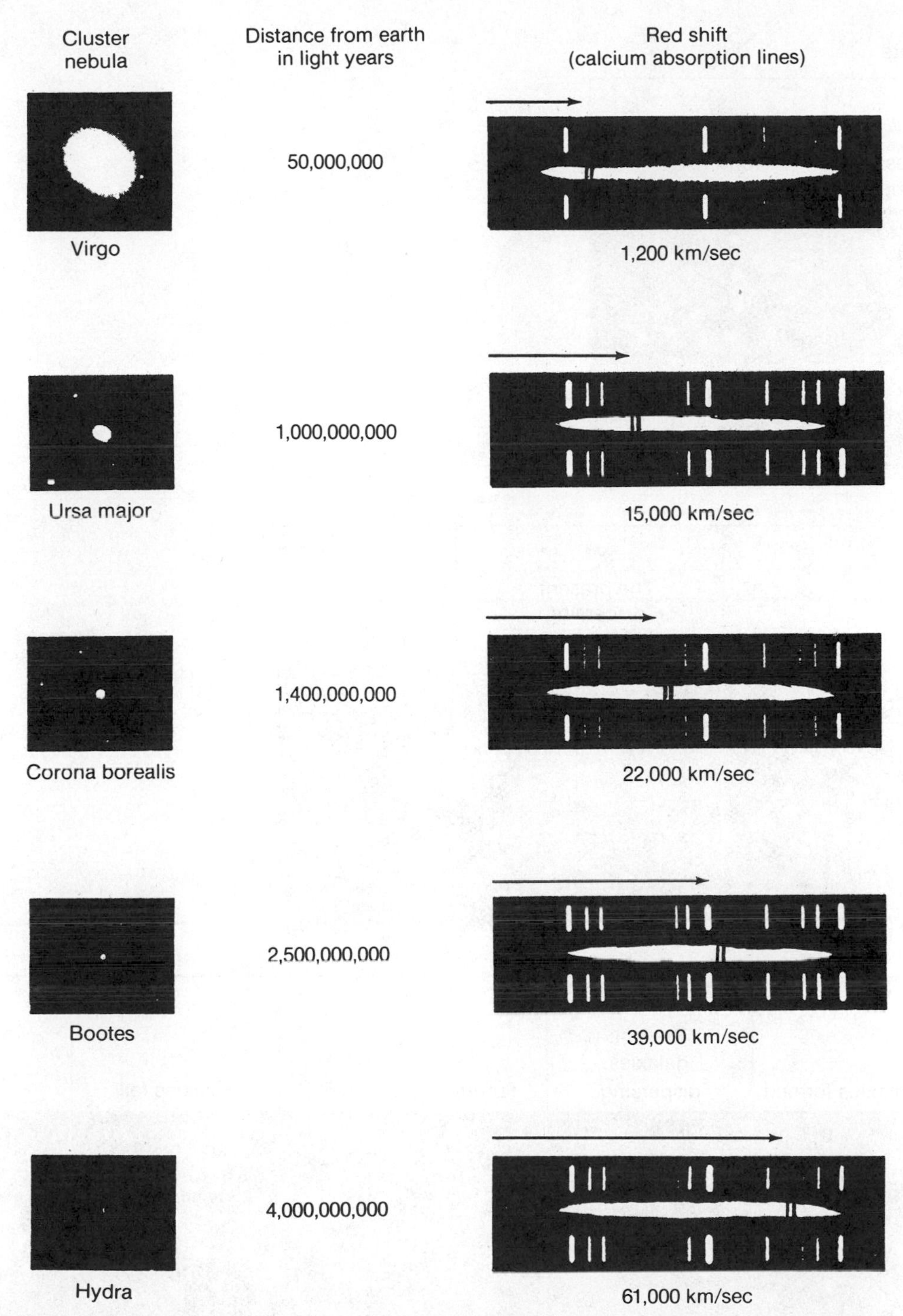

FIGURE 5-1 The shift (arrows) in calcium absorption line spectra towards the red for galaxies at various distances from the earth. There is a correlation between these distances and the speed at which a galaxy is receding, a relationship described by Hubble's Law. The present rate of recession is believed to lie in the range between 12 to 30 kilometers per second per million light years. (Adapted from Jastrow and Thompson with distances according to Kutter.)

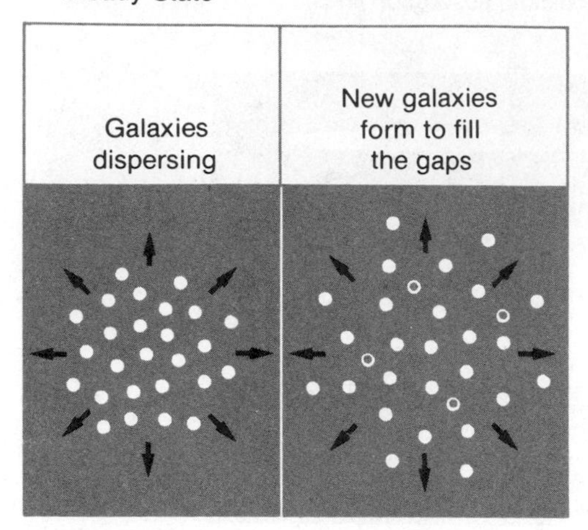

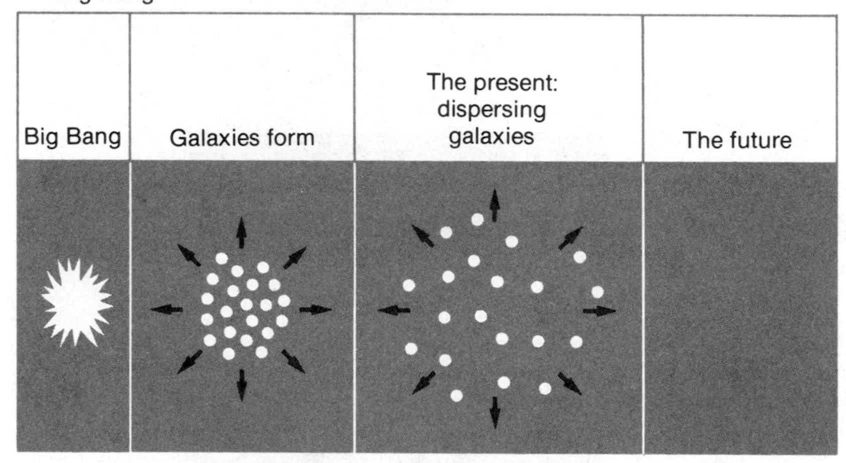

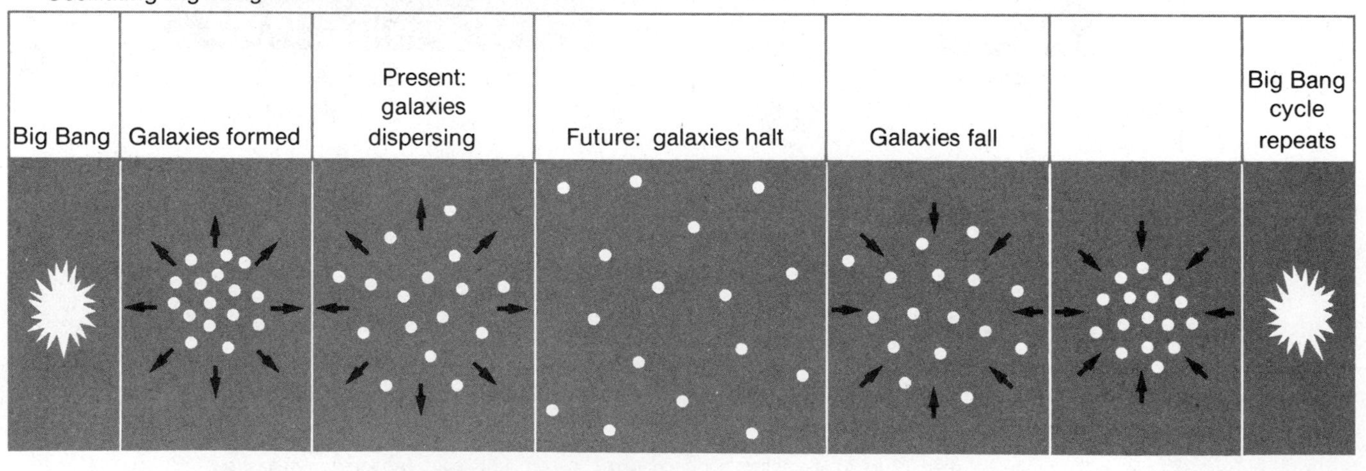

FIGURE 5-2 Schematic diagrams of three major cosmological theories. (Adapted from Jastrow and Thompson.)

curacy of measurements, although there now seems to be sufficient evidence to exclude the Steady State theory.

Whether Big Bang or Oscillating Big Bang, the evolution of the universe would in both cases follow for a time similar pathways. The question has therefore become, What was the sequence of events that followed the (last) Big Bang? In their attempts to answer this question, astronomers have helped provide us with an understanding of the origin of the elements, the development of the solar system, and a concept of the immense reaches of time that were essential to our existence.

According to modern reconstructions of the event, the Big Bang occurred about 15 billion years ago at a time when all of the matter in the universe was compacted into a density which may have equalled that of the matter in the nucleus of an atom. This density was more than one million billion times that on earth, sufficient to compress the entire universe into an area no larger than the present solar system. The temperature of this fireball mass must have been very high (10^{12}°K or higher) and may have lasted for a period of about 300,000 years. During this interval the Big Bang explosion is believed to have occurred along with the emission of hard gamma wave radiation. Expansion of the fireball enabled a decrease in temperature to take place during which individual atoms (mostly hydrogen, some helium) could begin to be formed.

When the universe reached an age of perhaps 100 million years, large masses of hydrogen separated from each other to form **protogalaxies**. These bodies of matter then gradually began to collapse inwardly to produce the giant galaxies in which individual stars evolved. About 100 billion galaxies are believed to exist, each galaxy containing perhaps 100 billion or more stars. Our own galaxy, the Milky Way, for example, may have as many as 150 to 200 billion stars and is organized in the shape of a flattened disc with spiral arms, about 100,000 light years in diameter. The galaxy is in constant motion, and the stars within it rotate around the galactic center at fairly rapid speeds. Thus our sun, which is approximately 30,000 light years from the galactic center, moves at a rate of more than 200 kilometers per second through space to complete its galactic orbit in about 250 million years. Spherical and elliptical galaxies are also common, the exact shape probably arising from the initial distribution of matter in the galaxy and the degree of spin imposed on this mass when it was formed.

Evolution of a Star and Its Elements

Among the main forces that act upon matter and participate in the evolution of a star or sun are:

1. A nuclear force which pulls particles of the atomic nucleus (protons, neutrons, etc.) together into densities of one billion tons per cubic inch. This force acts only over very small distances, no greater than one ten-trillionth of an inch.
2. An electromagnetic force which is 100 times weaker than the nuclear force. Electromagnetism serves to bind electrons to nuclei in the formation of atoms but weakens with distance.
3. Gravity, a force which is 10^{38} times weaker than electricity but can nevertheless act over long distances, such as between the earth and its moon, the sun and its planets, and the galaxy and its suns.

Initially, the first step in the evolution of a star is the gravitational condensation of fragments derived from the galactic cloud of gas and dust. Some of these condensing masses, or **protostars,** can be seen in the heavens as dark globules which have not yet reached the temperature necessary to emit intense light of their own. With the passage of time, the protostar continues to contract by gravity, releasing energy in the form of heat and light. This arises because atoms moving inwards from gravitational attraction pick up speed as the center increases in mass and density, and the greater the speed, the higher the temperature. As the protostar becomes smaller, its temperature therefore increases, and within 500,000 years its interior may reach a temperature of 100,000°K. A temperature of this magnitude causes the ionization of atoms (loss of outer electrons) but does not yet enable the **thermonuclear reactions** to occur that are necessary for this mass to start burning its own material. A much greater temperature, 10 million°K, is required in order to overcome the repulsive forces between hydrogen nuclei (protons) and to allow fusion reactions to occur.

Continued contraction of the initial gravitational mass keeps increasing the protostar's temperature until the ionized atoms now begin to absorb much of the radiation. This serves to trap the heat generated inside the mass so that it cannot escape to the exterior. A critical stage arrives when the 10 million-degree temperature for hydrogen fusion is reached and the subsequent movement of particles in the interior of the protostar is fast enough to prevent further contraction. At that point, approximately 10 million years from its origin for a star the size of our sun, the radiant energy of the star is maintained by thermonuclear reactions leading to the conversion of hydrogen (H) into helium (He^4):

proton (H nucleus) + proton (H nucleus) $\rightarrow$ deuteron (proton + neutron) + positron + neutron

deuteron + proton $\rightarrow$ He^3 (2 protons + 1 neutron) + γ rays

He^3 + He^3 $\rightarrow$ He^4 (2 protons + 2 neutrons) + 2 protons

Hydrogen burning enables the star to become hotter and more luminous, and it subsequently moves into what is called the **main sequence** or **main line** of stellar evolution.

As shown in Figure 5-3, when many stars are measured in terms of their external temperature and degree of luminosity, they generate a graph called the **Hertzsprung-Russell diagram** in which hydrogen-burning stars occupy positions along the main sequence (diagonal line). A star's position on this main sequence depends primarily on its size. Large stars

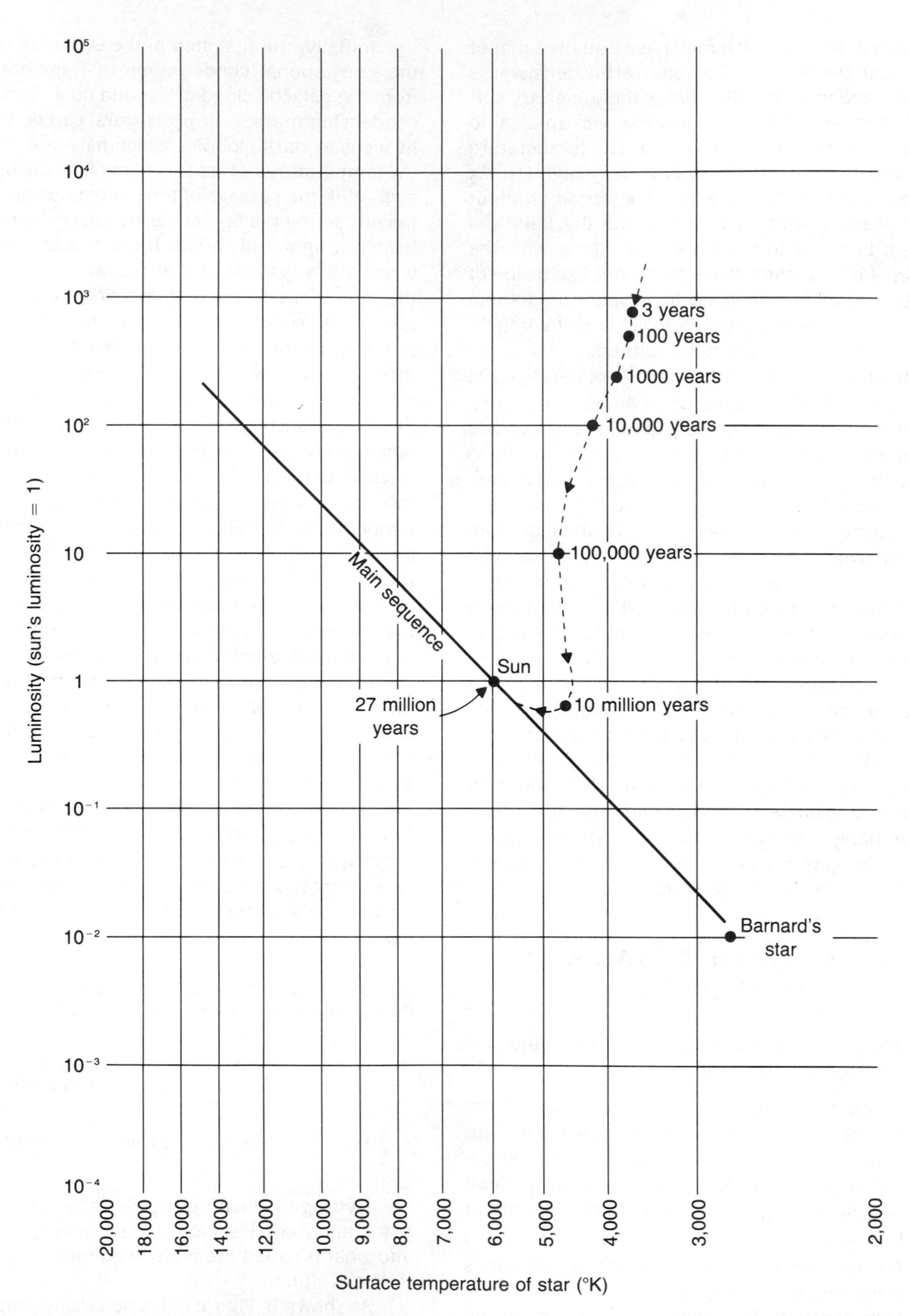

FIGURE 5-3 The relationship between surface temperature and luminosity for hydrogen-burning stars, shown as a diagonal line (the main sequence). The dashed lines and arrows show the path of a protostar about the size of our sun as it approaches the main sequence, and the number of years indicates the time from its origin until it reaches a particular stage. (From Jastrow and Thompson.)

with masses many times greater than our sun occupy positions on the main sequence at high luminosities and high surface temperatures (upper left). Small stars, such as Barnard's star, with a mass about one tenth of our sun, have relatively low luminosity and surface temperatures (lower right). Interestingly, because of the high temperatures generated, the large stars burn up their hydrogen more rapidly than the small stars. For example, the main-sequence lifetime of a star 20 to 30 times greater than our sun is only a few million years, whereas Barnard's star has the potential to continue burning hydrogen for trillions of years. Our sun, which is of intermediate mass, has already spent about half its lifetime, or 4.6 billion years, on the main sequence.

When a star has burned up a considerable portion of its hydrogen, a helium core accumulates and nuclear fusion diminishes. Because the interior of the star now has fewer thermonuclear reactions to prevent gravitational collapse, the star begins to shrink. This compression acts to heat the core rapidly and increase the rate of burning of the hydrogen shell around it. The collapse thus leads to an increase in the release of nuclear energy from the interior of the star which expands its outer layers. Expansion of the outer mass, in turn, absorbs most of the star's increased energy until the surface temperature drops to between 3,000°K and 4,000°K, and appears to have a reddish glow. The contraction of the helium core meanwhile continues, increasing the core temperature and increasing the burning rate of the enveloping hydrogen shell. This increases the luminosity of the star so that it now appears as a Red Giant, with features in the upper right of the Hertzprung-Russell diagram, far off the main line. At the time when our sun becomes a Red Giant it will be approximately 100 times its present size and its diameter will reach the earth's orbit.

The increased heating of the core of this expanded star eventually reaches the 100 million-degree temperature at which helium nuclei can fuse. This leads to an explosion of the core (the **helium flash**) followed by a further contraction, a process that may repeat itself a few times, during which various atomic products of helium-burning are produced, including beryllium (Be^8) and carbon (C^{12}):

$$He^4 + He^4 \rightarrow Be^8 \text{ (unstable)}$$
$$He^4 + Be^8 \rightarrow C^{12} + \gamma \text{ rays}$$
$$\text{or } 3\,He^4 \rightarrow C^{12} + \gamma \text{ rays}$$

The burning of helium, however, cannot continue for too long. There is a gradual accumulation of carbon in the core which then slows down the burning of helium just as the helium core, with its higher fusion temperature, had previously slowed down hydrogen burning. With the reduction in thermonuclear reactions, the star again begins to collapse. Should the star have a relatively small or medium-sized mass like our sun, the heat of gravitational collapse will be insufficient for the carbon core to reach the 600 million°K temperature at which carbon nuclei fuse. The star's outer envelope, however, will become greatly heated and expand. It may even separate from the core to form a ghostly planetary nebula. In such medium-sized stars, contraction of the core will eventually diminish, although considerable heat continues to be emitted. At that point, the contracted state of the star is such that if it were initially the size of our sun, that is, one million miles in diameter, it would now occupy a diameter of only 20,000 miles. The gravitational force at the surface of this compacted mass is high, one million times that on earth, and it is now called a White Dwarf. In time, the White Dwarf gradually cools, forming a dead, cold, lump of matter.

However, if the star is massive rather than small, its collapse, when helium burning begins to decline, may enable the carbon core to reach the critical 600 million-degree temperature. The burning of carbon would then take place, halting further collapse and producing various new elements, among which are oxygen (O^{16}), neon (Ne^{20}), sodium (Na^{23}), magnesium (Mg^{24}), silicon (Si^{28}) and sulfur (S^{32}):

$$C^{12} + C^{12} \rightarrow Mg^{24} + \gamma \text{ rays}$$
$$C^{12} + C^{12} \rightarrow Na^{23} + \text{proton}$$
$$C^{12} + C^{12} \rightarrow Ne^{20} + He^4$$
$$He^4 + C^{12} \rightarrow O^{16} + \gamma \text{ rays}$$
$$C^{12} + O^{16} \rightarrow Si^{28}$$
$$He^4 + Si^{28} \rightarrow S^{32} + \gamma \text{ rays}$$

As occurred previously in the helium core, the exhaustion of carbon burning then leads to a further gravitational collapse and further contraction. New nuclear fuels begin to burn, and new elements continue to be created until an iron (Fe^{56}) core appears.

The nuclear fusion of iron, in contrast to that of previous elements, absorbs energy rather than radiates it. The **iron core** thus acts as a "heat sink," leading to a gradual extinction of thermonuclear reactions. As nuclear fusion stops, collapse of the core begins again, but because no further reactions are taking place, the collapse of this large mass continues until its matter has reached a density of 10^{24} tons per cubic inch and a temperature of a trillion degrees. According to one view of subsequent events, the combination of heat and pressure causes this collapsing material to rebound rapidly like a giant spring in an explosion of immense proportions. Another view suggests that energy from the rapid gravitational collapse of the core

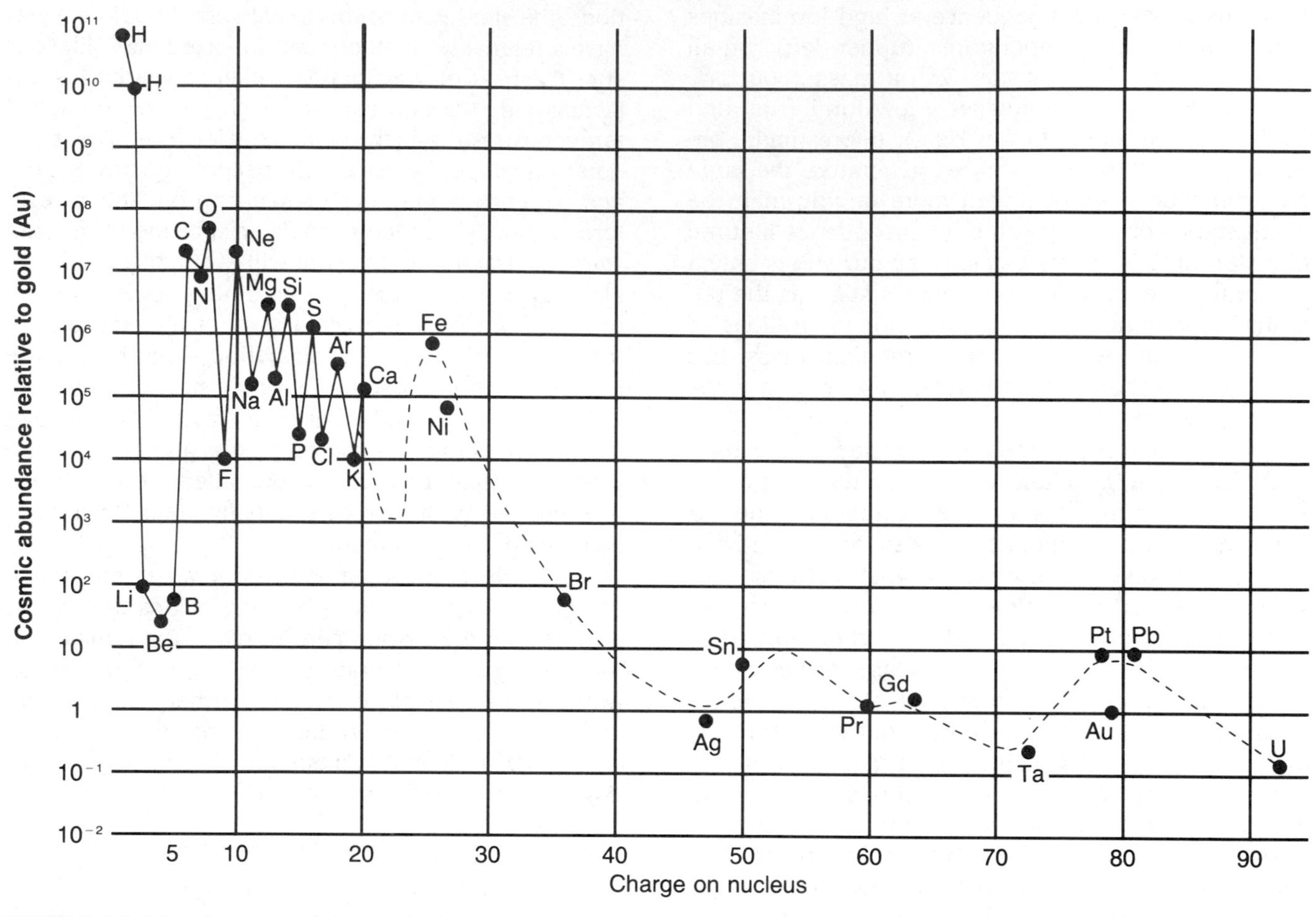

FIGURE 5-4 Relative abundance of elements in the universe, using gold (Au) as a standard of 1.00. Note that elements with atomic weights less than iron are relatively abundant since they are produced in many stars. Elements higher on the periodic table than iron, such as silver (Ag), lead (Pb), and uranium (U), are about 100,000 times less common because they are produced in the very brief interval of a supernova explosion. (Adapted from Jastrow and Thompson.)

causes an implosion which disperses the outer layers of the star at very high velocities. In both cases, a large part of the star's material is thrown out into space, and the sky may then light up to form the brightest object ever seen, a **supernova**. Supernovae now occur at a frequency of about one per 100 years in our galaxy but were probably more common in the past when star formation was more frequent. An example of the remnants of a supernova can be found in the Crab Nebula, a large, rapidly expanding cloud of matter in the constellation of Taurus which arose from a supernova explosion 920 years ago that was visible on earth during daylight. It is estimated that there have been about one billion supernovae during our galactic history.[2]

The supernova material, distributed widely through space, has important effects on later-born stars which use this material as part of their own formation. One consequence of a supernova is the creation of very high fluxes of free neutrons which are then captured by various atomic nuclei to form elements such as gold and uranium. It is primarily by this means that elements heavier than iron (the twenty-sixth element in the periodic table) can appear. Also, it is the relatively short duration of the supernova effect which accounts for the fact that these heavier elements are rare (Fig. 5-4). It is therefore obvious that our solar system, which contains samples of all elements, must have utilized the remnants of at least one supernova explosion in its formation, and some astronomers have already mapped some likely positions where such nearby supernovae may have occurred. Our sun may therefore be considered a second- or even third-generation star.

[2]After the supernova, the final core remnants of the massive star contract to become the densest material known—a neutron star—in which all protons and electrons have combined into a pure ball of neutrons about 10 miles in diameter. These bodies are now believed to be the pulsars which emit radiowaves from surface storms as they spin on their axes a few times a second. It has been suggested that, because of its heavy mass, a neutron star can continue to contract until its gravitational force is strong enough to prevent electromagnetic waves such as light from leaving its surface. When this immense density is reached, the collapsing star becomes a "black hole" in space.

SUMMARY

In order for evolution to occur, the history of the earth had to extend far beyond human comprehension. Although we now know that the universe is immensely old, when and how it came into being is still highly speculative. Two views of the origin of the universe have been propounded, both based on the universal use of hydrogen as the stellar fuel. The first, and less probable, theory (Steady State) is that hydrogen is continually replenished from outside the universe. The second (Big Bang) is that the universe orginated about 15 million years ago in an explosion of a small volume of extremely dense energy/matter. Support for this theory comes from the apparent continuing expansion of the universe, celestial radio waves associated with galaxies of a certain age, and the uniform distribution of helium and radiation (black body radiation) throughout the universe. There may have been only one big bang or the universe may oscillate between expansions and contractions.

After the big bang, the temperature of matter was enormously high. As it cooled, hydrogen and helium atoms formed. Perhaps 100 million years after the formative explosion, masses of hydrogen began to condense into galaxies, each galaxy producing many billions of stars in a manner still occurring today. Stars forming from galactic matter condense and increase in temperature. Ionization of atoms within the mass trap heat within the star, and the interior temperature reaches that necessary for the fusion of hydrogen atoms to form helium. The combustion of hydrogen raises the temperature still more, and the star moves into the main line of its evolution. The larger the star, the more rapidly it consumes its hydrogen, and nuclear fusion declines. It undergoes a complex series of events which cause it to expand and luminesce, at which point it is known as a red giant. The internal temperature continues to increase until helium nuclei can fuse. Eventually medium-sized stars will contract and cool, forming a white dwarf, which is the end of its evolution. Large stars may attain extremely high core temperatures, enabling carbon to burn and producing many new elements. This continues until the core of the star consists of iron, the combustion of which absorbs energy and eventually extinguishes further thermonuclear reactions. The star collapses, heats further, and explodes into a supernova. When supernovae occur, heavy elements are formed. All of these elements are found in our solar system and appear to be remnants of nearby supernovae formations.

KEY TERMS

Big Bang theory
carbon burning
helium flash
Hertzsprung-Russell diagram
hydrogen burning
iron core
main line
main sequence
Oscillating Big Bang theory
protogalaxies
protostars
Steady State theory
supernova
thermonuclear reactions

REFERENCES

Barrow, J. D., and J. Silk, 1983. *The Left Hand of Creation: Origin and Evolution of the Expanding Universe.* Heinemann, London.

Field, G. B., G. L. Verschuur, and C. Ponnamperuma, 1978. *Cosmic Evolution: An Introduction to Astronomy*. Houghton Mifflin Co., Boston.

Harrison, E. R., 1981. *Cosmology: The Science of the Universe*. Cambridge Univ. Press, Cambridge.

Hawking, S. W., 1988. *A Brief History of Time: From the Big Bang to Black Holes.* Bantam Books, Toronto.

Jastrow, R., and M. H. Thompson, 1972. *Astronomy: Fundamentals and Frontiers*. John Wiley & Sons, New York.

Kutter, G. S., 1987. *The Universe and Life: Origins and Evolution*. Jones & Bartlett, Boston.

Parker, B., 1988. *Creation: The Story of the Origin and Evolution of the Universe.* Plenum Press, New York.

Shklovskii, I. S., and C. Sagan, 1966. *Intelligent Life in the Universe*. Holden-Day, San Francisco.

Silk, J., 1980. *The Big Bang: The Creation and Evolution of the Universe.* W. H. Freeman, San Francisco.

Trefil, J. S., 1983. *The Moment of Creation*. Scribner's, New York.

Weinberg, S., 1977. *The First Three Minutes*. Basic Books, New York.

THE EARTH

There are two main theories for the evolution of planets in our solar system. The first, proposed originally by Buffon and later by Jeans and Jeffries, can be called the **collision theory**. It suggests that another star passing close to our sun pulled out, through gravity, material which became the planets. The most serious of many objections to this theory is the extreme rarity of such events: collisions or near-collisions between stars are estimated to have occurred in our galaxy only ten times in the last five billion years, whereas more than a billion stars in our galaxy are believed to have planets.

The second theory is held in various forms by most astronomers today and was first suggested by Kant and later by Laplace. It is called the **condensation theory** or nebular hypothesis, and proposes that the large, whirling mass of matter out of which our solar system initially condensed about 4.6 billion years ago did not have a uniform distribution of material. According to this theory, the large condensing mass at the center of this cloud became the sun when it reached thermonuclear reaction temperatures, whereas the smaller peripheral condensations never reached such critical temperatures and therefore became the **protoplanets** (Fig. 6-1). These peripheral masses remained tied to the solar orbit and formed the planets, although some, such as Earth, captured subplanets or moons of their own.

The condensation theory thus helps explain both the motion and location of the planets, although a number of questions still remain. For example, the rotation of the sun is about 100 times slower than would be expected from its large mass and the relatively smaller mass of its planets. (The sun has only two percent of the angular momentum of the solar system, rotating around its axis only once every 26 days, yet has 99.9 percent of the mass of the solar system.) According to some calculations, this reduced rotation rate would only occur if the mass of planetary material were several 100 times greater than observed. Also unexplained is the source for the difference in angle between the equatorial plane of the sun and the plane along which the planets revolve, a difference of about 7 degrees (Hughes).

However, in spite of these difficulties, most astronomers have little doubt that gravitational condensations must have occurred in the formation of the planets. In fact, the condensation theory implies that each time a star condenses out of the gaseous matter of space, the opportunity, or even likelihood, exists that

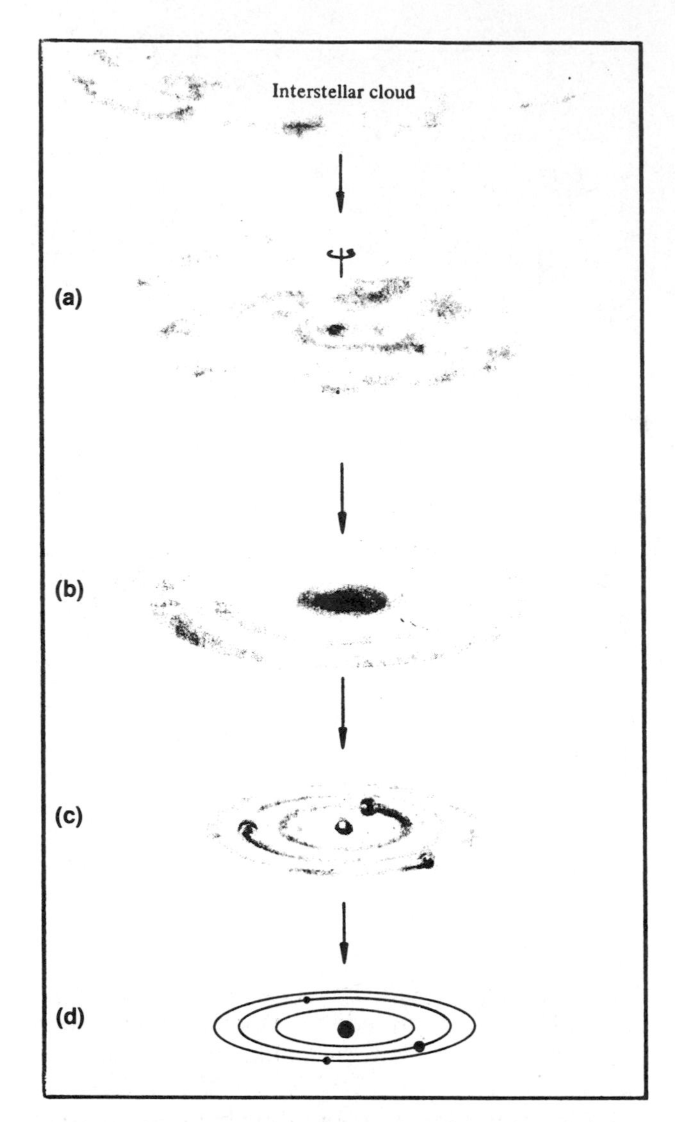

FIGURE 6-1 Stages during the condensation of the solar nebula into the solar planetary system. (a) Fragmentation of an interstellar cloud. (b) Contraction and flattening of the solar nebula. (c) Condensation of nebular material into meteorites and protoplanetary bodies. (d) Solidification of planets. (Adapted from Field, G. B., G. L. Verschuur, and C. Ponnamperuma, 1978. *Cosmic Evolution: An Introduction to Astronomy.* Houghton Mifflin, Boston.)

satellite planets will be formed. Unfortunately, aside from our own solar system, planets are almost impossible to discern elsewhere since they emit no radiation of their own and reflect only the light of their suns. Even if a planet of a nearby star were relatively large, such reflections would be much too feeble to be identified. Indirect observations have therefore been attempted by noting whether the motion of a star through the galaxy is affected by interaction with unseen planets. Through these means, several nearby stars are believed to maintain planets, and it has been reported that one star (Barnard's) has at least two large planets.

THE EARTH'S ATMOSPHERE

The planetary distribution of elements as our solar system condensed was not apparently uniform. According to some astronomers, a density gradient was established with many heavier elements condensing into the "earthlike" planets nearest the sun (Mercury, Venus, Earth, Mars) and relatively large amounts of the lighter, more volatile elements condensing into planets farthest from the sun (Saturn, Jupiter, Uranus, Neptune). It is suggested that these differences in condensation as well as the pressure of heat and solar radiation on nearby planetary atmospheres caused the earthlike planets to lose their initial hydrogen and helium atmospheres. Loss of this primary atmosphere then left these planets with the rocky materials so characteristic of them today.

Various gases, nevertheless, remained in the interior of the earthlike planets and gradually escaped to form a secondary atmosphere. On Earth, the outgassing of hydrogen, the most prevalent of cosmic elements, enabled the formation of three essential hydrogen-bearing compounds: methane (CH_4), water (H_2O), and ammonia (NH_3). Other gases that were present at the time probably included carbon monoxide (CO) and nitrogen (N_2) and some that can even now be observed to issue from volcanoes and hot springs, such as carbon dioxide (CO_2), hydrochloric acid (HCl), and hydrogen sulfide (H_2S). A number of the noble gases such as neon, argon, and xenon may also have been prevalent and should have persisted to this day in relatively high quantities since they are chemically inert. Their almost complete absence is so far unexplained.

In any case, there is now little question that the early atmosphere of the earth was either strongly or mildly **reducing** because of the prevalence of hydrogen compounds capable of providing electrons to **oxidizing** agents capable of accepting them. Evidence that supports this view exists in the finding of deposits laid down in South Africa and other places two or more billion years ago which became inaccessible to the earth's later atmosphere. Such deposits include sulfides of iron (FeS), lead (PbS), and zinc (ZnS), compounds which are highly unstable in the presence of oxygen. Were oxygen present in the atmosphere at the time these compounds were formed, they would have been deposited in the form of sulfates (e.g., $FeSO_4$) rather than sulfides.

Where, then, did our present oxygen atmosphere (Table 6-1) come from?

The answer to this question is not clear, although there seems to be general agreement among geochemists that the proportion of free oxygen in the atmosphere began to increase about 2 or 3 billion years

TABLE 6-1 Present Composition of the Earth's Atmosphere

Gas	Percent by Volume
Nitrogen (N_2)	78.09
Oxygen (O_2)	20.95
Argon (Ar)	0.93
Water (H_2O)	Variable (up to 1.00)
Carbon dioxide (CO_2)	0.03
Neon (Ne)	0.002
Helium (He), methane (CH_4), carbon monoxide (CO), krypton (Kr), nitrous oxide (N_2O), hydrogen (H), ozone (O_3), xenon (Xe)	Less than 0.001

ago. It has been suggested that ultraviolet irradiation of water in the upper atmosphere can produce free hydrogen ($2H_2O \xrightarrow{UV} 2H_2 + O_2$) which can then escape the earth's gravity and leave behind increasing amounts of molecular oxygen. Another proposal relies on the apparent correlation between increase in oxygen and increased domination of the earth's surface by plant life. As will be discussed in Chapter 9, electron transfer in the plant photosynthetic process involves the removal of hydrogen atoms from water molecules, producing free oxygen which then diffuses to the atmosphere. In whatever manner it first appeared, it is now generally agreed that the proportion of oxygen presently maintained in the atmosphere is related to photosynthesis in plants.

THE EARTH'S STRUCTURE

The formation of the earth from the wide band of material in its original orbit is believed to have been a process in which many subsidiary condensations first occurred. These subsidiary planetesimals were then drawn into the condensing earth, probably along with uncondensed orbital material, to form a structure which probably had some degree of differentiation. That is, different compounds and minerals probably occupied different positions in the earth depending on the temperature at which they condensed, the temperature of the condensing earth, and other variables. Within the first billion years of the earth's history, differentiation of its structure is believed to have proceeded at a fairly steady rate until relationships were achieved similar to those which exist today. Radioactive elements trapped within the earth during its condensation gave off small but incremental amounts of heat which gradually increased the temperature of surrounding material. Along with the heat of condensation and pressure, the center of the earth probably soon developed temperatures high enough to melt iron.

Present information concerning the interior of the earth is primarily derived from vibrational waves generated by earthquakes. These **seismic waves** can be detected by sensitive seismographs, and their paths and velocities can be shown to depend upon the composition, fluidity, and thickness of the materials through which they travel. Combined with studies of the earth's magnetic, electric, and gravity fields, seismic information indicates that the interior of the earth is divided into a number of concentric layers which undoubtedly differ in temperature, pressure, composition, and degree of crystallization (Fig. 6-2). At the center is a **core** about 2,100 miles in diameter that consists primarily of molten iron and nickel, mixed with sulfur or silicon. It is believed that shifts in this molten iron core are responsible for changes in the earth's magnetic field.

Surrounding the iron core is a hot **mantle** layer of rock, about 1,800 miles thick, that comprises approximately four fifths of the earth's volume. Because of radioactivity, pressure, and localized heating or cooling, the mantle has experienced repeated meltings and crystallizations, and geologists now characterize it as a partly molten plastic structure whose density increases in accord with its proximity to the core. Floating on the surface of the mantle is a thin **crust** of rock with a thickness of about 20 miles for the lighter continental land masses and about 5 to 7 miles for the heavier oceanic basins (Fig. 6-3). We know most about the crust, and three basic types of crustal rocks can be distinguished:

- **Igneous rocks:** These are rocks which have crystallized out of the molten liquid magma pushed up through cracks in the crust by the mantle. When deposited under existing rocks, igneous intrusions may be detected by the erosion of covering strata. Magma may also be deposited directly on the surface in the form of lava. The granites are a common example of igneous rocks, as are the dark, fine-grained basalts which appear often as the solidified lava of volcanoes.

- **Sedimentary rocks:** The erosion of igneous rocks by water, wind, chemical reactions, and volcanic activity produces small particles which can then be transported and reformed into new arrangements. Thus, a stream may deposit its sediments at the bottom of a lake, or wind, waves, and ice can shift sand, pebbles, and other

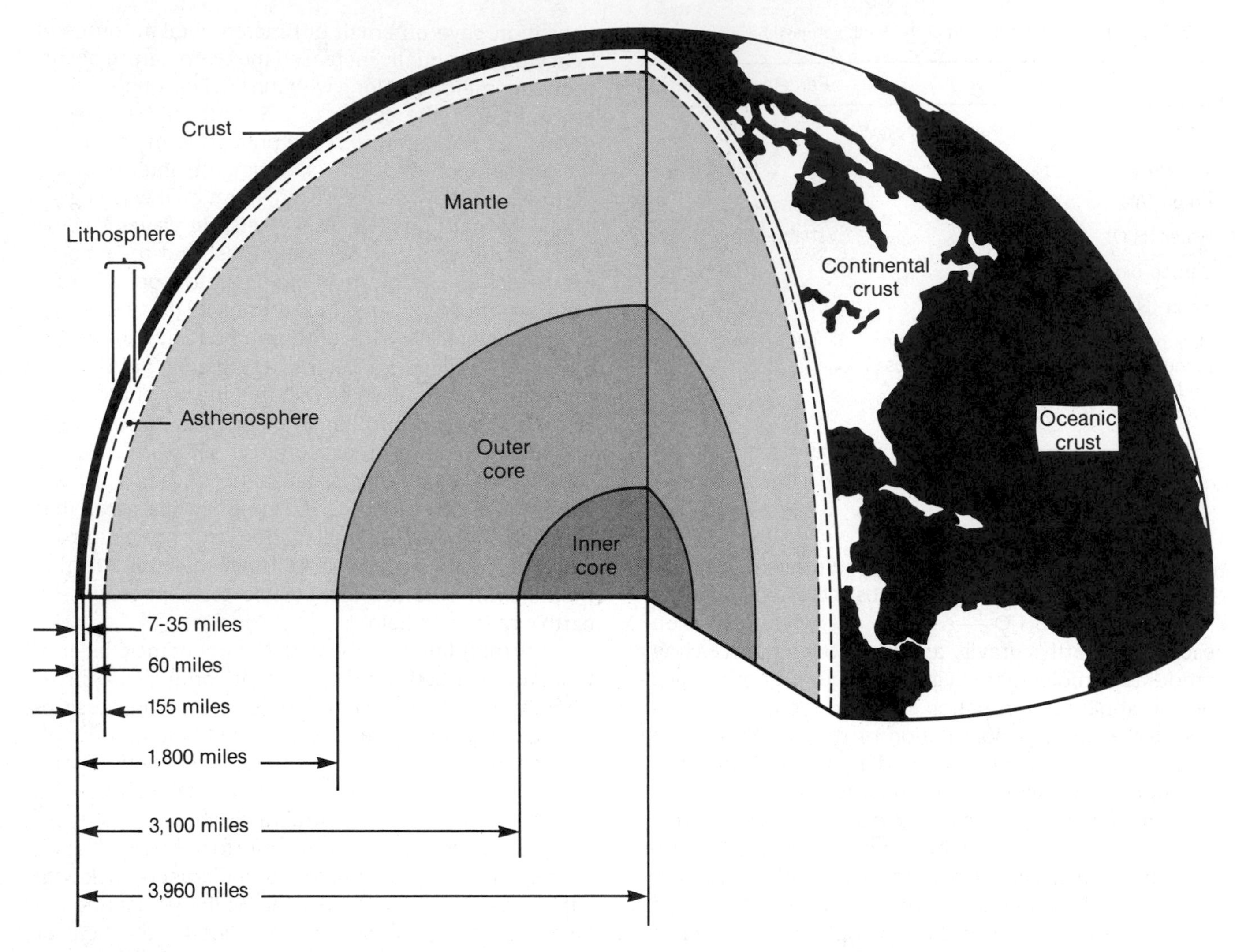

FIGURE 6-2 Section through the earth's interior, which possesses a radius of 3,948 miles (6,357 kilometers) at the poles and 3,960 miles (6,378 kilometers) at the equator. The lithosphere consists of relatively rigid plates composed of the rocklike crust plus a portion of the underlying mantle that reaches to a depth of about 50 miles at the oceanic basins and 60 to 90 miles at the continents. Below the lithosphere is a more fluid, deformable material, the asthenosphere, that allows the lithospheric plates to move about. (Adapted from Wyllie.)

geological debris into layers that settle out on various surfaces. Should such layers harden, either by the pressure of other layers upon them or by chemical means, a sedimentary rock is formed. In this process, gravity is the primary force accounting for the settling and layering observed. Sandstone (sand origin), shale (mud origin), and limestone (calcium carbonate) are examples of sedimentary rocks. In the case of limestone, it is most often found associated with the remains of organisms such as corals, molluscs, and other such organisms that lived in marine reefs and shallow seas and used calcium carbonate for their skeletal and habitat structures.

- **Metamorphic rocks:** These rocks were originally either igneous or sedimentary but later underwent significant changes because of heat, pressure, and/or chemical interactions. Marble, for example, is a metamorphic rock which was originally limestone, and slate is a metamorphic rock which was originally shale. According to some geologists, some forms of granite are also metamorphic rocks.

As shown in Figure 6-4, a rock cycle exists in which these three major types of rock, given enough time, are transformed from one to the other, although not necessarily in equal proportions. At present, the earth's crust is believed to consist, by volume, of 65 percent igneous rocks, 8 percent sedimentary rocks, and 27 percent metamorphic rocks. The surfaces of continental land masses, however, are covered mostly by a layer of sedimentary rocks.

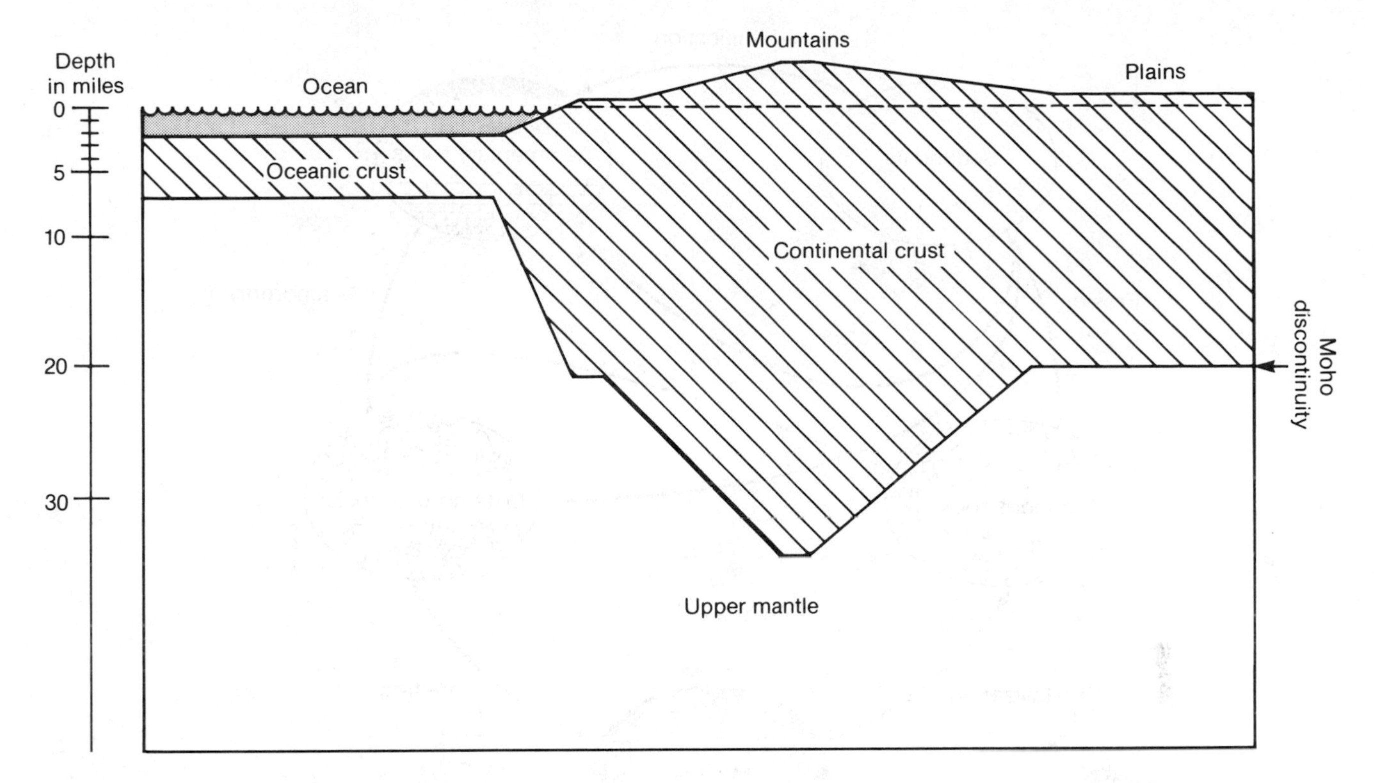

FIGURE 6-3 Section of the earth's crust showing differences in thickness. Since the crust has less density than the earth's subsurface material, it floats on the mantle, and different thicknesses of the crust float at different levels. Thus the thicker, and therefore more buoyant, continental land masses float higher compared to the thinner, less buoyant, oceanic basins. At the Moho discontinuity, a sharp change occurs in the velocity of certain seismic waves, and this area is considered to represent the boundary between crust and mantle. (Adapted from Sawkins et al.)

GEOLOGICAL DATING

Beginning in the seventeenth and eighteenth centuries, geologists became aware that the relative positions of different rocks could be used to determine their relative ages. Steno, an early proponent of the validity of fossils, was among the first to establish the **Law of Superposition**, which states that if a series of sedimentary rocks have not been overturned, the oldest layers or strata are at the bottom of the series and the youngest stratum is at the top. More than a century later, William Smith (1769–1839) discovered that different strata may be identified by the unique kinds of fossils found within them. As shown by Cuvier and others, the relative ages of the fossils seemed to correspond closely to the relative ages of the strata in which they were discovered. That is, fossils from the uppermost strata appeared to be more like modern organisms than fossils from lower strata (Fig. 6-5).

Fossils thus became a primary means by which a particular geological stratum or group of strata (system) could be traced in various localities. For example, the Cambrian system, named after a Welsh tribe by Sedgwick and Murchison in 1835, represents strata in which many marine invertebrate skeletons such as trilobites, brachiopods, and simple molluscs first appear. Cambrian strata can be found on all continents and occupy the same relative positions; that is, they are above Precambrian strata (absence of fossil shells) and below Ordovician and Silurian strata (true corals, echinoderms, small primitive fishes, etc.).

Unfortunately, not all geological strata are heavily marked by fossils, since fossils represent only a partial sampling of organisms and are mostly limited to those with shells, skeletons, or hard parts deposited in appropriate sediments. Soft-bodied organisms which could perhaps also identify strata are extremely rare in the fossil record. Furthermore, the same fossils are not always present in all locations of a stratum, since their distribution may have been limited to special habitats or areas. Nevertheless, fossils can usually serve to identify a particular stratum by the fact that all of its areas will generally contain at least some fossils characteristic of that period.

By these means a **Phanerozoic time scale** (or eon) has been defined as the period in which abundant visible (*phanero*) life (*zoon*) appears. It consists of three major eras of geological strata, beginning with the Paleozoic—the first in which significant numbers of hard-bodied fossils are found. As shown in Table 6-2, each

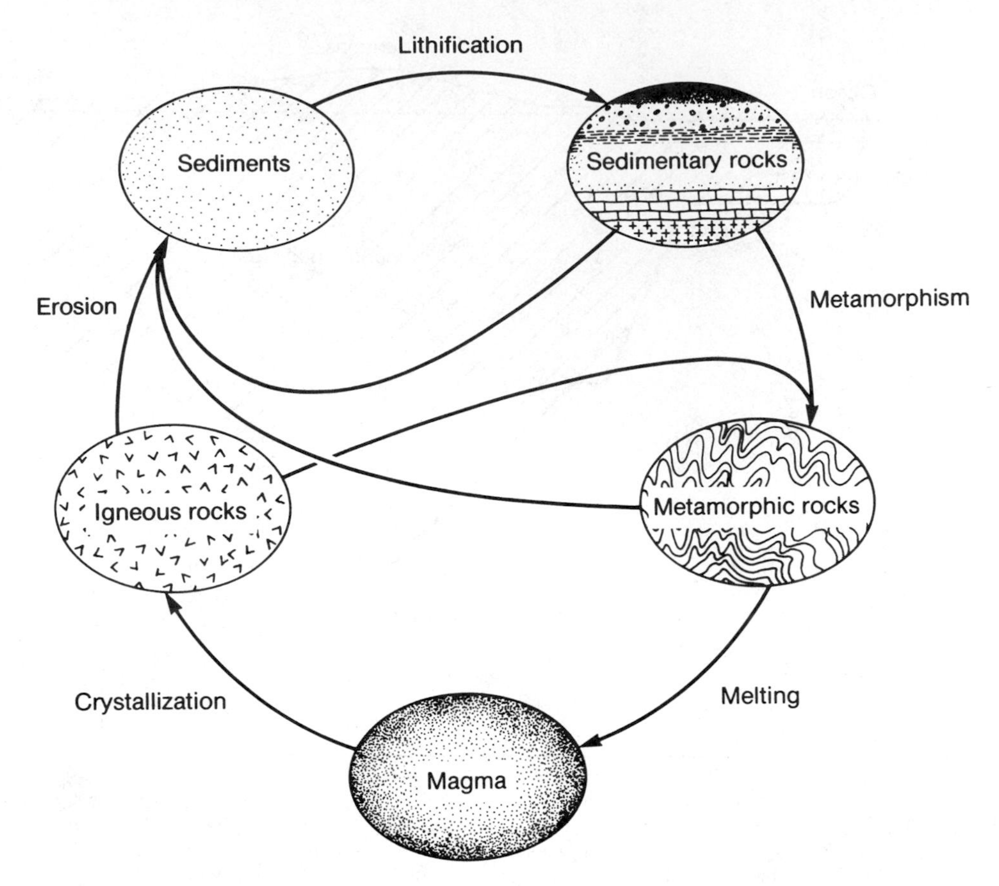

FIGURE 6-4 Diagrammatic representation of transitional events in the rock cycle. Some crustal rocks have been recycled many times, whereas others have persisted with little change from the initial formation of crustal rocks about 3.7 billion years ago. It is estimated that about half of all crustal rocks were formed during the last 600 million years.

era contains a number of subsidiary systems or periods, often further subdivided into series or epochs.[1]

However, although relative dating by stratigraphic methods usually enables establishment of a sequential relationship between different rocks and between different fossils, stratigraphy does not offer information on the time lengths involved. Sediments are not deposited in identical thicknesses from time to time or from place to place. Furthermore, in all localities large sections of the geological record have been worn away by erosions or destroyed by new rock formations and earth movement. Therefore, nowhere does the geological record offer a complete sequence that can be traced continuously, year by year to the present time.

DATING WITH RADIOACTIVE ELEMENTS

Fortunately, dating methods using radioactivity have been discovered which permit rocks even billions of years old to be dated with a fair degree of accuracy. All these methods rely on three main factors: (1) the ease with which many radioactive elements can be detected; (2) the known isotopes into which their atoms disintegrate; and (3) the known rates at which this disintegration occurs. For example, the radioactive element uranium 238 (^{238}U) is present in the mineral zircon found in most igneous rocks and disintegrates to

[1]The system of geological classification adopted in the eighteenth and early nineteenth centuries (initially suggested by the Italian geologist Arduino) followed the practice of designating primary rocks as those without fossils. These were believed to date from the origin of the earth's crust and appeared as typical non-stratified, ore-bearing outcroppings in mountainous areas. Stratified fossiliferous rocks, such as sandstone and limestone, were called secondary, and believers in the Judeo-Christian bible attributed their origin to the Noachian deluge. These secondary strata contained obviously ancient molluscan fossils such as ammonites and belemnites (Chapter 15) as well as early fish and reptiles that were considerably different from present forms. Tertiary sedimentary rocks were believed to be derived from secondary strata by flooding, erosion, volcanic action, and so on and contained ancient representatives of more recent forms such as mammals. Quaternary rocks represented the glacial and alluvial deposits that had occurred in relatively recent times. Since not all mountains nor all strata were of the same age, these divisions were difficult to apply universally, and the terms were eventually abandoned with the exception of Tertiary and Quaternary. Tertiary came to mean the period of preglacial deposits corresponding to most of the Cenozoic; Quaternary is used for the period dating from the Pleistocene ice age deposits to the present.

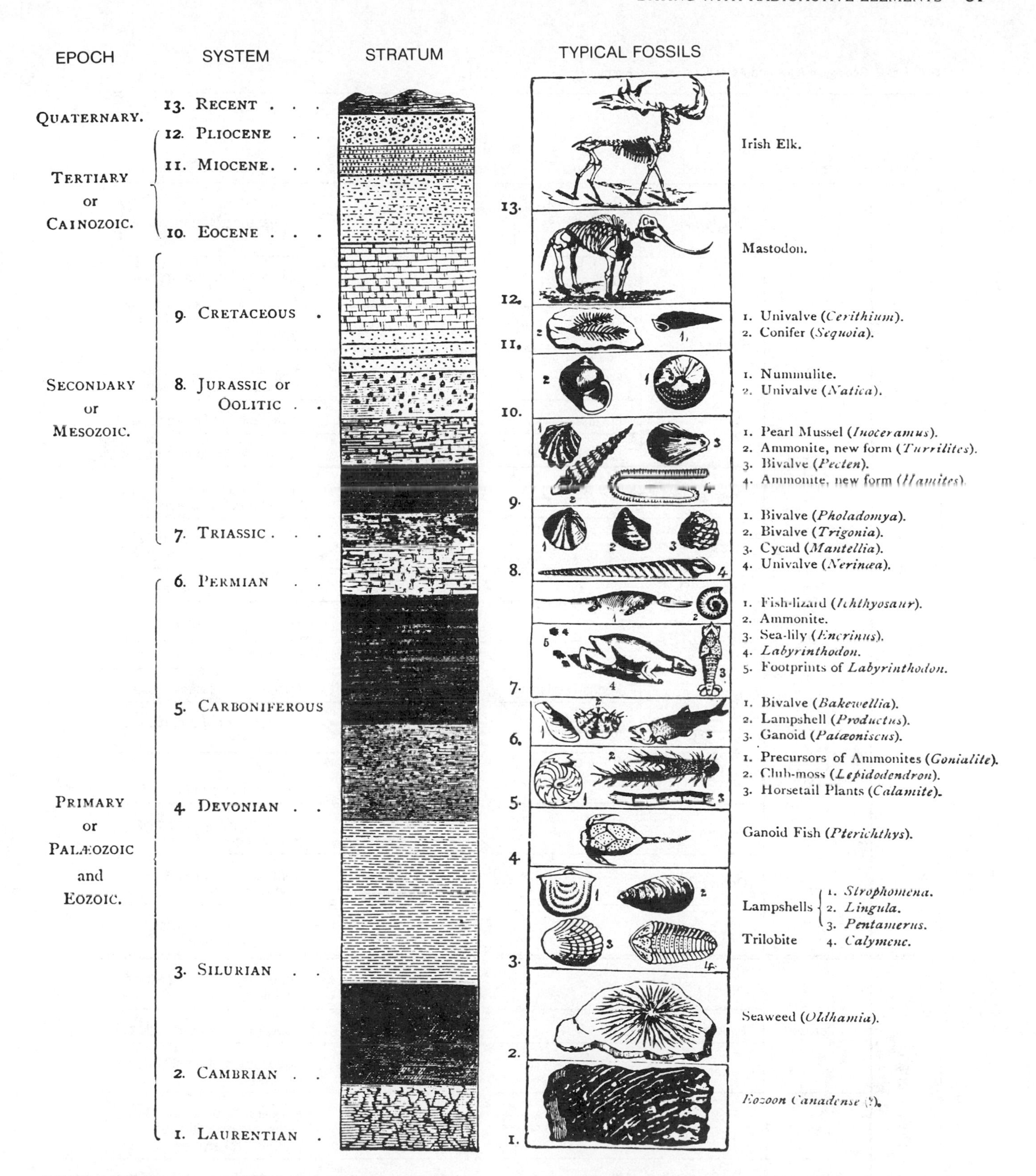

FIGURE 6-5 Nineteenth-century illustration of a table of stratified rocks that classifies geological strata according to their relative age and shows some of the fossils associated with each period. (From E. Clodd, 1888. *Story of Creation.* Longmans Green & Co., London.)

TABLE 6-2 Geological Ages and Associated Organic Events*

Time Scale (eon)	Era	Period	Epoch	Millions of Years Before Present (Approx.)	Duration in Millions of Years (Approx.)	Organic Events
Phanerozoic	Cenozoic	Quarternary	Recent (last 5,000 years)		2	Appearance of humans
			Pleistocene	2		
		Tertiary	Pliocene	5	3	Dominance of mammals and birds
			Miocene	24	19	Proliferation of bony fishes (teleosts)
			Oligocene	37	13	Rise of modern groups of mammals and invertebrates
			Eocene	54	17	Dominance of flowering plants
			Paleocene	65	11	Radiation of primitive mammals
	Mesozoic	Cretaceous		144	79	First flowering plants Extinction of dinosaurs
		Jurassic		213	69	Rise of giant dinosaurs Appearance of first birds
		Triassic		248	35	Development of conifer plants
	Paleozoic	Permian		286	38	Proliferation of reptiles Extinction of many early forms (invertebrates)
		Carboniferous	Pennsylvanian	320	34	Appearance of early reptiles
			Mississippian	360	40	Development of amphibians and insects
		Devonian		408	48	Rise of fishes First land vertebrates
		Silurian		438	30	First land plants and land invertebrates
		Ordovician		505	67	Dominance of invertebrates First vertebrates
		Cambrian		590	85	Sharp increase in fossils of invertebrate phyla
Precambrian	Proterozoic	Upper		900	310	Appearance of multicellular organisms
		Middle		1600	700	Appearance of eukaryotic cells
		Lower		2500	900	Appearance of planktonic prokaryotes
	Archean			3900	1400	Appearance of sedimentary rocks, stromatolites, and benthic prokaryotes
	Hadean			4500	600	From the formation of earth until first appearance of sedimentary rocks; no observable fossil organisms

*Dates derived mostly from Harland et al. Some geologists divide the Precambrian eon into only two major eras, Proterozic and Archean and then denote the Hadean as the first Archean period (see Fig. 9-13). However, the exact dates that mark each geological period are often only approximate, and other authors provide somewhat different time spans.

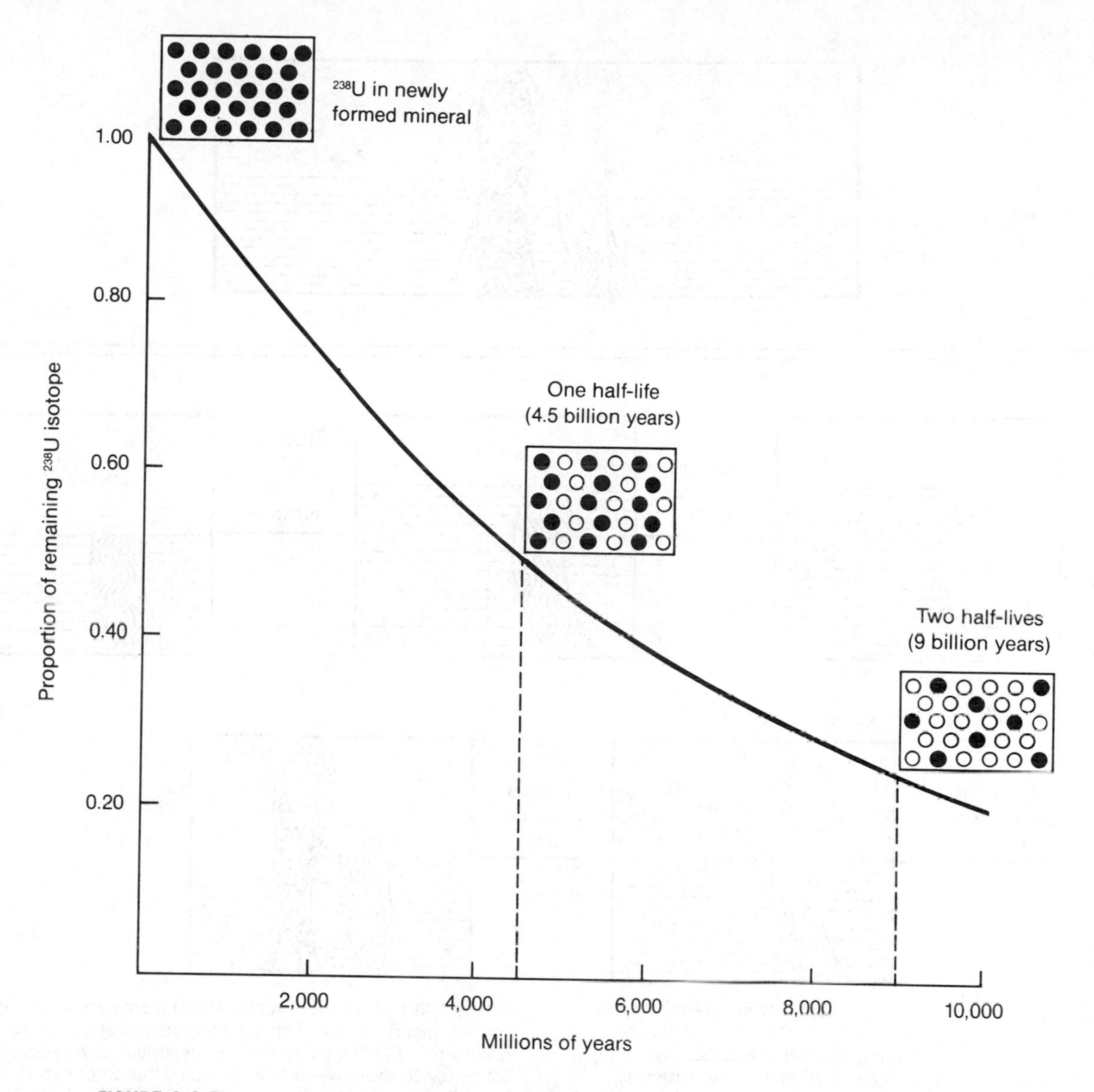

FIGURE 6-6 Theoretical relationship between duration of time in millions of years and the proportion of original ^{238}U isotope that remains in a rock, given a half-life of about 4.5 billion years. Note that the line is curved, not straight, and never quite reaches zero, since each half-life period reduces the amount of ^{238}U by 50 percent and some of the original isotope will always remain if the initial amount is large. (The fraction of ^{238}U that remains for any given period, x, is calculated as $(1/2)^y$ where $y = x/(4.5 \times 10^9)$.)

form the lead isotope ^{206}Pb at a rate which transforms half of the uranium into lead over a period (the half-life) of about 4.5 billion years (Fig. 6-6). Thus, after allowances are made for the presence of lead which was not produced by the uranium disintegration, and assuming these two isotopes have been fully retained, their relative amounts in a particular rock provide a fairly accurate dating method for older rocks.

A somewhat simplified formula that can be used for this purpose is

$$t = 1/\lambda \; ln \; (^{206}Pb/^{238}U + 1)$$

where t is time in years, λ is decay rate per year (1.537×10^{-10} for ^{238}U), and ln is the natural logarithm (base e). Thus, for example, a $^{206}Pb/^{238}U$ ratio of 0.360 in a particular sample would indicate that

$$t = 1/(1.537 \times 10^{-10}) \; ln \; (1.360)$$
$$= (6.508 \times 10^9)\,(.307) = 1.998 \times 10^9$$

that is, approximately two billion years have elapsed since the ^{238}U was first incorporated into this sample. Dates determined in this fashion can be checked by the disintegration rates of other radioactive elements present in the same material such as the decay of ^{235}U to ^{207}Pb (half-life of about 0.7 billion years). Additional radioactive elements used in dating include rubidium 87 which disintegrates to strontium 87, with a half-life of 47 billion years, and potassium 40 which disinte-

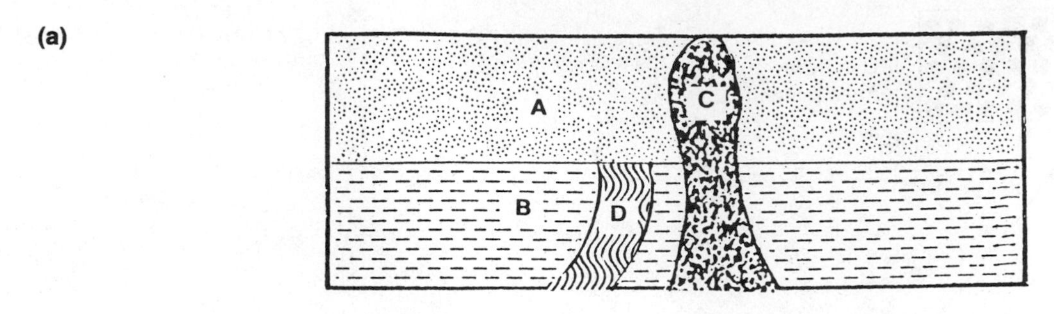

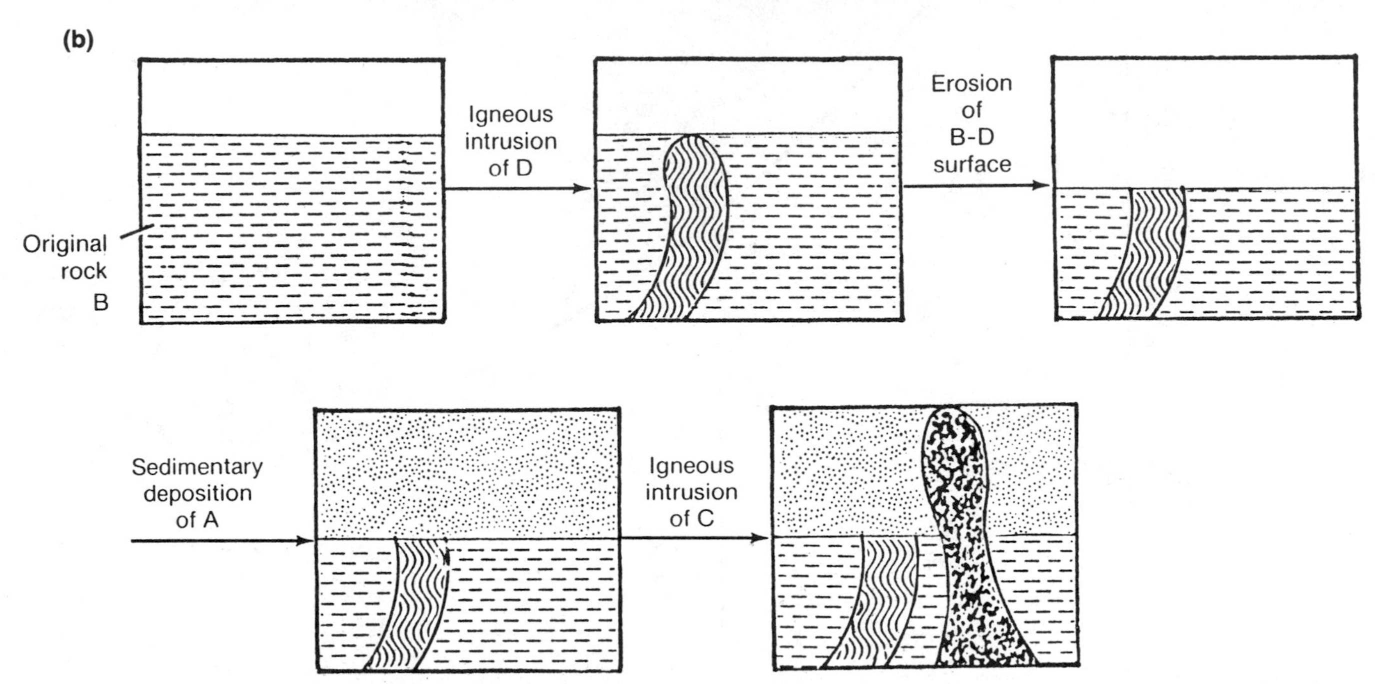

FIGURE 6-7 Use of relative and absolute dating in determining the ages of sedimentary and igneous rocks. (a) Diagram showing the observed relationships among four geological assemblies, two sedimentary (A and B) and two igneous (C and D). (b) Historical interpretation of the arrangement of these rocks based on the rules of superposition (younger sediments lie above older sediments) and crosscutting relationships (igneous rocks are younger than the rocks through which they cross). According to these principles, the B sediments are the oldest of rocks, and D represents a later igneous intrusion into B. Erosion then occurred, removing part of the intrusive igneous rock B, followed by the later deposition of A sediments. The last geological event was a new igneous intrusion of rock C into both A and B. The age relationships are therefore B>D>A>C. Thus, if the absolute ages of the two intrusions, C and D, can be determined by radioactive dating techniques, the upper and lower limits for the age of the A sediments can be determined.

grates to argon 40, with a half-life of 1.3 billion years. For dating fairly recent events, carbon 14 is now commonly used, which disintegrates into nitrogen 14 with a half-life of only 5,730 years.

So far, radioactive dating methods have been mostly applied to igneous rocks and the dates extended to sedimentary rocks by the relative positions of the two kinds of rock (Fig. 6-7). Thus, igneous rocks which coincide with the age of the Cambrian sediments are approximately 600 million years old. Later sedimentary rocks, as shown in Table 6-2, can be dated fairly precisely up to the recent period. As we go further back in time, the oldest terrestrial rocks are somewhat more than 3.5 billion years old, whereas estimates based on the combined isotope composition of lead in all earth materials point to an overall terrestrial age of about 4.6 billion years. This 4.6 billion-year estimate accords with the ages of moon rocks brought to earth by the Apollo lunar missions, as well as with similar estimates made for meteorites believed to have originated at the birth of the solar system. It is therefore now generally accepted that the earth was probably formed at low temperatures 4.6 billion years ago, and it took about a billion years for high pressures, radioactive heating, and surface cooling to generate continental masses and their igneous rocks. During the Archean period, 3.5 to 3.7 billion years ago, the presence of water and other weathering conditions were sufficient to enable the first of the presently observed sedimentary rocks to appear.

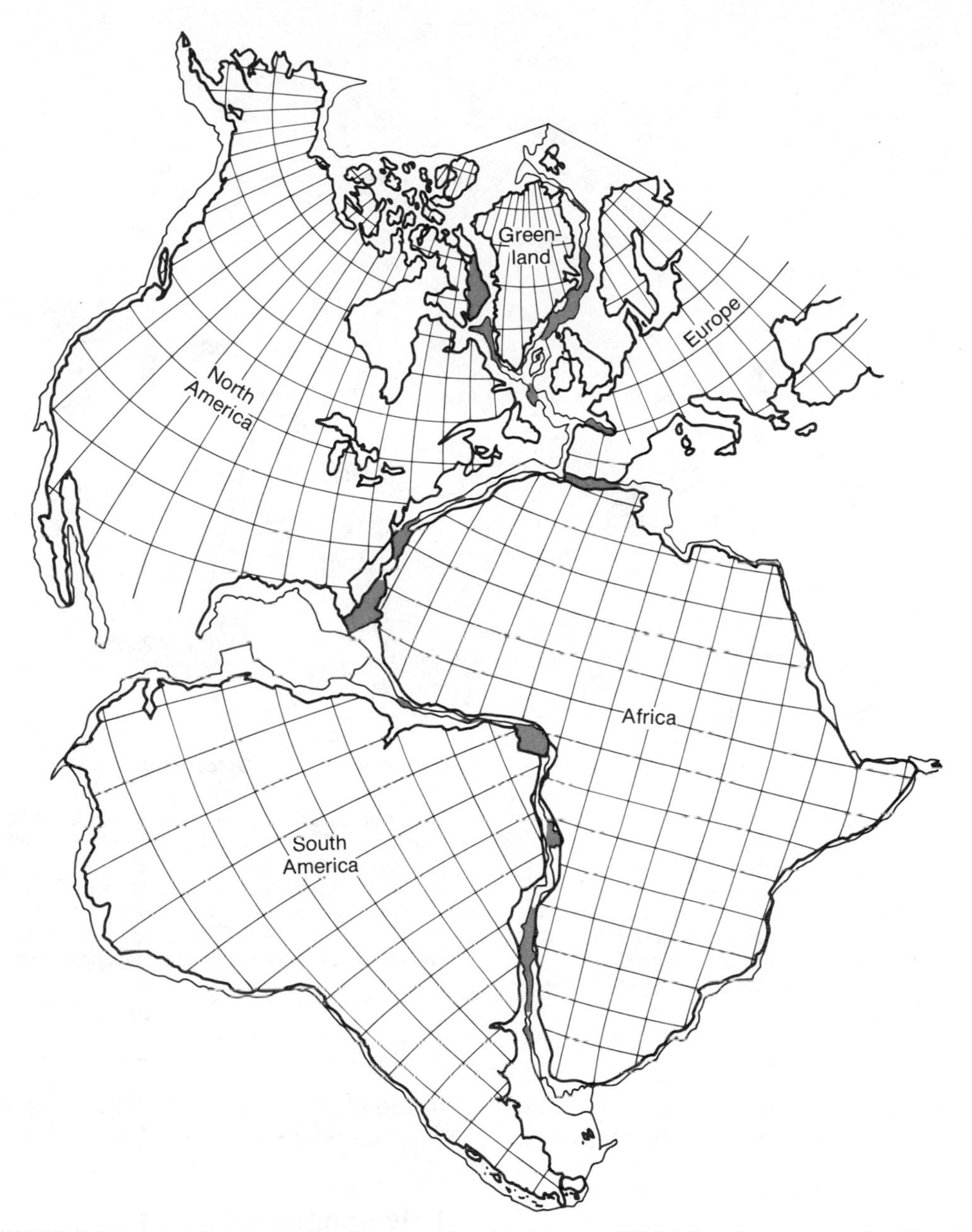

FIGURE 6-8 Matched fit between the offshore continental shelves at 500 fathoms deep on opposite sides of the Atlantic Ocean. (From Eicher, adapted from Bullard.)

Continental Drift

In the period between 1912 and 1930 a meteorologist, Alfred Wegener (1880-1930), developed the concept that all of the continents were at one time combined into a single land mass which he called **Pangaea**. He suggested that fissures occurred within this mass and the resulting fragments drifted apart to form the present continents. According to Wegener, drifting was caused by gravitational forces moving the continents through the viscous sea-floor material.

For the next few decades most geologists considered Wegener's theory little more than an imaginative fantasy until the evidence for continental drift became so overwhelming that it could no longer be ignored. This evidence is based on observations made of the fit between continents; similarity of rocks, fossils, and glaciation; paleomagnetism; and ocean floor spreading.

Fit of the Continents

As shown in Figure 6-8, one of the most striking geographic correlations is the exact match between the east coast of South America and the west coast of Africa. Not quite so obvious but nevertheless observable is the match between the east coast of North America and the northwest coast of Africa. These and other

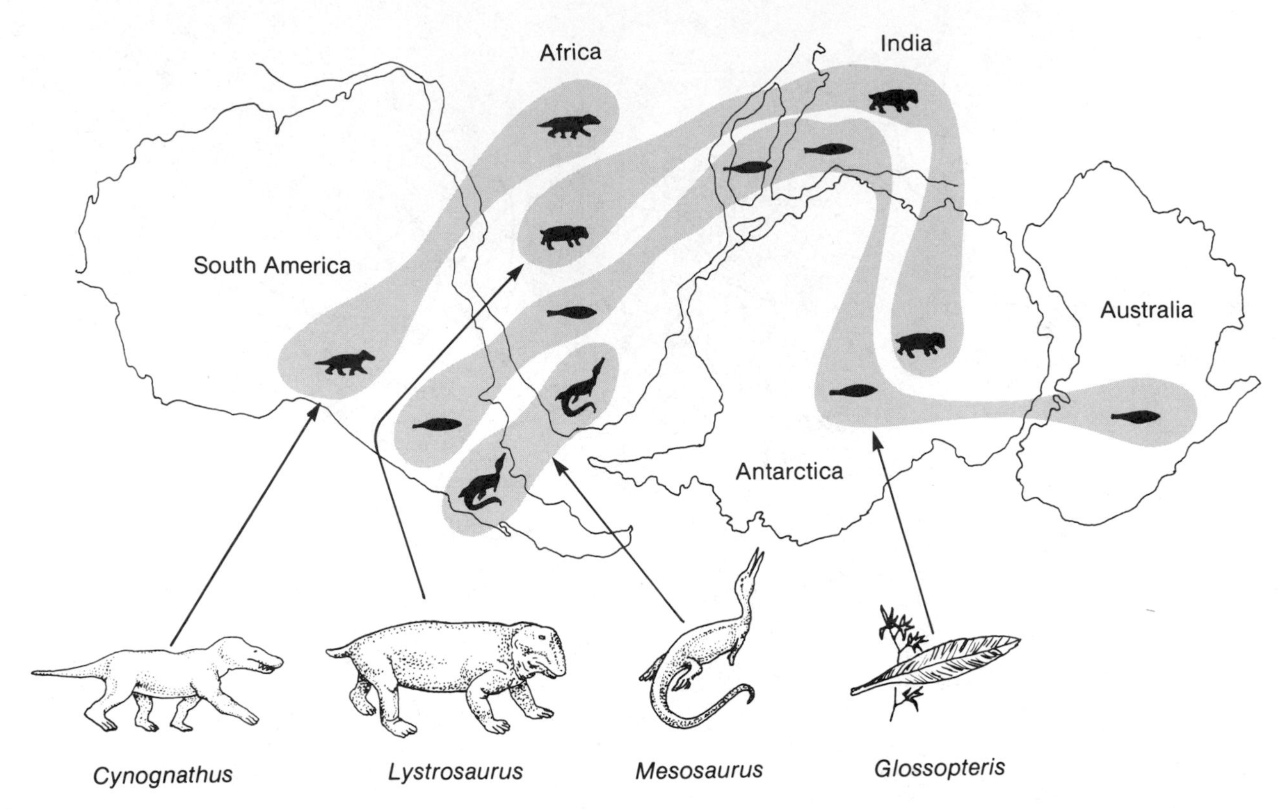

FIGURE 6-9 Distribution of various fossil plants and animals among the Gondwanaland continents. The presumed fit of the continental margins during the Permian-Triassic period is also shown. *Cynognathus* was a carnivorous mammal-like reptile, (therapsid, Chapter 17) with a distinctive doglike skull, found in Triassic-period deposits in South America and Africa. *Lystrosaurus* was also a Triassic mammal-like reptile but larger than *Cynognathus* and probably herbivorous, with beaklike jaws and two large tusks. The genus *Mesosaurus* represents a fossil order of freshwater reptiles restricted to Permian deposits in Brazil and South America. This reptile was about 1 1/2 feet in length with distinctive features of skull and limbs. *Glossopteris* was a fossil seed fern plant (gymnosperm) with large tongue-shaped leaves bearing a pattern of many reticulate veins. These fossil leaves are found in all of the Gondwana formations and date back to the early Permian period. (Adapted from Colbert.)

geographical juxtapositions indicate that the continents were at one time either joined together or extremely close.

Similarity of Rocks, Fossils, and Glaciations

A bed of rocks found in India called the Gondwana System dates from the late Carboniferous to the early Cretaceous period. Formations of extremely similar nature and composition are found in South Africa, South America, Antarctica, the Falkland Islands, and Madagascar. As shown in Figure 6-9, associated with a few of these Gondwana formations are unique types of fossil plants (*Glossopteris*) and animals (*Mesosaurus*, *Lystrosaurus*). Furthermore, all of the areas bearing Gondwana formations, along with Australia, were apparently covered by the same glaciation event during a Paleozoic ice age. To account for these observations geologists have suggested the existence of a massive southern continent, **Gondwanaland**, which included the areas that now carry the Gondwana formations and Australia. These land areas centered much closer to the South Pole than their present tropical locations and were therefore more easily glaciated.

Paleomagnetism

As new rocks arise from the cooling of magma, ferrous material within them is magnetized in a direction that depends upon the location and strength of the earth's magnetic field prevailing at the time. Should this magnetic field change for any reason, the magnetic field of newly formed rocks would also be expected to change. Thus rocks from all eras and all continents can be studied for their fossilized magnetism, or paleomagnetism, and the direction and distance of the earth's magnetic poles relative to these rocks can then be deduced.

Although paleomagnetic studies would be expected to show slight shifts in the magnetic poles, it was strange to find that these poles had shifted during past ages over thousands of miles and that the magnetic poles of different continents did not coincide for long periods of time. For example, as shown in Figure 6-10(a), the magnetic poles derived from analyzing

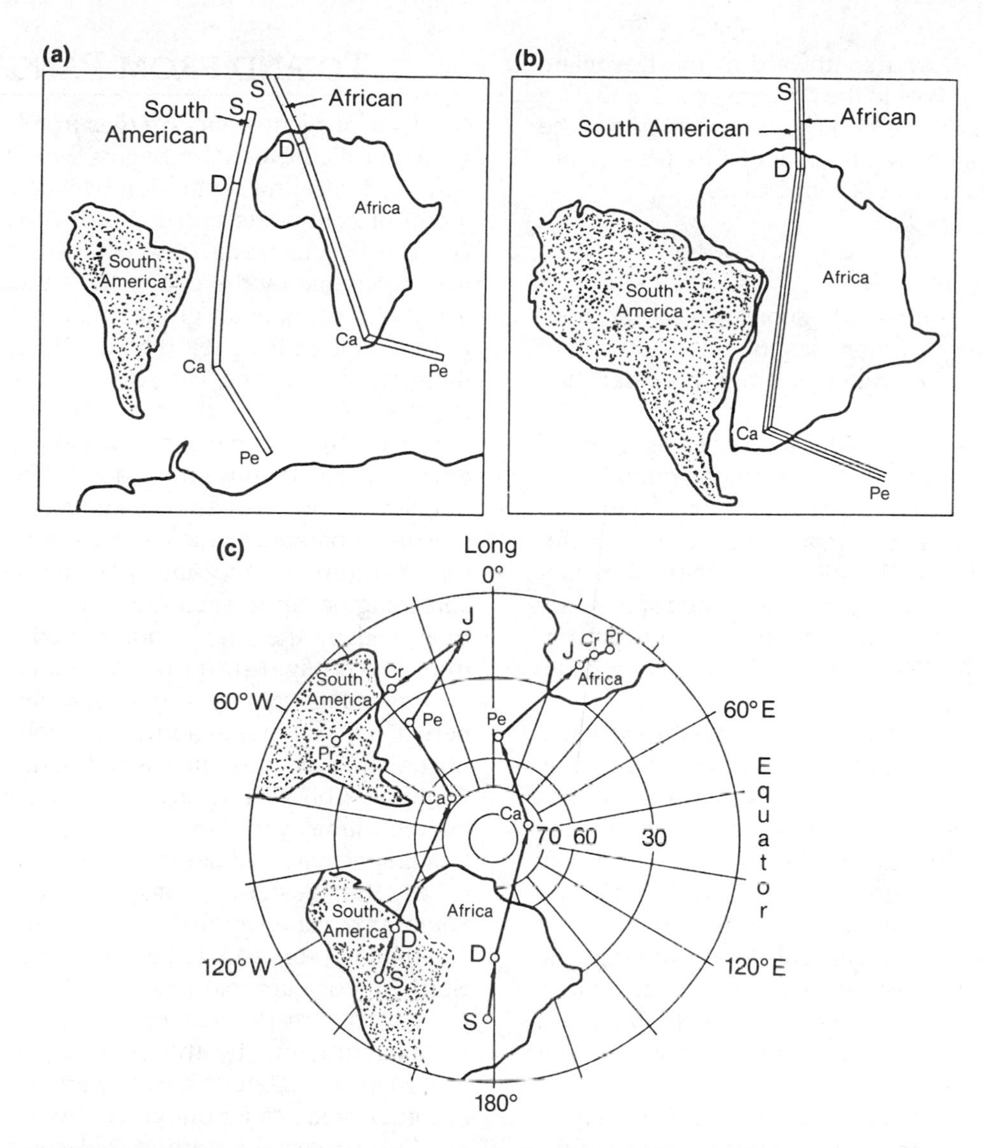

FIGURE 6-10 Magnetic pole wanderings of South America and Africa. (a) The two continents in their present relative positions, showing paleomagnetically determined locations of the south magnetic pole for the Silurian (S), Devonian (D), Carboniferous (Ca), and Permian (Pe) periods. The magnetic pole for these four periods is apparently in a different location for each continent, an anomaly that would be difficult to explain if each continent had always occupied its present relative position. (b) This reconstruction demonstrates that the two seemingly independent polar pathways shown in (a) coincide when the two continents are fitted together according to Figure 6-8. (c) View of the South Pole and Southern Hemisphere showing the geographic movement of the two continents from the Silurian period when they were juxtaposed to their present separated positions. The paleomagnetic data indicate that they moved as a unit across the South Pole through the Paleozoic era and began to separate during the Mesozoic (see also Fig. 6-13). The small circles in (c) indicate the geographic positions of a particular locality in each continent in various periods, from the Silurian (S) to the Jurassic (J), Cretaceous (Cr), and present (Pr). (Adapted from Cox and Moore.)

South American and African rocks that date between the Silurian and Permian periods are quite independent of each other. Since it is inconceivable that there were different magnetic poles for each continent, these different polar wanderings can best be explained as arising from the movement of continents relative to each other as well as from their movements relative to the poles. Figure 6-10(b) shows that the South American and African poles coincide for the Silurian-Permian period if the positions of these two continents are juxtaposed. Fig. 6-10(c) shows the geographical movements of these continents that are necessary to explain the apparent pathways of the magnetic poles if the magnetic poles are to be kept in their usual proximity to the geographic poles. Thus, a particular location in Africa during the Silurian period was at 30° latitude and 180° longitude in the Southern Hemisphere. This point (along with its adjoining continental land mass) then moved southward in the Devonian, almost to the South Pole in the Carboniferous, and then to further successive positions on the other side of the Southern Hemisphere until it reached its present position of 10° latitude and 30° longitude.

The Ocean Floor

A puzzling observation made soon after ocean floor samplings became common was the relative youth of the ocean floor. Its sediments seemed no older than 100 to 200 million years, and about 50 percent of its rock composition was no older than the beginning of the Tertiary period. Also, in contrast to the often folded and compressed sedimentary rocks in continental mountains, the oceanic mountains were almost exclusively made of igneous basalts. Since there was considerable evidence that oceans have existed since early geological history, the relative youth of the present ocean basins clearly indicated that older ocean floors must have been replaced.

Another unusual oceanic feature was the existence of magnetized belts that parallel the long midoceanic ridges found in almost all ocean basins. The measurement of the magnetic direction on both sides of such ocean ridges showed that each belt was symmetrically paired with a belt of approximately equal width and of the same magnetic orientation on the other side of the ridge. Belts that were adjacent to each other on the same slope, however, were usually magnetized differently.[2] Radioactive dating showed that the youngest belts were closest to the crest of the ridge and the older belts were farther away.

All these observations can best be explained if we assume that the midoceanic ridges represent fissures out of which new ocean floor emerges and then spreads to either side. Thus, molten rock emerging on each side of the oceanic ridge is magnetized upon cooling and is then displaced from the ridge by newly emerging material (Fig. 6-11). Somewhat like the annual growth rings of a tree, the ocean floor therefore retains its history in a series of parallel bands of rocks marked by magnetic fields prevailing at the time of their origin. However, **sea-floor spreading** is not uniform, and its annual rates vary from about 1 centimeter in the North Atlantic Ridge to 3 centimeters in the South Atlantic Ridge to as much as 9 centimeters in some portions of the Eastern Pacific Ridge.

TO AND FROM PANGAEA

One view of events that emerges from these studies is shown in Figure 6-12. It begins with a Devonian geography indicating separation between the Gondwana group of continents and a North American-Eurasian group called Laurasia (Fig. 6-12(a)). Most geologists now agree that by the end of the Paleozoic era, these two major continental groups had united to form the giant landmass Pangaea (Fig. 6-12(b,c)). According to these reconstructions, Pangaea began to break up during the Triassic period, about 225 million years ago (Fig. 6-12(d)). An oceanic rift developed between Western Gondwana (South America and Africa) and Eastern Gondwana (India, Antarctica, and Australia), and a further rift separated Laurasia from Western Gondwana (Fig. 6-12(e)). By the Late Jurassic period, sea-floor spreading began to separate North America from Europe, and by the Cretaceous period, South America from Africa (Fig. 6-12(f,g)).[3] The Indian subcontinent, moving independently from about the mid-Cretaceous period on, continued northward from the Antarctic-Australian mass until it reached Southern Asia in the Cenozoic. This Indian-Asian juncture is now marked by the Himalayan Mountains, in which mountain-building activity still seems to be going on.

In the Western Hemisphere, the rapid drift of South America away from Africa, which began about 100 million years ago, led eventually to a reunion with North America approximately 4 or 5 million years ago. In the Southern Hemisphere, Australia remained tied to Antarctica, but by the Eocene period, 45 million years ago, it had begun a northward journey which will eventually lead to its union with Asia.

The process of drifting and colliding continents has been extended even to Precambrian times. For example, it can be shown that the magnetic poles of the North American continent and the Gondwana group shared a common pathway for a period lasting more than a billion years during the Proterozoic era (Fig. 6-13). Although this indicates the existence of a giant continent containing most of the earth's surface, Northern Europe may have remained independent of North America until the middle of the Paleozoic era. Some geologists have also suggested that a number of Asian subsections were not actually united with each other and with Europe until the Mesozoic era (see Ziegler et al).

[2]Changes in magnetic orientation between adjacent belts are called reversals since they are caused by a 180-degree reversal in the polarity of the earth's magnetic pole. That is, the degree of magnetism weakens with time, and at some point it becomes reversed so that the south-pointing needle on a compass now points north. Using data from ocean floors and other sources, it has been shown that the duration of a particular magnetic polarity before it is reversed may vary from several thousand to 700,000 years.

[3]Some paleogeologists propose that sea-floor spreading is a consequence of an expanding earth. Owen, for example, suggests that the wide gap generally proposed to exist about 200 million years ago between Eastern Gondwana and Asia (occupied by the Tethys Ocean) would be eliminated if the earth's diameter at the time was 80 percent of its present value.

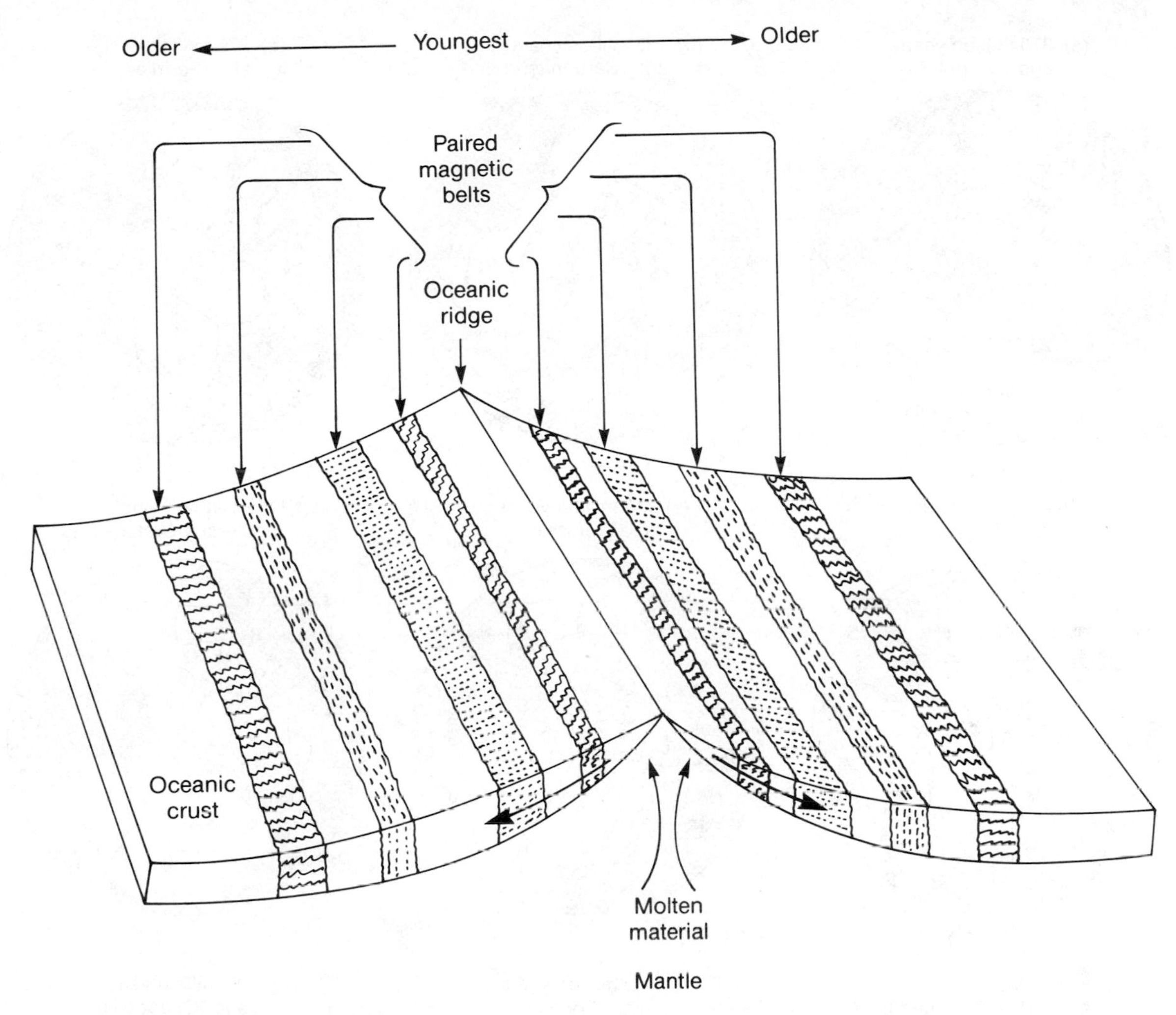

FIGURE 6-11 Diagrammatic section through an oceanic ridge showing sea-floor spreading. Hot molten material is added to the ridge from the mantle, falls away on both sides, and is magnetized in the orientation of the prevailing magnetic field as it cools. With time, the magnetic field changes in strength and/or direction, and new material added to the ridge forms a pair of belts distinctly different from adjacent belts.

TECTONIC PLATES

These seemingly varied and intricate continental movements are now known to be tied to the movements (tectonics) of gigantic plates of which the earth's crust seems to be composed. So far, six major plates as well as several minor ones have been delineated, marked mostly by earthquake belts that accompany plate movements (Fig. 6-14). In general, three major types of plate boundary events have been observed:

1. Plates can separate from each other by the addition of new lava material to their adjoining boundary. Such events occur in the oceanic ridges and account for sea-floor spreading and an increase in size of some oceanic basins (see Fig. 6-11).
2. Two adjoining plates can slide past each other at a common boundary, or fault, without any significant change in size. An example is the motion of the Pacific plate carrying a section of western California past the North American plate at the San Andreas fault. The speed at which this is occurring (6 centimeters per year) will bring Los Angeles to the same latitude as San Francisco in about 10 million years and to the Aleutian Islands near Alaska in about 60 million years.
3. One plate can move towards another causing a convergent boundary. When one such plate carries oceanic crust, the convergent event is often marked by the loss of plate material as this crustal mass plunges into the mantle. For example, the Pacific plate in its motion northward meets a border of the North American

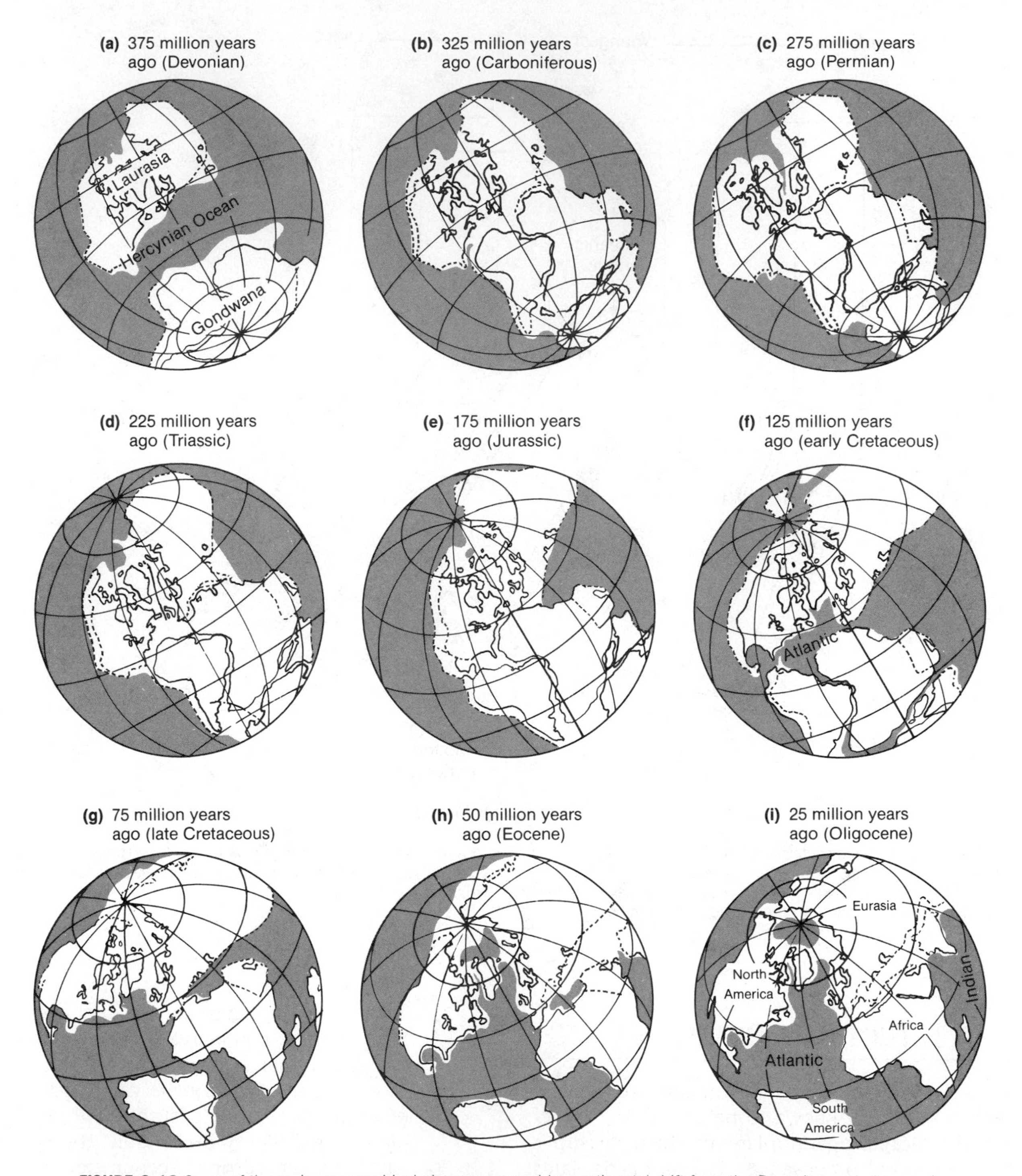

FIGURE 6-12 Some of the major geographical changes caused by continental drift from the Devonian period onward. (Adapted from Irving.)

plate at the Aleutian Islands and is then subducted under this trench (Fig. 6-15(a)). The descending plate changes or deforms the mantle which then produces volcanic activity and mountain formation in the crustal region above. Such processes account for some belts of volcanic islands, such as those located near the Aleutian and Java trenches, and also explain the origin of the South American Andes Mountains that lie near the Chilean trench, where the Pacific (Nazca) plate undercuts the westward-moving American plate (Fig. 6-15(b)) Once subduction has begun, a plate can continue to descend into the mantle, pushing oceanic sediments along with it until a continental mass meets the convergent boundary. Because the rocks of the continents are lighter and thicker than the ocean floor, they cannot apparently be

FIGURE 6-13 A possible reconstruction of the Gondwanaland-North American land mass for the approximate period between 1.0 and 2.2 billion years ago (bya). The heavy dark line shows the pathway of the magnetic pole determined for this land mass for given years (e.g., the magnetic pole was at the northwest tip of Africa 1.2 billion years ago). The exact position of Antarctica is uncertain. (Adapted from Piper, 1974.)

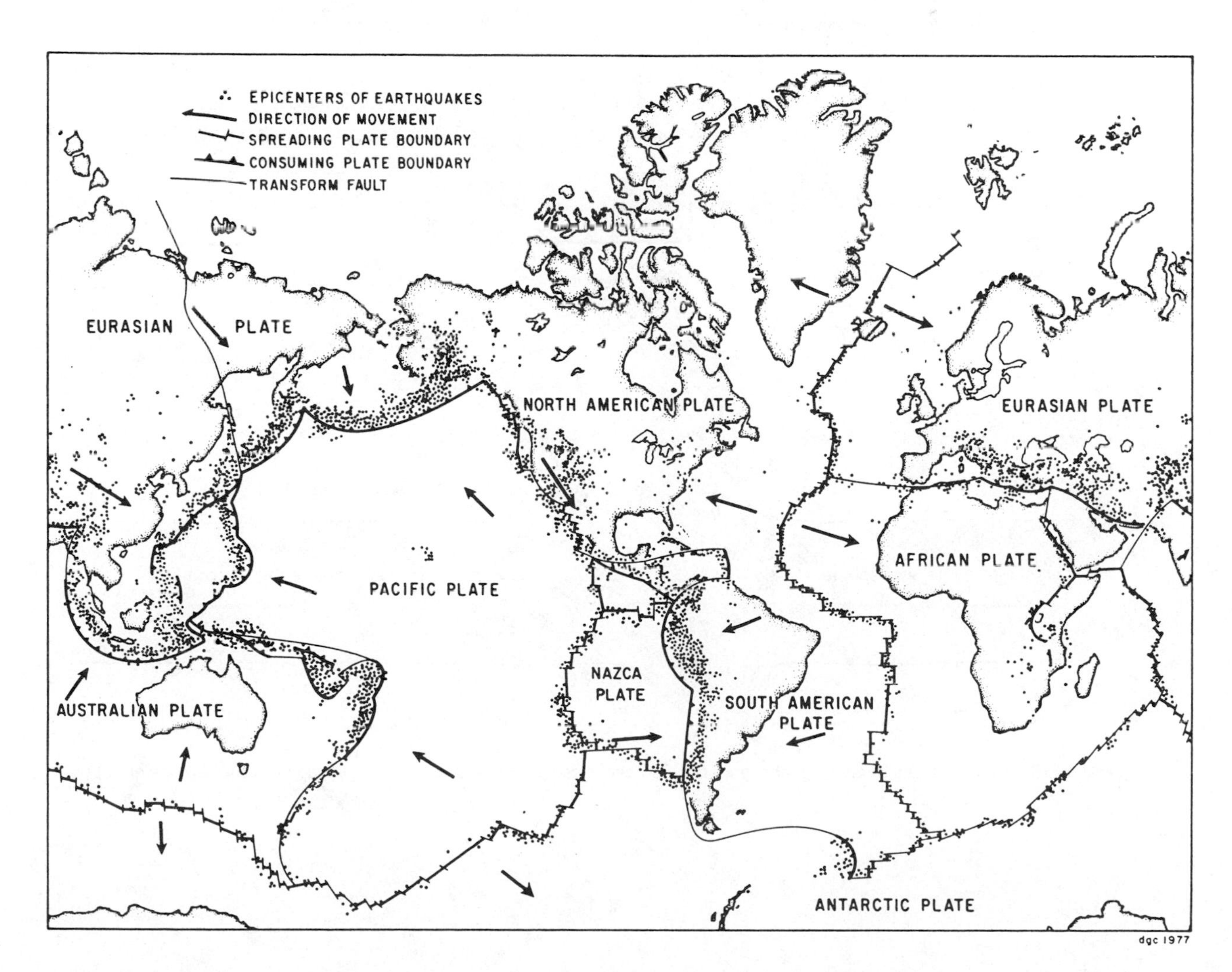

FIGURE 6-14 The major geological plates (and their boundaries) that account for many of the crustal movements. (From Cloud.)

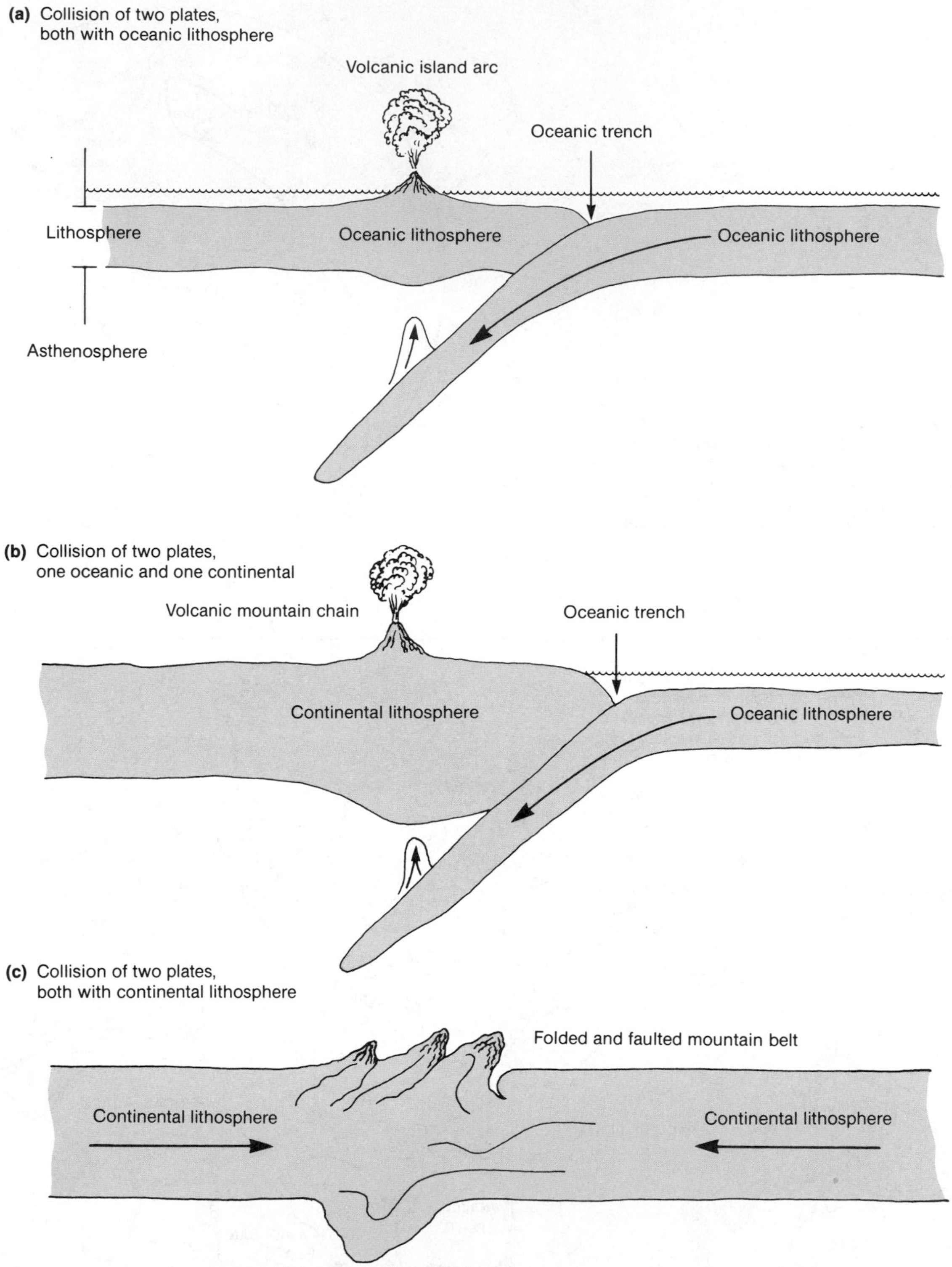

FIGURE 6-15 Three types of convergent events that can occur between lithospheric plates. (Adapted from Sawkins et al.)

forced under another plate, and, as a consequence, mountains such as the Himalayas are formed through the foldings and pressures of these colliding land masses (Fig. 6-15(c)).

Although the exact causes for plate tectonics are not fully known, plate tectonics has helped, more than any other geological theory, to explain the relative motion of continents since the Mesozoic era and clarified the localization of Gondwana deposits, island arcs, earthquake belts, and so on. Before the Mesozoic era, and certainly before the Paleozoic, mountain building and continental drift must also have occurred, but it is not yet clear whether such events were caused by the motions of the same rigid plates that are presently believed to exist or by the deformation of a more plastic crust. In any case, the separation and joining of continents has had important biological effects since it often determined the distribution of organisms and thus influenced subsequent evolution.

Biological Effects of Drift

One of the most prominent examples of the effect of continental drift on the distribution and evolution of organisms is the unique collection of primitive mammals found in Australia and South America. If we divide existing mammals (hairy skin, mammalian glands, special auditory skull bones; Chapter 18) into three groups, **Prototheria** (egg-laying monotremes such as the duckbilled platypus; Fig. 6-16(a)), **Metatheria** (marsupials, which undergo part of their development in an external female pouch; Fig. 6-16(b)), and **Eutheria** (placental mammals, which have their entire fetal development *in utero*; Fig. 6-16(c)), then we find that by the Miocene and Pliocene epochs, the placentals had completely replaced the more primitive monotremes and marsupials in every major locality except these two continents. In Australia there still exist two families of monotremes and thirteen of marsupials, but, with the possible exception of bats, no placentals were found on the continent until relatively recent times. The South American mammalian fauna seems to have been somewhat more advanced than that of Australia, and included, until the mid-Tertiary period, a number of placental families in addition to five families of marsupials. However, even the native South American placentals were generally primitive, as evidenced by mammals that persist there such as armadillos, anteaters, and tree sloths.

The picture of mammalian evolution that emerges from these studies is that primitive prototherians and metatherians had probably entered southern parts of Pangaea by the Upper Jurassic and Lower Cretaceous periods (Fig. 6-17(a)). The subsequent rifting of Australia isolated its primitive mammalian fauna from later competition with more advanced eutherian groups that evolved in Western Pangaea during the late Cretaceous and early Tertiary periods (Fig. 6-17(b)). In South America, the metatherians and primitive eutherians had probably replaced the prototherian groups by the early and mid-Tertiary period, but by that time the South American continent had drifted considerably from Africa and was also separated from North America (Fig. 6-17(c)). The evolution of mammals on the isolated island of South America was therefore largely independent of mammalian evolution elsewhere,[4] until South America rejoined North America via the Panama Isthmus during the Pliocene epoch (Fig. 6-17(d)). By the Pliocene epoch, however, considerable evolution towards more advanced eutherian forms had occurred either in Africa or Laurasia, whereas most of the South American mammalian fauna were still relatively primitive. When the Pleistocene epoch began, massive invasions of advanced northern eutherians were making their way south across the Central American land bridge, causing the rapid extinction of many South American mammalian families (Chapter 18). Only rarely, as in the case of opossums, did primitive South American mammals manage to successfully invade North America.

Supporting this view of continental drift and mammalian evolution is abundant fossil evidence of South American extinctions in the Pliocene and Pleistocene epochs, and the absence of any eutherian fossils in Australia up to recent times. As pointed out by Fooden, this view would also lead us to expect to find fossil remains of prototherians and metatherians in upper Triassic deposits of other land masses that separated from Pangaea concurrently with Australia, such as India and Antarctica. Should such fossils be discovered, they would add considerably to the predictive value of the continental drift theory.

In summary, continental drift must have had profound biological effects. It subjected moving land masses to new climatic conditions and geographical relationships, enabling their inhabitants to be selected for different evolutionary adaptations. Because of the breakup of land masses, drift must have also separated groups of organisms that were formerly associated, setting each such isolated species or group on its own evolutionary pathway. On the other hand, the joining of land masses because of continental drift led to com-

[4]During the Oligocene epoch, however, some island-hopping combined perhaps with transport on floating debris ("rafting") seems to have occurred, and a number of North American monkeys and caviomorph rodents made their way to South America.

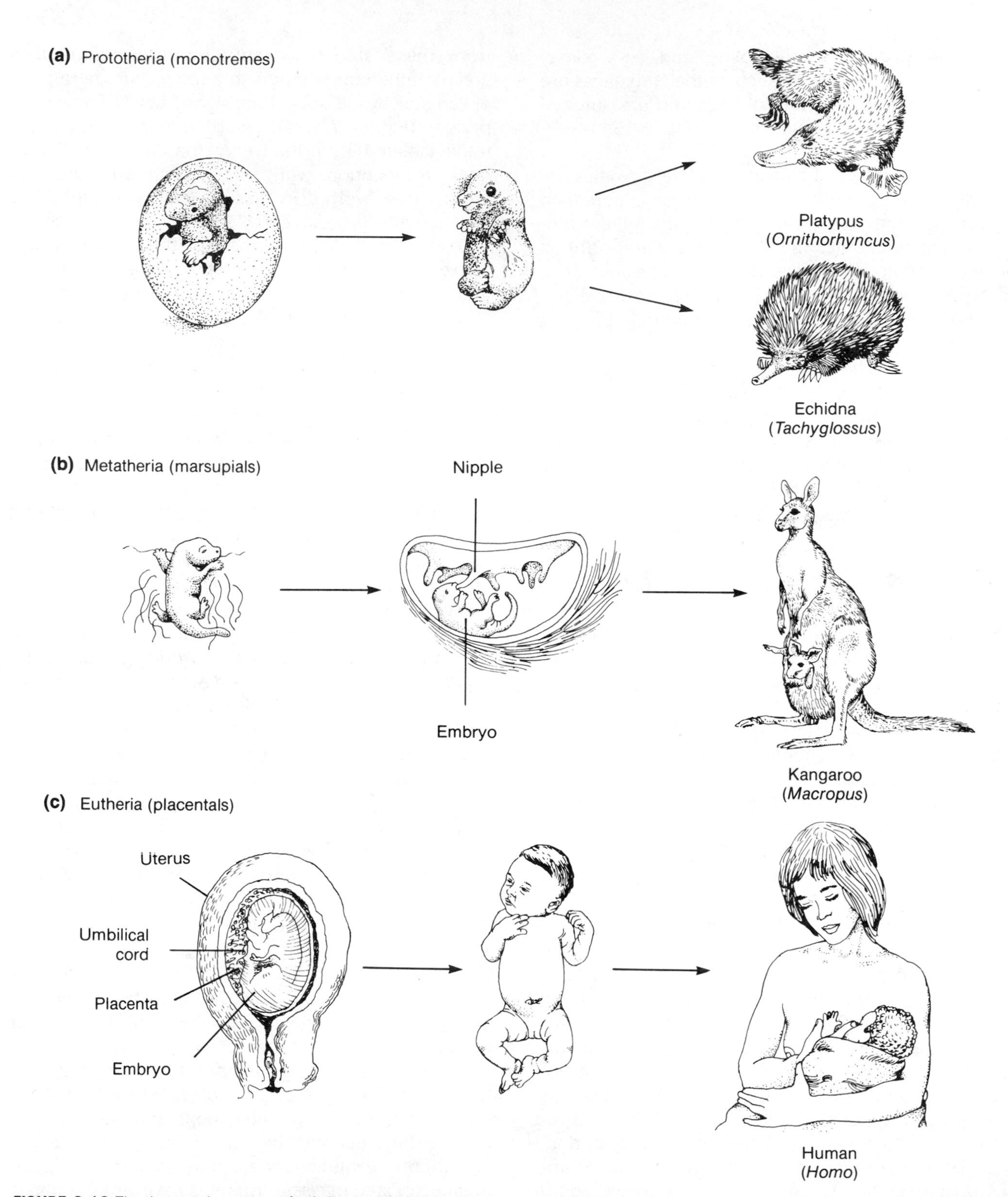

FIGURE 6-16 The three major groups of existing mammals, showing early developmental stages on the left and adult forms on the right. They all possess hair, mammary glands, and other features that distinguish them from reptiles but differ from each other in their pre-nursing development. Prototheria (a) lay eggs whose embryos hatch out to attach onto maternal abdominal hairs connected to mammary glands. In Metatheria (b), developing eggs remain in utero, and the emerging offspring then climb into the maternal pouch and attach themselves to mammary nipples. Eutherians (c) maintain the fetus in utero until a relatively late stage of development.

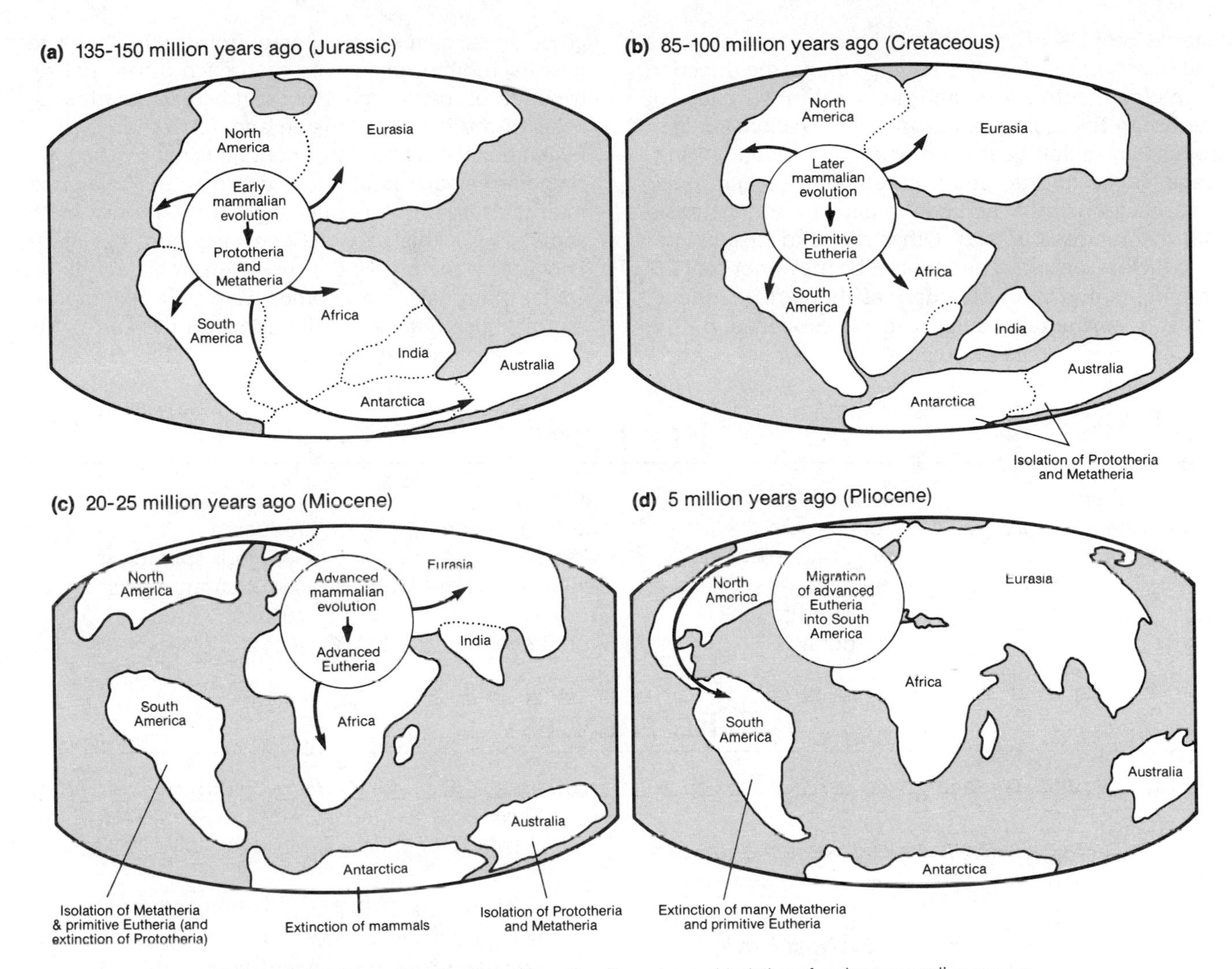

FIGURE 6-17 Effect of continental drift on the dispersion and isolation of major mammalian groups.

petition between previously separated groups of plants and animals that had evolved unique adaptations. These could then interact to produce increased complexity as well as extinction.

SUMMARY

Our solar system most probably originated by condensation from a rotating mass of gas about 4.6 billion years ago. The matter from which it condensed was unevenly distributed, and the central mass became the sun; the peripheral masses, the planets.

During this condensation elements were distributed according to a density gradient, with the outer planets receiving the lighter elements and the inner ones capturing the heavier ones. On earth an atmosphere was engendered containing many gaseous compounds such as ammonia, water, methane, carbon dioxide, and nitrogen, resulting in a reducing atmosphere with little or no oxygen. The high proportions of oxygen in the earth's present atmosphere probably arose from the later actions of photosynthetic organisms.

The interior of the earth is comprised of many concentric layers differing in many physical properties. The core is mainly molten iron and nickel and is surrounded by a thick, partially molten mantle. A thin crust of rock blankets the mantle and is composed of various types of rock: igneous, sedimentary, and metamorphic. The distribution of rocks in layers or strata accords with the age of the rock, and in the eighteenth and nineteenth centuries it became clear that different strata could be recognized by the types of fossils each contains. However, exact dating of strata had to await the discovery of radioactive elements, such as uranium, whose distintegration rates could be used for this purpose. The age of the earth has been determined to be 4.6 billon years by using such techniques.

The proposal by Wegener early in the twentieth century that the present land masses of the earth had once been united as one huge continent, Pangaea has since garnered much support in the form of the compatible profiles of continents, the similarity of rocks and fossils in previously conjoined areas, the direction of magnetism in rocks, and the youthful structure of the ocean floor. The disintegration of Pangaea began about 225 millon years ago when one section, Gondwana, broke up into South America-Africa and India-Antarctica-Australia, and both separated from Laurasia (North America-Eurasia). Other rifts and fusions created the present-day continents. The movements of the continents and many features of the earth's surface, such as earthquake belts, can be explained by the movement of at least six "plates" comprising the earth's crust. These plates can separate from each other, slide past each other, or converge on each other. The distribution of many organisms can be explained on the basis of the historical fusion and separation of these huge crustal masses. Most notable is the restriction of monotreme mammals to Australia and marsupial mammals both to Australia and South America by the separation of this part of Gondwana from the rest of Pangaea, while elsewhere they were replaced by placental mammals. Many other examples also indicate the biological effects of alterations in the earth's surface on evolutionary patterns.

KEY TERMS

Collision theory	Gondwanaland	oxidizing	radioactive dating
Condensation theory	igneous rocks	paleomagnetism	reducing
continental drift	Law of Superposition	Pangaea	sea-floor spreading
core	mantle	Phanerozoic time scale	sedimentary rocks
crust	metamorphic rocks	protoplanets	seismic waves
Eutheria	Metatheria	Prototheria	tectonic plates

REFERENCES

Badash, L., 1989. The age of the debate. *Sci. Amer.*, **261**(2), 90-96.

Cloud, P., 1978. *Cosmos, Earth, and Man: A Short History of the Universe.* Yale Univ. Press, New Haven, Conn.

Cocks, L. R. M. (ed.), 1981. *The Evolving Earth.* Cambridge Univ. Press, Cambridge.

Colbert, E. H., 1973. *Wandering Lands and Animals.* Hutchinson, London.

Dott, R. H., Jr., and R. L. Batten, 1981. *Evolution of The Earth*, 3rd ed. McGraw-Hill, New York.

Eicher, D. L., and A. L. McAlester, 1980. *History of the Earth.* Prentice-Hall, Englewood Cliffs, N.J.

Fooden, J., 1972. Breakup of Pangaea and isolation of relict mammals in Australia, South America, and Madagascar. *Science*, **175**, 894-898.

Harland, W. B., A. V. Cox, P. G. Llewellyn, C. A. G. Pickton, A. G. Smith, and R. Walters, 1982. *A Geologic Time Scale.* Cambridge Univ. Press, Cambridge.

Hughes, D. W., 1989. Planetary planarity. *Nature*, **337**, 113.

Irving, E., 1977. Drift of the major continental blocks since the Devonian. *Nature*, **270**, 304-309.

Jardine, N., and D. McKenzie, 1972. Continental drift and the dispersal and evolution of organisms. *Nature*, **234**, 20-24.

LeGrand, H. E., 1988. *Drifting Continents and Shifting Theories: The Modern Revolution in Geology and Scientific Change.* Cambridge Univ. Press, Cambridge.

Owen, H. G., 1983. Some principles of physical palaeogeography. In *Evolution, Time and Space: The Emergence of the Biosphere*, R. W. Sims, J. H. Price, and P. E. S. Whaley (eds.). Academic Press, London, pp. 85-115.

Piper, J. D. A., 1974, Proterozoic crustal distribution, mobile belts and apparent polar movements. *Nature*, **251**, 381-384.

__________, 1987. *Paleomagnetism and the Continental Crust.* Wiley & Sons, New York.

Sawkins, F. J., C. G. Chase, D. G. Darby, and G. Rapp, Jr., 1974. *The Evolving Earth.* Macmillan, New York.

Wetherill, G. W., 1988. Formation of the Earth. In *Origins and Extinctions,* D. E. Osterbrock and P. H. Raven (eds.). Yale Univ. Press, New Haven, Conn. pp. 43-82.

Wyllie, P. J., 1975. The earth's mantle. In *Continents Adrift and Continents Aground*, Scientific American (Introduction by J. T. Wilson.). W. H. Freeman, San Francisco, pp. 46-57.

Ziegler, A. M., C. R. Scotese, W. S. McKerrow, M. E. Johnson, and R. K. Bombach, 1979. Paleozoic paleogeography. *Ann. Rev. Earth Planet. Sci.*, **7**, 473-502.

MOLECULES AND THE ORIGIN OF LIFE

The problem of how life could have originated by ordinary chemical means long seemed insuperable. This is hardly surprising if we consider the present complexity of even the most elementary unit of life, the cell. At its simplest, a modern cell (Fig. 7-1) is surrounded by a highly selective permeable membrane composed of lipids and proteins which regulates the kinds of substances that pass through. Within the cell, the cytoplasm is composed of a multitude of structures and substructures involved in the synthesis, storage, and breakdown of a large variety of chemical compounds.

AMINO ACIDS

Foremost among the metabolic agents that enable the cell to function are the many different proteins that catalyze and regulate practically all living chemical reactions. In basic structure, proteins are composed of subunits, called amino acids, which have the following features:

$$\begin{array}{ccccc} & & H & & O \\ & & | & & \| \\ H_2N & - & C^* & - & C-OH \\ & & | & & \\ & & R & & \end{array}$$

1. An alpha carbon atom (C*) to which all other parts are attached.
2. An amino NH_2 group with a potential positive charge (NH_3^+).
3. A carboxyl COOH group with a potential negative charge (COO^-).
4. An H atom.
5. An R side chain that varies in structure among the different amino acids (Fig. 7-2).

These amino acids are linked together by chemical bonds called **peptide linkages** (see Fig. 7-10) into linear **polypeptide chains** that are the constituents of proteins. The highly specific structure of any protein molecule, whether it functions as an **enzyme (catalyst)** or for some other purpose, is derived from the exact linear placement of its various component amino acids. These specific amino acid sequences enable polypeptide chains to fold into specific three-dimensional forms that confer specific properties on proteins. It is, in fact, reasonable to claim that most present living phenomena, whether absorption, sensation, motion, structure, or whatever, derive from the enzymatic and regulatory activities of these long sequences of amino acids. Complexity, however, is not limited to proteins, since the amino acid sequences of

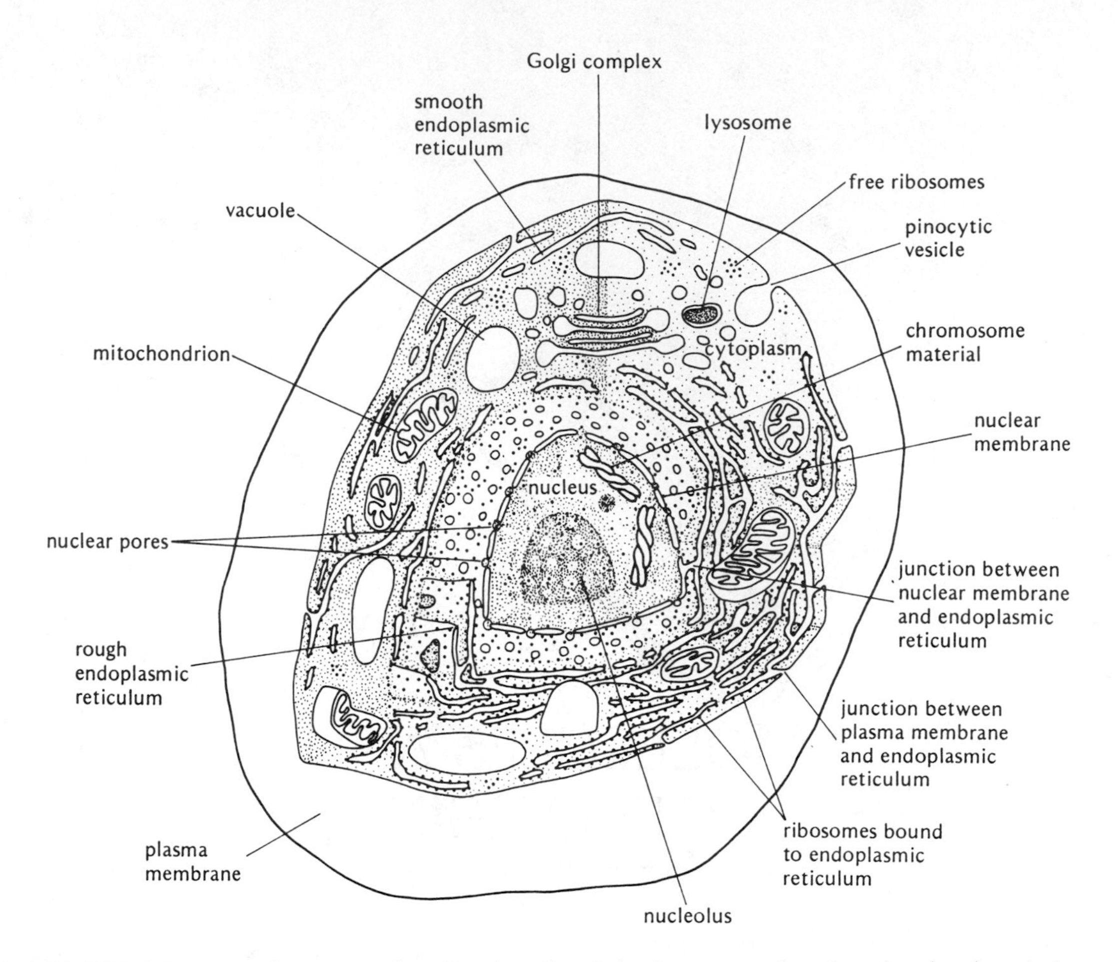

FIGURE 7-1 Diagrammatic representation of a eukaryotic cell showing cross sections through various important cellular organelles. (Adapted from Strickberger.)

proteins are actually determined by the nucleotide sequences in another group of basic molecules, nucleic acids.

Nucleic Acids

Nucleic acids, as shown in Figure 7-3, are long-chained molecules composed of **nucleotide** subunits, each containing a pentose (5-carbon) sugar, a monophosphate group, and a nitrogenous base. The two kinds of sugar used in nucleic acids, ribose (hydroxylated at the 2′ carbon position) and deoxyribose (lacks 2′ hydroxyl) provide the names for the two kinds of nucleic acids, ribonucleic acid (RNA) and deoxyribonucleic acid (DNA). In both these nucleic acids, the phosphate groups occupy the same position, serving to tie the 3′ carbon of one sugar to the 5′ carbon of its neighbor via a phosphodiester bond. Connected to the 1′ carbon of each sugar is one of four kinds of nitrogenous heterocyclic bases, of which two are purines [adenine (A) and guanine (G) in both DNA and RNA] and two are pyrimidines [cytosine (C) and thymine (T) in DNA and cytosine (C) and uracil (U) in RNA].

Since the complexity of proteins derives from the complexity of nucleic acids, it would seem that the restriction of nucleic acid composition to only four different kinds of bases would limit the message-bearing capacity of these molecules to only four kinds of messages, but this is not so. The facts that nucleic acid molecules may be many thousands or millions of nucleotides long and that each message can be encoded by a unique linear sequence of nucleotides endows these molecules with the capacity to bear an immense variety of highly complex messages. For any one nucleotide position, 4 different messages are possible (A, G, C, or T); for two nucleotides in tandem, 4^2 or 16 different messages are possible (AA, AG, AC, AT, GG, GC, ...); and so on: the rule being simply that for a linear sequence of n nucleotides, 4^n different possible messages can be encoded. Thus a linear sequence of only 10 nucleotides can be used to discriminate between more than 1 million (4^{10}) potentially different messages.

All of this helps explain the information-carrying role of nucleic acids but does not explain how this information is replicated and transmitted. The model presently accepted for nucleic acid replication is derived from the now-familiar **double helix** structure first

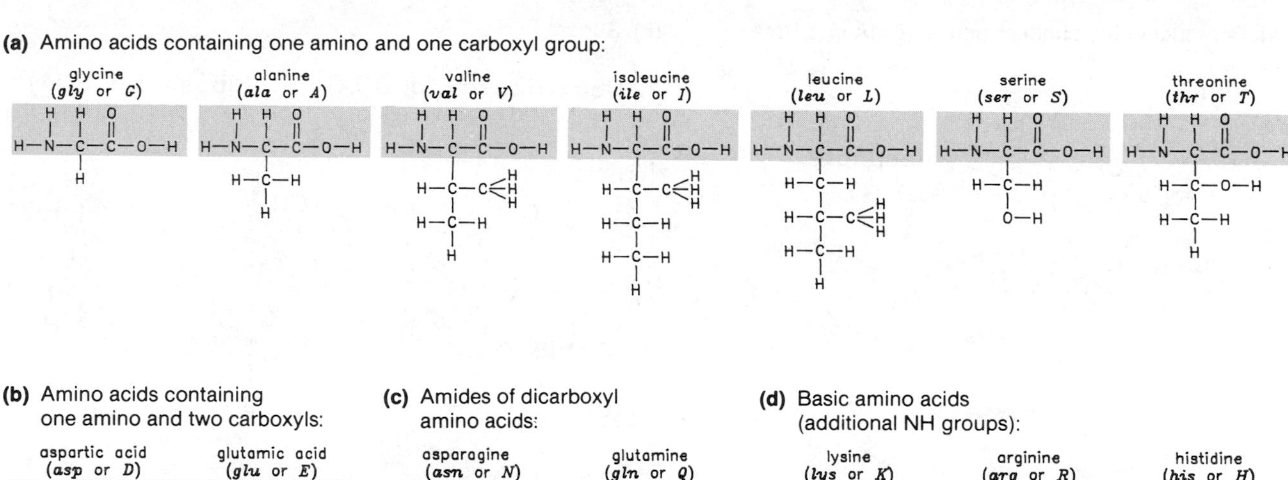

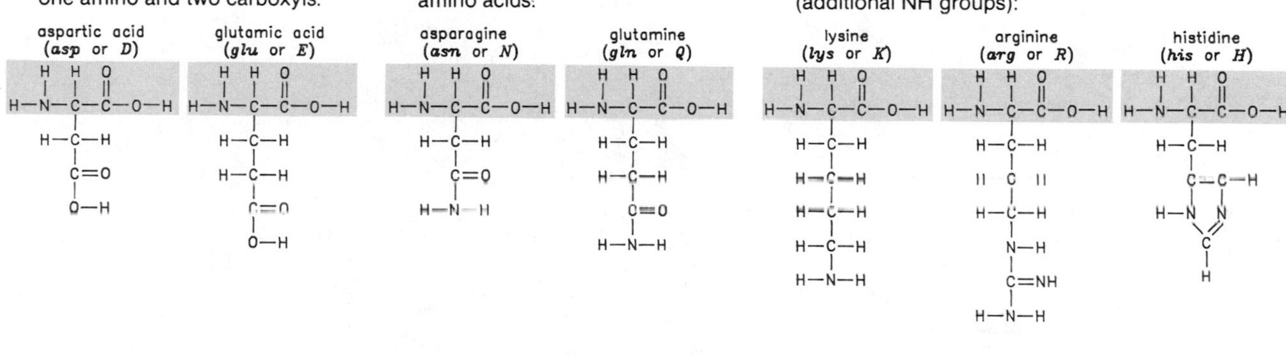

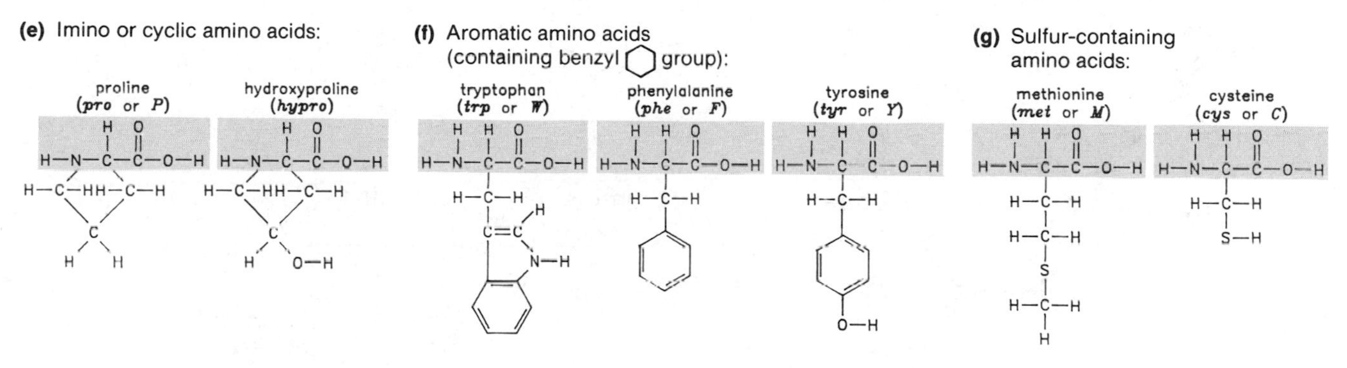

FIGURE 7-2 Structures of the common amino acids found in proteins.

offered by Watson and Crick. The DNA double helix is composed of two anti-parallel strands coiled around each other in the form of a right-handed screw, with complementary pairing between purine bases on one strand and pyrimidine bases on the other (A-T, G-C). In the familiar B form of DNA, diagrammed in Figure 7-4, there are approximately ten base pairs for each complete turn of the helix, and the bases are stacked almost perpendicularly to the helical axis.[1] The replicatory power of the DNA double helix obviously derives from the ability of each of the two strands to serve as a template for a newly complementary strand, so that two new double helices can be formed bearing nucleotide sequences which are identical to each other as well as to that of the parental molecule (Fig. 7-5). It is this unique quality of exact molecular replication, enabling similar messages to be transmitted from generation to generation, that confer upon nucleic acids their function as "genetic material."

Basic to our understanding of the relationship between genetic material and protein, therefore, is an important concept: that the three-dimensional structure of a protein—its form, shape, and subsequent function—is primarily determined by the linear se-

[1]DNA takes various structural forms depending on relative humidity, salt concentration, and so on. The first two DNA forms investigated, crystalline (dry) and wet (hydrated), were given the respective names A and B by Rosalind Franklin, an x-ray crystallographer whose photographs provided Watson and Crick essential information in devising their model. In the A form, 11 bases are reported per turn of the helix, and the bases are noticeably tilted (20°) relative to the helical axis.

(a) Polynucleotide chain structure (DNA and RNA)

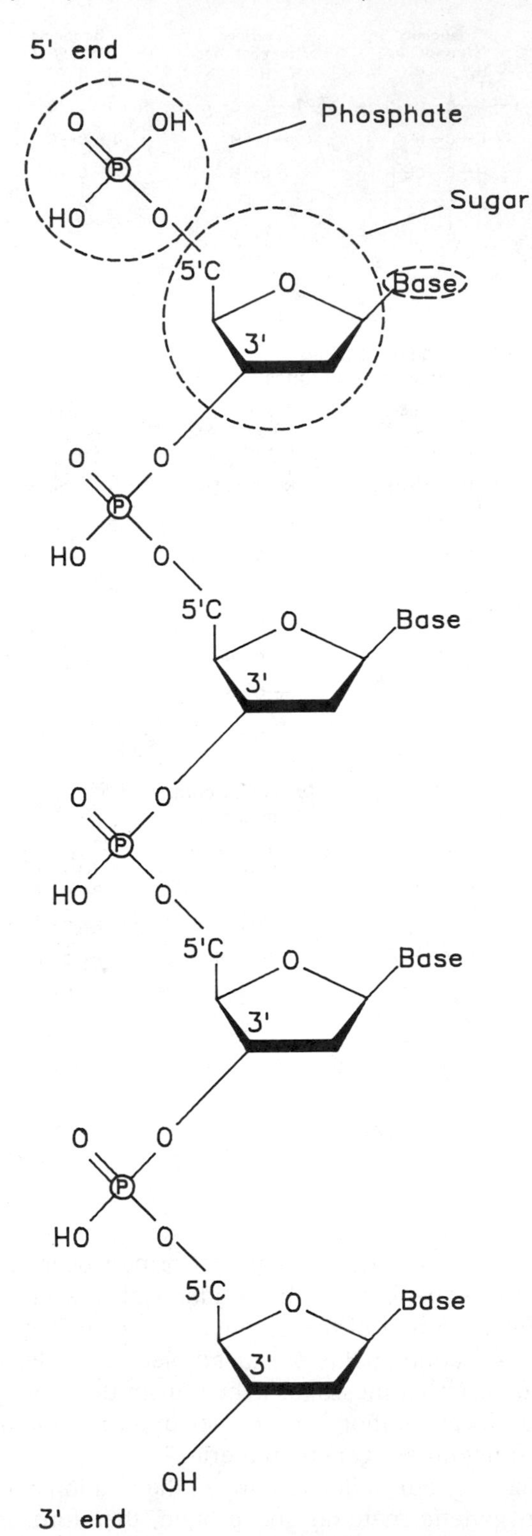

(b) Sugars

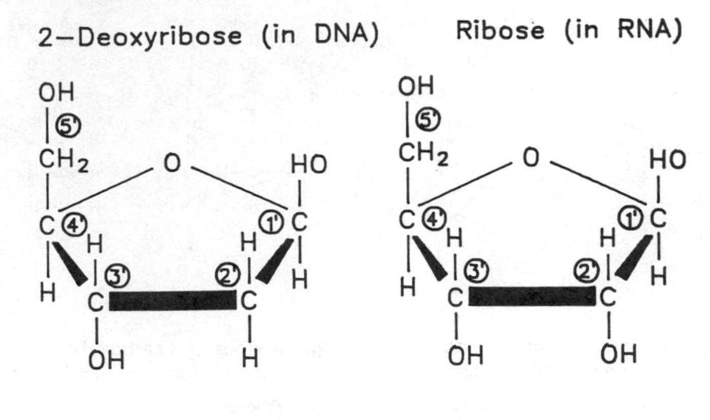

(c) Bases

Pyrimidines, one–ring bases:

thymine (T) cytosine (C) uracil (U)

in DNA

in RNA

Purines, two–ring bases (DNA and RNA):

adenine (A) guanine (G)

FIGURE 7-3 (a) General structure for DNA and RNA chains. The chains, or strands, may be of considerable length and are composed of a linear sequence of nucleotides. Each nucleotide consists of a phosphate group, a sugar, and a nitrogenous base, linked together in the manner shown. (b) Differences between the sugars found in DNA (deoxyribose) and RNA (ribose). (c) The basic kinds of nitrogenous bases found in DNA (T, C, A, G) and RNA (C, U, A, G).

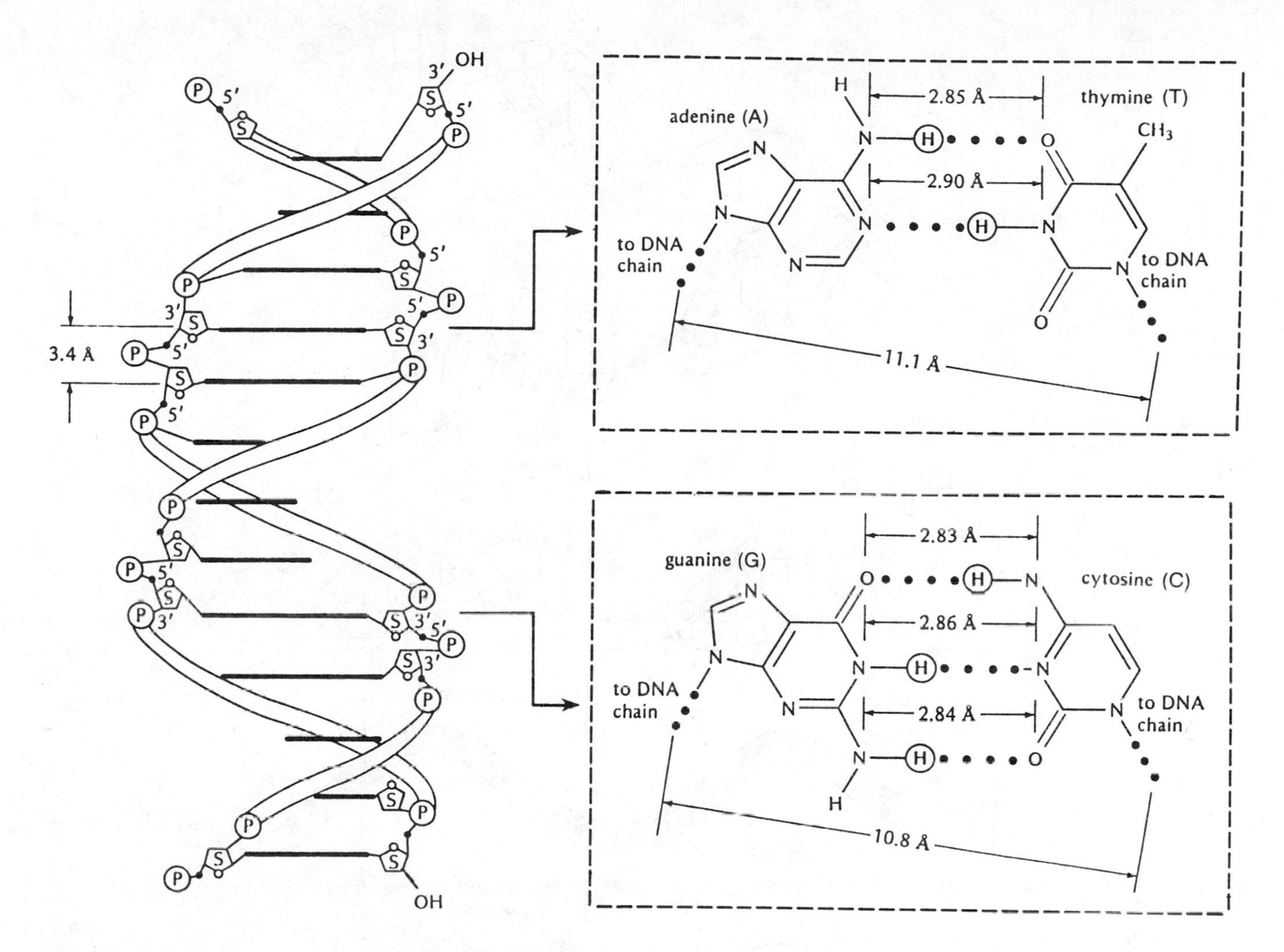

FIGURE 7-4 The Watson-Crick model of the standard (B form) DNA double helix on the left, with examples of hydrogen-bond pairing between bases on the right (A-T, G-C). The two strands of the double helix are bridged by parallel rows of such paired nucleotide bases stacked at regular 3.4-angstrom intervals [1 angstrom = 0.0001 micrometer (μm) = 0.0000001 millimeter (mm)]. (Adapted from Strickberger.)

quence of amino acids of which it is composed. This linear sequence of amino acids in turn derives from the linear sequence of bases in nucleic acids by means of a protein-synthesizing apparatus involving three different kinds of RNA.

In brief, as diagrammed in Figure 7-5, the genetic material, through the process of transcription, produces a molecule of **messenger RNA (mRNA)** which is, base for base, a complement to the bases on one of the DNA strands. Through the mediation of ribosomes, which are themselves composed of **ribosomal RNA (rRNA)** and protein, a sequence of bases in mRNA is then translated into a sequence of amino acids. This translation follows the triplet rule that a sequence of three mRNA bases designates one of the 20 different kinds of amino acids used in protein synthesis.

Note that during the process of translation, no physical material is actually inserted by the mRNA into the protein; only information is transferred. That is, the mRNA only designates the linear position in which each amino acid is to be placed by special molecules of **transfer RNA (tRNA)** that bring the amino acids to the messenger (Fig. 7-6). With the aid of the ribosome and various enzymes, the amino acids are then connected in sequence through peptide linkages. Thus a polypeptide chain of amino acids is formed in which the precise position of each component has been ultimately designated by the genetic material. DNA (or RNA in some viruses) thus provides the **genotype**, or genetic endowment of an organism. The expression of this nucleic acid information, via transcription and/or translation, provides the various aspects of an organism's appearance, or **phenotype**.

The presently observed circular interdependency of all these events, such as the replication of nucleotides because of the presence of appropriate enzymes and the determination of enzyme structure because of the presence of appropriate nucleotide sequences,

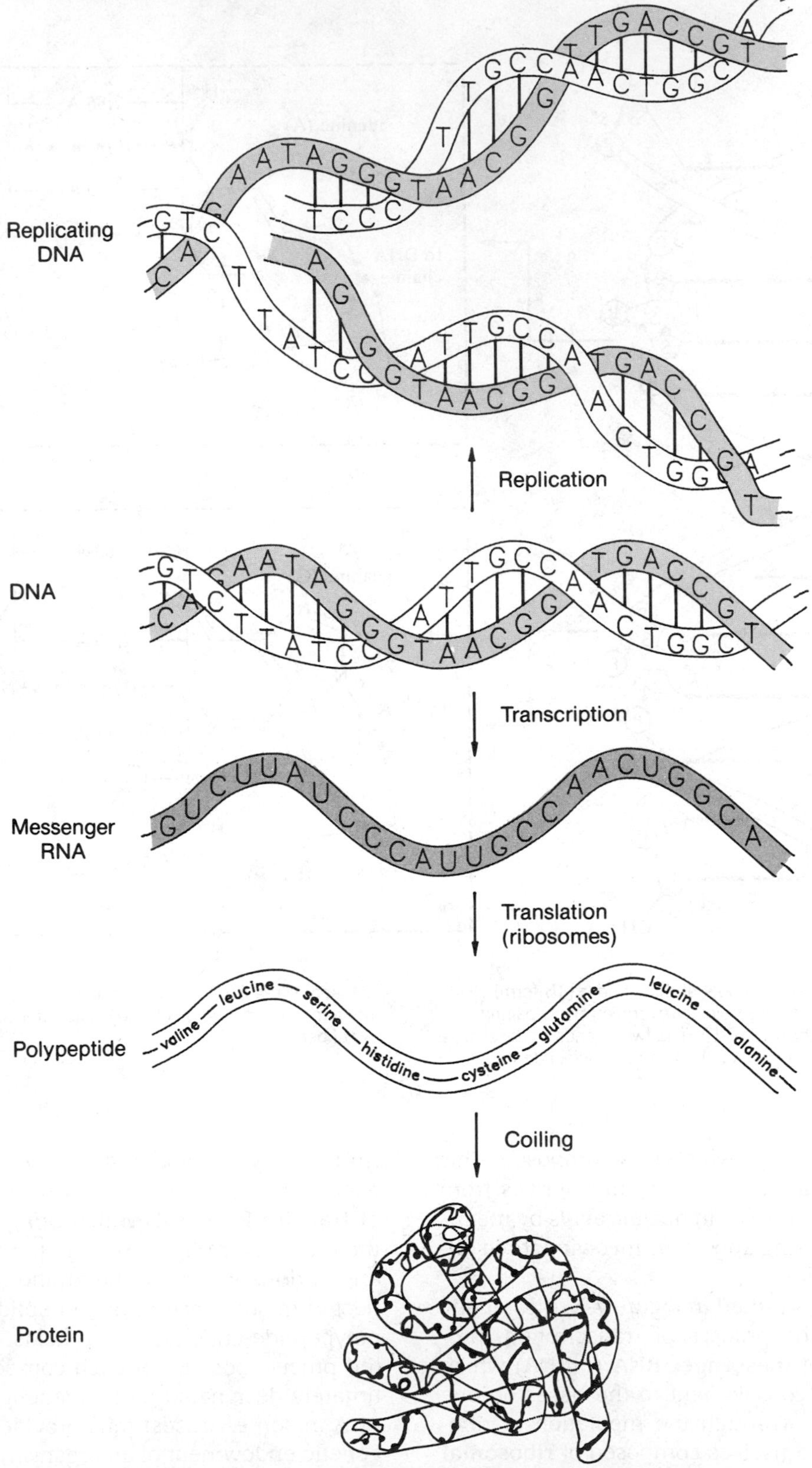

FIGURE 7-5 Diagrammatic illustration of how DNA is replicated and how information is transferred from DNA to RNA to protein. In DNA replication, special proteins break the hydrogen bonds between paired bases, allowing the two strands to unwind. Each unwound strand then acts as a template enabling a new complementary strand to be formed through base pairing catalyzed by a DNA polymerase enzyme. As a result, two double-stranded DNA molecules are produced, each an exact replica of the original parental double helix. In transcription, one of the two DNA strands, called the "sense" strand (lightly shaded), serves as a template upon which a molecule of messenger RNA is transcribed (darker shading). This messenger then serves in turn as a template upon which a molecule of protein is translated. Note that the messenger RNA is exactly complementary to its DNA template and that a sequence of three nucleotides (a triplet codon) on the messenger specifies one amino acid (e.g., 24 nucleotides = 8 amino acids). The amino acids in the illustrated polypeptide chain are coiled into a right-handed α-helix which enables this particular molecule to fold into a special protein called myoglobin used in the cellular transport of oxygen. Other proteins have, of course, different amino acid sequences, which may form different kinds of polypeptide structures such as β-pleated sheets and which may be composed of two or more polypeptide chains.

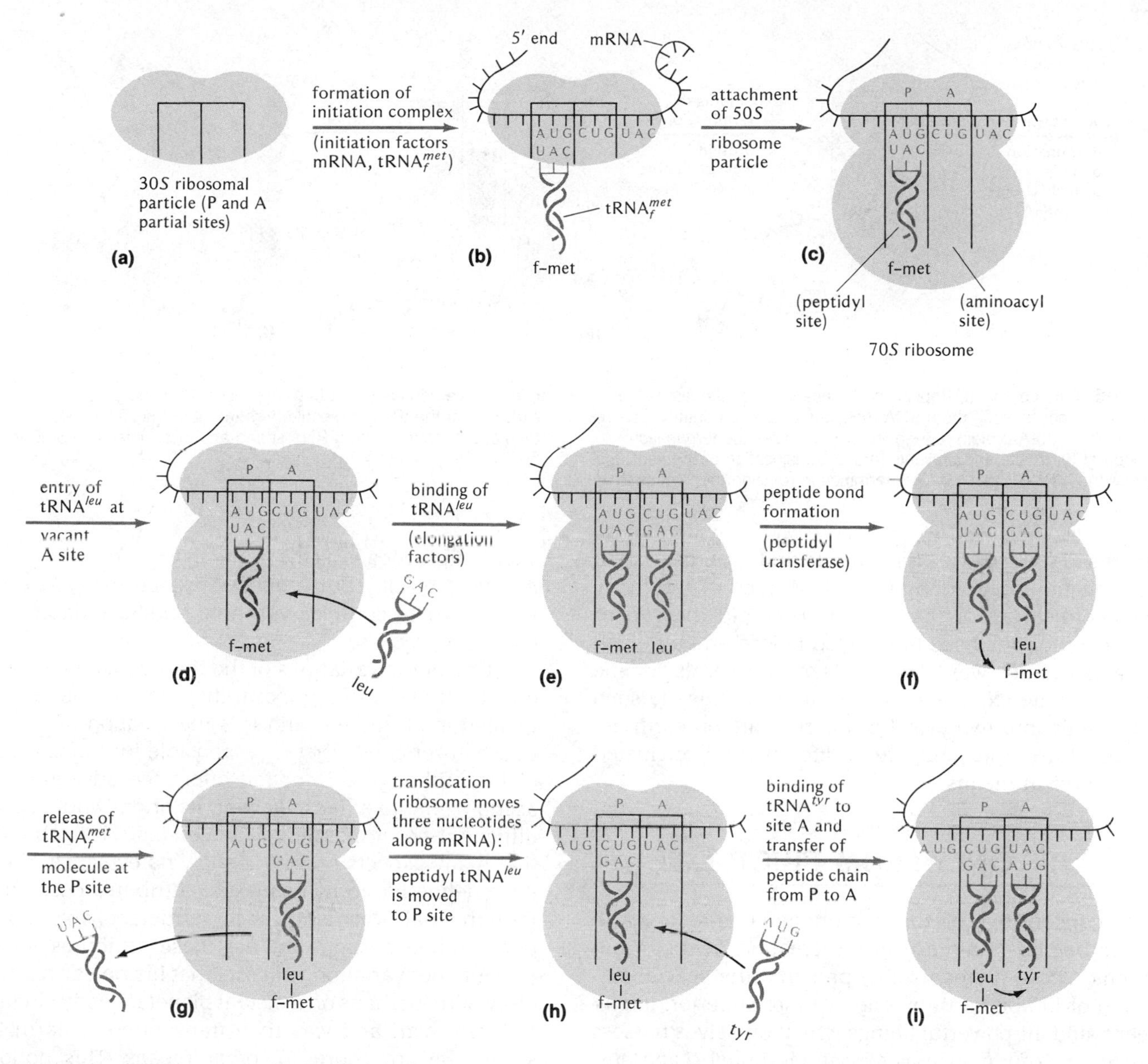

FIGURE 7-6 General scheme of protein synthesis in the bacterium *Escherichia coli.* The ribosome is composed of two subunits designated by their rates of sedimentation in a centrifuge as 30*S* and 50*S*. Each subunit is in turn composed of various proteins and molecules of ribosomal RNA. (a) The 30*S* ribosomal subunit bears two partial sites, peptidyl (P) and acceptor (A), which become functionally complete only when combined with the larger 50*S* subunit. (b) In the presence of mRNA and necessary protein initiation factors, only the special initiator transfer RNA, $tRNA_f^{met}$, bound to a formyl-methionine amino acid (f-met) occupies the partial P site opposite the mRNA initiation codon. (c) Joining of the 30*S* and 50*S* subunits then occurs, accompanied by cleavage of the phosphate-bond energy donor, guanosine triphosphate (GTP), releasing guanosine diphosphate (GDP) and inorganic phosphate (P_i). (d) Once the 70*S* ribosome is formed, the completed A site of the ribosome can then be entered by aminoacyl-charged tRNAs (tRNAs that carry amino acids such as leucine (leu), tyrosine (tyr), and so on) whose anticodons can match the mRNA codon at that site. (e) Special protein elongation factors allow the binding of the appropriate aminoacyl-charged tRNA to the A site, and this step is also accompanied by cleavage of GTP to GDP and P_i. (f) The amino acid (or peptide) at the P site is transferred to the amino acid at the A site through peptide bond formation (see Fig. 7-10(a)) catalyzed by a peptidyl transferase enzyme located on the 50*S* subunit. As a result, the lowermost amino acid in this diagram is at the amino (NH_2) end of the chain, and the uppermost amino acid is at the carboxyl (COOH) end. (g, h) Once the peptide bond is formed, the tRNA molecule that has donated its amino acid or peptide to the A site is released from the P site. Simultaneously, the ribosome is translocated along the mRNA molecule for a distance of three nucleotides, thereby placing the former A-site tRNA (with its newly elongated peptide chain) into the P site. An additional protein elongation factor is necessary for this to occur, and energy is again provided by the reaction GTP $\rightarrow$ GDP + P_i. This translocation step thus allows a new mRNA codon to appear at the now-vacant A site. (i) Binding of aminoacyl-charged tRNA to the A site is again followed by transfer of the peptide chain from P-site tRNA to A-site tRNA.

Continued.

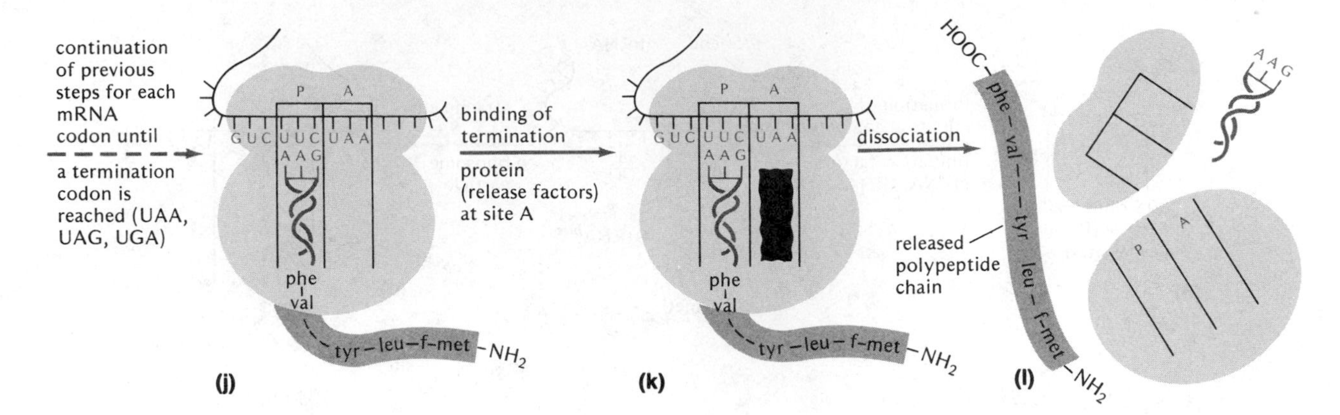

FIGURE 7-6, cont'd. (j) Repetition of these steps continues as the ribosome moves along the mRNA molecule until a termination codon is reached. (k) A protein release factor recognizes the termination codon at the A site and prevents further translocation of the ribosome along the mRNA molecule. (l) A termination reaction then occurs that releases the peptide chain from the P-site tRNA along with the expulsion of the tRNA molecule from the ribosome. The ribosome then separates from the mRNA strand and dissociates into 30*S* and 50*S* subunits. (From Strickberger.)

points to some of the difficulties in finding a reasonable explanation for the origin of life. Which of the many biochemical agents came first? How did they arise? How could they have functioned before an entire cellular structure was formed? Many proposals for the origin of life exist, and we can generally divide such proposals into two broad categories: life on earth developed from previous life, or life on earth originated by chemical means.

LIFE ONLY FROM PRIOR LIFE

The concept that life did not originally arise on earth is embodied in almost all the creation myths of humans. These myths usually presume the special creation of life on earth by one or more superior, intelligent, and all-powerful beings who themselves possess attributes of life such as sensation, thought, and purposive movement (Chapter 4). This concept has the advantage that it explains the origin of terrestrial life in a simple fashion that is especially attractive to those who believe that natural events are governed by conscious agents. It has the disadvantage that it does not explain the source of the initial creator and therefore permits the mystery of life's origin to remain.

Another ancient concept is that life can arise spontaneously at any time (Chapter 1), such as the presumed origin of insects from sweat and crocodiles from mud. Strangely enough, this view was often held simultaneously with the view that life derives from a conscious creator and was popular in Europe throughout medieval times. Spontaneous generation, however, was primarily put to rest by Pasteur and others in the 1860's and has not since been revived in its original form. As we shall see below, the modern concept of the spontaneous origin of life on earth does not include such simple means as the immediate action of sunlight on liquid or clay but proposes instead the past existence of more complex yet more understandable biochemical processes.

One of the variations of the theory that life comes only from life is the proposal that life is somehow ingrained in all matter, and it is the creation of matter by whatever cause that is responsible for the creation of life. Although this notion offers the advantage of ascribing life to a natural event, its origin would seem difficult if not impossible to understand. It is obvious that many aspects of matter show no evidence of life if we define life to include those attributes possessed by terrestrial organisms, such as metabolism, reproduction, and so on. How did these attributes arise?

Another variation suggests that life on earth arose elsewhere, perhaps on a distant planetary body circling a distant star, and was then transported to earth by space-resistant spores or other means. This notion, called **panspermia**, was fostered by the chemist Arrhenius (1859-1927) in the early part of this century and still has some adherents among scientists today (see, for example, Brooks and Shaw; Crick; and Hoyle and Wickramasinghe). It overcomes the difficulty of seeking a chemical explanation for the origin of life on earth but does not, of course, explain the distant origin of life.

Proponents of the panspermia hypothesis point to the fact that a number of different organic compounds have been discovered in carbon-containing meteorites (**carbonaceous chondrites**), and these compounds range from carbohydrates to amino acids. When looked at closely, however, the structural forms or isomers of amino acids found in these carbonaceous chondrites possess the two different kinds of **optical activity** (dextro- and levorotary) in equal amounts, therefore comprising a **racemic mixture** that shows

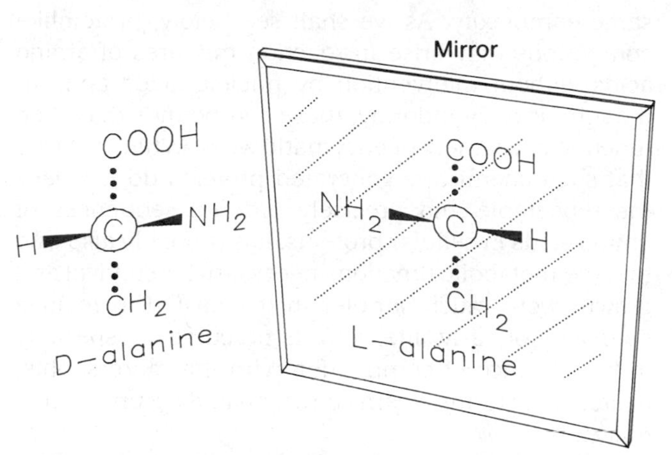

FIGURE 7-7 Structures of an L-form and a D-form amino acid. The dark, wedge-shaped bonds indicate that the attached NH_2 and H groups project above the plane of the paper. Note that although D-alanine is a mirror image of L-alanine, it is not identical to it, since the two molecules cannot be superimposed on each other. Each of these stereoisomers has optical activity that can be measured by the extent to which it rotates polarized light. However, the two forms rotate light in opposite directions, and an equimolar mixture of the two is a racemate which is not optically active.

no optical activity. On the other hand, the amino acids of living forms generally show optical activity of only one type, levorotary (Fig. 7-7). Furthermore, a number of the amino acids found in meteorites do not appear in proteins. Along with other observations, these findings indicate that organic compounds were probably formed through random chemical reactions in the meteorite itself or in its parent body rather than through ordered living processes.

Opposition to the panspermia hypothesis also comes from the difficulty of envisaging how spores or "bugs" from outer space can be transported to earth without the assistance of conscious agents in spaceships. If the bug is too large, it will not be easily ejected from the planet on which it occurs nor subsequently pushed out by radiation from the sun in its particular solar system. If it is too small, it will be kept from entering the earth's field by radiation pressure from our sun. Shklovskii and Sagan suggest that bugs that are larger than 0.6 μm cannot escape from earth, while bugs that are smaller than 0.6 μm will be pushed away from the earth. This means that the donor planet must have been capable of ejecting bugs which we cannot eject (e.g., about 1 μm); that is, the sun of the donor planet must have been very hot (high radiation pressure). But if such a sun were hot enough for this purpose, it would destroy ejected particles by radiation.

In addition, the hazards of interstellar travel are many. For example, ultraviolet radiation (UV) from our sun will kill ejected particles from earth within about one day of interplanetary travel. On the other hand, if the bugs were shielded from UV, they would be too heavy to be ejected. Among other space hazards are the hot ionized gases which surround early-type stars, the presence of cosmic rays, and the absorption of bugs into passing suns by gravitational attraction.

However, even assuming that bugs survived all other obstacles, the vastness of space would disperse these spores so widely that their chance of reaching earth would be infinitesimal. Shklovskii and Sagan have calculated that 100 million life-bearing planets in our galaxy would each have had to eject about 1,000 tons of spores in order for earth to have received a single microorganism during its first billion years of history. Such difficulties make it questionable whether an event as seemingly rare as panspermia is more probable than the chemical origin of life on this planet: Why should the origin of life on other bodies have had a greater probability than its origin on earth? Such misgivings, along with a rapid increase in our understanding of molecular biology and biochemistry, have influenced most scientists to concentrate their attention on a terrestrial origin of life.

THE TERRESTRIAL ORIGIN OF LIFE

The difficulty in visualizing life originating on earth is essentially one of visualizing the molecular environment and events that occurred in a long-distant past. Unfortunately, we possess as yet no certainties about the details of our molecular past, and we may never have such knowledge since molecular fossils, when found at all, are indeed sparse. At best, we can try to deduce the general nature of some of the original molecular events from present living structures and reactions and attempt to reconstruct such events experimentally under controlled conditions. However, before undertaking such a molecular review, it is important to consider the framework in which the question of the origin of life from a terrestrial source is usually posed. That is, we must attempt to deal with the problem of whether a highly complex, ordered phenomenon such as life could have arisen at all from the molecular chaos assumed to have existed during the earth's early history. To recapitulate a question discussed in Chapter 4, How can order arise from disorder?

Obviously, the probability for a modern, self-reproducing cell to arise from complete disorder is embarrassingly small. To use an oft-quoted example, can a monkey, given even billions of years, produce the works of Shakespeare by randomly pressing the keys of a typewriter? Even if we restrict Shakespeare's writings to 1 million (10^6) alphabetical letters and limit the typewriter to 26 keys, the chance for such an event is $(1/26)^{10^6}$. This means that even if a monkey could type 1 million words a second, such an event would be expected to occur only once in $7 \times 10^{1,414,965}$ years! By similar reasoning, the chances for most complex organic structures to arise spontaneously are infinitesimally small. Even a small enzymatic sequence of 100 amino acids would have only one chance in 20^{100} ($= 10^{130}$) to arise randomly, since there are 20 possible kinds of different amino acids for each position in the sequence. Thus, if we were to randomly generate a new 100-amino acid-long sequence each second, such a given enzyme would be expected to appear only once in 4×10^{122} years! In terms of the volume necessary to generate all such possibilities, the difficulty appears just as immense: were an entire universe 10 billion light years in diameter to be densely packed with randomly produced polypeptides, each 100 amino acids long, the number of such molecules, 10^{103}, would not equal their 10^{130} possibilities.

These arguments long seemed formidable and were further strengthened by the suggestion that nature itself would deteriorate any complex organization of matter even if such complexity were to arise accidentally. It was often pointed out that according to the **Second Law of Thermodynamics**, the energy in a system tends towards diffusion rather than concentration; that is, **entropy** (disorder) increases rather than decreases. Thus, there appeared to be only a negative answer to the question that Pasteur had posed in the nineteenth century: "Can matter organize itself?" It seemed either that the living organization of matter must be explained as arising from a mystical non-natural source, or that this event, if it were of natural origin, was of such low probability that any attempt at its comprehension or reconstruction would be meaningless.

In answer to these apparent difficulties, many scientists today point to two important considerations:

1. The likelihood that primeval "living" organisms did not possess many of their present complexities.
2. The formation of organic molecules and subsequent organic structures was not the result of completely random events, although such events were nevertheless natural in the sense that they followed chemical and physical laws.

The first consideration, that early organisms were more primitive than those of today, is extremely important. Perhaps the most basic quality of life we would recognize in even a primitive living organism is its ability to perform those reactions necessary for it to grow and replicate. Certainly the endless loop of metabolism and information transfer that we now see embodied in the intricate relationship between proteins and nucleic acids need not always have been of the same complexity. As we shall see below, proteinlike compounds may arise in reaction mixtures of amino acids without intervention by nucleic acids and, although formed randomly, these compounds may then function in a variety of enzymatic ways. While it is true that such chemically generated proteins do not have the repeatable and precisely ordered sequences of amino acids in cellular proteins, it is probably also true that the metabolic functions necessary for survival and growth were much simpler in the past. The relative simplicity of early "life" and its precursors, especially in the absence of competition with the more sophisticated later forms, seems a reasonable assumption to make.

The second consideration, that life did not arise from absolute chaos, has been supported in many ways. Many scientists have pointed out that the evolution of our solar system offered a number of essential prerequisites that enabled the development and sustenance of life:

1. Our planet possessed a sun of moderate size that was on the main sequence of stellar evolution. This provided a steady rate of emitted radiation over a long enough period of time for life to develop. As shown in Table 7-1, the amount of energy available from solar radiation seems to have always been far greater than that provided by any other source, although energy from other sources may also have been important in initiating particular chemical reactions.

TABLE 7-1 Present Energy Sources (Calories per Square Centimeter per Year) that Were Probably Available for Organic Synthesis on the Primitive Earth

Source	Energy
Total solar radiation (all wavelengths)	260,000
Ultraviolet light wavelengths (in angstroms)	
Below 3,000	3,400
Below 2,500	563
Below 2,000	41
Below 1,500	1.7
Electrical discharges (lightning, corona discharges)	4
Shock waves (meteorite impacts, lightning bolt pressure waves)	1.1
Radioactivity (to depth of 1 km)	0.8
Volcanoes (heat)	0.13
Cosmic rays	0.0015

From Miller and Orgel.

2. There existed a fairly large sampling of different elements such as H, O, C, N, S, P, Ca, and others. These elements must have provided considerable chemical diversity, enabling reactions to occur that were necessary for the formation of organic molecules involving carbon. The important chemical attributes of carbon, its ability to form four covalent bonds and the tetrahedral arrangement of its outer electrons, provided the opportunity for the formation of a large number of different kinds of stable molecules with considerable three-dimensional variety and complexity. Interestingly, the terrestrial presence of such molecules is not unique: by means of spectroscopy it is now possible to observe a variety of organic molecules in the dense interstellar clouds that give rise to stars and planets (Fig. 7-8). Such observations indicate that a number of compounds necessary for the origin of life were already present both before and during the formation of our solar system (see, for example, Hoyle and Wickramasinghe).
3. The earth followed a nearly circular orbit at a fairly uniform distance from the sun. Such an even orbit would eliminate temperature extremes in which organic molecules would be incapable of forming or functioning.
4. There was present on earth large amounts of an excellent solvent, water, which is stable in liquid form over a relatively wide range of temperatures and enables both acids and bases to ionize and react. Water also has the advantage that it floats in its crystalline frozen form (ice), so that bodies of water containing organic matter may remain liquid under a surface of ice, rather than freezing because of the subsurface accumulation of ice. It is now believed that water must have been present early in earth's history in the cold planetesimal condensations and appeared in liquid form as soon as the lithosphere reached appropriate temperatures. Additional water has been continually emitted into the atmosphere through volcanic activity, which was probably greater in the past than at present. Since water causes crustal erosion that leads to sedimentary rock formation, the presence of significant amounts of water dates back to the beginning of the geological record.
5. Hydrogen-containing gases existed for a long initial period in the earth's history, derived from the high cosmic abundance of hydrogen and the consequent abundance of its compounds. Even though free hydrogen was probably lost early in the earth's history, outgassing from the earth's interior would have led to at least a par-

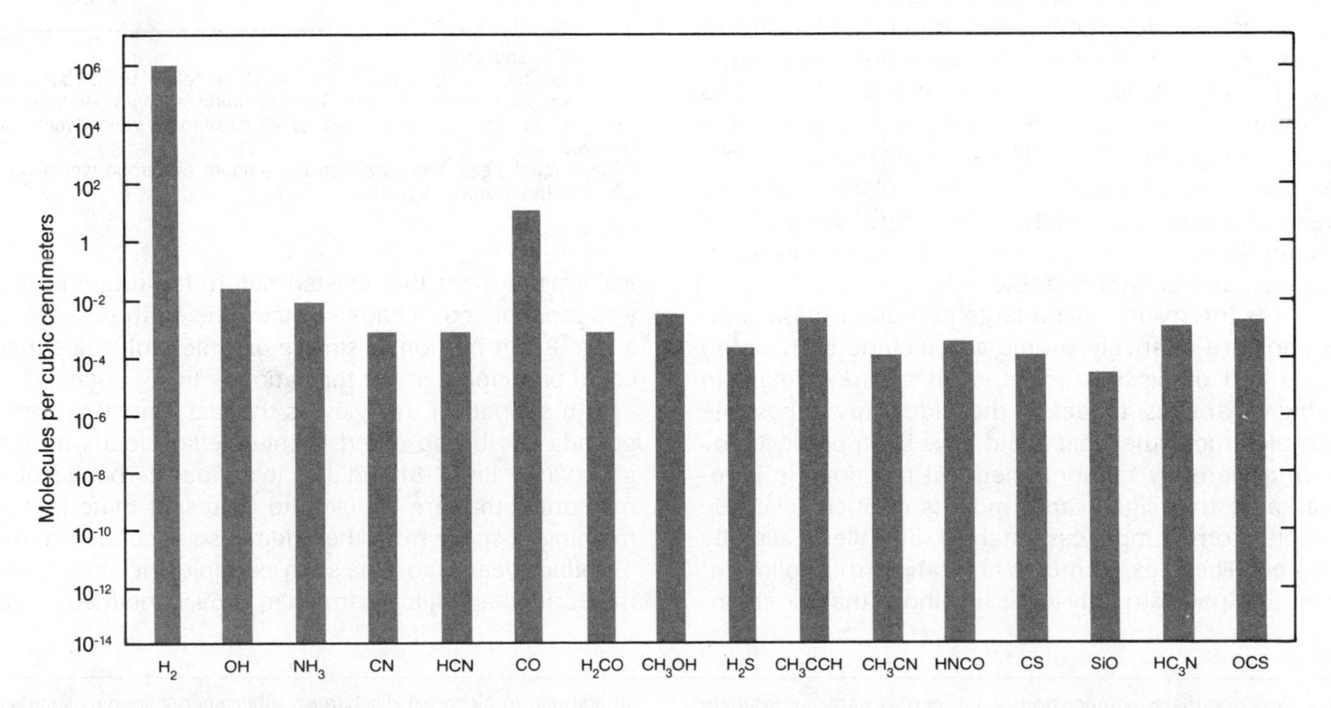

FIGURE 7-8 Densities of various molecules observed in molecular clouds within our galaxy. H_2 = hydrogen, OH = hydroxyl, NH_3 = ammonia, CN = cyanogen, HCN = hydrogen cyanide, CO = carbon monoxide, H_2CO = formaldehyde, CH_3OH = methyl alcohol, H_2S = hydrogen sulfide, CH_3CCH = methylacetylene, CH_3CN = methyl cyanide, HNCO = isocyanic acid, CS = carbon monosulfide, SiO = silicon monoxide, HC_3N = cyanoacetylene, OCS = carbonyl sulfide. (Adapted from Buhl.)

tially reducing atmosphere in some or many localities in which hydrogen protons could be donated to a variety of proton-accepting elements, especially carbon.[2] Such gases, along with the energy from solar and ultraviolet radiation, therefore enabled the formation of a variety of organic molecules that could, in turn, be used to provide both structure and energy for living processes, for example, amino acids, sugars, fatty acids, purines, and pyrimidines.

In summary, we can say that there was considerable **molecular preadaptation** for the biochemical events leading to the formation of life even before life appeared. That is, there were present appropriate energy sources, chemicals, temperature, and solvent. What kinds of reactions would then have taken place?

ORIGIN OF BASIC BIOLOGICAL MOLECULES

In the 1920's, Oparin, a Russian biochemist, and Haldane, an English geneticist, independently suggested that the primitive atmosphere of the earth was reducing and that organic compounds formed in such an atmosphere might be similar to those presently used by living organisms. It took, however, almost 30 years before an experimental test of this hypothesis was undertaken. In 1953, Miller placed together in a glass apparatus (Fig. 7-9) methane, ammonia, and hydrogen gases. An electric spark was generated in a large 5-liter flask, and water was boiled in a smaller flask to provide vapor to the spark as well as to circulate the gases. Compounds formed by sparking were then condensed, or recirculated if they were volatile. After one week of continuous electrical discharge, the products accumulated in the aqueous phase were chromatographed and analyzed (Table 7-2).

It is interesting that a large portion of these compounds are relatively simple and include both amino acids and other substances, such as urea, found in living organisms. In fact, of the wide array of possible complex molecules that could have been produced by such apparently random chemical reactions, it is remarkable that significant amounts of such relatively simple compounds essential to life were actually formed. These experiments and others that followed[3] therefore point strongly to the likelihood that the chemical environment that existed before the origin of life was probably not "chaos." Rather, the earth possessed a significant portion of simple organic molecules that could participate in the formation of living organisms.

TABLE 7-2 Yields of Various Organic Compounds Obtained from a Mixture of Water, Hydrogen, Methane, and Ammonia Exposed to Electrical Sparking*

Compound	Yield (percent)†
Glycine	2.1
Glycolic acid	1.9
Sarcosine	0.25
Alanine	1.7
Lactic acid	1.6
N-methylalanine	0.07
α-amino-n-butyric acid	0.34
α-aminoisobutyric acid	0.007
α-hydroxybutyric acid	0.34
β-alanine	0.76
Succinic acid	0.27
Aspartic acid	0.024
Glutamic acid	0.051
Iminodiacetic acid	0.37
Iminoaceticpropionic acid	0.13
Formic acid	4.0
Acetic acid	0.51
Propionic acid	0.66
Urea	0.034
N-methyl urea	0.051

From Miller and Orgel.
*These products represented only about 15 percent of the carbon that had been added to the apparatus. The remaining carbon products were mostly in the form of polymerized tarlike substances which were not analyzed.
†The percent yields are based on the amount of carbon which was added to the mixture as methane.

In support of this view is the fact that such compounds can be observed in interstellar clouds in our galaxy (see Fig. 7-8) and also in various carbonaceous meteorites that are believed to represent material remaining in space from the original solar condensation 4.6 billion years ago. One such example, the Murchison meteorite that fell in Australia in 1969, contained more

[2]Theories about the composition of the earth's early atmosphere range from strongly reducing to mildly reducing to neutral and nonreducing (see Chang et al.). However, there is no dispute that gases presently emitted from the earth's interior include hydrogen (H_2) and methane (CH_4), and it seems reasonable to assume that such gases were also emitted in the past.

[3]In addition to electrical discharges, other energy sources such as β-rays, γ-rays, x-rays, thermal heating, and ultraviolet light have been shown to produce amino acids and other organic compounds from simple gases. It is therefore not entirely surprising that some amino acids from nonbiological origin have been identified in rain and snow.

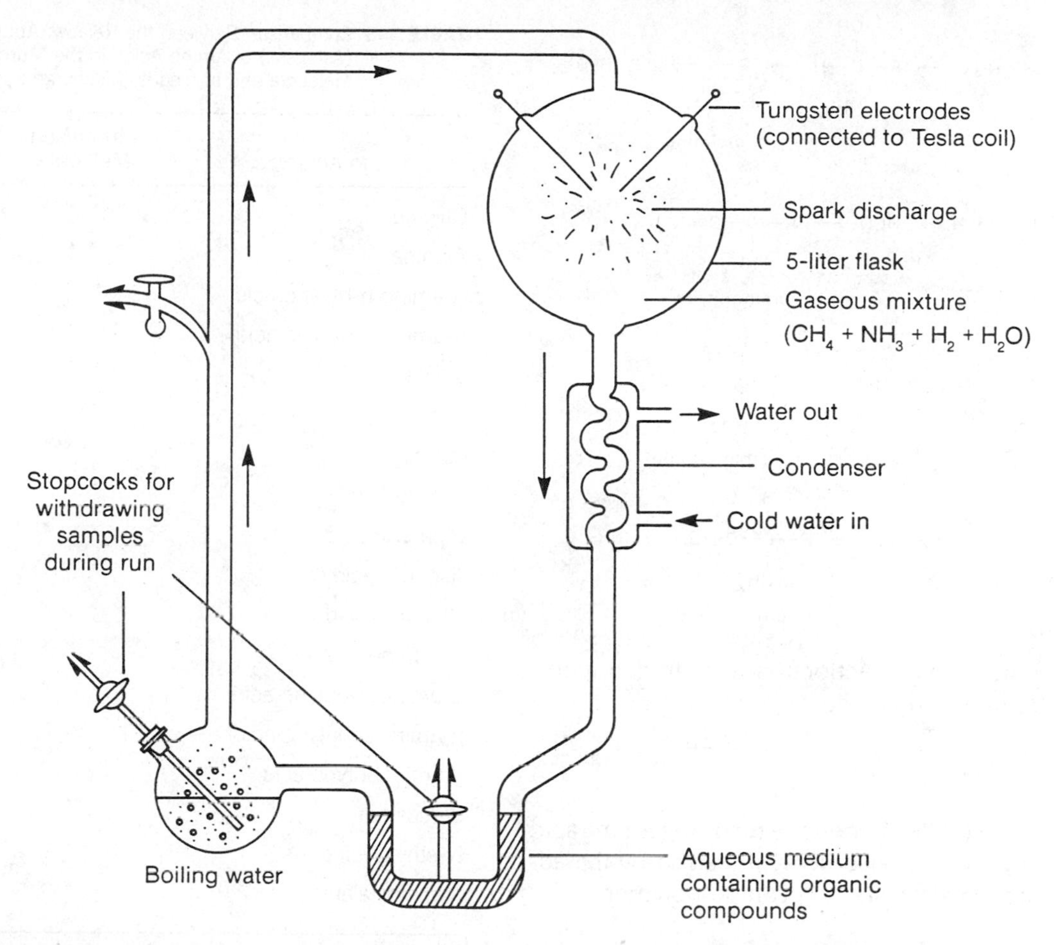

FIGURE 7-9 Apparatus used by Miller to demonstrate the synthesis of organic compounds by electrical discharge in a reducing atmosphere.

than 80 kilograms of carbonaceous material, of which about one percent was organic carbon. Upon analysis, this material included a greater concentration of amino acids normally used in proteins than nonbiological amino acids. Since the meteorite was analyzed by a number of different laboratories almost immediately after it landed and the results were consistent overall, it is believed doubtful that the protein-type amino acids could have originated from terrestrial contamination. Also remarkable is the fact that practically all of the amino acids found in the meteorite, both protein and nonprotein, are identical to amino acids produced by sparking mixtures in laboratory experiments (Table 7-3).

These observations strongly indicate that the laboratory experiments probably reflect actual chemical processes that also occurred in the synthesis of prebiotic organic compounds. However, the cause for these impressive correlations is not clear: What chemical mechanisms restrict or bias the production of organic compounds to those that are observed?

It has been suggested that the amino acids synthesized under primitive earth conditions arise primarily from the formation of aldehydes,

$$R-\overset{\overset{\Large O}{\|}}{C}-H$$

(where R may represent any group), which then interact with ammonia and cyanide compounds. These reactive chemicals may have been produced from a variety of simple gases or from their further interactions:

$$N_2 + H_2,\quad N_2 + H_2O \longrightarrow NH_3 \text{ (ammonia)}$$

$$CH_4 \longrightarrow C_2H_2 \text{ (HC}\equiv\text{CH, acetylene)}$$

$$CH_4 + N_2,\quad CH_4 + NH_3,\quad CO + NH_3,\quad C_2H_2 + N_2 \longrightarrow HCN \text{ (HC}\equiv\text{N, hydrogen cyanide)}$$

$$CH_4 + H_2O,\quad CH_4 + CO_2,\quad CO_2 + H_2,\quad CO_2 + H_2O \longrightarrow HCHO \text{ (H}\overset{O}{\overset{\|}{C}}\text{H, formaldehyde)}$$

$$CH_4 + H_2O \longrightarrow CH_3CHO \text{ (H}_3\text{C}-\overset{O}{\overset{\|}{C}}\text{H, acetaldehyde)}$$

According to one of the possible pathways of amino acid synthesis (the Strecker synthesis), subsequent steps are as follows:

$$R{-}\overset{O}{\overset{\|}{C}}{-}H + NH_3 \rightarrow R{-}\underset{NH_2}{\underset{|}{\overset{OH}{\overset{|}{C}}}}{-}H \rightarrow R{-}\underset{NH}{\underset{\|}{C}}{-}H + H_2O$$

aldimine

$$R{-}\underset{NH}{\underset{\|}{C}}{-}H + HCN \rightarrow R{-}\underset{NH_2}{\underset{|}{\overset{H}{\overset{|}{C}}}}{-}C{\equiv}N$$

aminonitrile

$$R{-}\underset{NH_2}{\underset{|}{\overset{H}{\overset{|}{C}}}}{-}C{\equiv}N + H_2O \rightarrow R{-}\underset{NH_2}{\underset{|}{\overset{H}{\overset{|}{C}}}}{-}\overset{O}{\overset{\|}{C}}{-}NH_2$$

aminoamide

$$R{-}\underset{NH_2}{\underset{|}{\overset{H}{\overset{|}{C}}}}{-}\overset{O}{\overset{\|}{C}}{-}NH_2 + H_2O \rightarrow R{-}\underset{NH_2}{\underset{|}{\overset{H}{\overset{|}{C}}}}{-}\overset{O}{\overset{\|}{C}}{-}OH + NH_3$$

α amino acid

If R in the above reactions is a hydrogen atom, that is, if the initial molecule,

$$R{-}\overset{O}{\overset{\|}{C}}{-}H$$

is formaldehyde (HCHO), then the resulting amino acid is glycine. Glycine can also be synthesized by the addition of water (hydrolysis) to cyanide polymers:

$$H{-}C{\equiv}N + H{-}C{\equiv}N \rightarrow H{-}\overset{NH}{\overset{\|}{C}}{-}C{\equiv}N$$

cyanide monomers dimer

$$H{-}\overset{NH}{\overset{\|}{C}}{-}C{\equiv}N + H{-}C{\equiv}N \rightarrow N{\equiv}C{-}\underset{NH_2}{\underset{|}{\overset{H}{\overset{|}{C}}}}{-}C{\equiv}N$$

trimer

$$N{\equiv}C{-}\underset{NH_2}{\underset{|}{\overset{H}{\overset{|}{C}}}}{-}C{\equiv}N + H{-}C{\equiv}N \rightarrow (H_2N)(N{\equiv}C)C{=}C(NH_2)(C{\equiv}N)$$

tetramer

$$(H_2N)(N{\equiv}C)C{=}C(NH_2)(C{\equiv}N) + H_2O \rightarrow H_2N{-}\underset{H}{\underset{|}{\overset{H}{\overset{|}{C}}}}{-}C{\equiv}N + H_2N{-}\overset{O}{\overset{\|}{C}}{-}C{\equiv}N$$

$$H_2N{-}\underset{H}{\underset{|}{\overset{H}{\overset{|}{C}}}}{-}C{\equiv}N + 2H_2O \rightarrow H_2N{-}\underset{H}{\underset{|}{\overset{H}{\overset{|}{C}}}}{-}\overset{O}{\overset{\|}{C}}{-}OH + NH_3$$

glycine

TABLE 7-3 Comparison Between the Relative Abundances (Asterisks) of Amino Acids in the Murchison Meteorite and in Electric Discharge Synthesis*

Amino Acid†	Murchison Meteorite	Electric Discharge
Glycine	* * * *	* * * *
Alanine	* * * *	* * * *
α-amino-n-butyric acid	* * *	* * * *
α-aminoisobutyric acid	* * * *	* *
Valine	* * *	* *
Norvaline	* * *	* * *
Isovaline	* *	* *
Proline	* * *	*
Pipecolic acid	*	*
Aspartic acid	* * *	* * *
Glutamic acid	* * *	* *
β-alanine	* *	* *
β-amino-n-butyric acid	*	*
β-aminoisobutyric acid	*	*
γ-aminobutyric acid	*	* *
Sarcosine	* *	* * *
N-ethylglycine	* *	* * *
N-methylalanine	* *	* *

From Miller (1987).
* Purine and pyrimidine compounds found in the nucleic acids of living organisms were not observed in the meteorite, although traces of nonbiological pyrimidines (e.g., 4-hydroxypyrimidine) were noted.
† Indicative of their nonbiological origin was the fact that the meteorite amino acids were optically inactive (racemic) mixtures of both D and L forms, rather than consisting exclusively of the levorotary forms produced biologically on earth.

The addition of formaldehyde to glycine under alkaline conditions can then produce serine:

$$H_2N{-}\underset{H}{\underset{|}{\overset{H}{\overset{|}{C}}}}{-}\overset{O}{\overset{\|}{C}}{-}OH + H{-}\underset{O}{\underset{\|}{C}}{-}H \rightarrow H_2N{-}\underset{H{-}\underset{OH}{\underset{|}{C}}{-}H}{\underset{|}{\overset{H}{\overset{|}{C}}}}{-}\overset{O}{\overset{\|}{C}}{-}OH$$

glycine formaldehyde serine

All of the 20 different amino acids presently used in protein synthesis possess a similar structural pattern, as shown in Figure 7-2, although many of them are synthesized by different biochemical pathways.

Among other basic organic molecules that could easily be synthesized under fairly simple conditions are the **sugars**. Glucose, ribose, and deoxyribose, for example, can be shown to occur in significant yields from the condensation of formaldehyde:

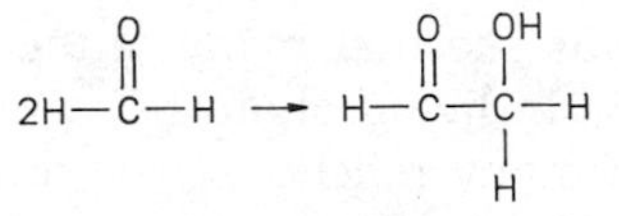

$$2H{-}\overset{\displaystyle O}{\overset{\|}{C}}{-}H \longrightarrow H{-}\overset{\displaystyle O}{\overset{\|}{C}}{-}\overset{\displaystyle OH}{\overset{|}{\underset{\underset{\displaystyle H}{|}}{C}}}{-}H$$

formaldehyde glycoaldehyde

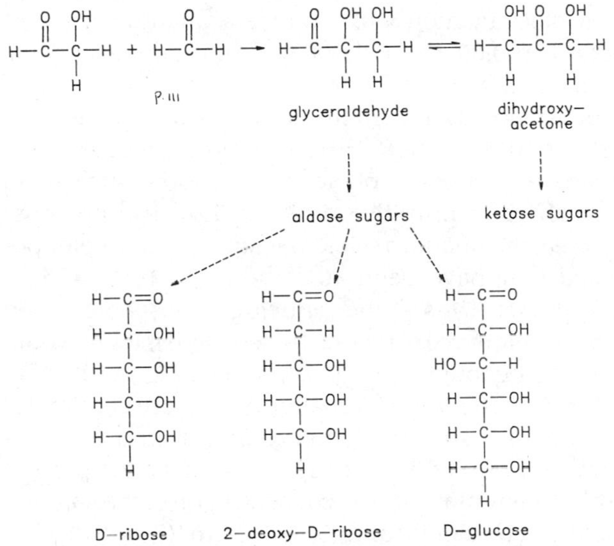

$$H{-}\overset{O}{\overset{\|}{C}}{-}\overset{OH}{\overset{|}{\underset{\underset{H}{|}}{C}}}{-}H + H{-}\overset{O}{\overset{\|}{C}}{-}H \longrightarrow H{-}\overset{O}{\overset{\|}{C}}{-}\overset{OH}{\overset{|}{\underset{\underset{H}{|}}{C}}}{-}\overset{OH}{\overset{|}{\underset{\underset{H}{|}}{C}}}{-}H \rightleftharpoons H{-}\overset{OH}{\overset{|}{\underset{\underset{H}{|}}{C}}}{-}\overset{O}{\overset{\|}{C}}{-}\overset{OH}{\overset{|}{\underset{\underset{H}{|}}{C}}}{-}H$$

glyceraldehyde dihydroxy-acetone

aldose sugars ketose sugars

D-ribose 2-deoxy-D-ribose D-glucose

The **purine** and **pyrimidine** bases that are essential components of nucleic acids can also be synthesized under prebiotic conditions. For example, Oró and co-workers have shown that the heating of aqueous solutions of ammonium cyanide (prepared by the reaction of HCN with NH_4OH) will produce up to 0.5 percent yield of adenine. Similarly, the action of ultraviolet radiation on a hydrogen cyanide solution will produce a number of purines, including adenine and guanine. The condensation reactions in the formation of adenine have been studied in some detail, and it has been suggested that one sequence may be as follows:

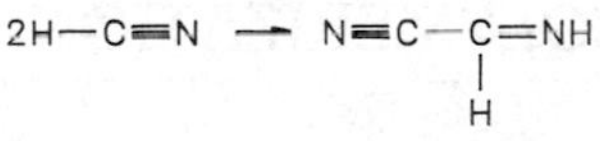

$$2H{-}C{\equiv}N \longrightarrow N{\equiv}C{-}\underset{\underset{H}{|}}{C}{=}NH$$

iminoacetonitrile

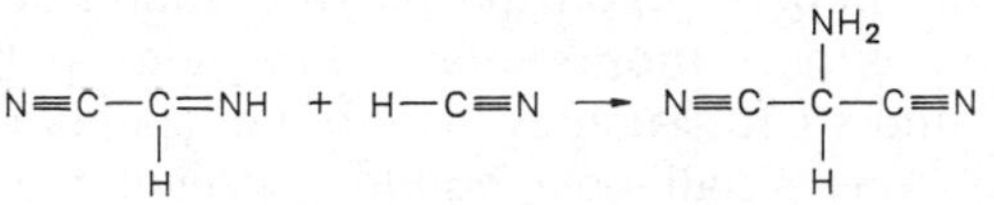

$$N{\equiv}C{-}\underset{\underset{H}{|}}{C}{=}NH + H{-}C{\equiv}N \longrightarrow N{\equiv}C{-}\overset{NH_2}{\overset{|}{\underset{\underset{H}{|}}{C}}}{-}C{\equiv}N$$

aminomalononitrile

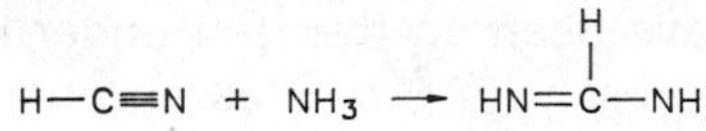

$$H{-}C{\equiv}N + NH_3 \longrightarrow HN{=}\overset{H}{\overset{|}{C}}{-}NH_2$$

formamidine

$$N{\equiv}C{-}\overset{NH_2}{\overset{|}{\underset{\underset{H}{|}}{C}}}{-}C{\equiv}N + 2NH_3 \longrightarrow$$

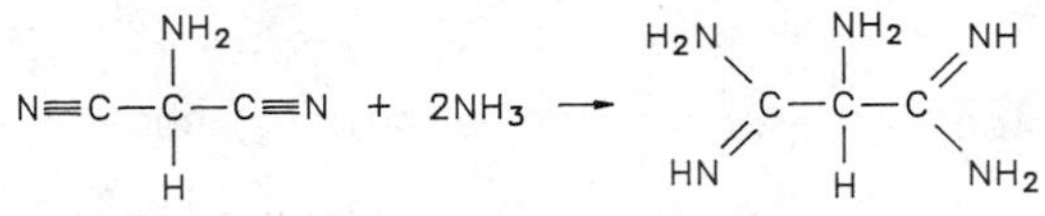

aminomalonodiamidine

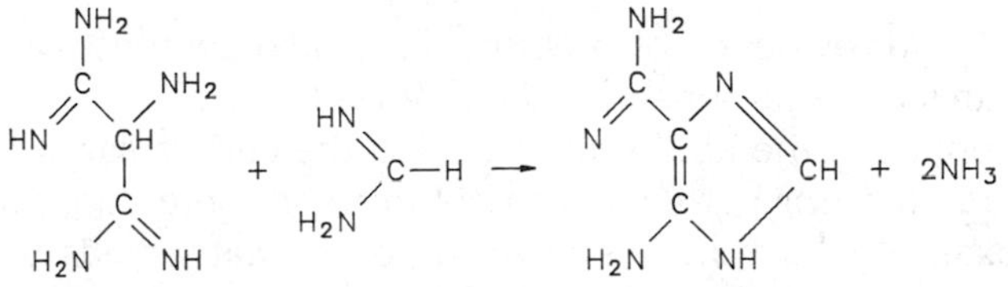

+ 2NH₃

4-aminoimidazole-5-carboxamidine

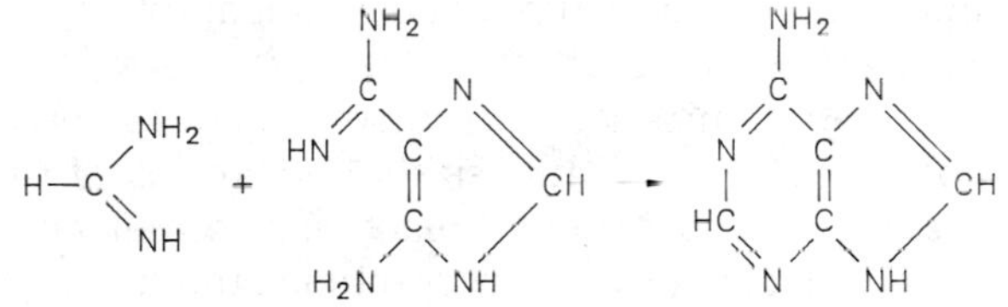

adenine

The pyrimidine cytosine has been produced by the reaction of cyanoacetylene with cyanates such as urea, as shown below, and similar synthetic procedures have been proposed for the other pyrimidines uracil and thymine:[4]

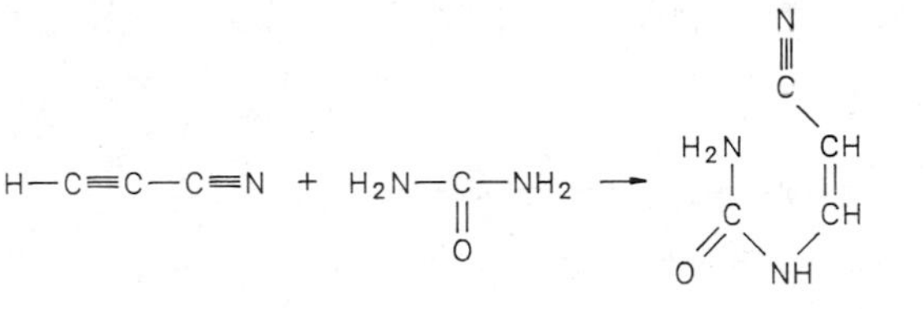

$$H{-}C{\equiv}C{-}C{\equiv}N + H_2N{-}\underset{\underset{O}{\|}}{C}{-}NH_2 \longrightarrow$$

cyanoacetylene urea β ureidoacrylonitrile

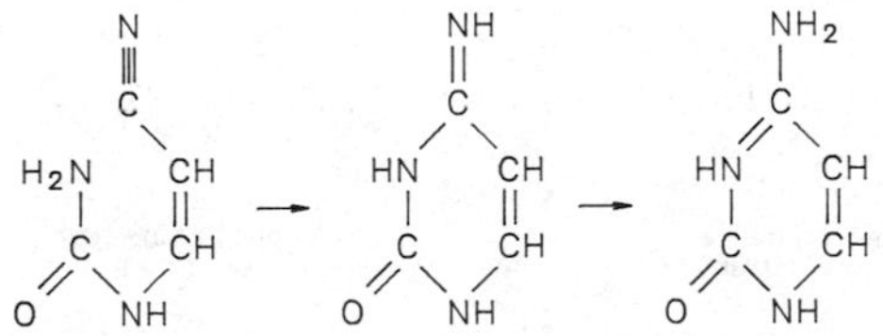

cytosine

Fatty acids, now used in membranes and storage tissues of living organisms, are among other basic mol-

[4]Ferris and Joshi have shown that orotic acid, which is a pyrimidine produced biologically as a precursor to uracil and also produced abiologically by hydrolysis of HCN compounds, can be relatively efficiently decarboxylated to uracil by ultraviolet light. Because the reactions are so alike in both of these systems, they suggest that early biological synthesis of uracil derivatives could easily have followed the pattern of abiological synthesis, but using catalytic enzymes rather than ultraviolet light.

ecules that have been synthesized under high atmospheric pressures, with γ-rays as an energy source:

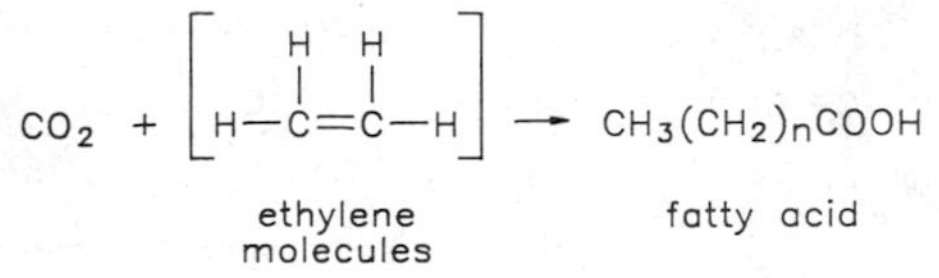

$$CO_2 + \left[H-\overset{H}{C}=\overset{H}{C}-H \right] \rightarrow CH_3(CH_2)_nCOOH$$

ethylene molecules — fatty acid

For evidence of prebiotic fatty acid synthesis we can look to carbonaceous chondrites that contain compounds of the kind synthesized in the early solar system. Interior portions of the Murchison meteorite, for example, have been shown to possess fatty acids up to eight carbons long (Lawless and Yuen). Moreover, experiments by Deamer indicate that a portion of uncontaminated Murchison meteorite compounds can produce fattylike structures and boundaried vesicles that resemble membranes.

Pyrroles, which are precursors of porphinelike compounds, can be synthesized in mixtures of CH_4, NH_3, and H_2O and can then react with formaldehyde (also benzaldehyde) to form porphine structures. Oxidation of these structures then yields the **porphyrin** rings found in heme, chlorophyll, and other pigments:

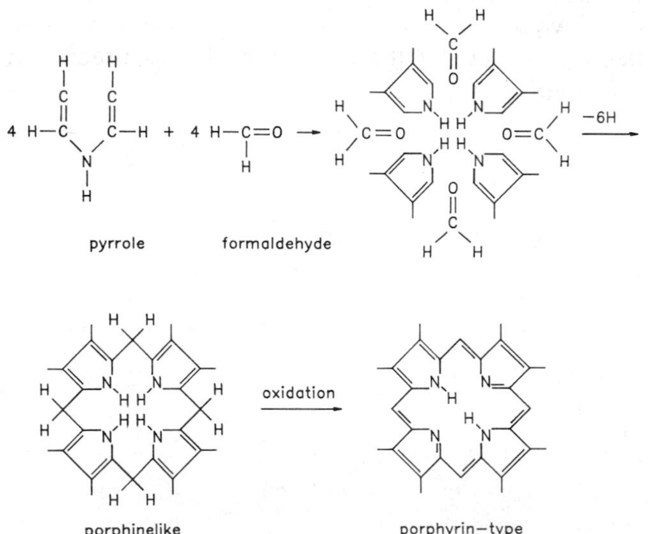

4 pyrrole + 4 formaldehyde → (−6H)

porphinelike structure —oxidation→ porphyrin–type ring

The porphyrin structure possesses alternating double and single bonds which can "resonate" by assuming a variety of different configurations without a change in the position of their constituent atoms. Such resonance confers stability upon porphyrins, enabling them to hold extra electrons and thus to function as electron acceptors (oxidation) or electron donors (reduction). Similar oxidative and reductive functions can be performed by nucleotide derivatives such as nicotinamide adenine dinucleotide (NAD), called **coenzymes** because they act in union with protein enzymes to catalyze a wide variety of chemical reactions.

We can thus see that many of the basic organic molecules used in living organisms are formed relatively easily in many reactions. The amounts per reaction are usually small, but the overall quantities of such substances may have been quite large. Shklovskii and Sagan, for example, point out that one photon of ultraviolet radiation will produce a quantum yield of about 1/100,000 to 1/1,000,000 of a simple organic molecule. If we take 10^{-22} grams as the average mass of such a molecule, then the quantum yield per photon is about $10^{-5} \times 10^{-22} = 10^{-27}$ grams. Since they estimate the number of photons at the top of the earth's atmosphere in primitive times at 3×10^{14} photons/cm^2/sec, the quantum yield per square centimeter per second may have been $10^{-27} \times (3 \times 10^{14}) = 3 \times 10^{-13}$ grams. Thus, if the reducing atmosphere lasted for 300 million years (about 10^{16} seconds), there would have been enough energy to produce $(3 \times 10^{-13}) \times 10^{16} = 3 \times 10^3$ grams of matter per square centimeter of the earth's surface. Furthermore, even if this material were diluted in as deep an ocean as the present (3×10^5 centimeters), the concentration of the solution would still be significant: (3×10^3)gm/(3×10^5)cc, or 0.01 gram per cubic centimeter.

Of course, ultraviolet radiation also causes decomposition of organic material, and such degradative effects may have been considerable. Nevertheless, once organic material was formed, it would undoubtedly have had many opportunities to accumulate in relatively protected localities, such as the fissures of rocks and the depths of pools inaccessible to decomposition by ultraviolet rays. In such places, the concentrations of organic materials may therefore have been quite high.

CONDENSATION AND POLYMERIZATION

Given localized concentrations of amino acids, sugars, and other organic molecules, further chemical evolution would depend upon the polymerization or condensation of these **monomers** into peptides, polysaccarides, and so on. Such events are not spontaneous: to obtain one small polypeptide (molecular weight 12,000) from a one molar aqueous amino acid solution in the absence of any other chemical forces would necessitate a volume of amino acids 10^{50} times that of the earth. How then could such polymerizations occur?

As shown in Figure 7-10, most polymerizations depend upon the removal of water molecules from the monomers that are to be condensed. Peptide bond

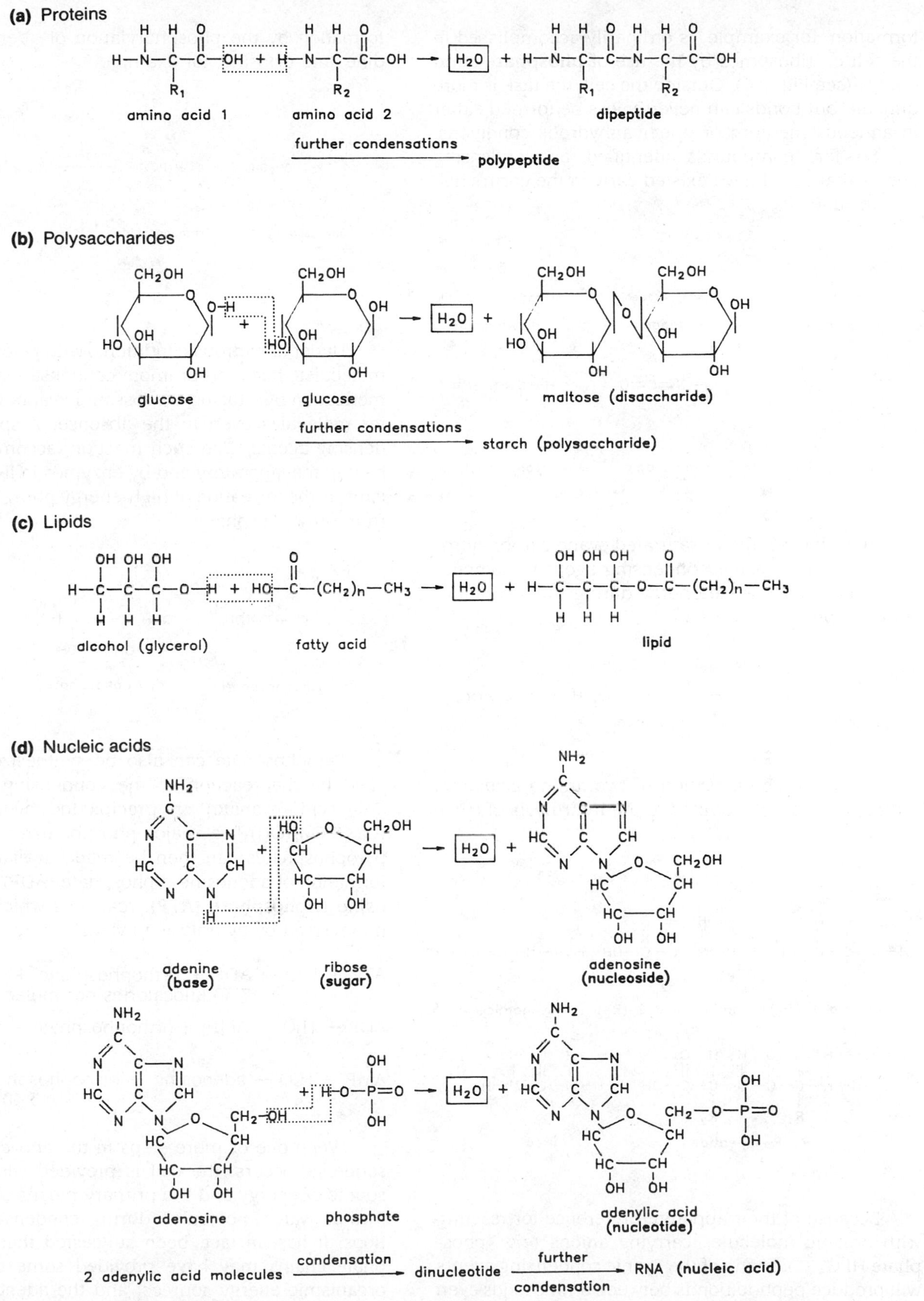

(a) Proteins

amino acid 1 + amino acid 2 → H_2O + dipeptide

further condensations → polypeptide

(b) Polysaccharides

glucose + glucose → H_2O + maltose (disaccharide)

further condensations → starch (polysaccharide)

(c) Lipids

alcohol (glycerol) + fatty acid → H_2O + lipid

$$H{-}C(OH)H{-}C(OH)H{-}C(OH)H{-}O{-}H + HO{-}C(=O){-}(CH_2)_n{-}CH_3 \rightarrow H_2O + H{-}C(OH)H{-}C(OH)H{-}C(OH)H{-}O{-}C(=O){-}(CH_2)_n{-}CH_3$$

(d) Nucleic acids

adenine (base) + ribose (sugar) → H_2O + adenosine (nucleoside)

adenosine + phosphate → H_2O + adenylic acid (nucleotide)

2 adenylic acid molecules —condensation→ dinucleotide —further condensation→ RNA (nucleic acid)

FIGURE 7-10 Examples of condensation reactions leading to the formation of peptides, polysaccharides, lipids, and nucleic acids. In (d) the sugar unit is ribose and the nucleic acid produced is therefore ribonucleic acid (RNA), whereas the sugar unit in deoxyribonucleic acid (DNA) lacks the oxygen atom at the 2′ carbon position. (Adapted from Calvin.)

formation, for example, is ordinarily accomplished in the cell on ribosomes by the use of **phosphate-bond** energy (see Fig. 7-6). Outside the cell, the task is more difficult, but bonds can nevertheless be formed either in aqueous medium or under anhydrous conditions.

So far, compounds identified as condensing agents that could have existed early in the earth's history include:

$$\mathrm{H{-}N(H){-}C{\equiv}N} \quad \text{cyanamide}$$

$$\mathrm{N{\equiv}C{-}C{\equiv}N} \quad \text{cyanogen}$$

$$\mathrm{N{\equiv}C{-}N(H){-}C{\equiv}N} \quad \text{dicyanamide}$$

$$\mathrm{(H{-}N{=})(H_2N{-})C{-}N(H){-}C{\equiv}N} \quad \text{dicyandiamide}$$

$$\mathrm{H{-}N{=}C{=}O} \quad \text{cyanic acid}$$

$$\mathrm{H{-}C{\equiv}C{-}C{\equiv}N} \quad \text{cyanoacetylene}$$

In each case, the unsaturated cyano carbon-nitrogen bonds enable the condensing agent to combine with water and release energy during this hydration. For example,

$$\underset{\text{cyanamide}}{\mathrm{H{-}N(H){-}C{\equiv}N}} + \mathrm{H_2O} \rightarrow \underset{\text{urea}}{\mathrm{H{-}N(H){-}C({=}O){-}N(H){-}H}} + \text{free energy}$$

Thus, the condensation of two amino acids into a dipeptide can be coupled to the **hydrolysis** of cyanamide:

$$\underset{\text{amino acid 1 }(R_1)}{\mathrm{H{-}N(H){-}C(H)(R_1){-}C({=}O){-}OH}} + \underset{\text{amino acid 2 }(R_2)}{\mathrm{H{-}N(H){-}C(H)(R_2){-}C({=}O){-}OH}} + \underset{\text{cyanamide}}{\mathrm{H{-}N(H){-}C{\equiv}N}} \rightarrow$$

$$\underset{R_1\text{–}R_2\text{ dipeptide}}{\mathrm{H{-}N(H){-}C(H)(R_1){-}C({=}O){-}N(H){-}C(H)(R_2){-}C({=}O){-}OH}} + \underset{\text{urea}}{\mathrm{H{-}N(H){-}C({=}O){-}N(H){-}H}}$$

Because of their apparent preference for reacting with organic molecules carrying anions (e.g., phosphate HPO_4^{--}), many of the cyanic condensing agents will produce peptide bonds between amino acids even in aqueous solutions. Some, such as cyanogen and cyanamides, have also been shown to cause nucleotide formation by the phosphorylation of adenosine, uridine, and cytosine, for example:

$$\underset{\text{adenosine}}{\text{adenosine }(\mathrm{NH_2},\ \mathrm{CH_2OH})} + \underset{\text{orthophosphate}}{\mathrm{H_3PO_4}} \xrightarrow{\text{dicyandiamide}} \underset{\text{adenylic acid (adenosine monophosphate or AMP)}}{(\mathrm{NH_2},\ \mathrm{CH_2OH_2PO_3})}$$

Under anhydrous conditions, with no or few water molecules, heat can promote condensation and polymerization by causing the loss and evaporation of water molecules even in the absence of specific condensing agents. One such reaction, accomplished by heat in the laboratory and by enzymes in living organisms, is the formation of high-energy phosphate bonds from orthophosphate:

$$2\left[\underset{\text{orthophosphate}}{\mathrm{O^-{-}P({=}O)(OH){-}OH}}\right] \rightarrow \underset{\text{pyrophosphate}}{\mathrm{O^-{-}P({=}O)(OH){-}O{-}P({=}O)(OH){-}O^-}} + \mathrm{H_2O}$$

Pyrophosphate can also be synthesized in high yield by the reaction of the condensing agent cyanic acid (cyanate) on precipitated hydroxyapatite $[Ca_{10}(PO_4)_6(OH)_2]$, a major phosphate mineral. Such pyrophosphates can then be made available for the formation of adenosine diphosphate (ADP) and **adenosine triphosphate (ATP)**, reactions which can then be reversed by hydrolysis to yield energy:

ATP + $H_2O \rightarrow$ ADP + orthophosphate + 7.30 kilocalories per molecular weight

ADP + $H_2O \rightarrow$ AMP + orthophosphate + 7.30 kcal/mole

AMP + $H_2O \rightarrow$ adenosine + orthophosphate + 3.40 kcal/mole

When one or more steps in the above hydrolytic sequence occurs, the cell is provided with its main source of energy and the primary means of removing further water molecules during condensation reactions. It has, in fact, been suggested that polyphosphate chains may have provided some of the first organismic energy sources, and the adenosine component in ATP was added later to act as a label that would allow enzymatic recognition.

TABLE 7-4 Amino Acid Compositions in Molar Percentages of Two Proteinoids Compared to the Initial Reaction Mixtures

Amino Acid	Initial Mixture	Proteinoid Product	Initial Mixture	Proteinoid Product
Aspartic acid	42.0	66.0	30.0	51.1
Glutamic acid	38.0	15.8	27.0	12.0
Alanine	1.25	2.36	2.72	5.46
Lysine	1.25	1.64	2.72	5.38
Semi-cystine	1.25	0.94	2.72	3.37
Glycine	1.25	1.32	2.72	2.79
Arginine	1.25	1.32	2.72	2.44
Histidine	1.25	0.95	2.72	2.03
Methionine	1.25	0.94	2.72	1.73
Tyrosine	1.25	0.94	2.72	1.66
Phenylalanine	1.25	1.84	2.72	1.48
Valine	1.25	0.85	2.72	1.16
Leucine	1.25	0.88	2.72	1.06
Isoleucine	125	0.86	2.72	0.90
Proline	1.25	0.28	2.72	0.59
Serine*	1.25	0.6	0.0	0.0
Threonine*	1.25	0.1	0.0	0.0

From Fox et al, 1963, *Arch. Biochem. & Biophys.* **102:**439.
*Serine and threonine were omitted from the 2:2:3 proteinoid. Tryptophan was present in the 2:2:1 proteinoid.

PROTEINOIDS

In the 1950's, Fox and co-workers developed a technique in which heat could also be used to produce peptides from dry mixtures of amino acids. Depending on the kinds of amino acids in the mixture, they found that temperatures of 150° to 180°C could produce as much as 40 percent yield of peptidelike products with molecular weights between 4,000 and 10,000 daltons. Fox called these polymers proteinoids (also thermal proteins), and he and his group have shown that these compounds bear remarkable proteinlike features. According to their analyses, the proteinoids possess nonrandom proportions of amino acids; that is, their compositions are not simply based on the frequency of the different amino acids in the initial mixture (Table 7-4).

They also suggest that the positions of the amino acids in the polymer are not based on their overall frequencies in the chain since some amino acids preferentially occupy the N- and C-terminals of the proteinoids.[5] The non-randomness of proteinoid structure also appears to be supported by the fact that these polymers all show similar properties as tested by sedimentation rates, electrophoretic techniques, column fractionation, and other measurements. Thus, there appears to be some preferential interaction between amino acids in proteinoid formation which dictates their position and frequency and leads to some degree of uniformity in the kinds of molecules produced.

Although not all of the amino acid bonds formed in such proteinoids are of the peptide variety, nor do the shapes of these molecules follow the familiar α-helix of protein structure, there still seem to be a sufficient number of peptide linkages to characterize them as proteins in many tests. Thus, proteinoids give positive color tests with the same reagents that proteins do; their solubilities resemble proteins; they are

[5]Since the interior amino acid sequences have not yet been analyzed, this point can be disputed.

TABLE 7-5 Properties Common to Thermally Produced Proteinoids and Biologically Produced Proteins

Qualitative amino acid composition
Range of quantitative amino acid composition (except serine and threonine)
Limited heterogeneity
Range of molecular weights (4,000-10,000)
Reaction in color tests (including biuret reaction)
Inclusion of non-amino acid groups (iron, heme)
Range of solubilities
Lipid quality
Salting-in and salting-out properties
Precipitability by protein reagents
Some optical activity (for polymers of L amino acids)
Hypochromicity
Infrared absorption patterns
Recoverability of amino acids with mineral acid hydrolysis
Susceptibility to proteolytic enzymes
Various enzymelike properties
Inactivation of catalysis by heating in aqueous buffer
Nutritive qualities
Hormonal activity (melanocyte stimulation)
Tendency to assemble into microparticle systems

From Fox and Dose.

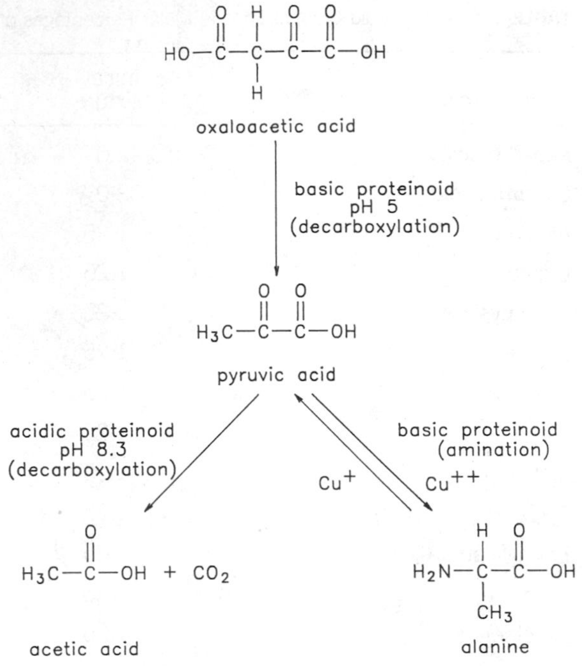

oxaloacetic acid: HO—C(=O)—C(H)(H)—C(=O)—C(=O)—OH

basic proteinoid pH 5 (decarboxylation)

pyruvic acid: H_3C—C(=O)—C(=O)—OH

acidic proteinoid pH 8.3 (decarboxylation)

H_3C—C(=O)—OH + CO_2

acetic acid

basic proteinoid (amination) Cu^{+} Cu^{++}

H_2N—C(H)(CH_3)—C(=O)—OH

alanine

FIGURE 7-11 Some sequential reactions catalyzed by different proteinoids or proteinoid complexes. (Adapted from Fox and Dose.)

precipitable with similar reagents and possess other proteinlike traits listed in Table 7-5.

Most importantly, proteinoids demonstrate a number of enzymelike activities which can increase the rates of various organic reactions. For example, they help to split apart certain molecules by addition of water (hydrolysis), they can be shown to catalyze the condensation of nucleotides such as ATP into di- and trinucleotides, and they help to remove carboxyl groups or amino groups from various structures. Moreover, they can improve catalytic activity of molecules such as heme that aid hydrogen peroxide in removing hydrogen from reduced compounds in oxidation reactions.[6]

Fox and Dose have suggested that some of these reactions, when combined into a particular sequence, may have served as the beginnings of metabolic systems that were to come later. Thus, decarboxylation of oxaloacetic acid can be followed by decarboxylation of its product, pyruvic acid, leading to acetic acid and carbon dioxide; or amination of pyruvic acid can lead to alanine (Fig. 7-11). Furthermore, some proteinoids even show relatively sophisticated hormonal activity and can stimulate the production of melanin-producing cells.

Although it has been debated whether the thermal synthesis of proteins could occur extensively in present natural surroundings (see, for example, Miller and Orgel), the exact conditions encountered on the primitive earth are certainly not known. It is possible that surfaces near some volcanic regions may have maintained appropriate temperatures for the condensation of amino acids, and cooling rains may have dispersed such thermally produced proteinoids to places where further interactions could take place.

In any case, a wide enough array of condensation mechanisms have been established, one or more of which most probably occurred in the past. Paecht-Horowitz and co-workers, for example, have shown that phosphate-activated amino acids such as aminoacyl adenylates will condense to form high yields of

[6]Fox, Jungck, and Nakashima have reported that ATP can also help to polymerize phenylalanine to the di- and tripeptide state when added to microparticles of polyadenine and lysine-rich proteinoids. They called this a protosome, or protoribosome—that is, a system involved in the synthesis of protoproteins from protonucleic acids.

polypeptide chains on layered clays such as montmorillonite:

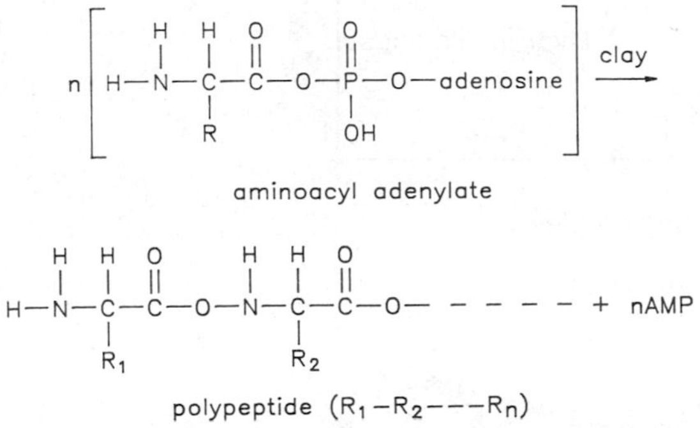

$$n\left[\mathrm{H{-}N(H){-}C(H)(R){-}C({=}O){-}O{-}P({=}O)(OH){-}O{-}adenosine}\right] \xrightarrow{\text{clay}}$$

aminoacyl adenylate

$$\mathrm{H{-}N(H){-}C(H)(R_1){-}C({=}O){-}O{-}N(H){-}C(H)(R_2){-}C({=}O){-}O{-}\;-\,-\,-\,-\,-} + n\mathrm{AMP}$$

polypeptide ($R_1-R_2---R_n$)

The amino acid ends of the adenylates apparently penetrate the narrow layers of the clay, and the condensation reactions take place there. Some clays may also polymerize nucleotides, and Burton and coworkers report that certain nucleotide compounds that are strongly absorbed by montmorillonite clays will form small amounts of dinucleotides. Thus, the early availability of cyanamides, heat, clays, and other condensing agents make it highly probable that polypeptides, polysaccharides, lipids, and perhaps even polynucleotides were present early in the earth's history.

THE ORIGIN OF ORGANIZED STRUCTURES

The presence of appropriate organic monomers and polymers is only a first step in the origin of life. Living processes of metabolism and function occur because the materials of which organisms are composed are highly organized. How did such organization come about?

At its earliest, interactions between molecules must have led them to assume relative positions based on forces such as hydrogen bonding, ionization, solubility, adhesion, surface tension, and so on. **Membranous droplets** or vesicles composed of lipids, polypeptides, or other molecules undoubtedly formed in great quantities, produced by the mechanical agitation of molecular films on liquid surfaces (Fig. 7-12) or even spontaneously (Deamer). The attainment of such droplet levels of organization was an important step in the origin of life for a number of reasons:

1. Depending on its structure and permeability, the membrane surrounding the droplet can selectively choose which compounds can enter from the environment and exit from the droplet.
2. Because of such **selective permeability**, concentrations of particular compounds can differ across the membrane, enabling reactions to occur within the droplet which would not have occurred outside the droplet.
3. There is evidence that the presence of a basic protein causes a 100-fold increase in the entrapment of nucleic acids into such droplets. As pointed out by Jay and Gilbert: "Protein-mediated encapsulation creates high local concentrations of protein and nucleic acids within the vesicular volume.... This would enhance the interaction of molecules with low affinities, potentiating the formation of aggregates with biological function."
4. The small size of the droplet can permit a chain or network of reactions to occur, the products of one reaction being available to serve as the substrates for another reaction.
5. Both the small size of the droplet and the concentration of various materials within it would permit localized precipitation to occur as well as the organization of compartments and substructures.
6. "Advanced" droplets of this kind may therefore be considered as unique subsystems that were able to preserve their organizational framework by partially separating themselves from the entropy or disorder in their surrounding environment. That is, although it is true that entropy tends to increase in the universe according to the Second Law of Thermodynamics, it can nevertheless decrease in such subsystems during their life spans. Because of their semipermeable membranes they can use entering energy and matter to retain, and even enhance, their organizational and informational structures as long as they can continue to perform biological processes (see Van Holde).[7]

It seems presumptuous and unrealistic to assume that the only kinds of organization capable of growth,

[7]One way of defining life is therefore as a system that prevents the attainment of a mass action equilibrium which has increased positive entropy. Biological organisms accomplish this by obtaining energy from external sources which they convert by metabolic processes into negative entropy (biological structures that sustain metabolism) for a period of time. Death for an organism is therefore simply the end of its ability to continue its metabolic functions, accompanied by a slide into mass action equilibrium, or positive entropy.

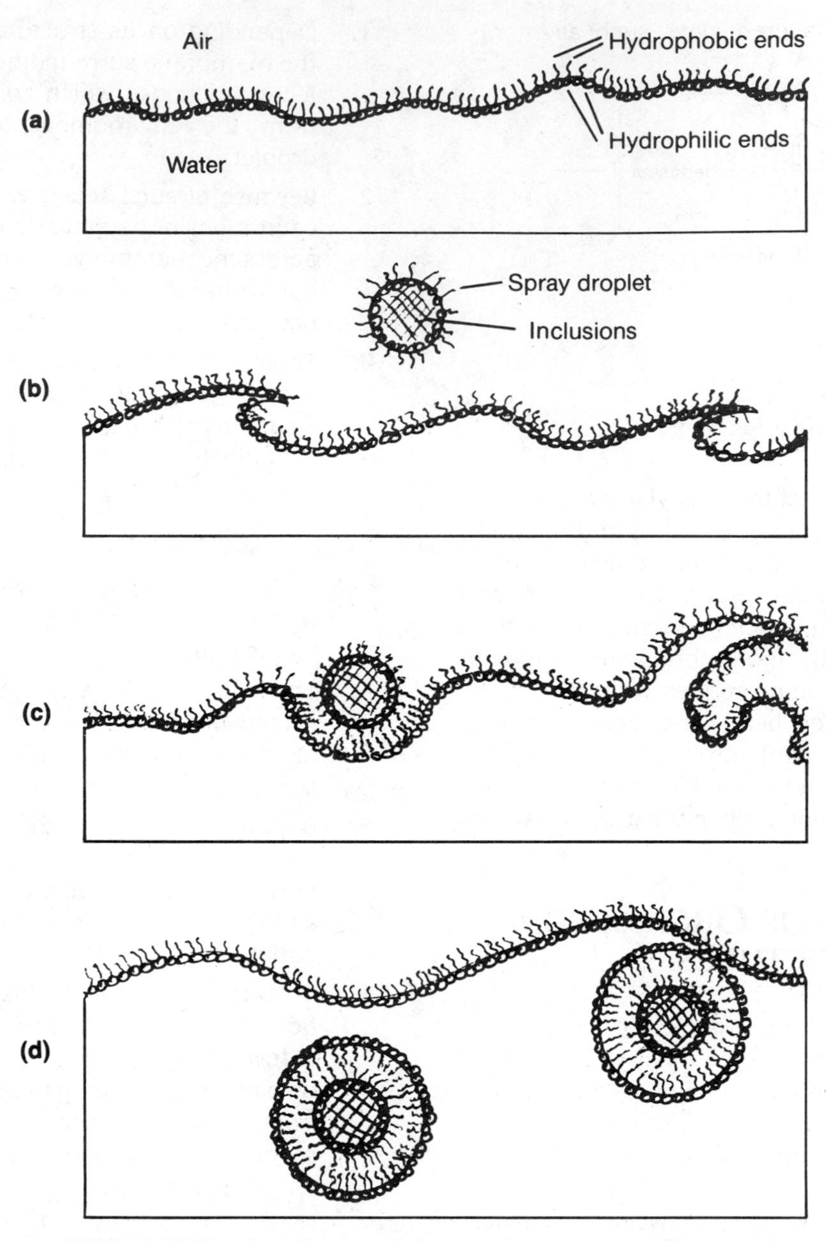

FIGURE 7-12 The effect of mechanical wave action on a surface film containing molecules oriented with one end pointed away from water (hydrophobic) and the other end towards water (hydrophilic). In organic molecules such as lipids the hydrocarbon chain is hydrophobic and the carboxyl end is hydrophilic. Wave action causes the formation of droplets and the bilayered vesicles shown in (d).

metabolism, and reproduction are the structures found in present-day organisms. Although present forms are highly efficient, early forms could have functioned at a much lower level of efficiency since they were not then competing with the more advanced forms. For a primitive form to show some (but certainly not all) "living" attributes it would have been sufficient if it could merely grow (e.g., increase in size), maintain its individuality, and divide. It is therefore interesting to note that there are at least two types of fairly simple laboratory-produced structures that appear to possess some aspects of these basic prerequisites: Oparin's coacervates and Fox's microspheres.

Coacervates

Coacervates have long been known to occur when dispersed colloidal particles separate spontaneously out of solution into droplets because of special conditions of acidity, temperature, and so on. (Fig. 7-13). If there is more than one type of macromolecular particle in the colloid, complex coacervates can be formed which show a number of interesting properties:

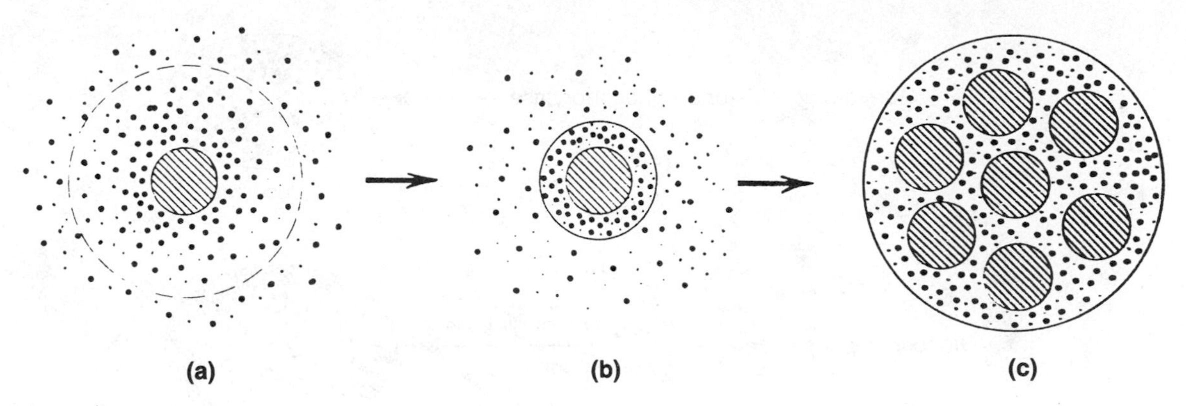

FIGURE 7-13 Formation of coacervates by the exclusion of water molecules (dots) from associated colloidal particles (hatched circles). The intervening water molecules can be removed through dehydration (e.g., increased salt concentration), or when colloidal particles are attracted to each other because they have opposite charges (e.g., negatively charged gum arabic and positively charged gelatin), or because some colloids are basic (e.g., histones) and others are acidic (e.g., nucleic acids). (From Kenyon and Steinman, adapted from Booij and Bungenburg de Jong.)

1. They possess a simple but persistent organization.
2. Although they are mostly unstable, some coacervates can maintain themselves in solution for extended periods.
3. They can increase in size.

Oparin, the first to draw serious attention to these droplets, developed artificial coacervate systems that could incorporate enzymes which performed functions such as the synthesis and hydrolysis of starch (Fig. 7-14) as well as synthesizing polynucleotides. In a coacervate system containing chlorophyll irradiated with visible light, Oparin and co-workers showed that there can be a constant inflow of reduced ascorbic acid and oxidized methylene red, which is then converted into a constant outflow of oxidized ascorbic acid and reduced methylene red. The chlorophyll picks up electrons from the ascorbic acid and then supplies these for the methylene red reduction—a process similar to common non-cyclic photosynthesis in which water molecules supply electrons for the reduction of the coenzyme $NADP^+$ to NADPH (Chapter 9).

Microspheres

These small spheres were shown by Fox to be formed when the thermally produced proteinoids were boiled in water and allowed to cool. The microspheres are uniform in size, stable, bounded by double membranes that appear somewhat cell-like, and can undergo fission and budding (Fig. 7-15). They are found in large numbers: 1 gram of proteinoid material can produce 10^8 or 10^9 microspheres. Among microsphere qualities indicating active internal processes are their selective absorption and diffusion of certain chemicals but not others, their growth in size and mass, and observations demonstrating osmosis, movement, and rotation. Moreover, microspheres show the potentiality for the transfer of information in that proteinoid particles can be seen to pass through junctions between them.

The spontaneous **self-assembly** of macromolecules into coacervates and microspheres indicates that their occurrence under primitive conditions would probably not have been an unusual event. They are not cells, of course, and it may well have taken considerable time before more elegant structures with more complex metabolic capabilities could have developed. Nevertheless, there is considerable evidence that the component materials of even more complex structures are capable of self-assembly without the immediate presence of a prior pattern. For example, the protein and nucleic acid components of tobacco mosaic virus will spontaneously aggregate into the exact configuration necessary to produce an active virus. A more complex virus such as T4, containing many different kinds of protein, also possesses a significant number of steps in which self-assembly occurs. Even cellular organelles such as ribosomes have been shown by Nomura and co-workers to be capable of being formed by self-assembly from component materials. Each level of self-assembly, from monomers to polymers to coacervates to various cellular organelles, may thus have been derived from non-random events, in that certain combinations are formed more quickly and easily than others.

Non-random self-assembly, however, is hardly sufficient by itself to account for more than a few of the various complexities of life. In their most essential aspects, such as the precisely ordered monomer se-

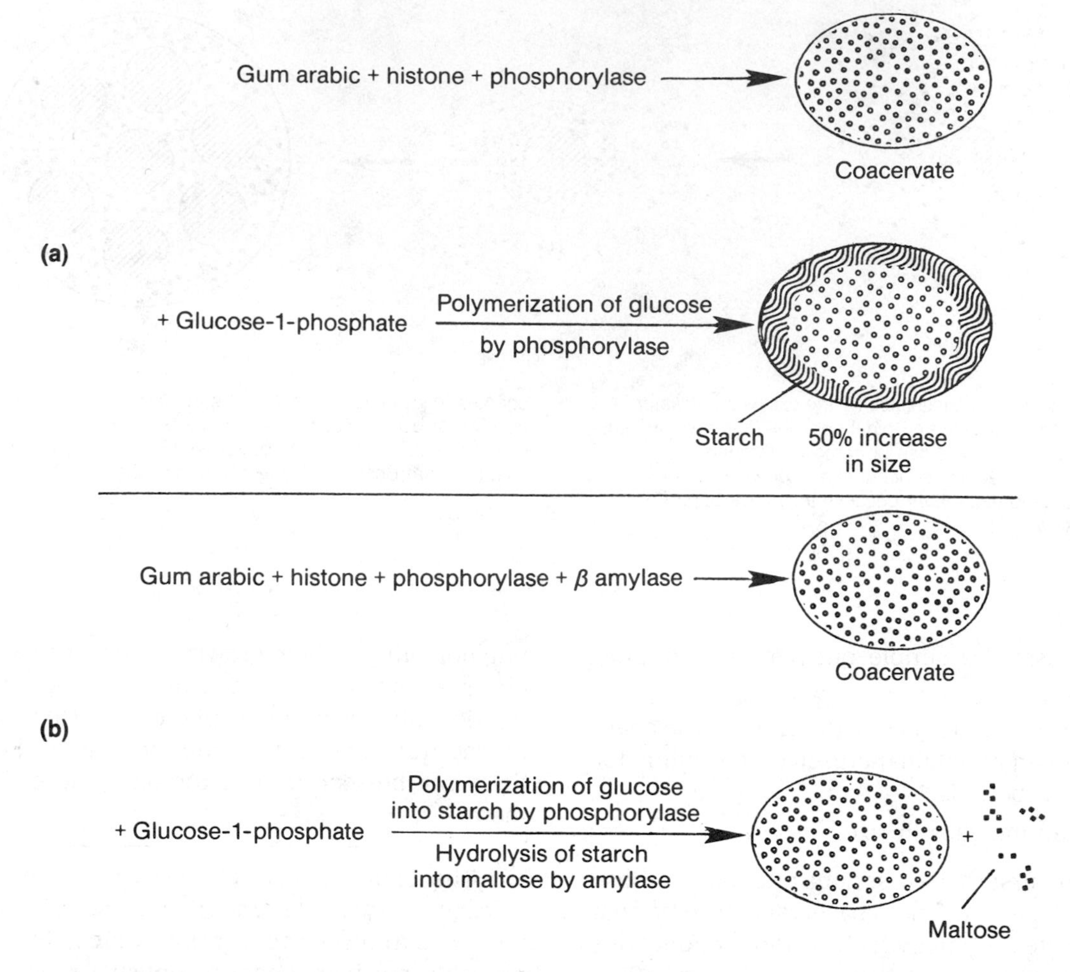

FIGURE 7-14 Synthesis and hydrolysis of starch in coacervate systems in which enzymes have been included in the droplets. In (a) the phosphorylase enzyme acts to polymerize phosphorylated glucose into starch, while in (b) the starch formed this way is hydrolyzed into maltose by the amylase enzyme.

quences found in proteins and nucleic acids, present forms of life are certainly not the result of mere chemical attractions between component amino acids or nucleotides. At the same time, we also have good reason to believe that, because of their specificity, the positioning of monomers in these precisely ordered sequences may nevertheless have had a non-random basis. To recapitulate a theme raised previously: How can the non-random biological order of amino acid and nucleotide sequences arise from disorder? The answer lies in selection.

THE ORIGIN OF SELECTION

Primitive structures, whether coacervates, microspheres, or other localized organizations, would have one important evolutionary feature: they would serve as the first distinctive, multi-chemical **individuals**, or **protocells**, that could interact as units with their environment. Together with their various neighbors and progenies, such individuals would form a group or **population** upon which selection could act. That is, protocells incorporating those organizations and metabolic activities most successful in growth and division would increase most in relative frequency or in area occupied.

Selection can thus be said to arise when the following conditions are reached:

1. A population of individuals exists.
2. The properties of these individuals are governed by reactions in which environmental material is absorbed and transformed into their own material.
3. There are differences between individuals in the efficiency with which these processes take place.
4. Limitations exist in the availability of materials and energy so that not all types of individuals can be formed, nor can all types of individuals that are formed survive.

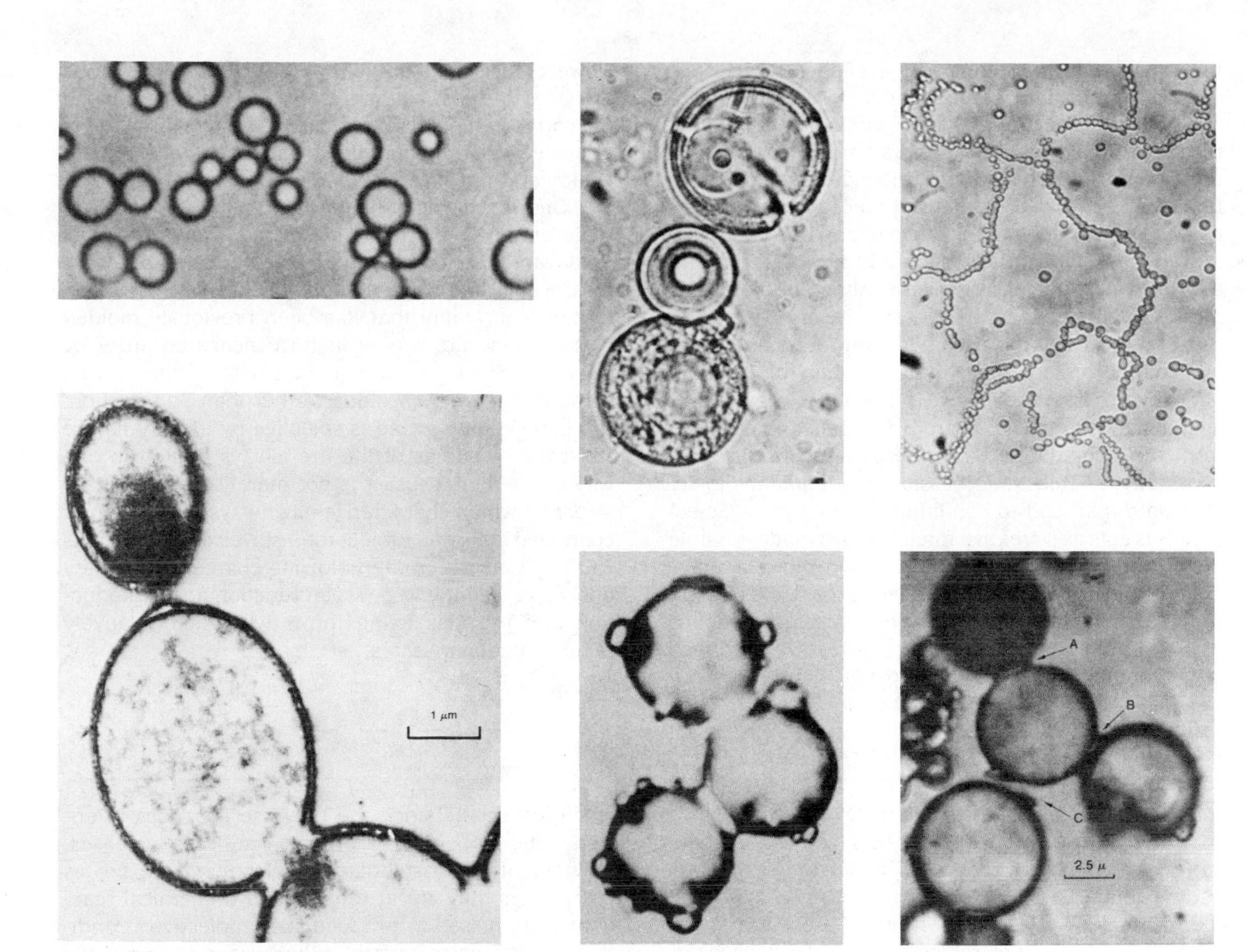

FIGURE 7-15 Various forms of proteinoid microspheres. (a) From a chilled solution of proteinoids. (b) Structured microspheres. (c) Streptococcus-like assemblages. (d) Electron micrograph showing double-layer membrane structure. (e) Budding. (f) Junctions formed between pairs of microspheres (arrows). (From Fox and Dose.)

The mechanism that enabled the formation of protocells is a crucial issue in understanding selection itself. If protocells could have originated only by self-replication, this would indicate that selection had always operated on the same efficient basis as it does now, with advantageous traits rapidly transmitted to succeeding generations by fairly exact replicative mechanisms. On the other hand, if protocells were initially formed only by acts of prevailing environmental chemistry, as many biologists now believe, this would mean that selection was not very efficient in the past and would itself have undergone evolution from chemical non-reproductive selection to the more modern biological natural selection. That is, early selection would have been confined to the survival of non-reproductive individuals who could wrest the most material from their environment and transform it for their own benefit with the least expenditure of energy. Although differences between such individuals could not be precisely transmitted, the fact that some such individuals survived and others did not would undoubtedly have had an effect on the composition and further interactions of succeeding groups.

Inheritance would therefore, at first, have been mostly a matter of transmitting molecular "things" that permit survival, rather than transmitting the nucleic acid patterns that produce the "things." As will be discussed in the following chapter, the earliest forms of "living" individuals may well have been protein-containing enclosures that replicated themselves poorly, yet passed on some of their metabolic and enzymatic properties, which continued to be selected and improved.

From a materialistic point of view, unless one postulates an accident of immense proportions and infinitely low probability, the gap between chemical evolution (changes in the composition of non-reproductive or poorly reproductive molecules, coacervates, microspheres, and so on) and biological evolution (changes in inherited differences among reproductive organisms) must have been bridged by selection. So far, selection is the only natural mechanism we know that can account for the creative changes among non-reproductive individuals that could have led them in the direction of living organisms. Although the events may be complex, the device is simple: organisms that react to their environment with improved functional information replace those that lack such information.

Nevertheless, selection is not merely a passive agent that sifts the good from the bad, the adaptive from the non-adaptive, but, because of its historical continuity, enables a succession of adaptations to accumulate that lead to something entirely new. Selection thus acts as a creative force that has made possible biological organizations that would otherwise have been highly improbable. To use a previous example, a polypeptide chain composed of a specific sequence of 100 amino acids has an extremely low probability (10^{-130}) of occurring spontaneously without selection. However, as explained in Chapter 4, if each step in the growth of the chain attains a selective advantage when the correct amino acid is inserted, then the probability of achieving a functionally advantageous polypeptide is almost immeasurably increased.[8]

Once the game of life has begun, the evolutionary replacement of the players, whether they are genes, organisms, races, species, or other entities, becomes inextricably bound to their ability to play the game further—an ability that has been previously molded by selection and is now in turn measured anew by selection. Some random replacement of the players certainly occurs by accident rather than by selection, but participation in life is selective by its very nature since the resources of life are always limited in one way or another. Thus, it is not merely their origin by selection which characterizes living systems but their continued ability to subject themselves to selection. As will be discussed in the following chapter, this ability has led to a coupling between function and reproduction that provides living forms with their relatively rapid evolutionary rates.

SUMMARY

For many years even those who believed in the reality of evolution could not explain how cells, with their presently complex membranes and compartmentalized systems, could have arisen. Recently, however, chemical pathways that may have led to the formation of the most important molecules in cells have been identified.

Proteins, composed of specific linear sequences of amino acids linked together by peptide bonds, are crucial to organisms because of their catalytic functions. Their structure is specified by the sequence of nucleotides in nucleic acids, which act as information storage molecules in cells. The probelm is to explain the origin of these now totally interdependent molecules.

If life arose on earth by chemical means, as seems more likely than by special creation or panspermia ("seeding" from outer space), the origins of complex organic molecules from simple substances present on the primitive earth must be explained. The probability of this occurring randomly is almost incalculably small, especially since the universe increases in entropy. The probability of organic syntheses increases, however, if the first organic molecules were not as complex as they are at present and if chemical reactions were biased to produce such molecules. Conditions on the earth several billion years ago probably offered a most favorable chemical environment: sufficient and continuous energy, availability of carbon and other important elements, an abundance of water (an excellent solvent), and vast amounts of hydrogen (and its compounds) providing a reducing or partially reducing atmosphere.

Some years ago Miller demonstrated that amino acids and other organic compounds could be spontaneously synthesized from hydrogen, ammonia, methane, and water in the presence of an energy source. Amino acids can also be synthesized by the interaction of aldehydes with nitrogenous compounds and by cyanide-cyanide reactions. Furthermore, formaldehyde can condense to form the pentose sugars used in nucleic acids. The nitrogenous bases of nucleic acids can

[8]For example, assuming that each of the 20 different amino acids are present in the surrounding medium in equal frequency, there is a probability of 1/20 that random chance will supply the correct, functionally advantageous amino acid to any position of the chain. Thus $100 \times 1/20 = 5$ positions that will be correctly occupied by chance alone without selection. For the remaining 95 positions, selection will operate so that each position, once occupied by its correct amino acid, enables further selection to occur at the next succeeding position. This stepwise procedure would entail perhaps 20 trials to achieve the correct amino acid at one position, another 20 trials to achieve the correct amino acid at the next position, and so on. In sum, a succession of only $95 \times 20 = 1{,}900$ trials may be necessary for selection to provide a functional amino acid sequence for a chain 100 amino acids long—a probability of $1/1900 \approx 2 \times 10^{-3}$ compared to the 10^{-130} probability in the absence of selection.

be produced from cyanide compounds; fatty acids and porphyrins, from simple chemicals. These simple subunits of complex molecules can be polymerized by the removal of water, and polymerization can be expedited by heat or with condensing agents. Fox showed that amino acid mixtures subjected to high temperatures will polymerize to form proteinoids, structures which have some peptidelike properties such as non-random amino acid frequencies and low-level catalytic activity. Polymerization of amino acids and even nucleotides can also occur in certain types of clays.

If complex molecules were produced from simpler ones long ago, how might they have become organized into cells? Intermolecular forces can bind macromolecules into membrane-enclosed droplets (coacervates and microspheres), which exhibit some features of living systems such as organization, selective permeability, and energy utilization. That these droplets, like primitive biological systems such as viruses, can self-assemble in non-random ways suggests that similar structures might have occurred on the pathway to living organisms. The droplet systems, or protocells, that could best maintain themselves would be perpetuated. Later, protocells that could reproduce, however inefficiently, would be acted on by natural selection, which would act as a creative force successively enhancing the probability of advantageous molecules, reactions, and structures.

KEY TERMS

adenosine triphosphate (ATP)
amino acids
carbonaceous chondrites
coacervates
coenzymes
condensation
double helix
entropy
enzyme
fatty acids
genotype
hydrolysis
individuals
membranous droplets
messenger RNA (mRNA)
microspheres
molecular preadaptation
monomers
nucleic acids
nucleotides
optical activity
panspermia
peptide linkages
phenotype
phosphate bond
polymerization
polypeptide chains
population
porphyrins
proteinoids
protocells
purines
pyrimidines
racemic mixture
ribosomal RNA (rRNA)
Second Law of Thermodynamics
selective permeability
self-assembly
sugars
transfer RNA (tRNA)

REFERENCES

Brooks, J., and G. Shaw, 1973. *Origin and Development of Living Systems.* Academic Press, New York.

Buhl, D., 1974. Galactic clouds of organic molecules. *Origins of Life*, **5**, 29-40.

Burton, F. G., R. Lohrmann, and L. E. Orgel, 1974. On the possible role of crystals in the origins of life. VII. The adsorption and polymerization of phosphoaramidates by montmorillonite clay. *Jour. Mol. Evol.*, **3**, 141-150.

Cairns-Smith, A. G., 1982. *Genetic Takeover and the Mineral Origins of Life.* Cambridge Univ. Press, Cambridge.

Calvin, M., 1969. *Chemical Evolution.* Oxford Univ. Press, Oxford.

Chang, S., D. DesMarais, R. Mack, S. L. Miller, and G. E. Strathearn, 1983. Prebiotic organic synthesis and the origin of life. In *Earth's Earliest Biosphere: Its Origin and Evolution*, J. W. Schopf (ed.). Princeton Univ. Press, Princeton, N.J., pp. 53-92.

Crick, F., 1981. *Life Itself: Its Origin and Nature.* Simon and Schuster, New York.

Deamer, D. W., 1986. Role of amphiphilic compounds in the evolution of membrane structure on the early earth. *Origins of Life*, **17**, 3-25.

Eigen, M., 1971. Self-organization of matter and the evolution of biological macromolecules. *Naturwiss.*, **58**, 465-523.

———, 1983. Self-replication and molecular evolution. In *Evolution From Molecules to Man,* D. S. Bendall (ed.). Cambridge Univ. Press, Cambridge, pp. 105-130.

Eigen, M., and P. Schuster, 1979. *The Hypercycle.* Springer-Verlag, Berlin.

Ferris, J. P., and P. C. Joshi, 1978. Chemical evolution from hydrogen cyanide: photochemical decarboxylation of orotic acid and orotate derivatives. *Science,* **201,** 361-362.

Fox, S. W., 1984. Proteinoid experiments and evolutionary theory. In *Beyond Neo-Darwinism*, M.-W. Ho and P. T. Saunders (eds.). Academic Press, London, pp. 15-60.

Fox, S. W., and K. Dose, 1972. *Molecular Evolution and The Origin of Life*. W. H. Freeman & Co., San Francisco.

Fox, S. W., J. R. Jungck, and T. Nakashima, 1974. From proteinoid microsphere to contemporary cell: formation of internucleotide and peptide bonds by proteinoid particles. *Origins of Life*, **5**, 227-237.

Haldane, J. B. S., 1929. The origin of life. *The Rationalist Annual*, **148**, 3-10.

Hoyle, F., and N. C. Wickramasinghe, 1978. *Lifecloud: The Origin of Life in the Universe*. Dent & Sons, London.

Jay, D. G., and W. Gilbert, 1987. Basic protein enhances the incorporation of DNA into lipid vesicles: model for the formation primordial cells. *Proc. Nat. Acad. Sci.*, **84**, 1978-1980.

Kenyon, D. H., and G. Steinman, 1969. *Biochemical Predestination*. McGraw-Hill, New York.

Lawless, J. G., and G. U. Yuen, 1979. Quantification of monocarboxylic acids in the Murchison carbonaceous meteorite. *Nature*, **251**, 40-42.

Matsuno, K., K. Dose, K. Harada, and D. L. Rohlfing (eds.), 1984. *Molecular Evolution and Protobiology*. Plenum Press, New York.

Miller, S. L., 1953. A production of amino acids under possible primitive earth conditions. *Science*, **117**, 528-529.

——, 1987. Which organic compounds could have occurred on the prebiotic earth? In Cold Spring Harbor Symposium on Quantitative Biology, Vol. LII, Cold Spring Harbor Laboratory, Cold Spring Harbor, N.Y.

Miller, S. L., and L. E. Orgel, 1974. *The Origins of Life on the Earth*. Prentice-Hall, Englewood Cliffs, N. J.

Nomura, M., 1973. Assembly of bacterial ribosomes. *Science*, **179**, 864-873.

Oparin, A. I., 1924. *Proiskhozhdenie Zhizny* ("The Origin of Life"). Moscovsky Robotschii, Moscow. (Original Russian edition of Oparin's theory; a revised edition was published in English 1938, reprinted 1953 by Dover Publications.)

Oró, J., and A. P. Kimball, 1961. Synthesis of purines under possible primitive earth conditions. I. Adenine from hydrogen cyanide. *Arch. Biochem. Biophys.*, **94**, 217-227.

Paecht-Horowitz, M., J. Berger, and A. Katchalsky, 1970. Prebiotic synthesis of polypeptides by heterogeneous polycondensation of amino acid adenylates. *Nature*, **228**, 636-639.

Schopf, J. W. (ed.), 1983. *Earth's Earliest Biosphere: Its Origin and Evolution*. Princeton Univ. Press, Princeton, N.J.

Shklovskii, I. S., and C. Sagan, 1966. *Intelligent Life in the Universe*. Holden-Day, San Francisco.

Strickberger, M. W., 1985. *Genetics*, 3rd ed. Macmillan, New York.

Van Holde, K. E., 1980. The origin of life: a thermodynamic critique. In *The Origins of Life and Evolution*, H. O. Halvorson and K. E. Van Holde (eds.). A. R. Liss, New York, pp. 31-46.

Wald, G., 1974. Fitness in the universe: choices and necessities. *Origins of Life*, **5**, 7-27.

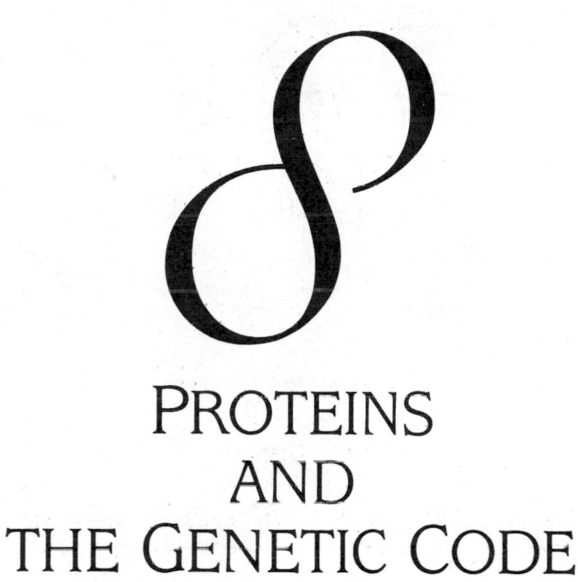

PROTEINS AND THE GENETIC CODE

In general, function in living organisms depends upon the transformation of material and energy outside the organism into processes that take place within it. These living processes must have originally existed on a fairly simple level not much different from some of the processes we have seen occurring in coacervates and microspheres. For a considerable period of time they may have depended, at least in part, on heat to provide some of the reactions that produce various cellular constituents. However, thermal energy, which may vary in place and in time, would hardly have been a reliable and consistent source for cellular necessities. At best, heat derived from the environment or from rapid oxidations would have provided only an explosive, uncontrolled release of energy. More valuable to the cell was the development of chemical energy providers, such as adenosine triphosphate (ATP), which can release small but significant amounts of phosphate-bond energy that can be both controlled and localized to specific reactions by the enzyme apparatus of the cell.[1]

The shift to chemical systems of energy therefore meant the elaboration of **organic catalysts** (enzymes) which could restrict chemical reactions to the most opportune times and places. Accompanying this shift, without doubt, must have been a remarkable increase in the efficiency with which particular reactions could take place. For example, inorganic ferric ion (Fe^{3+}) can be shown to have some catalytic activity in a variety of reactions including the decomposition of hydrogen peroxide into water and oxygen (Fig. 8-1(a)). However, when such ions are incorporated into porphyrin molecules to form **heme** (Fig. 8-1(b)), the molecules are about 1,000 times more effective than Fe^{3+} alone. If the protein component of the enzyme catalase is then added to the heme unit, catalytic efficiency is increased by a further factor of 1 billion (Fig. 8-1(c)). How did such proteins evolve?

[1]Phosphorylated nucleotides such as adenosine (ATP, ADP) or guanosine (GTP, GDP) are not the sole agents capable of transferring energy-rich phosphate bonds. In a reaction discovered by Siu and Wood, inorganic pyrophosphate (PP_i was found to serve as a phosphate donor and inorganic phosphate (P_i) as an acceptor: oxaloacetate + $PP_i \rightleftharpoons$ phosphoenolpyruvate (PEP) + CO_2 + P_i. Lipmann (1965) suggested that this reaction represents the metabolic fossil of a primitive form of energy transfer.

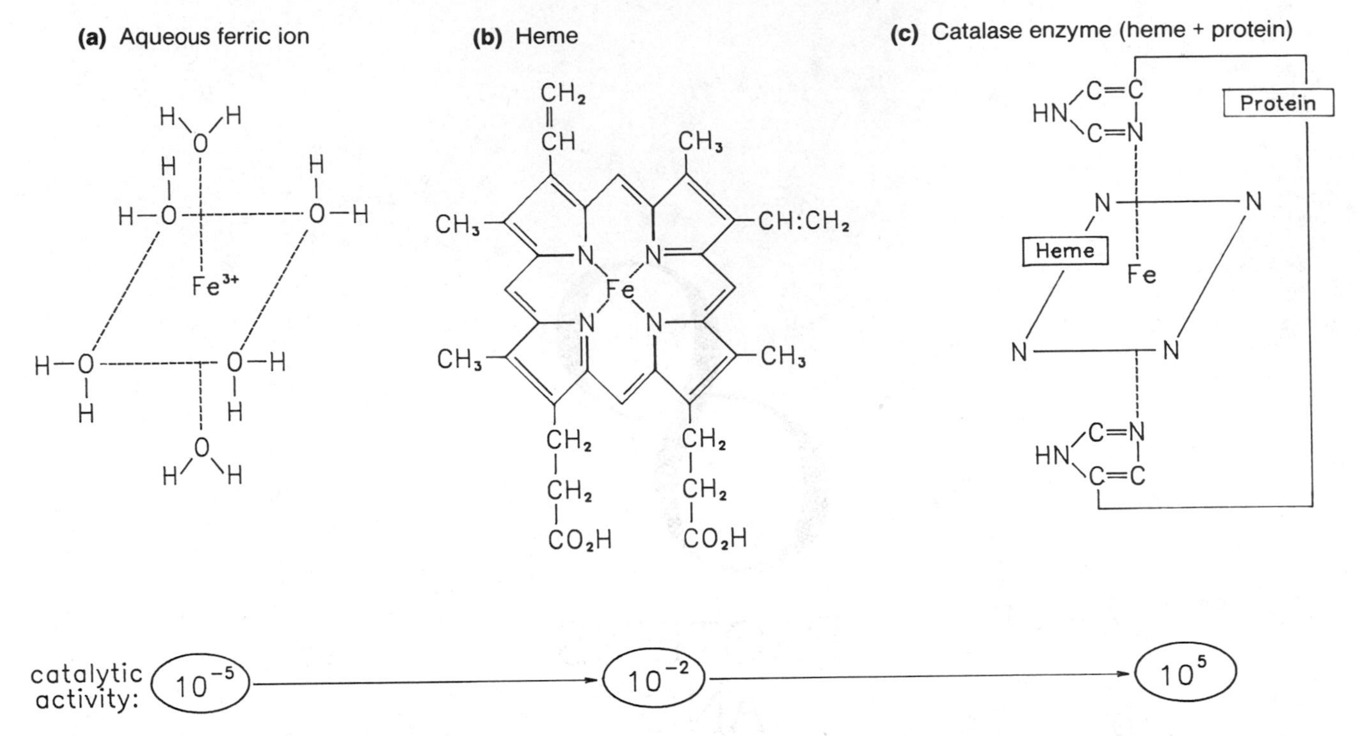

FIGURE 8-1 Change in catalytic activity for the reaction $2H_2O_2 \rightarrow 2H_2O + O_2$ when the iron atom is used by itself (a), or in different molecular combinations (b, c). (Adapted from M. Calvin, 1969. *Chemical Evolution.* Oxford Univ. Press, Oxford.)

PROTEINS OR NUCLEIC ACIDS FIRST?

At present, the amino acid sequences of the enzymes that serve as catalysts in the cell derive entirely from the nucleotide sequences in ribonucleic acid (RNA), which in turn derive from the nucleotide sequences of deoxyribonucleic acid (DNA) as discussed in Chapter 7. The fact that the entire chain of information transfer—replication (DNA → DNA), transcription (DNA → RNA), translation (RNA → protein)—is itself dependent on appropriate enzymes (Fig. 8-2) poses the serious question of how these functional and informational systems could possibly have evolved independently of each other.

One answer to this problem has been that nucleic acids arose first, and their self-replicating power then enabled selection to develop protein systems which would support further **self-replication**. The geneticist Hermann Muller long ago suggested that since the basic appearance, or phenotype, of an organism derives essentially from its genetic material, or genotype, this relationship must also have existed in the past. That is, the genotype was probably first in the evolutionary sequence. **Protein synthesis** might then have evolved by the direct interaction of specific amino acids with specific nucleotide sequences, or perhaps by the indirect placement of amino acids into such sequences through use of intermediary **adaptor molecules** that brought amino acids to the nucleotide chain.

Supporters of the view that nucleic acid replication arose first have generally used the argument that only a self-replicating system can provide the basis upon which selection can build a cooperative functional unit. In the absence of self-replication, function would presumably be quickly lost and the "organism" would be unable to maintain itself. The search has therefore been for an **autocatalytic** process that would explain the origin of a "naked gene" which could replicate itself without the help of proteins.

The possibility of autocatalytic **nucleic acid replication** has recently received some support from the discovery that some RNA molecules do possess catalytic properties even in the absence of proteins (Cech and Bass). Presumably, by making use of such **RNA catalysts**, short RNA sequences could theoretically replicate themselves without protein enzymes by forming templates for complementary RNA sequences (Loomis). However, in opposition to these proposals, Joyce et al. have pointed out that such RNA replication would have been strongly inhibited in pre-biological times because of the presence of different stereoiso-

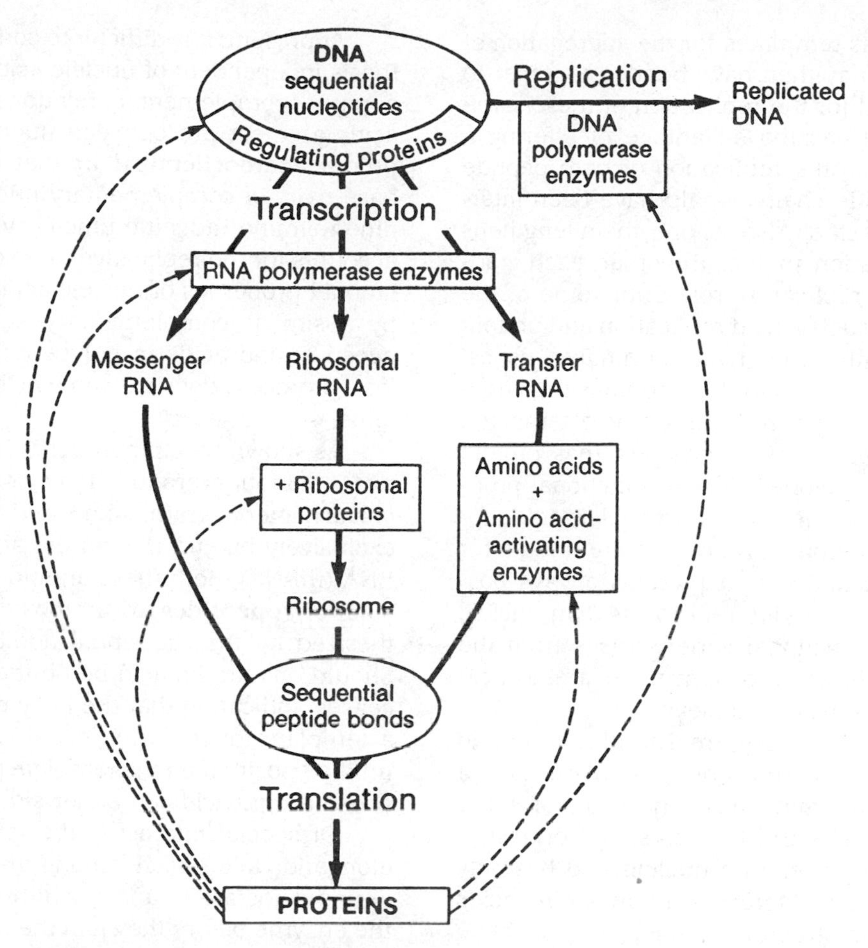

FIGURE 8-2 Schematic diagram showing the mutual dependence of information carried by nucleotide sequences and function governed by proteins. Heavy arrows indicate the general directions of information transfer, and dashed arrows point to proteins synthesized by this process. It is clear that the nucleotide sequence information determines the amino acid sequences of proteins, and the proteins in turn regulate and catalyze the transfer of nucleotide information.

mers (see Fig. 7-7) in the ribose-sugar backbone that would have prevented complementary base pairing. They therefore suggest that early nucleic acid genetic material was not based on ribose-containing nucleotides but on riboselike analogues in which such pairing difficulties were minimized and which were more easily synthesized than ribose. One such system has been experimentally demonstrated by Zielinski and Orgel. Nevertheless, numerous questions still remain as to the range of reactions that RNA could catalyze (its present range is quite limited) and how such systems, even if they occurred, would have made the transition to RNA synthesis and replication.

Instead of nucleic acids, Cairns-Smith suggests that the primitive self-replicating unit may have been composed of organic, claylike silicate crystals or layered clay mixtures (e.g., montmorillonite, see p. 117). Crystals can grow by the addition of subunits into their highly ordered structures and can be viewed as possessing some self-replicatory powers. According to Cairns-Smith, nucleic acids such as RNA could have been incorporated into such an assembly, followed by the formation of peptides and the evolution of a protein-synthesizing system.

Although such events cannot be excluded, they are difficult to envisage,[2] and some biologists have therefore placed more emphasis on the possibility that either proteins themselves or protein-nucleic acid combinations were the first self-replicating systems. Black, for example, suggests that some early peptides

[2]According to Joyce et al., "the accumulation of substantial quantities of relatively pure mononucleotides on the primitive earth is highly implausible."

might have served as templates for the aggregation of nucleotides which may then have bonded together to form a precise mold for the replication of these same peptides. This view has the advantage of offering a mechanism by which the replication of both peptide chains and nucleotide chains would have been interdependent from the start. Also, as one chain lengthens during further evolution so does the other, each gradually improving its replicatory role until some of the present features of nucleic acid replication and protein synthesis evolve. Unfortunately, as with naked genes, it is still difficult to conceive of the spontaneous origin of such a precisely organized nucleic acid template.

Various theories therefore suggest that protein systems probably developed diverse functional properties before they became coupled to nucleic acid replicative systems. Foremost among the arguments for this view is the likelihood that proteins always possessed more functionally different forms than nucleic acids. Part of this functional variety arises from the almost inexhaustible array of permuted amino acid sequences that proteins can achieve.

For example, since there are 20 different amino acid alternatives for each position in a polypeptide, a sequence of only 5 amino acids has 20^5, or more than 3 million possible arrangements. By contrast, a sequence of 5 nucleotides in a nucleic acid has only 4^5 or 1,024 possibilities. Moreover, many amino acids are quite different in structure (see Fig. 7-2) enabling them to interact in many different ways both with each other and with molecules such as water, metal ions, and various monomers and polymers. This confers an astronomical variety of possible three-dimensional configurations on a protein, in contrast to the relatively more rigid shapes assumed by many nucleic acids.[3]

Chemical experiments under presumed prebiological conditions also point to the difficulty of producing polymerized nucleic acids spontaneously, whereas long-chained polypeptides are produced in such experiments with relative ease. A number of authors (e.g., Fox) therefore make the claim that the development of a protein catalytic system must have occurred before the development of a nucleic acid replicative system. To many others, however, this view has a serious shortcoming, since if proteins arose first and were used by primitive cells or particles for functional purposes, how could they have been replicated without a nucleic acid translational system? Could proteins alone have synthesized proteins?

At present, it is difficult to conceive of protein synthesis independent of nucleic acids since there are no apparent complementary relationships between amino acids as there are between nucleotides. That is, the precise **stereochemical fit** that occurs between the base pairs of complementary nucleotide chains (adenine-thymine, adenine-uracil, cytosine-guanine) and accounts for the replicative, transcriptional, and translational properties of nucleic acids is nowhere echoed by a similar complementary stereochemical fit between amino acids in polypeptide chains. Nevertheless, a process does exist in which proteins make proteins.

As shown by Lipmann (1971) and others, at least two antibiotics produced by spore-forming *Bacillus brevis* bacteria, gramicidin S and tyrocidin, are formed exclusively by enzymes in the absence of messenger RNA (mRNA). Both these antibiotic molecules are circular **oligopeptides** ten amino acids long and are synthesized by the sequential addition of amino acids. Should one amino acid be omitted, peptide synthesis ceases, indicating that the enzyme involved serves as a template for the amino acid sequence. That is, an unfilled position on the template prevents bonding between amino acids on either side.

Of special interest is the form in which peptide elongation takes place among these antibiotics. Single amino acids are bound to sulfhydryl (-SH) groups on the enzyme before they join the peptide chain and are then connected by the sequential removal of their sulfur (thiol) groups and the formation of peptide bonds. That is, the chain maintains a thiol at its "head" end to furnish the connection for sequential growth. Lipmann called this process **head-growth polymerization** and pointed out that it is strikingly similar to the polymerization of carbon groups during fatty acid synthesis (also polymerized by use of sulfhydryl bonds) and to the polymerization of amino acids during ribosomal peptide synthesis (polymerized by phosphate bonds). He suggested that these similarities indicate a common underlying polymerization process that may have arisen early during chemical evolution. Of course, the enzymes now involved in antibiotic peptide synthesis are themselves synthesized via information transferred from genetic material, but the fact that proteins can produce proteins in these systems points to the possibility that repeatable copies of short-chained but functional peptides, 15 to 20 amino acids long, may have been produced in the past in the absence of nucleic acids.

[3]Nevertheless, as mentioned previously, some RNA molecules may well have functioned as catalysts in the absence of proteins. One such presently known example is an "intron" portion of transcribed RNA (p. 157, Chapter 9) in the protozoan ciliate *Tetrahymena* which splices itself out of the RNA molecule and helps form a chemical bond between the RNA sections on either side ("exons"). It has also been suggested (Woese [1980] and Gerbi) that molecular interactions presently observed between the various RNA molecules during protein synthesis (transfer RNA, messenger RNA, ribosomal RNA) may derive from a primitive translational system which had only RNA components.

EVOLUTION OF PROTEIN SYNTHESIS

Whatever the early composition of the templates used in the condensation of amino acids, the polymerization process itself must have been one of the first functions for which selection occurred, since it is only in peptide form that amino acids attain their catalytic properties. These early templates, however, were probably inefficient, producing peptide chains with only vague similarities to each other. Thus, as time went on, selection for improved polymerization must have led to the production of more efficient templates which were perhaps themselves polymerized by more efficient **polymerization enzymes**. The circular, autocatalytic tautology of life—to make more of those substances that can interact with the environment to make more such substances—can therefore be said to have become firmly established during this early period.[4]

Given a template for the synthesis of peptides and a polymerization enzyme, various sequences of events might have followed. The value of considering any of these is simply to show that enough molecular information has been accumulating to permit the development of different step-by-step hypotheses that can be used to solve the origin of life puzzle. If we continue with the notion that the first templates were probably not pure nucleic acids but perhaps some combination of peptides and other materials, one possible sequence is as follows:

1. Once peptide polymerization was established, selection for its improvement must have led to an increased number of polymerase enzymes per protocellular unit, as well as improved template efficiency of these polymerases and their ability to increase the rapidity of amino acid condensation.
2. Improvement in polymerase function was probably accompanied by improvements in the mechanism that brought amino acids to the template. That is, adaptor molecules evolved that could be bound to amino acids in the medium, and then bind these amino acids to the polymerase enzyme or the polymerizing template. In all likelihood these adaptor molecules were probably not very large, and their evolution might have meant simply new uses for some of the templates that had previously produced polymerase enzymes.
3. Further improvement probably involved selection for small, nucleotidelike coenzymes that were used in the binding and release of phosphate energy groups, similar perhaps to present coenzymes such as nicotinamide adenine dinucleotide (NAD) and flavine adenine dinucleotide (FAD), discussed in the following chapter. Such nucleotides were not necessarily RNA or DNA but could have consisted of combinations of various sugars, nucleic acid bases, and phosphates.
4. Both the template and the adaptor molecules may have incorporated nucleotide sequences which could then recognize each other by complementary base pairing via hydrogen bonding similar to the type presently seen between adenine and uracil or guanine and cytosine. These new adaptors would therefore be capable of bringing their amino acid to the nucleotide portions of the polymerization complex rather than to the peptide portions. Protein synthesis would thus gain the advantage of precise stereochemical pairing between complementary nucleotide bases as compared to the more inefficient amino acid-polypeptide interactions that may have been previously used. Events of this kind would therefore lead directly to improving the placement of different amino acids into specific positions on polypeptide chains and to a significant increase in the number of amino acids that could be incorporated into such chains.
5. If we consider the polymerization enzyme complex to be a primitive ribosome of sorts, some of its nucleotide components thus began to act as a template specifying the sequence of amino acid incorporation by base pairing with adaptor molecules. At the same time, these nucleotide sequences themselves could be replicated with considerable precision by base pairing with available nucleotides.
6. Replication of nucleotides led to the formation of master nucleic acid molecules that could be stored as genetic material, yet which could also serve in the translation of nucleotides to amino acids or produce complementary messenger strands for this purpose. Three separate functional classes of nucleotide sequences thus eventually arose: storage, messenger and

[4]Black suggests that the original force responsible for polymerizing organic molecules arose because of hydrophobic interactions between various amino acids ("a search by organic compounds for a means to separate from water"). The most stable of such interactions that would dissipate hydrophobic forces and produce the lowest free energy level presumably lies in the folded globular organization of polymerized protein. Black proposes that the energy to attain these polymerizations was derived from simultaneous degradation of other organic compounds. The coupling of these two processes, degradation and polymerization, would then provide the framework enabling the selection of polymers that could catalyze their own polymerization. Once such autocatalytic polymers arise, evolutionary refinements would place emphasis on selection for other attributes such as a code for polymer replication, metabolic pathways, response to environmental substances, homeostasis, and development.

translational (ribosomal and transfer), all probably RNA, since this nucleic acid is still used for two of these purposes in all organisms (messenger and adaptor) and is also used for genetic storage purposes in some viruses.[5]

7. Since RNA was probably used primitively for both information storage and protein translation, difficulties in separating these two functions must have offered advantages to organisms that could utilize a different nucleic acid, DNA, for storage purposes. The enzymes that translate RNA into protein do not function with DNA, thus permitting the more uniformly structured, double-helical DNA to be restricted exclusively to the storage of information and to the transcription of one of its strands to form mRNA.[6] The fact that deoxyribonucleotides are synthesized from ribonucleotides in cellular pathways supports the notion that DNA arose later in cellular evolution than RNA.

The evolution of protein synthesis may thus have had its start in a primitive polymerase enzyme that could be replicated inefficiently on a template composed of polypeptides and other materials. Evolution then proceeded to a self-contained ribosome that contained its own stored genetic information and also mRNA committed exclusively to the translation of a limited number of polypeptides. In later stages these functions were sequestered to different parts of the cell, with mRNA moving from its new site of transcription, where genetic information was now stored, to the ribosome for translation. The evolution of new kinds of ribosomes, no longer committed to the production of particular proteins, enabled different mRNAs to be translated by the same ribosome. This transferred the burden of regulating which proteins were to be made from the ribosome to the transcriptional process. That is, selection of the particular proteins to be produced could now be determined by regulating which mRNA molecules were to be transcribed from the stored genetic material. By these means, some of the basic modern features of protein synthesis may have come into being. (For an example of another possible scenario, see Loomis.)

EVOLUTION OF THE GENETIC CODE

Information transfer between nucleic acids and proteins is based on a genetic code which determines the placement in a protein of a particular type of amino acid from the placement of a particular trinucleotide sequence in mRNA. The terminology and characteristics of the code are listed in Table 8-1, and the code itself is given in Table 8-2.

As for any other biological trait, it is believed that this code must have evolved from a more primitive form, although no ancestral codes have so far been discovered. Attempts at an evolutionary reconstruction of the code have therefore generally relied on a detailed analysis of the features that characterize the present code. These features are as follows:

1. Messenger RNA molecules consist of only four kinds of nucleotide bases, adenine (A), guanine (G), uracil (U), and cytosine (C). These are arranged in chains of varying lengths and varying sequence.
2. An mRNA codon that specifies a particular amino acid is a triplet composed of a chain of three nucleotides.
3. The code is commaless and non-overlapping: each codon is translated in a continuous sequence, three successive nucleotides at a time, from one end of an mRNA reading frame to the other.
4. The codon sequence is complementary to an anticodon sequence on the adaptor or transfer (tRNA) molecule (Fig. 8-3) that carries a particular amino acid to the mRNA codon.
5. The coding dictionary of Table 8-2 is universal, shared by all living organisms, with some codon differences in mycoplasmas (bacteria lack-

[5]Dyson suggests two separate origins of life: one based on proteins and the other based on some form of replicating RNA nucleotides. According to him, RNA first entered protein-based cells as a replicating "infection," which later became incorporated as a more helpful symbiont that improved host cell replication. The likelihood that eukaryotic mitochondria and chloroplasts originated from symbiotic infections by prokaryotes (Chapter 9) is indicative to him that such symbioses have a long history which may extend back to nucleic acids themselves.

[6]It has also been proposed that DNA genetic material would offer a more easily protected molecule than RNA because the 2′ hydroxyl group in the ribose sugar of RNA causes it to undergo more rapid hydrolytic cleavage than the 2′ deoxyribose of DNA. A further suggested reason for a change in genetic material is that RNA replication is more error-prone than DNA replication because RNA polymerases do not possess the editing functions of DNA polymerase. In addition, the transition from RNA to DNA as the genetic material may have simply entailed the presence of special **reverse transcriptase** enzymes which can transcribe RNA sequences into DNA sequences. Such enzymes, perhaps originally used only for RNA replication (they can function as RNA polymerases), could later have been used to transfer genetic information from RNA to DNA.

TABLE 8-1 Definitions of Common Terms Used in Describing the Present Genetic Code

Term	Meaning
Code letter	Nucleotide, e.g., A, U, G, C (in messenger RNA) or A, T, G, C (in DNA)
Codon, or code word	Sequence of nucleotides specifying an amino acid, e.g., the RNA codon for leucine = CUG (or GAC in DNA)
Anticodon	Sequence of nucleotides on transfer RNA that complements the codon, e.g. GAC = anticodon for leucine (see Fig. 8-3)
Genetic code, or **coding dictionary**	Table of all the codons, each designating the specific amino acid into which it is translated (see Table 8-2)
Codon length, or word size	Number of letters in a codon, e.g., three letters in a **triplet code** (these are the same as coding ratio in a non-overlapping code)
Non-overlapping code	Code in which only as many amino acids are coded as there are codons in end-to-end sequence, e.g., for a triplet code, UUUCCC = phenylalanine (UUU) + proline (CCC)
Degenerate code	Presence of more than one codon for a particular amino acid, e.g., UUU, UUC = phenylalanine, or 20 different amino acids having a total of more than 20 codons
Synonymous codons	Different codons that specify the same amino acid in the degenerate code, e.g., UUU = UUC = phenylalanine
Ambiguous code	When one codon can code for more than one amino acid, e.g., GGA = glycine, glutamic acid (there are no ambiguities in the present code although there may have been such ambiguities in the past)
Commaless code	Absence of intermediary nucleotides (spacers) between codons, e.g., UUUCCC = two amino acids in triplet non-overlapping code
Reading frame	Particular nucleotide sequence coding for a polypeptide that starts at a specific point and is then partitioned into codons until the final codon of that sequence is reached
Frameshift mutation	Change in the reading frame because of the insertion or deletion of nucleotides in numbers other than multiples of the codon length. This changes the previous partitioning of codons in the reading frame and causes a new sequence of codons to be read.
Sense word	Codon that specifies an amino acid normally present at that position in a protein
Replacement mutation	Change in nucleotide sequence, either by deletion, insertion, or substitution, resulting in the appearance of a codon that produces a different amino acid **(missense)** in a particular protein, e.g., UUU (phenylalanine) mutates to UGU (cysteine)
Stop mutation	Mutation that results in a codon that does not produce an amino acid, e.g., UAG (also called a **chain-terminating** codon or **nonsense** codon)
Universality	Utilization of the same genetic code in all organisms, e.g., UUU = phenylalanine in bacteria, mice, humans, and tobacco (with some exceptions, e.g., mitochondria, see Table 8-2)

From Strickberger.

ing polysaccharide cell walls) and ciliate protozoa. Mitochondrial organelles also show a few codon differences from those used by cellular nuclei.

6. Ambiguities have not been found in the code: that is, two or more different amino acids are not specified by the same codon.
7. With the exception of methionine and tryptophan, all amino acids are each designated by more than one codon.
8. The pattern of degeneracy of the code is mostly in the third codon position. For example, eight amino acids, including valine, threonine and alanine, use quartets of codons, each member of a quartet varying only at the third position.
9. When an amino acid uses only a duet (two) of the codons in a quartet, the third codon positions in this duet are both pyrimidines (U and C) or both purines (A and G), never one pyrimidine and one purine.

Explanations for some features of code degeneracy have not been difficult to find. In part, degeneracy derives from the presence of more than one kind of tRNA for a single amino acid. The tRNAs used for leucine, for example, may have the anticodons AAU, AAC, and GAG. The fact that the pattern of degeneracy is

TABLE 8-2 Nucleotide Sequences in Messenger RNA Codons Specifying Particular Amino Acids*

UUU	Phe	UCU	Ser	UAU	Tyr	UGU	Cys
UUC	Phe	UCC	Ser	UAC	Tyr	UGC	Cys
UUA	Leu	UCA	Ser	UAA	STOP†	UGA	STOP†
UUG	Leu	UCG	Ser	UAG	STOP†	UGG	Trp
CUU	Leu	CCU	Pro	CAU	His	CGU	Arg
CUC	Leu	CCC	Pro	CAC	His	CGC	Arg
CUA	Leu	CCA	Pro	CAA	Gln	CGA	Arg
CUG	Leu	CCG	Pro	CAG	Gln	CGG	Arg
AUU	Ile	ACU	Thr	AAU	Asn	AGU	Ser
AUC	Ile	ACC	Thr	AAC	Asn	AGC	Ser
AUA	Ile	ACA	Thr	AAA	Lys	AGA	Arg
AUG‡	Met	ACG	Thr	AAG	Lys	AGG	Arg
GUU	Val	GCU	Ala	GAU	Asp	GGU	Gly
GUC	Val	GCC	Ala	GAC	Asp	GGC	Gly
GUA	Val	GCA	Ala	GAA	Glu	GGA	Gly
GUG	Val	GCG	Ala	GAG	Glu	GGG	Gly

*Some mitochondrial codons differ from those in this table, e.g., both AUA and AUG specify methionine. Such mitochondrial changes seem to be in the direction of economizing in the kinds of transfer RNA that are produced in a small organelle which makes relatively few polypeptides.
†These codons are also called chain-terminating codons or, in the past, nonsense codons.
‡This is the common codon used to initiate protein synthesis.

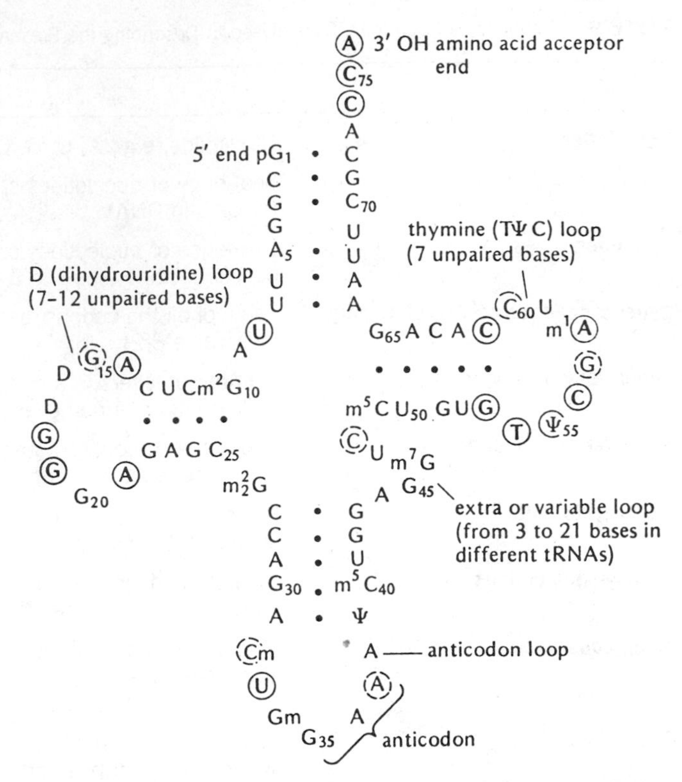

FIGURE 8-3 Sequence of the 76 nucleotides in phenylalanine tRNA of yeast shown in the commonly portrayed two-dimensional cloverleaf form. So far the more than 100 different kinds of tRNA molecules sequenced in a variety of organisms can be fitted into this same cloverleaf pattern, offering a maximum of pairing (dots) between complementary bases. The four major tRNA loops are indicated, including the anticodon loop which contains, in this case, the special sequence AAG that matches the mRNA codon for phenylalanine UUC (and also UUU because of wobble pairing; see also Fig. 8-5). Transfer RNA molecules that are specific for other amino acids bear, of course, different anticodon sequences as well as different nucleotides in some of the other positions. In the present diagram, bases encircled with solid lines occupy the same positions in all the different tRNAs examined so far, whereas those with dashed circles are more variable, indicating base pair positions that are always occupied by either purines or pyrimidines. Numbering of the bases, shown as subscripts, begins at the 5′ end of the nucleotide sequence. Unusual nucleotides found in tRNA include: D = dihydrouridine, Ψ = pseudouridine, mX = methylated nucleotide, and T = thymine. (Adapted from Strickberger.)

confined mostly to the third codon position was explained by Crick as arising from **wobble pairing** between certain bases of the tRNA anticodon and the mRNA codon in this position. Crick's wobble hypothesis, since proved correct, was that anticodons bearing inosine (I) at this position can pair with either U, A, or C;[7] anticodons bearing G can pair with either of the pyrimidines U and C; and anticodons bearing U can pair with either of the purines A and G. Third codon position degeneracy thus points to the importance of the first two codon positions in specifying amino acids; the eight amino acids that are coded by quartets of codons can be specified by nucleotide bases at only these positions.

The striking universality of the code—the fact that no other code has been found in any independent living organism—makes it reasonable to ask why this particular code? Since there are at least 10^{70} possible different codes using 64 codons to code for 21 entities (20 amino acids + chain termination), either the exclusive use of this particular code must have been derived from accidental causes, or there must be some relationship between amino acids and their codons (or anticodons) that excludes large numbers of other possible codes, or perhaps both factors operated. Some authors have suggested that the amino acids originally associated with their codons or anticodons through stereochemical fitting or shared other complementary properties such as hydrophilic and hydrophobic qualities. For example, some kind of pairing may have

[7]Inosinic acid is a nucleotide that uses the purine hypoxanthine as a base. Hypoxanthine in turn derives from the loss of an amino group from adenine. Other modified tRNA nucleotides that show wobble pairing have also been found (e.g., 5-methoxyuridine pairs with both A and G and less efficiently with U).

occurred between an amino acid such as phenylalanine and a codon such as UUC or its anticodon AAG (Jungck). On the other hand, other authors maintain that, in spite of considerable search, few, if any, examples of preferential affinity between amino acids and their codons or anticodons have been discovered. Furthermore, were such a hypothesis consistently applied, it would be grossly inadequate in explaining why serine uses two such markedly dissimilar sets of codons as the UCN quartet (where N designates any base) and the AGU-AGC duet.

An alternative hypothesis has therefore been offered suggesting that the universality of the code derives from the survival of only one of the possible codes that may actually have existed in the past. That is, early amino acid-codon relationships arose largely by chance rather than by restricted stereochemical pairing and may therefore have produced a number of different primordial genetic codes, each used by different groups. As time went on, however, only one of the groups carrying a particular code became successful enough to continue evolving, and the others became extinct.[8] The code thus became frozen in the form that it has since possessed.

The reason that a **frozen accident** of this kind will restrict any further evolution of the code is because protein synthesis in its mature form precisely positions each particular amino acid in all the various long-chained polypeptides in which it is found. This means that any change in the genetic code for even a single amino acid would cause a significant change in many different proteins. For example, if the code for phenylalanine were changed to include the cysteine codons UGU and UGC, then tRNA molecules carrying phenylalanine would now insert into all polypeptide positions formerly occupied by cysteine. Since these two amino acids differ considerably in structure and function, the effect of such a sudden change would undoubtedly be lethal.[9]

However, before being frozen, the genetic code must itself have undergone evolution to accompany some of the changes taking place in protein synthesis. That is, early proteins were probably much shorter than present proteins, composed of fewer kinds of amino acids, and produced by a much less accurate translation mechanism. These proteins may therefore have had only minimal functional specificity, and those produced by a single mRNA molecule could best be described as statistically alike rather than exactly alike. Thus, there would likely have been room for changes to take place between some of the codons of early amino acids, and not all of the present amino acids would have been incorporated into the primeval code(s).

A very early hypothesis on the evolution of the code was that it derived from a prior code that used fewer than three bases per codon. This idea seemed to be supported by the fact that the amino acids using quartets are still primarily specified by the first two nucleotide bases in each codon. Thus it seemed possible to suggest that an even earlier code may have been **singlet**, with four mononucleotide codons (A, U, G, C) specifying perhaps four kinds of amino acids. This code evolved later into the **doublet** form with 16 dinucleotide codons (AA, AU, AG, . . .) specifying a maximum of 15 amino acids and one chain-termination codon. Only after undergoing this doublet experience did the code finally achieve its present trinucleotide form of 64 codons.

Although the idea is superficially attractive, evolution of the code in this fashion seems extremely doubtful since each change in the size of the codon would change the meaning of practically all former codons. For example, a mRNA sequence being translated by a doublet code, AU CG UU GU AG CG . . ., would produce entirely different numbers and kinds of amino acids when translated by a triplet code, AUC GUU GUA GCG . . . . In the face of a radical transition of this kind an organism could hardly retain the function of most, if not all, of its genetic material.

The genetic code is therefore believed to have been triplet even during its beginnings, or perhaps doublet with single **nucleotide spacers**. Mechanical considerations support this view, since anything less than a triplet codon would probably not provide a stable pairing relationship between a tRNA anticodon and an mRNA codon. At the same time, quadruplet or quintuplet codons are probably too "sticky" to be used in the face of selection for an optimum turnover rate that would allow rapid dissociation between codons and anticodons. It has also been suggested that the trinucleotide width of a triplet anticodon provides a minimal space enabling tRNA molecules to lie close enough together for peptide bonding between their amino acids.

Given a small group of amino acids coded by such triplets, further evolution of the code would have probably proceeded under three selective conditions:

---

[8]According to Wong, individuals carrying the successful code may have possessed one or many unique advantages that would have provided them with important competitive superiority, such as the phased coupling of DNA replication and cell division.

[9]A frozen accident could also help to explain the optical rotation found in organic molecules produced by existing organisms (p. 105): that all proteins are composed of L-amino acids and all nucleic acids are composed of nucleotides with D-sugars. Perhaps once organisms developed enzymes that used subunits with a specific optical rotation, it would have been detrimental to change these specificities.

1. Nucleotide substitutions caused by errors in replication (mutations) should produce as few amino acid changes as possible.
2. The number of different codons per amino acid should generally be proportional to the frequency in which the amino acid occurs in proteins.
3. Errors occurring during the mRNA-tRNA translation process should lead to as few drastic protein changes as possible.

In respect to the first condition, selection would tend to expand the number of different codons used by an amino acid so that random base changes would still produce the same amino acid. Increasing the number of different codons for any particular amino acid would also be advantageous in diminishing the number of stop or nonsense codons that do not code for any amino acid at all, thereby ensuring that most random mutations do not terminate protein production. On the other hand, there would be a limit to the number of different codons that could be used by a single amino acid since a variety of amino acids have to be accommodated in the coding dictionary. Unfortunately, the extent to which these selective forces operated in the past is presently difficult to determine.

In respect to the numbers of codon assignments, there does appear to be a proportional relationship between the relative number of codons possessed by most amino acids and their frequencies in proteins (Fig. 8-4). However, the source for this relationship is still not clear. It may indicate a selective relationship (more codons are selected for the use of those amino acids that occur more frequently) or an accidental relationship (the overall composition of proteins derives from the frequency of amino acid codons), or perhaps both factors prevailed.[10]

The third level of codon evolution, selection for minimizing translational errors, may have had a considerable effect on codon assignments. For example, the prevailing degeneracy at the third codon position is precisely what we would expect if this position were the one most involved in translational errors. That is, the various quartets (e.g., GUU, GUC, GUA, GUG) and duets (e.g., UUU, UUC) arise from the selective advantage of assigning to the same amino acid codons that could most easily be mistaken for one another. The next most error-prone translational event occurs at the first codon position, and the code again shows some degeneracy at this position (e.g., UUA [leucine] → CUA [leucine]), or is so constructed that an error may occasionally enable the substitution of an amino acid with related function (e.g., UUA [leucine] → GUA [valine]). In general, however, the first codon position is considerably less error prone than the third, apparently because of modifications of the 40th tRNA nucleotide immediately adjacent to this position. These modifications include the addition of methyl or even bulkier groups to nucleotide no. 40, thereby preventing wobble at the first codon position. For example, one such mod-

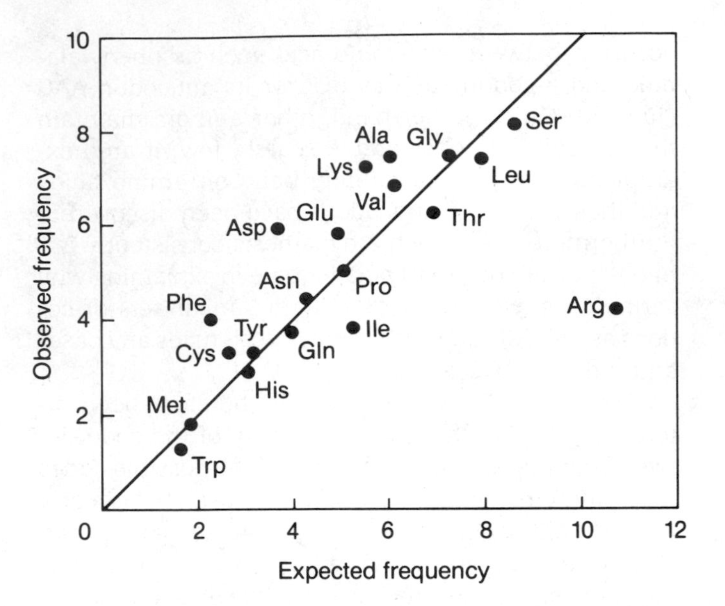

FIGURE 8-4 Comparison between observed and expected frequencies of amino acids at 5,492 positions in 53 different vertebrate polypeptides. If arginine is excluded, the correlation coefficient for these data is 0.89, which represents a high positive correlation. (The expected frequency of each amino acid is based on a technique of calculating the nucleotide composition of the mRNA from the frequency of bases in the first two codon positions used by the various amino acids that compose the proteins: U = 0.220, A = 0.303, C = 0.217, G = 0.261. Randomized triplet codons for this nucleotide composition then furnish the expected amino acids.) (From King and Jukes.)

[10]One anomalous feature of the code is the disproportionate number of arginine codons in the genetic code (6 codons = 9.8 percent of the total) compared to the actual frequency of this amino acid in most proteins (usually less than 5 percent). By contrast, lysine has only two codons (3.3 percent of the total) but is found more frequently (6.6 percent) in proteins. One reason for this may be that mutations leading to the substitution of arginine for other amino acids are usually more disadvantageous than mutations leading to the substitution of lysine. Jukes (1974) suggests that arginine is a relatively new amino acid in contrast to most others, although it has so many codons. According to him, ornithine was probably the amino acid that originally used the codons presently assigned to arginine. Presumably, the evolution of the urea cycle for nitrogen excretion

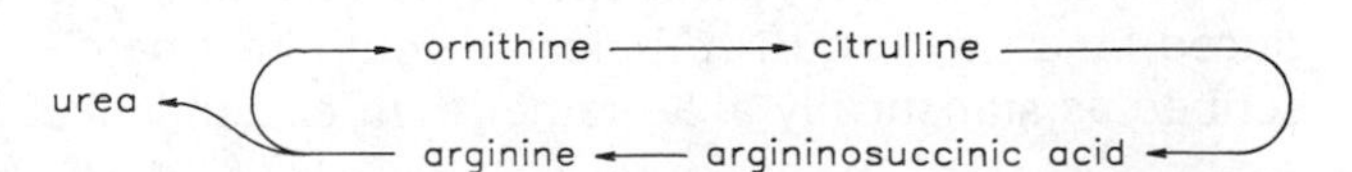

led to the production of arginine, which had a strong affinity for the tRNA molecules used by ornithine. Because arginine is a more basic amino acid than ornithine, it may therefore have also conferred new advantageous properties on some proteins. In general, however, when arginine usurped the ornithine codons, previous ornithine amino acid functions were probably assumed by lysine, which is somewhat like ornithine. Thus, Jukes claims that lysine with only two codons is now presumably used for many amino acid positions formerly occupied by ornithine, which had perhaps six codons.

TABLE 8-3 A Classification of Amino Acids According to the Nucleoside Found at Their Second Codon Positions and the Hydration Potentials of Their Side Chains

Amino Acid	Second Codon Letter	Hydration Potential (kcal/mole)	
Gly	G	+2.39	Most hydrophobic
Leu	U	+2.28	↑
Ile	U	+2.15	
Val	U	+1.99	
Ala	C	+1.94	
Phe	U	−0.76	
Cys	G	−1.24	
Met	U	−1.48	
Thr	C	−4.88	
Ser	C(G)	−5.06	
Trp	G	−5.89	
Tyr	A	−6.11	
Gln	A	−9.38	
Lys	A	−9.52	
Asn	A	−9.68	
Glu	A	−10.19	
His	A	−10.23	↓
Asp	A	−10.92	Most hydrophilic

Data from Wolfenden et al.

ified nucleotide is threonyl-6-adenine, which prevents the wobble pairing of U in the adjacent position with G in the first codon position.

It has been suggested that the second codon position, the least error prone during translation, may have been used at one time to separate entire classes of amino acids with unique functions. This system would offer a selective advantage by ensuring that amino acids are rarely substituted between classes. A possible remnant of such a grouping may be the assignment of U to the second codon position for leucine, isoleucine, and valine, all of which are **hydrophobic amino acids** found mostly in the interior of proteins, and assignment of A to the second codon position for glutamic acid, histidine, aspartic acid, and other **hydrophilic amino acids** found commonly on the protein surface (Table 8-3).

In summary, it is possible to claim that the triplet genetic code may have initially coded for relatively few amino acids, or, based primarily on the second codon position, been able to make distinctions only between general classes of amino acids. As time went on, the first codon position also came into use for amino acid positioning because of the modifications that could be made to the immediately adjacent tRNA nucleotide. With the advent of two accurately translated codon positions, the genetic code could then expand its repertoire of amino acids. Thus, until translation accuracy was firmly achieved, some shuffling of codons may have taken place between different amino acids, or entirely new amino acids may have been introduced using some or all of the codons of older amino acids. The retention of third codon position degeneracy or wobble may have served the purpose of economizing on the number of tRNA molecules needed for the translation of amino acids such as valine, threonine, alanine, and others.

In accord with these considerations, Jukes (1983) has proposed that an early stage in the evolution of the genetic code may have been somewhat like Figure 8-5(a). Each **quartet**, or **family**, of four codons specified perhaps one amino acid, and each amino acid was represented by only one kind of tRNA molecule whose anticodon could pair with all four codons in the family. (This type of code is commonly found in present-day mitochondrial organelles which, because of their small size and limited function, economize in the kinds of both tRNA and proteins they produce.) Such extreme wobble thus limited this code to no more than 15 or 16 amino acids. One family or partial family of codons was used to terminate translation since there were no amino acid-bearing tRNA molecules whose anticodons could pair with these stop codons. In subsequent evolution (Fig. 8-5(b), (c)), tRNA gene duplications enabled new anticodons to evolve, some of which could now mutate so that they were activated by new amino acids.

For example, the gene for an early tRNA molecule with anticodon AAU that specifies phenylalanine could form an additional tRNA gene by duplication which then mutates to produce anticodon AAG. The AAG anticodon becomes restricted to pairing with codons UUU and UUC, whereas the AAU anticodon pairs only with UUA and UUG. Although both tRNAs originally specify phenylalanine, the gene producing the tRNA with anticodon AAU may now mutate so that its tRNA product now binds to leucine rather than to phenylalanine. Thus leucine would now be specified by UUA and UUG codons.[11] Jukes points out that since such aminoacylation changes have arisen in the genetic code of mi-

[11]The process by which particular amino acids are bound to particular tRNA molecules is dependent on special **amino acid–activating enzymes** which recognize unique features of both the amino acids and the tRNAs. Thus, for example, some mutations of a tRNA gene may allow an amino acid–activating enzyme that had exclusively attached amino acid X to $tRNA^X$ to attach amino acid X to this new tRNA, e.g., $tRNA^Y$. A mutation of the amino acid–activating enzyme itself may then enable a new amino acid, Z, to attach to $tRNA^Y$, which can be considered as $tRNA^Z$ when it loses its prior function with amino acid X.

(a) An early code: 16 anticodons for perhaps 15 amino acids

Codons	Anticodon	Amino acid	Codons	Anticodon	Amino acid	Codons	Anticodon	Amino acid	Codons	Anticodon	Amino acid
UUU UUC UUA UUG	**AAU**	phe?	UCU UCC UCA UCG	**AGU**	ser	UAU UAC	**AUG**	tyr	UGU UGC UGA UGG	**ACU**	cys?
						UAA UAG	STOP				
CUU CUC CUA CUG	**GAU**	leu	CCU CCC CCA CCG	**GGU**	pro	CAU CAC CAA CAG	**GUU**	his?	CGU CGC CGA CGG	**GCU**	arg
AUU AUC AUA AUG	**UAU**	ile	ACU ACC ACA ACG	**UGU**	thr	AAU AAC AAA AAG	**UUU**	asn?	AGU AGC AGA AGG	**UCU**	ser?
GUU GUC GUA GUG	**CAU**	val	GCU GCC GCA GCG	**CGU**	ala	GAU GAC GAA GAG	**CUU**	asp?	GGU GGC GGA GGG	**CCU**	gly

tRNA gene duplications and mutations
- - - - →
evolution of new anticodons

(b) A later code: 31 anticodons for perhaps 18 amino acids

Codons	Anticodon	Amino acid	Codons	Anticodon	Amino acid	Codons	Anticodon	Amino acid	Codons	Anticodon	Amino acid
UUU UUC	**AAG**	phe	UCU UCC	**AGG**	ser	UAU UAC	**AUG**	tyr	UGU UGC	**ACG**	cys
UUA UUG	**AAU**	leu	UCA UCG	**AGU**	ser	UAA UAG	STOP		UGA UGG	**ACU**	trp
CUU CUC	**GAG**	leu	CCU CCC	**GGG**	pro	CAU CAC	**GUG**	his	CGU CGC	**GCG**	arg
CUA CUG	**GAU**	leu	CCA CCG	**GGU**	pro	CAA CAG	**GUU**	glu	CGA CGG	**GCU**	arg
AUU AUC	**UAG**	ile	ACU ACC	**UGG**	thr	AAU AAC	**UUG**	asn	AGU AGC	**UCG**	ser
AUA AUG	**UAU**	met	ACA ACG	**UGU**	thr	AAA AAG	**UUU**	lys	AGA AGG	**UCU**	arg
GUU GUC	**CAG**	val	GCU GCC	**CGG**	ala	GAU GAC	**CUG**	asp	GGU GGC	**CCG**	gly
GUA GUG	**CAU**	val	GCA GCG	**CGU**	ala	GAA GAG	**CUU**	glu	GGA GGG	**CCU**	gly

further tRNA gene duplications and mutations
- - - - →
evolution of new anticodons (and deletion of ACU anticodon to produce a UGA stop codon)

(c) The modern code: 43 known anticodons for 20 amino acids

Codon	Anticodon	Amino acid	Codon	Anticodon	Amino acid	Codon	Anticodon	Amino acid	Codon	Anticodon	Amino acid
UUU	**AAG**	phe	UCU	**AGI**	ser	UAU	**AUG**	tyr	UGU	**ACG**	cys
UUC		phe	UCC		ser	UAC		tyr	UGC		cys
UUA	**AAU**	leu	UCA	**AGU**	ser	UAA	STOP		UGA	STOP	
UUG	**AAC**	leu	UCG	**AGC**	ser	UAG			UGG	**ACC**	trp
CUU	**GAG**	leu	CCU	**GGI**	pro	CAU	**GUG**	his	CGU	**GCI**	arg
CUC		leu	CCC		pro	CAC		his	CGC	**GCG**	arg
CUA	**GAU**	leu	CCA	**GGU**	pro	CAA	**GUU**	gln	CGA		arg
CUG	**GAC**	leu	CCG		pro	CAG	**GUC**	gln	CGG		arg
AUU	**UAI**	ile	ACU	**UCI**	thr	AAU	**UUG**	asn	AGU	**UCG**	ser
AUC	**UAG**	ile	ACC	**UGG**	thr	AAC		asn	AGC		ser
AUA	**UAC***	ile	ACA	**UGU**	thr	AAA	**UUU**	lys	AGA	**UCU**	arg
AUG	**UAC**	met	ACG		thr	AAG	**UUC**	lys	AGG		arg
GUU	**CAI**	val	GCU	**CGI**	ala	GAU	**CUG**	asp	GGU	**CCG**	gly
GUC	**CAG**	val	GCC	**CGU**	ala	GAC		asp	GGC		gly
GUA	**CAU**	val	GCA		ala	GAA	**CUU**	glu	GGA	**CCU**	gly
GUG	**CAC**	val	GCG		ala	GAG	**CUC**	glu	GGG	**CCC**	gly

FIGURE 8-5 Some possible stages in the evolution of the genetic code based on a scheme suggested by Jukes. The mRNA codons are at the left of each box, and the tRNA anticodons are in bold capital letters to their right. The cytidine nucleotide in the UAC anticodon marked with an asterisk is acetylated, restricting this tRNA molecule to AUA (isoleucine) codons on mRNA. (From Strickberger.)

tochondria which produce relatively few proteins, they could also have occurred in early primitive organisms.

By such means the kinds of tRNA molecules could increase and new amino acids could be added to the code. Apparently, when 20 different amino acids were incorporated into the code, a sufficiently large number of proteins were being produced by these ancestral organisms so that codon changes necessary to include any further amino acids would lead to widespread protein malfunction and widespread lethality. At that point the code was frozen, and all organisms that were derived from this primitive ancestor shared this common code. The fact that, with the exception of a few codons in mitochondria, the genetic code is common to all organisms indicates that only the bearers of this particular code successfully survived the early evolutionary period.

SUMMARY

At present, cells are totally dependent upon protein catalysts for the production of both proteins and nucleic acids from nucleic acid templates. Although it is possible that nucleic acids appeared first, it seems more plausible to assume that proteins were the earliest polymers. If so, these proteins must have had numerous primary functions, such as catalysis, and only secondarily served as templates for nucleotide polymerization. Presumably these early proteins could self-replicate to some degree, a process which today occurs only in the case of a few bacterial antibiotics synthesized using the polymerizing enzyme as a template.

Because of the importance of lengthened molecular chains for both structure and function, polymerization was probably one of the earliest cellular processes to evolve. If one assumes the existence of some kind of template (perhaps a peptide–nucleic acid chimera) as well as a polymerizing enzyme, selection could have led to the origin of adaptor molecules carrying amino acids to the template, the appearance of coenzymes for the utilization of high-enery molecules, the formation of a base-pairing system by which adaptor and nucleotide portions of the template could recognize one another, and the formation of some primitive type of ribosome encompassing all of these functions. Subsequently a separate nucleic acid molecule developed for information storage, and this was added to the panoply of other nucleic acid sequences acting as messengers, adaptors, and regulators. Finally, there was a separation of storage and production functions into distinct molecules, that is, DNA and RNA.

At present the coded message from the storage molecule, DNA, to the synthesizing machinery is carried via RNA in the form of three-nucleotide units known as codons. From an evolutionary perspective, the code's most interesting features are its universality and its degeneracy (the use of several codons for the same amino acid). It seems likely that this particular code became fixed in its present form because of the success of the organisms bearing it and the disappearance of organisms carrying alternative codes.

It is likely that an early code, even then composed of three nucleotides, provided information for fewer amino acids than at present or for general classes of amino acids. To minimize errors in the replication of the master code or in its translation into protein, the third nucleotide in each codon was relatively unspecific (degenerate). The first and second nucleotides, less flexible, perhaps specified particular amino acids or discriminated among different groups of amino acids on the basis of certain features, such as their hydrophobic or hydrophilic properties. Having two accurately translated nucleotides increased the number of amino acids which could be determined and at the same time regularized the primary structure of proteins. The code ceased evolving once 20 amino acids were specified and a large number and variety of proteins were synthesized. At that time the code was "frozen" in our ancestral group, and it has since remained virtually invariable in every cellular organism.

KEY TERMS

adaptor molecules
ambiguous code
amino acid–activating enzymes
anticodon
autocatalytic
chain-terminating codon
code letter
coding dictionary
codon
codon family
codon length
codon quartet
commaless code
degenerate code
doublet code
evolution of protein synthesis
frameshift mutation
frozen accident
genetic code
head-growth polymerization
heme
hydrophilic amino acids
hydrophobic amino acids
missense mutation
nonsense codon
non-overlapping code
nucleic acid replication
nucleotide spacers
oligopeptides
organic catalysts
polymerization enzymes
protein synthesis
reading frame
replacement mutation
reverse transcriptase
RNA catalysts
self-replication
sense word
singlet code
stereochemical fit
stop mutation
synonymous codons
triplet code
universal code
wobble pairing

REFERENCES

Black, S., 1973. A theory on the origin of life. *Adv. in Enzymol.*, **38**, 193-234.

Cairns-Smith, A. G., 1982. *Genetic Takeover and the Mineral Origins of Life.* Cambridge Univ. Press, Cambridge.

Cech, T. R., and B. L. Bass, 1986. Biological catalysis by RNA. *Ann. Rev. Biochem.*, **55**, 599-629.

Crick, F. H. C., 1968. The origin of the genetic code. *Jour. Mol. Biol.*, **38**, 367-379.

Dyson, F., 1985. *Origins of Life*. Cambridge Univ. Press, Cambridge.

Eigen, M., W. Gardiner, P. Schuster, and R. Winkler-Oswatitsch, 1981. The origin of genetic information. *Sci. Amer.*, **244,** 88-118.

Fox, S. W., 1978. The origin and nature of protolife. In *The Nature of Life*, W. H. Heidecamp (ed.). University Park Press, Baltimore, pp. 23-92.

Gerbi, S. A., 1985. Evolution of ribosomal RNA. In *Molecular Evolutionary Genetics*, R. J. MacIntyre (ed.). Plenum Press, New York, pp. 419-517.

Joyce, G. F., 1989. RNA evolution and the origins of life. *Nature,* **338,** 217-224.

Joyce, G. F., A. W. Schwartz, S. L. Miller, and L. E. Orgel, 1987. The case for an ancestral genetic system involving simple analogues of the nucleotides. *Proc. Nat. Acad. Sci.*, **84**, 4398-4402.

Jukes, T. H., 1974. On the possible origin and evolution of the genetic code. *Origins of Life*, **5**, 331-350.

——, 1983. Evolution of the amino acid code. In *Evolution of Genes and Proteins,* M. Nei and R. K. Koehn (eds.). Sinauer Associates, Sunderland, Mass., pp. 191-207.

Jungck, J. R., 1978. The genetic code as a periodic table. *Jour. Mol. Evol.*, **11,** 211-224.

King, J. L., and T. H. Jukes, 1969. Non-Darwinian evolution. *Science*, **164,** 788-798.

Küppers, B-O., 1983. *Molecular Theory of Evolution.* Springer-Verlag, Berlin.

Lipmann, F., 1965. Projecting backward from the present stage of evolution of biosynthesis. In *The Origins of Prebiological Systems*, S. W. Fox (ed.). Academic Press, New York, pp. 259-280.

——, 1971. Attempts to map a process evolution of peptide biosynthesis. *Science*, **173,** 875-884.

Loomis, W. F., 1988. *Four Billion Years: An Essay on the Evolution of Genes and Organisms*. Sinauer Associates, Sunderland, Mass.

Siu, P. M. L., and H. G. Wood, 1962. Phosphoenolpyruvic carboxytransphorylase, a carbon-dioxide fixation enzyme from propionic acid bacteria. *Jour. Biol. Chem.*, **237,** 3044-3051.

Strickberger, M. W., 1985. *Genetics*, 3rd ed. Macmillan, New York.

Woese, C. R., 1973. Evolution of the genetic code. *Naturwiss.*, **60,** 447-459.

——, 1980. Just So Stories and Rube Goldberg machines: speculations on the origin of the protein synthetic machinery. In *Ribosomes, Structure, Function, and Genetics*, G. Chamblis, G. R. Craven, J. Davies, K. Davis, L. Kahan, and M. Nomura (eds.). University Park Press, Baltimore, pp. 357-373.

Wolfenden, R. V., P. M. Cullis, and C. C. F. Southgate, 1979. Water, protein folding and the genetic code. *Science*, **206,** 575-577.

Wong, J. T. F., 1976. The evolution of a universal genetic code. *Proc. Nat. Acad. Sci.*, **73,** 2336-2340.

Zielinski, W. S., and L. E. Orgel, 1987. Autocatalytic synthesis of a tetranucleotide analogue. *Nature*, **327**, 346-347.

FROM METABOLISM TO CELLS

For cellular protein and nucleic acid synthesis to evolve, there must also have been selection for biochemical pathways in which components of these and other polymers could be produced and chemical energy utilized. The results of such selection we now see everywhere: cellular metabolism is highly organized in time and space so that each metabolic step in a sequence occurs in fairly exact repeatable order governed by specific enzyme activity. Moreover, different metabolic sequences are often precisely coupled and regulated so that the products of one sequence (e.g., ATP) are available for use in other sequences at the appropriate time. Unfortunately, since there are no existing relics of ancient uncoordinated metabolic pathways, direct evidence of pre-cellular or early cellular metabolism has so far been impossible to obtain. Nevertheless, one approach towards understanding metabolic evolution has become available through **comparative biochemistry**: that is, to discover **metabolic pathways** or sequences within such pathways that are shared by different organisms and to consider such shared pathways as ancestral to these organisms.

ANAEROBIC METABOLISM

Anaerobic glycolysis, the breakdown of glucose in the absence of oxygen, is perhaps the most elemental of metabolic pathways, and various sections of this pathway are shared by all living creatures. This universality seems to depend upon the fact that all existing organisms derive their free energy from the chemical breakdown of such monosaccharides. In **heterotrophic** organisms, monosaccharides or organic materials that can be converted to monosaccharides are derived from sources outside the organism, whereas in **autotrophic** organisms such organic materials are made within the organism itself, usually by the reduction of carbon dioxide.

In both types of organisms glycolytic pathways may begin directly with glucose or with almost any organic material, such as sugars, fats, or amino acids, that can be formed into glucose. As shown in Figure 9-1, the **Embden-Meyerhof glycolytic pathway** leads to the formation of pyruvic acid (pyruvate), providing a net yield of two high-energy phosphate bonds in ATP, the basic currency for cellular chemical energy:

$$C_6H_{12}O_6 + 2\ ADP + 2\ P_i + 2\ NAD^+ \rightarrow 2\ C_3H_4O_3 + 2\ ATP + 2\ NADH + 2\ H^+ + 2\ H_2O$$

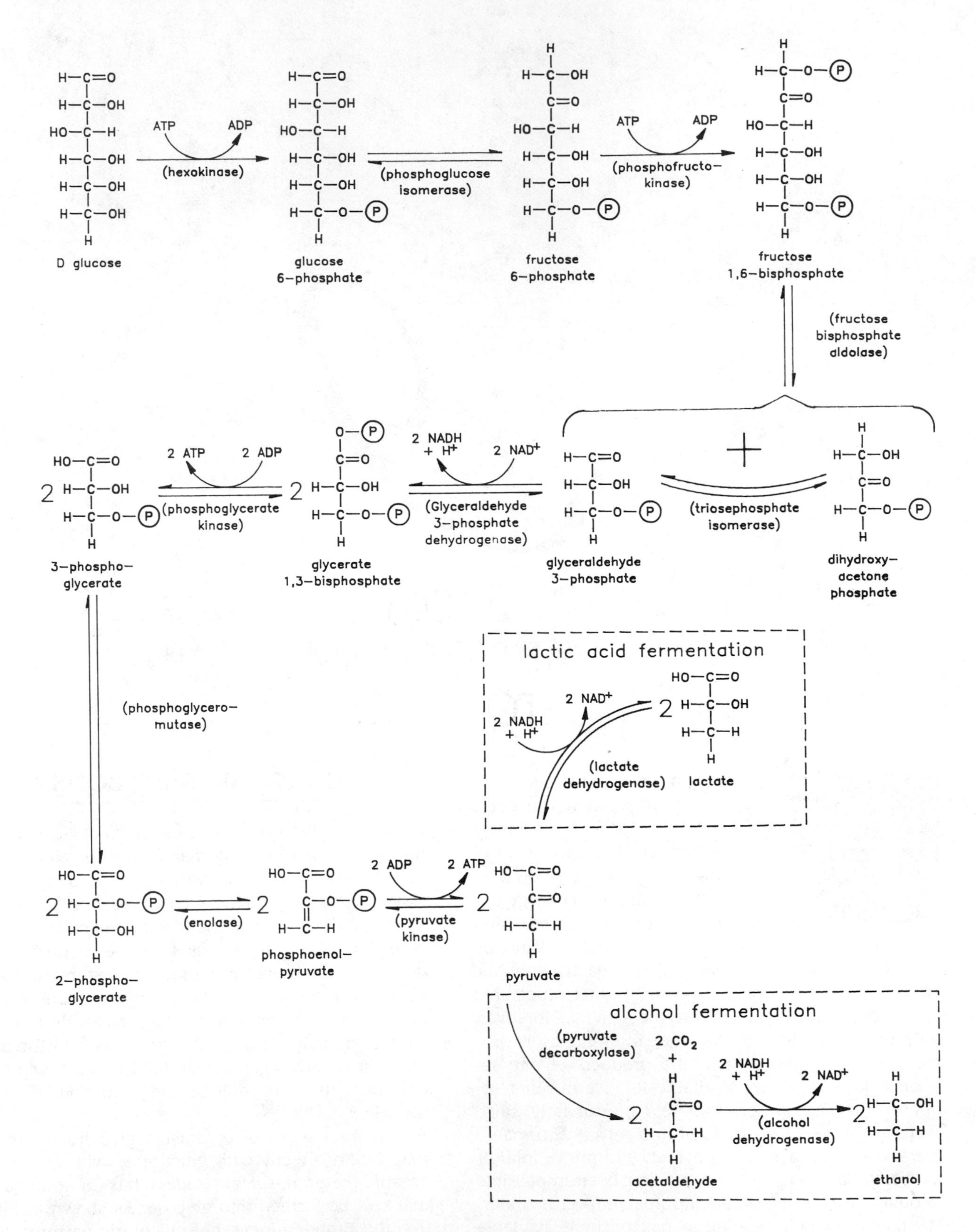

FIGURE 9-1 For legend see opposite page.

FIGURE 9-1 Steps in the anaerobic Embden-Meyerhof glycolytic pathway from glucose to pyruvate. Beginning with one molecule of glucose, the pathway degrades two ATP molecules to ADP but phosphorylates four ADP molecules to ATP. The overall advantage of this pathway to the cell therefore derives from the net formation of two high-energy phosphate bonds. Also indicated is the reduction of the pyridine nucleotide coenzyme, NAD^+. The reduced form of this compound (NADH) can then be oxidized to regenerate NAD^+ by reactions that donate hydrogens and electrons to form either lactic acid or ethanol:

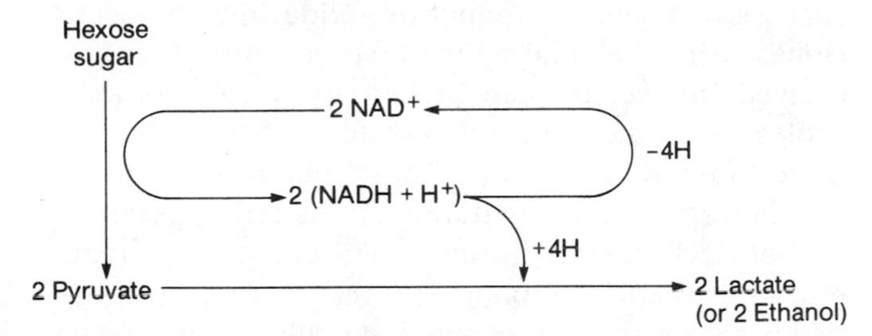

The pathway is actually more complex than illustrated here since the intermediate compounds formed during glycolysis can also be used as substrates in the synthesis of amino acids and nucleic acids. For example, 3-phosphoglycerate can serve as a substrate leading to the synthesis of serine, glycine, or cysteine amino acids, or purine bases. [For the illustrations in this chapter, the various carboxylic acids are shown in protonated form (HO—C=O or COOH) but their names are usually given as though they were unprotonated ($^-$O—C=O or COO^-), e.g., pyruvic acid = pyruvate. Phosphate groups (H_2PO_3) are indicated by circled P's, and the specific enzyme for each reaction is given in parentheses.]

During this process two molecules of the coenzyme NAD^+ (nicotinamide adenine dinucleotide, a carrier of protons [H^+] and electrons [e^-]) are reduced to NADH, but these can then be reoxidized in reactions forming either lactic acid (lactate) or ethanol.

Since only two of the reactions furnish ATP, the other steps in the pathway may be considered preparatory to these primary reactions. Nevertheless, the length of the pathway appears to be quite extended, and it has been suggested that some of these steps may partially recapitulate the succession of biochemical events that furnished energy to organisms in the past. For example, smaller molecules such as glyceraldehyde, or its phosphorylated form, may have been available in sufficiently high quantity in the past to allow relatively simple energy conversion in only a few steps. As these molecules were depleted, organisms that could metabolize some of the more available larger molecules to the glyceraldehyde stage could continue to use the same, but now extended, pathway.[1] By these means, succeeding organisms would need only to add or modify one or a few enzymes for each additional step as it occurred rather than to continually elaborate entirely new metabolic pathways.

The hypothesis that the gradual depletion of a necessary molecule causes biochemical pathways to lengthen in order to allow use of an available related compound was first formally proposed by Horowitz. It has been called **retrograde** or **backward evolution** and is believed to account for many intermediate steps in biochemical pathways that lead to the synthesis of compounds such as amino acids. For example, let us assume that A is a product essential for cellular function, B is a molecular precursor of A, and C is a molecular precursor of B. Obviously, there is no need for the organism to develop a pathway for the synthesis of A when A itself is present in the environment. But as A is depleted, a considerable selective advantage occurs for catalysis of the available B precursor molecules into A (e.g., by action of enzyme *1*). Similarly, as the supply of B is exhausted, selection occurs for the conversion of precursor C into B (e.g., by enzyme *2*). By these means a chain of metabolic reactions becomes established, $\ldots \xrightarrow{3} C \xrightarrow{2} B \xrightarrow{1} A$, which represents the evolution of enzyme *1* first, *2* second, *3* third, and so on.

However, even if the development of metabolic pathways did not depend on the depletion of necessary compounds but evolved more rapidly than Horowitz suggested, it is still clear that such sequential pathways derive from the chemical relatedness between compounds and the convertibility of one compound into another. This is most obvious in the simple stepwise changes noted in the glycolytic pathway and indicates that biochemical pathways, as is true of many other biological chemical phenomena, did not arise randomly but represent a form of pre-determinism based on biochemical structures.

Once such pathways were established, their survival must have depended on their ability to cope with persistent chemical problems. Glucose, for example, is now a common sugar present in various forms in both plants and animals, and its ready availability leads to the importance of glycolysis in virtually all organisms. In fact, each enzyme in the glycolytic pathway is found in all living multicellular organisms examined so far and in most single-celled organisms. Furthermore, the amino acid sequences in many of these enzymes are remarkably similar in organisms that have been evolving separately for at least a billion years. For example, the amino acid sequence of an **active site** of catalytic activity in the enzyme triosephosphate

[1]By similar reasoning, it is also possible that the glycolytic pathway may have been lengthened by beginning with large molecules and then proceeding to smaller ones. That is, the chemical energy first provided by the molecular breakdown of large molecules such as monosaccharides or polysaccharides was successfully increased by further metabolizing their smaller component products such as glyceraldehyde. The aerobic metabolic pathways discussed later may have offered, among other qualities, this type of advantage.

isomerase has apparently been conserved in organisms ranging from bacteria (*Escherichia coli*) to corn (*Zea mays*) (see Fig. 7-2 for a listing of amino acids and their letter codes):

E. coli	QGAAA F EGAVI AYEPVWAIGTGKSATPAQ
Yeast	EEVKDWTNVVVAYEPVWAIGTGLAATPED
Fish	DDVKDWSKVVLAYEPVWAIGTGKTASPQQ
Chicken	DNVKDWSKVVLAYEPVWAIGTGKTATPQQ
Rabbit	DNVKDWSKVVLAYEPVWAIGTGKTATPQQ
Corn	EKIKDWSNVVVATEPVWAIGTGKVATPAQ

Since it is highly improbable for so many amino acid sequences from different sources to have become similar by accident, it seems most likely that a single ancestral sequence existed for this catalytic purpose in the common progenitor of all these organisms. Sequence similarity was then preserved by continuous selection for the same enzymatic function (i.e., an essential step in glycolysis) in these various lineages. Gest and Schopf suggest that it was only "sugar-based cellular systems that provided the successful starting point for biochemical and cellular evolution."

Other anaerobic pathways have been discovered which are probably also of ancient lineage, dating back to the early anaerobic atmosphere. One example is the Entner-Doudoroff pathway, which circumvents some early steps of the Embden-Meyerhof pathway by allowing a more direct conversion of glucose-6-phosphate to three-carbon compounds, such as pyruvate, but produces a net gain of one less ATP molecule per metabolized glucose:

$$\text{glucose} + 2\,NAD^+ + ADP + P_i \rightarrow 2\,\text{pyruvate} + NADPH + ATP$$

By whatever means anaerobic metabolism occurred, organisms that relied upon it must have eventually faced a depletion of reduced carbon compounds that could be used as substrates. A change from such **organotrophic nutrition** to an ability to use simpler and more readily available carbon sources such as CO_2 (**lithotrophic nutrition**) must therefore have offered considerable evolutionary advantage. However, in order to be effectively used, CO_2 must be reduced, or fixed, by a process which provides electrons and hydrogen ions. Among the reducing compounds available for this purpose were probably H_2, H_2S, NH_3, and others. As can be observed among some organisms today, it is believed that these inorganic compounds were broken down with the aid of catalysts:

$$H_2(S) \xrightarrow[\text{enzyme}]{\text{hydrogenase}} 2\,H^+ + 2\,e^- + (S)$$

furnishing electrons and hydrogens that could then be utilized for energy production and for hydrogenation of carbon.

There is considerable agreement that many of these early reactions took place with the aid of membrane-bound enzymatic systems that allowed a successive chain of **reduction-oxidations** to occur (Jones). In the energy-production pathways that evolved, the electrons are picked up by acceptor molecules which then become transformed from the oxidized to the reduced form. These **electron acceptors** can then act as **electron donors**, transferring electrons farther down the chain until they reach the ultimate electron acceptor. Among electron carriers in these pathways are the iron-containing polypeptides known as ferredoxins (Fig. 9-2). These are sufficiently small and widespread in nature to believe that they, along with the porphyrins (p. 112), were among the first cellular oxidative-reductive agents. During the process of electron transfer, which now involves other agents such as cytochromes, energy released in specific oxidation-reduction steps is consumed through the transfer of protons across the cell membrane, thereby creating a **proton gradient** (see Fig. 9-6). The potential energy available in this protein gradient can be converted into chemical energy by the **phosphorylation** of ADP to ATP. Using the proton gradient (or low-potential reductants such as H_2) as an energy source, electrons can also be transferred via a hydride ion (H^-) to the coenzyme NAD^+ (or in some cases, to another coenzyme, FAD, flavine adenine dinucleotide), which then becomes reduced NADH. By means of the reduced coenzyme, a hydrogen is then transferred to a carbon recipient, leading to a reduced carbon compound that can then be used either structurally or metabolically.

PHOTOSYNTHESIS

In spite of advantages offered by electron transport systems in these early stages, reliance upon chemical energy sources undoubtedly restricted organisms to those specific localities or conditions where these organic and inorganic compounds could be found. Perhaps the most important step towards environmental independence therefore occurred when a mechanism evolved that could use light as a means to generate ATP (**photophosphorylation**).[2] The order in which

[2]Some authors have suggested that the photosynthetic production of ATP occurred before heterotrophic phosphorylations of the kind energized by anaerobic glycolysis (Broda). Photophosphorylation, however, appears more complex in terms of membrane organization than does anaerobic phosphorylation, and the electron transfer mechanisms associated with ATP production are therefore generally believed to have begun on the non-photosynthetic level.

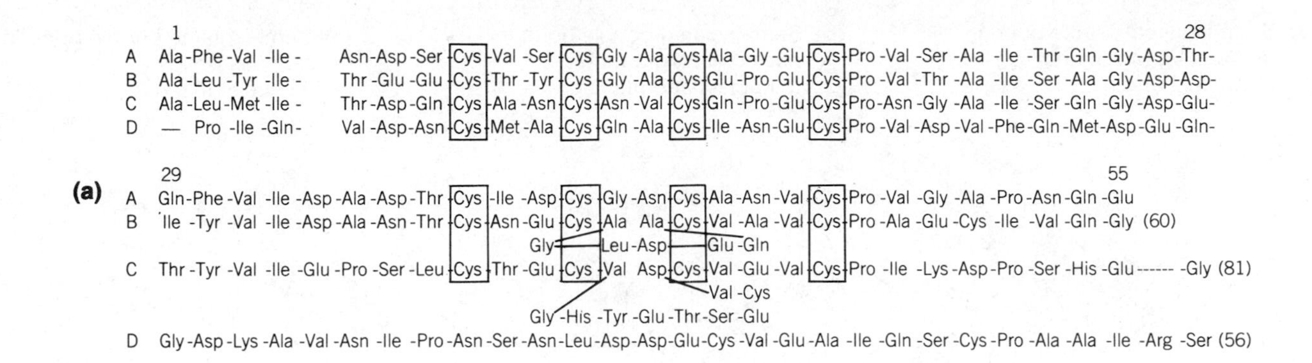

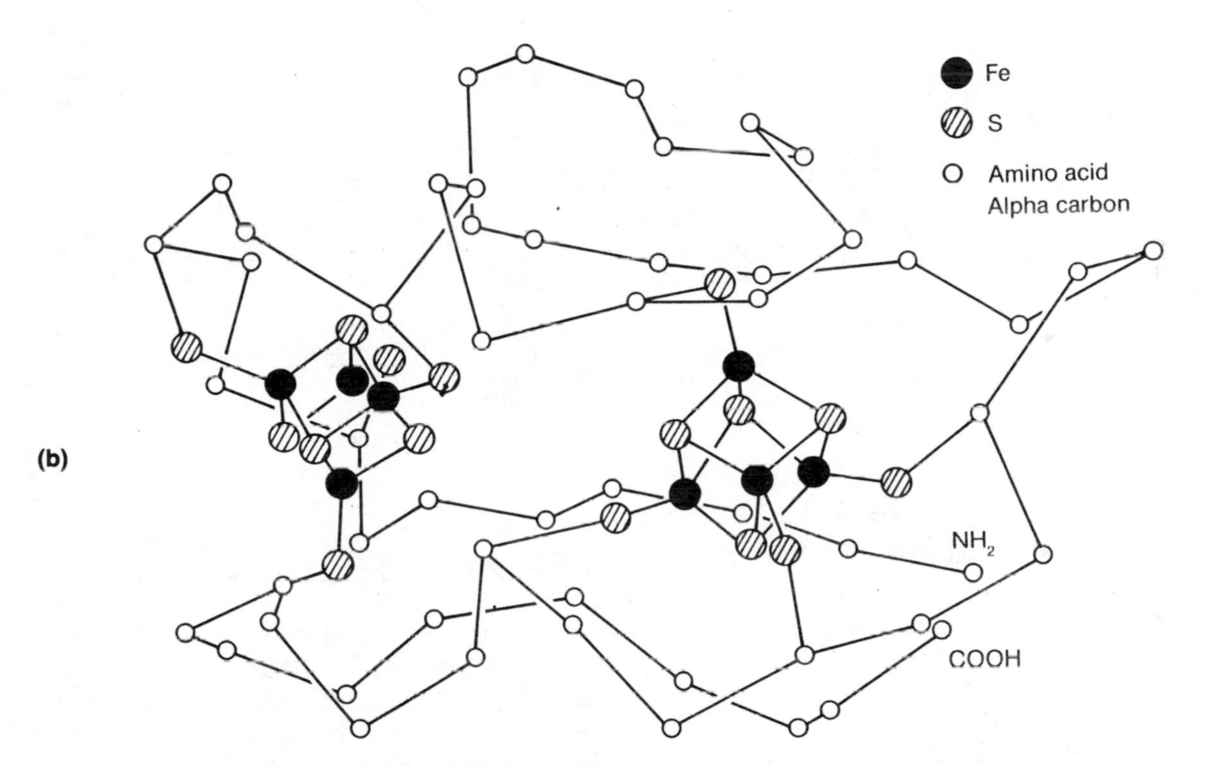

FIGURE 9-2 (a) Amino acid sequences of ferredoxin in four bacterial species: A, *Clostridium butyricum*, an anaerobic fermenting bacterium; B, *Chlorobium limicola*, a green photosynthetic bacterium; C, *Chromatium vinosum*, a purple photosynthetic bacterium; D, *Desulfovibrio gigas*, a sulfate-reducing bacterium. (From Hall et al.) Note that there is considerable similarity between the amino acid sequence in the first halves of these molecules (nos. 1-28) and the sequence in the second halves, an observation that initially prompted Eck and Dayhoff to suggest that this protein originated from a genetic duplication. (b) Folding of a ferredoxin molecule from *Peptococcus aerogenes* as revealed by x-ray analysis. There are two identically distorted cubes, each containing four iron atoms that are held by sulfur bonds arising from the cysteine residues enclosed in rectangular boxes in part (a). (Adapted from Adman et al.)

such light-sensitive mechanisms appeared is not known, but virtually all those observed today are dependent upon the presence of **chlorophyll**, and this was probably also true in the past. In fact, Woese and co-workers have pointed out that five of the major bacterial groups (gram-positive, purple, green sulfur, green nonsulfur, and cyanobacteria) possess photosynthetic species, and they therefore suggest the likelihood that most, perhaps all, modern bacterial groups may have been derived from a common photosynthetic ancestor.

As with other porphyrins, chlorophyll can assume a number of resonance forms in which double and single bonds shift while the molecule remains rigid and stable (Fig. 9-3). This resonating structure enables chlorophyll to maintain light-absorbed energy, as well as transfer it to similar molecules or receive it from pigments such as carotenoids which absorb light energy at other wavelengths.

A photosynthetic pathway believed to be quite ancient is shown in Figure 9-4. It is a **cyclic** one in which solar energy acting upon light-sensitive chlorophyll excites the molecule to a high-energy state, allowing an electron to be passed on to other electron transfer agents. At a very early evolutionary stage, this system must have been bound to the membrane of a primitive

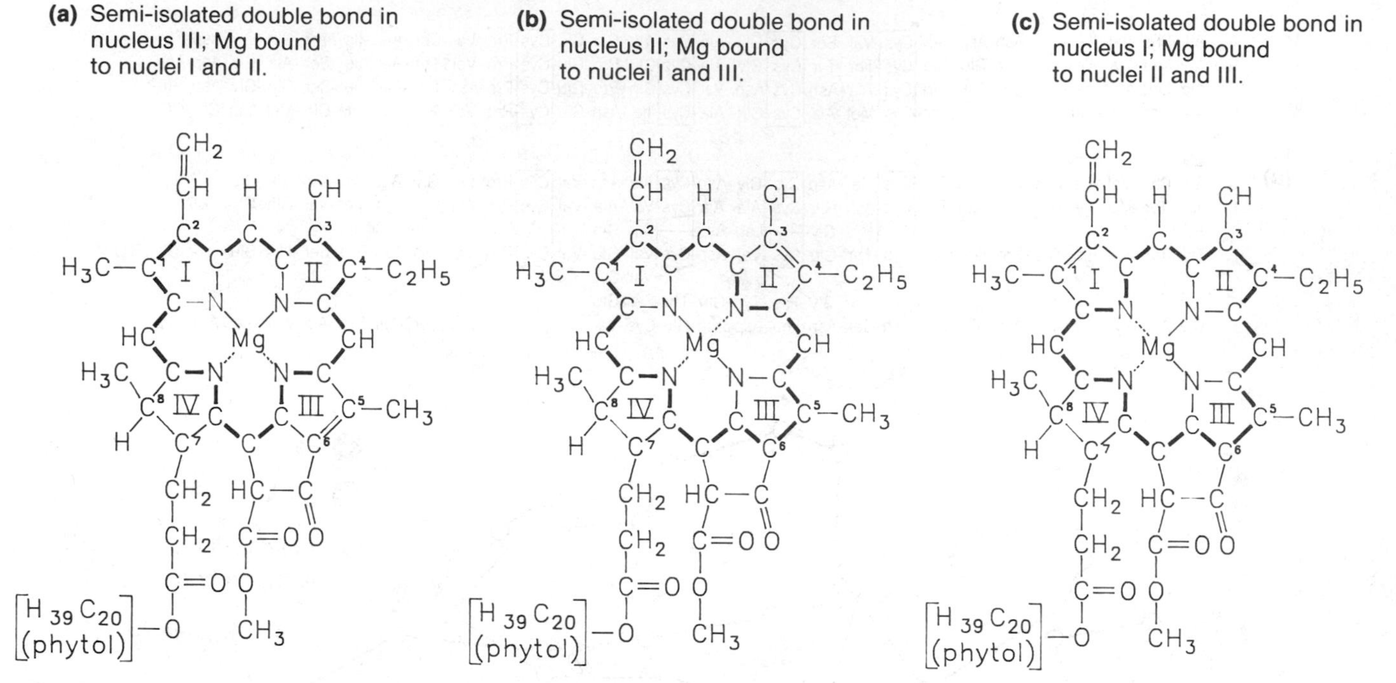

(a) Semi-isolated double bond in nucleus III; Mg bound to nuclei I and II.

(b) Semi-isolated double bond in nucleus II; Mg bound to nuclei I and III.

(c) Semi-isolated double bond in nucleus I; Mg bound to nuclei II and III.

FIGURE 9-3 Three resonance forms of chlorophyll a, showing stability of the molecule as its double and single bonds shift around the ring system in various ways (heavy lines). This ability to resonate enables chlorophyll to temporarily retain a high electron energy level resulting either from the excitation of electrons exposed to appropriate wavelengths of light or from electrons transferred to chlorophyll from other pigments such as carotenoids, phycobilins, or flavines. Chlorophyll can also transmit such energy to other molecules used in photosynthetic reactions. (Adapted from Wald.)

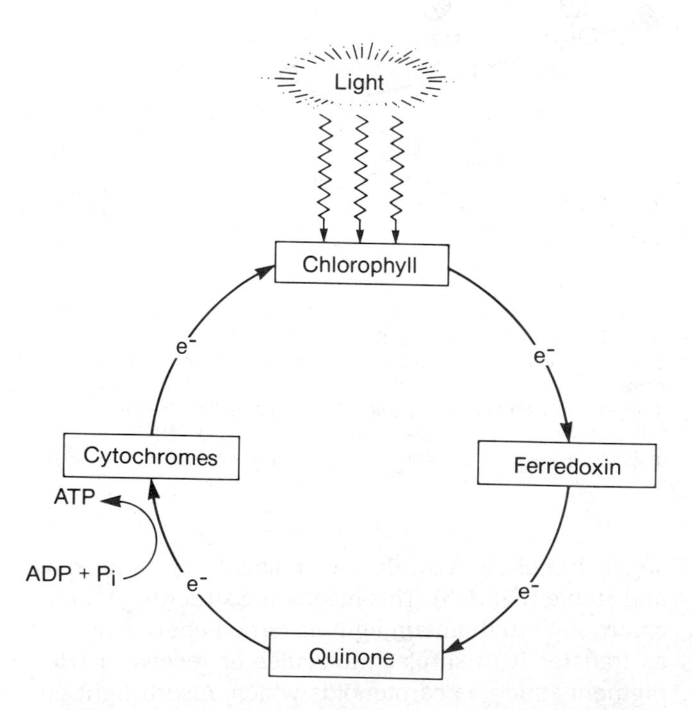

FIGURE 9-4 Diagrammatic view of a cyclic photophosphorylation pathway. Excited chlorophyll molecules force electrons to flow towards the more electronegative direction occupied by the electron acceptor, ferredoxin. Electrons then flow back to the now positively charged chlorophyll through electron carriers that include quinones and various cytochrome pigment proteins. During the transfer of electrons towards electropositive components, a proton gradient is formed across the membrane, and this gradient is used to supply energy for phosphorylation of ADP to ATP (see Fig. 9-6).

cell or droplet, allowing photoactive energy to be coupled to existing membrane-contained systems that could phosphorylate ADP to ATP.[3] ATP generated by this coupled system would be available for metabolic needs and would enable growth to occur independently of environmental chemical energy sources such as glucose intake.

The source of carbon, however, is as important to autotrophs as it is to heterotrophs. Some bacterial photophosphorylators such as the purple nonsulfur bacteria (*Rhodospirillum*) seem to be at least partly dependent upon complex organic molecules for their carbon sources and also use organic substrates as electron donors (e.g., the oxidation of succinic acid to fumaric acid), whereas others, such as the purple sulfur bacteria (*Chromatium*) use CO_2 exclusively and can obtain electrons from inorganic material such as H_2S.

[3]One such system probably originated as a membrane complex able to deacidify the cell interior by transporting protons (H^+) out of the cell using the energy provided from ATP → ADP breakdown. It is thought that incorporation of anaerobic oxidation-reduction enzymes and electron transfer components into the cell membrane provided an opportunity for evolving a system that could also reverse the ATP → ADP reaction: that is, evolution of an ATP synthesizing system energized by re-entry of protons into the cell. The association of chlorophyll with such membrane components thus offered a pathway for powering proton-gradient formation by photoactivity.

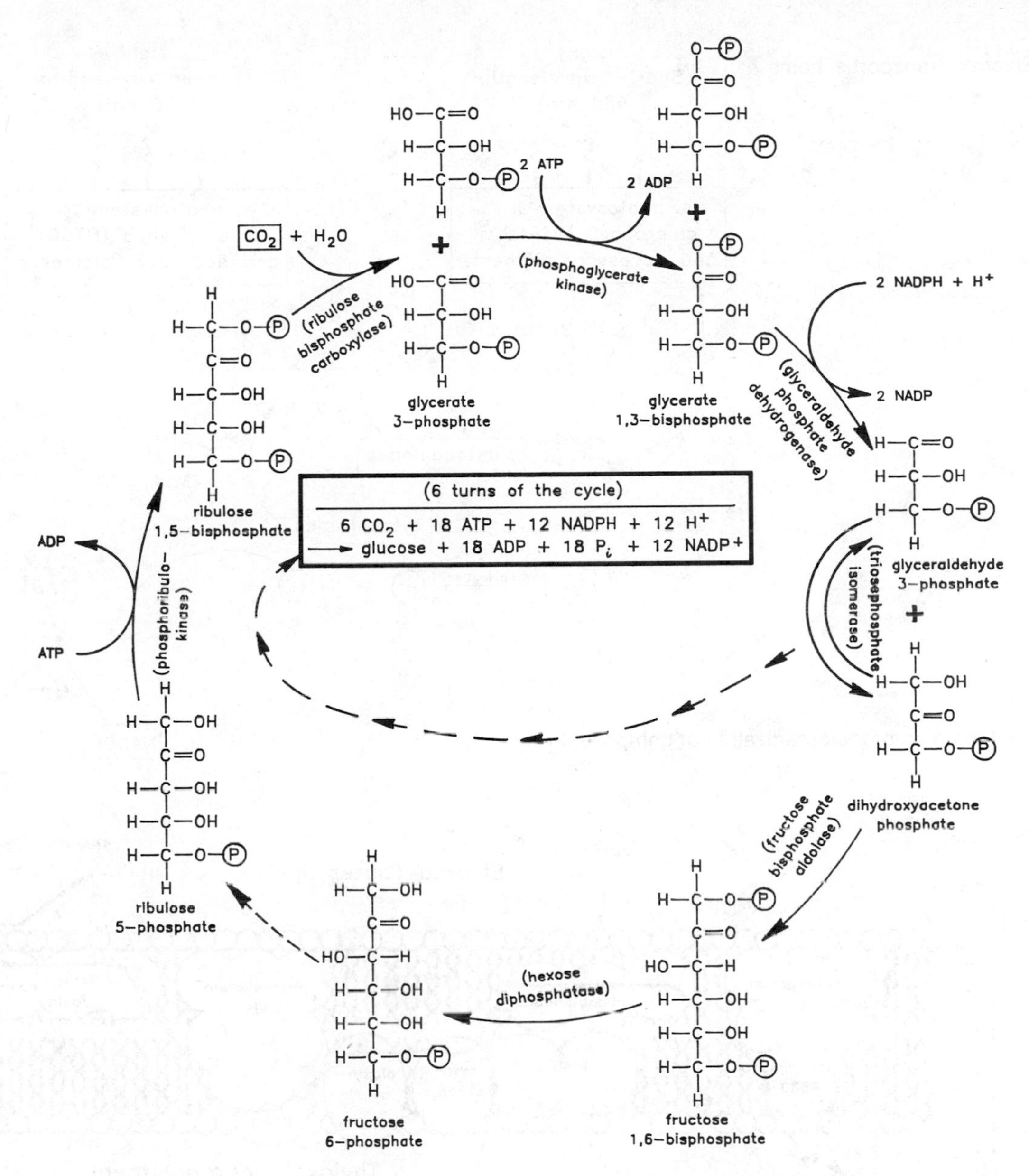

FIGURE 9-5 Simplified diagram of the Calvin cycle, the main metabolic pathway for carbon dioxide fixation in photosynthetic organisms. It relies upon reducing power contributed by NADPH formed during photosynthetic reactions and on energy provided by ATP. ($NADP^+$ is the oxidized form of the coenzyme nicotinamide adenine dinucleotide phosphate, and NADPH is the reduced form.) Between glyceraldehyde 3-phosphate and ribulose 5-phosphate in the pathway are a number of reactions, only two products of which are illustrated here (fructose 1,6-bisphosphate and fructose 6-phosphate).

Although it is not clear as yet which of these conditions was more primitive, the ability to use CO_2 as a source of carbon must have offered early photosynthesizers an important opportunity to expand their distribution. In fact, the **Calvin cycle**, which is the most common pathway for the reduction of CO_2, has been detected in practically all observed photosynthetic organisms. As shown in Figure 9-5, one carbon dioxide molecule is incorporated for each turn of this cycle, and one molecule of ribulose bisphosphate is regenerated for each CO_2 molecule incorporated. Six turns of the cycle are therefore necessary to produce one glucose molecule, and the overall reaction is:

$$6\,CO_2 + 18\,ATP + 12\,NADPH + 12\,H^+ \rightarrow \text{glucose} + 18\,ADP + 18\,P_i + 12\,NADP^+$$

Nevertheless, in spite of the advantage of utilizing readily available CO_2 and easily obtainable photosynthetic energy, the distribution of early photosynthesizers probably remained restricted because of their de-

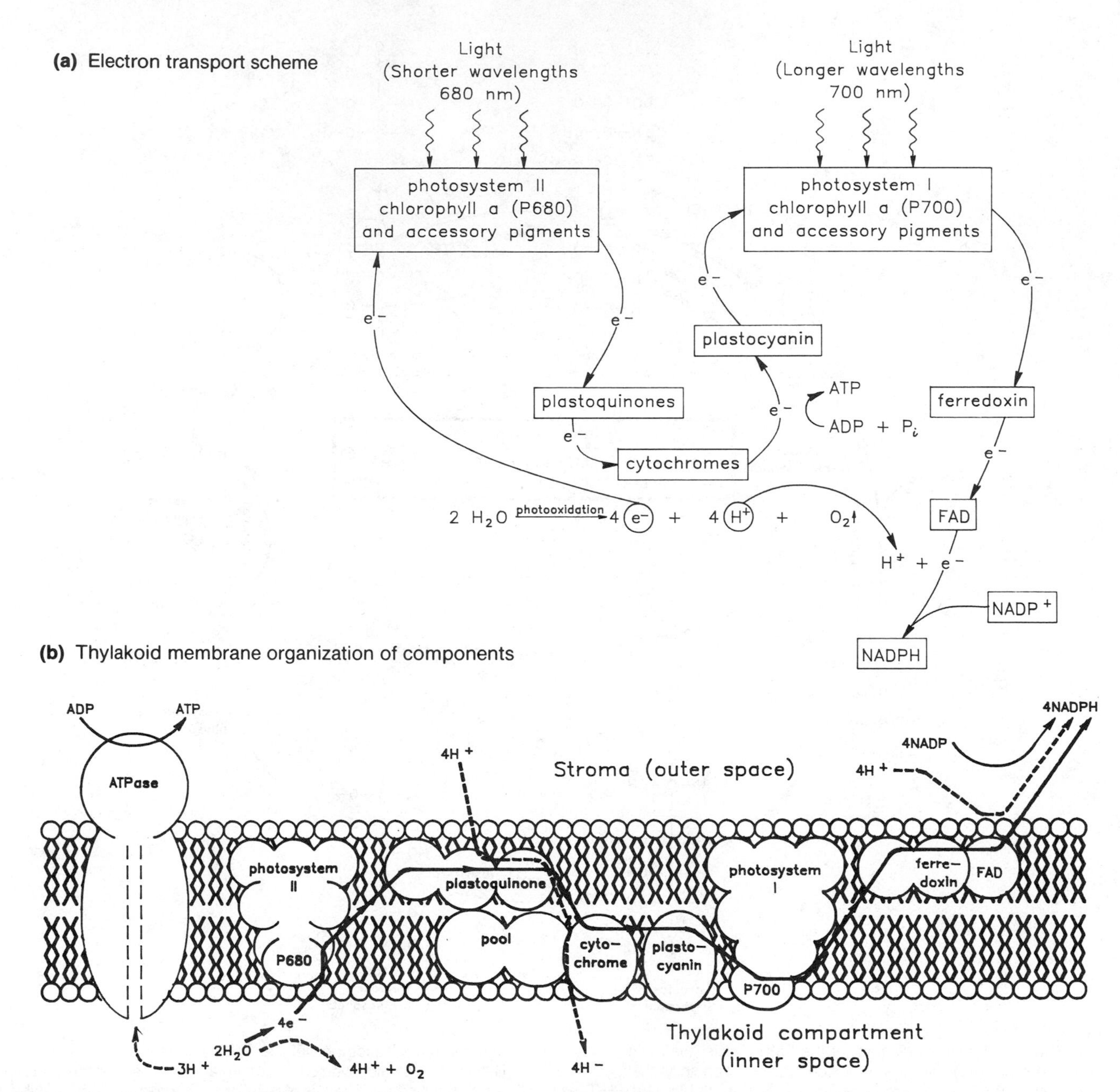

FIGURE 9-6 Diagram of non-cyclic electron flow in which electrons obtained from water molecules are transferred to the electron acceptor $NADP^+$. (a) Two photosystems, I and II, each sensitive to slightly different wavelengths of light, can be activated to high energy levels so that electrons are passed along the chain. The loss of electrons to $NADP^+$ by photosystem I is compensated by the transfer of electrons from photosystem II. Photosystem II electrons, in turn, are replaced by electrons derived from the photo-oxidation of water molecules. As in cyclic photophosphorylation, the flow of electrons produces a proton gradient that can supply energy for the phosphorylation of ADP to ATP. This is shown more clearly in the thylakoid membrane structure diagrammed in (b) where the ATP-synthesizing enzyme, ATPase, generates ATP by using the H^+ gradient caused by the flow of H^+ ions into the thylakoid compartment. Also shown are the general locations of the two photosystem components and some of the electron transport chain proteins. The presumed flow of electrons is indicated by a heavy line, and the flow of H^+ ions by dashed lines. (Part (b) adapted from Wolfe and also Zubay.)

pendence on compounds such as H_2S for hydrogen sources. Such dependence can still be noted today among some bacterial photosynthesizers (purple and green sulphur bacteria) in which hydrogen sulfide provides the electrons that allow the hydrogenation of carbon:

$$2\ H_2S + CO_2 \xrightarrow{\text{light}} \underset{\text{carbohydrate}}{(CH_2O)} + H_2O + 2\ S$$

A primary revolution in the distribution of living organisms therefore occurred when photosynthetic mechanisms evolved that could derive their electrons from readily available water molecules. This involved two chlorophyll systems (**non-cyclic photosynthesis**) and their union into distinctively specialized photosynthetic membranes (**thylakoids**) found today only in prokaryotic cyanobacteria and the chloroplasts of eukaryotic algae and plants. As shown in Figure 9-6, the source of electrons for the photosystem II chlorophyll component comes from the oxidation (dissociation) of water into electrons and H^+ ions and the release of molecular oxygen: $2\ H_2O \rightarrow 4\ e^- + 4H^+ + O_2 \uparrow$. The light-excited, positively-charged chlorophyll component of photosystem II readily accepts electrons released during this dissociation, and the coenzyme $NADP^+$ (nicotinamide adenine dinucleotide phosphate) is readily reduced to NADPH by hydrogens and electrons passed on to it through the electron transfer chain. The NADPH and ATP formed in the solar-illuminated **light reactions** are then used in **dark reactions** to reduce carbon by the Calvin cycle.

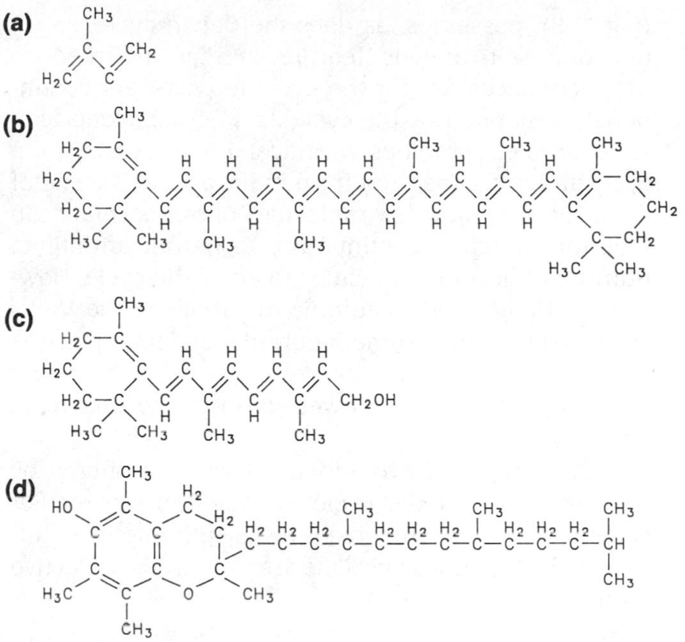

OXYGEN

The consequences of using water as an electron and hydrogen donor in photosynthesis were profound since the liberation of molecular oxygen began to produce an oxidizing, aerobic, environment whose chemical effects were quite different from the relatively more reducing environment previously encountered. The speed at which oxygen accumulated because of photosynthesis is not known, but organisms found in South African Bulawayan limestone and gunflint strata date back to about 3 billion years ago and are believed to have been oxygen-generating cyanobacteria. It has been suggested that oxygen concentration in the atmosphere may have remained at 1 percent of the present level until about 2 billion years ago and then gradually increased to its present concentration with increased success of photosynthetic forms. Although these estimates are conjectural, it seems clear that the initiation of an oxygen atmosphere led to an increase in the number and kinds of organisms capable of utilizing aerobic metabolic pathways. By the Cambrian period or somewhat earlier, oxygen levels had apparently become high enough to permit rapid evolution of large aerophilic multicellular organisms (Chapter 14).

Since oxygen is a highly reactive element that can rapidly oxidize organic material, one of the immediate selective effects of an oxygen atmosphere would be to increase the frequency of cellular **anti-oxidant compounds**. These anti-oxidants were probably not much different from some isoprenoid derivatives that can be found today, such as vitamin K, coenzyme Q, the phytol groups attached to chlorophyll, the carotenes, vitamin E, and others. (Fig. 9-7). Enzymes which neutralize superoxide radicals ($O_2^{\bullet -}$) and peroxide products (H_2O_2) formed by oxygen within cells must also have evolved at that time, such as the superoxide dismutases, catalases, and peroxidases (Fridovich). Interestingly, one possible early attempt to detoxify oxygen is still believed to be used in some present organisms such as luminescent bacteria, and that is to localize oxidation to special reactions that emit fluorescent light. The peroxides formed by oxidation in such organisms react with luciferase enzymes to produce organic acids and water, radiating light in the process.

FIGURE 9-7 Examples of isoprenoid compounds. (a) The basic 5-carbon isoprenoid unit. (b) β-carotene, a plant isoprenoid offering protection against oxidation in visible light. (c) Vitamin A_1 (retinol 1), a fat-soluble vitamin found only in animals, formed by the cleavage of carotenes. (d) Vitamin E (α-tocopherol), a plant anti-oxidant for which nutritional dependence has been demonstrated in rodents.

AEROBIC METABOLISM

Perhaps the most significant change in metabolism that accompanied the new aerobic environment was the evolution of a respiratory pathway by which oxygen is utilized to produce much more energy from the breakdown of glucose than can be produced by anaerobic glycolysis. This system involved the elaboration of a series of enzymes that transform pyruvic acid into an activated acetic acid group (acetyl-coenzyme A or one of its evolutionary precursors), then carry these small acetyl groups along a special cycle in which they are converted into carbon dioxide and their hydrogens removed:

$$CH_3COOH\text{-coenzyme A} + 2\ H_2O \rightarrow 2\ CO_2 + 8\ (e^- + H^+) + \text{coenzyme A}$$

This part of the pathway, variously called the **Krebs cycle**, citric acid cycle, or tricarboxylic acid cycle

(Fig. 9-8), possesses, as does the Calvin cycle, an interesting self-catalytic feature in that intermediate products necessary for the cycle to occur are continuously generated by the cycle itself. Oxaloacetic acid, for example, combines with acetic acid to begin the cycle and is regenerated from malic acid at the end of the cycle. One or a few molecules of oxaloacetate can therefore function continuously to permit an infinite number of acetate molecules to enter the cycle. However, although tied to aerobic metabolism, the Krebs cycle itself is not immediately dependent upon the presence of molecular oxygen, and it has therefore been suggested that it owes its origin to anaerobic pathways.

One proposed evolutionary sequence offered by Weitzman is that the cycle evolved from an earlier stage in anaerobes in which it was split into two metabolically different arms. One arm followed a reductive pathway (see also Gest):

pyruvate → oxaloacetate → malate →
fumarate → succinate → succinyl-coenzyme A

The other arm also began with an initial pyruvate but was engaged in oxidative metabolism:

pyruvate → acetyl-coenzyme A → citrate →
cis-iconitate → isocitrate → α-oxoglutarate

Coupled to reactions in the reductional arm was the oxidation of NADH to NAD^+, necessary for the replacement of NAD^+ used in glycolysis, which, in turn, was necessary for ATP formation. (The final product of the reductional arm, succinyl-coenzyme A could also be used in the synthesis of porphyrins and various amino acids.) Reactions in the oxidative arm had the obvious function of providing reduced nucleotides (NADH, NADPH) for carbohydrate synthesis as well as providing compounds such as α-oxogluterate, a precursor of glutamic amino acid. Both these arms apparently still function in some organisms, such as cyanobacteria and anaerobically grown *E. coli* bacteria.

According to Weitzman, the gap between these two arms (between α-oxogluterate coming from the oxidative arm to succinyl-coenzyme A coming from the reductive arm) was bridged by an enzyme complex (α-oxoglutarate dehydrogenase) that can be considered a genetic variant of the pyruvate dehydrogenase complex which helps oxidize pyruvate to acetyl-coenzyme A. He suggests that selection to change the remaining reductive steps of the cycle in an oxidative direction was associated with the increase of atmospheric oxygen. Thus, although the Krebs cycle itself does not use molecular oxygen, its evolution and adoption by aerobic organisms seems based on a membrane-bound electron transport system where oxygen serves as the final electron acceptor in the chain: one more stage in the evolution of membrane-bound systems.

In this electron transport process, the pyridine nucleotide coenzyme NAD^+ picks up electrons and associated protons (e^- and H^+) and transfers them into a respiratory chain composed of various electron carriers. The final electron transfer to oxygen, the ultimate electron acceptor, causes the production of water: $O_2 + 4\,e^- + 4\,H^+ \rightarrow 2\,H_2O$.[4] As diagrammed in Figure 9-9, such electron transfers occur along an electrical potential gradient that provides sufficient energy exchange at three steps (coupling sites) to allow an ADP molecule to be phosphorylated to ATP. In sum, complete aerobic oxidation of a molecule of glucose, including **oxidative phosphorylation**, produces maximally about 38 molecules of ATP (compared to only two molecules of ATP formed by anaerobic glycolysis):

$$C_6H_{12}O_6 + 6\,H_2O + 6\,O_2 + 38\,ADP + 38\,P_i \rightarrow 6\,CO_2 + 12\,H_2O + 38\,ATP$$

If we consider that the hydrolysis of a mole of ATP (507 gm) to ADP probably provides at least 7 kilocalories, then cellular oxidation of a mole of glucose (180 gm) can produce at its theoretical maximum about 38 x 7 = 266 kilocalories. This amount is 39 percent of the 686 kilocalories produced by burning a mole of glucose in air and indicates that the efficiency of oxidative phosphorylation can be perhaps four times higher than human-designed mechanical energy-conversion systems, which rarely provide more than 10 percent efficiency.

In addition to its metabolic effects, the oxygen atmosphere enabled a stratospheric **ozone** (O_3) layer to be formed that screens out shortwave ultraviolet radiation from reaching the earth's surface. Since absorption of such wavelengths by organic ring structures (e.g., nucleotides) can cause lethality by deteriorating or modifying vital molecules, ozone screening must have been essential in enabling the expansion of life to ocean surfaces as well as to land.

[4]It has been suggested (Dickerson) that the terminal components of aerobic respiratory mechanisms must have evolved more than once in bacteria by adapting different heme proteins to carry electrons from cytochromes to oxygen. Earlier in evolutionary history, membrane-bound anaerobic respiratory chains apparently used fumarate as a terminal electron acceptor (Jones).

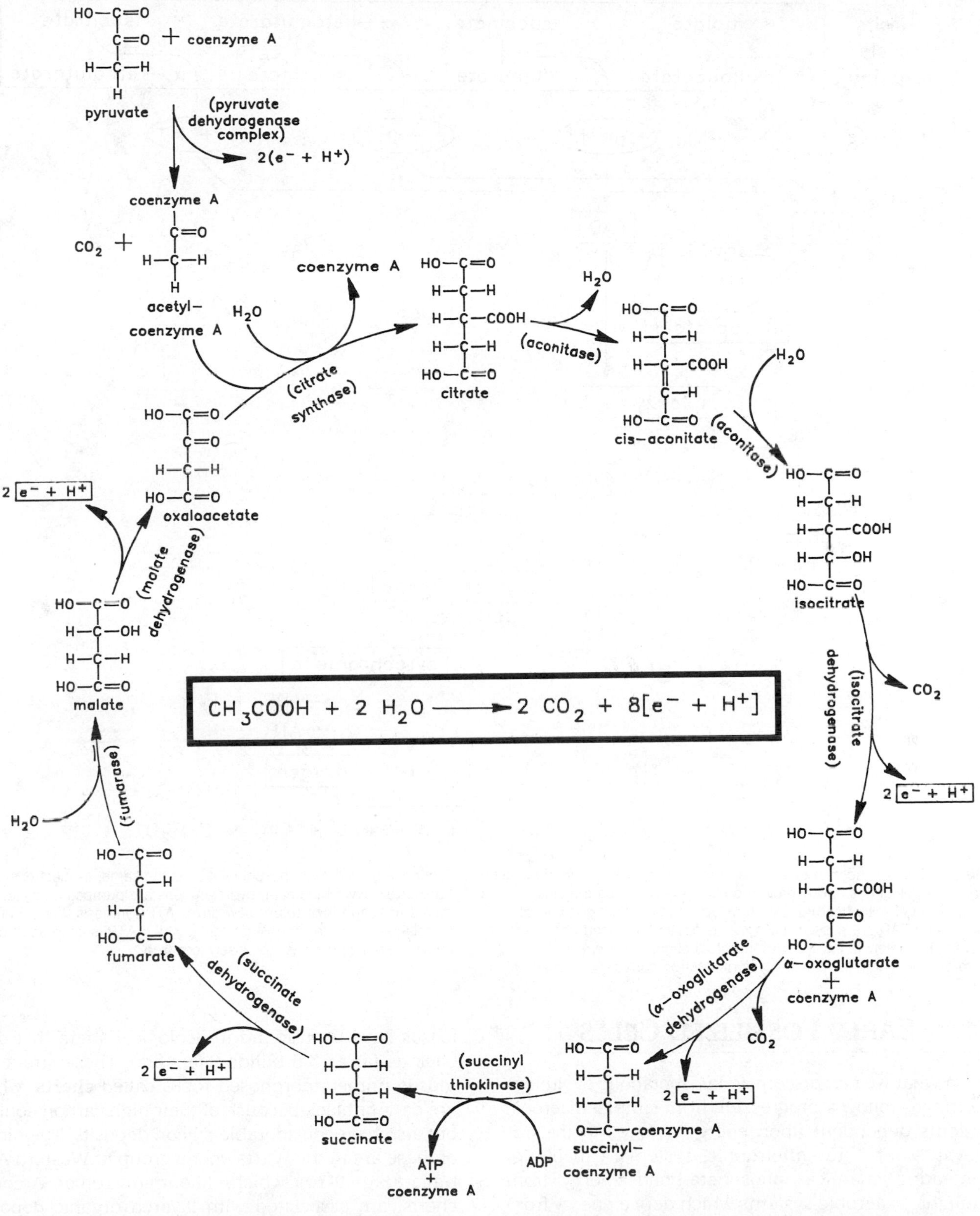

$$CH_3COOH + 2\ H_2O \longrightarrow 2\ CO_2 + 8[e^- + H^+]$$

FIGURE 9-8 Diagram of Krebs cycle reactions for the conversion of a 2-carbon acetate molecule (entering the cycle as acetyl-coenzyme A) into two carbon dioxides and eight $[e^- + H^+]$ units used for reduction. In three instances, the cofactor NAD^+ is used as the hydrogen acceptor ($NADP^+$ can be used in one of these steps) and FAD is used in the succinate → fumarate reaction. These reduced coenzymes are then passed on to the respiratory chain described later in the text, where they are reoxidized by the transfer of electrons to molecular oxygen. This reoxidation process releases large amounts of energy that can be used to form phosphate bonds of ATP molecules. The Krebs cycle itself possesses only one reaction, succinyl-coenzyme A → succinic acid, in which a component molecular bond transfers its energy directly to a phosphate bond.

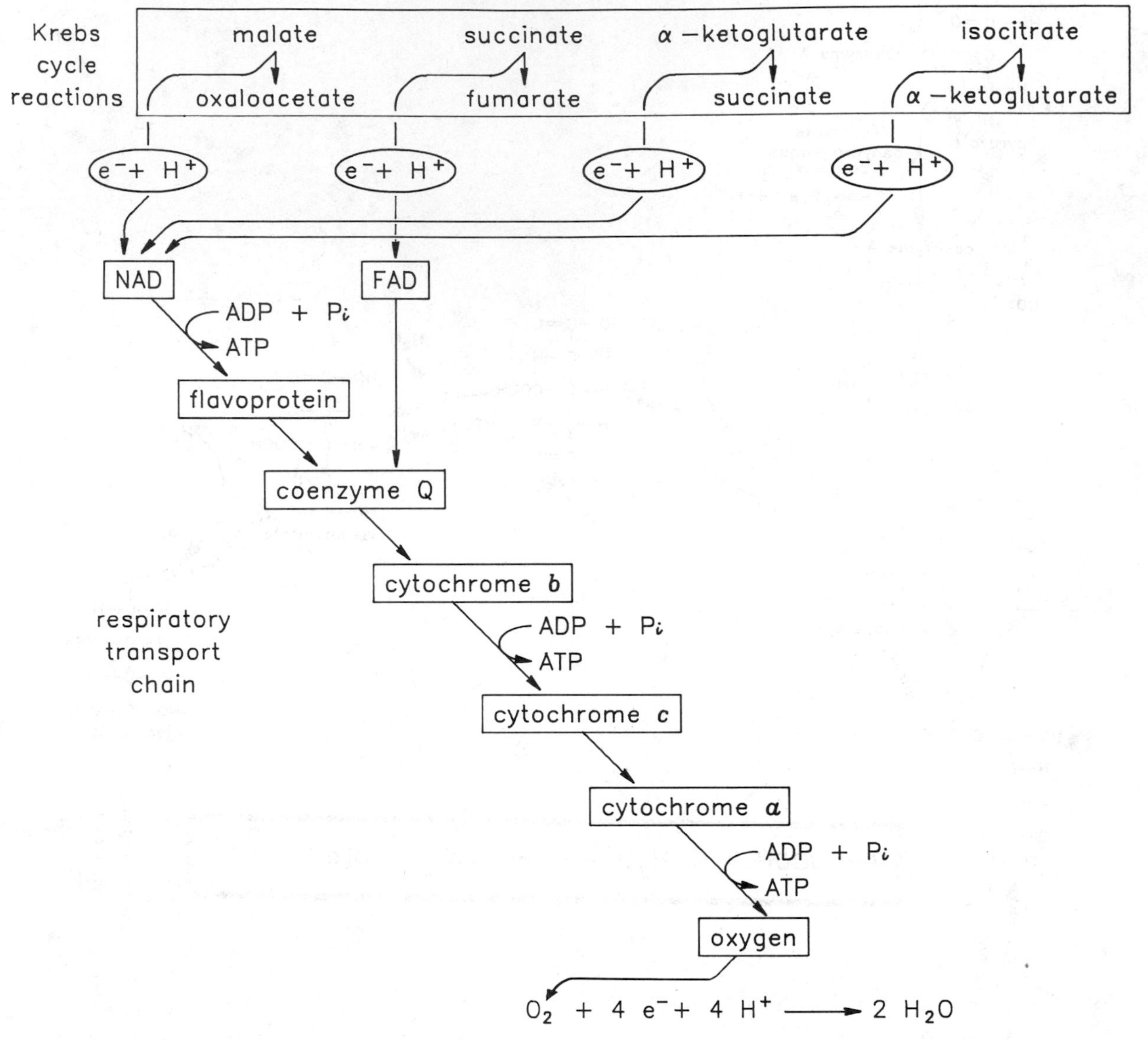

FIGURE 9-9 A simplified diagram of a respiratory pathway for the transfer of electrons and hydrogen protons occurring on the inner membrane of mitochondria. Electrons and protons from coenzymes NAD^+ and FAD are passed down the respiratory transport chain leading ultimately to oxygen and the production of water. Although different microorganisms use different electron carriers, the general sequence of oxidative phosphorylation is the same: as electrons are transferred down the chain, hydrogen ions are pumped across the membrane, and their return flow drives ATP synthesis. For simplicity, the diagram provides coupling sites at which ATP is generated, but these sites have not been precisely localized.

EARLY FOSSILIZED CELLS

From what we can discern so far, metabolic evolutionary stages follow a progression from simple anaerobic systems dependent upon energy sources in the primeval "soup," to autotrophic systems capable of generating chemical phosphate-bond energy from sunlight, to aerobic systems which derive energy from the transfer of electrons to oxygen. (A proposed phylogeny of prokaryotes based on sequencing amino acids in proteins and nucleotides in RNA is given in Fig. 12-17.)

The exact periods during which these evolutionary steps took place are not known, but various cell-like fossils have been found in geological strata that date back as far as 3.5 billion years ago. These strata are mostly unmetamorphosed rocks called **cherts**, which are dark or black because of their high carbon content but also bear considerable silicon deposits. The oldest of these are in the Warrawoona group of Western Australia (Fig. 9-10) which, like many other Archean cherts, are associated with layered organic deposits called **stromatolites**.

In their modern form, stromatolites are composed of mats of photosynthetic cyanobacteria whose filaments trap various aqueous sediments which are then cemented together to form characteristic laminated structures shaped like giant knobs. As shown in Figure

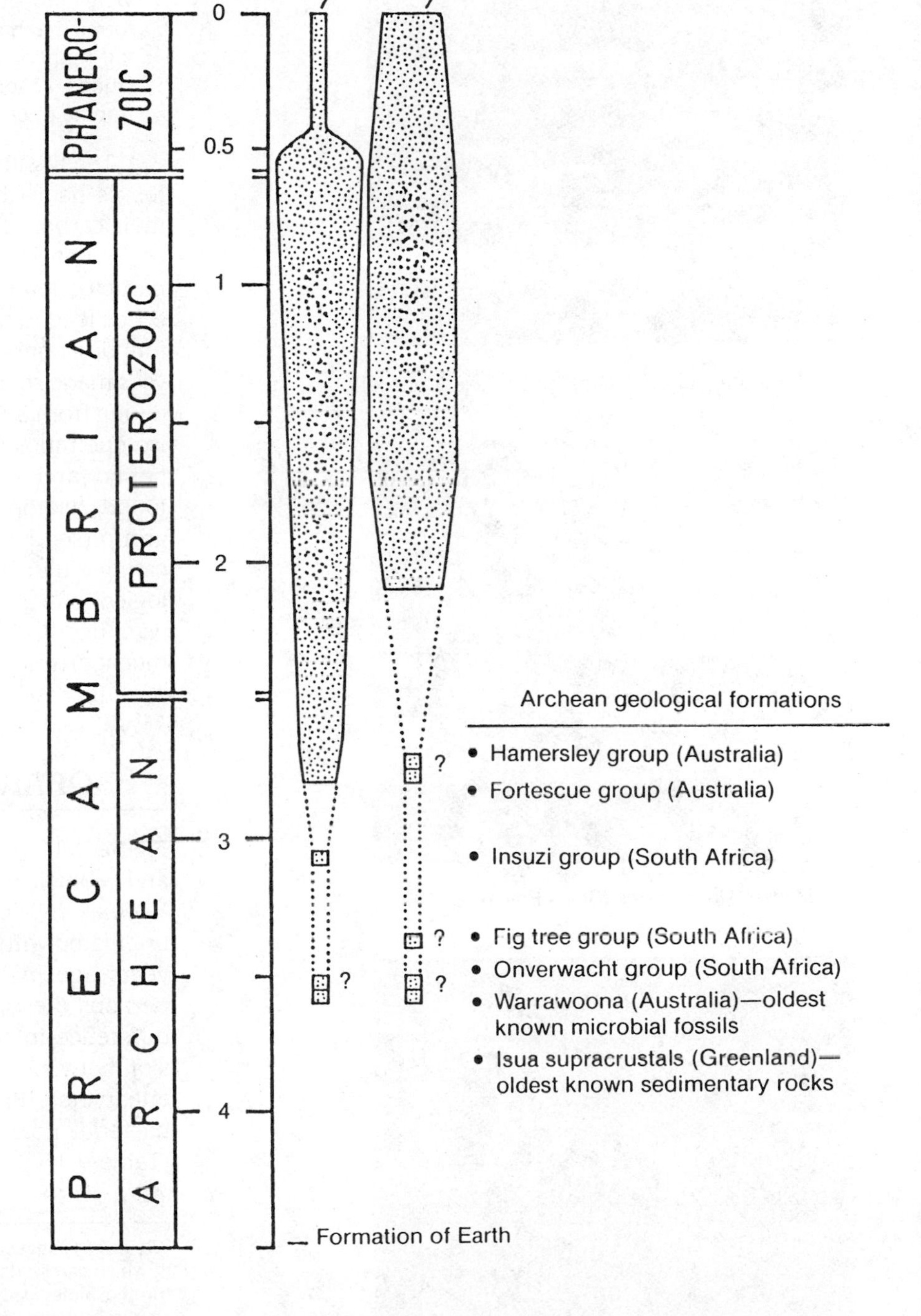

FIGURE 9-10 The chronological record of stromatolite deposits and microbial fossils, indicating their presence in some Archean geological formations. Stromatolites became abundant before the end of the Archean era and decreased only at the close of the Precambrian eon, apparently because of grazing metazoa and competition from eukaryotic algae. The early microbial fossil record generally parallels stromatolite abundances, although not all microbial fossils are found in stromatolite deposits. (Adapted from Schopf and Walter.)

9-11, these modern structures are remarkably similar to ancient stromatolites, which must also have arisen from the deposition of biological organisms. There are nevertheless differences in distribution between modern and ancient forms, since modern stromatolites are found only in extremely inhospitable environments protected from grazing metazoans such as snails and sea urchins (e.g., they occur in salinities ten times that of sea water and at temperatures greater than 65°C), whereas ancient stromatolites were undoubtedly more widely dispersed because of the absence of such herbivores. Interestingly, many of the fossil organisms found in stromatolite deposits are remarkably similar to modern prokaryotes (Fig. 9-12). In the words of Schopf and Walter,

> It seems reasonable to conclude that: (1) shallow water and intermittently exposed environments (and possibly also, land surfaces and open oceanic waters) were habitable by prokaryotic microorganisms at least as early as 3.5 billion years ago; (2) such organisms com-

(a) Fossil stromatolites (Great Slave Lake, Canada)

(b) Modern stromatolites (Shark Bay, Australia)

(c) Stromatolite internal laminated structure

FIGURE 9-11 Comparison between (a) fossil stromatolites approximately two billion years old and (b) modern stromatolites. (c) A cross section of a stromatolitic knob, which can have a diameter of several feet across, showing laminated layers of algal mats. (From Margulis.)

prised finely laminated, multi-component, stromatolitic communities of the sediment-water interface, biocenoses where the principal surficial mat-building taxa were probably filamentous and photo-responsive, forms that may have been capable of phototactic, gliding motility; and (3) such communities probably included anaerobes and both autotrophic and heterotrophic microorganisms.

The likelihood that many of these ancient cell-like fossils had a biological origin can be surmised from their carbon isotope ratios ($^{13}C/^{12}C$). There are differences between these two isotopes in respect to their participation in cellular metabolism, and these differences lead to ratios that are unique compared to those found in nonbiological material. According to the analyses made so far, almost all of the stromatolite deposits dating from 3.5 billion years ago and later have carbon isotope ratios similar to rocks from the Carboniferous period and other strata in which living forms are found. Isotope ratios in earlier rocks of the Isua group (see Fig. 9-10) are considerably different, indicating either high-temperature environmental or geological effects or perhaps conditions in which carbon was fixed by non-photosynthetic or nonbiological mechanisms.[5]

PROKARYOTES AND EUKARYOTES

Figure 9-13 summarizes some of the major biological and geological evolutionary events from the lowermost Hadean division to the present Phanerozoic eon that originated with the Cambrian period about 600 million years ago. Aside from the origin of photosynthesis, perhaps the most significant biological change is the difference in cellular complexity that marks the division between organisms classified as prokaryotes and eukaryotes. **Prokaryotes**, a term used to include all bacteria, now generally distinguishes two kingdoms (Table 9-1):

[5]Cloud has suggested that **banded iron formations** (BIF), the oldest of which can be dated to 3.76 billion years ago, may also have been formed biologically. According to him, internal sources of oxygen were responsible for the change from ferrous to ferric ion, and these sources were probably living protocyanobacteria that were splitting water molecules and releasing O_2:

$$\underset{\text{(ferrous oxide)}}{4\,FeO} + \underset{\text{(from photoautotrophs)}}{O_2} \rightarrow \underset{\text{(ferric oxide)}}{2\,Fe_2O_3}$$

The ferrous ion thus served as a "sink" for molecular oxygen generated during photosynthesis and would have protected the anaerobic metabolic systems of these early photoautotrophs. The oxidized BIF bands may have resulted from episodic supplies of ferrous ion which enabled growing anaerobic photoautotrophs to precipitate iron in the form of ferric oxide.

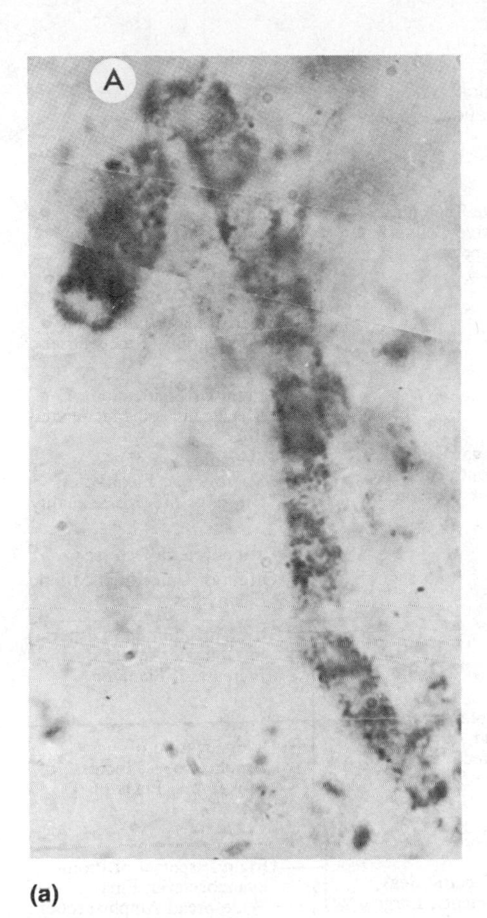

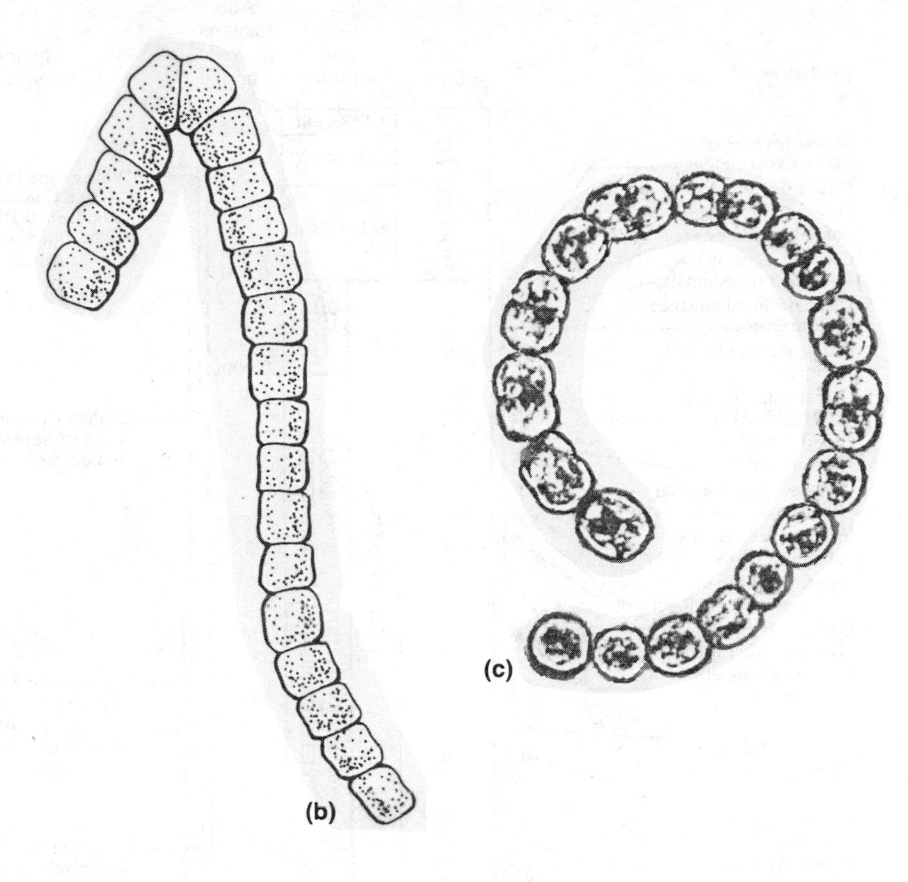

FIGURE 9-12 (a) Thin section of a filamentous prokaryotic fossil found in stromatolite chert in the 3.5 billion-year old Warrawoona formation in Western Australia. (b) Reconstruction of this fossil. (c) Diagram of a phase-contrast microphotograph of a filament of cells of a modern cyanobacterium (*Anaebaena* species). Among other modern prokaryotic groups which appear similar to the microfossil in (a) are colorless sulfur-gliding bacteria (e.g., *Thioplaca schmidlei*) and green sulfur bacteria (*Oscillochloris chrysea*). (Parts (a) and (b) from Schopf and Walter.)

- **Eubacteria**, encompassing the major forms of bacteria as well as the cyanobacteria, practically all possessing unique peptidoglycan or murein cell walls (chains of sugars cross-linked with short peptides, some of which contain D-amino acids)
- **Archaebacteria**, which use other materials for their cell walls and live under more rigorous environmental conditions than eubacteria, such as hot sulfur springs and extreme salt concentrations[6]

Classified under the term **eukaryotes** are many single-celled **protistan** organisms such as photosynthetic algae and non-photosynthetic protozoans, as well as multicellular plants (metaphyta), animals (metazoa), and fungi.

The difference between prokaryotes and eukaryotes is most obvious in the absence of a **nuclear envelope** in the generally smaller prokaryotic cells (1-10 μm) and its presence in the generally larger eukaryote cells (10-100 μm). In addition, prokaryotes divide mostly by **binary fission** rather than by the **mitotic mechanisms** of nearly all eukaryotes. Prokaryotes consequently lack organelles such as mitotic spindles and centrioles and do not possess the histone proteins used to structurally organize the relatively larger and more numerous eukaryotic chromosomes. Prokaryotes also lack some of the membrane structures and **organelles**

[6]In addition to eubacteria and archaebacteria, Lake and co-workers have suggested subdividing the archaebacteria into two groups, one of which is to be considered a third prokaryotic kingdom called eocyta. The eocyte bacteria are sulfur-dependent and, according to Lake, their ribosomes are more similar to those of eukaryotes than to the other prokaryotes.

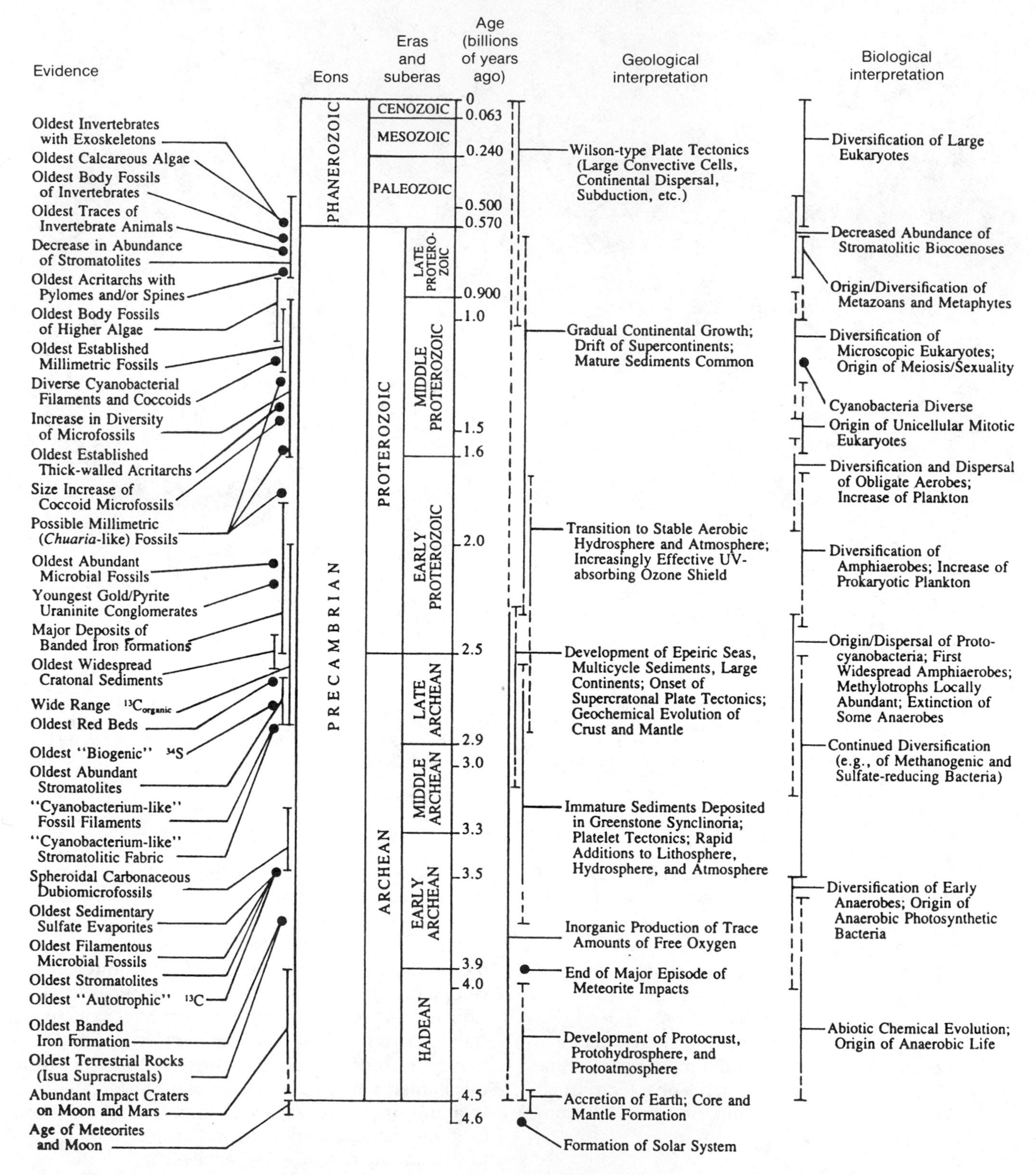

FIGURE 9-13 A summary of the evidence for Precambrian evolution along with the geological and biological interpretations of these observations. (From Schopf et al.)

TABLE 9-1 Some Molecular Biological Characteristics that Distinguish Eubacteria, Archaebacteria, and Eukaryotes

	Prokaryotes		Eukaryotes
	Eubacteria	Archaebacteria	
Cell wall	All incorporate muramic acid	None incorporate muramic acid	Absent (except in plants—but no muramic acid)
Genome organization	One circular chromosome	Single chromosome?	Usually several linear chromosomes
Introns (transcribed intervening sequences)	None	In tRNA genes	Yes
RNA polymerases	One	Probably one	Three (for mRNA, rRNA, and tRNA)
Capping of 5' mRNA terminal	No	Probably absent	Yes
3' mRNA poly-A tails	Absent	Present	Present
Ribosome unit sizes	$30S + 50S = 70S$	$30S + 50S = 70S$	$40S + 60S = 80S$
Ribosome RNA sizes	23*S*, 16*S*, 5*S*	23*S*, 16*S*, 5*S*	28*S*, 18*S*, 5.8*S*, 5*S*
Ribosome sensitivity to:			
Anisomycin	No	Yes	Yes
Chloramphenicol	Yes	No	No
Kanamycin	Yes	No	No
Protein translation sensitivity to diphtheria toxin	No	Yes	Yes
Protein initiator tRNA amino acid	Formylmethionine	Methionine	Methionine

Adapted from Woese (1983), and Doolittle and Daniels.

commonly found in eukaryotes, such as endoplasmic reticulum (usually associated with ribosomes in protein synthesis), Golgi apparatus (secretory bodies), and mitochondria (used in aerobic respiration). In fact, almost all eukaryotes are aerobic with the exception of some forms such as yeast that can also function anaerobically (amphiaerobes) and some amoeban-type protists such as *Pelomyxa* (members of a lineage that may always have been anaerobic).

Among the further attributes of eukaryotes which are almost entirely absent in prokaryotes are **split genes** in which adjacent amino acids in a polypeptide product are coded by nucleotide sequences that may be hundreds of bases apart. The nucleotide sequences that code for amino acids in such polypeptides are called **exons** ("expressed sequences"), and the intermediate non-coding nucleotide sequences are called **introns** ("intervening sequences"). These split genes transcribe their nucleotide sequences from DNA to RNA, but the RNA is then processed so that the introns are removed and the exons spliced together (Fig. 9-14). Such processed mature mRNA molecules are then transported through pores of the nuclear envelope to the cytoplasm to be translated into polypeptides.

The advantages of intron-exon structures seem to be sufficiently great to account for the fact that split genes occur in almost all vertebrate protein-coding genes so far examined, as well as in many similar genes in other eukaryotes. The reasons for these advantages, however, are not entirely clear. Gilbert suggested that each exon may have originally coded for a single **polypeptide domain** that could be used for a specific function, and others such as Blake suggested a relationship between the average size of an exon (coding for about 20 to 80 amino acids) and the size of the smallest polypeptide sequence that could be folded into a stable structure (about 20 to 40 amino acids). Because of the flexibility of exon arrangement and intron removal, the exons coding for these polypeptide subunits could then be combined in various ways to form genes that produced optimally functional polypeptides. Even some single genes can produce several different functional proteins by arranging their exons in several different ways through alternate splic-

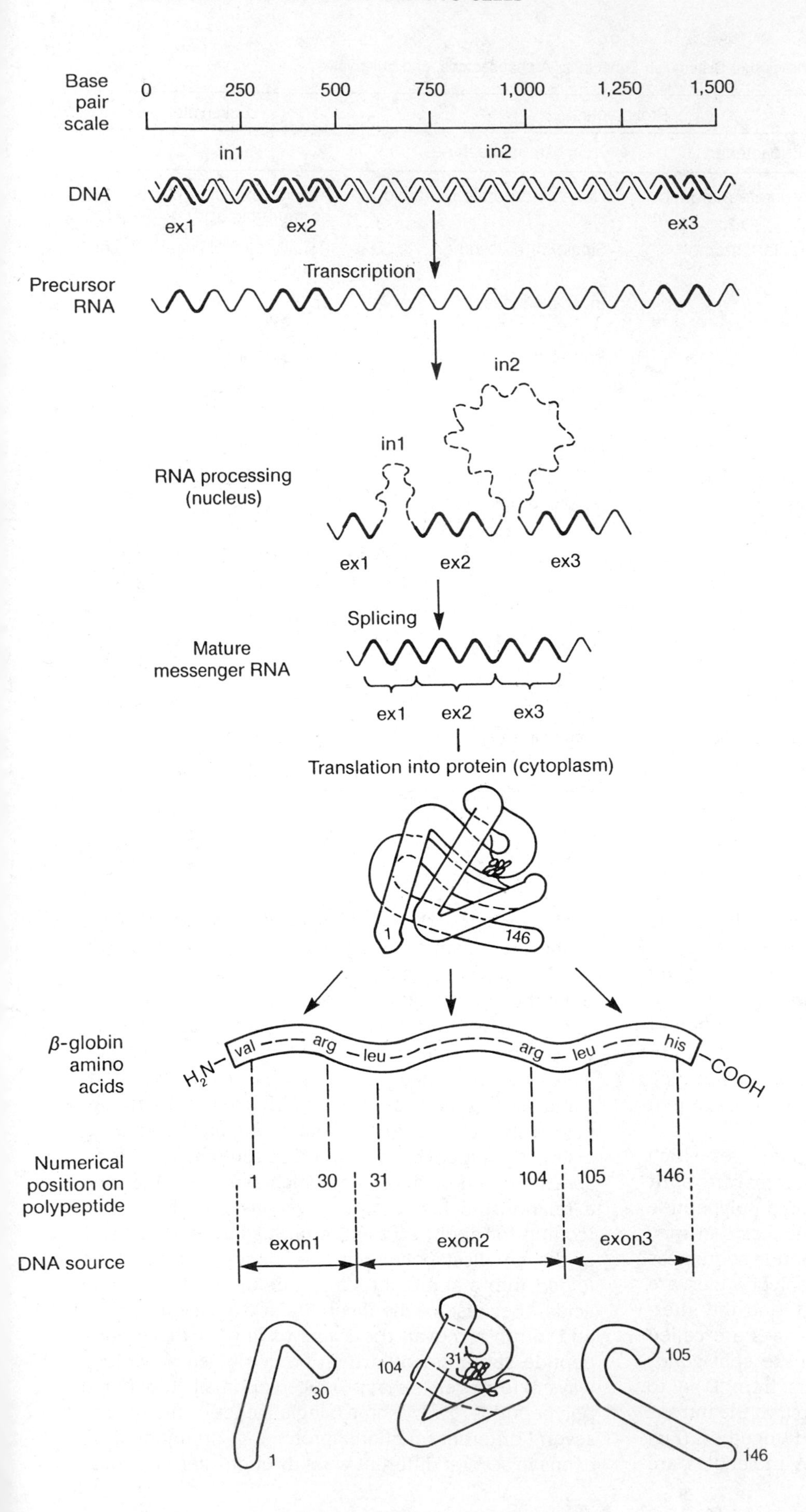

FIGURE 9-14 The intron-exon structure for nucleic acid sequences involved in the production of the human β-globin polypeptide chain, 146 amino acids long, one of the components of normal adult hemoglobin. The top portion of the diagram indicates the approximate 1,500 nucleotide base-pair length in the β-globin DNA sequence and the structure of this sequence in terms of two introns (in1, in2) and three exons (ex1, ex2, ex3). Intron 1 is 130 nucleotides long and separates codons that will later be translated into amino acids numbered 30 and 31 on the β-globin chain. Intron 2, 850 nucleotides long, separates codons for β-globin amino acids numbered 104 and 105. Although the entire DNA sequence is transcribed into mRNA, the introns are precisely removed by special RNA processing in the nucleus, and the exon sequences are spliced together and then translated into a continuous β-globin amino acid sequence. At the bottom of this figure, a single β-globin polypeptide chain has been diagrammatically split to indicate the three subcomponents (domains) respectively produced by the three exon nucleotide sequences. (Adapted from Strickberger.)

ing patterns (Andreadis et al.). According to Doolittle and others, cellular organisms ancestral to both prokaryotes and eukaryotes probably possessed such intron-exon structures, but these were mostly abandoned in prokaryotic lines because of an increased intensity of selection for streamlining DNA replication and improving transcription efficiency.

Other characteristics that distinguish eubacteria, archaebacteria, and eukaryotes are shown in Table 9-1. Woese and co-workers have also catalogued the presence and frequencies of various sequences in the 16*S* RNA component of ribosomes in the three groups and shown that there are distinctive differences among them (Table 9-2).

TABLE 9-2 Presence and Frequencies of Some Oligonucleotide Sequences Found in the 16*S* Ribosomal RNA Component of Eubacteria, Archaebacteria, and Eukaryotes

	Percent Occurrence		
Sequence*	Eubacteria	Archaebacteria	Eukaryotes
CYUAAYACAUG	83	0	0
AYUAAG	1	62	100
ACUCCUACG	97	0	0
CCCUACG	0	97	0
ACNUCYANG	0	0	100
YYUAAAG	3	97	0
AUACCCYG	93	3	0
AAACUUAAAG	0	100	100
UYAAUYG	1	100	0
CAACCYUYR	91	0	0
CCCCG	0	100	0
UCCCUG	0	97	100
AUCACCUC	91	100	0

Abridged from Woese (1985).
* R designates either purine (adenine or guanine); Y designates either pyrimidine (uracil or cytosine); N designates any of the four bases.

Although these tables show that archaebacteria occupy an intermediate evolutionary position between eubacteria and eukaryotes, these data are not yet sufficient to determine the exact phylogenetic relationship among these groups. It has therefore been suggested that no one group has been derived from any other, but rather they all diverged from a common cellular ancestor, called the **progenote** (Woese, 1987). This view is supported by the observation that all three of these groups possess common basic attributes such as a similar mode of DNA replication (new nucleotides are added to the 3′ end of the molecule), a common genetic code, a similar protein-synthesizing system, many similar metabolic pathways, a similar cell membrane structure composed of a phospholipid bilayer, and a similar mechanism of molecular transfer across membranes (**active transport**). However, subsequent events are still unclear: hypotheses have been proposed that eubacteria and eukaryotes arose from archaebacteria (Woese), and that prokaryotes arose from eukaryotes (Darnell), and that eukaryotes arose directly from prokaryotes (e.g., Cavalier-Smith, 1981). Future studies will undoubtedly help in deciding between these various views, and it is even possible that some distinctions between these major groups may have to be reevaluated (Lake).

EVOLUTION OF EUKARYOTIC ORGANELLES

More information has been accumulating in respect to the origin of the mitochondria and chloroplast organelles found within eukaryotic cells. The most popular hypothesis, offered in its modern form by Stanier, Margulis, and others, proposes that eukaryotic cells evolved by physically incorporating prokaryotic organisms into their cytoplasms. According to this theory, the ancestral anaerobic eukaryotic cell achieved the ability to perform **endocytosis**, the ingestion of supramolecular particles, because of a change in surface membrane properties. This lineage then became actively predaceous so that selection led to increased predatory abilities such as an increase in cell size as well as to other innovations that affect movement, capture of prey, and digestion. Among the various prokaryotes upon which this animal fed were bacterialike aerobes capable of performing oxidative phosphorylation as well as cyanobacteria capable of photosynthesis. At various times such ingested photosynthetic and aerobic prokaryotes assumed mutually advantageous symbiotic relationships with their hosts, not much different from those that are even now occasionally observed (Fig. 9-15).

According to one scheme, symbiotic relationships were first established with mitochondrionlike aerobic bacteria that improved eukaryotic metabolism and broadened eukaryotic predatory activity. Later, one or more lines of these new aerobic eukaryotes began similar symbiotic relationships with photosynthetic cyanobacteria, eventually evolving into the various eukary-

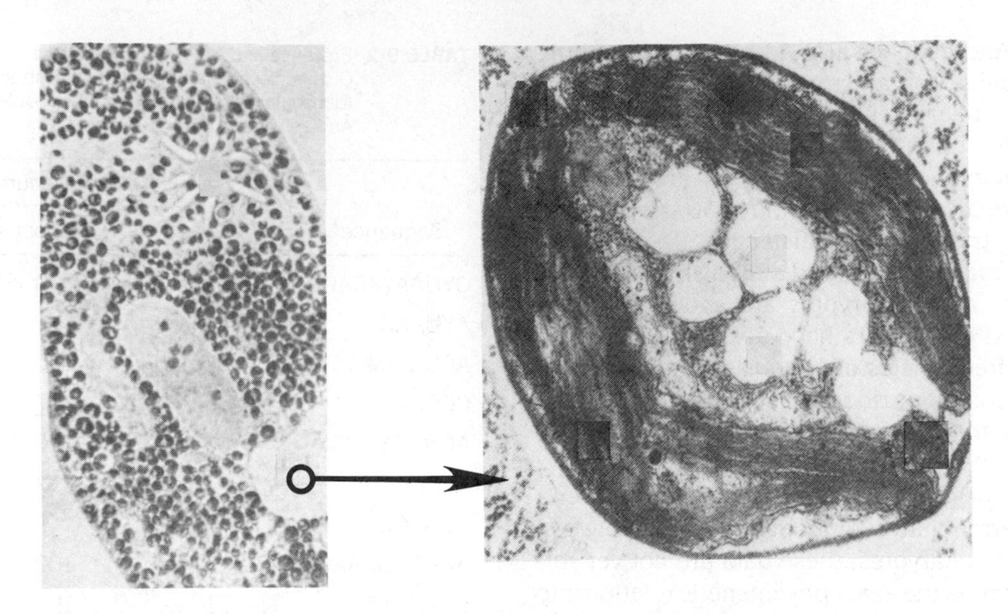

FIGURE 9-15 Symbiotic relationships between a eukaryote and its photosynthetic organelles. The protozoan ciliate *Paramecium bursaria* (left) harbors hundreds of symbiotic algae (right) which may be released from the cell and cultured independently. (From Margulis.)

otic algae and plants (Fig. 9-16).[7] In support of this view is the fact that mitochondria and chloroplasts are known to have their own genetic material (DNA), and their ribosomes are more similar to those of prokaryotes than to host ribosomes in respect to size, sensitivity to antibiotics, and nucleotide sequences of ribosomal RNA components. Whatley and co-workers have also pointed out that the aerobic bacterium *Paracoccus denitrificans* appears closest to what may have been the ancestral mitochondrion symbiont because of its many mitochondrionlike enzymatic, respiratory, and membranous features.[8]

Whatever the origin of their mitochondrial and chloroplast organelles, it seems clear that by 1.5 billion years ago or even earlier, eukaryotic cells had appeared, all of them in the form of single-celled organisms (Fig. 9-17). Because of their relatively large size and complexity, these protistans possessed a much larger amount of genetic material (DNA) than is traditionally found among prokaryotes. The difficulty of manipulating and replicating one or more large DNA molecules possessing circular prokaryotic forms, each with only a single point (origin) of replication, apparently led to the division of eukaryotic DNA into one or more linear chromosomes, each with multiple origins of replication. The near-universal presence of microtubules in eukaryotes can be explained as enabling nuclear chromosomal division to supplant the usual prokaryotic non-nuclear method, which depends on separating dividing chromosomes by their individual attachment to a lengthening cell membrane. Instead, each of the relatively larger, more complex, and often more numerous eukaryotic chromosomes possesses a **centromere** (or **kinetochore**) that can attach to a microtubular network of sliding spindle fibers during division, thus allowing the daughter chromosomes to move to opposite poles fairly efficiently. So far, the exact sequence of steps in the evolution of eukaryotic chromosome structure and cell division is unknown, although some hypotheses have been offered (Cavalier-Smith, 1987).

It seems very likely that once eukaryotic mitosis evolved, sex-cell division, or **meiosis**, must have quickly followed, since meiosis and sexuality are al-

[7]Since flagella, cilia, centrioles, and similar eukaryotic organelles are all composed of microtubules and seem to be structurally homologous, some proponents of the symbiotic theory suggest that these also arose by capture of prokaryotes followed by subsequent symbiosis. According to Margulis (see also Margulis and Sagan), an original cilium-type symbiont was a spirochetelike organism as can now be observed in the protozoan *Myxotricha paradoxa*. She proposes that all eukaryotic mitotic mechanisms owe their origin to the microtubular proteins that were initially involved in flagellar movements.

[8]A different hypothesis, less popular at present, suggested that these organelles arose as genetic duplications, each duplicate genome eventually enclosed by a separate set of membranes (Raff and Mahler). Mitochondria, now present in both eukaryotic plants and animals, presumably originated in a eukaryotic plantlike ancestor similar perhaps to some existing eukaryotic green algae, such as *Euglena*. One phylogenetic line of these photosynthetic organisms then led to the higher plants, whereas another line lost their chloroplasts and gave rise to eukaryotic fungi and animals. Arguments against this hypothesis mainly reside in the striking similarities between chloroplasts and cyanobacteria and between mitochondria and other eubacteria in their ribosomal RNA sequences (e.g., Palmer).

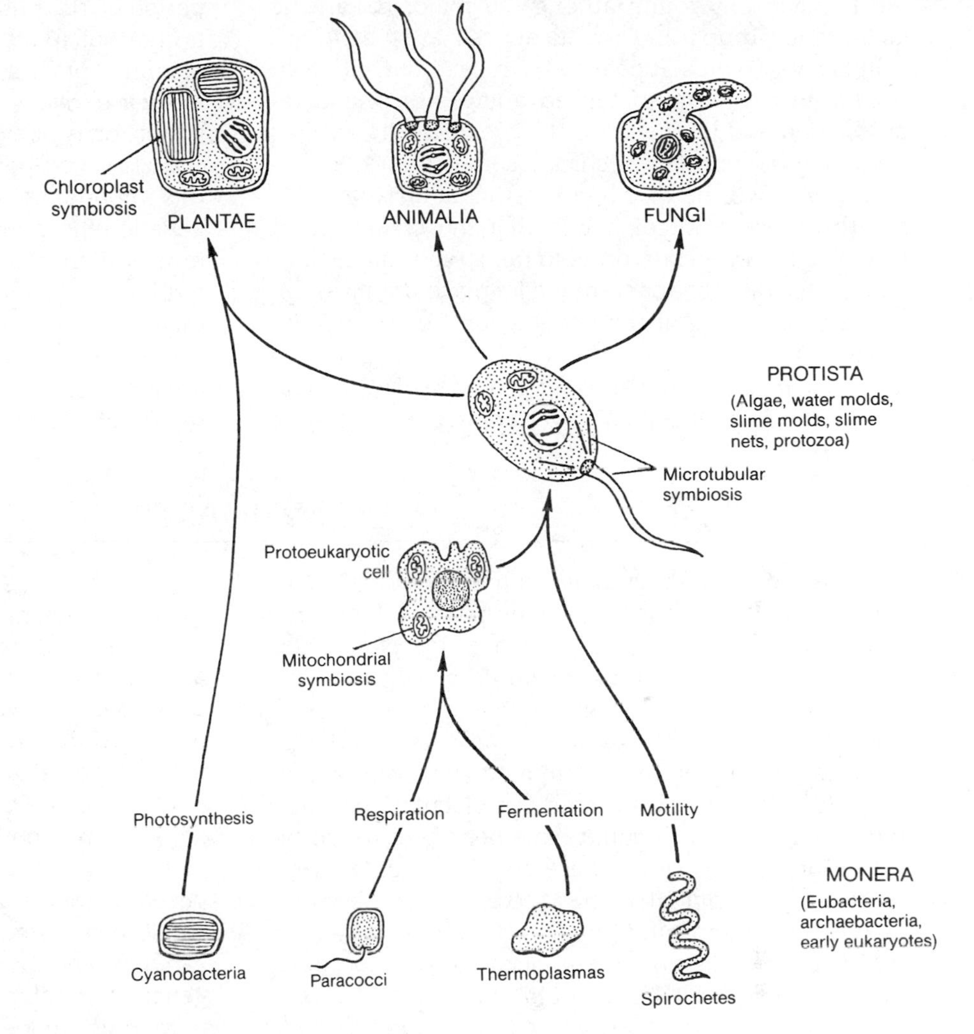

FIGURE 9-16 Symbiotic events during the evolution of eukaryotic cells according to Margulis. Mitochondrial symbioses arose from an early invasion of aerobic respiring prokaryotes such as *Paracoccus*. Eukaryotic motility as well as the microtubular structures involved in mitosis and meiosis was then gained by symbiosis with a prokaryotic spirochetelike form. In a later event, some eukaryotic cells were invaded by photosynthetic cyanobacteria, thus giving rise to plant chloroplasts. (Adapted from Margulis.)

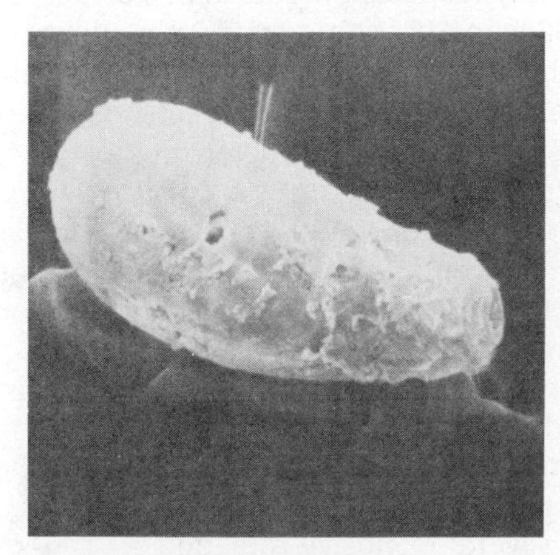

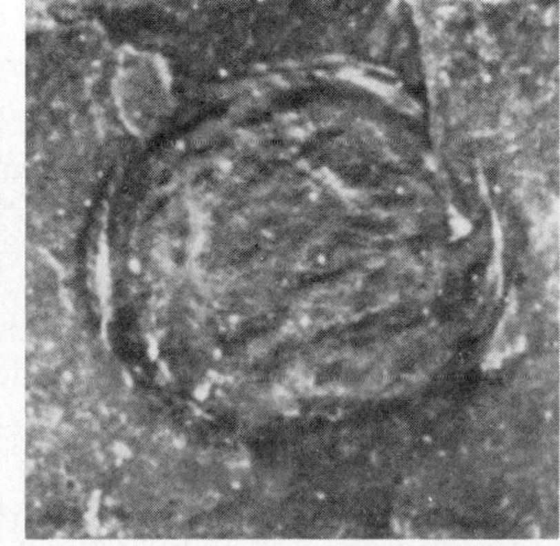

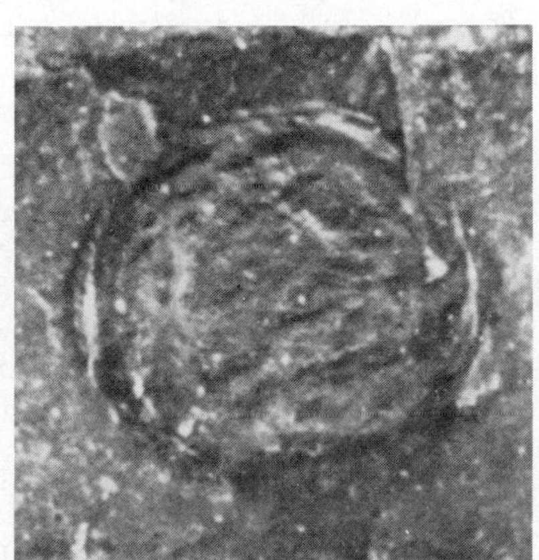

(a) Grand Canyon shales (800 million years old)

(c) Central Australian sediments (850 million years old)

FIGURE 9-17 Microfossils of probable eukaryotic cells that date back to the Proterozoic era. The cells in (a) and (b) are many times larger than any known prokaryotic cells and are considerably more complex. The group of cells in (c) are in a characteristic tetrahedral arrangement suggesting they were formed either through mitotic or meiotic eukaryotic cell division mechanisms. (From Schopf.)

most universally found among the major eukaryotic taxonomic groups and practically all asexual multicellular eukaryotes appear to have been derived from sexual forms. Some of the advantages of sexual reproduction and the need for the meiotic cell divisions to reduce the number of chromosomes prior to cross-fertilization will be discussed in Chapters 10 and 13. For the present we can note that the change from single-celled protistan forms to the large multicellular eukaryotes did not apparently take place until just before the beginning of the Phanerozoic eon. When the Cambrian period began about 600 million years ago there was the sudden marked appearance of numerous new forms of multicellular invertebrates. Perhaps in a period of 100 million years or less, an explosive radiation of eukaryotes occurred in which a large number of animal phyla appeared in the fossil record. The story of their derivations and relationships has long been a major focus of evolutionary biologists.

However, before dealing with these phenomena, it is important to obtain a clear concept of some basic genetic mechanisms that provide fuel and substance to evolution: that is, to understand the origin of hereditary variability between organisms and its transmission between generations. What kinds of genetic variability are there? What are their causes, and how are they transmitted? The next chapter provides a brief review of this topic.

SUMMARY

The prevalence of metabolic pathways common to many organisms suggests that comparative biochemistry would be a fruitful approach to investigate the evolution of metabolism. Anaerobic glycolysis is an almost universal pathway in which energy is released as glucose is degraded to pyruvic acid, and the reactions comprising it may illuminate its evolutionary past. For instance, when small, molecular reactants were abundant, they would have been the first to be utilized for energy. After these small molecules were depleted, the original pathway could have been extended by metabolizing larger compounds. Whatever the correct explanation, metabolic pathways, because of their sequential nature, could not have arisen randomly but were "predetermined" by available preexisting compounds. Many pathways, and enzymes controlling individual reactions, have been retained because of their selective advantages and are commonly found in living systems. This astonishing similarity cannot be due to chance but to evolution from a common ancestral system.

Eventually, high-energy carbon compounds became less plentiful and some organisms were able to switch to reducing more copious carbon compounds, such as CO_2 by means of a membrane-bound oxidation-reduction system. A revolution occurred when some organisms began to reduce carbon by using light as an energy source (photosynthesis). Such reduction requires a source of electrons, for which substances such as H_2S were first used. It later became advantageous to split water, which was plentiful. Oxygen, a by-product of this process, increased to such an extent that it was exploited to enhance energy yields from glucose breakdown. A new pathway, the Krebs cycle, arose from anaerobic pathways, the entire process being contingent upon the presence of oxygen. Electrons removed from oxidized compounds in the cycle are transferred to a membrane-bound electron transport system for which oxygen acts as the ultimate electron acceptor molecule. Much more energy is released in this process than is released in anaerobic glycolysis.

Prokaryote-type cells arose about 3.5 billion years ago, and by about 2 billion years later some had diversified into eukaryotes. Present prokaryotes composed of the Eubacteria and the Archaebacteria, are small, contain no nuclear membrane or complex organelles, and divide by binary fission. Eukaryotes are generally aerobic and more complex, as they contain numerous organelles and a microtubular apparatus for mitotic cell division. Eukaryotic genes, unlike those of prokaryotes, often have nucleotide sequences within them (introns) which are not translated into peptide sequences. In many respects the Archaebacteria seem intermediate between Eubacteria and eukaryotic cells, but it may be that all three types of cells arose from a single common ancestor, the progenote.

Because of the cell-like characteristics of mitochondria and chloroplasts, many biologists now explain the presence of these eukaryotic organelles as remnants of an ancient symbiotic relationship. The large amount of DNA in eukaryotic cells led to its fission into several chromosomes and to the appearance of microtubule-dependent cell division. Meiosis must also have arisen early, since it is almost universal in eukaryotes. Somewhat later, perhaps just prior to the Phanerozoic era, multicellular eukaryotes appeared and underwent several explosive radiations.

KEY TERMS

active site
active transport
aerobic metabolism
anaerobic glycolysis
anti-oxidant compounds
archaebacteria

autotrophic
backward evolution
banded iron formations
binary fission
Calvin cycle
centromere
cherts
chlorophyll
comparative biochemistry
cyclic photosynthesis
dark reactions
electron acceptors
electron donors
Embden-Meyerhof pathway
endocytosis
eubacteria
eukaryotes
exons
heterotrophic
introns
kinetochore
Krebs cycle
light reactions
lithotrophic nutrition
meiosis
metabolic pathways
mitotic mechanisms
non-cyclic photosynthesis
nuclear envelope
organelles
organotrophic nutrition
oxidative phosphorylation
ozone
phosphorylation
photophosphorylation
polypeptide domain
progenote
prokaryotes
protistans
proton gradient
reduction-oxidation
retrograde evolution
split genes
stromatolites
thylakoids

REFERENCES

Adman, E. T., L. C. Sieker, and L. H. Jensen, 1973. The structure of a bacterial ferredoxin. *J. Biol. Chem.*, **248,** 3987-3996.

Andreadis A., M. E. Gallago, and B. Nadal-Ginard, 1987. Generation of protein isoform diversity by alternative splicing: mechanistic and biological implications. *Ann. Rev. Cell Biol.*, **3**, 207-242.

Barker, H.A., 1972. ATP formation by anaerobic bacteria. In *Horizons of Bioenergetics*, A. San Pietro and H. Gest (eds.). Academic Press, New York, pp. 7-31.

Blake, C. C. F., 1985. Exons and the evolution of proteins. *Int. Rev. Cytol.*, **93**, 149-185.

Broda, E., 1975. *The Evolution of the Bioenergetic Processes*. Pergamon, Oxford.

Cavalier-Smith, T., 1981. The origin and early evolution of the eukaryotic cell. In *Molecular and Cellular Aspects of Microbial Evolution (32nd Symposium of the Society for General Microbiology)*. M. J. Carlile, J. F. Collins, and B. E. B. Moseley (eds.). Cambridge Univ. Press, Cambridge, pp. 33-84.

——, 1987. The origin of cells: A symbiosis between genes, catalysts, and membranes. *Cold Sp. Harb. Symp. Quant. Biol.*, **52**, 805-824.

Cloud, P., 1974. Evolution of ecosystems. *Amer. Sci.*, **62,** 54-66.

Darnell, J. E., 1978. Implications of RNA-RNA splicing in evolution of eukaryotic cells. *Science*, **202**, 1257-1260.

Dickerson, R. E., 1980. The cytochromes: an exercise in scientific serendipity. In *The Evolution of Protein Structure and Function*. D. S. Sigma and M. A. B. Brazier (eds.). Academic Press, New York.

Doolittle, W. F., 1978. Genes-in-pieces, were they ever together? *Nature*, **272**, 581.

Doolittle, W. F., and C. J. Daniels, 1985. Prokaryotic genome evolution: what we may learn from the archaebacteria. In *Evolution of Prokaryotes*, K. H. Schleifer and E. Stackebrandt (eds.). Academic Press, London, pp. 31-44.

Eck, R. V., and M. O. Dayhoff, 1966. Evolution of the structure of ferredoxin based on living relics of primitive amino acid sequences. *Science*, **152,** 363-366.

Fridovich, I., 1975. Oxygen: boon and bane. *Amer. Sci.*, **63,** 54-59.

Gest, H., 1981. Evolution of the citric acid cycle and respiratory energy conversion in prokaryotes. *FEMS Microbiol. Lett.*, **12,** 209-215.

Gest, H., and J. W. Schopf, 1983. Biochemical evolution of anaerobic energy conversion: the transition from fermentation to anoxygenic photosynthesis. In *Earth's Earliest Biosphere: Its Origin and Evolution*, J. W. Schopf, (ed.). Princeton Univ. Press, Princeton, N.J., pp. 135-148.

Gilbert, W., 1979. Introns and exons: playgrounds of evolution. In *Eucaryotic Gene Regulation*, R. Axel, T. Maniatis, and C. F. Fox (eds.). Academic Press, New York, pp. 1-12.

Hall, D. O., J. Lumsden, and E. Tel-Or, 1977. Iron-sulfur proteins and superoxide dismutases in the evolution of photosynthetic bacteria and algae. In *Chemical Evolution of the Early Precambrian*, C. Ponnamperuma (ed.). Academic Press, New York, pp. 191-210.

Herdman, M., and R. Y. Stanier, 1977. The cyanelle: chloroplast or endosymbiotic prokaryote? *FEMS Microbiol. Lett.*, **1,** 7-12.

Horowitz, N. H., 1945. On the evolution of biochemical synthesis. *Proc. Nat. Acad. Sci.*, **31**, 153-157.

Jones, C. W., 1985. The evolution of bacterial respiration. In *Evolution of Prokaryotes*, K. H. Schleifer and E. Stackebrandt (eds.). Academic Press, London, pp. 175-204.

Knoll, A. H., and E. S. Barghoorn, 1977. Archean microfossils showing cell division from the Swaziland system of South Africa. *Science*, **198,** 396-398.

Krebs, H., 1981. The evolution of metabolic pathways. In *32nd Symposium of the Society for General Microbiology*, M. J. Carlile, J. F. Collins, and B. E. B. Moseley (eds.). Cambridge Univ. Press, Cambridge, pp. 215-228.

Lake, J. A., 1987. Prokaryotes and archaebacteria are not monophyletic: rate invariant analysis of rRNA genes indicates that eukaryotes and eocytes form a monophyletic taxon. *Cold Sp. Harb. Symp. Quant. Biol.*, **52**, 839-846.

Lake, J. A., E. Henderson, M. Oakes, and M. W. Clark, 1984. Eocytes: a new ribosome structure indicates a kingdom with a close relationship to eukaryotes. *Proc. Nat. Acad. Sci.*, **81,** 3786-3790.

Lehninger, A. L., 1984. *Biochemistry*, 3rd ed. Worth, New York.

Margulis, L., 1981. *Symbiosis in Cell Evolution.* W. H. Freeman, San Francisco.

Margulis, L., and D. Sagan, 1986. *Origins of Sex: Three Billion Years of Genetic Recombination.* Yale Univ. Press, New Haven, Conn..

Mitchell, P., 1979. Keilin's respiratory chain concept and its chemiosmotic consequences. *Science*, **206,** 1148-1149.

Molecular Origins and Evolution of Photosynthesis, 1981. A special issue of *BioSystems* (Vol. 14, No. 1).

Nagy, L. A., 1974. Transvaal stromatolite: first evidence for the diversification of cells about 2.2×10^9 years ago. *Science*, **183,** 514-515.

Nes, W. R., and W. D. Nes, 1980. *Lipids in Evolution*. Plenum Press, New York.

Palmer, J. D., 1985. Evolution of chloroplast and mitochondrial DNA in plants and algae. In *Molecular Evolutionary Genetics*, R. J. MacIntyre (ed.). Plenum Press, New York, pp. 131-240.

Raff, R. A., and H. R. Mahler, 1975. The symbiont that never was: an inquiry into the evolutionary origin of the mitochondrion. In *Symbiosis (29th Symposium of the Society for Experimental Biology)*. D. H. Jennings and D. L. Lee (eds.). Cambridge Univ. Press, Cambridge, pp. 41-92.

Rao, K. K., D. O. Hall, and R. Cammack, 1981. The photosynthetic apparatus. In *Biochemical Evolution*, H. Gutfreund (ed.). Cambridge Univ. Press, Cambridge, pp. 150-202.

Schopf, J. W., 1978. The evolution of the earliest cells. *Sci. Amer.*, **239** (3), 110-134.

Schopf, J. W., J. M. Hayes, and M. R. Walter, 1983. Evolution of earth's earliest ecosystems: recent progress and unsolved problems. In *Earth's Earliest Biosphere*, J. W. Schopf (ed.). Princeton Univ. Press, Princeton, N.J., pp. 361-384.

Schopf, J. W., and M. R. Walter, 1983. Archean microfossils: new evidence of ancient microbes. In *Earth's Earliest Biosphere: Its Origin and Evolution*, J. W. Schopf (ed.). Princeton Univ. Press, Princeton, N.J., pp. 214-239.

Stanier, R. Y., 1970. Some aspects of the biology of cells and their possible evolutionary significance. *Symp. Soc. Gen. Microbiol.*, **20**, 1-38.

Strickberger, M. W., 1985. *Genetics*, 3rd ed. Macmillan, New York.

Tribe, A., A. J. Morgan, and P. A. Whittaker, 1981. *The Evolution of Eukaryotic Cells*. Edward Arnold, London.

Wald, G., 1974. Fitness in the universe: choices and necessities. In *Cosmochemical Evolution and the Origins of Life*, J. Oró, S. L. Miller, C. Ponnamperuma, and R. S. Young (eds.). D. Reidel Publ. Co., Dordrecht, Netherlands, pp. 7-27.

Weitzman, P. D. J., 1985. Evolution in the citric acid cycle. In *Evolution of Prokaryotes*, K. H. Schleifer and E. Stackebrandt (eds.). Academic Press, London, pp. 253-275.

Whatley, J. M., P. John, and F. R. Whatley, 1979. From extracellular to intracellular: the establishment of mitochondria and chloroplasts. *Proc. Roy. Soc. London Series B*, **204,** 165-187.

Woese, C. R., 1983. The primary lines of descent and the universal ancestor. In *Evolution from Molecules to Men*. D. S. Bendall (ed.). Cambridge Univ. Press, Cambridge, pp. 209-233.

——, 1985. Why study evolutionary relationships among bacteria? In *Evolution of Prokaryotes*, K. H. Schleifer and E. Stackebrandt (eds.). Academic Press, London, pp. 1-30.

——, 1987. Bacterial evolution. *Microbiol. Rev.*, **51**, 221-271.

Woese, C. R., B. A. Debrunner-Vossbrinck, H. Oyaizu, E. Stackebrandt, and W. Ludwig, 1985. Gram-positive bacteria: possible photosynthetic ancestry. *Science*, **229,** 762-765.

Wolfe, S. L., 1981. *Biology of the Cell*, 2nd ed. Wadsworth Publ. Co., Belmont, Calif.

Zubay, G., 1988. *Biochemistry*, 2nd ed. Macmillan, New York.

PART 3

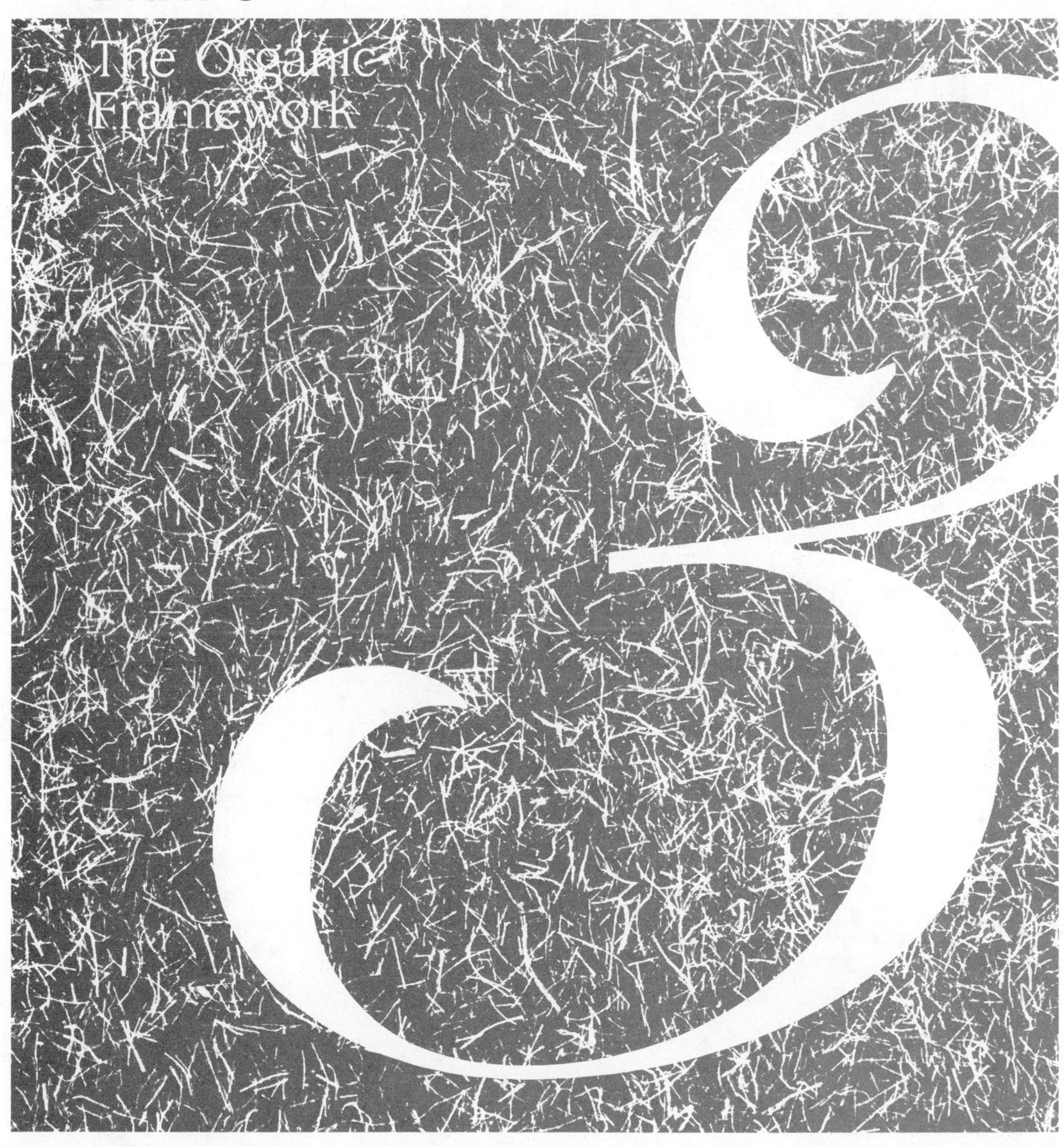

GENETIC CONSTANCY AND VARIABILITY

Evolutionary changes are based on two fundamental aspects of genetics or biological inheritance: **constancy** and **variability.** Constancy resides in the genetic observation that like produces like and derives from the ability of nucleic acid macromolecules to replicate (Chapter 7). The evolutionary significance of constancy is that all living processes, from biochemistry to behavior, depend upon the transmission of constant or reliable information from previous generations. On the other hand, variability resides in the genetic observation that like also produces unlike and derives from the fact that the replication of biological information is not always constant or exact, thus producing inherited biological changes (**mutations**). The obvious evolutionary significance of genetic variability is that it provides the fuel for evolution by enabling organisms to become different from their ancestors. This chapter offers a brief review of some fundamental concepts of genetic constancy and variability.

CELL DIVISION

In modern organisms the transmission of biological information is coordinated with cellular division so that both parental and daughter cells carry copies of the same information. Such cell division processes must have originated early in the history of life as a solution to the problem of how membrane-enclosed organisms could grow and expand without enlarging themselves to the point of endangering their existence.[1] Since biological information is coded in the form of long-chained nucleic acid molecules, it is basic to cell division that the structures that bear these molecules, the **chromosomes,** must replicate and divide.

In prokaryotes possessing a single chromosome, the two products of chromosomal replication attach to different points on the cell membrane. As the cell elongates to form two daughter cells, the two chromosome products separate, each becoming enclosed in a separate daughter cell. In eukaryotes, more than one chromosome is usually present, and more complex cell division processes occur, divided into **mitosis** (cell di-

[1]As a cell grows in the absence of cell division, its volume, which is dependent on the cube of its radius, increases proportionately greater than its surface area, which is dependent on the square of the radius. The inner constituents of non-dividing cells with increasing radius would thus develop considerable difficulties in both obtaining and excreting metabolic substances because they have relatively less surface area for each increase in volume.

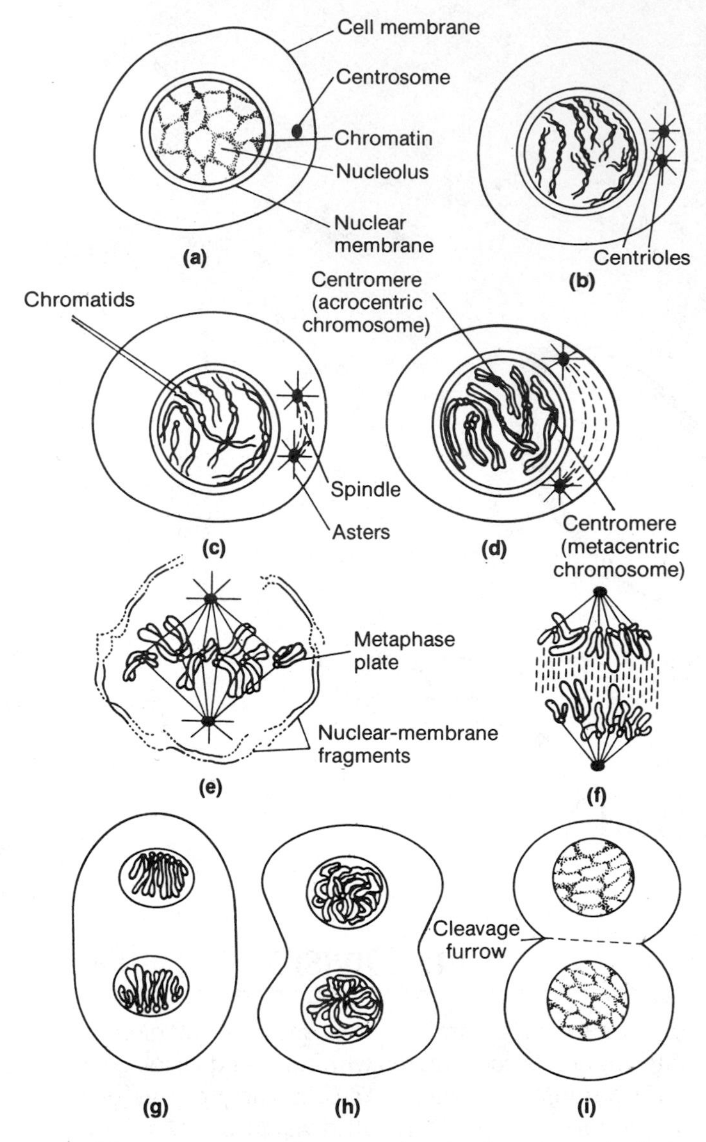

FIGURE 10-1 Diagrammatic presentation of various stages of mitosis in a somatic cell. (a) Interphase; (b, c, and d) early, medium, and late prophase stages showing increased thickening and condensation of chromosomes; (e) metaphase; (f) anaphase. During the telophase stages (g, h, and i) a nuclear membrane re-forms around each polar group of daughter chromosomes, and these chromosomes then revert to the more extended interphase state. Division of the cytoplasm (cytokinesis) is also completed during this final mitotic stage. Note in (d) that each chromosome has its centromere placed at a constant linear position: at the center for **metacentric** chromosomes, off-center for **acrocentric** chromosomes, and at or near the tip for **telocentric** chromosomes.

vision of somatic or body tissues) and **meiosis** (cell division of gamete-producing tissues).

In mitosis, the parental and daughter cells have exactly the same numbers and kinds of chromosomes, and it is this form of cell division that provides all the body cells of an organism with the same chromosome constitution, or **karyotype.** As illustrated in Figure 10-1, mitosis first becomes obvious during the **prophase** stage in which the gradually condensing chromosomes have replicated so that each chromosome is now composed of two **chromatids** connected together at their centromeres (Fig. 10-1(b), (c), (d)). These chromosomes next arrange themselves on a **metaphase** plate in which the two members of a pair of chromatids are connected to spindle fibers that go to opposite poles (Fig. 10-1(e)). Thus, when chromatid separation occurs at the **anaphase** stage (Fig. 10-1(f)), each daughter cell receives one of the chromatid replicates (which can now be considered an individual chromosome) derived from each of the original chromosomes in the parental cell. If, for example, a parental cell has four chromosomes, A^1, A^2, B^1, and B^2, mitosis will divide these into chromatids A^1-A^1, A^2-A^2, B^1-B^1, and B^2-B^2, to produce two daughter cells, each of the chromosome constitution (karyotype) A^1, A^2, B^1, and B^2.

In meiosis, the cell division process performed by eukaryotes engaged in sexual reproduction, gametes are formed that contain only one representative of each kind of chromosome. For example, a gamete containing two kinds of chromosome, A^1 and B^1, will fertilize a gamete containing the same two kinds of chromosome, A^2 and B^2, to form a zygote that now has one pair of each, A^1A^2 B^1B^2. If these events occur in an organism whose life cycle is primarily **diploid** (i.e., cells possessing two sets of chromosomes), mitosis will replicate the A^1A^2 B^1B^2 karyotype in every somatic cell, until gametes are formed by meiosis in the male and female gonadal tissues. Each such gamete will now contain one of the **haploid** karyotypes (i.e., only one set of chromosomes), either A^1B^1, A^1B^2, A^2B^1, or A^2B^2. If the zygote is formed in an organism whose life cycle is primarily haploid, the meiotic process will produce non-gametic cells carrying one of these four haploid karyotypes. In both cases, whether haploid or diploid, meiosis generally follows the stages illustrated in Figure 10-2 and is based primarily on close **homologous pairing** between similar kinds of chromosomes called homologues, such as between $A^1 \leftrightarrow A^2$ and between $B^1 \leftrightarrow B^2$. Pairing between such homologues enables them to separate (disjoin) from each other at the end of the first meiotic division and provides each gamete with a representative of each kind of chromosome when **disjunction** is normal. A comparison between mitosis and meiosis is diagrammed in Figure 10-3.

Sexual reproduction and its accompanying meiotic divisions provide two important sources of variability. First, because of the phenomenon of **recombination** or **genetic exchange** (also called **crossing over**), sections of homologous chromosomes can exchange material, thereby forming different linear arrays of nucleotides. For example, chromosome A^1, bearing the hypothetical nucleotide sequence AGC AGC AGC . . . , can exchange material during meiosis with chromosome A^2, AGC TCG TCG . . . , to yield products that are part A^1 and part A^2, such as AGC TCG AGC. . . .

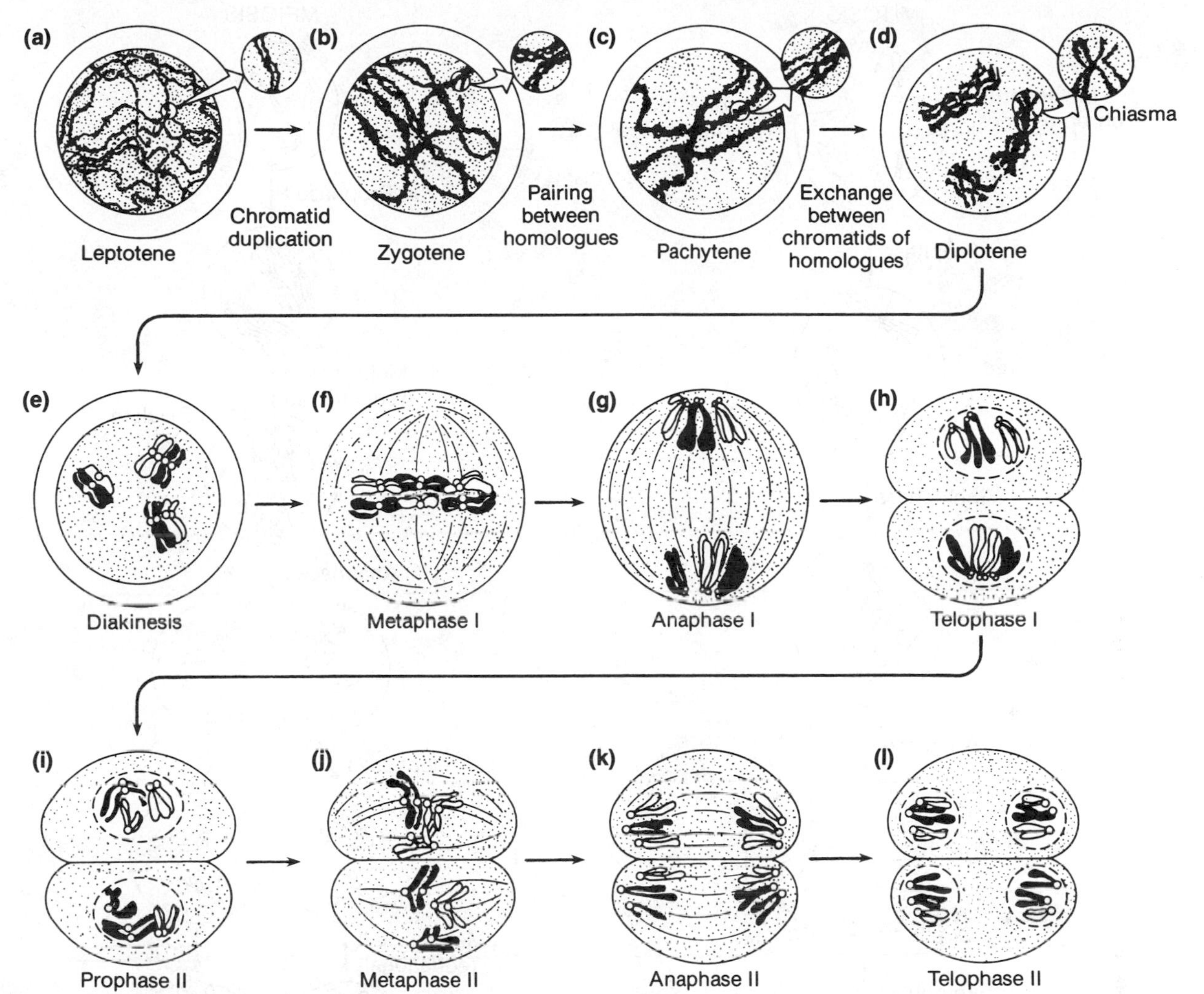

FIGURE 10-2 Principle stages of meiosis, showing the first (I) and second (II) meiotic divisions. (a - e) Pairing between homologous chromosomes during the various prophase I stages. Each chromosome has replicated to form two sister chromatids, which are then paired with the two sister chromatids of their homologue to form a group of four chromatids, or **bivalent**. At metaphase I (f) these paired groups are arranged on an equatorial plate preparatory to their separation. At anaphase I (g) the two homologous chromosomes separate, each taking their chromatids with them to opposite poles. A telophase stage (h) may then follow, which involves some form of cytokinesis. In many organisms this first meiotic division is a reduction division because it reduces the number of homologues in a nucleus from two to one. Thus, if there are differences between homologous chromosomes A^1 and A^2, the resultant nuclei after meiosis I will not be identical, since each will have only one of these chromosomes. The next meiotic division (II) that follows in such organisms (i - l) separates the two chromatids of each chromosome, yielding two nuclei for each nucleus formed during division I. A diploid cell undergoing meiosis will therefore produce four haploid cells, each of which carries one representative of a homologous pair of chromosomes. The four haploid products of a sperm-producing cell (spermatogonium) can all function as gametes, while only one of the haploid products of an egg-producing cell (oogonium) functions as a gamete. (Adapted from Wolfe.)

In addition, as the number of pairs of homologous chromosomes increases, the meiotic process assorts these into increasingly numerous varieties of chromosome combinations. Thus, as previously described, two homologous pairs of chromosomes (A^1 A^2, B^1 B^2) segregating in a cross can yield four different kinds of gametes (A^1B^1, A^1B^2, A^2B^1, A^2B^2), whereas segregation of three pairs of homologous chromosomes (A^1 A^2, B^1 B^2, C^1 C^2) can yield eight different gametic combinations ($A^1B^1C^1$, $A^1B^1C^2$, $A^1B^2C^1$, $A^1B^2C^2$, $A^2B^1C^1$, $A^2B^1C^2$, $A^2B^2C^1$, $A^2B^2C^2$). If we extend this to four pairs of chromosomes, there are 16 different possible gametic combinations, the rule being that the number of possible different kinds of gametes is equal to 2^n, where n represents the number of pairs of chromosomes undergoing meiosis. In humans with normal karyotypes of 23 chromosome pairs, the different kinds of gametes that can be formed by this simple chromosome assortment process alone is 2^{23}, or more than 8 million varieties.

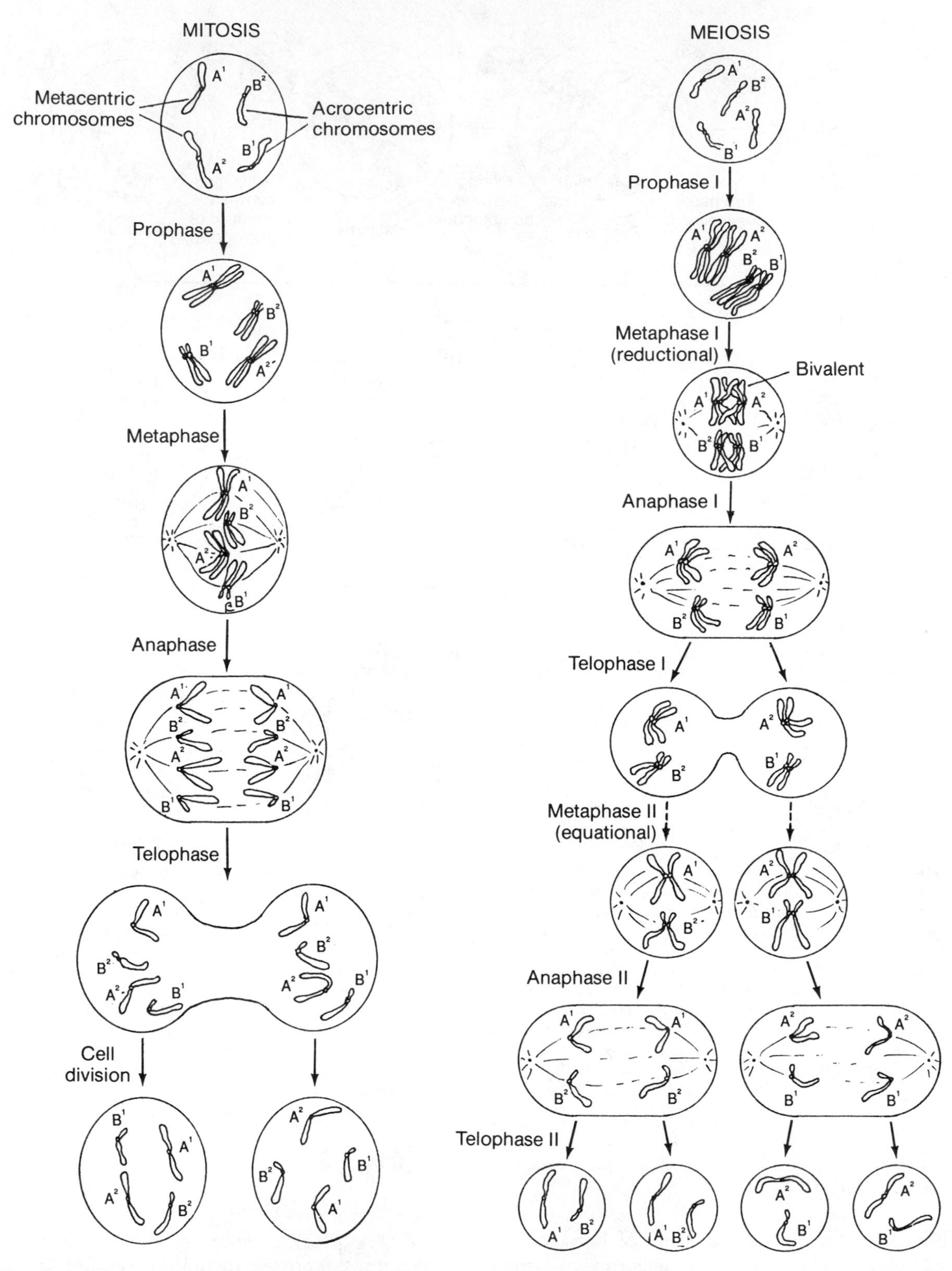

FIGURE 10-3 Diagrammatic comparison of mitosis and meiosis, using an organism which has two pairs of homologous chromosomes, A^1 and A^2, and B^1 and B^2. Replication of each chromosome into two chromatids occurs during prophase of each type of cell division, but they differ in respect to their time of separation. In mitosis, the four chromosomes arrange themselves individually on the metaphase plate, and the anaphase division separates the two sister chromatids of each chromosome so that each pole receives an identical group of four chromosomes. In meiosis, the four chromosomes pair up as two homologous pairs (rather than as individual chromosomes) which appear during metaphase I as two bivalents. Anaphase I of meiosis separates each chromosome member of a homologous pair, but the two sister chromatids of each chromosome are not separated from each other until anaphase II.

MENDELIAN SEGREGATION AND ASSORTMENT

Because an understanding of the molecular basis of genetics was long difficult to achieve, fundamental genetic principles were primarily derived from observations on transmission of more visible and obvious biological characteristics. The earliest genetic laws derived from such observations were discovered by Gregor Mendel in Brno, Czechoslovakia, in the early 1860's, although his work did not become generally known until the beginning of the twentieth century.

As described in Chapter 3, the pre-mendelian view of heredity that prevailed throughout the nineteenth century was that heredity followed a blending process in which offspring inherited a dilution, or blend, of parental characteristics. It was Mendel's exceptional contribution to prove that biological characteristics were inherited by means of discrete units, later called **genes**, that remained undiluted in the presence of other genes.

Mendel demonstrated this concept by using a number of characters in the pea plant, *Pisum sativum*, in which each character possessed two alternative appearances, or traits, such as smooth or wrinkled seeds, yellow or green seeds, tall or short plants, and so on. In his experiments, he counted the appearance of each different trait among the individuals in every generation, then analyzed these numerical results in terms of ratios that led to two fundamental principles of heredity.

1. The Principle of **Segregation.** When two pure-breeding parental stocks are intercrossed which differ in respect to a character such as seed shape, for example, smooth × wrinkled seeds, the first filial generation, or F_1, will carry the genes for each of these traits. Since the pea plant is diploid, an individual possesses a pair of such genes for seed shape, as well as pairs of genes that govern other characters such as seed color, plant size, and so on. In modern terminology, the two individual members of a particular gene pair are known as **alleles**. For simplicity, we can designate alleles by alphabetical letters, for example, the smooth and wrinkled alleles are, respectively, *S* and *s*. These alleles then segregate in the gametes of the F_1 hybrid which then unite to form a second filial generation (F_2) in predictable proportions. As shown in Figure 10-4, these F_2 proportions arise from the fact that the allele for one trait (e.g., *S*) has **dominant** effects over the other **recessive** allele (*s*) when both are present together in *Ss* individuals. That is, *Ss* peas appear smooth (as are *SS* peas), and the only wrinkled peas are *ss*. Individuals that carry two different alleles for any particular character (e.g., *Ss*) are designated as **heterozygotes**, and individuals who carry two identical alleles (e.g., *SS*, *ss*) are **homozygotes**. Mendel's experiments showed that these alleles had obviously not been changed or blended in the heterozygote but segregated from each other to be transmitted as discrete and constant particles between generations.

2. The Principle of **Independent Assortment.** In crosses involving two different characters, such as seed shape and seed color, Mendel found that the results were predictable if genes that determined one character (e.g., seed shape) had no effect on the segregation of genes for the other character (e.g., seed color). That is, genes for different characters segregated independently of each other. As shown in Figure 10-5, the F_1 of a cross differing in two such characters, smooth yellow × wrinkled green, shows only the dominant phenotypes, but the F_2 proportions are predictably 9/16 smooth yellow, 3/16 smooth green, 3/16 wrinkled yellow, and 1/16 wrinkled green. These predictions are based on the fact that each of the four kinds of gametes produced by the F_1 arise from an independent association between a seed shape allele (probability 1/2) and a seed color allele (probability 1/2), so that each different gamete has a frequency of $1/2 \times 1/2 = 1/4$. The explanation for such independent assortment turned out to be the localization of the genes for each of the two characters to a different nonhomologous pair of chromosomes (Fig. 10-6): how one pair of homologous chromosomes assort themselves towards opposite poles during meiosis has no effect on how a nonhomologous pair assort themselves.

DOMINANCE RELATIONS AND MULTIPLE ALLELES

The particular ratios that Mendel observed in his diploid pea plants (3:1, 9:3:3:1, etc.) held true only as long as the genes involved showed complete dominance and recessiveness and possessed only two alleles for each character. After 1900 it was quickly realized that dominance relations between alleles could range quite widely, and a gene determining a particular character could be represented by three or more alleles. For example, the condition of **incomplete dominance** can be expressed in flower color of some plants in which the homozygote G^1G^1 produces red

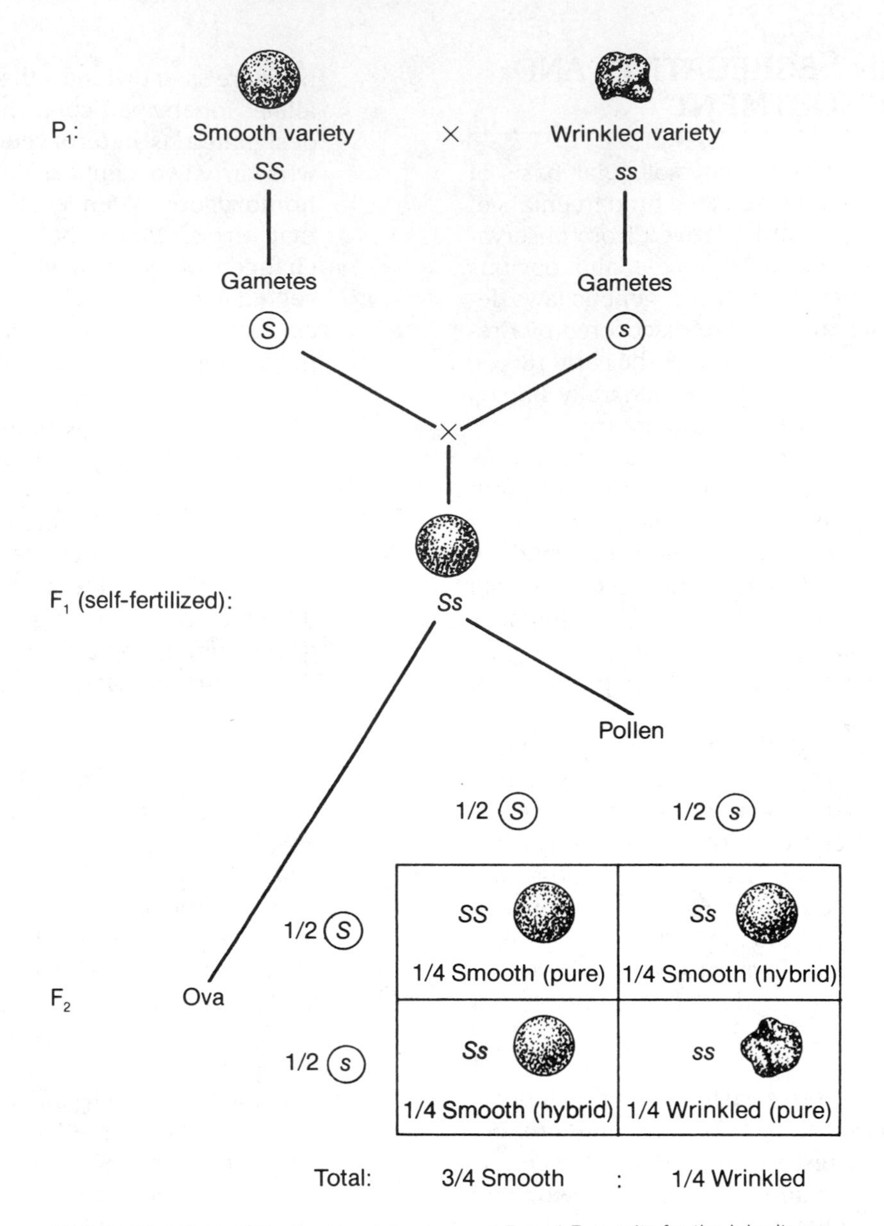

FIGURE 10-4 Explanation for Mendel's observed F_1 and F_2 results for the inheritance of the seed-shape character in garden peas, based on the segregation of alleles *S* (dominant) and *s* (recessive).

flowers, the homozygote G^2G^2 produces white flowers, and the heterozygote G^1G^2 produces an intermediate shade of pink flowers. Or, on the other hand, both alleles G^1 and G^2 are considered **codominant** when they have distinguishable effects in the heterozygote; that is, they each produce a uniquely recognizable substance, such as a special sugar compound, so that the heterozygote G^1G^2 possesses both compounds.

Any gene may also possess more than two alleles, $G^1, G^2, G^3, \ldots$ producing, for example, a different color or different compound in each different kind of homozygote, G^1G^1, G^2G^2, $G^3G^3, \ldots$. The source for such **multiple allelic** systems arises from the fact that a gene is composed of hundreds or thousands of nucleotides, and an allelic difference may be caused by only a single nucleotide change. Moreover, there may be different dominance relationships between the alleles in such a system so that G^1, for example, produces a codominant effect with G^2, but alleles G^1 and G^2 act as though they are each completely dominant over G^3. In addition, interactions between genes in different allelic systems may occur so that the expression of allele G^1, for example, is changed because of the presence of alleles at a different gene pair, such as H^1 or H^2. Among such interactions are those which can modify the dominance relations of a particular allele so that the effect of G^1, for example, is dominant over G^2 in certain genotypes (e.g., $H^1I^2J^1K^3$) but is recessive when the background genotype is changed (e.g., $H^2I^1J^3K^2$). Complications of this kind have been given the name

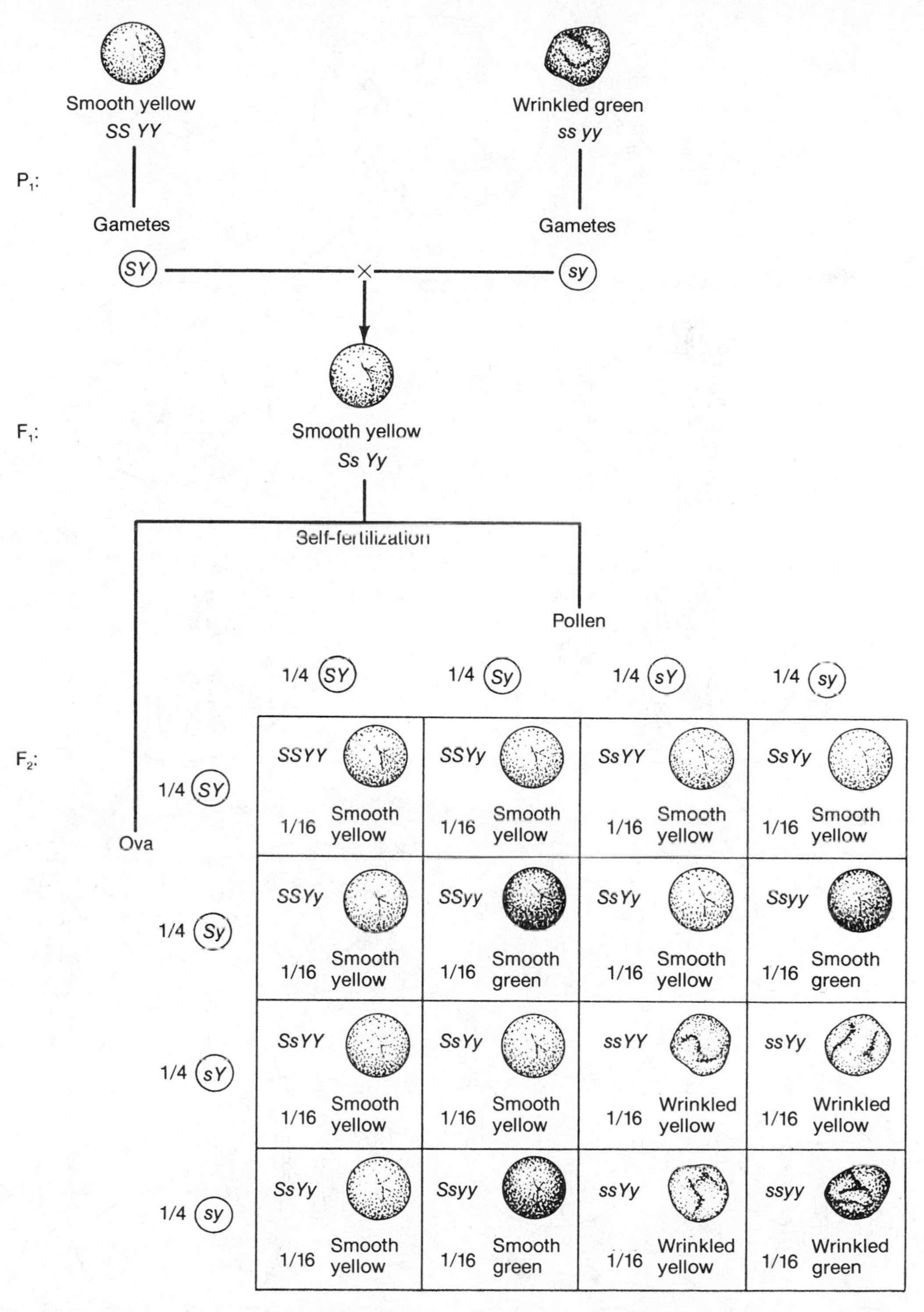

FIGURE 10-5 Explanation of Mendel's results for the segregation and assortment of alleles at two pairs of genes; *S* and *s* for seed shape, and *Y* and *y* for seed color.

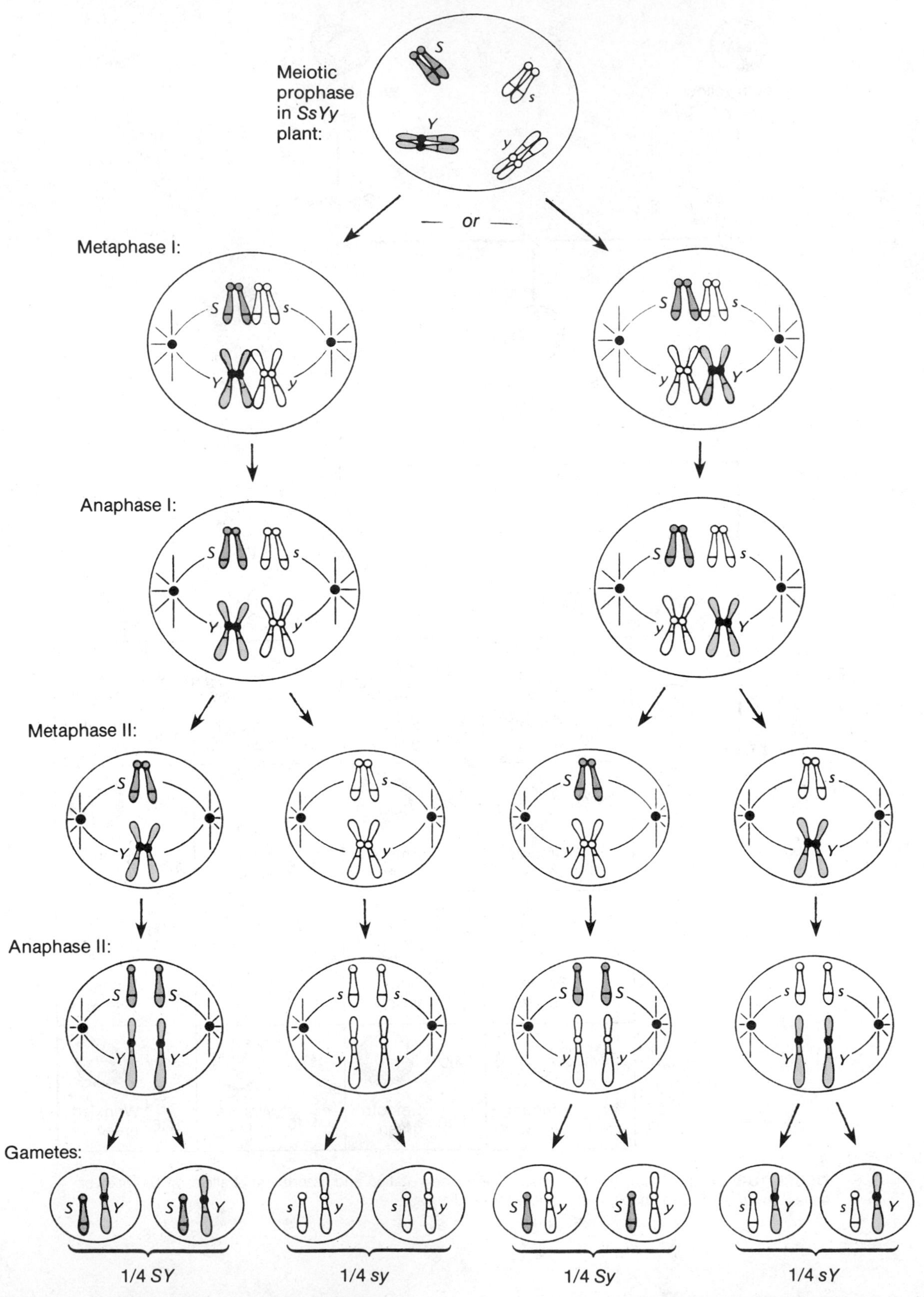

FIGURE 10-6 Explanation for the segregation and independent assortment of seed shape and seed color in Mendel's experiments in terms of factors or genes (*S*, *s*, and *Y*, *y*) localized on different chromosomes. Because of independent assortment, all four possible combinations of chromosomes in the gametes correspond to all four possible combinations of genetic factors. (Adapted from Strickberger.)

epistatic interactions, to describe when one or more gene pairs change the effect caused by other gene pairs. Another term, **modifiers**, is used when the effects of genes on other genes can be measured quantitatively.

In general, most common, or **wild type**, alleles have been evolutionarily selected to be dominant in diploid organisms because they produce advantageous products in the presence of other alleles whose products may not be as advantageous or may even be deleterious. On the molecular level, such wild type allelic products are mostly functional proteins such as enzymes, whereas the products of mutant alleles often appear to be non-functional, only partly functional, or even absent (**null alleles**).

Because there are so many possible allelic differences, as well as different dominance and interaction effects, it is easy to see that the variability generated by these means is far beyond that described previously for the independent assortment of chromosomes during meiosis. To take a simple example, a gene with four alleles (G^1, G^2, G^3, G^4) can produce ten different possible diploid genotypes (G^1G^1, G^1G^2, G^1G^3, G^2G^1, G^2G^2, etc.), and 100 such genes can produce 10^{100} possible genotypic combinations, a figure larger than the estimated number of protons and neutrons in the universe. Since any cellular organism carries more than 100 genes, each with many possible alleles, its potential genetic variability is probably numerically greater than any conceivable aspect of reality.

SEX LINKAGE

Early in the twentieth century the discovery was made that various genes could be localized to a particular chromosome associated with sex determination. A prominent example of this is the "bleeder" disease, hemophilia, localized to the X chromosome of humans and other mammals. In these organisms males and females differ in respect to a pair of sex chromosomes so that males are XY and females are XX. Thus males (the **heterogametic sex**) produce sperm that contain either X or Y chromosomes, whereas females (the **homogametic sex**) produce only X-bearing eggs. If the proportion of X- and Y-bearing sperm are equal, fertilization restores the two sexes in equal frequency: ½ X-bearing sperm x X-bearing eggs → ½ XX females; ½ Y-bearing sperm x X-bearing eggs → ½ XY males. However, since males carry only a single X chromosome, and the Y chromosome is mostly inactive, alleles present on the X chromosome will express their effects in males, although such alleles may be recessive when in females. Thus a single hemophilia-producing allele on the X chromosome of a **hemizygous** male causes the classic hemophilia disease, whereas two such alleles are necessary to cause hemophilia in XX females.

Interestingly, of the more than 100 such sex-linked genes now identified in humans, many are also known to be sex-linked in other mammals (Table 10-1). The conservation of the same genes on different mammalian X chromosomes indicates that a large part of this chromosome has been conserved throughout mammalian evolution, at least for 90 million years. Ohno, in fact, has proposed that any sex-linked gene found in one mammalian species can be expected to be sex-linked in other mammals as well.

TABLE 10-1 Some Gene Products and Diseases Linked to the X Chromosome in Both Humans and Other Mammals

X-linked in Humans	X-linked in Other Mammals
α-galactosidase deficiency	Chimpanzee, gorilla, sheep, cattle, pig, rabbit, hamster, mouse
Anhidrotic ectodermal dysplasia	Cattle
Bruton-type agammaglobulinemia	Mouse
Copper transport deficiency	Mouse
Duchenne/Becker muscular dystrophy	Mouse, dog
Glucose-6-phosphate dehydrogenase deficiency	Chimpanzee, gorilla, sheep, cattle, pig, horse, donkey, hare, hamster, mouse, kangaroo
Hemophilia A (factor VIII defiency)	Dog
Hemophilia B (factor IX defiency)	Dog
Hypoxanthine-guanine phosphoribosyl transferase (Lesch-Nyhan syndrome)	Chimpanzee, gorilla, horse, hamster, dog, mouse
Ornithine transcarbamylase defiency	Mouse
Phosphoglycerate kinase	Chimpanzee, gorilla, horse, hamster, mouse, kangaroo
Steroid sulfatase deficiency (ichthyosis)	Mouse
Testicular feminization syndrome	Cattle, dog, mouse, rat
Vitamin D–resistant rickets	Mouse
Xg blood cell antigen	Gibbon

Adapted from Strickberger.

LINKAGE AND RECOMBINATION

The localization of different genes to the X chromosome, first demonstrated in the fruit fly, *Drosophila melanogaster*, provided the opportunity for establishing distances between such genes, or **linkage** relationships. Such determinations arose from crossing over or recombinational events in which exchanges occurred between the chromatids of paired homologous chromosomes. For example, the experiment diagrammed in Figure 10-7 showed that two *Drosophila* sex-linked genes, those involved in *white* eyes (w^+ and w) and *miniature* wings (m^+ and m), recombined with a frequency of about 38 percent to produce new chromosomal combinations.

This frequency of recombination provided a measure of the **linkage distance** between genes on the same chromosome: the greater the recombination frequency, the greater the distance. Thus, since the recombination frequency between the genes for *white* eyes and *cut* wings was about half the frequency of that between *white* and *miniature*, the linkage distance between *white* and *miniature* could be assumed to be about twice that between *white* and *cut*. Such experiments with both sex-linked and nonsex-linked (**autosomal**) genes provided linkage maps such as that shown in Figure 10-8, indicating that chromosomes are composed of linear arrays of genes, or **loci**, whose relative positions are additive (if three genes are linked in the order *H-G-I*, then the *H-I* linkage distance is the sum of *H-G* + *G-I* distances).

In humans and other mammals, techniques involving somatic cell fusions rather than mating between sex cells have been used to obtain linkage relationships. These somatic hybridization techniques have been extraordinarily successful, and a large number of genes are now localized to many of the 23 human chromosomes. Moreover, like the finding of similar sex-linked genes in mammals, genes localized to particular human chromosomes (identified by special banding patterns; see Fig. 10-17) can be found localized to similar chromosomes in other mammals (Fig. 10-9).

In bacteria and viruses, recombinational techniques are considerably more advanced than in eukaryotes because of the very small size of these organisms, their rapid generation times, and the ease with which biochemical mutant genes can be identified. These advantages have led to elaborately detailed linkage maps such as that shown in Figure 10-10. In addition, new methods that permit exact determinations of nucleotide sequences in both prokaryotes and eukaryotes are now providing information beyond anything conceived even a decade ago. All these studies indicate the immense variability that can be generated through recombination: any chromosome may come to differ from its homologues by carrying a distinctive combination of alleles, for example, ...$G^1H^2I^1J^3K^2$... vs. ...$G^2H^1I^3J^2K^1$... vs. ...$G^1H^1I^3J^3K^1$..., and so on.

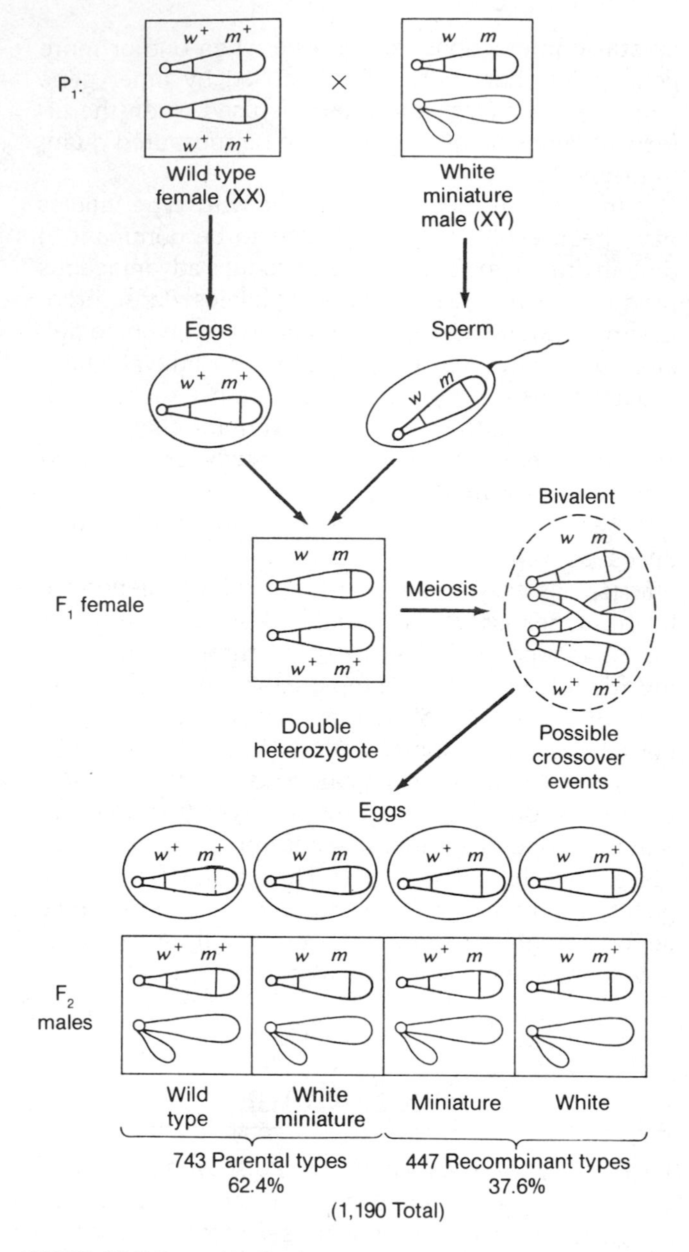

FIGURE 10-7 Recombination between the sex-linked genes *white* and *miniature* in *D. melanogaster* as evidenced by recombinant classes among F_2 males. Since the only sex chromosome contributed by the paternal parent to male offspring is the Y, the phenotypes of F_2 males in respect to *white* and *miniature* are strictly determined by the meiotic events that occur in the double heterozygous F_1 female. As shown in this illustration, these meiotic events involve crossovers between chromatids at the four-chromatid, or bivalent, stage (see Fig. 10-3). The frequency of observed recombinants for any two given linked genes is thus dependent upon the frequency in which such meiotic crossovers occur between the two genes. (Adapted from Strickberger.)

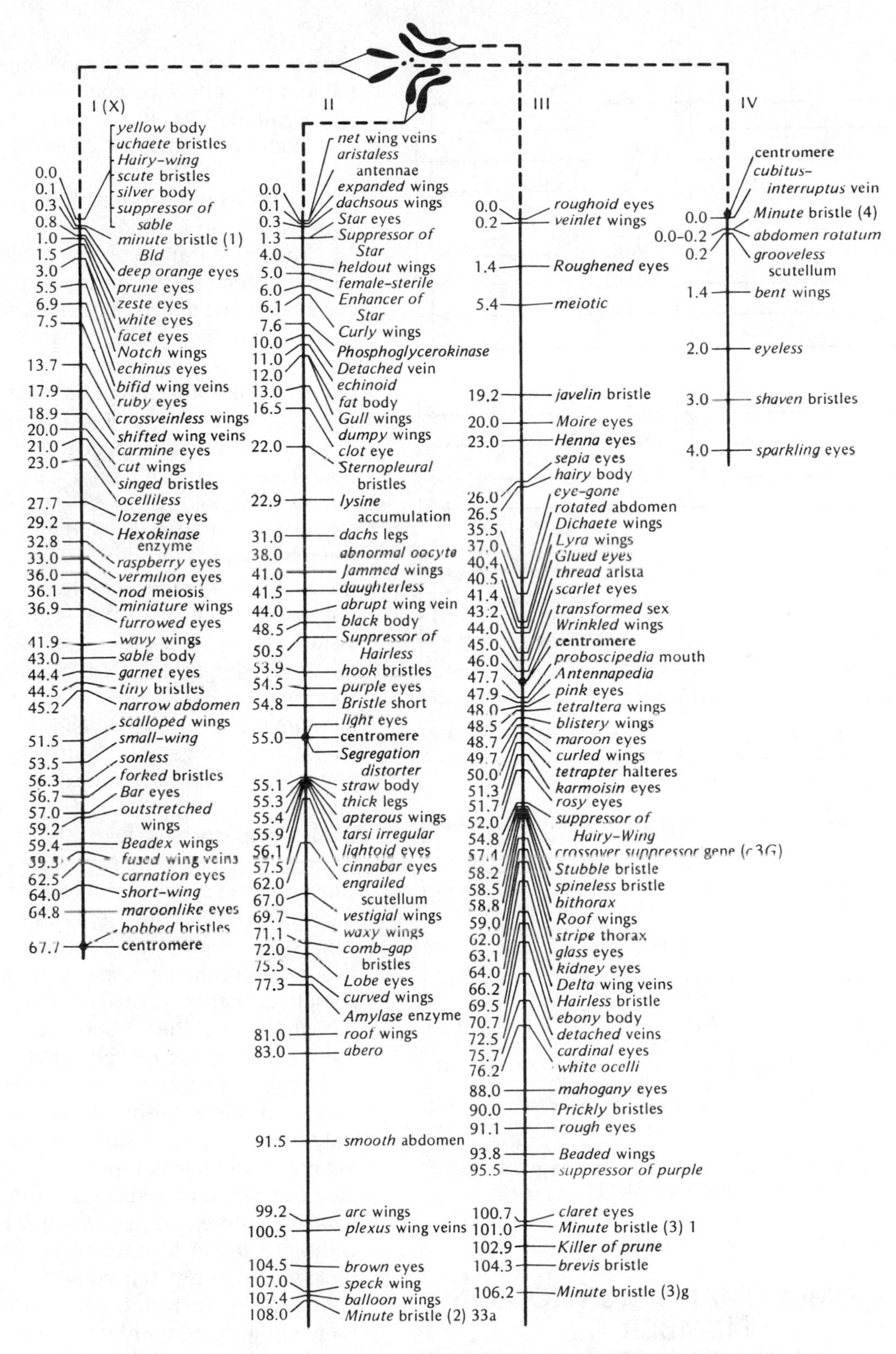

FIGURE 10-8 Linkage map of some of the important genes in the four chromosomes of *D. melanogaster*. Note that a variety of different genes may affect a specific character such as eye color, wing shape, and bristles, indicating that many steps exist in the development of a particular function, each step governed or capable of being modified by separate and different genes. (From Strickberger.)

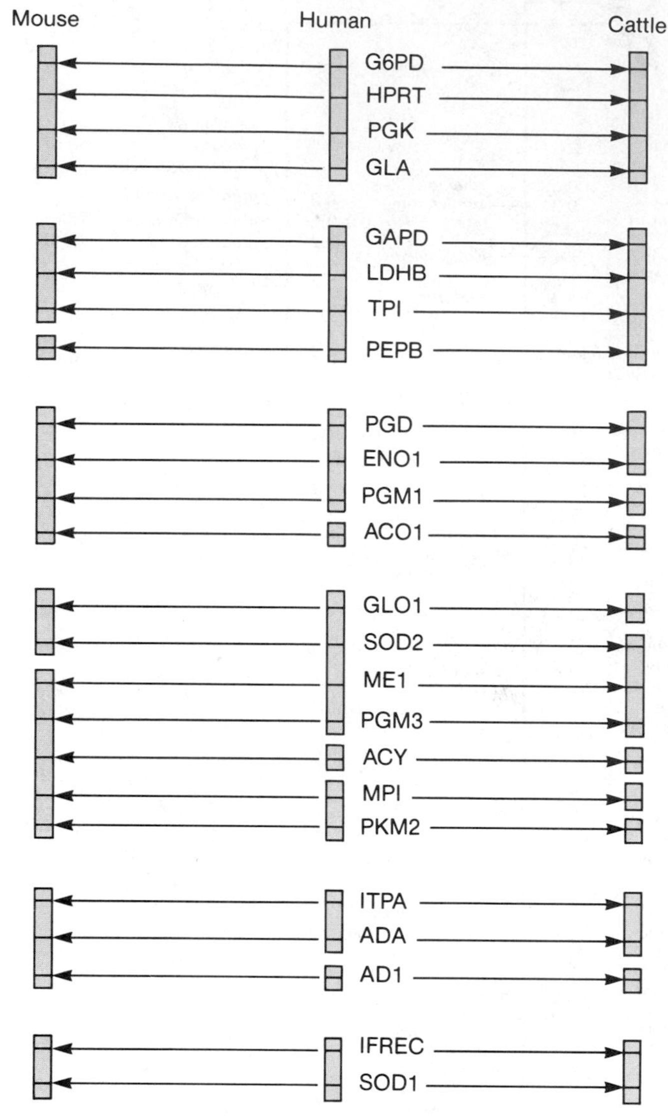

FIGURE 10-9 Linkage relationships between 24 human gene loci (center column) compared to those of similar genes in mice (left) and cattle (right). Each vertical rectangular block of genes indicates a chromosome in the respective species. For example, the top rectangles in each column give the relative positions of four genes on the X chromosome of each species. Note that the linkage order for many gene loci are the same in each of the three species, indicating that a large number of linkage relationships have been evolutionarily conserved. (Adapted from Womack and Moll.)

CHROMOSOMAL VARIATIONS IN NUMBER

Forms of variation that are often microscopically observable are changes in the number or structure of chromosomes. The numerical chromosome variations are of two major kinds: changes in the number of entire sets of chromosomes, **euploid variations**; and changes in the number of single chromosomes within a set, **aneuploid variations**. Since most sexually reproducing eukaryotes are diploids with two sets of chromosomes (2n), euploid variations may extend from the haploid or monoploid condition (1n) to various levels of **polyploidy** (3n, 4n, ...) as shown in Table 10-2.

Such events may be caused by the fertilization of an egg by more than one sperm, or by cell division failures in which, for example, a diploid rather than haploid gamete is produced.

The appearance of extra sets of chromosomes within a species itself, **autopolyploidy**, seems to be a common mode of evolution in many plant groups such as mosses, apples, pears, bananas, tomatoes, and corn. Polyploids that originate from crossing between different species, **allopolyploids**, are also found in some plants such as wheat. For example, a gamete from species *A* may fertilize species *B* to produce the diploid hybrid *AB* which can then undergo polyploidy to form the allotetraploid *AABB*, or other variations shown in Figure 10-11. Both auto- and allopolyploids have been produced with chemicals such as colchicine which break down spindle fiber microtubules and thereby interfere with chromosome segregation during cell division.

Interestingly, the first laboratory-created species was the product of a cross between two tobacco plants, *Nicotiana tabacum* (a diploid with 48 chromosomes whose haploid gametic number, n, was 24) and *N. glutinosa* (24 chromosomes, n = 12). As shown in Figure 10-12, the hybrid of this cross was sterile although it possessed two sets of chromosomes, one from each species. In such cases, sterility arises because an even division of chromosomes during meiosis depends upon pairing between homologous chromosomes, but these homologues are absent when the two cross-fertilizing species have evolved chromosomal differences between them. Nevertheless, although sterile, the hybrid plant could continue to grow and increase by vegetative cuttings. Eventually, a chromosome-doubling event occurred that produced an allopolyploid (also called amphidiploid) that was fully fertile because each chromosome now had a pairing mate, and normal meiosis could take place. Thus, at one stroke, a new species with 72 chromosomes, *N. digluta*, was created that was only fertile when crossed with itself (all of its chromosomes could undergo homologous pairing) but could not produce fertile offspring when crossed to either of the parental species (only some of its chromosomes could pair homologously with those of either parent).

In animals, polyploidy is a much rarer event than in plants because most animals show much greater developmental sensitivity to even a small change in chromosome number. This sensitivity extends also to chromosomal sex-determining mechanisms, so that animal polyploids face difficulties in maintaining the same proportions of X and Y chromosomes present in normal diploids. For example,

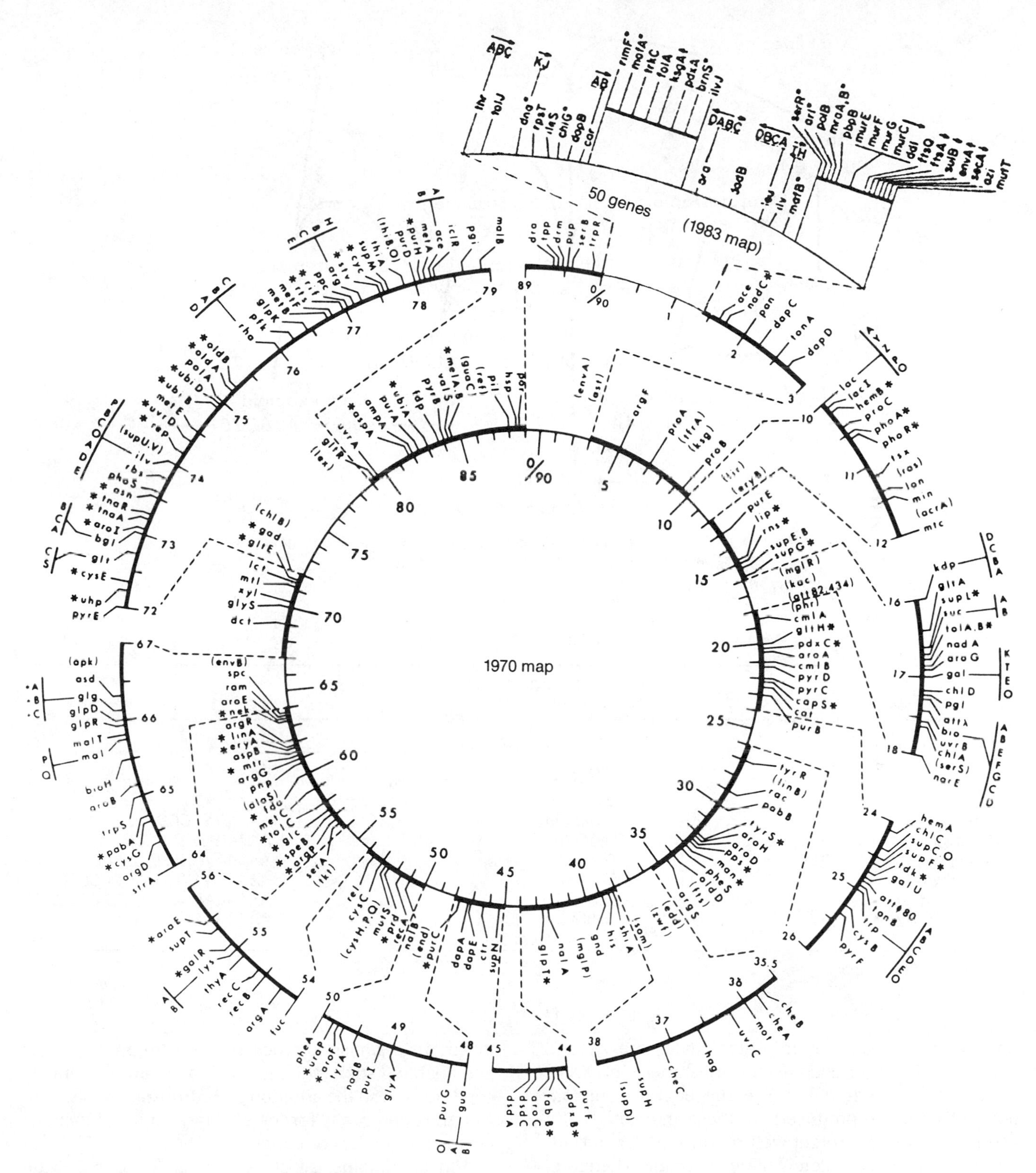

FIGURE 10-10 An early (1970) linkage map of the *E. coli* bacterial chromosome. This map has been continually enlarged as new genes have been discovered and localized. For example, in the upper right corner is a small section, about one-fiftieth of the genome, to which 18 genes were localized in 1970 and 50 genes localized in 1983. (From Bachmann.)

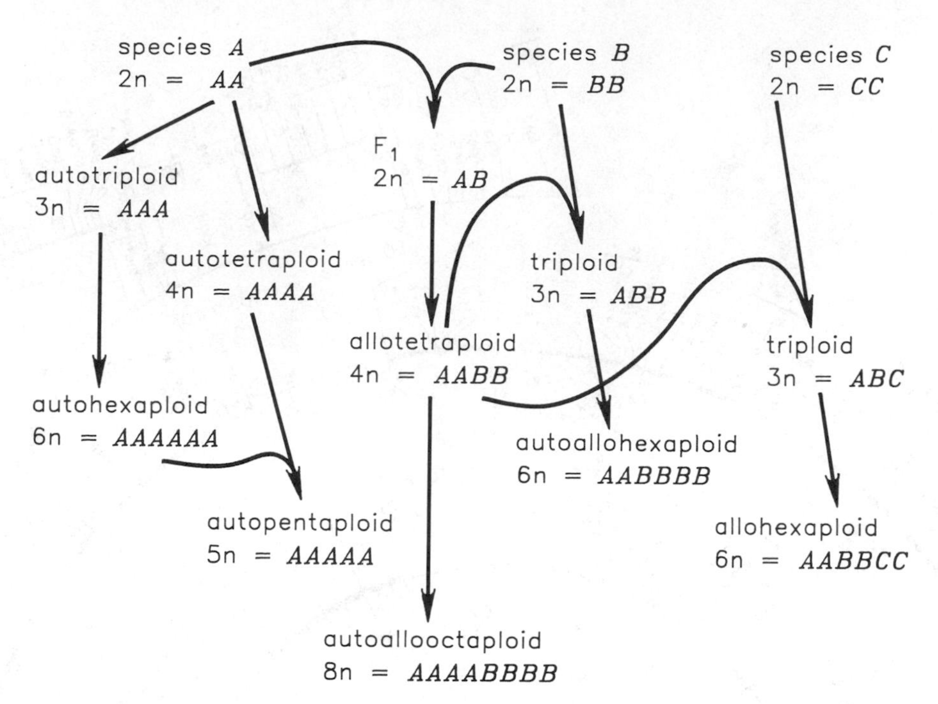

FIGURE 10-11 Terminologies used for different polyploids and some of the pathways by which they originate. (Adapted from Strickberger.)

TABLE 10-2 Euploid Variations, Involving Entire Sets of Chromosomes

Euploid Type	Number of Homologues Present for Each Chromosome	Example
Haploid or monoploid	One (1n)	*A B C*
Diploid	Two (2n)	*AA BB CC*
Polyploid	More than two	
Triploid	Three (3n)	*AAA BBB CCC*
Tetraploid	Four (4n)	*AAAA BBBB CCCC*
Pentaploid	Five (5n)	*AAAAA BBBBB CCCCC*
Hexaploid	Six (6n)	*AAAAAA BBBBBB CCCCCC*
Septaploid	Seven (7n)	*AAAAAAA BBBBBBB CCCCCCC*
Octaploid, etc.	Eight (8n), etc.	*AAAAAAAA* . . . , etc.

Adapted from Strickberger.

polyploidy in a species in which males are XY and females XX could lead to establishment of XXYY males and XXXX females. The subsequent combinations of gametes produced by these individuals (XY sperm + XX eggs) might well produce XXXY individuals that are not completely male or female. Therefore, when animal polyploid species do occur, they usually possess some form of **asexual reproduction** such as **parthenogenesis** (embryonic development of eggs without fertilization).

Nevertheless, there are probably some advantages in polyploidy. In both plants and animals, the extra polyploid chromosomes may act as multiple buffers in various organismic processes, thus improving vigor and enabling such individuals to face new and drastic conditions. Also the additional chromosomes may provide the opportunity for the evolution of new functions for their extra sets of genes.

In aneuploids, as shown in Table 10-3, a wide range of variations may result when chromosomes are either added or subtracted from a normal set. Such events may occur because meiotic disjunction between homologous chromosomes is abnormal (**nondisjunction**). Again, plants seem to be more tolerant of such chromosomal variations than animals, but aneuploids are found in both groups.

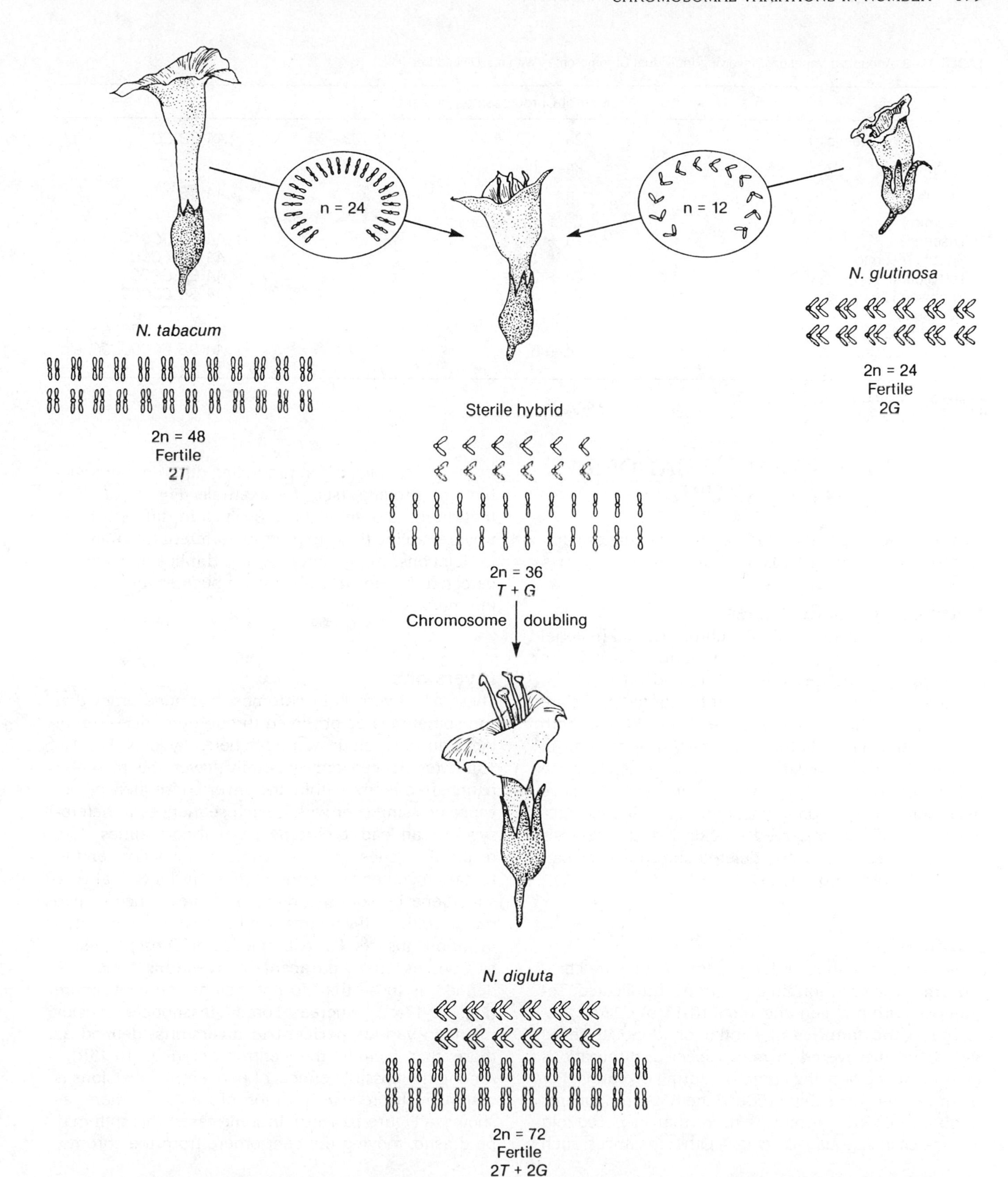

FIGURE 10-12 Flowers and karyotypes of *N. tabacum*, *N. glutinosa*, their sterile diploid hybrid, and the allotetraploid (amphidiploid) formed by chromosome doubling in the sterile hybrid.

TABLE 10-3 Aneuploid Variations, Involving Individual Chromosomes Within a Diploid Set

Type	Number of Chromosomes Present	Example
Disomic (normal diploid)	2n	*AA BB CC*
Monosomic	2n − 1	*AA BB C*
Nullisomic	2n − 2	*AA BB*
Polysomic	Extra chromosomes	
Trisomic	2n + 1	*AA BB CCC*
Double trisomic	2n + 1 + 1	*AA BBB CCC*
Tetrasomic	2n + 2	*AA BB CCCC*
Pentasomic	2n + 3	*AA BB CCCCC*
Hexasomic	2n + 4	*AA BB CCCCCC*
Septasomic	2n + 5	*AA BB CCCCCCC*
Octasomic, etc.	2n + 6, etc.	*AA BB CCCCCCCC*, etc.

Adapted from Strickberger.

CHROMOSOMAL VARIATIONS IN STRUCTURE

A variety of changes in chromosome structure are diagrammed in Figure 10-13.

Deletions or Deficiencies

These terms describe losses of chromosomal material (Fig. 10-13(a)). In general, the severity of a deletion depends upon how extensive it is and on what nucleotides or genes are missing. If functional genes are involved, deletions can be quite deleterious in both diploids and haploids but not necessarily harmful in polyploids or aneuploids where such genes may be present in extra chromosomes. Various deletions have been detected in populations, and many of these produce cytologically observed "buckles" during meiosis when the deleted and non-deleted chromosomes pair up closely in heterozygotes.

Duplications

These are either short or long segments of extra chromosome material originating from duplicated sequences within a genome (Fig. 10-13(b)). Since numerous **gene families** of similar or identical genes have been discovered in many species, duplications show evidence of being common during evolution. For example, the genes that produce the RNA components of ribosomes are present as more than one 100-fold duplicates in various eukaryotes. Other instances, such as the genes involved in producing different hemoglobin-type proteins (see, for example, Fig. 12-6) show that duplicated genes have evolved in different pathways, enabling them to perform different functions. As in deletions, many chromosome duplications can be detected in heterozygotes by the buckles formed during meiotic pairing.[2]

Inversions

These are reversals in chromosomal gene order that can sometimes be observed through the formation of loops during meiotic pairing in heterozygotes (Fig. 10-13(c), (d)). Inversions generally lower the recombination frequency within the inverted sequence, because crossing over within such sequences in heterozygotes can lead to chromosomal abnormalities. As a result, the genes included within an inversion tend to remain together as a nonrecombinant block, called a supergene by some workers. The prevalence of two major kinds of inversions has been well documented in various insects. For example, some *Drosophila* species possess many **paracentric inversions**, which are defined as those that do not include the centromere (Fig. 10-13(c)); whereas some grasshopper species maintain various **pericentric inversions,** defined as those that include the centromere (Fig. 10-13(d)). Among the possible effects of pericentric inversions is a shift in the relative position of the centromere, as shown in Figure 10-13(d). In some cases, this shift may be drastic, moving the centromere from the chromo-

[2]Among the mechanisms that produce duplications is **unequal crossing over**, in which homologous pairing during recombination is slightly askew, thereby producing a crossover product that contains extra chromosomal material (see Fig. 12-5).

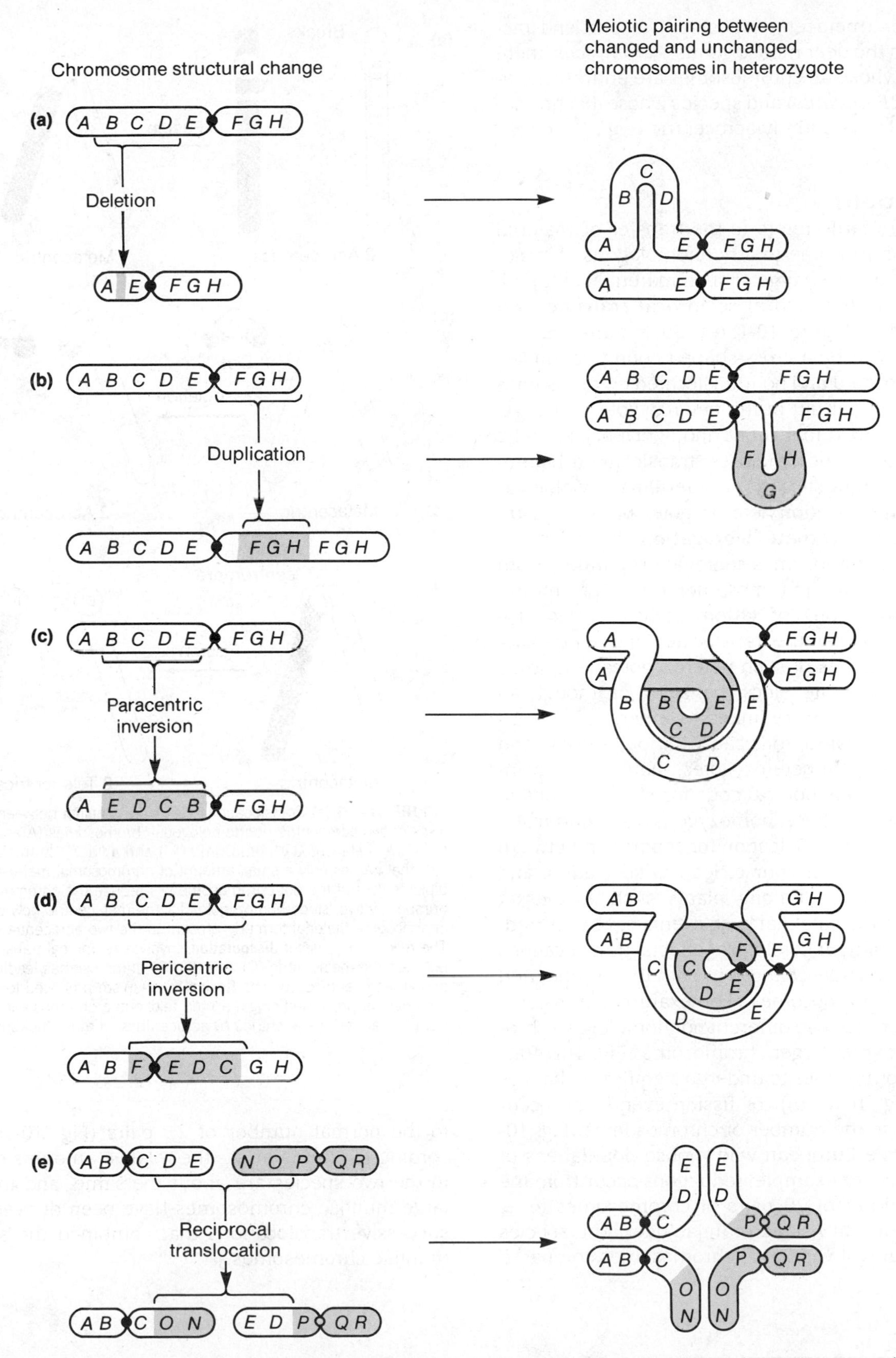

FIGURE 10-13 Major kinds of structural chromosomal changes and their effects on meiotic pairing in heterozygotes who carry both changed and unchanged homologues. (For diagrammatic simplicity, the double chromatid structure of each chromosome is omitted, although it is understood that meiotic pairing between two homologues actually involves four chromatids.)

somal center (metacentric position) to one end (acrocentric). In the deer mouse genus, *Peromyscus*, there are species whose 48 chromosomes are entirely metacentric (e.g., *P. collatus*) and species whose 48 chromosomes are almost entirely acrocentric (e.g., *P. boylei*).

Translocations

This term primarily refers to the transfer of material from one chromosome to a nonhomologous chromosome. When the exchange of such material is mutual, it can result in the kind of **reciprocal translocation** diagrammed in Figure 10-13(e). Such translocations are recognizable by a cross-shaped configuration between translocated and nontranslocated chromosomes during meiotic pairing in the heterozygote. Since gametes containing duplications and deficiencies can be produced during such meioses, translocation heterozygotes often show sterility.[3] The fertile and viable gametes that translocation heterozygotes produce result primarily from **alternate segregation**, in which the translocated chromosomes segregate separately from the nontranslocated chromosomes (these are the diagonal combinations of chromosomes in Fig. 10-13(e)). Thus genes in the translocated and nontranslocated chromosomes of such heterozygotes tend to be inherited as separate blocs, behaving as though all genes on each bloc were linked together.

Because translocations cause various sterility and fertility problems in heterozygotes, there is an advantage for individuals homozygous for translocations to be isolated from those homozygous for nontranslocated chromosomes. Selection for separation between these groups may, of course, lead to speciation, and there are examples among plants such as *Clarkia* (Lewis) which indicate that such events have occurred.

In terms of their cytological effects, translocations may obviously cause changes in both the number and structure of chromosomes. For example, translocations may combine two different nonhomologous chromosomes into one larger chromosome (Fig. 10-14(a)) or cause chromosomes to undergo significant changes in shape (Fig. 10-14(b)), or fission events can occur which increase the number of chromosomes (Fig. 10-14(c)). In some European wild mouse populations of *Mus musculus*, for example, reductions occur from the standard number of 20 pairs of chromosomes to as few as 11 pairs. In Asiatic muntjac deer, one species has only 3 pairs of very large chromosomes compared to the normal number of 25 pairs (Fig. 10-15). According to Liming et al., the relative amounts of DNA in the two species are about the same, and the very large muntjac chromosomes have been derived from successive translocations that combined the smaller muntjac chromosomes.

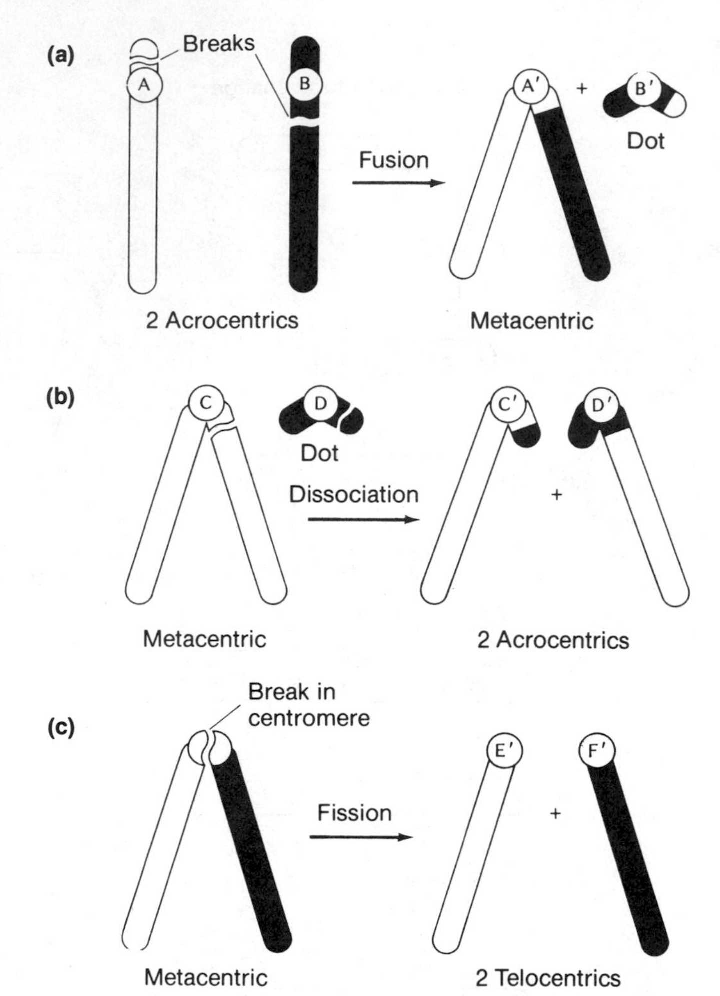

FIGURE 10-14 (a) Translocations that lead to **fusion** between the arms of two acrocentric nonhomologous chromosomes (A and B) to form one metacentric chromosome (A′) and a "dot" chromosome (B′) that carries only a small amount of chromosomal material. In diploids, loss of the dot chromosome will reduce the chromosome number by two, since homozygotes for the metacentric now carry the chromosome material formerly present in the two acrocentrics. (b) The reverse process of **dissociation**, involves reciprocal translocations between the metacentric (C) and dot (D) chromosomes leading to C′ and D′ acrocentrics. (c) The **fission** mechanism proposed to explain the origin of presumed single-armed telocentric chromosomes (E′, F′) from two-armed metacentrics or acrocentrics. (From Strickberger.)

[3]Note that the gametes produced by the translocation heterozygote illustrated in Figure 10-13(e) would have such duplications and deficiencies if the two upper chromosomes went to one pole and the two lower chromosomes went to the other pole.

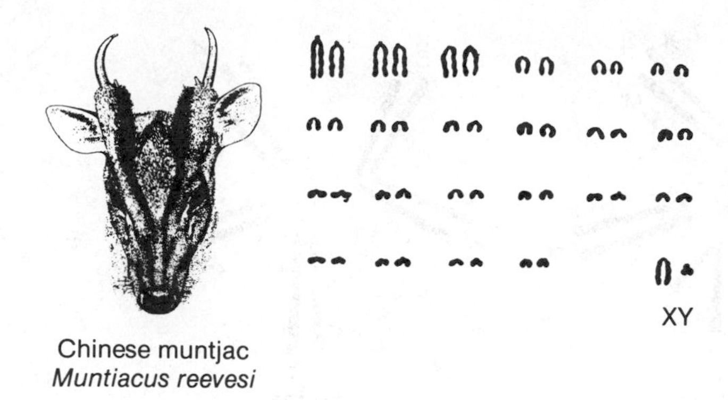

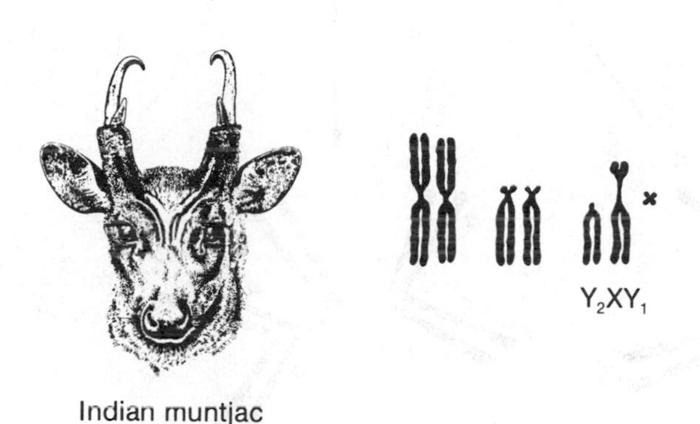

FIGURE 10-15 The Chinese and Indian muntjac deer and their karyotypes. The Indian muntjacs, with two pairs of autosomes and three sex chromosomes, have the lowest known chromosome number of any mammal. (Adapted from Austin and Short.)

CHROMOSOMAL EVOLUTION IN *DROSOPHILA* AND PRIMATES

In those instances where linear sections of chromosomes can be identified through distinctive bandings, chromosomal evolution can be charted in great detail. In *Drosophila* species and other dipteran insects, the chromosomes of salivary gland cells and other tissues have been replicated many times over, yet each replicate still remains closely apposed to other such replicates within the same nucleus. As a result, such **polytene chromosomes** are tremendously enlarged and show highly detailed banding configurations along their lengths. These distinctive banding arrangements enable even minor chromosomal changes to be traced. Thus practically all of the chromosomal changes that have occurred in the evolution of hundreds of species of these flies have been described, a sample of which is shown in Figure 10-16.

However, although polytene chromosomes are absent in many organisms, new chromosomal staining techniques have been devised which enable detailed comparisons between even relatively small mammalian chromosomes. For example, the G-banding technique illustrated in Figure 10-17 allows a comparison of human, chimpanzee, gorilla, and orangutan chromosomes. These bandings show that some chromosomes (nos. 6, 13, 19, 21, 22, and X) are practically identical in all four species, and various arms or sections of other chromosomes are homologous throughout. The changes that have occurred during the evolution of these primates can be accounted for by the simple chromosomal variations described above. Thus the difference in number between humans (n = 23) and apes (n = 24) derives from a fusion event that combined the two indicated chimpanzee-type chromosomes to form the no. 2 human-type chromosome.

GENE MUTATIONS

Gene mutations, or **point mutations,** have been defined as mutations that are not observable at the chromosomal levels discussed previously; that is, they are presumed to affect the nucleotide structure of the gene itself. At present, many gene mutations can be discerned either directly by modern procedures of nucleotide sequencing or indirectly by their resultant effects on the amino acid sequences of proteins.

Among the various kinds of mutational changes at the molecular level are **base substitutions**, a term meaning simply nucleotide changes that involve substitution of one base for another. As shown in Figure 10-18, these are called **transitions** when exchanges occur either between purines (A↔G) or between pyrimidines (T↔C) and **transversions** when purines are exchanged for pyrimidines or vice versa (A, G↔T, C). Some such changes may occur spontaneously through copying errors caused, for example, by rare tautomeric nucleotide base changes that enable complementary pairing between adenine and cytosine or between guanine and thymine (Fig. 10-19). Other base substitutions may arise from the action of mutagenic agents such as hydroxylamine, nitrous acid, and nitrogen mustards. In addition, there are instances in which specific nucleotide sequences cause increased mutations among adjacent nucleotides. Such "hot spots of mutation" may act by coiling the DNA molecule in ways which influence DNA polymerase enzymes to produce replication errors. There are also differences in the replication accuracy of polymerase enzymes: some strains carry enzymes that are apparently more prone to produce mutational errors than others.

Other mutational events at the molecular level include nucleotide deletions, duplications, and insertions. These can arise spontaneously within the cell or be evoked by externally applied mutagenic agents such as acridine dyes. Rearrangements of nucleotides also occur, such as inversions (reversals in nucleotide

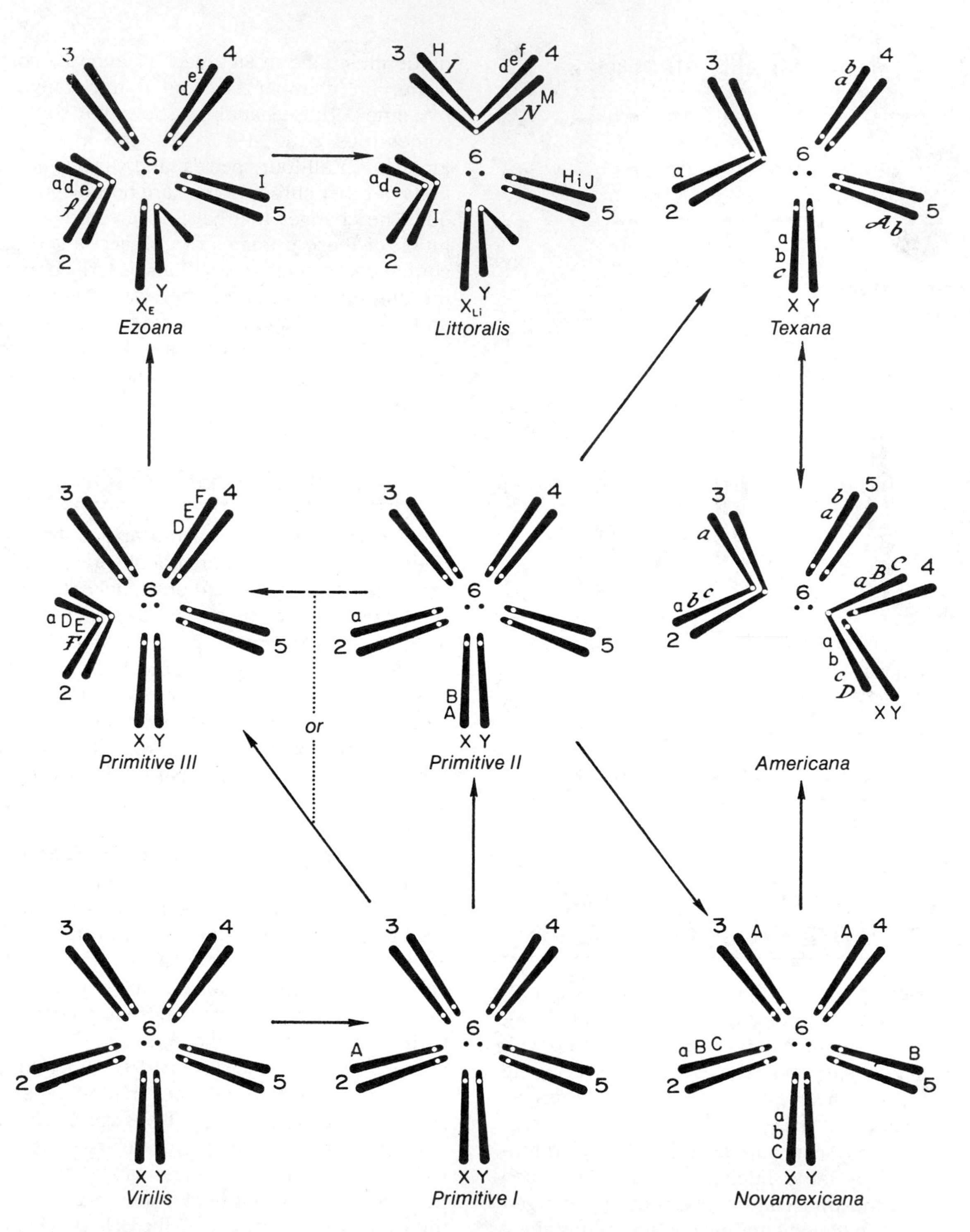

FIGURE 10-16 Paths of chromosomal evolution in some species of the *virilis* group of *Drosophila*. The chromosomes of what was probably the original karyotype of the genus *Drosophila* (lower left) are numbered from 1(X) to 6, and specific chromosomal banding arrangements are indicated by letters. (Adapted from Stone.)

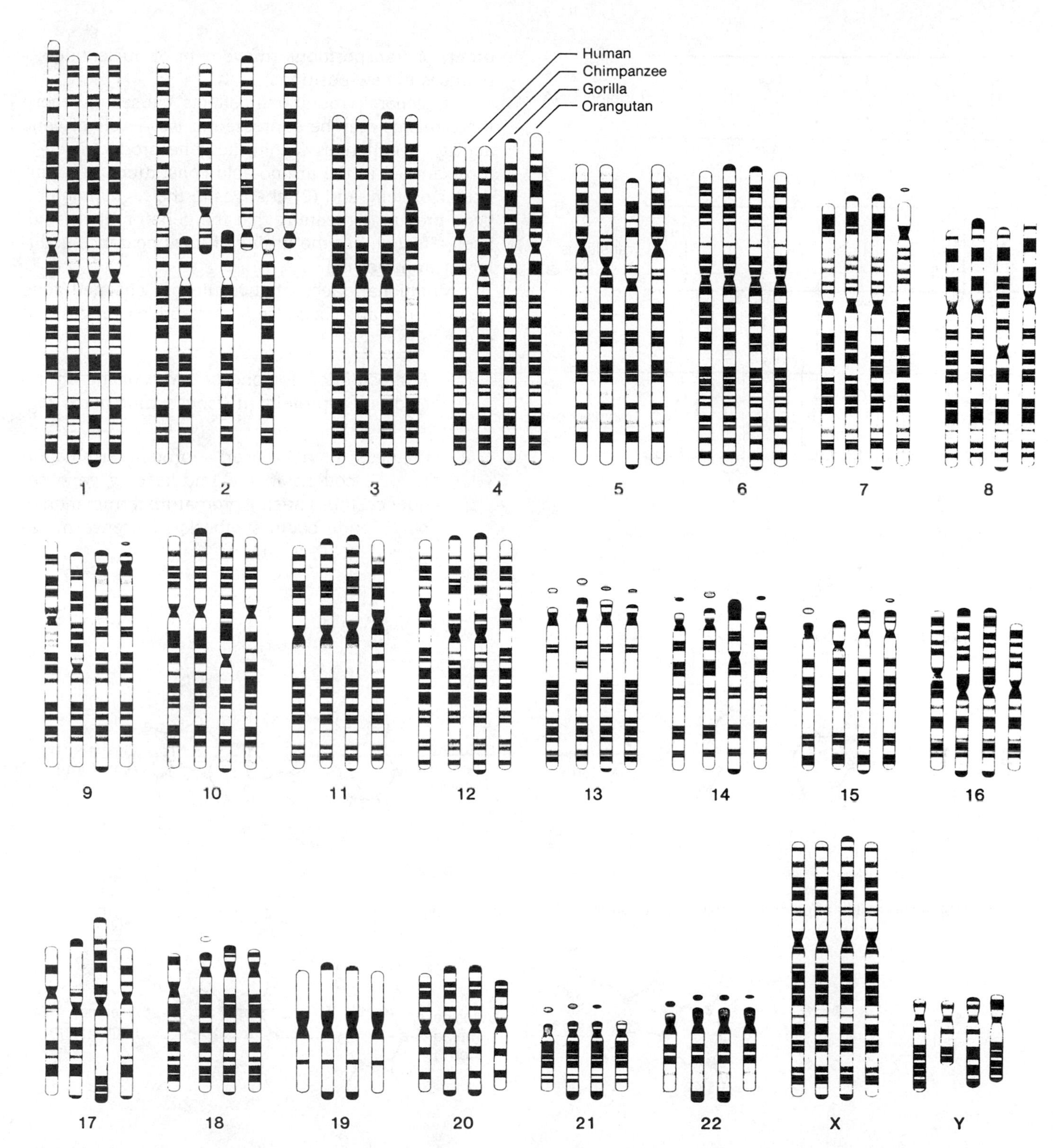

FIGURE 10-17 Banding arrangements of the chromosomes of humans, chimpanzees, gorillas, and orangutans, in respective order from left to right for each chromosome. Note that the 24 pairs of chromosomes in the great apes are reduced to 23 pairs in humans (nos. 1 to 22 + XY) because of fusion of two different chromosomes into a single no. 2 human chromosome. This fusion event along with some other changes (e.g., inversions in chromosomes 1 and 18) must have taken place some time after the human line separated from a human-chimpanzee common ancestor. On the whole, these banding arrangements indicate that humans have a closer evolutionary relationship to chimpanzees than to gorillas and a more distant one to orangutans. (From Yunis and Prakash.)

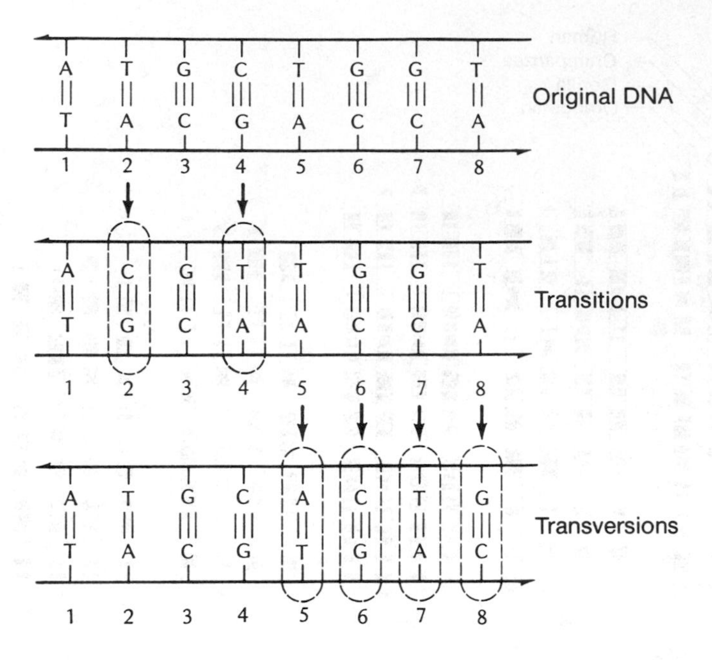

FIGURE 10-18 Examples of specific base-pair changes in a section of double-stranded DNA. (Adapted from Strickberger.)

order) or transpositions (movement of nucleotide sequences to new positions).

In general, mutational effects caused by these mechanisms may be expressed at two levels of gene activity: (1) changes within the gene product itself—for example, in the amino acid constitution of a particular protein; and (2) changes in the regulation of a gene product—meaning that the gene product is not itself affected, but the timing of its appearance is different from normal.

Mutational effects that result in a changed gene product may arise because of nucleotide changes that cause:

1. A substitution for one or more of the amino acids in a protein (missense mutations, Fig. 10-20(a)).
2. Changes that insert protein-termination ("stop") codons in the middle of a gene sequence, thus causing premature termination of polypeptide chain synthesis (nonsense mutations, Fig. 10-20(b)).

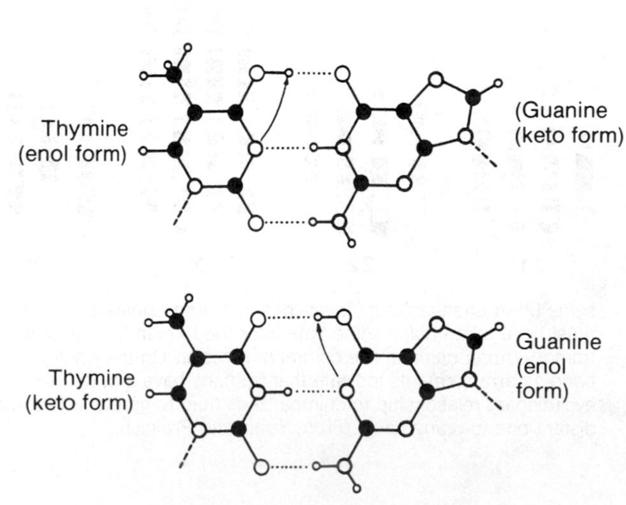

FIGURE 10-19 (a) Normal complementary pairing between nucleotide bases during DNA replication. (b) Modified base-pairing relationships that result from *tautomeric* molecular changes. Because of such changes, base substitutions can occur which produce, for example, transitions from T-A base pairs to C-G base pairs. (Adapted from Drake.)

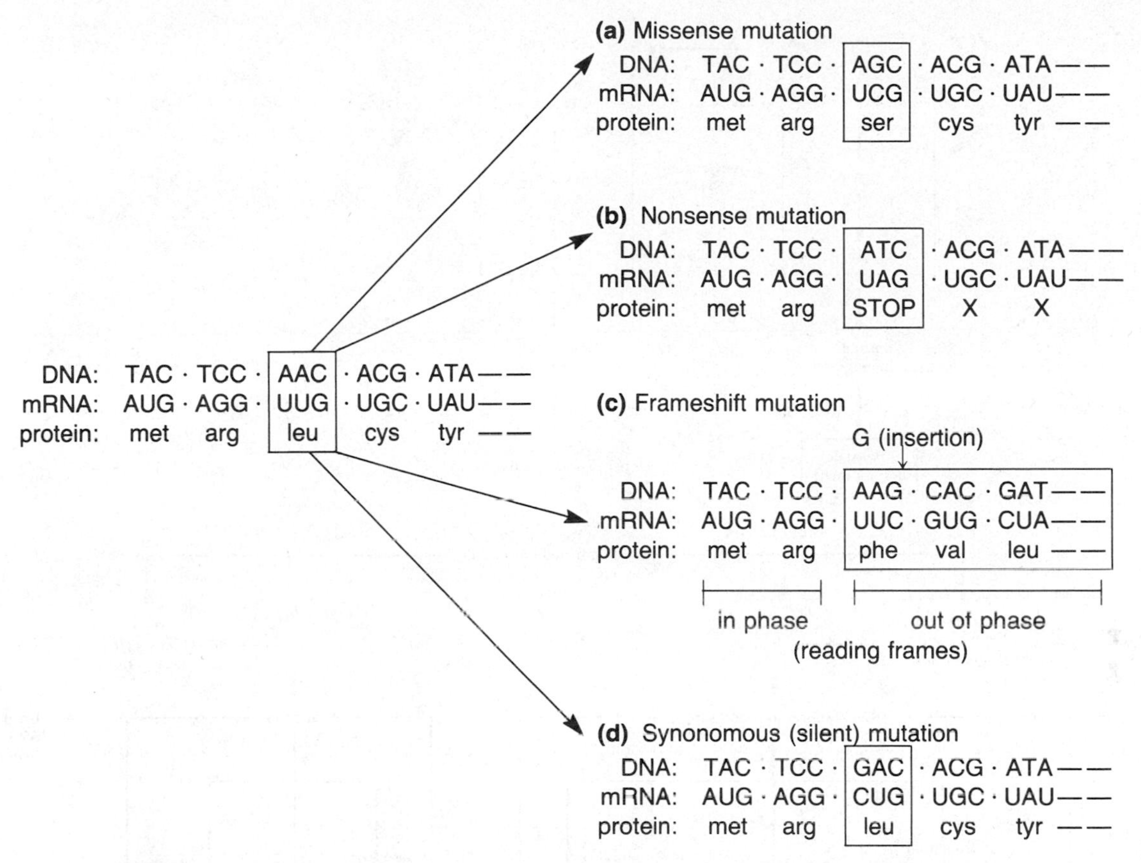

FIGURE 10-20 Different kinds of mutations produced by the various indicated nucleotide changes.

3. Nucleotide insertions or deletions that modify the messenger RNA (mRNA) protein-translation reading frame so that a new and different sequence of codons appears (frameshift mutations, Fig. 10-20(c)).
4. In addition, since practically all amino acids are coded by more than one kind of codon, synonymous mutations can change an amino acid codon without producing an amino acid substitution (Fig. 10-20(d)).

Depending on its position, a nucleotide mutation may have important consequences on the organism though it causes only a single amino acid substitution in a long-chained protein. A prominent example of such effect is that caused by the **sickle-cell** mutation in humans, a gene which, in the United States, is almost entirely confined to blacks. Normally, the adult hemoglobin molecule in human blood cells is composed of four polypeptide globin chains, two αs and two βs, each about 140 amino acids long with its own specific sequence. However, in homozygotes for the sickle-cell gene (Hb^S/Hb^S) all β-globin chains differ from normal βs at the no. 6 position because of a transversion which changed the glutamic acid codon GAA to the sickle-cell valine codon GUA. As shown in Figure 10-21, the effects of this single genetic mutation are profound, causing a variety of phenotypic changes which often lead to inviability. (**Pleiotropy** is the name given to multiple phenotypic effects of a single gene.) Sickle-cell disease is known to kill more than 10 percent of American black homozygotes before the age of 20, and probably has even more lethal effects in Africa where medical facilities are limited. The reasons for the high frequency of this gene in black populations is related to the selective advantage of sickle-cell heterozygotes in malarial regions, a topic that will be discussed in Chapter 21.

REGULATORY MUTATIONS

Regulatory mutations are those that affect the rates at which gene products are produced, although the products themselves may be unaffected. Among such examples are the **thalassemias**, genetic diseases in which the production of either α or β chains is absent or diminished. Although such mutations, like the sickle-cell allele, often cause lethality in homozygotes, they apparently also offer protection against the dread

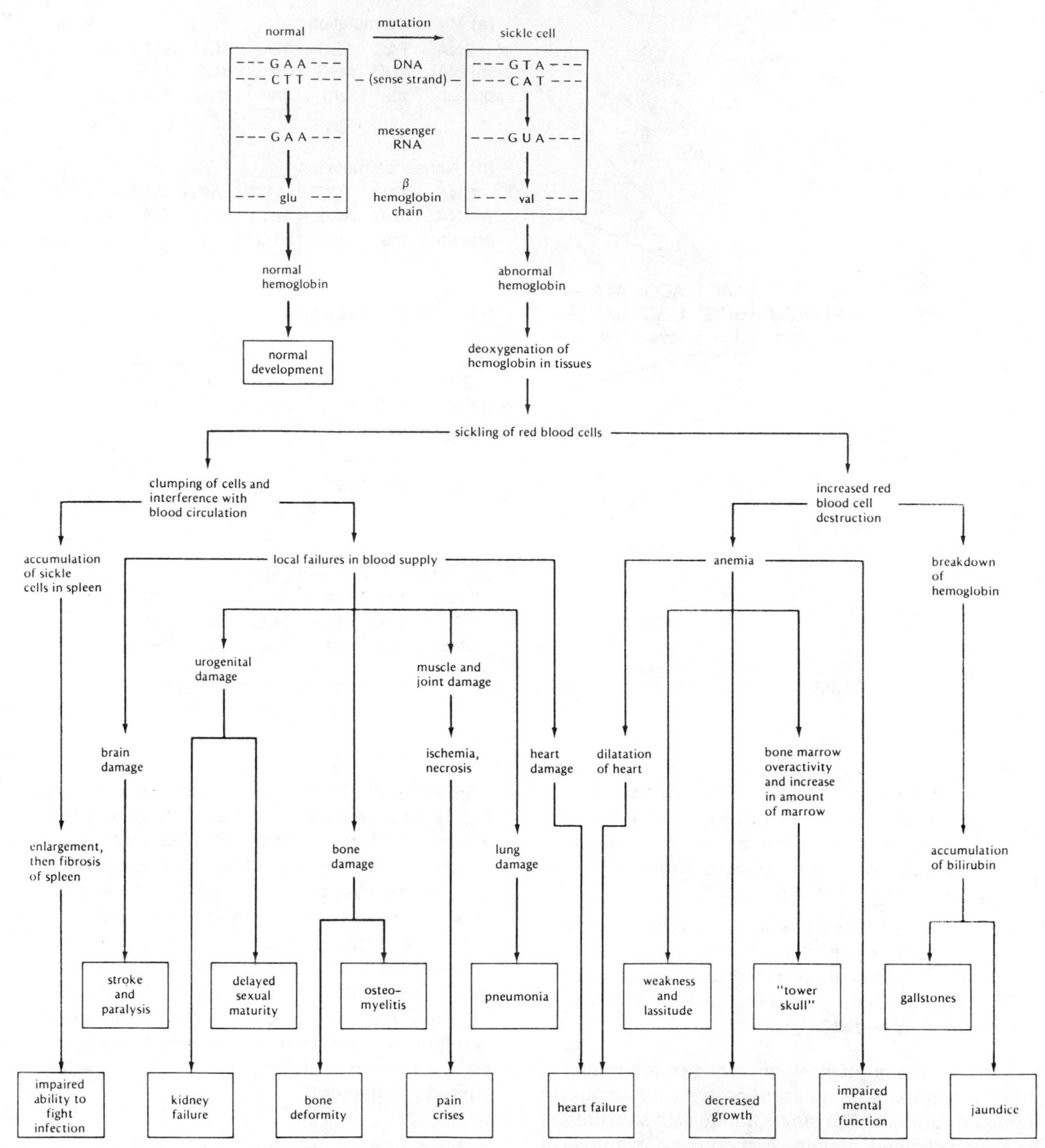

FIGURE 10-21 Varied (pleiotropic) effects of the sickle-cell mutation beginning with the transversion that changed a thymine nucleotide to an adenine nucleotide on the DNA sense strand of the β-hemoglobin gene. The resultant GUA trinucleotide (triplet) coding sequence is then translated into a valine amino acid instead of the normal glutamic acid, producing developmental consequences that can obviously be serious for sickle-cell homozygotes. (From Strickberger.)

malarial parasites. It is presumably this special advantage of some thalassemia heterozygotes which explains the frequency of these genes in human populations.

Viruses and prokaryotes have, so far, provided much more information on regulatory mechanisms than have eukaryotes. The prokaryotic system in *Escherichia coli* bacteria (Fig. 10-22) governs the production of enzymes involved in lactose sugar metabolism (Fig. 10-22(a)). Since *E. coli* bacteria are not commonly exposed to lactose, the genes used in *lac* enzyme synthesis (**structural genes**) are normally prevented from

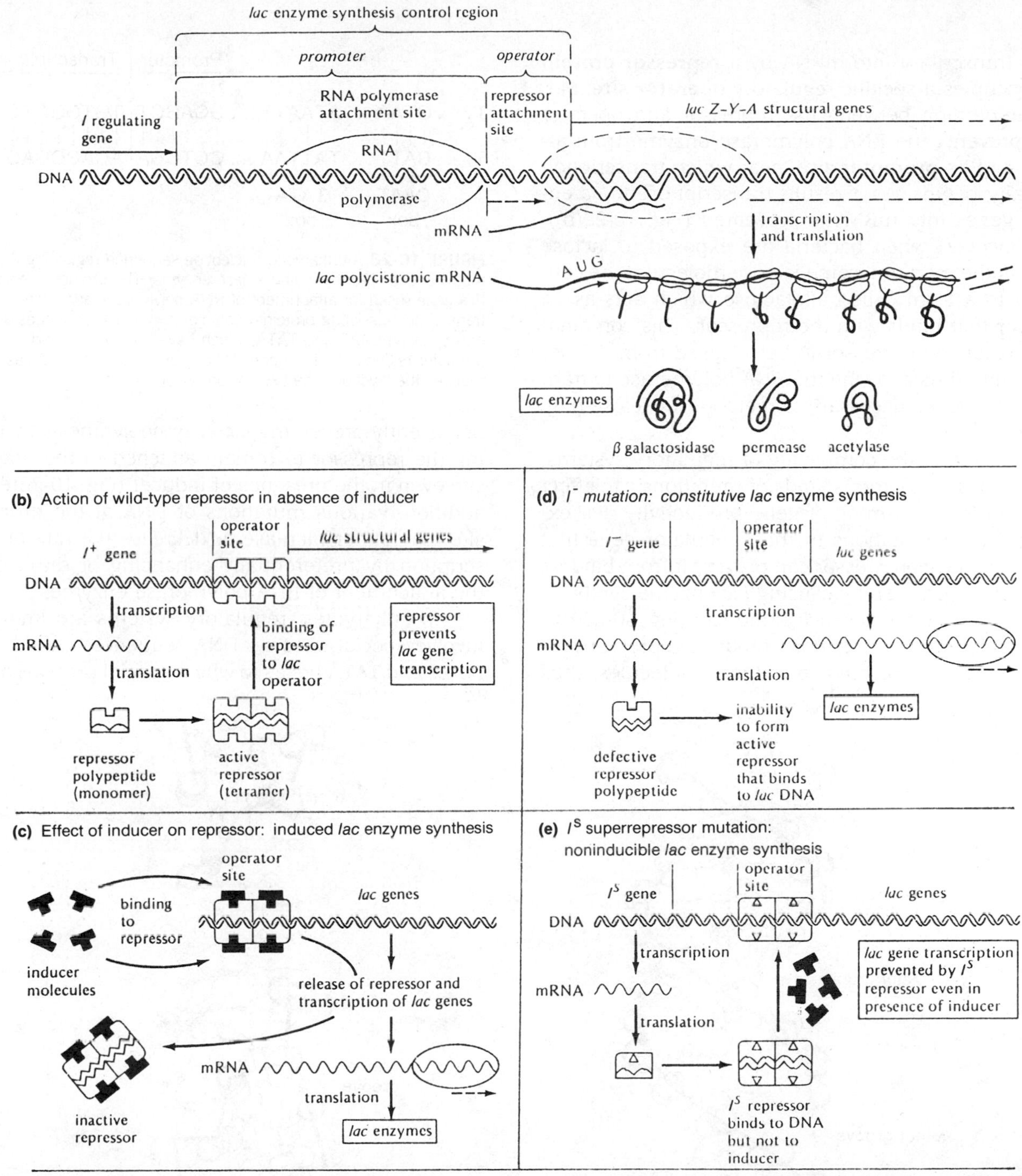

FIGURE 10-22 General scheme of *lac* enzyme synthesis in *E. coli* and the effects of repressor function or dysfunction on this inducible system. (a) The DNA region involved in controlling transcription of the *lac* structural genes *Z*, *Y*, and *A* is composed of two major regulatory sites, the operator and promoter, each of which serves to bind specific proteins. In the absence of the repressor protein, RNA polymerase begins transcription at the operator, which is also the site to which the repressor attaches. (The *lac* repressor itself is coded at a regulator locus, *I*, adjacent to the *lac* locus, but such proximity is not necessarily true for all systems controlled by repressor genes.) Transcription of the *lac* genes, coupled with translation, leads, as shown, to synthesis of the three *lac* enzymes. (b) Transcription and translation of the I^+ normal gene produces a normal repressor protein which binds to the operator site of the *lac* locus, blocking the transcription of *lac* genes by RNA polymerase. This repressed state is found in normal *E. coli* cells that are not grown on lactose medium. (c) Transfer of cells to lactose medium leads to introduction of allolactose inducer molecules which causes the repressor to disassociate from its DNA binding site on the operator. This allows transcription of the *lac* structural genes to proceed, followed by their translation and synthesis of *lac* enzymes. As shown diagrammatically, the repressor is a tetramer (a molecule composed of four polypeptide chains) that acts as an **allosteric protein**, signifying that it has more than one binding site: in this case, one for DNA and one for the inducer. Binding of the inducer to the repressor changes the form or steric configuration of the DNA binding site, making the repressor inactive. (d) When an I^- mutation produces an inactive repressor that cannot bind preferentially to the *lac* operator, the transcription of *lac* genes is unimpeded by repressor. The synthesis of *lac* enzymes thus proceeds "constitutively" in the absence of inducers, that is, even when grown on non-lactose medium. (e) A superrepressor mutation at the *I* locus causes the production of a *lac* repressor that no longer recognizes inducers but maintains its site for normal *lac* operator attachment. The result is repression of *lac* enzyme synthesis even in the presence of inducer. (From Strickberger.)

being transcribed into mRNA by a **repressor protein** that occupies a specific regulatory **operator site**. Molecular binding between the repressor and operator DNA prevents the RNA polymerase enzyme from attaching to its **promoter site**, near which transcription normally begins. As a result, transcription of *lac* enzyme genes into mRNA is prevented (Fig. 10-22(b)).

However, when bacteria are exposed to lactose sugar in the medium, some lactose molecules are converted to a form called allolactose which acts as an **inducer** that binds with the repressor. This combination causes the repressor to be released from the operator site, thus allowing the RNA polymerase to transcribe the genes necessary to metabolize lactose (Fig. 10-22(c)).

Because of the complexity of regulatory systems, it is obvious that various kinds of mutations can affect the quantity and timing of gene productivity. For example, some mutations in the *I* regulator gene that produces the *lac* repressor can prevent it from binding to the operator, thereby causing *lac* enzyme synthesis to occur even in the absence of inducer (Fig. 10-22(d)). Conversely, other *I* mutations produce repressor proteins which cannot bind to inducer molecules, thus

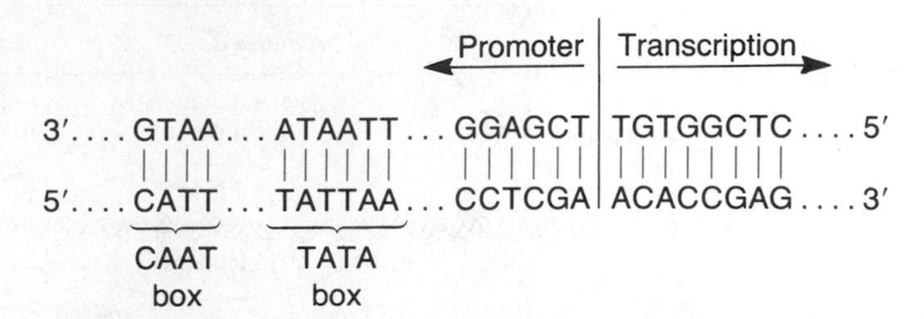

FIGURE 10-23 A eukaryotic nucleotide sequence regulating the gene that produces the thymidine kinase enzyme. The promoter region of this gene (used for attachment of RNA polymerase and other transcription-assisting proteins) contains two short sequences or boxes, called CAAT and TATA, which have also been found in promoter regions of other genes. Alteration of these sequences reduces the level of gene expression (Felsenfeld).

persistently preventing *lac* enzyme synthesis by causing the repressor to remain attached to the operator site even in the presence of inducer (Fig. 10-22(e)). In addition, various mutations of DNA at the promoter site may either increase or decrease the rate of transcription by preferentially enhancing or diminishing the attachment of RNA polymerase enzymes.

In eukaryotes, regulatory systems are known to involve special sites on DNA sequences called CAAT boxes and TATA boxes to which special proteins attach

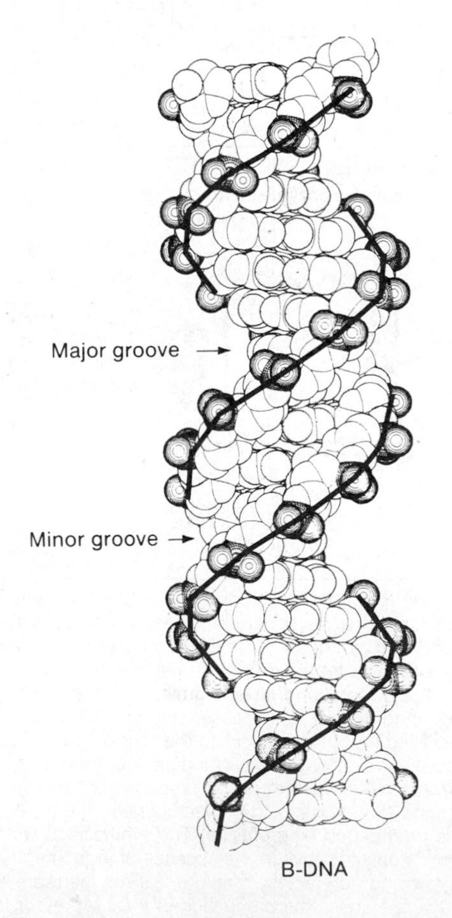

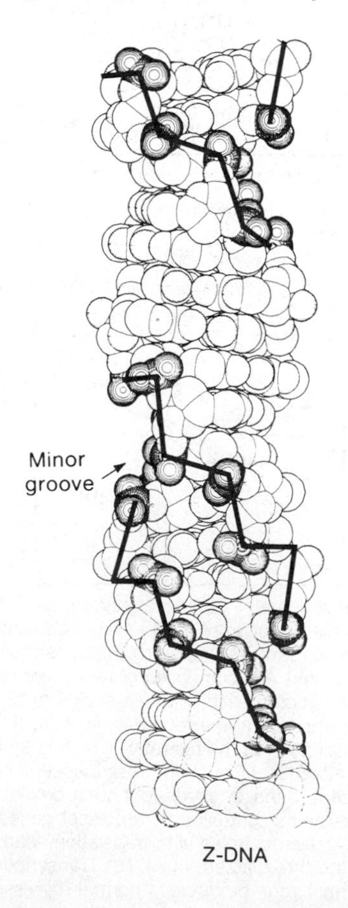

FIGURE 10-24 Space-filling models of B-DNA and Z-DNA double helical molecules. The heavy lines used to connect the phosphate groups in each chain show the zig-zag placement of phosphates in Z-DNA in contrast to the smoother curve of their relationship in B-DNA. The major and minor grooves in B-DNA differ in depth but do not extend to the central axis of the molecule, whereas the indicated Z-DNA groove penetrates the axis of the double helix. (From Rich et al.)

that allow transcription (Fig. 10-23). Some investigators have also demonstrated that the DNA double helix at some eukaryotic regulatory sites changes from a right-handed to a left-handed form, accompanied by a zig-zag placement of phosphate groups (Fig. 10-24).

Since regulation plays an essential role in the timing and placement of all metabolic reactions, regulatory mutations can easily affect both the morphology and function of any organism. For example, the *bithorax* locus in *Drosophila melanogaster*, which governs the placement of structures in various segments, may have mutations that produce an extra set of wings (Fig. 10-25). It has long been suggested that changes in the shapes of various related fishes, as well as in many other species groups, can be ascribed to simple regulatory changes that affect their developmental growth coordinates (Fig. 10-26).

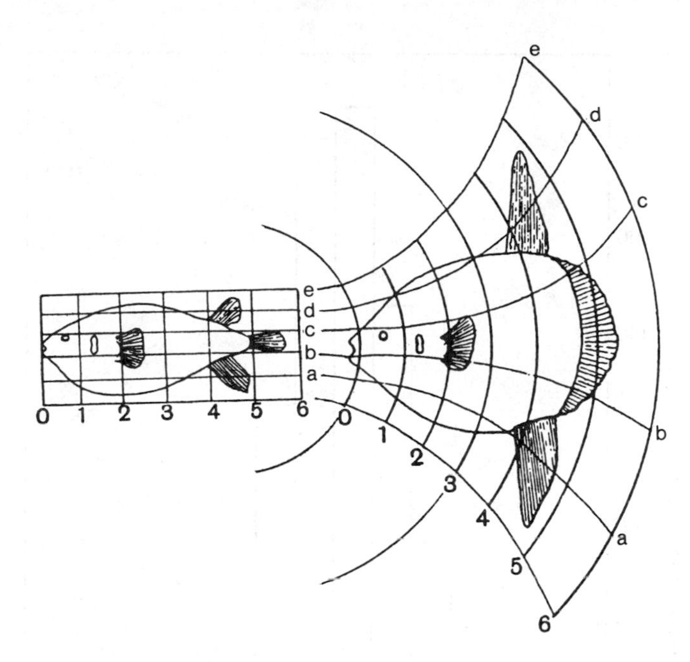

FIGURE 10-26 D'Arcy Thompson's demonstration how completely different organismic shapes can be generated by simple developmental changes in geometric coordinates. Thus, if the vertical coordinates of the puffer fish, *Diodon* (left), are changed into concentric circles, and its horizontal coordinates into hyperbolas, the resultant animal has the shape of the sunfish, *Orthagoriscus* (right). (Adapted from Thompson.)

Quantitative Variation

Although large regulatory changes have been used to explain some major differences between groups (Chapter 12), the extent to which they account for most other evolutionary events is still unclear. Beginning with Darwin himself, many evolutionists suggested that it is rather small heritable changes that provide most of the variation upon which natural selection acts. In Darwin's words (*On the Origin of Species*),

> Extremely slight modifications in the structure and habits of one species would often give it an advantage over others; and still further modifications of the same kind would often still further increase the advantage.... Under nature, the slightest differences of structure or constitution may well turn the nicely-balanced scale in the struggle for life, and so be preserved.

These small changes are most obvious in respect to measurable characters such as size and yield and produce the **continuous variation** often seen for characters distributed in bell-shaped curves (normal distributions), such as human heights (Fig. 10-27).

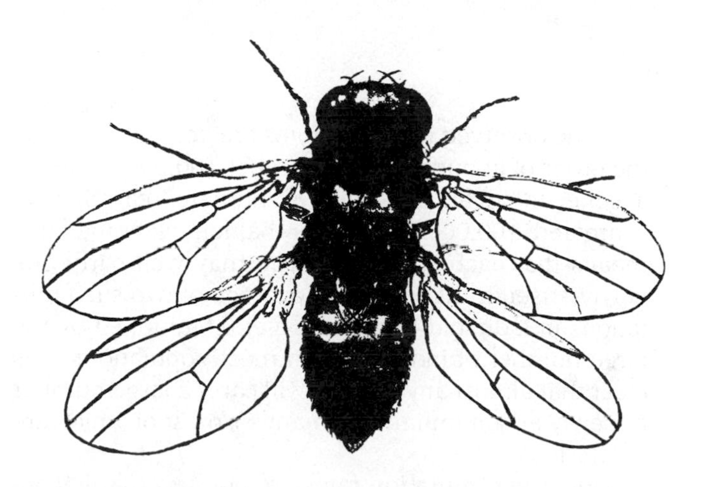

FIGURE 10-25 A four-winged *Drosophila* caused by mutations at the *bithorax* locus. Normally, like all dipteran insects, in *Drosophila* there is only a single pair of wings which arise from the second of the three thoracic segments. As shown here, certain *bithorax* mutations will cause the third thoracic segment to produce its own pair of wings, a condition which simulates that of the ancestral four-winged fly. (From a photograph by E. B. Lewis; see Bender et al.)

Support for the importance of continuous variation in evolution was garnered from various studies which showed that small differences can accumulate through selection to give large differences. For example, Figure 10-28 shows that selection for the presence or absence of white spotting in Dutch rabbits can lead to completely colored or completely white strains. The genetic cause for these changes are genes with small phenotypic effect, called **multiple factors** or **polygenes**, that can produce the familiar normal distributions when they assort independently. Thus, a mating between heterozygotes for three pairs of genes, each with two alleles—one colored, one white—produces the phenotypes shown in Figure 10-29, ranging from all colored to all white. Experiments and analyses of this kind therefore demonstrated that selection for quantitative characters could be explained on the basis of the seg-

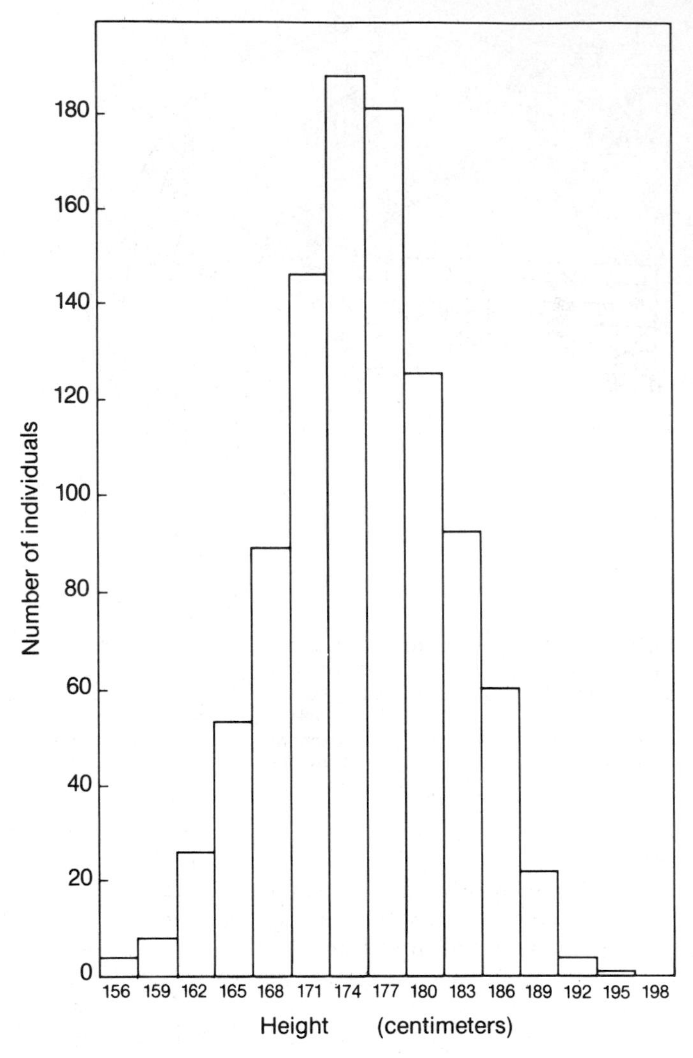

FIGURE 10-27 Distribution of heights among one thousand Harvard students aged 18 to 25. (Adapted from Castle.)

FIGURE 10-28 White spotting in Dutch rabbits, ranging from almost no spotting (grade 1) to complete spotting (grade 18). Selection experiments on animals with intermediate spotting (e.g., grades 7 to 12) showed that spotting can be increased or decreased. (From Castle.)

regation and assortment of simple mendelian genes whose individual small effects may add up to large phenotypic differences.

MUTATION RATES

Based on everything considered so far, the opportunity for all kinds of mutations obviously derives from various sources and may have various phenotypic effects. For newly arisen mutations, these effects will most likely be harmful since prevailing genotypes are generally well adapted for their particular environments, and most changes are unlikely to improve them further. The detection of new mutations in most organisms has therefore usually been associated with the observation of newly inherited deleterious effects, some of which are given in Table 10-4.

The observed mutation rates are generally low, on the order of about one mutation per 100,000 copies of a gene. In organisms such as humans, carrying an estimated 100,000 genes per haploid genome, this means that each sperm and egg may well carry one newly arisen mutation, or an average of two such mutations in a diploid fertilized zygote. Multiplied by the large number of individuals in most populations, it is clear that almost any species will carry a large number of newly arisen mutations, many or most of which are deleterious.

However, mutation rates are not necessarily constant. Among the causes that can modify mutation rates are genes for polymerase enzymes that replicate DNA. Some alleles of these genes have been found to act as **mutator genes** that can increase mutation rates manyfold, whereas other alleles act as **antimutators** to decrease mutation rates. Mutation rates, like other essential traits, seem therefore adjusted to optimum values in most, if not all, organisms. Nevertheless, mutation rates may be considerably affected by external causes, including, surprisingly, infectious elements that can be transmitted from other individuals. For example, some viruses, such as herpes simplex, rubella (German measles), and chicken pox, can cause breaks and deletions in chromosomes because they release nuclease enzymes that attack host DNA.[4]

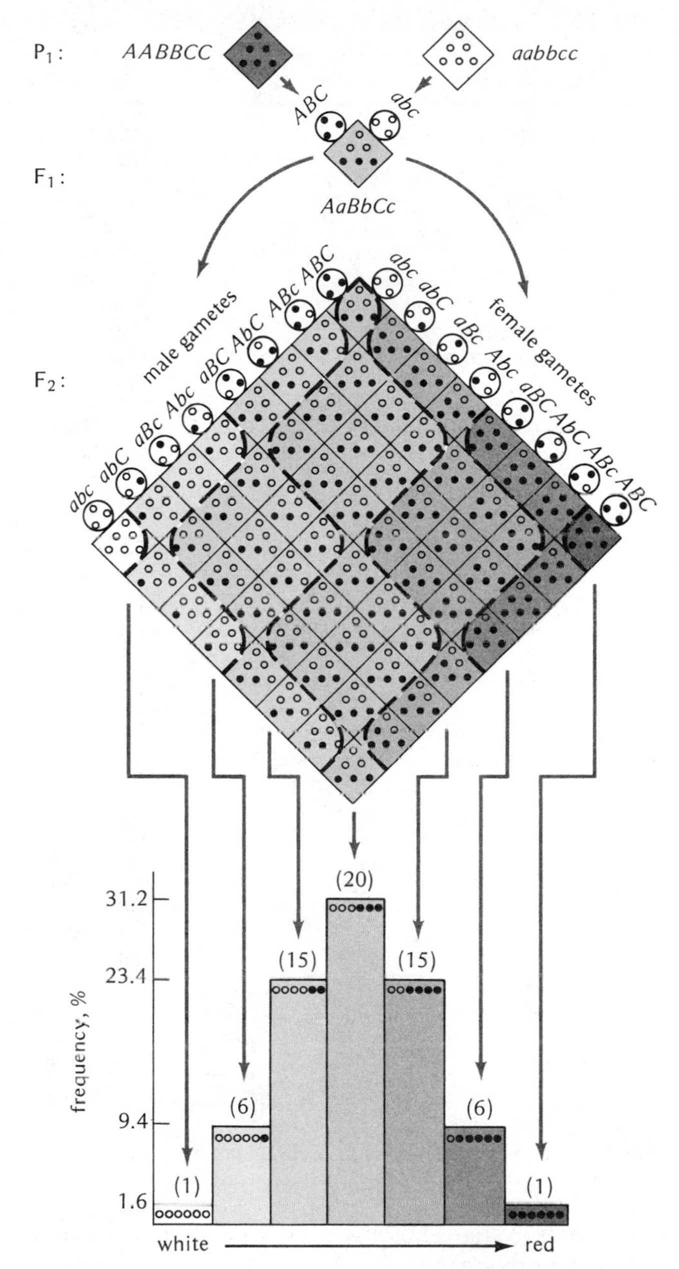

FIGURE 10-29 The results of crosses between two strains of wheat differing in three gene pairs that determine grain color. Each gene pair assorts independently of the others, and the alleles at each gene pair lack dominance, so that *Aa*, for example, has a color intermediate between *AA* and *aa*. The F_1, carrying three color (*ABC*) and three non-color alleles (*abc*) is therefore intermediate in color to the parental stocks, and the F_2 produces a range of colors in the frequencies shown in the histogram. (From Strickberger.)

TRANSPOSONS, REPEATED SEQUENCES, AND SELFISH DNA

Other sources of mutational change in both prokaryotes and eukaryotes are **transposons:** nucleotide sequences that can promote their own transposition between different genetic loci. This is accomplished by special transposase enzymes produced by the transposon which enable copies of the transposon to be inserted into various target sites. For example, the IS1 transposon illustrated in Figure 10-30 makes staggered cuts at each side of a nine-nucleotide base pair sequence, and a copy of IS1 is then inserted within the gap produced by these cuts. Depending on where transposons are inserted, mutations of all kinds may be produced, marked by target site repeats, in which there are similar sequences of nucleotide bases at each end of the insertion but in inverted order.

Inverted repeats have therefore been used to detect the presence of transposable elements in various species and indicate that DNA sequences can be picked up by a transposon and transferred to other DNA locations. This has been shown for antibiotic resistance genes that can be transmitted between bacterial strains by small circular DNA particles called **plasmids** which have received transposon insertions. Examples of this kind indicate that some hereditary traits carried by transposons may have been passed horizontally between individuals of the same generation, rather than through normal vertical transmission between generations.

Also of interest is the fact that there may be many such sequences within any species. In primates, for example, a sequence with transposonlike features called *Alu* is represented by perhaps more than one million copies per human diploid cell. Smaller repetitive sequences of the type discussed in Chapter 12 have also been shown to be widely prevalent in various

[4]Important factors that act to correct nuclear damage from these and other causes, including ultraviolet radiation, are a variety of **DNA repair mechanisms.** These enzyme systems are found in practically all cells, indicating that all forms of life have faced common problems of DNA damage. In fact, the inability to repair DNA damage is often lethal: in humans, such deficiencies are found in genetic diseases such as xeroderma pigmentosum which often causes lethality because of increased incidence of cancer.

TABLE 10-4 Spontaneous Mutation Rates at Specific Loci for Various Organisms

Organism*	Trait	Mutation per 100,000 Gametes
Virus: T4 bacteriophage	Rapid lysis ($r^+ \rightarrow r$)	7.0
	New host range ($h^+ \rightarrow h$)	0.001
Bacteria: *E. coli*	Streptomycin resistance	0.00004
	Phage T1 resistance	0.003
	Leucine independence	0.00007
	Arginine independence	0.0004
	Arabinose dependence	0.2
Salmonella typhimurium	Threonine resistance	0.41
	Histidine dependence	0.2
	Tryptophan independence	0.005
Fungus: *Neurospora crassa*	Adenine independence	0.0008-0.029
	Inositol independence	0.001-0.010
Insect: *D. melanogaster*	y^+ to *yellow*	12.0
	bw^+ to *brown*	3.0
	e^+ to *ebony*	2.0
	ey^+ to *eyeless*	6.0
Plant: corn (*Zea mays*)	*Sh* to *shrunken*	0.12
	C to *colorless*	0.23
	Su to *sugary*	0.24
	Pr to *purple*	1.10
	I to *i*	10.60
Rodent: *Mus musculus*	a^+ to *nonagouti*	2.97
	b^+ to *brown*	0.39
	c^+ to *albino*	1.02
	d^+ to *dilute*	1.25
Primate: *Homo sapiens*	Achondroplasia	0.6-1.3
	Aniridia	0.3-0.5
	Dystrophia myotonica	0.8-1.1
	Epiloia	0.4-1.0
	Huntington's chorea	0.5
	Intestinal polyposis	1.3
	Neurofibromatosis	5.0-10.0
	Retinoblastoma	0.5-1.2

Adapted from Strickberger.
*Mutation rate estimates in viruses, bacteria, and fungi are based on particle or cell counts rather than gametes.

eukaryotes. What explains their widespread distribution and persistence?

According to some writers (Orgel and Crick, Doolittle and Sapienza) many transposable elements and other forms of repeated sequences contribute little, if any, function to their host cells. Since the DNA replication process cannot discriminate between functional and non-functional sequences, any introduced DNA sequence will be replicated. Transposon DNA may therefore be perpetuated parasitically as either "junk" or "selfish" DNA. Other explanations propose that certain repeating sequences may function as essential elements in regulating gene activity, or serve as origins of DNA replication, or act as mutator genes that occasionally provide new adaptive mutations. There is as yet no agreement on how to weigh the selfishness or unselfishness of such sequences, and it is possible that some repeated sequences fulfill both roles.[5]

[5]Although "selfishness" indicates that a unit of life is primarily concerned with its own replication, it is debatable whether, as argued by Dawkins, such selfishness is the exclusive property of DNA simply because DNA replicates itself so well. According to this argument, all biological entities such as cells and organisms are merely the means for the replication of DNA. It would seem that such attempts to reduce explanation of life to such simple levels could lead easily to a meaningless chain of arguments: organisms are the means enabling cells to replicate; cells enable chromosomes to replicate; chromosomes enable gene replication; genes enable codon replication; codons enable nucleotide replication; nucleotides enable the replication of nitrogenous bases, sugars, phosphates, and so on; until statements can be developed asserting that all organisms are the means enabling the perpetuation of various atomic or even subatomic particles. Selfishness may have some meaning in terms of one component of life competing with similar components at the same level of organization (selfish organisms, selfish cells, selfish DNA, etc.), but there seems little to be gained by trying to establish which level is most selfish.

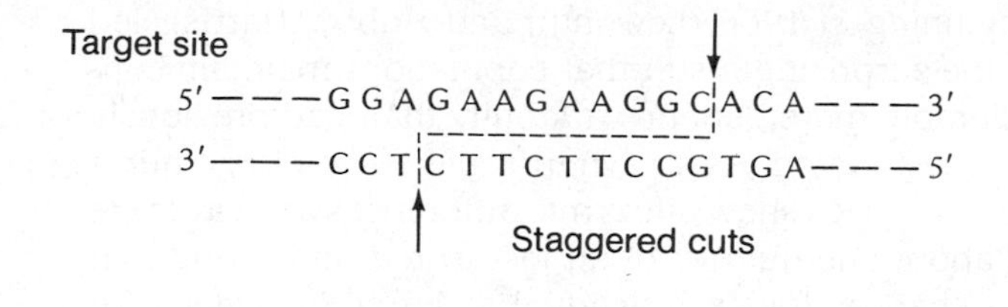

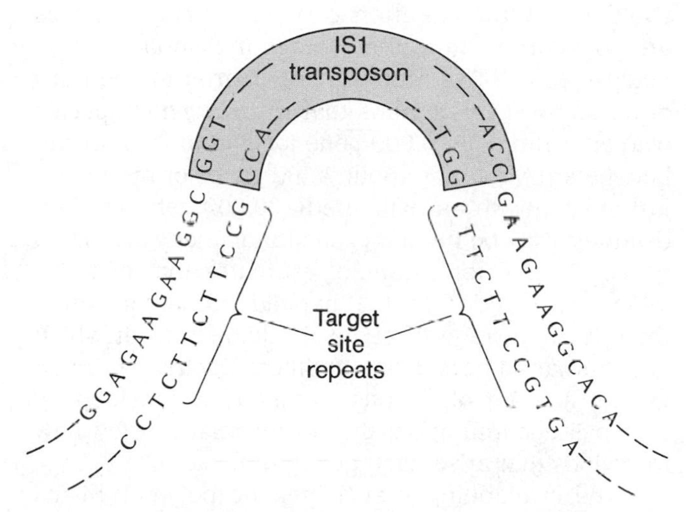

FIGURE 10-30 Mode of insertion of the IS1 transposon. The nine-base pair DNA sequence at the top of the diagram is recognized as a target site by the transposon and is cleaved at the indicated arrows. IS1 is then inserted into the resulting gap, and DNA sequences are synthesized complementary to the former single-strand sections of the target site. This process therefore produces identical nine-base pair repeats at each end of the transposon. (From Strickberger.)

THE RANDOMNESS OF MUTATION

Until the 1950's many bacteriologists felt that bacteria possessed a unique "plastic heredity" in which appropriate mutations arise as an immediate response to the needs of the environment. This concept seemed supported by observations in which bacteria exposed to some virus or antibiotic would quickly develop a resistant form. The explanation offered was that mutations do not originate on a random basis prior to exposure to some selective agent (**preadaptive** mutations), but rather that appropriate mutations are stimulated to arise only after bacteria have been exposed to a selective agent (**postadaptive** mutations). Like Lamarck, the postadaptive mutation concept had to rely on some unknown, perhaps mystical, agency which allowed the environment to directly cause the appearance of new adaptive hereditary factors instead of employing Darwin's natural selection to choose among those preadaptive hereditary factors that are already present.

One of the impressive tests that enabled a decision to be made between these pre- and postadaptive models was performed by J. and E. Lederberg using a novel technique called replica-plating. In their experiment, samples from a "lawn" of bacteria growing on a petri dish (master plate) were transferred to other petri dishes (replica plates) which contained selective media such as viruses (bacteriophages) or antibiotics (streptomycin). These transfers were performed in a manner enabling bacteria derived from specific clones on the master plate to be localized to the same positions on the replica plates (Fig. 10-31). The finding that resistant bacteria occupied identical positions on various replica plates indicated that they arose from the same clone on the master plate, a clone which must have been present before exposure to the selective medium on the replica plates. In other words, these mutants shared a common preadaptive origin on the master plate and did not arise postadaptively because of a Lamarckian response to a stimulus by the selective medium.[6]

GENETIC POLYMORPHISM: THE WIDESPREAD NATURE OF VARIABILITY

New mutations that have an immediate beneficial effect on the organism seem generally to be quite rare, although there are some that are either neutral in their effect or deleterious only when they occur in relatively rare homozygotes. Such neutral and deleterious but recessive mutations can therefore often accumulate in a population without any immediately serious unfavorable effects, thereby furnishing a reservoir of genetic variability.

Genetic variability, or **polymorphisms,** may include the maintenance of different kinds of chromosomal anomalies such as inversions, translocations, and extra chromosomes. In *Drosophila pseudoobscura*, for example, populations in different localities in the western United States are polymorphic for a wide variety of third chromosome gene arrangements (Fig. 10-32(a)), and the frequencies of such arrangements may

[6]Both Cairns et al. and Hall have performed experiments which suggest that some selective environments can induce adaptive bacterial mutations at a frequency higher than expected were such mutations strictly random. These experiments and their interpretation have generated considerable controversy (see, for example, Symonds and Lenski et al.), and the matter has not yet been resolved. So far, the disputed phenomena appear to be limited to particular genes or gene combinations and to particular environments.

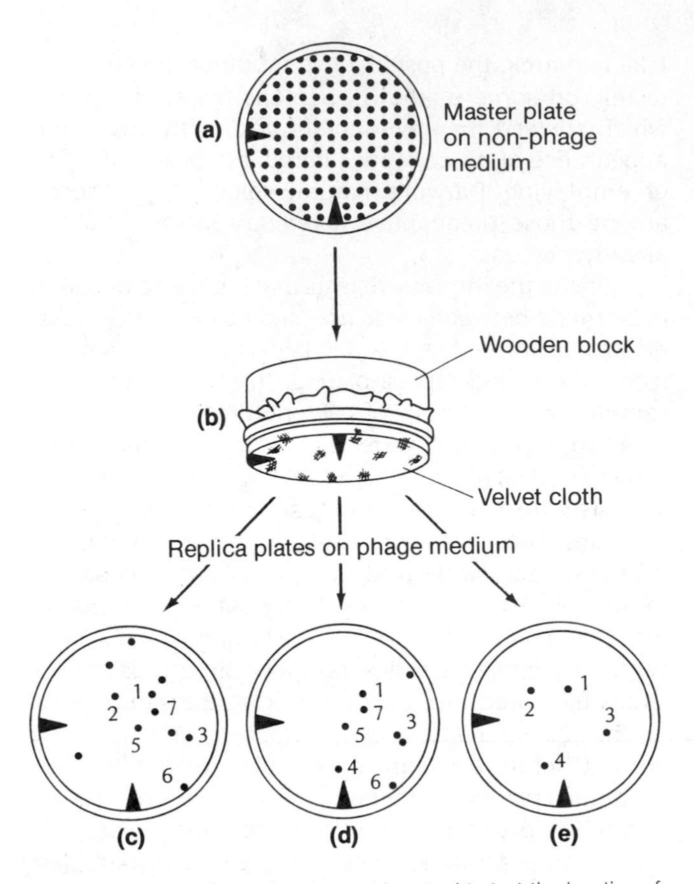

FIGURE 10-31 Replica-plating technique used to test the location of clones of *E. coli* resistant to T1 bacteriophage (virus). (a) Master plate showing diffuse bacterial growth of phage-sensitive *E. coli* on a non-phage medium. Replicas are made by pressing a velvet-covered wooden block (b) against the master plate, then pressing this, oriented in the same direction, to the surface of petri dishes containing culture medium mixed with phage T1. One master plate possesses sufficient bacteria to start colonies on a number of replica plates. (c - e) Replica plates showing occurrence of resistant colonies (e.g., 1, 3, 4) in identical locations, indicating that resistance to phage T1 must have been present at each of these positions in the master plate. (Based on Strickberger, adapted from Lederberg and Lederberg.)

change seasonally (Fig. 10-32(b)). This indicates not only that chromosomal polymorphism is generally adaptive in this species, but that certain polymorphic variations are preferentially adaptive in helping populations adjust to their specific environment at specific times.

Polymorphism on the gene level itself is certainly also present, and its magnitude can be discerned by techniques which make allelic differences visible. Among such studies are electrophoretic methods that measure the mobility of a protein in an electric field, distinguishing even slight variations in electric charge (Fig. 10-33). Since each different electrophoretic form of a protein usually indicates a different amino acid sequence (and therefore a different nucleotide sequence in the gene that produced it) they are each considered to signify an allelic difference, or **allozyme**. Application of this technique to natural populations, beginning in 1966 (Lewontin and Hubby; Harris), led to the surprising result that populations maintain considerably more genetic variability than had previously been estimated. As shown in Table 10-5, a large number of species show allozymic differences at an average of about one quarter of all loci tested, indicating that the chances for an individual to be heterozygous for any particular tested locus is more than 7 percent.

Since it is estimated that only perhaps one third of amino acid changes in proteins are detectable by electrophoretic techniques, these observed values should probably be tripled. Thus one can say that about two thirds or three quarters of all loci in many species are polymorphic, and the average individual may be heterozygous for as much as one quarter to one third of all its loci. This means that in *Drosophila* species with an estimated 10,000 gene loci, an individual can be a heterozygote for about 2,500 genes or more; and in humans, with an estimated 100,000 gene loci, individuals may be heterozygous for as many as 25,000 genes! Britten, for example, estimates that of the 3 billion nucleotides in the haploid human genome, there is an average of about 5 million sites at which one human differs from another. Clearly, a much greater amount of genetic variability is provided by such past accumulations than by the relatively few new mutations that arise each generation.

The availability of such genetic polymorphisms thus allows many populations to confront new environmental challenges with a large variety of mutations, some of which may be advantageous. For example, the exposure of insect populations to DDT pesticides has caused a widespread increase in the frequency of various DDT-resistant mechanisms such as

1. An increase in lipid content that enables the fat-soluble DDT to be separated from other parts of the organism.
2. The presence of enzymes that break down DDT into relatively less toxic products.
3. A reduction in toxic response of the nervous system to DDT.
4. Changes in the permeability of the insect cuticle to DDT absorption.
5. A behavioral response that reduces contact with DDT.

It is therefore not surprising that insecticide resistance has been shown to be associated with numerous genes. For example, the genes responsible for DDT resistance in *Drosophila* are located on all major chromosomes, each gene acting as a polygene with a small incremental effect (Fig. 10-34).

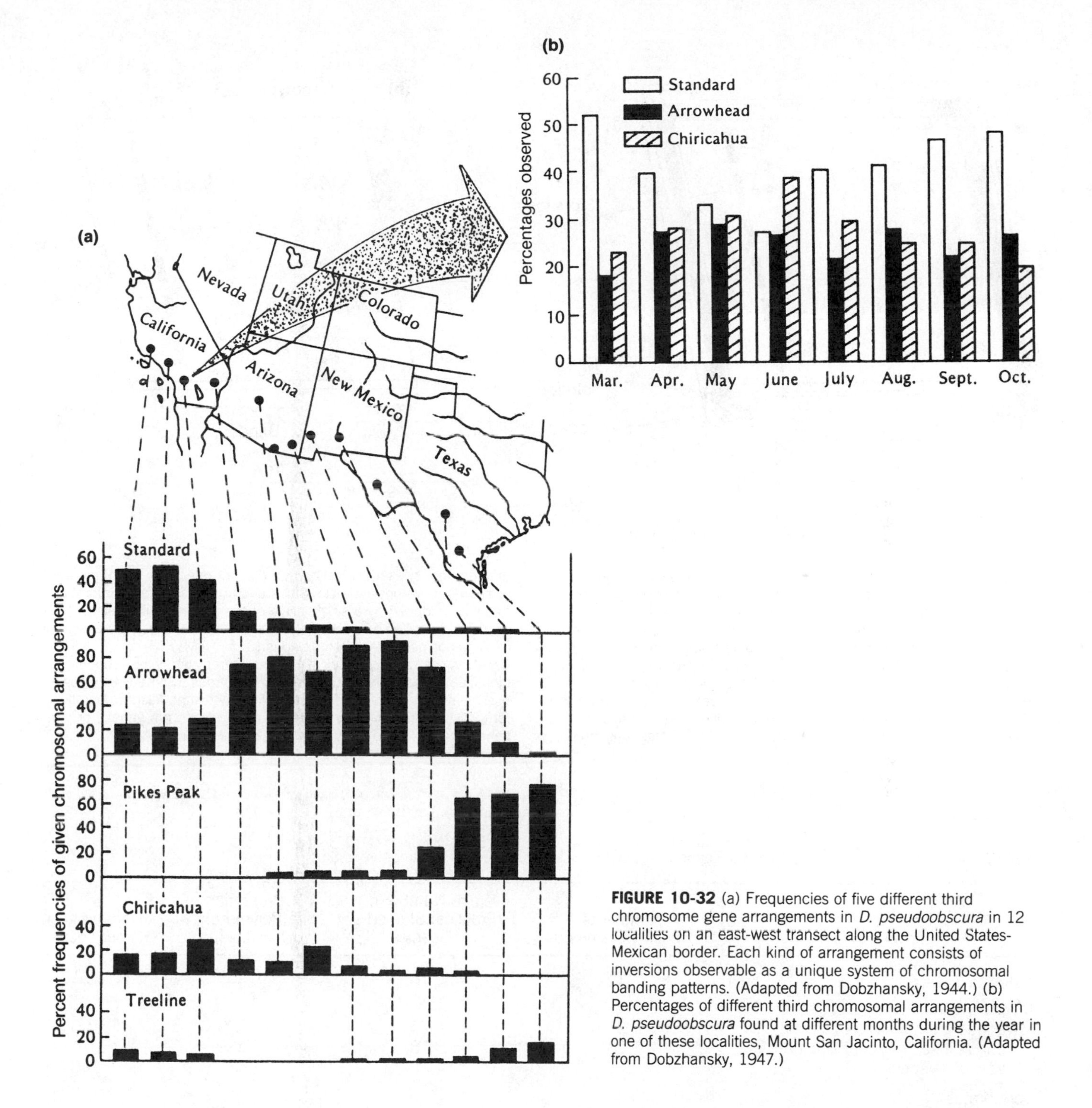

FIGURE 10-32 (a) Frequencies of five different third chromosome gene arrangements in *D. pseudoobscura* in 12 localities on an east-west transect along the United States-Mexican border. Each kind of arrangement consists of inversions observable as a unique system of chromosomal banding patterns. (Adapted from Dobzhansky, 1944.) (b) Percentages of different third chromosomal arrangements in *D. pseudoobscura* found at different months during the year in one of these localities, Mount San Jacinto, California. (Adapted from Dobzhansky, 1947.)

On the whole, it seems clear that most populations do not await the lucky incidence of new favorable mutations to provide for their evolutionary needs. Instead, populations tend to use their reservoir of genetic variability, consisting of many historically accumulated mutations. Evolutionary potential and genetic variability can therefore be considered as two sides of the same evolutionary coin.

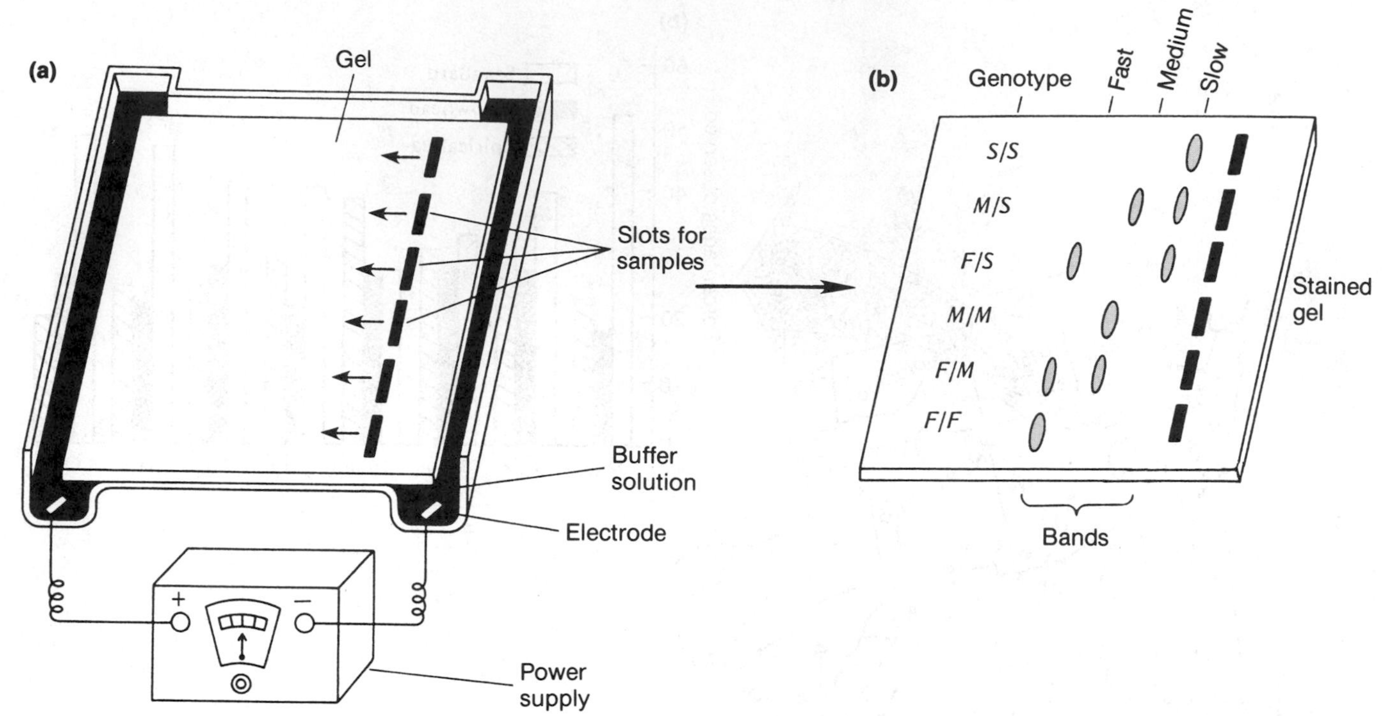

FIGURE 10-33 (a) General scheme for electrophoresis, using a gel (starch or polyacrylamide) in which samples are placed along a row and subjected to an electric current carried through an aqueous buffered solution. Depending on their size and electrical charge, molecules in the samples will separate, going toward either the negative or positive pole. The position they occupy on the electrical gradient can be identified as bands when the gel is treated with agents that can assay molecular or enzymatic activity or is exposed to ultraviolet light. (b) Treatment of the gel with dyes sensitive to a specific enzyme activity shows that the enzyme on this gel has three different forms, each migrating at a distinct rate, slow, medium, and fast, toward the positive pole. Since each of these three enzymatic forms is produced by a single allele of the gene for this protein, *S, M, F*, it is often called an allozyme. Thus an individual may possess one of six genotypes, either homozygous (*S/S, M/M, F/F*) or heterozygous (*M/S, F/S, F/M*), each different genotype producing an identifiable electrophoretic pattern of allozymes. (Some terminologies use the more general term isozyme for any distinct electrophoretic form of a protein, whether its uniqueness arises from genetic or non-genetic causes.) (Adapted from Strickberger.)

TABLE 10-5 Estimates of Genetic Variability Found in Natural Populations, Based on Electrophoretic Studies

Organisms Tested	Number of Species Examined	Average Number of Loci (Proteins) Studied per Species	Proportion of Polymorphic Loci	Heterozygosity per Locus
Plants	15	18	.259	.071
Invertebrates				
Various groups except insects	27	25	.399	.100
Various insects except Drosophilidae	23	18	.329	.074
Drosophila species	43	22	.431	.140
Vertebrates				
Fish	51	22	.152	.051
Amphibia	13	22	.269	.079
Reptiles	17	23	.219	.047
Birds	7	21	.150	.047
Mammals (except primates)	43	26	.148	.036
Primates				
Humans	1	71	.28	.067
Chimpanzees	1	43	.05	—
Macaque monkeys	1	29	.10	.014
TOTALS AND AVERAGES	242	23	.263	.074

From Strickberger, derived from data collected by Nevo.

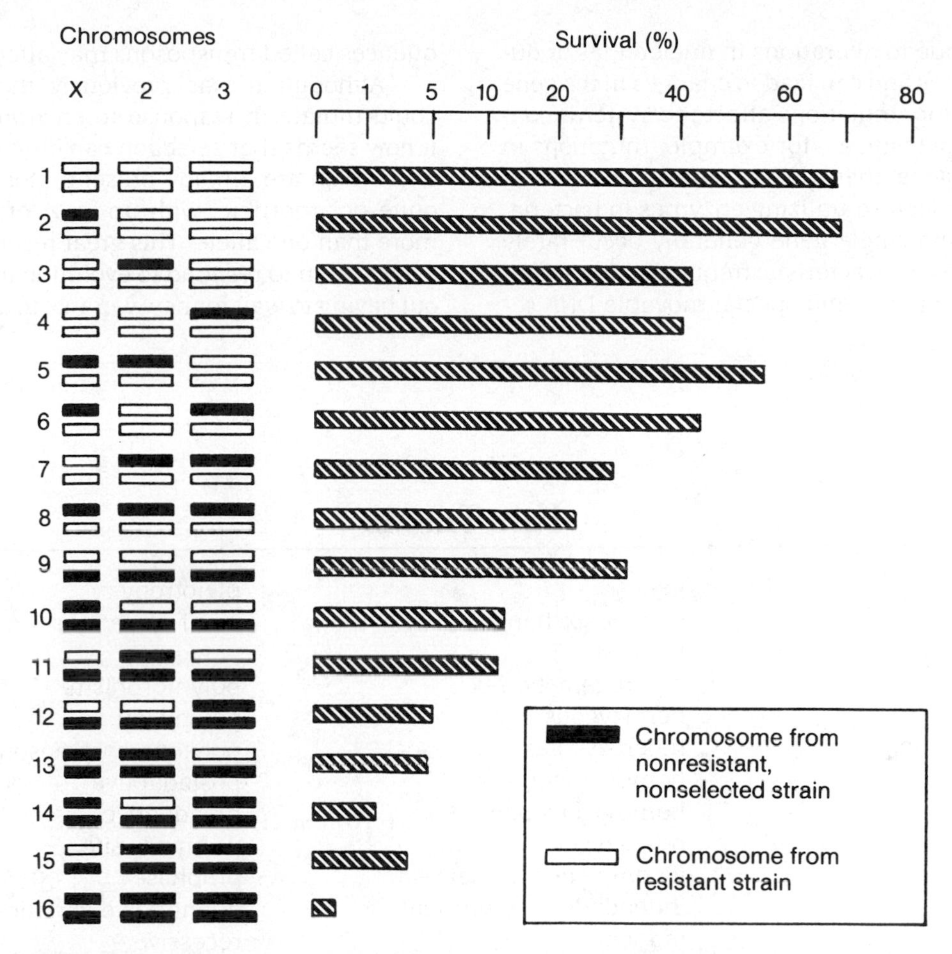

FIGURE 10-34 Percent survival of 16 different types of *D. melanogaster* flies exposed to a uniform dose of the insecticide DDT. Each type of fly carries a unique set of chromosomes derived from DDT-resistant and DDT-nonresistant strains. It is obvious that DDT resistance increases with the increased number of chromosomes from resistant strains. (Adapted from Crow.)

SUMMARY

Life depends on the genetic constancy with which organisms transmit information to their offspring, but evolution cannot occur without genetic variability. Genetic traits, changed or unchanged, are transmitted from one generation to the next by some type of cell division: by binary fission in prokaryotes and, in eukaryotes, by mitosis (somatic cell division) or meiosis (gamete-producing division). In mitosis all of the daughter cells are genetically equivalent; while in meiosis, variability is provided by recombination between homologous chromosomes and by random assortment of chromosomes into the gametes.

In the nineteenth century most people believed in blending inheritance. However, in the 1860's, Gregor Mendel broached two principles of inheritance which contradicted this idea. The first, the principle of segregation, states that alleles of a single gene will segregate from each other into the gametes as discrete particles. The second, the principle of independent assortment, involves the independent segregation into gametes of genes on different chromosomes.

Among the genetic sources of variability are incomplete dominance or codominance of alleles, the presence of multiple alleles for a gene, and the many instances when one gene locus affects the phenotypic expression of another (epistasis). In sex linkage the heterogametic sex (XY) will express all alleles lying on the X chromosome without regard to dominance relationships. In linkage genes located on the same chromosome are constrained to varying degrees (recombination frequencies, linkage distances) to remain together during meiosis.

Variability will be enhanced when there are alterations in chromosome number, either changes in entire sets (euploidy) or in individual chromosomes (aneuploidy), or when chromosomes undergo modifications in their structure, such as deletions, duplications, inversions, or translocations. Translocations can modify the size, composition, and even the number of chromosomes. Small localized chromosomal changes, or mutations, are due to alterations in nucleotides or nucleotide sequences and can lead to changes in the gene product. Regulatory mutations affect the systems controlling genetic activity, as, for example, mutations in the regulatory gene that governs the expression of genes coding for lactose-utilizing enzymes in bacteria. Mutations within a single gene generally occur rarely and usually have a characteristic frequency. But mutator genes, environment, and special movable DNA sequences called transposons may affect mutation rates.

Although it was previously thought that genes could mutate in response to environmental demand, it now seems that selection can choose only from variants which are already present. Most populations are quite polymorphic, with up to ¾ of gene loci having more than one allele. This great reservoir of variability allows them to respond to evolutionary pressures without having to wait for new variants to arise by mutation.

KEY TERMS

acrocentric
alleles
allopolyploids
allosteric protein
allozyme
alternate segregation
anaphase
aneuploid variations
antimutators
asexual reproduction
autopolyploidy
autosomal
base substitutions
bivalent
chromatids
chromosomes
codominant
constancy
continuous variation
crossing over
deficiencies
deletions
diploid
disjunction
dissociation
DNA repair mechanisms
dominant
duplications
epistatic interactions
euploid variations
fission
fusion
gene families
gene mutations
genes
genetic exchange
haploid
heterogametic sex
hemizygous
heterozygotes
homogametic sex
homologous pairing
homozygotes
incomplete dominance
independent assortment
inducer
inversions
karyotype
linkage
linkage distance
loci (plural; singular = locus)
meiosis
metacentric
metaphase
mitosis
modifiers
multiple alleles
multiple factors
mutations
mutator genes
nondisjunction
null alleles
operator site
paracentric inversions
parthenogenesis
pericentric inversions
plasmids
pleiotropy
point mutations
polygenes
polymorphism
polyploidy
polytene chromosomes
postadaptive
preadaptive
promoter site
prophase
quantitative variation
recessive
reciprocal translocations
recombination
regulatory mutations
repeated sequences
repressor protein
segregation
selfish DNA
sex linkage
sickle-cell disease
structural genes
telocentric
thalassemia
transitions
translocations
transposons
transversions
unequal crossing over
variability
wild type

REFERENCES

Austin, C. R., and R. V. Short, 1976. *The Evolution of Reproduction*. Cambridge Univ. Press, Cambridge.

Bachmann, B. J., 1983. Linkage map of *Escherichia coli* K-12, edition 7. *Microbiol. Rev.*, 47, 180-230.

Bender, W., M. Akam, F. Karch, P. A. Beachy, M. Peifer, P. Spierer, E. B. Lewis, and D. S. Hogness, 1983. Molecular genetics of the bithorax complex in *Drosophila melanogaster*. *Science*, **221**, 23-29.

Britten, R. J., 1986. Rates of DNA sequence evolution differ between taxonomic groups. *Science*, **231**, 1393-1398.

Cairns, J., J. Overbaugh, and S. Miller, 1988. The origin of mutants. *Nature*, **335**, 142-145.

Castle, W. E., 1932. *Genetics and Eugenics*, 4th ed. Harvard Univ. Press, Cambridge, Mass.

Crow, J. F., 1957. Genetics of insect resistance to chemicals. *Ann. Rev. Entomol.*, **2**, 227-246.

Dawkins, R., 1976. *The Selfish Gene*. Oxford Univ. Press, New York.

Dobzhansky, Th., 1944. Chromosomal races in *Drosophila pseudoobscura* and *D. persimilis*. *Carnegie Inst. Wash. Publ. No. 554*, Washington, D. C., pp. 47-144.

———, 1947. A directional change in the genetic constitution of a natural population of *Drosophila pseudoobscura*. *Heredity*, **1**, 53-64.

Doolittle, W. F., and C. Sapienza, 1980. Selfish genes, the phenotype paradigm, and genome evolution. *Nature*, **284**, 601-603.

Drake, J. W., 1970. *The Molecular Basis of Mutation*. Holden-Day, San Francisco.

Felsenfeld, G., 1985. DNA. *Sci. Amer.*, **253** (4), 58-67.

Hall, B. G., 1988. Adaptive evolution that requires multiple spontaneous mutations. I. Mutations involving an insertion sequence. *Genetics*, **120**, 887-897.

Harris, H., 1966. Enzyme polymorphisms in man. *Proc. Roy. Soc. London B*, 164, 298-310.

Lederberg, J., and E. M. Lederberg, 1952. Replica plating and indirect selection of bacterial mutants. *Jour. Bact.*, **63**, 399-406.

Lenski, R. E., M. Slatkin, and F. J. Ayala, 1989. Another alternative to directed mutation. *Nature*, **337**, 123-124.

Lewis, H., 1973. The origin of diploid neospecies in *Clarkia*. *Amer. Nat.*, **107**, 161-170.

Lewontin, R. C., and J. L. Hubby, 1966. A molecular approach to the study of genic heterozygosity in natural populations. II. Amount of variation and degree of heterozygosity in natural populations of *Drosophila pseudoobscura*. *Genetics*, **54**, 595-609.

Liming, S., Y. Yingying, and D. Xingsheng, 1980. Comparative cytogenetic studies on the red muntjac, Chinese muntjac, and their F_1 hybrids. *Cytogenet. and Cell Genet.*, **26**, 22-27.

Mendel, G., 1866. Versuch über Pflanzen-Hybriden. This is Mendel's classic paper, originally published in the Proceedings of the Brünn Natural History Society. It has been translated into English and reprinted under the title *Experiments in Plant Hybridization*.

Nevo, E., 1978. Genetic variation in natural populations: patterns and theory. *Theor. Pop. Biol.*, **13**, 121-177.

Ohno, S., 1979. *Major Sex Determining Genes*. Springer-Verlag, Berlin.

Orgel, L. E., and F. H. C. Crick, 1980. Selfish DNA: the ultimate parasite. *Nature*, **284**, 604-607.

Rich, A., A. Nordheim, and A. H.-J. Wang, 1984. The chemistry and biology of left-handed Z-DNA. *Ann. Rev. Biochem.*, **53**, 791-846.

Stone, W. S., 1962. The dominance of natural selection and the reality of superspecies (species groups) in the evolution of *Drosophila*. *Univ. of Texas Publ.*, **6205**, 507-537.

Strickberger, M. W., 1985. *Genetics*, 3rd ed. Macmillan, New York.

Symonds, N., 1989. Anticipatory mutagenesis? *Nature*, **337**, 119-120.

Taylor, A. L., 1970. Current linkage map of *Escherichia coli*. *Bacteriol. Rev.*, **34**, 155-175.

Thompson, D. W., 1942. *On Growth and Form*, 2nd ed. Cambridge Univ. Press, Cambridge.

Wolfe, S. L., 1981. *Biology of the Cell*. Wadsworth Publ. Co., Belmont, Calif.

Womack, J. E., and Y. D. Moll, 1986. Gene map of the cow: conservation of linkage with mouse and man. *Jour. Hered.*, **77**, 2-7.

Yunis, J. J., and O. Prakash, 1982. The origin of man: A chromosomal pictorial legacy. *Science*, **215**, 1525-1529.

REFERENCES

SYSTEMATICS AND CLASSIFICATION

he chapters that follow seek to describe, in a general way, the probable course of events that took place in the evolution of various groups of organisms. Much of this effort derives from morphological and functional descriptions of these organisms and is based on enumerating and comparing their similarities and differences—areas that are the traditional province of **systematics, or classification**.[1]

As we have seen in Chapter 1, techniques of classifying organisms were formulated much earlier than acceptance of the concept that their similarities and differences arose from evolutionary causes. It is, in fact, fairly easy to classify organisms in a variety of ways that can obscure their common origins. For example, fish and whales can be classified in one group, flies and birds in another, frogs and crocodiles in a third, and squirrels and monkeys in a fourth. Of course, by the eighteenth and nineteenth centuries the criteria of classification were less arbitrary, and taxonomists such as Linnaeus used a multitude of features in their descriptions rather than the single character of whether an organism swims in water, flies in air, crawls in mud, or climbs trees. Nevertheless, there is still dispute on how many characters must be compared in order to obtain a "natural" classification and which characters are to be given greater consideration than others (weighting).

With the advent of the Darwinian revolution an additional consideration entered into the thinking of at least some systematists, and that was whether classification could or should be used to reflect evolutionary relationships. There were certainly strong indications that many organisms grouped together because they possessed a large number of similar features could also be said to descend from a common ancestor (p. 35ff). But such determinations revealed only one

[1]The terms systematics, classification, and taxonomy are often used interchangeably, although some taxonomists such as Simpson consider systematics a much broader field: the study of the diversity of organisms and all their comparative and evolutionary relationships, including such topics as comparative anatomy, comparative ecology, comparative physiology, comparative biochemistry. As a subtopic of systematics, classification is defined by Simpson (1961) as the ordering of organisms into groups, and taxonomy is considered to be the study of the principles and procedures of classification.

aspect of evolutionary classification. Another aspect was to discern lines of descent between groups that shared perhaps only a few features in common. As Darwin put it,

> Our classification will come to be, as far as they can be so made, genealogies . . . we have to discover and trace the many diverging lines of descent in our natural genealogies, by characters of any kind which have long been inherited.

This question of exact genealogy, or **phylogeny**, between the many different groups of organisms was not, and is often still not, easily soluble.

A primary reason for the difficulty in determining phylogenetic relationships is the difficulty in finding an unbroken line of ancestors that connects different groups. The fossil record may be fairly complete for some groups such as horses (pp. 45-46), but it is quite meager for birds, whales, insects, early angiosperms (flowering plants), and many other organisms. These fossil-record inadequacies arise from a number of causes:

1. The organisms themselves may be destroyed before, as well as after, their deposition in a sedimentary layer.[2]
2. The formation of sedimentary layers is usually limited to areas which allow them to accumulate without the great disturbances caused by wind, wave action, or other forces.
3. Even when formed, sedimentary layers may later be eroded or moved about by various geological events, causing discontinuities in the record.
4. Only a small portion of fossil-bearing sedimentary rocks are accessible to paleontologists.

For the most part, evolutionists are therefore forced to hypothesize phylogenies by considering and weighting all the known relationships among the different groups, rather than by directly observing common ancestors in the fossil record (although some fossils may be quite similar to ancestral forms). Also, it has seemed to many evolutionists that how organisms are grouped and classified should somehow coincide with their phylogenetic relationships.

Unfortunately, taxonomists are not always agreed on methods of classification and therefore on which groups of organisms are to be classified together. As shown in Figure 11-1, five different schemes of classification have been offered for the same groups of insects. Moreover, even when similarities between organisms are agreed upon, it is not always clear how many subgroups are involved. For example, the single genus *Rubus* (blackberries, raspberries, loganberries) has been divided into 500 species by one botanist, 200 by another, and 25 by a third. To the evolutionist, taxonomic problems seem to be at least twofold:

- How to recognize the basic unit of classification, the species, and then to identify this unit, if possible, with a fundamental evolutionary unit.
- How to order species into systems that will connect them all into a reasonably accurate phylogenetic scheme.

SPECIES

Among the variety of species definitions that have been offered, taxonomists have generally used morphological criteria, since it is primarily in this form that most individuals have been compared. Thus, Davis and Heywood define species as "assemblages of individuals with morphological features in common and separable from other such assemblages by correlated morphological discontinuities in a number of features." Simple as this procedure may seem, it relies heavily on the personal predilections of the taxonomist involved, and as described above, leads in numerous instances to different numbers of species for the same groups. "Lumpers" tend to combine populations into single species or groups, whereas "splitters" tend to separate the same populations into different species or groups. The distinction between extreme proponents of these two points of view seems to lie in whether taxonomy should be used to unite those organisms that share any common features at all (lumpers) or to separate organisms that differ in any respect at all (splitters).

In order to eliminate at least some such arbitrary distinctions, numerical methods have been proposed in which taxonomic distinctions are based on the size of the statistical correlation for as large a number of characters as possible. These characters, all given equal numerical weight, are presumably the ultimate

[2]Newell estimates that, after organic decomposition, only the remains of individuals from 10 or 15 species that died on a tropical river bank can be identified out of the 10,000 or so species that live in the area. The proportion of preserved species is greater in some ocean environments such as limestone reefs, but even then the proportion of identifiably preserved species is no greater than 1 or 2 percent.

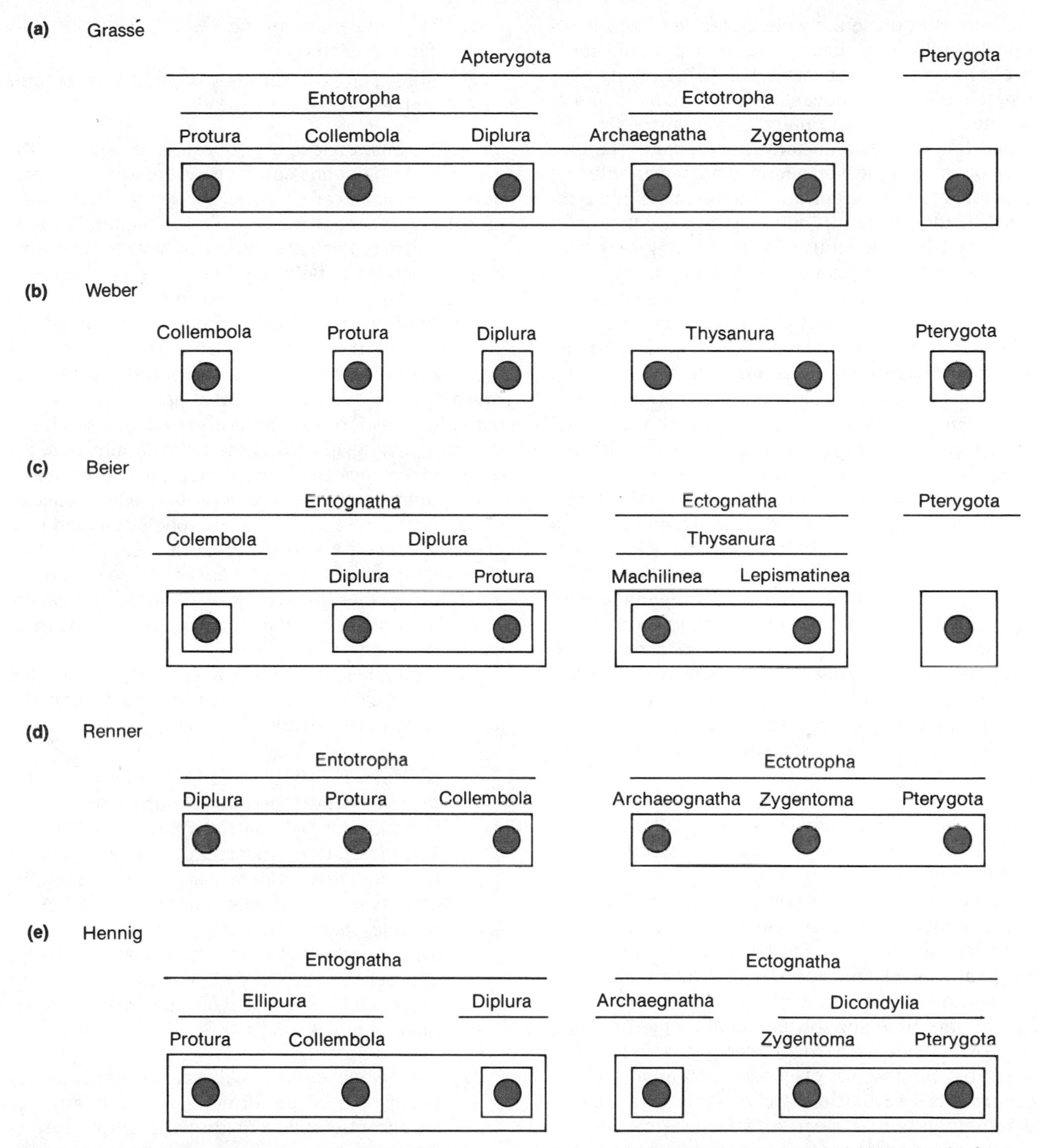

FIGURE 11-1 Five different classifications presented by different authors (a-e) for the same six groups of insects. These groups, indicated by filled circles, have been ranked in categories from subclass to suborder. (From Hennig 1975.)

distinctive qualities of the organisms involved and are not further divisible. A high statistical correlation between individuals for a large number of such characters would indicate their membership in the same species or groups, and a low correlation would point to their separation into different species or groups. To a large degree, this method, called **numerical** or **phenetic taxonomy**, formalizes some of the processes used intuitively by taxonomists but adds to these a degree of quantitative numerical consistency which pheneticists present as a more exact alternative to the usual **classical taxonomy**.

There is at present a wide literature in which numerical studies have been used to help clarify some taxonomic problems (Sneath and Sokal). Although this approach offers the advantage that many different characters can be evaluated simultaneously, the method has not been universally accepted. The presumption in numerical taxonomy that measurable unit characters are not further divisible is considered a serious difficulty, since it is certainly possible that characters used by taxonomists, such as wing length or body measurements, can be divided into further components and therefore lead to different statistics and different relationships. Also, to give equal weight in classification to every character, no matter what its complexity, may hardly be appropriate for characters which have greater evolutionary importance than others. For example, differences in chromosomal structure or in homologous pairing, characteristics that may easily isolate different groups because of cell division abnormalities (see pp. 176 and 182), are certainly more frequently significant in evolution than a character such as petal color. Some practicing taxonomists have also pointed out that the statistical results obtained from numerical methods still depend on subjective choices as to how they are to be taxonomically evaluated. For example, there is still a choice as to which value of a correlation coefficient is to be used for classifying individuals into a single species.

We can see that whether the approach is traditionally morphological or statistically numerical, both are based to some degree on subjective elements in the minds of taxonomists. That is, they both exemplify arbitrary choice, or, as John Locke stated long ago, "the boundaries of the species, whereby men sort them, are made by men."[3]

Many biologists have not been satisfied to allow species definitions to rest primarily on a subjective morphological approach and have instead adopted a **biological species concept**. Derived from Buffon (pp. 9-10) and others, this concept defines a species as a sexually interbreeding or potentially interbreeding group of individuals normally separated from other species by the absence of genetic exchange, that is, by **reproductive isolation**. The obvious advantage of this definition is that species distinctions can be objectively tested by two relatively simple criteria:

1. Do populations in the same locality normally fertilize each other?
2. Should cross-fertilization occur, are the hybrids viable and fertile?

If the answer to either question is no, then the evaluated populations can be considered to be reproductively isolated or separate species. (A description of isolation mechanisms is given in Chapter 23.)

It is such biological criteria that have enabled species distinctions to be made between similar-appearing populations that could not formerly be separated on the basis of the usual morphological taxonomic criteria. Thus, the fruit fly species *Drosophila pseudoobscura* and *D. persimilis*, called **sibling species** because they are almost identical in appearance, do not normally cross-fertilize; this is also true for some leafy-stemmed sibling species in the phlox family, *Gilia tricolor* and *G. angelensis*. Such biological tests have also led to unifying different groups into single species which had been separated by morphological and geographical criteria into distinct species (e.g., the union of various species of North American sparrows into one **polytypic species** consisting of multiple geographic races or subspecies, the song sparrow, *Passarella melodia*).

However, in spite of these advantages, application of the biological species concept universally encounters considerable difficulties:

1. Although it may be possible to observe reproductive barriers between groups found in the same locality (**sympatric populations**; Chapter 23), there are many practical limitations in attempting crosses between groups that are ordinarily separated (**allopatric populations**). To provide space and appropriate environments for the enormous numbers of possible crosses between all allopatric combinations of similar organisms appears beyond the present capability of biologists.[4]

2. Even when crosses between allopatric populations can be performed, some arbitrary decisions must still be made. For example, results from interbreeding experiments may range

[3]"An Essay Concerning Human Understanding" (1689).

[4]For plants, Baker writes:

> The number of hybridization attempts which must be made, the number of plants which must be raised in the first hybrid generation and the number of subsequent generations which must be grown to see whether fertility is maintained and whether segregation occurs, all place limitations on the comprehensiveness of the experiments.... The total task, for naturally occurring plants, is beyond human capacity for achievement.

from no genetic exchange at all for certain attempted crosses to a fairly large degree of genetic exchange for others (Fig. 11-2). The question then arises, At which point on this scale of interbreeding values shall we separate species?[5]

3. Fossil populations cannot be tested as to whether they can or cannot exchange genes either among themselves or with present populations.

4. In asexual organisms, no matter whether reproduction occurs by fission or by parthenogenesis, each clone of individuals is essentially genetically isolated from every other clone, yet few, if any, biologists would consider describing each clone as a separate species.

In order to salvage the objective criteria used in characterizing the biological species but to broaden its definition to those populations of the type listed above, various authors have proposed an **evolutionary species concept**. That is, a species is defined in terms of differences that are not dependent on sexual isolation but rather on its evolutionary isolation, of which sexual isolation is only one aspect. In Simpson's words, "an evolutionary species is a lineage (an ancestral-descendant sequence of populations) evolving separately from others and with its own unitary evolutionary role and tendencies." This definition may also be considered to lay the groundwork for the changes that result from competition and interaction between species: the existence of separate evolutionary lineages implies that an important factor affecting their success and survival may be the success and survival of other such lineages.

However, again, the problem of an evolutionary species definition is that such distinctions may be difficult to make in practice; taxonomists faced with a large variety of specimens often have few techniques to distinguish between them, other than the purely morphological. Also, since evolutionary speciation is a process, defining the point at which groups of organisms have reached complete separation still has, of necessity, some arbitrary elements of choice. (To define a species as an "individual" clearly separated from other "individuals" can therefore be a gross oversimplification: see the discussion in Ruse on whether species are individuals, classes or populations.) Nevertheless, it seems clear that the evolutionary species concept justifies bringing to bear ecological, behavioral, genetic, and morphological evidence as a reflection of evolutionary separation or distance. It is, after all, evolutionary separation between populations that accounts for species differences: without evolution there are no species.

In summary, the difficulties in species taxonomy are, to a large extent, inherent in the process of speciation itself. That is, the differences between populations that makes some of them difficult to classify as varieties, subspecies, or species arise from the fact that they undergo evolutionary changes that can differ in intensity and sequence. Phenotypic and genotypic distinctions do not therefore evolve in a uniform fashion, but rather comparisons between populations generally show different degrees of change in various characters. In sexual forms, the overall result of such differences is different degrees of reproductive isolation and morphological distinctiveness, whereas in asexual forms evolutionary distinctions are reflected in differences other than reproductive isolation.[6] In both cases, members of a species share a community of descent that explains many of their common features. The task of the taxonomist is not always made easier because of this evolutionary explanation, but it has the virtue of providing an understanding for some of the difficulties in separating certain groups of organisms into species.

PHYLOGENY

Given the existence of a primitive ancestor, the question of the origin of species is basically the question of the origin of new species. Darwin mostly devoted himself to explaining how, under natural selection, a single species can change through time. Successional changes within a single lineage have been called by various authors **phyletic evolution** or **anagenesis**, an

[5]Interestingly, even when genetic exchange is completely uninhibited between some allopatric populations, it may still seem desirable to consider them as separate species since they do not hybridize under normal conditions. One well-known example is the discovery of two widely separate populations of trees occupying similar habitats, one in China (*Catalpa ovata*) and the other in the eastern United States (*C. bignoides*). Although they can cross with each other to produce hybrids that are as viable and fertile as the parents, these populations have probably been separated for many millions of years, and botanists have therefore generally agreed to continue their separate species identifications.

[6]According to some authors, the magnitude of the differences between recognized species in sexual forms, other than reproductive isolation, should also be used to distinguish species in asexual forms.

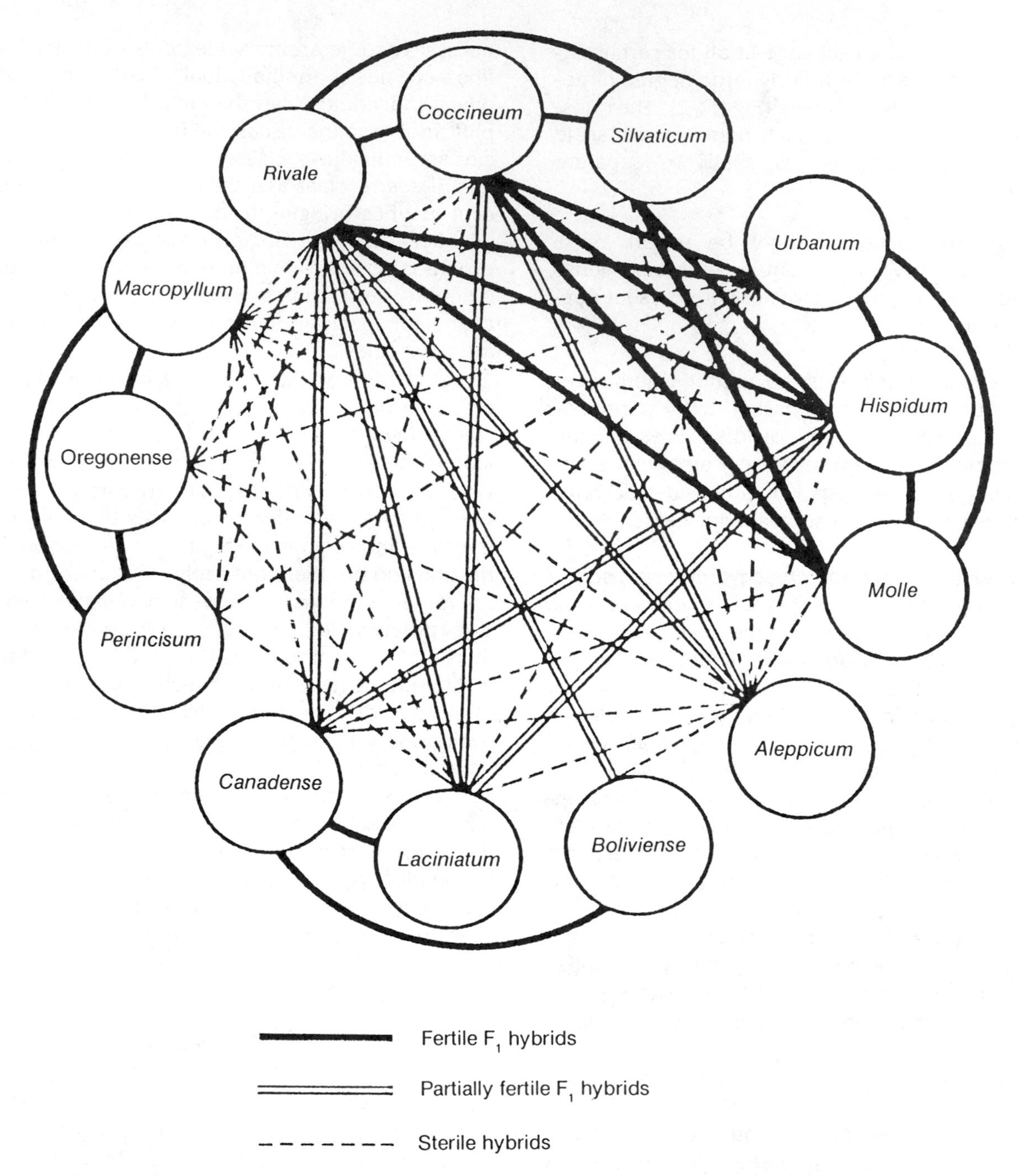

FIGURE 11-2 Differences in fertility observed for F_1 hybrids derived from crosses between 13 species of the plant genus *Geum*, a perennial herb. (From Gajewski.)

example of which is shown in Figure 11-3. However, an important concern of many evolutionists was the problem of multiplication of species (p. 33): to explain the splits and divisions within an ancestral line that cause appearance of more than one species—a cluster of species or **clade**. This pattern, known as **phylogenetic branching**, or **cladogenesis**, was first offered by Lamarck (see Fig. 1-5), but it was Haeckel, beginning in the 1860's, who popularized this form of evolutionary description.

As explained previously, the determination of phylogenetic trees is often difficult since the common ancestors of different groups of organisms are usually long extinct and the fossil record is usually inadequate. The absence of complete fossil information thus puts a great deal of emphasis on constructing phylogenies by comparisons between known organisms, whether existing or fossil. In general, the more a group of species shares common inherited attributes, the more likely their descent from a common ancestor. All avail-

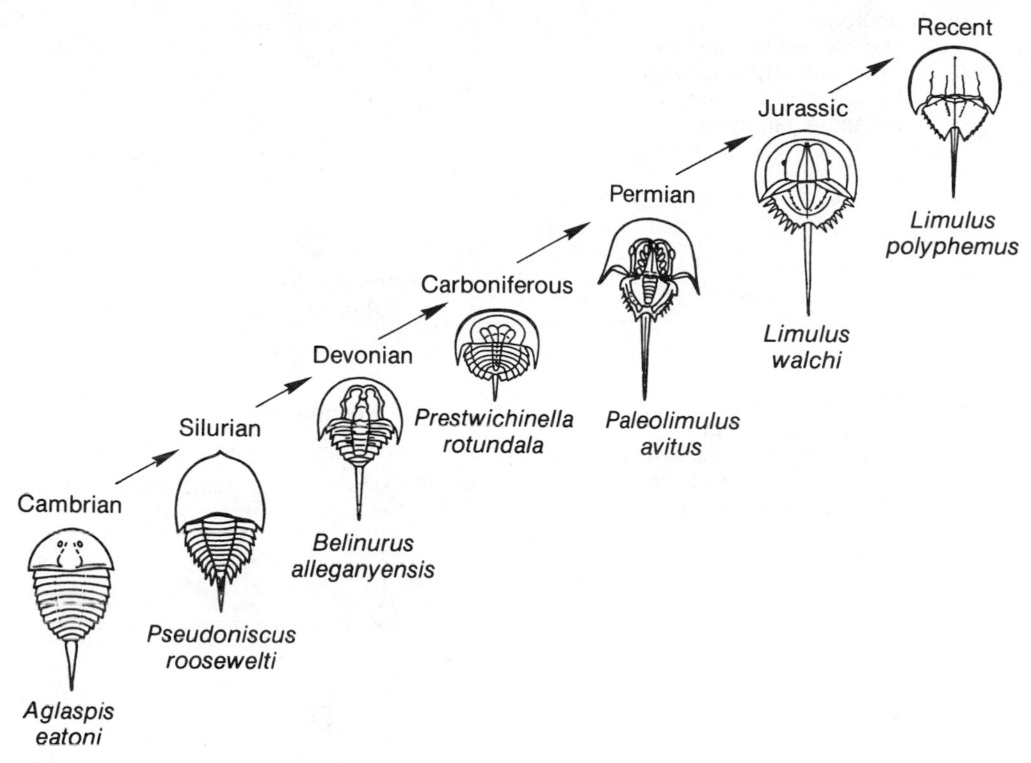

FIGURE 11-3 Phyletic evolution in merostome arthropods (horseshoe crabs), indicating relatively small phenotypic changes over long periods of time. (Adapted from Newell.)

able heritable characteristics are therefore enlisted in making comparison between species: morphological, embryological, behavioral, physiological, biochemical, genetic, and chromosomal. It is assumed that the more two species are alike in these respects, the more they share common hereditary characteristics and the closer at hand their common ancestor. Nevertheless, even when species share common features, the genotypic basis for this may derive from various evolutionary causes, as shown in Figure 11-4:

Homology The same feature occurs in different species because it derives directly from a common ancestor that bore the same characteristic. For example, the many similar features in the forelimb structure of vertebrates indicate derivation from a common vertebrate ancestor (see Fig. 3-6).[7]

Parallelism A similar feature occurs in different species, but their immediate common ancestor appeared differently. For example, anteaterlike features have appeared in different lines of mammals that were descended from non-anteater mammalian groups (Fig. 11-5).

Convergence A similar feature occurs in different species whose ancestral lineages differed from each other in this respect for a considerable period of time. We have seen an example of this in the similarity of marine hydrodynamic forms among the widely separated fish, reptile, and mammalian classes of vertebrates (see Fig. 3-7).

It is obvious that it is homology that is the basis for establishing phylogenetic lineages since the sharing of phenotypes because of common descent signifies a closer relationship between organisms than any other cause for phenotypic similarity. However, in comparing organisms, many characters are usually considered, and it is necessary to realize that some similarities may be caused by homology and others by parallelism or convergence. Perhaps an extreme ex-

[7]The term **serial homology** has been used to describe similarities between parts of the same organism, such as homologies between the various vertebrae in a vertebrate, or between the different feathers in a bird, or between the different kinds of hemoglobin molecules (α, β, γ, etc.) produced by a particular individual. The genetic basis for serial homology often comes from the duplication of a gene responsible for producing or affecting a particular structure. Such duplicates may originally possess similar features, but they may also evolve differently from each other (see Chapter 12).

(a) Homology:
two species bearing the same phenotype caused by common ancestry for the same genotype

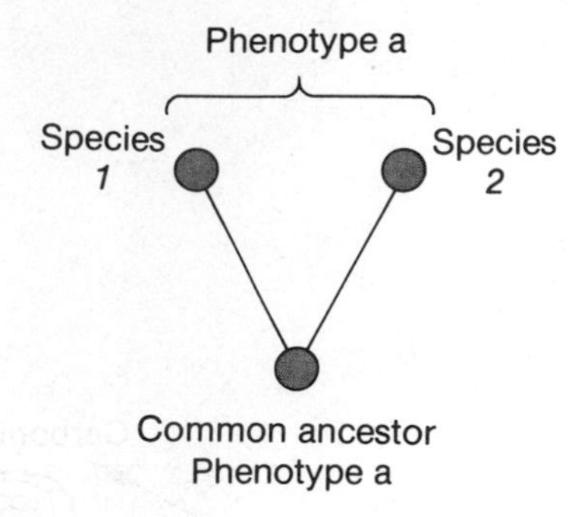

(b) Parallelism:
two species with the same phenotype descended from a common ancestor with a different phenotype

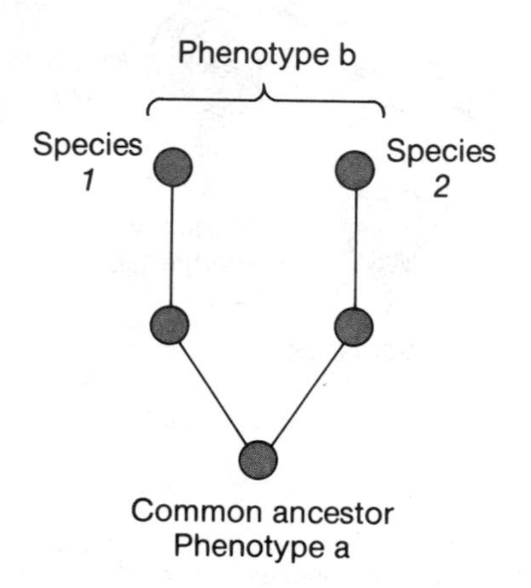

(c) Convergence:
two species with the same phenotype whose common ancestor is very far in the distant past

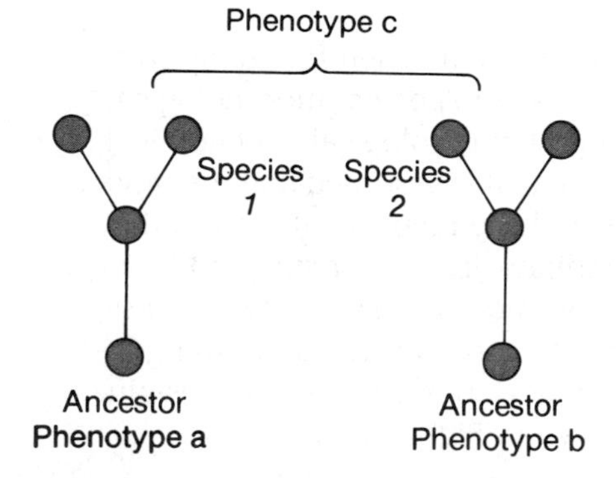

FIGURE 11-4 The phenomena of homology, parallelism, and convergence diagrammed for two species, labeled *1* and *2*, that are phenotypically similar for a particular character. Note that the distinction between parallelism and convergence may be arbitrary since there are no rules that restrict how far in the past one can establish a common ancestor for parallel evolution, and even convergent lineages have common ancestors, albeit distant ones.

ample of this is to note that the similarity between the shape of the pectoral flipper in mammalian whales and the shape of the pectoral flipper in ancient reptilian ichthyosaurs undoubtedly arises from convergence caused by selection in the two groups for swimming efficiency, yet there are many structural features of both these limbs that are homologous because of their relationship to the forelimb of a common land vertebrate ancestor. Determining the phylogenies of these organisms obviously depends upon separating the homologies (they are both vertebrates) from the convergences or parallelisms (they are not members of the same vertebrate class).

In some related lines, however, such distinctions are not always easy to make. Mammalian fossils, for example, can be defined by various criteria (lower jaw composed of a single bone [dentary], a dentary-squamosal jaw joint, and ear with three ossicles; Chapter 18) and their ancestry traced back to an ancient order of reptiles, the Therapsida. Using these criteria,

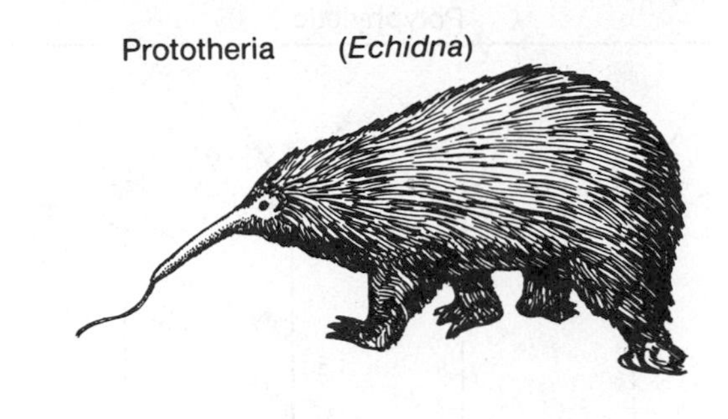

FIGURE 11-5 Similar phenotypic features among anteaters (long snout and tongue, powerful claws) that evolved independently within each of the three major groups of existing mammals.

it is quite possible to claim that different groups among the therapsids independently evolved the mammalian level, or **grade**, of organization and function because of parallel evolution (Kermack and Kermack); that is, similar mammalian features shared by some lines of early mammals are not the result of homology but of parallelism. Such instances have been given the name **polyphyletic evolution** to define cases in which a particular grade of organization is arrived at independently by different groups of organisms, in contrast to **monophyletic evolution** in which all organisms included within a particular grade derive from only a single ancestral population (Fig. 11-6). According to Simpson (1961), one extreme polyphyletic view suggested that each mammalian species arose separately from a single ancestral protozoan species.

PROBLEMS OF CLASSIFICATION

Ideally, the most descriptive phylogenetic picture of a particular population of organisms would be to show it as a portion of a multi-limbed tree which has branched connections to all present and ancestral populations and which indicates, through these connections, its degree of relationship to all other populations (Fig. 11-7). Since there may be as many as 5 to 10 million existing species of organisms, both known and unknown, and there undoubtedly existed some hundreds of millions of extinct species in the past, mostly unknown, it is obvious that such a complete picture is impossible to achieve. Nevertheless, a picture of some sort is desirable, and evolutionists have usually placed much of this descriptive burden on how organisms are classified.

Traditional classification, as exemplified in Linnaeus's system (p. 9), was not based on evolutionary criteria but on what seemed to be "natural." The attempt to use natural relationships to organize the large numbers of groups of organisms that were being discovered into simpler but fewer categories prompted the introduction of hierarchical classifications in which an organism was placed not only in a particular species but also in ranked categories that included other species (the genus), other genera (the family, the order), and so forth. This system, somewhat extended from the time of Linnaeus, is still in use today.

Each unit of classification, whether it be a particular species, genus, order, or whatever, is called a **taxon**, and is given a distinctive name. Taxa are arranged in categories so that a taxon in a "higher" category includes one or more taxa in "lower" categories. For example, Figure 11-8 shows some of the categories and taxa often used in the classification of humans (*Homo sapiens*) and fruit flies (*D. melanogaster*).[8]

It is clear that this mode of classification offers a simple scheme for identifying and cataloging large numbers of species. For example, it is sufficient to use phyla such as Arthropoda and Chordata to distinguish many animals and to use mammalian orders such as Primates and Rodentia to distinguish many mammals. From an evolutionary point of view, this classification

[8]Some classifications introduce additional categories to those shown in Figure 11-8 either by adding new terms (e.g., tribe, a rank between family and genus) or by adding the prefixes super-, sub-, and infra-, to the given categories. Thus, the class Mammalia is usually included within the subphylum Vertebrata and the order Primates in the infraclass Eutheria.

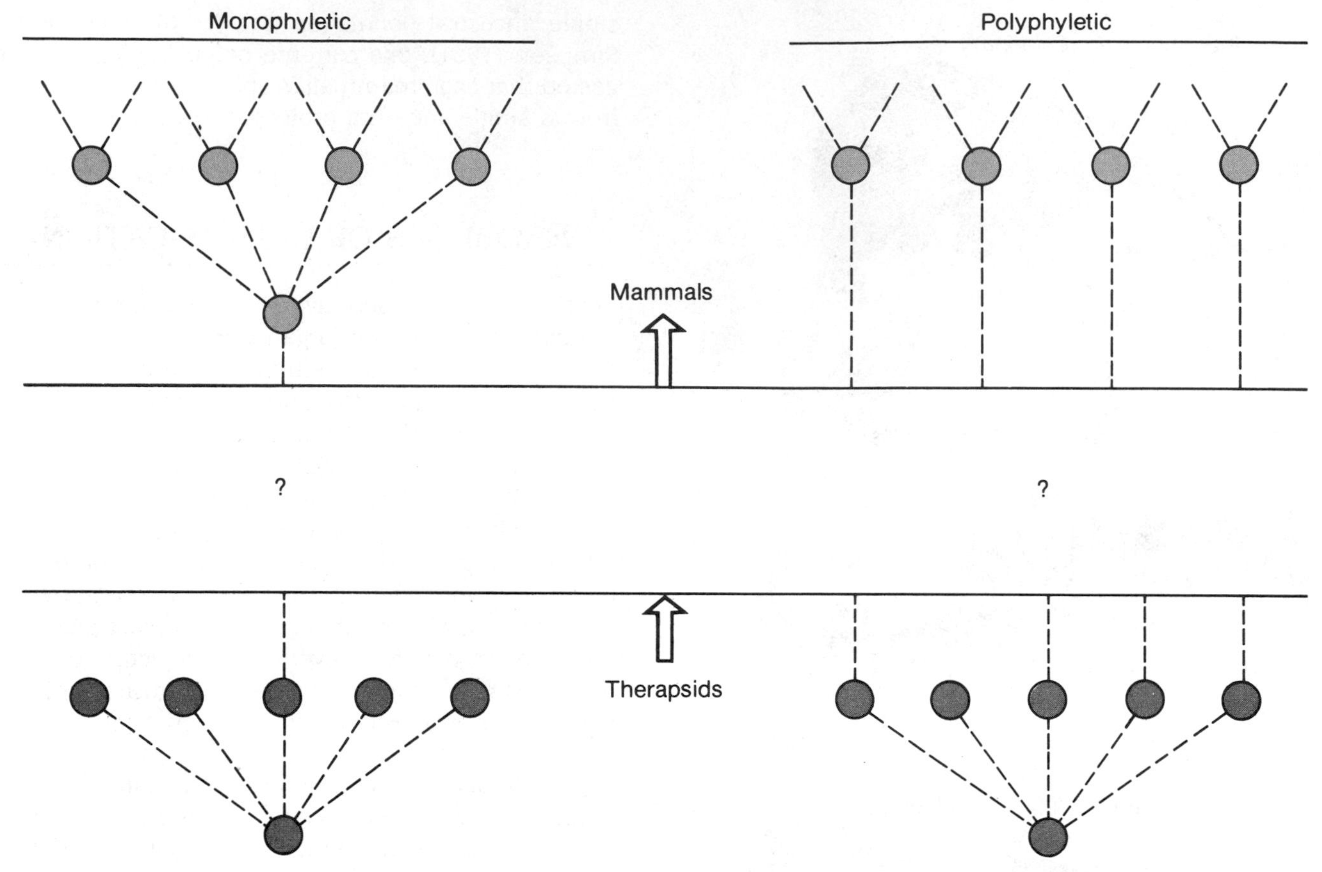

FIGURE 11-6 Monophyletic and polyphyletic schemes that can be used to explain the evolution of mammals from therapsid reptiles. In monophyletic evolution only a single therapsid group served as ancestor to the mammalian radiation, whereas two or more therapsid groups gave rise to mammals in the polyphyletic scheme. (Instead of monophyletic, some authors use the term holophyletic to describe that portion of a phylogenetic tree that includes a common ancestor and all of its descendants.)

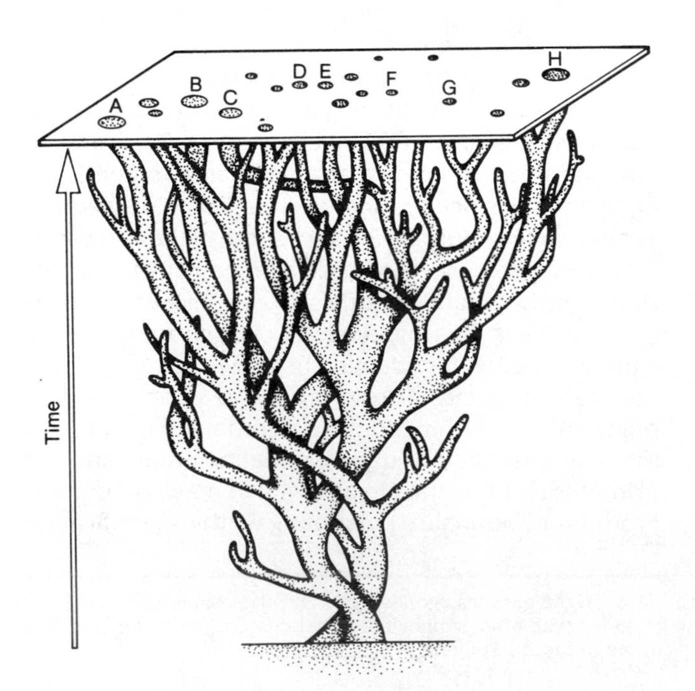

FIGURE 11-7 Diagram of a phylogenetic tree of related populations shown as continuous branches undergoing evolutionary changes through time. Some populations have become extinct, and others have merged or diverged to produce new and different forms. If we consider time as the vertical axis in this illustration, distances along the x and y axes might indicate measurements of different genetic traits. Thus, the differences between some populations, such as A and H at the present time level (top of figure), may be sufficiently great to warrant separate taxonomic designations, whereas others (e.g., D and E) may not yet be taxonomically distinct. Note also that there may be convergences between two separate lineages (e.g., B and C) in respect to the measured traits, which can conceal their evolutionary separation. (Adapted from Levin.)

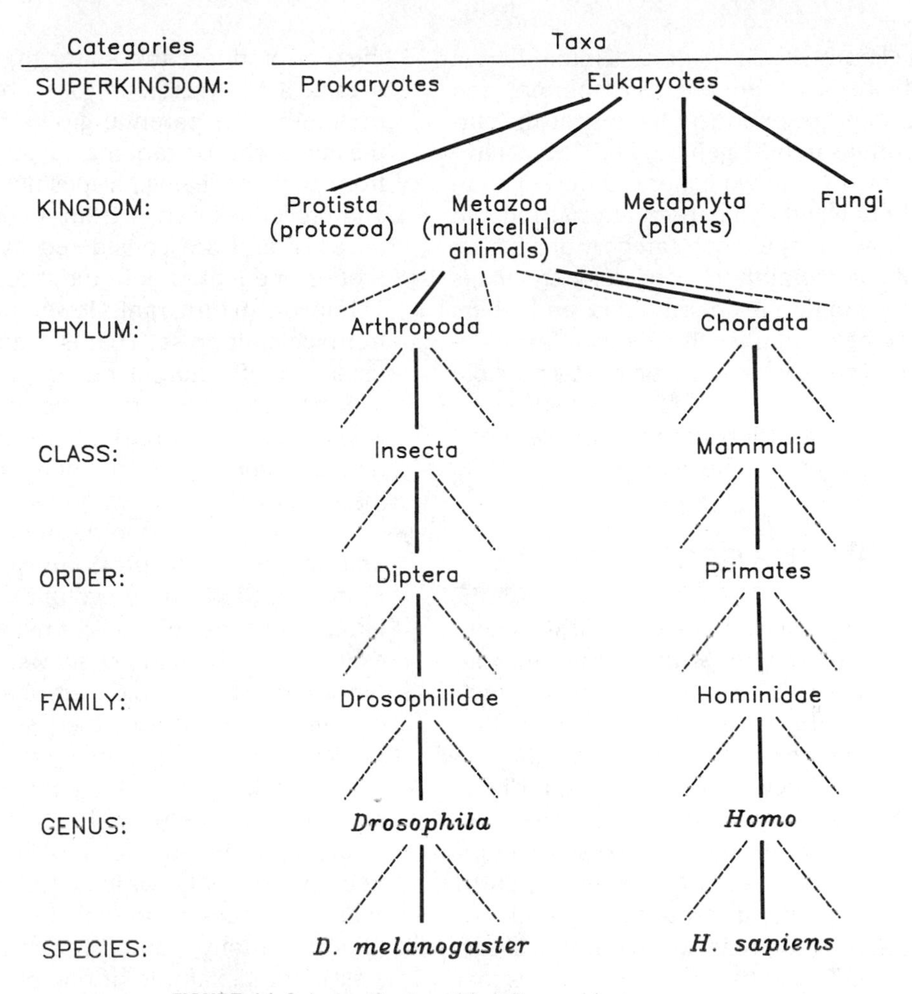

FIGURE 11-8 A classification of fruit flies and humans.

also has much to commend it, since there are obvious homologies between species within taxa such as Diptera or Primates. To a significant extent, these classification schemes reflect an underlying phylogenetic pattern in which each taxon seems to have originated from the one in which it is included.

Unfortunately, each traditional taxon is not always clearly monophyletic or derived from a single common ancestor. Mammals, as explained previously, may have had a polyphyletic origin. Also, the arthropod grade of organization, characterized by an exoskeleton with jointed appendages, is now believed to have been achieved by a variety of segmented annelid-type organisms undergoing parallel evolution; some gave rise to the ancient trilobites and modern crustaceans, while others served as ancestors to groups such as insects (Chapter 15). Claims have also been made by some botanists that some major plant taxa—bryophytes, tracheophytes, gymnosperms, angiosperms—may have had multiple origins (see, for example, discussions in Stewart and in Thomas and Spicer). If these taxa, as well as others, are polyphyletic, then the correlation between classification and phylogeny can hardly be exact: the taxon Arthropoda is not really a single phylogenetic lineage, nor is this necessarily true for the taxon Mammalia, or some others.[9]

A further serious problem is that classification at this level, like species concepts, has arbitrary elements in it. It is common, for example, for different taxonomists working with organisms such as arthropods or

[9]Simpson attempted to deal with this difficulty by defining monophyly as derivation of a group from an ancestral taxon of equal or lower rank, even if more than one population in this taxon are the direct ancestors. Thus, he considered the class Mammalia as monophyletic since the different reptilian populations from which it was derived were all in the same order (Therapsida). Ghiselin points out that a consistent application of this view could lead to the absurdity that the vertebrate phylum is monophyletic even if birds came from echinoids and mammals from crinoids—since both these hypothetical ancestral groups are members of the same echinoderm phylum. Perhaps an alternative solution for eliminating the problem of polyphyletic origin of mammals would be to remove the taxon Synapsida (a group that includes therapsids) from the class Reptilia and place it in the class Mammalia. Adoption of this approach depends, of course, on how prone taxonomists are to accept the early, obviously reptilelike synapsids as mammals.

angiosperms to classify the same groups into different families. As with species, there are also "splitters" and "lumpers"; e.g., some taxonomists have split the various catlike mammals into 28 genera, and others have combined them into one or two genera. Different opinions also exist about whether a particular group should be considered to have attained one category or a higher one. For example, some authors consider nematodes as a class within a phylum Aschelminthes, and others consider them to have attained the rank of an independent phylum. The existence of so many possible arbitrary decisions for classifying organisms above the species level has led to a variety of proposals designed to make classification more objective.

PHENETICS

One proposal, usually called phenetic classification, applies numerical taxonomy to arranging groups into genera and higher ranks. It has the advantage, previously described, that similarities can be evaluated on an objective numerical basis, thereby overcoming the personal prejudices of individual taxonomists. Unfortunately, it has the disadvantage that phylogenetic relationships between different taxa can be obscured if some or many of the characters used in the numerical analysis have become similar because of parallel or convergent evolution. In such cases, the numerical correlations are considerably enhanced between taxa that from an evolutionary point of view should really be separated. Also, the significance of correlations between taxa is based on presuming that the analyzed characters have evolved independently; yet it is often difficult, if not impossible, to show that all of the characters used in numerical taxonomy are independent of each other, since a character such as wing length may be dependent on a character such as body length. Also, some characters may have evolved at a different pace than others if they are tied to unique functions or to particular parts of the life cycle. For example, Michener has reported that a numerical taxonomic analysis of allodapine bees (a group whose larvae are raised in common burrows rather than isolated cells) gives one classification when larval characteristics are used and quite another when adult characteristics are used.

CLADISTICS

A more evolutionary approach to classification than numerical taxonomy is **cladistics**. In this system, whose chief proponent has been Hennig, every significant evolutionary step marks a dichotomous branching that produces two **sister taxa** equal to each other in rank. Since the ranking of such sister groups is below the rank of the parental group that gave rise to them, the hierarchy or ranking of groups derives logically from their genealogical position. For example, birds and crocodiles derive from a common reptilian stem ancestor and are considered by cladists to be sister groups of equal rank in the taxon Archosauromorpha. This taxon, in turn, ranks lower than the class of reptiles from which it arose. That is, cladists do not follow the usual classification of ranking birds (Aves) as a class separate from the class of reptiles (Reptilia), since that would imply that birds arose from a primitive prereptilian stem ancestor rather than from the reptiles themselves.

In cladistic taxonomy importance is put on separating ancestral (**plesiomorphic**) characters from those later derived by evolution and emphasizing the latter in order to establish phylogenies. As the simple example in Figure 11-9 shows, it is whether derived (**apomorphic**) characters are shared (**synapomorphic**) or unique that dictates the phylogeny. When this approach is extended to many taxa and more than one character, fairly complex phylogenies can be inferred by various mathematical methods (e.g., Felsenstein). One popular technique is to choose that phylogenetic tree for the data which minimizes the number of changes necessary to explain its evolutionary history (the "parsimony method"; see also Chapter 12). To use a simplified example, if one phylogenetic tree necessitates five convergent events to explain similarities between some of its species and another tree necessitates only two such events, the latter would be preferred since it minimizes extraneous evolutionary explanations.

Because of its strict genealogical consistency, cladistic classification has been espoused by an increasing number of taxonomists, although it still excites considerable controversy. Among the objections raised against cladism are the following (see also Mayr [1974], Sokal [1975], Ghiselin, and the historical account by Hull):

1. It seems unwarranted to assign a completely new taxonomic designation to each branch if one of the branches remains identical to the previous ancestral population (Fig. 11-10). Certainly, a single ancestral taxon giving rise to new offshoot taxa may be no different after the split than it was prior to the split.

2. Although anagenetic changes in a species can occur through time (see Fig. 11-3), cladistic classification proposes to use the same taxon name throughout evolution as long as a taxon

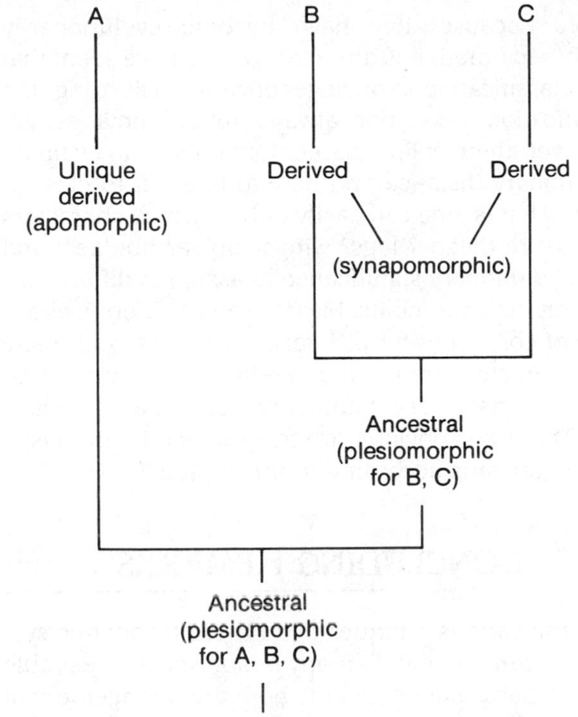

FIGURE 11-9 A cladogram for three taxa, A, B, C, in which a particular character is designated as either ancestral (plesiomorphic, according to cladists) or derived (apomorphic). The sharing of the same character derived from an ancestral group (synapomorphy) by B and C in this diagram indicates that they are sister taxa.

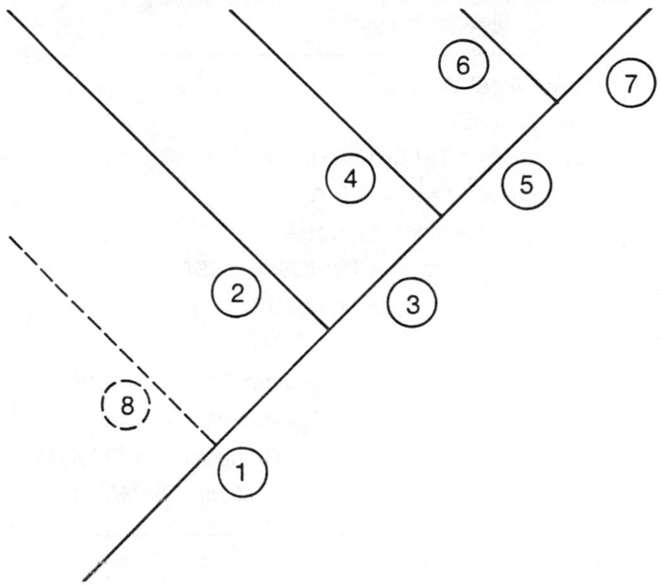

FIGURE 11-10 Illustration of the kind of diagram commonly used in cladistic classification (called a Hennigian comb), where each lineage that extends from the node of a branched fork represents a distinct taxon. Thus populations 1, 3, 5, and 7 are given different taxonomic designations although they may not be recognizably different genetically. If a new lineage of organisms (indicated by the dotted line 8) is discovered to have split from population 1, cladists will change the taxonomic designation for 1 so that it differs before and after this split.

has produced no discernible branches. Obviously, phyletic evolution is not recognized in this method of classification.

3. The cladistic dichotomous scheme is restricted to only two new species at each fork, yet a taxon may certainly produce more than two offshoot taxa (e.g., by multiple budding at the periphery of its distribution).

4. The ranking of taxa into groupings such as families, orders, classes, has been tied by various cladists to their chronological age. That is, branching during the Precambrian eon produced phyla; classes arose during the Cambrian-Devonian period; orders stem from the Mississippian-Permian, and so on. According to this reckoning, taxa that arose at an early age but were limited to one or very few species (such as the brachiopod genus *Lingula*, a living fossil dating back to pre-Devonian times) should be higher in rank than some taxa that are produced at a later stage but include large numbers of genera and species (such as mammals and birds).[10]

5. The rule that each genealogical branching point must be ranked lower than the one that occurred previously has led some cladistic taxonomists to offer an unwieldy proliferation of categories for previously simple classifications. For example, most non-cladist classifications consider the order Primates as ranked below the class Mammalia by two intermediate categories, the subclass Theria (mammals that produce live young) and the infraclass Eutheria (placental mammals). On the other hand, one cladistic classification of primates offers ten intermediate categories (Table 11-1), which may well be further expanded if more genealogical

[10]There appears to be an underlying assumption held by such cladistic proponents that branching is a uniform process with respect to time and that an older group has necessarily branched into a larger number of taxa than a younger group. This is, of course, not always true, since even modern groups can branch extensively, as evidenced by the large numbers of Hawaiian Drosophilidae (650 to 700 species) that are descended from perhaps one or two ancestral species that came to these islands no earlier than perhaps 20 or so million years ago.

TABLE 11-1 Classification of Primates According to One Cladistic Proposal

- Class MAMMALIA
  - Subclass THERIA
    - Superlegion TRECHNOTHERIA
      - Legion CLADOTHERIA
        - Sublegion ZATHERIA
          - Infraclass TRIBOSPHENIDA
            - Supercohort EUTHERIA
              - Cohort EPITHERIA
                - Magnorder PREPTOTHERIA
                  - Superorder TOKOTHERIA
                    - Grandorder ARCHONTA
                      - Order PRIMATES

Adapted from McKenna.

branches between primates and early mammals are discovered. Thus although genealogical accuracy can be gained by this method, it is clear that informational simplicity can be lost.

EVOLUTIONARY CLASSIFICATION

A long-standing alternative to the taxonomic systems described above dates from the post-Darwin period and has been named **evolutionary classification**. According to its major modern proponents (Simpson, Mayr) it uses the traditional taxonomic categories but gives special consideration to evolutionary relationships and biological attributes rather than to strict morphological relationships. That is, incorporated into group taxonomy are factors such as functional and morphological innovations, adaptive range, and numbers of species. Evolutionary classification therefore attempts to provide both the genealogical relationship between groups and the amount of evolutionary change or distance between them.

Unfortunately, attempts to provide a classification based on evolutionary distances can distort phylogenetic relationships and vice versa. Thus, although birds and reptiles are designated as distinct vertebrate classes because they have become evolutionarily widely separated, it is also clear, as we have seen, that such classification is phylogenetically misleading: the classification does not always reflect phylogeny.[11] Moreover, there is little agreement on how to evaluate evolutionary distance and how to relate it to classification. That is, one can easily debate which characters to measure (heads? legs? wings? tibiae? fibulae?) and what evolutionary significance to assign to differences between measurements. Measurement differences are also not consistent for different characters, and many such characters are limited to relatively few groups so that comparisons of evolutionary distance are also limited. There are obvious arbitrary elements in classifying organisms according to this system.

CONCLUDING REMARKS

From the various critiques of all these taxonomic systems we can see that, to many biologists, the desirable goals of classification include both the arrangement of groups into a pattern that accurately reflects their evolutionary relationships and the placement of groups into a reference system so their major features are easily and efficiently described and identified (information storage and retrieval). These two purposes are apparently not fully accomplished by any single classification system reviewed so far. Traditional morphological classification and numerical taxonomy may simplify the placement of organisms into a classification scheme but can ignore their evolutionary relationships. Cladistic classification may offer advantages in clarifying some evolutionary patterns but overlooks others or seriously complicates the use of taxonomic information. Evolutionary classification can be more arbitrary than cladistics by classifying a group such as birds outside its strict phylogenetic sequence, but it can also offer descriptive and evolutionary information in a more useful form. The fairly common acceptance of evolutionary classification probably lies in the fact that it is a synthesis incorporating some of the essential attributes of other systems. The flexibility afforded by the eclecticism of this system enables it to offer more evolutionary features than traditional or numerical taxonomy, as well as more simplicity, and even, at times, more evolutionary information than cladistic classification. Nevertheless, it seems clear that the perfection of a biological taxonomic system is yet to be achieved.

[11] As Simpson (1980) explains this,

> The fundamental difference is that phylogeny is something that happened and classification is an arrangement of its results. Although phylogeny cannot be observed as such over periods long enough to be really significant, it existed as a sequence of factual events among real things (organisms) and in a philosophical or logical sense it is objective or realistic in nature. Classification is not. It is an artifice with no objective reality. It arises and exists only in the minds of its devisers, learners, and users.

SUMMARY

In the earliest stages, classification involved observations of similarities and differences among organisms, without regard to their origins. Since Darwin's time, many taxonomists have attempted to construct a system which would reflect phylogenetic (genealogical) relationships, but fossil evidence of exact ancestral relationships, the ideal basis for such schemes, is often lacking. The main goals of the evolutionary taxonomist are to recognize the basic unit of classification, species, and to order them into as realistic a phylogenetic system as possible.

Various methods have been used to define the species. Until recently morphology had been used almost exclusively, but to increase exactitude, newer systems have been devised. In phenetics or numerical taxonomy, numerical values are assigned to characters, and a cluster of these values is used to define a species. The biological species concept is intended to overcome the subjectivity inherent in the above methods, and the ability of populations to interbreed and produce viable offspring are used as the primary criteria for determining species boundaries. However, because of the enormous number of species, their geographic dispersal, and limitations in space and manpower, and because most are extinct, it is impossible to differentiate many species by these criteria.

The evolutionary species concept represents an effort to resolve these difficulties by defining species based on their evolutionary isolation from each other. Ideally, this method utilizes morphological, genetic, behavioral, and ecological variables, although it too does not resolve all the problems intrinsic to species taxonomy since not all traits evolve at the same rate or in the same sequence.

Reconstructing phylogenies is even more difficult than defining species. Phenotypes may be alike because of common origin (homology), because of similar evolutionary patterns arising separately in different lines from not-too-distant common acestors (parallelism), or because of the development of similar characteristics in groups originating from completely different ancestors (convergence). Traditional classification schemes utilize hierarchical schemes based on natural criteria to order organisms into taxa. But these taxa are not necessarily monophyletic (coming from a common ancestor), and their designation may be arbitary. Phenetics endeavors to make classification more objective by applying numerical methods to taxa, while cladistics assumes that evolution occurs in a series of dichotomous branchings, each evolutionary branchpoint giving rise to taxa of equal rank. Major difficulties arise with phenetics if convergent or parallel evolution has occurred, and cladistics does not deal adequately with phyletic evolution or multiple branching.

In evolutionary classification, traditional taxonomic categories and morphological criteria are employed, but many other biological factors are incorporated as well. In this way it is hoped that genealogical relationships as well as evolutionary distance can be depicted. However, resolution of these issues within a single classification system remains elusive.

KEY TERMS

allopatric populations
anagenesis
apomorphic
biological species concept
clade
cladistics
cladogenesis
classical taxonomy
classification
convergence
evolutionary classification
evolutionary species concept
grade
homology
monophyletic evolution
numerical taxonomy
parallelism
phenetic taxonomy
phyletic evolution
phylogenetic branching
phylogeny
plesiomorphic
polyphyletic evolution
polytypic species
reproductive isolation
serial homology
sibling species
sister taxa
species
sympatric populations
synapomorphic
systematics
taxon

REFERENCES

Baker, H. G., 1970. Taxonomy and the biological species concept in cultivated plants. In *Genetic Resources in Plants*, O. H. Frankel and E. Bennett (eds.). Blackwell, Oxford, pp. 47-68.

Brooks, D. R., and E. O. Wiley, 1985. Theories and methods in different approaches to phylogenetic systematics. *Cladistics*, **1**, 1-11.

Cracraft, J., 1983. The significance of phylogenetic classifications for systematic and evolutionary biology. In *Numerical Taxonomy*, J. Felsenstein (ed.). Springer-Verlag, Berlin, pp. 1-17.

Davis, P. H., and V. H. Heywood, 1963. *Principles of Angiosperm Taxonomy*. Van Nostrand, Princeton, N. J.

Eldredge, N., and J. Cracraft, 1980. *Phylogenetic Patterns and the Evolutionary Process*. Columbia Univ. Press, New York.

Felsenstein, J., 1982. Numerical methods for inferring phylogenetic trees. *Quart. Rev. Biol.*, **57**, 379-404.

Gajewski, W., 1959. Evolution in the genus *Geum*. *Evolution*, **13**, 378-388.

Ghiselin, M. T., 1984. Narrow approaches to phylogeny: a review of nine books on cladism. In *Oxford Surveys in Evolutionary Biology*, R. Dawkins and M. Ridley (eds.). Oxford Univ. Press, Oxford, pp. 209-222.

Hennig, W., 1966. *Phylogenetic Systematics*. Univ. of Illinois Press, Urbana.

——, 1975. "Cladistic analysis or cladistic classification?": A reply to Ernst Mayr. *Systematic Zoology*, **24**, 244-256.

Hull, D. L., 1988. *Science as a Process: An Evolutionary Account of the Social and Conceptual Development of Science*. Univ. of Chicago Press, Chicago.

Janvier, P., 1984. Cladistics: theory, purpose, and evolutionary implications. In *Evolutionary Theory: Paths into the Future*, J. W. Pollard (ed.). John Wiley & Sons, Chichester, pp. 39-75.

Kermack, D. M., and K. A. Kermack, 1984. *The Evolution of Mammalian Characters*. Croom Helm, London.

Levin, L., 1983. *The Earth Through Time*. Saunders, Philadelphia.

Mayr, E., 1969. *Principles of Systematic Zoology*. McGraw-Hill, New York.

——, 1974. Cladistic analysis or cladistic classification? *Zeitschrift f. Zoologische Systematik u. Evolutionsforschung*, **12**, 94-128.

——, 1981. Biological classification: Toward a synthesis of methodologies. *Science*, **214**, 510-516.

McKenna, M. C., 1975. Towards a phylogenetic classification of the mammalia. In *Phylogeny of the Primates*, W. P. Luckett and F. S. Szalay (eds.). Plenum Press, New York, pp. 21-46.

Michener, C. D., 1970. Diverse approaches to systematics. *Evol. Biol.*, 4, 1-38.

——, 1977. Discordant evolution and the classification of allodapine bees. *Systematic Zoology*, **26**, 32-56.

Nelson, G., and N. Platnick, 1981. *Systematics and Biogeography: Cladistics and Vicariance*. Columbia Univ. Press, New York.

Newell, N. D., 1959. The nature of fossil record. *Proc. Amer. Philos. Soc.*, **103**, 264-285.

Patterson, C., 1981. Significance of fossils in determining evolutionary relationships. *Ann. Rev. Ecology and Systematics*, **12**, 195-223.

Ross, H. H., 1974. *Biological Systematics*. Addison-Wesley, Reading, Mass.

Ruse, M. (ed.), 1987. *Biology and Philosophy*, **2** (2), 127-225.

Simpson, G. G., 1961. *Principles of Animal Taxonomy*. Columbia Univ. Press, New York.

——, 1980. *Why and How: Some Problems and Methods in Historical Biology*. Pergamon Press, Oxford.

Slobodchikoff, C. N. (ed.), 1976. *Concepts of Species*. Dowden, Hutchison & Ross, Stroudsberg, Penn.

Sneath, P. H., and R. R. Sokal, 1973. *Numerical Taxonomy*. W. H. Freeman, San Francisco.

Sokal, R. R., 1975. Mayr on cladism — and his critics. *Systematic Zoology*, **24**, 257-262.

——, 1985. The continuing search for order. *Amer. Nat.*, **126**, 729-749.

——, 1986. Phenetic taxonomy: theory and methods. *Ann. Rev. Ecol. Syst.*, **17**, 423-442.

Stewart, W. N., 1983. *Paleobotany and the Evolution of Plants*. Cambridge Univ. Press, Cambridge.

Systematic Biology, 1969. (Proceedings of an International Conference sponsored by The National Research Council.) National Academy of Sciences, Washington, D. C.

Thomas, B. A., and R. A. Spicer, 1986. *The Evolution and Palaeobiology of Land Plants*. Croom Helm, London.

Wiley, E. O., 1978. The evolutionary species concept reconsidered. *Systematic Zoology*, **27**, 17-26.

——, 1981. *Phylogenetics: Theory and Practice of Phylogenetic Systematics*. John Wiley & Sons, New York.

12 MOLECULAR PHYLOGENIES

Among modern attempts at overcoming some of the usual phylogenetic problems has been to use molecular information rather than rely exclusively on morphological information. On the molecular level, information can be obtained by comparing sequences of nucleotides in various DNA and RNA molecules as well as by comparing sequences of amino acids (and their molecular configurations) in different proteins. These techniques are obviously restricted to recent organisms from which such compounds can be extracted, and the information gathered so far has generally been limited to only some proteins and some nucleic acid sequences. Nevertheless, the advantage of a molecular approach offsets many of its sampling limitations, since evolutionary changes on the level of amino acid and nucleotide substitutions can be measured and compared between existing organisms no matter how greatly they differ in other phenotypic features. That is, differences between sampled molecules from different organisms may be compared on a unit scale of amino acids or nucleotides when there may be no simple comparative units of morphology, behavior, ecology, physiology, and so on. For example, as shown previously on page 142, an amino acid sequence in a particular protein can be compared between corn plants, rabbits, chickens, fish, yeast, and bacteria.

IMMUNOLOGICAL TECHNIQUES

The earliest and still among the simplest of comparative molecular methods makes use of immunological techniques in which antibodies produced in a particular host (usually a rabbit) against proteins (antigens) of one species are measured as to their activity against proteins of other species. For example, if antibodies against species A precipitate much of the protein in species C but little of the protein in species B, then the A and C proteins are presumed to possess more similar molecular configurations (antigenic components) and are therefore more evolutionarily alike than those of A and B.

In the **immunodiffusion technique** practiced by Goodman and co-workers, antibodies produced against the blood serum (plasma) of one species is placed in a center well on an agar plate with the serum of the two species to be compared placed in nearby wells forming a trefoil arrangement. As can be seen in Figure 12-1, the antibodies and antigens diffuse

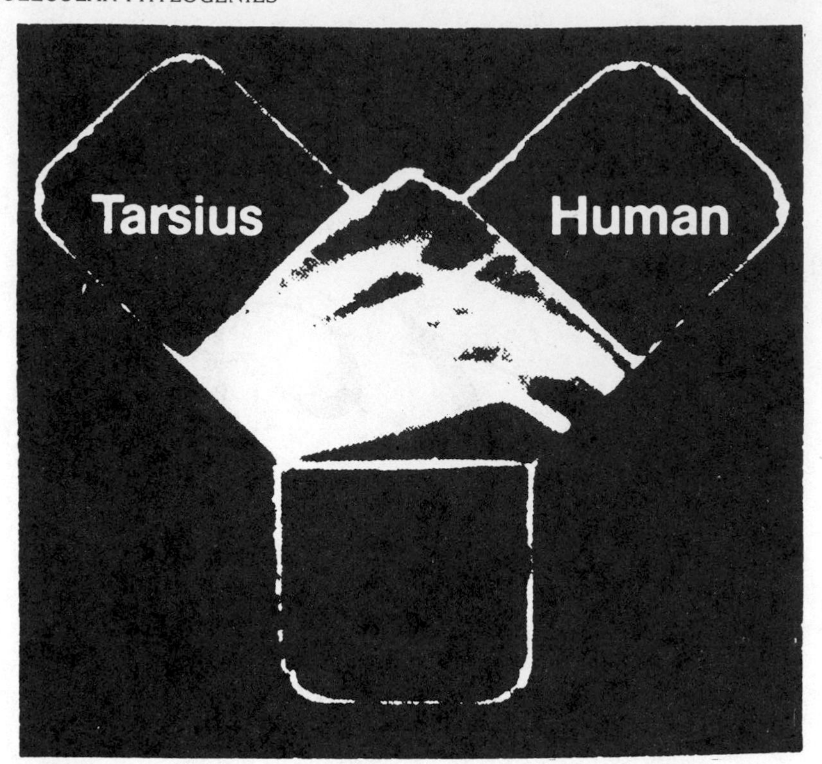

FIGURE 12-1 Reaction occurring on an agar plate when rabbit antibodies (lower well), produced against antigens of the *Tarsius* primate interact with *Tarsius* serum (left) and with human serum (right). Although not distinctly visible in this reproduction, eight separate precipitin bands are formed against *Tarsius* serum, each producing a spur against human serum. The number of spurs is a measure of the differences between the two sera. (From Goodman [1975].)

through the agar and form precipitation bands where the antibodies react with specific antigens. Because the blood serum contains a variety of proteins with different rates of diffusion, a number of separate precipitation reactions occur, each recognized by a separate band. If the proteins of the two tested species are alike, the bands will fuse; if they differ, spurs are formed, indicating greater antibody reaction with one species than the other. The number and length of the spurs are then converted into **antigenic distances** by special tables: the higher the number, the greater the distance between the species being compared.

Using a variety of antibodies, the information gathered by this technique can then be analyzed by various mathematical rules (algorithms) to obtain a phylogenetic tree that best correlates the antigenic distance between species with the length of time since they shared a common ancestor. Thus, if two species are antigenically closer to each other than to a third species, the third species is presumed to have broken off earlier from the common stem that all three originally shared. Successive comparisons are made between all possible combinations of species until the entire phylogenetic tree is obtained. A picture of such a tree for anthropoids (humans, apes, monkeys) is shown in Figure 12-2. In contrast to traditional taxonomy of the time, this technique showed that humans (*Homo*), gorillas (*Gorilla*), and chimpanzees (*Pan*) are antigenically closer to each other than to the Asian orangutan (*Pongo*). Based on this method, one can therefore separate the first three genera from the orangutans and place them in a separate group.

Other immunological techniques, such as **microcomplement fixation** used by Sarich and Wilson, involve the production of antibodies against specific proteins found in blood serum (albumin and transferrins) or enzymes such as lysozyme. Antigenic distances detected by measuring the amount of antigen-antibody reactions[1] then provide data that generally support the phylogenies obtained by other taxonomic methods, although some differences occur. One of the immunological problems is **nonreciprocity**: the degree to

[1]In microcomplement fixation, rabbits immunized with a protein antigen from one species will produce antiserum that gives a strong reaction against that antigen (homologous antigen) but not against the same protein from another species (heterologous antigen). The degree of antigenic difference between the two species is then measured by the concentration to which the antiserum must be increased for the heterologous antigen to react similarly as the homologous antigen.

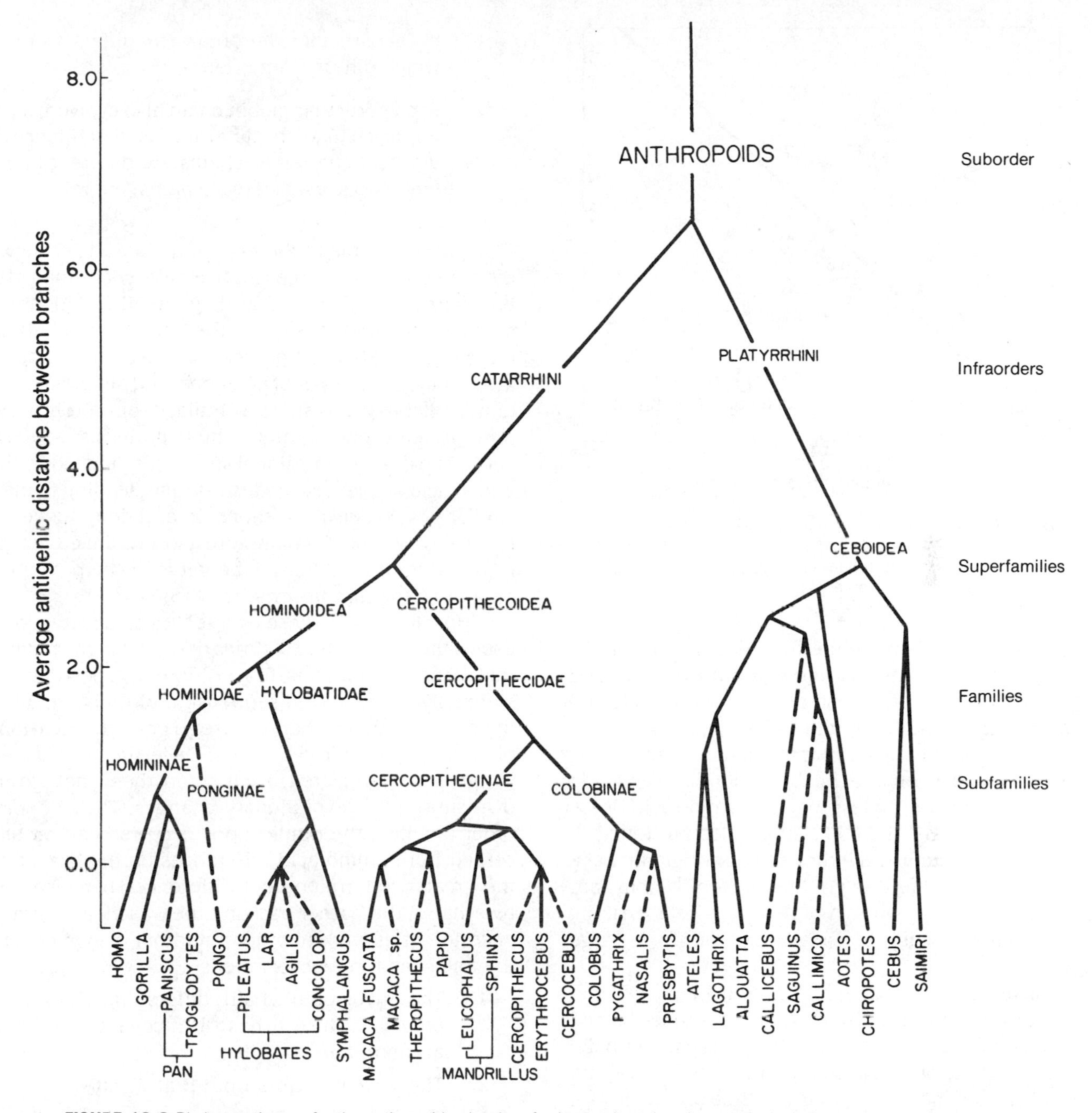

FIGURE 12-2 Phylogenetic tree for the anthropoid suborder of primates based on immunological distances derived from blood serum antigens. (From Dene et al.)

which antibodies produced against antigen A will react with antigen B is not always the same as the degree to which antibodies formed against B will react with A. An example of this is the finding that antigenic distance between humans and baboons is almost doubled when human lysozyme antibodies are used. However, nonreciprocal antigenic reactions are usually taken into account in suggesting phylogenies (Sarich and Cronin), and there seem to be, in most cases, fairly constant relationships between antigenic distances and amino acid differences. As Figure 12-3 shows, antigenic distances for three proteins including lysozyme are reflected in a linear increase in amino acid differences (a high correlation coefficient of 0.9).

Amino Acid Sequences

A more direct but more laborious approach to molecular phylogeny has been to actually sequence the amino acids in proteins by biochemical methods and to compare such sequences for the same protein in

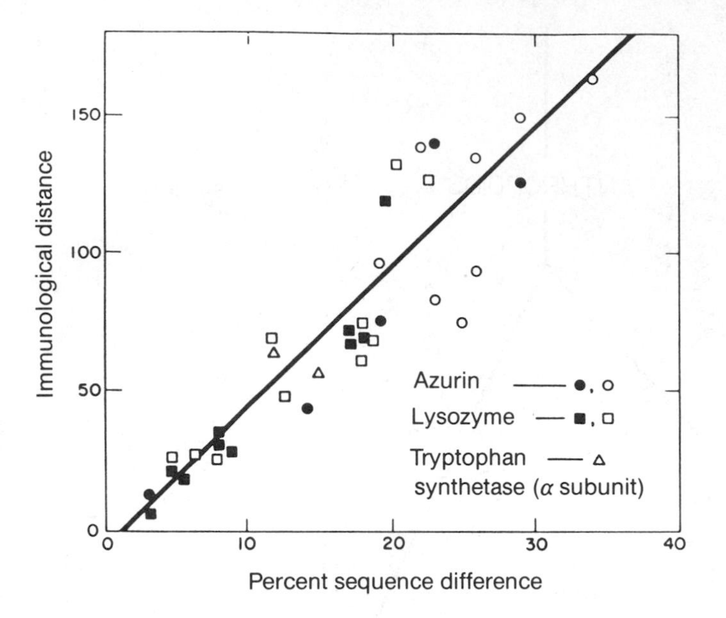

FIGURE 12-3 Correlation between immunological distance and the percent difference in amino acid sequence among three known proteins, bacterial azurins, α subunits of tryptophan synthetase, and bird lysozymes. (From Champion et al.)

different species. Among the first proteins to yield its amino acid sequence was **hemoglobin** (Ingram), and it probably still remains one of the most investigated of all proteins. The basic unit of hemoglobin consists of an iron-containing porphyrin (heme) that can reversibly bind oxygen attached to a globin polypeptide chain that is usually no less than 140 amino acids long. The fact that hemoglobinlike molecules are found in both plants and animals indicates their origin far back in the history of life. In vertebrates, hemoglobins are usually the primary protein of red blood cells, making them relatively easy to isolate and purify in large amounts.

As explained in Chapter 10, red blood cell hemoglobin of normal human adults is a four-chained molecule or tetramer, consisting of two pairs of polypeptide chains, one pair bearing the α sequence and the other pair mostly bearing the β sequence ($\alpha_2\beta_2$). Some adult hemoglobin utilizes δ chains instead of βs ($\alpha_2\delta_2$), and a common form of embryonic hemoglobin has two αs and two γs ($\alpha_2\gamma_2$). Other types of hemoglobin chains have also been found (e.g., ϵ), and hemoglobinlike molecules such as **myoglobin** occur in other tissues. All these chains are distinguished by their somewhat different properties and different amino acid sequences.

The facts that a species can possess different kinds of globin molecules and each of these in turn can differ between different species points to two major kinds of globin evolution:

1. Evolution occurred between the different kinds of globin chains to produce the variety carried by a particular vertebrate (α differs from β, which differs from γ, etc.).
2. Each particular globin chain also evolved, leading to changes in its amino acid sequence in different species (α chains are different in different species, as are β chains, etc.).

As an example of the first kind, Table 12-1 shows the amino acid sequences for five different human globin chains. It is obvious that they are all variations on a single homologous theme: the chains are all about the same length; they possess identical amino acids at a significant number of positions; and all functioning human globin genes share a similar exon-intron structure—three exons separated by two introns (see Fig. 9-14). The three-dimensional structures dictated by the amino acid sequences are also similar, leading to their similar physiological functions. In addition, the genes for the β, γ, and δ chains are closely linked (chromosome no. 11), although the gene for the α chain is on a different chromosome (no. 16).

How to explain these events? These observations, especially the sequence similarities, strongly suggest that rather than arising from different genes that accidentally converged in sequence and function, all of the different globin chains arose as **gene duplications** of an original globin-type gene. Once such duplicated globin genes appeared, each could then undergo its own subsequent evolutionary changes. The temporal order in which the duplications occurred can be discerned from amino acid differences on the basis that the greater the amino acid differences between any two chains, the further back in time was their common ancestor. Thus, we can order a number of events from the following three observations:

1. The myoglobin chain differs most from all others because it has distinctive amino acids at more than 100 sites.
2. The α chain differs from β at 77 sites.
3. The β chain differs from γ at 39 sites but differs from δ at only 10 sites.

These findings mean that the gene for myoglobin must have been formed from a very early duplication, which was then followed by a later duplication that separated the α and β genes. Because they differ least, the separation between the β and δ chains derives from a fairly recent duplication.

Figure 12-4 portrays the genetic phylogeny of the five globins in terms of the numbers of nucleotides necessary to account for the amino acid differences, along with the chronological periods in which each duplication is presumed to have occurred. We can see that the duplication events led to the early coexistence of myoglobin with an α-like chain, the former probably

TABLE 12-1 Amino Acid Sequences for Human Myoglobin and Four Human Hemoglobin Chains (α, β, γ, and δ)*

	1		2	3	4	5	6	7	8	9	10	11	12	13	14	15	16	17	18	19
Myoglobin	G	—	L	S	D	G	E	W	Q	L	V	L	N	V	W	G	K	V	E	A
α chain	V	—	L	S	P	A	D	K	T	N	V	K	A	A	W	G	K	V	G	A
β chain	V	H	L	T	P	E	E	K	S	A	V	T	A	L	W	G	K	V	—	—
γ chain	G	H	F	T	E	E	D	K	A	T	I	T	S	L	W	G	K	V	—	—
δ chain	V	H	L	T	P	E	E	K	T	A	V	N	A	L	W	G	K	V	—	—

	20	21	22	23	24	25	26	27	28	29	30	31	32	33	34	35	36	37	38	39
Myoglobin	D	I	P	G	H	G	Q	E	V	L	I	R	L	F	K	G	H	P	E	T
α chain	H	A	G	E	Y	G	A	E	A	L	E	R	M	F	L	S	F	P	T	T
β chain	N	V	D	E	V	G	G	E	A	L	G	R	L	L	V	V	Y	P	W	T
γ chain	N	V	E	D	A	G	G	E	T	L	G	R	L	L	V	V	Y	P	W	T
δ chain	N	V	D	A	V	G	G	E	A	L	G	R	L	L	V	V	Y	P	W	T

	40	41	42	43	44	45	46	47	48	49	50	51	52	53	54	55	56	57	58
Myoglobin	L	E	K	F	D	K	F	K	H	L	K	S	E	D	E	M	K	A	S
α chain	K	T	Y	F	P	H	F	—	D	L	S	H	—	—	—	—	—	G	S
β chain	Q	R	F	F	E	S	F	G	D	L	S	T	P	D	A	V	M	G	N
γ chain	Q	R	F	F	D	S	F	G	N	L	S	S	A	S	A	I	M	G	N
δ chain	Q	R	F	F	E	S	F	G	D	L	S	S	P	D	A	V	M	G	N

	59	60	61	62	63	64	65	66	67	68	69	70	71	72	73	74	75	76	77
Myoglobin	E	D	L	K	K	H	G	A	T	V	L	T	A	L	G	G	I	L	K
α chain	A	Q	V	K	G	H	G	K	K	V	A	D	A	L	T	N	A	V	A
β chain	P	K	V	K	A	H	G	K	K	V	L	G	A	F	S	D	G	L	A
γ chain	P	K	V	K	A	H	G	K	K	V	L	T	S	L	G	D	A	I	K
δ chain	P	K	V	K	A	H	G	K	K	V	L	G	A	F	S	D	G	L	A

	78	79	80	81	82	83	84	85	86	87	88	89	90	91	92	93	94	95	96
Myoglobin	K	K	G	H	H	E	A	E	I	K	P	L	A	Q	S	H	A	T	K
α chain	H	V	D	D	M	P	N	A	L	S	A	L	S	D	L	H	A	H	K
β chain	H	L	D	N	L	K	G	T	F	A	T	L	S	E	L	H	C	D	K
γ chain	H	L	D	D	L	K	G	T	F	A	Q	L	S	E	L	H	C	D	K
δ chain	H	L	D	N	L	K	G	T	F	S	Q	L	S	E	L	H	C	D	K

	97	98	99	100		102		104		106		108		110		112		114	
Myoglobin	H	K	I	P	V	K	Y	L	E	F	I	S	E	C	I	I	Q	V	L
α chain	L	R	V	D	P	V	N	F	K	L	L	S	H	C	L	L	V	T	L
β chain	L	H	V	D	P	E	N	F	R	L	L	G	N	V	L	V	C	V	L
γ chain	L	H	V	D	P	E	N	F	K	L	L	G	N	V	L	V	T	V	L
δ chain	L	H	V	D	P	E	N	F	R	L	L	G	N	V	L	V	C	V	L

	116		118		120		122		124		126		128		130		132		134
Myoglobin	Q	S	K	H	P	G	D	F	G	A	D	A	Q	G	A	M	N	K	A
α chain	A	A	H	L	P	A	E	F	T	P	A	V	H	A	S	L	D	K	F
β chain	A	H	H	F	G	K	E	F	T	P	P	V	Q	A	A	Y	Q	K	V
γ chain	A	I	H	F	G	K	E	F	T	P	E	V	Q	A	S	W	Q	K	M
δ chain	A	R	N	F	G	K	E	F	T	P	Q	M	Q	A	A	Y	Q	K	V

		136		138		140		142		144		146		148		150		152	
Myoglobin	L	E	L	F	R	K	D	M	A	S	N	Y	K	E	L	G	F	Q	G
α chain	L	A	S	V	S	T	V	L	T	S	K	Y	R						
β chain	V	A	G	V	A	N	A	L	A	H	K	Y	H						
γ chain	V	T	G	V	A	S	A	L	S	S	R	Y	H						
δ chain	V	A	G	V	A	N	A	L	A	H	K	Y	H						

*The amino acids are abbreviated by single capital letters as shown in Figure 7-2 and Table 12-2. The chains are aligned with the 153 amino acids in myoglobin, and the boxes indicate identical amino acids found in all chains at the designated numbered positions.

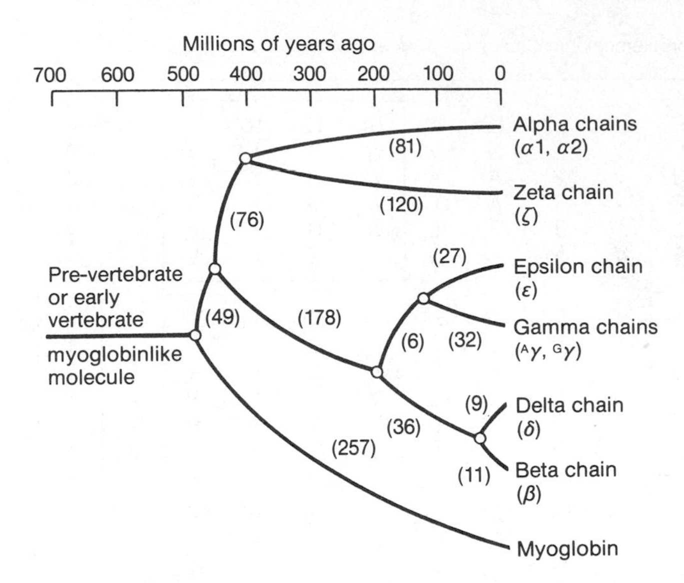

FIGURE 12-4 Phylogenetic relationships between globin-type proteins found in humans, showing the estimated times at which they originally diverged from each other. The estimated number of nucleotide replacements necessary to cause the observed amino acid changes in each branch of the lineage is given in parentheses. (From Strickberger, data from Goodman and co-workers.)

assuming (or maintaining) an intracellular function and the latter probably assuming a circulatory function. When a duplication of the α-like gene further evolved into a β-like gene, the advantage of having two pairs of different chains in a tetramer hemoglobin molecule must have been sufficiently great to account for the fact that all subsequent hemoglobins in the circulating blood of most vertebrates preserve this tetramer organization. After the β-like gene was formed, a translocation separated it from α and transferred it to a different chromosome. Duplications then occurred in the β-like gene, yielding eventually the modern β, γ, and δ genes.

It is now clear that such gene duplications are not unusual and can easily arise from an **unequal crossing over** event that produces a recombinant product possessing increased chromosome material (Fig. 12-5).[2] The human α gene, for example, is known to have two side-by-side (tandem) duplicates, α1 and α2, and the **β gene cluster**, also called a **gene** or **multigene family**, is composed of a sequence of seven such genes (Fig. 12-6).

In some cases, such as globin chains, serine proteases (chymotrypsin, trypsin), and lactate dehydrogenases, duplicated genes have preserved similar although not identical functions. In other instances, duplicated genes appear to have evolved in completely different functional directions, although they share enough common amino acid sequences to indicate their common origin. Thus the vertebrate nerve growth factor protein that enhances the outgrowth of neural cells from sympathetic and sensory ganglia has amino acid sequences similar to insulin and may share some functional similarities as well (Frazier et al.). Also, the α-lactalbumin protein, which is part of an enzyme used in the synthesis of lactose in mammalian milk, has an amino acid sequence remarkably similar to that of the lysozyme enzyme found in tears that is used to degrade the mucopolysaccharides of bacterial cell walls (Hill et al.). This similarity is further reflected in the fact that both proteins are the products of tissues (mammary, tear duct) that were at one time probably sebaceous glands, and they both use sugar molecules as their substrates.

Duplications also exist within genes themselves, as shown by the presence of three homologous amino acid regions (polypeptide domains, p. 155) within the γ heavy chain of the immunoglobulin G antibody. In the human haptoglobin α-2 protein found in blood serum, a segment of 59 amino acids (positions 13 to 71) is an almost exact repetition of an adjacent segment (positions 72 to 130). Repeated amino acid sequences have also been discerned within the ferredoxin protein (used as an electron carrier in various biochemical processes; see Fig. 9-2),[3] in the glutamate dehydrogenase enzyme, and various other proteins (Li).

In spite of these and other fairly conspicuous examples, ancestral homologies based on similarities between amino acid sequences are not always obvious. Various tests for detecting amino acid sequence similarity have therefore been devised, some more sensitive in distinguishing homologous relationships than others (Fitch [1973], Dayhoff et al., Nei). In general, the most sensitive of these tests depend upon comparing fairly long amino acid sequences, since comparisons between short sequences can be misleading; for example, similar sequences only three or four amino acids long may often arise in non-related pro-

[2]Unequal crossing over may also cause gene fusion by eliminating chromosomal material between two formerly separate genes. The union in fungi between the A and B components of the tryptophan synthetase enzyme, normally separate in bacteria, is believed to have occurred through such fusion events.

[3]The small size of ferredoxin, its limited sampling of amino acids, and the simple positioning of its iron atoms indicates that it is one of the most primitive electron transport agents in the cell. These features, along with its apparent duplicated structure, have suggested that ferredoxin may owe its origin to an even earlier peptide formed on one of the then-primitive templates. (The length of ferredoxin's basic amino acid sequence before duplication is only about 28 amino acids long.) Perhaps only when greater accuracy of reproduction and translation could be achieved was it possible for such small early enzymes to increase in size and thereby improve their reaction specificity, catalytic activity, and structural stability.

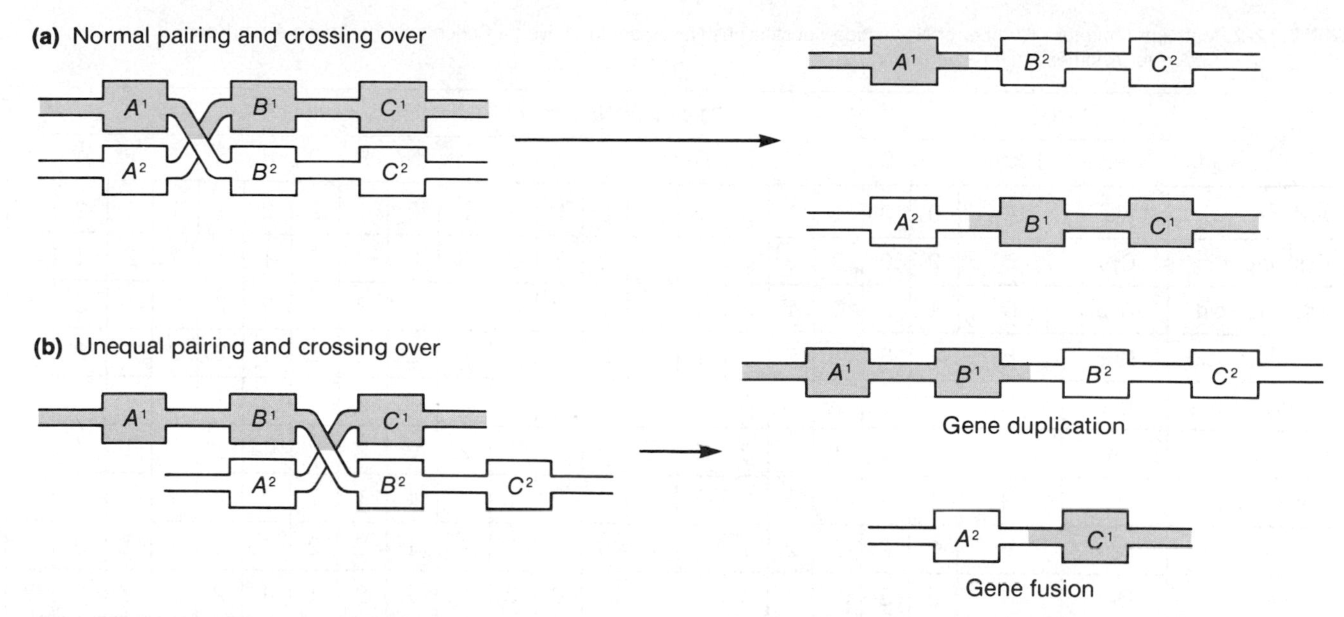

FIGURE 12-5 Results of equal and unequal crossing over for three genes on a chromosome. (a) When pairing between homologous sections on two chromosomes is equal, the crossover products have the same amounts of chromosomal material (e.g., they both have an *A*, *B*, and *C* gene). (b) When pairing between the two chromosomes is unequal, one of the crossover products carries a gene duplication (the *B* gene in this illustration), and the other product shows a fusion between genes (*A* - *C*) that were formerly separated by the intervening segment.

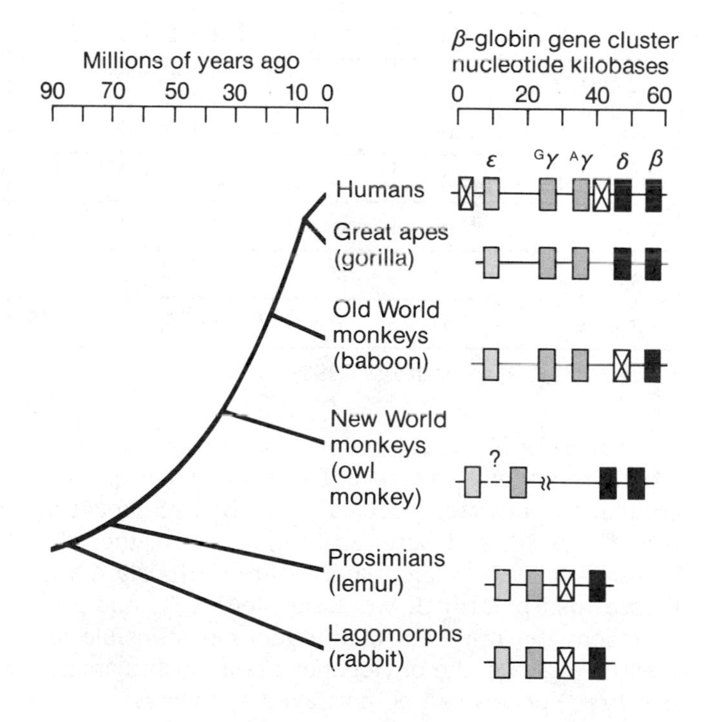

FIGURE 12-6 The clustered organization of the β-globin-type genes in five different primates and in the rabbit (right side), along with a proposed evolutionary tree (left side). The genes in each of these clusters are linked together on the same chromosome, for example, the human β-globin cluster is localized to a span of 60,000 nucleotides on chromosome 11. Each gene, denoted by a small rectangle, is transcribed from left to right, with the genes responsible for embryonic and fetal development on the left (lighter shadings) and the genes that produce adult β-globins on the right (darker shading). Genes marked with crosses indicate **pseudogenes**, duplicates which have become non-functional for various reasons. (From Strickberger, adapted from Jeffreys et al.)

teins. Methods to overcome difficulties caused by gaps or insertions have also been proposed (e.g., Elleman).

The second aspect of protein evolution, comparisons between their amino acid sequences in different species, has been used to produce phylogenies involving a large number of organisms. For such comparisons, Fitch has suggested that duplicate genes in a species, although homologous to each other in the sense that they derive from a common ancestor, be termed differently from homologies for the same gene in different species. He called the former **paralogous**, and the latter **orthologous**. It is therefore comparisons between orthologous genes that are valid for determining the phylogenetic relationships of the species that carry them.

One basic technique for detecting orthologous amino acid sequences is not essentially different from that used in determining phylogenies based on immunological distances, except that it can furnish more precise phylogenetic positioning because it begins with an alignment of amino acid sequences from the same known protein in different species. Once aligned, the minimum number of mutations necessary to transform one amino acid in one sequence to that of a different amino acid in the same position in the other sequence can then provide an estimate of mutational distances, such as those shown in Table 12-2.

TABLE 12-2 Matrix of Minimum Number of Nucleotide Substitutions Necessary to Convert a Codon for One Amino Acid (Rows) into a Codon for Another Amino Acid (Columns)

	Abbreviations		Minimum Number of Nucleotide Substitutions																			
Amino Acid	3-letter	1-letter	A	C	D	E	F	G	H	I	K	L	M	N	P	Q	R	S	T	V	W	Y
Alanine	Ala	A	0	2	1	1	2	1	2	2	2	2	2	2	1	2	2	1	1	1	2	2
Cysteine	Cys	C	2	0	2	3	1	1	2	2	3	2	3	2	2	3	1	1	2	2	1	1
Aspartic acid	Asp	D	1	2	0	1	2	1	1	2	2	2	3	1	2	2	2	2	2	1	3	1
Glutamic acid	Glu	E	1	3	1	0	3	1	2	2	1	2	2	2	2	1	2	2	2	1	2	2
Phenylalanine	Phe	F	2	1	2	3	0	2	2	1	3	1	2	2	2	3	2	1	2	1	2	1
Glycine	Gly	G	1	1	1	1	2	0	2	2	2	2	2	2	2	2	1	1	2	1	1	2
Histidine	His	H	2	2	1	2	2	2	0	2	2	1	3	1	1	1	1	2	2	2	3	1
Isoleucine	Ile	I	2	2	2	2	1	2	2	0	1	1	1	1	2	2	1	1	1	1	3	2
Lysine	Lys	K	2	3	2	1	3	2	2	1	0	2	1	1	2	1	1	2	1	2	2	2
Leucine	Leu	L	2	2	2	2	1	2	1	1	2	0	1	2	1	1	1	1	2	1	1	2
Methionine	Met	M	2	3	3	2	2	2	3	1	1	1	0	2	2	2	1	2	1	1	2	3
Asparagine	Asn	N	2	2	1	2	2	2	1	1	1	2	2	0	2	2	2	1	1	2	3	1
Proline	Pro	P	1	2	2	2	2	2	1	2	2	1	2	2	0	1	1	1	1	2	2	2
Glutamine	Gln	Q	2	3	2	1	3	2	1	2	1	1	2	2	1	0	1	2	2	2	2	2
Arginine	Arg	R	2	1	2	2	2	1	1	1	1	1	1	2	1	1	0	1	1	2	1	2
Serine	Ser	S	1	1	2	2	1	1	2	1	2	1	2	1	1	2	1	0	1	2	1	1
Threonine	Thr	T	1	2	2	2	2	2	2	1	1	2	1	1	1	2	1	1	0	2	2	2
Valine	Val	V	1	2	1	1	1	1	2	1	2	1	1	2	2	2	2	2	2	0	2	2
Tryptophan	Trp	W	2	1	3	2	2	1	3	3	2	1	2	3	2	2	1	1	2	2	0	2
Tyrosine	Tyr	Y	2	1	1	2	1	2	1	2	2	2	3	1	2	2	2	1	2	2	2	0

Adapted from Fitch and Margoliash (1967).

For example, if the minimum, or most parsimonious, mutational distance for a particular protein comparison between species A and B is 25, between A and C is 20, and between B and C is 30, then the two most closely related species are obviously A and C (see also p. 214). If we assign legs x, y, and z to represent the numbers of mutations responsible for their divergence, the phylogenetic relationship among the three species can be portrayed as follows:

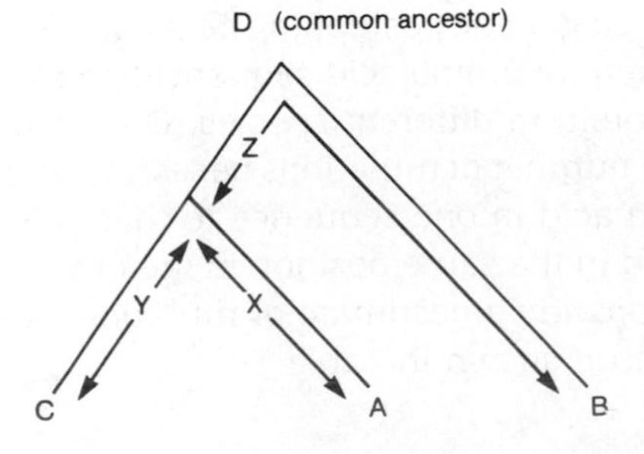

The lengths of these legs can be determined by noting that the A-B distance (25) is less than the C-B distance (30), hence the leg x must be 5 mutations less than y. Since $x + y = 20$ and $y - x = 5$, one equation can be subtracted from the other and solved for x:

$$\begin{array}{r} x + y = 20 \\ -\ (-x + y = \ \ 5) \\ \hline 2x = 15 \\ x = 7.5,\ y = 12.5 \end{array}$$

The leg z must therefore be equal to the A-B distance minus x (or the C-B distance minus y); that is, $z = 25 - 7.5$ (or $30 - 12.5$) $= 17.5$.

As increased numbers of species are used to provide mutational data for a particular protein, a **phylogenetic tree** can be established for these species by different numerical and algorithmic methods (Felsen-

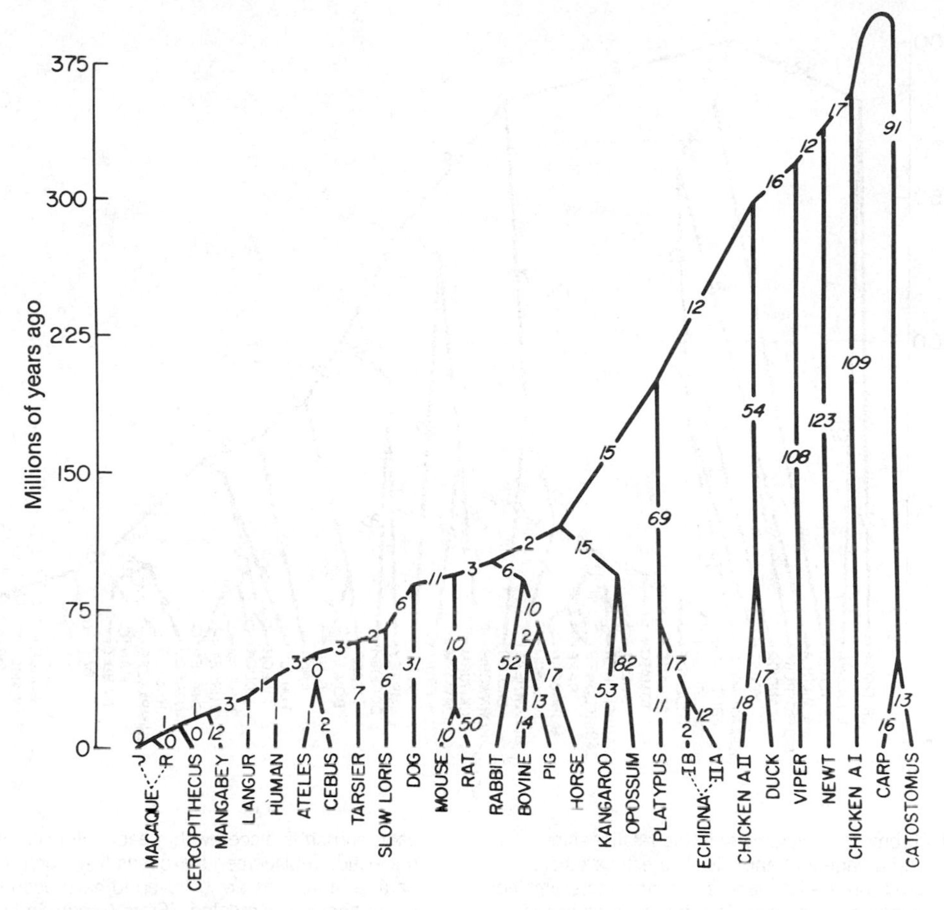

FIGURE 12-7 A phylogeny of α-hemoglobin chains in a variety of vertebrate species determined by calculating the number of nucleotide replacements that account for the number of observed amino acid substitutions presumed to have occurred during each evolutionary interval. The vertical scale, given in millions of years, is based on paleontological estimates for the age of the common ancestor of each branch. (From Goodman [1976].)

stein, Nei).[4] One example is shown in Figure 12-7 for the sequence of α-hemoglobin chains in 29 different vertebrates involving a total of 630 nucleotide replacements. Because hemoglobin phylogenies exclude invertebrates, plants, and fungi, other more ubiquitous proteins have been used that can be compared between more widely varied groups of organisms. One such protein, cytochrome *c*, also contains heme but generally functions in aerobic organisms as part of the respiratory electron transport chain (see Fig. 9-9). A phylogenetic tree based on 53 different amino acid sequences of cytochrome *c* is shown in Figure 12-8.

Not surprisingly, phylogenies derived from studies of only one or two proteins can hardly be expected to always reflect true evolutionary history nor discriminate accurately even between different species. The fact that there are no differences between human and chimpanzee α-hemoglobins marks such lack of discrimination. Similarly, the fact that camels and whales share the same cytochrome *c* amino acid sequence hardly reflects their obvious evolutionary divergence. (Note also, for example, the similar sequence in cows, pigs, and sheep.) Nevertheless, it is remarkable how closely the information obtained from this limited sample of only two proteins out of many thousands approximates the phylogenetic relationships obtained from extensive studies in comparative anatomy and paleontology. A more faithful reflection of evolutionary

[4]An important condition of the **parsimony method** used above is that the observed mutational distance between two species must be less than or equal to the sum of the distances from their common ancestor to each species in the projected phylogenetic tree. In the illustrated example, the observed distance between A and B is 25, and the distances from their hypothesized common ancestor at the triangular apex (e.g., D) to these two species must total at least 25, if not more. The reason that the sum of the (D-A) + (D-B) legs may be more than 25 is that undiscovered **back mutations** (e.g., adenine → guanine → adenine) may exist in these two legs, or **parallel mutations** may have occurred in each leg (e.g., cytosine → adenine; guanine → adenine), so that the difference observed between A and B is really an underestimate of the actual mutational distance. This triangle inequality condition therefore signifies the evolutionary presence of back and parallel mutations and is often taken into account by proportionally augmenting the number of mutations in an ancestral leg in accord with its length. That is, not only is the A-D-B distance, for example, increased numerically over the the measured A-B distance, but longer distances are even further augmented.

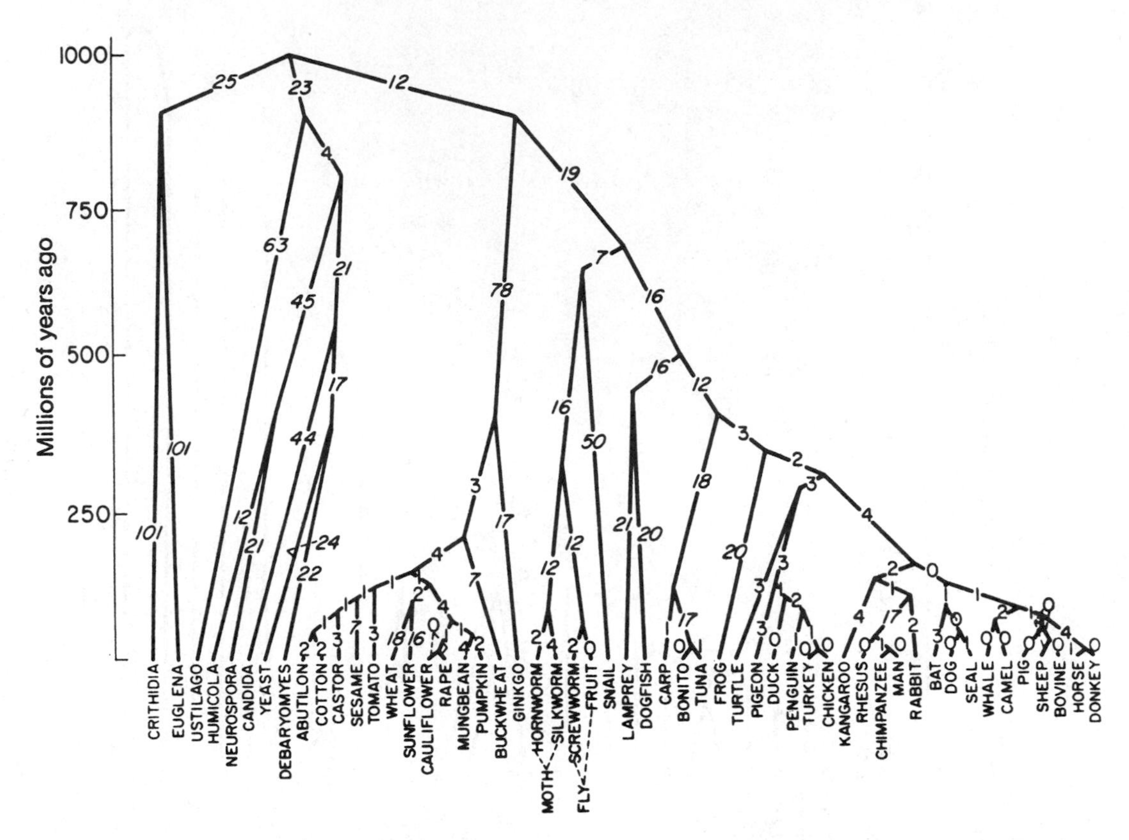

FIGURE 12-8 A cytochrome *c* phylogeny showing relationships among plants, fungi, and a variety of animals. The estimated ages of branching points are given on the left-hand scale, and the estimated number of nucleotide replacements necessary for the evolution of each branch is placed within each interval. When the number of nucleotide replacements exceeds five, some undetectable back, or parallel, mutations are believed to have occurred, and that number is proportionally augmented. (From Goodman [1976].)

history might therefore be expected if we could combine studies from many different proteins, allowing comparisons among many different amino acid positions. Although there are difficulties in such procedures, some attempts have been made, and the results of one such study is diagrammed in Figure 12-9. (Another example, involving both proteins and nucleic acids, will be discussed below.)

Quantitative DNA Measurements

It seems likely that the first self-replicating organisms had no more than a small amount of genetic material, sufficient only for the maintenance of those relatively few functions necessary for their preservation. As time passed and these organisms continually faced more challenges, there must have been considerable advantage in possessing greater amounts of genetic material that could be used to increase the number of functions and their regulation. Competition among organisms for successful adaptation to their environments therefore rapidly became dependent on the numbers and kinds of genes they possessed, and evolutionary changes have obviously proceeded on both of these levels. Present techniques for evaluating some of the consequences of these evolutionary factors include measurements of the amounts of genetic material in different organisms and the detection of nucleic acid homologies between them.

Measurement of the amounts of DNA in different organisms began with the studies of Mirsky and Ris in the early 1950's and presently extends to more than 1,000 species. The techniques include: (1) measuring the amount of stained DNA in cells (Feulgen staining), (2) isolating DNA chemically and deriving an average DNA amount from a known number of cells, and (3) indirectly estimating DNA content from chromosomal size or nuclear volume. Based on results from these techniques, Figure 12-10 shows the observed range of cellular DNA content in numbers of nucleotides (1 picogram = 10^{-12} gm = 2.01×10^9 nucleotides). Interestingly, these data do indicate a relationship between the amount of DNA and general evolutionary status, going from small amounts in viruses and bacteria to relatively large amounts in the higher eukaryotes. However, this progression is certainly not uniform, since some fairly "primitive" representatives of various groups (e.g., ferns, psilopsids, lungfishes, salamanders) show relatively large amounts of DNA. Comparative amounts of DNA are thus not much of a clue to phylogenetic relationship, especially since the cells of some very different organisms, such as mammals and gymnosperms, have exactly the same DNA content, whereas obviously related species among amphibians and other groups vary widely in DNA content.

Nevertheless, some authors have attempted to derive a few generalities from these data. Hinegardner has pointed out that specialized species within some groups have less DNA than the more generalized species. For example, the more generalized fish species such as salmon and cod have cellular DNA contents ranging from 1.2 to 4.4 picograms, whereas the DNA of specialized forms such as sea horses and angler fish range from 0.45 to 0.80 picograms. At the same time, however, various species of algae, protozoa, ferns, and amphibia may also be considered quite specialized yet contain relatively large amounts of DNA. Relationship between DNA amount and chromosome number is another feature whose consistency varies: DNA content correlates well with chromosome number in plants and fishes but poorly in mammals. In general, it seems clear that knowledge of DNA amount alone is insufficient to derive the fine textural patterns of evolutionary history, and a more complete analysis is necessary.

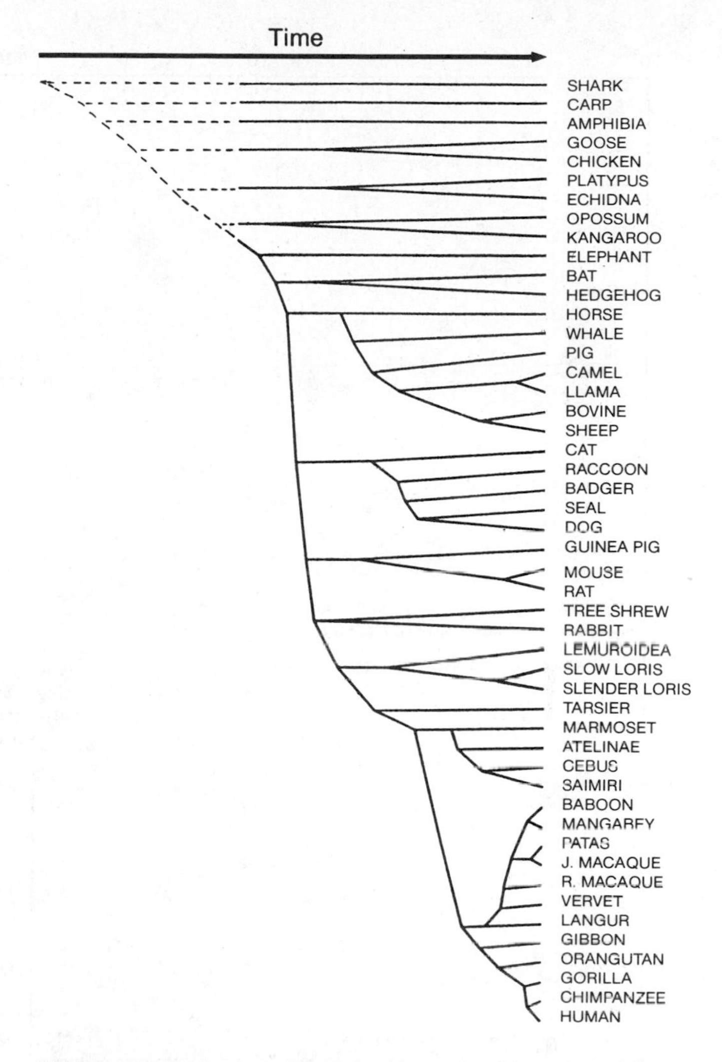

FIGURE 12-9 A phylogenetic tree for 49 vertebrate taxa derived from aligning the amino acid sequences in up to seven different polypeptide chains: α-hemoglobin, β-hemoglobin, myoglobin, lens α crystallin A, fibrinopeptide A, fibrinopeptide B, and cytochrome *c*. (Adapted from Goodman et al.)

REPETITIVE DNA SEQUENCES

One approach towards analyzing DNA in greater detail than simply measuring its amount has been to determine differences among the kinds of DNA present in a single organism. These studies began with the discovery by Britten that DNA sheared to specific sizes then separated into single strands would reassociate into double-stranded molecules at rates based on the nature of their nucleotide sequences. For example, a single strand of DNA bearing a sequence repeated many times over throughout a genome would find a complementary "mate" and form a double-stranded molecule much more rapidly than a rare complex sequence. (Reassociation rate can be detected by measuring optical changes that occur in the transition from single- to double-stranded DNA.) By these means, Britten and co-workers were able to classify DNA as either **repetitive** or **unique**, referring to sequences that occurred frequently and those that occurred only as single copies.

As Table 12-3 shows, a significant fraction of DNA in tested eukaryotic organisms is repetitive, some sequences present in 200 copies or less and others repeated more than 1 million times.

The lengths of these repetitive sequences may vary from less than 100 to more than 2,000 nucleotides long. With the exception of *Drosophila*, the shorter length repetitive sequences seem to be interspersed among sequences of nonrepetitive unique DNA, although the function of this arrangement is unknown. According to some authors (Davidson and Britten) these interspersed sequences act as regulatory genes similar to prokaryotic operators and promoters (Chapter 10), whereas others have suggested that they are involved in packaging long transcribed strands of heterogeneous nuclear RNA (HnRNA) so they can be further processed into messenger RNA. The fact that many eukaryotic messenger RNA molecules are shorter than their parental HnRNA molecules certainly indicates that there are extra DNA nucleotides in each gene that are not translated into protein: some are at the gene termini, and others, known as introns (Fig. 9-14), are distributed throughout eukaryotic protein-coding genes.

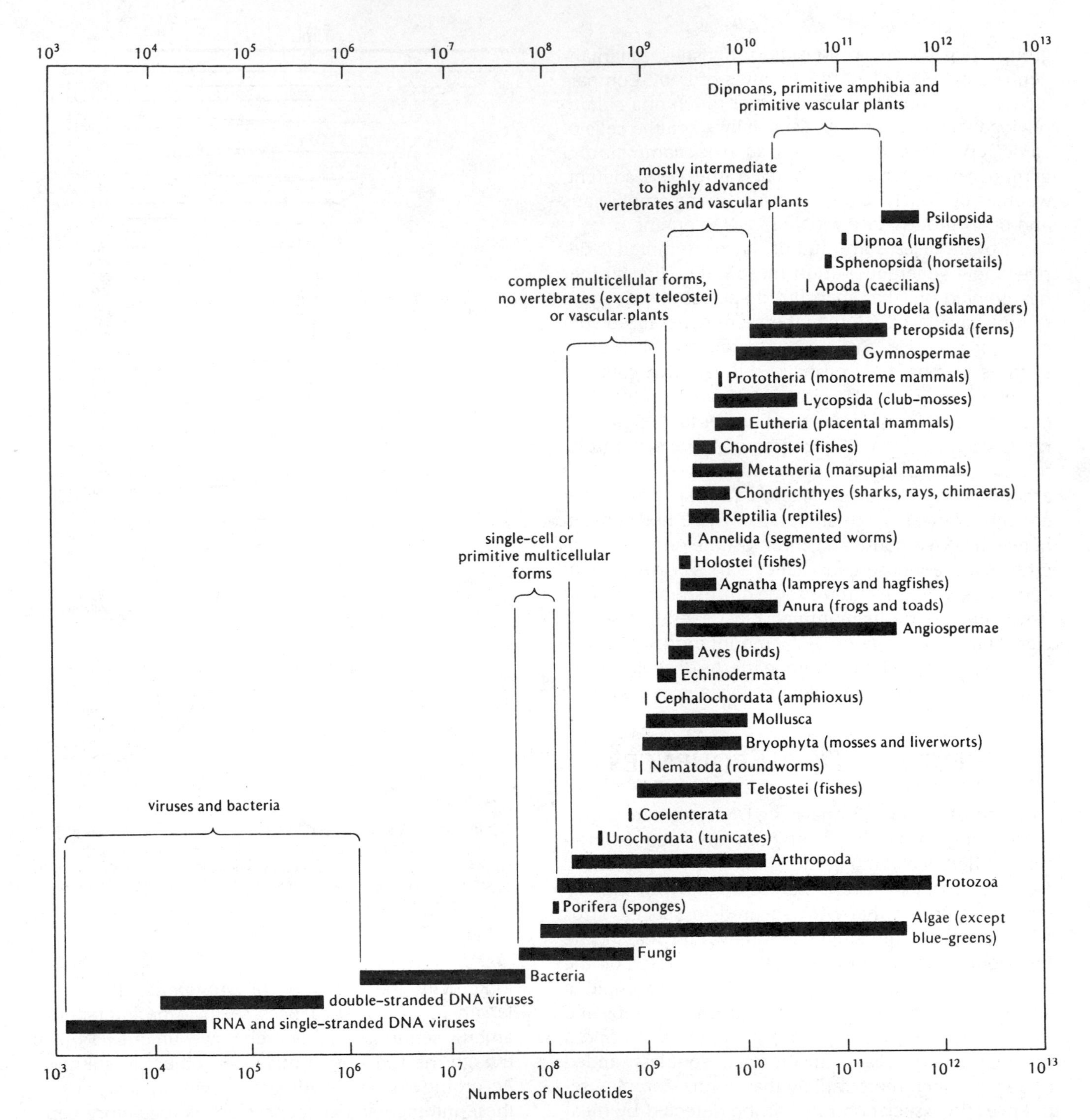

FIGURE 12-10 Comparison of the numbers of nucleotides in the genetic material of different types of organisms. Each bar in the figure represents the range of nucleotide numbers found among the species sampled in a designated group. The nucleotide values given are largely derived from estimates of the weight of nucleic acid in the haploid complement of an organism according to the formula, 1 gm of nucleic acid = 2.0×10^{21} nucleotides. Thus the DNA in the diploid complement of human chromosomes (6.4×10^{-12} gm) contains $(2.0 \times 10^{21}) \times (6.4 \times 10^{-12}) = 12.8 \times 10^9$ nucleotides, or 6.4×10^9 nucleotide pairs of double-stranded DNA. This is equal to a length of more than 2 meters (2.9×10^6 nucleotide pairs = 1 millimeter). (From Strickberger, adapted from Sparrow, Price, and Underbrink.)

Among the long-length repetitive DNA sequences, a number undoubtedly code for multigene families such as the many duplicate, tandemly arranged ribosomal RNA genes, transfer RNA genes, and histone genes. A function for the more repetitive but shorter nucleotide sequences found in **satellite DNA** (recognized by its separation from the main portion of DNA after centrifugation) seems more obscure, although these have often been localized to distinctively staining chromosome sections and centromere regions (**heterochromatin.**) In the house mouse (*Mus musculus*) a satellite DNA that comprises about 10 percent of

TABLE 12-3 Estimates of the Frequencies of Nonrepetitive DNA Sequences and Three Classes of Repetitive DNA Sequences in Various Genomes

Organism	Nonrepetitive (Single Copy)	Partially Repetitive (to about 200 Copies per Genome)	Intermediately Repetitive (250-60,000 Copies per Genome)	Highly Repetitive (70,000-1,000,000 or More Copies per Genome)
Chlamydomonas reinhardtii (algae)	.70		.30	
Physarum polycephalum (fungus)	.58		.42	
Ascaris lumbricoides (annelid)	.77		.23	
Drosophila melanogaster (fruit fly)	.78	.15	.07	
Strongylocentratus purpuratus (sea urchin)	.38	.25	.34	.03
Xenopus laevis (clawed toad)	.54	.06	.37	.03
Gallus domesticus (chicken)	.70	.24		.06
Bos taurus (cattle)	.55		.38	.05
Homo sapiens (human)	.64	.13	.12	.10

Data from Straus.

the genome interestingly shows no homology to the DNA of related rodents such as rat, field mouse, and hamster. On the other hand, a number of satellite sequences have been found to be widely conserved in different species groups such as crustaceans (crabs) and insects (*Drosophila*). From such studies and others it seems clear that at least some satellite DNA can arise quite rapidly during the evolution of a species by the addition of many copies of a new DNA sequence or by the amplification of DNA sequences present in ancestral species. The different kinds of satellite DNA, their different amounts, and their possible different origins indicate that they probably possess different functions, or perhaps no function at all (e.g., Miklos). As discussed in Chapter 10, it has been tempting to various biologists to consider some or many such sequences as forms of "selfish DNA."

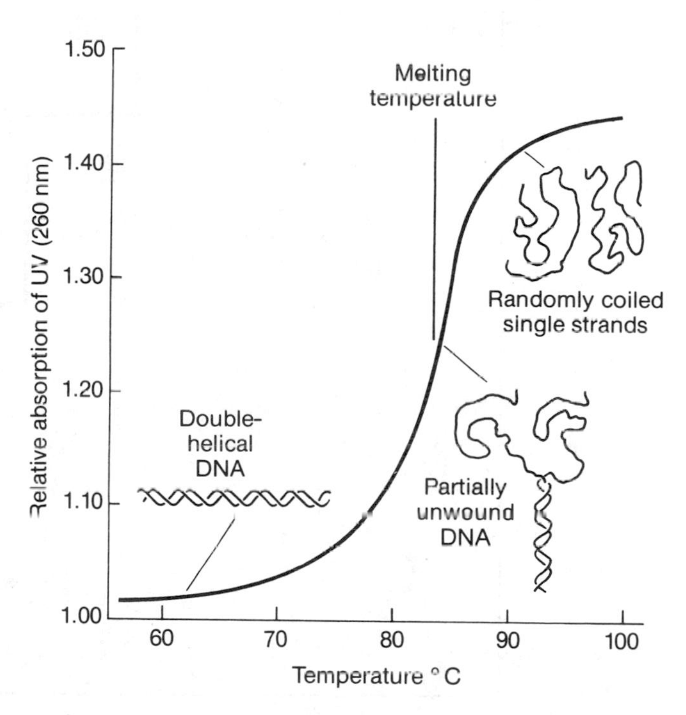

FIGURE 12-11 Melting temperature curve for DNA of T4 bacteriophage showing the marked change in ultraviolet absorption that occurs at approximately 84°C when about half the DNA has changed from double helical form to single-stranded form. (From Strickberger.)

NUCLEIC ACID PHYLOGENIES BASED ON DNA-DNA HYBRIDIZATIONS

In order to estimate the extent of homology between nucleic acids of different sources, measurements can be taken of the degree to which homologous nucleotide sequences in different single strands pair up to form double-stranded sections. In one of the techniques presently used, DNA is extracted from two organisms, X and Y, dissociated into single strands, then given the opportunity to form X-Y hybrid double-stranded DNA by incubating the different DNA molecules together at appropriate temperatures. To enable the separation of interspecific X-Y DNA from intraspecific X-X or Y-Y DNA, the nonrepetitive DNA of one species, X, is radioactively labeled and only relatively small amounts of it are used in the incubation mixture. Because of its rarity the DNA of X will therefore have very little chance of forming X-X double strands, and all radioactively labeled double-stranded DNA can be assumed to be X-Y. This double-stranded DNA can then be extracted (on hydroxyapatite crystals) and its properties examined. If the DNA from X and Y are perfectly homologous and possess no nucleotide differences at all (i.e., they are of the same species), then the **melting temperature** at which the hybrid X-Y DNA dissociates into single strands (Fig. 12-11) will, of course, be the same as either X-X or Y-Y. However, should there be nucleotide differences between X and Y, then the X-Y hybrid will dissociate more easily because of nucleotide mismatching; that is, its stability is reduced and its melting

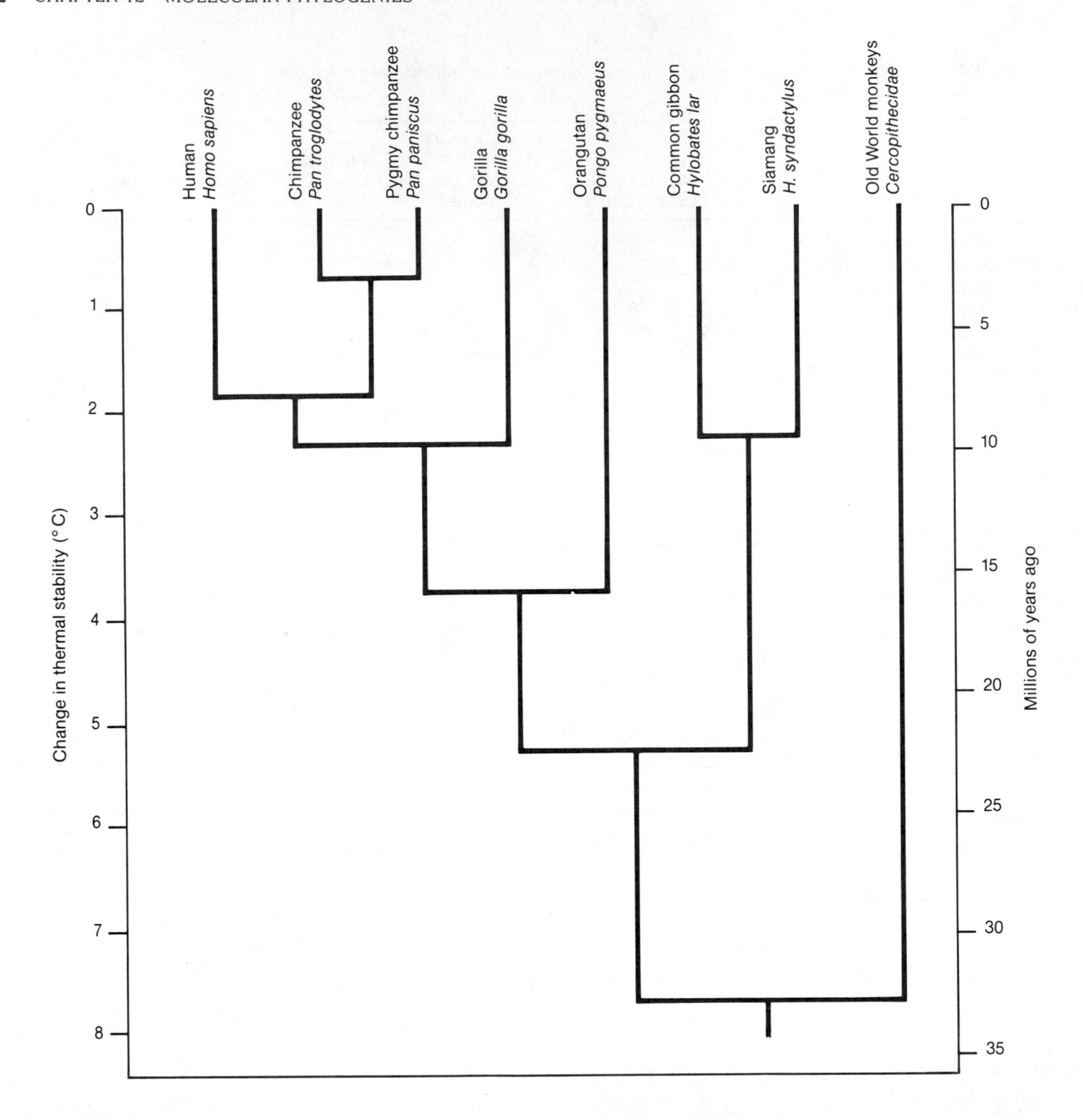

FIGURE 12-12 Phylogenetic tree and dates of divergence for humans, apes, and Old World monkeys based on DNA-DNA hybridization studies. According to Templeton, it is still difficult to distinguish between this illustrated phylogeny and one in which chimpanzees and gorillas are placed together in a lineage separate from humans, a view that seems to be supported by some recent morphological comparisons (Andrews). On the other hand, recent papers by Miyamoto et al., Maeda et al., and Holmquist et al. using methods other than DNA hybridization support Sibley and Ahlquist's conclusions. (Adapted from Sibley and Ahlquist [1984].)

temperature will be lowered. Various experiments show that for each 1 percent difference in nucleotide composition between X and Y, the thermal stability of the X-Y hybrid DNA molecule is lowered by about 1°C.

Sibley and Ahlquist point out that such techniques enable comparisons between perhaps a billion or more nucleotides simultaneously and can provide considerably more information than is usually obtained from comparing a few characters at a time. For example, they have produced detailed bird order phylogenies that once seemed difficult or impossible to determine. They have also suggested that primate phylogenies based on such DNA-DNA comparisons can specify relationships that formerly seemed obscure, such as human-chimpanzee-gorilla (Fig. 12-12). Moreover, they propose that since billions of nucleotides are involved in these determinations and changes are registered over millions of years, average rates of change over time can be estimated with a fair degree of confidence. Thus, assuming from paleontological evi-

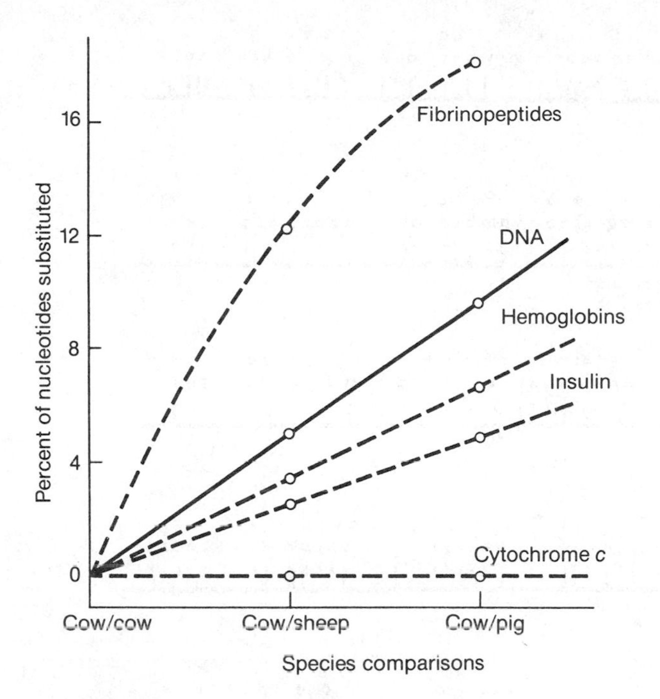

FIGURE 12-13 Nucleotide differences observed among three species of artiodactyls using the DNA hybridization technique described in the text (solid line) and estimates of nucleotide substitutions derived from amino acid analysis of various proteins (dashed lines). (Adapted from McCarthy and Farquhar.)

dence that the divergence between the lineages of Old World monkeys and apes-humans occurred about 33 million years ago (Chapter 19), and observing an approximate 7.7°C change in thermal stability between these groups and their common ancestor, there is an average of 1°C change for each 33/7.7 = 4.3 million-year interval. Thus the right hand scale of Figure 12-12 provides estimates of the dates at which these various primate taxa diverged.

These techniques also make possible comparisons between **nucleotide substitution rates** obtained from DNA-DNA hybridization data and substitution rates obtained from amino acid changes in known proteins. When such comparisons are undertaken, as in Figure 12-13, the rate of change in DNA seems more rapid than the rate of change obtained from most proteins except for fibrinopeptides.[5] It therefore seems likely that considerable portions of these tested DNA sequences do not code for essential proteins such as cytochrome and insulin, and perhaps code for no proteins at all. Such sequences, perhaps largely "junk" DNA or "selfish" DNA (see p. 194), can consequently accumulate changes more rapidly than can genes which code for stringently selected proteins.

However, despite their value in detecting divergences between entire genomes, objections to DNA-DNA hybridization techniques have nevertheless been made. One objection repeatedly raised (e.g., Templeton) is that DNA hybridization experiments compress all divergence information into a single distance measurement, thereby losing information on specific nucleotide sequence changes that could provide a more statistically supported choice among the different possible phylogenetic trees. Therefore DNA comparisons have also been explored by other techniques as discussed below.

Nucleic Acid Phylogenies Based on Restriction Enzyme Sites

One approach to comparative DNA analysis is to use **restriction enzymes**, which recognize specific short nucleotide sequences and cleave the molecule at these sites. For example, the enzyme *Eco*RI, isolated from *E. coli* bacteria, recognizes the hexanucleotide sequence (N represents non-specified nucleotides):

```
                 Recognition site
                  ↓
5' .......N N G A A T T C N N .... 3'
          | | | | | | | | | |
3' .......N N C T T A A G N N .... 3'
                          ↑
```

and produces cleavage products at the points between G and A indicated by the arrows:

```
                    ↓
5' .... N N G3' :               : 5'A A T T C N N .... 3'
        | | |   :........ + .......:        | | |
3' .... N N C T T A A5' :          :       3'G N N .... 3'
                                   ↑
```

Since DNA molecules can differ from each other in nucleotide sequence, and therefore differ in the number and placement of sites recognized by *Eco*RI, each particular kind of DNA will have fragments of characteristic length when subjected to the enzyme. Also, since there are different kinds of restriction enzymes, many of which recognize target sites different from other such enzymes, a DNA molecule subjected to a battery of different enzymes will produce cleavage products that are unique for that particular kind of DNA. By such means, **restriction maps** have been obtained for a variety of DNAs, ranging from relatively small repetitive DNA sequences to mitochondrial DNA and the DNA of even larger chromosomes.

[5]The fact that the fibrinopeptides accumulate many changes over relatively short periods of time is known to be related to their function: they are sections of fibrinogen molecules that are removed during blood clot formation, and most amino acid changes in this sequence have relatively little effect on this function.

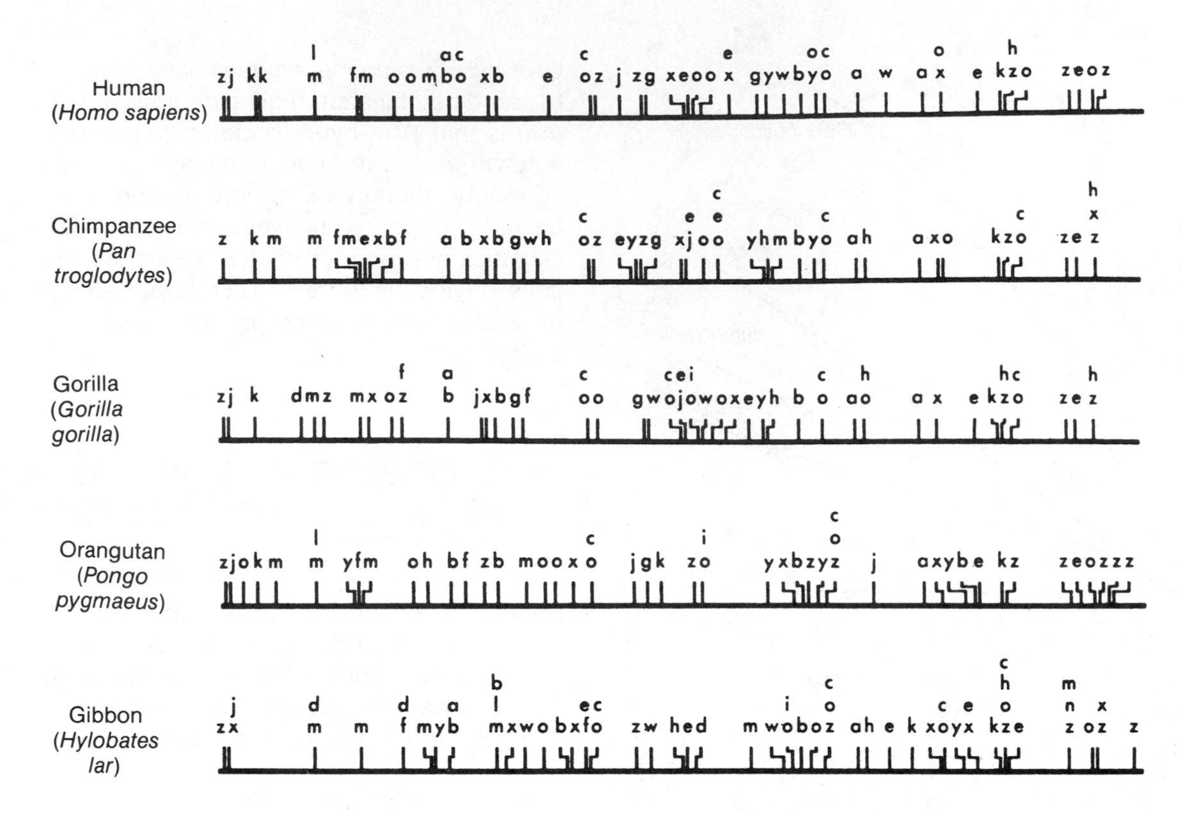

FIGURE 12-14 Cleavage maps of mitochondrial DNA from humans and four other species of higher primates, derived from the use of 19 restriction enzymes. Cleavage sites for each enzyme are designated by small letters: a, *Eco*RI; b, *Hin*dIII; c, *Hpa*I; d, *Bgl*II; e, *Xba*I; f, *Bam*HI; g, *Pst*I; h, *Pvu*II; i, *Sal*I; j, *Sac*I; k, *Kpn*I; l, *Xho*I; m, *Ava*I; n, *Sma*I; o, *Hin*cII; w, *Bst*EII; x, *Bcl*I; y, *Bgl*I; and z, *Fnu*DII. The position at the left of each map is the replication origin of the mitochondrial chromosome. (Adapted from Ferris et al.)

In one example analyzed by Ferris et al., mitochondrial DNA from humans and apes was subjected to 19 different restriction enzymes. As shown in Figure 12-14, approximately 50 sites were cleaved by these enzymes for each mitochondrial chromosome, allowing a detailed comparison of target site sequences among the five species. Thus, in accord with previously determined phylogenies, humans share more such sites with chimpanzees and gorillas than with orangutans and gibbons. Unfortunately, the exact branching order among humans, chimpanzees, and gorillas is not obvious from these data (Smouse and Li), and further studies using other techniques (e.g., Maeda et al.) are being used to test the conclusions reached by Sibley and Ahlquist. Among the more fruitful mitochondrial DNA studies have been those showing relationships among various present human races; these will be discussed in Chapter 19.

Nucleic Acid Phylogenies Based on Nucleotide Sequence Comparisons and Homologies

A more precise method of phylogenetic determination is to compare actually known nucleotide sequences from different organisms rather than infer relationships from hybridization studies or restriction enzyme maps. This offers advantages in comparing changes between protein coding and noncoding DNA sequences and in determining the extent of synonymous and nonsynonymous nucleotide substitutions in the amino acid coding regions (Li et al.). However, since accumulation of nucleotide sequence information has begun only fairly recently, complete sequence data have been obtained for only some small viruses and for relatively few cellular genes and mitochondrial organelles, although the number is growing rapidly.

One phylogeny derived from characterizing DNA sequences in considerable detail was shown in Figure 12-6 for the β-globin gene clusters of various primates. Some other examples come from nucleic acid structures that are more widely distributed, such as 5*S* RNA. This ribonucleotide chain, a component of the larger of the two ribosomal subunits, is believed to function in ribosome binding of the various transfer RNA molecules that carry the different amino acids. Since this cornerstone of the basic protein-synthesizing apparatus would undoubtedly be difficult, if not impossible, to change, it has therefore been evolutionarily conserved in all organisms. Thus the secondary structure of 5*S* RNA appears to be universally the same (Fig. 12-15), a feature that enables all the various 5*S* RNAs to be aligned for every nucleotide position. When such alignments are effected, differences between 5*S* RNAs can be used to generate a phylogenetic tree in which

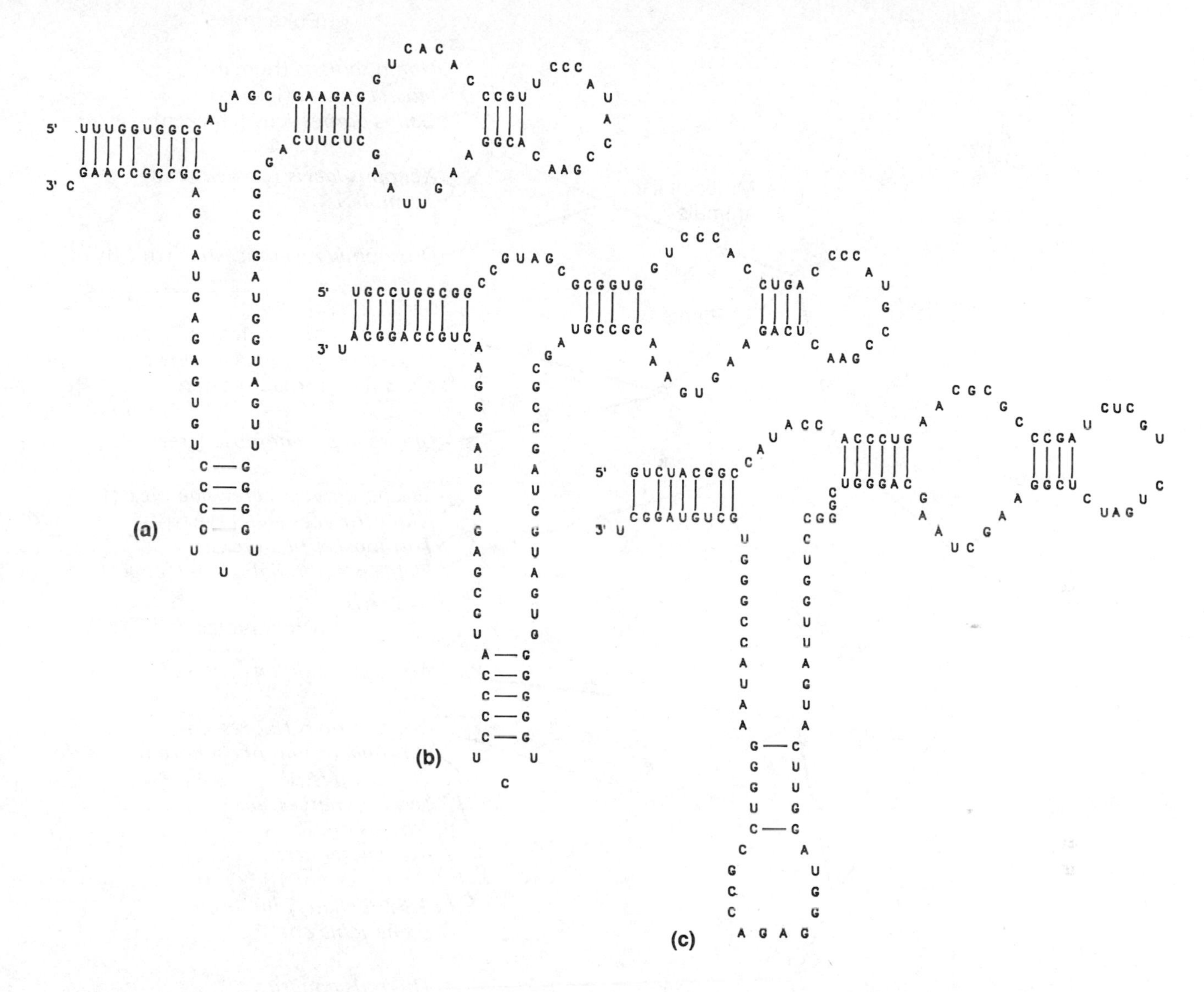

FIGURE 12-15 Models of the secondary structures of 5*S* ribosomal RNA molecules from (a) *Escherichia coli*, (b) *Bacillus subtilis*, and (c) humans. (Adapted from Hori and Osawa.)

nucleotide changes measure evolutionary distance. As shown in Figure 12-16, this tree shows a divergence between primitive prokaryotes and primitive eukaryotes at a time perhaps 50 percent earlier than the divergence between fungi (e.g., yeast) and eukaryotic plants and animals (see also Gouy and Li). Since the latter divergence is estimated to have occurred about 1.2 billion years ago, the earlier prokaryote-eukaryote separation may well have occurred 1.8 billion years ago.[6]

COMBINED NUCLEIC ACID-AMINO ACID PHYLOGENIES

A comprehensive phylogeny that takes into account both nucleotide sequences from 5*S* RNA and amino acid sequences from ferredoxin and the *c*-type cytochromes has been offered by Barnabas and co-workers and is diagrammed in Figure 12-17. Since ferredoxin is a very primitive iron-containing protein used in a

[6]Because 5*S* RNA molecules have not been found in animal mitochondria and are considered too small by some experimenters (e.g., Hasegawa et al.), other ribosomal RNA sequences are being used for nucleotide comparisons. For example, the distinction between prokaryotes and eukaryotes is also marked by differences in the frequencies of certain sequences of the 16*S* RNA ribosomal component, as shown in Table 9-2. A study of this kind by Gray et al. (1984) using mitochondrial and chloroplast RNAs provides added support for the endosymbiotic theory of organelle evolution (Chapter 9) by showing that both these organelles can be traced to a eubacterial origin. Interestingly, this study also proposes that animal and fungal mitochondria originated from non-photosynthetic aerobic bacteria, whereas plant mitochondria originated separately from cyanobacteria. (In a later study [1989] these authors suggest that plant mitochondria may be mosaics resulting from two different symbiotic events.) The fact that mitochondria derive from a bacterial origin which possessed a genetic code shared by all other organisms (the universal code) supports the concept that changes in amino acid associations with mitochondrial codons (see Table 8-2) came after the universal code was established.

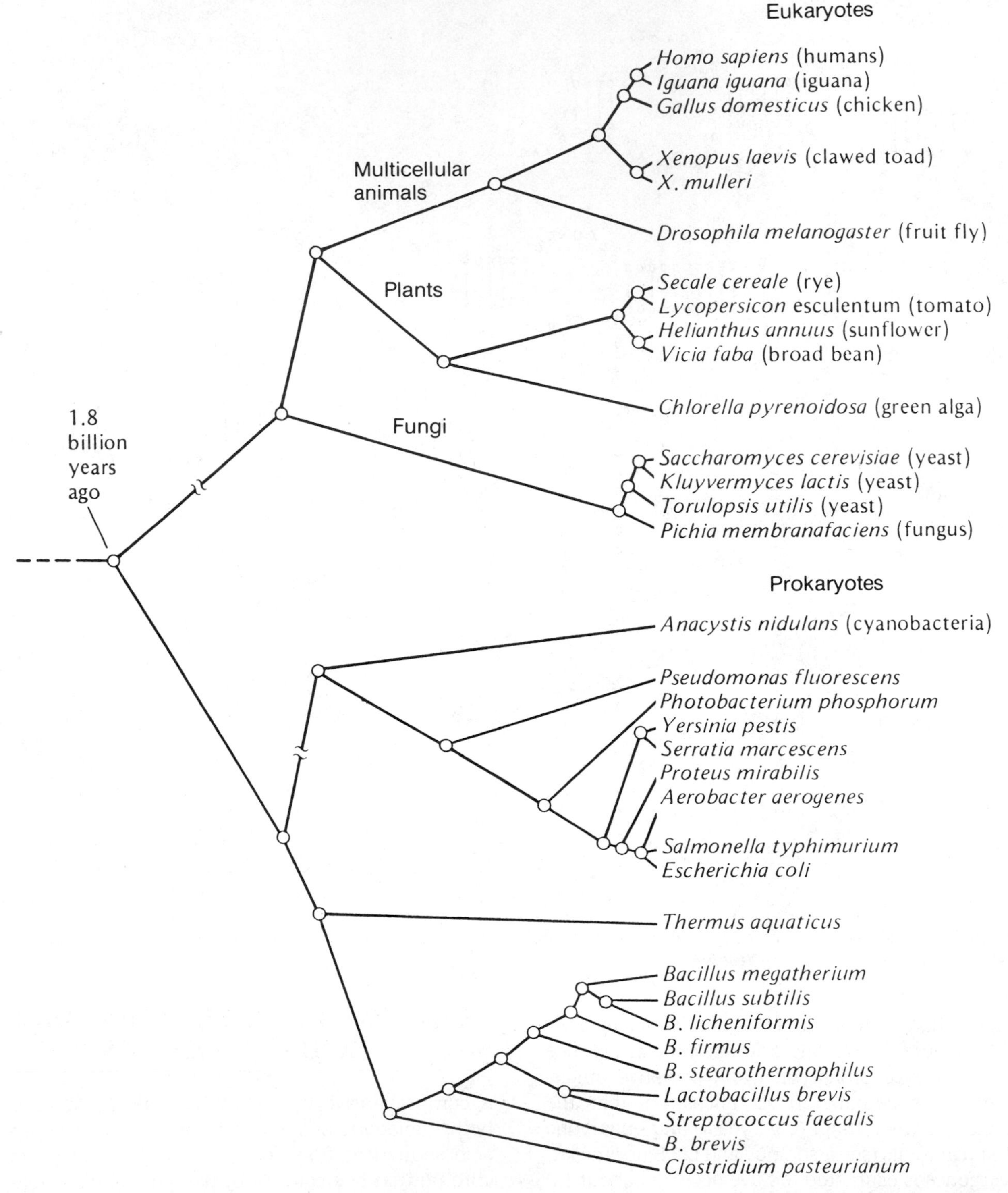

FIGURE 12-16 A phylogenetic tree that best explains the data gathered from comparing 5*S* RNA nucleotide sequences among many different species. The separation between eukaryotes and prokaryotes appears to be close to about 2 billion years ago. Since these studies, the 5*S* RNA sequences of many other species have also been determined; for example, a phylogenetic tree was constructed for 28 5*S* RNA plant sequences by Hori and Osawa, 1985. (From Strickberger, adapted from Hori and Osawa.)

number of basic oxidative-reductive pathways, the doubling event that occurred early in its evolution provides a base line for the phylogenetic tree. Organisms whose ferredoxins most resemble the inferred sequence of the primitive duplicated molecule are anaerobic and heterotrophic bacteria such as *Clostridium*. Divergence subsequently occurred, leading to anaerobic photosynthetic bacteria such as *Chromatium* and to the main line of aerobic respiratory organisms.

The cytochrome *c* analysis used by Barnabas et al. shows a phylogeny in which the eukaryotic sequences are most similar to the cytochrome *c*2 sequences of the nonsulfur purple photosynthetic bacteria (*Rhodospirillaceae*). Since cytochrome *c* functions exclusively

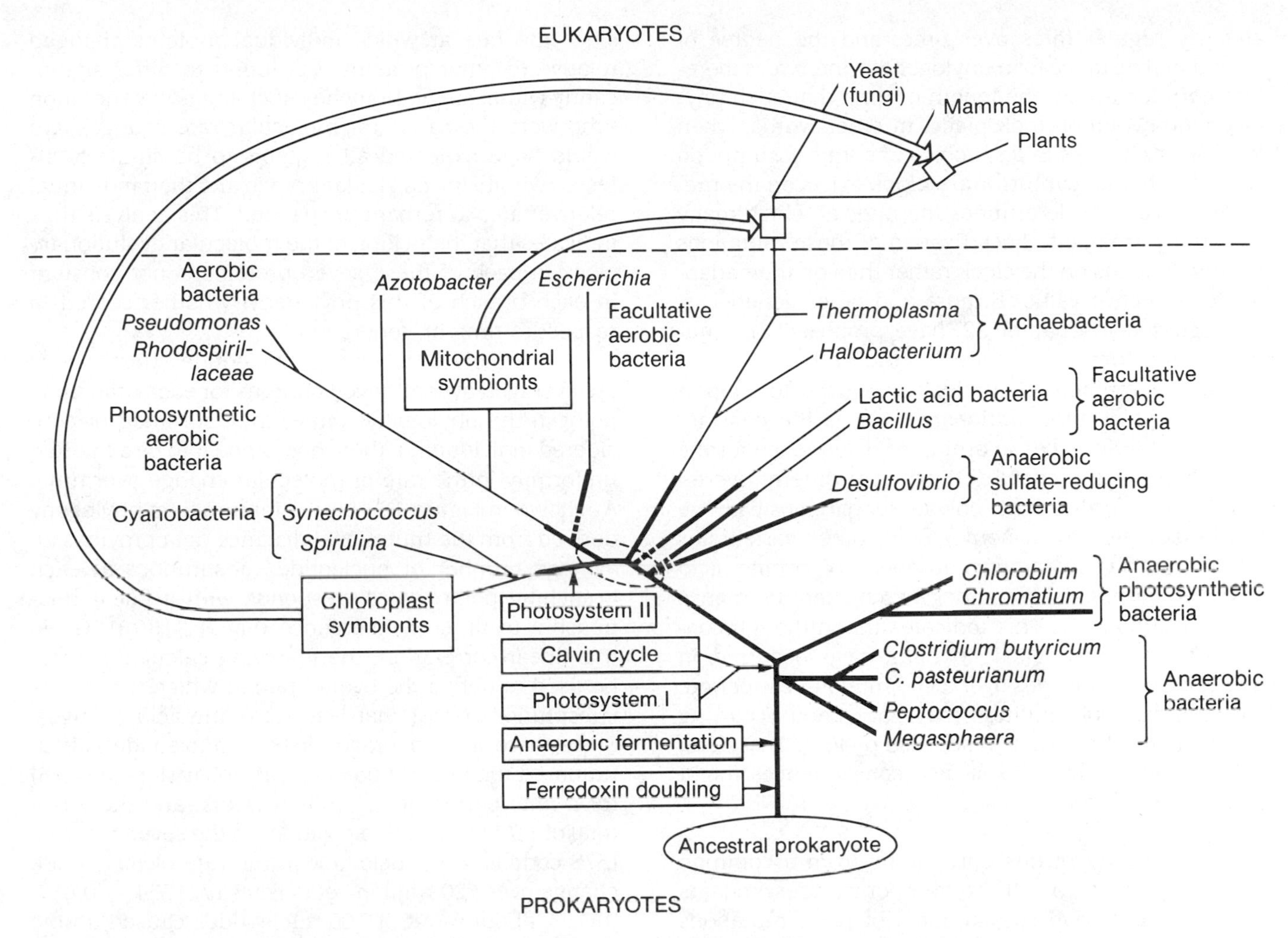

FIGURE 12-17 A composite evolutionary tree based on sequence analyses of ferredoxin, *c*-type cytochromes, and 5*S* ribosomal RNAs. The heavy lines indicate segments of the tree dependent on anaerobic metabolism, and the thin lines indicate groups using aerobic respiration. Not all branching relationships have been clearly resolved (e.g., the central dotted region). Note that photosynthesis evolved fairly early in prokaryotic phylogeny and is then presumed to have been independently lost in a number of derived lines such as *Bacillus* and the eukaryotes but was regained later in eukaryotic plants through a chloroplast invasion from a lineage close to the cyanobacteria. According to this tree, early eukaryotic-type cells were probably facultatively aerobic since they stemmed from a line close to *Escherichia* and *Bacillus* bacteria, and it was apparently only later that they replaced these pathways with a more advanced aerobic system introduced by the mitochondrial symbiosis. The dashed line across the illustration separates the eukaryotes (above) from the prokaryotes (below). (Adapted from Barnabas et al.)

in the eukaryotic mitochondrion (although coded by DNA in the nucleus), it seems most likely that it is the mitochondrion organelle itself (rather than the eukaryotic cell) that derives from this bacterial line. This view is supported by 5*S* RNA studies (see Fig. 12-17) showing that eukaryotes actually diverged from an earlier prokaryotic form and not from the later *Rhodopseudomonas* line. It was apparently after the inclusion of the mitochondrion organelle into the eukaryotic cell that the gene for cytochrome *c* was incorporated into the eukaryotic nucleus.

Similarly, the fact that there are strong homologies between cyanobacteria and the chloroplasts of higher plants for their cytochrome and ferredoxin sequences indicates that eukaryotic plant cells must have incorporated a chloroplast organelle that may at one time have been a prokaryotic cyanobacterium. This phylogeny therefore offers very strong support to the symbiotic theory of organelle function (Chapter 9).

RATES OF MOLECULAR CHANGE: EVOLUTIONARY CLOCKS

Inherent in all the various phylogenies is the concept that evolutionary differences between organisms arise from mutational differences, and therefore, in general, the greater the number of mutational differences between organisms, the greater their evolutionary distance. In some of the phylogenies considered so far (e.g., Fig. 12-12) this notion has been used to furnish evolutionary time scales; that is, the assumption has been made that mutations are incorporated (or **fixed**) at fairly regular rates over time, and the degree of mutational distance for a phylogenetic interval is therefore correlated with the length of time that such phylogenetic evolution took place. In other words, when looking at changes in a specific gene, this assumption suggests that an **evolutionary clock** exists on the molecular level that determines the rates at which many mutations are fixed. Since fixation of these mutations mostly depends on the clock rather than on their adaptive or selective value, Kimura and other geneticists, as discussed in Chapter 22, have proposed that mutations are primarily "neutral" in their effect.

One prominent example that is used to support the concept of an evolutionary clock is the constant number of differences in amino acid sequence for the same hemoglobin chain derived from different vertebrates. Specifically, if we look at comparisons with the shark sequence for α-hemoglobin, other vertebrates differ from it by similar numbers of amino acid changes: carp 85, salamander 84, chicken 83, mouse 79, and human 79. This indicates that although considerable morphological changes have occurred in these different lineages over a 400 million-year period, constant rates of mutation may have been occurring for at least some proteins (see also p. 460). Therefore, if evolutionary clocks exist, two consequences might be expected to follow:

1. The lines of descent leading from a common ancestor to all contemporary descendants should have similar rates of fixed mutations since they have experienced similar durations.

2. The proportional rate of fixation that occurs in one gene relative to the rates of fixation in other genes should be preserved throughout any line of descent.

A classical study by Fitch and Langley tested these attributes of the evolutionary clock hypothesis for seven proteins whose amino acid sequences were examined in 18 vertebrate taxa. Using commonly accepted dates of divergence for the various common ancestors of these taxa, the temporal length of each separate line of descent was obtained, thus allowing comparisons to be made between the number of nucleotide substitutions that occurred within a given time period for all proteins together and for each protein individually. The results of the test showed that the rate at which all proteins have changed together varies significantly among the branches in the different lines of descent, indicating that molecular changes are not uniform for these geological periods.

Moreover, these differences in rates of protein change cannot be simply explained as arising from different generation times in different lines of descent, since the rate at which individual proteins changed relative to other proteins was found to differ significantly within single branches. If changes in generation time were the cause for molecular rate changes, we would have expected all proteins to be similarly affected within any particular branch and their individual relative rates to remain unchanged. This analysis thus indicates that the ticking of the molecular evolutionary clock in each of these seven proteins is not constant in each branch of this phylogeny, whether scored in respect to time or generation.

Nevertheless, when the nucleotide substitutions are averaged over all seven proteins for each branching point in the phylogeny (rather than summed or considered individually), there does appear to be a marked uniformity in the rate of molecular change over time. As shown in Figure 12-18(a), a mammalian phylogeny derived from the mutational distance data provides an average number of nucleotide substitutions at each branching point that corresponds with a linear relationship to time of divergence (Fig. 12-18(b)). Given this linear correlation, the following calculations can be used to derive the overall rate at which nucleotide substitutions occur that lead to amino acid changes. Since there is an average of 98.17 nucleotide substitutions at the farthest point of this linear slope (no. 16) for a time period of 120 million years, and there is a total of 1,734 nucleotide positions in the seven proteins (578 codons $\times$ 3 nucleotides), the rate of nucleotide change over 120 million years is $98.17/1734 = 0.057$; that is, about 6 out of 100 nucleotides caused amino acid substitutions during this interval. The annual rate of amino acid-changing nucleotide substitutions in this lineage is therefore $0.057/(120 \times 10^6) = 0.47 \times 10^{-9}$.

The need for averaging changes among different genes in order to obtain an annual rate of nucleotide substitutions indicates that there is no single evolutionary molecular clock applicable to every nucleotide sequence. The most probable reason for this is that selection intensity varies in different parts of the genome, causing different rates at which mutations are fixed. In addition, as Britten points out (see also Li et al.), there appear to be significant differences between taxonomic groups in nucleotide substitutions that have neutral effects on the phenotype, such as synonymous codon changes [e.g., UUU (phenylalanine) → UUC (phenylalanine)]. The data so far show two different rates at which such substitutions have been incorporated: a slow rate of divergence for humans, apes, and birds, and a faster rate for rodents, lower primates, *Drosophila*, and sea urchins. Britten offers a possible reason for slower rates of divergence in the lower mutation rate that would result from improved DNA repair systems. He suggests that such lower mutation rates would be advantageous in groups which evolved toward increased parental investment

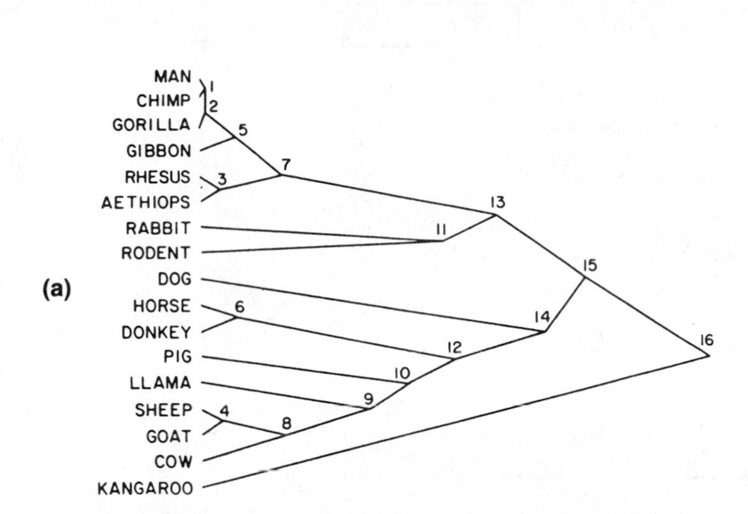

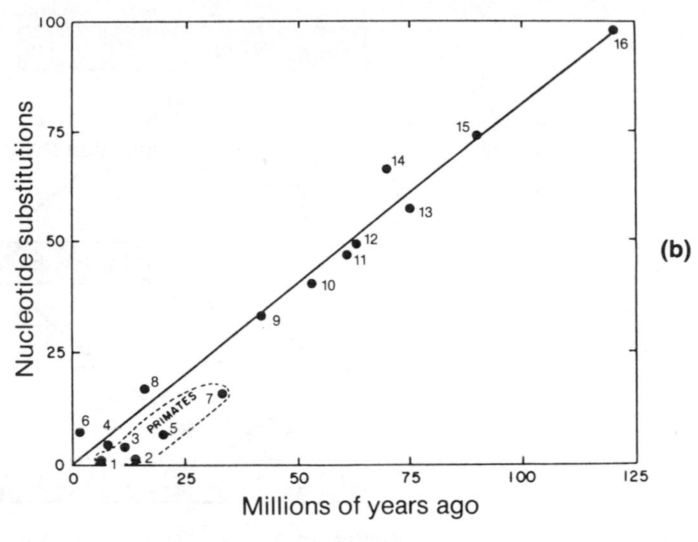

FIGURE 12-18 (a) A phylogeny of 17 vertebrate species determined from amino acid sequences examined in seven different proteins. Each numbered nodal point represents a common ancestor for its two diverging branches. (b) The relationship between the average number of nucleotide substitutions at each nodal point to the estimated time at which its branches diverged. A straight line has been drawn from the origin to the farthest nodal point (no. 16), which represents the placental-marsupial divergence approximately 120 million years ago. The slope of this line indicates a substitution rate of 0.47 nucleotides per billion years, and, with the exception of primate rates of evolution (nos. 1, 2, 3, 5, 7), most nodal points fall fairly close to this value. The source for the decreased nucleotide substitution rate in primates is not known, although it has been suggested that the number of nucleotide substitutions is more variable in this group, or estimated dates for primate divergences are less than those commonly accepted, or some primate mutation rates have decreased, or that a combination of these and possibly other factors applies. This pattern of "primate slowdown" is discussed by Britten (see also Maeda et al.). (Adapted from Fitch and Langley.)

in their offspring—that is, reduced birth rate and greater postnatal care.

Wilson and co-workers, on the other hand, propose that the differences in evolutionary rates among taxonomic groups cited by Britten are probably exceptional. Instead, they suggest that an examination of many different genes in both bacteria and mammals show fairly similar rates of nucleotide substitutions at synonymous codon sites (Fig. 12-19). This view has been disputed for such sites in *Drosophila* species (Riley), so all one can say at present is that although there may be evidence for an evolutionary clock in various lineages, it apparently does not tick at the same rate in all taxonomic groups.

REGULATORY GENES AND SOME EVOLUTIONARY CONSEQUENCES

One of the frequent observations that has emerged from comparing different organisms on the molecular level is the fact that so many of them share the same kinds of proteins. For example, whether organisms are prokaryotes or eukaryotes, they all share similar enzymes involved in basic biochemical processes such as amino acid synthesis, DNA replication, and protein synthesis. These polymerases, proteases, ferredoxins, cytochromes, nucleases, and numerous other gene

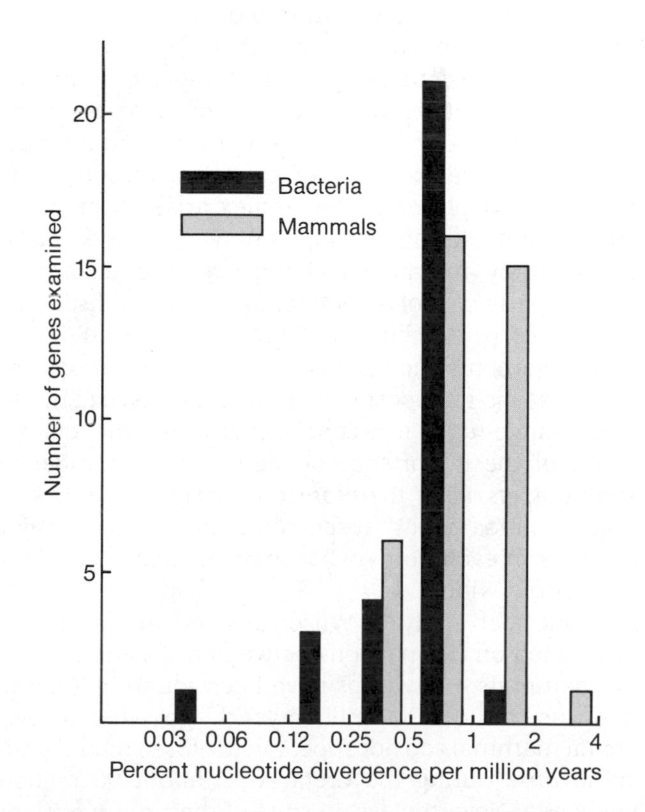

FIGURE 12-19 The rates of nucleotide substitutions per million years at synonymous codon sites in 30 bacterial and 38 mammalian genes, according to Wilson et al.

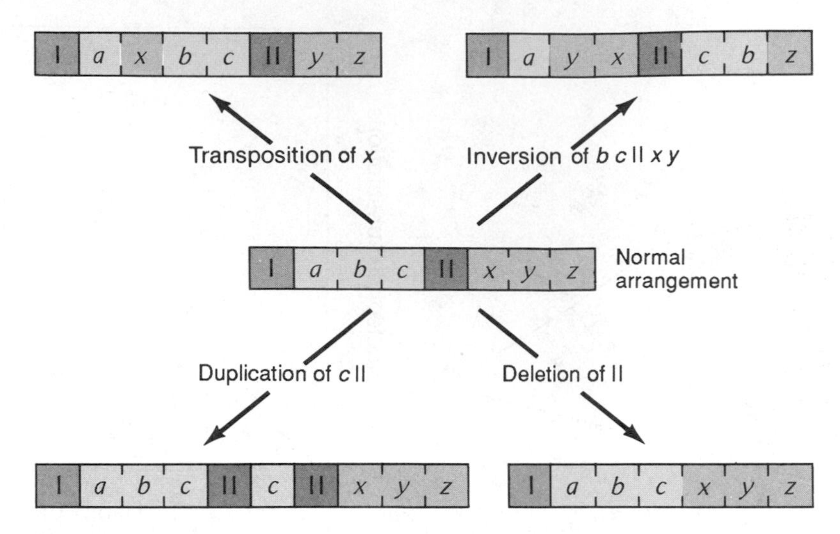

FIGURE 12-20 Some of the possible rearrangements that can occur in a gene sequence containing two sets of structural genes (*a, b, c* and *x, y, z*), each set controlled by separate regulator genes (*I, II*) and each change capable of causing striking mutant effects. For example, the deletion of regulator gene *II* places control of structural genes *x, y,* and *z* under regulator gene *I*, and the indicated inversion reverses the previous regulatory controls over genes *b, c* and *x, y*. (From Strickberger, adapted from Wilson.)

products are common enzymes that are universally distributed. When examined closely, the distinctive structural features of different organisms within any group, such as vertebrates, seem less dependent on differences among the kinds of proteins they possess than on how various common proteins such as actin, myosin, collagen, and albumin are organized and regulated. Thus in comparing the anatomy of cat with dog or cat with mouse, differences seem to be primarily based on the extent and location of their common tissues (e.g., bone, muscle, nerve). This control over the quantity and placement of tissues arises from regulatory events during development which can easily be modified by mutational changes in the eukaryotic counterparts of prokaryotic regulator genes discussed in Chapter 10. As shown in Figure 12-20, simple gene rearrangements such as deficiencies, duplications, inversions, and transpositions (translocations) can markedly change regulatory control over gene function. Because of the importance of regulation, a number of evolutionists have therefore emphasized changes in regulation as being responsible for various major changes in evolution—what may be called "new bottles for old wine."

One such study, by Wilson and co-workers (1974), was based on comparisons between frogs and placental mammals. Early frogs have been found in Triassic deposits of about 200 million years ago, whereas placental mammals do not appear in the fossil record until some time during the Cretaceous, about 90 million years ago. Nevertheless, in spite of their more ancient fossil history and their numerous array of species, frogs have undergone few phenotypic changes in evolution compared to the enormous adaptive radiation of placental mammals. The more than 3,000 species of frogs look very much alike (Fig. 12-21(a)) and have been relegated by taxonomists into a single order (Anura), whereas the 3,800 species of placental mammals (of which about 1,700 are rodents) diverged widely and have usually been classified into 15 or 16 orders ranging from bats to primates to whales (Fig. 12-21(b)). Since the amino acid sequences in proteins of both of these groups seem to have evolved at approximately the same rate, Wilson suggests that the phenotypic similarity among frogs indicates that relatively few regulatory mutations have become established in frogs compared to mammals.

The importance of **regulatory changes** seems obvious in the sharp phenotypic contrast between related species such as humans and African apes. The differences between these two groups (brain size, facial structure, bipedal locomotion, etc.) are sufficient for many anthropologists to place them in different taxonomic families (Hominidae, Pongidae), yet they are strikingly similar in composition of structural proteins: they possess almost identical α- and β-hemoglobin chains, cytochrome *c* proteins, and even fibrinopeptides. In fact, comparisons for any given protein between these two groups show an average of about 99 percent identity in amino acid sequence (King and Wilson).

It seems clear from these and other examples that regulatory mutations can play a larger role in the morphological and functional differentiation of species than many structural gene changes. This observation has been used by some evolutionists to support the notion that new species, new genera, or new orders (**macroevolutionary** events) can arise from regulatory changes with large phenotypic effect occurring over relatively short periods of time. That is, a population undergoing only small gradual changes (**microevolution**) may persist that way for long periods until pronounced regulatory changes are incorporated that allow one or more of its small isolated groups to evolve into a higher taxonomic category.

Since some of the fossil data for vertebrates, molluscs, and other animals show such periodic bursts of evolutionary activity, paleontologists such as Eldredge and Gould have given the name **punctuated equilibria** to what they see as long-term fossil uniformity of populations punctuated by geologically short periods of rapid speciation. Others argue that because the fossil record has gaps, punctuated equilibria are only apparent, not real: what seems rapid in geological time may actually involve many thousands of generations. However, although this dispute has generated many arguments and counter-arguments (see discussion in Chapter 23), all evolutionists agree that both gradual and rapid changes occur during evolution. What has not yet been resolved is the relative importance of these changes in explaining speciation and the evolution of higher taxonomic categories. On the phenotypic level, some speciation events certainly seem to be gradual, so that sibling species of *Drosophila*, for example, look very much alike, whereas other related species that are probably no older than some *Drosophila* sibling species (e.g., chimpanzees and humans) look very different. It would obviously be helpful if the genetic and evolutionary events in each lineage could be analyzed.

MOLECULAR EVOLUTION IN THE TEST TUBE

The phenomenal growth of molecular information in biology has sparked various attempts to demonstrate evolutionary processes in the laboratory where detailed analysis of successive changes is more possible than in nature. Since this approach usually demands considerable biochemical analyses and rapid generation times, as well as rigid control over genetic and environmental conditions, most of these studies have been performed with microbial organisms. Among the common techniques of **test-tube evolution** is to subject a strain of bacteria to a new carbon source (e.g., xylitol) or nitrogen source (e.g., butyramide) which the cells are not able to metabolize properly. When such cells are simultaneously exposed to a mutagenic agent, mutations increase in frequency, and some adaptive mutations may then arise. Natural selection then proceeds by permitting the survival of those bacterial

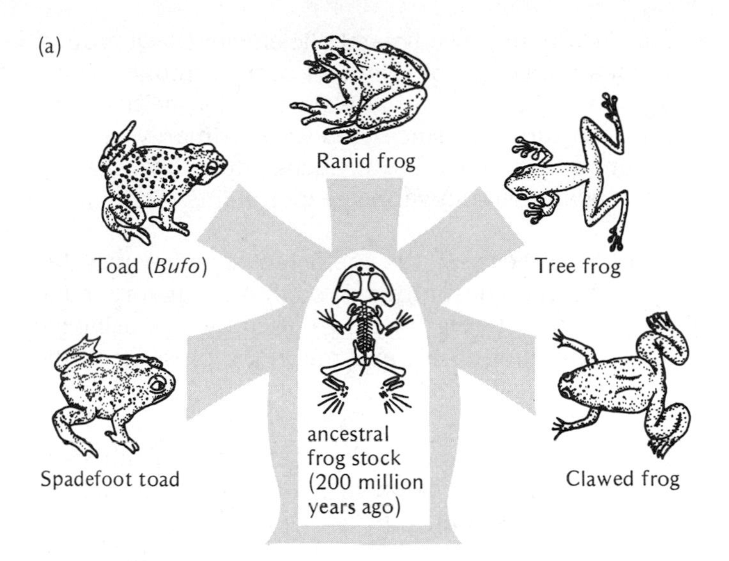

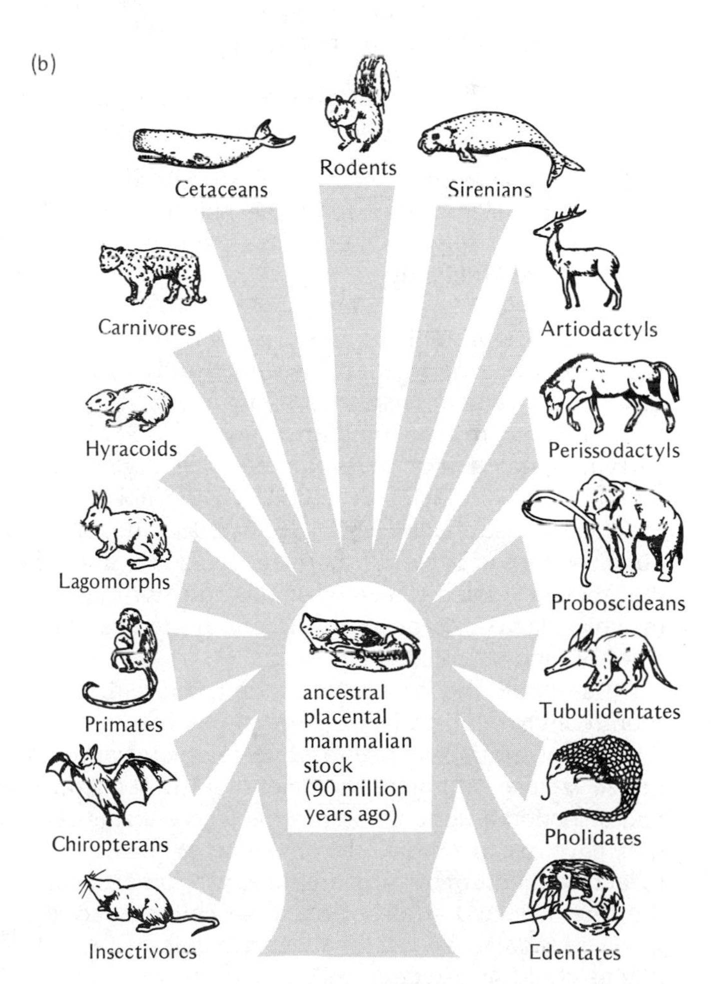

FIGURE 12-21 (a) Some of the major forms of present-day frogs along with a Triassic froglike fossil dating back to about 200 million years ago. (b) Representatives of 15 present-day orders of placental mammals and the fossil skull of a Cretaceous placental-like mammal dating back to about 90 million years ago. (From Strickberger.)

strains with improved metabolic efficiency. Often, evolutionary changes of this kind occur through the adaptation of enzymes that were initially inefficient on the new substrate since they were primarily used for other purposes. A variety of such adaptational changes have been noted in various experiments (Clarke):

1. Synthesis of the inefficient enzyme may become constitutive through a regulatory mutation (see Fig. 10-22(d)), thereby increasing the amount of this enzyme in the presence of the new substrate.
2. A regulatory mutation may enable synthesis of the inefficient enzyme to become inducible by the new substrate.
3. Mutations may occur that enable the substrate to enter the cell more easily.
4. A gene duplication may occur that enables increased production of the inefficient enzyme.
5. Mutations may occur in the structural gene for the enzyme itself to enable the former inefficient enzyme to metabolize the new substrate more efficiently.

A striking demonstration of enzymatic evolution is shown in an experiment by Campbell and co-workers on an *E. coli* protein. Instead of trying to adapt bacteria to an unusual artificial medium, they exposed a strain carrying a deletion of the β-galactosidase *Z* gene (see Fig. 10-22(a)) to lactose medium. In the absence of β-galactosidase, hydrolysis of lactose does not occur. Such colonies of bacteria can be recognized by the fact that they develop a red color on an indicator medium in which lactose is present, in contrast to the white color of lactose-utilizing colonies.

Campbell and co-workers found that within one month of growth on a lactose-containing medium, the *Z*-deficient strain gave rise to white colonies that could utilize lactose, although inefficiently. Further growth and selection among these new lactose-utilizing cells then gave rise to a more efficient strain. When exposed to lactose medium unsupplemented with other sugars, the final selected strain of bacterial cells, called *ebg* (evolved β-galactosidase), could form colonies as rapidly as could wild-type *E. coli*.

Various tests then showed that the *ebg* strain had evolved a lactose-hydrolyzing enzyme (EBG) completely different from β-galactosidase. This new enzyme had a greater molecular weight, different immunological properties, and different ionic sensitivities. Interestingly, as shown by Hall and Hartl, its appearance could be regulated by lactose, and regulatory mutations were found similar to those that control β-galactosidase. After nucleotide sequencing of the genes involved, Stokes and Hall concluded that the EBG system is a remnant of an ancient duplication of the *E. coli lac* enzyme system. In sum, these experiments demonstrate that a protein with only vague affinities for a particular function can assume that function with remarkable efficiency by a stepwise evolutionary process of mutation and selection (Hall, Hartle).

On the nucleotide level, Mills and co-workers were able to narrow test-tube evolutionary experiments down to some of the smallest self-replicating molecules known so far. They began with the RNA nucleic acid of a Qβ virus that was about 4,220 nucleotides long, a molecule which could be replicated in test tubes upon the addition of a replicase enzyme and various chemical components. Selection for rapid replication was then practiced by successive transfers of only the earliest replicating molecules to new cultures. Under these conditions successful Qβ molecules need only retain those sequences that enable them to be recognized by the replicase enzyme provided in the culture. That is, they no longer have need for genes that formerly coded for what are now unnecessary proteins—the coat protein and the replicase enzyme. Selection was then even further intensified by placing fitness advantages on RNA molecules that could self-replicate when only a single such molecule is present in an entire culture. This single-stranded molecule, or plus strand, must rapidly attract a replicase enzyme to form a complement, or minus, strand which then forms a new plus strand, and so on. By these means, short, independently replicating RNA molecules were selected, including one type called midivariant which was only about 220 nucleotides long (Fig. 12-22)!

Eigen and his group then showed that when these experiments were reversed—that is, when such mixtures begin without any Qβ RNA sequences at all—the Qβ replicase will nevertheless splice together nucleotides on its own without a template. Moreover, evolution in such mixtures can produce a variety of *de novo* RNA sequences capable of adapting to different environmental conditions, including sequences that reach by accretion that optimal self-replicating midivariant length of 220 nucleotides! These and other experiments have led Eigen and co-workers to conclude that the number of nucleotides (information content) in an RNA strand determines the frequency at which mutant sequences arise, as well as the number of reproductive cycles necessary for the selection of an optimal mutant with high reproductive rate. The prevailing (wild type) genotype achieves stability when its selective advantage is sufficiently great to overcome the error rate of replication for that nucleotide sequence. If the error rate is too high, adaptive information is lost. If the error rate is too low, the capacity for further adaptation is reduced. Such factors therefore govern the nucleotide length of a molecule in these experiments. For example, Eigen points

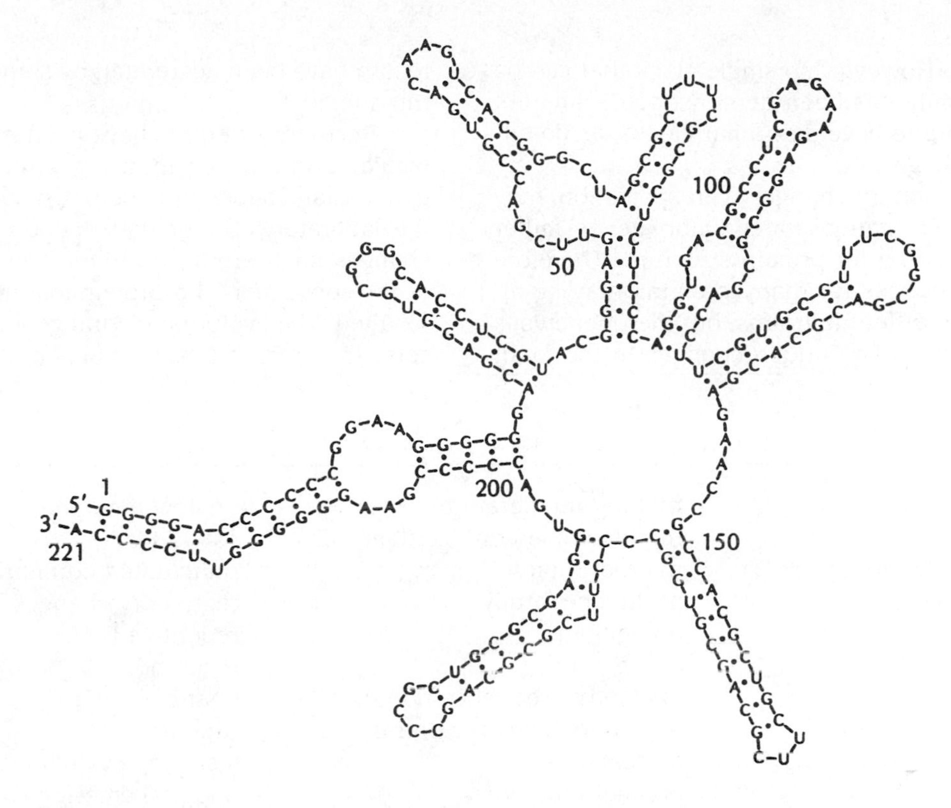

FIGURE 12-22 Nucleotide sequence and secondary structure of the smallest RNA molecule that can replicate independently in a mixture containing the replicase enzyme of Qβ virus. (Adapted from Miele et al.)

out that since RNA polymerases do not replicate nucleotides as accurately as DNA polymerases, the RNA molecules in single-stranded RNA viruses are usually no longer than 10^4 nucleotides, a value that can be theoretically calculated from error rates and selective advantages.

SUMMARY

In the search for methods to determine phylogenetic relationships, many new techniques have been utilized that allow molecular comparisons among organisms, even among those without common morphological features. Among these techniques are immunological methods that test proteins from different organisms for antigenic similarities. Another useful technique surveys amino acid sequences in individual proteins and can be informative about the evolution of individual genes. For example, amino acid sequencing indicates that myoglobin and the many different forms of hemoglobin arose from one ancestral globin gene which subsequently underwent a series of duplications. Such events were apparently common, with many of the duplicated genes evolving along different lines and carrying out disparate functions. Phylogenetic distances also may be determined by comparing amino acid changes in the same protein in different species and calculating the number of mutations necessary to convert one amino acid to another. To construct a realistic phylogenetic tree, however, information from many proteins is required. Techniques involving measurement of DNA levels generally show that the quantity of DNA is not always proportional to the taxonomic status of the group. Analyses of the types of DNA found in organisms indicate that much DNA is repetitive, sometimes highly so. A method used by some investigators is to measure the degree to which DNAs from different organisms hybridize in order to detect homologies that can be useful in determining phylogenetic lineages and in dating points of divergence of various taxa.

Restriction enzymes, which cleave DNA into fragments at particular sites, also allow comparisons among DNAs of different species. The most difficult, but most informative, of these techniques is a comparison of exact nucleotide sequences of DNAs from different sources. For example, data from 5*S* ribosomal RNA has been used to compose a phylogentic tree in which nucleotide changes are used as a measure of evolutionary distance. Phylogenies may also be constructed utilizing both DNA and protein sequencing.

It has been assumed by some workers that mutations become fixed into genotypes at a fairly regular rate (the evolutionary clock), and there does appear to be constancy in the rate of molecular change over time for some genes. However, no single clock that can be applied universally has been demonstrated; different parts of the genome have dissimilar clocks, as do different taxonomic groups.

Major evolutionary change, even speciation, may occur because of mutations in regulatory genes, rather than in genes coding for protein structure. Therefore amino acid sequences for many proteins may be almost identical in different groups, but the phenotypes may vary considerably. Sudden changes in the fossil record have been accredited by some workers to such mutations.

Recently, attempts have been made to study molecular evolution by inducing mutations in microorganisms and subjecting them to selection pressures in the laboratory. Under these conditions, evolutionary changes, such as enzyme adaptation to a new substrate or the conversion of a protein to a new function, have resulted. The evolution of viral genomes by molecular selection pressures has also been demonstrated.

KEY TERMS

antigenic distances
back mutations
DNA-DNA hybridization
evolutionary clock
fixed mutations
gene cluster
gene duplications
gene family
hemoglobin
heterochromatin
immunodiffusion technique
macroevolution
melting temperature
microcomplement fixation
microevolution
multigene family
myoglobin
nonreciprocity
nucleotide sequence homologies
nucleotide substitution rates
orthologous
parallel mutations
paralogous
parsimony method
phylogenetic tree
pseudogenes
punctuated equilibria
regulatory changes
repetitive DNA
restriction enzymes
restriction maps
satellite DNA
test-tube evolution
unequal crossing over
unique DNA

REFERENCES

Andrews, P., 1987. Aspects of hominoid phylogeny. In *Molecules and Morphology in Evolution: Conflict or Compromise?* C. Patterson (ed.). Cambridge Univ. Press, Cambridge, pp. 23-53.

Barnabas, J., R. M. Schwartz, and M. O. Dayhoff, 1982. Evolution of major metabolic innovations in the Precambrian. *Origins of Life*, **12**, 81-91.

Britten, R. J., 1986. Rates of DNA sequence evolution differ between taxonomic groups. *Science*, **231**, 1393-1398.

Britten, R. J., and E. H. Davidson, 1971. Repetitive and non-repetitive DNA sequences and a speculation on the origins of evolutionary novelty. *Quart. Rev. Biol.*, **46**, 111-138.

Britten, R. J., and D. E. Kohne, 1968. Repeated sequences in DNA. *Science*, **161**, 259-540.

Campbell, J. H., J. A. Lengyel, and J. Langridge, 1973. Evolution of a second gene for β-galactosidase in *Escherichia coli*. *Proc. Nat. Acad. Sci.*, **70**, 1841-1845.

Champion, A. B., K. L. Soderberg, A. C. Wilson, and R. P. Ambler, 1975. Immunological comparison of azurins of known amino acid sequence: dependence of cross-reactivity upon sequence resemblance. *Jour. Mol. Evol.*, **5**, 291-305.

Clarke, P. H., 1978. Experiments in microbial evolution. In *The Bacteria, Vol. 6: Bacterial Diversity*, L. N. Ornston and J. R. Sokatch (eds.). Academic Press, New York, pp. 137-218.

Davidson, E. H., and R. J. Britten, 1973. Organization, transcription, and regulation in the animal genome. *Quart. Rev. Biol.*, **48**, 565-613.

Dayhoff, M. O. (ed.), 1978. *Atlas of Protein Sequence and Structure*. Vol. 5, Supplement 3. National Biomedical Research Foundation, Washington, D.C.

Dayhoff, M. O., R. M. Schwartz, and B. C. Orcutt, 1978. A model of evolutionary change in proteins. In *Atlas of Protein Sequence and Structure*, Vol. 5, Supplement 3, M. O. Dayhoff (ed.). National Biomedical Research Foundation, Washington, D.C., pp. 345-352.

Dene, H. T., M. Goodman, and W. Prychodko, 1976. Immunodiffusion evidence on the phylogeny of primates. In *Molecular Anthropology*, M. Goodman and R. E. Tashian (eds.). Plenum Press, New York, pp. 171-195.

Eigen, M. 1983. Self-replication and molecular evolution. In *Evolution From Molecules To Man*, D. S. Bendall (ed.). Cambridge Univ. Press, Cambridge, pp. 105-130.

Eldredge, N., and S. J. Gould, 1972. Punctuated equilibria: an alternative to phyletic gradualism. In *Models In Paleobiology*, T. J. M. Schopf (ed.). Freeman, Cooper, San Francisco, pp. 82-115.

Elleman, T. C. 1978. A method for detecting distant evolutionary relationships between proteins or nucleic acid sequences in the presence of deletions or insertions. *Jour. Mol. Evol.*, **11**, 143-161.

Felsenstein, J., 1988. Phylogenies from molecular sequences: inference and reliability. *Ann. Rev. Genet.*, **22**, 521-565.

Ferguson, A., 1980. *Biochemical Systematics and Evolution*. John Wiley, New York.

Ferris, S. D., A. C. Wilson, and W. M. Brown, 1981. Evolutionary tree for apes and humans based on cleavage maps of mitochondrial DNA. *Proc. Nat. Acad. Sci.*, **78**, 2432-2436.

Fitch, W. M., 1973. Aspects of molecular evolution. *Ann. Rev. Genet.*, **7**, 343-380.

——, 1976. Molecular evolutionary clocks. In *Molecular Evolution*, F. J. Ayala (ed.). Sinauer Associates, Sunderland, Mass., pp. 160-178.

Fitch, W. M., and C. J. Langley, 1976. Protein evolution and the molecular clock. *Fed. Proc.*, **35**, 2092-2097.

Fitch W. M., and E. Margoliash, 1967. Construction of phylogenetic trees. *Science*, **155**, 279-284.

Fitch W. M., and E. Margoliash, 1970. The usefulness of amino acid and nucleotide sequences in evolutionary studies. *Evol. Biol.*, **4**, 67-109.

Frazier, W. A., R. H. Angeletti, and R. A. Bradshaw, 1972. Nerve growth factor and insulin. *Science*, **176**, 482-488.

Galau, G. A., M. E. Chamberlin, B. R. Hough, R. J. Britten, and E. H. Davidson, 1976. Evolution of repetitive and nonrepetitive DNA. In *Molecular Evolution*, F. J. Ayala, ed. Sinauer Associates, Sunderland, Mass., pp. 200-224.

Goodman, M., 1975. Protein sequence and immunological specificity. In *Phylogeny of the Primates*, W. P. Luckett and F. S. Szalay (eds.). Plenum Press, New York, pp. 219-248.

——, 1976. Toward a genealogical description of the primates. In *Molecular Anthropology*, M. Goodman and R. E. Tashian (eds.). Plenum Press, New York, pp. 321-353.

Goodman, M., A. E. Romero-Herrera, H. Dene, J. Czelusniak, and R. E. Tashian, 1982. Amino acid sequence evidence on the phylogeny of primates and other eutherians. In *Macromolecular Sequences in Systematic and Evolutionary Biology*, M. Goodman (ed.). Plenum Press, New York, pp. 115-191.

Gould, S. J., 1982. The meaning of punctuated equilibrium and its role in validating a hierarchical approach to macroevolution. In *Perspectives On Evolution*, R. Milkman (ed.). Sinauer Associates, Sunderland, Mass., pp. 83-104.

Gouy, M., and W-H. Li, 1989. Molecular phylogeny of the kingdoms Animalia, Plantae, and Fungi. *Mol. Biol. and Evol.*, **6**, 109-122.

Gray, M. W., D. Sankoff, and R. J. Cedergren, 1984. On the evolutionary descent of organisms and organelles: a global phylogeny based on a highly conserved structural core in small subunit ribosomal RNA. *Nuc. Acids Res.*, **12**, 5837-5852.

———, 1989. On the evolutionary origin of the plant mitochondrion and its genome. *Proc. Nat. Acad. Sci.*, **86**, 2267-2271.

Hall, B. G., 1983. Evolution of new metabolic functions in laboratory organisms. In *Evolution of Genes and Proteins*, M. Nei and R. K. Koehn (eds.). Sinauer Associates, Sunderland, Mass., pp. 234-257.

Hall, B. G., and D. L. Hartl, 1974. Regulation of newly evolved enzymes. I. Selection of a novel lactase regulated by lactose in *Escherichia coli*. *Genetics*, **76**, 391-400.

Hartl, D. L., 1989. Evolving theories of enzyme evolution. *Genetics*, **122**, 1-6.

Hasegawa, M., Y. Iida, T. Yano, F. Takaiwa, and M. Iwabuchi, 1985. Phylogenetic relationships among eukaryotic kingdoms inferred from ribosomal RNA sequences. *Jour. Mol. Biol.*, **22**, 32-38.

Hill, R. L., K. Brew, T. C. Vanaman, I. P. Trayer, and J. P. Mattock, 1969. The structure, function, and evolution of α-lactalbumin. *Brookhaven Symp. Biol.*, **21**, 139-152.

Hinegardner, R., 1976. Evolution of genome size. In *Molecular Evolution*, F. J. Ayala, ed. Sinauer Associates, Sunderland, Mass., pp. 179-199.

Holmquist, R., M. M. Miyamoto, and M. Goodman, 1988. Analysis of higher-primate phylogeny from transversion differences in nuclear and mitochondrial DNA by Lake's methods of evolutionary parsimony and operator metrics. *Mol. Biol. and Evol.*, **5**, 217-236.

Hori, H., and S. Osawa, 1979. Evolutionary change in RNA secondary structure and a phylogenetic tree of 54 5S RNA species. *Proc. Nat. Acad. Sci.*, **76**:381-385.

Hori, H., and S. Osawa, 1985. Evolution of green plants as deduced from 5S rRNA sequences. *Proc. Nat. Acad. Sci.*, **82**, 820-823.

Ingram, V. M., 1963. *The Hemoglobins in Genetics and Evolution*. Columbia Univ. Press, New York.

Jeffreys, A. J., S. Harris, P. A. Barrie, D. Wood, A. Blanchetot, and S. M. Adams, 1983. Evolution of gene families: the globin genes. In *Evolution From Molecules to Men*, D. S. Bendall (ed.). Cambridge Univ. Press., Cambridge, pp. 175-195.

Jukes, T. H., 1966. *Molecules and Evolution*. Columbia Univ. Press, New York.

Kimura, M., 1979. The neutral theory of molecular evolution. *Sci. Amer.,* **241** (5), 94-104.

King, M. C., and A. C. Wilson, 1975. Evolution at two levels: molecular similarities and biological differences between humans and chimpanzees. *Science*, **188**, 107-116.

Li, W-H., 1983. Evolution of duplicate genes and pseudogenes. In *Evolution of Genes and Proteins*, M. Nei and R. K. Koehne (eds.). Sinauer Associates, Sunderland, Mass., pp. 14-37.

Li, W-H., C-C. Luo, and C-I. Wu, 1985. Evolution of DNA sequences. In *Molecular Evolutionary Genetics*, R. J. MacIntyre (ed.). Plenum Press, New York, pp. 1-94.

Maeda, N., C. Wu, J. Bliska, and J. Reneke, 1988. Molecular evolution of intergenic DNA in higher primates: pattern of DNA changes, molecular clock, and evolution of repetitive sequences. *Mol. Biol. and Evol.*, **5**, 1-20.

McCarthy, B. J., and M. N. Farquhar, 1972. The rate of change of DNA in evolution. *Brookhaven Symp. Biol.*, **23**, 1-41.

Miele, E. A., D. R. Mills, and F. R. Kramer, 1983. Autocatalytic replication of a recombinant RNA. *Jour. Mol. Biol.*, **171**, 281-295.

Miklos, G. L. G., 1985. Localized highly repetitive DNA sequences in vertebrate and invertebrate genomes. In *Molecular Evolutionary Genetics*, R. J. MacIntyre (ed.). Plenum Press, New York, pp. 241-321.

Mills, D. R., F. R. Kramer, and S. Spiegelman, 1973. Complete nucleotide sequence of a replicating RNA molecule. *Science,* **180**, 916-927.

Mirsky, A. E., and H. Ris, 1951. The deoxyribonucleic acid content of animal cells and its evolutionary significance. *Jour. Gen. Physiol.*, **34**, 451-462.

Miyamoto, M. M., B. F. Koop, J. L. Slightom, M. Goodman, and M. R. Tennant, 1988. Molecular systematics of higher primates: genealogical relations and classification. *Proc. Nat. Acad. Sci.*, **85**, 7627-7631.

Nei, M., 1987. *Molecular Evolutionary Genetics.* Columbia Univ. Press, New York.

Ohno, S., 1970. *Evolution by Gene Duplication*. Springer-Verlag, New York.

Riley, M. A., 1989. Nucleotide sequence of the *Xdh* region in *Drosophila pseudoobscura* and an analysis of the evolution of synonomous codons. *Mol. Biol. and Evol.*, **6**, 33-52.

Sarich, V. M., and J. E. Cronin, 1976. Molecular systematics of the primates. In *Molecular Anthropology*, M. Goodman and R. E. Tashian (eds.), Plenum Press, New York, pp. 141-170.

Sarich, V. M., and A. C. Wilson, 1966. Quantitative immunochemistry and the evolution of primate albumins: microcomplement fixation. *Science*, **154**, 1563-1566.

Sibley, C. G., and J. E. Ahlquist, 1984. The phylogeny of primates as indicated by DNA-DNA hybridization. *Jour. Mol. Evol.*, **20,** 2-15.

——, 1987. Avian phylogeny reconstructed from comparisons of the genetic material, DNA. In *Molecules and Morphology in Evolution: Conflict or Compromise?* C. Patterson (ed.). Cambridge Univ. Press, Cambridge, pp. 95-121.

Smouse, P. E., and W.-H. Li, 1987. Likelihood analysis of mitochondrial restriction-cleavage patterns for the human-chimpanzee-gorilla trichotomy. *Evolution*, **41**, 1162-1176.

Stokes, H. W., and B. G. Hall, 1985. Sequence of the *ebgR* gene of *Escherichia coli*: evidence that the EBG and LAC operons are descended from a common ancestor. *Mol. Biol. Evol.*, **2**, 478-483.

Straus, N. A., 1976. Repeated DNA in eukaryotes. In *Handbook of Genetics,* Vol. 5, R. C. King (ed.). Plenum Press, New York, pp. 3-29.

Strickberger, M. W., 1985. *Genetics*, 3rd ed. Macmillan, New York.

Templeton, A., 1986. Relations of humans to African apes: a statistical appraisal of diverse types of data. In *Evolutionary Processes and Theory,* S. Karlin and E. Nevo (eds.). Academic Press, Orlando, Fl., pp. 365-388.

Wilson, A. C., 1975. Evolutionary importance of gene regulation. *Stadler Symp.*, **7**, 117-133.

Wilson, A. C., L. R. Maxson, and V. M. Sarich, 1974. Two types of molecular evolution: evidence from studies of interspecific hybridization. *Proc. Nat. Acad. Sci.*, **71**, 2843-2847.

Wilson, A. C., H. Ochman, and E. M. Prager, 1987. Molecular time scale for evolution. *Trends in Genetics*, **3,** 241-247.

13

EVOLUTION IN PLANTS AND FUNGI

The photosynthetic eukaryotic organisms that are most commonly believed to have been ancestral to the vascular land plants were **algae** similar to the present Chlorophyta (green algae) and Charophyceae (stoneworts). These algae are presumed to have originated from single-celled flagellated organisms somewhat like *Chlamydomonas* which, in turn, evolved from eukaryotic cells that had been invaded by prokaryotic chloroplastlike symbionts (Chapter 9). In time, multicellular photosynthetic colonial organisms appeared, probably aided by the ease of association between their cell division products and the advantages that accrue to larger structures whose component cells can undertake a division of labor. Some of the steps in this evolutionary sequence may be echoed in the presently observed series that extends from *Chlamydomonas* to *Volvox* (Fig. 13-1); that is, a progression from organisms in which most or all cells can be reproductive to organisms in which most cells are somatic and only a few are reproductive. At some unknown historical point, one or more groups of algal organisms lost their motility and began to follow an evolutionary direction toward higher plants.

TERRESTRIAL ALGAE

The present sessile forms of algae are, of course, not necessarily direct relics of the ancient progenitors of land plants, although, like motile forms, they may also follow unicellular or multicellular organization. Many of these plants are found growing terrestrially on soil or as epiphytes on trees. Some land-dwelling algal species such as *Fritschiella tuberosa* possess rhizoids that penetrate the ground as well as branched, multicellular filaments that are both prostrate and erect (Fig. 13-2). According to some reports the *Fritschiella* life cycle alternates between the haploid **gametophyte** (n) and diploid **sporophyte** (2n) phases common to so-called higher plants (Fig. 13-3). That is, although both phases are multicellular and grow through regular mitotic cell division, the transition from diploid to haploid is achieved through a meiotic reductional division in the sporophyte. Haploid **spores** produced by the sporophyte develop into the gametophyte phase, which then produces sexual gametes mitotically. These unite, in turn, to form again the diploid zygote and subsequent sporophyte. In a process that is also similar to that of higher plants, *Fritschiella* undergoes the cell-splitting phase of cell division (cytokinesis) by means of cell-

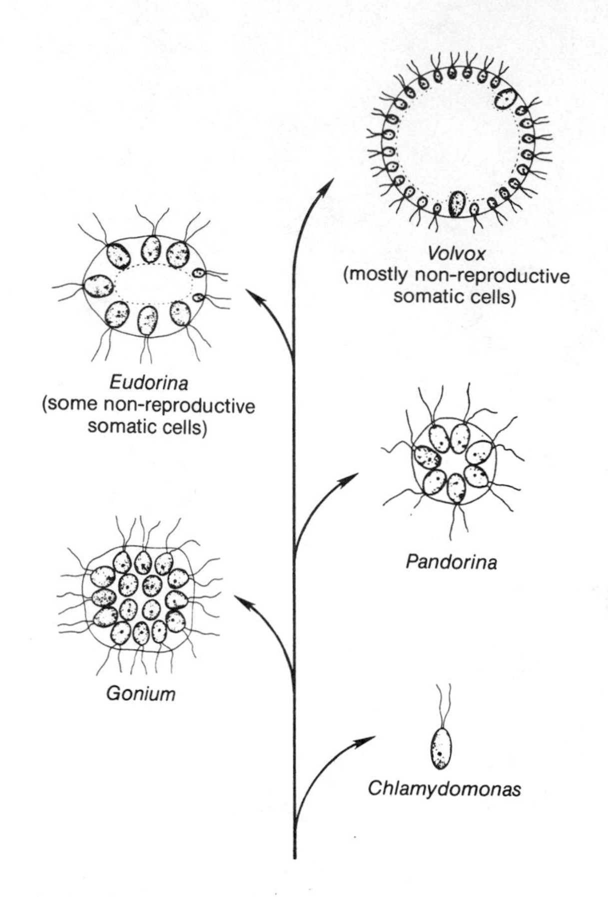

FIGURE 13-1 Possible phylogenetic sequence showing the origin of some multicellular algal aggregates such as *Volvox*. Land plants may have had a similar origin although the intermediary types were probably different. (Adapted from Pickett-Heaps.)

plate formation rather than by cytoplasmic constriction or cell furrowing.

Among further similarities between green algae and higher plants is the fact that green algae store their carbohydrate reserves in the form of starch and many have rigid, cellulose-reinforced cell walls. In addition, both green algae and higher plants use similar types of chlorophyll (a and b) and carotenoids (α and β). So far, a number of green algae are known, such as *Ulva* and *Caulerpa* (Fig. 13-4), whose membranous, leaflike forms simulate the appearance of some higher vascular plants yet show their evolutionary ancestry in that they pass through an algalike, filamentous stage.

Perhaps the most significant aspect of adaptation to land was the prevention of water loss because of cell surface evaporation, a problem that obviously does not exist in most aquatic algae. In those instances where dehydration (**desiccation**) can occur in algae, two major mechanisms of coping with this difficulty have evolved. One mechanism used in algae such as *Trentepohlia* has been simply to confine cellular growth to aquatic conditions and to become dormant under dry conditions. The absence of large, watery vacuoles in *Trentepohlia* cells, as in air-dispersed spores of other plants, enables such cells to suffer relatively little change in shape and volume during dehydration compared to those with vacuoles. On the other hand, some species with water-filled vacuolated cells, such as *Cladophorella* and *Fritschiella*, seem to maintain a waxy **cuticle** on their airborne parts which retards water loss. The vacuoles, in turn, enable cells to continue their metabolic activity as though under a constant marine environment and also provide mechanical rigidity, or turgor, that prevents cellular collapse. The large volume occupied by vacuoles also forces the cytoplasm into a relatively thin sheet along the perimeter of the cell, thereby maximizing the available photosynthetic surface. In a number of important qualities, mud-dwelling algae such as *Fritschiella* therefore seem eminently preadapted to begin the journey to land.

When this journey began is unknown. Tiffney and others point out that a fall in sea level during Ordovician glaciations would have caused aquatic plants in shoreline communities to undergo selection for resistance to desiccation. Other environmental conditions contributing to land plant evolution have been reviewed by Chapman. These include an increase in at-

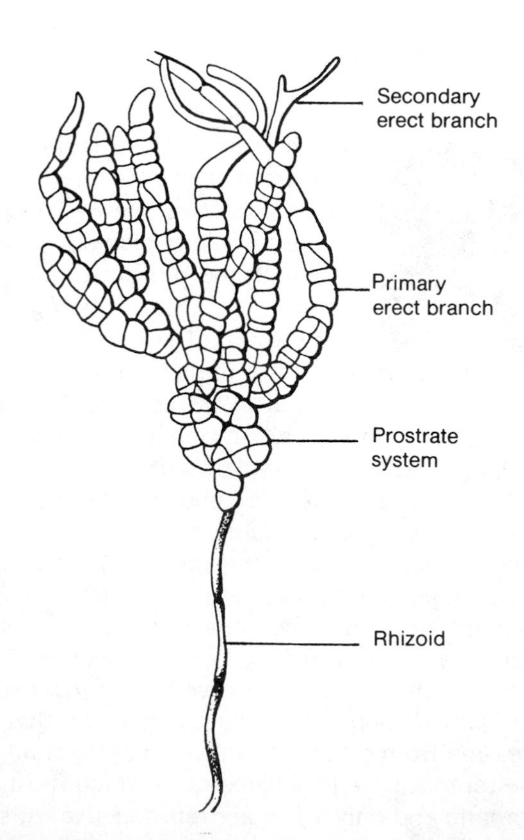

FIGURE 13-2 Sketch of *Fritschiella tuberosa*, a soil alga, showing the branching, filamentous, erect system and the more three-dimensional prostrate system that resembles the parenchyma of higher plants. (Adapted from Delevoryas.)

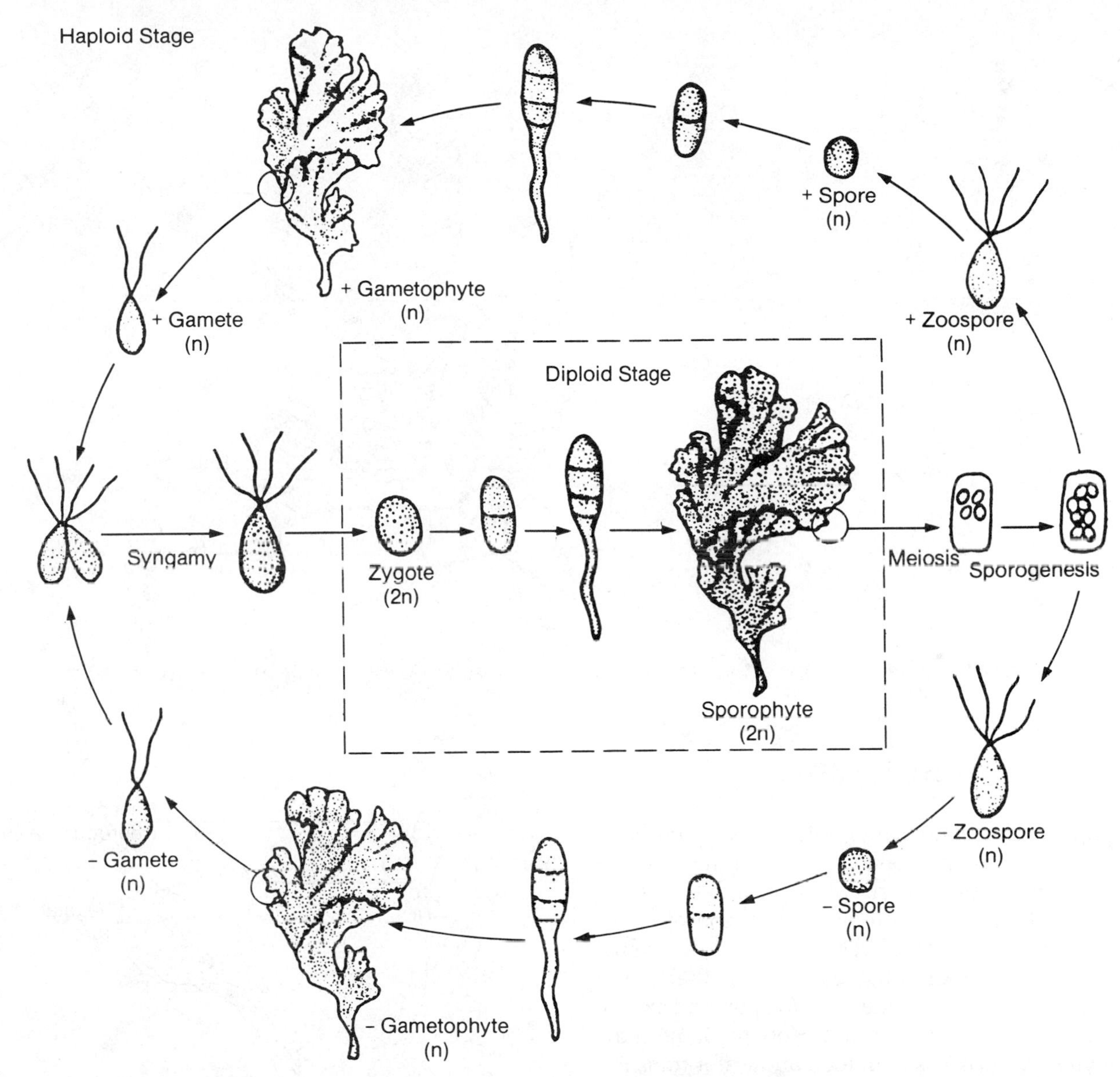

FIGURE 13-3 Mode in which haploid and diploid generations alternate in some green aquatic algae (e.g., *Ulva*), showing similar-appearing (isomorphic) gametophytes (n) and sporophytes (2n). In land plants the sporophyte embryo is retained and nourished within the gametophyte tissue and matures into a different form from the gametophyte (heteromorphic alternation of generations). In bryophytes, as explained in the text, the sporophyte remains dependent on the gametophyte, whereas the sporophyte is independent in higher vascular plants (Tracheophyta).

mospheric oxygen that aids in the formation of highly oxygenated polymers, such as cutin (waxy cuticle material used in waterproofing) and lignin (a stiffening polymer used for mechanical support and water-conducting tissues), and produces an ozone screen against harmful ultraviolet rays (Chapter 9).

Among the algae themselves, evolutionary relationships are not entirely clear, although most botanists recognize that the golden brown algae (Chrysophyta) and brown algae (Phaeophyta) show relatively advanced features, especially in respect to differentiated structures. Both these types of algae possess **planktonic** (motile) and **benthic** (nonmotile) forms, the latter attaching to the sea floor in shallow areas. In some brown algae such as *Fucus* (Fig. 13-5) cell division appears localized, as in higher plants, to a specific meristematic growth area below the elongating tip of the plant. Also, the differentiated gonadic organs of *Fucus* produce sperm and eggs, and the relatively complex body tissues include specialized conducting cells that appear similar to the sieve tubes in the phloem of higher plants. Nevertheless, these algae are generally not considered to be ancestral to land plants since their pigmentation (chlorophyll c, fucoxanthins, etc.) and storage products are so different. Also, among factors shared with other algae, they lack the waterproof cuticles that would prevent desiccation on land.

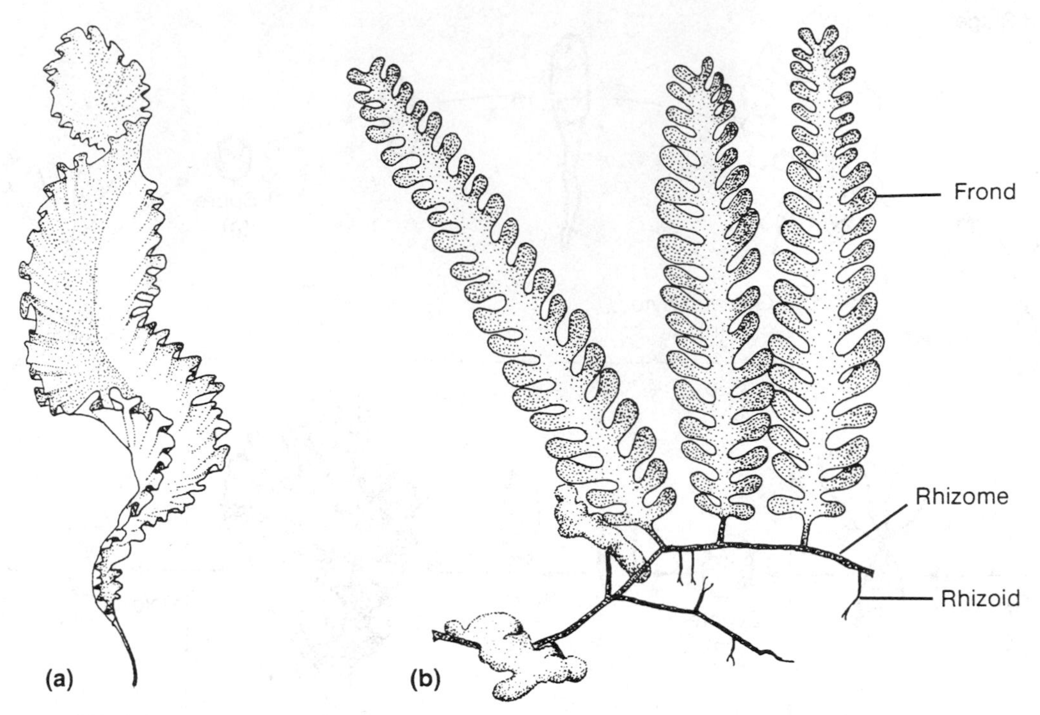

FIGURE 13-4 Green algae with leaflike forms. (a) *Ulva*. (b) *Caulerpa*. (Adapted from Delevoryas, based on other sources.)

BRYOPHYTES

The simplest of the land plants—liverworts, hornworts, and mosses—are traditionally classified into a single group, **Bryophytes**.[1] These plants exhibit features common to land plants through possession of multicellular reproductive structures, a cuticle in their aerial parts, and many epidermal pores (stomata) that permit the transfer of carbon dioxide, water vapor, and oxygen between their tissues and the atmosphere. Some of the bryophytes also possess both food and water transport tissues, although these do not seem to be as efficient as the phloem and xylem of the more advanced vascular plants. Because of limitations in food and water transport, bryophytes are apparently limited to small stature and live mostly in moist environments where they can transport water along their surfaces. When found in arctic or arid environments, their growth is usually suspended until the warm, moist season begins.

Among characteristics bryophytes bear in common is an **alternation of generations** in which the haploid gametophyte generation is free-living and the diploid sporophyte generation remains permanently attached to it. In liverworts such as *Riccia* (Fig. 13-6) and *Sphaerocarps*, the sporophyte is relatively undif-

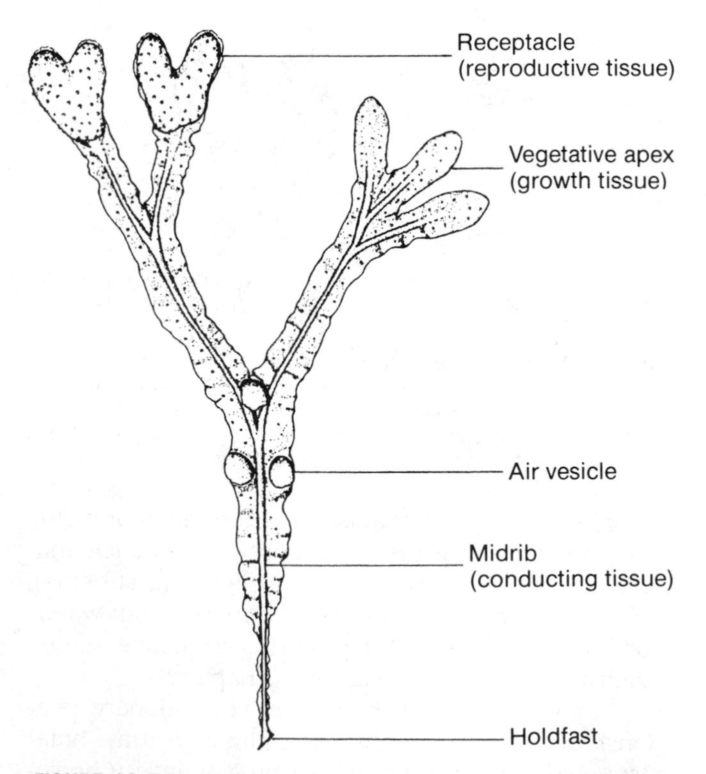

FIGURE 13-5 The plant body of *Fucus vesiculosis*, a brown alga that commonly grows in the intertidal zone. (Adapted from Bold et al.)

[1]Some authors restrict this name to the mosses and call the liverworts and hornworts Hepatophytes.

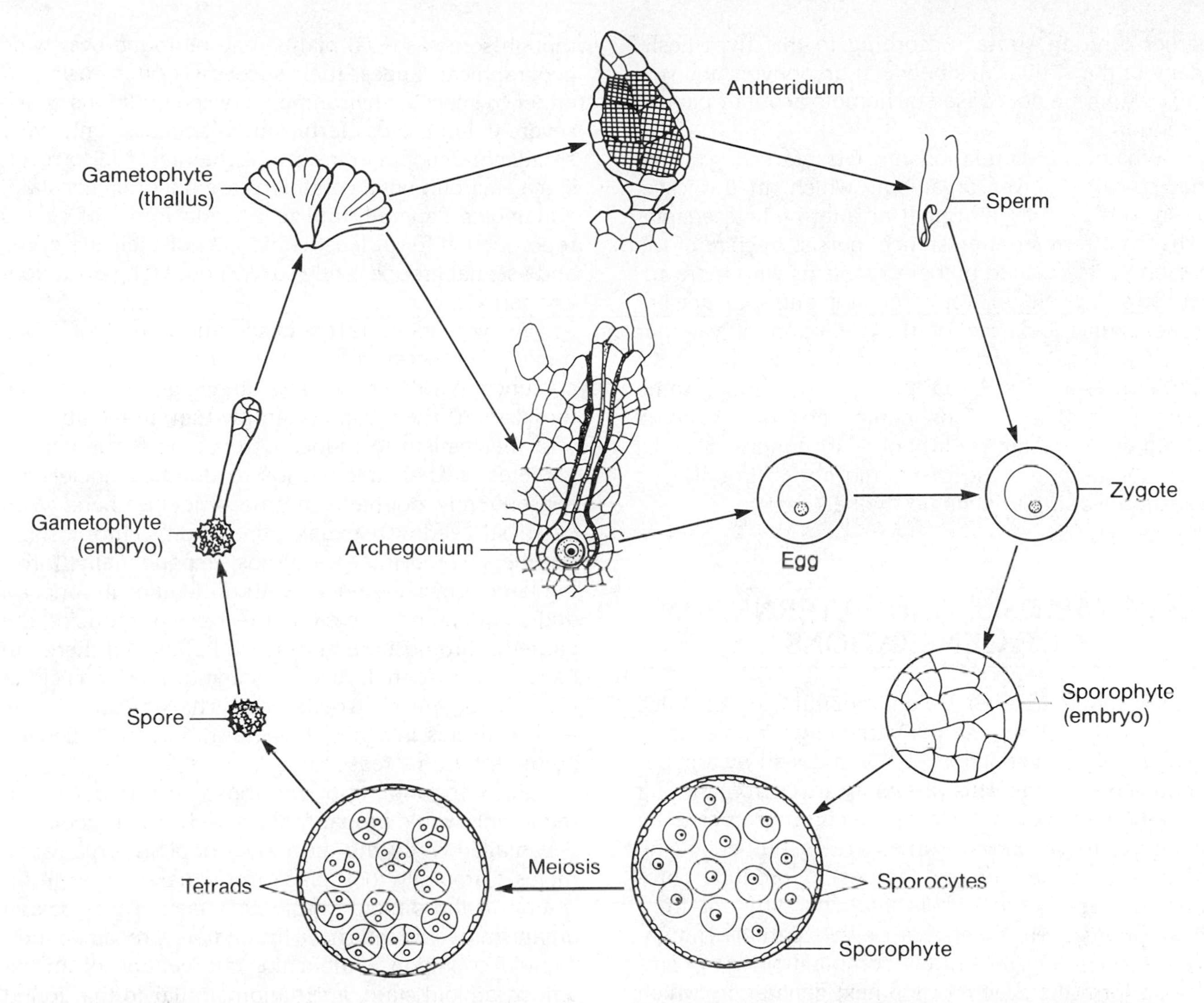

FIGURE 13-6 Life cycle of a liverwort species in the genus *Riccia*. The gametophyte is a small, flattened, chlorophyll-bearing thallus with rootlike rhizoids on its undersurface. The spherical sporophyte grows on the tissue of the gametophyte, and each of its interior sporocytes divides meiotically to form a tetrad composed of four haploid spores. Germination of the spore leads to the gametophyte, and the cycle then continues as shown. (Adapted from Holman and Robbins.)

ferentiated, whereas it is considerably more complex in the hornwort *Anthoceros*.

Although there is little clue as to the direct ancestry of bryophytes or even the phylogenetic relationships between their major groups, many botanists believe these plants had an algal origin. They point to the presence of both spore-forming and gamete-forming tissues in various algae and to the similarity between the filamentous growth pattern of some green algae and the branching filamentous protonema stage observed in many mosses. The aquatic environment of most bryophytes and their dependence on water for fertilization also points to an aquatic origin.

One evolutionary sequence, often accepted in the past, is that algae evolved into bryophytes, which then evolved into vascular plants (e.g., ferns). However, other views have also recently been set forth, including the suggestion that bryophytes evolved from land plants that had already achieved terrestrial adaptations and their aquatic features are only secondary. According to this view, the bryophyte sporophyte may represent a reduced stage of the dominant, more independent sporophyte found in higher plants.

A popular concept is that bryophytes and the higher plants may have each had an independent algal origin because of the notable differences between them. For example, the sporophyte generation of bryophytes is dependent for nutrients and support on the gametophyte, whereas the sporophyte of higher plants is completely independent. Furthermore, the earliest unequivocal appearance of bryophytes in the fossil record is in the Devonian period for liverworts and in the Carboniferous for mosses, whereas recognizable fossils of vascular land plants have been discovered in earlier Silurian strata. According to this hypothesis, many of the similarities between bryophytes and vascular plants are not caused by homology but by parallel evolution.

Whatever their relationship, it is interesting to note that so-called lower organisms which, in this case, seem to have only marginal or intermediate adaptations for a terrestrial existence, persist in spite of the presence of so-called higher organisms with more advanced adaptations. Nonvascular plants did not become extinct because of the evolution of vascular plants, and non-seed plants such as ferns still survive in the presence of seed plants. Apparently, some environmental conditions confer no overwhelming advantage on later evolutionary inventions; that is, evolution among these plants did not go in only one direction leading to a single "higher" form.

SEX, MEIOSIS, AND ALTERNATION OF GENERATIONS

As indicated in Chapter 10, an important consequence of sex is to help provide the variability which enables a wide array of genotypes to be produced by a population of organisms. This variability arises from the fact that when two parents contribute chromosomes to an offspring, these chromosomes are reshuffled in the offspring's sexual meiotic tissues to produce chromosomal and genetic combinations different from those originally received from either parent. The gametes containing these new combinations can combine to form the zygote of the next generation, which then has new genetic combinations; these are reshuffled in the following generation, and the process continues (Chapter 22).

For populations continually encountering different environments, sex is therefore an obvious advantage to a lineage in producing new combinations of genes, some of which may be adaptive and allow the lineage to survive. On the other hand, lineages without the variability introduced by sex can more easily become extinct under changing circumstances. However, in a long-standing population that continually endures the same environment, it is reasonable to expect that genotypes will evolve that are eminently adapted to that environment and that most, if not all, new genetic combinations will have lower adaptive value than the parentals. Under such constant circumstances, the advantages of sex are not apparent. In fact, there are numerous examples in plants where sexual reproduction has been abandoned and replaced by asexual methods such as the spread of vegetative somatic tissues or by parthenogenesis (reproduction through unfertilized eggs—found also in some animals). Although some asexual plants may be found over wide geographical ranges, their success is often restricted either to specific environments or to conditions which severely limit cross-fertilization because only very small inbred populations can be maintained. However, since environmental conditions are not often constant, eukaryotes generally utilize meiotic forms of sexual reproduction for at least some part of their life cycle, and asexual groups rarely survive over long evolutionary periods.

As yet, there is no exact knowledge of when meiosis originated, although it must have appeared in conjunction with, or soon after, the beginning of sexual fertilization. The reason is simply that, in the absence of a mechanism to reduce chromosome numbers in gametes, sexual union leads to doubled nuclei and consequently doubled chromosome numbers. With each succeeding sexual generation, chromosome numbers would increase almost exponentially, forming large, unwieldy nuclei with difficulties in function and coordination. A meiotic mechanism reducing the gametic chromosome number to half would therefore have been essential. At what stage of the life cycle of primitive organisms would meiosis have occurred? The answer to this is again conjectural, but the following argument seems reasonable.

Since a doubling of chromosome number in somatic cells would probably not have been immediately advantageous to the primitive haploid organisms, meiosis probably took place immediately after fertilization in the diploid zygote cell itself. These sexual organisms would thus have immediately regained their haploid condition without the intervention of an extended diploid state, a situation similar to that found in algae such as *Chlamydomonas* and fungi such as *Neurospora*.

There are, however, advantages to lengthening the diploid stage, not the least being that such cells may now possess two kinds of genetic information, one from each parent, enabling a single organism to use different developmental pathways in responding to different environmental conditions. In addition, the two alleles of a gene in a diploid may each produce unique products which can then buffer each other to ensure developmental uniformity in any particular environment ("heterozygote advantage"; Chapter 21). Diploidy also provides the opportunity for dominant genetic relationships which can enhance the evolutionary potential of a population by helping to retain deleterious recessive alleles that may be advantageous under future conditions.

It can also be argued that when meiosis takes place immediately after fertilization the gametes possess relatively limited genetic variability because only one reductional division has occurred. For example, a single diploid cell with three pairs of chromosomes (or three pairs of genes), A^1, A^2, B^1, B^2, and C^1, C^2, might produce four haploid gametes from a meiotic division that are of constitutions $A^1B^1C^2$, $A^1B^1C^2$, $A^2B^2C^1$, and $A^2B^2C^1$. A multicellular diploid organism, on the other hand, could produce a greater variety of gametes, since numerous kinds of reduction divisions can take place in a large number of parental cells undergoing meiosis (meiocytes). Thus some meiocytes of such an organism could produce $A^1B^1C^2$ and $A^2B^2C^1$ gametes, others could produce $A^1B^2C^1$ and $A^2B^1C^2$ gametes or $A^2B^1C^1$ and $A^1B^2C^2$, and so forth. To the extent that the production of genetic variability is advantageous, a population of organisms whose diploid meiotic tissues are multicellular would have greater potential for evolutionary change than a population containing a similar number of organisms in which the diploid meiotic stage is unicellular. (Further hypotheses to explain the origin and persistence of sex can be found in Michod and Levin.)

In animals, the lengthened diploid stage became the dominant feature of the life cycle, and the haploid stage is now mostly restricted to the gametes themselves. In plants, the lengthened diploid stage, or sporophyte, also produces meiotic products as in animals, but these are spores rather than gametes. The meiotically produced spores develop into haploid gametophytes which only later produce gametes by mitosis.

The sporophyte-gametophyte alternation of generations in plants (Fig. 13-7) has long been puzzling, and various explanations can be offered. For example, aside from the advantages of maintaining a diploid state, the sporophyte produces dispersible, encapsulated spores that are resistant to desiccation. By contrast, plant as well as animal gametes are relatively unprotected and generally depend upon an aqueous environment for dispersion and fertilization. In fact, it is probably the vulnerability of sexual gametes to terrestrial conditions that accounts for the persistence of the gametophyte stage in plants, a stage that is easily eliminated in animals. That is, animals are sufficiently mobile to enable the transfer of gametes by direct contact between organisms in the diploid stage, whereas plants are sessile, and the transfer of gametes between them is restricted to moist environmental conditions. Thus, the features of plant sporophytes that enable resistance to desiccation and conquest of the land are apparently quite different from gametophyte features necessary for the aqueous transfer of plant gametes.

The evolutionary development of the plant sporophyte from dependence on the gametophyte stage to independence is therefore a reflection of the advantages of diploidy as well as spore production, and the persistence of the gametophyte stage can be considered a reflection of the advantage for immobile gamete-producing individuals to grow in aqueous proximity to each other.

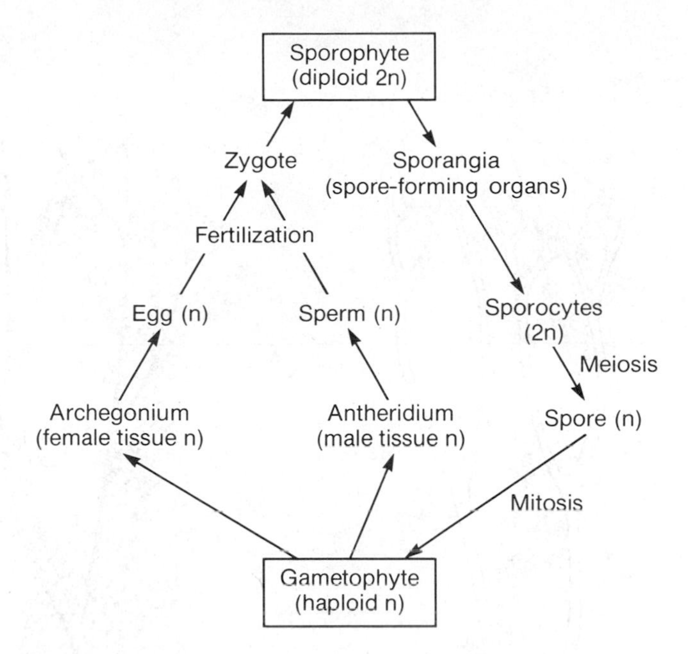

FIGURE 13-7 Alternation of gametophyte and sporophyte generations in the plant life cycle (see also Fig. 13-3). In some plants, the gametophyte is unisexual, either male or female. In others, the gametophyte is bisexual or hermaphroditic, producing both male and female tissues.

Traditionally, botanists have concentrated their disputes about the alternation of generations on whether the two generations were initially similar or different. According to the **antithetic**, or interpolation, theory of sporophyte origin, the initial sporophyte differed both functionally and morphologically from the parental gametophyte and soon became even more distinct as an increasing proportion of its vegetative tissue became converted to vegetative purposes such as photosynthesis. The notion that the sporophyte originated as a fertilized single diploid cell which gradually became increasingly multicellular by delaying meiosis and spore formation to later stages accords with this view.

A contrasting **homologous**, or transformation, theory suggests that the sporophyte showed little initial difference from the gametophyte, since they both share the same genetic constitution derived from the same organism and should therefore have shared the same patterns of growth. In support of the homologous theory is the similarity between sporophyte and gametophyte in many algae, as well as structural similarities between their stems in the "living fossil" fern *Psilotum* and in some other primitive ferns such as *Stromatopteris*. There also appear to be some basic developmental similarities: sporophyte tissue can arise in some gametophytes without the intervention of gametic fertilization (apogamy), and some sporophytes may produce gametophytes without spore formation (apospory). As yet, the issue of sporophyte origin is not resolved, although there is general agreement that

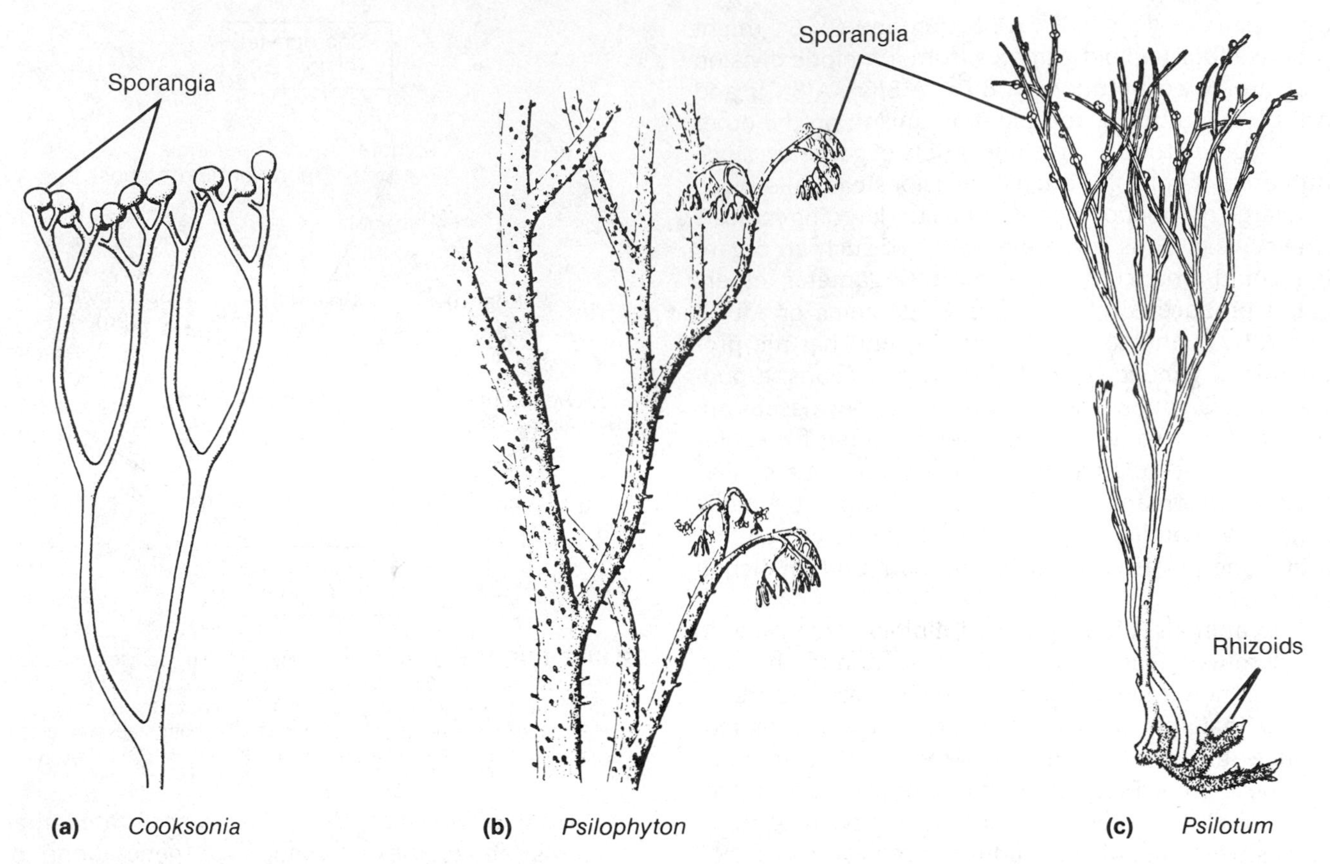

FIGURE 13-8 Reconstructions of the sporophytes of two early fossil plants and a modern representative of the Psilopsida. (a) An Upper Silurian plant, *Cooksonia caledonica,* about an inch or so high, which had no distinctive leaves or roots but showed naked, dichotomously branched axes with terminal sporangia. Similar fossils were long known from Devonian rocks in Rhynie, Scotland, and have been classified together in a group called the **Rhyniophyta,** or Rhynia-type plants. (b) *Psilophyton princeps,* section of a spiny, leafless fossil plant that first appears less than 10 million years after *Cooksonia.* It had a main stem axis with lateral branches terminating in sporangia and a vascular structure that seems to have been larger than *Cooksonia,* so may have grown taller. (c) The modern plant *Psilotum nudum* has a number of features that resemble the fossil forms: simple stems, non-discernable leaves, and absence of a modern root system. It is not clear whether this plant is a "living fossil" or a secondary descendant of a more advanced form (Stewart). (a and b adapted from Taylor, c adapted from Bold et al.)

land plant evolution can best be characterized by increased sporophyte importance. It is, in fact, to sporophytes that the vascular tissues of land plants are generally restricted.

EARLY VASCULAR PLANTS

Whatever the origin of the land plants, the fossil record shows a rapid evolutionary radiation from the time of their first appearance in the late Silurian period more than 400 million years ago. By the end of the Devonian, 50 million years later, forests containing woody trees of relatively great variety had become well established. These successful land plants were **vascular**, bearing conductive tissue (xylem) that enables water to reach the erect parts of the plant, associated with tissue (phloem) that enables food to be distributed. They have often been given the name **Tracheophyta** because of the presence of tracheids, fluid-conducting tubes impregnated with an organic substance (e.g., lignin) that also provides mechanical support for erect growth.

The earliest of the Silurian vascular fossils includes a number of simple plants with leafless stems classified in the genus *Cooksonia* (Fig. 13-8(a)), some of which may have had terminal spore-bearing organs (**sporangia**). Together with other leafless and rootless fossil plants, these are somewhat similar to plants in the modern genus *Psilotum* (Fig. 13-8(c)). According to Banks and others, early plants such as these were ancestors of multibranched plants that rapidly evolved into taxa such as *Psilophyton* (Fig. 13-8(b)).

Found also in the Devonian period are primitive leafless plants that differ from *Cooksonia* types in carrying their sporangia laterally along branches rather than terminally. These plants, of which *Zosterophyllum* is an example (Fig. 13-9(a)), have been presumed to be the ancestors of the early **club mosses**, or **lycopods**,

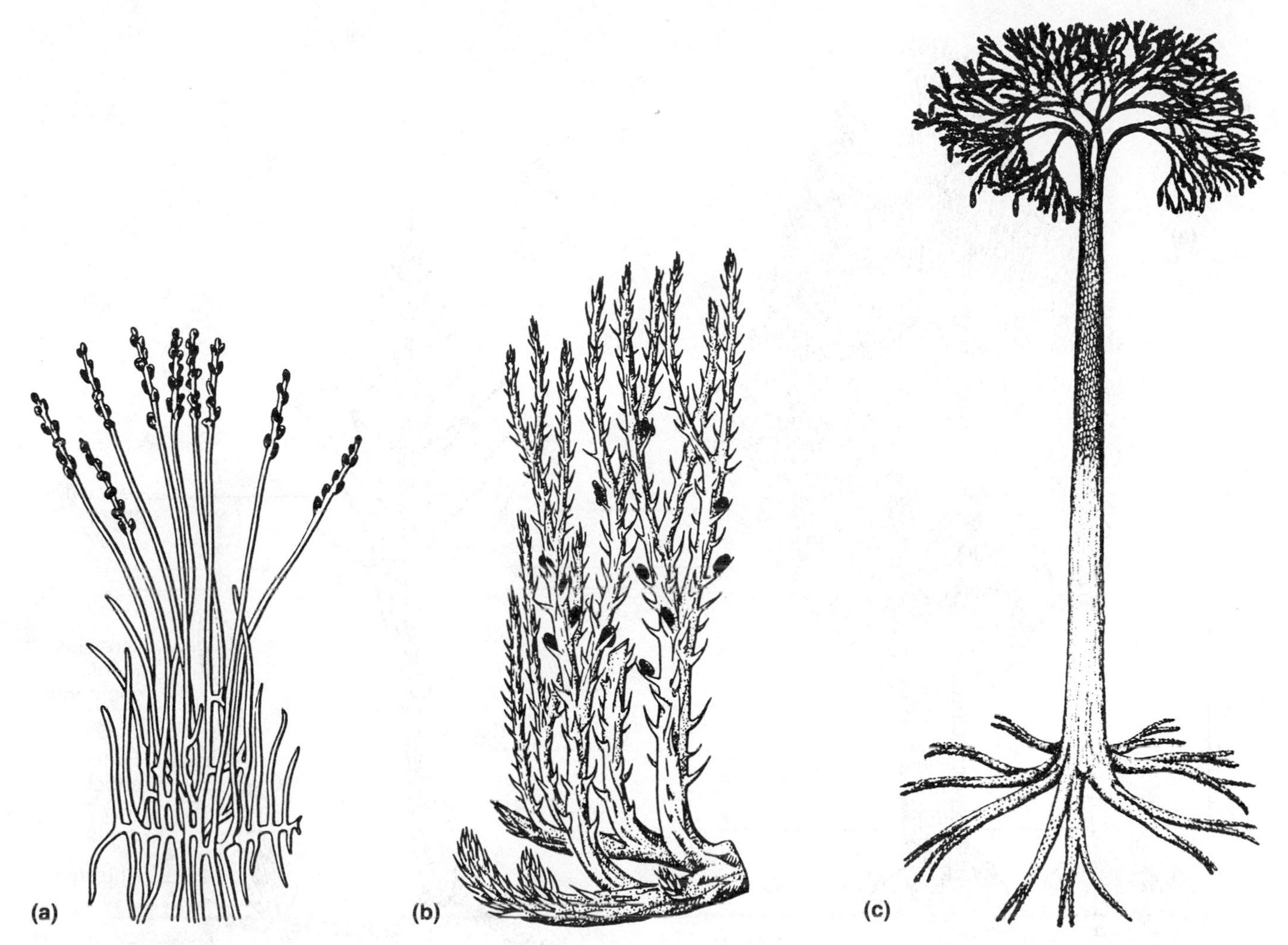

FIGURE 13-9 Reconstructions of several Devonian plants. (a) *Zosterophyllum myretonianum* (about 7 inches tall). (b) *Asteroxylon mackiei* (about 2 feet tall). (c) *Lepidodendron* species (about 150 feet tall. The scars along the upper stem of *Lepidodendron* are leaf cushions, where the long, filamentous leaves had been attached during earlier growth. The sporangia are carried in the heavy, pendulous cones. (a and b adapted from Foster and Gifford, c adapted from Stewart.)

such as *Asteroxylon* (Fig. 13-9(b)), which in turn led to arborescent lycopods such as *Lepidodendron* (Fig. 13-9(c)), so abundant during the Carboniferous age. Herbaceous lycopods are found even to the present time.

The Devonian period also saw the origin of **sphenopsids**, horsetail plants with segmented stems and whirled leaves and branches. These plants were common until the Mesozoic period, contributing huge trees to the Carboniferous coal forests (Fig. 13-10(a)), but are now represented only by the genus *Equisetum* (Fig. 13-10(b)), composed of a group of about 25 herbaceous species.

In terms of evolutionary persistence, an enduring group among these early spore-bearing plants were the **ferns**, Pterophyta, now numbering about 10,000 species in which the sporangia are carried directly on the leaves (Fig. 13-11). These include, then and now, both small, herblike forms as well as large tree ferns (Fig. 13-12). Ferns were apparently the first plants to exploit the use of large, prominent leaves, megaphylls, in contrast to the smaller leaves, microphylls, used by the lycopods and sphenopsids. The origin of either leaf type is not clear, although various theories have been offered.

The **telome theory**, as developed by Zimmermann, suggests that primitive psilophyte branches, called telomes, evolved in two major alternative directions: the first towards greater complexity and vascularization, leading eventually to the leaves and branches of ferns and higher plants; the second in a retrogressive direction toward a single unbranched form, leading eventually to bryophytes such as the hornwort *Anthoceros*. According to this hypothesis,

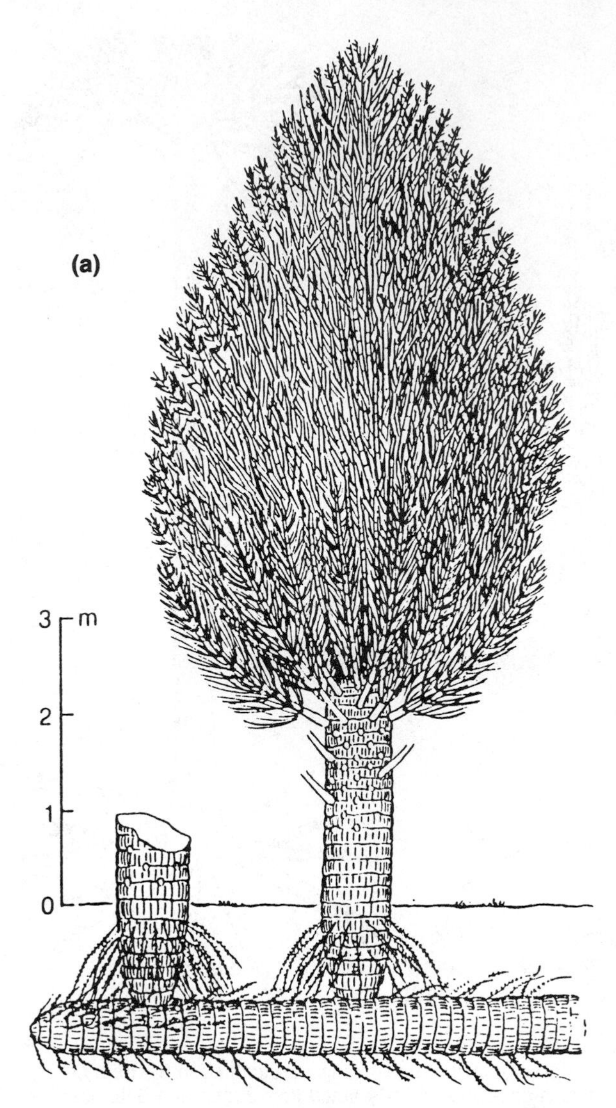

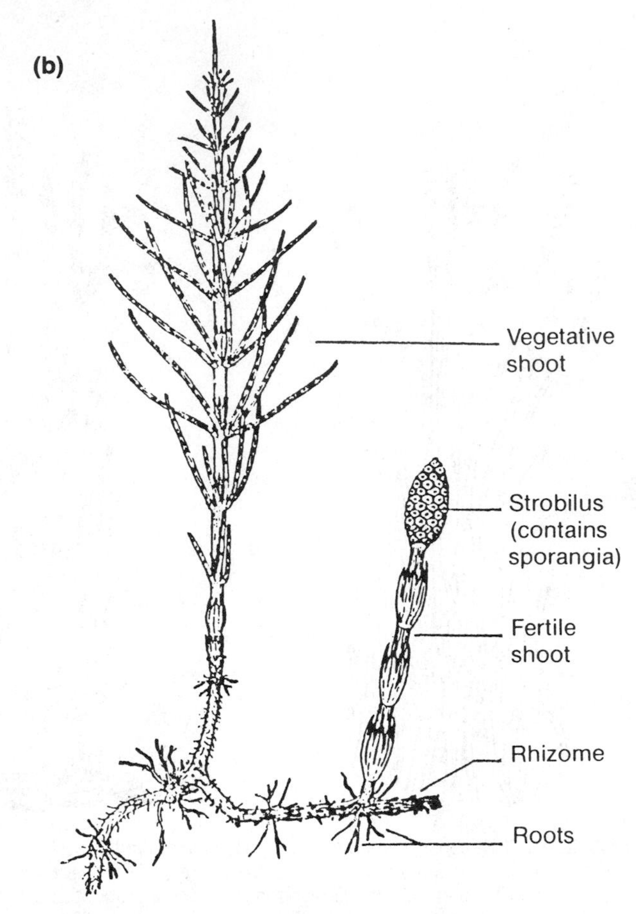

FIGURE 13-10 Ancient and modern representatives of Sphenopsida. (a) Reconstruction of *Calamites*, a common tree during the Carboniferous period, that reached heights of 90 to 100 feet with trunks 2 feet thick. (Adapted from Foster and Gifford.) (b) Modern *Equisetum arvense*, showing both vegetative and fertile shoots. Its reproductive system is homosporous: the strobili bear sporangia that produce spores alike in size. (Adapted from Bold et al.)

leaves originated from small branches that lay in the same plane. As shown in Figure 13-13(a), webs that formed between such planated branches could have produced leaflike structures.

A different proposal, the **enation theory** (Fig. 13-13(b)), suggests primitive leaves arose from small outfoldings along the stem, somewhat like microphylls in the fossil lycopod *Asteroxylon*. Only later were these leaves vascularized.

As yet, the correctness of either theory is not fully determined, although it is clear that evolutionary changes affected almost every aspect of these early plants. For example, beginning with the first spore-bearing plants, there is a progressive change from **homospory**, in which all spores are alike, to **heterospory**, in which sporophytes produce both large-diameter **megaspores** (200 μm) and smaller diameter **microspores**. By the Upper Devonian period the heterosporous lines had evolved megaspores more than 2,000 μm in diameter. This increase in megaspore size apparently resulted from a reduction in the number of cells in the megasporangium so that only a single tetrad of spores is produced, of which three spores abort and one enlarges.

Complexities of the vascular bundles also originated from a primitive form, the **protostele**, in which the conductive tissues were simply arranged in a solid circular core. Later in evolution these divided into the

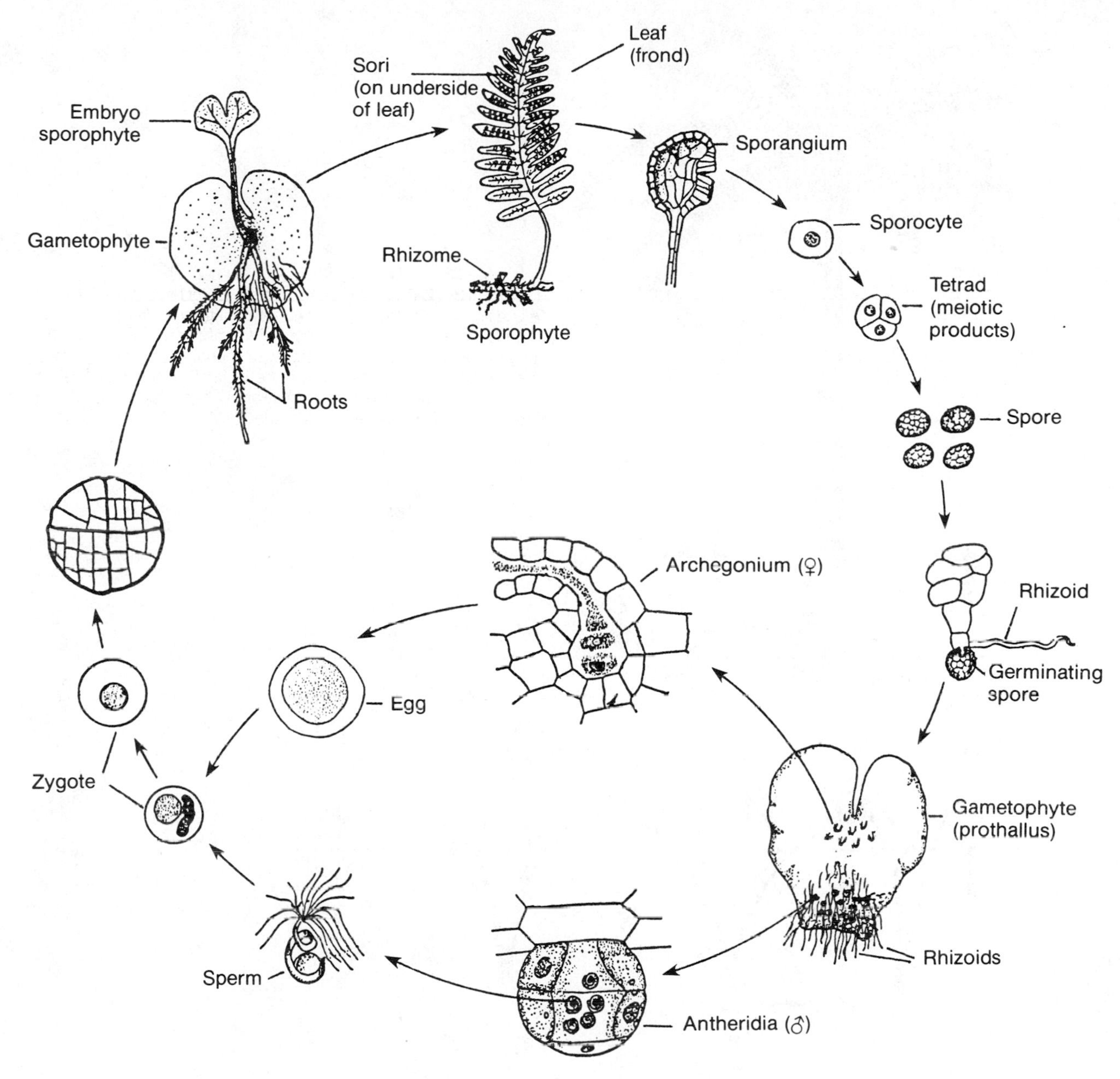

FIGURE 13-11 Life cycle of a common fern *Polypodium vulgare.* (Adapted from Holman and Robbins.)

various lobed and concentric arrangements shown in Figure 13-14.

The structural and functional advantages provided by many of these innovations bolstered the vertical development of plants and enabled large trees and shrubs of all types to evolve. By the Carboniferous period, lush and extensive forests had developed in vast swamps along the eastern coast of North America and similar coastal regions of Europe and North Africa. The absence of annual growth rings in many tree trunks of this period indicates that the climate was mostly tropical and growth was rapid. In this environment, decay of many fallen trees and shrubs was inhibited by their rapid submergence below the watery swamp surface, making them immune to attack by all but anaerobic bacteria. As the sea level fluctuated in these areas, successive generations of swamps were formed and submerged, and thick layers of organic strata were compressed into peat. Further sedimentation and compression led to the escape of volatile hydrocarbons, enabling formation of the enormously thick and extensive coal seams.

FIGURE 13-12 Reconstruction of the Carboniferous tree fern *Psaronius,* about 25 feet tall. Leaf scars left by earlier fronds that have fallen away are visible near the top of the trunk, and the trunk's long pyramidal shape is caused by surrounding adventitious roots that increase in thickness toward the base. The root structure suggests that these trees grew in swampy habitats. (Adapted from Foster and Gifford, based on Morgan.)

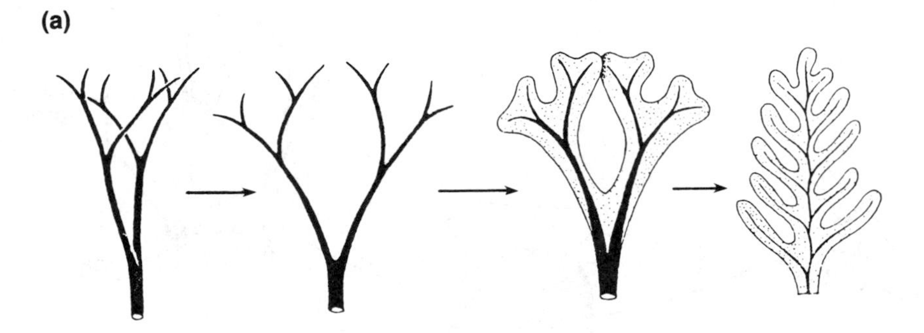

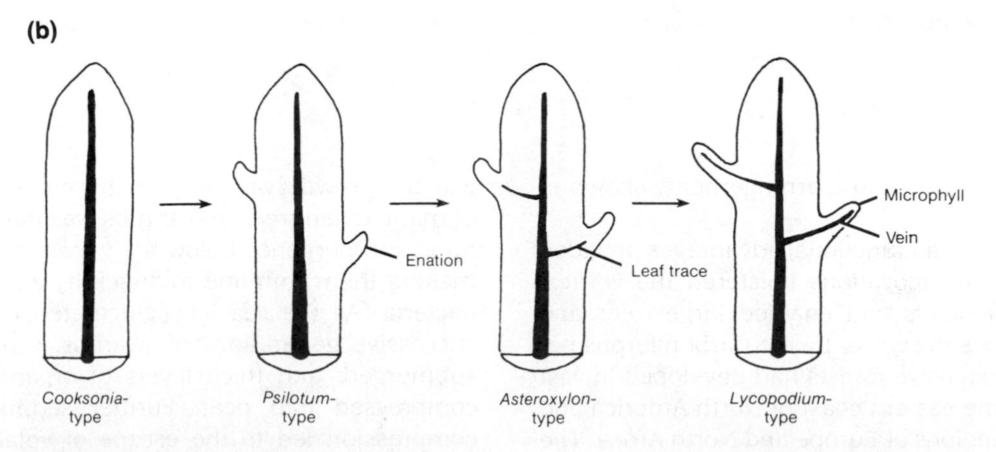

FIGURE 13-13 Diagrammatic representations of two theories explaining the origin of leaves. (a) Flattening (planation) of a branch system according to the telome theory, followed by webbing between the branches to form a flat, veined megaphyll. (b) Evolution of microphylls according to the enation theory. (Adapted from Bold et al., from Stewart.)

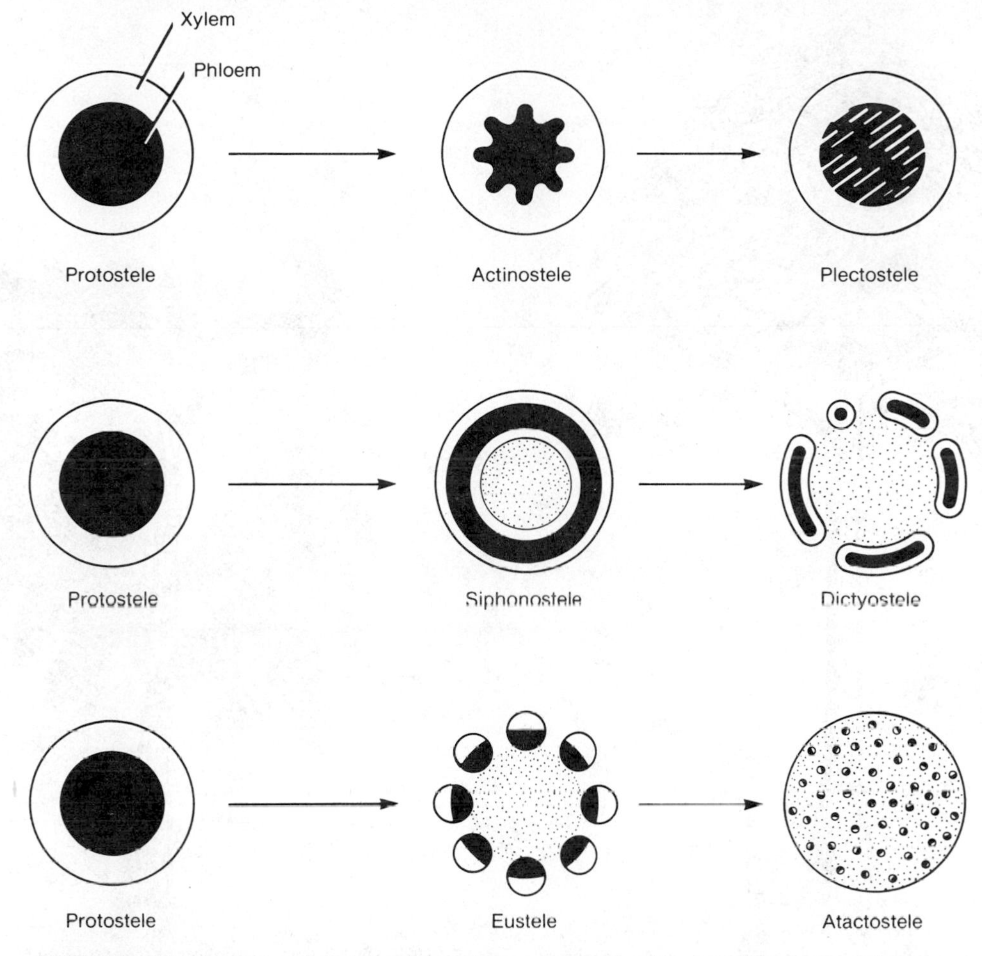

FIGURE 13-14 Proposed evolutionary relationships among some of the vascular cylinders (steles) found in plants. The various protosteles (top row) are considered primitive and occurred in Rhynia-type plants; siphonosteles and dictyosteles are characteristic of many ferns; various seed plants possess eusteles; and some of the complex atactosteles are found in flowering plants. (Adapted from Bold et al.)

FROM SWAMPS TO THE UPLANDS

Successful as they were, Carboniferous spore-bearing plants were limited to a moisture-laden environment because the motile male gamete was dependent on aqueous transmission to the female gametophyte. In order to extend their range onto dry land, plants therefore had to await the evolution of enclosed desiccation-protected gametophytes in which cross-fertilization could occur by non-aqueous devices such as wind dispersal. Similarly, there was considerable advantage in the evolution of protected sporophyte embryos whose distribution could be independent of their parental gametophytes. In essence, it was the reduction in size of the gametophyte and the evolution of easily dispersible **pollen** (male gametophytes) and **seeds** (sporophyte embryos) that helped further the conquest of dry land.

Unfortunately, the evolutionary progression from early vascular plants to pollen-producing, seed-bearing plants is not yet apparent. According to the fossil record, **gymnosperms** (naked seeds) appear in notable frequency during the Carboniferous period, eventually giving rise to modern representatives that include ginkgos and cycads as well as conifers such as pines, cedars, and sequoias. The **angiosperms** (covered seeds), now the dominant land plants, accounting for about 96 percent of all plant species, have no identifiable fossils earlier than the Cretaceous period of the Mesozoic. Transitional fossil forms leading directly to either gymnosperms or angiosperms have not yet been discovered, although there are fossils such as *Archaeopteris*, dating to the Devonian period, that may represent a group ancestral to all seed-bearing plants. As shown in Figure 13-15, *Archaeopteris* was a tall tree resembling a modern conifer with a crown of leafy branches. Its stem is also considered to bear a number of features in common with gymnosperms, although it produced free spores rather than protected seeds. According to Beck, Banks, and others, it was Paleozoic plants of this kind, called **progymnosperms**, combin-

FIGURE 13-15 Reconstruction of the progymnosperm *Archaeopteris.* (Adapted from Foster and Gifford, from Beck.)

FIGURE 13-17 Reconstruction of a seed fern, *Medullosa,* about 12 to 15 feet tall. (Adapted from Andrews, from Stewart and Delevoryas.)

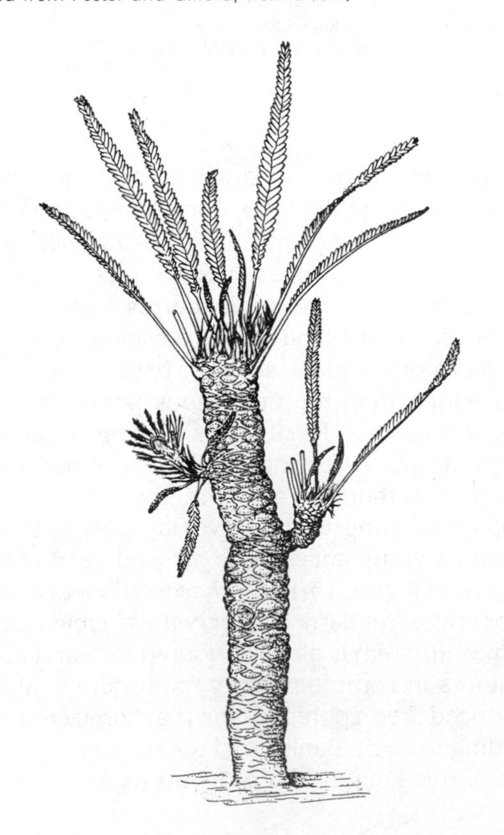

FIGURE 13-16 Reconstruction of a cycad gymnosperm, *Williamsonia sewardiana,* from Jurassic period rocks in India. The cycads were highly abundant contemporaries of the dinosaurs, and this period is also known as the Age of the Cycads. Like the dinosaurs, most of this group became extinct, although about 100 species of cycads exist, mostly in the tropics. Another "living fossil" is the *Ginkgo biloba* tree, sole remnant of a gymnosperm class Ginkgopsida, also common during the Mesozoic era. (From Andrews, adapted from Sahni.)

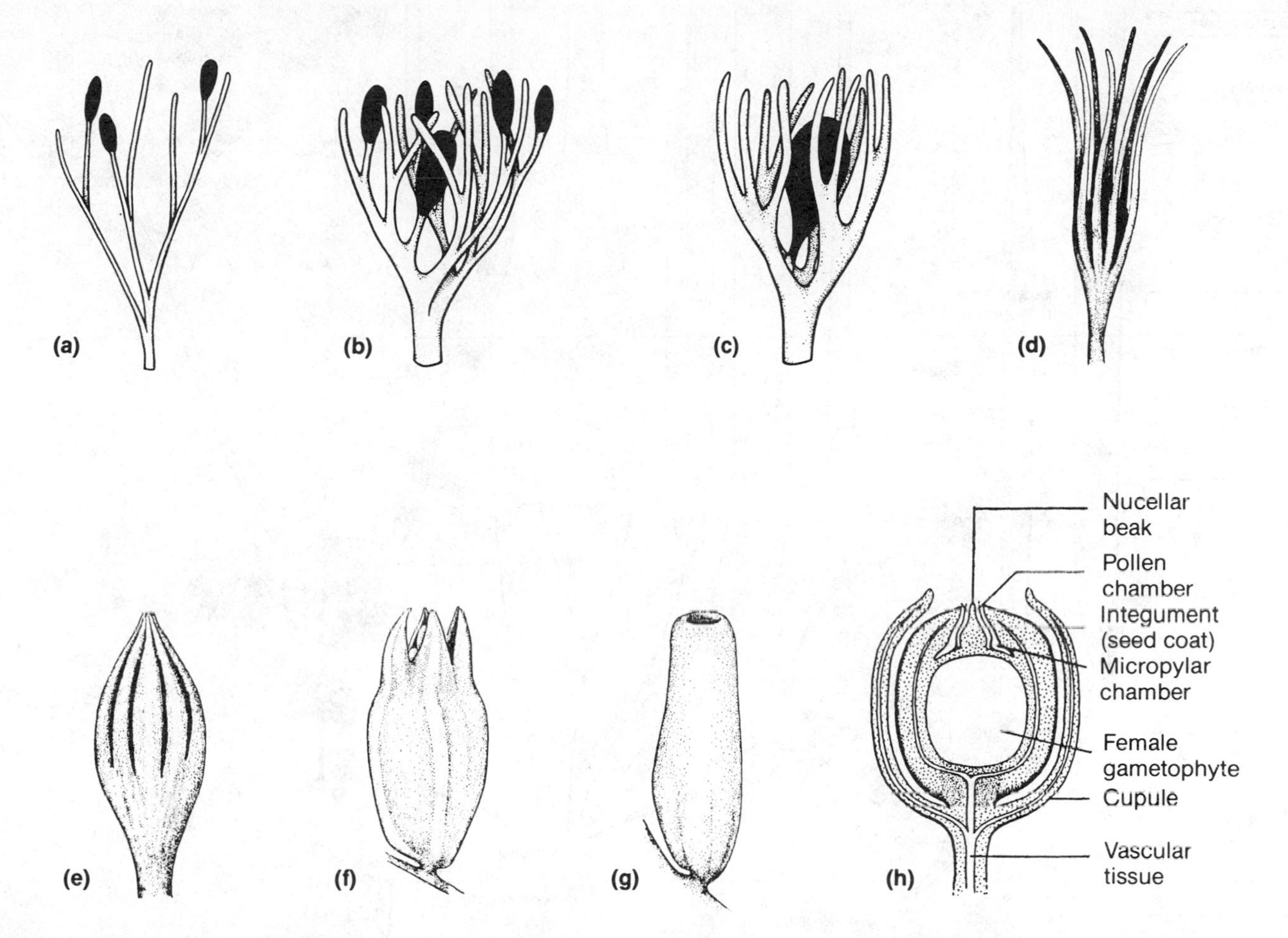

FIGURE 13-18 (a-g) One of the possible sequences in the evolution of the pteridosperm seed. The exposed sporangium that produces megasporocytes is gradually enclosed, enabling the female gametophyte (produced by the megaspore) to be completely protected within sporophyte tissue (nucellus). Fertilization takes place when male gametophytes in the form of pollen tubes grow through the micropyle, thus enabling sperm to reach the female gametophytic egg. The complete seed that envelops the zygote and developing embryo is coated with an integument produced by the parental sporophyte. (d-g) Seeds of fossil pteridosperms. (h) Section of a pteridosperm ovule. (Adapted from Foster and Gifford.)

ing pteridophytic, sporulating reproductive modes with more advanced anatomical structures such as large trunks, that gave rise to the gymnosperms which became so successful during the relatively dry Mesozoic (Fig. 13-16).

Another fossil group, perhaps also arising from the progymnosperms, were fernlike plants that bore seeds rather than spores (Fig. 13-17). These **seed ferns** (Pteridospermales) show a variety of seed forms that suggest a progression in the method by which the female gametophyte is enclosed in the seed integument (Fig. 13-18). However, whether seeds originated in only one group of progymnosperms (monophyletic origin) or more than one group (polyphyletic origin) is not yet discernible (Fig. 13-19). In any event, such evolution may have occurred early: Pettitt and Beck, for example, have described a fossil seed that dates back to the Upper Devonian, 350 million years ago.

ANGIOSPERMS

Angiosperms were the last major plant group to have evolved, appearing first in the early Cretaceous and reaching considerable abundance and variation by the late Cretaceous. Among other features, they are characterized by unique **flower** structures which enable many of them to be insect- or bird-pollinated, and they also bear unique seeds that are often adapted to dispersal by animals.

The adaptive advantage of pollination by animals is the simple one of ensuring cross-fertilization with other members of the same species by utilizing only a relatively small amount of pollen compared to the large amounts of pollen necessary in random wind pollination. As a result, angiosperm flowers, derived from leaves modified into petals, sepals, and related

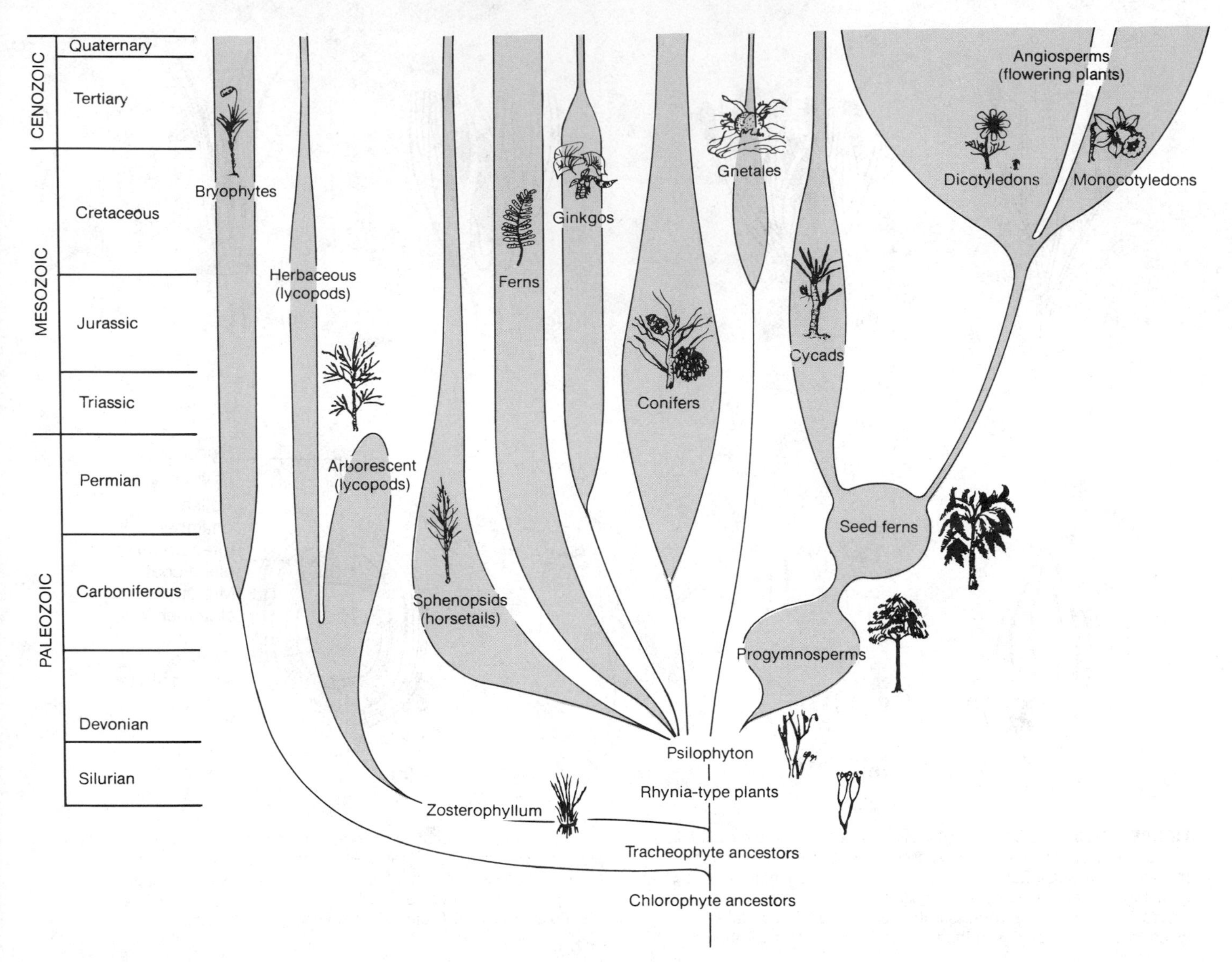

FIGURE 13-19 Possible evolutionary phylogeny stemming from some of the early plant forms. Widths of shaded sections provide very general estimates of the relative abundance of particular plant groups.

structures, are among the most intricate and attractive organs that ever appeared in plants (Fig. 13-20). They possess size, color, and odor differences that can attract specific animal pollinators, an advantage that can spread so rapidly that even some closely related plants have evolved different kinds of flowers that attract faithful pollinators (Fig. 13-21). Such pollinators must undoubtedly have had an effect on the sexual organization of the angiosperm flower since it would also be to the advantage of a plant to contribute its pollen to a mobile animal pollinator at the same time that its own ova were fertilized by pollen from another plant. That is, flowers would be selected in which pollen transfer and fertilization occurred in a single visit. The flowers of early angiosperms relying on insect pollination were therefore probably bisexual, in contrast to their wind-pollinated ancestors that would mostly have used unisexual flowers in order to help prevent self-fertilization between pollen and ova of the same flower.[2]

Also distinctive in angiosperms is **double fertilization**: two gametic nuclei of the pollen tube fertilize

[2]The fact that some groups of flowering plants possess species that are also wind pollinated indicates that evolutionary reversals can occur from one form of pollination to the other. In the case of the evolution of figs, such reversals may well have occurred more than once: the first reversal being in the order Urticules from insect pollination to wind pollination, followed by a second subsequent reversal to insect pollination in the family Moraceae, in which the genus *Ficus* is almost exclusively pollinated by species of chalcid wasps.

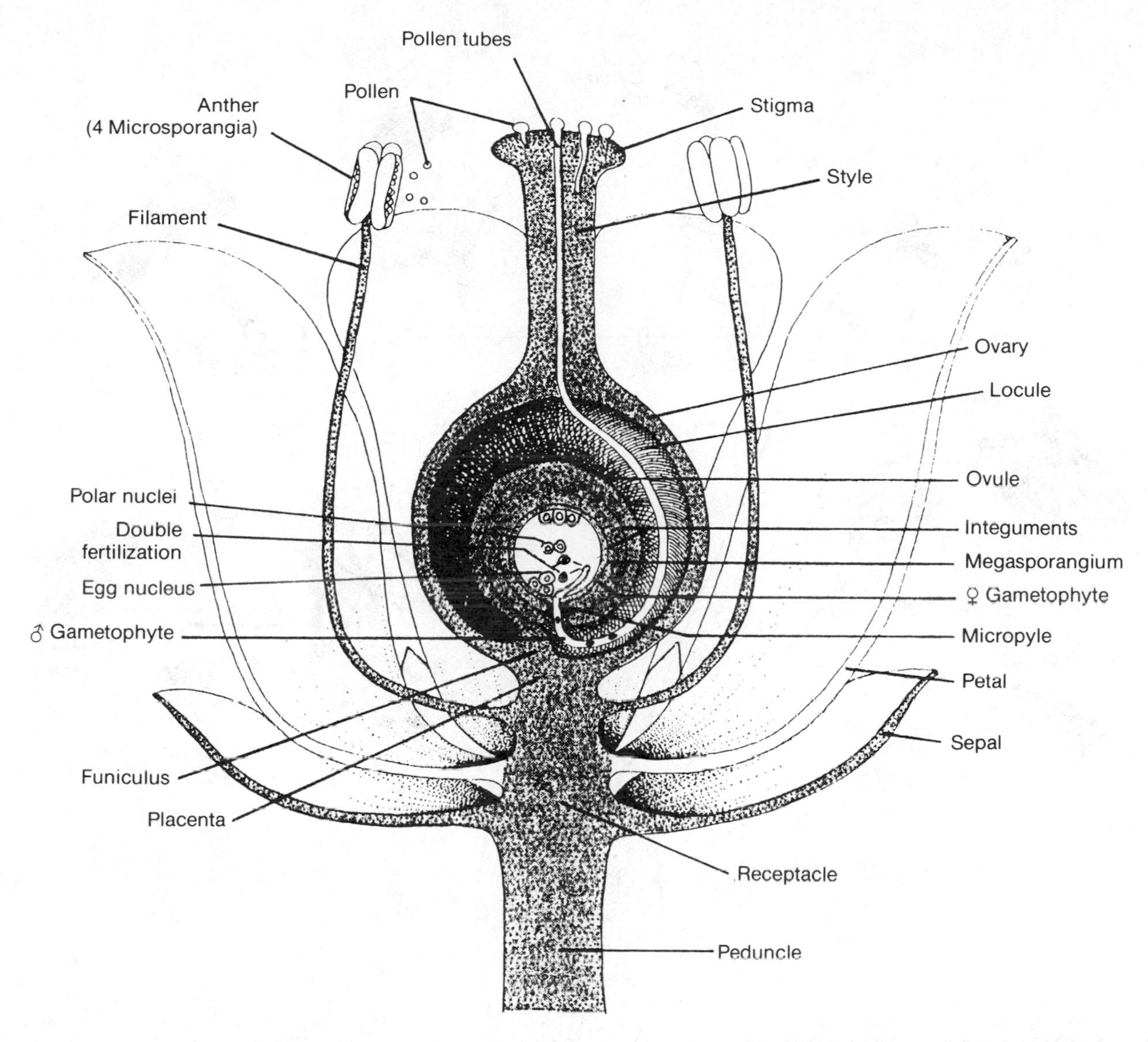

FIGURE 13-20 Diagram of a generalized angiosperm flower after fertilization. The female gametophyte is within the ovule, and the male gametophyte is the pollen tube. Two sperm nuclei are produced by the male gametophyte; one fertilizes the egg nucleus to produce the zygote (2n), and the other fertilizes the two polar nuclei to produce the endosperm (3n). (Adapted from Neushul.)

the female gametophyte, one producing the diploid (2n) embryonic nucleus, and the other often producing a polyploid [commonly triploid (3n)] **endosperm** nucleus used for embryonic nutrition. Maturation of the fertilized ovule thus leads to an angiosperm seed which has two integuments rather than the single integument found in gymnosperms. As the name angiosperm ("seed vessel") implies, these seeds are often covered, either with fleshy, fruity tissues, adhesive burs, feathery parachutes, or devices that enable the seed to be dispersed either by animals or the elements.

Dispersal ability is only one of the selective forces acting on seeds. Among others are the ability of the seed coats to protect the seed against predators and the elements, the necessity of adequate food storage for embryonic development, and the programming of seed germination to coincide with the developmental period available. All of these factors lead to specific anatomical and physiological adaptations, although there have been noticeable quantitative adaptations as well. For example, some of these forces, such as selection for wide dispersal, put a premium on small size and large numbers, whereas others, such as selection for vigorous competitive embryos, emphasize large seeds and smaller numbers (discussed in Chapter 22 as *r* and *K* selection). On the whole, the size and number of seeds produced by a particular species is therefore a compromise between these various factors and the physical limitations of the plant.

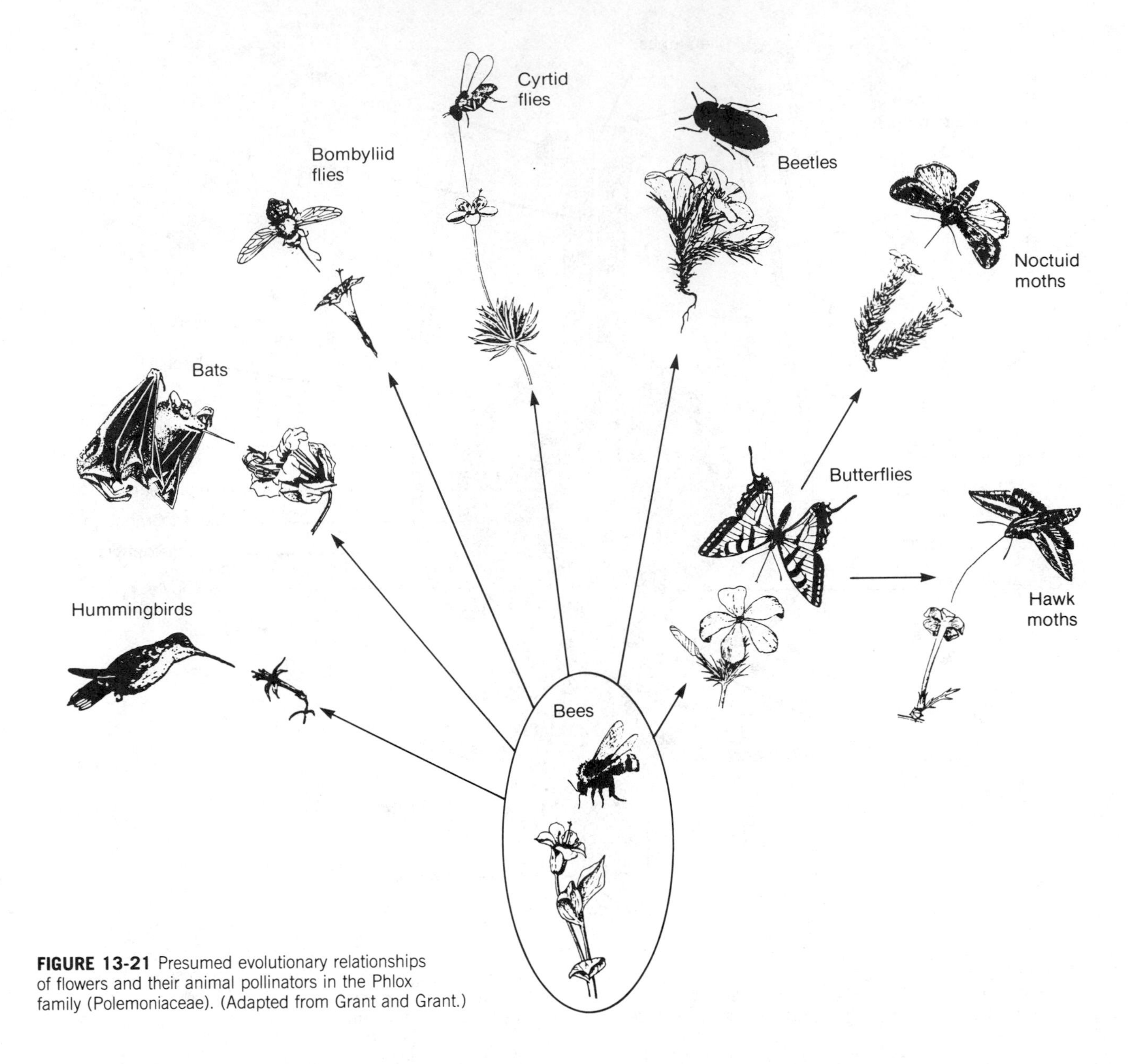

FIGURE 13-21 Presumed evolutionary relationships of flowers and their animal pollinators in the Phlox family (Polemoniaceae). (Adapted from Grant and Grant.)

EVOLUTION OF ANGIOSPERMS

A century ago, Darwin termed angiosperm origin "an abominable mystery"—a mystery that is still unsolved today. What is disputed is not only the source but also the time of angiosperm origin, with estimates ranging from an unknown period during the Paleozoic (Martin et al.) to a date that corresponds with the earliest of the angiosperm fossil records in the early Cretaceous period of the Mesozoic. Since most discovered plant fossils appear to have been associated with wet lowland areas where organic decomposition could be inhibited by silt and mud, those who propose an earlier Mesozoic or Paleozoic origin for angiosperms have suggested that they first arose in upland mountainous areas where deposition of fossils would have been rare because of active erosion. Justification for this view lies in the finding that the first angiosperm fossil leaves already show considerable differentiation, as though preceded by a lengthy evolutionary period (Fig. 13-22). On the other hand, the earliest fossil pollen that can confidently be ascribed to angiosperms is in the early Cretaceous period in the form of single-furrowed (monocolpate) grains, followed soon after by new pollen types. Doyle and Hickey therefore propose that angiosperm evolution was rapid during the early Cretaceous, and it is this rapid diversification which accounts for the variety of fossil forms found in this period, rather than a much earlier unobserved origin.

Based on summarizing a large amount of information, Stebbins has hypothesized that ancestral angiosperms were shrubs with spirally arranged, simple leaves and woody tissues formed from a single vascular cylinder. These progenitors bore bisexual flowers

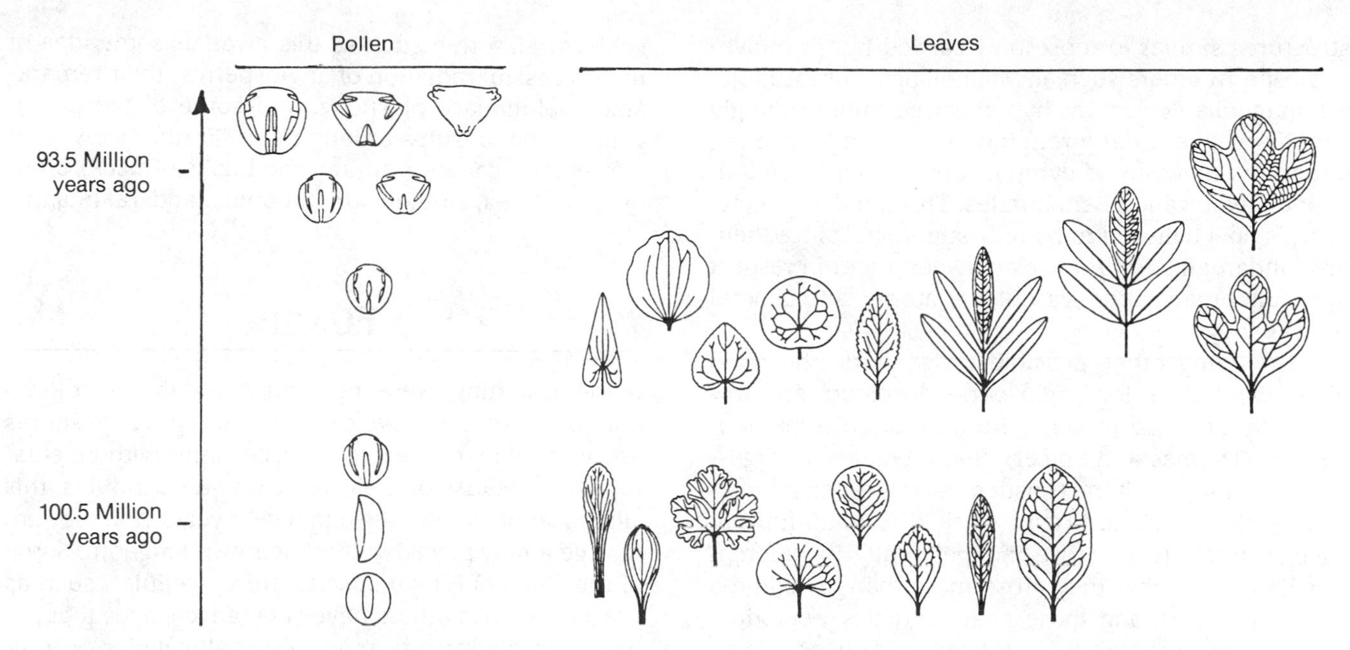

FIGURE 13-22 Appearance of angiosperm pollen and leaves at different time levels in mideastern United States Cretaceous outcroppings (Potomac Group). The earliest pollens possess single furrows (monocolpate), but these become more sculpted as time goes on. The leaves are generally characterized by a hierarchy of vein complexities: primary veins, secondary veins, intercostal veins, and so on. (Adapted from Doyle and Hickey.)

at the ends of branches, with the short male stamens lying in peripheral bundles and producing monocolpate pollen. The infolded female carpels had terminal stigmas and bore their ovules near the folded margins. After fertilization, development of the embryo and endosperm proceeded rapidly by nuclear division without cell wall formation (coenocytic), producing a two-leaved (dicotyledonous) embryo surrounded by considerable endosperm.

According to Stebbins, the ecological impact which led to the evolution of angiosperms was the alternation of dry and wet seasons with its emphasis on rapid gametophyte and embryonic development. Rainy periods followed by calms after storms also would provide an opportune time for flowering and insect pollination, as well as promoting selection for protective seed structures such as closed carpels. It is not clear where such conditions might have appeared, but Stebbins suggests that semi-arid mountainous regions with annual droughts would have offered early evolutionary opportunities for angiosperms, similar perhaps to those inferred from the rapid evolutionary rates observed among angiosperms that now inhabit mountainous regions in South Africa, Ethiopia, Ecuador, and Mexico.[3] This view is supported by Doyle and Hickey and by others.

Raven, on the other hand, points out that tropical lowland conditions with their large insect populations are more favorable for plants dependent on insect pollination than on wind pollination. Since there was a significant expansion of tropical climates during the Cretaceous period, he suggests it was this environment which allowed the early angiosperms to disperse and become dominant.

Their place of origin is only one aspect of the "abominable mystery" of angiosperms; the other is, of course, their ancestry. One approach to angiosperm phylogeny has been to search for groups that possess

[3]Data collected along the U. S. Pacific Coast by Stebbins for more than 8,000 plant species belonging to more than 800 genera show that some ecologically specialized regions such as alpine areas, deserts, lakes, swamps, and bogs have fewer plant species per genus than can be found in more ecologically variable habitats such as fields, meadows, and open woods. Numerically, there are fewer than four or five species in each genus in ecologically specialized regions compared to about ten species per genus in ecologically variable regions. Apparently the latter provides greater evolutionary opportunities than the former. Stebbins points out that, by these criteria, tropical flora with their large numbers of different species may nevertheless be considered to exist in evolutionarily restricted areas since the number of species per tropical genus is probably not much more than five. It is rather the large number of genera in the tropics, along with the large number of families, that accounts for tropical floral diversity. Presumably much of this diversity reflects the continued persistence of species that are relics of genera and families that underwent evolutionary radiation into the tropics in the past, rather than examples of new rapid speciation. Stebbins, in fact, calls tropical plant communities "museums"—"plant communities that have suffered the least disturbance during the past 50 to 100 million years and so have preserved the highest proportion of archaic forms in an essentially unchanged condition." Tropical communities would thus mostly represent geographical depositories for ancient plant groups rather than sites of origin. That is, this argument suggests that angiosperm diversity in tropical flora reflects dispersion into the tropics rather than origin from the tropics.

structures similar to those now carried by "primitive" angiosperm orders such as Magnoliales. For example, the magnolia flower was hypothesized to be strikingly similar to the axial grouping of sporangia-bearing structures (strobili) of gymnosperms, cycads, and an extinct taxon called Bennetitales. This similarity, however, is now believed to be only superficial since there are fundamental differences between them in respect to sexual organization, vascular anatomy, and general morphology.

Excluding other possible progenitors, the most likely candidates for angiosperm ancestors are now generally believed to come from among the pteridosperms (Thomas and Spicer). Some botanists suggest that the unique characteristics of angiosperms indicate a monophyletic origin, since it is doubtful that such characteristics arose independently in different groups or even that they arose more than once in the same group. Among these characteristics, reproductive mechanisms stand out: mitosis is reduced to only two cell divisions between formation of the haploid microspore and production of the male gamete, and to only three cell divisions between the megaspore and the multinucleate embryo sac; also, only angiosperms utilize double fertilization to produce simultaneously both the diploid sporophyte zygote and the commonly triploid nutritionally supporting endosperm.

On the other hand, characteristics such as the sepals and petals of flowers, xylem vessels, and other traits (Stewart) are not universally found in all angiosperms and may be considered similar to structures in gymnosperms and other vascular plants. Therefore some botanists propose that the combinations of characteristics that place plants in the angiosperm taxon may have arisen in more than one ancestral group, and the angiosperm taxon may have had a polyphyletic origin.

However this matter will be resolved, it seems clear that angiosperm advantages in rapid gametogenesis, biparental contributions to the endosperm, improved pollination, and fruity seed coverings enabled this group to radiate into widely different ecological habitats and become the dominant group in many of them. Angiosperms, with their protected and nutritionally endowed seeds, like mammals with their fetuses, developed forms adapted to dry climates, wet climates, and various types of terrain. Some reinvaded the sea, others became parasitic, and some such as sundews and Venus flytraps are even carnivorous.

Figure 13-23 shows a cross section of a hypothetical evolutionary tree with the various orders of angiosperms arranged as branches around an ancestral complex that served as the primeval trunk. The figure is drawn so that orders close to the ancestral complex are more primitive in respect to early angiosperm characteristics than those farther away. Although much is conjectural, a tree such as this gives us some idea of the successful radiation of angiosperms, their remarkable evolutionary plasticity, and some of the phylogenetic relationships among them. Further views and information can be found in the books of Beck, Cronquist, Hughes, Hutchinson, Stebbins, and Takhtajan.

FUNGI

In the past, fungi were included within the plant kingdom because they have cell walls and produce spores and were often defined as "simple plants without chlorophyll." Because of their many unique attributes, this classification has changed in recent years, and they are now generally placed within their own kingdom. Some of the 120,000 fungal species are unicellular, such as yeasts, whereas others have vegetative stages that are mostly in the form of branched multicellular or multinuclear filaments called **hyphae** which aggregate into a mass called the **mycelium**. Restricted by their growth form and absence of chlorophyll, they are, as a result, heterotrophic, deriving their nutrition either **parasitically** (live hosts) or **saprophytically** (dead organic material). In most classifications, fungi are separated from the various slime molds that possess amoeboid stages, such as *Myxomycetes,* which form plasmodial, acellular aggregates, and the cellular *Acrasiales.*

Prior to Darwin, it was often suggested that fungi were a form of algae, and were grouped with them into a single division, Thallophyta, that possessed branched, threadlike filaments and produced motile, algalike zoospores. With the appearance of *The Origin of Species,* the ancestry of fungi was therefore sought among the algae, especially the red algae. During the course of their evolution, it was believed that fungi lost the algal chloroplasts responsible for their former photosynthetic mode of nutrition and consequently became exclusively parasitic or saprophytic. Because of the possibility that resemblances between fungi and plant algae (presence of cell walls, non-motile habit) arise from convergence rather than ancestry, this view is no longer universally held, and other hypotheses have been put forth.

One suggestion is that early fungi shared a common ancestry with chemotrophic flagellated cells using inorganic sulfur or nitrogen for energy. Those chemotrophs that gave rise to fungi then evolved saprophytic forms dependent on organic materials synthesized by previous organisms. The fact that fungal cells are eukaryotic has led some workers to propose that some of the intermediary steps occurred through protozoanlike forms. The adoption of a parasitic existence on live hosts would then have offered the opportunity for early aquatic fungi (perhaps related to the present

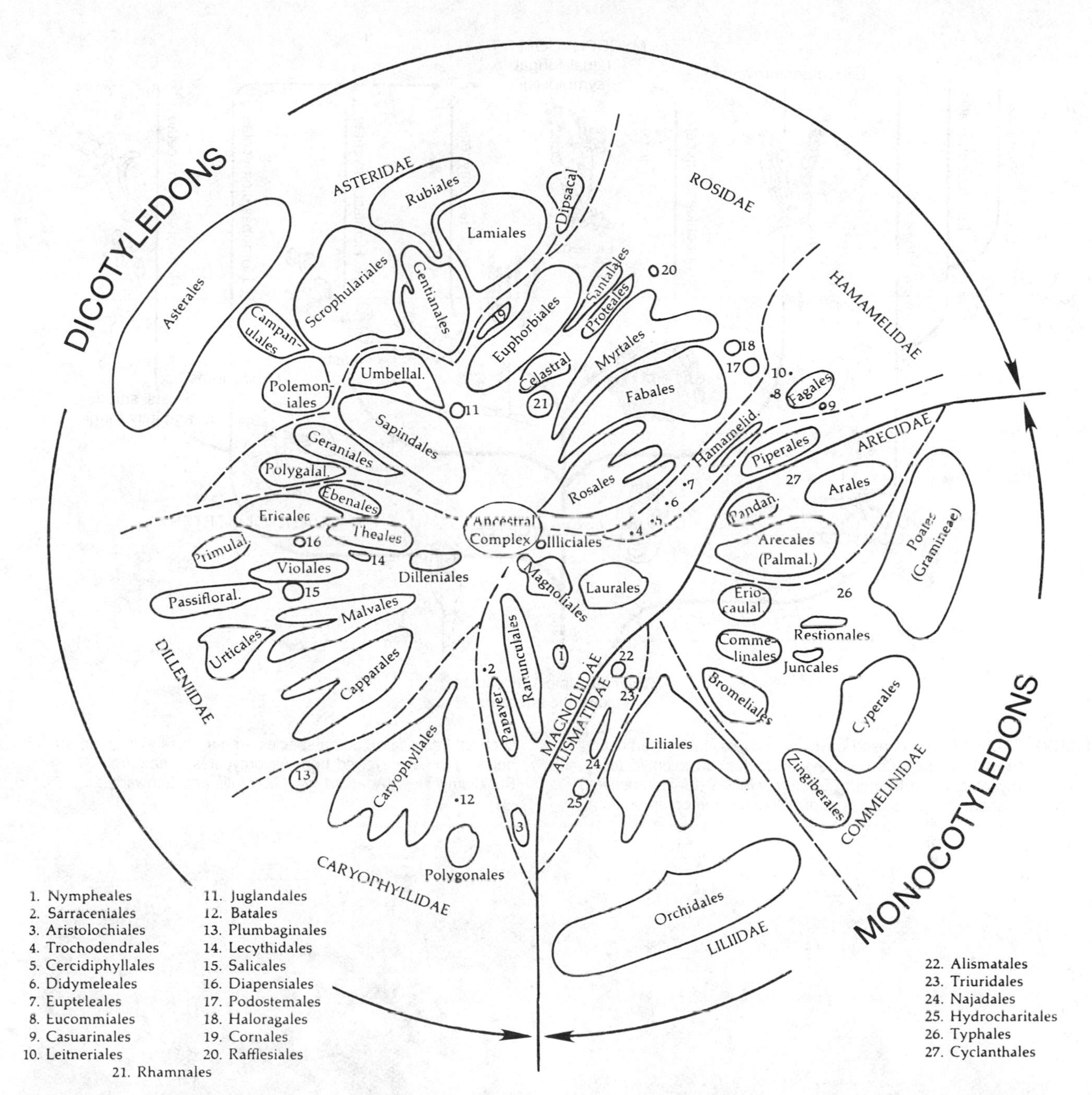

FIGURE 13-23 Proposed evolutionary relationships among major groups of angiosperms according to Stebbins. The class of **dicotyledons** (two embryonic leaves) are presumed to have evolved before the **monocotyledons** (one embryonic leaf). The phylogenetic relationships between subclasses and between orders are shown by their physical proximities in the figure (e.g., the subclass Hamamelidae may derive from the subclass Magnoliidae). Within each subclass, each order occupies an area indicative of its relative population size. (Similar diagrams, but with different placements of subclasses and orders, can be found in Sporne and also Thorne.) (Adapted from Stebbins.)

forms, Oomycetes and Chytridiomycetes) to resist desiccation in the host tissues of land-borne plants and evolve subsequently into more modern forms that use aerially dispersed spores (Zygomycetes, Ascomycetes, Basidiomycetes, and the asexual Deuteromycetes; Fig. 13-24).

Another view suggests that evolution proceeded in the direction from obligate parasitism on host-provided nutrients to a more independent saprophytic habit capable of synthesizing simple protoplasmic products into more complex useful compounds. According to Raper,

> The eventual competence of an escaping parasite to continue to flourish on the remains of its dead host inevitably increased its reproductive and dispersal potential and conferred enhanced fitness in competition with related forms that remained totally dependent upon living host tissue.

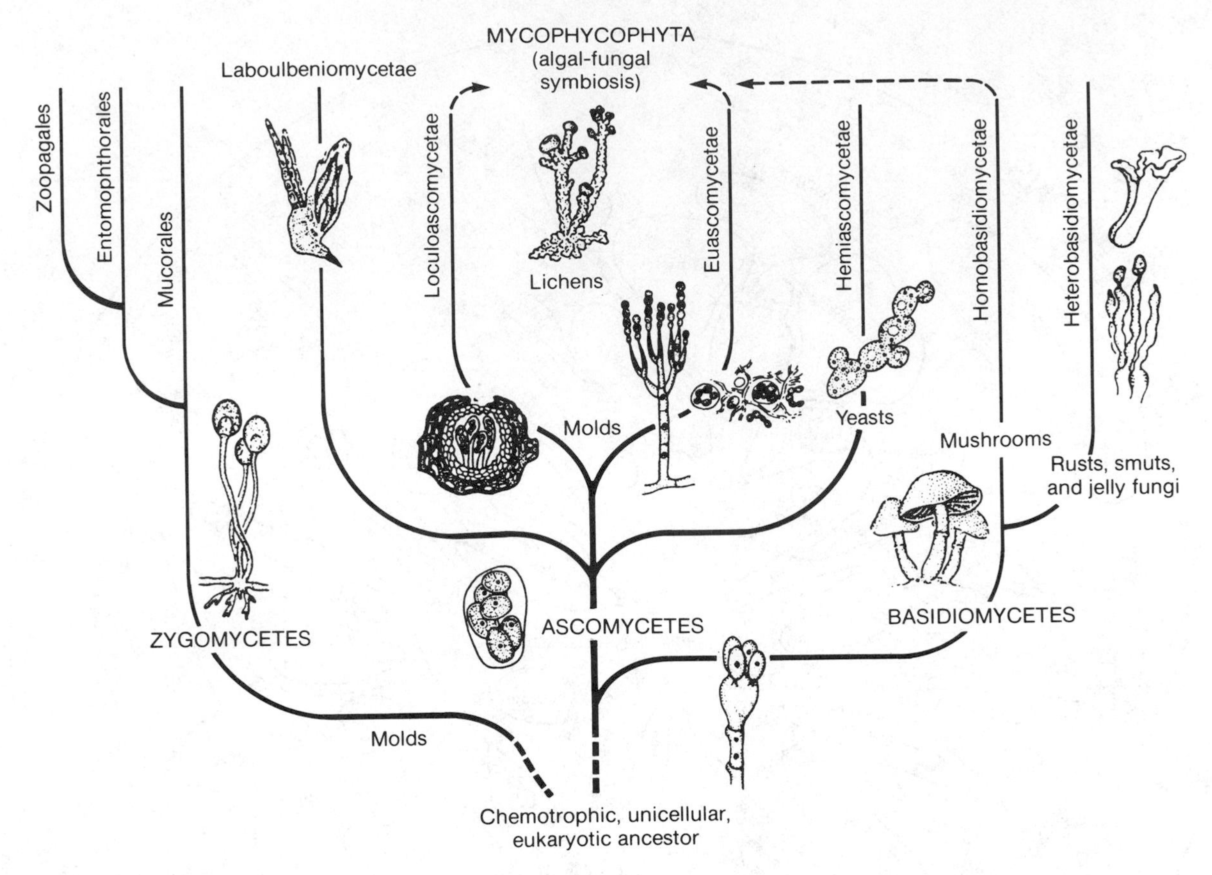

FIGURE 13-24 Some proposed phylogenetic relationships among fungi, showing three major groups, Ascomycetes, Basidiomycetes, and Zygomycetes. A fourth group, Deuteromycetes (fungi imperfecti), consists of about 25,000 species and includes some common fungi such as *Penicillium*. Some species among this latter group are believed to have evolved from Ascomycetes, others from Basidiomycetes. (Adapted from Margulis and Schwartz.)

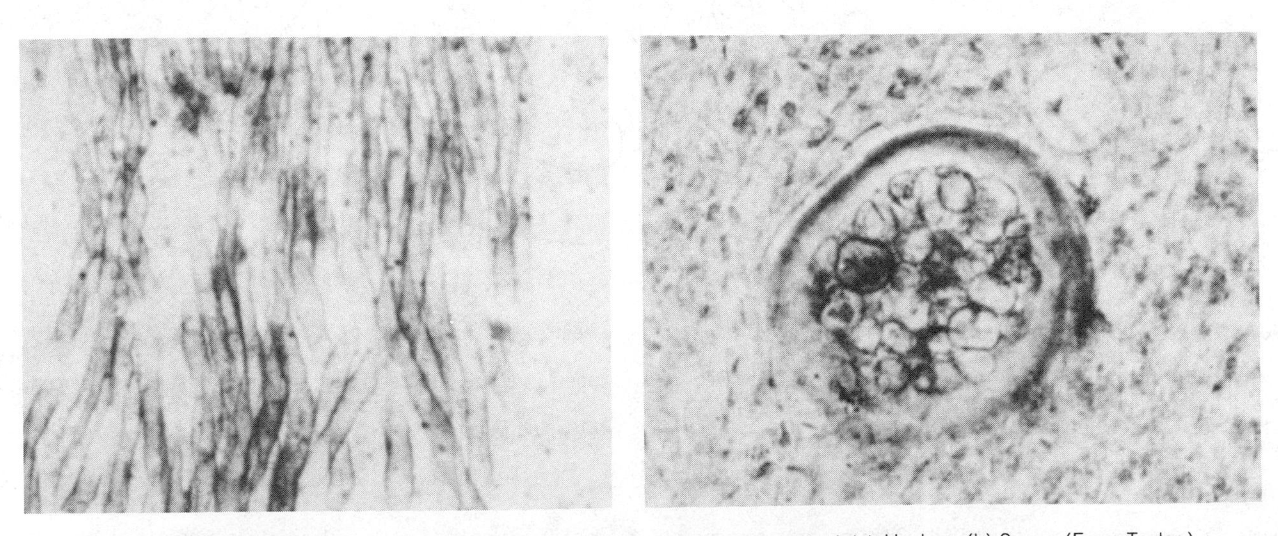

FIGURE 13-25 Fossils of different parts of a Devonian fungus, *Paleomyces gordoni*. (a) Hyphae. (b) Spore. (From Taylor.)

Whatever their origin, it is likely that the parasitic-saprophytic transition occurred a number of times in fungal evolution and in both directions. According to some interpretations of the fossil record, their success as a group stems from the Precambrian era, and they appear to have evolved into most of their presently observed forms by the end of the Paleozoic (Fig. 13-25). At present, relationships among the various major groups of fungi is still speculative, although it is clear that the parasitic fungi, at least, are still actively evolving: each new genetic variant of a host which provides it with resistance against a fungal parasite is often overcome by an increase in frequency of a fungal genetic variant that permits host susceptibility (see p. 472).

SUMMARY

Vascular land plants share certain features with green algae, which are believed similar to ancestral forms from which land plants arose. Among these characteristics are starch storage, cell-plate formation during cytokinesis, and the use of chlorophylls a and b in photosynthesis. Other land plant characteristics that may have been present in their algal ancestors were alternation of diploid and haploid generations and meiotic spore production. The progenitors evolved into colonial and then into multicellular organisms in which the cells became differentiated but lost individual motility. When plants became terrestrial, they developed techniques to minimize water loss.

The evolutionary origins of bryophytes, the simplest land plants, are obscure. They live in moist areas because of their simple water- and food-distributing tissues and are characterized by a distinctive life cycle in which the diploid generation, the sporophyte, is parasitic on the haploid plant, the gametophyte.

According to many biologists, the increased variability among offspring produced by meiosis has caused it to become entrenched in the reproductive processes of most eukaryotes. Whatever its origin, it may have served as a means of reestablishing the prevailing haploid condition in early sexually reproducing organisms. Probably because of the advantages in having two sets of genetic information, the initially brief diploid stage has been considerably lengthened in many forms.

In plants, the products of meiosis are resistant spores that germinate into a haploid structure, the gametophyte. Fertilization of gametes produced by these structures yields the diploid sporophyte. Because of the need to provide the vulnerable gametes with an aqueous environment, the gametophyte generation has persisted in all land plants, but the sporophyte has become dominant and independent of the gametophyte.

The earliest vascular plants were leafless and rootless, and they are the ancestors of the club mosses, horsetails, and ferns. The evolutionary trends in these plants included the development of two types of spores (heterospory), the formation of large leaves (megaphylls), and the differentiation of increasingly elaborate vascular tissues.

When embryos gained independence from water and became enclosed in seeds, the gametophyte was greatly reduced. Plants, in the form of the gymnosperms, which bear naked seeds, and the angiosperms, with covered seeds, became truly terrestrial. The angiosperms are characterized by elaborate flower structures, which attract pollinators, and by double fertilization, in which a triploid nutritional endosperm as well as a diploid embryo is produced.

Little is known of angiosperm origins although fossil pollen is found in early Cretaceous deposits. Angiosperms may have first appeared as woody shrubs in climatically unstable highland areas, or they may have been lowland tropical plants. Because of their unique characteristics, particularly reproductive features such as extreme truncation of gametogenesis, efficient pollination, and protected seeds, the angiosperms have become the dominant group in many environments.

The fungi, an ancient group but one which is still actively evolving, are unicellular or multicellular saprophytic or parasitic organisms. They may have been derived from chemotropic protista or from obligate parasites, some of which became saprophytic.

KEY TERMS

algae
alternation of generations
angiosperms
antithetic theory
benthic
bryophytes
club mosses
cuticle
desiccation
dicotyledons
double fertilization
enation theory
endosperm
ferns
flowers
fungi
gametophyte
gymnosperms
heterospory
homologous theory
homospory
hyphae
lycopods
megaspores
microspores
monocotyledons
mycelium
parasitic
planktonic
pollen
progymnosperms
protostele
rhyniophyta
saprophytic
seed ferns
seeds
sphenopsids
sporangia
spores
sporophyte
telome theory
tracheophyta
vascular plants

REFERENCES

Andrews, H. N., Jr., 1961. *Studies in Paleobotany*. John Wiley, New York.

Banks, H. P., 1970. *Evolution and Plants of the Past*. Macmillan, London.

Beck, C. B. (ed.), 1976. *Origin and Early Evolution of Angiosperms*. Columbia Univ. Press, New York.

Bold, H. C., C. J. Alexopoulos, and T. Delevoryas, 1980. *Morphology of Plants and Fungi*, 4th ed. Harper & Row, New York.

Chapman, D. J., 1985. Geological factors and biochemical aspects of the origin of land plants. In *Geological Factors and the Evolution of Plants*, B. H. Tiffney (ed.). Yale Univ. Press, New Haven, Conn., pp. 23-45.

Cronquist, A., 1968. *The Evolution and Classification of Flowering Plants*. Nelson, London.

Delevoryas, T., 1977. *Plant Diversification*, 2nd ed. Holt, Rinehart & Winston, New York.

Doyle, J. A., and L. J. Hickey, 1976. Pollen and leaves from the mid-Cretaceous Potomac Group and their bearing on early angiosperm evolution. In *Origin and Early Evolution of Angiosperms*, C. B. Beck (ed.). Columbia Univ. Press, New York, pp. 139-206.

Foster, A. S., and E. M. Gifford, Jr., 1974. *Comparative Morphology of Vascular Plants*, 2nd ed. W. H. Freeman, San Francisco.

Gensel, P. G., and H. N. Andrews, 1984. *Plant Life in the Devonian*, Praeger, New York.

Grant, V., and K. A. Grant, 1965. *Flower Pollination in the Phlox Family*. Columbia Univ. Press, New York.

Holman, R. M., and W. F. Robbins, 1940. *Elements of Botany*, 3rd ed. John Wiley, New York.

Hughes, N. F., 1976. *Palaeobiology of Angiosperm Origins*. Cambridge Univ. Press, Cambridge.

Hutchinson, J., 1969. *Evolution and Phylogeny of Flowering Plants: Dicotyledons; Facts and Theory*. Academic Press, London.

Margulis, L., and K. V. Schwartz, 1982. *Five Kingdoms*. W. H. Freeman, San Francisco.

Martin, W., A. Gierls, and H. Saedler, 1989. Molecular evidence for pre-Cretaceous angiosperm origins. *Nature*, **339**, 46-48.

Michod, R. E., and B. R. Levin, 1988. The *Evolution of Sex*. Sinauer Associates, Sunderland, Mass.

Neushul, M., 1974. *Botany*. Hamilton Pub. Co., Santa Barbara, Calif.

Pettitt, J. M., and C. B. Beck, 1968. *Archaeosperma arnoldii*—a cupulate seed from the Upper Devonian of North America. *Contr. Mus. Paleontol. Univ. Michigan*, **22**, 139-154.

Pickett-Heaps, J. D., 1975. *Green Algae: Structure, Reproduction, and Evolution of Selected Genera*. Sinauer Associates, Sunderland, Mass.

Raper, J. R., 1968. On the evolution of fungi. In *The Fungi, An Advanced Treatise*, Vol. III, G. C. Ainsworth and A. S. Sussman (eds.). Academic Press, New York, pp. 677-693.

Raven, P. H., 1977. A suggestion concerning the Cretaceous rise to dominance of the angiosperms. *Evolution*, **31**:451-452.

Sporne, K. R., 1976. Character correlations among angiosperms and the importance of fossil evidence in assessing their significance. In *Origin and Early Evolution of Angiosperms*, C. B. Beck (ed.). Columbia Univ. Press, New York, pp. 312-329.

Stebbins, G. L., 1974. *Flowering Plants: Evolution Above The Species Level*. Harvard Univ. Press, Cambridge, Mass.

Stewart, W. N., 1983. *Paleobotany and the Evolution of Plants*. Cambridge Univ. Press, Cambridge.

Takhtajan, A., 1969. *Flowering Plants: Origin and Dispersal*. Oliver & Boyd, Edinburgh.

Taylor, T. N., 1981. *Paleobotany*. McGraw-Hill, New York.

Thomas, B. A., and R. A. Spicer, 1987. *The Evolution and Palaeobiology of Land Plants*. Croom Helm, London.

Thorne, R. F., 1976. A phylogenetic classification of the Angiospermae. *Evol. Biol.*, **9**:35-106.

Tiffney, B. H., 1985. Geological factors and the evolution of plants. In *Geological Factors and the Evolution of Plants*, B. H. Tiffney (ed.). Yale Univ. Press, New Haven, Conn., pp. 1-21.

Zimmermann, W., 1952. Main results of the "Telome Theory." *Paleobotanist*, **1**, 456-470.

FROM PROTOZOA TO METAZOA

When the Cambrian period began 590 million years ago there was the sudden, marked appearance of numerous new skeletonized forms of multicellular invertebrates. In a time span of 100 million years or less an explosive radiation of eukaryotes occurred, signified by the appearance of most major animal phyla in the fossil record (Fig. 14-1). Many of these skeletonized forms probably represent entirely new adaptive radiations of Precambrian forms. That is, the acquisition of hard parts by some of the older forms undoubtedly initiated changes in the location and attachment of their soft tissues, leading to entirely different animals than previously existed. Perhaps the Cambrian period marks a warming trend, allowing the mineralization of tissues. Or perhaps the atmospheric accumulation of oxygen through photosynthesis had reached sufficient levels to permit mineralization, as well as to form a protective blanket of ozone that facilitated the rapid expansion of multicellular animals.[1]

Among other explanations for the **Precambrian-Cambrian discontinuity** is a formerly popular concept that most signs of life were obliterated in strata earlier than the Phanerozoic because of the heat and pressure involved in geological processes such as mountain building. A further notion was that living forms evolved mostly in freshwater areas, and their fossils are therefore absent in Precambrian sediments, which are primarily of marine origin. However, since there is now considerable evidence that both prokaryotic and eukaryotic Precambrian organisms can be clearly identified, the Cambrian discontinuity seems to be real and not merely the result of geological metamorphism or imperfect fossilization.

A hypothesis offered by Stanley suggests a biological cause for the sudden increase of Cambrian forms, based on the principal of **cropping**. In cropping, predators feed on the most abundant prey species, thereby reducing their numbers and allowing other species to utilize resources formerly monopolized by the dominant prey. For example, cropping of a field

[1]It was Berkner and Marshall who suggested that metazoan evolution, dependent on aerobic metabolism, had to await an oxygen atmosphere sufficient to sustain it at a level that was perhaps 1 percent of present atmospheric oxygen. Unfortunately, data on Cambrian and Precambrian oxygen pressures are not available, and it is even possible that plentiful supplies of free oxygen produced by algae may have been present during a billion-year interval before the Cambrian (e.g., see Cloud).

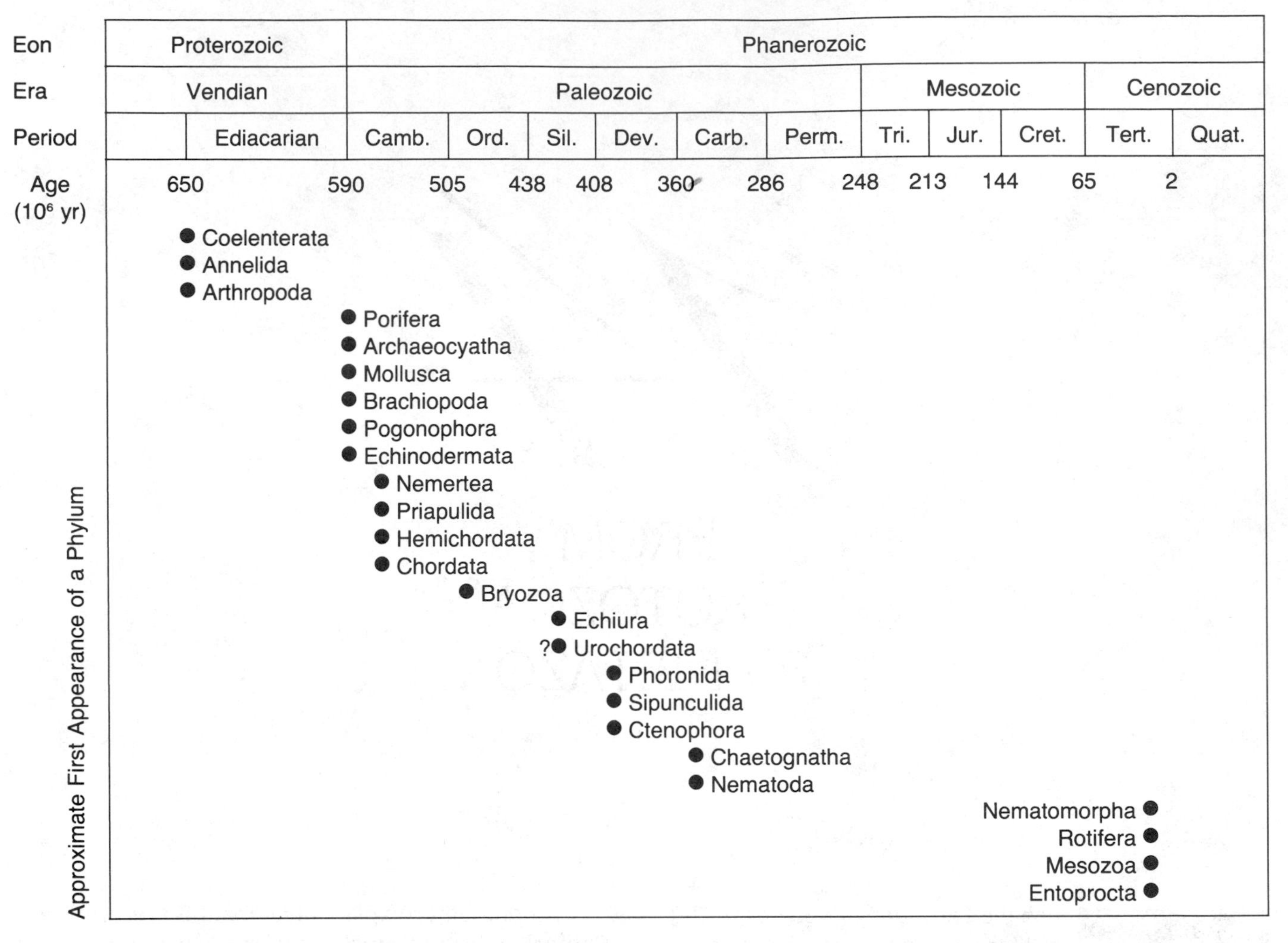

FIGURE 14-1 Approximate times in which various well-described metazoan phyla first appear in the fossil record. (Adapted from Dobzhansky et al.; see also Brasier.)

that had been restricted to a single dominant plant species soon opens numerous niches for other plant species that can now grow in an area where they were formerly excluded. The evolutionary value of cropping extends also to predators through a **feedback cycle**, since the diversification of prey species leads in turn to the diversification of predator species.

The Precambrian fossil record indicates the widespread existence of algal stromatolites (p. 150) growing under conditions unhampered by limitations other than those of available light and nutrients. In the words of Stanley, "We can envision an all-producer Precambrian world that was generally saturated with [autotrophic] producers and biologically monotonous." In the absence of cropping, new algal prokaryotic species and even new algal eukaryotes must have been preempted from occupying these areas. In fact, the finding by Brock that prokaryotic algae do not exist in acid conditions (perhaps because their chlorophyll molecules are relatively unprotected in their sites on the plasma membrane) suggests that such environmentally restricted areas may have provided the first opportunity for colonization by eukaryotic algae.

In any case, until eukaryotic cropping animals appeared, significant evolutionary changes may have been relatively minor, since biological habitats that could have potentially been occupied by many newly evolved species would have been limited. The occurrence of heterotrophic croppers in the Precambrian period may thus have allowed diversification of their autotrophic prey which then caused, in turn, diversification among these new **herbivores**. According to Sleigh, it was the increase in marine autotrophic eukaryotes (photosynthetic plankton) during the Late Precambrian that provided the basic food supply for Cambrian herbivores. Since there may have been only a few steps from being a herbivore to feeding on herbivores, new classes of **carnivores** could now also evolve, leading to the explosive adaptive radiation of the Cambrian period. The appearance of hard exoskeletons may therefore have provided a common function to a wide variety of Cambrian organisms in offering varying degrees of protection from predation. The fact that eukaryotes had by this time evolved sexual reproductive methods undoubtedly also enhanced their ability to diversify into new functions and new available habitats.

Whatever the reasons for the Cambrian proliferation of skeletonized metazoans, it is now clear that their origin as soft-bodied animals must have occurred much earlier. The fossils of a variety of soft-bodied animals have been found in the Precambrian **Ediacarian strata** of South Australia as well as in tidal deposits at ten other places around the globe dating back to a period about 600 to 700 million years ago (Fig. 14-2). Some reports suggest the existence of animal remains, burrows, or fecal pellets that are about 1 billion years old (Glaessner [1983]). Most of these Precambrian fossils appear to be either coelenterates or worms of various kinds, including annelids. Because their organization seems relatively advanced, with intricate surface structures and dimensions that may have been more than 1 meter long and 1 to 2 centimeters wide, it is clear that these fossil organisms must have undergone considerable prior evolution. What were the first steps in multicellular animal evolution?

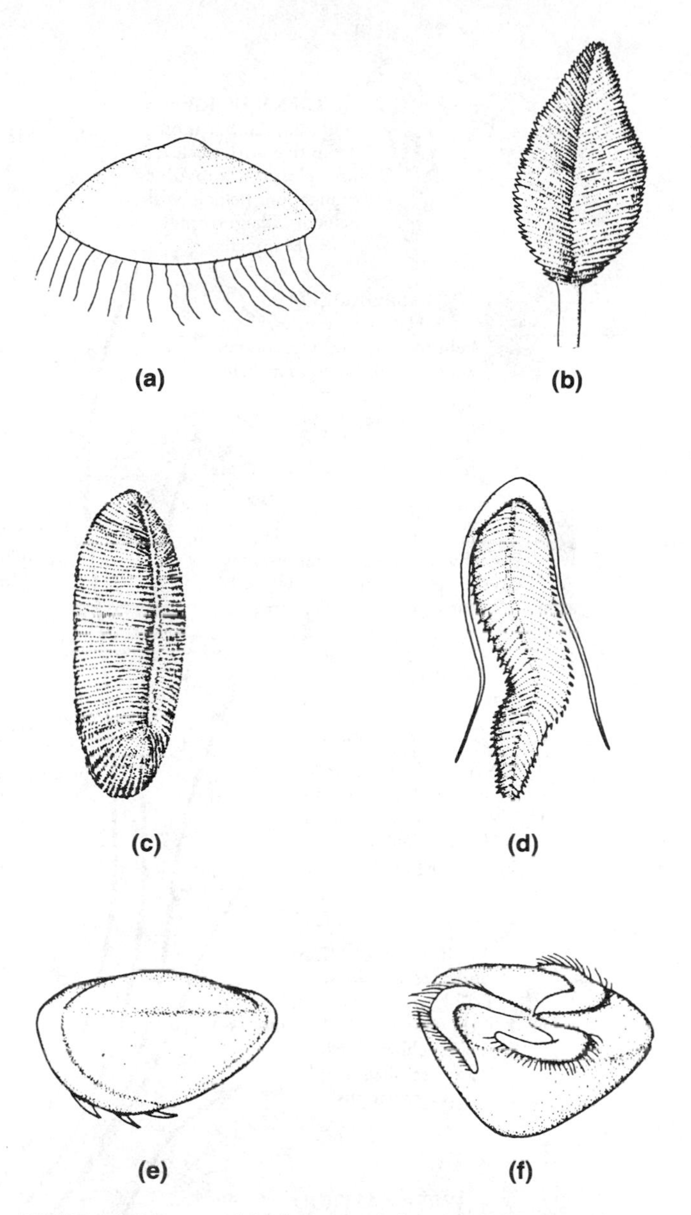

FIGURE 14-2 Some soft-bodied animals found in Ediacarian tidal flat deposits of South Australia and about ten other places throughout the globe, all occurring prior to the Cambrian period approximately 590 to 650 million years ago. Some of these fossils (a and b) resemble modern coelenterates such as jellyfish and sea pens, whereas others (c and d) bear likenesses to segmented annelids and perhaps also (e and f) to molluscs and echinoderms. (Adapted from Olson and Robinson, from Glaessner.)

Protistan Ancestry

Most recent taxonomic schemes classify all unicellular eukaryotes into an exclusive kingdom, **Protista.** These organisms include (1) protozoa, which ingest their food directly, (2) photosynthetic algae, and (3) some saprophytic fungi. According to the prevailing five-kingdom classification system first proposed by Whittaker, the Protista stand between the prokaryotic Monera (Archaebacteria and Eubacteria) and the three multicellular kingdoms of Plantae, Fungi, and Animalia. Thus, aside from their diverse forms and the many ways in which they affect other forms of life, protistans are believed to be the essential link between the early progenote cells (p. 157) and all multicellular eukaryotes.

Evolutionarily, protistans are the first eukaryotes with fossils that date back about a billion years ago (see Fig. 9-17). As discussed in Chapter 9, it is now generally presumed that endosymbiotic events provided them with mitochondria, chloroplasts, and perhaps even cilia and flagella. In addition to these organelles, protistans share with other eukaryotes many features, including a nuclear membrane as well as **cilia** and **flagella** whose substructures are organized into nine pairs of microtubules circling two microtubules in the axial center ("9 + 2" arrangement).

Because not all photosynthetic protistans contain the same light-gathering pigments, Sleigh and others suggest that photosynthetic prokaryotes invaded eukaryotic cells more than once. Some of these endosymbiotic events resulted in protistan algae, completely reliant on autotrophic nutrition, whereas others produced protistans that can alternate from autotrophic to heterotrophic nutrition, such as *Euglena*. The lack of chloroplasts, whether caused by their initial absence or later loss, gave rise to the large diversity of protistan heterotrophs, or protozoa.

Figure 14-3 offers a very general scheme for protistan radiation. It begins with a **protoflagellate**, which

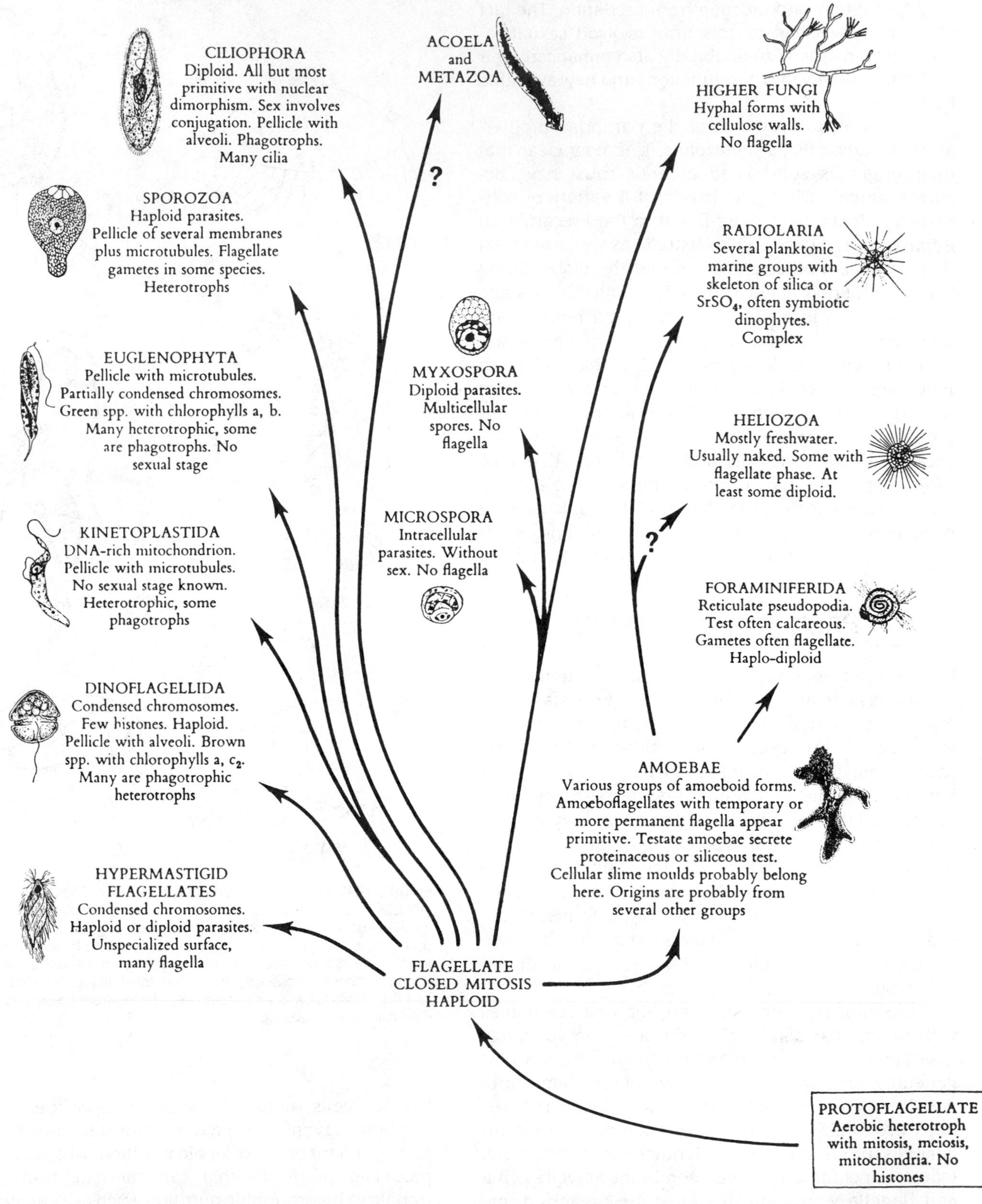

FIGURE 14-3 One proposed phylogeny for various protistan groups leading to metazoans, higher fungi, and land plants. Although this scheme provides a separate evolutionary origin for sponges and indicates a possible polyphyletic origin for other metazoa (Cnidaria

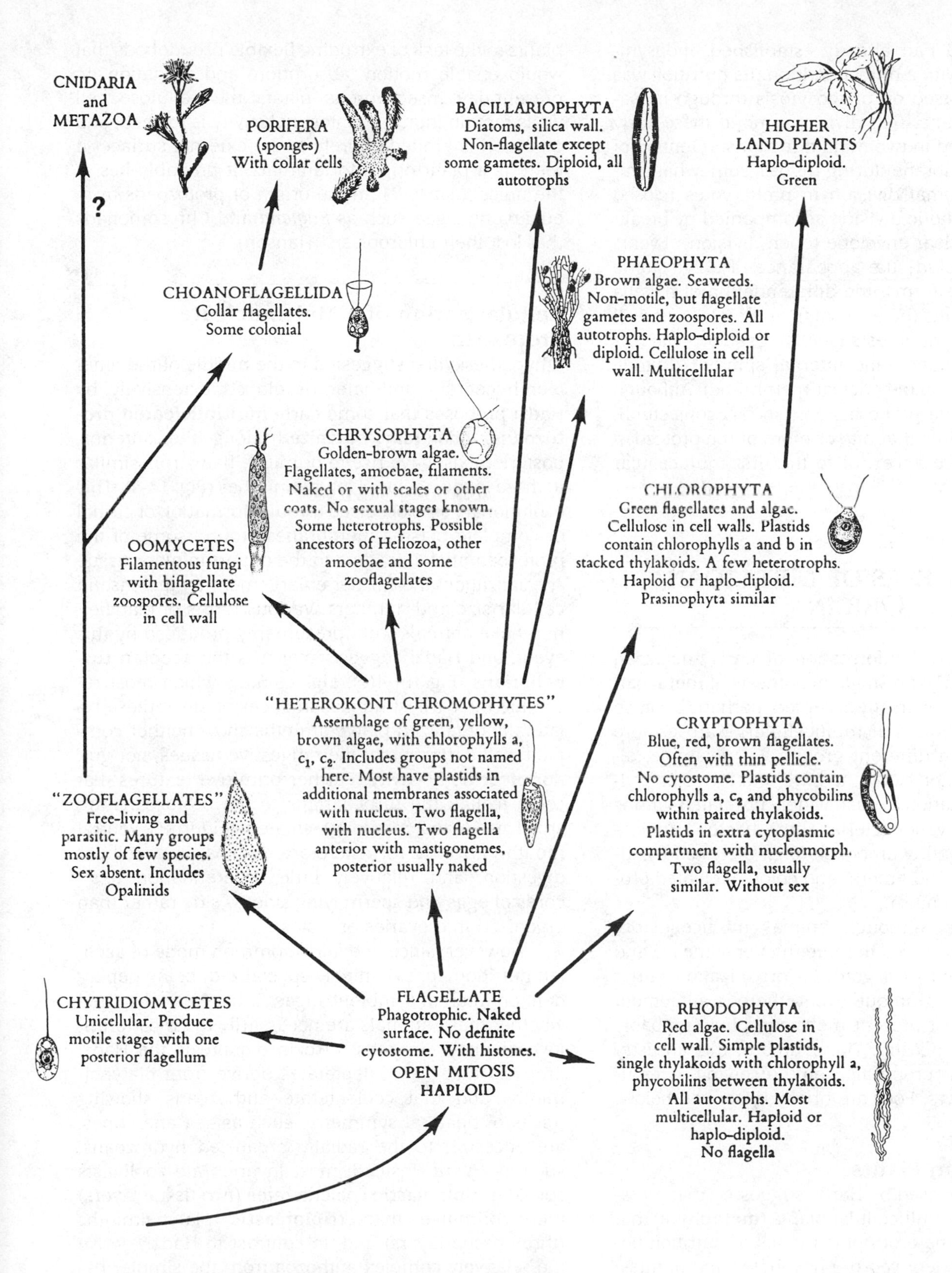

and acoelans), monophyletic evolutionary schemes are still popular (e.g., see Barnes). Nucleotide sequencing data derived from some of these protistan groups (e.g., Baroin et al.) are now beginning to furnish more exact phylogenetic information. (Adapted from Sleigh.)

had flagella and had already established endosymbiotic relations with a mitochondrion. Its nutrition was heterotrophic, based on phagocytosis through its naked cell membrane. Cell division among these early protistans evolved in two main directions: retention of the nuclear membrane during mitosis somewhat similar to chromosomal division in prokaryotes (closed division); and mitotic division accompanied by breakdown of the nuclear envelope (open division). Events that followed include the appearance of a cytostome (mouth) in some forms, the differentiation of flagella into uneven lengths (heterokonts), or their loss, as well as the development of tests (shells), pseudopodia (cytoplasmic extensions), and internal spindles (mitotic microtubules within the nuclear membrane). Although exact phylogenetic proposals are, so far, conjectural, it seems reasonable that one or more of the protozoan heterotrophs were ancestral to the first multicellular animals (metazoa).

HYPOTHESES OF METAZOAN ORIGIN

Since detailed fossil information of early metazoan evolution is lacking, no single hypothesis of **metazoan origin** has been universally accepted, perhaps because it is highly probable that multicellularity arose more than once and in different groups. For example, although they are not metazoans, we find forms of multicellular organization in such widely unrelated groups as filamentous cyanobacteria, slime bacteria (myxobacteria), aggregating amoebae (e.g., dictystelids), algae (brown, red, and green), and colonial ciliated protozoans (Zoothamnium).

Nevertheless, although animal multicellularity may have been achieved in more than one lineage and can be considered as a grade of organization rather than the result of a unique event, there is agreement that many or most present metazoan phyla probably had a common ancestry. The source of this ancestry has led to various proposals, some receiving more attention than others. Four are briefly described below.

Evolution from Plants

A hypothesis proposed by Hardy suggested that metazoans arose from multicellular plants (**metaphyta**) that were forced into heterotrophic modes of nutrition because they were deprived of phosphates and nitrates. As these carnivorous plants presumably became successful in absorbing and capturing other organisms, they lost their chloroplasts and turned into metazoans. Objections against this concept point mostly to the difficulty of adapting the rigid cellulose cell wall of plants to the task of extruding flexible pseudopods that would enable motion, absorption, and predation. In present-day insectivorous plants the cellulose cell walls remain intact, no intestinal cavity is formed, and digestion is limited entirely to the external surface. If there is a plant origin for animals, it probably lies at the unicellular level, in the origin of protozoans from eukaryotic algae such as *Euglena* and Chrysomonads that lost their chloroplasts (Hanson).

Cellularization of a Multinucleate Protozoan

A hypothesis first suggested in the middle of the nineteenth century and later developed extensively by Hadzi proposes that some early **multinucleated protozoans**, bilaterally organized along the anterior-posterior axis, gave rise to primitive flatworms similar to those in the phylum Platyhelminthes (Fig. 14-4). This evolutionary step occurred through formation of partial or complete plasma membranes around some of the protozoan nuclei, leading to the opportunity for tissue specialization and further enlargement by increase in cellular size and numbers. Various forms of platyhelminthlike animals were presumably produced by this event, and Hadzi suggests that it is the **acoelan turbellarians** (Fig. 14-4(c), Fig. 15-6(a)) which most resemble the earliest metazoan ancestor since these bilaterally organized platyhelminths show neither complete cellularization of their digestive tissues, nor gut, nor other body cavity. Further primitive features that seem to link the turbellarians to the protozoans are small size (1 to 2 mm long), ciliated epidermis, ventral mouth, absence of excretory organs, intracellular digestion, and relatively little differentiation (e.g., cords of eggs and sperm lying side by side rather than organized into ovaries or testes).

However, since one fairly common mode of acoelan nutrition appears highly specialized, being dependent on internal symbiotic algae, it has been disputed whether these animals are necessarily the most primitive of platyhelminths. Also in dispute is the Hadzi proposal that the coelenterates derive from platyhelminths and that coelenterate anthozoans showing traces of bilateral symmetry, such as sea anemones, are ancestral to the radially organized hydrozoans, such as *Hydra*. Instead, most invertebrate zoologists consider **diploblastic** coelenterates (two tissue layers) more primitive than **triploblastic** platyhelminths (three tissue layers) and, in contrast to Hadzi, derive the relatively complex anthozoa from the simpler hydrozoa.

An attempt to salvage at least part of the hypothesis of protozoan-platyhelminth evolution was the proposal that coelenterates may have originated from protozoans independently of platyhelminths and that the

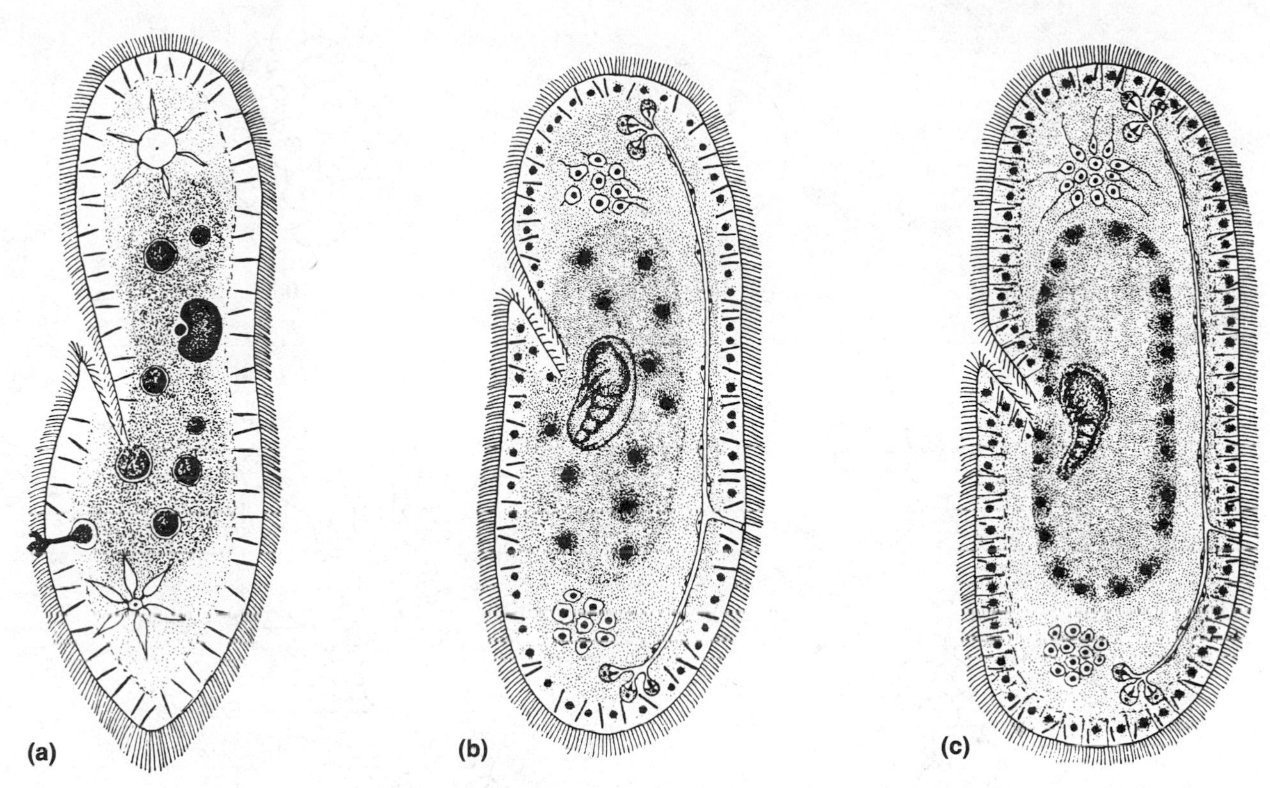

FIGURE 14-4 Hypothetical transformation of a ciliated protozoan (a) into a primitive turbellarian metazoan (c). (Adapted from Hadzi.)

two phyla are therefore not directly related (e.g., see top of Fig. 14-3). Support for such a polyphyletic origin of metazoans can be gathered from the likelihood that sponges (Porifera) may also have had an independent protozoan origin, arising from choanoflagellates (collared zooflagellates, see Chapter 15). This view appears supported by the study of Field et al. comparing various regions of 18*S* RNA molecules in a wide variety of eukaryotes. These 18*S* molecules are essential for ribosomal protein synthesis, and, like 5*S* RNA (p. 234), portions of their sequences are mostly or partly conserved in all eukaryotes. According to Field et al., sequence comparisons show that coelenterates, fungi, and plants derive from a distinctly separate protistan origin than do platyhelminthes and other metazoan groups (see Fig. 15-12).

Gastrulation of a Colonial Protozoan

In the 1870's Haeckel proposed that hollow-balled colonies of flagellated protozoans, not unlike the modern alga *Volvox* (see Fig. 13-1) but lacking chloroplasts, developed an anterior-posterior axis as they swam through primitive waters. Food particles in this primitive **blastula** (presumably recapitulated in the blastula embryonic stage of many metazoa) were swept toward the posterior pole by ciliary action, and cells at that end became specialized for digestive functions. According to Haeckel and his followers, these digestive cells then invaginated through a circular **blastopore** into the hollow interior of the organism to form an internal digestive tract or **archenteron** (Fig. 14-5). This new bilayered, cuplike organism with **ectoderm** on the outside and **endoderm** on the inside was believed to be similar to one of the developmental stages in some present-day metazoa, the **gastrula**. The **gastraea hypothesis**, then, suggests that the primitive nature of sponges and coelenterates lies in the fact that they have remained at this diploblastic gastrula level.

A further corollary of the gastraea hypothesis is that an important body cavity of most metazoans, the coelom, originated from lateral pockets formed in the archenteron. There are different views as to the number of pockets involved, but it is agreed among supporters of this hypothesis that the consequence of coelomic formation was to allow the development of a third tissue layer, the **mesoderm**, lying between the ectoderm and endoderm. Further evolution then proceeded to form all the various triploblastic phyla.

A major objection to the gastraea hypothesis is that gastrulation by invagination is not common in the embryological development of many metazoans. Even in hydrozoan coelenterates, presumed by Haeckel to exemplify the gastrula stage of evolution, endodermal tissues are formed by ectodermal cells that appear to wander in from an intact epithelial surface rather than by a cuplike folding process. To the extent that developmental patterns are conserved in evolution (pp. 41-

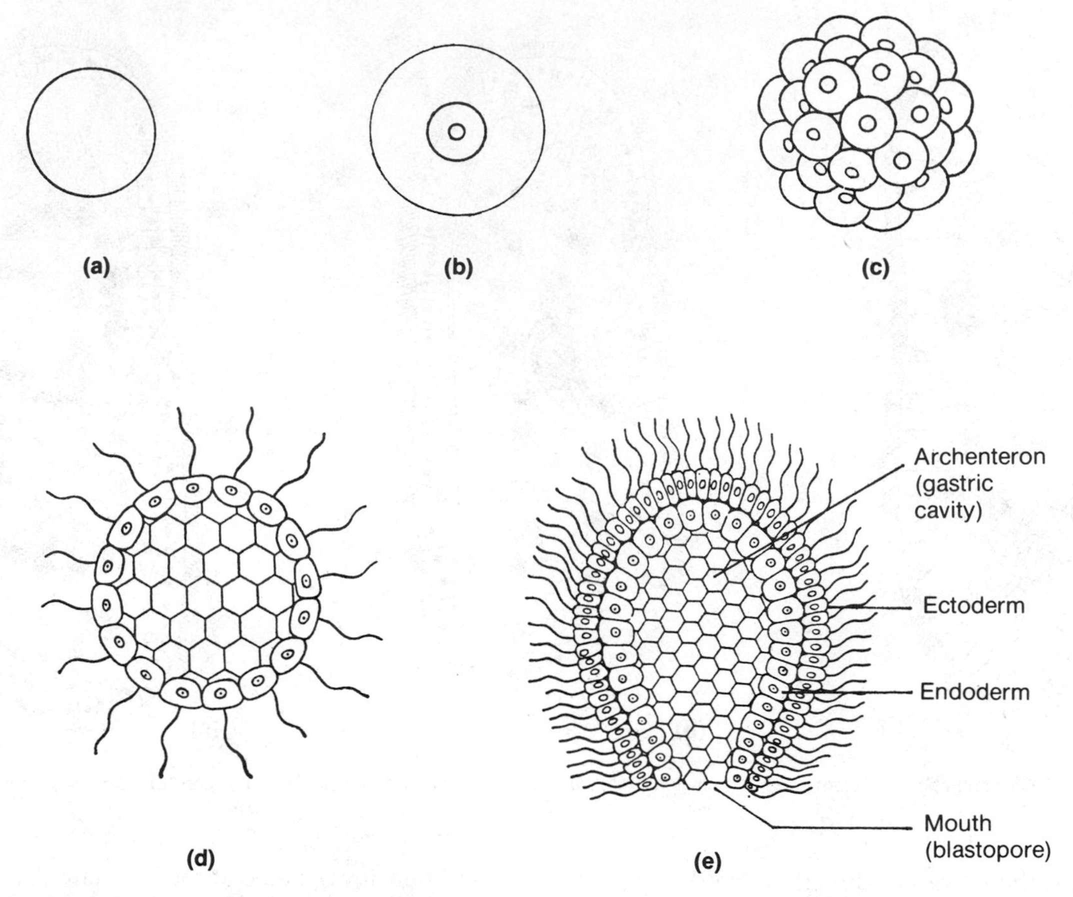

FIGURE 14-5 Stages in the evolution of a multicellular organism according to Haeckel. The monerula (a) has no nucleus; the cytula (b) is nucleated; the morula (c) is a compacted solid ball of cells; the blastula (d) is a hollow, single-layered cellular sphere; and the gastrula (e) is a bilayered organism with an exterior opening. (Adapted from Kerkut, based on Haeckel.)

44), gastrulation by invagination does not appear to be a primitive pattern among such coelenterates.

Many workers in this field are also unwilling to go along with the implication that the coelomic sacs formed in early gastrulalike organisms gave rise directly to the segmented coelomic cavities found in invertebrate animals such as phoronids and pterobranchs. These segmented coelomates would presumably have been ancestral to acoelomates such as platyhelminths and nemerteans as well as to nonsegmented coelomates such as nematodes and sipunculids. As most zoologists have come to believe that the platyhelminths are more primitive than segmented coelomic animals and that the coelom is probably a feature that arose in fairly large, nonciliated animals to aid in burrowing and swimming (see below), the phylogeny derived from the gastrulation hypothesis has been seriously disputed.

Planula Hypothesis

A fourth, more popular hypothesis at present is that Haeckel's blastula was followed not by gastrulation but by the formation of a solid ball of cells (**planula**) in which the ectodermal cells became specialized for locomotion and the endodermal cells for digestion (Fig. 14-6). As shown by Metschnikoff and others in the nineteenth century, many lower metazoans do not make use of a mouth and digestive tube since digestion is phagocytic and intracellular. The finding of planula-type larvae in primitive metazoan phyla such as sponges and coelenterates and the observation that various groups among the platyhelminths and pogonophorans have a solid gut filled with endodermal cells indicate that primitive planula-type organisms would have been viable. According to this hypothesis, the formation of a hollow archenteron and open blastopore would have occurred during later evolutionary stages (Fig. 14-6(e)). "Contrary to Haeckel's opinion it is probable that entoderm formation by invagination is a derived rather than the original method, and represents one of those short cuts common in embryology" (Hyman [1940]).

In spite of their differences, the advantages of multicellularity are assumed by all these hypotheses.

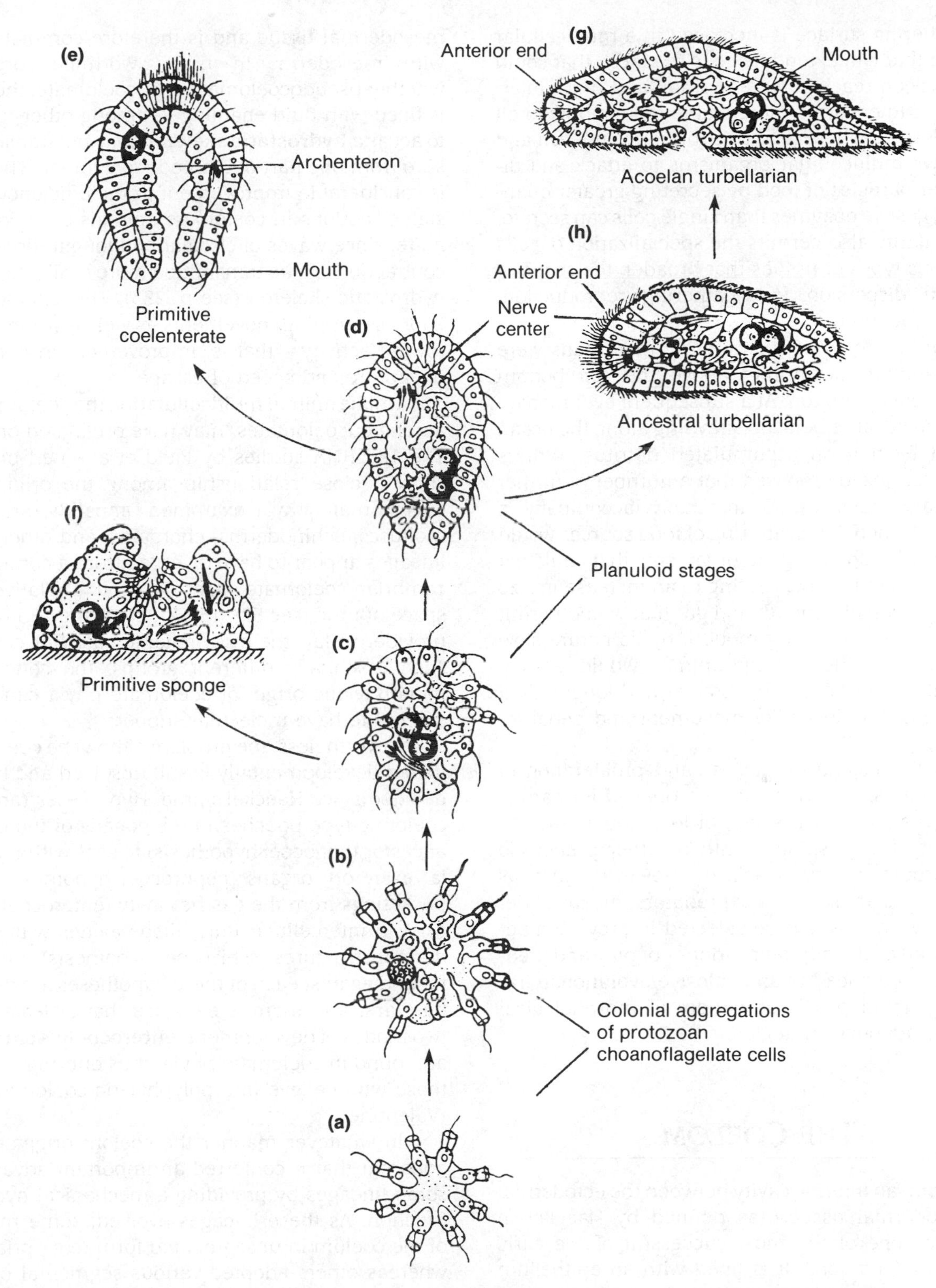

FIGURE 14-6 Illustration of one version of the planula hypothesis, beginning with a colonial choanoflagellate that evolved into planuloid organisms. These, in turn, led to sponges, coelenterates, and bilateral creeping turbellarians. (Adapted from Glaessner [1984], from Ivanov.)

Food-gathering surface is increased in a multicellular organism that can extend its cells to places that could not have been reached were it small and unicellular. This serves to ensure a more stable food supply to all of its cells even where food distribution is uneven and also allows multicellular organisms to attack and digest larger particles of food by secreting greater quantities of digestive enzymes than single cells can secrete. Multicellularity also permits the specialization of cells into various types of tissues that broaden the scope of protection, dispersion, food gathering, reproduction, and other functions.

According to these views, early metazoans were **pelagic** animals swimming above the sea bottom mainly by ciliary motion. At a subsequent evolutionary stage some became **benthic**, crawling along the ocean floor and feeding on accumulated detritus. Writers such as Clark have proposed that a number of further evolutionary steps would inevitably accompany a benthic existence. The scattering of food sources would give a selective advantage to organisms that could eat more food more rapidly, leading to an increase in size and to evolution of a mouth and gut that would permit selective digestion. Ciliary motion, by its nature slow and cumbersome for a large animal, would, as discussed later, be replaced by circular and longitudinal muscles enabling leechlike movement and pedal locomotion.

With the increased success and proliferation of bottom feeders, a niche would be opened for carnivorous animals in which speed of locomotion and development of a grasping mouth or other prehensile organs would be emphasized. However, just as improvements are selected in predators, means of defense and escape would be selected in prey. Competition among all the different varieties of prey and predators would thus lead to an explosive evolutionary radiation, generating a large variety of morphological forms and adaptive strategies.

THE COELOM

The **coelom**, an internal cavity between the ectodermal and endodermal tissues (as defined by Haeckel in 1872), was one of the most successful of the early metazoan adaptations. It is lined with an epithelium which often contains testes or ovaries and possesses ducts to the exterior used in the transmission of gametes or waste products. As shown in Figure 14-7, there are two types of coelomate phyla: **pseudocoelomates** (false coelomates), in which the coelom is derived from a persistent blastocoele and only partially lined with mesoderm, and **eucoelomates** (true coelomates), in which the coelom arises as a cavity within mesodermal tissue and is therefore completely lined with mesoderm. In many wormlike organisms, whether pseudocoelomate or eucoelomate, the coelom is filled with fluid enabling it, among other functions, to act as a **hydrostatic skeleton** that can transmit pressure from one part of the body to another. Thus, there is considerable improvement in the efficiency of peristaltic motions in coelomate animals over non-coelomate, since waves of circular and longitudinal muscle contraction can be transmitted more easily through the hydrostatic skeleton (see p. 283). This enables undulatory swimming movements as well as improved burrowing activity—that is, improvement in both speed of capture and speed of escape.

Unlike animal multicellularity, the coelom, at least among eucoelomates, may have originated only once. The 18*S* RNA studies by Field et al. cited previously show a close relationship among the origins of all eucoelomate phyla examined: annelids, arthropods, molluscs, echinoderms, chordates, and others. These lineages appear to have diverged from a common Precambrian coelomate ancestor within a relatively short space of time (see Fig. 15-12). Since the 18*S* RNA data provide, so far, the most reliable basis for comparing such essentially different groups, the concept of a monophyletic origin of coelomate phyla can be considered to have molecular support.

Nevertheless, the problem of how the coelom originated developmentally is still unsolved and has been debated since Haeckel's time. Hypotheses range from coelomic-type pouches in the gonads of the coelomic ancestor (gonocoel hypothesis) to sacs within nephridial excretory organs (nephrocoel hypothesis) to outpocketings from the gastric cavity (enterocoel hypothesis) to intercellular fluid-filled cavities within mesodermal structures (schizocoel hypothesis). Arguments for and against each of these hypotheses are discussed by Clark, and there is evidence that at least the last two modes of development (**enterocoely**, **schizocoely**) are found in coelomate phyla, thus offering support to those who believe in a polyphyletic coelomate origin (Valentine).

In whatever manner the coelom originated, it is apparent that it conferred an important advantage in some lineages by providing a mechanical hydrostatic function. As these lineages evolved, some made use of the coelom in unsegmented form (e.g., priapulids), whereas others adopted various segmental organizations (see Table 15-1) that had profound effects on their future evolution.

METAMERISM

The serial segmentation of the body along an anterior-posterior axis, called **metamerism**, is found

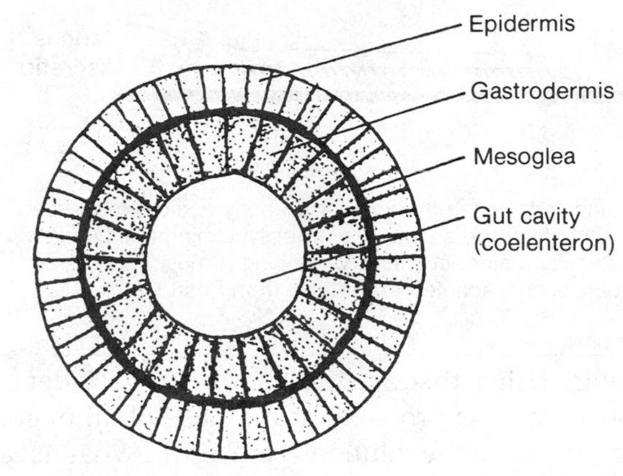

(a) Diploblastic coelenterate

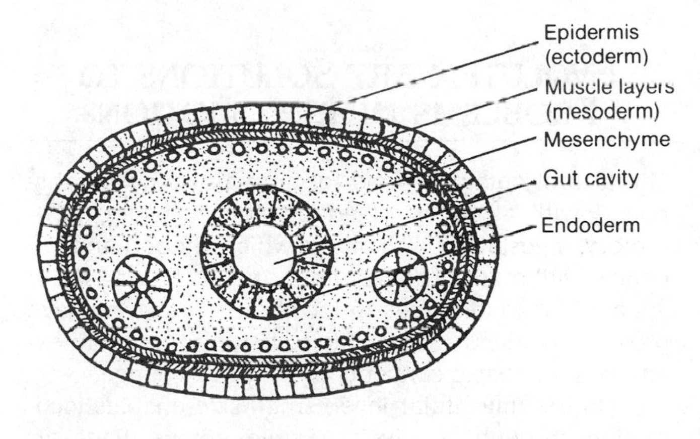

(b) Triploblastic acoelomate

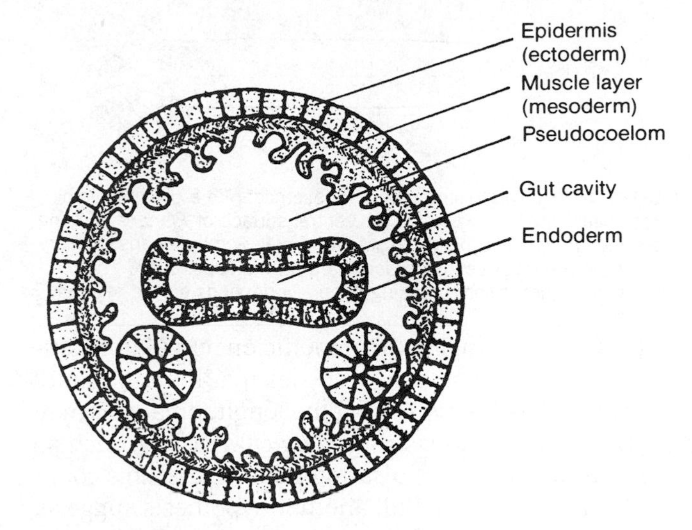

(c) Pseudocoelomate

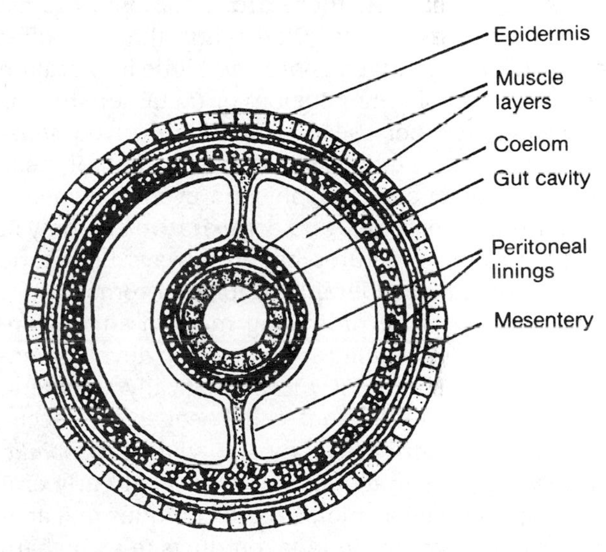

(d) Eucoelomate

FIGURE 14-7 Diagrammatic illustrations of general kinds of metazoan body cavities. (a) Diploblastic body plan found in coelenterates such as *Hydra*. (b) Triploblastic plan in which the coelom is absent (acoelomate), as can be observed in platyhelminths. (c) Pseudocoelomate plan observed in various aschelminth phyla in which the body cavity is only partially lined with mesoderm. (d) Eucoelomate body plan found in phyla such as arthropods, annelids, chordates, and echinoderms. (Adapted from Kershaw.)

in a variety of metazoan phyla including some coelenterates, platyhelminths, annelids, arthropods, chordates, and others. Often organ systems such as nephridia, gonads, and nerve ganglia are repeated within each segment, or metamere, and the segments are also marked by constrictions of the body wall musculature and by the repetition of coelomic cavities. Although there are no existing animals which possess identical segments throughout, since the head and anal segments differ from other metameres in all known cases, some animals such as the polychaete annelids show remarkable identity among many of their segments. There are two general kinds of segmentation: mesodermal, beginning in the mesoderm and proceeding from the interior of the animal outwards, such as in annelids, arthropods, and chordates; and superficial, which begins externally from the cuticular surface and then proceeds inwardly, often involving only the body wall musculature, as found in the Acanthocephala and other Aschelminthes phyla.

Like the origin of the coelom, a variety of hypotheses have attempted to explain the origin of metamerism. For example, one hypothesis points to the fact that many internal organs such as nerve ganglia, gonads, and excretory organs are serially repeated in some pseudometameric animals in which body wall segmentation is absent (platyhelminth turbellarians and nemerteans). The hypothesis suggests that the development of metameric organization in such animals would allow simple organ replacement if the animal is injured; that is, a nearby intact organ could replicate itself and thereby replace an adjacent injured or missing organ. On the other hand, the advantage of pseudometamerism might be to provide multiple excretory

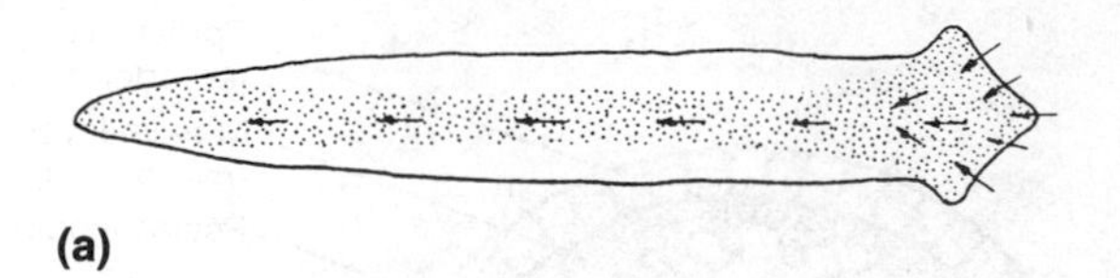
(a)

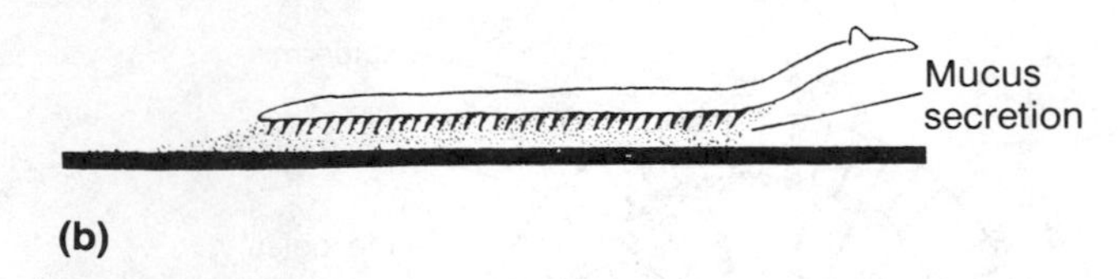

(b)

FIGURE 14-8 Ciliary distribution and locomotion in a platyhelminth turbellarian flatworm, *Planaria.* (a) Ventral surface of *Planaria* showing the direction of ciliary beats. (b) Mode of creeping by means of ciliary beating on a mucus secretion deposited on the substratum. The efficiency of such ciliary creeping generally depends on the relatively small size and flattening of the turbellarian body in order to present as large a ventral surface as possible. In platyhelminths with larger and more circular dimensions, ciliary creeping is mostly, if not entirely, abandoned. (Adapted from Clark, from Pearl.)

organs to an animal with an inefficient circulatory system. Another hypothesis proposes that the uniformity of mesodermal growth along the longitudinal axis may be broken for various embryological reasons, such as the introduction of a pulsating pattern, and this could lead to metamerism. Still another hypothesis suggests that the segmentation of muscular tissue originated from improved undulatory swimming motions conferred upon a flexible animal.

For the most part, there are arguments against each of these hypotheses (Clark), and they all suffer from a fossil record that has provided little information on the evolution of either coelom or metamerism. It is still unclear whether these structures evolved separately or together, and it is still debated whether these were monophyletic or polyphyletic events. Evidence derived from embryology and comparative anatomy in respect to these structures cannot always be clearly interpreted, and considerable ambiguity remains.

Among the more promising modern studies are those which have revealed segmentation developmental patterns that are not morphologically apparent, based on tracing individual cell lineages in insects. These studies, performed mostly with *Drosophila*, make use of genetic techniques that label early embryonic cells as well as molecular techniques that analyze DNA and its transcriptional products (e.g., Ingham et al.). Thus, it is now known that specific genes exercise control over segmental patterns, segmental borders, and the ability of segments to differentiate into particular structures such as wings and legs (see Fig. 10-25). Interestingly, a fairly long DNA sequence of about 180 nucleotides (called the homeo box) recurs in various *Drosophila* genes that regulate segmentation, and homologous sequences have also been found in annelids, vertebrates, and some other segmented animals (McGinnis et al.). Whether such findings indicate a common segmental ancestry for these metazoans has yet to be determined.

A somewhat different approach towards gaining an understanding of at least some aspects of the evolution of various body structures is to study their selective advantages in terms of locomotion and function. Such studies, although theoretical, are applicable to a wide variety of organisms which face similar environments and occupy similar roles within these environments. Using this approach, we can consider a body form that was probably at the base of all major forms of metazoan evolution—a worm. What is a worm, and how did it evolve?

EVOLUTIONARY SOLUTIONS TO PROBLEMS OF LOCOMOTION

The task of obtaining food for animals is inextricably bound with a variety of adaptations; sensory, locomotory, ingestatory, and others which support and enhance fulfillment of this primary need. Among these adaptations, differences in locomotory behavior have provided workers such as Clark the opportunity to examine some basic concepts of adaptive change.

On the unicellular level, small size enables locomotion through relatively simple ciliary, flagellar, pseudopodial, or even Brownian motion. Once the metazoan grade of organization is reached, locomotory cells face the problem of moving relatively large masses in concerted activity. The earliest of metazoan animals, perhaps planula-like organisms, probably moved by ciliary activity not unlike the motion of the small acoelan turbellarians. The acoelans, many of a size no larger than a millimeter, use ciliary cells on their ventral surface for creeping, and some degree of swimming can also be attained this way. Motion in a directional fashion quickly confers an anterior-posterior orientation upon the animal, making the anterior portion more concerned with those adaptations necessary for both sensing and confronting the environment being entered. **Bilaterality**, or the distinction between right and left sides, is an immediate consequence of a dorso-ventral, anterior-posterior anatomy and leads to opportunities for organizational complexity. However, as animals grow larger, the use of cilia alone limits more rapid locomotion because of the relatively small forces cilia can generate (Fig. 14-8), and a range of other methods are employed, all dependent on organized muscular tissues.

In its simplest form, tissue organization in triploblastic animals takes the shape of a **worm**, which can be defined as a long, flexible tube of constant volume enclosed by a muscular body wall. To allow coordinated activity, the muscle tissue is organized into two major groups: circular muscles whose contraction reduces the diameter of the animal and increases its length, and longitudinal muscles that contract with opposite effect by reducing length and increasing diameter. The opposed activity of these muscle tissues means essentially that for one type of muscle to extend, the other must contract. Also, the effect caused by a localized muscle contraction depends strongly upon what other muscles do: if a circular muscle contracts, then a neighboring area will expand unless its circular muscles also contract. If a longitudinal muscle contracts, then the animal will flex in that direction should the longitudinal muscles on the other side of the body be relaxed. Limitations on the extent of movement depend upon the size of the muscles, their locations, attachments, and the degree to which the body wall can be distorted (e.g., the deformability of the cuticular basement membrane).

In many of the platyhelminth turbellarians that have reached sizes much larger than the acoelans, locomotory movements are almost entirely transmitted through a pedal longitudinal "foot." Pedal locomotory waves arise by contraction and relaxation of those ventral longitudinal muscles that are in contact with the surface. This somewhat inefficient creeping mechanism allows the locomotion of animals whose bodies are mostly solidly filled with cells, since muscular effects on the body wall are restricted to relatively short distances.

As fluid accumulates in either cells or sinuses within the wormlike body, effects of muscular changes can be transmitted through greater distances and improve locomotion. For example, although a coelom is lacking in the ribbon worms (phylum Nemertea), some of these animals possess a gelatinous parenchyma that furnishes a relatively long, fluid skeleton, allowing the effects of muscular contractions to be transmitted more easily than through solid tissue. Undulatory swimming movements can then occur, produced by contraction of longitudinal muscles on opposite sides of the body. Furthermore, in some nemerteans we see the early signs of **peristaltic movements** that are to become an important feature of animals possessing a true fluid-filled coelom.

Essentially, the coelomate condition of a continuous body cavity represents a significant evolutionary advance in providing a fluid skeleton that eliminates cellular barriers to hydrostatic pressure. Because of this, the effects of contractions in one part of the body can be immediately transmitted to other parts. Peristaltic motion, defined as a wave of circular muscle contraction followed by longitudinal muscle contraction, can then generate much larger forces than in acoelomates since the entire coelomic hydrostatic skeleton and all of the body wall musculature is involved. For example, peristalsis adds adaptive value to a burrowing animal by enabling the entire circumference of the body to be used in thrusting through the substrate. Furthermore, a fluid-filled coelom also enables the rapid eversion of a proboscis or lophophore by simple hydrostatic pressure, as can be seen in the rapid burrowing movements of priapulids (Fig. 14-9) and the extension and withdrawal of the tentacular polypide in animals such as ectoprocts (Fig. 14-10).

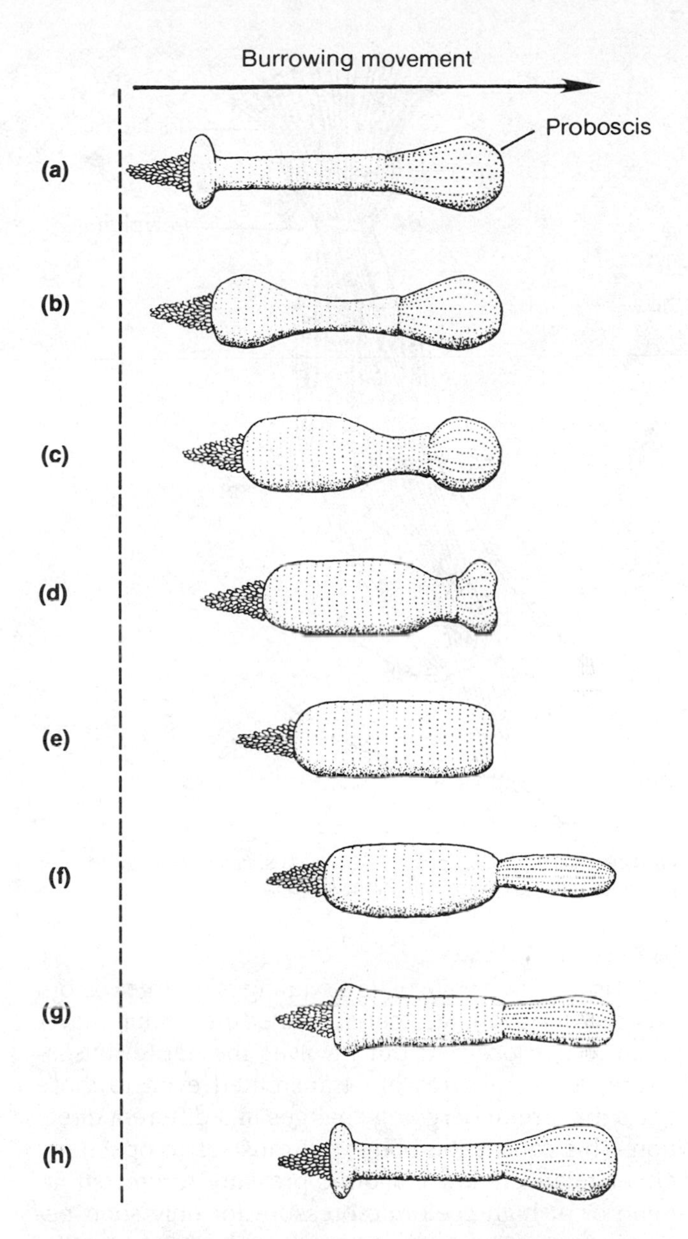

FIGURE 14-9 Stages in the burrowing activity of a *Priapulus* worm beginning with (a) lengthwise extension of its body and enlargement of its proboscis which serves as an anchor, allowing the animal to move anteriorly (to the right). The length of the animal then contracts by increasing in diameter (b-e), and the proboscis becomes enclosed. In stages (f-h) the proboscis is extended again, and the animal elongates to repeat the cycle. (Adapted from Clark, from Friedrich and Langeloh.)

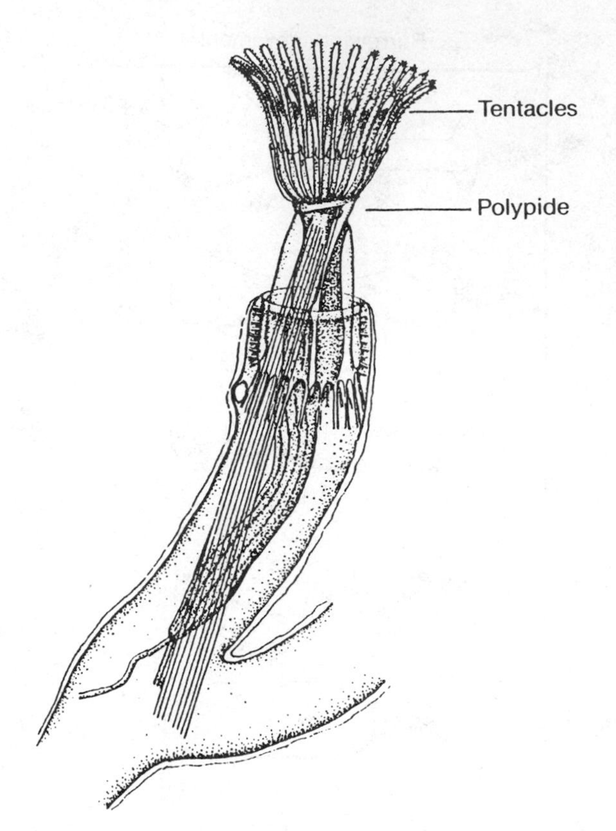

FIGURE 14-10 The ectoproct, *Fredericella sultana*, with everted polypide. (Adapted from Clark, from Allman.)

However, in spite of its advantages, a large coelom has the disadvantage that sustained peristaltic movement is not localized but involves the entire musculature, and pressures are transmitted even to those parts that are undergoing changes in a different direction. This causes the body wall muscles to operate at either relatively low coelomic pressure to prevent fatigue or at high coelomic pressure for only short periods of time. Thus unsegmented coelomic worms like sipunculids and echiuroids that cannot localize their hydrostatic pressures are generally slow-moving and relatively sedentary. Segmentation of the body by metamerism is therefore a mechanism which allows the localized establishment of pressure gradients and overcomes the generalized coelomic pressures that affect the entire musculature at the same time.

Selection for segmentation of the coelom appears to have taken at least two major directions. One was towards the **oligomerous** animals which have three main coelomic areas, the most anterior being an unpaired pocket, the protocoel, followed by a pair of mesocoels and a posterior pair of metacoels. Hemichordates, such as *Balanoglossus* (the acorn worm) and the pterobranchs (Fig. 14-11), as well as echinoderms and pogonophorans, have all three coelomic sacs to various degrees, whereas the protocoels mostly disappear in phoronids, ectoprocts, and brachiopods. In another evolutionary direction are metameric animals such as annelids which possess multiple body wall divisions that provide a smooth transition of the peristaltic wave. The sustained and efficient burrowing by oligochaete annelids such as earthworms is a direct consequence of their numerous segments (Fig. 14-12).

Once segmentation appeared, further locomotory adaptations rapidly evolved. In many of the polychaete class of annelids, motion occurs primarily by means of oarlike **parapodia** (Fig. 14-13), with consequent reduction in the circular muscles. The polychaete septa provide rigid attachment points for the parapodial muscles, enabling turgor in the parapodium so that it can be moved as a single unit. Nevertheless, in spite of its locomotory advantages, segmentation has drawbacks, since each segment, separated from its neighbor by a septum, must have its own set of organs, such as nerve ganglia, nephridia, gonads, and musculature. It is therefore inevitable that the numbers and kinds of segments, as well as the coelomic cavities, become modified or reduced with changes in habit or function. For example, although arthropods obviously originated

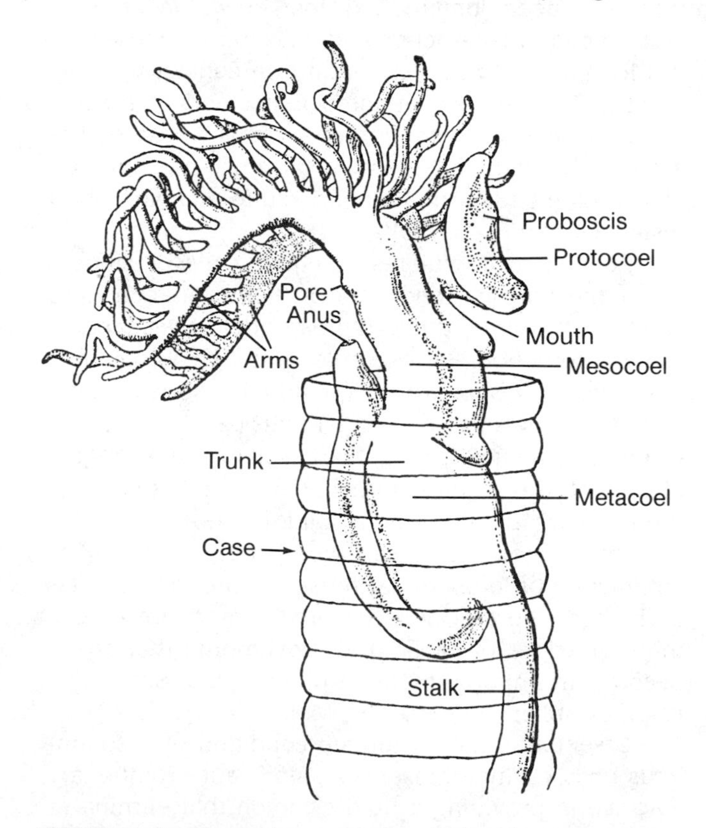

FIGURE 14-11 *Rhabdopleura*, a species of colony formers (pterobranchs) of the phylum Hemichordata, animals that are believed to be evolutionarily related to chordates. There are three coelomic regions in these individuals: the proboscis, or cephalic shield, contains the protocoel; the collar possesses a pair of mesocoels that extend into the tentacle-bearing arms; and the trunk contains the paired metacoel cavities. (Adapted from Borradaile et al.)

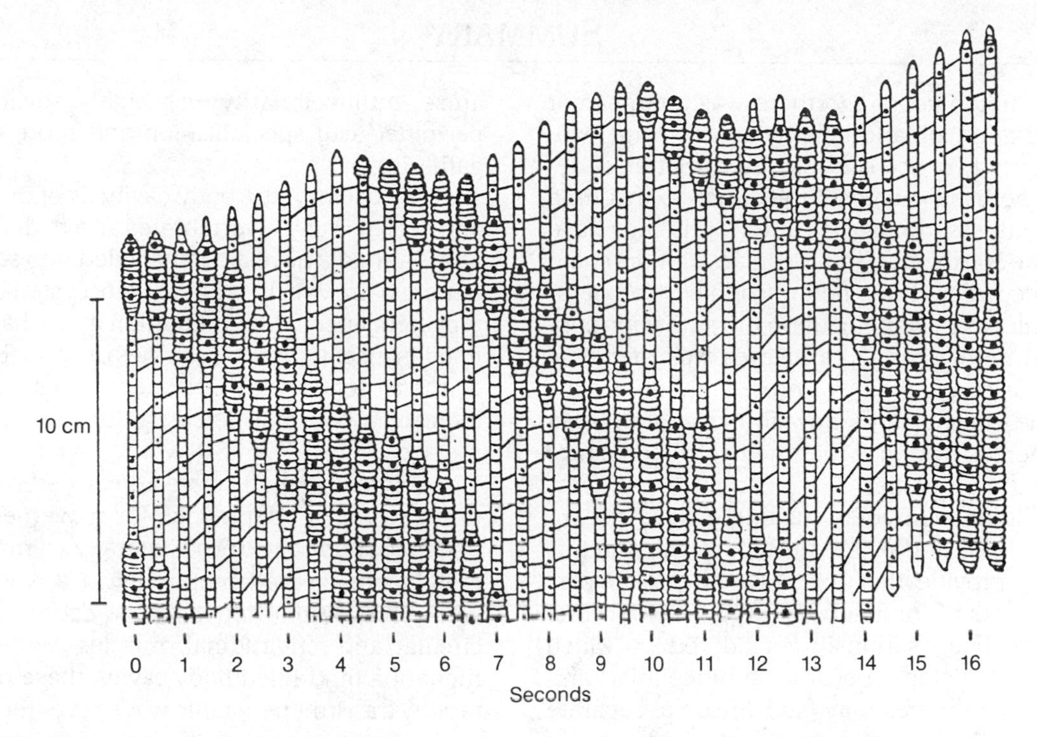

FIGURE 14-12 Peristaltic motion of burrowing earthworms as observed by changes in their segmental diameters. Contraction of longitudinal muscles in a group of segments causes widening of these segments and enables the animal to wedge against the sides of the burrow. Segments behind this group are then pulled up by further longitudinal muscle contraction, shortening the body and increasing the number of segments in the "anchor." Some of the widened segments in the anchor now undergo circular muscle contraction, causing elongation of the body and extension of these segments in a forward direction. These elongated segments contract and widen in turn, and the peristaltic cycle is repeated. Connecting lines indicate the relative motion of particular segments. (Adapted from Clark, from Gray and Lissmann.)

from annelidlike ancestors, their partitioned skeleton has become rigid, and localized muscular movement can now occur in the absence of either coelom or septa.[2]

Evolution among a variety of invertebrates, ranging from acoelomates to coelomates, will be discussed in the following chapter.

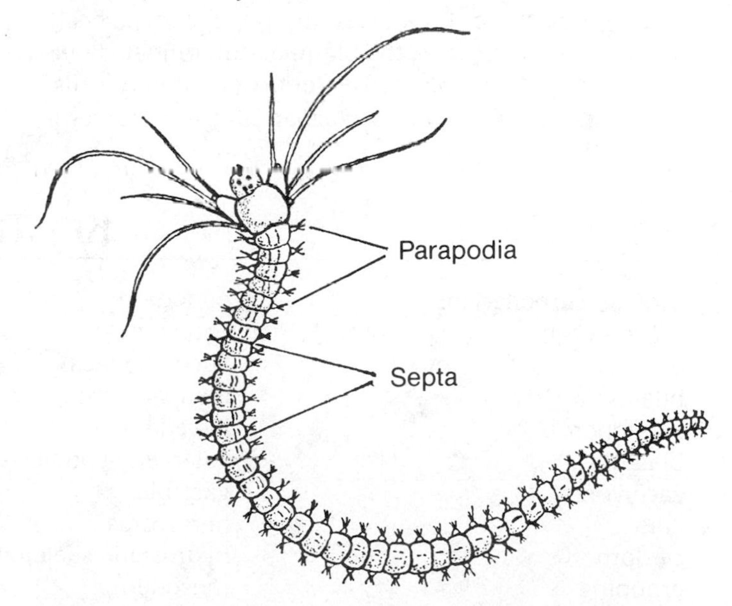

FIGURE 14-13 One of the free-swimming nereid polychaete annelids, *Platynereis.* (Adapted from Smith.)

[2]In vertebrates, a group that also possesses a metameric body plan (Chapter 16), the selective value of segmentation appears to derive from their increased swimming efficiency, given an axial skeleton. That is, vertebrates with an axial notochord can best achieve undulatory swimming movements by applying longitudinal muscular contractions to small sections of the axial skeleton. In early chordates whose spinal columns were most likely similar to the notochords of tunicate larvae (see p. 323), these muscles probably inserted directly on the notochord itself. This was then followed by segmentation of the spine through the occurrence of vertical septa at repeated intervals (as seen in the ammocoete larva of the lamprey *Petromyzon*), since such organization would provide increased mechanical advantage for longitudinal muscles. At some point in the evolution of fish, these vertical septa were replaced or became transformed into inclined septa (myocommata) which allow transmission of both longitudinal and lateral forces to the long axis of the animal. Because of this evolutionary history it is important to note that the segmentation produced by the myocommata does not exactly correspond to the segmentation of the axial skeleton (the vertebrae): to enable smooth locomotory movement each myocomma has insertions on a number of adjacent vertebrae, and each vertebra bears insertions from two or more myocommata.

SUMMARY

Six hundred million years ago there was an apparent explosive radiation of animal phyla with hard exoskeletons, but why it occurred is still a matter of debate. One hypothesis suggests that prior to this time a homogeneous assemblage of autotrophs may have dominated the earth, but significant evolution could not have occurred until heterotrophic "croppers" opened new niches by preying upon the autotroph population. From herbivores to carnivores was then only a small succession of steps. For protection, organisms developed the exoskeletons now found as Cambrian fossils, but there is evidence that their soft-bodied precursors arose much earlier.

All unicellular eukaryotes, including algae, the animal-like protozoa, and a few fungi, are classified as Protista. They provide the link between prokaryotes and the kingdoms of multicellular organisms. Several pathways have been seriously considered by which heterotrophic protistans became multicellular animals. 1. Cell membranes may have arisen to separate the nuclei in a multinucleate protistan, producing an acoelous flatworm. 2. A proposal by Haeckel that flagellated colonial protistans became bilaterally symmetrical and developed a gut by means of invagination of "digestive" cells, thus producing a gastrulalike structure, the gastraea. 3. A presently popular hypothesis that a solid ball of cells, the planula, differentiated into interior digestive cells and exterior locomotory cells, as occurs in some lower metazoan larvae. However it arose, multicellularity was highly advantageous and permitted cell specialization and more efficient food gathering.

The coelom, the body cavity, is of uncertain origin and acts in many invertebrates as a hydraulic skeleton. Some animals have a body divided into segments (metameres) in which many organ systems are serially repeated. Specific segmentation genes have been identified in insects, and similarities of parts (homeo boxes) of these genes with DNA sequences in other phyla indicate a common ancestry for metamerism in the metazoa.

Originally small ciliated creatures, metazoans had to make many adaptations for movement when they enlarged and became bilaterally symmetrical. The early, three-layered metazoa had a wormlike shape, whose movement depended on coordinated layers of circular and longitudinal muscles. With the development of a fluid-filled body cavity, these muscles could rapidly transmit peristaltic waves over the entire length of the body, enabling effective burrowing activity. In segmented animals, hydrostatic pressure can be apportioned more efficiently to localized areas. Few animals are completely metameric, however, because it then becomes necessary to provide a set of organs for each segment. Further adaptations, especially in hard-bodied forms, led to changes in body plan, and both coelom and segments have become modified in most animal phyla.

KEY TERMS

acoelan turbellarians
archenteron
benthic
bilaterality
blastopore
blastula
carnivores
cilia
coelom
cropping
diploblastic
ectoderm
Ediacarian strata
endoderm
enterocoely
eucoelomates
feedback cycle
flagella
gastraea hypothesis
gastrula
herbivores
hydrostatic skeleton
mesoderm
metamerism
metaphyta
metazoan origin
multinucleate protozoans
oligomerous
parapodia
pelagic
peristaltic movements
planula hypothesis
Precambrian-Cambrian discontinuity
protista
protoflagellate
pseudocoelomates
schizocoely
triploblastic
worm

REFERENCES

Barnes, R. D., 1980. *Invertebrate Zoology*, 4th ed. W. B. Saunders, Philadelphia.

_____, 1985. Current perspectives on the origins and relationships of lower invertebrates. In *The Origins and Relationships of Lower Invertebrates*, S. Conway Morris, J. D. George, R. Gibson, and H. M. Platt (eds.). Clarendon Press, Oxford, pp. 360-367.

Baroin, A., R. Perasso, L-H. Qu, G. Brugerolle, J-P. Bachellerie, and A. Adoutte, 1988. Partial phylogeny of the unicellular eukaryotes based on rapid sequencing of a portion of 28*S* ribosomal RNA. *Proc. Nat. Acad. Sci.*, **85**, 3474-3478.

Berkner, L.V., and L.C. Marshall, 1965. On the origin of oxygen concentration in the earth's atmosphere. *Jour. Atmosph. Sci.*, **22**, 225-261.

Borradaile, L. A., F. A. Potts, L. E. S. Eastham, and J. T. Saunders, 1959. *The Invertebrata*, 3rd ed., revised by G. A. Kerkut. Cambridge Univ. Press, Cambridge.

Brasier, M. D., 1979. The Cambrian radiation event. In *The Origin of Major Invertebrate Groups*, M. R. House (ed.). Academic Press, London, pp. 103-159.

Brock, T. D., 1973. Lower pH limit for the existence of blue-green algae: evolutionary and ecological implications. *Science*, **179**, 480-483.

Clark, R. B., 1964. *Dynamics in Metazoan Evolution*. Clarendon Press, Oxford.

Clarkson, E. N. K., 1979. *Invertebrate Palaeontology and Evolution*. Allen & Unwin, London.

Cloud, P., 1976. Beginnings of biospheric evolution and their biogeochemical consequences. *Paleobiology*, **2**, 351-357.

Cloud, P., J. Wright, and L. Glover III, 1976. Traces of animal life from 620-million-year-old rocks in North Carolina. *Amer. Sci.*, **64**, 396-406.

Dobzhansky, Th., F. J. Ayala, G. L. Stebbins, and J. W. Valentine, 1977. *Evolution*. W. H. Freeman, San Francisco.

Dougherty, E. C. (ed.), 1963. *The Lower Metazoa*. Univ. Calif. Press, Berkeley.

Field, K. G., G. J. Olsen, D. J. Lane, S. J. Giovannoni, M. T. Ghiselin, E. C. Raff, N. R. Pace, and R. A. Raff, 1988. Molecular phylogeny of the animal kingdom. *Science*, **239**, 748-753.

Glaessner, M. F., 1983. The emergence of metazoa in the early history of life. *Precambrian Res.*, **20**, 427-441.

_____, 1984. *The Dawn of Animal Life*. Cambridge Univ. Press, Cambridge.

Gray, J., and H. W. Lissmann, 1938. Studies in animal locomotion. VIII. The earthworm. *Jour. Exp. Biol.*, **15**, 506-517.

Hadzi, J., 1963. *The Evolution of the Metazoa*. Macmillan, New York.

Hanson, E. D., 1977. *The Origin and Early Evolution of Animals*. Wesleyan Univ. Press, Middletown, Conn.

Hardy, A. C., 1953. On the origin of the metazoa. *Jour. Microscop. Sci.*, **94**, 441-443.

House, M. R. (ed.), 1979. *The Origin of Major Invertebrate Groups*. Academic Press, London.

Hyman, L., 1940. *The Invertebrates: Protozoa Through Ctenophora*. McGraw-Hill, New York.

_____, 1951. *The Invertebrates: Platyhelminthes and Rhynchocoela, the Acoelomate Bilateria*. McGraw-Hill, New York.

Ingham, P. W., K. R. Howard, and D. Ish-Horowicz, 1985. Transcription pattern of the *Drosophila* segmentation gene *hairy*. *Nature*, **318**, 439-445.

Ivanov, A. V., 1968. *The Origin of Multicellular Animals* (in Russian). Nauka, Leningrad.

Jagersten, G., 1972. *Evolution of the Metazoan Life Cycle*. Academic Press, London.

Kerkut, G. A., 1960. *Implications of Evolution*. Pergamon Press, Oxford.

Kershaw, D. R., 1983. *Animal Diversity*. University Tutorial Press, Slough, Great Britain.

McGinnis, W., R. L. Garber, J. Wirz, A. Kuroiwa, and W. J. Gehring, 1984. A homologous protein-coding sequence in *Drosophila* homeotic genes and its conservation in other metazoans. *Cell*, **37**, 403-408.

Olson, E. C., and J. Robinson, 1975. *Concepts of Evolution*. C. E. Merrill, Columbus, Ohio.

Sleigh, M. A., 1979. Radiation of the eukaryote Protista. In *The Origin of Major Invertebrate Groups*, M. R. House (ed.). Academic Press, London, pp. 23-54.

Smith, J. E. (ed.), 1971. *The Invertebrate Panorama*. Universe Books, New York.

Stanley, S. M., 1973. An ecological theory for the sudden origin of multicellular life in the late Precambrian. *Proc. Nat. Acad. Sci.*, **70**, 1486-1489.

Trueman, E. R., 1975. *The Locomotion of Soft-Bodied Animals*. Arnold, London.

Valentine, J. W., 1989. Bilaterians of the Precambrian-Cambrian transition and the annelid-arthropod relationship. *Proc. Nat. Acad. Sci.*, **86**, 2272-2275.

Whittaker, R. H., 1969. New concepts of kingdoms of organisms. *Science*, **163**, 150-160.

15
Evolution Among Invertebrates

Perhaps 90 percent or more of all present metazoan species do not possess vertebral axial skeletons. These **invertebrates** range from jellyfish to worms, squids, starfish, shrimp, flies, and myriad other forms and adaptations. In general, aside from parasitic invertebrates which have tended to become smaller and less differentiated during evolution, free-living invertebrates have become more complicated, with each innovation often helping the animal gain further control over its particular environment by enabling it to become a better burrower, crawler, swimmer, or food selector. Such adaptations often involve many organs (integument, muscles, nerves, etc.) that act in concert and form part of the architectural framework, or **body plan**, of the animal in dealing with its specialized problems of survival and reproduction.

It is difference in body plans which provides the basis for separating invertebrates into more than 30 different present phyla, some of which are also grouped together because they share developmental as well as morphological features. Thus, as discussed in Chapter 14, the absence of a coelom separates the Platyhelminthes from many other phyla. Similarly, the radial organization of the coelenterates (Cnidaria, Ctenophora) separates them from bilaterally organized phyla. On the embryological level, as described later, the fate of the blastopore, whether it becomes mouth or anus, separates the **protostome superphylum** from the **deuterostome superphylum**. Using such criteria, one approach towards classifying metazoan phyla is shown in Table 15-1.

From an evolutionary point of view, these and other morphological and developmental differences undoubtedly reflect some progressive stages in the ancestry of these organisms. Figure 15-1 therefore provides only one of a sampling of proposals offered to relate various phyla evolutionarily. Since so much is still conjectural and a detailed review of possible phylogenetic schemes for all 30-odd invertebrate phyla would be beyond the scope of this book, only some aspects are discussed here.

Porifera (Sponges), Placozoa, and Mesozoa

Sponges are among the most primitive metazoan phyla. They possess a simple body plan (Fig. 15-2) which extracts food particles from water currents and

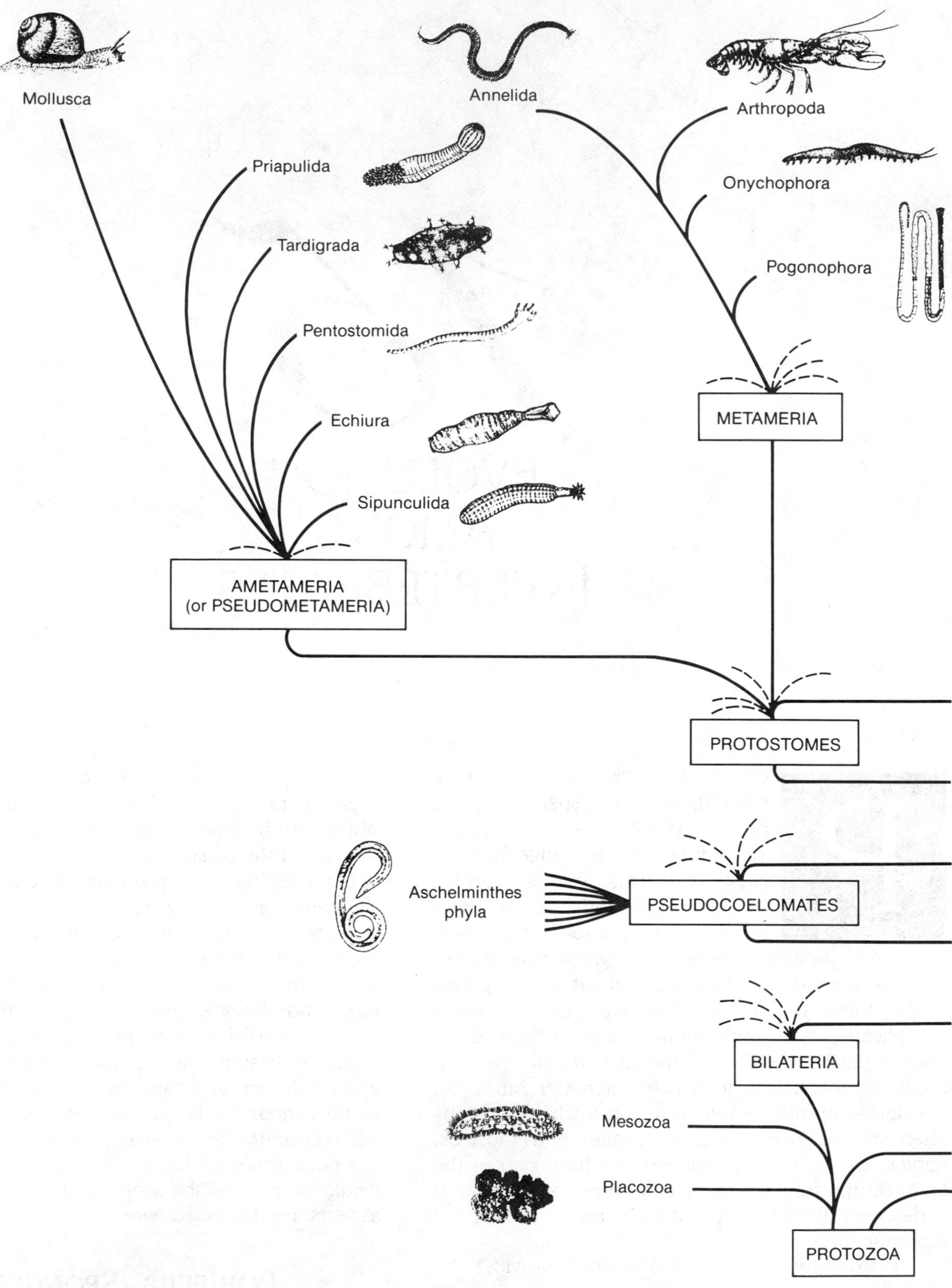

FIGURE 15-1 A diagrammatic arrangement of metazoan phyla showing possible phylogenetic relationships, along with an illustrated sample species for each phylum.

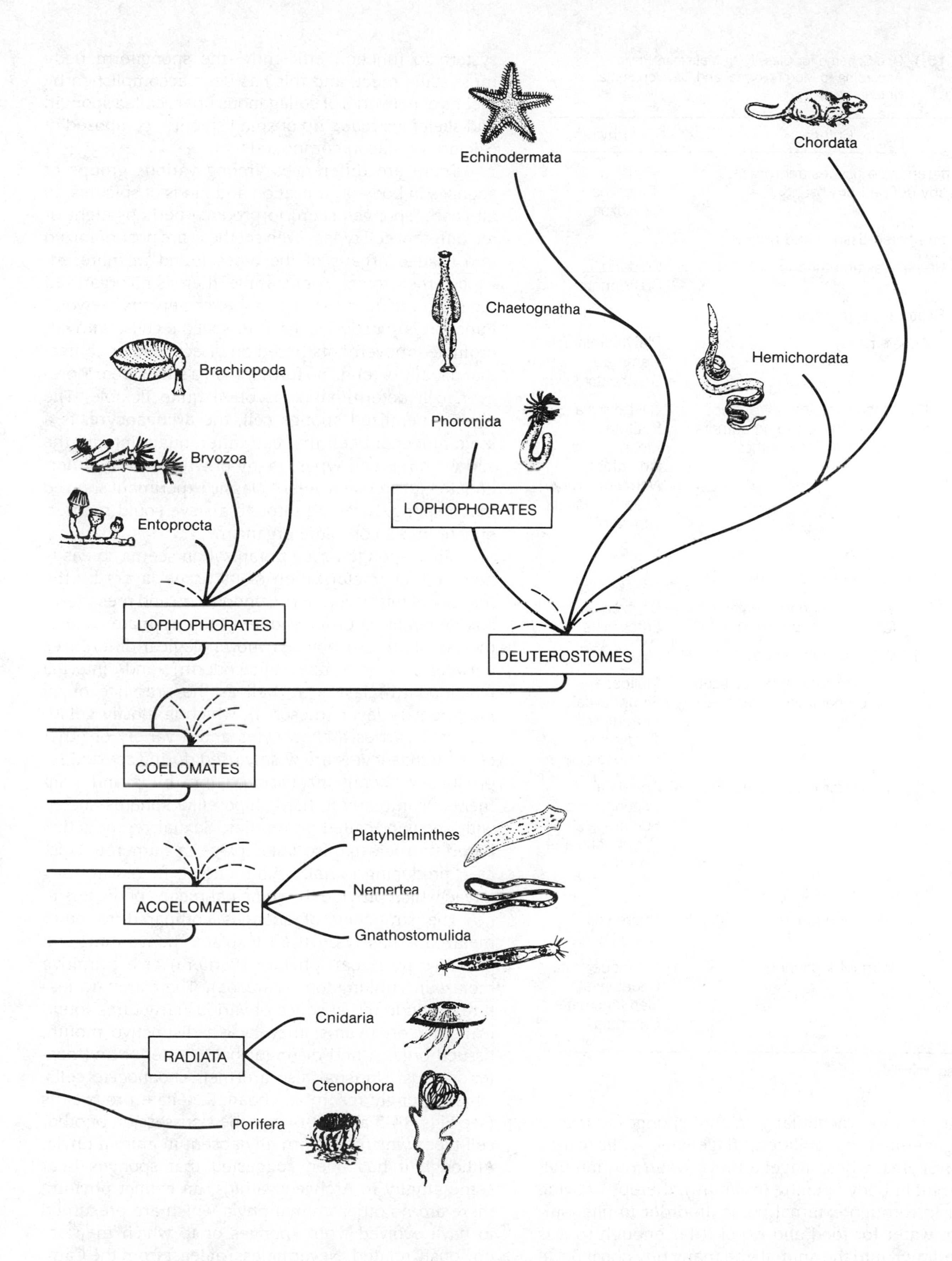
Echinodermata
Chordata
Chaetognatha
Hemichordata
Brachiopoda
Phoronida
Bryozoa
LOPHOPHORATES
Entoprocta
LOPHOPHORATES
DEUTEROSTOMES
COELOMATES
Platyhelminthes
Nemertea
ACOELOMATES
Gnathostomulida
Cnidaria
RADIATA
Ctenophora
Porifera

TABLE 15-1 One Scheme for Classifying Metazoan Phyla According to Morphological and Developmental Criteria

Criteria	Phylum
I. Differentiated tissues and organs poorly defined or absent	Porifera Placozoa Mesozoa
II. Differentiated tissues and organs	
A. Radially symmetrical	Cnidaria Ctenophora
B. Bilaterally symmetrical	
1. Acoelomates	Platyhelminthes Nemertea Gnathostomulida
2. Pseudocoelomates (grouped together by some authors as the phylum Aschelminthes)	Gastrotricha Rotifera Kinorhyncha Nematoda Nematomorpha Acanthocephala Loricifera
3. Coelomates	
a. Protostomes	
i. With lophophore (tentacled food-gathering crown)	Bryozoa Entoprocta
ii. Without lophophore	
(a) Non-metameric or pseudometameric organization	Mollusca Sipunculida Priapulida Tardigrada Pentastomida
(b) Metameric organization	Annelida Pogonophora Echiura Onychophora Arthropod
b. Deuterostomes	
i. With lophophore	Phoronida Brachiopoda
ii. Without lophophore	Echinodermata Chaetognatha Hemichordata Chordata

Adapted from Lutz.

digests them intracellularly. Within sponges, currents are generated by collared, flagellated cells called **choanocytes**, whose flagella move water out through an exhalant body opening (osculum), thereby drawing water in through external pores. In order to filter sufficient water for food and expel it far enough so it is not redrawn into the animal, the many tiny choanocyte flagella combine to produce a forceful exhalant current that exceeds more than 6 inches a second. For this system to function efficiently, the spongiform body must stand erect, and this has been accomplished by means of networks of collagenous fibers called spongin and skeletons made up of small spicules composed of calcium or silicon compounds.

There are differences among various groups of sponges in body organization and kinds of spicules. In all cases, sponges seem to produce perhaps eight or ten different cell types; even so, these are not organized into tissued organs of the types found in more advanced metazoans. For example, there is no organized digestive organ, muscular tissue, or nervous network. Function is mostly localized to specific cells, with coordinated movements based on direct cellular contact that usually extends no further than a small area. Moreover, cell determination is often quite flexible. The most generalized sponge cell, the **archeaocyte**, is a large amoeboid cell that can differentiate into all the various other cell types, many of which can dedifferentiate into archaeocytes. A classic experiment showed that a sponge strained through a sieve could redifferentiate into a complete organism.

Since sponge tissue organization seems so easily modified, characterization of its tissue layers by the traditional terms ectoderm, endoderm, and mesoderm has generally been considered inappropriate. Nevertheless, there are obvious morphological differences between its external (pinacoderm) and internal (choanoderm) layers as well as the presence of an intermediary layer (mesohyl), which is mostly gelatinous but carries archaeocytes and a variety of other cells. Archaeocytes are widely used during asexual reproduction, being incorporated into buds and fragments or into small, hardy, sporelike spheres coated with spongin, called gemmules. Sexual reproduction based on meiosis also takes place in numerous species, producing radially symmetrical, free-swimming larvae which provide the principal means of dispersal.

The simplicity of sponges compared to other metazoans has been used to place them in either a primitive metazoan phylum (Porifera) or a primitive metazoan subkingdom (Parazoa). Their archaic features include the absence of various structures found in higher organisms: they lack a distinctive mouth, tissued organs, and distinguishable anterior and posterior ends. Unusual also are their choanocyte cells, which strongly resemble choanoflagellate protozoans (see Figs. 14-3 and 14-6) and their possession of other cell types which are normally absent in animal phyla. Although it has been suggested that sponges bear some affinity to Archaeocyathids, an extinct phylum, there are no other animal phyla which are presumed to have evolved from sponges or to which they are obviously related. Nevertheless, at least from the Cambrian period onward, their simple body plan has apparently been a successful way of generating water

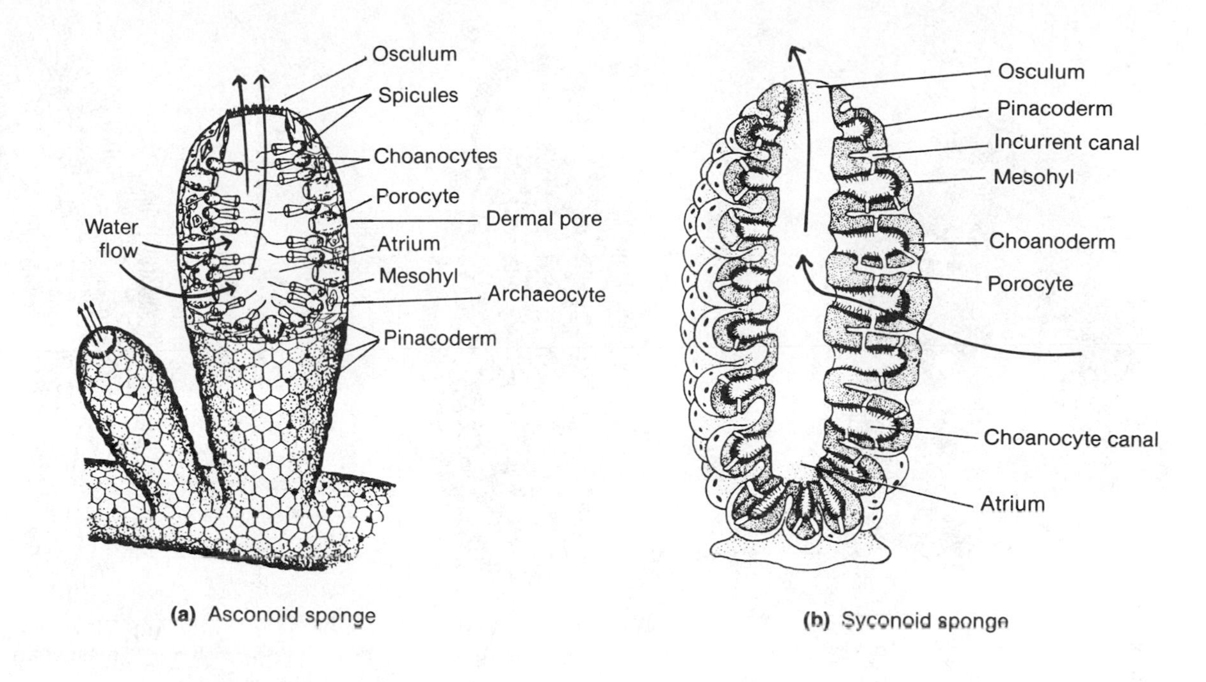

(a) Asconoid sponge

(b) Syconoid sponge

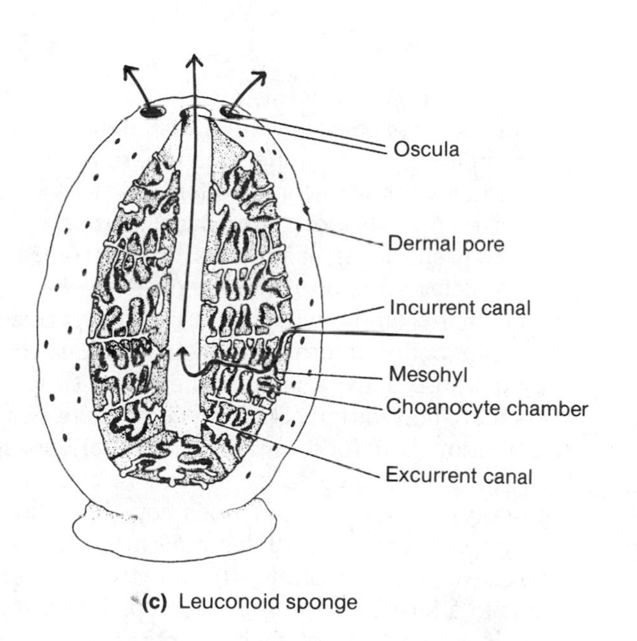

(c) Leuconoid sponge

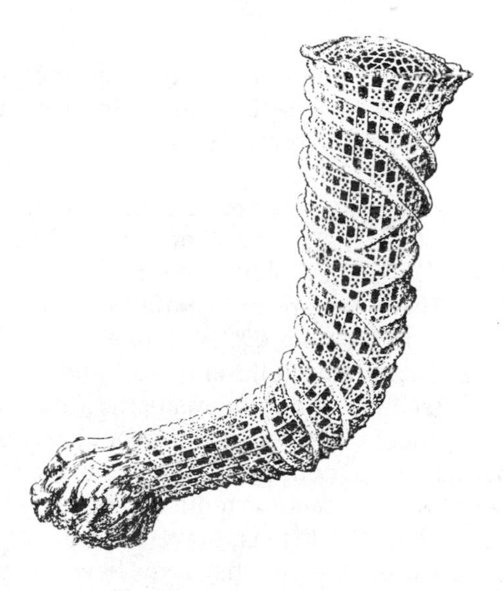

(d) Skeleton of hexactinellid "glass" sponge

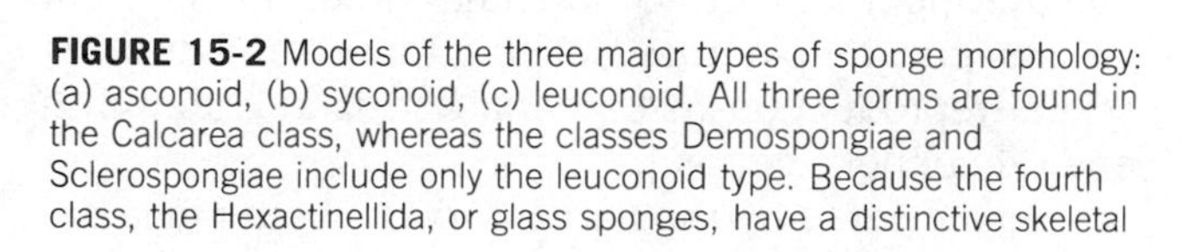

FIGURE 15-2 Models of the three major types of sponge morphology: (a) asconoid, (b) syconoid, (c) leuconoid. All three forms are found in the Calcarea class, whereas the classes Demospongiae and Sclerospongiae include only the leuconoid type. Because the fourth class, the Hexactinellida, or glass sponges, have a distinctive skeletal framework (d), do not possess the same type of pinacoderm found in other groups, and lack cell wall separation between many of their cells (syncytium), proposals have been offered that they be considered as a separate phylum (Bergquist).

currents for feeding, and they are still successful today, with about 5,000 species distributed widely from freshwater to marine areas and from shallow regions to great depths.

Morphologically more allied to other metazoa than the sponges are two phyla of very simple multicellular animals. One, called **Placozoa**, consists of a single known species, *Trichoplax adhaerans*, a flattened, free-living marine organism only a few millimeters in diameter (Fig. 15-3(a)). It has an upper and lower layer of flagellated epithelial cells enclosing a sheet of loose, fibrous mesenchymal cells. Its body can assume irregular shapes as it creeps along the substratum like an oversized amoeba, enveloping food particles and digesting them through its lower surface. Asexual reproduction occurs by fission and budding, and sexually

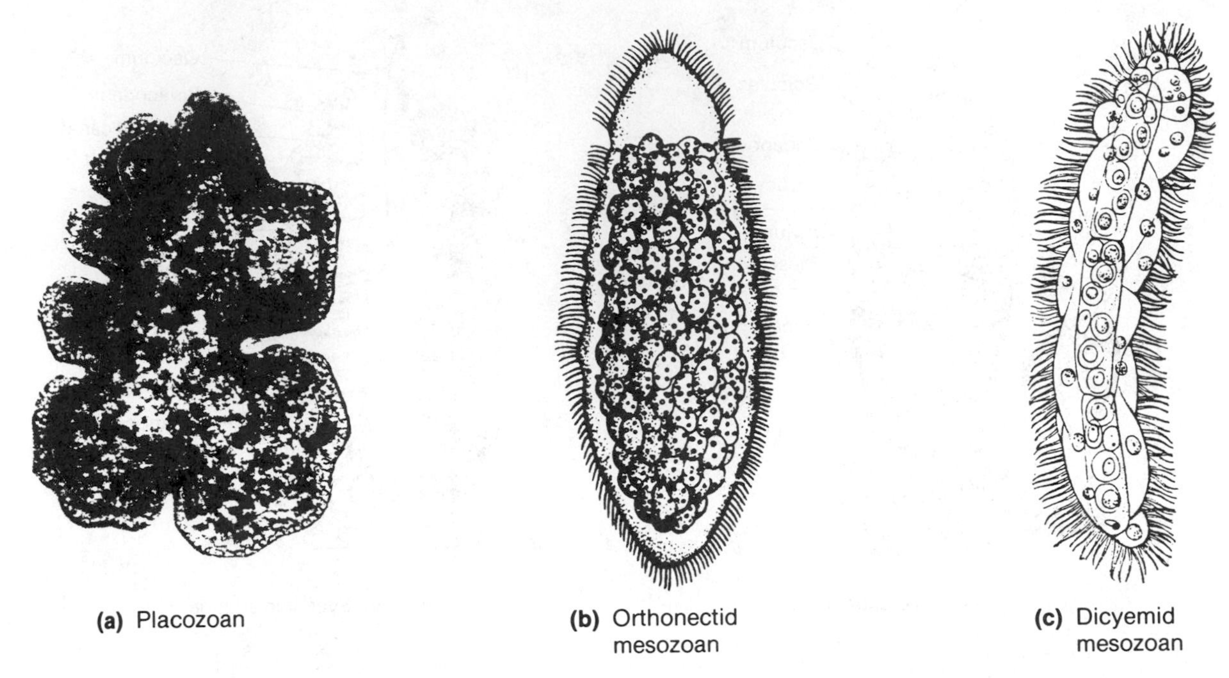

FIGURE 15-3 (a) *Trichoplax adhaerans*, the only known species in the phylum Placozoa. Adult stages of orthonectid (b) and dicyemid (c) members of the phylum Mesozoa.

produced eggs have been found in the mesenchyme. Although little else is known about this primitive metazoan, it is undoubtedly a remnant of a very early metazoan offshoot.

Also of simple body plan are members of the phylum **Mesozoa** (Fig. 15-3(b), (c)), of which about 50 species are known, all of which are parasites of marine invertebrates. These too are quite small but have apparently more complex life cycles than Placozoa, involving male and female differentiation and various larval stages. Because of their parasitism, it has been proposed that mesozoans are really degenerate platyhelminths that abandoned a free-living life-style, causing their tissues to become reduced and simplified in the process. Most biologists, however, still classify them as a separate phylum that may have become parasitic very early in its evolution. As yet, the relationship among Placozoa, Mesozoa, and other metazoans has not been determined.

RADIATA

The two **coelenterate** phyla, Cnidaria and Ctenophora, collectively called **Radiata**, show a major step forward in metazoan organization through possession of a mouth and a specialized gastrovascular digestive cavity (coelenteron). These expandable organs allow the ingestion of much larger food particles than can be filtered by sponges, permitting even entire prey organisms to be broken down extracellularly before they are absorbed intracellularly.

Tissue organization is also more advanced in radiates than in sponges, with a distinctively organized outer epidermis and inner gastrodermis, believed to be homologous with the ectodermal and endodermal layers, respectively, of more advanced metazoans. The middle tissue layer of radiates, the mesoglea, is primarily gelatinous, and most body functions are confined to the other two layers. Radiates possess, for example, both epidermal and gastrodermal muscular tissue whose activity is coordinated by simple nerve nets that allow various body movements to be used in locomotion and food capture. However, no specific cells or tissues devoted exclusively to circulatory, respiratory, or excretory purposes have been found.

Reproductively, radiates may utilize both sexual and asexual modes. Gamete formation, when it occurs, leads to a fertilized egg that develops into a solid, externally ciliated ball of cells, the **planula** (p. 278). Further development of the planula varies in different groups, but the planula itself is a universal feature of sexual reproduction in Cnidaria and is also found in one Ctenophora genus.

As indicated by the name Radiata, both phyla are radially organized so that almost any plane through the central oral axis of the animal cuts it into two approximate mirror-image halves. Also, both phyla are soft-bodied and use flexible tentacles to bring food to their extensible oral cavity. However, although this soft-bodied structure enables varied changes in shape—in some stages by use of the gastric cavity as a hydrostatic organ (see p. 280)—there are no hard parts upon which antagonistic muscles can operate; that is, there are no levers or fulcra which can amplify

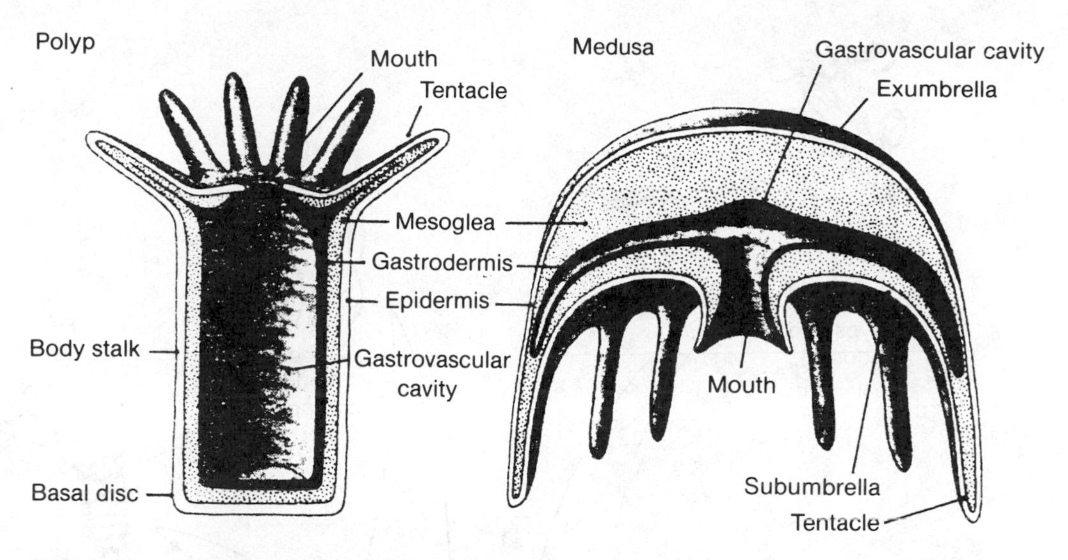

FIGURE 15-4 The two primary body forms found in Cnidaria. Note that these shapes are essentially inversions of each other, except that the intermediary mesogleal layer is usually thicker in the medusa form. (Adapted from Barnes.)

movements through the use of flexor and extensor muscles. Thus the speed of movement among coelenterates is relatively slow, and their muscles must function over a broader and less efficient range of contraction and expansion than in organisms with hard skeletal structures. Moreover, as the animals grow larger, mesogleal tissues increase, and body wall contractions must be transmitted through bulkier layers. Added to these limitations is the fact that the gastric cavity has an external oral opening which prevents its use as a hydrostatic organ when the animal is feeding.

Nevertheless, the Radiata have been successful throughout metazoan history, and they include about 10,000 described species. Present radiates are all carnivorous, although the method of food capture varies between the two phyla. In the more common of the two, the Cnidaria, the tentacular epidermis (and often also sections of the gastrodermis) is armed with specialized cells, cnidocytes, which contain miniature stinging, harpoon-like organelles called **nematocysts** that immobilize prey and allow them to be brought to the gastric cavity. This cnidarian feature, used for both offense and defense, apparently dates back to their early history and may have originated from glandular secretory cells since there are no obvious protozoan nematocyst homologues (Robson). However it arose, this unusually effective mode of food capture certainly helps account for cnidarian evolutionary persistence.

Many cnidarians also undergo developmental changes that produce either one or two different body forms, the **polyp** and the **medusa** (Fig. 15-4). The polyp is mostly a stationary (sessile) form with a tubular body. Tentacles surround the polypoid mouth at its oral end, and it is often attached to the substratum by a basal disc at its aboral end. The medusa, on the other hand, is usually a free-swimming form resembling an inverted umbrella-shaped polyp. Its concave undersurface bears a centrally located mouth surrounded by tentacles that hang down from the umbrellar margin. The close relationship between these two body forms can be seen in the fact that the medusa can become a polyp in some cnidarians when it attaches to a solid substrate. Generally, the polyp is used for stationary food-gathering and the medusa for dispersion, although they may each assume different importance in the various cnidarian classes. Thus, medusa are entirely absent in the Anthozoa (sea anemones and corals), polyps are absent in some of the Scyphozoa (jellyfish) and inconspicuous in others, but both forms can occur in some species of Hydrozoa (hydra and its various solitary and colonial derivatives) and Cubozoa (sea wasps).

Different views have been offered as to evolutionary relationships between these classes, but most zoologists agree that a Precambrian group capable of producing larval polyps and adult medusa gave rise to the various Cnidaria and also most probably to the Ctenophora (Fig. 15-5). The latter phylum, whose most typical forms are known as comb jellies, uses rows of ciliated plates (combs) for locomotion and mostly uses special adhesive cells (collocytes) for food capture. Although ctenophorans also differ from cnidarians in tentacle attachment and some other traits, many zoologists believe there is a close phylogenetic relationship between them.

PLATYHELMINTHES AND OTHER ACOELOMATES

Crossing the boundary from diploblastic to triploblastic animals meant an increase in the number and organizational complexity of mesodermal cells. These ad-

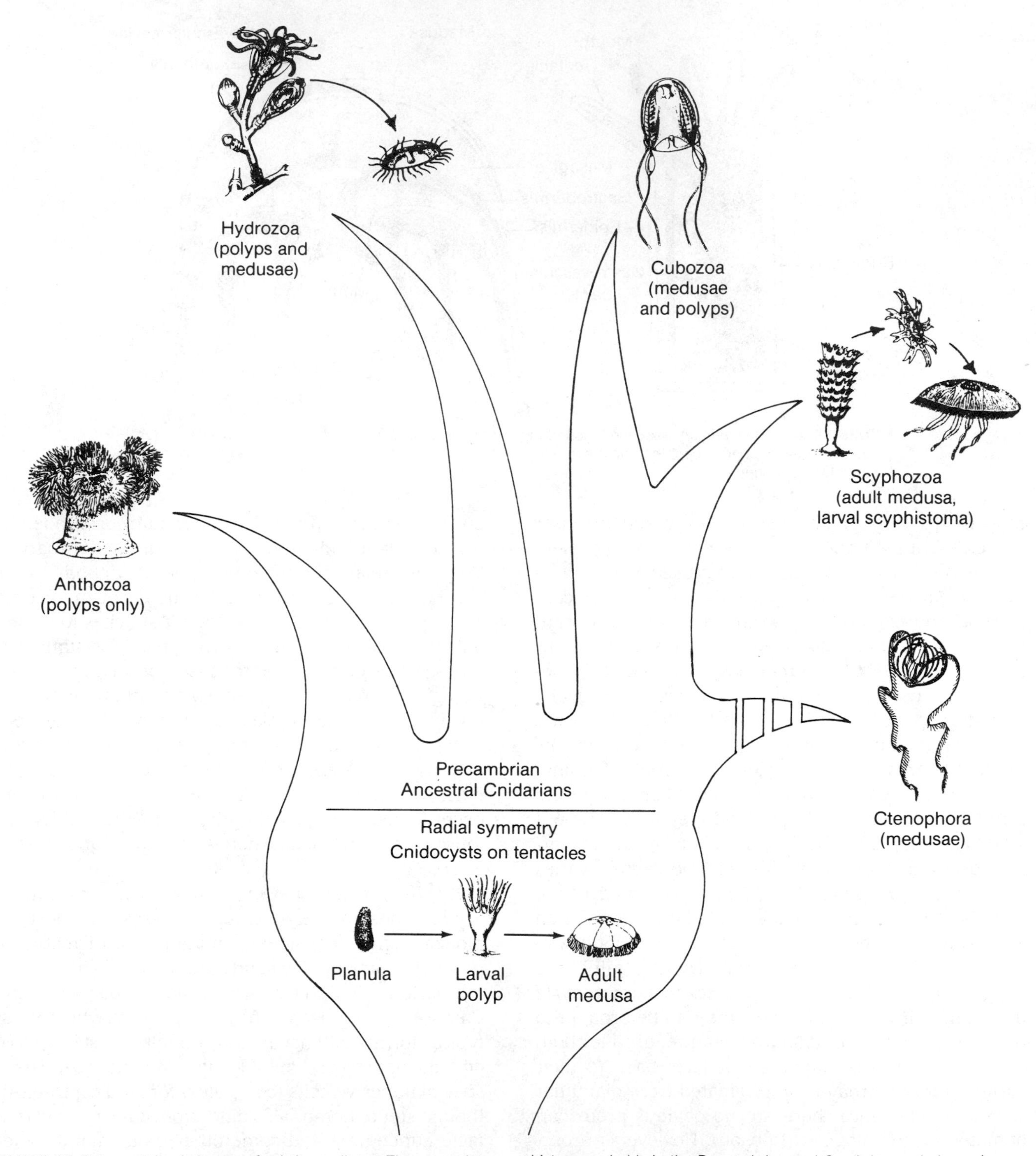

FIGURE 15-5 A possible phylogeny of existing radiates. The present classes of Cnidaria are believed to have arisen early in evolutionary history, probably in the Precambrian and Cambrian periods, and almost in parallel with origin of the phylum Ctenophora.

vances enabled a wide range of novel tissues and organs to be generated, such as bundles of circular and longitudinal muscles, excretory organs, circulatory channels and tissues, and complex reproductive systems. Along with these changes went an increase in organismal size and, together with active locomotion, an anterior-posterior orientation, or polarity, that aided food gathering and provided various animal groups with a bilateral symmetry

The phylum **Platyhelminthes** represents an early but successful stage in the triploblastic progression, comprising at present about 15,000 species. Platyhelminthes, also called flatworms, have a permanent mesodermal layer from which is derived muscular tis-

(a) Turbellaria: Order Acoela

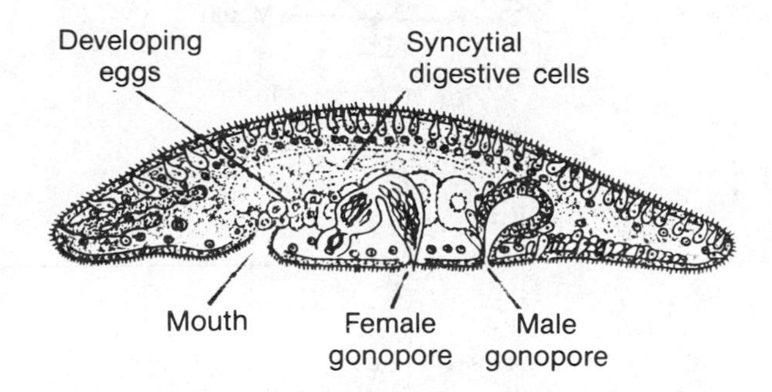

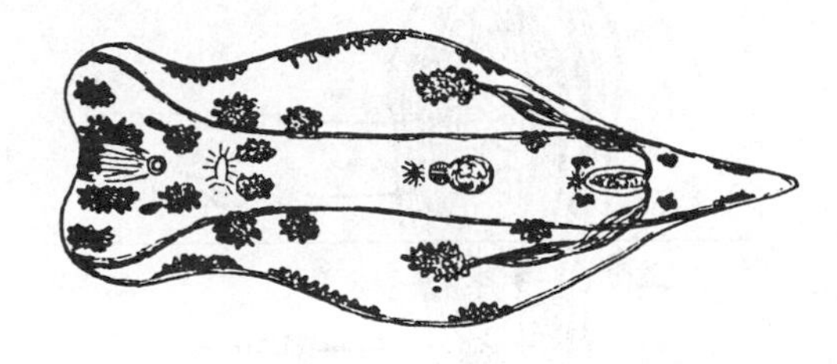

(b) Turbellaria: Order Prolecithophora

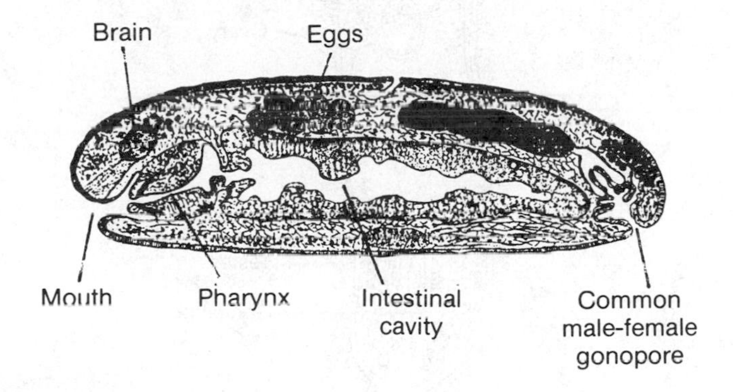

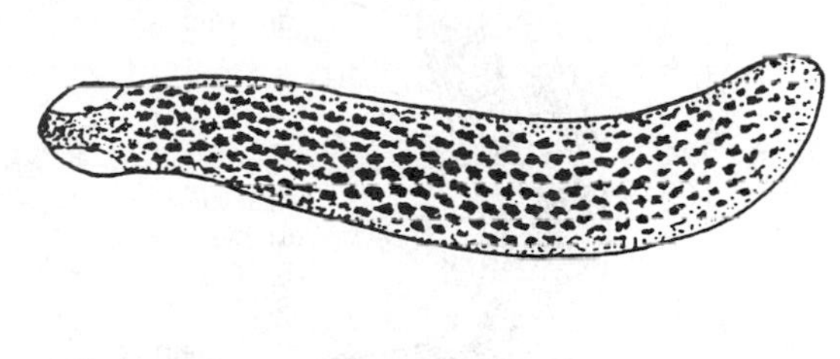

FIGURE 15-6 Median sagittal sections (left side) and dorsal views (right side) of genera from two different turbellarian orders. (a) An acoelan, *Convoluta*, showing the syncytial mass of digestive cells. (Adapted from Barnes.) (b) A freshwater turbellarian with muscular pharynx and gut cavity, *Hydrolimax*. (Adapted from Hyman, 1951.)

sue, an extensive hermaphroditic reproductive system, and relatively simple osmoregulatory organs (protonephridia). A circulatory system for gas exchange and excretion is absent but apparently not essential in these dorsoventrally flattened animals, whose interior cells are generally near either the external surface or an internal gut surface. Morphologically significant is the platyhelminth anterior-posterior organization with nervous and sensory structures concentrated at the cephalic end.

In the class Turbellaria, which includes the free-living flatworms, the mouth serves as both entrance and exit for the digestive organ. In the acoelan turbellarians this organ is a syncytial mass of cells (Fig. 15-6(a)), whereas in other turbellarian orders it is composed of one or more blind sacs similar to the coelenterate digestive cavity (Fig. 15-6(b)). In all cases spaces between the internal organs and body wall are filled with tissue, and there is no coelomic cavity. Generally, turbellarian locomotion is restricted to cilial movement and/or ventral (pedal) muscular creeping (see Fig. 14-8).

The other two classes of platyhelminths are entirely parasitic: the Trematoda (flukes) are like turbellarians in possessing a mouth and digestive cavity, whereas the Cestoda (tapeworms) depend entirely on absorption of host nutrients through the body wall. Apparently, once relatively large potential host organisms evolved, **parasitism** became a successful way of life for many platyhelminths because their small, flattened bodies do not seriously or immediately hinder host functions. Flatworm parasitic adaptations involve devices that fasten onto host tissues, such as hooks and suckers, as well as the reduction or loss of sensory and digestive organs that are no longer necessary for a dependent existence. Most important, in coping with host defense systems, parasites have evolved integuments that protect them against host enzymes and antibodies.

Interestingly, although parasitism may have simplified various organs, it also led in many cases to increased complexity of the parasite's life cycle. Some tapeworms, for example, may pass through a few different intermediate hosts ranging from arthropods to fish before the adult stage develops in the primary mammalian host. Such developmental networks must often have followed opportunities provided by other evolutionary events. Thus, Hyman (1951) has sug-

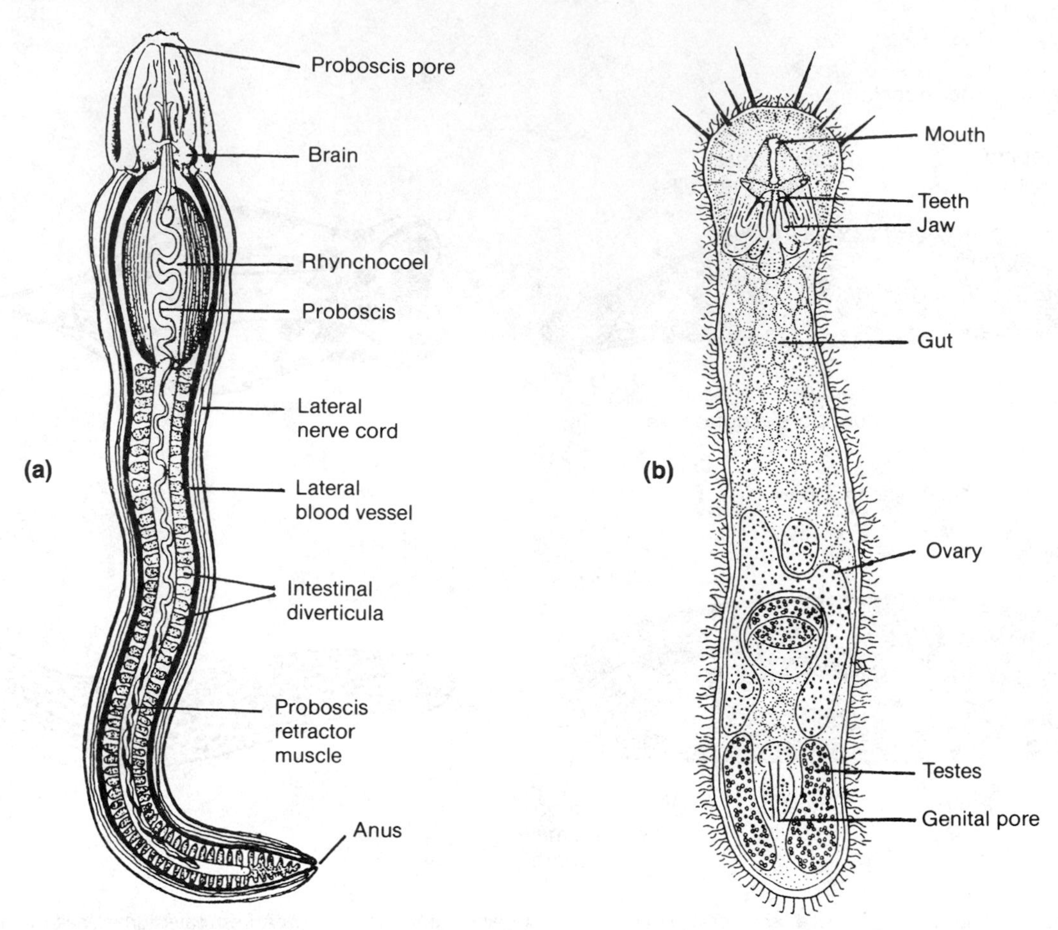

FIGURE 15-7 (a) Diagrammatic structure of a nemertean, *Cerebratulus*, showing lateral blood vessels and nerve cords. The rhynchocoel cavity and proboscis extends almost through the entire length of the animal. (Adapted from Kershaw.) (b) A gnathostomulid, *Problognathia*. (Adapted from Margulis and Schwartz.)

gested that some members of one group of trematodes (order Digenea) were originally infective only in molluscs. After fish and other vertebrates evolved, many digenetic trematodes invaded these newer groups, and molluscs were then retained as the intermediate host.

The adaptive advantage of such life cycle complexity allows the parasite to build up population numbers in intermediate hosts to improve its chances for infection of a primary host. Also, spreading its early stages among intermediate hosts does not exhaust primary host resources and allows the adult parasite to remain productive for relatively long periods. For example, some adult *Schistosoma* trematodes, the source for the widespread tropical disease schistosomiasis, may live for 30 years, and some human tapeworms are active for 20 years or more.

Characteristic of all these parasites is their enormous reproductive powers. Asexual reproductive stages often supplement sexual reproduction so that a single adult in some species can potentially produce hundreds of thousands, if not millions, of offspring. In various trematodes and cestodes almost their entire anatomy and physiology seems to be devoted to reproduction. Since their offspring have extremely low survival rates because of the many chance factors and hazards in parasite distribution and infection, such features must obviously have been long selected. Limited survival opportunities explain why usually only a minority of an appropriate host species is infected by a particular parasite at any one time.

Evolutionarily, it is generally believed that the parasitic platyhelminth classes were derived from a turbellarian ancestor, since the turbellarians appear to be the most primitive group in the phyla. Whether the ancestral turbellarian was an acoelan or had a simple gut of the type shown in Figure 15-6(b) is not yet clear. In any case, turbellarians undoubtedly stand close to the evolutionary base of all metazoans, including the two other acoelomate phyla, Nemertea and Gnathostomulida. These phyla share features with free-living flatworms but also show distinctive characteristics of their own. The nemerteans, known as ribbon worms, have an eversible proboscis and are perhaps the most primitive metazoans with a circulatory blood system and a digestive tract that has both mouth and anus (Fig. 15-7(a)). The gnathostomulids (Fig. 15-7(b)) are like flatworms in lacking an anal opening but have, among other unique characteristics, a pair of comblike, serrated jaws and reversible ciliary propulsion. (Some workers have placed them among phyla in the aschelminthes group as a pseudocoelomate offshoot.)

PSEUDOCOELOMATE (ASCHELMINTHES) PHYLA

Although their form and structure may vary considerably, the most primitive metazoans to show a distinctive, fluid-filled body cavity include phyla often grouped together under the name **Aschelminthes** (Table 15-2). Characteristic of the coelom among these phyla is that it encloses a thin-walled digestive cavity that lacks peritoneal linings, muscles, and supporting mesenteries (see Fig. 14-7). Since animals with a true coelom have such structures, the aschelminth phyla are generally called **pseudocoelomates**. Many of the aschelminths also possess an epidermal cuticle, adhesive organs, constant cell numbers (eutely), and a digestive tract with mouth, anus, and a muscular pharynx that pumps food into the flaccid gut cavity.

The most common and perhaps most representative of these phyla, **nematodes**, have a tubular shape maintained by high internal pressures which distend the animal to the extent permitted by its thick cuticle (almost like an overstuffed sausage). Overall changes in length are therefore slight since the animal is almost always fully extended by its high internal pressure. Nematodes are therefore not highly adapted for burrowing, which demands peristaltic activity. Instead, they function as undulatory swimmers and coilers by means of antagonistic longitudinal muscle contractions.

Some authors have proposed that the features shared by the aschelminths are sufficient to unite these groups into one phylum. However, for most zoologists, phylogenetic relationships among the aschelminths are still difficult to discern, and their distinctions seem to be sufficiently great to justify classifying them separately.

It has also been proposed that although some of these phyla are probably related, the pseudocoelomate features of the others may be the result of convergence. For example, many zoologists suggest that the gastrotrichs, nematodes, and nematomorphs share a common heritage in their derivation from a single group of acoelan turbellarians, whereas the four remaining phyla were derived independently from other acoelan groups. Generally, gastrotrichs are considered the most primitive of these phyla because they are all aquatic and ventrally ciliated. In any case, judging from the numbers of species in these phyla, the pseudocoelomate condition and its various adaptations have certainly endowed many of its bearers with continued evolutionary persistence. It seems clear that a fluid-filled tube of whatever nature, provided with circular and longitudinal muscles, offered significant advantages both as a hydrostatic organ for locomotion and as a carrier of metabolites, wastes, and gases throughout the body.

COELOMATES

In terms of known numbers of species, distribution of habitats, and total mass, the so-called higher or true coelomates can be considered the most successful of metazoans. They can be classified into 16 different phyla, all characterized to varying degrees by possession of a coelom surrounded by mesodermal tissues (see Fig. 14-7). Further divisions commonly separate these phyla into two major groups, mostly distinguished by the embryonic location of the mouth. In protostomes the mouth develops at or near the blastopore, which is the blastula slit (see Fig. 14-5) that invaginates to form the primitive gut. In deuterostomes the blastopore develops into the anus, and the mouth develops elsewhere. In addition, coelomic development is mostly schizocoelous in protostomes and enterocoelous in deuterostomes.

Early embryonic cleavage patterns also often differ between the two groups, with **spiral cleavage** in protostomes and **radial cleavage** in deuterostomes (Fig. 15-8). Also, in protostomes, development is mostly **de-**

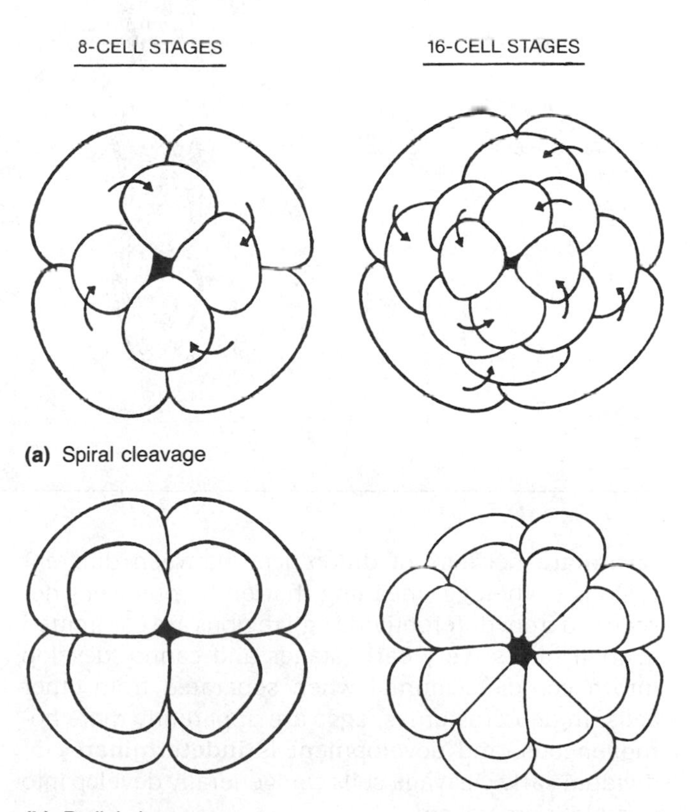

FIGURE 15-8 Modes of spiral and radial cleavages at 8- and 16-cell stages.

TABLE 15-2 Some Characteristics of the Pseudocoelomate Aschelminthes Phyla

Phylum*	Approximate Number of Described Species	Adult of a Sample Species	Life Styles	Habitat	Features
Nematoda (round worms)	10,000-30,000		Both free-living and parasitic	Marine, freshwater and soil	Complex flexible cuticle; lack flagella or cilia; tubular excretory system; mostly dioecious sexual reproduction
Nematomorpha (horsehair worms)	230		Adults free-living, but larvae parasitic in arthropods	Mostly freshwater and damp soil	Thick flexible cuticle; digestive tract absent in adults; dioecious
Gastrotricha (gastrotrichs)	400		All free-living	Marine and freshwater	Ciliated ventral surface; pseudocoel is diminished or absent; mostly hermaphroditic, but some parthenogenetic reproduction
Rotifera (wheel animals)	1,800		Mostly free-living; some sessile and colonial forms; a few are parasitic	Mostly freshwater; others in marine habitats and in bryophytes	Ciliated crown; grinding pharynx(mastax); sexually dioecious, but some parthenogenesis

terminate because of differences between different regions of the egg, meaning that embryonic cells descended from differentiated egg regions are committed to their fate at very early stages and cannot develop into a complete animal when separated from other cells. In deuterostomes, eggs are apparently more homogeneous, and development is **indeterminate**: individual early cleavage cells can generally develop into complete organisms.

Further distinctions can also be made among protostome phyla in respect to metamerism: some are obviously segmented, whereas others show no or little segmentation and have been designated ametameric or pseudometameric. (All deuterostome phyla show some degree of segmentation.) Also, both protostomes and deuterostomes possess phyla which are primarily sessile and feed mainly by capturing food particles using ciliated tentacular arms called **lophophores**.

Although the fossil record is incomplete for small soft-bodied organisms, most coelomate phyla appear

TABLE 15-2 Some Characteristics of the Pseudocoelomate Aschelminthes Phyla *continued*

Phylum*	Approximate Number of Described Species	Adult of a Sample Species	Life Styles	Habitat	Features
Kinorhyncha (kinorhynchs)	100		All free-living	Burrow in marine sediments	Segmented cuticle; lack external cilia; have large movable spines on trunk; dioecious
Acanthocephala (spiny-headed worms)	500		Parasitic	Larval stages in arthropods; vertebrates are primary hosts	Both the retractable proboscis and body wall covered with short spines; digestive tract absent in adults; dioecious
Loricifera	1		Free-living	Marine sediments	Spiny head; telescoping mouth; abdomen enveloped by plates (lorica); two oar-like tail appendages in larvae for swimming and climbing; dioecious

*Some authors also include an eighth phylum, Priapulida, in this group, but others claim that priapulids have a true coelom and should therefore be placed among more advanced metazoa.

to be quite ancient, probably of Precambrian origin. Many of the large soft-bodied specimens found in the Ediacarian fossil strata are obviously coelomates (see Fig. 14-2) and may well have had a history that extends 100 million or more years before the Ediacarian period. Certainly by the Cambrian period deuterostomes had already completed their evolution from protostomes, and most metazoan body plans seemed to have become fairly well established during the Paleozoic era.

Since a discussion of each of these phyla is beyond the scope of this book, only some general evolutionary trends in some common coelomate invertebrates will be discussed. Further information can be found in textbooks such as those of Barnes and Lutz, as well as in various collections of articles (e.g., those edited by Conway Morris et al. and House).

MOLLUSCA

Molluscs, a phylum which now numbers over 100,000 species, were perhaps among the earliest of metazoan herbivores and possessed a body plan based on creeping over shallow marine substrates (Fig. 15-9). One of their distinctive features is the possession of a **radula**, a rasplike organ bearing chitinous teeth which unrolls from the mouth and is used for scraping algae from rocks. In some members of the phylum this herbivous tool has been abandoned because of new feeding habits, but its vestiges often remain. Because of their early herbivorous habits, many molluscs also have long intestinal tracts which are dorsally located so that the ventral creeping surface, the **foot**, can remain free. Correlated with protecting and providing respiration

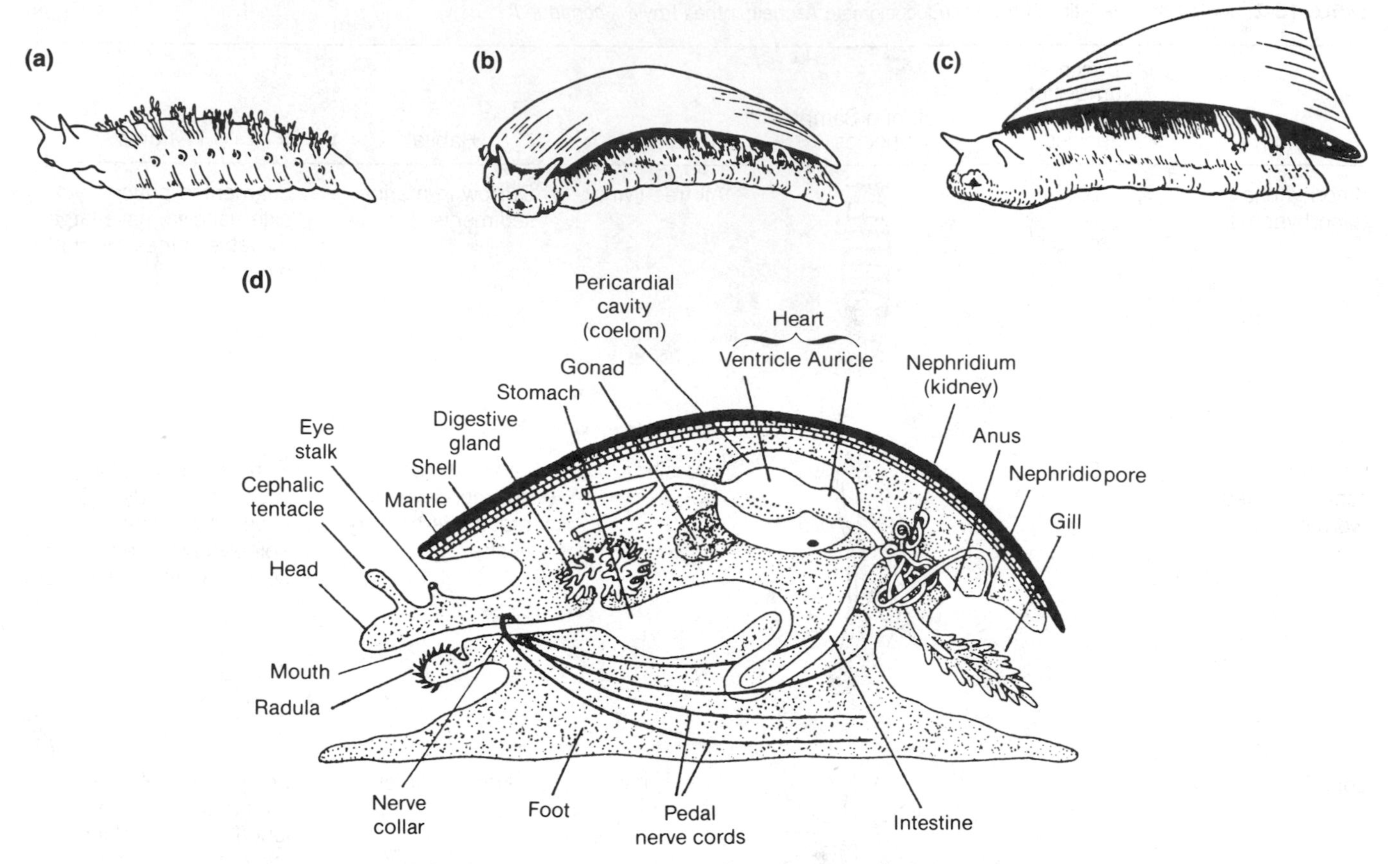

FIGURE 15-9 Hypothetical stages in the early evolution of molluscs. (a) A wormlike benthic creeper which may have possessed repeated sets of external gills and excretory pores. (b) Evolution of a protective calcified dorsal shield. (c) Development of the shield into a shell that moves forward to cover the head. (d) A lateral view of a later evolutionary stage showing the basic body plan and organs of a hypothetical ancestral mollusc. (Adapted from Lutz and from Solem.)

for this visceral bulk is a hard shell and a set of active gills enclosed in a body fold called the **mantle**. Various groups of molluscs using this initial architecture have evolved subsidiary changes in structures which are often recognizably molluscan but considerably different from each other (Fig. 15-10). A few are described briefly below.

Gastropods, which include snails, whelks, and limpets, have two pairs of tentacles, one small and one bearing the eyes. They often have a cone-shaped shell which was originally straight and served not merely as a dorsal shield but as a protected retreat for enclosure of the entire animal. As these animals evolved into larger forms, their shells were apparently difficult to balance in longer lengths, and asymmetric, spirally coiled shells were selected in various lines. At some point in this evolution, the gastropod stage was reached when the mantle bearing the gills was rotated from facing posteriorly to facing anteriorly (**torsion**) with an accompanying rotation of the viscera. Among the explanations for torsion is that it permitted additional room for withdrawal of the head and allowed water to enter the mantle cavity from the head end rather than from the rear. The problem of sanitation of the mantle apparently became important since it contains excretory organs whose products can foul the respiratory gills. One solution found in keyhole limpets was to have a small hole in the mantle and shell immediately above the anus, positioned so that feces can be excreted instead of being passed down the gills. A more common solution was to eliminate one member of the original pair of gills so that incoming water is circulated through the remaining gill and then propelled outward along the other side, which now contains the anal and kidney duct orifices. Interestingly, species in one of the gastropod subclasses (opistobranchs) show "detorsion," with a tendency towards reduction of the remaining gill and the formation of new bilateral gills and bilateral symmetry.

Bivalvia (also called Pelecypods, or hatchet foot) are the hinge-shelled **bivalves** such as clams and mussels that are flattened from side to side. In these ani-

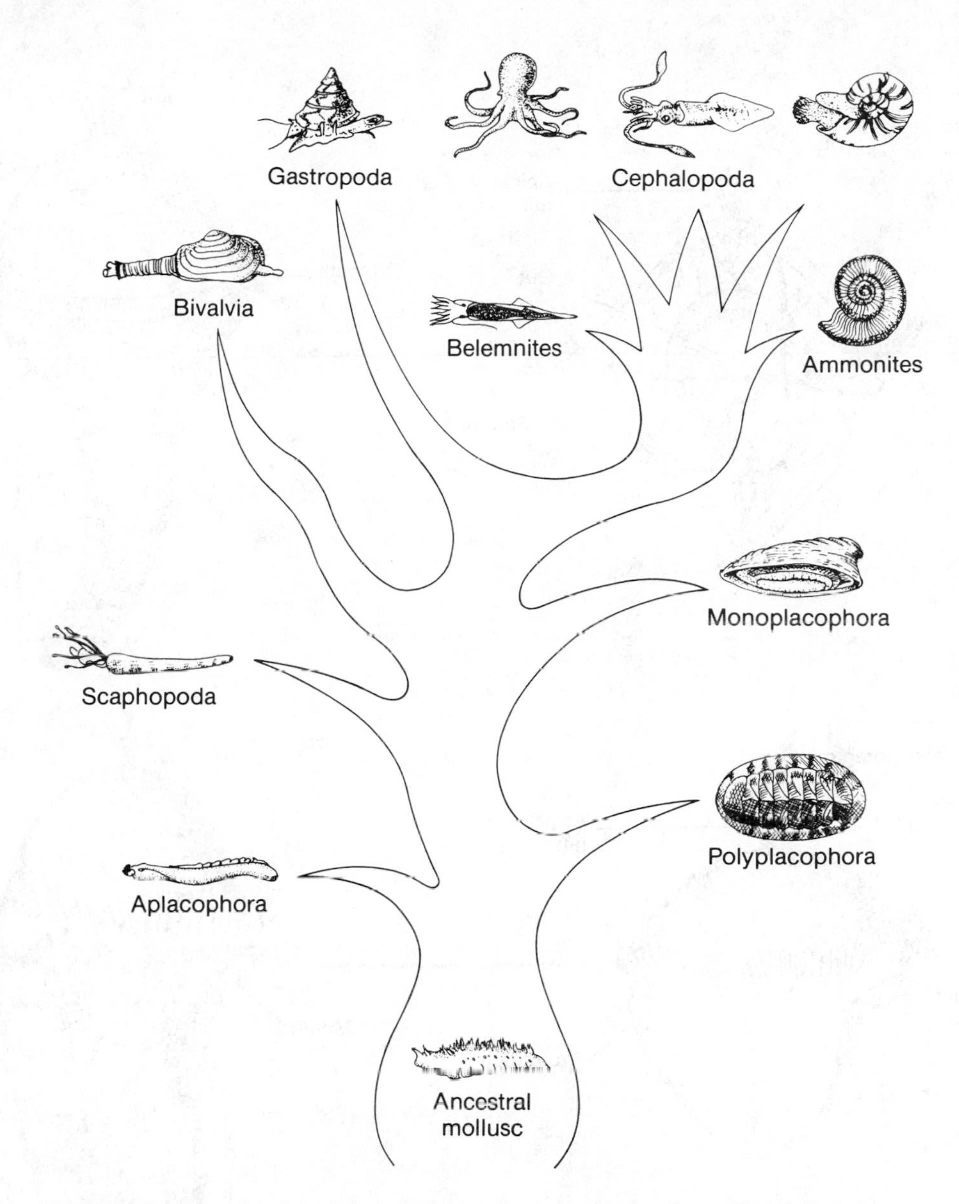

FIGURE 15-10 One possible phylogeny of the various classes of molluscs. The shelled classes are found in many fossil strata, some of which can be traced back to the Cambrian period, such as the existing *Nautilus* group (upper right figure). Similar cephalopods with coiled external shells, the ammonites, became extinct at the end of the Cretaceous period. The belemnites, marked by internal shells, also became extinct at that time and are allied with modern squids and octopoid forms.

mals the gills are much larger than in gastropods, since they are also used for food gathering by means of ciliary tracts which carry captured food particles to the mouth. Because these particles are relatively small and dispersed, the radula is no longer needed and the animals can now collect food in a sedentary fashion. The head is greatly reduced since sensory orientation is no longer necessary, and the foot is either reduced or converted into a burrowing tool. In some bivalves evolution has led to a complete separation between inhalant and exhalant water currents.

Cephalopods (head-foot) include the most mobile of molluscs such as squids and octopi. The head, bearing the largest and most complex of invertebrate brains, now occupies the main locomotory position formerly occupied by the foot, and the foot has been reduced to a ring of tentacles around the head. This style of life represents a transition from relative passivity to rapid mobile aggression and was apparently very successful through the Mesozoic era, as witnessed by the large numbers of fossil ammonites and belemnites. The mantle is heavily muscularized in this group, and water is drawn around the sides and squirted out like a syringe through a tubelike opening called the **siphon**. The siphon can be aimed in any direction, allowing jet-propelled movement in the opposite direction. As a concomitant to hunting, both nervous system and vision are highly developed, and the eyes

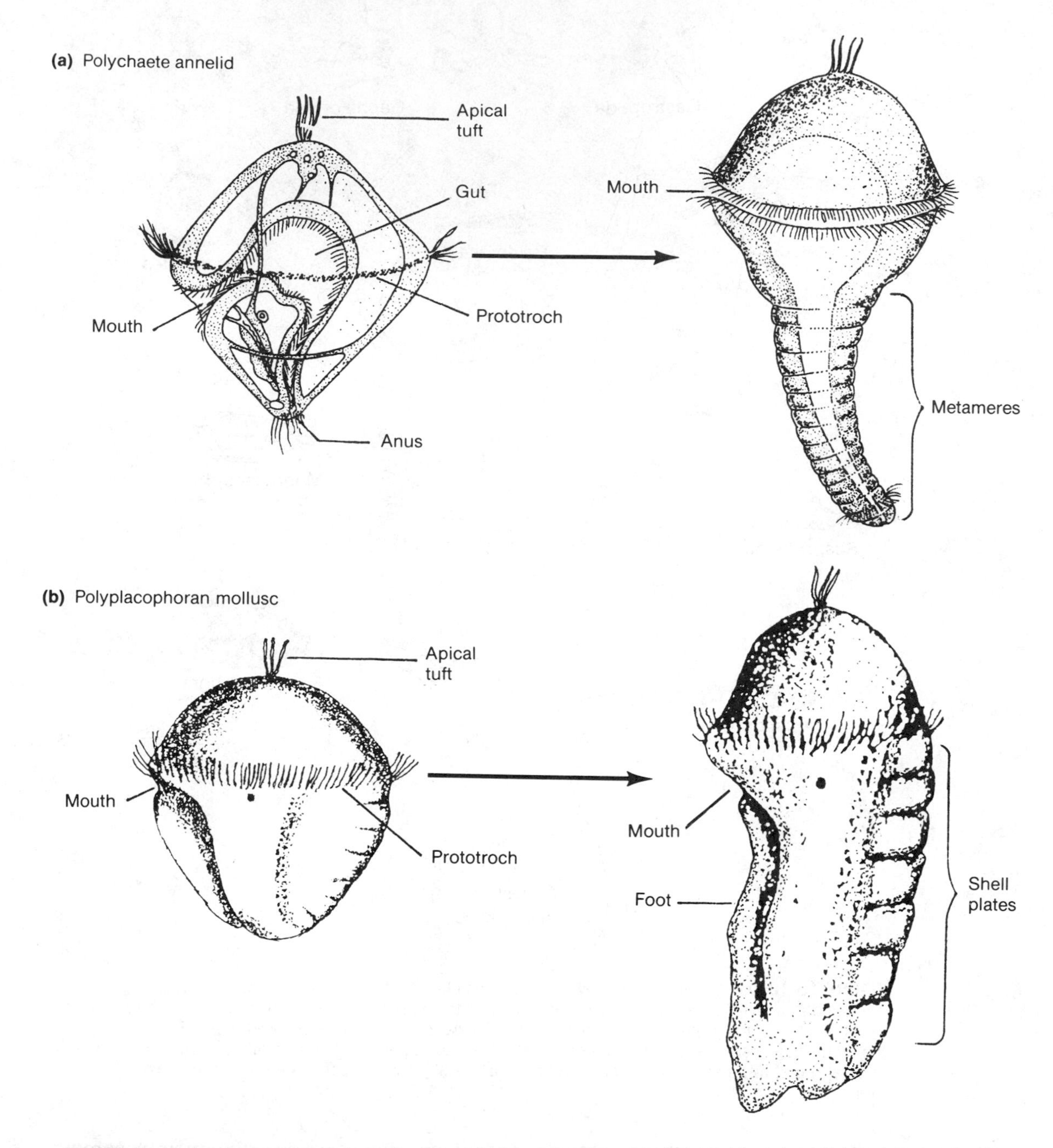

FIGURE 15-11 Trochophore larval stages in an annelid (a) and mollusc (b). (Adapted from Lutz and from Barnes.)

form images which are probably as clear as those formed by the similar and evolutionarily convergent eyes of vertebrates (see Fig. 3-1). Modern cephalopods, with the exception of *Nautilus* and *Spirula*, show a reduction in shell size and complexity; in squids the shell, reduced to a chitinous plate, has now adopted the function of an internal skeleton. In the octopus, only tiny shell remnants remain.

Other molluscan groups include two which show some degree of metameric organization, the chitons, or polyplacophorans (multi-plated shell), and the monoplacophorans (single flat shell). The latter group was believed to have become extinct at the end of the Devonian period until a modern member, *Neopilina*, was discovered in the 1950's. Many of its organs, such as gills, muscles, and nervous system show serial repetition, and some zoologists have therefore proposed that molluscs have had an annelidlike origin. In support of this view is the occurrence of a **trochophore larva** in species of both phyla (Fig. 15-11). Other zoologists point out that serial repetition of parts is not necessarily the same as annelid segmentation, al-

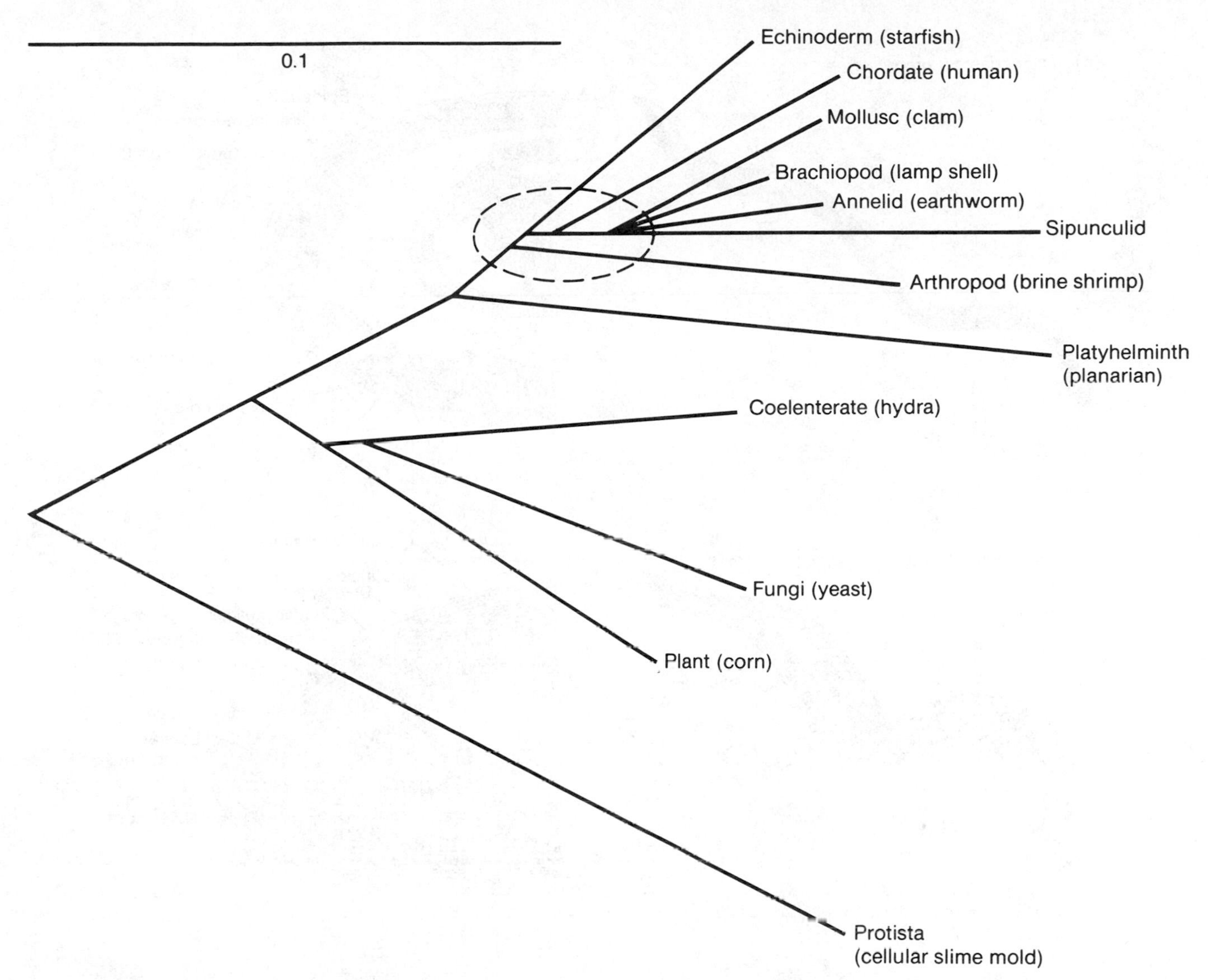

FIGURE 15-12 An evolutionary tree based on comparisons between 18*S* RNA sequences from the different groups shown. As indicated, the platyhelminths and later eucoelomate groups appear to have diverged from a protistan ancestry separate from the protistan ancestry of coelenterates, fungi, and plants. Unfortunately, the exact branching order within the dotted ellipse cannot be resolved from the data, although it seems clear that these different groups all radiated relatively rapidly from what may well have been a common coelomate ancestor that was also metameric. The scale bar at the upper left represents an evolutionary distance calculated as 0.1 substitutions for each nucleotide position in the sequences that were compared (a total of more than 800 nucleotides). (Adapted from Field et al., Figs. 2 and 5.)

though both phyla may have shared a common coelomate ancestor. Clark and others carry the argument further and suggest that the presumed molluscan coelom, its pericardial cavity, probably originated independently of the annelid-type coelom. Since there is no other evidence of a molluscan coelom, they propose that molluscs arose from benthic acoelomate animals that may have been similar to turbellarians and nemerteans. How this question will finally be resolved is still unclear, although the 18*S* RNA data mentioned previously (p. 277) suggest that the ancestral group from which coelomic metameric arthropods are derived probably also gave rise to obviously coelomic metameric annelids as well as to molluscs (Fig. 15-12).

As in many other instances, the difficulty in discovering intermediate forms between one phylum and another probably stems from the fact that many such transitional events occurred in Precambrian times among very small soft-bodied organisms that were poorly fossilized, if at all. Moreover, different phyla represent different major types of organization adapted to entirely different ways of life, each phylum often utilizing a distinct mechanism of obtaining food along with distinct metabolic needs, reproductive modes, and so on. Forms that may have been transitional between two such distinct adaptive zones would most likely have suffered considerable competitive disadvantages, once better-adapted forms had been derived. That is, a transitional form produced by phylum A lead-

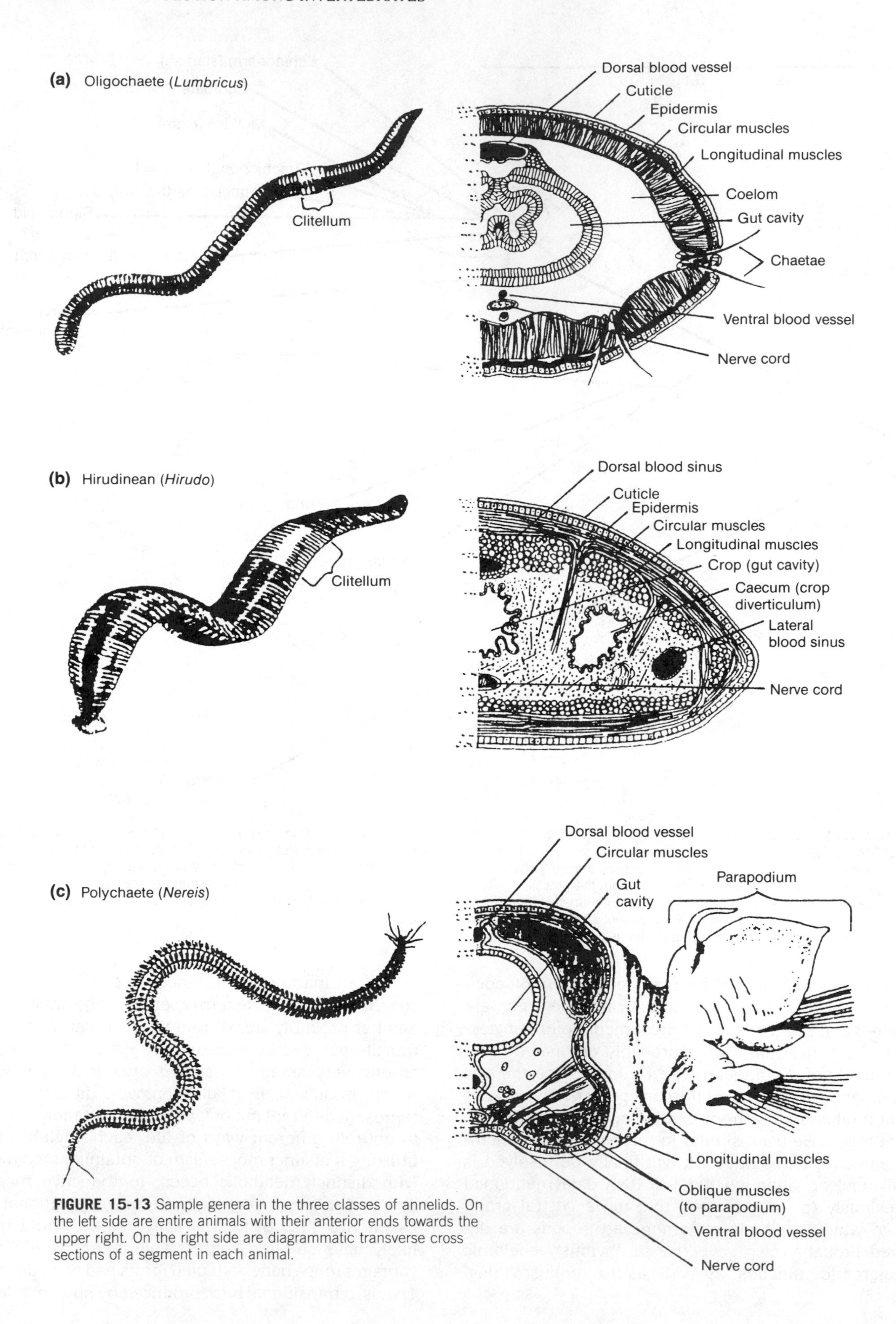

FIGURE 15-13 Sample genera in the three classes of annelids. On the left side are entire animals with their anterior ends towards the upper right. On the right side are diagrammatic transverse cross sections of a segment in each animal.

ing to phylum B would probably be rapidly replaced by newly evolved forms of phylum B. The scarcity of most transitional forms, either among living species or in the fossil record, is therefore a reflection of their typical short-term survival and helps explain the relatively wide evolutionary gaps between major groups that occupy different adaptive zones.

ANNELIDA

Although unsegmented coelomate worms are burrowers, their habits are relatively sedentary, and none of them engage in continuous burrowing. The sausage-shaped Sipunculida (peanut worms) and the similarly shaped Echiura (proboscis worms) are two such coelomate phyla that use peristalsis to move slowly through marine substrata. As discussed in Chapter 14, it is only in segmented coelomates such as **annelids** that hydrostatic pressures can be localized to specific segments and active sustained burrowing makes its appearance.

Annelids are a group of about 9,000 species possessing a soft, wormlike body with various numbers of segmented coelomic compartments separated by transverse septa. A prostomium and pygidium, each of which has a unique structure different from other segments, cap, respectively, the anterior and posterior ends of the trunk. Of the three existing annelid classes, Oligochaeta, Polychaeta, and Hirudinea (Fig. 15-13), the first represent an offshoot of what was probably a very early annelid benthic stock. **Oligochaetes** include the familiar earthworms characterized by their spine-like chaetae (or setae) used for traction during the continuous burrowing that so enhances soil fertility.

Allied to Oligochaeta are the Hirudinea (**leeches**). Both these groups are hermaphroditic and possess a glandular organ called the **clitellum** that usually covers five to ten segments near the anterior end. The clitellum secretes a mucous coat to help bind two copulating animals together and also produces a "cocoon" for the deposition of fertilized eggs. Because this organ appears to be homologous in both groups, oligochaetes and hirudineans are considered by some zoologists as subgroups within a "Clitellata" class or subphylum. Leeches, however, have abandoned burrowing in favor of predation or blood-sucking parasitism, and their locomotory habits are mostly based on looping, in which the entire circular and longitudinal musculature contracts as a unit and peristalsis is not needed at all (Fig. 15-14). Since peristaltic locomotion is absent, septa are no longer needed, and even the coelom is reduced, since a hydrostatic organ for the maintenance of circularity and transmission of muscular pressure need not be as great as in a burrowing animal.

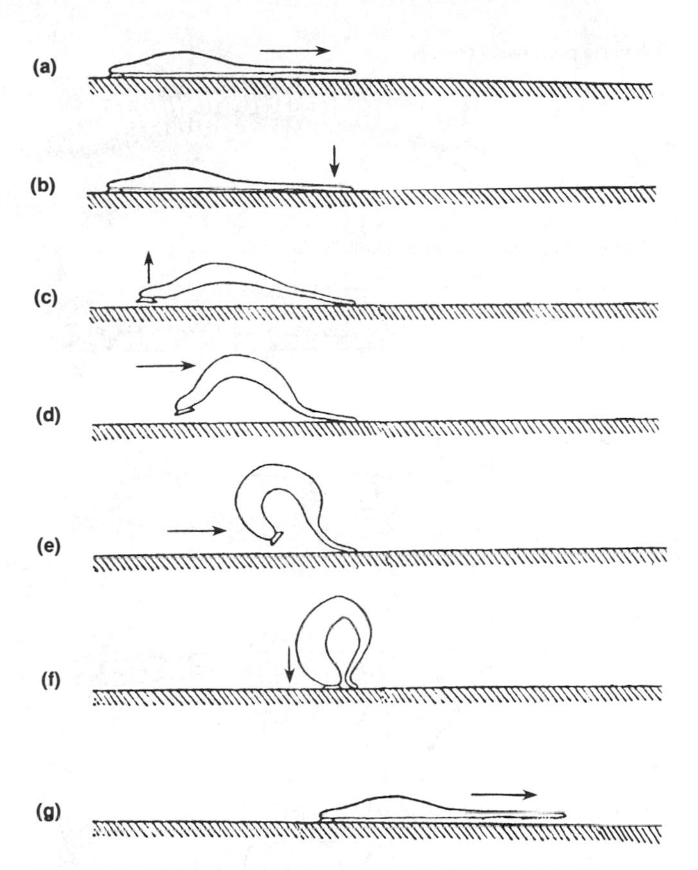

FIGURE 15-14 Stages in the locomotion of a leech, *Helobdella stagnalis.* (a) Posterior sucker (left side) is fixed to the substrate, and anterior end (right side) is extended in direction of arrow. (b) Anterior sucker is fixed. (c-e) Posterior sucker is released, and animal contracts. (f) Posterior sucker is fixed immediately behind the anterior sucker, and the cycle begins again. (Adapted from Clark [1964].)

By contrast to oligochaetes and leeches, many free-living **polychaetes** such as the marine worm *Nereis* evolved lateral appendages or **parapodia** (see Fig. 14-13). This enabled them to mostly abandon burrowing and appears to have been associated with adaptation to locomotion in loose, dispersed material rather than compacted substrates. Under those circumstances, parapodia are much more effective locomotory agents than peristaltic waves, and such animals would have had increased opportunities to move rapidly and feed directly on the surface bottom rather than hide below it. Once parapodia appeared, swimming above the substratum must have become available to numerous groups of animals by further development of parapodial musculature.

The use of appendages, however, breaks up the body's circular and longitudinal muscles since the parapodial muscles necessary to move them must be inserted across the internal coelom (in polychaetes, into the midventral line). Thus, the longitudinal muscles do not cover the parapodial region, and the circular muscles are confined to regions between parapodia. These limitations in circular muscles makes peristalsis

Annelid polychaete (*Nereis*)

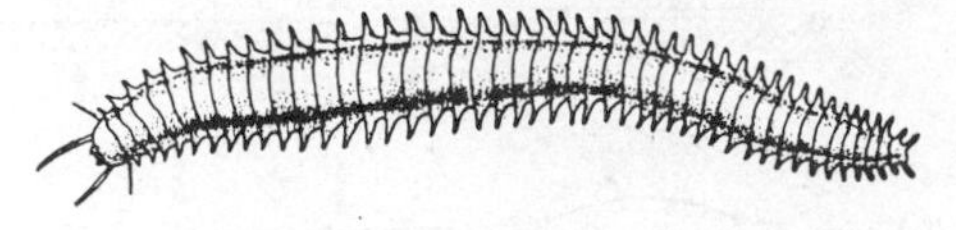

Protonychophora (*Aysheaia*, Cambrian period)

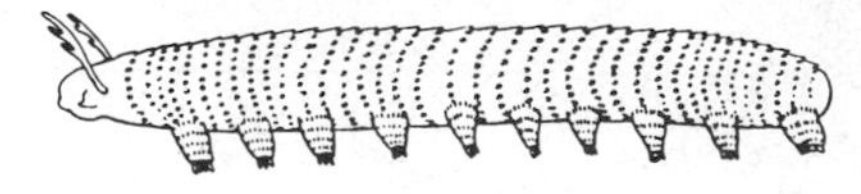

Onychophora (*Peripatus*)

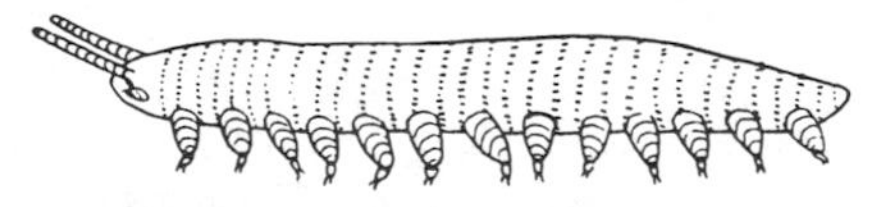

Arthropoda (myriapod)

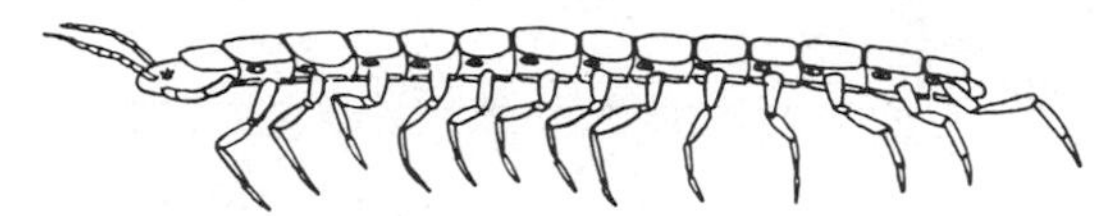

Arthropoda (insect, Carboniferous period)

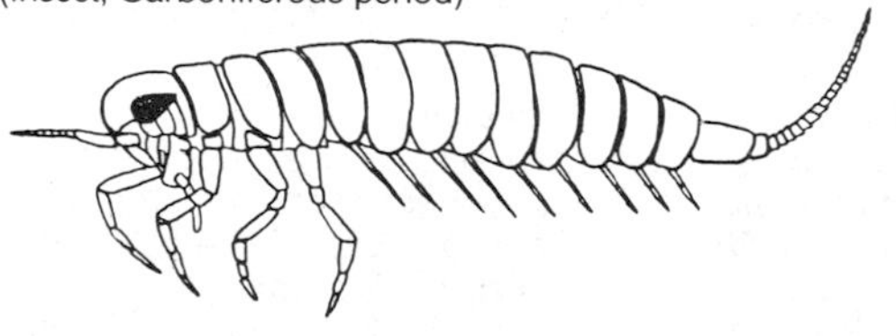

FIGURE 15-15 Evolutionary progression from polychaete- or oligochaetelike annelids to arthropods through intermediary forms similar to those found in the phylum Onychophora. Valentine (1989), on the other hand, suggests that arthropods and annelids share an ancestry among noncoelomate segmented worms from which they evolved independently. (Adapted from Clarke.)

difficult or impossible to perform, and the parapodial animal is now an effective crawler or swimmer, but no longer a good burrower.

Parapodia in a soft-bodied organism such as the annelid polychaete, however, is only a partial answer to the development of appendages since the "hinge" of the appendage cannot be very firmly fixed in the animal's soft body and the animal cannot gain maximum leverage for its limbs. This lack of fixity also causes the muscles that move the parapodia to have variable lengths, and they are therefore not as efficient as they would be in a hard-bodied organism. The next great evolutionary advance in rapid locomotion therefore came with the development of hard-bodied skeletons, as can be seen in arthropods.

ARTHROPODA

In **arthropods** a tough, chitinous cuticle provides an exoskeleton with fixed hinges for the jointed appendages. Except for some arthropod larvae which are burrowers and soft-bodied, this hardened skeleton eliminates the need for a hydrostatic scaffold, and the coelom is now restricted to the excretory organs. Although internal septa are lost, various degrees of segmentation remain that betray a common ancestry with annelids. As shown in Figure 15-15, the link between the two phyla, at least in the case of some arthropods, may have progressed from an annelid or pre-annelid ancestor through animals similar to those in the phylum Onychophora. Among other features indicative of their relationship is that both annelids and arthropods share a similarly structured central nervous system with ventral nerve cords and a contractile, dorsal, tubular "heart."

Hardened, chitinous exoskeletons with their inner projections (apodemes, or endophragma) for muscle attachments, jointed appendages, and the development of new kinds of cephalic structures provided many opportunities for arthropod evolutionary radiation: 80 percent of all known animal species today are arthropods, and these are found in almost every conceivable ecological habitat. Because phylogenetic relationships among major groups are not yet clear, the diversity of arthropod species has been classified in various ways. One common classification system, shown in Table 15-3, divides the phylum into four subphyla.

A different view promoted by Manton and others is that each of these subphyla should be classified as a separate phylum since they each probably originated from a different annelidlike ancestor, probably in Precambrian times (Fig. 15-16). That is, arthropods evolved polyphyletically, and the group should be considered a grade or **superphylum** rather than a single monophyletic phylum.

Whatever their origins, it is clear that their body plan provided them with almost immediate success. During the Cambrian period, trilobite and merostome arthropods became dominant marine animals in terms of size, mobility, and predatory powers. (Some fossil eurypterids reached almost 10 feet long!) In the later Paleozoic era these particular groups declined because of competition with more advanced arthropods and other invertebrates, as well as with vertebrates (Chapter 16).

Following their marine origins, perhaps the most significant arthropod advance was the invasion of land which gave rise to terrestrial arachnids and uniramians in addition to some terrestrial crustaceans. Among these, **insects** (class Hexapoda) underwent what is

TABLE 15-3 Some Characteristics of the Four Subphyla of Arthropoda

Subphylum	Approximate Number of Existing Described Species	Adult of a Sample Species	Habitat	Pairs of Antennae	Pairs of Legs	Characteristics Features
TRILOBITOMORPHA						
Trilobites	Extinct		Marine	1	Many	Body with variable numbers of segments organized into three longitudinal lobes; biramous (two-lobed) legs
CHELICERATA						
Merostomes (horse-shoe crabs; eurypterids—extinct)	4		Marine	0	5	Anterior appendages are chelicerae (pincers or fangs); uniramous (one-lobed) legs; book gills (merostomes); book lungs, waxy cuticle, and muscular pumping pharynx (arachnids)
Arachnids (scorpions, spiders, ticks, mites, and others)	65,000		Mostly terrestrial	0	4	
CRUSTACEA						
Crustaceans (crabs, lobsters, shrimp, copepods, isopods, barnacles, and others)	35,000		Mostly marine, some freshwater and terrestrial	2	Many	Biramous legs; carapace (dorsal cover); paired compound eyes; larvae when present are of nauplius type
UNIRAMIA						
Myriapods (centipedes, millipedes)	10,500		Terrestrial	1	Many	Uniramous legs; 2 pairs of maxillae; malpighian tubule excretory system; paired compound eyes and complete metamorphosis through pupal stages (many insects); respiration via air tubules (trachea)
Insects (flies, beetles, bugs, bees, locusts, etc.)	1,000,000		Terrestrial, aerial, and aquatic	1	3	

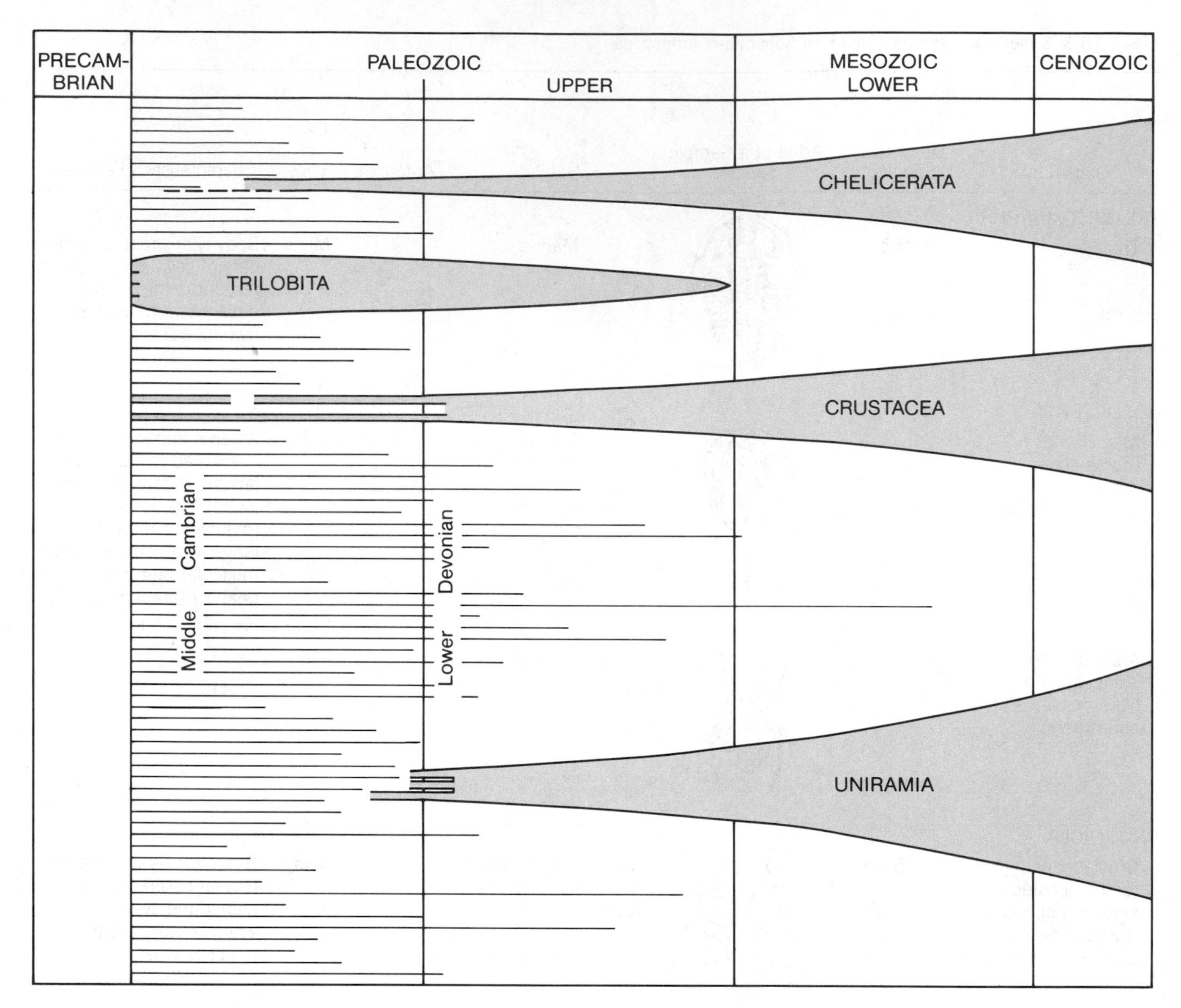

FIGURE 15-16 General evolutionary patterns and fossil frequencies of arthropod groups. The thin parallel lines indicate families or genera of arthropods that had a mineralized exoskeleton. (Adapted from Whittington.)

probably the most explosive radiation of any metazoan group, occupying practically all of the many varied terrestrial habitats. There are approximately 1 million described insect species, and there are probably more than twice that number undescribed!

The causes for such relatively rapid evolutionary changes and widespread radiation are difficult to trace at present, but must undoubtedly be connected to the small size of insects, their rapid generation times, their evolution of winged forms, and their apparently endlessly malleable structures. That is, practically all sections of insect anatomy—such as legs, mouth parts, wings, and eyes—have shown a capacity for evolving new structures and functions that seems superficially like the ease with which interchangeable parts are inserted into a child's modular building toy (Fig. 15-17). The specialized larval stages of many insect orders live and feed differently from the adult forms and thus provided a further advantage in allowing these groups, especially those with winged adults, to finely partition their resources and exploit a wide range of environments and diets. Beetles, with more than 300,000 described species, possess an additional adaptation in the extra-thick chitinous armor which protects even their vulnerable wings against predators and parasites.

It is also among insects that the only social animal organizations are found until we reach the vertebrates. **Social organization** means that there is a division of labor among different members of a group, a phenomenon which is far beyond the kinds of simple aggregation found in swarms of migrating locusts or in cooperative "tents" built by tent caterpillars. The truly

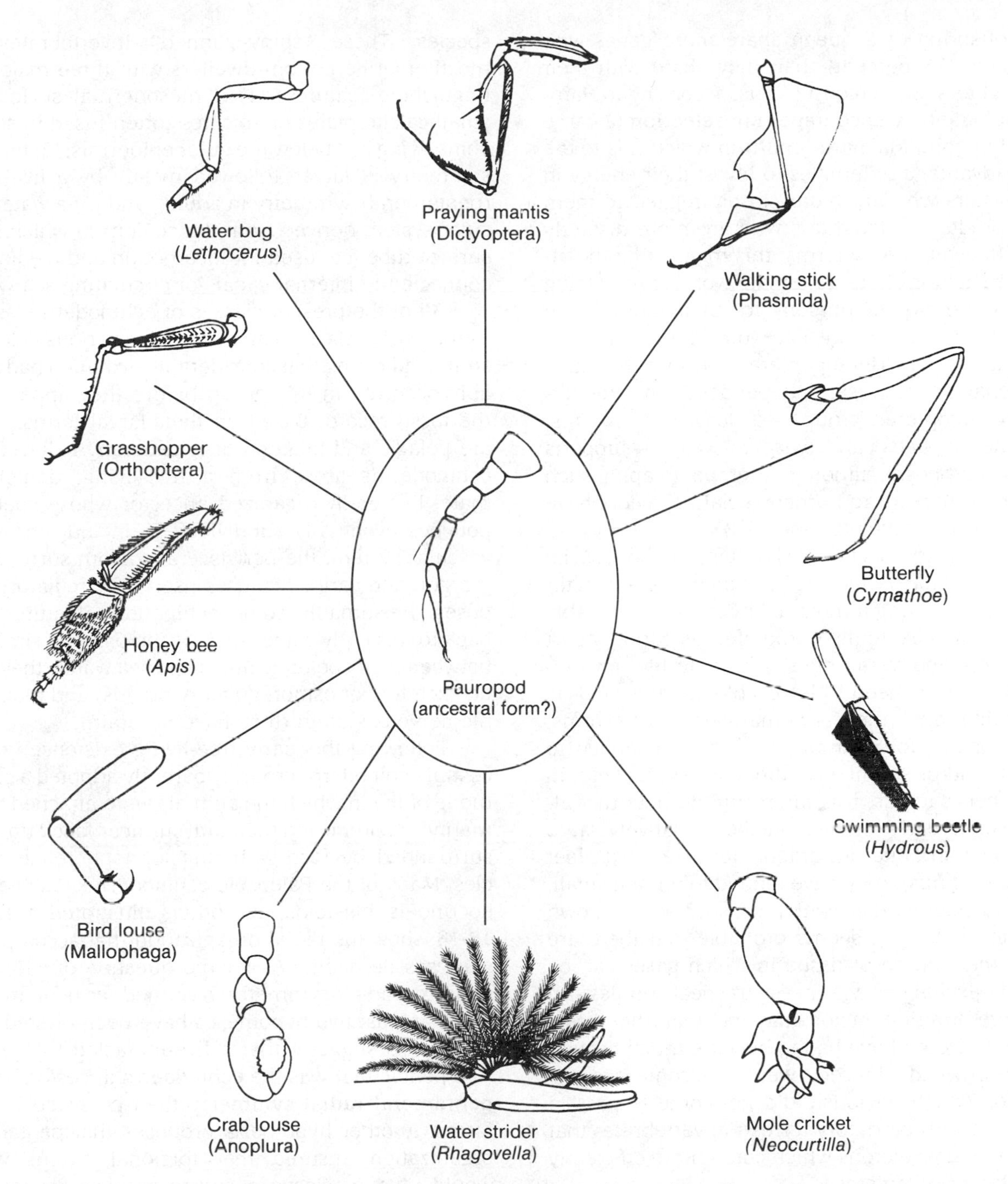

FIGURE 15-17 Legs in different insect groups. (Adapted from Fox and Fox.)

social insect societies include the order Isoptera (termites) and various members of the order Hymenoptera (ants, bees, and wasps). In these groups different morphological types (castes) assume different social functions. Each colony, for example, usually has only one fertile queen engaged in egg production, one or more fertile males for egg fertilization, and one or more castes of sterile workers that are exclusively engaged in food gathering, cooperative brood care, nest maintenance, and defense of the colony. Differentiation between fertile queens and sterile workers is based on the production of substances by the queen that suppress fertility in workers but which can be overcome by various environmental influences such as special foods (e.g., royal jelly in honeybees).

In termites both sexes derive from fertilized eggs, but in hymenopterans the distinction between sexes derives from development of males from unfertilized eggs and females from fertilized eggs (haploid males and diploid females). It has been pointed out that this hymenopteran system of **haplodiploidy** is especially prone to the evolution of sociality since the diploid female offspring of a queen share more genes with their sisters (75 percent) than they share with their own daughters (50 percent).[1] Thus, according to Hamilton, haplodiploidy encourages **kin selection** (a name invented by John Maynard Smith) in which it is to the genetic advantage of females to invest their energy in raising sisters who are more closely related to them than in producing daughters who are more distantly related. In evolutionary terms, this means that the **altruistic behavior** of sterile female workers in helping to raise more worker progeny for their mother (the queen) is a phenomenon that would be expected to arise frequently in Hymenoptera. The fact that such social behavior has arisen independently in three different hymenopteran groups—at least once in ants, eight times in bees, and twice in wasps—supports Hamilton's proposal, although attempts to apply such genetic determinism to human social behaviors have led to serious disputes (Chapter 24).

In summary, insect success extends to both social and non-social forms and has been an important factor in almost every living terrestrial environment probably dating back at least to the Carboniferous period. Apart from insects and vertebrates, relatively few invertebrate phyla have been able to colonize the land as successfully, and these—nematodes, earthworms, and gastropods, for example—are often confined to special humid or soil-like conditions. Nevertheless, in spite of their success, it is interesting to note that although some early insects reached relatively large sizes (some Carboniferous dragonflies measured 2 feet between wing tips), they have tended to remain small, especially by comparison with most species in our own vertebrate phylum. It seems probable that there are limits to the volume of tissue in which gases can be effectively exchanged via insect tracheal tubules. An even more limiting factor may be that the insect exoskeleton would have had to become much heavier and more unwieldy if insects were to become larger—a situation which obviously did not come to pass in competition with predatory terrestrial vertebrates that possessed endoskeletons which could more efficiently support larger organisms.

ECHINODERMATA

Aside from chordates, **echinoderms** are the largest of the deuterostome phyla, containing at present 6,500 species. These "spiny-skinned" invertebrates are mostly marine bottom-dwellers with three major distinguishing features: (1) a mesodermal skeleton of small calcite plates or spicules (often fused in sea urchins) lying just below the outer epidermis; (2) bilateral symmetry in larvae followed mostly by a five-rayed (pentameral) symmetry in adults; and (3) a water vascular system derived from the coelom in which small surface tube feet used for locomotion and feeding are connected to internal canals of circulating sea water.

All of the present classes of echinoderms as well as the extinct classes can be found in various Paleozoic strata, indicating that considerable evolution had probably occurred in this group before they appeared in the fossil record. Based on their larval forms, which are pelagic and bilateral, it is generally believed that echinoderms arose from a free-living, deuterostomate, bilaterally-organized ancestor whose coelomic pouches eventually subdivided to include the water vascular system, the perivisceral coelom surrounding the gut, and various sinuses used for circulatory purposes. Presumably some echinoderm features date back to this early stage, such as the osmotic similarity between their coelomic fluids and sea water (they have no excretory or osmoregulatory organs) and their simple nervous system (they have no brain).

Following this early free-living existence, an ancestral echinoderm group apparently adopted a sessile mode of life in which the animals were attached to the marine substrate and their oral surfaces faced upwards surrounded by food-gathering lophophorelike tentacles. Many of the Paleozoic echinoderms, such as the eocrinoids, blastoids, and others illustrated in Figure 15-18, show the radial organization that accompanies such sessile habits. As for the question of why echinoderm radial symmetry assumed a pentamerous form, at least two hypotheses have been offered. One hypothesis suggests that a five-tentacled lophophore (the pentactula) was the echinoderm ancestor, and its **pentameral radial symmetry** then persisted in later forms. Another hypothesis proposes that pentameral organization ensured that torsional strains which would tend to cleave a suture between plates surrounding an essential central area of the animal would not easily be transmitted to a suture that was immediately opposite. That is, a structure such as  is less likely to break apart than  . In the absence of fossilized echinoderm ancestors, it is difficult to decide which of these hypotheses is correct.

[1] Fifty percent of the genes shared by sisters come from their haploid father who gives the same set of genes to all of his daughters, which also share half of their remaining genes because genes in different haploid eggs produced by their diploid mother have the probability of being 50 percent alike [.50 + (.50 x .50) = .75]. On the other hand, a female shares only 50 percent of her genes with her female offspring since each of her daughters gets 50 percent of its genes from the father.

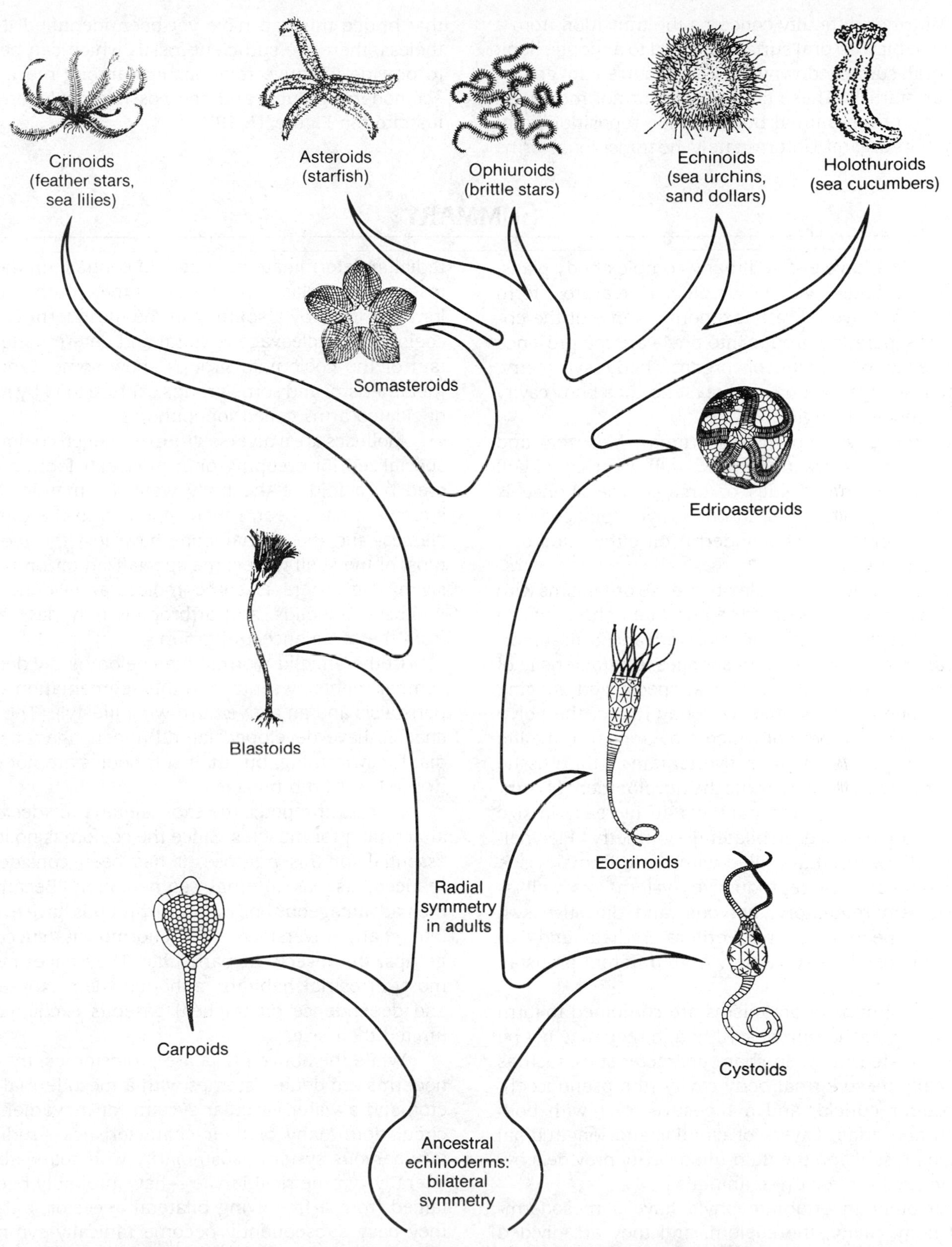

FIGURE 15-18 A possible phylogeny of echinoderm groups, some of which have been classified as subphyla or as classes. The five groups at the top are contemporary, and those below are extinct. A somewhat different phylogeny is offered by Paul and Smith. (Individual illustrations are from Nichols and from Weller.)

A further difficulty concerns the transition from a sessile habit with oral surface upward to a mobile habit with oral surface downward, as occurred in groups such as starfish. These changes meant not merely an inversion of the animal but a drastic repositioning of many of its organs. Unfortunately, no transitional forms that bridge this gap have yet been identified. Nevertheless, there are sufficient fossils which can be used to provide tentative relationships among most major echinoderm groups, and one possible phylogeny is illustrated in Figure 15-18.

SUMMARY

The metazoa have evolved many complex body plans, which, combined with embryological features, form the basis for invertebrate taxonomy. Some of the criteria for separating groups into phyla are the presence or absence of tissues or organs, body symmetry, metameric organization, and presence of a body cavity and its mode of origin.

Of the phyla, sponges (Porifera), Mesozoa, and Placozoa are the least complex, with a variety of cell types but no true tissues or organs. The sponge is generally organized into an outer layer (pinacoderm) and an inner layer (choanoderm) on either side of a gelatinous mesohyl.

Next in order of complexity are the organisms with radial symmetry, the Cnidaria and Ctenophora, which have a digestive cavity and more advanced tissue organization than that seen in sponges. Characteristic of cnidarians are the planula larva, specialized stinging cells or nematocytes, and two body forms, the polyp and the medusa, both of which may appear in the life cycle of a species. In other invertebrates a third tissue layer, the mesoderm, permits the development of complex organ systems and parallels an increase in size and the appearance of bilateral symmetry. Platyhelminths (flatworms) are filled with mesodermal tissue and have complex reproductive systems, as well as simple osmoregulatory, nervous, and digestive systems. It appears likely that Porifera, Radiata, and Platyhelminthes arose separately from different protistan ancestors.

Many groups of organisms are combined to form the phylum Aschelminthes. Although they may not be closely related, they do share characteristics such as a partially mesodermal body cavity (the pseudocoel), an exterior cuticle, and a digestive tract with both mouth and anus. Layers of circular and longitudinal muscle pressing on the fluid-filled cavity provide a hydraulic skeleton for these animals.

All other invertebrate phyla have a mesoderm-lined body cavity, the coelom, and they are divided into two groups mainly on embryological criteria. The deuterostomes are characterized by the development of the blastopore into the anus, a coelom formed by outpocketing of the mesoderm (enterocoelous), and radial, indeterminate cleavage. In contrast, in the protostomes, the blastopore becomes the mouth, the coelom is formed by a splitting of the mesoderm (schizocoelous), and cleavage is spiral and determinate. Certain of the coelomate groups show some degree of metamerism, and some groups capture food by means of ciliated arms called lophophores.

Molluscs are a diverse group of shelled coelomates specialized for creeping on a muscular foot and covered by a fold of the body wall, the mantle. These structures have been much modified in the different classes, and the cephalopods have lost the foot and most of the shell to become specialized for propulsive swimming. There is some molecular evidence that molluscs, annelids, and arthropods may have arisen from the same ancestral group.

In the annelid worms, the coelom is divided into compartments by septa, and this segmentation allows peristalsis and an active burrowing lifestyle. The polychaetes have developed lateral appendages, parapodia, for swimming, but their soft bodies are not effective foils for limb muscles.

In the arthropods, the exoskeleton provides a firm attachment for muscles. Since the coelom is no longer essential for this purpose, it has been considerably reduced, as has internal segmentation. Because of their advantageous body plan, arthropods have evolved into many diversified and enormously successful groups, the insects in particular. They have invaded most terrestrial habitats, although their exoskeleton and dependence on tracheal gaseous exchange has limited their size.

While the above phyla are protostomes, the echinoderms are deuterostomes with a mesodermal skeleton and a water vascular system for movement and circulation. Many of their characteristics—rudimentary nervous system, isosmolarity with sea water, bilaterally symmetrical larvae—have probably been retained from a free-living bilateral ancestor, although they have subsequently become radically symmetrical with a pentamerous organization. Alternatively they may have been derived from a pentamerous lophophore.

KEY TERMS

altruistic behavior
annelids
archaeocyte
arthropods
Aschelminthes
bivalves
body plan
cephalopods
choanocytes
clitellum
coelenterates
determinate development
deuterostome superphylum
echinoderms
foot (mollusc)
gastropods
haplodiploidy
indeterminate development
insects
invertebrates
kin selection
leeches
lophophores
mantle (mollusc)
medusa
Mesozoa
molluscs
nematocysts
nematodes
oligochaetes
parapodia
parasitism
pentameral radial symmetry
Placozoa
planula
Platyhelminthes
polychaetes
polyp
protostome superphylum
pseudocoelomates
radial cleavage
Radiata
radula
siphon
social organization
spiral cleavage
sponges
superphylum
torsion
trochophore larva

REFERENCES

Barnes, R. D., 1980. *Invertebrate Zoology*, 4th ed. Saunders, Philadelphia.

Bergquist, P. R., 1985. Poriferan relationships. In *The Origin and Relationships of Lower Invertebrates*, S. Conway Morris, J. D. George, R. Gibson, and H. M. Platt (eds.). Clarendon Press, Oxford, pp. 14-27.

Boudreaux, H. B., 1979. *Arthropod Phylogeny With Special Reference to Insects*. Wiley-Interscience, New York.

Clark, R. B., 1964. *Dynamics in Metazoan Evolution*. Clarendon Press, Oxford.

——, 1979. Radiation of the metazoa. In *The Origin of the Major Invertebrate Groups*, M. R. House (ed.). Academic Press, London, pp. 55-102.

Clarke, K. U., 1973. *The Biology of Arthropods*. Edward Arnold, London.

Clarkson, E. N. K., 1979. *Invertebrate Palaeontology and Evolution*. Allen & Unwin, London.

Conway Morris, S., J. D. George, R. Gibson, and H. M. Platt (eds.), 1985. *The Origins and Relationships of Lower Invertebrates*. Clarendon Press, Oxford.

Dobzhansky, Th., F. J. Ayala, G. L. Stebbins, and J. W. Valentine, 1977. *Evolution*. W. H. Freeman, San Francisco.

Dougherty, E. C. (ed.), 1963. *The Lower Metazoa*. Univ. Calif. Press, Berkeley.

Field, K. G., G. J. Olsen, D. J. Lane, S. J. Giovannoni, M. T. Ghiselin, E. C. Raff, N. R. Pace, and R. A. Raff, 1988. Molecular phylogeny of the animal kingdom. *Science*, **239**, 748-753.

Fox, R. M., and J. W. Fox, 1964. *Introduction to Comparative Entomology*. Reinhold, New York.

Hamilton, W. D., 1964. The evolution of social behavior. *Jour. Theoret. Biol.*, **1**, 1-52.

House, M. R. (ed.), 1979. *The Origin of Major Invertebrate Groups*. Academic Press, London.

Hyman, L., 1940. *The Invertebrates: Protozoa Through Ctenophora*. McGraw-Hill, New York.

——, 1951. *The Invertebrates: Platyhelminthes and Rhynchocoela, the Acoelomate Bilateria*. McGraw-Hill, New York.

Kershaw, D. R., 1983. *Animal Diversity*. University Tutorial Press, Slough, Great Britain.

Lutz, P. E., 1986. *Invertebrate Zoology*. Addison-Wesley, Reading, Mass.

Manton, S. M., 1977. *The Arthropoda: Habits, Functional Morphology, and Evolution*. Clarendon Press, Oxford.

Margulis, L., and K. V. Schwartz, 1982. *Five Kingdoms*. W. H. Freeman, San Francisco.

Nichols, D., 1969. *Echinoderms*, 4th ed. Hutchinson, London.

Olson, E. C., and J. Robinson, 1975. *Concepts of Evolution*. C. E. Merrill, Columbus, Ohio.

Paul, C. R. C., and A. B. Smith, 1984. The early radiation and phylogeny of echinoderms. *Biol. Rev.*, **59**, 443-481.

Robson, E. A., 1985. Speculations on coelenterates. In *The Origin and Relationships of Lower Invertebrates*, S. Conway Morris, J. D. George, R. Gibson, and H. M. Platt (eds.). Clarendon Press, Oxford, pp. 60-77.

Smith, J. E. (ed.), 1971. *The Invertebrate Panorama*. Universe Books, New York.

Solem, G. A., 1974. *The Shell Makers*. John Wiley, New York.

Trueman, E. R., and M. R. Clarke (eds.), 1985. *The Mollusca: Vol. 10, Evolution*. Academic Press, Orlando, Fla.

Valentine, J. W., 1977. General patterns of metazoan evolution. In *Patterns of Evolution, as Illustrated by the Fossil Record*, A. Hallam (ed.). Elsevier, Amsterdam, pp. 27-57.

———, 1989. Bilaterians of the Precambrian-Cambrian transition and the annelid-arthropod relationship. *Proc. Nat. Acad. Sci.*, **86**, 2272-2275.

Weller, J. M., 1969. *The Course of Evolution*. McGraw-Hill, New York.

Whittington, H. B., 1979. Early arthropods, their appendages and relationships. In *The Origin of Major Invertebrate Groups*, M. R. House (ed.). Academic Press, London, pp. 253-268.

16

ORIGIN OF VERTEBRATES

The separation of vertebrates as a group from invertebrates as a group is a reflection of the significance we attach to separating ourselves and those organisms that resemble us from all organisms that do not. There are also, of course, other factors in making this distinction since vertebrates are members of a unique phylum, **Chordata**, and among the most varied and successful of all animals. They number about 40,000 species, ranging in size from minuscule fish to giant whales, and have invaded a wide variety of habitats from oceanic depths to soaring heights above the earth. In what characters are vertebrates unusual?

Although Figure 16-1 is a composite of two different groups, fish and human, widely separated along the vertebrate spectrum, it portrays some special morphological features possessed by this phylum:

1. A paired series of clefts, or **gills**, that lead outward from the pharynx and are present in the embryo as well as in the adults of some vertebrate groups.
2. An internal skeletal structure oriented along the anterior-posterior axis which derives from the embryonic presence of a flexible rod, the **notochord**. Although the notochord is replaced by a column of cartilaginous or bony **vertebrae** in the adults of many groups (subphylum Vertebrata), it is this structure which gives the name Chordata to this phylum.
3. A **tail** which may be quite prominent and extends beyond the anus in embryos of all groups, although not in all adults.
4. A single **hollow nerve cord** that runs **dorsally** above the notochord.

These characters seem unusual compared to those of other phyla, and the question arises whether we can find the basis for such organization elsewhere, or perhaps find a direct fossil connection to another phylum. The answers are not yet encouraging.

HYPOTHESES ON THE ORIGIN OF VERTEBRATES

By the time vertebrates appear in the fossil record they are quite distinctive, showing all of the major attributes that characterize them as chordates. Since reliable fossil connections between vertebrates and ancestral

FIGURE 16-1 Composite drawing of man and fish, showing various vertebrate characteristics. (Adapted from S. Ohno, 1970. *Evolution by Gene Duplication.* Springer-Verlag, New York.)

forms are absent, hypotheses about vertebrate origins have been based on comparative methods that seek to discover whatever homologies with other phyla can be found. For the most part, these approaches have relied on morphological and embryological comparisons and have produced a variety of hypotheses.

Among these proposals were suggestions for an annelid-type origin based on some general similarities between the two groups (Table 16-1). To enable the transition from annelids to vertebrates, it was suggested that the dorsoventral position of the annelid nerve cord and the annelid direction of blood vessel flow could be reversed to coincide with the dorsoventral orientation in vertebrates. However, the replacement of both mouth and anus that would accompany such transformations was difficult to visualize, and, as listed in the lower part of Table 16-1, sufficient additional differences were noted between the two groups to abandon that particular hypothesis.

Suggestions for a vertebrate origin from arthropods raised similar objections and were generally abandoned because, like those for an annelid origin, they necessitated what seemed to be a blatant manipulation of organs and tissues—radical changes that would probably have had deleterious effects on the animals involved.

Probably the most popular of present hypotheses is the concept of a common ancestry shared by echinoderms and vertebrates. This concept rests, for the most part, on a variety of traits presumed to be strongly conservative and generally used to distinguish the

TABLE 16-1 Summary of the Main Morphological Arguments For and Against an Annelid Origin for Vertebrates in Respect to Designated Characteristics

	Annelids	Vertebrates
ARGUMENTS FOR AN ANNELID ORIGIN		
Bilateral symmetry	Yes	Yes
Presence of coelom	Yes	Yes
Metameric organization	Yes	Yes
Terminal growth	Yes	Yes
Dorsal and ventral longitudinal blood vessels	Yes	Yes
Anterior "brain"	Yes	Yes
ARGUMENTS AGAINST AN ANNELID ORIGIN		
Complete segmentation through body wall	Yes	No
Coelom embryology	Schizocoelous	Enterocoelous
Position of nerve cord	Ventral	Dorsal
Skeleton	External	Internal
Gill slits	No	Yes
Flow of dorsal blood vessel	Anteriorly	Posteriorly
Flow of ventral blood vessel	Posteriorly	Anteriorly
Fate of blastopore	Mouth (protostome)	Anus (deuterostome)

Adapted from Neal and Rand.

echinoderm superphylum (**deuterostomes**) from the annelid superphylum (**protostomes**):

1. The fate of the blastopore in echinoderms and vertebrates is to produce the adult anus (deuterostome), whereas the blastopore becomes the mouth in the annelid group (protostome).
2. Both echinoderms and vertebrates show radial cleavage of early zygotic cells (blastomeres) rather than the spiral cleavage presumed to be the rule in the annelid group (see Fig. 15-8). Furthermore, since isolated blastomeres of echinoderms and amphibian vertebrates have been shown to develop into normal embryos, their developmental process is considered indeterminate, in contrast to the abnormal embryos produced by isolated blastomeres undergoing spiral cleavage (determinate development).
3. The origin of the coelom in echinoderms and vertebrates is enterocoelous, whereas it is schizocoelous in the annelid superphylum.

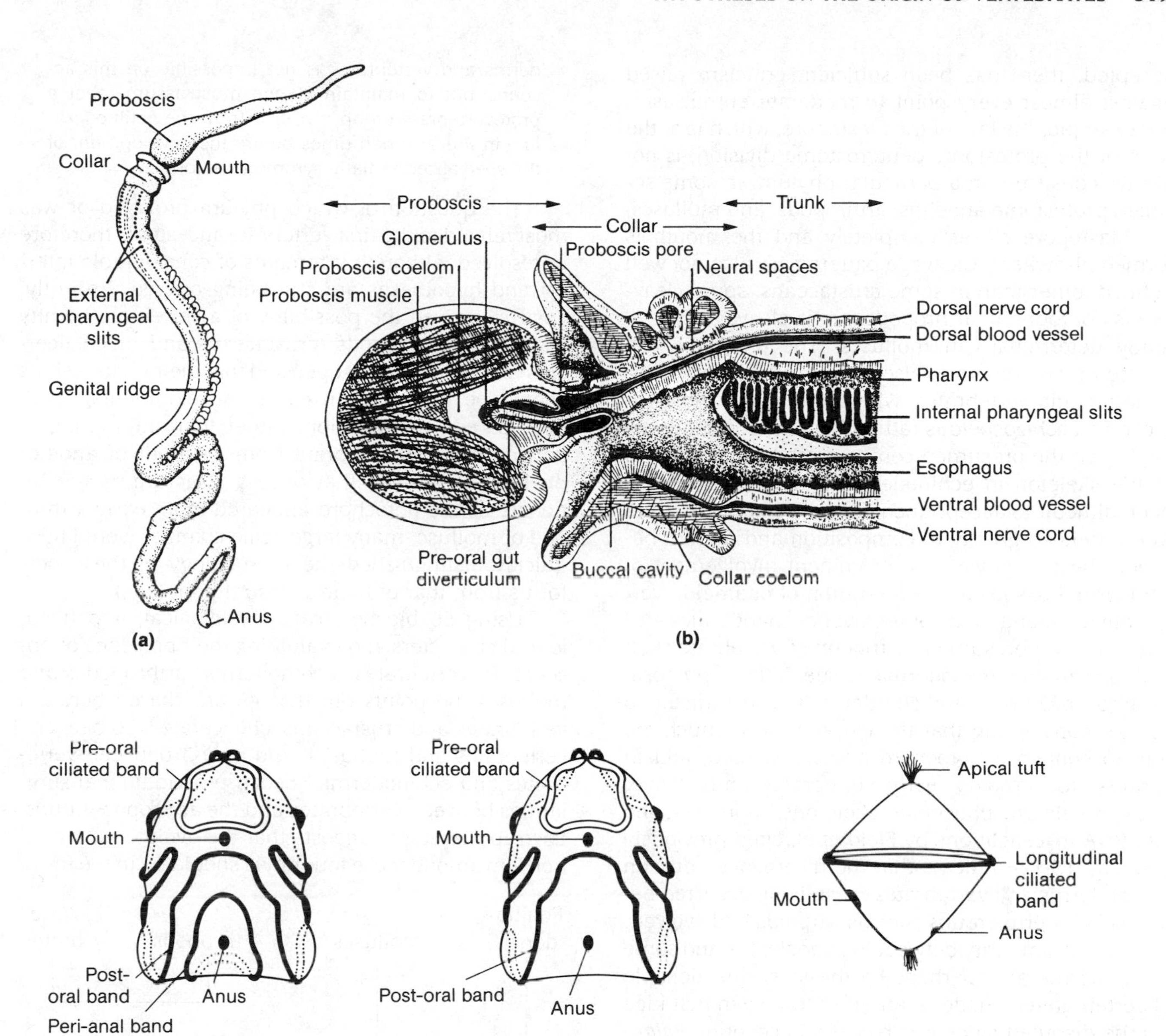

FIGURE 16-2 (a) Adult acorn worm of the genus *Balanoglossus.* (b) Internal structure of the anterior region of a *Balanoglossus* species. The pre-oral gut diverticulum is considered by some zoologists as a primitive notochord. (c) The tornaria larva of *Balanoglossus.* (d) The auricularia larva of echinoderm holothurians. The major difference between (c) and (d) is in the presence or absence of a peri-anal ciliated band. (e) The trochophore larva found in annelids and molluscs (see also Fig. 15-11).

4. The skeleton of vertebrates and echinoderms originates embryonically from mesodermal tissue, but originates from ectodermal tissue in the annelid superphylum.
5. Jeffries and co-workers have proposed that some groups among the echinoderm carpoids (see Fig. 15-18) have lateral openings that resemble gill slits. These and other suggested vertebratelike features, such as tails, may indicate a direct connection between echinoderms and chordates.
6. The larvae in echinoderms are either of the pluteus-type or a variation of it, such as the auricularia larvae in holothurians (sea cucumbers). By contrast, the annelid superphylum produces a trochophore larvae. Since the acorn worm, *Balanoglossus* (Fig. 16-2(a)), has an auricularia-type larva (called tornaria) and this animal is generally believed to be a chordate, although primitive (**hemichordate**; subphylum Hemichordata), a strong link between echinoderms and vertebrates logically follows (Fig. 16-2(c), (d)).

Although some of these basic arguments for echinoderm affinity to vertebrates have been widely accepted, there has been sufficient criticism raised against almost every point to moderate enthusiasm. For example, the fate of the blastopore, which is at the base of the protostome-deuterostome division, is not always consistent in a particular phylum: in some so-called protostome annelids, arthropods, and molluscs, the blastopore closes completely and the mouth is formed elsewhere. Cleavage patterns are also not well defined: other than in some crustaceans, spiral cleavage is not found in arthropods, nor is cleavage consistently determinate in molluscs. The embryological source of the coelom is also in dispute for both echinoderms and vertebrates, with some observations indicating a schizocoelous rather than enterocoelous origin. Even the presumed common mesodermal origin of the skeleton in echinoderms and vertebrates has been difficult to accept unequivocally since this structure differs so greatly in composition and pattern between the two phyla. The prominent involvement of ectodermal tissues in the formation of enamel in vertebrate teeth and in the outer layer of shark scales also questions the presumed restriction of vertebrate skeletal formation to mesodermal tissue. Jeffries' proposal of a carpoid origin for chordates is also difficult to accept, considering that the carpoid tail is much too thin to contain a notochord and nerve cord, and in contrast to carpoids, early vertebrates had already evolved calcium phosphate skeletons. Moreover, the 18*S* RNA investigations by Field et al. cited previously (see Fig. 15-12) indicate that the divergence between echinoderms and vertebrates most likely occurred before echinoderm groups such as carpoids had evolved.

Larval similarity between *Balanoglossus* and some echinoderms may perhaps be the least questionable of vertebrate-echinoderm affinities, but even that idea can be disputed since it is based on accepting *Balanoglossus* as a chordate-like invertebrate—a point that has been argued from the time it was first proposed by Bateson in 1886. According to Bateson, the *Balanoglossus* pre-oral gut diverticulum was a notochord of sorts (see Fig. 16-2(b)), although no one has yet found anything rodlike or notochordlike in it. Nevertheless, many zoologists still consider *Balanoglossus* to have chordate similarities, basing this association primarily on the presence of gill slits.

From these considerations the hypothesis of a vertebrate-echinoderm relationship seems only slightly supported. As Stahl pointed out:

> All extant echinoderms develop a radial symmetry and a semisessile habit that are far distant from the vertebrate condition. A connection between echinoderms and vertebrates is not impossible on this account, but to maintain it, one must assume that a protovertebrate group diverged from the echinoderm line in Precambrian times before the development of the specialized radially symmetrical forms.

The question of which phylum provided or was most related to the first vertebrate ancestor is therefore unresolved, although it remains of considerable interest and hypotheses are still being offered. Recently, Løvtrup revived the possibility of a vertebrate affinity with either arthropods (crustaceans and/or chelicerates) or molluscs and suggested that their relationships should be traced to an earlier developmental level, perhaps egg organization, rather than attempting to derive an adult vertebrate from an adult of another phylum. He argues that once a structure as revolutionary as the notochord appeared in an early arthropod or mollusc, many large-scale changes would have quickly distinguished the morphology of these new forms from that of their ancestral population.

Using 56 biochemical, physiological, and histological characters and evaluating their presence or absence in vertebrates, echinoderms, arthropods, and molluscs, he points out that 45 are shared between vertebrates and crustaceans/chelicerates, 50 between vertebrates and molluscs, and only 3 between vertebrates and echinoderms.[1] Because of additional similarities between vertebrates and the arthropod groups, Løvtrup therefore suggests that the evolutionary relationship among these four phyla should be in the form:

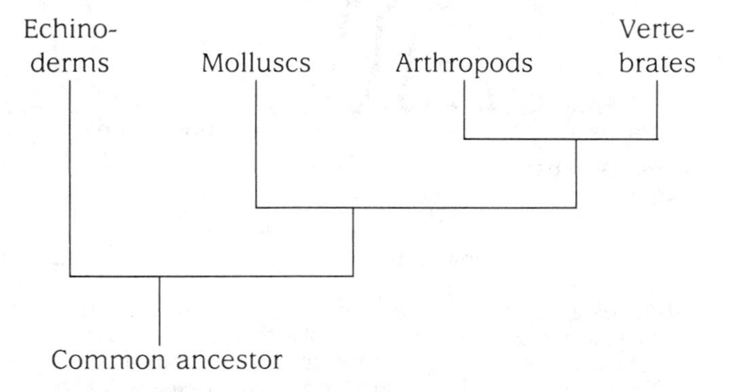

Whichever phylogeny seems more convincing, that of Løvtrup or the commonly accepted hypothesis of echinoderm origin or derivation from still another phylum, it is clear that there are still many hidden steps in the origin of chordates. Since so much is unknown, it may be more tractable to simply accept the existence of a primitive chordatelike animal and ask rather general questions of how it may have lived and under what conditions it may have evolved further.

[1]The sharing of some of these characters between vertebrates and the arthropod-molluscan groups undoubtedly arises from convergence, but it seems unlikely that all of the similarities are convergent.

CEPHALOCHORDATES AND UROCHORDATES

In addition to vertebrates, a number of living chordatelike animals are also usually classified as subphyla of Chordata. Although none of these possess vertebrae or some of the other characteristics of vertebrates, they all have features that have prompted zoologists to offer hypotheses on what early chordates were like. Among these, the most similar to vertebrates are **cephalochordates** such as the marine lancelet, *Branchiostoma*, known also by the common name of **amphioxus**. These 1- to 2-inch long fishlike marine animals swim by contracting metamerically organized muscles (myotomes) placed alongside a semi-rigid notochord that runs from tip to tip (Fig. 16-3). As an adult, the animal is mostly sedentary, burrowing in the sea bottom, then extending its anterior end into the waters above to filter-feed on passing food particles. They have a large pharyngeal "branchial basket," penetrated by up to 200 gill slits, that filters water drawn through the mouth by ciliary currents and then passes extracted food along a mucosal strand into the digestive tract.

It has often been pointed out that cephalochordates are probably not in the direct line of vertebrate ancestry because (1) their notochord does not end at a "brain" but extends to the very anterior tip of the animal, (2) they possess excretory cells more related to those of platyhelminths and annelids, and (3) they have no sense organs that can be related to those of vertebrates. However, their possession of notochord, gill slits, dorsal nerve cord, metamerically organized myotomes, posterior direction of blood flow in the dorsal vessels and anterior direction in the ventral vessels, as well as vertebratelike organs such as the thyroid indicates that they are certainly close relatives to vertebrates. An accidental convergence of so many basic characters in two unrelated groups would be difficult to accept. It seems therefore reasonable to propose that cephalochordates, although perhaps not directly ancestral to modern vertebrates, represent a mode of life that early chordatelike animals probably shared, that is, swimming and **filter feeding**.[2]

At least two points of evidence support this view. One is the fact that lampreys, a primitive group of jawless and boneless vertebrates but vertebrates nevertheless, possess an **ammocoete** larval form that is also a swimming filter feeder remarkably similar to cephalochordates. Another point that will be discussed

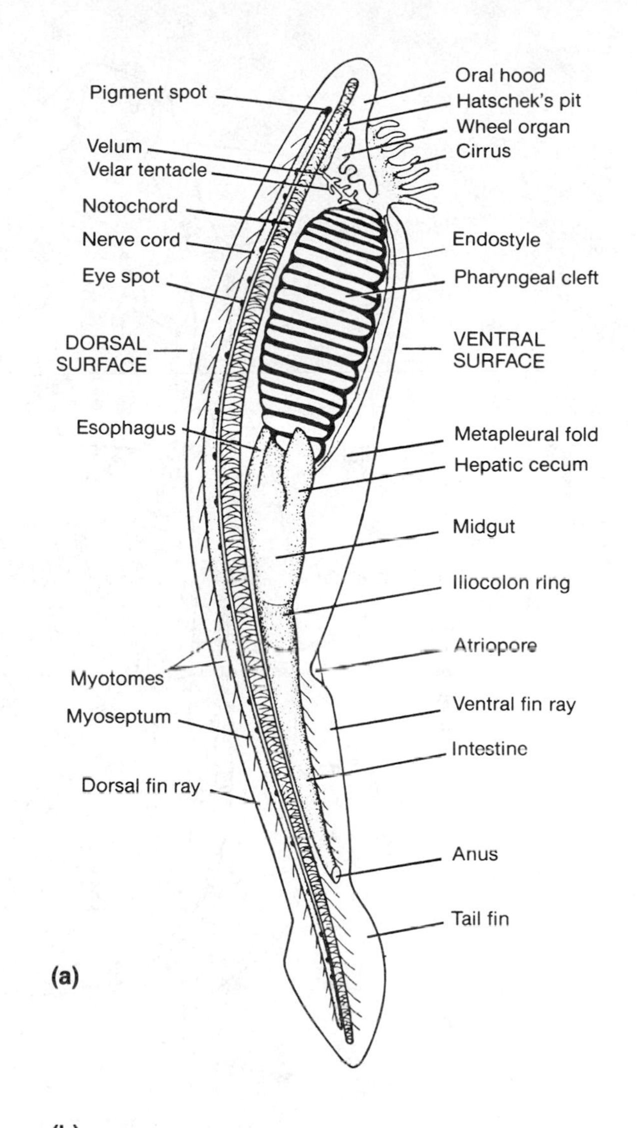

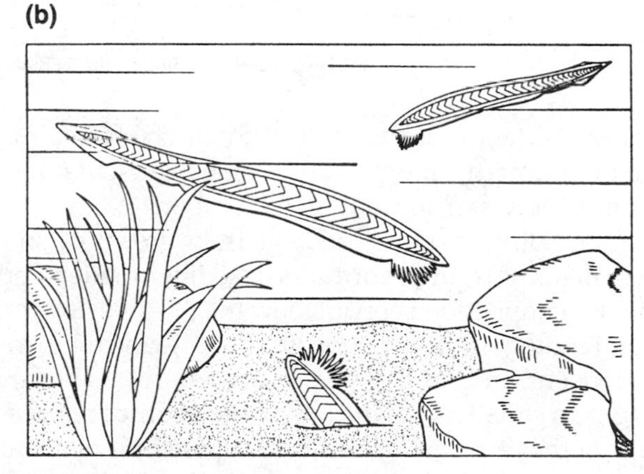

FIGURE 16-3 Anatomical features (a) and habitat (b) of the cephalochordate *Branchiostoma*.

[2]A song popular among summer students who attended or worked at the Marine Biological Laboratory at Woods Hole on Cape Cod, followed the World War I tune "It's A Long Way To Tipperary":

It's a long way from Amphioxus
It's a long way to us;
It's a long way from Amphioxus
To the meanest human cuss;
Goodbye fins and gill slits,
Welcome teeth and hair;
It's a long long way from Amphioxus
But we came from there!

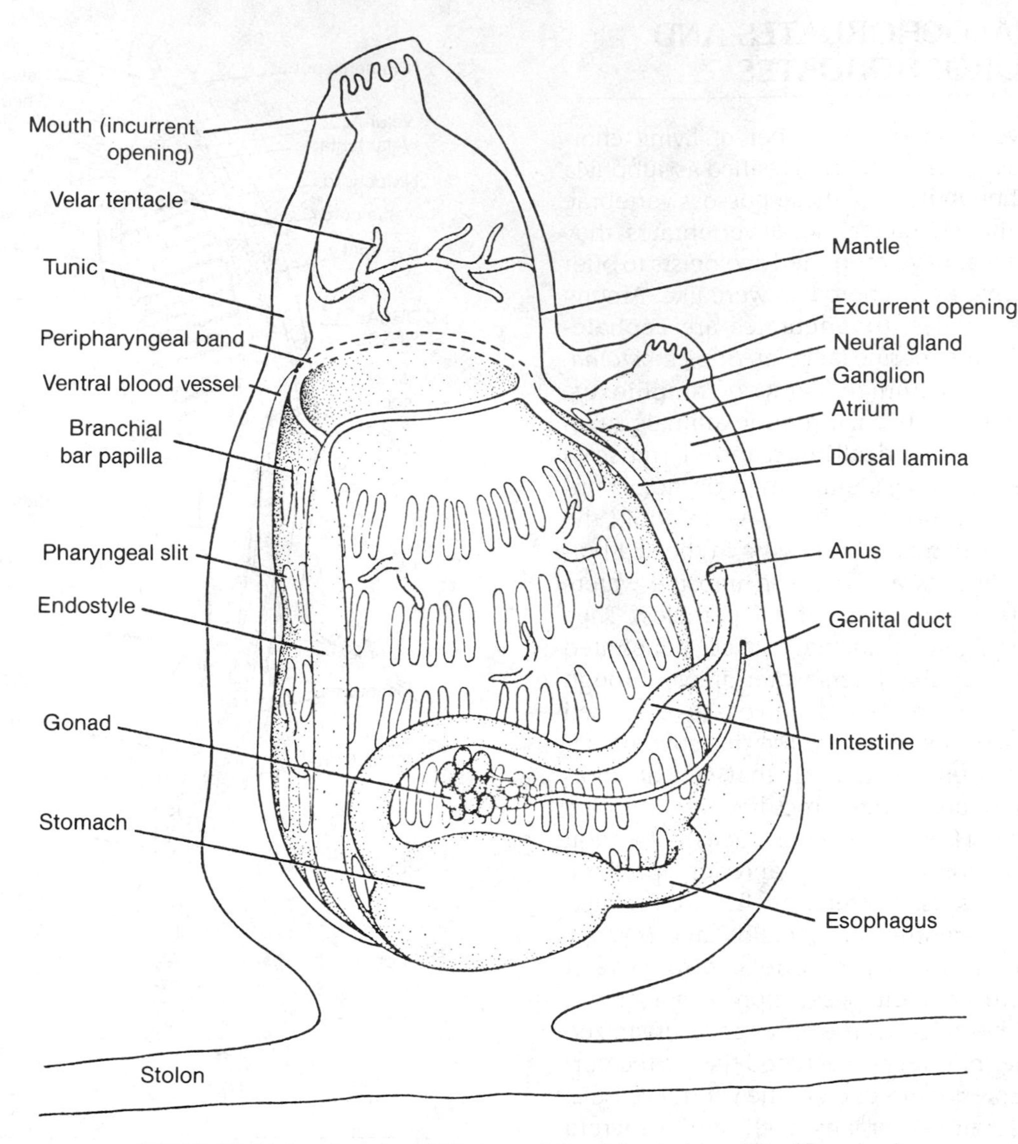

FIGURE 16-4 Internal anatomical features of an adult colonial ascidian (*Perophora*).

later is the fact that the earliest fossil chordates were also undoubtedly filter feeders, using a muscular pharynx to suck water into the gill chamber.

According to many zoologists, a group of small marine animals, **urochordates**, also filter feeders, provide the connection showing how the progression from filter feeding to active notochordial swimming may have occurred. Among the urochordates is a large subgroup called ascidians (also called **tunicates** or sea squirts) that are sessile in the adult, feeding by means of a large pharyngeal basket through which water is drawn, filtered, and then expelled through many gill slits (Fig. 16-4). Although aside from gill slits there is no immediate evidence of chordate structure, the organization of the ascidian larvae is remarkably different from the adult. As shown in Figure 16-5(a) this larva has a notochord and dorsal nerve cord as well as gill slits. Its function is obviously to actively seek out suitable habitats and distribute itself as widely as possible before metamorphosing into the sessile adult form (Fig. 16-5(b), (c)).

The possibility that this metamorphosis could be increasingly delayed with each successive generation so that the swimming larval form itself becomes sexually mature was developed by Garstang and quickly accepted by many biologists. This process, termed **paedomorphosis** (shaped like a child), involved the incorporation of adult sexual features into earlier immature stages. Garstang's hypothesis was that some urochordate groups such as the Larvacea arose by paedomorphosis and can now be considered as sexually mature larvae. Extending this argument, early free-swimming chordates were believed to have passed through this same process of paedomorphosis in descending from ancestors that had chordatelike larval stages: that is, the swimming, filter-feeding larvae of

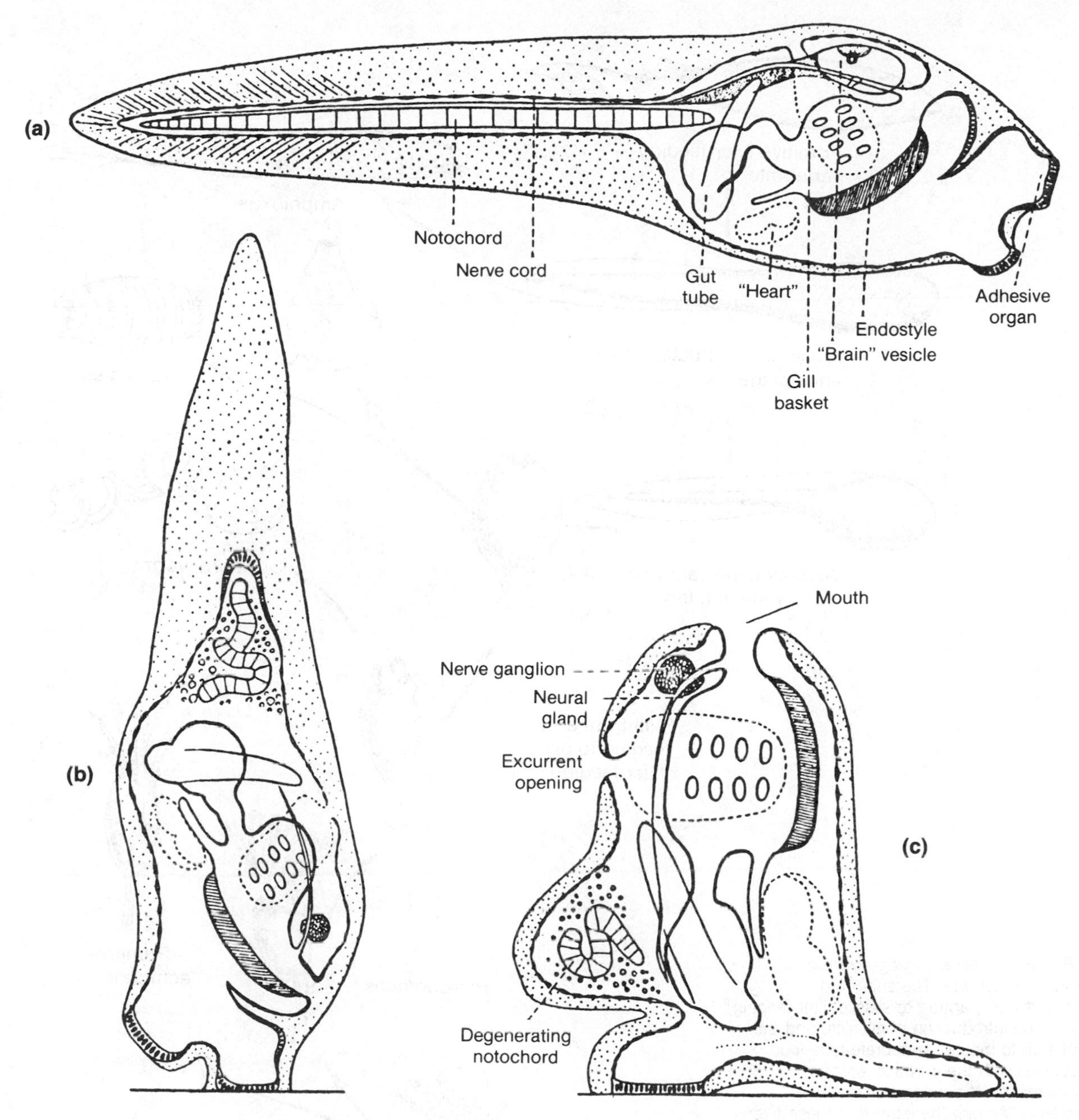

FIGURE 16-5 (a) Free-swimming ascidian larva before metamorphosis. (b) Attachment of the larva to a substrate by its anterior suckers is accompanied by degeneration of the notochord and rotation of its internal structures. (c) Stage at which the resorption of the tail is almost complete and the body parts are assuming their adult positions.

prechordate animals became more successful than their adult sessile forms, perhaps because they could follow and search out new food supplies as well as escape predators. Selection for early sexual maturity then occurred among these larval prechordates yielding adult "chordates."[3]

In general, the relationship of vertebrates to cephalochordates and urochordates makes it likely that the earliest of all these chordate groups were actively swimming filter feeders. It is certainly possible that early chordates arose from sessile filter feeders by the incorporation of adult sexual traits into motile juvenile

[3]For those who believed that the similarity between the tornaria larva of the hemichordate *Balanoglossus* and the auricularia larva of echinoderms stems from a phylogenetic relationship, Garstang's proposal had the added attraction of explaining that the transition between the two phyla occurred through a larval form rather than through changes in the considerably different adult forms. Evolution of echinoderms into chordates was visualized as the development of dorsal and neural folds from ciliated bands of the auricularia accompanied by the introduction of gill slits and a notochord. The origin of the latter structures was not explained, nor was it clear how this unusual larva reached the adult sexual stage.

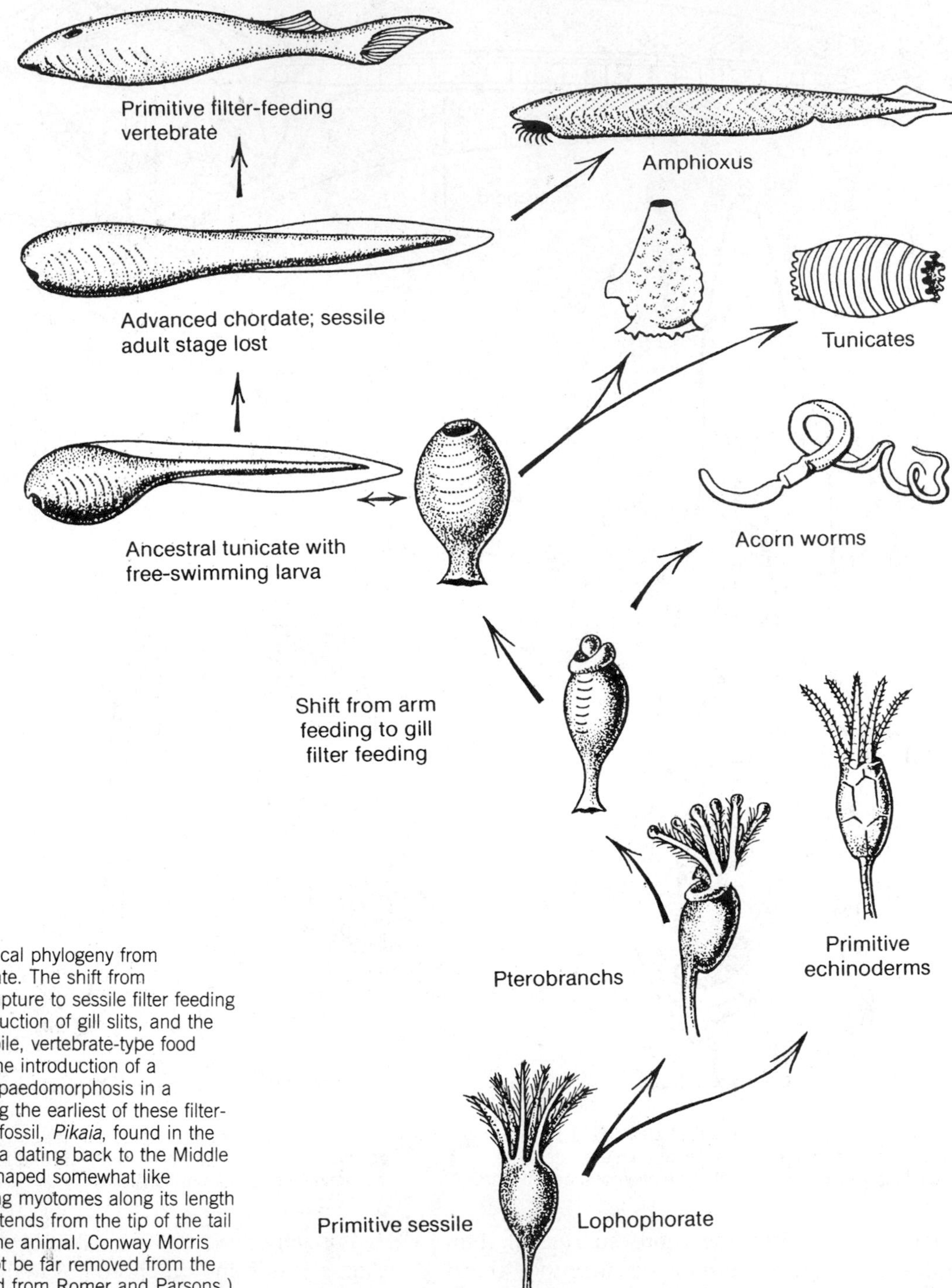

FIGURE 16-6 Hypothetical phylogeny from lophophorate to vertebrate. The shift from lophophore-type food capture to sessile filter feeding occurred with the introduction of gill slits, and the subsequent shift to mobile, vertebrate-type food capture occurred with the introduction of a notochord by means of paedomorphosis in a tunicatelike larva. Among the earliest of these filter-feeding vertebrates is a fossil, *Pikaia*, found in the Burgess Shale of Canada dating back to the Middle Cambrian period. It is shaped somewhat like amphioxus with repeating myotomes along its length and a notochord that extends from the tip of the tail to the anterior third of the animal. Conway Morris suggests that "it may not be far removed from the ancestral fish." (Adapted from Romer and Parsons.)

forms through paedomorphosis. This process and its converse, **neoteny** (the retention of some immature morphological traits into adult stages), are not unusual events, as witness, for example, the observation that some salamanders such as the mudpuppy and axolotl reproduce while in their gilled, immature forms.

Since there are hemichordates such as the pterobranch *Rhabdopleura* (see Fig. 14-11) which use a lophophore (group of ciliated tentacles) for food capture, it has been suggested that links in the chordate chain extend back to a lophophorate-type ancestry which may have been shared with other lophophorates such as phoronids, brachiopods, and ectoprocts, as well as with the crinoidlike ancestors of the echinoderms. A proposed phylogeny of this type, extending from lophophore ancestor to vertebrates, is shown in Figure 16-6.

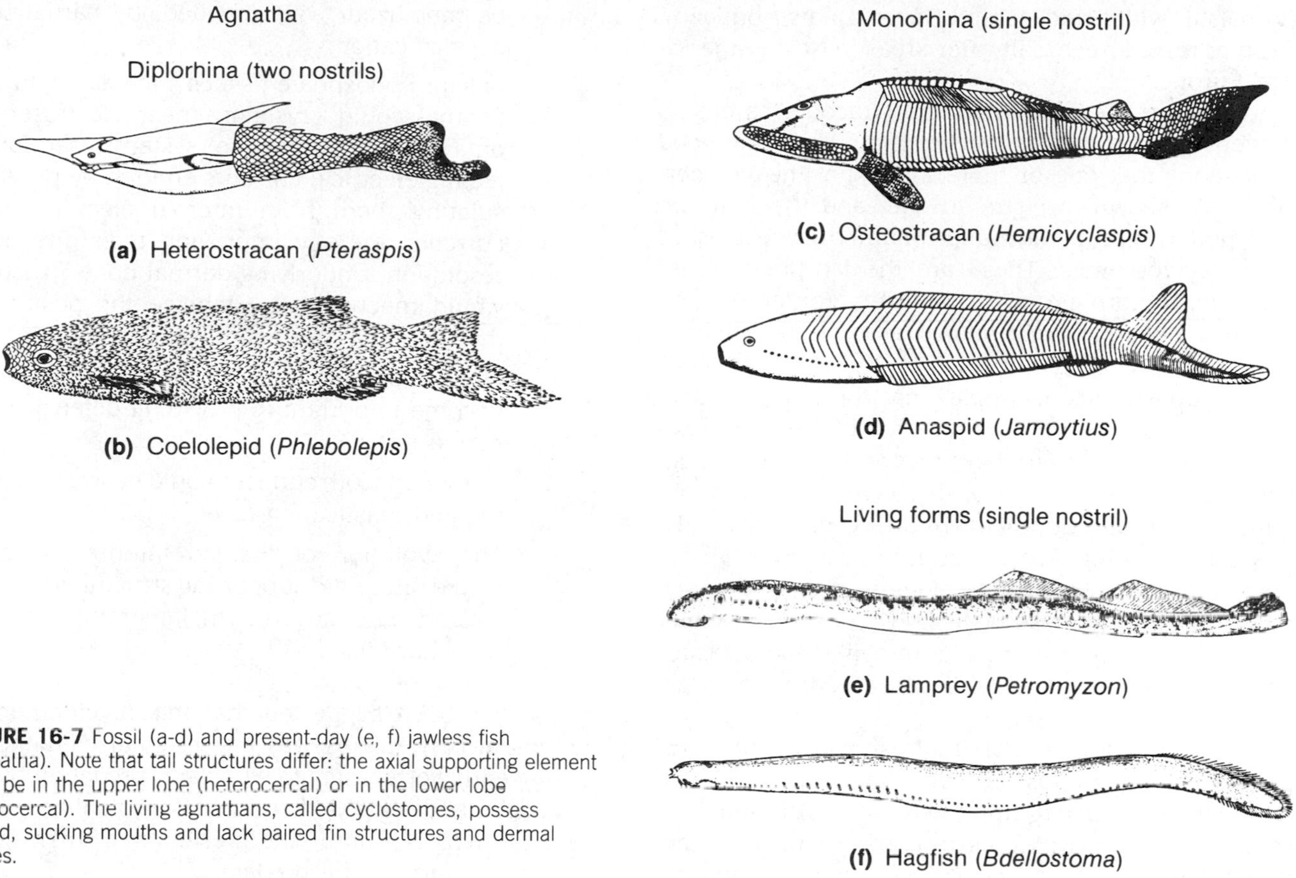

Agnatha

Diplorhina (two nostrils)

(a) Heterostracan (*Pteraspis*)

(b) Coelolepid (*Phlebolepis*)

Monorhina (single nostril)

(c) Osteostracan (*Hemicyclaspis*)

(d) Anaspid (*Jamoytius*)

Living forms (single nostril)

(e) Lamprey (*Petromyzon*)

(f) Hagfish (*Bdellostoma*)

FIGURE 16-7 Fossil (a-d) and present-day (e, f) jawless fish (Agnatha). Note that tail structures differ: the axial supporting element may be in the upper lobe (heterocercal) or in the lower lobe (hypocercal). The living agnathans, called cyclostomes, possess round, sucking mouths and lack paired fin structures and dermal bones.

FOSSIL JAWLESS FISH (AGNATHA)

The earliest known vertebrates are a group of **jawless fishes**, the heterostracans, whose armored skin plates (dermal bones) appear in marine deposits of the late Cambrian period and whose fossils extend into the late Devonian. They were usually small fish, 8 to 12 inches long, encased in large dermal plates anteriorly and smaller plates or scales posteriorly (Fig. 16-7(a)). In addition to two laterally placed eyes they also possessed a mid-dorsal opening for a median eye or pineal organ. Various forms of these fishes are found, some obviously bottom feeders, mostly with ventrally placed mouths, but others with mouths on the dorsal surface that may have fed on plankton. Both the heterostracans and an associated group that were about half their size, the coelolepids (hollow-scaled, Fig. 16-7(b)) had two nasal openings (subclass Diplorhina) distinguishing them from other jawless fishes with single nostrils (Monorhina). The latter include the heavily armored osteostracans (bony shells, Fig. 16-7(c)) and the lightly armored but apparently more maneuverable "shieldless" anaspids (Fig. 16-7(d)), both forms appearing in the mid-Silurian and becoming extinct during the Devonian.

It has been suggested (Moy-Thomas and Miles) that all four of these jawless groups, often called **ostracoderms** (bony skins), made use of a large, **muscular pharynx** that enabled them to suck up food-laden water more rapidly and in much larger amounts than could be achieved by the ciliary activity of invertebrate lophophorates and filter feeders. On the other hand, Northcutt and Gans propose that although the early vertebrate pharynx may have originally been used for pumping water and filter feeding, it evolved into an active predatory organ in adult ostracoderms. According to them, jawless fish used the pharynx for scooping up small soft-bodied or lightly armored bottom-dwelling animals. However, whether through improved filter feeding or active benthic predation, most paleontologists agree that pharyngeal adaptations were important in accounting for ostracoderm early success. Recent reports indicate they inhabited late Cambrian and early Ordovician marine environments

(Repetski) with a widespread Devonian distribution in fresh or mixed fresh/saltwater areas in North America and Europe.[4]

After the Devonian period, ostracoderms are no longer found in the fossil record, but there are good indications that one or more agnathan lineages persisted. As shown in Figure 16-7(e) and (f), there are two modern groups of round-mouthed jawless fishes called **cyclostomes**. These are the lampreys (order Petromyzontiformes) and hagfishes (order Myxiniformes), both probably related to ostracoderms, although the precise pattern of ancestry is unclear. It has been proposed, for example, that the fossil anaspid *Jamoytius* may be in the direct line of descent of ostracoderms and modern lampreys and perhaps of hagfishes as well. In any case, the advent of jawed and finned vertebrates during the mid-Devonian eventually pushed remaining jawless ostracoderms and their descendants into restricted ecological niches in which they have nevertheless been able to maintain themselves to this day, either as lampreylike ectoparasites or as burrowing hagfishlike detritus feeders and scavengers.

It is interesting to note that fish with **bony plates** appear first in the fossil record, and only in the late Devonian are **cartilaginous fish** (sharks) found. This sequence obviously seems to conflict with the long-popular Haeckelian concept ("ontogeny recapitulates phylogeny," p. 42) that since cartilage development precedes bone development in ontogeny, the phylogeny being recapitulated is one in which cartilaginous fish evolved first and bony fish later. Since cartilage does not fossilize well, it is difficult to prove which of these tissues evolved first, but the relatively early occurrence of bone must have provided various immediate and subsequent advantages, such as the following.

1. Excess calcium ions, diffusing through the skin and gills, can be deposited in the skin as an osmotically inert substance, thus conserving energy that would otherwise be expended in excreting these ions.
2. Tissue deposits of calcium and phosphate would provide a metabolic reserve that could be mobilized, when needed, by partial bone decalcification.
3. Calcium phosphate tissues such as dentin and enamel could crystallize near electrosensory organs (e.g., "lateral line" systems used for detecting electrical currents emitted by prey), insulating them from internal electrical body currents, thereby improving their directional resolution. Underlying dermal bone structures would mechanically stabilize the position of these organs (Northcutt and Gans).
4. The elaboration of such bone structures would become important in providing defensive dermal armor.
5. Hardened tooth surfaces could evolve for grasping and masticating food.
6. The evolution of ossified internal skeletons would offer rigid supporting structures for muscle and organ attachment, far stronger than the notochord itself.

It has been suggested that the development of defensive bony plates was an essential element enabling early vertebrates to withstand predation by the voracious and widely distributed scorpionlike eurypterids—that is, until the vertebrates themselves evolved into important predators.

EVOLUTION OF JAWED FISHES (GNATHOSTOMATA)

The first jawed fossil vertebrates appeared during the Silurian period and are divided into two groups. The earliest of these, the **acanthodians** (spiny sharks), are generally represented by *Climatius*, which was only a few inches long and characterized by both paired and unpaired spiny fins (Fig. 16-8). Although the earliest known forms are in marine sediments, they were, for the most part, freshwater animals found in river, lake, and swamp deposits, many surviving up to late Paleozoic times.

Sometime after the first appearance of these spiny fishes, towards the end of the Silurian period, another

[4]Smith, from physiological evidence, suggested that early vertebrates originated in freshwater streams and lakes and only later entered the saltwater marine environment. He proposed that the vertebrate kidney arose as an organ regulating osmosis in a fresh-water environment in which the concentration of ions is much lower than in cellular tissues. To prevent "swamping" of body tissues by incoming water, the kidney glomeruli pump out excess water, while the kidney tubules resorb necessary ions and small molecules back into the circulatory system. In contrast to these views, it has been pointed out that the presence of vertebrate kidney glomeruli do not necessitate a freshwater origin but may have arisen in areas of dilute sea water, such as the brackish coastal estuaries where continental rivers empty their contents. The osmoregulatory function of the kidney emphasized by Smith may also be secondary to its excretory function; that is, the adaptive value of the kidney was primarily to get rid of waste products accumulated by an animal with a high rate of metabolism. The likelihood of a marine and estuarine origin for early vertebrates is also emphasized by the observation that presumed ancestral or closely related forms (cephalochordates, urochordates, and hemichordates) are all marine fauna, found mostly in shallow waters.

FIGURE 16-8 The acanthodian *Climatius* showing the broad-based spiny fins running mid-dorsally and (in two rows) ventrally. These fins were completely covered by small armored scales. (Adapted from Colbert.)

group of jawed fishes evolved called **placoderms** (plate skinned), which flourished during the Devonian, then rapidly became extinct. Some placoderms were bottom dwellers, such as the skatelike rhenanids (Fig. 16-9(a)) and the antiarchs with their stiltlike, jointed pectoral appendages (Fig. 16-9(b)). Others were predators of gigantic proportions such as the arthrodire *Dunkleostus*, with a length of more than 30 feet (Fig. 16-9(c)).

Significant in both acanthodians and placoderms were the presence of **jaws** and further development of **paired fins**. Jaws revolutionized the way of life of these early vertebrates by offering them new food resources which had previously been excluded because of the limitations in filter feeding and sucking. Carnivorous behavior could now be extended to all sizes of prey through grabbing, tearing, and chewing. Hard or armored food materials that were formerly inaccessible, such as molluscs, could now be ground and milled by flattened, opposed teeth. In addition, jaws provided defensive and aggressive behaviors that could be used both intra- and interspecifically and offered greater opportunities to manipulate the environment in the building of nests or grasping of mates. The teeth that provided primitive jaws with their cutting function came either from the evolution of skin "denticles" that had initially served as armor plating in these early vertebrates or from cutting edges on the jaws themselves, as could be seen in some of the placoderms.

The intermediary steps between jawless and jawed fish are not known from the fossil record, although views have been expressed that jaws arose by the transformation of pharyngeal **gill arches** previously used in filter feeding and perhaps respiration as well (Romer and Parsons). These gill arches are paired on each side and supported by V-shaped hinged structures whose apices point posteriorly. According to proponents of the gill arch-jaw transformation the fate of the anterior pairs of gill arches is unclear; they may have disappeared, been incorporated into the base of the cranium, or used to form one or more of the mouth structures. Whatever happened to them, the first gill arch posterior to these anterior pairs underwent changes so that the upper part of the hinge became the upper jaw, or palatoquadrate bone, and the lower part became the mandible (Fig. 16-10). Behind these structures, an arch called the hyoid was incorporated into the complex by contributing its dorsal portion, the hyomandibular, to anchor the hinge of the jaw to the brain case. Among the evidence to support this view is the archlike appearance of palatoquadrate and mandible in acanthodians as well as in later sharks and bony fish (Fig. 16-11). A further jaw/gill slit connection is the observation that the trigeminal cranial nerve in the shark (and some of the other cranial nerves) runs one of its branches down to the lower jaw and another anteriorly to the upper jaw as though a gill slit had been enclosed at one time (Fig. 16-12). The fact that there are cranial nerves anterior to the trigeminal has been used to indicate there once were gill slits anterior to those involved in jaw formation.

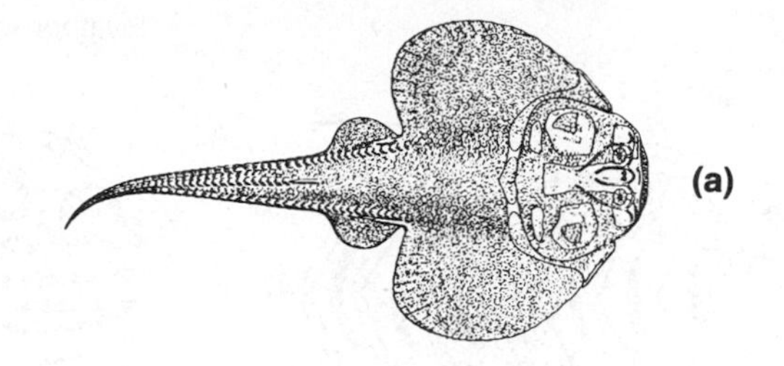

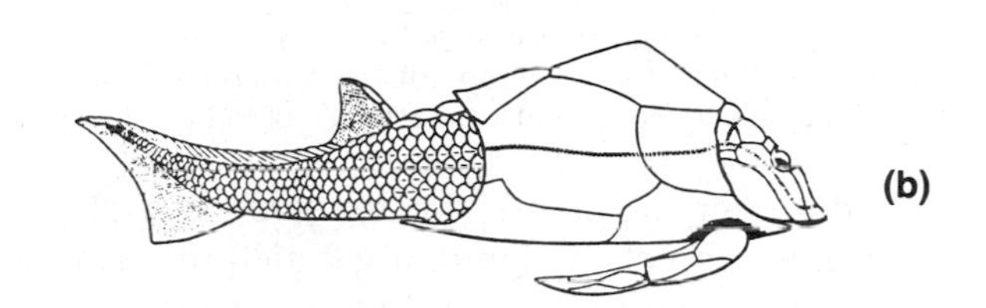

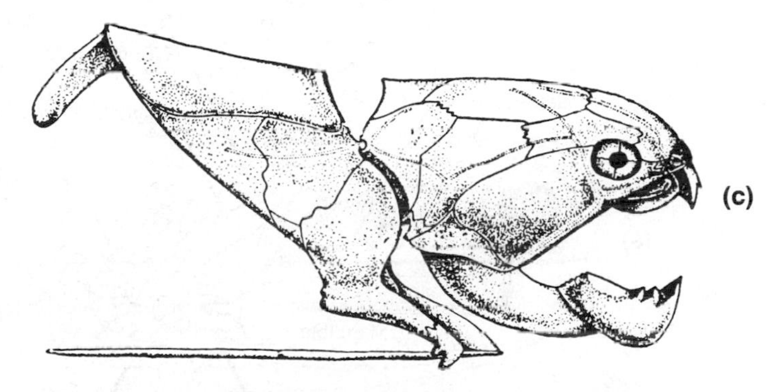

FIGURE 16-9 Placoderms. (a) Skatelike rhenanid (*Gemuendina*) with large lateral fins. (b) An antiarch (*Pterichthyodes*) showing the scaled posterior portion and the heavily armored anterior trunk and head regions. The pectoral "fins" of antiarchs were encased in bony plates that may have enabled them to crawl along the sea bottom. (c) Ten-foot-long anterior bony shield of the arthrodire *Dunkleostus*. The animal was about 30 feet long. (Adapted from Romer.)

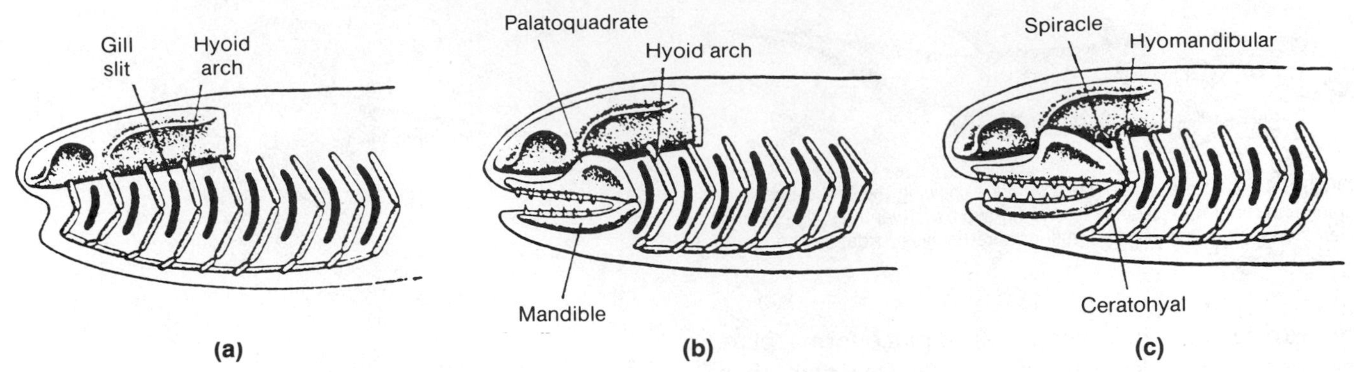

FIGURE 16-10 Stages in the evolution of jaws according to the hypothesis that the vertebrate jaw is derived from one of the anterior gill arches. (a) The jawless condition. (b) The conversion of an anterior gill arch into jaws. (c) Incorporation of bones from the hyoid arch to support the hinge of the jaw. In this progression, the gill slit anterior to the hyoid arch became reduced to the spiracle. (Adapted from Romer and Parsons.)

A different hypothesis for the evolution of jaws proposes that no clear sign of a gill arch remnant anterior to the mouth has been found in either ancient or modern vertebrates. According to this view, the palatoquadrate and mandibular bones may never have served as gill supports, and the mouth, as well as its supporting structures, was always distinctly separate from the pharynx (Carroll). Although it is not certain which of these hypotheses is correct, there is general agreement that, because their jaw structures are so similar, the major modern forms of fish must have derived their jaw pattern from a common ancestral group.

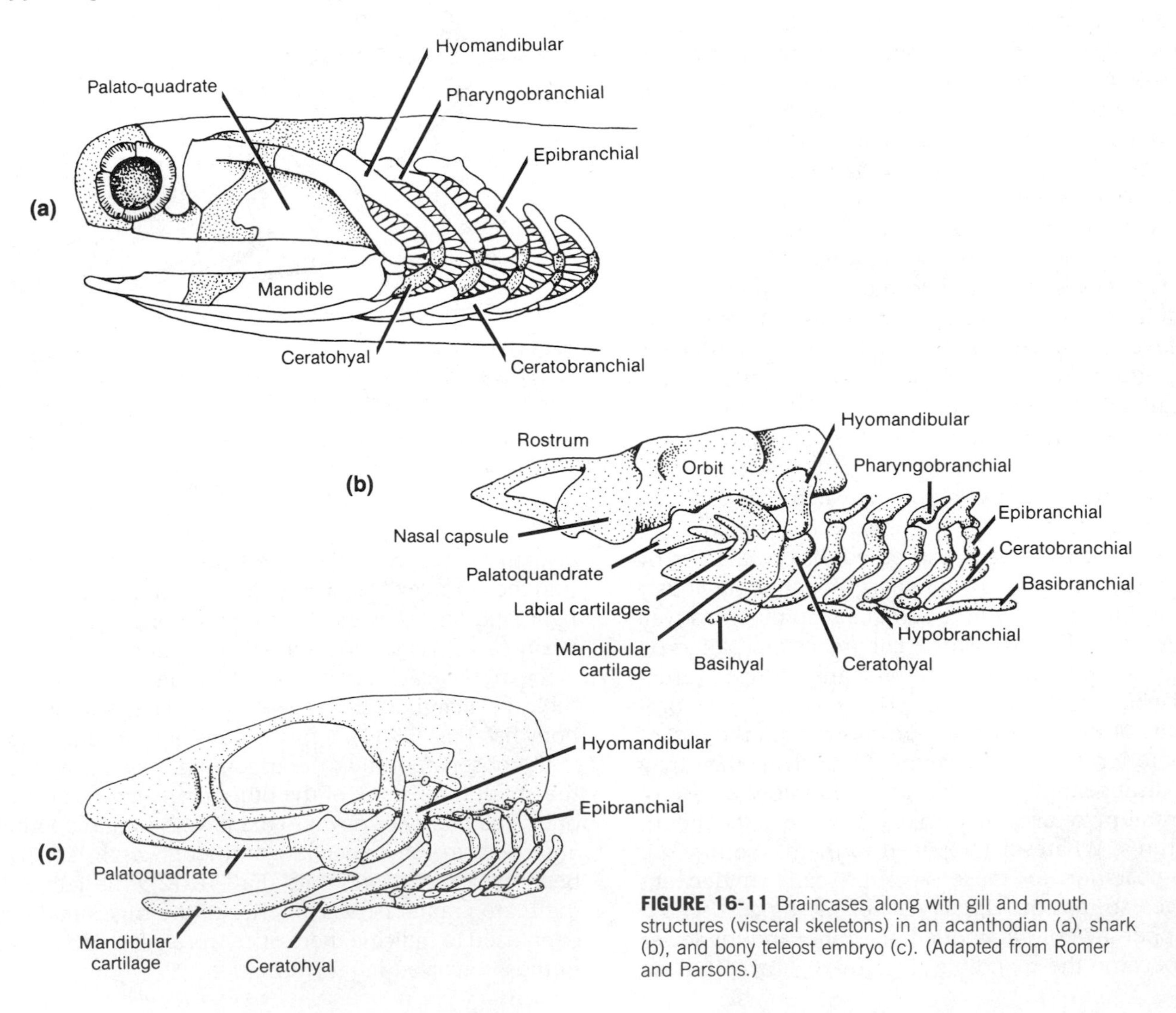

FIGURE 16-11 Braincases along with gill and mouth structures (visceral skeletons) in an acanthodian (a), shark (b), and bony teleost embryo (c). (Adapted from Romer and Parsons.)

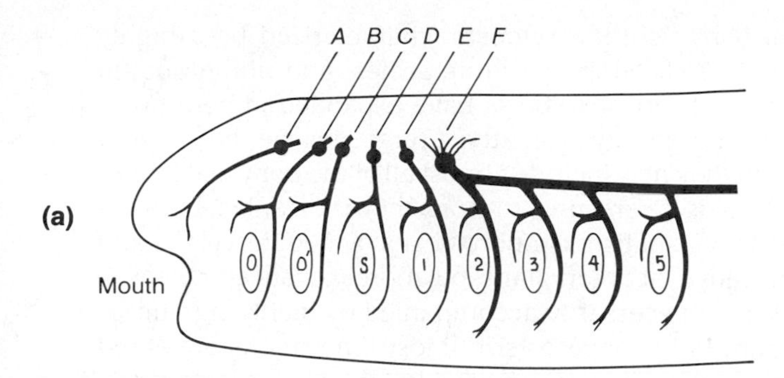

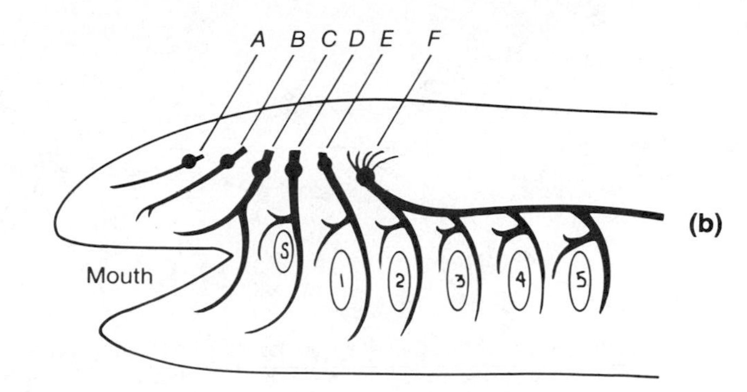

FIGURE 16-12 Diagrammatic views of the dorsal root cranial nerves that innervated the gill arches (a) in a hypothetical primitive fish, and (b) in a later jaw-bearing fish such as the shark. The letters *o* and *o'* represent gill slits lost with the acquisition of jaws, *s* represents the gill slit later used as the spiracle in sharks, and *1* to *5* are gill slits posterior to the spiracle. *A* to *F* are cranial nerves. *A* represents the terminal cranial nerve found in numerous vertebrates that may have innervated the anteriormost member of the gill series. *B* represents a nerve that is independent in lower vertebrates but combines with the trigeminal nerve in mammals. *C* is the trigeminal nerve that innervates the upper and lower jaws in all present vertebrates. *D, E,* and *F* represent, respectively, the facial, glossopharyngeal, and vagus nerves. (See Fig. 3-11 for a comparison between innervations of the vagus nerve in fish and in mammals.) (Adapted from Romer and Parsons.)

CHONDRICHTHYES AND OSTEICHTHYES

Evolving from a group that was most probably separate from that of placoderms, new groups of jawed fishes began to appear during the Devonian period. Their success was probably influenced by improved swimming efficiency resulting from increased nervous and muscular coordination and progressively streamlined body forms. Cartilaginous fishes among these groups are classified as members of the class **Chondrichthyes** and bony fishes in the class **Osteichthyes**.

Although there is a variety of subgroups in each of these classes, there are sufficient common features within each class that indicate some degree of homology. The Chondrichthyes are cartilaginous, with no identifiable bone tissue, although some parts can become calcified. Almost all of these cartilaginous fish have heterocercal tails, a feature also common in early bony fish. Exact genealogical lines of descent to modern chondrichthyians such as sharks, skates, and rays (and to a strange group called chimaeras) are not known, although these modern forms are all unquestionably in the same group. The Osteichthyes are characterized by the bony composition of skull, jaws, gill cover (operculum), scales, vertebrae, and ribs. Also, in contrast to cartilaginous fish, most bony fish possess a hydrostatic organ, the **swim bladder**, used for buoyancy control.

Both groups show an improved use and arrangement of fins compared to the older acanthodians and placoderms. To varying degrees, the fin patterns include (1) a caudal (tail) fin used for propulsory motion, (2) usually stationary dorsal and ventral fins that act as keels to prevent rolling and side-slipping, and (3) mobile, paired pectoral and pelvic fins that provide vertical controls, "brakes" and "bilge rudders" (Fig. 16-13).

Some workers have suggested that the dorsal, caudal, and anal fins are derived from a continuous fold of skin that was originally present along the dorsal midline of the body. In the fin-fold theory, the paired pectoral and pelvic fins presumably arose from similar folds along the flanks. There is no paleontological evidence yet to support this view, although it is clear that fin structures have undergone considerable evolution: that is, there have been changes in the placement of fins, in their supporting structures, and in their mobility and flexibility. For example, the bases of the paired fins in Devonian cartilaginous sharks, such as *Cladoselache*, are quite wide (Fig. 16-14), but these become narrower and more mobile in later forms. Among the bony fishes, the dorsal fin in some forms (*Dorypterus*) is almost as long as the fish itself.

Since it is through lineages of Osteichthyes that land vertebrates evolved, much attention has been paid to them. The middle Devonian strata in which

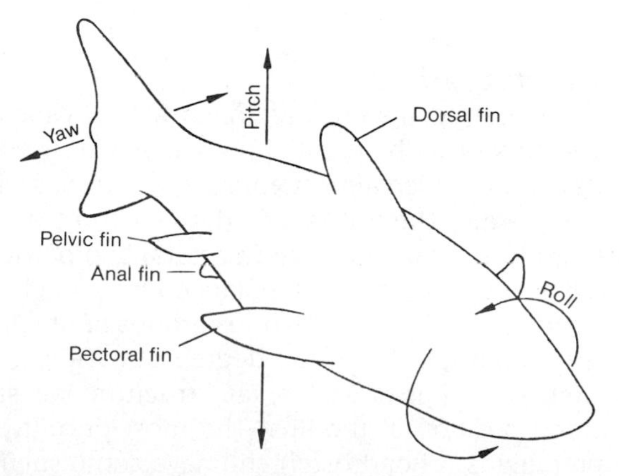

FIGURE 16-13 Generalized design of a modern fish that enables it to cleave through the water rapidly with considerable control and cope efficiently with disturbing forces such as yaw, pitch, and roll. (Adapted from Waterman et al.)

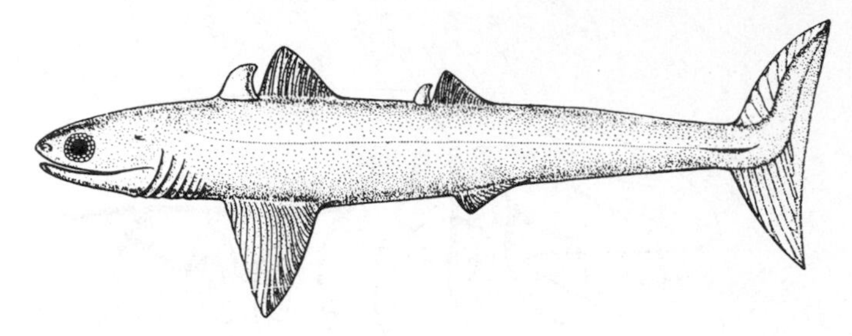

FIGURE 16-14 *Cladoselache*, a late Devonian shark ranging from 1.5 to 4 feet long. It had paired pectoral and pelvic fins as well as fins on the mid-dorsal line, each supported by a broadened row of unconnected rods of cartilage. (Adapted from Romer and Parsons.)

early bony fish are found indicates that they evolved from fish that had entered freshwater areas but did not reenter the ocean in any great numbers until the Mesozoic. Because of this freshwater history, according to Romer and Parsons, most or all of these early forms appear to have carried a new organ, **lungs**, that enabled them to breathe atmospheric oxygen directly. It seems likely that such lungs served initially as accessory organs to gills when dissolved oxygen diminished in stagnant waters because of either high temperatures or the growth of algae and microorganisms (eutrophication). They suggest that later changes in climatic conditions and the movement of many bony fish to the sea led, in many cases, to transformation of lungs into the homologous saclike organ, the swim bladder.

ACTINOPTERYGII AND SARCOPTERYGII

The first appearances of bony fish in Devonian sediments already show their division into two subclasses based on fin structure and other features: the **actinopterygians** and **sarcopterygians**.

Actinopterygii

The actinopterygians are ray-finned fish in which the fins are supported by parallel bony rays whose movements are controlled almost entirely by muscles within the body wall. They all seem derived from a basic ancestral form that had paired pectoral and pelvic fins as well as a single dorsal fin balanced by a single anal fin on the ventral surface. Various groups of ray-finned fish evolved, differing in the degree of ossification of the skeleton, types of scales, tail structure, jaw structure, and position of the fins. The most primitive actinopterygians (Chondrostei) still have some surviving species today such as the sturgeon and the Mississippi paddlefish. More complex forms, the Neopterygei, gave rise to a number of groups including the modern bowfin and garpike, as well as to the most recent of all bony fish, the **teleosts**, characterized by a highly ossified skeleton, very thin scales, and numerous fin specializations. Teleosts have expanded in both number and variety from the Cretaceous period onward until they now include almost all of the bony fish found.

It is interesting to note that the replacement of each type of fish, chondrosteans by neopterygians and primitive neopterygians by more advanced teleosts, was not necessarily accompanied by highly significant changes in shape or size. Most innovations appeared to be relatively minor, yet within the competitive struggle for existence, it is clear that even minor changes that improve locomotion and feeding adaptations (Carroll) can significantly increase survival ability.

Sarcopterygii

Within the sarcopterygians are grouped the flesh-finned fish which supported the fin with small individual bones, arranged either along the fin axis or in rows parallel to the body. Fin movements were apparently mostly controlled by muscles within the fin itself. Early flesh-finned fish were distinguished by having two dorsal fins and a pineal opening at the top of the skull which bore a median eye. Some also had internal nostrils which functioned in air breathing and may have been directly connected to the lungs. Descendants of one group of these fish still persists in the form of the African, Australian, and South American **lungfishes**, called Dipnoi (dipnoans), that are air breathers during dry seasons when their pools stagnate. In contrast to the single lung of the Australian form, the African and South American lungfish have a pair of lungs and can encyst themselves in mud during the dry season, breathing air through openings in their burrows. It has been claimed that the present distribution of lungfish in Africa, South America, and Australia derives from the proximity between these continents in Gondwanaland during the Mesozoic era (Chapter 6).

Another group of Sarcopterygii are the Crossopterygii or tassel-finned (also lobe-finned) fish which differ significantly from the dipnoans in respect to ossification of the skeleton, skull structure, and other features. The **crossopterygians** are classified into two main groups that trace back to the Devonian period, the **rhipidistians** and **coelacanths**, the former being primarily freshwater fish while the latter became primarily marine. The internal nostrils of the coelacanths were lost, and the lung was often converted to a calcified swim bladder. During the Permian period the rhipidistians became extinct, whereas the coelacanths were believed to have disappeared later during the Mesozoic. This view prevailed until a living coelacanth, *Latimeria chalumnae*, was found in 1938 (see Fig. 3-14). Since then, many of this species have been caught in the Indian Ocean between Africa and Madagascar. A general phylogeny of major groups of fish mentioned in this chapter is given in Figure 16-15.

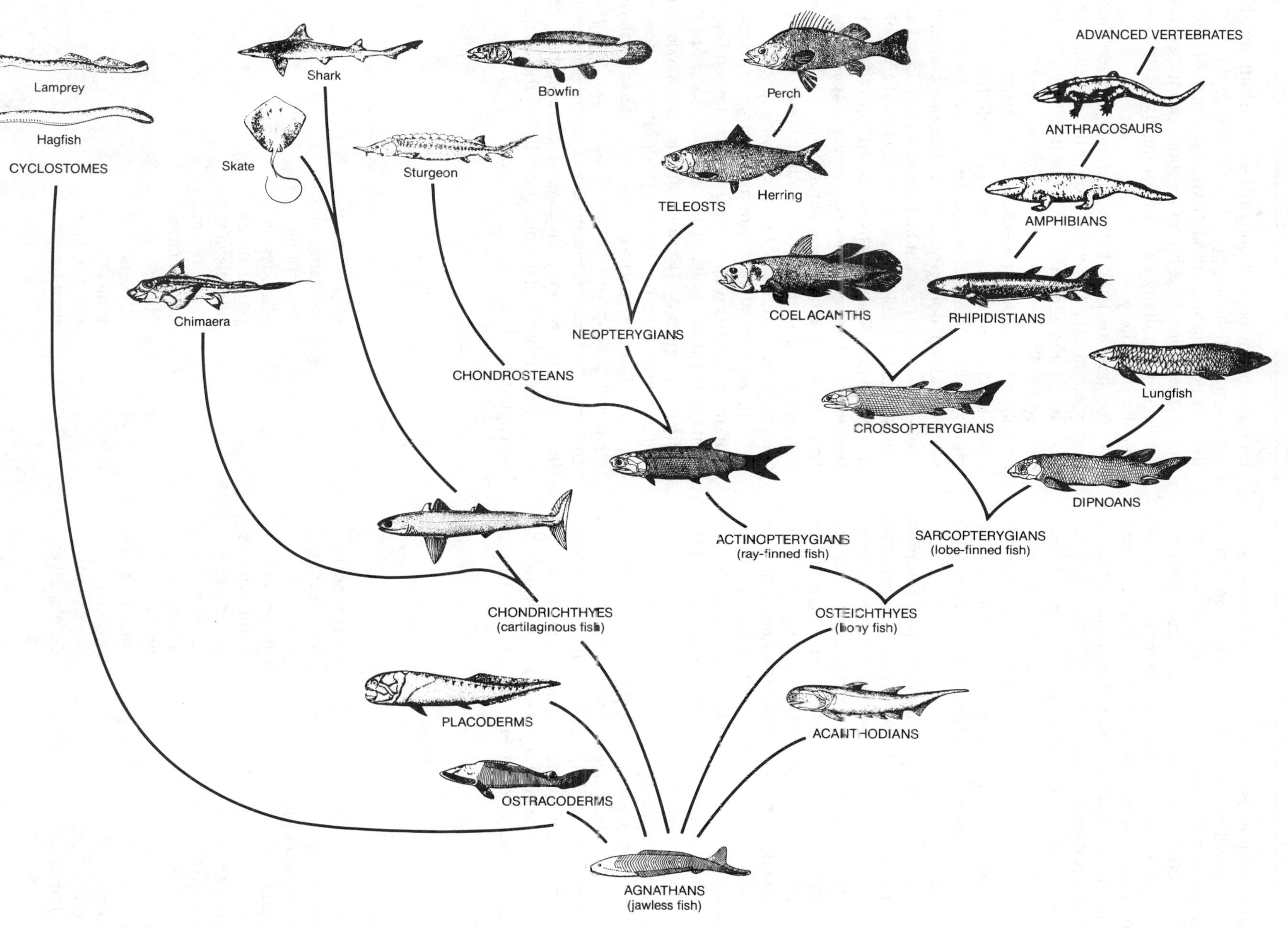

FIGURE 16-15 Phylogenetic relationships among different lines of fish. The ancestor of each given lineage is not necessarily the species illustrated but is believed to have been similar.

SUMMARY

The vertebrates possess several singular features such as pharyngeal gill clefts, an internal skeleton derived from a mesodermal notochord, a tail, and a hollow dorsal nerve cord. Although linking ancestors have not been located in the fossil record, attempts have been made to draw lines of descent from annelids, arthropods, and, more plausibly, from a common ancestor with the echinoderms. Embryological similarities between echinoderms and vertebrates include deuterostomy, radial indeterminate cleavage, enterocoelous origin of the coelom, and mesodermally derived skeletons. However, there are many inconsistencies in the data, and the relationship between the two groups is quite distant.

Some living chordate subgroups may resemble the primitive vertebrate ancestor. Although probably not ancestral to vertebrates, cephalochordates are obviously related and have a life style—swimming and filter feeding—characteristic of the putative protovertebrates. Many larval urochordates have gill slits, notochords, and dorsal nerve cords, structures which are lost in the sessile adults. If such larvae underwent paedomorphosis and reproduced, they might have evolved into free-swimming chordates and, eventually, vertebrates.

The heterostracans, the jawless fishes, are the most ancient fossil vertebrates known and were covered with dermal bony plates and scales. They had a muscular pharynx, perhaps at first engaged in filter-feeding and pumping water but later converted into a predatory organ. The lampreys and hagfishes are modern representatives of these fishes, most of which were replaced by jawed and bony fishes. Interestingly, the cartilaginous fishes appear later in the fossil record than the bony fishes, although which one actually arose first is debatable. Bone has numerous advantages over cartilage, particularly as a supporting tissue and as a calcium depository.

The jaws of later fishes probably evolved from bony pharyngeal gill arches used for filter feeding, and because of this adaptation fish became very effective carnivores. The earliest jawed groups, acanthodians and placoderms, became extinct and were replaced by more efficient swimmers, the Chondrichthyes (cartilaginous fishes) and Osteichthyes (bony fishes with swim bladders). Both groups have elaborate and highly evolved fin patterns including a caudal fin, dorsal and ventral fins for stability, and paired pectoral and pelvic fins for control. The bony fish evolved from freshwater fishes most of which had lungs as gill accessories. In fish reentering a marine environment, the lung was converted to a swim bladder, but in other cases the lung allowed vertebrates to colonize the land.

The early bony fishes are divided into Actinopterygii (ray-finned fish) and Sarcopterygii (fleshed-finned fish). Living derivations of the former group are the sturgeon, garpike, and the teleosts, the most successful of the bony fishes but not much different in size or structure from the earliest bony fish. The sarcopterygians had fins supported by small bones, a pineal eye, and often had internal nostrils and lungs. The lungfishes and the recently discovered coelocanths, previously thought to be extinct, are relics of the ancient sarcopterygians.

KEY TERMS

acanthodians
actinopterygians
ammocoete larva
amphioxus
bony fish
cartilaginous fish
cephalochordate
Chondrichthyes
chordates
coelacanths
crossopterygians
cyclostomes
deuterostomes
dorsal hollow nerve cord
filter feeding
gills
gill arches
hemichordate
jawed fish
jawless fishes
lungfishes
lungs
muscular pharynx
neoteny
notochord
Osteichthyes
ostracoderms
paedomorphosis
paired fins
placoderms
protostomes
rhipidistians
sarcopterygians
swim bladder
tail
teleosts
tunicates
urochordates
vertebrae

REFERENCES

Berrill, N. J., 1955. *The Origin of Vertebrates*. Clarendon Press, Oxford.

Carroll, R. L., 1988. *Vertebrate Paleontology and Evolution*. W. H. Freeman, New York.

Colbert, E. H., 1981. *Evolution of the Vertebrates*, 3rd ed. John Wiley, New York.

Conway Morris, S., 1979. The Burgess Shale (Middle Cambrian) fauna. *Ann. Rev. Ecol. and Syst.*, **10**, 327-349.

DeBeer, G., 1958. *Embryos and Ancestors*, 3rd ed. Clarendon Press, Oxford.

Garstang, W., 1928. The morphology of the Tunicata, and its bearings on the phylogeny of the Chordata. *Quart. Jour. Microscop. Sci.*, **72**, 51-187.

Gould, S. J., 1977. *Ontogeny and Phylogeny*. Harvard Univ. Press, Cambridge, Mass.

Jarvik, E., 1977. The systematic position of acanthodian fishes. In *Problems of Vertebrate Evolution*. S. M. Andrews, R. S. Miles, and A. D. Walker, eds. Academic Press, London, pp. 199-225.

____, 1980. *Basic Structure and Evolution of Vertebrates*, Vols. 1 and 2. Academic Press, New York.

Jeffries, R. P. S., 1986. *The Ancestry of Vertebrates*. British Museum (Natural History), London.

Kerkut, G. A., 1960. *Implications of Evolution*. Pergamon Press, Oxford.

Løvtrup, S., 1977. *The Phylogeny of Vertebrata*. John Wiley, London.

McFarland, W. N., F. H. Pough, T. J. Cade, and J. B. Heiser, 1985. *Vertebrate Life*, 2nd ed. Macmillan, New York.

Moy-Thomas, J. A., and R. S. Miles, 1971. *Palaeozoic Fishes*, 2nd ed. Saunders, Philadelphia.

Neal, H. V., and H. W. Rand, 1939. *Comparative Anatomy*. Blakiston, Philadelphia.

Northcutt, R. G., and C. Gans, 1983. The genesis of neural crest and epidermal placodes: a reinterpretation of vertebrate origins. *Quart. Rev. Biol.*, **58**, 1-28.

Olson, E. C., 1971. *Vertebrate Paleozoology*. Wiley-Interscience, New York.

Repetski, J. E., 1978. A fish from the Upper Cambrian of North America. *Science*, **200**, 529-531.

Romer, A. S., 1966. *Vertebrate Paleontology*, 3rd ed. Univ. of Chicago Press, Chicago.

Romer, A. S., and T. S. Parsons, 1977. *The Vertebrate Body*, 5th ed. W. B. Saunders, Philadelphia.

Schmalhausen, I. I., 1968. *The Origin of Terrestrial Vertebrates*. Academic Press, New York.

Smith, H. W., 1961. *From Fish to Philosopher*. Doubleday, New York.

Stahl, B. J., 1974. *Vertebrate History: Problems in Evolution*. McGraw-Hill, New York.

Waterman, A. J., B. E. Frye, K. Johansen, A. G. Kluge, M. L. Ross, C. R. Noback, I. D. Olsen, and G. R. Zug, 1971. *Chordate Structure and Function*. Macmillan, New York.

FROM WATER TO AIR: AMPHIBIANS, REPTILES, AND BIRDS

Crossopterygians were apparently preadapted for moving out of water onto land since they had functioning lungs and two pairs of bone-strengthened muscular fins on which they could move their bodies and support themselves terrestrially without depending on the buoyancy of water. For many of these fish, it would seem that relatively few further changes were necessary to provide them with at least a primitive terrestrial existence (Fig. 17-1). Nevertheless, the question of why some of them abandoned their shallow-water habitats and went onto land is difficult to answer with certainty.

A classic hypothesis popularized by Romer suggests that when the shallow, hypoxic habitats of ancient crossopterygians dried up or stagnated further, some varieties that were preadapted to breathing atmospheric oxygen would have searched for new pools of water and probably survived on land for short periods of time. According to Romer, seasonal droughts were common in the Devonian, and selection during such periods would lead to increased intervals of terrestrial exploration until some groups could eventually maintain themselves out of water for significant parts of their life cycle.

Another hypothesis, more commonly accepted at present (McFarland et al.), suggests that these aquatic forms escaped to land because of population pressures resulting from predation (probably other crossopterygians) as well as from competition for space, food, and breeding sites in these warm, swampy habitats. The transition to land in a moist tropical climate might have produced relatively little stress in such terrestrially preadapted crossopterygians, and some invertebrate food sources on land may not have been much different than in the swamps themselves.

Whether because of drought or competition or both, once existence on land was established as an important stage in survival, further selection would operate on many levels to improve air breathing, eliminate carbon dioxide, increase resistance to desiccation, increase head mobility, and enhance further transformations. The possibility of such changes can be seen in the fact that some present-day fish such as mudskippers, climbing perch, and walking catfish have developed various terrestrial adaptations, even to the point of climbing trees and capturing food.

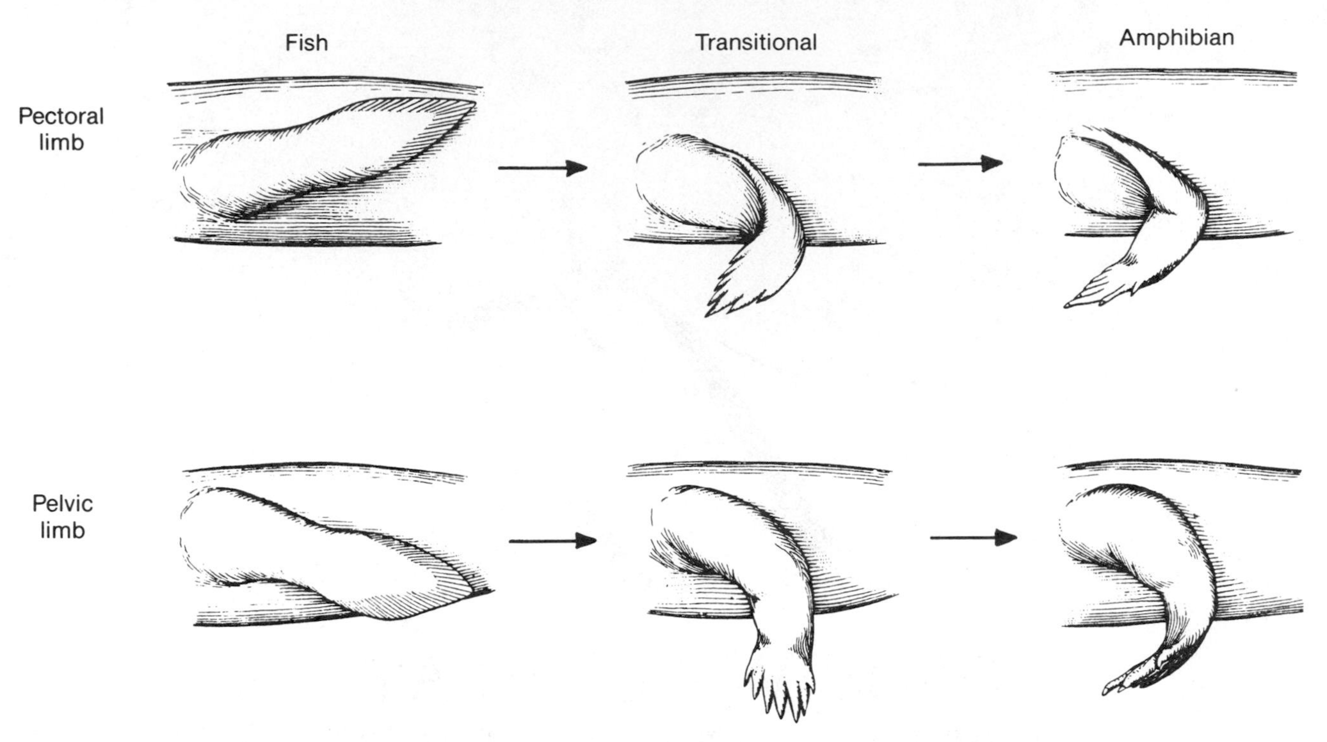

FIGURE 17-1 Shifts in the positions taken by pectoral and pelvic limbs in the transition from fish to amphibian. (Adapted from Romer and Parsons.)

EARLY AMPHIBIANS

However land vertebrates first evolved, paleontologists generally agree that they were related to lobe-finned fishes of the rhipidistian type, and this transition from fish to crawling four-legged land **tetrapod** is believed to have occurred by the end of the Devonian period, about 350 million years ago. The earliest of such identified **amphibians** in the fossil record, called **ichthyostegids** (Fig. 17-2), show their relationship to rhipidistians in a number of features:

1. Although there are some differences, the arrangement of dermal bones in the skulls of rhipidistians and ichthyostegids are quite similar. As shown in Figure 17-3, it is fairly easy to identify a large number of bones occupying relatively similar positions. Even the remnant of a preopercular bone is present in these primitive amphibia although they possessed no operculum (gill cover).
2. The fins of rhipidistians and their supporting girdles possess bones that can easily be considered homologous to those of early tetrapods (Fig. 17-4).
3. The tooth structure of both rhipidistians and ichthyostegids possesses similar complex labyrinthine foldings of the pulp cavity (Fig. 17-5). In fact, the prevalence of these unusual teeth has given the name Labyrinthodontia to the ichthyostegids and also to two other orders of fossil amphibians, the anthracosaurs and temnospondyls. (It is from early **anthracosaurs** that reptiles are believed to have been derived.)
4. The sensory lateral line system of rhipidistians which extended into the skull appears homologous to a similar pattern of sensory canals embedded in the ichthyostegid skull (see Fig. 17-3).
5. The ichthyostegids possessed a fin-rayed caudal tail that showed obvious fishlike ancestry.
6. The structure of the vertebrae of ichthyostegids had changed relatively little from the vertebral structure of rhipidistians (Fig. 17-6).

Unfortunately, almost nothing is known of the soft-body structures of ichthyostegids, and there is no obvious clue as to how they solved the problem of preventing desiccation. Very probably, like many modern amphibians, a considerable part of their development was spent in water, and the adult form never wandered too far from moist surroundings. Nevertheless, their many anatomical innovations indicate that they and their contemporaries had evolved successful solutions to at least some mechanical problems faced by land-dwelling vertebrates.

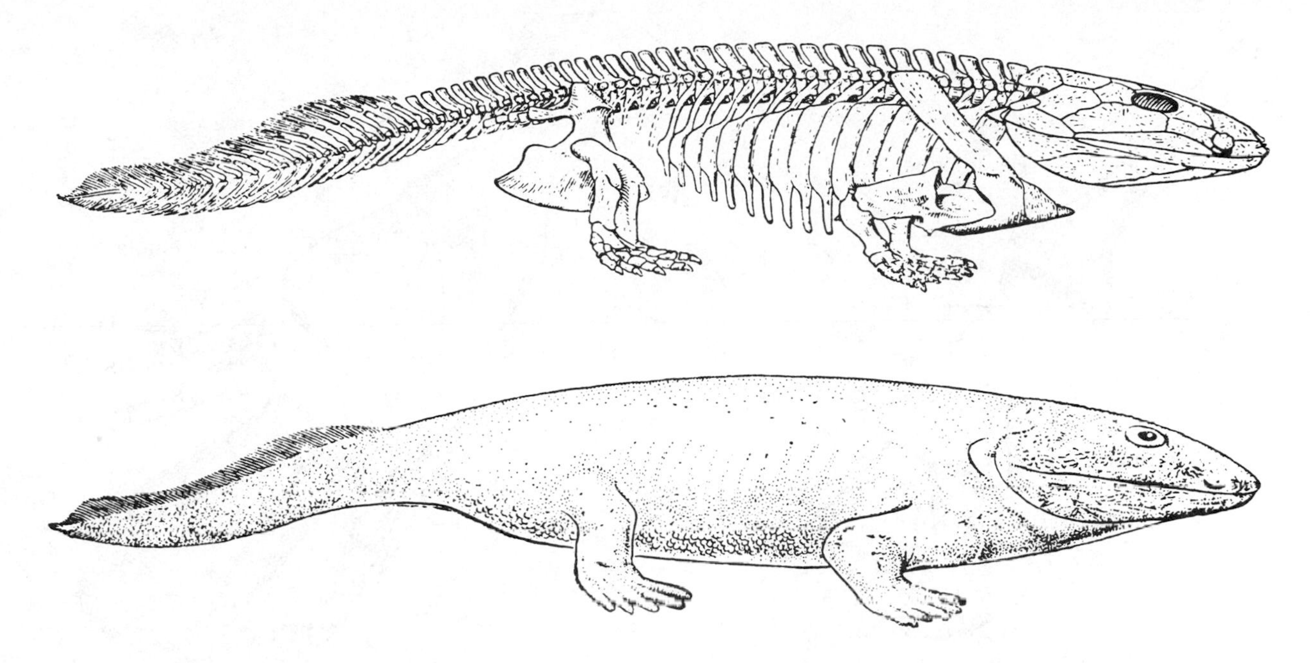

FIGURE 17-2 Reconstruction of the skeleton and external appearance of the earliest complete amphibian fossil found, *Ichthyostega*. This specimen dates back to the late Devonian period, and was about 3 feet long. Although it had stout limbs, it still had many fishlike features including scales, dorsal tail fin, and a hydrodynamic shape. (Adapted from Schmalhausen, from Jarvik.)

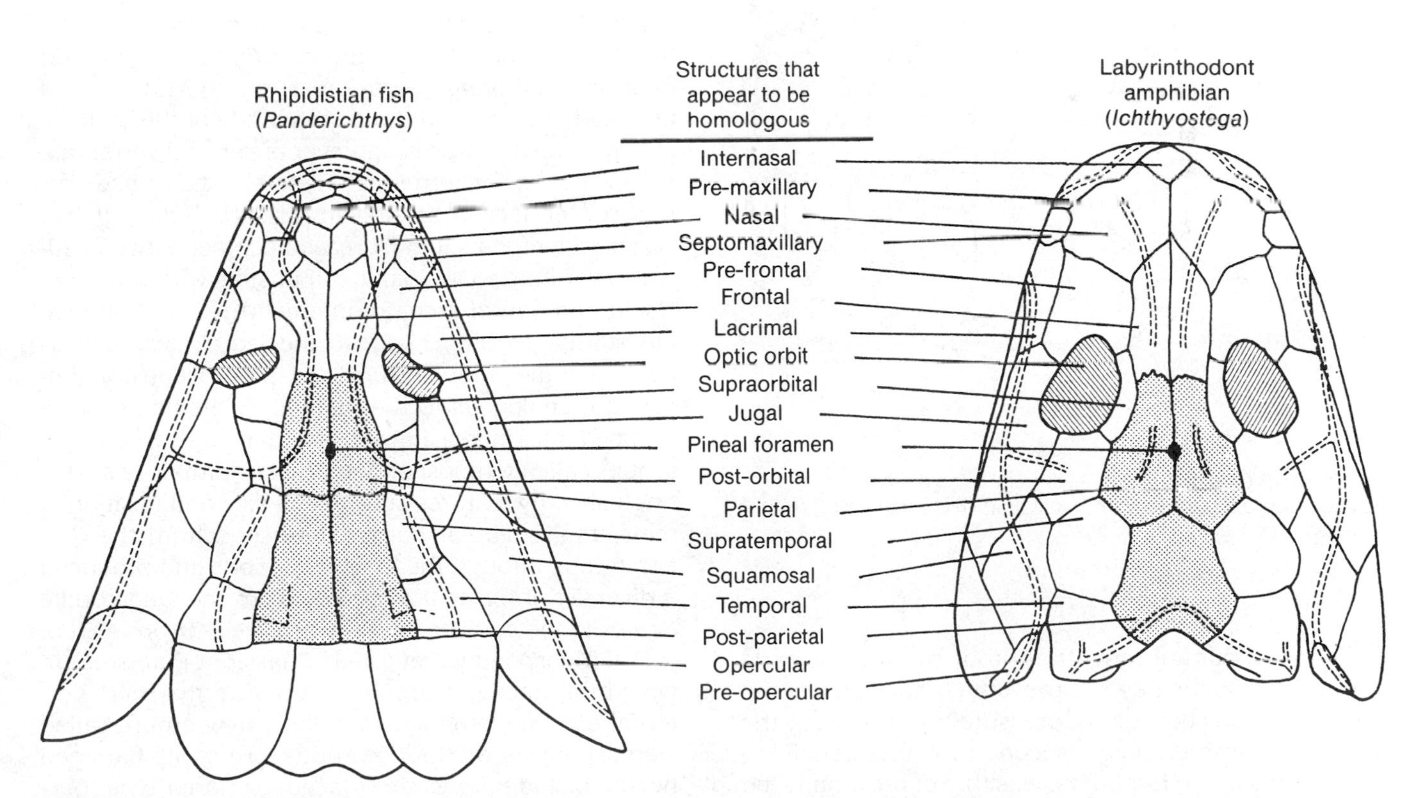

FIGURE 17-3 Dorsal views of the tabular bones in skulls of a rhipidistian and ichthyostegid compared in terms of likely homologous structures. Dashed lines indicate sensory canals. (Adapted from Duellman and Trueb.)

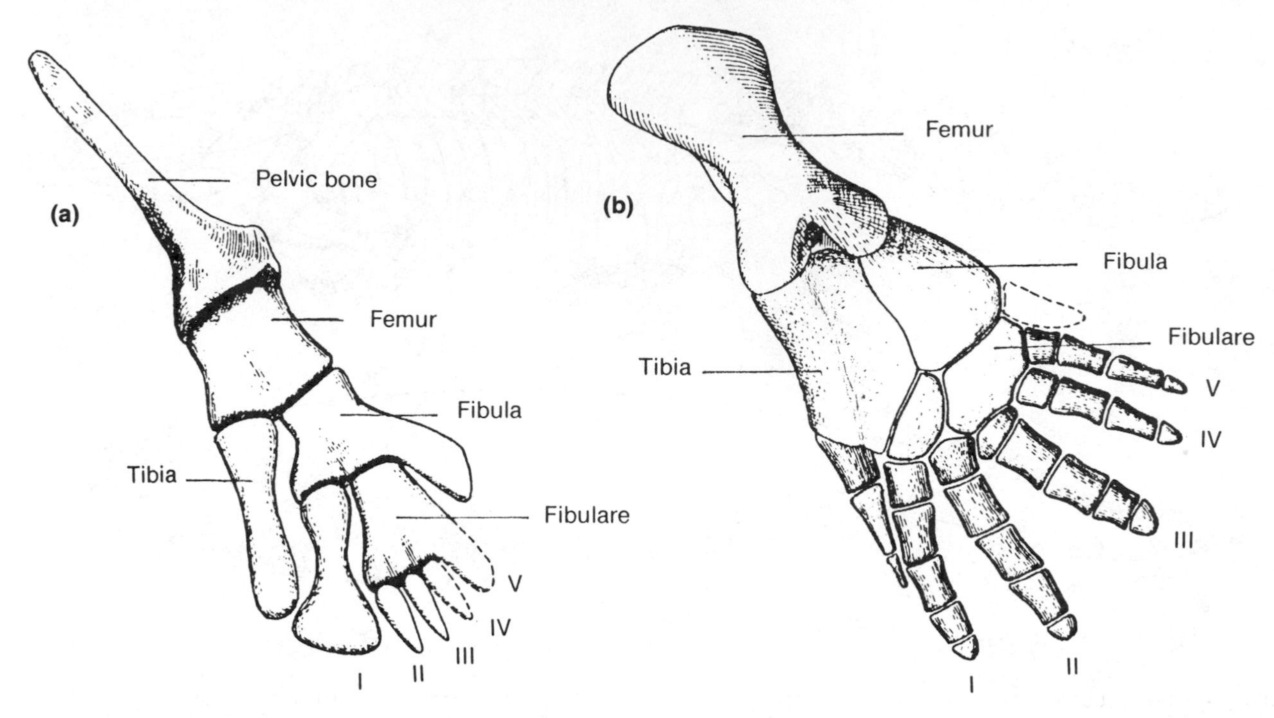

FIGURE 17-4 Comparison between bones in the pelvic fin of a rhipidistian (a) and those in the hind limb of a fossil amphibian (b). Roman numerals indicate presumed homologies between some rhipidistian bones and amphibian digits. (Adapted from Jarvik.)

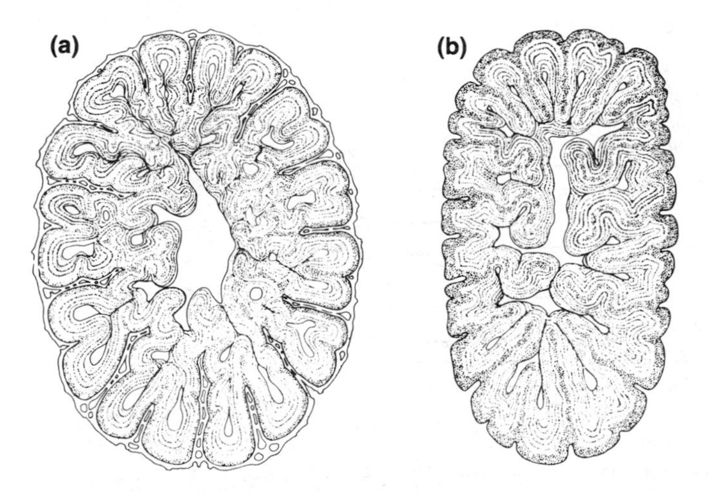

FIGURE 17-5 Cross sections of teeth from a rhipidistian, *Polyplocodus* (a), and a labyrinthodont amphibian, *Benthosuchus* (b). (Adapted from Stahl, from Bystrow.)

One important terrestrial problem for these early tetrapods was the need to prevent compression of the internal organs because of pressure transmitted by the limbs and limb girdles. A second problem, pertaining specifically to the forelimbs, was that of preventing the impact of terrestrial locomotion from being transmitted to the braincase since the rhipidistian pectoral fins are connected to the skull. As shown in Figure 17-7, these difficulties were solved by attaching first the pelvic and then the pectoral girdles to the main axis, the spine. The spine thus became a "suspension bridge" that (1) absorbed the impact of terrestrial motion, (2) freed internal organs from pressure, and (3) enabled the head to turn and lift independently of the body by placing it on a forward cantilever of cervical vertebrae.

Selection for terrestrial skeletal rigidity had various consequences, placing emphasis on strengthened vertebral elements and increased contact between adjacent vertebrae. Originally, rhipidistian vertebrae consisted of a neural arch, an intercentrum, and intercalary cartilages (Fig. 17-8). In the later rhachitomous condition the intercentrum and pleurocentrum (believed homologous to the intercalary elements) were retained but increasingly ossified in the amphibian groups called temnospondyls. These animals (also distinguished by features such as the size and position of their cranial tabular bones) flourished from the Carboniferous through the Triassic periods and produced a diversity of forms that ranged from the alligatorlike *Eryops* to the smaller, dorsally armored *Cacops*. Out of the temnospondyls evolved Triassic organisms in which the intercentrum alone became the main vertebral central element. Among this new group, called stereospondyls, are found various forms with flattened heads, including one short-faced, armored type, *Gerrothrax*, which had obviously returned to an aquatic existence by maintaining gills in the adult stage.

In the anthracosaur line of amphibians, vertebral structure evolved in a direction opposite to that of temnospondyls; that is, it was the pleurocentrum rather

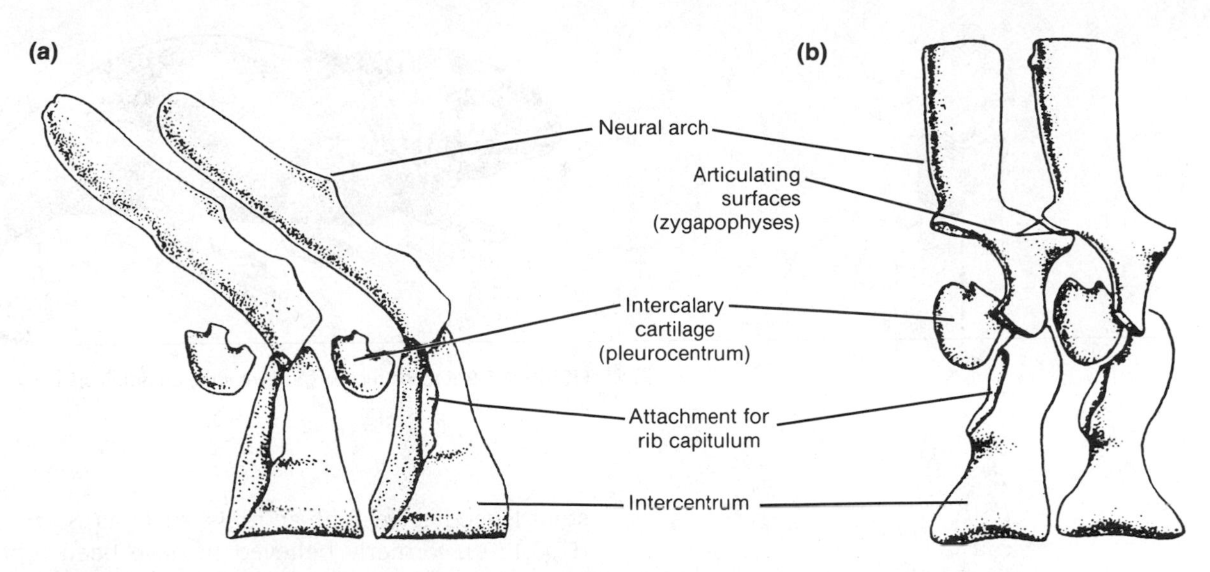

FIGURE 17-6 Lateral view of vertebrae from a rhipidistian, *Eusthenopteron* (a), and from the fossil amphibian, *Ichthyostega* (b). (Adapted from Romer [1966].)

(a) *Eusthenopteron*

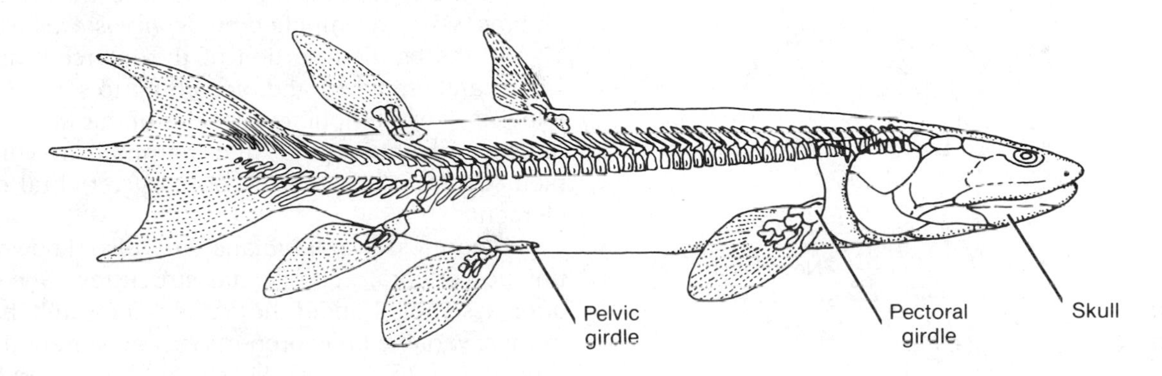

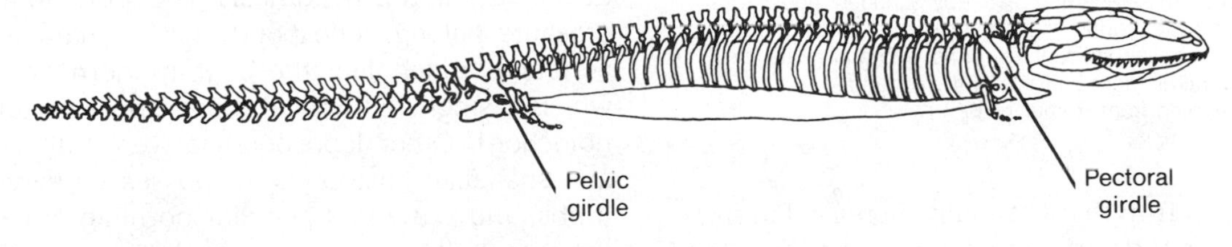

(c) *Hylonomus*

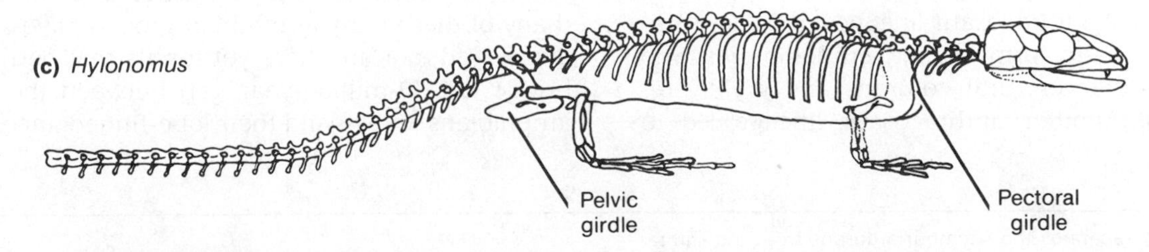

FIGURE 17-7 Attachments of the pectoral and pelvic girdles in a rhipidistian crossopterygian (a), an early labyrinthodont amphibian (b), and a very early reptile (c). (Adapted from Romer [1967].)

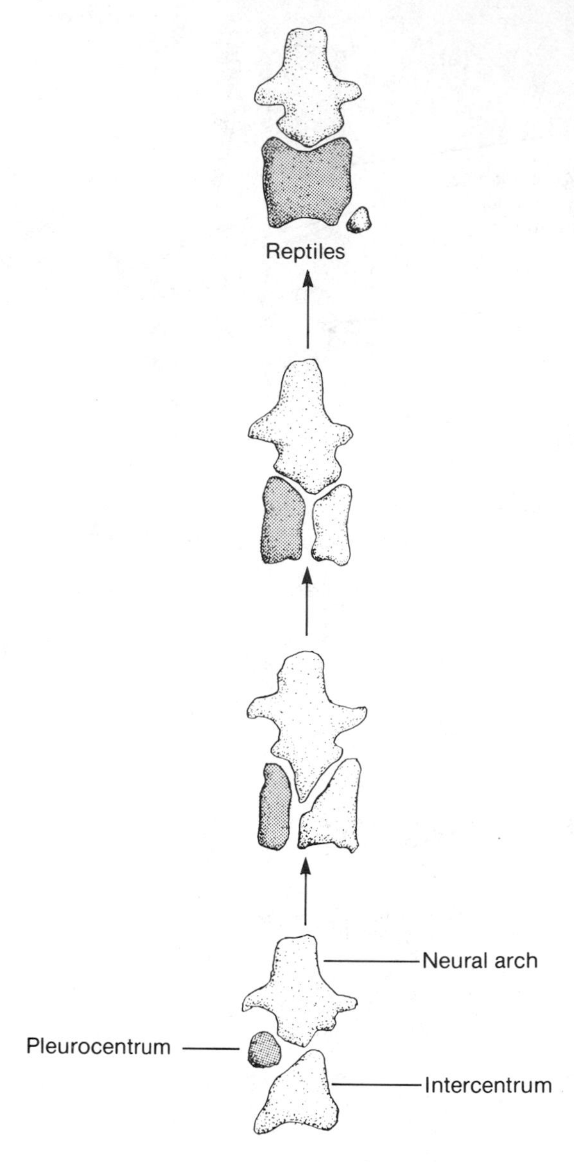

FIGURE 17-8 The evolution of vertebral central elements (lateral views) beginning with an early amphibian labyrinthodont and proceeding through stages found in amphibian anthracosaurs up to the reptilian grade. In advanced reptiles, birds, and mammals, the entire vertebral central element consists of the expanded pleurocentrum. (Adapted from Romer [1966].)

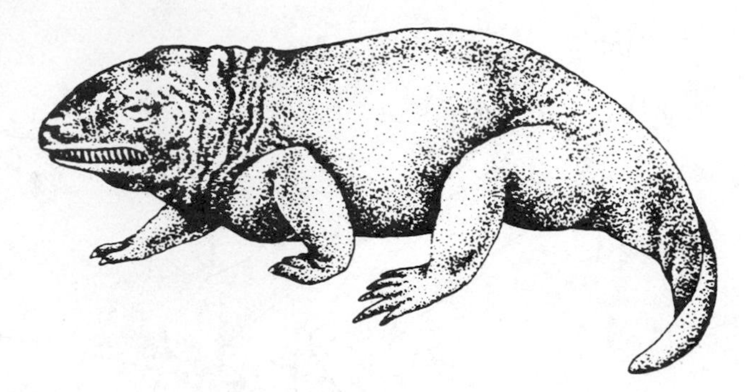

FIGURE 17-9 Reconstruction of *Seymouria*, an amphibian anthracosaur.

than the intercentrum that gradually increased in size. Fewer in number and variety than temnospondyls, the anthracosaurs had also a shorter fossil history, first appearing during the early Carboniferous period and becoming extinct by the end of the Permian. During this interval, one anthracosaur lineage led back to predominantly aquatic forms called embolomeres who had very flexible vertebral columns and greatly reduced limbs. Another anthracosaur lineage led to stout-legged terrestrial animals such as *Seymouria*, (Fig. 17-9), formerly believed to have been reptilian until gilled amphibian larvae were discovered among some of their group. One of the seymouriamorphs, *Diadectes*, was probably one of the earliest, if not the first, of the tetrapod herbivores.[1] In a third group of anthracosaurs, exemplified by *Gephryostegus* (see Fig. 17-15), the ossified portion of the intercentrum was significantly reduced and, according to some paleontologists, it was amphibian species of this kind that led directly to higher reptiles and mammals in which the pleurocentrum became the primary vertebral central element.

Other fossil amphibians are also known with unique vertebral and skeletal structures. One order, often assembled under the name microsaurs, had extremely variable body proportions and vertebral structures that appear to parallel those found in anthracosaurs. Although they, too, have been proposed as possible reptilian ancestors, Carroll and others have pointed to their many peculiar specializations as evidence against this hypothesis. The group to which microsaurs belong, lepospondyls, are characterized by vertebrae much like those of modern amphibia in which a single cylindrical, bony spool surrounds the notochord. Other lepospondyls were either limbless and snakelike (aistopods) or possessed very reduced limbs; and some had peculiar hornlike projections of the rear skull bones (nectridians). These small aquatic forms flourished during the Carboniferous period, but disappeared by the close of the Permian.

It is unfortunate that transitional forms between many of these early amphibian groups have not been discovered, nor are there yet fossils that permit us to bridge the 30-million-year gap between the earliest amphibians known and their lobe-finned ancestors. In

[1] Panchen (1977) separates the seymouriamorphs from the anthracosaurs and considers them as the two suborders of the order Batrachosauria.

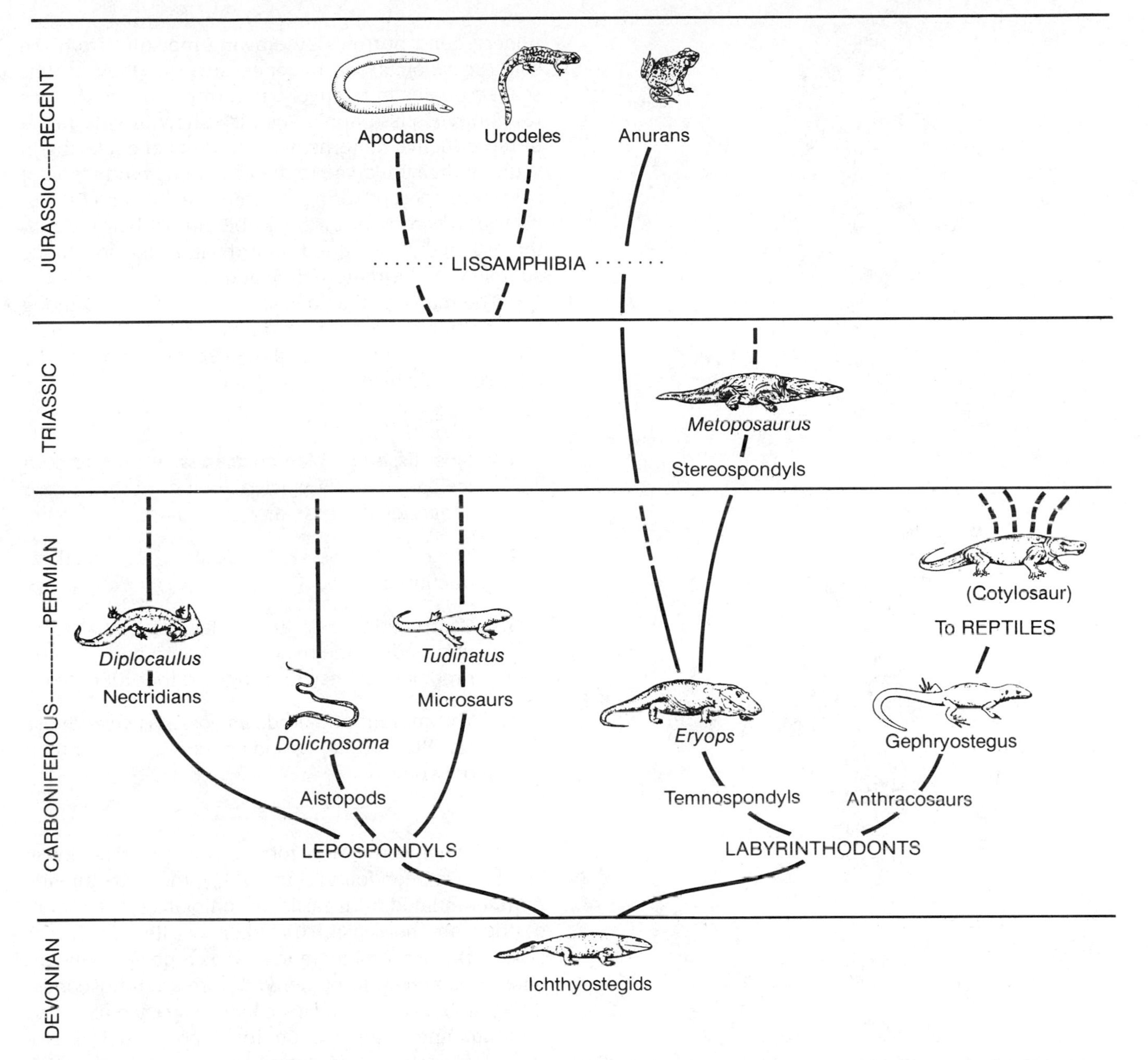

FIGURE 17-10 One possible phylogeny for some of the primary lineages among amphibians. Most major early forms of amphibians appear to have become extinct by the end of the Permian period, with the exception of the stereospondyls who survived until the late Triassic. None of the present-day amphibia (Lissamphibia) can yet be directly traced to any specific Paleozoic form.

the absence of a clear phylogeny, some paleontologists suggest that at least some of the amphibian diversity stemmed from a polyphyletic origin in which different groups of crossopterygians or rhipidistians served as ancestors for different groups of amphibians (Jarvik). Others argue that evolution of the characteristic bone arrangements in amphibian limbs and the five-toed foot would probably not have arisen more than once. The serious gaps in the fossil record seem to preclude a firm decision on this score, although present information does allow some phylogenetic schemes to be offered (Fig. 17-10).

MODERN AMPHIBIANS

Among the intriguing unanswered questions is the origin of modern Amphibia. These are usually classified together under the subclass **Lissamphibia** and separated into three orders: apodans (legless, wormlike caecilians), anurans (frogs and toads), and urodeles (newts and salamanders). In contrast to the scaled skins of early amphibians, modern lissamphibians possess a permeable, glandular skin allowing for considerable water and gaseous exchange. (Caecilian skins have, in some species, small embedded scales.)

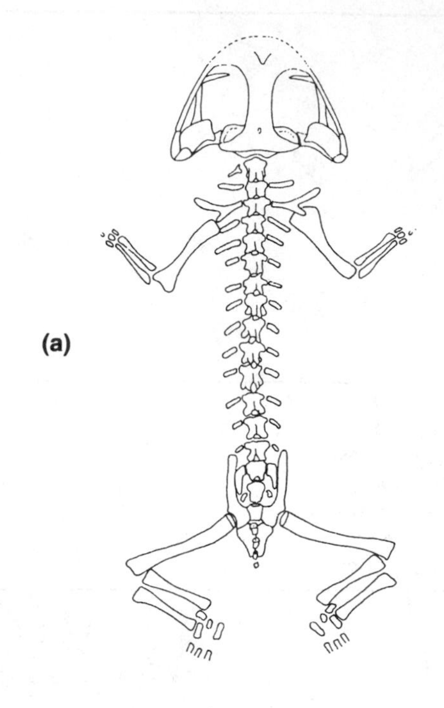

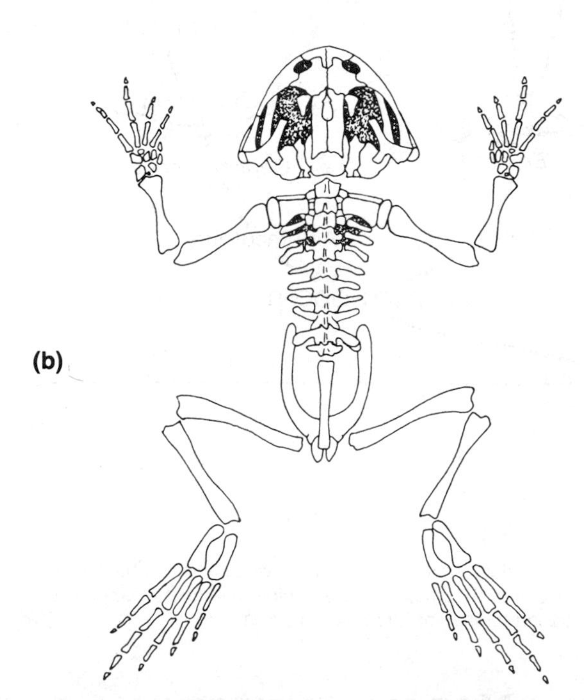

FIGURE 17-11 Skeletal reconstructions of two froglike Mesozoic fossils: *Triadobatrachus* (a), found in Triassic deposits, and the early Jurassic *Neobatrachus* (b). Modern urodeles have a fossil record dating back to the Cretaceous period, but there are no fossils as yet for the modern apodan amphibians. (Adapted from Estes and Reig, from Romer [1966].)

Although this feature limits important segments of their activity to moist or aqueous surroundings, lissamphibians have achieved a variety of remarkable adaptations. Some toads, for example, are desert inhabitants capable of surviving long periods of drought in underground burrows by drawing moisture from the soil (*Scaphiopus*) or by encapsulating their water-soaked bodies in relatively impermeable membranes (*Cyclorana*). Lissamphibians also show a wide range of reproductive patterns, from traditional egglaying in water and gilled larvae to the viviparous production of well-developed offspring. In some caecilians (*Typhlonectes*), offspring at birth can be almost half the maternal length, having fed on a thick, milky substance supplied by the maternal oviduct.

The many distinctions and specializations among lissamphibians make them difficult, as yet, to connect phylogenetically to any of the Paleozoic forms for the following reasons:

1. Most lissamphibian species have unique pedicellate teeth in which a zone of fibrous tissue separates the base and crown.
2. Their skulls and skeletons show a marked reduction of bone.
3. Compared to fossil amphibians other than temnospondyls, the hands of modern anurans and urodeles possess four digits rather than five.
4. The anurans and urodeles possess two auditory ossicles, the stapes and operculum, rather than only one.

Although some Mesozoic froglike and urodelelike fossils have been found (Fig. 17-11), these are already so differentiated from earlier amphibians that their ancestries are not easily traced. According to Carroll (1988), the lissamphibian taxon has a polyphyletic origin: frogs appear to be derived from a temnospondyl group, and urodeles and caecilians may each have had separate ancestries among the lepospondyl microsaurs. Perhaps, as Schmalhausen has suggested, some, if not most, lissamphibia evolved in isolated, poorly fossilized, mountainous ponds and streams that offered protection to these land vertebrates because the environment was cold and relatively inhospitable to other early tetrapods. Modern amphibia are even now more frequent in some cooler areas than are reptiles. The special jumping adaptations developed by anurans, enabling them to escape predators with a few giant leaps into or out of water and to swim rapidly by "frog kicking," were probably among the advantageous features that allowed them to reenter the tropics. In any case, when considered in terms of their continued persistence for more than 200 million years and the significant numbers of existing species (about 4,000), modern amphibians seem to be a well-adapted and successful group.

REPTILIAN GRADE OF EVOLUTION

A considerable amount of material has been written on reptilian evolution, which is particularly interesting for many reasons.

1. By evolving a shelled **amniotic egg** (see Fig. 17-14), reptiles freed land vertebrates from reproductive dependence on an aqueous environment and allowed them to enter a full terrestrial existence.
2. Reptiles have provided large numbers of fossils from the late Paleozoic era onward.
3. Reptilian phylogenetic relationships seem to be clearly delineated in a number of groups.
4. Reptiles represent, in terms of numbers, size, and mass, the ruling land vertebrates throughout the long Mesozoic era.
5. Reptiles clearly gave rise to mammals, our own vertebrate class.

The features which distinguish reptiles from present-day amphibians are fairly easy to note and include the following:

1. Skull and skeletal differences: Modern reptilian skulls possess one occipital condyle compared to two such condyles in modern amphibians; at least two vertebrae are incorporated into the reptilian sacrum compared to only one in amphibians; and there are five digits in the reptilian foreleg compared to four or fewer in amphibians.
2. Heart: Amphibians have a single ventricle, whereas the reptilian ventricle is at least partially divided—the left side sending oxygenated blood to the carotid artery, and the right side sending venous blood to the pulmonary artery (Fig. 17-12).
3. Epidermis: The amphibian epidermis is generally soft and moist, allowing some degree of gaseous and aqueous exchange, whereas the more heavily cornified reptilian epidermis acts as a barrier to such exchange.
4. Gonadic ducts and excretion: In many amphibians a single excretory duct system services both the gonads and the kidneys, whereas reptiles have separate ducts for each of these systems (Fig. 17-13). Also, the nitrogenous products of excretion in reptiles, urea and uric acid, can be concentrated and do not necessitate a large flow of water for their removal. In many amphibians, the urine is quite dilute and may contain a considerable amount of ammonia.

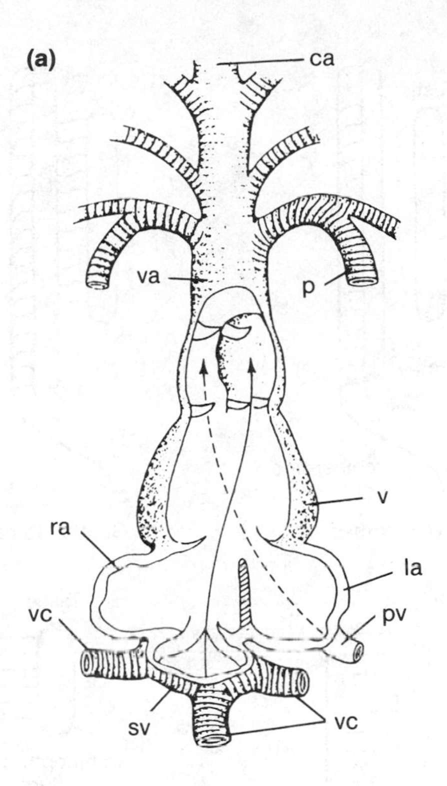

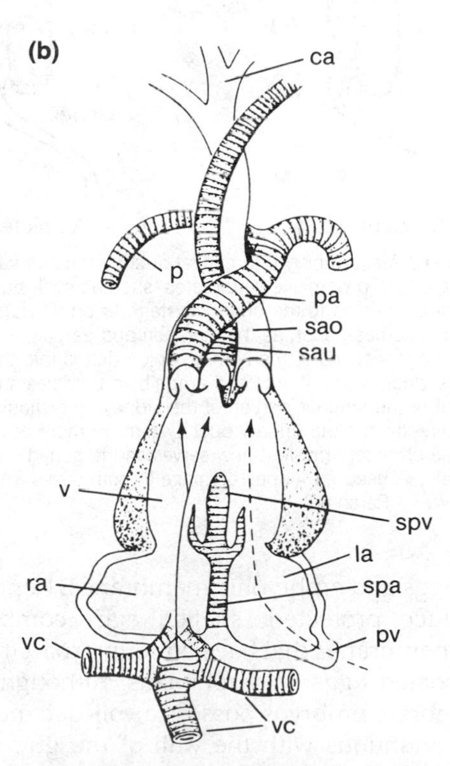

FIGURE 17-12 Ventral views of amphibian (a) and reptile (b) hearts, showing their anterior chambers and blood flow (arrows). Abbreviations: *ca*, carotid artery; *la*, left auricle; *p*, pulmonary artery; *pa*, pulmonary aorta; *pv*, pulmonary vein; *ra*, right auricle; *sao*, systemic aorta carrying oxygenated blood; *sau*, systemic aorta carrying unoxygenated blood; *spa*, interauricular septum; *spv*, interventricular septum; *sv*, sinus venosus; *v*, ventricle; *va*, ventral aorta; *vc*, systemic vein. (Adapted from Stahl, from Goodrich.)

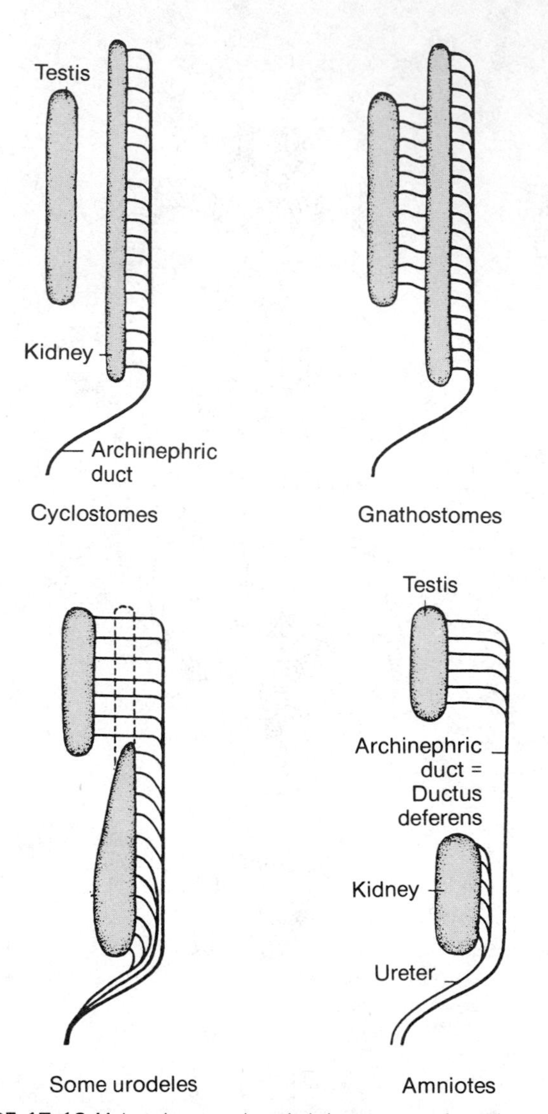

FIGURE 17-13 Male urinary and genital duct systems in various vertebrate groups. In primitive vertebrates, such as cyclostomes, the gonad is not connected to the urinary system. In primitive jawed fishes (gnathostomes), such as the sturgeon and garpike, the testis has multiple connections to the kidney, which drains into the archinephric duct. Many sharks and amphibian urodeles show replacement of the anterior portion of the kidney by testicular ducts that drain directly into the urinary duct system. In more advanced amniotes the archinephric duct now serves as the gonadic duct, and a single ureter is used for kidney drainage in both sexes. (Adapted from Romer and Parsons.)

5. Eggs and embryonic membranes: Reptiles produce protected, shelled eggs composed of membranes that have no counterpart in the gel-coated eggs of amphibians. Although all vertebrate embryos possess a yolk sac membrane continuous with the wall of the gut, reptilian embryos (as well as those of birds and mammals) produce an additional membrane continuous with the embryonic body wall that folds around it to yield an outer chorion and inner amnion (Fig. 17-14). The amniotic cavity prevents adhesions by isolating the embryo from direct contact with the shell and helps provide the embryo with protection against temperature fluctuations. In addition, amniotic eggs have a sac called the **allantois** that grows out of the embryonic hindgut and rapidly covers the inner surface of the chorion. The allanto-chorion membrane complex is well supplied with blood vessels and acts as a respiratory organ that allows inward diffusion of oxygen through the permeable shell as well as the outward passage of carbon dioxide. Nitrogenous wastes of the reptilian embryo are deposited into the allantoic cavity as relatively insoluble, non-toxic precipitates such as uric acid that need not be immediately eliminated.

Of all traits that characterize the reptilian advance, it is the amniotic egg that appears to be most significant. With the exception of some viviparous forms, the moisture-dependent amphibian egg is an important element in maintaining amphibian aquatic ties. Reptiles, on the other hand, can lay their eggs in a large variety of terrestrial environments, and in some lizards and snakes, embryonic development can proceed even with a loss of water. To many evolutionists, the complexity of the amniotic egg suggests that the transition from amphibians to reptiles did not occur more than once, and the reptilian grade of evolution is therefore most probably monophyletic. Unfortunately, since eggs are rarely fossilized, the point at which this transition occurred is unknown,[2] although it seems clear that it must have been preceded by the following steps:

1. Since a shelled amniotic egg can only be fertilized prior to egg laying, internal fertilization must have been an antecedent behavior in this evolutionary line.
2. The habit of laying eggs on land must also have been present, since the shelled reptilian embryo is dependent on gas exchange and would have been unable to obtain sufficient oxygen while immersed in water.
3. In order for the amniotic embryo to be born on land, the stage of an aquatic gilled larva was probably absent.
4. To the above attributes, Carroll adds small size, since the early eggs laid on land could not have been too large if a satisfactory rate of gas exchange was to be maintained before the amniotic membranes evolved.

[2]The earliest fossil purported to be a reptilian egg dates to the early Permian period.

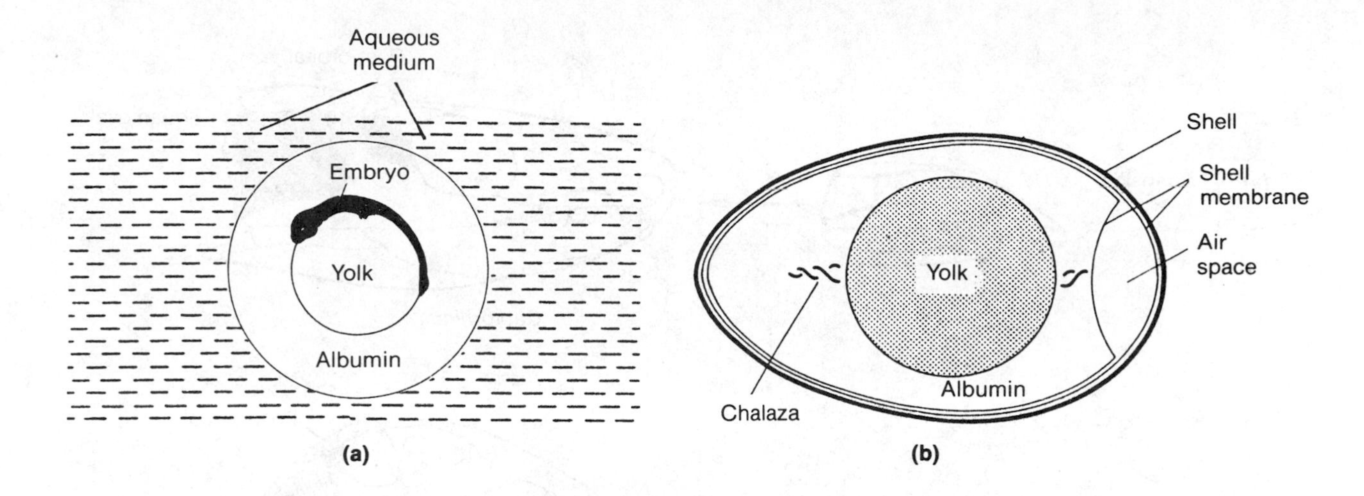

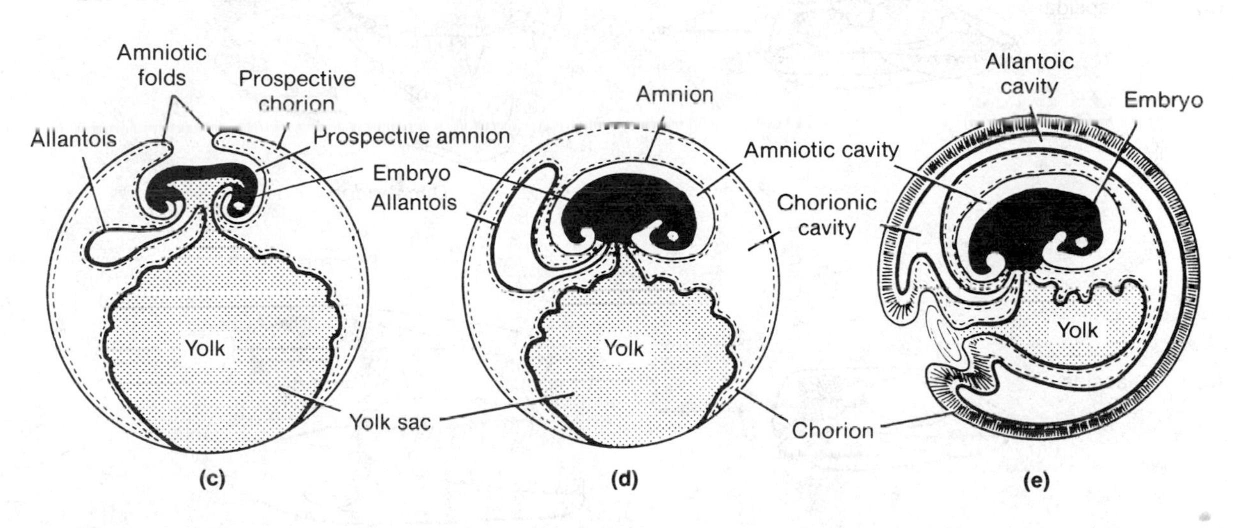

FIGURE 17-14 Diagrammatic views of a generalized anamniotic amphibian egg (a), an amniotic (bird) egg (b), and the embryonic membranes formed during a few developmental stages in avian development (c-e). (Adapted from Alexander, from McFarland et al.)

The achievement of reptilian organization is believed to have occurred during the Carboniferous period by a group called the **captorhinomorphs**, considered the earliest of the stem reptiles, or **cotylosaurs**. According to a number of paleontologists, it was through a line of small, lizardlike amphibian anthracosaurs that these reptiles evolved, and *Gephryostegus* (Fig. 17-15) is believed to be one such amphibian form close to the reptilian line.

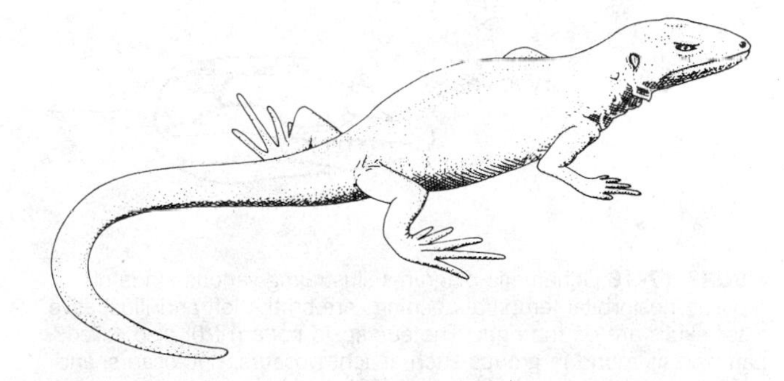

FIGURE 17-15 Reconstruction of *Gephryostegus*, an anthracosaur member of the amphibian labyrinthodont lineage that existed in the late Carboniferous period. It was about 20 inches long and was probably mostly terrestrial in its habits. Since fossils of early captorhinomorph reptiles are very similar to gephryostegids, many paleontologists suggest that these amphibians were close to the line from which reptiles arose. (Adapted from McFarland et al., from Carroll.)

Since many reptilian traits involve soft tissues that are rarely, if ever, preserved, paleontological criteria for crossing the amphibian-reptilian boundary concentrate on skeletal characteristics such as structure of the palate: in early reptiles the pterygoid bone in the skull begins to show a transverse flange associated with what becomes the largest jaw-closing muscle, the pterygoideus. The large palatal fangs of labyrinthodont amphibians are lost, and there is also a reduction in the postparietal, tabular, and supratemporal skull bones. Other changes include increased heterogeneity of rep-

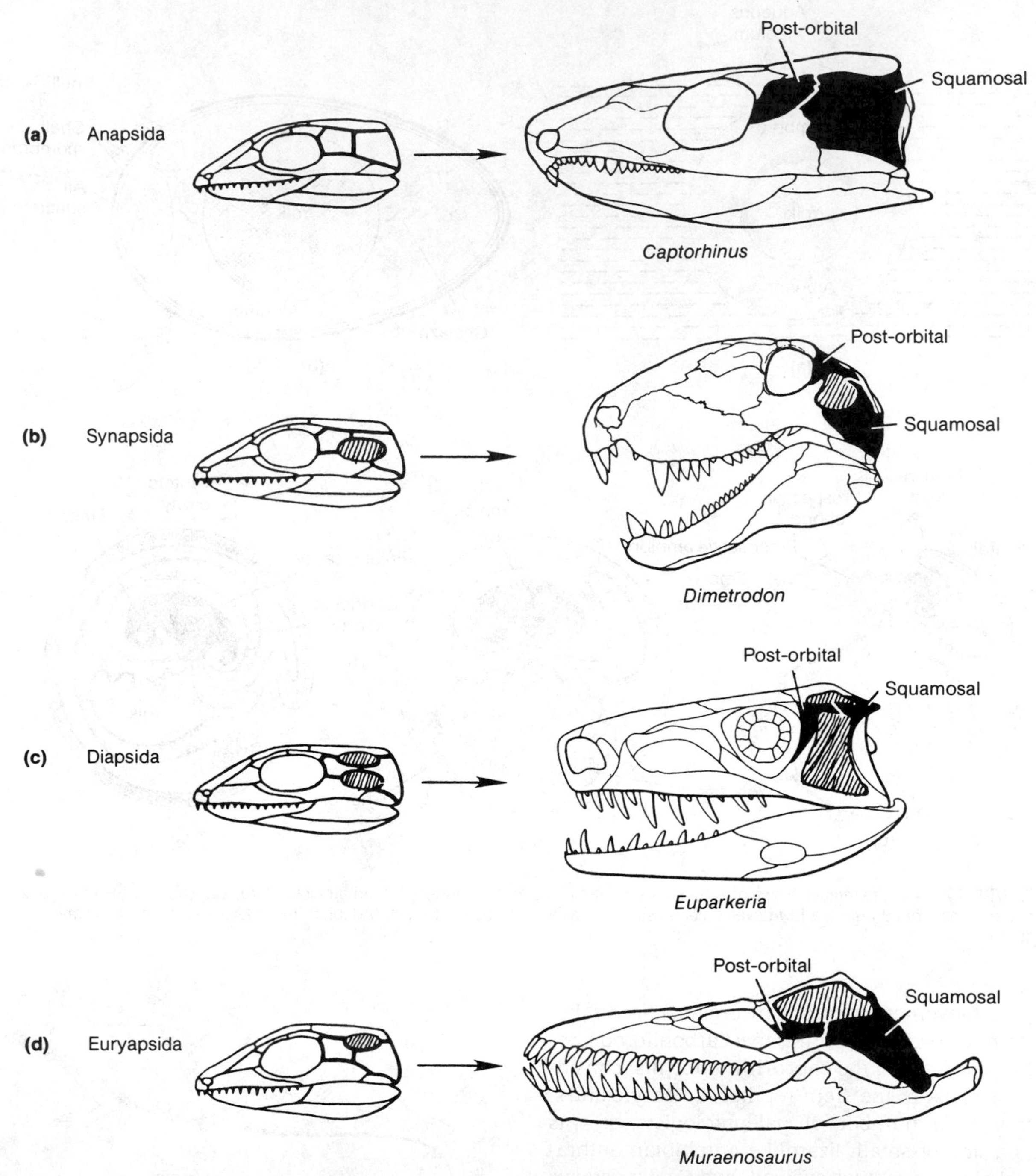

FIGURE 17-16 Schematic diagrams illustrating various kinds of reptilian postorbital temporal openings are on the left and illustrative fossil skulls are on the right. The euryapsid pattern (d), also called parapsid, is found in groups such as ichthyosaurs, nothosaurs, and plesiosaurs. According to Carroll (1988), the euryapsids are derived from early diapsids, and their fenestral pattern evolved by loss of the temporal bar beneath the lower fenestrum, accompanied by thickening of the postorbital and squamosal bones. (Adapted from Colbert.)

tilian teeth compared to the uniformly shaped amphibian teeth, as well as changes in the proportions and degree of ossification in the pectoral and pelvic limb girdles, probably associated with selection for improved support and locomotion in less aquatic habitats.

By the end of the Permian period a large variety of different reptilian lines appear, and, because connections between some of them are not yet clear, they have been grouped in various ways. One common method of classification makes use of openings, called **fenestra**, in the temporal region of the skull behind the optic orbits. In the earliest reptiles the skull is solidly roofed, and the temporalis muscles used to close the jaw run between the inside (medial) surface of the lower jaw and the braincase, within the outer bony layer of the skull. This unfenestrated condition defines the subclass **Anapsida** ("an" = without, "apse" =

arch), who are thereby provided with a relatively rigid skull structure (Fig. 17-16(a)). Nevertheless, this anapsid structure may have had disadvantages because the expansion of the temporalis muscle is restricted by the outer bony covering. According to this view, fenestral openings in the cheek region of the skull would have enabled jaw muscles to increase in size and allow a stronger bite and more efficient mastication. Frazetta has also suggested that the bony edges of fenestral openings serve as a much stronger anchorage for jaw muscles than do the internal surfaces of the skull bones.

Reptiles with single temporal openings, the **Synapsida** (Fig. 17-16(b)), are the first group to have diverged from the ancestral anapsid stocks and include the mammal-like reptiles. The presence of fenestra above and below a bar formed by joining the postorbital and squamosal bones defines the subclass **Diapsida** (Fig. 17-16(c)), now separated into the infraclasses Lepidosauria and Archosauriomorpha. Based on these views, the classification system given in Table 17-1 can be applied to all the reptilian orders, although some placements, such as the mesosaurs, placodonts, and ichthyosaurs are still unclear.

TABLE 17-1 Classification System for the Class Reptilia

Subclass Anapsida
- Order Captorhinida (Cotylosauria): stem reptiles
- Order Mesosauria: mesosaurs (aquatic freshwater reptiles)
- Order Testudinata: turtles

Subclass Synapsida
- Order Pelycosauria: pelycosaurs
- Order Therapsida: mammal-like reptiles

Subclass Diapsida
- Infraclass Lepidosauria
  - Order Eosuchia: early diapsids
  - Superorder Lepidosauria
    - Order Sphenodonta: sphenodontids
    - Order Squamata: lizards and snakes
  - Superorder Sauropterygia
    - Order Nothosauria: nothosaurs
    - Order Plesiosauria: plesiosaurs
    - Order Placodontia: placodonts
    - Order Ichthyopterygia: ichthyosaurs
- Infraclass Archosauromorpha
  - Order Protorosauria: protorosaurs
  - Order Trilophosauria: trilophosaurids
  - Order Rhynchosauria: rhynchosaurs
  - Superorder Archosauria
    - Order Thecodontia: early archosaurs
    - Order Crocodilia: crocodiles and alligators
    - Order Pterosauria: flying reptiles
    - Order Saurischia: lizard-hipped dinosaurs
    - Order Ornithischia: bird-hipped dinosaurs

Adapted from Carroll (1988).

REPTILIAN EVOLUTION

A general scheme showing some reptilian phylogenetic relationships is given in Figure 17-17, which also indicates approximate times during which various groups became extinct. In brief, the first reptiles are found in Pennsylvanian deposits of the Carboniferous period,[3] although paleontologists suggest that their evolution from anthracosaurs may well have occurred earlier, perhaps during the late Mississippian. By the beginning of the Permian, these early reptiles had undergone considerable evolutionary changes, and numerous new reptilian groups are found side by side with many varieties of amphibians. The end of the Permian and beginning of the Triassic records a decline in amphibian fossils (with the exception of stereospondyls) and also marks a striking expansion of the mammal-like reptiles, the **therapsids**. By the time of the Jurassic, almost all of the major reptilian groups have emerged, accompanied by a drastic fall in numbers of therapsids. From the Jurassic onward, reptilian adaptations enable widespread dispersion to many habitats including the aquatic, and for the next 100 million years or so there is a veritable age of dinosaurs, pterosaurs, and marine reptiles. This reptilian dominance lasted until the end of the Cretaceous, when almost all reptilian groups disappeared, except for lizards, snakes, turtles, crocodiles, and the New Zealand tuatara (*Sphenodon*). Although many explanations have been offered for this remarkable drama of reptilian radiation and decline, only a sample of these can be discussed.

The early reptilian captorhinomorphs appear first as small, slender animals, about 1 to 2 feet long, at a time in the Carboniferous during which many insects evolved terrestrial forms. The exact relationship between insects and reptiles is difficult to determine, but it has been suggested that captorhinomorphs functioned primarily as insectivores in the terrestrial food chain. Adaptation for a terrestrial existence seems also

[3]The unusual nature of their fossilization has been pointed out by Carroll (1988) and provides an illustration of the importance of accidental factors in such processes:

> These fossils are not found in normal coal-swamp deposits, such as those from which the majority of Carboniferous tetrapods have been found, but rather within the upright stumps of the giant lycopod *Sigillaria*. These trees grew in areas that were subject to periodic flooding, which resulted in the burial of the trees in several meters of sediments. The trees died and the central portion rotted out, but the bark was stronger and retained the cylindrical shape of the stump. After the withdrawal of the water, animals living on the newly developed land surface would occasionally fall into the hollow stumps. Eventually they died and were covered with sediments and fossilized.

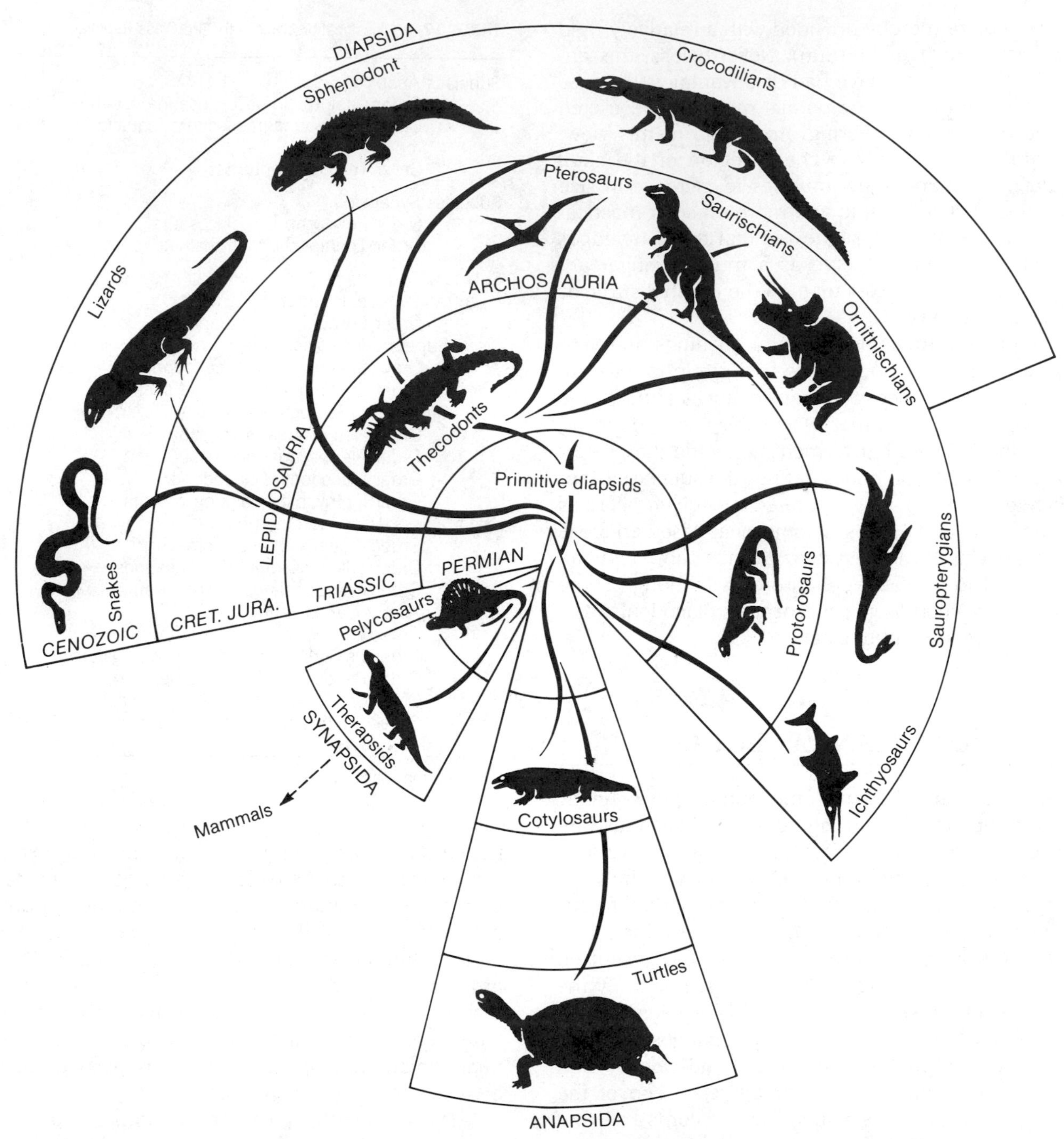

FIGURE 17-17 General evolutionary scheme showing relationships among the major reptilian groups, beginning with their origin in the Paleozoic era. Further major evolutionary events would include the origin of mammals from synapsids and the origin of birds from an archosaurian lineage. (Adapted from Colbert.)

to have been shared by other captorhinomorph-derived reptiles, such as small synapsids called **pelycosaurs** that also appear during the Pennsylvanian epoch.

The rapid radiation of these early forms led, by the end of the Carboniferous period, to exploitation of various environments. In aquatic habitats were found mesosaurs with long, toothy jaws (see Fig. 6-9). Among terrestrial forms were probably some of the larger pelycosaurs, whereas others of this group were more aquatic and also preyed upon fish and amphibious vertebrates. Early reptiles that were more exclusively dependent on non-aquatic food sources were either herbivores, such as the pareiasaurs and caseids, or were predators on various upland insectivorous and herbivorous forms, such as the carnivorous therapsids.

There are strong indications that pelycosaurs such as *Dimetrodon* (Fig. 17-18) are close to the line which gave rise to the therapsids, and these, in turn, later gave rise to the mammals. *Dimetrodon*, however, was a specialized animal with extremely long neural spines that supported a dorsal "sail" believed to have been used in temperature regulation. A sail of this kind, well supplied with blood vessels, would have enabled an animal that had cooled off at night to resume an optimum metabolic temperature at daybreak by placing its body perpendicular to the sun's rays. Further heating and cooling would have been accomplished by increasing or decreasing blood flow into this large, heat-exchanging dorsal surface. The animal may also have cooled off during very warm periods by moving into the shade or orienting itself parallel to the sun's rays. Since continuous enzymatic activity in muscle and other tissues depends upon the maintenance of optimum body temperatures, selection for such mechanisms would be an important force enabling pelycosaurs and their therapsid cousins to engage in longer periods of active predation or escape.

In the therapsids, however, temperature-regulating mechanisms other than dorsal sails are thought to have been used, and it is quite possible that some of these new forms were the first of the **endothermic** vertebrates; that is, optimum body temperatures were now sustained by internal metabolic heat-producing mechanisms rather than by external **ectothermic** means. As pointed out by Bennett and Ruben, higher metabolic rates in endothermic animals provide not only higher body temperatures, but (in association with greater numbers of mitochondria and increased aeration) also enable higher levels of oxygen utilization. Increased aerobic metabolism, in turn, enables more sustained activity and greater stamina than can be achieved by ectotherms who become rapidly exhausted because they rely mostly on anaerobic metabolism. Given the advantage of a high body temperature, selection for insulating mechanisms to help maintain it would have led to thicker layers of subcutaneous fat as well as the modification of scales into hair and feathers. By contrast, such insulation would be disadvantageous in ectotherms since they achieve their optimum body temperature through heat exchange with the environment.

FIGURE 17-18 Reconstruction of the early Permian pelycosaur, *Dimetrodon*. (Adapted from Romer [1968].)

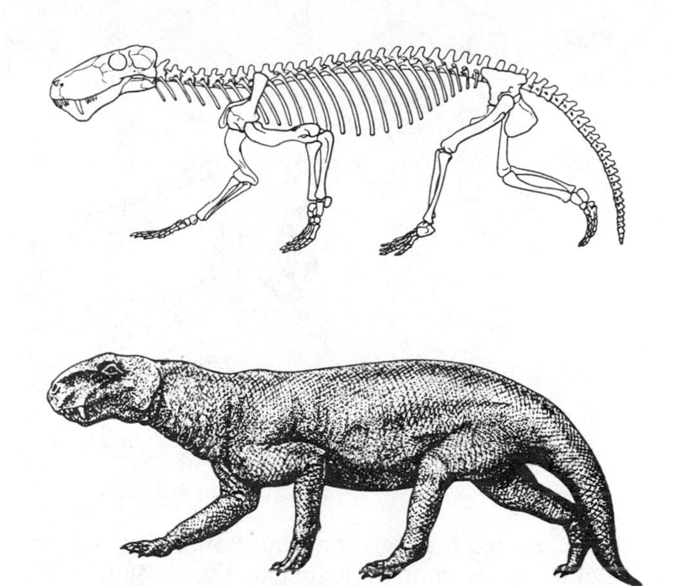

FIGURE 17-19 Skeleton and reconstruction of a carnivorous therapsid (*Lycaenops*) from the late Permian, showing a number of mammal-like features. It was about the size of a wolf, with large upper canines functioning as "saber teeth." (Adapted from Colbert, from Romer and Parsons.)

Because considerable additional amounts of food must be consumed to provide energy for high metabolic rates, the cost of endothermy is relatively high. For a given body weight, five to ten times more energy is required to maintain an endothermic mammal or bird at the same body temperature as an ectothermic reptile or amphibian. Also, despite their limitations in stamina, ectotherms are capable of short bursts of activity through anaerobic metabolism. Thus ectotherms can survive well under conditions which stress low energy expenditure and may even compete successfully with endotherms when predatory pursuit or escape is limited to short distances.

However, in environments that stress sustained activity, the race has gone to endotherms, of which therapsids may have been the earliest forms. Along with improved locomotory adaptations that moved the legs beneath the body, these forms seem to have been markedly successful both as herbivores and carnivores (Fig. 17-19), some reaching sizes 10 to 12 feet long. By the end of the Permian period about six out of seven reptilian fossils were therapsids.

Surprisingly, although many therapsids crossed the Permian boundary into the Mesozoic era, their numbers significantly diminished before the end of the Triassic. Among the reasons offered to explain this event is the possibility that the warm, constant climate

FIGURE 17-20 Reconstruction of *Euparkeria*, an early Triassic thecodont. It was about 2 to 3 feet long, with a short trunk counterbalanced by a heavy, muscular tail that would have enabled it to run bipedally.

of the Mesozoic reduced the importance of therapsid temperature-regulating advantages by enabling many ectothermal reptiles to maintain stable high body temperatures with less food intake. That is, increased numbers of active reptilian ectotherms could now be produced by the same amount of ingested energy necessary to maintain the high metabolic rate of many fewer endotherms.

Also interesting is the fact that the remainder of the Mesozoic, from the Jurassic to the close of the Cretaceous, was characterized by ruling reptiles from an entirely different subclass, the diapsid **archosaurs**. Numerous other reptilian groups also persisted and evolved such as turtles, lizards, and snakes, but it was among the archosaurs that the ruling **dinosaurs** are found, some of whom reached dimensions that still dwarf those of any other land vertebrate yet evolved.

EARLY ARCHOSAURS

The earliest diapsids are first known from upper Pennsylvanian deposits and most probably had a captorhinomorph ancestry (Reisz). These animals bore two fenestrae behind the optic orbit and an additional opening near the tip of the snout. During the radiation of these diapsids, two main infraclasses arose, lepidosauromorphs and archosauromorphs, differentiated by many traits, but most importantly by a unique ankle and foot structure in archosauromorphs that facilitates an upright posture. By the end of the Permian and beginning of the Triassic period, archosauromorph evolution had proceeded sufficiently far to produce a variety of groups, including **bipedal** forms in which the forelegs were shorter than the hind legs. Also, these animals had their teeth set in sockets (**thecodonts**) rather than fused to the jaw margins in lizard (lepidosauromorph) fashion.

The environmental pressures accounting for the innovation of archosaur bipedalism is still unknown,[4] but its development and persistence seem to be associated with selection for improved running speed as well as selection for large size. Even small early Triassic archosaurians begin to show the effects of selection toward bipedalism. For example, *Euparkeria*, a Triassic thecodont, had hind legs approximately one and a half times the length of the forelegs and bore two fenestrae (one in the lower jaw and the other anterior to the orbit) that anticipate those found in the later dinosaurs. It was obviously a carnivorous form, about 2 feet long, lightly built with hollow bones. As shown in Figure 17-20, it probably ran bipedally but rested or ambled slowly on all fours.

By the middle of the Triassic, bipedal innovations appeared to have developed further in a number of thecodont lines, whereas other lines such as phytosaurs and crocodiles maintained an obligate four-footed gait. In some bipedal forms, selection for increased length of stride by the hind legs had reached the point where the tibia was about as long as the femur, and these animals must have been quite speedy. (In the racehorse, the tibia/femur ratio is about 0.9.) However, in spite of various thecodont adaptations, the dinosaur radiation had already begun during the Triassic, and many tetrapod groups, including thecodonts and therapsids, became extinct before the close of that period.

[4]One conjecture put forth was that ancient archosaurs were swamp-dwellers that had developed strong hind legs for paddling, steering, and kicking, as well as a strong, muscular tail for propulsion. Thus when many swamps began to dry up during the Permian-Triassic, some of these forms found themselves preadapted for short-distance running on their hind legs, using their muscular tails as a balancing organ.

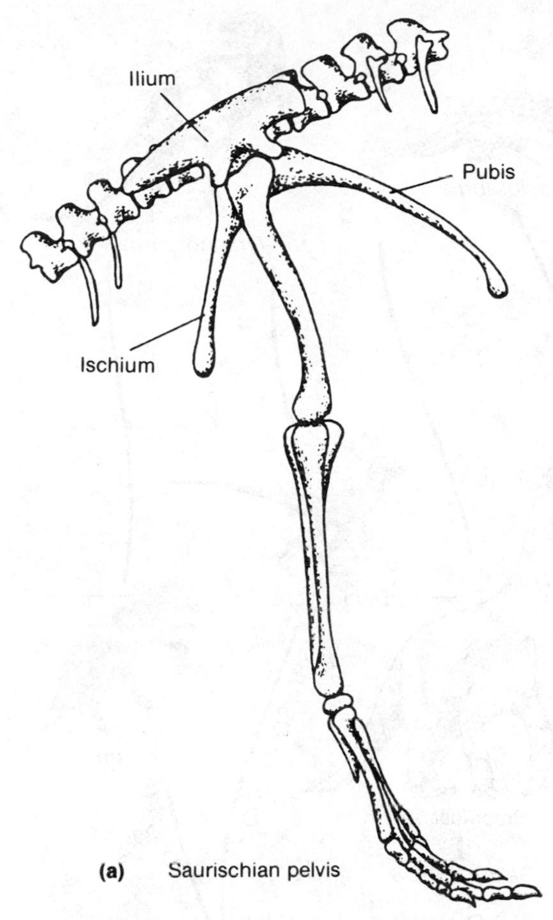

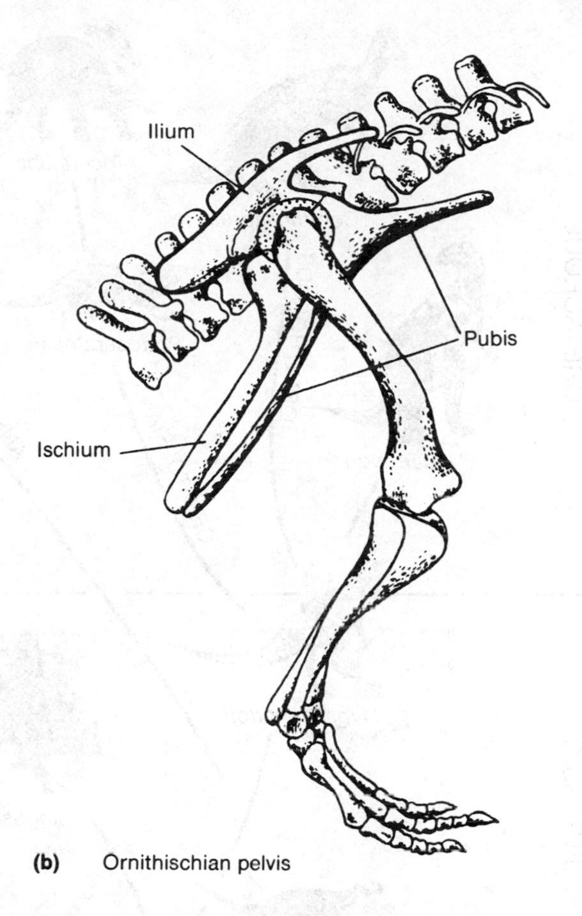

FIGURE 17-21 The two general types of pelves found in dinosaurs. In primitive reptiles, the pelvis is a solid, platelike structure from which the femur projects horizontally (parallel to the ground). With the evolution of bipedalism and a vertical femur, greater leverage for moving the hind legs arose by attaching the limb muscles to fore and aft extensions of the pelvis. In birds, the posterior extension of the pubis helps support the ischium, but the anterior pubis is not well developed. (Adapted from Romer [1966].)

THE DINOSAURS

The dinosaurs are popularly considered a single group but are actually composed of two major orders, **Saurischia** and **Ornithischia**, perhaps sharing a common thecodont ancestry during the Triassic. (Charig questions such common ancestry, and Carroll [1988] supports it.) A primary difference between the two orders in pelvic structure appears early in their evolutionary history: the Saurischia show the original thecodont triradiate structure in which the pubis extends anteriorly and the ischium posteriorly (Fig. 17-21(a)), while the Ornithischia possess a tetraradiate pelvis with the pubis usually parallel to the ischium (Fig. 17-21(b)). The two groups show their bipedal ancestry in shorter forelimbs than hind limbs, although some dinosaurs returned to a quadrupedal stance with lengthened forelegs.

As shown in Figure 17-22, phylogenetic branching among the dinosaurs proceeded throughout the Mesozoic, each of the two orders producing a variety of suborders as well as infraorders. Among the saurichians, the carnivorous **therapods** (four or fewer toes on the hind feet) have traditionally been classified into two groups based on presumably distinct body proportions, the smaller coelurosaurs and larger carnosaurs. Since some therapods appear to share traits from both groups, these distinctions no longer hold fast. However, these bipedal forms are eventually classified, it is clear that by the Cretaceous period they had radiated widely, evolving into forms such as the small, ostrichlike *Ornithomimus*, small to medium-sized deinonychosaurs such as *Deinonychus*, and large therapods such as *Tyrannosaurus* that may have weighed 6 to 8 tons and stood 20 feet above the ground.

The increase in body size in some lines of therapods appears to have paralleled the increase in body size in some lines of their herbivorous prey, probably because an **arms race** develops in which protection offered to prey by an increase in size selects in turn

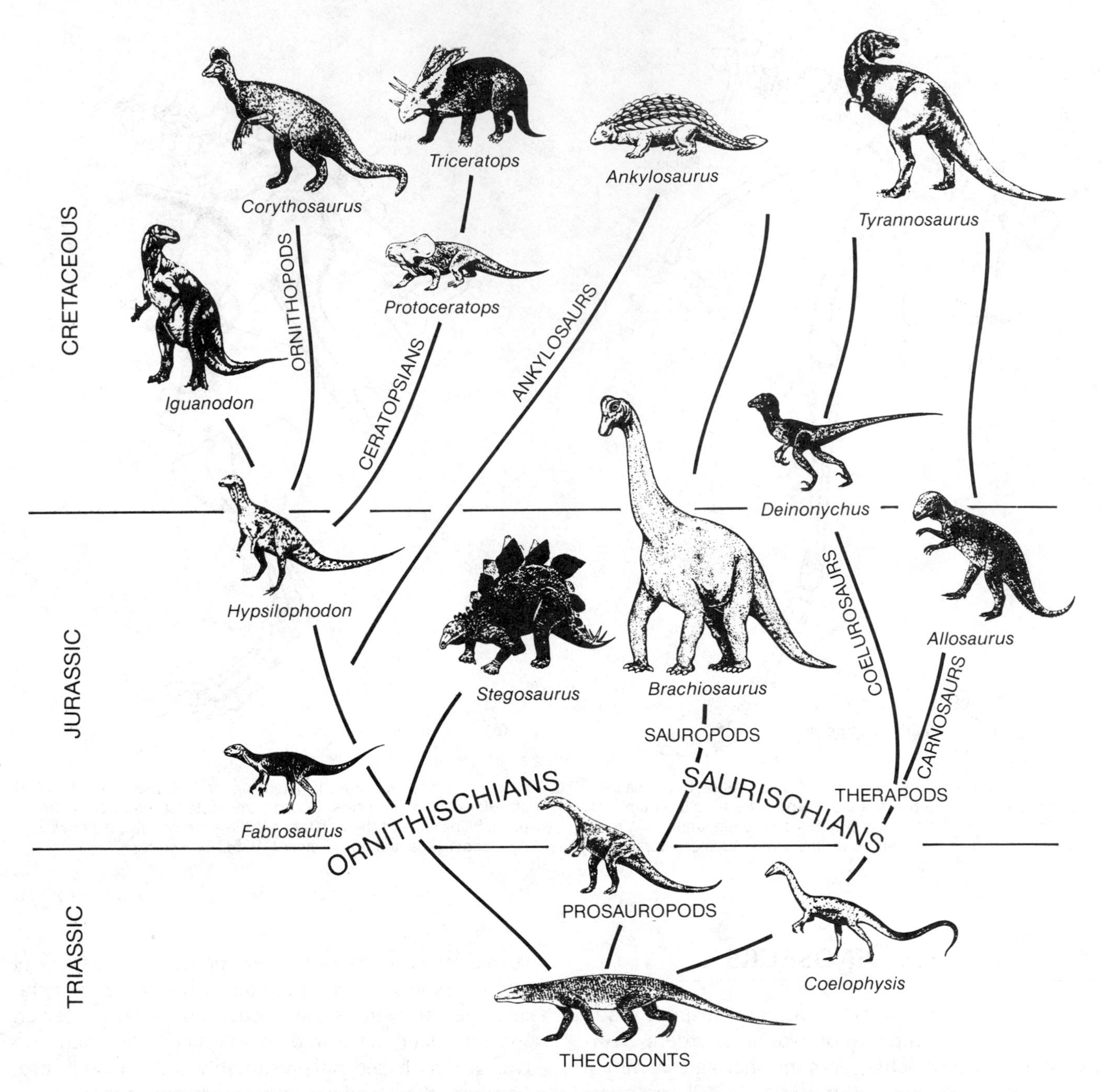

FIGURE 17-22 Dinosaur phylogeny from the Triassic to the end of the Cretaceous. (Adapted from Colbert.)

for increased size in predators. This cycle of selection for body size thus gave rise on one hand to the large therapods, and on the other hand to large herbivorous ornithischians and even larger herbivorous sauropods (five toes on the hind legs) such as *Apatosaurus* (also known as *Brontosaurus*) that were 70 to 80 feet long, and weighed 50 tons or more. The existence of very large fossil bones indicates that even these impressive dimensions were probably exceeded by some forms, and it is possible that terrestrial vertebrate body size had not yet reached its upper limit when the dinosaurs became extinct at the end of the Cretaceous.

The classic hypothesis that many of the giant sauropods were aquatic because only water could have buoyed up their immense weights is still debated and has been opposed by the view that they were primarily terrestrial. A terrestrial existence is supported by the discovery of fossilized stomach contents in one animal, indicating a diet of woody, leafy material, and an early Cretaceous trackway found in Texas indicating that they traveled in socially organized herds. Presumably, at least some of the long-necked sauropods fed upon the upper branches of tall Mesozoic conifers, using gizzard stones to macerate plant tissues. Their large

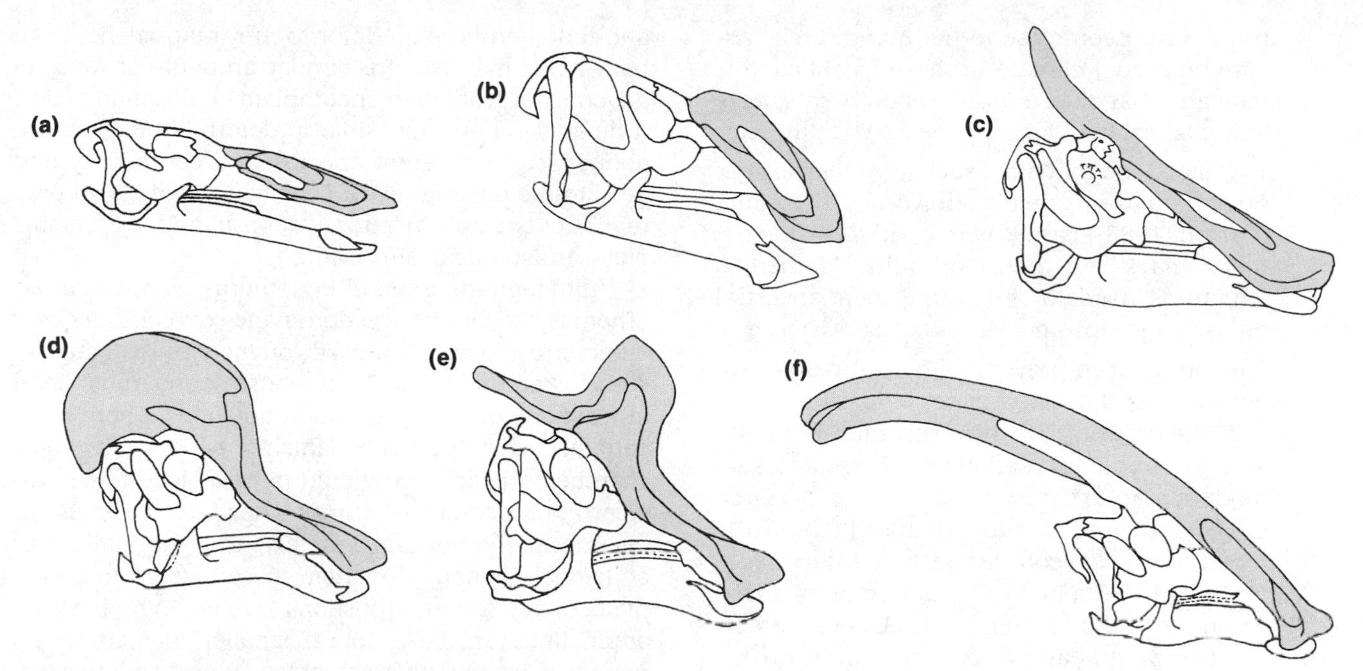

FIGURE 17-23 Skulls of various duck-billed hadrosaurs of the Upper Cretaceous. Behind a flattened, toothless beak lay rows of teeth (as many as 700 per jaw ramus!) used apparently to grind tough vegetable matter. As shown, the nasal and premaxillary bones (shaded) assumed unusual shapes in different groups, producing long loops of nasal passages for purposes which are still undetermined. Among suggestions for their use have been improvement in sense of smell, visual signals for species recognition, male weapons or shields during mating competition; and resonators for amplifying sound. (a) *Anatosaurus*; (b) *Kritosaurus*; (c) *Saurolophus*; (d) *Corythosaurus*; (e) *Lambeosaurus*; (f) *Parasaurolophus*. (From Romer [1966].)

size, powerful tails, and social organization probably gave them considerable protection against all but the largest predators.

Various lines of the exclusively herbivorous ornithischians did not possess the same degree of bipedalism and size as some lines of saurischians, but many were nevertheless quite large, reaching lengths of 30 feet and weights of 5 tons or more. Among the ornithischian groups are the duck-billed hadrosaurs, the armored stegosaurs and ankylosaurs, and the horned ceratopsians. There appears to have been considerable variation and specialization among some of these groups; hadrosaurs, for example, were the most diverse of the dinosaurs in terms of skull morphology, with a large variety of crested forms (Fig. 17-23). This diversity is reflected in their relative success, since hadrosaurs may have comprised as much as 75 percent of terrestrial vertebrate biomass in numerous localities by the end of the Cretaceous.

ENDOTHERMY VERSUS ECTOTHERMY

Clearly, dinosaurs were successful creatures throughout much of the Mesozoic, and no other group of reptiles or mammals approached their size or importance in terrestrial deposits until after the Cretaceous period. Some paleontologists such as Bakker and Ostrom have suggested that dinosaur success stemmed at least partially from the high metabolic rates provided by endothermy, enabling continuous high levels of activity. A large, sluggish, ectothermal reptile, whether herbivore or carnivore, could hardly have competed successfully with endothermal therapsids and their mammalian descendants. That is, if dinosaurs were only typical ectothermal reptiles, why did the mammals remain small and insignificant throughout the Mesozoic?

Marshalled in further support of dinosaur endothermy are a number of arguments:

1. The fully erect posture of many dinosaurs, with their limbs extending vertically downward beneath the body, is a posture which is found today only among endotherms (all modern ectotherms have a sprawling gait). The speed and agility that accompanies an erect posture would presumably have had its source and sustenance in high metabolic activity.
2. The microscopic bone structure of dinosaurs shows a high density of blood-carrying haversian canals similar to those found in mammals, whereas modern ectotherms show few such canals.
3. The distribution of dinosaur fossils during the Cretaceous extends to areas in Canada that

must have been close to the Arctic Circle. Assuming a cool climate in these localities, this indicates that dinosaurs, like endotherms, were animals that could supply their own body heat.

4. The likelihood that birds such as *Archaeopteryx* originated from small carnivorous dinosaurs (Carroll [1988]) carries with it the corollary that there may have been primitive featherlike structures used for insulating their ancestral dinosaur groups against loss of body heat.
5. The predator-to-prey ratio of carnivorous-to-herbivorous dinosaurs in some fossil deposits is of the order of 3 to 100. This ratio is similar to that found for endothermal predators in modern animal communities who need a large prey population to support their high-energy metabolism. By contrast, ectothermal predators with lower metabolic requirements can exist on a prey population ten times less numerous (e.g., a predator-to-prey ratio of 3 to 10).

Since these and other arguments suggest that the distinction between dinosaurs and reptiles may be as profound as the distinction between mammals and reptiles or birds and reptiles, Bakker and co-workers propose the removal of dinosaurs from the class Reptilia and their placement in a separate vertebrate class, Dinosauria.

However interesting these arguments may be, each has nevertheless been disputed by other paleontologists. Erect posture, for example, may have little to do with endothermy but may be instead the only stance a large, heavy, terrestrial animal can assume without causing undue bending of its supporting limbs. Similarly, haversian bone structures may be related to growth rate, body size, and other factors rather than to endothermy, especially since some ectothermal reptiles such as turtles show such structures and some small mammals and birds do not. Dinosaur radiation to northern latitudes is also not sufficient evidence for endothermy since Cretaceous climates were warmer than they are now, continental drift may have moved their ancient habitats northward, and such northern deposits also contain fossils of obviously ectothermal crocodiles and turtles. Also, whether feathers were originally used for insulation rather than for flight is still disputed (Feduccia [1985]), and there is, as yet, no evidence of feathers among dinosaurs. Some biologists also dispute the predator-to-prey ratios cited by Bakker for some localities. Others point out that it is difficult to discriminate between large ectotherms and endotherms on predator-to-prey ratios alone, since both types may require similar amounts of food. In general, the problem of incomplete fossilization makes it difficult if not impossible to determine the relative abundance of different species in a community and adds to the uncertainty of specifying predator-to-prey ratios; for example, one fossil deposit in Texas contains only carnivorous coelurosaurs.

Although the issue of endothermy is not resolved (Thomas and Olson), the dominance of dinosaurs over other terrestrial vertebrates throughout most of the Mesozoic is an undisputed fact whose source must lie in at least one or more special adaptations. There must certainly have been mechanisms, such as the four-chambered heart, that would permit blood to be efficiently pumped to the sauropod head many feet above heart level. If dinosaurs were also endothermal, many additional elements for their success seem to be explained, yet serious questions remain. Why have no adult dinosaurs been found smaller than about 20 pounds in weight, whereas practically all endothermal Mesozoic mammals (and many ectothermal reptiles) were below this size? What accounts for the extinction of all dinosaurs at the end of the Mesozoic, yet the survival of various other vertebrate groups including mammalian endotherms? In sum, why were presumed endothermal dinosaurs unable to adapt to ecological and climatological conditions to which endothermal mammals adapted?

For those who believe in dinosaur ectothermy, it is possible that large size alone played a significant role in dinosaur success as well as in their limitations. As discussed previously, large ectotherms in a warm climate would have been able to preserve body heat and perhaps attain fairly high rates of metabolism and activity without paying the high cost of endothermy. Given such dependence on high body temperatures, smaller dinosaurs, subject to greater temperature fluctuations because of their size and lack of insulation, would not have been as successful.[5] Also, if dinosaurs were primarily ectothermal, even large size would not have protected them against the more variable climate that is believed to have inaugurated the Cenozoic era. For example, a large ectotherm with its reduced ability to change body temperature rapidly would have had considerable difficulty losing heat during a hot summer as well as difficulty gaining enough heat during winter's prolonged cold periods. By contrast, endothermal mammals were able to survive the end of the Cretaceous and increase during the Cenozoic because their activity was not as dependent on external temperatures.

[5]Temperature stability in modern ectothermal reptiles is a function of size: the larger the animal, the more stable its body temperature (Spotila).

THE LATE CRETACEOUS EXTINCTIONS

There also remains considerable debate as to whether increased climatological activity was the primary cause for dinosaur extinction at the end of the Cretaceous period. Volcanic activity, epidemics of disease, changes in plant composition, shifting continental profiles, elevated carbon dioxide level (greenhouse effect), changes in sea level or ocean salinity, high doses of ultraviolet radiation, dust clouds caused by collisions with comets or asteroids, and ionizing radiation from supernova explosions or other sources are among the explanations offered for the Cretaceous extinctions (e.g., see discussions in Russell, Kerr, and Hallam).[6]

Although no dinosaurs are known to have survived the Cretaceous, other groups also suffered, and it is estimated that more than half of all animal species, classified into the various groups given in Table 17-2, became extinct during a relatively short geological period. Since the extinctions covered so many different kinds of organisms, the extent to which dinosaur ectothermy or endothermy affected their survival is unclear: no terrestrial vertebrate larger than 50 pounds is known to have survived the Cretaceous, and, with the exclusion of crocodiles and turtles, the extinctions embraced numerous marine organisms of varying sizes and metabolic features.

According to Raup and Sepkoski, these extinctions are not isolated events but seem tied to other extinctions that occurred throughout the Phanerozoic eon. They point out that mass extinctions of families and genera occur with a periodicity of about 26 million years, and an analysis of their data by Fox supports this view (Fig. 17-24). However, since there appears to be no clear cause for this periodicity, the Raup-Sepkoski hypothesis has been disputed, although extra-terrestrial factors such as the effects of a possible companion star to our sun (named by some, Nemesis), or the impacts of comets or asteroids, have been offered as explanations. Interestingly, an anomalous iridium-rich layer at the Mesozoic-Cenozoic boundary found in nearly 50 different localities on earth lends credence to a large meteorite impact at the end of the Cretaceous periods (Alvarez), but dinosaurs and other animal groups had already declined in numbers or disappeared before this impact layer was deposited (Sloan et al.). Knowledge of the cause(s) for mass extinctions and the role of extinction in evolution therefore still awaits more exact information (Raup).

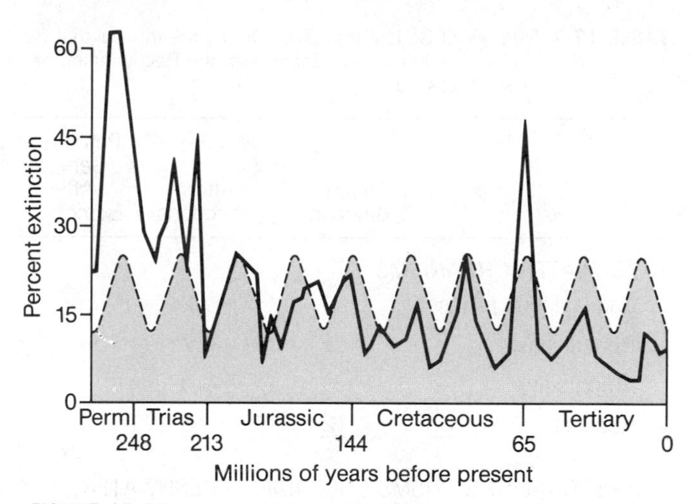

FIGURE 17-24 Heavy solid lines show the percent of marine animal extinctions during the approximate 260-million-year interval from the Permian period to the present. Dashed lines and shaded areas are based on a periodicity of 26 million years (calculated as the tenth Fourier harmonic). According to Fox, the relationship between the extinctions and 26-million-year periods is statistically significant. (Adapted from Fox.)

REPTILIAN FLIGHT: PTEROSAURS

Airborne flight is a highly adaptive form of locomotion enabling (1) rapid escape from terrestrial predators, (2) access to feeding and breeding grounds that would otherwise be difficult or impossible to reach, and (3) relatively swift transit between localities. Although gliding forms capable of parachuting for short distances appear in various vertebrate groups including fish, known adaptations for sustained ascending flights appeared only three times in two vertebrate classes: twice in reptiles (**pterosaurs** and **birds**) and once in mammals (bats). The two reptilian-derived flying forms differ in respect to mechanisms and accompanying adaptations: in pterosaurs a flight membrane was formed by a thin fold of skin stretched between the trunk and elongated fourth finger of each hand, while in birds the flying surface is composed of many stiff wing feathers that project posteriorly from the front limb.

[6]A once popular hypothesis sought to explain the extinction of dinosaurs by internal rather than external causes. Just as individuals are born, grow old, and die, it was suggested that races, species, and other taxonomic categories follow a similar life history. This concept, called **racial senescence**, seemed to be supported by many observed evolutionary successions, and its cause was believed to lie in the gradual decline of some unknown vital force in each group, leading ultimately to the appearance of bizarre and non-adaptive characters. This is certainly not true. Far from senescent, dinosaur groups were progressively adaptive for more than 100 million years, and even towards the end of the Cretaceous period new groups, such as the ceratopsians, appeared to be continually evolving in adaptive directions. There is no indication that the extinction and replacement of dinosaurs or any organism is caused by mechanisms other than the failure of their adaptations to cope with new environmental or competitive challenges.

TABLE 17-2 Number of Genera that Lived During the Interval that Began 20 Million Years Before the Extinctions at the End of the Cretaceous Period and Ended with the Beginning of the Cenozoic Era, Compared with the Number that Existed 10 Million Years Afterwards.*

	Before Extinctions	After Extinctions	Percent of Genera After Extinctions
FRESHWATER ORGANISMS			
Cartilaginous fishes	4	2	
Bony fishes	11	7	
Amphibians	9	10	
Reptiles	12	16	
	36	35	97
TERRESTRIAL ORGANISMS (INCLUDING FRESHWATER ORGANISMS)			
Higher plants	100	90	
Snails	16	18	
Bivalves	0	7	
Cartilaginous fishes	4	2	
Bony fishes	11	7	
Amphibians	9	10	
Reptiles	54	24	
Mammals	22	25	
	226	183	81
FLOATING MARINE MICROORGANISMS			
Acritarchs	28	10	
Coccoliths	43	4	
Dinoflagellates	57	43	
Diatoms	10	10	
Radiolarians	63	63	
Foraminifers	18	3	
Ostracods	79	40	
	298	173	58

	Before Extinctions	After Extinctions	Percent of Genera After Extinctions
BOTTOM-DWELLING MARINE ORGANISMS			
Calcareous algae	41	35	
Sponges	261	81	
Foraminifers	95	93	
Corals	87	31	
Bryozoans	337	204	
Brachiopods	28	22	
Snails	300	150	
Bivalves	399	193	
Barnacles	32	24	
Malacostracans	69	52	
Sea lilies	100	30	
Echinoids	190	69	
Asteroids	37	28	
	1,976	1,012	51
SWIMMING MARINE ORGANISMS			
Ammonites	34	0	
Nautiloids	10	7	
Belemnites	4	0	
Cartilaginous fishes	70	50	
Bony fishes	185	39	
Reptiles	29	3	
	332	99	30
OVERALL TOTALS	2,868	1,502	52

From Russell.
*The record for terrestrial organisms is limited to North America but is global for marine organisms.

Although primitive forms of pterosaurs appear in the Upper Triassic and birds in the Jurassic, earlier evolution of these two types has unfortunately left no obvious record, perhaps because they started out as small arboreal creatures in poorly fossilized highland habitats. Nevertheless, their skeletal features provide a number of fairly clear homologies with earlier reptiles. For pterosaurs, one phylogeny suggests they originated from early Triassic bipedal thecodonts (Fig. 17-25, left side). The ancestry of birds includes an additional dinosaur intermediate (Fig. 17-25, right side), since the similarity between the first fossil bird (*Archaeopteryx*) and dinosaurs is so striking that *Archaeopteryx* would have been classified as a dinosaur in the absence of its feathers. In general, it is clear that by the time of their first fossil appearance, both types were well adapted to active gliding and probably to sustained flying as well.

A well-described Jurassic pterosaur, *Rhamphorhynchus*, (Fig. 17-26(a)), was about 2 feet long with a typical diapsid archosaurian skull and an additional preorbital fenestra that helped lighten the head. The **rhamphorhyncoid** tail was long with a small, rudder-like flap of skin at the end, the bones were light and hollow, and the elongated jaws were armed with strong, pointed teeth. According to Padian, the sternum and its accessory bones provided sufficient surface to have allowed attachment of large flight muscles like

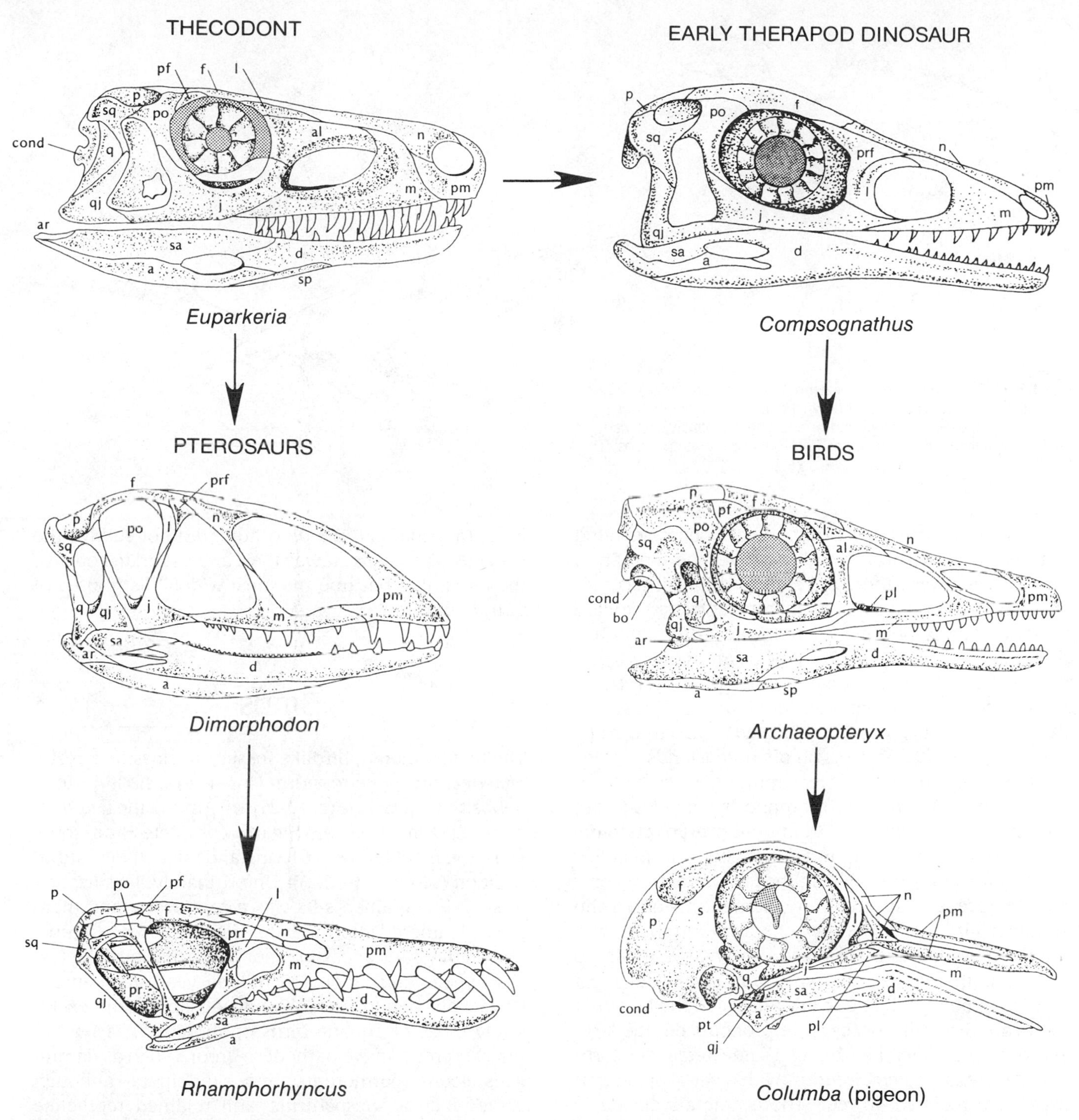

FIGURE 17-25 Similarities in skull structures among early bipedal archosaurs such as *Euparkeria* and the pterosaur and bird lineages that are believed to have arisen from them. Abbreviations: *a*, angular; *al*, adlacrimal; *ar*, articular; *bo*, basioccipital; *cond*, occipital condyle; *d*, dentary; *f*, frontal; *j*, jugal; *l*, lacrimal; *m*, maxilla; *n*, nasal; *p*, parietal; *pf*, postfrontal; *pl*, palatine; *pm*, premaxilla; *po*, postorbital; *pr*, prootic; *prf*, prefrontal; *pt*, pterygoid; *q*, quadrate; *qj*, quadratojugal; *sa*, surangular; *sp*, splenial; *sq*, squamosal. (Adapted from Stahl, from Romer and Heilmann.)

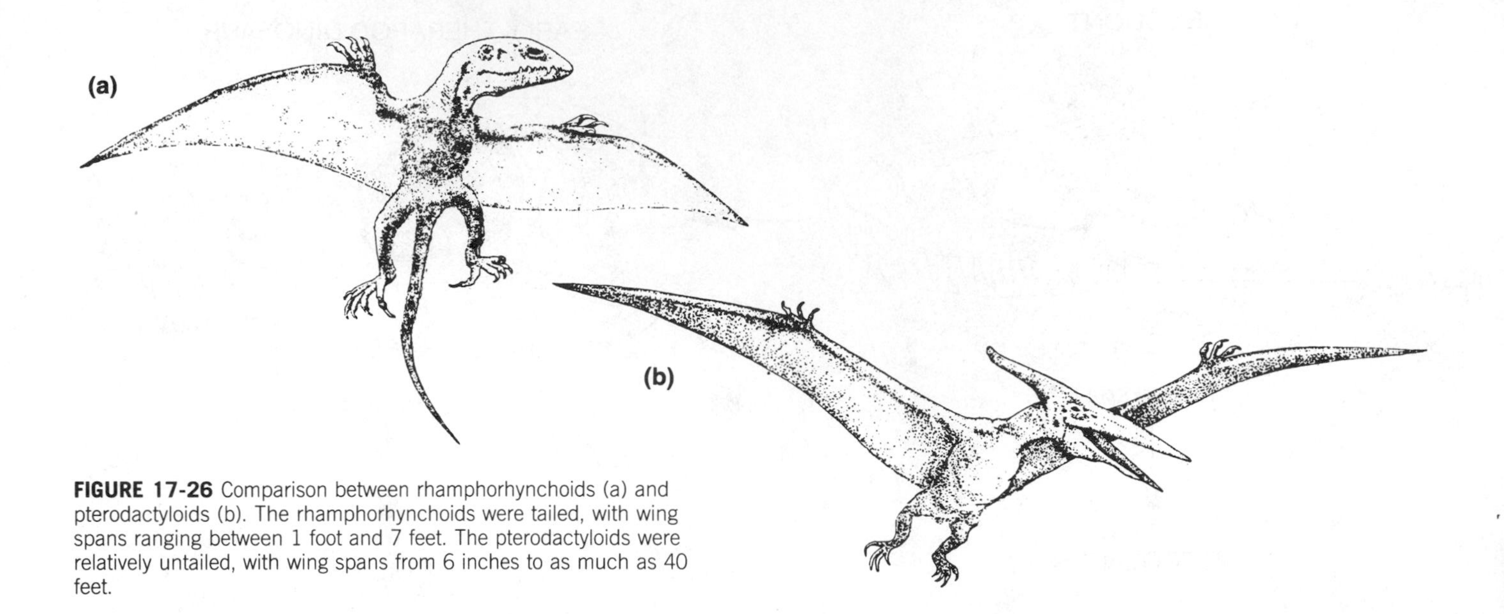

FIGURE 17-26 Comparison between rhamphorhynchoids (a) and pterodactyloids (b). The rhamphorhynchoids were tailed, with wing spans ranging between 1 foot and 7 feet. The pterodactyloids were relatively untailed, with wing spans from 6 inches to as much as 40 feet.

those of modern flying birds. He also demonstrated that the wing membrane did not cover the hind limb, and the legs were therefore free for bipedal locomotion. The traditional concept of pterosaurs as restricted to aerial gliding but feeble and awkward on land has therefore been replaced by that of an actively flying animal capable of reasonably efficient terrestrial bipedal locomotion. From their fossil locations, it seems likely that these pterosaurs roosted near large lakes and coastal areas, perhaps on offshore islands in trees or cliffs, and hunted for fish swimming at the surface.

By the late Jurassic, rhamphorhynchoid success had given rise to another group, the **pterydactyloids**, which continued well into the Cretaceous. In the pterydactyloids (Figure 17-26(b)), the tail was almost completely absent, a special bony element anchored the shoulder girdle to a number of spinal vertebrae, and the teeth tended to be reduced, leading eventually to a long, toothless beak. In *Pterandon*, a long skull crest extending behind the optic orbit doubled the length of the head. Size differences were pronounced, the largest fossil pterydactyloid being a late Cretaceous form found in Texas, *Quetzalcoatlus*, with a wingspread that may have reached 40 feet! These animals obviously must have been successful gliders, probably using sea or land thermals for lift, and may have been capable of powered flapping for short distances. Since pterosaurs could not have utilized the various environmental devices for temperature regulation that were available to terrestrial ectotherms, such as movement in and out of shady areas, it has been suggested that they must have been endothermal, and some pterosaur fossils do show evidence of hairlike scales that may have served as insulation. Perhaps, along with dinosaurs, pterosaurs also merit non-reptilian class status. In any case, they too, like the dinosaurs, did not survive the Cretaceous; instead, it was their archosaurian cousins, the birds, that became the most widely distributed of Cenozoic flyers.

BIRDS

The first feathered, birdlike fossils, all classified as ***Archaeopteryx***, were found in Upper Jurassic limestone deposits in Bavaria (Fig. 17-27(a)). Among the five skeletons known, three are nearly complete, and some show the flight feathers of wing and tail in their natural position (see Fig. 3-12(a)). Unfortunately, fossilization in such fine-grained silts was a rare event, and there are no feathered intermediates between *Archaeopteryx* and its dinosaur ancestors, nor do further birdlike fossils appear until about 10 million years later in the Cretaceous period. These Cretaceous fossils are exclusively those of aquatic birds or shorebirds, a few already representative of modern groups such as flamingoes, loons, cormorants, and sandpipers, although some, such as *Hesperornis*, still retained reptilelike teeth.

Among its various consequences, the absence of an adequate fossil record preserves the mystery of when and how feathers first evolved from reptilian scales and has made the origin of avian flight a matter of dispute. Were feathers primarily an adaptation for insulating the presumed endothermic reptilian ancestors of birds, or were they primarily associated with flight and their insulating qualities secondary? Were primitive ancestral birds originally **arboreal** reptiles who used their developing wings to glide from branch

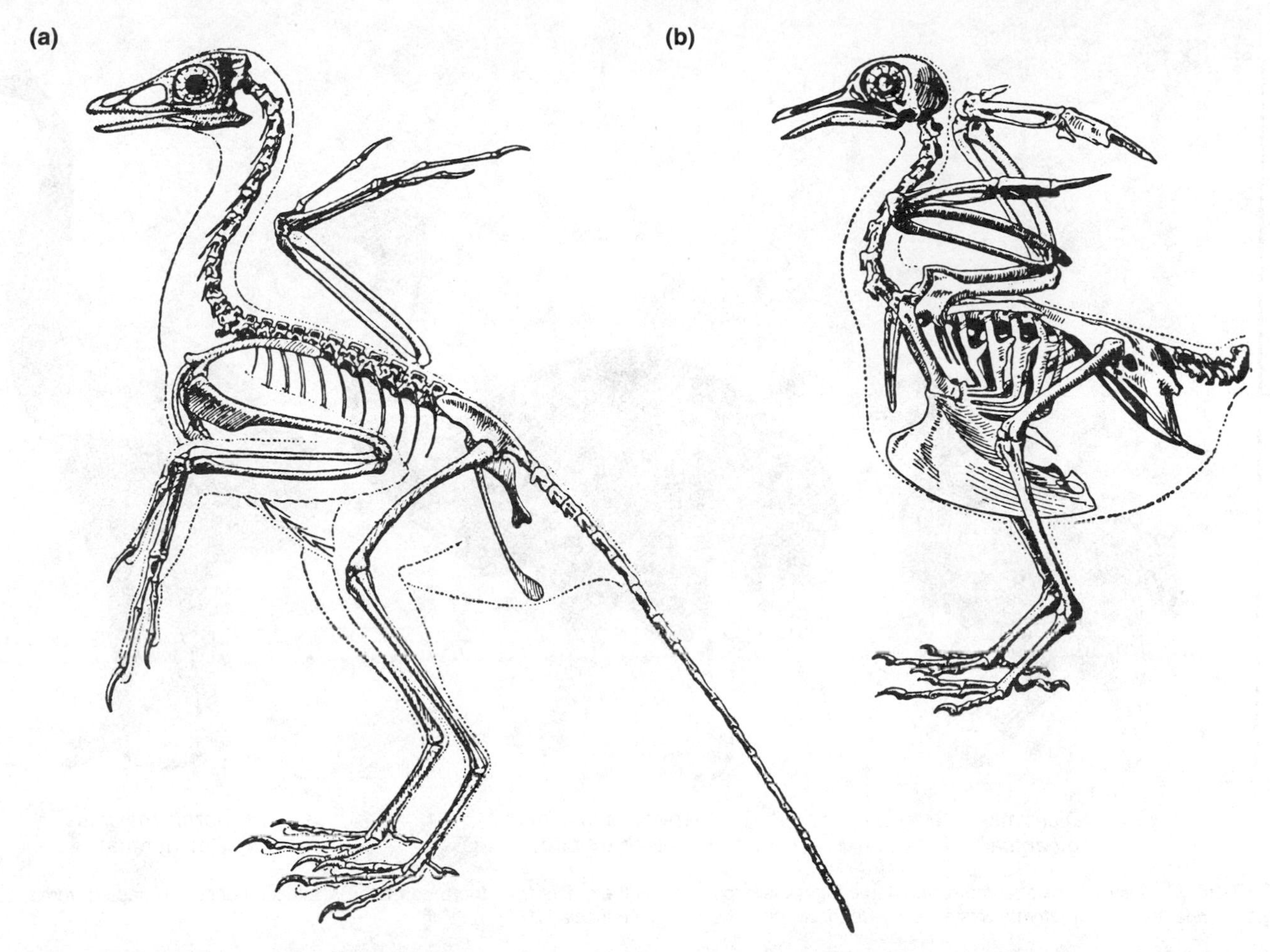

FIGURE 17-27 Skeletons of (a) *Archaeopteryx*, and (b) the comparably sized modern pigeon, *Columba*. Note the changes: the pelvis and sacrum have coalesced into a single structure, the sternum has become enlarged for the attachment of flight muscles, the hand bones have fused, and the long, bony tail has diminished. (Adapted from Colbert, from Heilmann.)

to branch, or were they **cursorial**, ground-dwelling creatures whose primitive feathers formed planing surfaces enabling them to increase running speed? Different opinions extend also to *Archaeopteryx*, although it is generally agreed that *Archaeopteryx* represents a fairly advanced stage in a long history of bird evolution that may have begun much earlier. The primary feathers of *Archaeopteryx* are remarkably similar in vane structure to the primary feathers of modern flying birds, whereas non-flying birds have feathers with different structures (Feduccia and Tordoff): *Archaeopteryx* could fly, if weakly.

Ostrom presents the view that feathers evolved primarily as a means of controlling heat loss in some endothermal dinosaurs, and these feathers, especially on the forelimbs, could then be used to help capture prey such as insects. A feathered "insect net" of this type, along with accompanying muscular adaptations such as enlarged pectorals, would then serve as an incipient **wing**, preadaptive for powered flight. Other paleontologists argue that the evolution of wings in a cursorial animal would have been a hindrance rather than an aid in gaining speed, since wing lift would have interfered with ground traction necessary for the hind legs. On the other hand, if *Archaeopteryx* had an exclusively gliding arboreal history, what would explain its bipedal stance, obviously derived from ground-dwelling forms?

There are no clear answers to these questions as yet, although there are some suggestions that bird evolution may not have been a straightforward process but endured a number of stages, beginning with a bipedal, cursorial reptile which later became arboreal. It was presumably in the trees that long-feathered gliding forms developed. Even if *Archaeopteryx* was primarily cursorial, this might mean only that one transitional ground-dwelling form was fossilized from an extensive avian adaptive radiation.

Once begun, bird evolution seems to have advanced in significant steps from the Cretaceous period

FIGURE 17-28 Reconstructions of some extinct large flightless birds showing their relative sizes. *Diatryma* was an early Cenozoic bird, the others date from the much later Pleistocene epoch. (Adapted from Feduccia [1980].)

through the beginning of the Tertiary. The Cretaceous birds show such modern features as an enlarged brain, fusion of the skull bones, and reduction of the temporal fenestrae. The sternum was greatly enlarged in some forms, indicating the attachment of powerful flight muscles, and there were also obvious skeletal changes such as fusion of the pelvis and sacrum. Although bird fossils are never too plentiful, there are sufficient numbers and kinds of Eocene and Oligocene deposits to indicate that almost all modern orders of birds had evolved by then.

One dramatic adaptive opportunity caused by the extinction of the dinosaurs at the end of the Cretaceous was to leave open terrestrial niches into which various large, flightless, ground birds evolved, such as the 7-foot-tall *Diatryma* (Fig. 17-28) and others (Phorusrhacidae) that may have reached 10 feet or more. Some of these giant forms were widely distributed until they became extinct later in the Cenozoic because of competition with advanced mammalian carnivores and predation by humans. Only relatively few flightless ground birds now survive, such as the ostriches of Africa and the rheas of South America; the smaller flightless species such as kiwis and island rails are mostly confined to gradually diminishing habitats.

At present, systematists classify birds into a total of about 35 orders subdivided into about 200 families. The distinctions among these groups are sometimes subtle, mostly based on external traits rather than on pronounced anatomical and skeletal differences used in separating groups among mammals and other vertebrate classes. Nevertheless, among the existing 8,800 species of birds are a variety of feeding and locomotor adaptations, ranging from flesh-eating to nectar-feeding, and from rapid running to gliding, swooping, diving, and swimming (Fig. 17-29). Although evolutionary changes that gave rise to some of these adaptations have been proposed, the absence of fossil intermediates makes relationships between avian groups uncertain, especially among those in arboreal habitats. The molecular approach to this problem by Sibley and Ahlquist, based on DNA studies (Chapter 12), is now providing considerable information and shows promise of clarifying many disputed issues.

FIGURE 17-29 Some of the many adaptations of bird bills (above) and feet (below). (Adapted from Feduccia [1980].)

SUMMARY

The sarcopterygian fish were preadapted to terrestrial life since they had lungs and fleshy fins. They may have invaded the land when their swampy habitats desiccated, or when competition and predation forced them to seek new habitats. Similarities of teeth, vertebrae, and other bones indicate that the earliest amphibia arose from these fish. A terrestrial environment favored certain structural modifications. A rigid spine provided suspension for limb girdles, freed internal organs from pressure, and acted as a cantilever for the head. In certain amphibia, the pleurocentrum rather than the intercentrum became the primary vertebral element, a trait retained by reptiles and mammals.

Modern amphibia live in many different habitats and are quite specialized. Their unique teeth, reduced skeletal bone, four-digit hands and unusual ear structure make their phylogenetic relationships difficult to trace as yet.

The reptiles, arising from small, lizardlike amphibia, have many novel attributes such as a partially-divided ventricle, a cornified epidermis, separate reproductive and excretory ducts, and a shelled egg containing a membrane-enclosed embryo. The development of the amniotic egg liberated reptiles from dependency upon water and was preceded by terrestrial egg laying and internal fertilization. The development of fenestrae in the temporal bone, a trait often used to classify fossil reptiles, provided better anchorage for jaw muscles.

Reptiles diversified into many terrestrial habitats and became the dominant vertebrate group throughout the Mesozoic era. Therapsid reptiles, ancestors of the mammals, appear to have had temperature-regulating mechanisms and may have been the first endotherms. Special limb bones allowed the archosaur reptiles to become bipedal, increasing their speed and size. Their descendants, the dinosaurs, radiated into almost every terrestrial habitat, and some became so enormous as to approach terrestrial size limits. Some paleontologists claim that dinosaurs were endothermic, based on bone structure, posture, and biogeographical distribution. According to these views, endothermy, in combination with their size and other favorable attributes, may have enabled dinosaurs to dominate all other land vertebrates. At the end of the Cretaceous period, dinosaurs along with large marine reptiles and various other groups became extinct. This extinction may have been caused by a unique single event or a succession of events or may have been only one in a cycle of periodic extinctions.

Adaptations for sustained flight appeared twice in the reptiles, giving rise to pterosaurs and birds. The pterosaurs evolved from dinosaurs be developing hollow bones and flight membranes between trunk and forelimbs and, in some cases, by losing the reptilian tail and teeth, leaving the jaw as a beak. Birds may have originated from bipedal, ground-dwelling endothermic reptiles, their feathers deriving from scales used as insulating mechanisms and/or as structures that aided gliding. Birds evolved rapidly with alterations of bone structure and enlargement of the sternum and brain. On the paleontological level, little is known of evolutionary relationships among birds since bird fossils are relatively rare.

KEY TERMS

allantois
amniotic egg
amphibians
Anapsida
anthracosaurs
arboreal
Archaeopteryx
archosaurs
arms race
bipedalism
birds
captorhinomorphs
cotylosaurs
cursorial
Diapsida
dinosaurs
ectothermic
endothermic
fenestra
ichthyostegids
Late Cretaceous extinctions
Lissamphibia
Ornithischia
pelycosaurs
pterosaurs
pterydactyloids
racial senescence
reptilian flight
rhamphorhyncoids
Saurischia
Synapsida
tetrapod
thecodonts
therapsids
therapods
wing

REFERENCES

Alexander, R. M., 1975. *The Chordates*. Cambridge Univ. Press, Cambridge.

Alvarez, L. W., 1983. Experimental evidence that an asteroid impact led to the extinction of many species 65 million years ago. *Proc. Nat. Acad. Sci.*, **80**, 627-642.

Archibald, J. D., and W. A. Clemens, 1982. Late Cretaceous extinctions. *Amer. Sci.*, **70**, 377-385.

Bakker, R. 1986. *The Dinosaur Heresies*. Longman, Harlow, England.

Bennett, A. F., and J. A. Ruben, 1979. Endothermy and activity in vertebrates. *Science*, **201**, 649-654.

Carroll, R. L., 1970. The ancestry of reptiles. *Phil. Trans. Royal Soc. London B*, **257**, 267-380.

——, 1977. Patterns of amphibian evolution: an extended example of the incompleteness of the fossil record. In *Patterns of Evolution, As Illustrated by the Fossil Record*, A. Hallam, (ed.). Elsevier, Amsterdam, pp. 405-437.

——, 1988. *Vertebrate Paleontology and Evolution*. W. H. Freeman, New York.

Charig, A., 1983. *A New Look at the Dinosaurs*. British Museum, London.

Colbert, E. H., 1981. *Evolution of the Vertebrates*, 3rd ed. John Wiley, New York.

Desmond, A. J., 1976. *The Hot-Blooded Dinosaurs: A Revolution in Paleontology*. Dial Press/James Wade, New York.

Duellman, W. E., and L. Trueb, 1986. *Biology of Amphibians*. McGraw-Hill, New York.

Estes, R., and O. A. Reig, 1973. The early fossil record of frogs: a review of the evidence. In *Evolutionary Biology of the Anurans*, J. L. Vial (ed.). Univ. of Missouri Press, Columbia, Mo., pp. 11-63.

Feduccia, A., 1980. *The Age of Birds*. Harvard Univ. Press, Cambridge, Mass.

——, 1985. On why dinosaurs lacked feathers. In *The Beginnings of Birds*, M. K. Hecht, J. H. Ostrom, G. Viohl, and P. Wellnhoffer (eds.). Freunde des Jura-Museums Eichstätt, Willibaldsburg, Eichstätt, West Germany pp. 75-79.

Feduccia, A., and H. B. Tordoff, 1979. Feathers of *Archaeopteryx*: asymmetric vanes indicate aerodynamic function. *Science*, **203**, 1021-1022.

Fox, W. T., 1987. Harmonic analysis of periodic extinctions. *Paleobiology*, **13**, 257-271.

Frazetta, T. H., 1969. Adaptive problems and possibilities in the temporal fenestration of tetrapod skulls. *Jour. Morphol.*, **125**, 145-158.

Hallam, A., 1987. End-Cretaceous mass extinction event: argument for terrestrial causation. *Science*, **238**, 1237-1242.

Heilmann, G., 1927. *The Origin of Birds*. Appleton, New York. (Reprinted 1972, Dover Books, New York.)

Jarvik, E., 1980. *Basic Structure and Evolution of Vertebrates*, Vols. 1 and 2. Academic Press, New York.

Kerr, R. A., 1987. Asteroid impact gets more support. *Science*, **236**, 666-668.

Langston, W., Jr., 1981. Pterosaurs. *Sci. Amer.*, **244** (2), 122-137.

McFarland, W. N., F. H. Pough, T. J. Cade, and J. B. Heiser, 1985. *Vertebrate Life*, 2nd ed. Macmillan, New York.

Olson, E. C., 1971. *Vertebrate Paleozoology*. Wiley-Interscience, New York.

Ostrom, J. H., 1974. *Archaeopteryx* and the origin of flight. *Quart. Rev. Biol.*, **49**, 27-47.

Padian, K., 1985. The origins and aerodynamics of flight in extinct vertebrates. *Paleontology*, **28**, 413-433.

Panchen, A. L., 1977. The origin and early evolution of tetrapod vertebrae. In *Problems in Vertebrate Evolution* (Linnaean Soc. Symp. Ser., Vol. 4), S. M. Andrews et al. (eds.). Academic Press, London, pp. 289-318.

——(ed.), 1980. *The Terrestrial Environment and the Origin of Land Vertebrates*. (Systematics Association Special Volume No. 15.) Academic Press, New York.

Randall, D. J., W. W. Burggren, A. P. Farrell, and M. S. Haswell, 1981. *The Evolution of Air Breathing in Vertebrates*. Cambridge Univ. Press, Cambridge.

Raup, D. M., 1986. Biological extinction in earth history. *Science*, **231**, 1528-1533.

Raup, D. M., and J. J. Sepkoski, Jr., 1986. Periodic extinction of families and genera. *Science*, **231**, 833-835.

Reisz, R. R., 1977. *Petrolacosaurus*, the oldest known diapsid reptile. *Science*, **196**, 1091-1093.

Romer, A. S., 1966. *Vertebrate Paleontology*, 3rd ed. Univ. of Chicago Press, Chicago.

——, 1967. Major steps in vertebrate evolution. *Science*, **158**, 1629-1637.

——, 1968. *The Procession of Life*. World Publ. Co., Cleveland.

Romer, A. S., and T. S. Parsons, 1977. *The Vertebrate Body*, 5th ed. W. B. Saunders, Philadelphia.

Russell, D. A., 1979. The enigma of the extinction of the dinosaurs. *Ann. Rev. Earth Planet Sci.*, **7**, 163-182.

Schmalhausen, I. I., 1968. *The Origin of Terrestrial Vertebrates*. Academic Press, New York.

Sibley, C. G., and J. E. Ahlquist, 1987. Avian phylogeny reconstructed from comparisons of the genetic material, DNA. In *Molecules and Morphology in Evolution: Conflict or Compromise?* C. Patterson (ed.). Cambridge Univ. Press, Cambridge, pp. 95-121.

Sloan, R. E., J. K. Rigby, Jr., L. M. Van Valen, and D. Gabriel, 1986. Gradual dinosaur extinction and simultaneous ungulate radiation in the Hell Creek Formation. *Science*, **232**, 629-633.

Spotila, J. R., 1980. Constraints of body size and environment on the temperature regulation of dinosaurs. In *A Cold Look at the Warm-Blooded Dinosaurs*, R. D. K. Thomas and E. C. Olson (eds.). AAAS Selected Symposium **28**. Westview Press, Boulder, Colo., pp. 233-252.

Stahl, B. J., 1974. *Vertebrate History: Problems in Evolution*. McGraw-Hill, New York.

Thomas, R. D. K., and E. C. Olson (eds.), 1980. *A Cold Look at the Warm-Blooded Dinosaurs.* AAAS Selected Symposium **28.** Westview Press, Boulder, Colo.

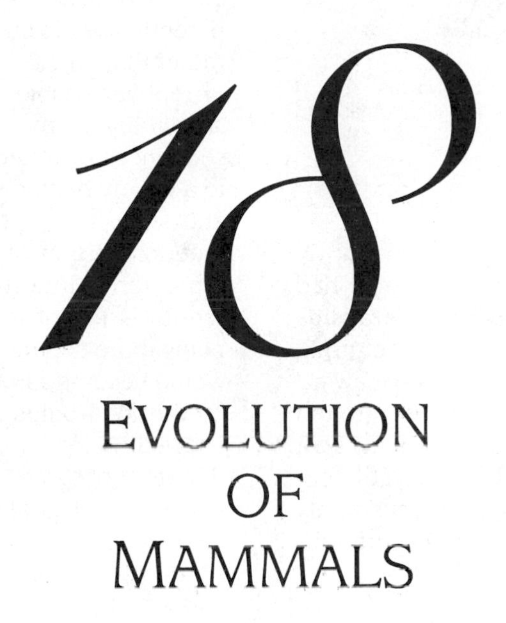

EVOLUTION OF MAMMALS

ammals derive their name from the maternal mammary glands used to suckle their young after birth. Mammalian uniqueness, however, extends also to numerous other anatomical, physiological, and behavioral traits that evolved throughout much of the Mesozoic era.

In addition to mammaries, the soft-body features of mammals are (1) **live birth** (except for monotremes); (2) endothermy along with adaptations such as hairy coverings to control heat loss and sweat glands to augment evaporation and cooling; (3) a **diaphragm** to increase the inspiration of oxygen and expiration of carbon dioxide, both necessary for high metabolic activity; (4) a **four-chambered heart** that completely separates oxygenated arterial blood from venous blood; and (5) greater intelligence derived from expansion of the **neocortex** of the brain.

Among the hard-body, or skeletal, differences that set living mammals apart from other vertebrates are (1) a **double occipital condyle** at the rear of the skull that articulates with the first cervical vertebra; (2) a **mandible** composed of a single bone (the dentary) with a condyle that articulates with the squamosal bone of the skull; (3) the transformation of the reptilian quadrate and articular bones, formerly used for jaw articulation, into the incus and malleus **ear ossicles** used for sound transmission; (4) a **bony secondary palate** separating the nasal passages from the mouth; (5) a **single nasal opening** in the skull; (6) a relatively **large braincase**; and (7) greater differentiation between teeth (**heterodont** dentition) characterized largely by **multi-rooted cheek teeth** (molars and premolars) with **multi-cusped crowns**.

Since mammalian soft-body features are, as yet, impossible to trace, hypotheses on the origin of mammals from reptilian forms are mainly derived from studies of fossilized skeletal materials. At present, the most prominent candidates for mammalian ancestors lie among extinct groups of **synapsid reptiles** that initially appeared in the Carboniferous period. Although at first these were ungainly looking pelycosaurs (see Fig. 17-18), by the Permian they had evolved into a variety of therapsid forms that were adapted primarily to a terrestrial existence (see Fig. 17-19). A number of paleontologists consider it likely that by the early Triassic some advanced therapsids had become the first of the vertebrate endotherms and may have possessed other soft-body features.

Skeletally, the therapsids were distinct from other reptiles and certainly evolving in a mammalian direc-

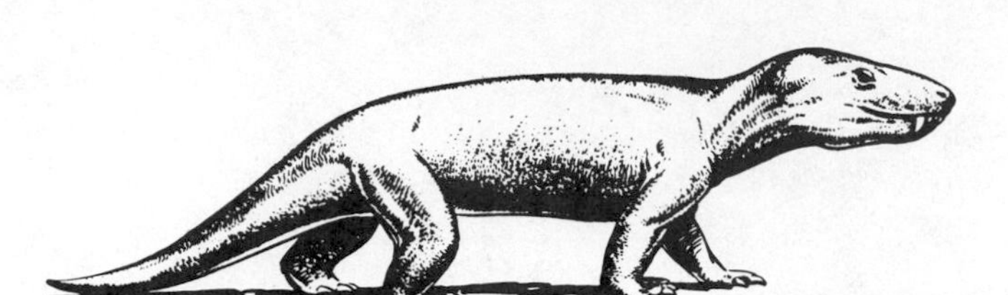

FIGURE 18-1 Reconstruction of *Cynognathus*, a carnivorous cynodont therapsid of the early Triassic, about 4 feet long. (Adapted from Colbert.)

tion. For example, the sprawled reptilian stance had changed in some of the doglike **cynodont therapsids** to a more vertical placement of limbs below the trunk, with the knees pointing forward and elbows backward (Fig. 18-1). This elevated the body from the ground and enhanced mobility by enabling direct fore and aft leg motion. Although the jaw articulation was still reptilian, the later therapsid mandible was almost entirely composed of the **dentary bone**, as in the therapsid-derived mammals.

TEETH

Among other highly significant mammal-like features in therapsids was the development of cusps on the crowns of cheek teeth. These extra surfaces were apparently used to cut and grind food into small particles rather than to gulp large chunks or swallow whole prey in reptilian fashion. Since continued breathing is essential for the metabolic needs of mammals, the consequences of retaining food orally while chewing led to a variety of innovative changes. In most amphibians and terrestrial reptiles the nasal openings are in the anterior portion of the mouth, and breathing can be temporarily interrupted without ill effect while the mouth is full of food. Mammals, on the other hand, being more dependent on constant aerobic respiration, would become asphyxiated if inspired air were blocked by the food bolus long enough for a mouthful to be chewed. Selection for an extended secondary palate therefore occurred in mammal-like lines, allowing air to be carried to and from a point beyond the mouth near the trachea (Fig. 18-2).

Tooth replacement was another area in which con-

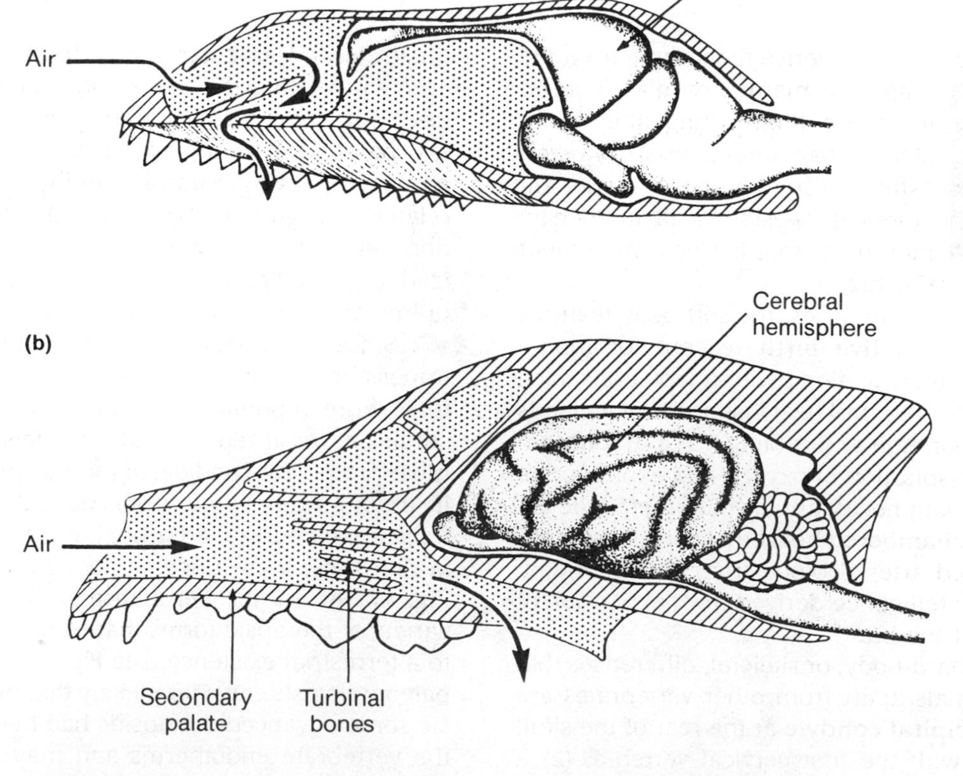

FIGURE 18-2 Air pathways in reptiles (a) and mammals (b), showing the long mammalian secondary palate that separates air entering the pharynx from food being retained in the mouth. The mammalian turbinal bones are covered with mucous membranes that warm and moisten the entering air. (Adapted from Romer [1968].)

siderable evolution had to occur before the mammalian grade was reached. In general, the addition and replacement of teeth is closely associated with relationships between the size of the teeth and the size of the skull during growth. There is obviously considerable value for a growing animal whose head and mouth are enlarging to keep pace with the increased size of its food by increasing the number of large teeth. In newborn reptiles, teeth along the jaw margins are small in accord with the size of the animal, and these are shed and replaced by larger teeth in an alternating pattern. That is, as each tooth matures and becomes more firmly fixed to the jaw, adjacent older teeth weaken and are replaced. Since a fully mature reptile may have a skull ten times longer than when it begins tooth replacement, this alternating cycle ensures that firm teeth of appropriate size are always present.

In mammals such continuous replacement of alternate teeth would interfere with the precise fit between the upper and lower teeth necessary for effective chewing action. Therefore, only one set of teeth is replaced in mammals, called the **deciduous** or milk teeth, which include incisors and canines as well as the post-canine deciduous molars (cheek teeth). In immature mammals, it is the deciduous molars that perform the chewing function, and these are replaced by permanent premolars when the more posterior adult molars have emerged. The need for only a single replacement of mammalian teeth probably derived from the fact that their young are born relatively large and can survive without teeth by suckling at maternal mammaries until their heads are even larger than at birth. By the time the deciduous teeth have erupted and the weaned young mammal is subsisting on adult food, their skulls have reached about 80 percent of postnatal growth. This means that only a relatively small amount of further skull growth is necessary before the permanent teeth replace the deciduous set.

JAWS AND HEARING

The precise fit between upper and lower mammalian cheek teeth is a consequence of selection for improved chewing activity and is correlated with changes in both jaw muscles and tooth shape. That is, in addition to the relatively limited grasping and puncturing functions of the reptilian jaw, mammalian evolution has emphasized shearing, grinding, and crushing activity in both premolar and molar regions. There are now muscles (masseter and pterygoid) that permit considerable force to be exerted in biting down in these regions, muscles (internal pterygoid) that move the jaw from side to side during chewing, and muscles (buccinator and tongue) that move the food within the mouth. It is believed that the rearrangement and change in mammalian jaw muscles, along with lengthening of the coronoid and angular processes of the dentary, reduced strain at the jaw articulation and correlated with the transition of former posterior bony elements of the jaw, the **articular** and **quadrate**, to assume auditory functions within the middle ear.

A major stimulus for these changes—improved hearing to capture prey and escape predators who were not in the direct line of vision—must have been one of the primary selective forces acting on early land vertebrates. This was especially true for early mammals, who survived the Mesozoic "reptilian tyranny" by entering a **nocturnal** environment which placed emphasis on auditory and olfactory perception. A possible evolutionary sequence for these events, diagrammed in Figure 18-3, can be described as follows.

The **tympanic membrane**, which functions as a taut, drumlike receptor for airborne sound, may well have been lacking in early land vertebrates and even perhaps in early synapsid reptiles, who relied mainly on ground-transmitted vibrations (Fig. 18-3(a)). In these therapsid ancestors, transmission of sound from ground to inner ear was mostly through bone conduction via the relatively thick stapes which maintained contact with both the quadrate in the skull and the articular in the lower jaw (Fig. 18-3(b)). Kermack and Mussett suggest that the evolution of a tympanum in mammal-like reptiles would have provided hearing for airborne sounds but would nevertheless have been inefficient in detecting a wide range of frequencies because of the relatively large mass and immobility of the bones between the tympanic membrane and inner ear (Fig. 18-3(c)). Since the difficulty in reducing the size of the articular and quadrate bones in these lineages derives from their use as the jaw hinge, one solution to this problem was to utilize other bones for this purpose and allow the articular and quadrate to diminish in size and confine themselves to sound conduction.[1]

This process is already apparent in some Triassic therapsids such as *Diarthrognatus* who show a new mammal-like jaw articulation involving the dentary bone in the lower jaw and the **squamosal** bone in the upper jaw, in addition to the old reptilian articular-quadrate joint. In early mammals such as *Morganu-*

[1] In lineages leading to modern reptiles (sauropsids), this difficulty was apparently not encountered, since the tympanic membrane was placed more posteriorly, allowing the stapes to serve as a single bone connecting the tympanum and inner ear without the intervention of the quadrate and articular (which remained as the reptilian jaw joint).

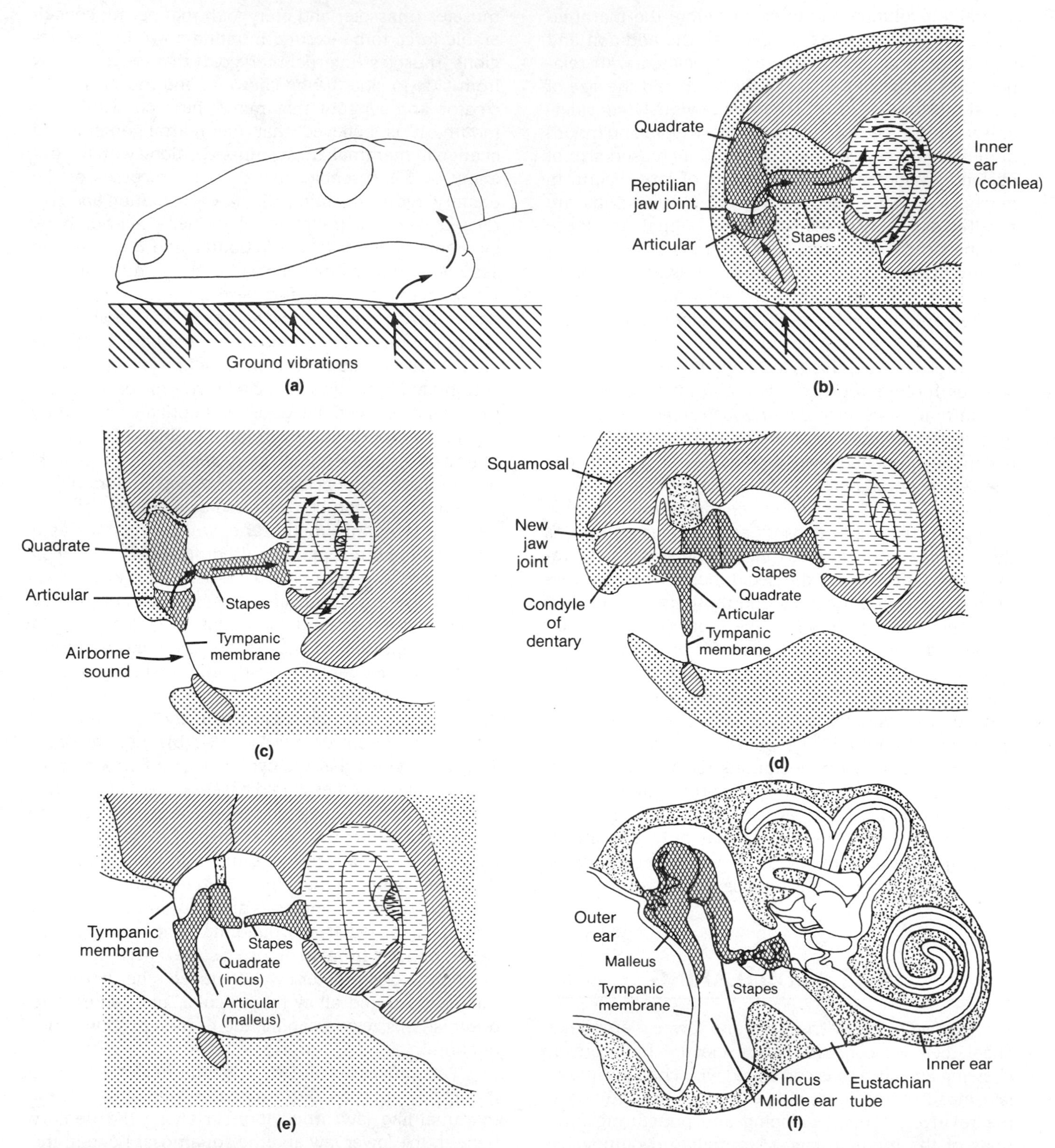

FIGURE 18-3 Proposed stages in the evolution of the ear apparatus, beginning with a land tetrapod that picks up ground vibrations through bone conduction (a, b). In reptilian synapsid lineages that lead to the mammal-like therapsids (c), a tympanic membrane picks up airborne sound and transmits it to the articular and quadrate bones of the jaw hinge and thence into the stapes that connects to the inner ear. As therapsid evolution proceeds, a new mammalian jaw joint evolves (squamosal-dentary) because of selection for improved molar chewing abilities. In *Morganucodon*, an early Triassic mammal (d), both jaw joints are present, although the size of the articular-quadrate-stapes bones have diminished. In late Triassic mammals (e), the squamosal-dentary joint has become the only jaw hinge, and the articular-quadrate-stapes bones are now entirely involved in hearing. The diagram in (f) presents a more anatomical view of the shape and positioning of these bones in the ear of a modern mammal. (Adapted from Kermack and Mussett.)

codon, reduction of the articular and quadrate had proceeded even further, although they still formed part of the jaw hinge (Fig. 18-3(d)). It was apparently only in the Jurassic that the quadrate and articular were freed from their function in jaw articulation and became incorporated into the mammalian middle ear as small ossicles called the **incus** (anvil) and **malleus** (hammer), respectively (Fig. 18-3(e, f)). In support of this view are embryological studies showing that the quadrate and articular in the mammalian fetus first occupy a reptilian position on the side of the jaw and are later transformed into the mammalian ear ossicles.

Unfortunately, because of a scarcity of very early mammalian fossils, the point of transition between therapsids and mammals is still unknown both in respect to time and place. Early mammals appear to have been small, often about the size of mice or rats, and their bodies were therefore usually rapidly disarticulated even in areas that were well fossilized. With some exceptions, the complete mammalian skeletons that have been discovered date no earlier than the late Cretaceous. As a result, possible evolutionary lineages among earlier Mesozoic mammals have been derived almost entirely from fossil teeth and jaws. Fortunately, hard enameled dentition is the most easily preserved part of the vertebrate body, and the cusps, ridges, and depressions on the surfaces of teeth follow heritable genetic patterns that can be used to point the way to phylogenetic relationships.

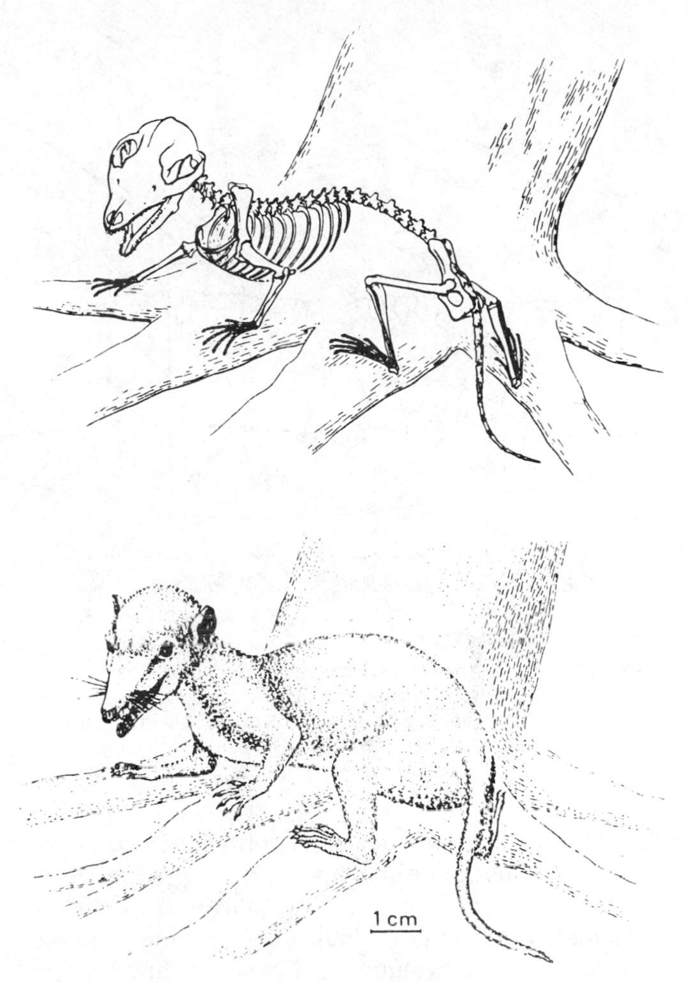

FIGURE 18-4 Proposed skeletal and full-body reconstructions of a late Triassic-early Jurassic mammal, *Megazostrodon*. It was about 4 inches (10 centimeters) long and weighed approximately 1 ounce. (Adapted from Crompton et al.)

EARLY MAMMALS

The earliest mammalian departure from the therapsid line is found in a geographically widespread group of late Triassic-early Jurassic fossils called **morganucodontids** (Fig. 18-4). For the first time there is a clear differentiation of the post-canine teeth into premolars and molars, with only the last premolars showing evidence of tooth replacement, whereas the molars followed the mammalian pattern of permanence. Among other distinctive morganucodont traits is the precise occlusion between upper and lower jaws, which produced a consistent pattern of molar wear facets (Fig. 18-5). In general, the structure of a morganucodont molar was three cusps aligned along the anterior-posterior axis of the tooth, an arrangement called **triconodont**. Fossil triconodonts with patterns similar to or derived from these have been found throughout the remainder of the Mesozoic era, and are among the variety of groups traditionally classified in the subclass **Prototheria**.

As mentioned previously (see p. 93 and Fig. 6-16), the modern remaining prototherian lines are the Australian and New Guinean **monotremes**, egg-laying mammals now represented by the grub- and shrimp-eating platypus (*Ornithorhyncus*) and ant-eating echidna (*Tachyglossus* and *Zaglossus*).[2] Although these animals are presently quite rare and their fossil record is meager, some of their features may nevertheless reflect those of their more numerous Mesozoic prototherian ancestors. For example, in addition to egg laying, the monotremes still use the reptilian cloaca as a common chamber for both the rectal and urogenital openings and show a number of reptilian skull characters. Also, in contrast to the more advanced marsupials and placentals, monotreme pectoral girdles lack a scapular spine and still contain an interclavicle,

[2]Based on an analysis of an early Cretaceous jaw fragment identified as monotreme, Kielan-Jaworowska and co-workers suggest that rather than being prototherians, monotremes originated from an ancestral therian mammal. However, other monotreme skeletal features are so primitive that this view is not readily accepted (Carroll).

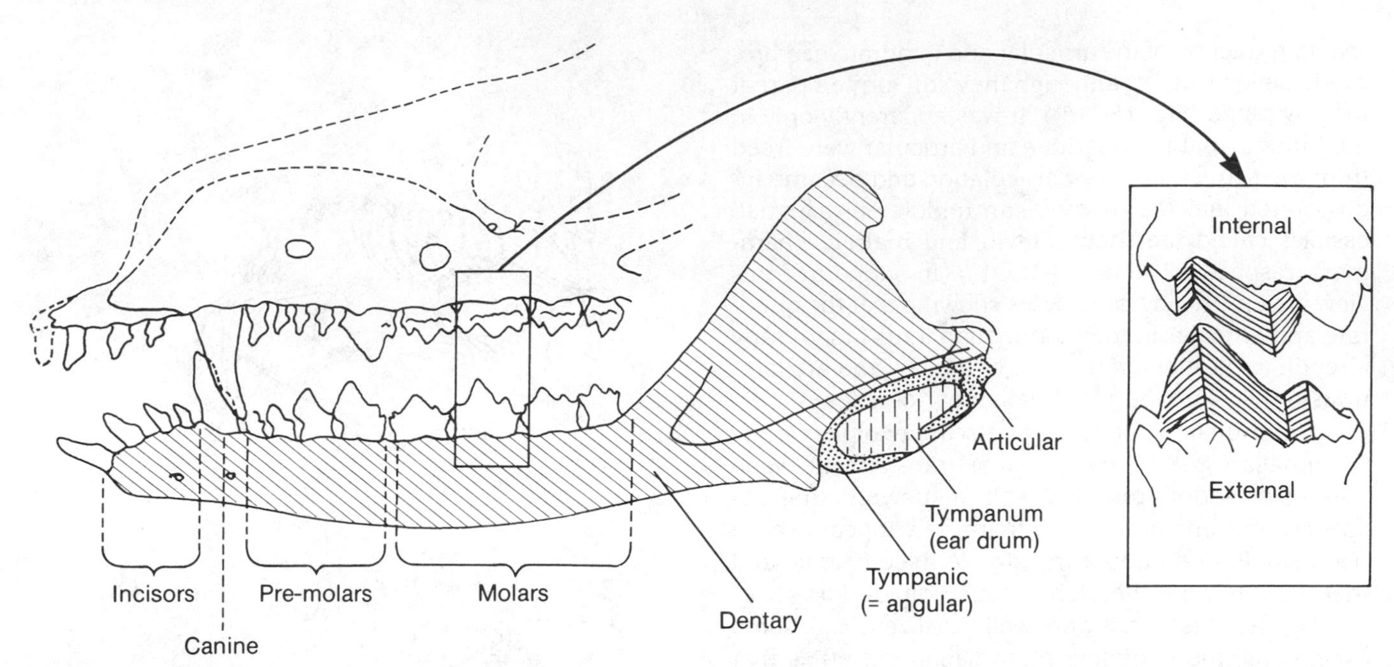

FIGURE 18-5 Lateral view of the inch-long jaws of a late Triassic morganucodontid, *Morganucodon.* The molar teeth occlude more precisely than in any reptile, and, as can be seen in the inset, show matching wear facets between the internal surface of the upper molars and the external surface of the lower molars. (To illustrate the internal surface of the upper molar, the tooth is drawn as though it were transparent.) (Adapted from Crompton and Jenkins.)

much like the Triassic-Jurassic mammalian docodonts, another primitive mammalian group. The absence of teeth in monotremes is a rare specialization that helps make their phylogeny difficult to determine but can be explained by an evolutionary history confined either to mud burrowing or ant eating. It is probably the persistence of such unique specializations in the relatively isolated Australian continent that enabled these Mesozoic relics to survive for such a long period.

Discoveries of a variety of fossil teeth indicate that by the end of the Triassic period an important mammalian division occurred, separating the morganucodontids and their subsequent prototherian lineages from a new group called the **therians**, marked by a more sophisticated molar structure. Therian molars, called **tribosphenic** or tritubercular, are distinguished from prototherian types by possessing a triangular arrangement of cusps in the upper molars, one of which (protocone) fitted closely into a lower molar basin (talonid), much like a pestle into a mortar (Fig. 18-6 (a, b)). The crushing action of this arrangement was supplemented by shearing and cutting surfaces which progressed in number from three to six as therian lineages evolved (Fig. 18-6 (c-g)). These activities were further enhanced by evolution of a narrower lower jaw suspended in a sling of muscle that enabled side-to-side grinding action, a feature whose beginnings are already observed in earlier mammals. It seems likely that these improved oral pulverizing mechanisms accompanied new dietary opportunities as well as increased digestive efficiency.

The earliest ancestor to the tribosphenic-type molar is found among contemporaries of the morganucodonts, called kuehnotheriids (Fig. 18-6(c)), and it is this evidence that has been used by some paleontologists to suggest a diphyletic or polyphyletic origin of mammals from different therapsid stocks. That is, it is presumed that more than one line of mammal-like therapsids, distinguished from other groups by possessing cusped molar teeth, gave rise separately to morganucodonts, kuehnotheriids, and perhaps also to some other early mammalian lineages. On the other hand, important braincase similarities between therian and atherian mammals have prompted different paleontologists to adopt a monophyletic position (Kemp [1988]). Other authors point to the apparent complexity of early mammalian evolution and the skimpiness of fossils still makes a clear decision difficult between monophyly, diphyly, and polyphyly. According to Kermack and Kermack, if one asks the question, "'Are the major taxonomic divisions of the [early] Mammalia more closely related to each other than to any other taxonomic group?' then the answer has to be at the moment that we simply do not know."

EARLY MAMMALIAN HABITATS

Even with the many gaps in the fossil evidence, some aspects of early mammalian lives and habitats can nevertheless be reasonably supposed. Certainly, improved dentition and mastication in Mesozoic mam-

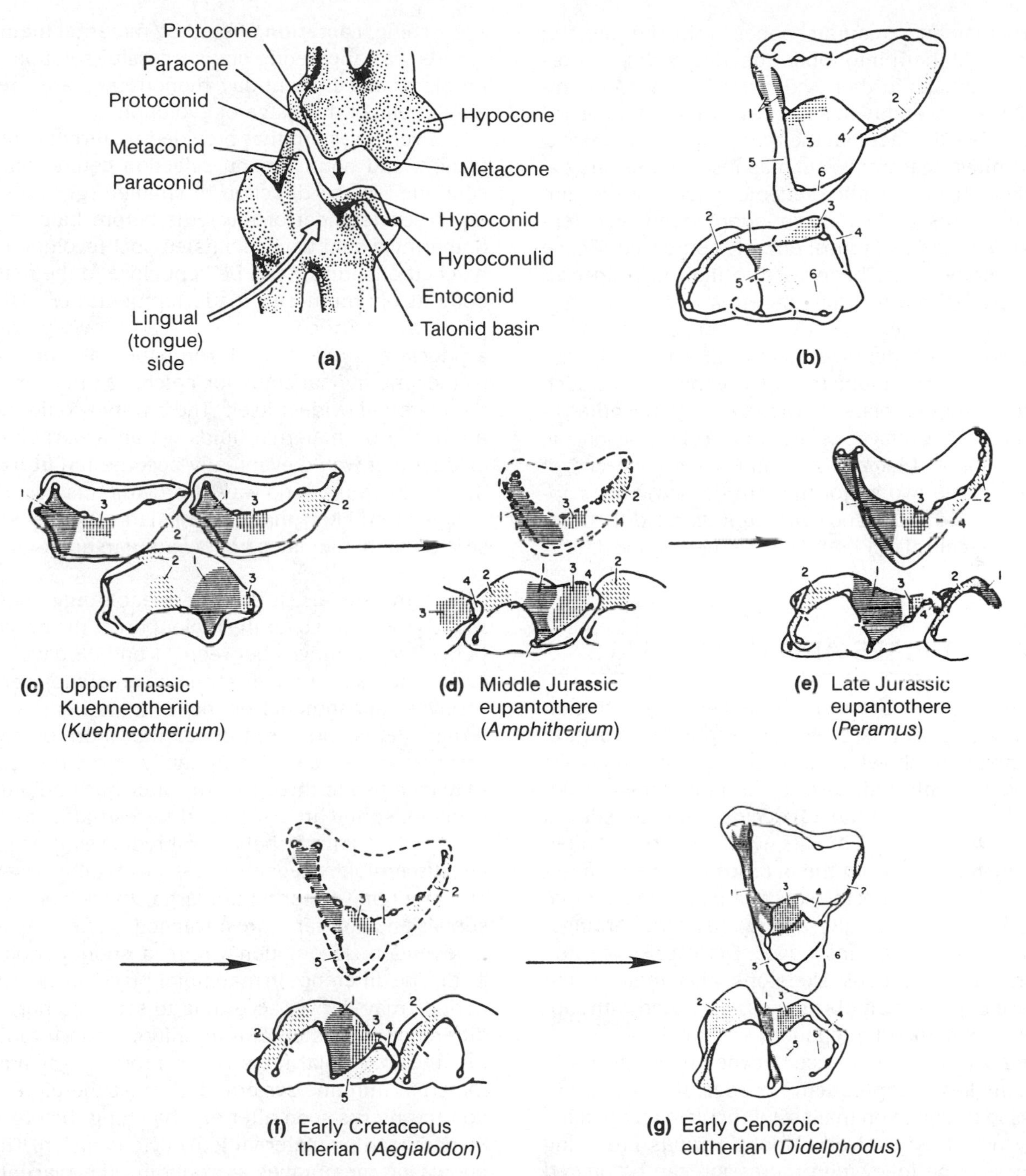

FIGURE 18-6 (a) Generalized upper and lower tribosphenic molars, based on those of a modern therian, the opposum *Didelphis* (oriented with the anterior of the animal to the left). (b) Crown (occlusal) view, with matching wear facets numbered and shaded alike. (c-g) Crown views of upper molars (above) and lower molars (below) representing general stages in the evolution of the therian tribosphenic molar (only lower molars are actually known for fossils d and f). (Adapted from McFarland et al., from Bown and Kraus.)

mals would have helped maintain constant body temperatures in these small animals by promoting the rapid absorption of food. Because of their endothermy, there have been suggestions that these primitive mammals were primarily nocturnal and insectivorous, functioning in the cool of the evening when their ectothermal reptilian predators were inactive. In support of this nocturnal role, Jerison and others point out that early mammalian brains were three or four times larger than those of even advanced therapsids, and that a significant portion of this increase can be attributed to neural connections for enhanced auditory (and perhaps also olfactory) acuity associated with adaptation to a nocturnal habitat.

The increased specialization of the mammalian auditory apparatus, accomplished in part by freeing the articular and quadrate bones from the jaw and transforming them into more effective middle ear ossicles, is perhaps further evidence of selection for improved sensory ability in a nocturnal, light-diminished habitat. Also, the fact that the retinae of many present-day primitive mammals, such as insectivores, are extremely rich in rod photoreceptors sensitive to dim light, in contrast to the daylight-adapted retinae of reptiles such as lizards (almost entirely composed of cone photoreceptors), has been used to indicate a specialized nocturnal mammalian ancestry.

Dental changes in early mammalians, probably prompted by a rodentlike, nocturnal way of life, appeared to have continued throughout the Jurassic; such changes in fossil molars are diagrammed in Figure 18-7. In the early Cretaceous, the first truly tribosphenic molar is found (*Aegialodon*), and toward the end of that period the two major modern therian groups appear in relative abundance: **marsupials** in North America and **placentals** in both North America and Asia.

MARSUPIALS AND PLACENTALS

Fossil differences between these two major therian groups can be seen in both skull and tooth structure, with marsupials showing the following distinctions: (1) a relatively small braincase; (2) unique bony composition of the auditory bulb; (3) deciduous teeth reduced to only the posterior premolars which, in turn, are often markedly different from the anterior molars; (4) a relatively large number of incisors (a total of eight or more in the complete upper jaw); and (5) distinctive arrangements of molar cusps and ridges. Discoveries of some post-cranial Cretaceous skeletons also indicate the presence of pelvic epipubic bones associated with support of a marsupial-type pouch.

Since a major difference between marsupials and placentals lies in reproductive modes, the absence of soft tissue fossilization makes it difficult to recapitulate exactly how they evolved. Among various prevailing hypotheses, the following progression can be offered (see Lillegraven). Very early mammals, distinguished by small size, endothermy, and heterodont dentition, were probably egg laying, but these eggs were probably small. Under such circumstances, a significant selective advantage would have accrued to animals that could raise their young past the immature stages of hatching. **Lactation** offered by maternal **mammary glands** was apparently one successful solution to the problem, and present-day monotremes are presumably a relic of this stage of evolution.

Given a system that provided maternal care, protection, and nourishment, selection could then have continued in the direction of smaller eggs and more rapid development of the fetus before hatching. Endothermy would have facilitated such evolution since hatched offspring could be kept close to the maternal body at optimum enzymatic temperatures. At some point, **oviparity** could be replaced by **viviparous** reproduction, since it would probably take only a few further mutational steps for hatching to occur within the maternal oviduct itself. The embryo could then be nourished on maternal fluids within a portion of the oviduct that would eventually become the **uterus**. Although there is yet no way of proving this hypothesis, it is believed likely that mammalian viviparity was restricted to therian lines, that is, to marsupials and placentals.

In marsupials, a thin, permeable egg shell surrounds the embryo for the major part of the pregnancy period, which ranges between 11 and 38 days for different species. During this time, the marsupial embryo receives nourishment from both egg nutrients and maternal uterus, but because of the short pregnancy, emerges from the vagina in highly immature form.[3] It may take two to three months after birth before marsupial offspring are capable of terrestrial locomotion.

In placentals, a shelled embryonic stage is no longer discernible, pregnancy is considerably extended, and emerging offspring are larger than those of marsupials and also far more advanced—often capable of independent locomotion within a short period after birth. The difference in marsupial-placental gestational periods may derive, according to some authors, from differences in maternal immunity response to the fetus: the marsupial fetus is not protected against the maternal immune system, and must therefore abandon the uterus soon after egg hatching, before it can be damaged by maternal leukocyte invasion. The trophoblastic membranes surrounding the placental fetus, on the other hand, ordinarily prevent exchange between maternal and fetal tissues and thus act as barriers that help keep the fetus from being immunologically rejected.

It seems therefore likely that mammalian ancestors to placentals would have undergone evolutionary

[3]This immaturity places significant restrictions on the directions that marsupial evolution can take. For example, it has been suggested that the absence of marsupial forms with front flippers, as are found among placental seals, or the absence of marsupial hooved forms, lies in the necessity for marsupial offspring to possess forelegs with claws in order to crawl from the vagina to the maternal mammary teat. A fully aquatic life, such as found among placental cetaceans, is also closed to marsupials since they would be unable to provide their very immature young with air during that considerable period of weeks or months in which the young remain continually fastened to the teat, nor could tiny marsupial offspring survive the temperature stress caused by complete immersion in water.

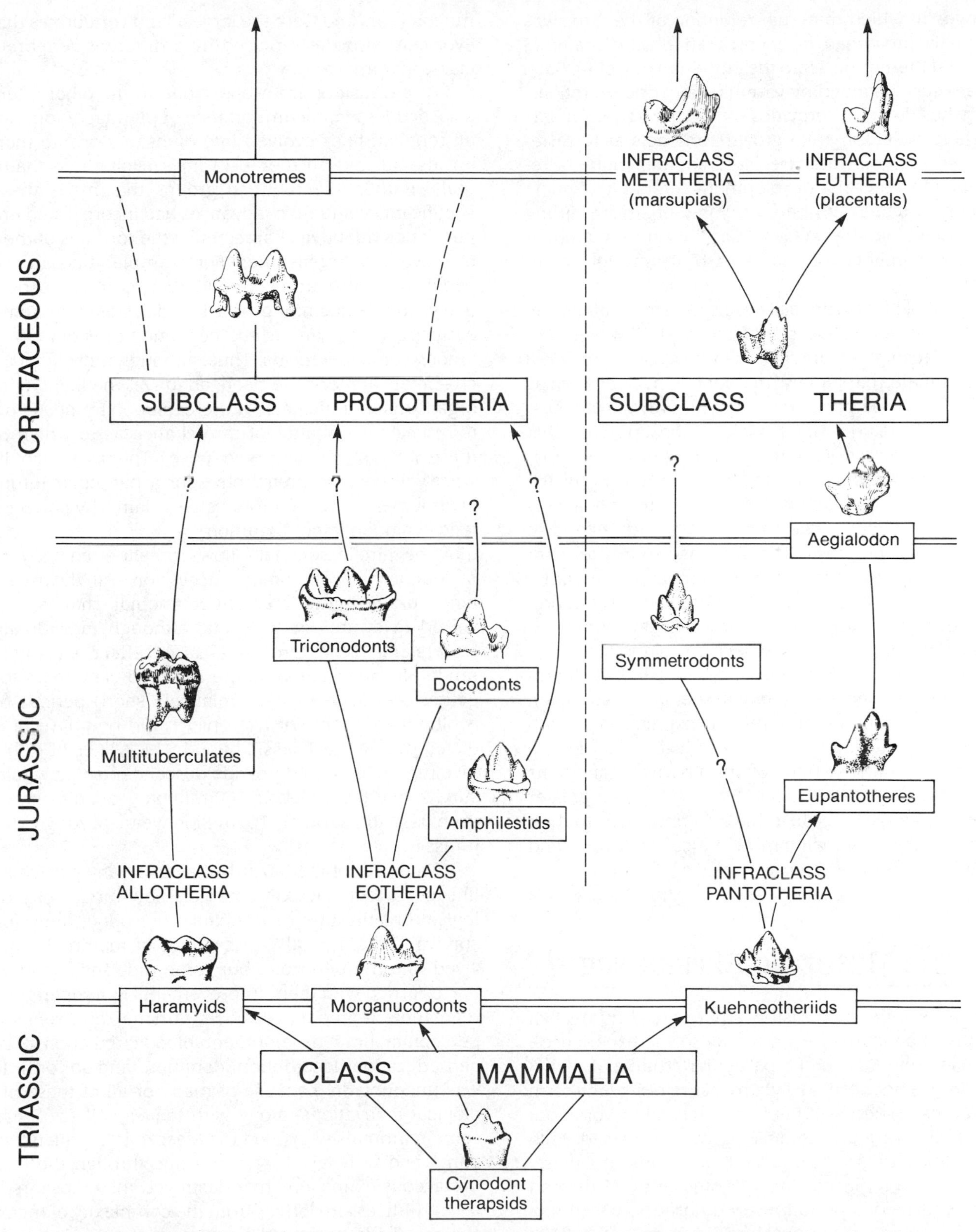

FIGURE 18-7 A phylogeny of mammals from the Triassic to the Cenozoic, shown in terms of changes in molar teeth.

changes in which maternal retention of the fetus was gradually prolonged by incorporating both maternal and fetal membranes into the eutherian placenta. Such a placenta, sustained by various endocrine secretions, nourishes the fetus, provides the oxygen requirements for rapid developmental growth, and acts as a waste-removal system. Compared to any other mode of reproduction, uterine development probably confers greater protection to the embryonic organism during its most vulnerable stages. Also, because maternal-child attachments continue past birth through mammary feeding, the stage is set for prolonged family relationships that emphasize learning and intelligence.

It is nevertheless significant that however profound placental advantages may have been, they did not eliminate marsupials: marsupials have at least as long a history as placentals, are still prevalent in Australia, and include forms such as the opossum that compete successfully with placentals in various placental-dominated localities. One basic reason for the persistence of marsupials derives from their relatively minor reproductive investment: their birth size is so small that offspring can be abandoned soon after birth without great maternal loss. Placentals, on the other hand, commit much greater resources to early reproductive stages, and a pregnancy will often continue in the face of serious maternal sacrifice. Thus marsupial reproduction can more easily be adjusted to appropriate environmental conditions; that is, reproduction and nursing are continued when conditions are advantageous, and there is little expense in discarding their minuscule newborn offspring when conditions turn poor. Placentals, on the other hand, take greater reproductive risks because their commitment to their offspring is greater, being more difficult to reverse and involving considerable cost.

THE MESOZOIC EXPERIENCE

As in all evolutionary history, there are a number of lessons to be learned from the Mesozoic experience.

First, dominance of a particular group at a particular time is not necessarily a measure of its long-term evolutionary success. This lesson has been borne out repeatedly: various therapsid groups replaced each other, dinosaur groups replaced therapsids, and these, in turn, were replaced by other dinosaurs. At the end of the Cretaceous period, even dinosaurs, which had been dominant in various forms for more than 100 million years and were the largest land vertebrates that ever existed, were replaced by a different vertebrate class, mammals.

The transition from one group to the other offers a second lesson, the importance of preadaptation. Not all reptilian lines evolved into dinosaurs or into therapsids, nor did all therapsid lines evolve into mammals. Rather, the reptilian groups that made these significant evolutionary advances had incorporated important preadaptive characters by the fortunes of their own evolutionary histories. For example, dinosaur bipedalism and mammalian endothermy trace, respectively, to the beginning of a bipedal stance in some early archosaurs and to the beginning of endothermy among some therapsids. Thus, although most if not all characters are, or have been, adaptive, the inability of organisms to anticipate future evolutionary needs often makes it a matter of rare chance as to which of these characters are preadaptive. The limitation in what will become preadaptive for a particular future environment may therefore help explain why polyphyletic evolution is not common.

The third lesson that follows is the unpredictability of long-term evolutionary succession. All the many kinds of biological and environmental changes involved in evolutionary events, although individually understandable in terms of cause and effect, act largely randomly when attempts are made to view them together over long (or even relatively short) periods of evolutionary time. For example, could one have predicted during the Triassic period which lines of primitive mammals would provide descendants that would survive into the Cenozoic 150 million years distant, or even last for another 70 million years through the Jurassic?

Thus, out of the hazards that await any particular lineage, a fourth lesson emerges: new modes of biological organization can enhance the opportunity for survival. True, not all Triassic lines of mammals survived into the Cenozoic, but some did, and many of these carried with them improvements in temperature regulation, reproductive mode, nursing care, sensory perception, brain development, blood circulation, oxygen utilization, locomotion, dentition, and so forth. It was undoubtedly because of many or all of these biological innovations, along with their small size, that some mammals withstood the Mesozoic "reptilian tyranny" and were given safe passage through the late Cretaceous extinctions that destroyed the dinosaurs.[4]

A fifth lesson derives from the complexity of major biological adaptations: the evolution of a new grade of

[4]Although one can claim that time provides the arrow which orients the direction of evolutionary change and that an evolutionary trend represents "progress," there is obviously more than one direction and many lines of progress. That is, if we consider the multitude of species that have evolved and their many different life styles, the only discernable evolutionary trend shared by all is the opportunism that became embedded in the earliest of their ancestors. Like many authors in the past, we can call the results of this opportunism "progress" (see Nitecki), but it is questionable whether this term or other value judgments help us understand evolutionary processes.

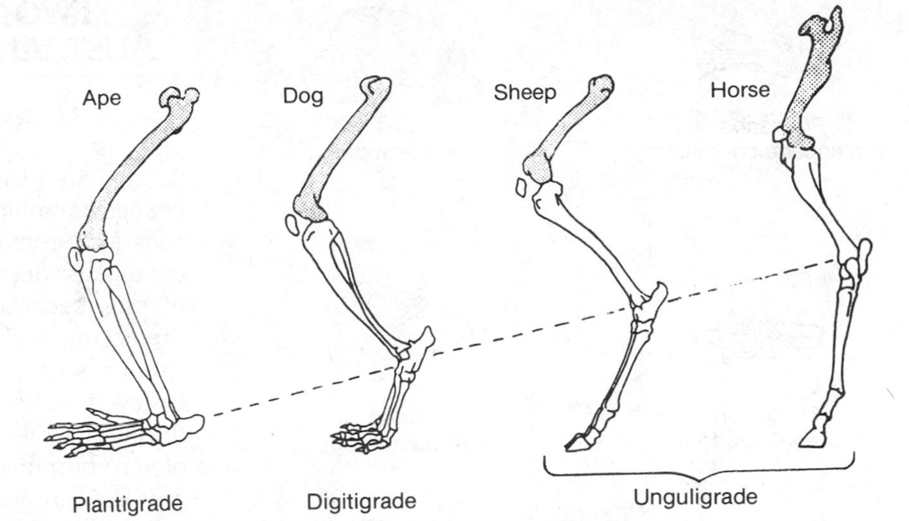

FIGURE 18-8 A series of modifications in the mammalian hind limb generally associated with increased running speeds (from left to right). These changes include positioning of the foot to run on the toe tips, lengthening of the foot bones (indicated by height of the diagonal line from ground), and fusion of the tibia and fibula (also the forelimb radius and ulna) to prevent rotation of the foot during running. (Adapted from Savage and Long.)

organization, like that of mammals, is marked by coordinated changes in many different traits occurring often over a considerable period of time. For example, although we can characterize a mammal by one or another of its unique traits, the trait itself, such as dentition or reproductive mode, is actually the result of a number of successive mutations, each of which must be coordinated with its entire genetic architecture. The transition from reptile to placental mammal may well have taken 75 to 100 million years because a wide range of **coadaptive mutations** had to be integrated into evolving organisms. Also, in support of monophyletic evolution, the complexity of this integrative process makes it unlikely that many different lines would have continued to undergo the same succession of identical genetic changes. If there is similarity in the evolutionary progress of different lines, as observed in therapsid-early mammalian transitions, such apparent polyphyletic events may really be caused by parallel or convergent evolution; that is, similar characters evolved through different genetic events, and their genetic lineages are therefore different. Furthermore, the "sweepstakes" nature of evolution makes it unlikely that even such parallel evolution could have continued in a variety of therapsid lines throughout the Mesozoic, with all such different lines reaching the placental mammalian grade in the Cretaceous.

A sixth lesson is that once a new adaptive innovation appears, or a new grade of adaptive organization is reached, opportunities for widespread radiation eventually follow. Mammalian endothermy must have opened a nocturnal niche into which many lines entered, just as improvements in mammalian dentition opened a dietary niche that few, if any, reptiles had ever fully exploited.

Perhaps a final lesson of the Mesozoic is that the survival or extinction of any group may be closely connected to the survival or extinction of other groups. That is, individuals of a particular group are dependent on the existence of entire constellations of associated organisms. Replacement of therapsids by dinosaurs as the new dominant land vertebrates was, for example, a mass phenomenon, involving many genera and families. Also, as mentioned previously and discussed below, a significant cause for radiation of mammals during the Cenozoic can be ascribed to many new functional roles and habitats made available to them by the extinction of the dinosaurs.

CENOZOIC ERA: THE AGE OF MAMMALS

However important the Mesozoic was to early mammalian evolution, it was in the Cenozoic that the full flowering of mammalian radiation burst forth. The extinction of the dinosaurs seemed to have removed many mammalian Mesozoic constraints: they could invade herbivorous and carnivorous niches formerly closed to them and were able to become active **diurnally** as well as nocturnally. New mammalian life styles thus began to appear during the Paleocene epoch, but observable morphological differences accumulated slowly.

However, by the middle and late Paleocene, radical evolutionary changes appear to have occurred in a few major centers, especially North America. By the middle and late Eocene, 20 to 30 million years after the Cretaceous, skeletal adaptations for creeping, running, digging, swimming, flying, and climbing had morphologically differentiated many major fossil groups. These and further changes led, for example, to modifications in the lower limbs of some terrestrially mobile animals from a flat-footed stance (**plantigrade**) to running on the digits (**digitigrade**) or on the tips of the toes (**unguligrade**). Increased speed in groups such as horses and other hoofed animals was accompanied by reduction in the number of toes and lengthening of the limbs and foot bones (Fig. 18-8). Cerebral

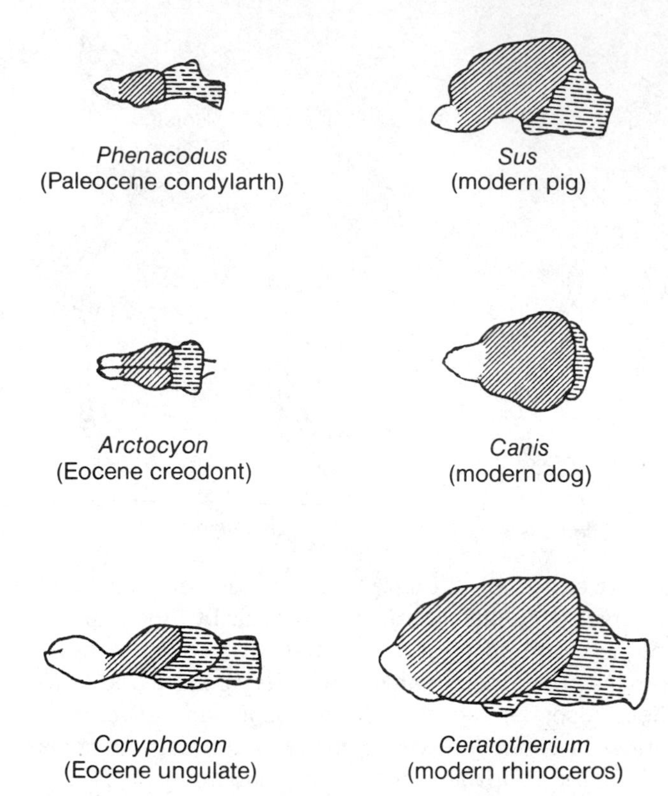

FIGURE 18-9 Comparisons between brains of ancient and modern mammals of similar size. The cerebral hemispheres are indicated by diagonal lines, olfactory lobes are blank, and the cerebellar and medullary areas are shown as horizontal broken lines. (Adapted from Lull, from Osborn.)

brain size, a mark of the ability to integrate sensory and motor information, increased in both mammalian prey and predators (Fig. 18-9). Such evolutionary changes, as well as many others, led to replacement of most of the primitive Mesozoic and early Cenozoic mammalian forms, a process that continued throughout the Tertiary period. Of more than 30 orders of mammals, about half are now extinct. Of those that remain, almost all can be traced back to the Eocene epoch (Fig. 18-10).

TWO ISLAND CONTINENTS: AUSTRALIA AND SOUTH AMERICA

Most of the new adaptive radiations occurred among placentals, although two continents, Australia and South America, showed significant evolutionary changes among marsupials as well. Some of the reasons for the marsupial radiation in the two southern continents derive from the isolation of these land masses because of continental drift during the late Cretaceous and early Cenozoic (Chapter 6). Although various scenarios have been offered, it is generally believed that marsupials originated in North America during the mid-Cretaceous, and, along with a couple of very primitive placental groups, migrated down an arc of Central American islands into South America before the end of that period.

From South America, marsupials dispersed into an Antarctican continent that was considerably warmer than at present and, unaccompanied by placentals, reached Australia during the early Eocene, about 50 million years ago. By mid-Eocene, perhaps 5 million years later, Australia separated from Antarctica and began its northern journey toward Asia, carrying along its isolated marsupial population.[5] New Zealand apparently separated from the Gondwanaland mass even earlier, probably during the Cretaceous, and neither native nor fossil terrestrial mammals have been found there.

The success and diversity of herbivorous and carnivorous marsupials produced during their Australian radiation (diagrammed in Fig. 3-4) are ascribed to the absence of placental rivals. The only other mammalian subclass present in Australia, the prototherian monotremes, were probably too primitive or too specialized to offer much competition.[6] As a result, by mid-Miocene times, at least 15 families of marsupials existed. Many of these were browsers of various kinds that probably fed on temperate rain forest vegetation, along with at least two groups that performed a carnivorous role.

By the late Miocene, drier conditions led to an expansion of grasslands, followed by the evolution of many different kinds of grazing kangaroos, including one Pleistocene species whose adults were 10 feet high. The invasion of Australia by humans both during the Pleistocene and more recently has been accom-

[5]The North American marsupials also threw off a colony to Europe during the early Eocene and short-lived lineages to Asia and Africa, probably through a North Atlantic-Greenland-Europe connection. During the Miocene, marsupial populations of both northern continents became extinct, and it was only when a North American-South American land bridge was re-established in the Pliocene that some marsupials reinvaded North America from the south.

[6]Fossil evidence for monotremes is extremely poor, although Pleistocene deposits do show monotremes in Australia at that time in the forms of both platypus and echidna genera, and a middle Miocene date has been given to teeth that may have belonged to a platypus-type animal. However, the mode of monotreme arrival in Australia, the time at which it occurred, and any subsequent radiations are unknown.

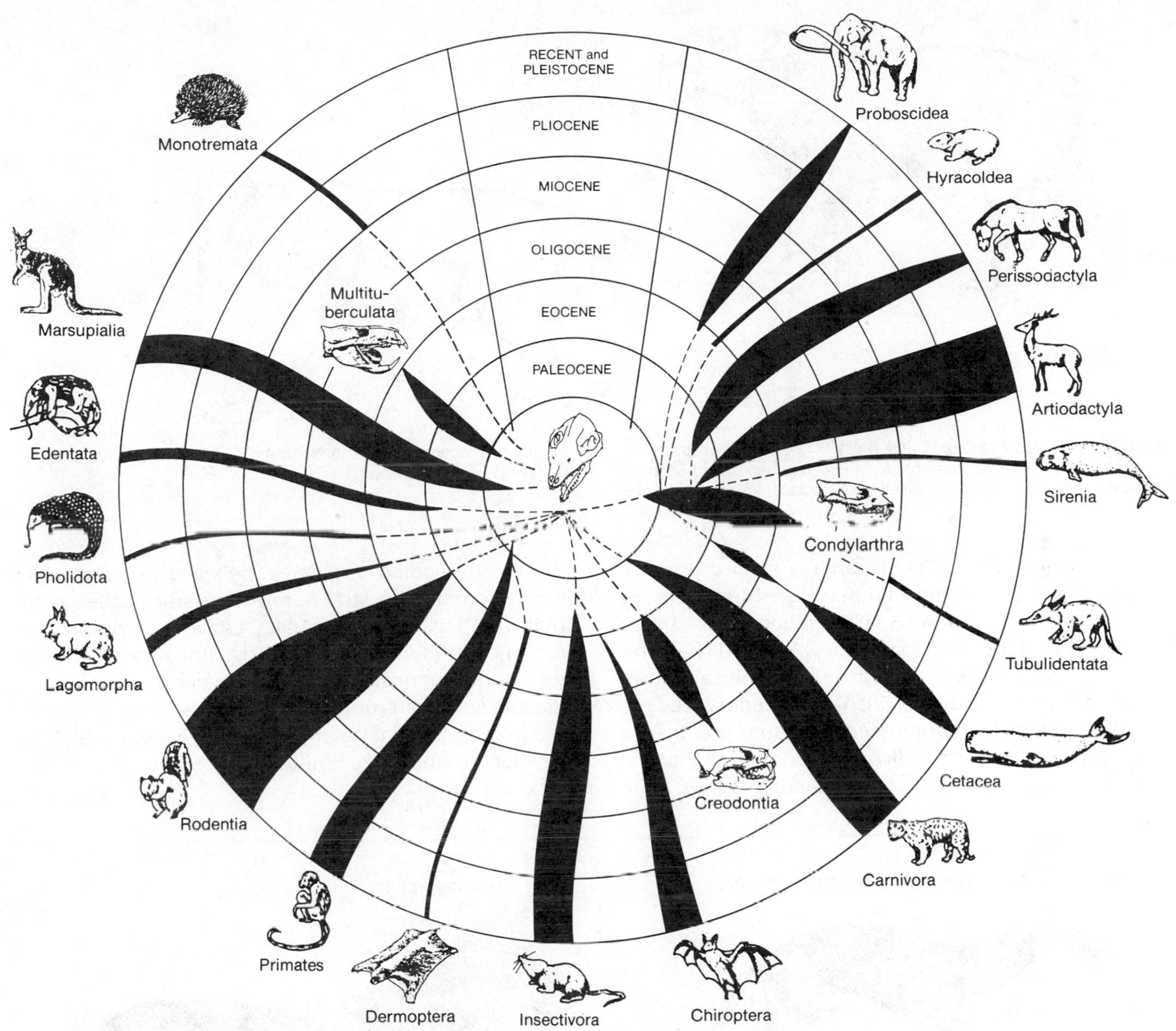

FIGURE 18-10 Radiation pattern of mammalian orders beginning with the Cretaceous period (center circle), including three extinct groups (multituberculates, condylarths, and creodonts). Widths of black bars indicate rough estimates of relative fossil abundances at various times. Exact phylogenetic relationships among many of these mammalian orders (dotted lines) are still disputed (Benton; Novacek et al.). (Data from Gingerich.)

panied by other placental groups, from dogs to rabbits, sheep, and rodents. Based on their vulnerability to placental competition, it is doubtful that many present Australian marsupial groups will survive without protection.

In South America, marsupial radiation followed a different pattern since the presence of placental herbivores and edentates channeled the marsupials into carnivorous and insectivorous niches. These animals ranged from numerous species of opossumlike didelphids to jumping, gnawing, and doglike forms. Some of the latter, members of the borhyaenid family, are believed to be ancestral to *Thylacosmilus*, the marsupial saber-toothed "tiger" of the South American Pliocene (Fig. 18-11(a)), a carnivore strikingly similar to the large placental saber-toothed cat, *Smilodon*, of the North American Pleistocene (Fig. 18-11(b)). The borhyaenids also show marked similarities to the Australian marsupial family of thylacines that included the Tasmanian wolf (Fig. 18-12). Obviously, evolutionary convergence or parallelism, caused by selection for a similar way of life, produced similar structures in the genetically different placentals and marsupials even in different continents.

The South American placentals, although beginning only with some ungulates and **xenarthrans**

FIGURE 8-11 Reconstructions and skulls of (a) *Thylacosmilus*, a saber-toothed marsupial carnivore from Pliocene deposits in South America; and (b) *Smilodon*, a saber-toothed placental carnivore found in late Pleistocene deposits in North America. (Adapted from Steel and Harvey, from Simpson.)

("strange-jointed"), radiated perhaps even more rapidly than did marsupials on that isolated continent. By the early Eocene, within 15 to 20 million years of their initial late Cretaceous colonization, placentals had produced 75 to 100 new genera divided into about 15 families. The xenarthrans (also called **edentates** because of reduced or suppressed dentition) produced a strange bestiary of armadillos, glyptodonts, sloths, and anteaters (Fig. 18-13(a)). Also radiating widely were the mostly hoofed ungulates, believed to have originated from an ancestral herbivorous stock called **condylarths** (Fig. 18-13(b)). Again, convergent or parallel evolution produced some striking similarities: some of the South American litopterns, apparently selected for grazing and rapid running, were, by the early Miocene, remarkably similar to the one-toed horses that first developed about 20 million years later in North America.

FIGURE 18-12 (a) *Prothylacynus patagonicus*, a borhyaenid marsupial from the early Miocene period in southern Argentina. (b) *Thylacinus cynocephalus*, the recently extinct marsupial Tasmanian wolf. (c) *Canis lupus*, the modern placental North American wolf. (From Marshall.)

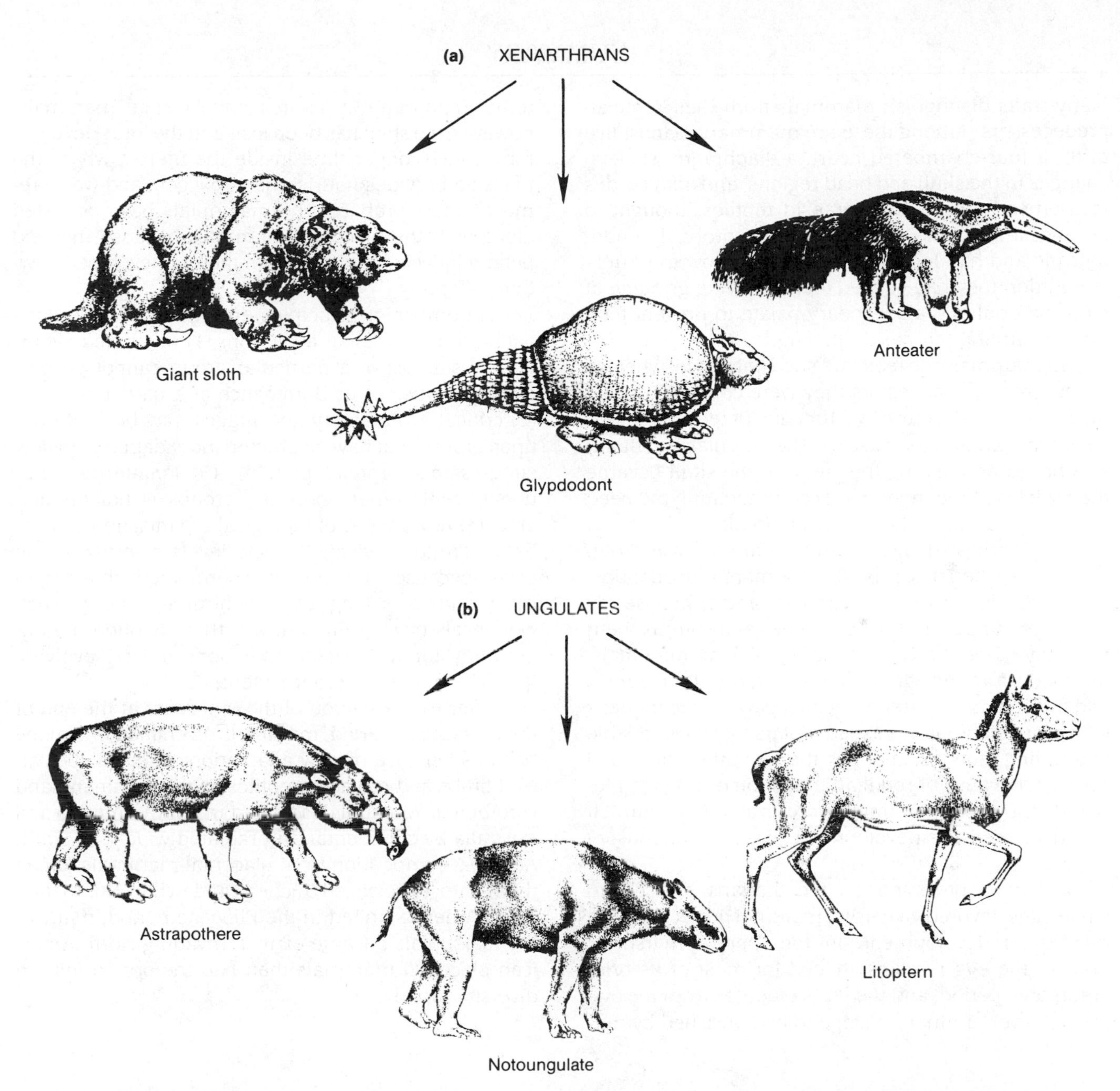

FIGURE 18-13 Reconstructions of some of the (a) xenarthrans and (b) ungulates produced by the South American placental mammalian radiation. (Adapted from Steel and Harvey.)

In the early or late Oligocene, a similar rapid radiation began among the rodents and primates that had reached South America at that time from either North America or Africa. The rodents produced a great diversity of caviomorphs (cavies) distinguished by special jaw muscle attachments. The primates, confined mostly to tropical areas, produced the wide array of New World monkeys collected in the superfamily Ceboidea.

When South America was next united with North America, 30 to 35 million years later in the Pliocene, an extensive interchange between the mammals of these two continents followed. Many South American groups became extinct, including marsupial carnivores and many placental ungulates, at least partially because of competition with more advanced North American placentals. Some of the successful invading North American groups also diversified quite rapidly. For example, cricetid rodents (field mice) evolved into more than 30 genera and 180 species within the 5- or 6-million-year period of their South American immigration. As elsewhere, extinction and radiation in South America seemed to go hand in hand, testifying again to the basic opportunism of evolutionary change.

SUMMARY

Many traits distinguish mammals from their reptilian predecessors. Among these are mammary glands, live birth, a four-chambered heart, a diaphragm, skeletal changes in the skull and head regions, and adaptations for homeothermy. The therapsid reptiles, thought to be ancestral to the mammals, were probably endothermic and had legs placed directly below the trunk. In addition they had cusped teeth allowing grinding of food material and a secondary palate to prevent food from occluding the nasal openings.

In mammals the teeth are shed only once, whereas in their reptilian ancestors they were continuously replaced. As the structure and function of teeth changed, so did the jaw and its muscles. The articular and quadrate bones articulating the jaw with the skull became the ossicles of the inner ear, accommodating the need for keen hearing in these small animals.

The earliest known mammals are a fossil group dating from the Triassic period, the morganucodontids, which developed tricusped molars and a precise occlusion of upper and lower cheek teeth. From them probably came the Prototheria, egg-laying mammals, the modern representatives of which are the platypus and the echidna. A later branching gave rise to therians with more elaborately cusped molars and jaws capable of grinding motions, enabling them to utilize new food sources. Because of reptilian predation during the Mesozoic era, most mammals were probably nocturnal, which favored the development of excellent sensory organs.

Selection pressures on the therians, which had small eggs, favored viviparity, maternal protection, and rapid fetal development. In the therian marsupial branch, the egg remains shelled for most of its brief embryonic period, and the fetus emerges from the oviduct at a very immature stage to be nourished by maternal mammary structures. In placental mammals, however, the shell has been lost, and the fetus develops for a much longer time inside the uterus, where the placenta provides nourishment, oxygen, and waste removal. Most probably the marsupials have persisted along with the placental mammals because they expend relatively little energy on their undeveloped newborn offspring.

A number of principles emerge from an examination of mammalian evolution: (1) continued evolutionary success of a particular group cannot be predicted because of its dominance at a particular time; (2) critical evolutionary advantages may be conferred upon groups that have characteristics adaptable to new circumstances (preadaptation); (3) long-term evolutionary replacement between groups is not predictable; (4) new modes of biological organization can enhance group survival; (5) new levels of organization occur because of complex, coordinated changes in many traits over long intervals of time; (6) once such new levels have been attained, there is often the opportunity for widespread radiation; and (7) evolving groups are often interdependent.

After the extinction of the dinosaurs at the end of the Cretaceous period, mammals diversified into many habitats because of new adaptations such as specialized limbs and replaced reptiles as the dominant land vertebrates. Marsupials isolated in South America and Australia by continental drift radiated widely. In South America, competition from placental mammals forced them into specialized niches, and when South and North America united in the Pliocene period, many of the marsupials became extinct. Invading North American placental mammals then had the opportunity to diversify rapidly.

KEY TERMS

articular
bony secondary palate
coadaptive mutations
condylarths
cynodont therapsids
deciduous teeth
dentary bone
diaphragm
digitigrade
diurnal
double occipital condyle
ear ossicles
edentates
four-chambered heart
heterodont
incus
lactation
large brain case
live birth
malleus
mammary glands
mandible
marsupials
monotremes

morganucodontids
multi-cusped teeth
neocortex
nocturnal
oviparity
placentals
plantigrade
Prototheria
quadrate
single nasal opening
squamosal
synapsid reptiles
therians
tribosphenic molar
triconodont
tympanic membrane
unguligrade
uterus
viviparity
xenarthrans

REFERENCES

Archer, M., and G. Clayton (eds.), 1984. *Vertebrate Zoogeography and Evolution in Australia*. Hesperian Press, Carlisle, Australia.

Benton, M. J., 1988. The relationships of the major group of mammals: new approaches. *Trends in Ecol. and Evol.*, **3**, 40-45.

Bown, T. M., and M. J. Kraus, 1979. Origin of the tribosphenic molar and metatherian and eutherian dental formulae. In *Mesozoic Mammals: The First Two-thirds of Mammalian History*, J. A. Lillegraven, Z. Kielan-Jaworowska, and W. A. Clemens (eds.). Univ. of California Press, Berkeley, pp. 172-181.

Carroll, R. L., 1988. *Vertebrate Paleontology and Evolution*. W. H. Freeman, New York.

Colbert, E. H., 1981. *Evolution of the Vertebrates*, 3rd ed. John Wiley, New York.

Crompton, A. W., and F. A. Jenkins, Jr., 1979. Origin of mammals. In *Mesozoic Mammals: The First Two-thirds of Mammalian History*, J. A. Lillegraven, Z. Kielan-Jaworowska, and W. A. Clemens (eds.). Univ. of California Press, Berkeley, pp. 59-73.

Crompton, A. W., and P. Parker, 1978. Evolution of the mammalian masticatory apparatus. *Amer. Sci.*, **66**, 192-201.

Crompton, A. W., C. R. Taylor, and J. A. Jagger, 1978. Evolution of homeothermy in mammals. *Nature*, **272**, 333-336.

Eisenberg, J. F., 1981. *The Mammalian Radiations: Evolution, Adaptation, and Behavior*. Univ. of Chicago Press, Chicago.

Gingerich, P. D., 1977. Patterns of evolution in the mammalian fossil record. In *Patterns of Evolution as Illustrated by the Fossil Record*, A. Hallam (ed.). Elsevier, Amsterdam, pp. 469-500.

Jerison, H. J., 1973. *Evolution of the Brain and Intelligence*. Academic Press, New York.

Kemp, T. S., 1982. *Mammal-like Reptiles and the Origin of Mammals*. Academic Press, London.

——, 1988. Interrelationships of the Synapsida. In *The Phylogeny and Classification of the Tetrapods*, Vol. 2, M. J. Benton (ed.). Oxford Univ. Press, Oxford, pp. 1-22.

Kermack, D. M., and K. A. Kermack, 1984. *The Evolution of Mammalian Characters*. Croom Helm, London.

Kermack, K. A., and F. Mussett, 1983. The ear in mammal-like reptiles and early mammals. *Acta Palaeontologica Polonica*, **28**, 147-158.

Kielan-Jaworowska, Z., A. W. Crompton, and F. A. Jenkins, 1987. The origin of egg-laying mammals. *Nature*, **326**, 871-873.

Lillegraven, J. A., 1979. Reproduction in Mesozoic mammals. In *Mesozoic Mammals: The First Two-thirds of Mammalian History*, J. A. Lillegraven, Z. Kielan-Jaworowska, and W. A. Clemens (eds.). Univ. of California Press, Berkeley, pp. 259-276.

Lillegraven, J. A., Z. Kielan-Jaworowska, and W. A. Clemens (eds.), 1979. *Mesozoic Mammals: The First Two-thirds of Mammalian History*. Univ. of California Press, Berkeley.

Lull, R. S., 1940. *Organic Evolution*. Macmillan, New York.

Marshall, L. G., 1980. Marsupial paleobiogeography. In *Aspects of Vertebrate History*, L. L. Jacobs (ed.). Museum of Northern Arizona Press, Flagstaff, Ariz., pp. 345-386.

Marshall, L. G., S. D. Webb, J. J. Sepkoski, Jr., and D. M. Raup, 1982. Mammalian evolution and the great American interchange. *Science*, **215**, 1351-1357.

McFarland, W. N., F. H. Pough, T. J. Cade, and J. B. Heiser, 1985. *Vertebrate Life*, 2nd ed. Macmillan, New York.

Nitecki, M. H. (ed.), 1988. *Evolutionary Progress*. Univ. of Chicago Press, Chicago.

Novacek, M. J., A. R. Wyss, and M. C. McKenna, 1988. The major groups of eutherian mammals. In *The*

Phylogeny and Classification of the Tetrapods, Vol. 2, M. J. Benton (ed.). Oxford Univ. Press, Oxford, pp. 31-71.

Olson, E. C., 1971. *Vertebrate Paleozoology*. Wiley-Interscience, New York.

Rich, P. V., and E. M. Thompson (eds.), 1982. *The Fossil Vertebrate Record of Australia*. Monash Univ. Press, Clayton, Australia.

Romer, A. S., 1966. *Vertebrate Paleontology*, 3rd ed. Univ. of Chicago Press, Chicago.

——, 1968. *The Procession of Life*. World Publ. Co., Cleveland.

Romer, A. S., and T. S. Parsons, 1977. The *Vertebrate Body*, 5th ed. W. B. Saunders, Philadelphia.

Savage, R. J. G., and M. R. Long, 1986. *Mammal Evolution: An Illustrated Guide*. British Museum (Natural History), London.

Simpson, G. G., 1980. *Splendid Isolation: The Curious History of South American Mammals*. Yale Univ. Press, New Haven.

Stahl, B. J., 1974. *Vertebrate History: Problems in Evolution*. McGraw-Hill, New York.

Steel, R., and A. P. Harvey, 1979. *The Encyclopaedia of Prehistoric Life*. Mitchell-Beazley, London.

Young, J. Z., 1981. *The Life of Vertebrates*, 3rd ed. Clarendon Press, Oxford.

19

PRIMATE EVOLUTION AND HUMAN ORIGINS

rimates, the mammalian order that includes humans, are species that possess a number of adaptations indicating an arboreal (tree-living) ancestry. Among others listed in Table 19-1, these adaptations include (1) ability to move the four limbs in various directions; (2) grasping power of the hands and feet; (3) slip-resistant cutaneous ridges (dermatoglyphs) on the ventral pads of these extremities, which also contain specialized tactile-sensitive organs (Meissner's corpuscles); (4) retention of the clavicle (collar bone) to support the pectoral girdle in positioning the forelimb; and (5) flexibility of the spine to allow twisting and turning.

In addition to possessing a highly developed brain in relation to body size, anthropoid primates (monkeys, apes, and humans) also undergo a relatively long postnatal growth period accompanied by considerable parental care for a relatively small number of offspring. The selective value of the last feature probably arises from the limited number of offspring that can be successfully born and carried by highly mobile primates, along with the long dependent learning period needed to cope with many complex environmental and social variables. Although all of the features mentioned in Table 19-1 are not characteristic of every primate, all existing primates possess a sufficient number to distinguish them from other mammalian arboreal groups such as shrews, squirrels, and raccoons.

PRIMATE CLASSIFICATION

There are presently about 185 primate species, all usually classified (Table 19-2) into two suborders, **prosimians** (or Strepsirhini) and **anthropoids** (or Haplorhini). The prosimians, often known as the lower primates, generally retain more early mammalian features (e.g., claws, long snout, lateral-facing eyes) than do the anthropoid higher primates. With the exception of Madagascar, an island that separated from Africa before anthropoids had evolved, prosimians found in other localities are small bodied and nocturnal. (The tree shrews [tupaiids], once thought to be a prosimian group, are now generally considered members of the most primitive of eutherian orders, Insectivora, or assigned to an order of their own [Tupaioidea].)

The anthropoids, which include monkeys, apes, and humans, are mostly larger than prosimians and are generally diurnal rather than nocturnal. Compared

TABLE 19-1 Traits and Tendencies Found in Primate Groups

- Independent mobility of the digits
- A wrap-around first digit in both hands and feet (thumb, big toe)
- Replacement of claws by nails to support the digital pads on the last phalanx of each finger and toe
- Teeth and digestive tract adapted to an omnivorous diet
- A semi-erect posture that enables hand manipulation and provides a favorable position preparatory to leaping
- Center of gravity positioned close to the hind legs
- Well-developed hand-eye motor coordination
- Optical adaptations that include overlap of the visual fields to gain precise three-dimensional information on the location of food objects and tree branches
- Bony orbits to help protect the eyes from arboreal hazards
- Shortening of the face accompanied by reduction of the snout
- Diminution of the olfactory apparatus in diurnal forms
- Compared to practically all other mammals, a very large and complex brain in relation to body size

to prosimians, anthropoids also possess more of the primate features enumerated previously, such as a shortened face, forwardly directed eyes, and a larger, more complex brain. Species members from the two primate suborders are illustrated in Figure 19-1 and can briefly be described as follows.

Lemurs

The lemurs are found exclusively in Madagascar, which was probably separated from the African continent sometime during the late Cretaceous period. In this relatively protected area they have produced a range of small and large species that often parallel the role of forest monkeys on the mainland. But lemurs are generally more primitive than monkeys, possessing a longer snout and moist philtrum between nose and upper lip that accentuates their sense of smell. They also possess a special toilet claw on the second toe, thick fur, sensitive facial hairs (vibrissae), and a dental comb formed by the nearly horizontal (procumbent) orientation of the lower incisors and canines that is used for both grooming and feeding. Because lemurs possess a "tapetum" (retinal layer that reflects incoming light back through the retina), some have become exclusively nocturnal (mouse and dwarf lemurs) and others are active during the dim crepuscular light of late dusk and early dawn as well as diurnally (true lemurs).

Lorises

The lorises are found in forests of both Africa (pottos and galagos [bush babies]) and Southeast Asia (slender and slow lorises). The snout is shorter than in lemurs, and the relatively large eyes face forward, indicating increased emphasis on visual predation. These and other adaptations, including a retinal tapetum, permit either nocturnal or crepuscular activity. Like lemurs, lorises possess dental tooth combs, a toilet claw, and a moist philtrum.

Tarsiers

These nocturnal Southeast Asian primates seem to stand between prosimians and anthropoids. Although they have two toilet claws on each foot and enormous eyes relative to head size, like anthropoids, they lack the retinal tapetum that characterizes lemurs and lorises. Tarsiers also show anthropoid characteristics in replacement of the moist philtrum by a dry, furry space between nose and lip, as well as upright lower incisors. Unique to tarsiers is the fusion of tibia and fibula in the lower leg, an adaptation that apparently aids their ability to make single leaps as long as 6 or 7 feet.

TABLE 19-2 Classification of Existing Subgroups in the Order Primates, with Common Names for Some Members in Each Group*

Suborder Prosimii
- Superfamily Lemuroidea: lemurs
- Superfamily Lorisoidea: lorises, galagos (bush babies)
- Superfamily Tarsioidea: tarsiers

Suborder Anthropoidea
- Infraorder Platyrrhini (New World)
  - Superfamily Ceboidea
    - Family Callitrichidae: marmosets, tamarins
    - Family Cebidae: capuchins, howler monkeys, spider monkeys
- Infraorder Catarrhini (Old World)
  - Superfamily Cercopithecoidea
    - Family Cercopithecidae: macaques, baboons, vervet monkeys
    - Family Colobidae: langurs
  - Superfamily Hominoidea
    - Family Hylobatidae: gibbons, siamangs
    - Family Pongidae†: orangutans, gorillas, chimpanzees
    - Family Hominidae: humans

*Primate taxonomy has generated considerable debate (Aiello). Some primatologists rename the prosimians as the suborder Strepsirhini (moist, doglike muzzle between nose and lip) and the anthropoids as the suborder Haplorhini (dry skin or fur between nose and lip) and include the tarsiers as a haplorhine infraorder (Tarsii).

†In some molecular classifications based on immunological studies (see Fig. 12-2) and DNA sequencing (see Miyamoto et al.), all the great apes are placed together with humans in a single family, Hominidae, which is then subdivided into two subfamilies: one containing the orangutans (Ponginae), and the other containing humans, chimpanzees, and gorillas (Homininae). (See also Tattersall et al.)

FIGURE 19-1 A tree shrew (order Insectivora) and various representatives of the order Primates.

Platyrrhines

The platyrrhine infraorder designates the New World monkeys found in Central and South America, all of which are arboreal. They are characterized by broad noses with widely spaced nostrils facing laterally and three premolars on each side of the jaw. In one family (Callitrichidae) are marmosets and tamarins, small animals that possess claws on all digits except the big toe. Species in the other platyrrhine family (Cebidae) have nails instead of claws, and some also possess prehensile tails.

Catarrhines

This infraorder includes two superfamilies, the Cercopithecoidea (Old World monkeys) and Hominoidea (apes and humans). They share narrowly spaced nostrils facing downwards and a dental formula of 2.1.2.3 (two incisors, one canine, two premolars, and three molars on each side of the centerline in both upper and lower jaws). The catarrhine monkeys are mostly larger than the New World monkeys, lack prehensile tails, and have produced terrestrial (baboons, mandrills) as well as arboreal forms.

Hominoids

In this catarrhine superfamily of apes and humans are found a number of adaptations to brachiating (arm-hanging and -swinging) arboreal locomotion along with different degrees of adaptation to a ground-dwelling existence. Perhaps because arboreal hominoids had to adapt to holding on to overhead tree branches, their posture is more erect than that of monkeys. Also, as an aid in brachiation, the arms and shoulders are more flexible, the wrists and elbows are more limber, and the spine is shorter and stiffer. Other hominoid attributes that distinguish them from monkeys are: a broader and larger pelvis to support more vertical weight; visceral attachments and arrangements that provide more vertical support for stomach, intestines, and liver; loss of the tail; five-cusped lower molars rather than the four cusps in monkeys; a broad but shallow thorax because of the change to a more vertical posture; scapulae placed dorsally on the thorax to position the shoulder joint so the arms can be extended laterally; and numerous other features.

Gibbons

Perhaps the most primitive of hominoids are the gibbons and siamangs, an almost entirely arboreal group confined to Southeast Asia. They share with many Old World monkeys a relatively small size (none are more than 25 pounds) and ischial callosities (cornified sitting pads fused to the ischial bones). Compared to other hominoids, they are superb acrobatic brachiators who swing with elongated arms through the trees, their legs often folded beneath them.

Orangutans

The orangutans are large apes (some males may weigh more than 200 pounds) restricted to Borneo and Sumatra. With the exception of adult males, they are mainly arboreal and on the ground move mostly quadrupedally with clenched fists to support the upper torso. However, like chimpanzees and gorillas, they lack ischial callosities. As discussed previously (Chapter 12), molecular data indicate that this group separated at an early stage in hominoid evolution from the group that includes chimpanzees, gorillas, and humans.

Chimpanzees

The chimpanzees are found in Equatorial Africa where they live in groups of about 40 to 50, socially organized in a dominance hierarchy. Although they sleep and do most of their feeding arboreally, they are less specialized for arboreal pursuits than Asiatic apes and more specialized for terrestrial locomotion. Like the gorillas, they travel on the ground by "knuckle-walking," using the second phalanx of four manual digits as forelimb support. Their diet is mainly frugivorous, but they have also been observed to eat termites as well as capture and eat young baboons, monkeys, and occasionally even young chimpanzees. Compared to all other primates except humans, chimpanzees show a remarkably wide array of expressions, postures, and gestures.

Gorillas

The gorillas are the largest of apes (some males may weigh 500 pounds or more) and are found in Equatorial Africa in two main distributions: lowland gorillas west of the Congo basin and mountain gorillas eastward. Adult males rarely climb in trees and make their sleeping nests on the ground. Their social groups of 10 to 20 individuals are organized around a single dominant male ("silverback"), with other adult males occasionally present. They do not seem to be as active as chimpanzees and have a diet which seems to be almost entirely herbivorous.

HUMAN-APE COMPARISONS

Compared to all hominoids, humans (**hominids**) present the greatest number of adaptations to bipedal terrestrial locomotion. Their hind limbs are longer rel-

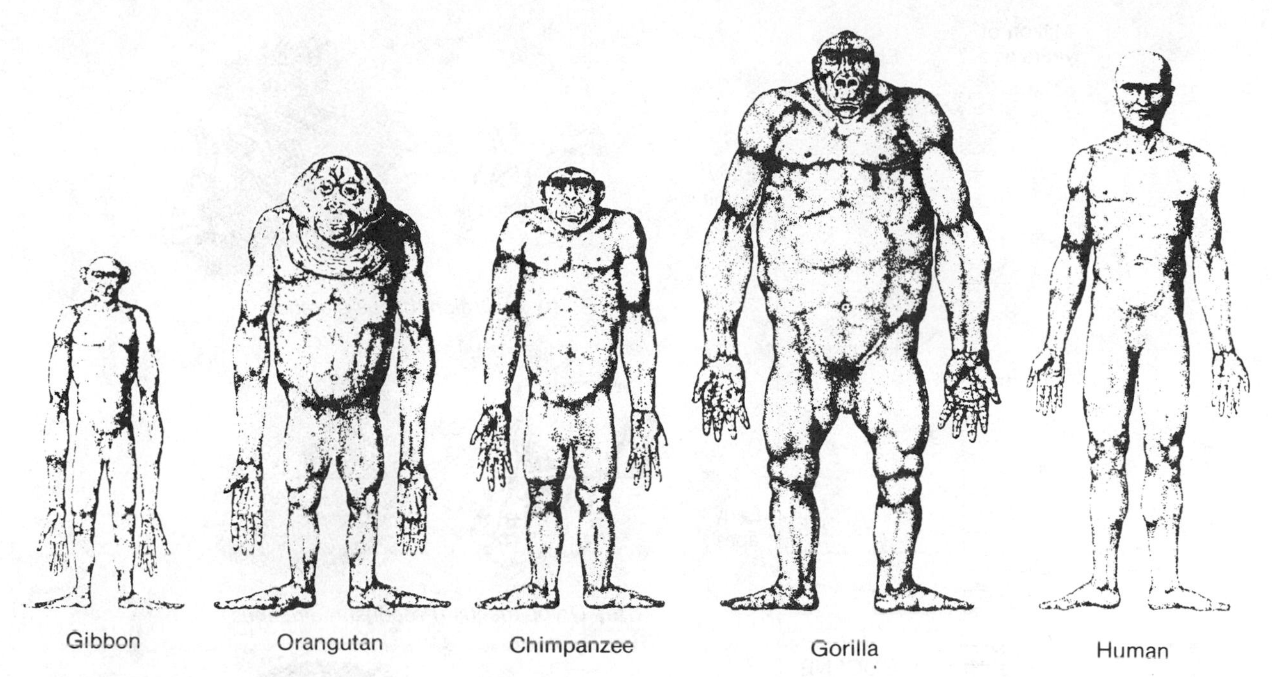

FIGURE 19-2 Body contours and proportions of adult apes and humans with all hair removed, drawn to the same scale. (Adapted from Schultz.)

ative to their forelimbs than in any of the apes, and their hands, freed from supporting the body, provide the most refined of manipulatory controls. Additional uniquely human anatomical traits include a relatively large brain and small face, reduction in length of the canines, reduced body hair covering, and numerous cranial, dental, skeletal, and other features. Nevertheless, in spite of the differences, if comparisons are made bone for bone, muscle for muscle, organ for organ, humans are strikingly similar to apes, although differing in proportions (Fig. 19-2). This similarity between apes and humans can also be seen in some of their motions and postures:

1. Because their arms are extended laterally in brachiator fashion and their elbows and shoulders are remarkably mobile, they both scratch the back of their head from the side rather from the front as do monkeys.
2. As can be seen in the sitting individuals in Figure 19-3, positioning of the limbs can be remarkably alike—both support the chin with their hands and cross their legs. Even some facial expressions seem similar (see Fig. 19-15).
3. In walking the heel is touched first to the ground, whereas in monkeys the metatarsals are touched first.
4. The knuckle-supporting stance of crouching American football players is the conventional stance for ground movement in chimpanzees and gorillas.

The Fossil Record

According to the fossil record, many Mesozoic mammals were very much like extant tree shrews (Fig. 19-4(a); see also Fig. 18-4), probably adapted to an insectivorous life-style that encompassed both the forest floor and trees and shrubs. Although many of these

FIGURE 19-3 A chimpanzee and movie actress (Dorothy Lamour) resting on a 1938 movie set of the motion picture *Jungle Love*. (From a photograph in W. M. Mann, 1938. *National Geographic*, **73**, 615-655.)

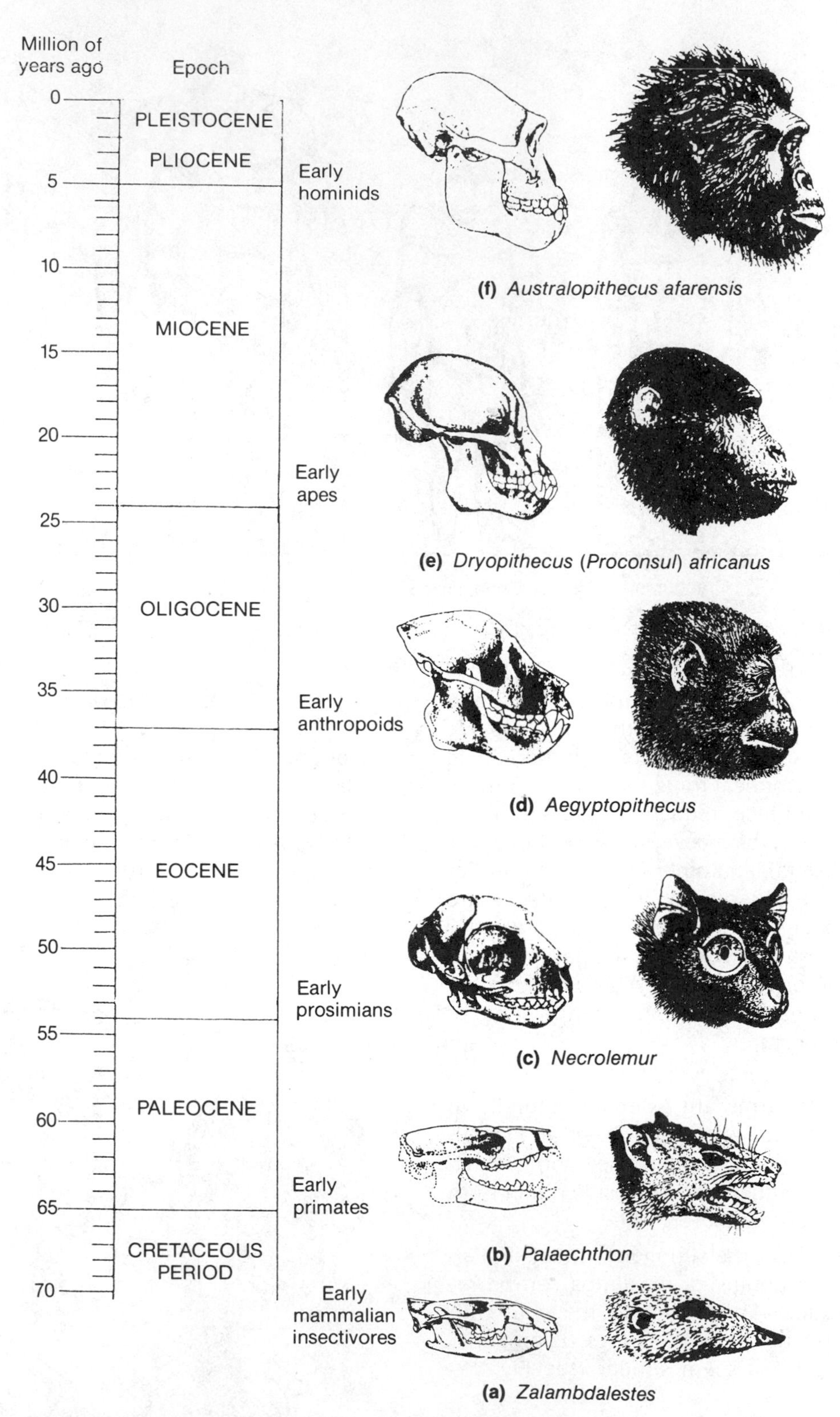

FIGURE 19-4 Beginning from an early insectivore of the Mesozoic era (a), are samples from branches of the primate evolutionary tree (b-f) showing the chronology, skulls, and reconstructions of some fossil species.

basic arboreal adaptations were further developed by the primates that evolved from these early forms, the point at which primate origin took place is still difficult to discern. Part of this difficulty lies in the relative absence of fossilized forest animals in sedimentary rocks. Generally, arboreal animals are usually completely disarticulated soon after they reach the forest floor, and only rare chance events will wash their skeletons down into rivers, lakes, or marine sediments where they can be more easily fossilized.

Nevertheless, Paleocene deposits do show the presence of some early archaic primates classed in a suborder Plesiadapiformes (Fig. 19-4(b)) which possess uniquely structured auditory regions and dentitions that differentiate them from related insectivorous forms (Fig. 19-4(a)). By the Eocene, some of these groups had evolved further changes such as a bony ring around their optical orbits and digital nails rather than claws. One Eocene family (adapids, Fig. 19-4(c)) then gave rise to lemurs and lorises and another (omomyids) to the tarsiers.

The next evolutionary stage, which led to the platyrrhine (New World) and catarrhine (Old World) anthropoids, has been the subject of considerable debate. Although most primatologists now consider both anthropoid groups to have had a monophyletic origin, questions remain as to whether they evolved from the same omomyid group as the tarsiers and whether the platyrrhine monkeys arrived in the New World via dispersal from North America or by island-hopping from Africa. Perhaps all one can say at present is that the similarities of anthropoids to Eocene omomyids seem greater than to adapids (Carroll). Also since there are no Cenozoic prosimian fossils in South America nor Cenozoic anthropoid fossils in North America, it seems unlikely that the platyrrhines originated in either South America or North America. Rather, the New World monkeys may well have come from Africa, which, through continental drift, was closer to South America during the early and middle Cenozoic than it is at present (see Fig. 6-12). During this time, the two continents were probably spanned by one or more chains of islands that were later submerged, some of which may have served as intermediate stations for the platyrrhine journey.

In any case, recent investigations of fossils found in the Faiyum Province of Egypt (e.g., *Aegyptopithecus*, Fig. 19-4(d)) suggest that fairly primitive anthropoids had already evolved by the Oligocene epoch more than 30 million years ago (Fleagle et al.). Over the next 10 million years these African groups probably diverged into both Old World monkeys (cercapithecoids) and apelike forms (hominoids) as well as into the New World forms that eventually colonized South America (ceboids). (The earliest known South American platyrrhine fossils occur about 5 million years after the early anthropoids in the Faiyum deposits.) Included among the hominoidlike fossils found later during the Miocene is a group called the dryopithecines, of which *Proconsul* is considered an early member (Fig. 19-4(e)).

It is interesting to note that compared to their relatively fewer numbers today, apelike forms were more common during the early and middle Miocene than those of monkeys. This situation, however, was reversed during the late Miocene when monkeys became much more numerous and widespread than apes. Andrews (1981) suggests that dietary changes in Old World monkeys during the Miocene enabled them to compete successfully against many of the arboreal apes, perhaps by developing the ability to eat and digest fruits before they became ripe enough for the hominoids.

One important consequence of the replacement of apes by monkeys in arboreal habitats was probably to increase the selective pressure among various ape (and some monkey) species for ground-dwelling adaptations. It seems likely that it was from a group of these terrestrial Miocene apes that the ancestral humans evolved. The exact timing of this event is nevertheless uncertain. One view, originally based on the presumed humanlike lower jaws of a fossil ape called *Ramapithecus* was that the ape-human split occurred about 12 to 14 million years ago or earlier. However, with the recent discovery of more complete ramapithecine fossils, this view has been discarded, and this group is now considered to be linked to a much earlier apelike lineage that may have been ancestral to orangutans (Wolpoff 1982, Andrews 1983, Pilbeam 1984). The divergence between apes and humans may therefore have occurred considerably after the ramapithecine fossilizations. In support are molecular studies cited in Chapter 12 indicating that this divergence took place either 3.5 to 5.5 million years ago (Sarich and Cronin) or 5.5 to 7.7 million years ago (Sibley and Ahlquist; see Fig. 12-12). The fact that no unequivocal hominidlike fossils (such as the australopithecines described below) have yet been found earlier than the middle Pliocene, about 4 million years ago, may indicate that the hominid lineage is not much older than that.

THE AUSTRALOPITHECINES

The earliest of recognizably hominid fossils have been ascribed to the genus *Australopithecus* (southern ape). The first such fossil was discovered by Dart in 1924 from a lime quarry at Taung in the Cape Province of South Africa. This fossil, named *A. africanus*, was of a 6-year-old (the "Taung child") and consisted of the front part of the skull and most of the lower jaw (Fig. 19-5(b)). All of the deciduous teeth were present as well

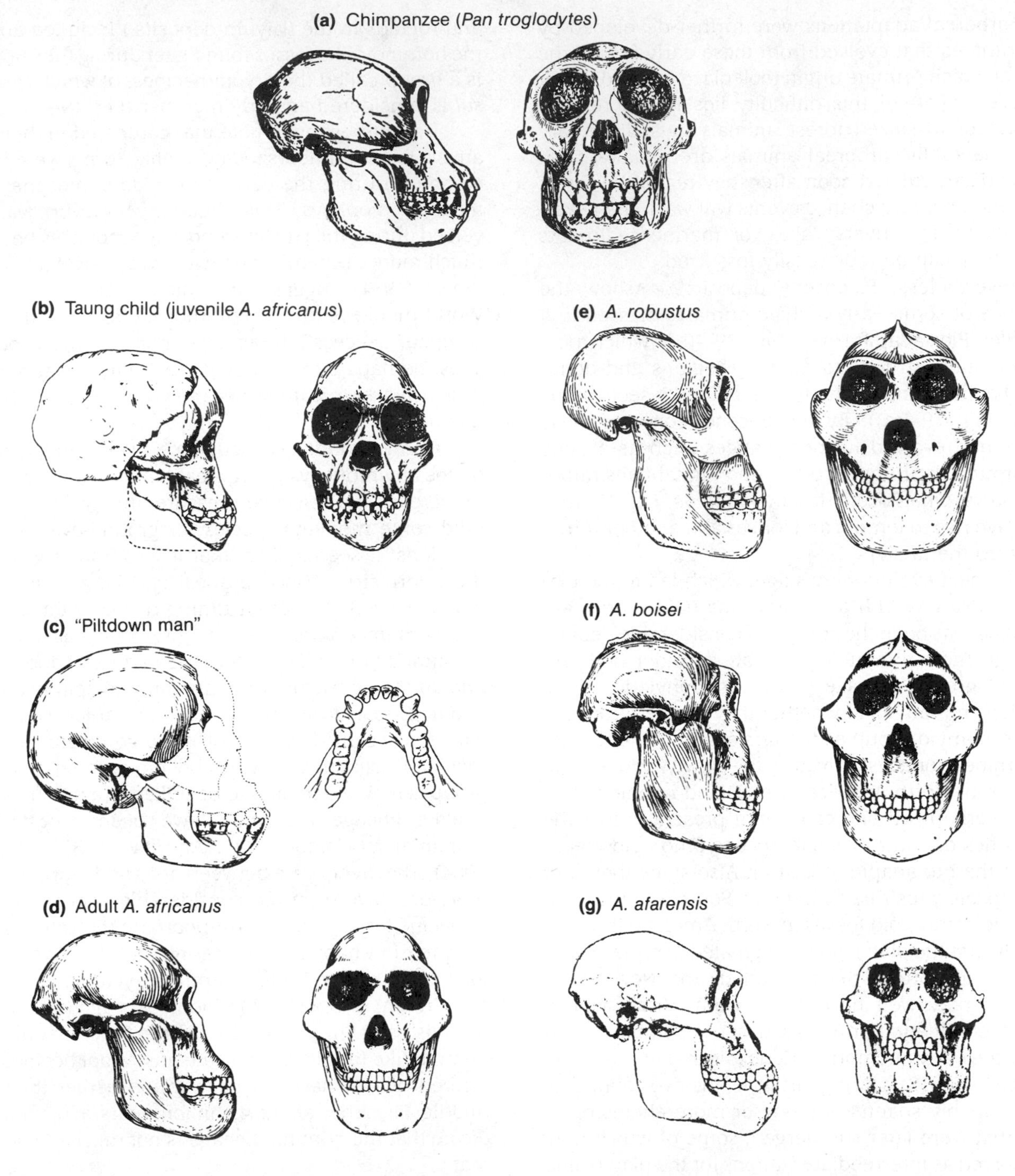

FIGURE 19-5 Skulls of a chimpanzee (a) and fossil hominids described in the text, including cranial and lower jaw fragments of the Piltdown forgery (c). (Adapted from Johanson and Edey.)

as the first of the permanent replacement molars. Although these teeth were generally larger than those of humans, they showed humanlike features in the multicusped nature of the anterior milk molar, which is single-cusped in apes. Also, the lunate sulcus (the anterior border of the visual area in the brain) observed in the endocranial cast was posterior to its usual position in apes, more like that of humans.[1] The Taung

[1]In humans this sulcus is pushed posteriorly largely because of expansion of the parietal cerebral areas responsible for symbolic associations connected with language and sequential reasoning.

skull indicated that the adult brain volume was about 450 cc (midway between chimpanzee and gorilla) and that the adult body was probably smaller than that of chimpanzees, with weights between 40 and 70 pounds.

At the time of the Taung discovery, however, most anthropologists believed that primitive humans had large braincases and apelike jaws with large canines. Evidence for this belief came from a so-called fossil cranium and lower jaw found in 1912 at Piltdown, England, that showed such features (Fig. 19-5(c)). Unfortunately, the Piltdown fossil was accepted as valid by many anthropologists for about 40 years until a number of anthropologists showed that the entire skull was a hoax: the teeth had been artificially ground down, the cranium was of a different age than the jaw, the bone color was caused by artificial pigmentation, and the molar teeth had long roots like those of apes. Moreover, the associated animal fossils at the Piltdown site had a large accumulation of radioactive salts whose origin could be traced to a site in Tunisia. The **Piltdown Man** therefore turned out to be a combination of a human cranium and the lower jaw of a female orangutan, a hoax perpetrated by someone who knew enough to destroy all obvious signs of the pseudofossil's true origin by removing the jaw joint and modifying other features.

Because of the Piltdown forgery, it was more than 20 years before most anthropologists began to accept the humanlike nature of the australopithecine fossils. By then, numerous such fossils had been found in other sites in Africa, such as the adult australopithecine skulls discovered at Sterkfontein not far from Taung (Fig. 19-5(d)). The thickly enamelled teeth of these fossils indicated heavy tooth usage that enabled the teeth to wear flat before the dentin was exposed and probably also indicated a fairly long life span compared to apes. The Sterkfontein fossils also showed that, although there were differences from human teeth, especially in respect to the larger australopithecine premolars and molars, their incisors and canines were reduced compared to apes. The date for these fossils was set at about 2½ to 3 million years ago.

Post-cranial australopithecine skeletal material, such as the pelvis and vertebrae, was shown by Robinson and others to be humanlike with a distinct lumbar curvature of the spine indicating erect posture. As in humans, vertical weight transmission occurred through the outer condyle of the knee, and *Australopithecus* was undoubtedly a bipedal walker.

Fossils of a larger hominid having a mature weight of about 80 to 140 pounds were found in other South African sites and called *A. robustus* (Fig. 19-5(e)). Accompanying the larger size of this group, often represented as an offshoot of *A. africanus*, were significantly larger teeth and jaws, which undoubtedly reflected a different, perhaps more herbivorous, diet with a need for more powerful grinding. The brain, too, was larger than *A. africanus*, with a volume of about 550 cc, but this may have been primarily associated with increased relative body size rather than increased intelligence.

In East Africa a Pliocene australopithecine, *A. boisei* (Fig. 19-5(f)), apparently underwent selection for even larger molars than *A. robustus*. It is the largest of the australopithecines and shows various cranial adaptations for powerful masticating jaw muscles that indicate a diet of tough plant food such as seeds and fruits with hard husks and pods. Interestingly, an early form of *A. boisei* found recently in 2½-million-year old deposits at Lake Turkana, Kenya, indicates that these massively built australopithecines may have been evolving separately but parallel to the *africanus-robustus* lineage (Lewin). If we follow both these lineages forward from the Pliocene into the Pleistocene less than 2 million years ago, it seems that the molars were becoming larger while the face was shortening and becoming more vertical and humanlike. The significance of these parallel changes in two different lineages is difficult to understand so far, since the selective forces acting upon these groups are as yet unknown.

Even more primitive than the above groups are a series of fossils found more recently in East African sites at Laetoli, Tanzania, and the Afar (Hadar) region of Ethiopia. These fossils, dated to between 3½ to 4 million years ago, have been included among the australopithecines under the species name *A. afarensis* (Fig. 19-5(g)). A reconstruction of the face of this oldest known hominid species is given in Figure 19-4(f), showing the heavy brow ridges, low forehead, and projecting (prognathous) mouth. However, in spite of some such similarities, *A. afarensis* displays a large number of important cranial, dental, and skeletal differences from apes. For example, although the canines are larger in *afarensis* males than in females, this sexual dimorphism is much less pronounced than in apes or early Miocene pongidlike fossils. Tooth positioning, enamel thicknesses, and the resulting wear facets have also been modified to allow greater transverse jaw movements that retain cutting functions but improve side-to-side grinding.

The fairly complete skeleton of an *afarensis* female ("Lucy") found at Afar shows a small, muscularly powerful body, perhaps only 3½ to 4 feet in height, with relatively longer arms than modern humans but possessing a habitual bipedal stance and bipedal locomotion. In support of such early bipedalism are footprints dated to about 3.7 million years ago, preserved under a layer of volcanic ash at Laetoli. These prints of two individuals who walked along the same path for a distance of more than 70 feet are of distinctly bipedal hominids, demonstrating that bipedalism must have preceded many other hominid adaptations such as increased brain size.

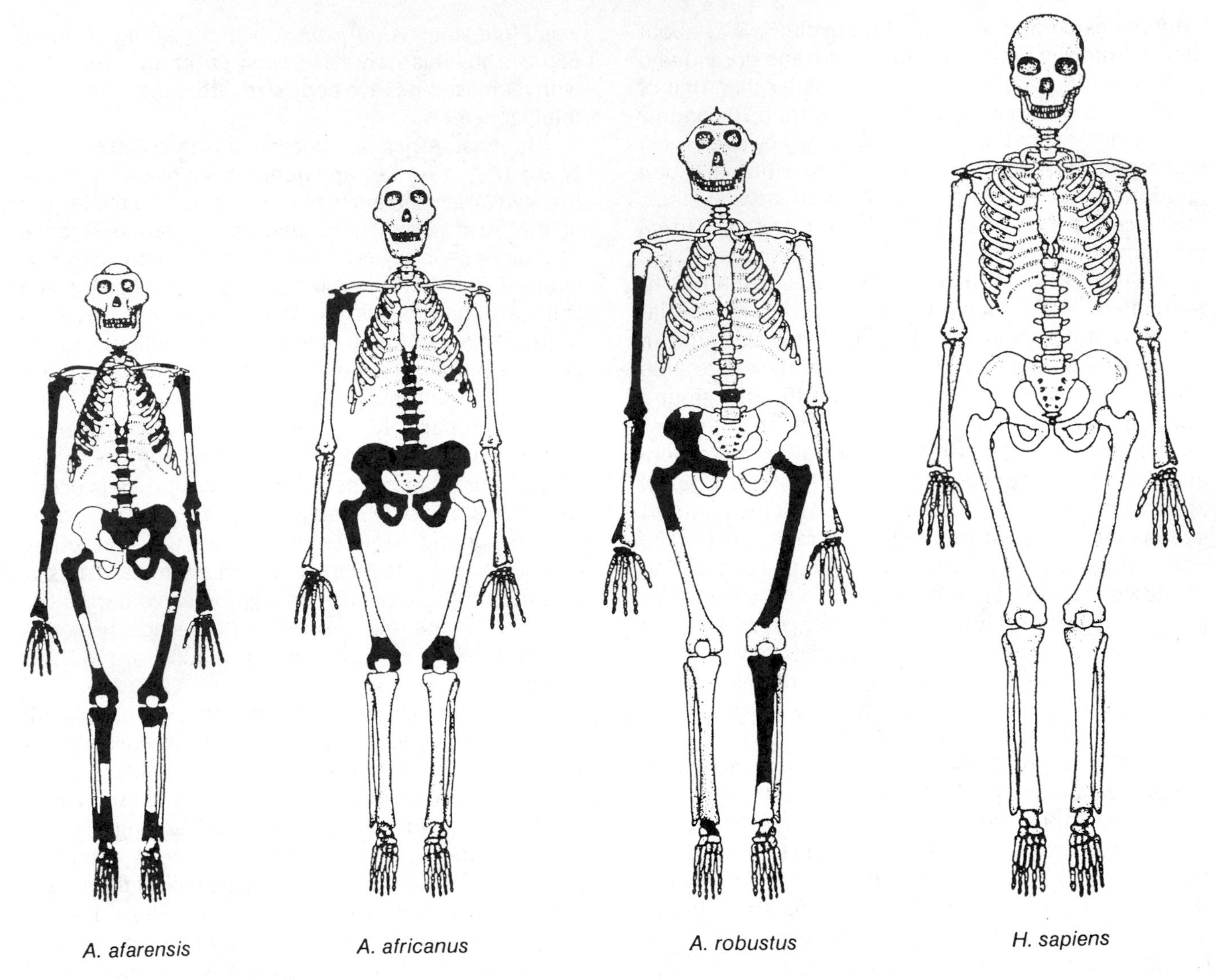

FIGURE 19-6 Skeletons of three australopithecines and a modern human. Black portions indicate the actual fossils found in the australopithecine skeletons. In all these hominids the pelvis is relatively shallow and rounded, and the femurs tilt towards the midline. This indicates that the trunk is being supported by the pelvis and body weight is being transmitted directly downwards through the knees when the individual stands erect. (From Johanson and Edey.)

Since all australopithecine skeletal reconstructions show the bipedal stance (Fig. 19-6), it must have been from a population within this group that the later human genus, *Homo*, evolved. What is not yet clear are the exact lineages within the australopithecines: Was *A. afarensis* actually an early form of *A. africanus*? Was *A. afarensis* so specialized in the direction of increased food grinding that its descendants could only have been heavy-jawed australopithecines such as *A. robustus* or *A. boisei*? Or, do the fossils ascribed to *A. africanus* represent two species, one allied to the "robust" australopithecines and the other ancestral to the *Homo* lineage? In any case, there is little question that *A. africanus* stands at or very near the base of hominid phylogeny and indicates that the australopithecines should be regarded as a group in which considerable evolutionary change was occurring.

BIPEDALISM

Although bipedalism has been established as a longstanding feature in the above lineages, why and how it originated has been disputed for over a century and is still a matter of controversy among paleoanthropologists. Its importance is obvious since it either accompanied or led to numerous adaptations that helped cast the future evolution of bipedal primates into a human framework. In a brief review of these controversies, Day (1986) has offered three arenas in which selective pressures might have enhanced bipedalism, each of which may have been influenced by the others:

- **Improved Food Acquisition.** Early hominids lived in a patchy environment consisting of mixed woodland and savannah (relatively dry grassland and bushland with occasional trees) that provided seasonal food supplies.[2] This placed emphasis on an omnivorous diet, demanding relatively more time spent searching for food over longer distances than in a more localized, homogeneous environment. An upright stance and bipedal striding would thus have enhanced **long-distance foraging** by enabling the manual transport of food gathered in different places. Tanner proposes that this was originally a female function, prompted primarily by food sharing with their offspring, and led to the invention and use of food-gathering tools. Unfortunately, the kind of food, plant or animal, acquired and carried in these early journeys is unknown. A primary diet of dispersed plant foods would seem to be in accord with the ape-like teeth that early hominids possessed. On the other hand, some workers suggest that early hominids relied heavily on scavenging carcasses from migratory herds of ungulates, and bipedalism became important both for terrestrial locomotion and for manual transport of immature offspring.

- **Improved Predator Avoidance.** Since bipedalism enhances height, it improves the ability of a hominid to see over tall grasses or obstructions and to wade in deeper water to pursue game or seek protection from predators. Day points out that the ability to climb trees would also have helped in escaping predators and increased the field of view in detecting danger. The curved hand and foot bones and relatively long arms of early australopithecines and humans (i.e., *Homo habilis*) point to persistent tree-climbing abilities.
- **Improved Reproductive Success.** Lovejoy proposed that bipedalism enabled adult males to carry food manually to their females and offspring who could then remain sequestered in a single locality, the **home base**. This mode of provisioning reduced the need for females to be continuously mobile in foraging both for themselves and their attached offspring as in other competing hominids, thereby offering three important advantages: (1) a relatively stable home base which provided more constant social relationships and perhaps closer mother-infant relationships that improved infant survival, (2) reduced infant injuries because infants no longer were attached to a continuously mobile mother, and (3) a reduction in the spacing between births by allowing more offspring to be cared for successfully.

Although the extent to which these proposals represent historical events is still debated (e.g., see the collection edited by Kinzey), Foley has pointed out that "evolution is as much about reproductive strategy as foraging behavior." It would certainly seem that a survival strategy dependent on bipedalism and a home base also makes use of other adaptations and preadaptations. For example, **sexual bonding** between males and females can motivate male foragers to continue to provision their family group because of their ties to particular females and can also extend male involvement into helping parent their offspring. Certainly, one important element that encourages human sexual bonding and year-round copulation is the absence of seasonal estrus cycles marked by specific, externally recognizable signals. Continued interest of both sexes in sexual bonding is also stimulated by human secondary sexual characteristics that persist from puberty onward. These traits include the relatively large penis in males, the enlarged mammaries and increased subcutaneous fat deposits in females (e.g., buttocks),[3] as well as hair adornments and apocrine scent glands displayed in both sexes.

Thus, the life-style introduced by bipedalism, long-distance foraging, continuous sexual activity, and their many corollaries probably improved survivorship by (1) intensifying the involvement of parents in their offspring, (2) extending the childrens' learning period, and (3) promoting a supporting and familial relationship among siblings whose births could be placed closer together. In an important sense, it was this evolutionary stage which led to strengthening of ties between related individuals, on the level of both the hominid nuclear family and more extended kinships. To these significant advantages can be added that bipedalism fosters the use of manual weapons such as stick wielding and stone throwing, which extends the reach

[2]There is evidence that beginning with the late Miocene and extending through Pliocene-Pleistocene times, periodic decreases in global temperature took place, marked by the onset of ice sheet formation in Antarctica and glaciation in the northern hemisphere. As water became locked up in ice, various terrestrial areas became relatively dry, and more open environments such as woodlands and grasslands replaced many rain forests in tropical regions.

[3]Cant points out that there is (or was) a relationship between reproductive success and the size of female breasts and buttocks in the sense that these anatomical parts provide an easily visible signal to males of the degree of feminine fat reserves. Minimal levels of fat reserves are essential for continued ovulation and lactation, and fat breasts and buttocks probably interfere less with bipedal locomotion than fat deposits placed elsewhere. He suggests that the evolution of such localized fat deposits therefore began as an expression of the feminine nutritional state and was probably reinforced by its use as an attractant in sexual bonding, helping reproductively successful choices to be made and maintained between sexual partners.

(a) *Pan troglodytes*

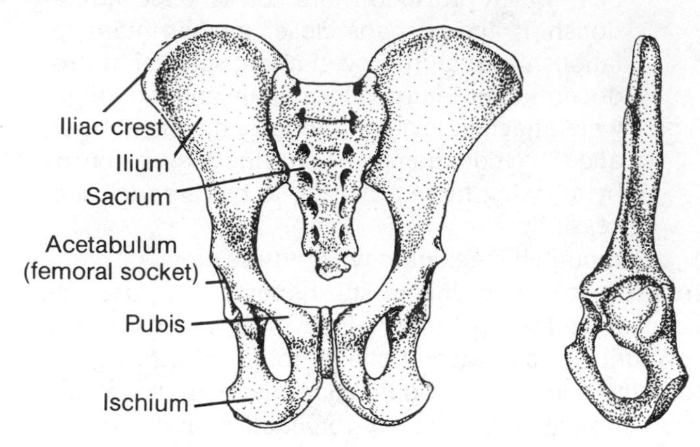

(b) *Australopithecus africanus*

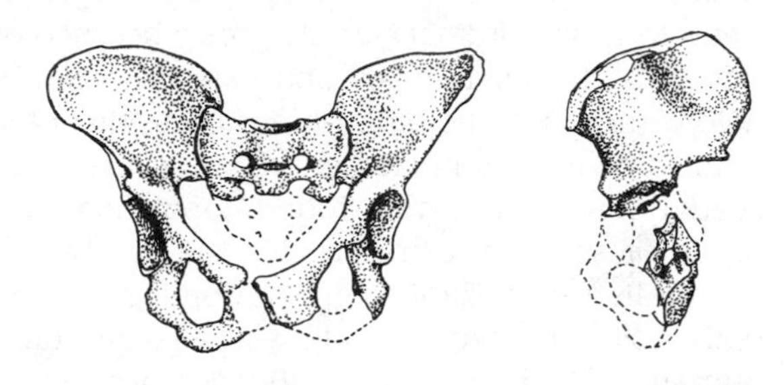

(c) *Homo sapiens*

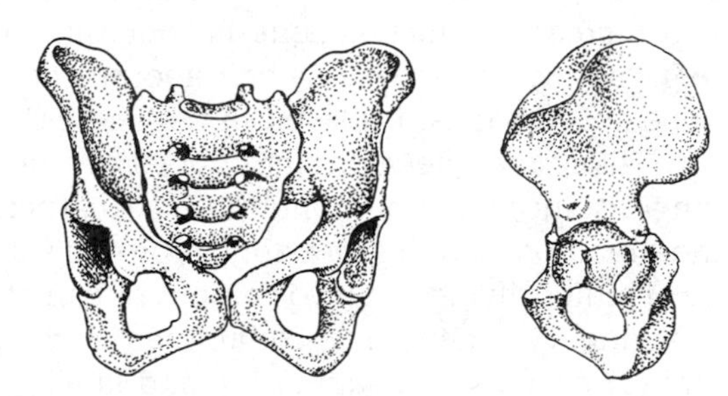

FIGURE 19-7 Comparison between the pelves of a chimpanzee (a), an australopithecine (b), and a modern human (c). Frontal views are on the left and lateral views on the right. Note that the distance between left and right acetabuli in *A. africanus* is less than in *H. sapiens.* This increase in interacetabular distance in humans was apparently the result of selection for a relatively large birth canal to permit the passage of newborn infants with larger crania than possessed by australopithecine newborns. (Adapted from LeGros Clark.)

of hominids beyond the teeth, claws, or defenses of animal competitors, predators, and prey. Bipedalism also allows even primitive tools and weapons to be carried from place to place, as well as offspring to be moved from one camp to another or from one food resource to another.

Anatomically, bipedalism is based upon broadening of the pelvis, changes in hind-limb muscle origins and insertions, and convergence of the femora towards the knee, conferring a knock-kneed stance compared to apes. In quadrupedal vertebrates, the hip bone, or innominate, of the pelvis has three components (ilium, ischium, and pubis) that link the hind limbs to the spine and help provide the propulsive force for quadrupedal motion. In animals like the tree shrew, the ilium is a long, narrow blade that lies alongside the sacral vertebrae. It lengthens further in primates as more hind-limb power is used to leap from place to place. The ischial tuberosities are also widened in many primates in association with selection for stability in sitting. The more erect anthropoids also used their pelvis for visceral support and, as a consequence of such selection, the ilium has become wider and more of the sacral vertebrae are fused into it (Fig. 19-7). In bipedal hominids, the pelvis assumes a broad, shallow, bowl-like shape with a widened but shortened ilium, thus bringing the sacrum closer to the acetabulum, the femoral socket. These changes, together with forward curvature of the lower spine and further flattening of the thoracic cage, help transmit weight of the trunk directly to the legs, producing a balanced center of gravity along a vertical axis (Fig. 19-8).

Normal hind-limb musculature in mammals involves flexors and extensors that move the femur forwards or backwards in relation to the pelvis, as well as abductors and adductors that control the lateral positioning of the trunk and pelvis upon the legs. Since humans walk while standing erect, the necessary leverages for these muscles have been mostly attained through attachments provided by the broadened pelvis and its various bony projections, such as the quadriceps flexor (rectus femoris) that swings the leg forward (Fig. 19-9).

On the dorsal surface, the rearward development of the iliac spine enables the gluteus maximus muscle, formerly an abductor of the thigh, to become an extensor that provides the important power stroke in running and climbing. Other changes from apelike bipedalism include transformation of the gluteus medius and minimus, formerly extensors, into abductors that balance the body laterally during walking. Moreover, since there is a straight vertical line going through the hominid leg—from hip joint to knee joint to ankle joint to foot surface—body weight is efficiently transmitted directly through the bones without tension rather than through the muscles. Convergence of the femora towards the knees ensures that the axis of weight remains close to the center of gravity. The pelvis rotates around this axis during walking, and the body can be kept in a forward plane by swinging the arms.

Below the knee, the tibia bears all the weight of walking, and little, if any, is transmitted through the fibula. Whereas the fibula in tree-dwelling primates is

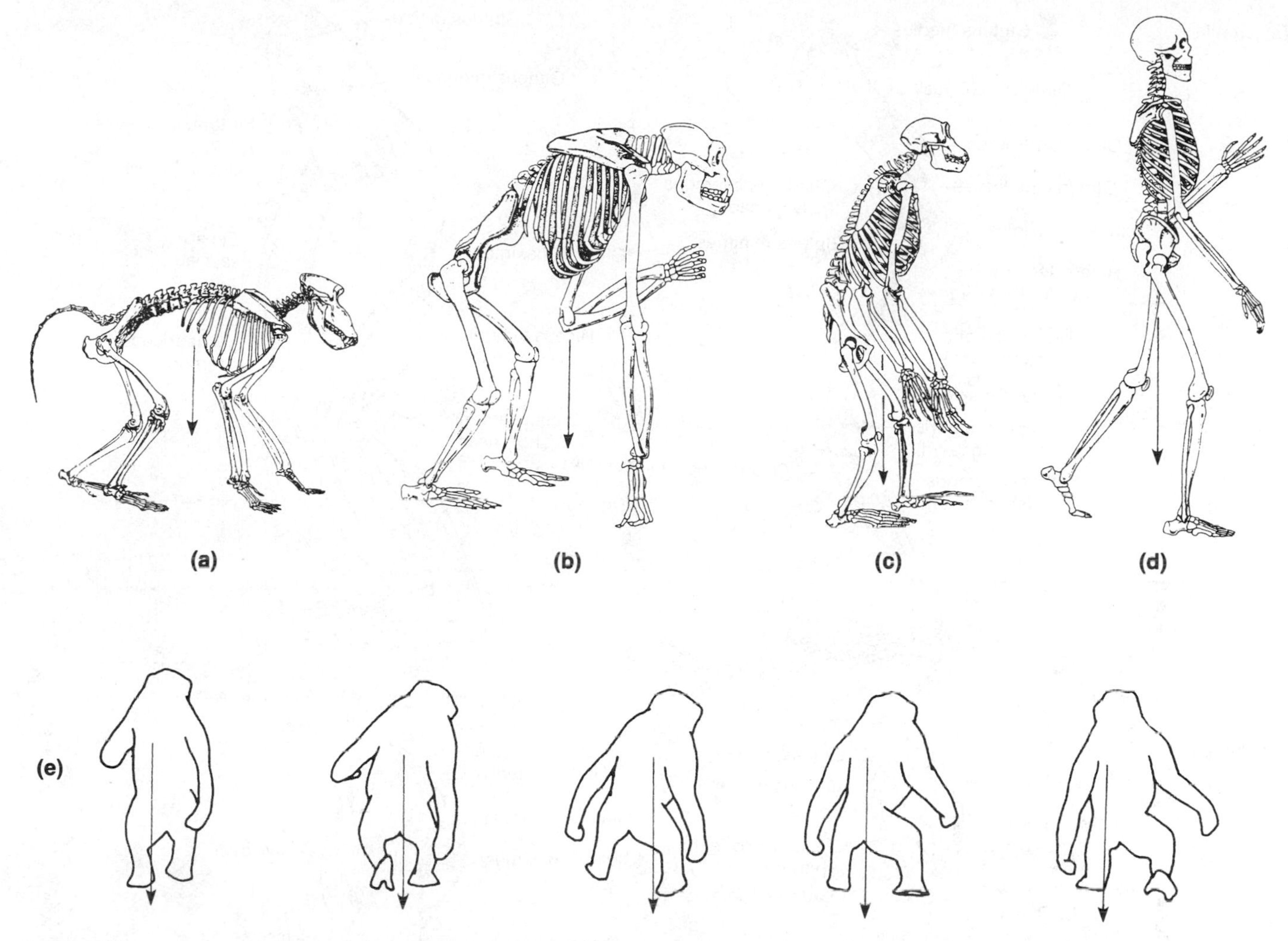

FIGURE 19-8 Centers of gravity (arrows) in the common stances of four ground-dwelling primates: baboon (a), gorilla (b), chimpanzee (c), and human (d). When humans walk bipedally the center of gravity remains in approximately the same vertical plane as the skeletal axis. When apes walk bipedally (e) the body tilts from side to side, causing shifts in the center of gravity that have to be corrected by muscular exertion.

useful in revolving the foot, such rotation has been reduced in humans since they walk and run on the ground, and the fibula is consequently reduced.

Walking has also involved changes in the foot to provide both upright balance and striding power. Balance is achieved by a three-point weight distribution between the heel and two points on the ball of the foot, the inner (first) and outer (fifth) metatarsals. Power is achieved by sequentially transferring weight from the heel to the fifth metatarsal to the first metatarsal to the big toe that serves as the "push off" (Fig. 19-10). In fact, the shape of the big toe alone is extremely helpful in determining whether an individual is capable of humanlike striding.

Taken together, these modifications allow sustained hominid walking to be accomplished without apelike bipedal shifting of the trunk from side to side and therefore with relatively minor expenditures of energy.

HOMO

The earliest fossils ascribed to the genus *Homo* have been found in both East and South Africa and date between 2.2 and 1.8 million years ago at the Pliocene-Pleistocene boundary. These early forms, first named *Homo habilis* by Louis Leakey (Fig. 19-11(a)), were about as short as *A. afarensis*, with males at 4½ feet tall and a recent fossil individual with a height of about 3 feet. Their cranial capacities, however, were at about 600 to 700 cc, pointing to a significant departure from the australopithecines. Associated with their fossils are artifacts, indicating clearly that these new hominids were engaged in making regularly patterned **stone tools**, products of the stone industry named Oldowan (Fig. 19-12(a)). The tools were apparently used in hunting and butchering animals, including small reptiles, rodents, pigs, and antelopes, and probably used also in scavenging carcasses of animals as large as ele-

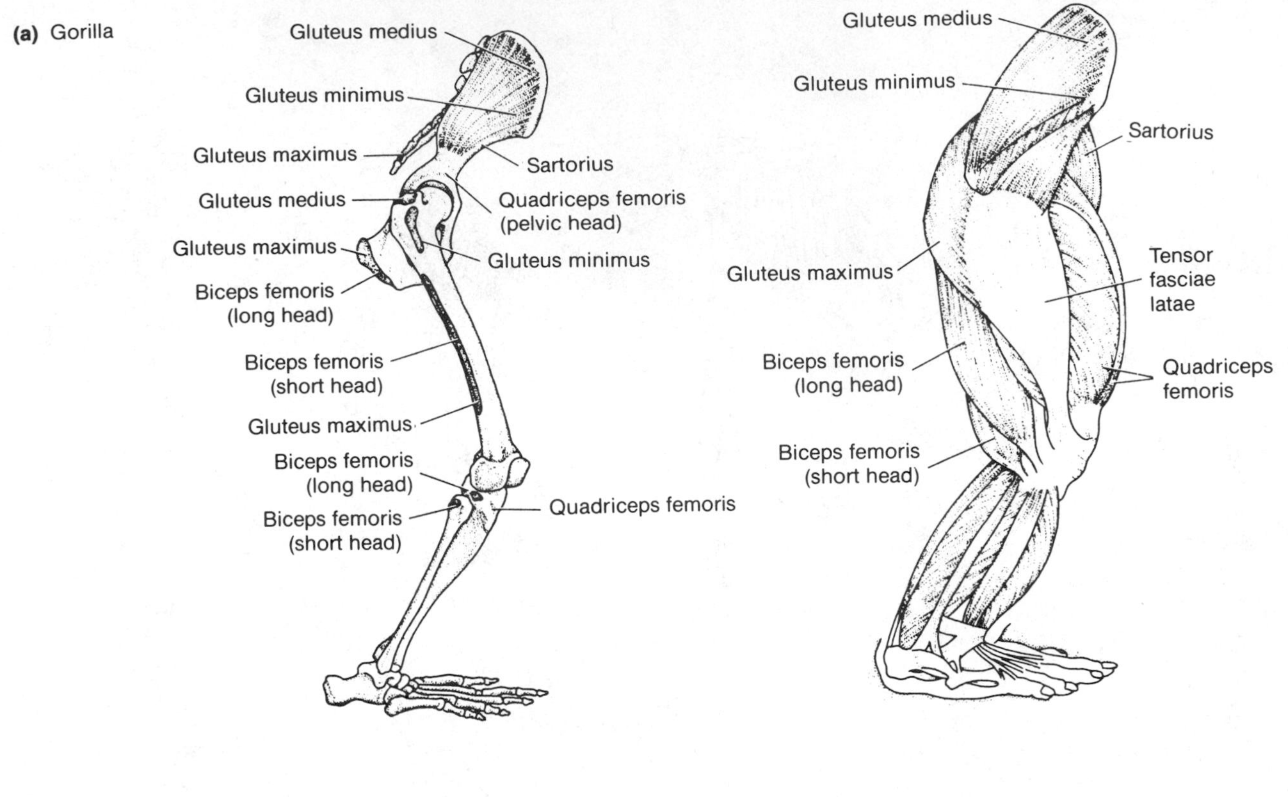

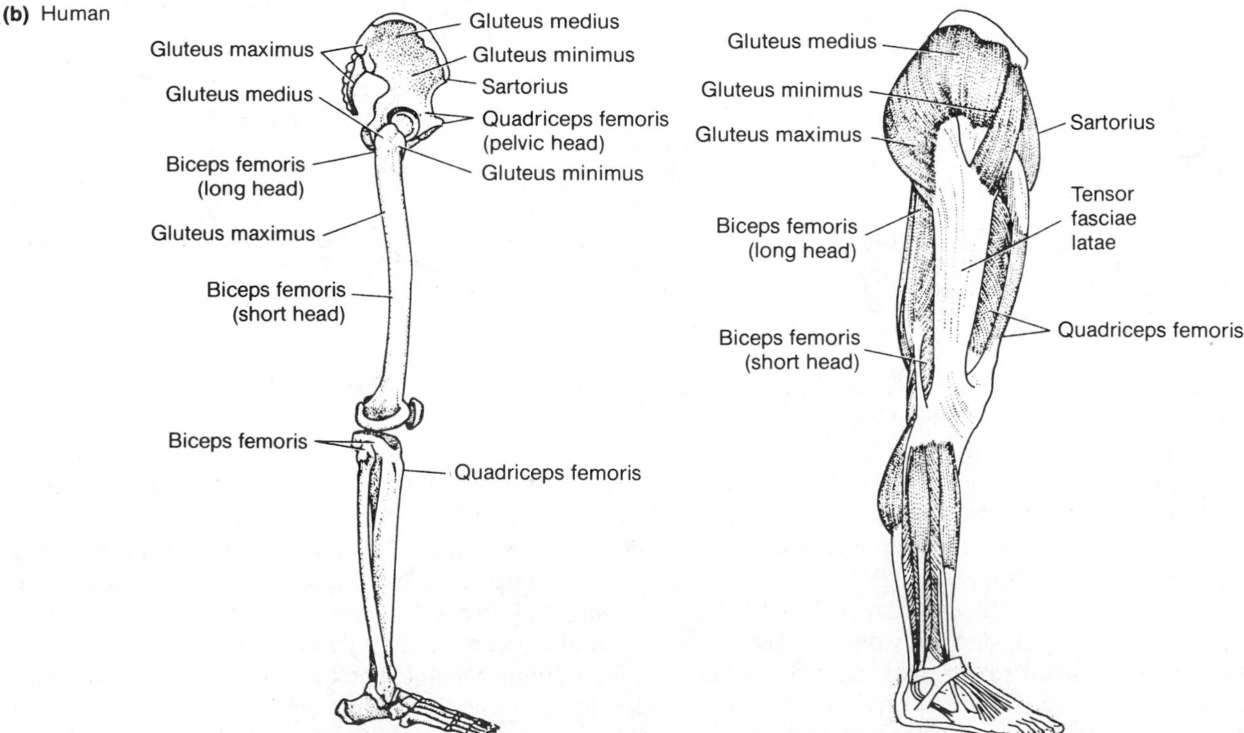

FIGURE 19-9 Bones of the hind limbs (left side) showing origins and insertions of the major muscles (right side) of a gorilla (a) and human (b). In both primates the muscles that cause the femur to swing forward in relation to the pelvis (flexion) are the sartorius and quadriceps. However, the broadened human pelvis provides leverage (pulling power) for these muscles when the individual is standing erect, whereas the gorilla can only gain such leverage in the bent position shown. Similarly, the human ability to swing the femur backwards (extension) can be accomplished in the erect position because the gluteus maximus muscle that extends the femur is attached to a rearward projection of the pelvis. Extension of the gorilla femur, on the other hand, is again dependent upon its bent position, utilizing the biceps femoris as an extensor. For side-to-side positioning of pelvis and femur during abduction (swinging the femur laterally outward in respect to the pelvis), the gorilla uses mainly the gluteus maximus muscle (converted to an extensor in humans), whereas humans rely on the gluteus medius and minimus. (From Wolpoff, 1980, adapted from Napier. See also Lovejoy, 1988.)

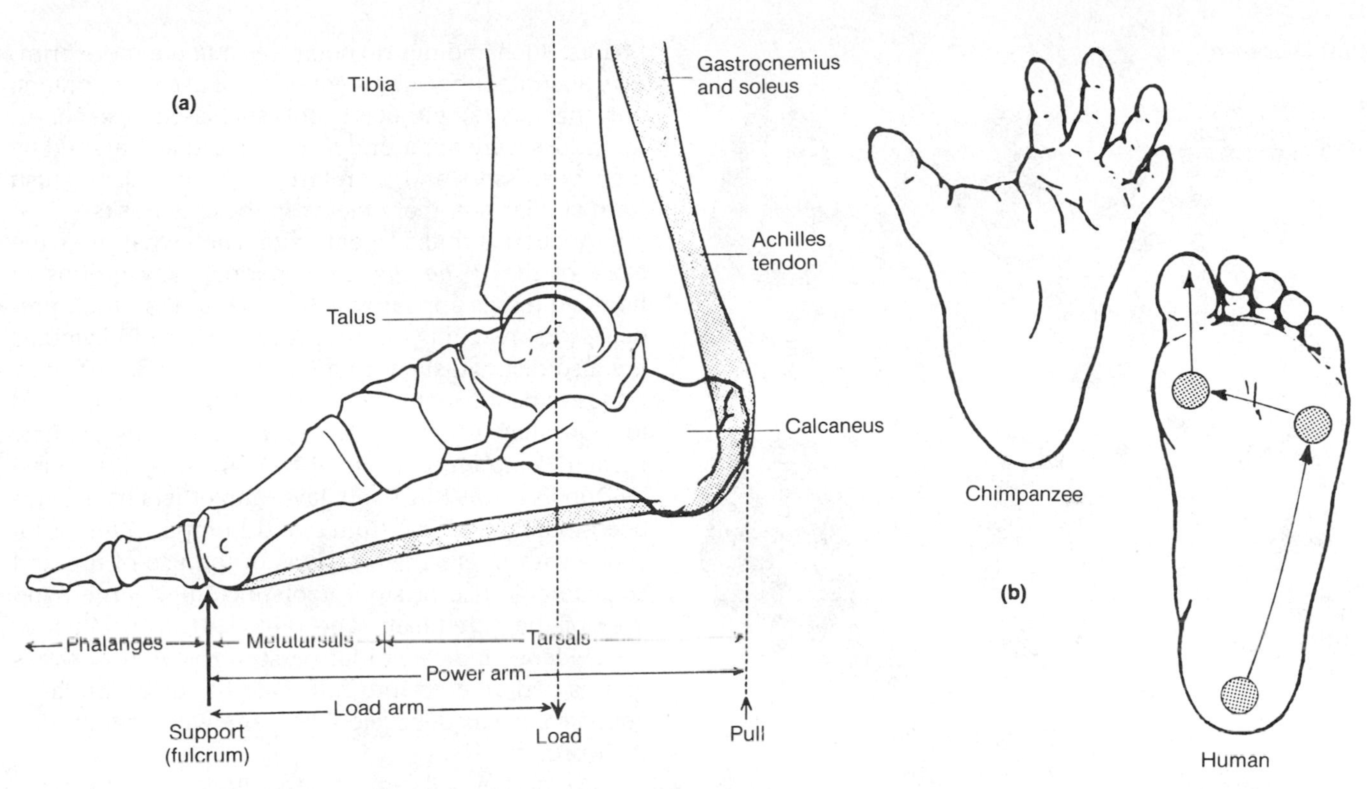

FIGURE 19-10 (a) Leverages in the human foot that provide propulsive forces for walking. Contraction of the gastrocnemius and soleus muscles in the calf of the leg pulls the Achilles tendon attached to the heel (calcaneus). This produces the power (power arm) that transfers the weight load (load arm) to the metatarsals. For light-bodied leaping primates the load arm is almost as long as the power arm, thus producing the springing jump, whereas in humans and heavier primates the load arm is relatively shortened, and power is increased at the expense of leaping. (b) Plantar view of chimpanzee and human feet. In the chimpanzee, the first metatarsal and its accompanying phalanx (big toe) are at a marked angle to the other metatarsals, the phalanges are relatively long and curved, and the foot can be used for grasping. In humans the phalanges are reduced, and the more robust first metatarsal and big toe are set parallel to the others, thus enhancing the ability to walk or run directly forwards. Note that although the human foot is narrow, it acts as a tripod upon which weight is stably distributed (indicated by the three circles), the arrows show how weight is transferred between these three centers to the "push off" on the big toe. (Adapted from Campbell.)

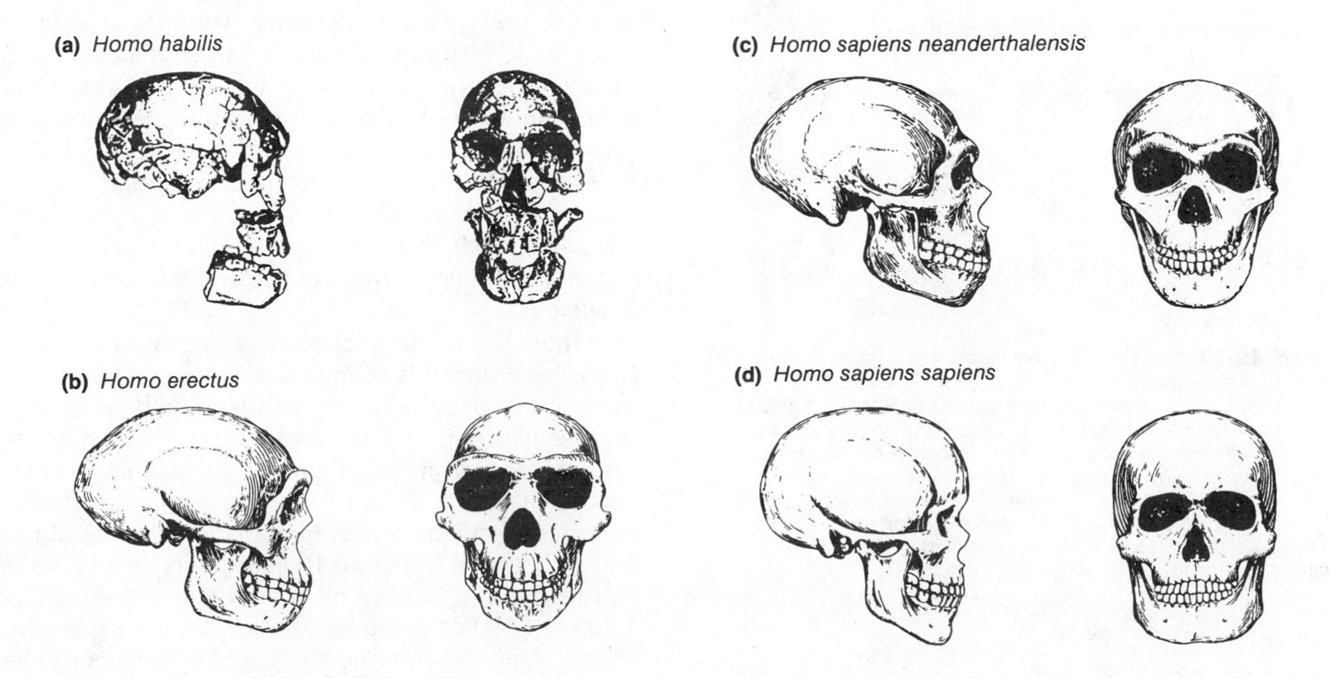

FIGURE 19-11 Comparisons among hominid fossil skulls. (Adapted from Johanson and Edey.)

(a) Oldowan

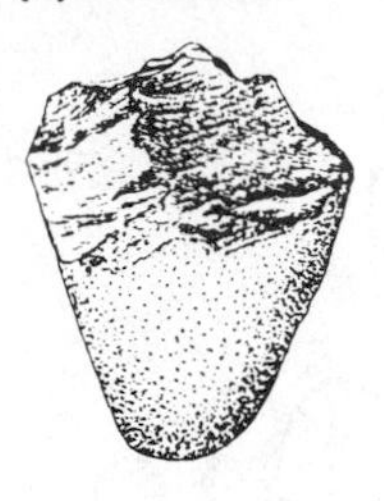

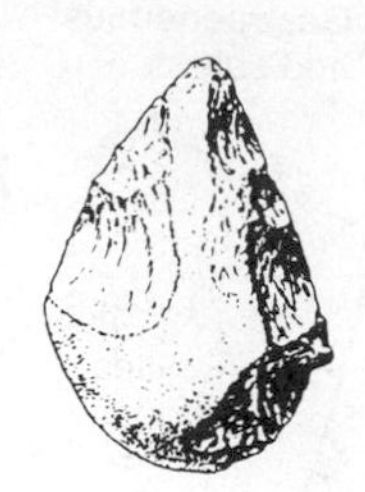

(b) Acheulean

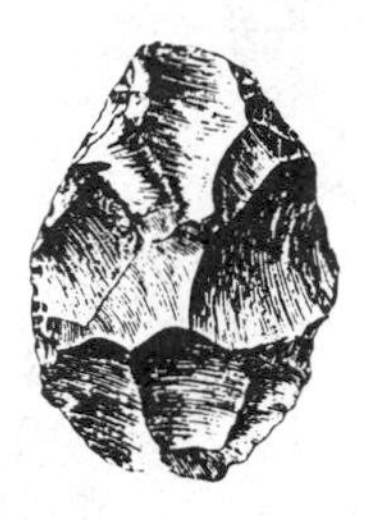

(c) Mousterian

(d) Aurignacian and upper Paleolithic

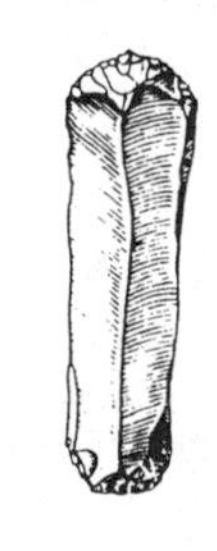

FIGURE 19-12 Stages in stone tool development beginning with the Oldowan stone industry (a) which is believed to have been associated with *Homo habilis*. Later stages were associated with other groups: (b) Acheulean (*Homo erectus*), (c) Mousterian (*Homo sapiens neanderthalensis*), (d) Aurignacian and Upper Paleolithic (*Homo sapiens sapiens*). These industries and cultures constitute the anthropologists' Paleolithic Stone Age. Following this period are the Mesolithic and Neolithic ages, the latter beginning about 10,000 years ago and marked by polished stone tools, pottery, domesticated animals, cultivated plants, and woven cloth.

phants. Such findings do not imply that the more primitive australopithecines were not tool users or hunters, only that australopithecine tools such as stones, bones, and sticks were apparently simpler and less altered by their possessors and therefore difficult to distinguish from similar non-tool objects in these deposits.

About 1.8 million years ago, somewhat after the close of the *H. habilis* fossil period, new groups of hominid fossils appear which are taller than their predecessors, reaching about 5½ feet. These individuals are also distinguished by thicker skulls, heavier brow ridges, smaller teeth, and larger brain volumes (750 to more than 1,000 cc). The first of these fossils, now named *Homo erectus* (Fig. 19-11(b)), was discovered by Dubois in 1891 in Trinil, Java, and others have since been found in Africa, China, and Europe. Often associated with *H. erectus* are signs of the use of fire and considerable use of stone tools including large hand axes of the Acheulean type (Fig. 19-12(b)). Much of the evidence indicates that most, if not all, *H. erectus* groups had entered into full-scale hunting with large animals such as deer, elephant, and wild boar among their prey.

Were they all known, the various distinctions among different *H. erectus* groups over time would probably be sufficient to mark off new evolutionary levels and perhaps even new species. These distinctions, however, encompass anatomical and behavioral traits that are not always perceptible on the fossil level, and it is therefore not yet clear at which points separations or transitions occurred. The distinctions are even more blurred between *H. erectus* and *Homo sapiens*, our own species. For example, hominid fossils found near the Solo River in Java, dated to less than 250,000 years ago, show brain volumes averaging 1,100 to 1,200 cc, significantly larger than those of middle Pleistocene *H. erectus* fossils from the same area; yet they are like older fossils in respect to prominence of brow ridges and some other features. European fossils (from Swanscombe in England, Steinheim in Germany), dated to about 200,000 years ago, also show such increased brain volumes as well as anatomical traits intermediate between *H. erectus* and *H. sapiens*.

Thus, if it is true that *H. sapiens* evolved from an *H. erectus* group, this change is not clear-cut but seems to be marked by gradual changes found in various intermediate populations. What is known is that by the time the European Neanderthals appear (*H. sapiens neanderthalensis*, Fig. 19-11(c)) along with similar types found in Zimbabwe (Africa) and Shanidar (Iraq), that is, between 50,000 to 100,000 years ago, humans had reached their present brain volume averages of 1,300 to 1,500 cc. Although these types are somewhat shorter than modern humans and show distinctive brow ridges, large jaws, small chins, and some other anatomical relics, they were socially and behaviorally quite advanced in many respects: they were apparently skillful hunters of large animals such as the cave bear and mammoth; they produced many complex stone tools; and they apparently performed ritualistic social ceremonies, including placing flowers in graves of their dead. Many anthropologists, although not all (see p. 407), feel that the Neanderthals probably deserve as full a membership in *H. sapiens* as the higher-skulled races, that is, *H. sapiens sapiens* (Fig. 19-11(d)) that began to replace the Neanderthals in various parts of the world about 40,000 years ago.

So far, among the earliest fossils of the more modern humans are those found in Mount Carmel in Israel, dated to about 90,000 years ago (Stringer et al.). Other transitional forms have been found in other localities in the Middle East, all associated with what is called the Mousterian stone age industry (Fig. 19-12(c)). Since the Mousterian culture is also associated with the Neanderthals, there must have been some important cultural overlap between these various *H. sapiens* groups. In any case, by about 35,000 years ago, an era named the Upper Paleolithic (last part of the Old Stone Age) began in Europe, characterized by new methods of flaking flint to form stone tools. These cultures, among which the earliest was the Aurignacian (Fig. 19-12(d)), were accompanied by a marked change in human fossils. The Neanderthals were apparently then replaced by new Cro-Magnon types in which the brow ridges were reduced, the skull vault was higher, and the face was smaller and less prognathous. Many examples of representational art now appeared, painted on cave walls and sculpted in clay or bone. In essence, modern humans had arrived in Europe and elsewhere, and these new forms can be considered among the present human races.[4]

Figure 19-13 shows one possible phylogenetic scheme which broadly traces relationships among various known hominid fossil groups, beginning with *A. afarensis*. Although exact transitional events between these groups remain unknown, the end point of this phylogeny—relations among presently existing human races—can be analyzed by molecular techniques discussed in Chapter 12. One such technique, phylogenies determined by differences between mitochondrial DNAs (mtDNA), has been used by Cann et al. to indicate that the ancestral mtDNA sequence from which all modern human mtDNA sequences derive probably had an African origin between 140,000 and 290,000 years ago.

The advantages of using mtDNA include the fact that it is a circular molecule, 16,569 base pairs long, whose complete nucleotide sequence is known. Also, it is inherited only through the maternal lineage as a haploid unit (males do not pass on their mitochondria) that does not recombine either with other mtDNA or with nuclear DNA. Therefore, unless modified by mutation (usually by single nucleotide substitutions), an mtDNA molecule remains unchanged from one generation to the next, is homogeneous within an individual, and can be isolated from human tissues such as the placenta and red blood cells (there are about 10^{16} molecules of mtDNA per individual). Furthermore, in contrast to nuclear DNA, mtDNA evolves about ten times faster in vertebrates, enabling comparisons between groups which would be more difficult to differentiate if slower and more complexly evolving nuclear sequences were used. More important, when mtDNA sequences are analyzed with restriction enzymes (p. 233), differences between individuals can be traced that help determine their evolutionary relationships.

As shown in Figure 19-14, the evolutionary tree of human mtDNA begins with a single ancestral sequence in Africa (*a*),[5] and its nine major descendant sequences (*b-j*) were then dispersed to races in Africa and other geographical regions in which further branching occurred. Since any geographical race outside Africa can be traced to more than one unique mtDNA branch, there were apparently numerous colonizations of each area by females carrying their particular mtDNA sequences. For example, Cann et al. suggest that mitochondrial genomes in their sample of New Guinea Highlanders had seven different maternal origins, most from Asia and the remainder probably from Australia.

[4]So far, little is known about the abrupt disappearance of the Neanderthals; some of their populations may have died out, whereas others may have been absorbed into the new dominant forms. One hypothesis suggests that the Neanderthals represent a separate offshoot of the human line, differing from both the *Homo sapiens* groups that preceded it and those that followed. According to Rak, the extraordinarily large Neanderthal face was the result of biological innovations that allowed strong biting forces to be exerted on their front teeth, which were much larger and had deeper roots than in other *Homo sapiens* groups. The fact that their incisors and canines often show heavy wear indicates that the Neanderthals may have used these teeth for processing tough foods or hides, or both. Presumably, the expanded nasal chamber produced by their large face may have served as a radiator, warming and humidifying inspired air in the dry, cold, glacial climates that many European Neanderthals inhabited.

[5]Although Cann et al. propose that all of our mtDNA can be traced to a single ancestral mitochondrial "Eve," we should keep in mind that her contribution of nuclear DNA (the major component of heredity) has been greatly diluted by the sexual process: in the absence of inbreeding (Chapter 20), an individual inherits only half its nuclear DNA from each parent, one quarter from each grandparent, one sixteenth from each great grandparent, and so on. The mitochondrial "Eve" may therefore have contributed relatively few of the 3 billion nucleotides in our nuclear genome. Furthermore, it is possible that there may have also been more than one mitochondrial "Eve" in the founding human population, since many females with *a*-type mtDNA sequences (Fig. 19-14) could have been present at the time the founding event occurred. Once phylogenies of the Y chromosome are obtained, it will be of considerable interest whether such Y-chromosome sequences, inherited only through males, show common paternal ancestors ("Adams?").

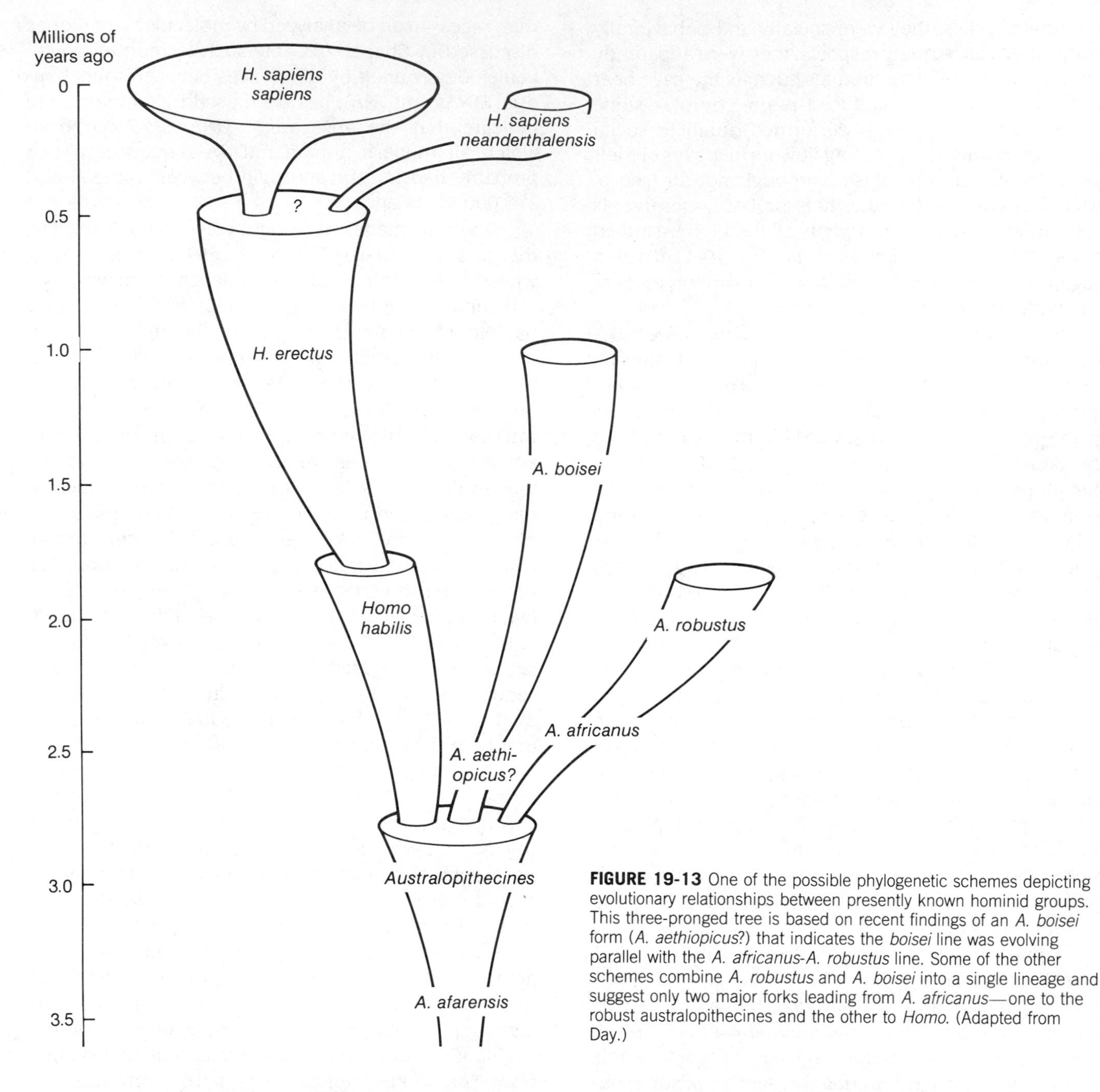

FIGURE 19-13 One of the possible phylogenetic schemes depicting evolutionary relationships between presently known hominid groups. This three-pronged tree is based on recent findings of an *A. boisei* form (*A. aethiopicus*?) that indicates the *boisei* line was evolving parallel with the *A. africanus-A. robustus* line. Some of the other schemes combine *A. robustus* and *A. boisei* into a single lineage and suggest only two major forks leading from *A. africanus*—one to the robust australopithecines and the other to *Homo*. (Adapted from Day.)

Although the timing of such migratory events is still unclear, it is possible to estimate the rate at which mtDNA sequences diverge using measurements of differences between mtDNA sequences whose common ancestry can be approximately dated. According to Cann and co-workers, this divergence rate is most likely between 2 and 4 percent nucleotide change per million years in vertebrates, thus giving the dates shown in Figure 19-14. In general, these findings support fossil evidence indicating that archaic forms of *Homo sapiens* were present in Africa between 100,000 and 200,000 years ago (Clarke; Stringer and Andrews). From these beginnings, *Homo sapiens* radiated outward to different localities and differentiated into the various races, apparently replacing the indigenous races of *H. erectus*. This view is supported by the fact that all *Homo sapiens* mtDNA sequences derive from a single African sequence, and there is no evidence of hybridization with other, more primitive native hominid forms (such as Asian *H. erectus*) which probably had much different mtDNA sequences.

HUNTING HOMINIDS

Apart from finding fossil and biochemical evidence of hominid origins, the concern of many anthropologists is also to understand how past environments affected

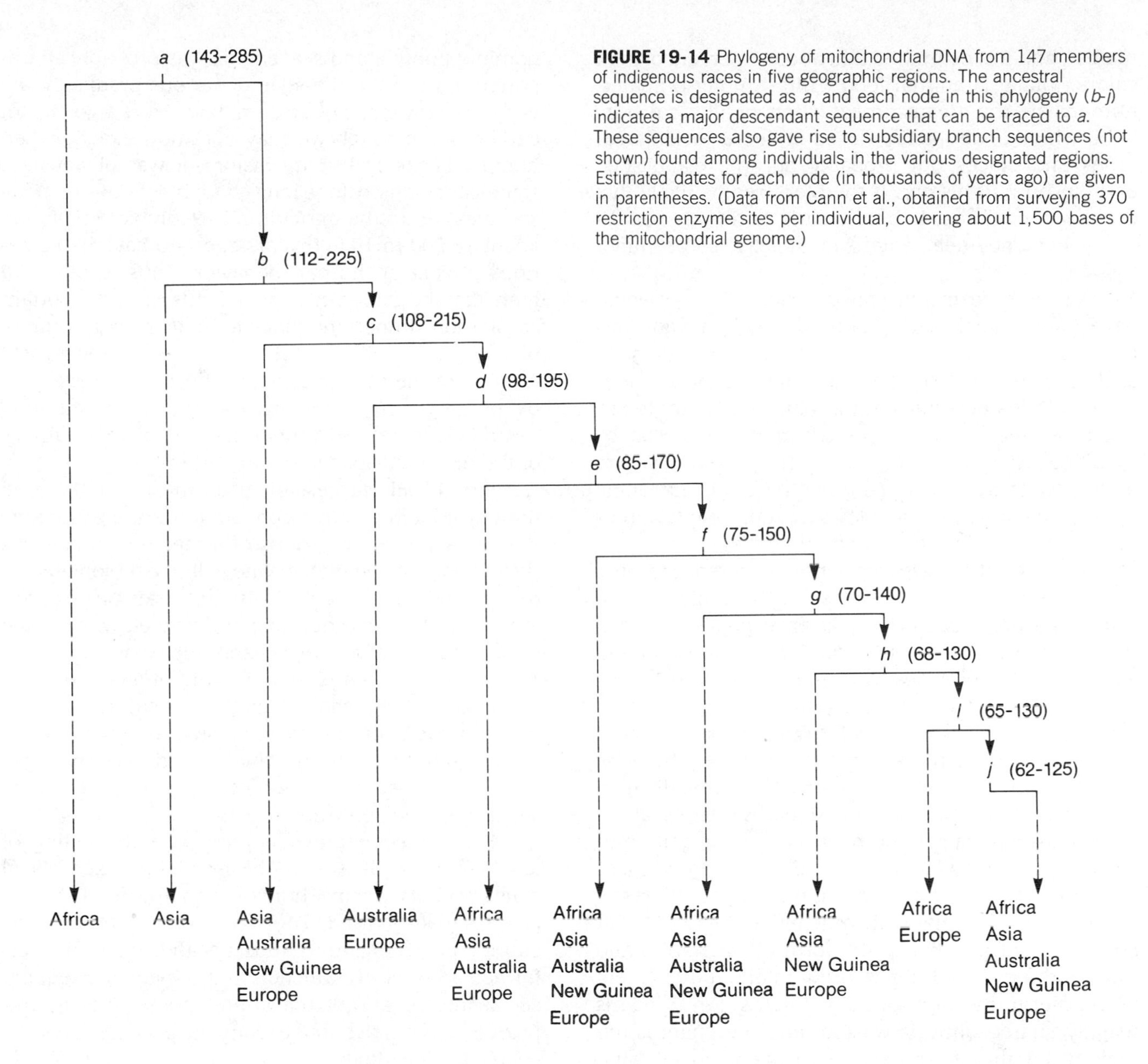

FIGURE 19-14 Phylogeny of mitochondrial DNA from 147 members of indigenous races in five geographic regions. The ancestral sequence is designated as *a*, and each node in this phylogeny (*b-j*) indicates a major descendant sequence that can be traced to *a*. These sequences also gave rise to subsidiary branch sequences (not shown) found among individuals in the various designated regions. Estimated dates for each node (in thousands of years ago) are given in parentheses. (Data from Cann et al., obtained from surveying 370 restriction enzyme sites per individual, covering about 1,500 bases of the mitochondrial genome.)

and selected among various human traits, and how humans, in turn, affected and selected their environments. As discussed previously, the introduction of bipedalism and the home base, however they occurred, would have had profound effects on the kinds of environments that these early hominids were able to exploit. They could now move from forest to savannah with greater ease than ever before and cover much larger areas in their search for food. Food resources in the savannah, however, are different from those in the forest, and the impact of this new environment placed emphasis on a host of new behaviors and adaptations.

The relatively low rainfall in the savannah provided fewer high-quality plant foods than in the forests and made the distribution of such resources patchy, that is, present in some places and not in others. These resource irregularities, combined with plant seasonality, would have engendered further selection for increased hominid mobility, broad dietary habits, and flexible strategies in searching for food. Since the savannah grasslands also supported various herbivores including migratory herds of large mammals, a selective advantage for **meat eating** may have appeared fairly early, including strategies for both avoiding and successfully competing with the large predators that preyed upon these mammals.

Certainly, a most important change in the life-style of these hominids would have been an increase in the relative amount of meat in their diet. **Animal hunting** is not a novel trait in primates, and numerous instances have been recorded of baboon and chimpanzee groups engaged in purposeful hunting of animals smaller than themselves (Harding and Teleki; Goodall). Various baboon troops are known to hunt small ungulates, other primates, and hares. Practically all chimpanzee groups studied to date engage in hunting (mostly other primates), and it is estimated that about 3 percent of their caloric intake is animal food (Hill). Among primates, humans are the greatest meat eaters of all—a dietary habit which undoubtedly varied in degree at different times and places but probably became established early in human history. It therefore seems likely that even in their forest habitats, early hominids had already become meat eaters to at least some minor degree.

An increase in the consumption of meat would have offered such early hominids many advantages: (1) meat is a rich source of essential amino acids such as lysine, tryptophan, and histidine; (2) it provides more calories per unit weight than most plant foods; (3) it is either packaged (small animals) or can be modified by cutting and tearing (large animals) into units easy to transport to a home base; (4) the killing of only one large animal allows a group of individuals to be fed, often for more than one day.

Although there are undoubted differences, some idea of early hominid life-styles may be obtained from **hunter-gatherer societies** that have persisted to the present day in places such as Central Africa (Mbuti Pygmies), South Africa (Kalahari Bushmen), South America (Yanamamo Indians), and Australia (Aborigines). These consist of social groups or bands where males are usually the hunters and females the plant gatherers. Since their omnivorous diet depends upon highly variable plant and animal food sources that are often seasonal, each band moves about several times a year over fairly wide ranges to different home bases or settlements. Among some groups of the Kalahari Bushmen, it is estimated that the females gather about 60 percent of the diet in the form of vegetables and fruit, and the male hunters bring in about 40 percent of the diet in the form of animal game. Since food is usually shared, little is wasted, and the chances are maximized that some will be available to all band members. As Silberbauer points out, whatever the proportion of meat in the diet, hunting is a "prestigious activity.... The [Bushmen] are hungry for meat, and any description of the 'good life' always includes mention of a plentiful supply of it."

Unfortunately, information on the proportions of plant and animal foods in the diets of ancient hominids is still unobtainable, especially since many plants that may have been used by such hominids have not been fossilized or are poorly preserved. Nevertheless, bone accumulations found in conjunction with traces of hominid settlements at East African sites indicate that hominid hunting and scavenging was probably an important part of the life-style of various groups by the early Pleistocene, about 2 million years ago, if not earlier. So, although we may not know exactly when hunting began in human history, it was obviously a significant industry for a long enough period—in many societies, up to the agricultural (Neolithic) revolution about 15,000 to 10,000 years ago—to have had a serious impact on human behavior. Thus, even if we grant that the gathering of plant foods was as important as or more important than hunting in early human history (Tanner; Shipman), it seems reasonable to ask, what were the selective forces and effects generated by hunting? As Tooby and DeVore stress, hunting "would elegantly and economically explain a number of the unusual aspects of hominid evolution."

First of all, successful medium and large game hunting requires active cooperation among hunters. We see this even in groups of foraging chimpanzees (usually two to five males) who will tree a monkey and then cut off its escape by assuming strategic positions around it.[6] Hominid hunters, empowered with simple weapons such as wooden spears, clubs, and hand axes, probably used such techniques and others, including tracking, stalking, chasing game into cul-de-sacs and swamps, over cliffs, or into ambushes, or by continuing the chase until the animal tired. Utilizing **cooperative** means, larger animals could be brought down than by single hunters alone.

Second, cooperative hunting and the killing of large animals placed emphasis on increased social cohesion both during hunting and the food-sharing process that followed. Transfer of information in executing successful hunts thus became vital and performed a necessary function in the many subsequent social interactions of the entire group. That is, improved communication became especially advantageous as individuals assumed more complex roles in planning, hunting, helping, food gathering, food sharing, infant care, child training, and other vital activities.

Third, successful hunting emphasized the perception and retention of information on migratory pathways, watering sites, and home base settlements, whose geographical positions extended over home ranges (regions habitually occupied by a group) probably greater that those occupied by most other primates or carnivores (Foley). The experiences and observations of hominid hunters thus had to be mentally dissected into component geographical and ecological features, prey behaviors, weather effects, and seasonal

[6]Goodall lists more than 200 observed incidents where red colobus monkeys were caught and/or eaten by Gombe chimpanzees near Lake Tanganyika, in addition to cannibalism and the capture and consumption of many other mammals.

changes, then stored and synthesized into communicable mental maps that enable prediction, planning, and modification.

Fourth, hominid hunting involved stresses that would have fostered increased locomotory adaptations such as persistence in the chase (humans can continue jogging for distances which are generally longer than many large animals can continue running), maneuverability in the kill, and long-distance traveling to or between home bases while carrying heavy burdens. Various writers have also pointed out that the need for increased diffusion of metabolic heat during these pursuits would have selected for the loss of body hair and increased numbers of sweat glands, features that among primates are unique to humans.

Fifth, the technological skills necessary for a clawless, canineless hominid to capture and butcher large prey led to a variety of snares, weapons, and tools, including the stone implements that date back at least to *H. habilis* (see Fig. 19-12). Such technologies, especially evident in the fossil tools, involve the shaping of material according to some preconceived notion of what its appearance should be after the process is completed. These toolmaking skills therefore involve not only manual dexterity, hand-eye coordination, and considerable concentration, but also the ability to plan and visualize an object which is not apparent in the raw material from which it is created. The final form of a tool must be conceptualized in its three dimensions, and such concepts had to be implemented by mastering a series of techniques. These included finding and recognizing appropriate, workable stones in outcrops that were often widely dispersed, carrying these stones back to a base, and shaping them into tools by a sequence of precise strokes. The considerable mental abilities used in toolmaking had to be supplemented with social and communicatory abilities enabling the toolmakers to transmit such skills to other individuals who could continue the industry.

Finally, hunting placed further social emphasis on the home base to allow food exchange among foraging subgroups, particularly when the food supply was irregular, as it often is in hunting. A home base also has value for nursing and pregnant females who could not always or easily cover the long distances necessary for large-scale hunting. The home base would thus have become a center for food sharing, shelter, hunting preparation, sexual bondings, child care, and other social exercises, in which all members had to be tied together by communicatory skills.

However, like the origin of bipedalism, the role of hunting among human ancestors has been considerably disputed, especially since the extent of its practice in early human groups is unknown (Harding and Teleki). Nevertheless, from what we can surmise from present hunting-gathering groups, and even from individuals in more modern societies who engage in hunting as a sport, the practice of hunting, whatever its role, was probably reinforced in various emotional ways: by the pleasures of seeking out and subduing prey; by the satisfactions of mastering the physical skills necessary for efficient aiming, throwing, and grappling; and by mastering the intellectual skills used in devising cunning offensive and defensive strategies. These behaviors arise early in human development, especially in play among juveniles and adolescents, and their perfection in adults has been socially approved and rewarded in every known historical culture. The fact that hunting is no longer necessary in most modern societies has not lessened interest in these behaviors, and athletes (and "warriors") who develop such skills are often greatly esteemed.

COMMUNICATION

Communication is the means through which a stimulus from one individual can trigger a response in others. Communication methods may include signals transmitted through any of the sensory channels: scent, touch, vision, and sound. Practically all animals which interact with each other make use of one or more communicatory methods, but they are especially well developed in social animals where information is essential in providing cues to other individuals about factors such as food sources, predator encounters, territorial boundaries, sexual readiness, social ranking (dominance), and emotional states. Examples of these can be found throughout the primate order (Jolly).

For instance, various prosimians and monkeys use urine or scents emitted by special glands to mark trails and territorial boundaries. **Olfactory cues** are also commonly used to attract sexual partners and signal the onset of ovulation. Such communication has its counterpart in humans who emit odors from their axillary and genital regions. Although these odors are now generally washed off or disguised by deodorants and perfumes in Western culture, tests have shown that many people can use such body scents to distinguish between the two sexes as well as among individuals.

Tactile communication assumes its most common primate form as grooming, or fur cleaning with fingers, lips, and teeth. It is one of the most obvious and frequent kinds of interaction observed in many mammalian groups and seems to serve as the main social cement that binds pairs of individuals or group members together. In chimpanzees grooming is supplemented by other tactile behaviors such as holding hands, patting, embracing, and kissing (Goodall). Since humans have relatively little fur, the traditional form

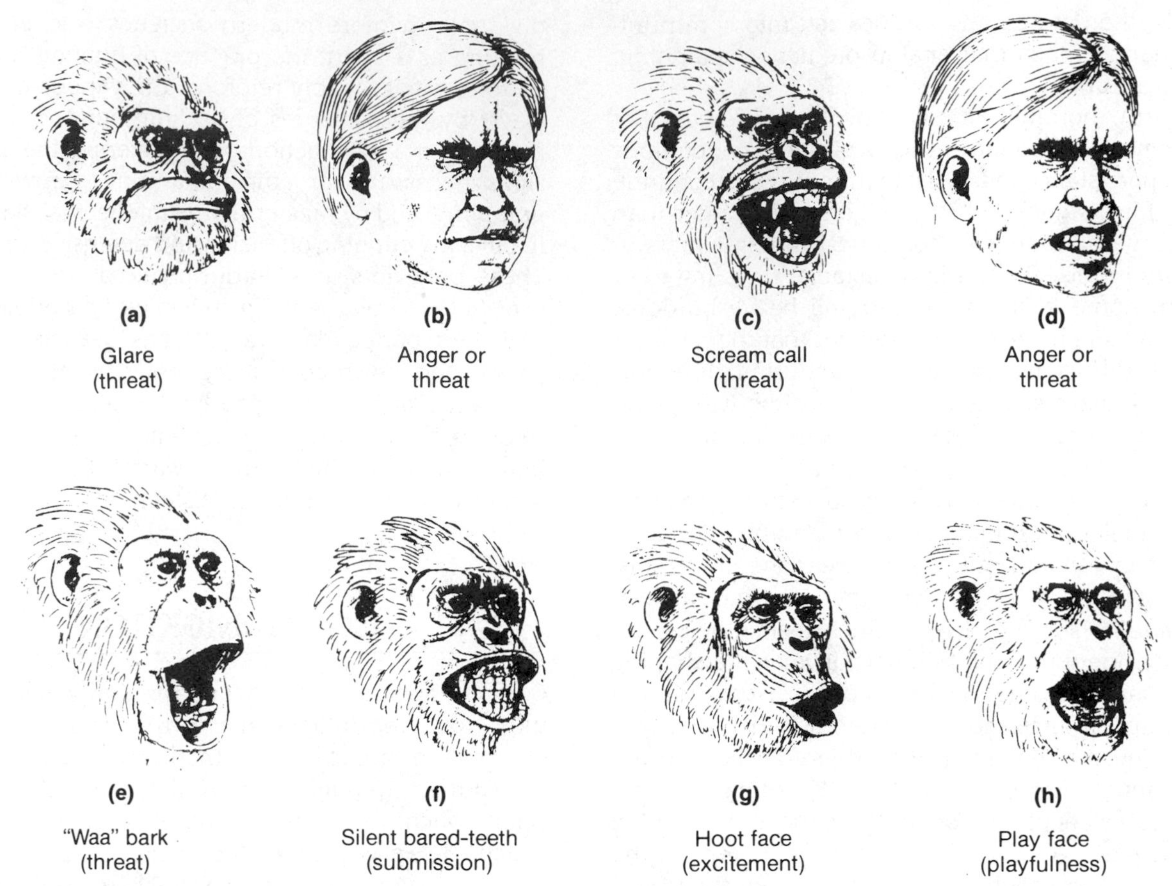

FIGURE 19-15 A small sampling of chimpanzee facial expressions indicating various emotional states, along with two seemingly related human expressions in (b) and (d). (Adapted from Chevalier-Skolnikoff.)

of primate grooming is excluded, and it is little surprise that tactile social reassurances are mostly confined to other tactile behaviors.

On the **visual** level, primate signals may include physical gestures or anatomical displays such as the postures, genital swellings, and colorations used to signal sexual receptivity. As shown in Figure 19-15, facial expressions may be quite varied and are easily visible in hairless faces. Some of these expressions, such as the glare and scream call (Fig. 19-15(a) and (c)), probably signal threat messages throughout the primate order and can be considered to mark an aggressive attitude even in human cultures (Fig. 19-15(b) and (d)).

Compared to visual displays, **vocalizations** have the advantage that they go from mouth to ear and leave the hands and body free for other activities. Oral sounds also have the advantage of providing feedback by allowing the vocalizer to hear its own vocalizations and thereby evaluate and control them while (or perhaps even before) they are uttered. Moreover, although sound fades rapidly, it can be transmitted over long or short distances, in all directions, even around obstacles which would interfere with visual communication. In general, because sound leaves no record, the message in most primates is usually short and simple, indicating, for example, predator alarms or territorial calls. On the other hand, some primate vocalizations have a variety of gradations, each providing a somewhat subtle meaning. Thus, Japanese macaques use particular variations of the "coo" sound in specific situations, such as a male separated from the group, females contacting their young, dominants contacting subordinates, and females in estrus (Green and Marler).

Other primate vocalizations not only reflect the emotional state of the vocalizer but also direct attention to specific external events. A prominent example is that of three different alarm calls given by vervet monkeys, each designating a specific kind of predator: (1) the "rraup" is emitted upon detecting a hawk and causes the troop to look up and then seek cover in lower branches; (2) the "chirp" is given on seeing a mammalian predator and causes the troop to ascend to the forest canopy; (3) the "chutter" is used when detecting a snake, and the troop may then adopt aggressive positions on the ground. Each of these calls is therefore **symbolic** in the sense that it denotes an object which has no direct relationship to the call itself (e.g., a chirp is not a leopard) and may be considered a primate preadaptation for human communication in which the sounds of words do not correspond to their meanings.

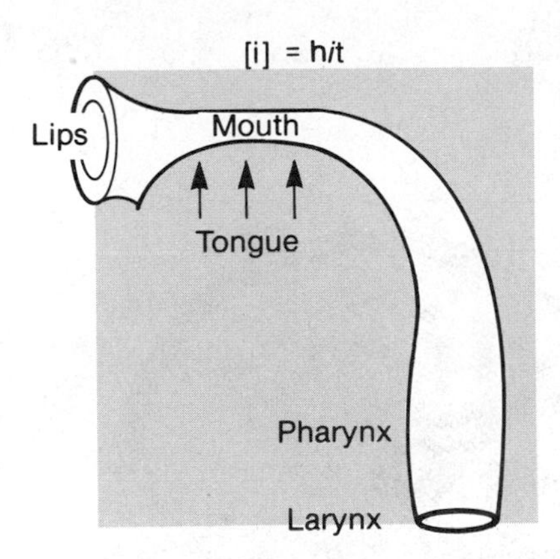

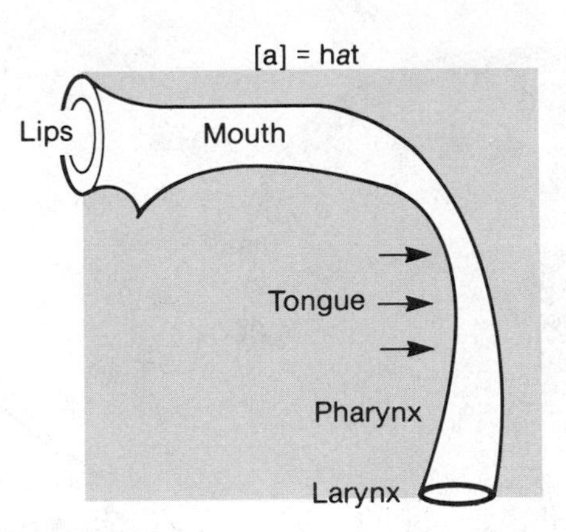

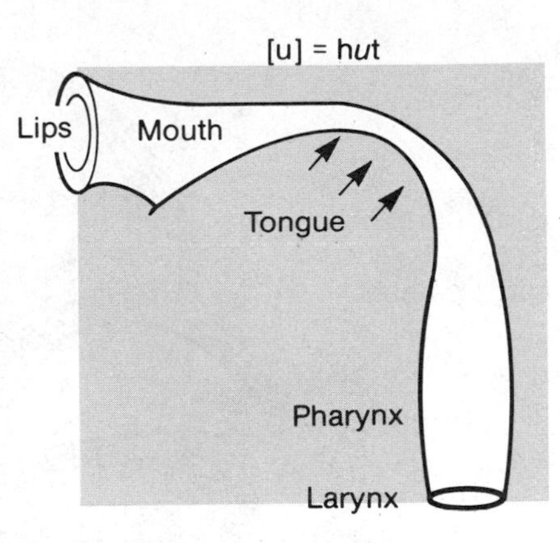

FIGURE 19-16 Diagrammatic views of how three vowel sounds are produced in adult humans by positioning the tongue (arrows) in different parts of the oral airway. Note that a sharp bend (formed by the hard palate above the mouth and rear wall of the pharynx) partitions this airway into right-angled mouth and pharyngeal sections that are essential for these vowel sounds. By contrast, the vocal airways of chimpanzees and newborn human infants are shorter and primarily in the shape of a slightly curved tube, making these vowels much less distinct.

SPEECH

The most symbolic of primate vocalizations is, of course, human **speech** and **language**. Here, new characteristics have been introduced through a wide range of different-sounding syllables that can be strung together in various ways to provide a vocabulary of different meanings (words). Compared to using only a sequence of sounds limited to single tones, human speech provides a rapid means of communication. For example, a sequence of dots and dashes, as used in the Morse code, can be interpreted at a rate that is often much less than 50 words a minute, whereas a sequence of spoken syllables can often easily be understood when delivered at a rate of 150 words a minute.

As in other mammalian vocalizations, it is the **larynx**, located in the upper part of the tracheal tube, which provides the basis for speech. Its origin, however, is not connected with sound but stems back to early air-breathing fish. These ancestral vertebrates, not unlike the lungfish of today, opened a valve in the floor of the pharynx to help swallow air into the lungs when out of the water and closed the valve when in the water. As selection for air breathing continued in terrestrial vertebrates, this laryngeal valve became imbued with fibers and cartilages that provided more precise control over laryngeal dilation and closure and allowed increased amounts of air to be aspirated when necessary. Thus, like so many other evolutionary features, the ability of the larynx to generate oral sounds was a preadaptation of an organ originally used for a different purpose.

In producing sound, the larynx acts like a woodwind reed that controls vocal pitch by opening and closing rapidly so that expired air from the lungs is interrupted to form puffs: the greater the frequency of puff formation, the higher the pitch. However, in order to produce the vowels of human-like speech the laryngeal puffs must pass through a tubelike airway (the pharynx) whose length and shape determine the eventual frequency patterns that are emitted and thus the quality of the different vowel sounds (Fig. 19-16).

This basic mode of vocalization, via a laryngeal-like output and a supralaryngeal "filter," is used by all terrestrial, air-breathing animals that produce oral sounds, from frogs to mammals. In addition, there seem to be neural auditory units in each such animal that respond with maximum sensitivity to specific ranges of frequencies. It is these specific neural sensitivities which allow bullfrogs, for example, to respond to the mating calls of their own species and not to those of others. In humans, there seem to be special neural brain circuits that can perceive and identify various categories of sound combinations and thus distinguish between different kinds of spoken syllables, an ability which has apparently evolved from primates with more limited powers of distinction (Lieberman).

Figure 19-17(a) shows a diagram of the adult human upper respiratory tract, with its sound-producing airway which begins at the larynx and proceeds through the pharynx and mouth. Note that compared to that of the chimpanzee (Fig. 19-17(b)), the adult human mandible extends forward for a relatively shorter distance. Among the consequences of reduced mandibular size[7] and lower positioning of the human

[7]Excessive tooth crowding is apparently one of the biological costs in reduction of the mandibular body.

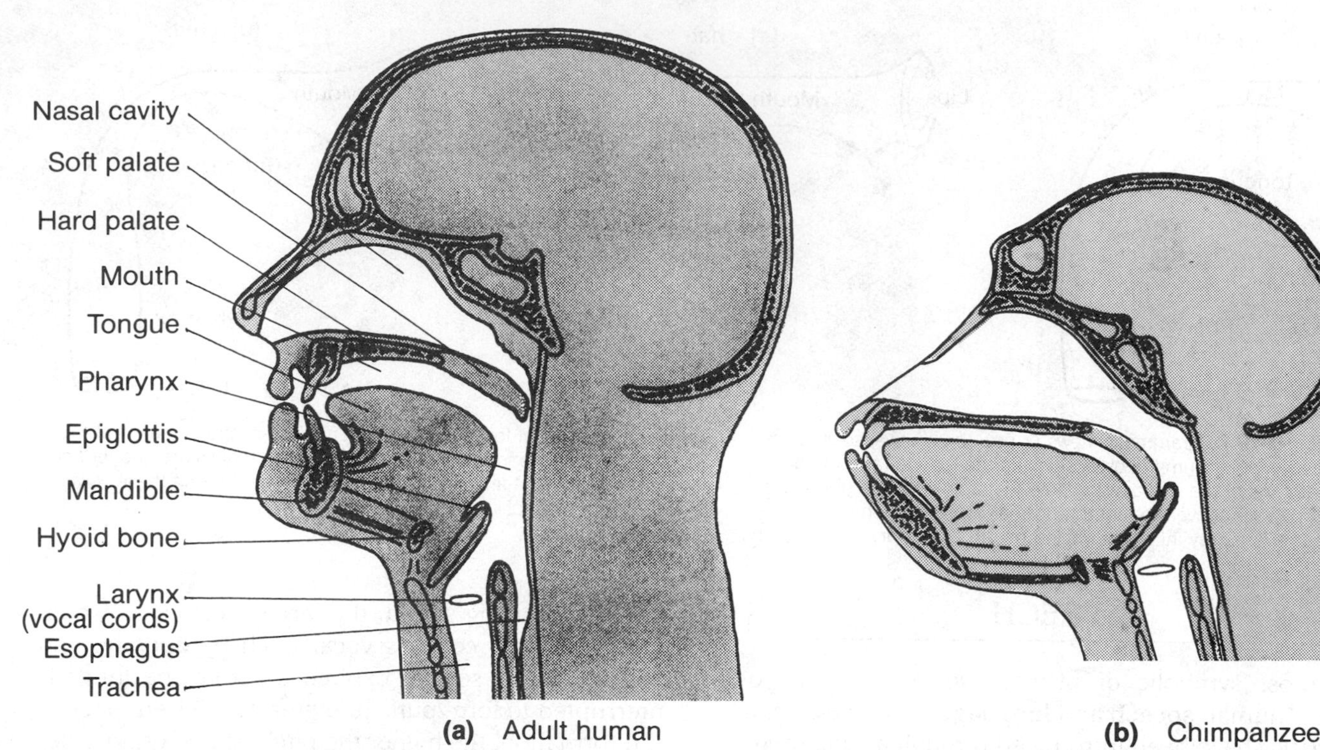

FIGURE 19-17 Upper respiratory systems of an adult human (a) and chimpanzee (b), showing important structures associated with vocalization. The pharynx is significantly lengthened in humans because the larynx is displaced downwards in the neck, and the bulging tongue formed by a reduction in the length of the mandible (note also the lower position of the hyoid bone and epiglottis) now forms the anterior wall of the pharynx. As a result, humans can enunciate vowels and syllables more clearly by positioning the tongue in both mouth and pharynx. (Newborn human infants show the same overlap between epiglottis and soft palate as non-human primates, but the human pharynx lengthens considerably during infancy and childhood, transforming humans from obligate nose breathers as infants to the adult condition of voluntary mouth breathers). (Adapted from Lieberman.)

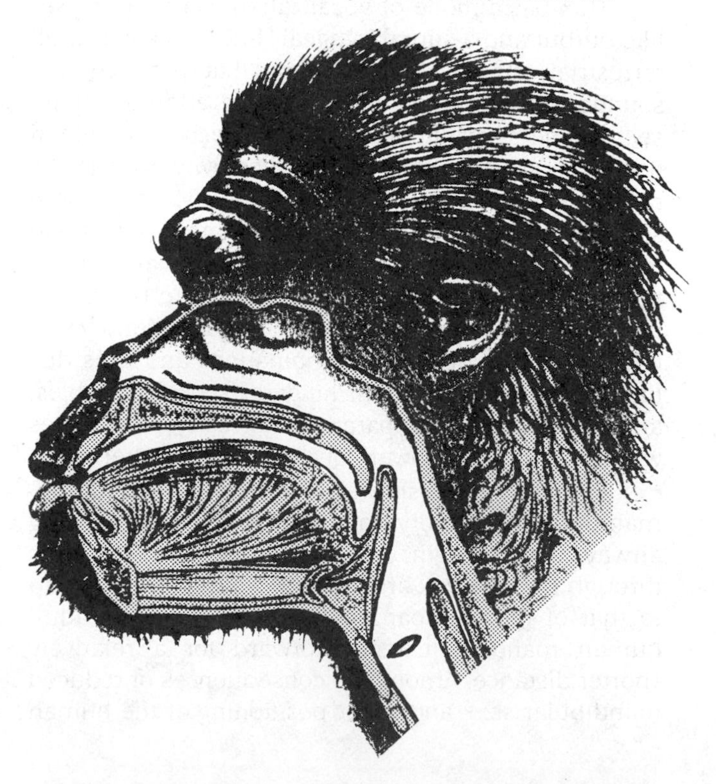

FIGURE 19-18 Reconstruction of the presumed upper respiratory system of an adult australopithecine. Like non-human primates, the epiglottis overlaps the soft palate, the back of the tongue does not reach the pharynx, and the larynx is positioned relatively high in the vocal tract. (Adapted from Lieberman.)

larynx in the vocal airway are the thickening and rounding of the tongue to form the anterior wall of the pharynx. In the chimpanzee (and newborn human infant) the pharyngeal section of the oral tract is shorter, and the epiglottis (used to cover the trachea during food intake) overlaps the soft palate. As a result, the tongue is isolated from the pharynx and respiration normally proceeds through the nasal cavity. Non-human primates and newborn humans can thus drink and breathe simultaneously because the nasal respiratory pathway to the lungs can remain open while liquid passes around it into the esophagus. Thus, although there is a better separation between breathing and swallowing in non-human primates and newborn human infants, adult human speech sounds produced by manipulating the tongue in the pharyngeal cavity are absent.

Because speech depends so much upon soft, non-fossilizable tissues, the problem of uncovering its phylogenetic history is not easy to solve. Nevertheless, various important correlations can be made between skull structure and the positioning of the larynx, size of the tongue, and length of the pharynx. According to such studies (Lieberman), it seems fairly certain that the vocal tract of the australopithecines (Fig. 19-18) was no different from that of non-human primates, and this was probably also true for most, if not all, of the lineages classified as *H. erectus*. In fact, Lieberman and others suggest that *H. sapiens neanderthalensis* had a chimpanzeelike vocal tract that would have considerably limited its speech patterns. Sounds dependent on pharyngeal shape and control, such as [i], [a], [u], [k], and [g], would probably have been considerably distorted. Like chimpanzees and newborn human infants, Neanderthal speech would have been nasalized because the nasal cavity cannot be closed off from the pharynx.[8] If this view is correct, a primary reason for the divergence between Neanderthals and modern humans may have been because of differences in phonetic ability and consequent differences in the kinds of languages that could be employed. On the other hand, a contrasting view has been proposed based on the discovery of a 60,000-year-old Neanderthal skeleton with an intact hyoid bone used for laryngeal muscle control. According to Arensburg and coworkers, this crucial bone shows little difference from that of present-day humans, and they suggest that it signifies Neanderthal capability of human speech. Since the actual position of the larynx and other soft tissues is not known in this fossil or in earlier ones, it is possible that there were various evolutionary experiments in hominid vocal systems over the last 200,000 or so years, perhaps only one of which led to that of modern humans.

However, whether these hypotheses are true or not, the sophistication of modern human language is now one of the crucial keys in separating humans from all other animals, and some of its features and possible evolutionary characteristics are worth considering.

LANGUAGE AND SELF-AWARENESS

In contrast to Darwin who stated in *The Descent of Man* that it would be "impossible to fix on any definite point when the term 'man' ought to be used," Max Müller, a linguist, soon laid down the challenge that "language is our Rubicon, and no brute [ape] will dare cross it." It seemed that Müller's barrier could only be overcome by teaching apes a human language. However, various trials made throughout the early twentieth century were unsuccessful, although it was often clear that apes understood far more than they could communicate.

Among the first serious, long-term attempts to bridge the language gap was the 1947 undertaking by Keith and Catherine Hayes who raised an infant chimpanzee (Viki) in a normal human environment and tried to teach her to articulate human speech. The attempt was a failure, and after 6 years Viki could not utter more than four distinguishable words: "papa," "mama," "cup," and "up." As pointed out previously, it is now evident that a chimpanzee is quite limited in the range of sounds that it can produce because the length of its vocal tract is relatively short and apes seem deficient in ability to control its shape. This is not surprising, since chimpanzees in their natural habitats are usually silent except when aroused.

A different and more successful approach was begun in 1965 by Allen and Beatrice Gardner who raised a chimpanzee (Washoe) from the age of 10 months in an environment where its human caretakers used American Sign Language (ASL) which does not involve speech. By the time Washoe was 5 years old, she had learned to use at least 132 different signs covering a variety of names, actions, modifiers, and functions. Her sentences, however, rarely extended to more than one or two words or repetitions of these, indicating that her language abilities stopped at about the level of a 2- to 3-year-old human child. Findings of this kind have generally also been true for other ASL-taught chimpanzees since Washoe.

Nevertheless, in spite of their limitations, there seems little question that chimpanzees do have linguistic abilities, although on a more elementary level than humans. For example, as shown in Table 19-3, chimpanzees trained in ASL by the Gardners, Roger Fouts, and others can use various appropriate and imaginative combinations of signs to signify items that are not in their vocabulary. They can also refer to events that are distant in time and place (**displacement**)—one of the important features of language. This ability was demonstrated in Washoe's ASL conversation with a human companion in which she repeatedly asked for an orange, then signed, "You go car gimme orange hurry," indicating that she could communicate an event that was to take place in the future at some other place. (Washoe had been in a car more than 2 years earlier and knew that one could get oranges in a store.) Even some simple word orders can be grasped by chimpanzees as shown in Roger Fouts' observation that Lucy could distinguish between "Roger tickle Lucy" and "Lucy tickle Roger." Another chimpanzee, Ally, showed that ape discourse can extend beyond their own persons to comment on the environment when she signed "George smell Roger" to her trainer after Roger Fouts lit a pipe. Chimpanzee communications also include deceptions, as when Booee asked for a tickle from his chimpanzee com-

[8]In adult modern humans, access from the pharynx to the nasal cavity can be opened or closed at will by lowering or raising a flap (velum) of the soft palate. However, although there are languages which use nasalized vowels (e.g., Portugese), there is more difficulty in distinguishing small differences between such vowels compared to distinguishing small differences between non-nasalized vowels. In general, therefore, most modern human languages tend to avoid nasalized vowels.

TABLE 19-3 Sign Sequences (Word Combinations) Created by Chimpanzees for Items Not in Their Ordinary Vocabulary

Item	Chimpanzee Sign Sequence
Onion, radish	Cry fruit, cry hurt fruit
Watermelon	Candy fruit, drink fruit
Alka Seltzer	Listen drink
Cigarette lighter	Metal hot
Swan	Water bird
Brazil nut	Rock berry
Hateful objects	Dirty- . . . (probably signifying fecal), e.g., dirty leash, dirty monkey, dirty Roger, etc.

panion, Bruno, while obviously trying to get to the raisins that Bruno possessed.

A somewhat more abstract two-dimensional language taught to chimpanzees by David Premack involves metal-backed plastic tokens of arbitrary shape (lexigrams) to represent words which are arranged in vertical sequence on a magnetized board. This approach was further developed by Duane Rumbaugh through use of computer-connected keys embossed with lexigrams that, when pressed, light up sequentially on a screen above the keyboard console. Lana, one of the computer-trained chimpanzees, demonstrated that she could use this system to ask "intelligent" questions. For example, having been taught the symbol "name-of," she used it to elicit the unknown name of the object ("box") that contained a desirable item, candy. First she asked, "Tim give Lana name-of this," and, when provided with the name, called for the "box." Perhaps a greater linguistic feat was that of Sarah, trained in Premack's plastic-token language, who used it to understand "if-then" causal relationships: for example, "If Sarah take apple—then Mary give chocolate Sarah; if Sarah take banana—then Mary no give chocolate Sarah."

These experiments show that although apes do not use and understand language in its adult human sophisticated forms (e.g., see the description of language by Jolly), they do possess a variety of important linguistic abilities:

1. They can use words (or signs) as arbitrary symbols for real objects and actions.
2. They can create combinations of words for objects that are not in their vocabulary.
3. They can refer to activities or objects distant in time or place.
4. They can use words to deceive.
5. They can use words to gain information.
6. They can use words to comment on the world about them.
7. They can understand word sequences which utilize the logic of cause and effect.

According to many observers (but not all), apes thus show they have gone beyond merely perceiving events and objects to also conceiving how they occur and how they are related. In their natural habitat, these accomplishments are especially obvious in chimpanzee toolmaking and use, such as the way they employ twigs and vines in fishing for termites. Termite fishing involves being aware of its most favorable months (October and November), locating the sealed termite tunnels, often importing the necessary supply of tools from distances as far away as half a mile, shaping some of the tools by removing leaves, biting off the ends of the tools to achieve an optimum length, inserting the tool with a proper twisting motion that can follow the curves of the termite tunnel, vibrating it gently to bait the termite soldiers, and retracting it carefully to avoid tearing off the termites. Learning these tasks takes years, and even an anthropologist (Geza Teleki) who studied the technique for months was no better at it than a chimpanzee novice.

Lieberman and others suggest that the "rules" involved in sequencing such motor-controlled operations are preadaptations for language, which also must follow **sequencing rules**, since a sentence is composed of a sequence of phrases made up of components such as noun, verb, adjective, and subject. The ability to devise and follow such rules apparently lies in the association centers of the brain, and glimmerings of this can even be found in simple vertebrates such as frogs who can perform a sequential set of actions such as fly catching.[9]

[9]According to Chomsky, the ability to structure words into meaningful sentences (**syntax**), no matter what the language, is determined by inherited neural structures. This does not mean that any human child will necessarily communicate by language without hearing or speaking one. In fact, children who by misfortune have been raised in isolation without the opportunity of communicating with other humans do not acquire language and remain seriously deficient in social and intellectual development. The neuronal components of language and speech apparently have to be exercised at a critical early stage in order to become functional, just as kittens temporarily deprived of vision in one eye between 4 and 12 weeks of age will not possess binocular vision even when full vision is restored. Thus, although the ability to acquire language appears to be innate in humans, the particular language used in communication has to be learned by experience and imitation. The need for learning seems especially true for producing understandable speech sounds and words, since it takes much longer for children to speak understandably than for them to comprehend what is spoken to them.

It seems clear that to enable a word (e.g., an auditory stimulus) to stand as a symbol for an object or action (e.g., a visual stimulus), there must be a neural associative center that allows such connections to be made. Such **cross-modal associations**, as they are called, are already present in many mammals since they obviously do understand and can act upon some of the words spoken to them. In the case of apes, and even monkeys, cross modality extends even to the ability to correlate objects that they cannot see but can feel to objects that they can see. Tests of apes, for example, show that even a glimpse of the photograph of an object will allow them to choose correctly by feeling for it among different objects hidden from their eyes.

The neurological areas of the brain that allow cross-modal associations and matching are not yet known in apes, but one such human center is believed to be the angular gyrus shown in Figure 19-19. Near this region, on the parietal lobe of the left cerebral hemisphere, is **Wernicke's area**, concerned with formulating and comprehending intelligent speech; persons who have lesions in this area emit informationless, wordy babble. Patients with lesions in **Broca's area** on the left frontal lobe have difficulty speaking, since this region apparently serves to coordinate vocal muscular movements.

Although these speech and language areas are most often on the left cerebral hemisphere, there is some variability, and left-handed individuals may have these areas localized on the right hemisphere. The existence of such functional laterality, with one cerebral hemisphere dominant over the other, seems to be present in other primates, even in rhesus monkeys, and may be considered a preadaptation for development of the associative centers used in human speech and language. That is, the cross-modal associations used in connecting various faculties, such as being able to visualize a predator upon hearing a particular alarm call, may have led to the dominance of one cerebral hemisphere, and these areas were then further enhanced in human lineages during selection for the acquisition of language.[10]

Interesting as these findings are, it is possible to claim, as many already have, that language can only develop in those organisms which have a concept of "I," or **self-awareness**, so they can intellectually separate themselves from the rest of the world. That is, to be capable of language one must analyze external events by taking them apart and putting them back together in a symbolic, thoughtful, manipulative con-

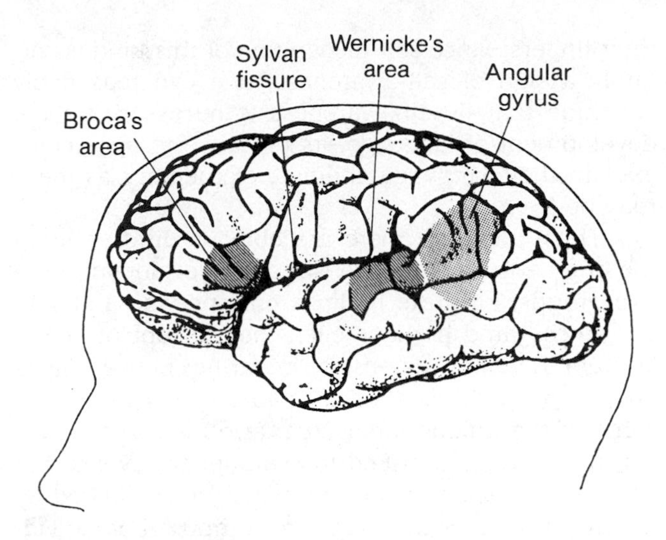

FIGURE 19-19 Centers associated with speech on the human left cerebral hemisphere.

text (the intellectual "I") different from one's own immediate, non-thoughtful, reflexive reactions. However, even from this challenging point of view, the kernel of self-awareness can be demonstrated in apes as shown by a psychologist, Gordon Gallup, who discovered that its existence could be tested by a relatively simple procedure—the reaction of an individual to its image in a mirror.

In human children, the realization that the mirror reflects themselves rather than another child becomes apparent at the age of about 20 months. This achievement indicates that the child has already attained a concept of itself as distinct from that of other children, a concept whose imaged reflection can be identified in the mirror. The child will then use the reflection to observe and examine itself rather than examine the image on the mirrored surface, signifying an awareness of its own individuality. Birds, dogs, and most other animals, including monkeys, are unable to recognize anything else in the mirror but a different member of the same species, no matter how long they are exposed to the mirror image. The "I" seems to be less developed or missing in these animals.

Most interestingly, when Gallup exposed wild-born chimpanzees to full-length mirrors he found the development of "self-directed" behavior within a few days: the chimpanzees used the mirrors to examine and manipulate different parts of themselves. Moreover, when such chimpanzees were anesthetized and painted with bright red spots which they could see only in the mirror, they used their reflections to detect these spots on themselves, to touch the spots and then smell

[10]On the intellectual plane, preadaptations are as important as they are on the physical plane: chimpanzees were certainly not selected in the past to learn American Sign Language nor humans to practice higher mathematics. The abilities which allow these primates to perform these intellectual feats are obvious preadaptations that evolved for other selected purposes, such as evaluating and responding to intricate social and environmental factors and relationships.

their fingers. Since self-recognition of this kind is certainly a sign of self-awareness, one can reasonably conclude that if self-awareness is necessary for the development of language, its presence in apes helps explain their language abilities, elementary as these may be.

Thus, although there are obvious differences in the degree to which these traits are developed in the two groups, all these findings on language and self-awareness (and perhaps even the concept of death; Desmond) point to extensive similarities between apes and humans. It seems reasonable to conclude that some of the unique intellectual attributes of humans can probably be ascribed to evolutionary events that occurred sufficiently early in history to be shared by the two groups. Both groups are composed of individuals who continually interact in complex and changing social patterns, and this was most likely true for their common ancestors as well. Since self-awareness and language depend almost entirely on social interactions (isolated young chimpanzees do not develop self-awareness, and isolated human children do not develop language), an environment of complex interpersonal relations probably played an essential role in the evolution of these traits.

ALTRUISM AND MORALITY

The importance of social interactions in the development of behavioral and communication skills can be seen throughout primate groups in the panoply of calls, grimaces, gestures, and activities they use to indicate social positions (e.g., dominance, subordination, group affiliation); needs (e.g., food, sex, reassurance); and changes in any of these areas (e.g., new social positions, alliances, sexual states, or dietary interests). Such behaviors range from transmitting only information on themselves as individuals to actions that may have immediate effects on the survival of other group members.

Information affecting the survival of other group members is most obvious, for example, when a monkey encounters a leopard and reacts with a loud scream, signifying that nearby listeners are to take refuge. Thus, although this warning signal may call the predator's attention to the screamer and diminish its own chances for survival, the effect can nevertheless be helpful in preserving its relatives or compatriots. Population geneticists beginning with Haldane and Wright (Chapter 22) suggested that there were genetic advantages in such **altruistic behavior** in which individuals may even go so far as to endanger their own genetic future for those who carry closely related genotypes. In 1964, Hamilton popularized this cooperative process under the name **kin selection**, and provided formulae by which some of its benefits could be evaluated (see p. 312). As Maynard Smith has pointed out, "the main reason for thinking that kin selection has been an important mechanism in the evolution of cooperation is that most animal societies are in fact composed of relatives."

Some years after Hamilton's proposals, Trivers introduced a concept of altruism that seemed to have special applicability to human social behavior. Trivers' theory of **reciprocal altruism** suggested that altruism can become established in a group where the frequency of interaction between individuals is high and the life span sufficiently long to enable recipients of altruistic acts to return favors to the altruists. The benefits to individuals who partake in such reciprocal altruism can far outweigh the costs, since even slight expenditures of altruistic energy (such as throwing a lifeline to a drowning individual) may have significant benefits to the altruist when it is reciprocated by the previous beneficiary or other group members. Frequent interaction and exchange of roles ("sometimes an altruist, sometimes a beneficiary") is necessary in order to recognize "cheaters" early on, who would otherwise continually try to act as beneficiaries and exploit the altruists.[11]

By refusing cheaters the benefit of future aid, through either punishment or exile, **moral sentiments** are developed and enhanced in such cooperative groups, and emphasis is put on precise accounting and balancing of exchanges between individuals. As pointed out by Trivers, the maintenance of such systems will therefore be supported by introducing or reinforcing a variety of **emotional traits**:

Friendship The emotional bonds established between individuals who behave altruistically towards each other.

Moral indignation and aggression The feelings of injustice and hostility towards cheaters that can lead to retribution against them.

[11]The other side of this social coin can be seen where **dominance relations** are allowed to hold sway, and cheating by one or more dominant individuals replaces cooperative relationships. Under such circumstances, reciprocal altruism would be absent or tend to diminish except in those areas in which social alliances are forged ("you help me dominate X, and I'll help you dominate Y"), or where dominance is excluded (chimpanzees who capture game are considered its proprietors whether they are dominant or subordinate, and the entire troop may then line up for handouts). However, even in human societies unclouded by overt dominance relations, cheating may be practiced where it can remain undetected, or where local conditions may allow it to occur (e.g., smuggling). In Trivers' words, there are differences among humans "in the degree of altruism they show and in the conditions under which they will cheat."

Gratitude The emotional responses of recipients to what they perceive as altruistic acts.

Sympathy, kindness, and generosity Emotional motivations that aid individuals in performing altruistic acts.

Guilt An emotion engendered by active cheating or its contemplation that either prevents cheating from happening or leads to reparation and thereby helps prevent the rupture of social bonds.

Although these emotions are not uniformly felt or expressed by all individuals under all circumstances, they, like other emotions such as love and security, obviously serve the purpose of helping to preserve social groupings based on intricate reciprocal relationships. It therefore seems very likely that human morality and its accompanying sentiments such as fairness and justice did not emerge full-blown from human thought but derive from an evolutionary history which probably traces back to early hominid groups or even earlier. As shown by Goodall and others, there are hints of the existence of some of these emotions in other primates, in addition to well-expressed emotions such as fear that are probably shared by all mammals. Their origin is no mystery: emotions are devices that aid the individual in performing tasks necessary for its own preservation or that of the group. The extent to which evolutionarily derived emotions and strategies guide our present social behavior has been the subject of considerable debate, and aspects of this topic, called sociobiology, will be discussed in Chapters 23 and 24.

SUMMARY

Primates have a number of features indicating an arboreal past, as well as more recent adaptations such as large brain size. They have few offspring with an extended childhood during which the young can learn and receive care from their parents. Claws and other early primate features characterize the prosimian suborder of lemurs, lorises, and tarsiers, many of which are nocturnal and have a retinal tapetum which permits vision in dim light. The anthropoids—the old and new-world monkeys, apes, and humans—are usually diurnal and have forward-directed eyes and larger brains than the prosimians. Among the hominoids are the gibbons, orangutans, chimpanzees, gorillas, and humans, all of which have some adaptations for bipedal locomotion.

It is generally assumed that the prosimians evolved from insectivorous mammals, and the tarsier line later gave rise to the anthropoids. Within the anthropoids the apes diverged from the monkeys about 20 to 30 million years ago, and approximately 4 or 5 million years ago an ape lineage split into the modern apes and the hominid line.

Fossils of the genus *Australopithecus*, with some human-like features of teeth, spine, and brain, represent the oldest known hominid remains, and there is good evidence that they were bipedal. Of the several species, *A. africanus*, with its large teeth and jaws, appears to be the most ancient and apelike.

Bipedalism may have been favored because of advantages in foraging, evading predators, carrying provisions, and using tools and weapons. Among the consequences of bipedalism were probably establishment of a "home base," sexual bonding, and parental involvement in offspring. Changes in the pelvis, legs, and feet accompanied the new body alignment.

The earliest fossils of the genus *Homo*, *H. habilis*, are about 2 million years old and have brains much larger than those of the australopithecines. *Homo erectus*, a widespread species which utilized distinguishable stone tools, appeared somewhat later and possibly gave rise to *Homo sapiens*. The neanderthals, early representatives of this new species, had a modern brain size, heavy jaws and brow ridges, and were socially advanced. They yielded about 35,000 years ago to Cro-Magnon man, who had an advanced culture and was modern in appearance.

Evidence from mitochondrial DNA evolution suggests that the ancestral *H. sapiens* mtDNA sequence arose roughly 200,000 years ago in Africa and that this species spread worldwide without mixing with other extant species.

At some point anthropoids began supplementing their foraging activities with hunting and its increasing importance enhanced social cohesion, cooperation, and communication, as well as favoring improvements in technology.

Primates can transmit signals to other individuals through any of the sensory mechanisms: scent, grooming behavior, gestures and facial expressions, displays, and vocalization. The development of the tongue, soft palate, and larynx allowed humans to produce a variety of sounds not available to other primates.

Although attempts to teach chimpanzees how to speak have failed, a number of investigators have shown that chimpanzees possess conceptual linguistic skills not dependent on the structure of the vocal tract. It may be that language can arise only in organisms, man and apes, which have awareness of their "self" or separateness from their environment, traits that arise because of the complexity of social interactions. Other traits, such as altruism, morality, friendship, and guilt may be based on frequent interactions between individuals in a group where such behaviors and emotions are adaptations that promote survival and social cohesion.

KEY TERMS

altruistic behavior
angular gyrus
animal hunting
anthropoids
australopithecines
bipedalism
Broca's area
catarrhines
chimpanzees
communication
cooperative hunting
cross-modal associations
displacement (language)
dominance relations
emotional traits
gibbons
gorillas
home base
hominids
hominoids
Homo
hunter-gatherer societies
kin selection
language
larynx
lemurs
long-distance foraging
lorises
meat eating
moral sentiments
olfactory cues
orangutans
Piltdown man
platyrrhines
prosimians
reciprocal altruism
self-awareness
sequencing rules
sexual bonding
speech
stone tools
symbolic calls
syntax
tactile communication
tarsiers
visual signals
vocalizations
Wernicke's area

REFERENCES

Aiello, L. C., 1986. The relationships of the Tarsiioformes: A review of the case for the Haplorhini. In *Major Topics in Primate and Human Evolution*, B. Wood, L. Martin, and P. Andrews (eds.). Cambridge Univ. Press, Cambridge, pp. 47-65.

Andrews, P., 1981. Species diversity and diet in monkeys and apes during the Miocene. In *Aspects of Human Evolution*, C. B. Stringer (ed.). Taylor & Francis, London, pp. 25-61.

———, 1983. The natural history of *Sivapithecus*. In *New Interpretations of Ape and Human Ancestry*, R. Ciochon and R. Corruccini (eds.). Plenum Press, New York, pp. 441-464.

Arensburg, B., A. M. Tillier, B. Vandermeersch, H. Duday, L. A. Schepartz, and Y. Rak, 1989. A middle Paleolithic human hyoid bone. *Nature*, **338**, 758-760.

Campbell, B., 1985. *Human Evolution*, 3rd ed. Aldine, New York.

Cann, R. L., M. Stoneking, and A. C. Wilson, 1987. Mitochondrial DNA and human evolution. *Nature*, **325**, 31-36.

Cant, J. G. H., 1981. Hypothesis for the evolution of human breasts and buttocks. *Amer. Naturalist*, **117**, 199-204.

Carroll, R. L., 1988. *Vertebrate Paleontology and Evolution*. W. H. Freeman, New York.

Chevalier-Skolnikoff, S., 1973. Facial expressions of emotion in nonhuman primates. In *Darwin and Facial Expression*, P. Ekman (ed.). Academic Press, New York, pp. 11-89.

Chomsky, N., 1972. *Language and Mind*. Harcourt, Brace, and Jovanovich, New York.

Ciochon, R. L., and J. G. Fleagle (eds.), 1985. *Primate Evolution and Human Origins*. Benjamin/Cummings, Menlo Park, Calif.

Clarke, R. J., 1985. A new reconstruction of the Florisbad cranium, with notes on the site. In *Ancestors: The Hard Evidence*, E. Delson (ed.). A. R. Liss, New York, pp. 301-305.

Day, M. H., 1986. *Guide to Fossil Man*, 4th ed. Cassell, London.

———, 1986. Bipedalism: pressures, origins and modes. In *Major Topics in Primate and Human Evolution*, B. Wood, L. Martin, and P. Andrews (eds.). Cambridge Univ. Press, Cambridge, pp. 188-202.

Delson, E. (ed.), 1985. *Ancestors: The Hard Evidence*. A. R. Liss, New York.

Desmond, A. J., 1979. *The Ape's Reflexion*. Dial Press, New York.

Fleagle, J. G., 1988. *Primate Adaptation and Evolution*. Academic Press, San Diego, Calif.

Fleagle, J. G., T. M. Bown, J. D. Obradovitch, and E. L. Simons, 1986. Age of the earliest African anthropoids. *Science*, **234**, 1247-1249.

Foley, R., 1987. *Another Unique Species: Patterns in Human Evolutionary Ecology*. Longman, Harlow, Great Britain.

Gallup, G. G., Jr., 1977. Self-recognition in primates: A comparative approach to the bidirectional properties of consciousness. *Amer. Psychologist*, **32**, 329-338.

Goodall, J., 1986. *The Chimpanzees of Gombe: Patterns of Behavior*. Harvard Univ. Press, Cambridge, Mass.

Green, S., and P. Marler, 1979. The analysis of animal communication. In *Handbook of Behavioral Neurology*, Vol. 3, P. Marler and J. G. Vandenbergh (eds.). Plenum Press, New York, pp. 73-158.

Hamilton, W. D., 1964. The evolution of social behavior. *Jour. Theoret. Biol.*, **1**, 1-52.

Harding, R. S. O., and G. Teleki (eds.), 1981. *Omnivorous Primates: Gathering and Hunting in Human Evolution*. Columbia Univ. Press, New York.

Hill, K., 1982. Hunting and human evolution. *Jour. Hum. Evol.*, **11**, 521-544.

Johanson, D., and M. Edey, 1981. *Lucy: The Beginnings of Mankind*. Simon & Schuster, New York.

Jolly, A., 1985. *The Evolution of Primate Behavior*. Macmillan, New York.

Kinzey, W. G. (ed.), 1987. *The Evolution of Human Behavior: Primate Models*. SUNY Press, Albany, N. Y.

LeGros Clark, W. E., 1978. *The Fossil Evidence for Human Evolution*, 3rd ed. Univ. of Chicago Press, Chicago.

Lewin, R., 1986. New fossil upsets human family. *Science*, **233**, 720-721.

Lieberman, P., 1984. *The Biology and Evolution of Language*. Harvard Univ. Press, Cambridge, Mass.

Lovejoy, C. O., 1981. The origin of man. *Science*, **211**, 341-350.

——, 1988. The evolution of human walking. *Sci. Amer.*, **259** (5), 118-125.

Maynard Smith, J., 1983. Game theory and the evolution of cooperation. In *Evolution From Molecules to Man*, D. S. Bendall (ed.). Cambridge Univ. Press, Cambridge, pp. 445-456.

Miyamoto, M. M., B. F. Koop, J. L. Slightom, M. Goodman, and M. R. Tennant, 1988. Molecular systematics of higher primates: genealogical relations and classification. *Proc. Nat. Acad. Sci.*, **85**, 7627-7631.

Napier, J. R., and P. H. Napier, 1985. *The Natural History of Primates*. British Museum, London.

Pilbeam, D., 1984. The descent of hominoids and hominids. *Sci. Amer.*, **250**, 84-96.

——, 1986. Hominoid evolution and hominoid origins. *Amer. Anthropol.*, **88**, 295-312.

Rak, Y., 1986. The Neanderthal: a new look at an old face. *Jour. Hum. Evol.*, **15**, 151-164.

Robinson, J. T., 1972. *Early Hominid Posture and Locomotion*. Univ. of Chicago Press, Chicago.

Schultz, A. H., 1933. Die körperproportionen der erwachsenen catarrhinen Primaten, mit spezieller Berücksichtigung der Menschenaffen. *Anthropol. Anz.*, **10**, 154-185.

Shipman, P., 1985. The ancestor that wasn't. *The Sciences*, **25** (2), 43-48.

Silberbauer, G., 1981. Hunter/gatherers of the Central Kalahari. In *Omnivorous Primates: Gathering and Hunting in Human Evolution*, R. S. O. Harding and G. Teleki (eds.). Columbia Univ. Press, New York, pp. 455-498.

Stringer, C. B., and P. Andrews, 1988. Genetic and fossil evidence for the origin of modern humans. *Science*, **239**, 1263-1268.

Stringer, C. B., R. Grün, H. P. Schwarcz, and P. Goldberg, 1989. ESR dates for the hominid burial site of Es Skhul in Israel. *Nature*, **338**, 756-758.

Szalay, F. S., and E. Delson, 1979. *Evolutionary History of the Primates*. Academic Press, New York.

Tanner, N. M., 1987. The chimpanzee model revisited and the gathering hypothesis. In *The Evolution of Human Behavior: Primate Models*, W. G. Kinzey (ed.). SUNY Press, Albany, N. Y., pp. 3-27.

Tattersall, I., E. Delson, and J. Van Couvering (eds.), 1988. *Encyclopedia of Human Evolution and Prehistory*. Garland Publ., New York.

Tooby, J., and I. DeVore, 1987. The reconstruction of hominid behavioral evolution through strategic modeling. In *The Evolution of Human Behavior: Primate Models*, W. G. Kinzey (ed.). SUNY Press, Albany, N. Y., pp. 183-237.

Trivers, R., 1985. *Social Evolution*. Benjamin/Cummings, Menlo Park, Calif.

Wolpoff, M. H., 1980. *Paleoanthropology*. A. A. Knopf, New York.

——, 1982. *Ramapithecus* and hominid origins. *Curr. Anthropol.*, **23**, 501-510.

Part 4

20

Populations, Gene Frequencies, and Equilibrium

At the center of Darwin's theory of evolution was the concept that small heritable changes provided the **continuous variation** upon which natural selection acted. However, soon after publication of *The Origin of Species*, Francis Galton, Darwin's cousin, became convinced that a mathematical approach to heredity showed that evolution must have proceeded in sharp, discontinuous steps. By 1871, Galton had already disproved Darwin's pangenesis hypothesis (p. 28) to his own satisfaction by showing that transfusion of blood between rabbit strains had no effect on heredity. Similar to August Weismann's later germ-plasm theory, Galton in 1875 suggested that instead of somatically acquired "gemmules," the hereditary material was passed on between generations with little or no change. It seemed apparent to Galton that parents who deviate significantly from the average for some continuous quantitative trait such as height tended to produce offspring that were closer to the average than themselves (Galton's Law of Regression). He surmised that continuous variation was therefore not the agency which leads to the origin of new species but that non-blending, **discontinuous variation** ("sports") provided the abrupt changes between species.

Mutationists and Selectionists

By the end of the nineteenth century, two schools of thought had established themselves in England based on their preference for continuous or for discontinuous variation. Upholding the importance of continuous variation in evolution were the mathematically oriented **biometricians**, Weldon and Pearson, and in opposition to them was Bateson and his supporters. When Mendel's 1865 paper on the genetics of peas was discovered in 1900, these two camps polarized further. Bateson and the **mendelians** proposed that most hereditary characteristics were discontinuous and could be explained by the segregation of mendelian factors, whereas the biometricians insisted that most characteristics were continuous and mendelian factors were only involved in exceptional traits. To mendelians such as De Vries and others, evolution could only be effective if selection operated on large mutations of the kind discovered in the evening primrose, *Oenothera lamarckiana*, whereas the biometricians allied themselves with Darwin's original concept, that selection acting on small differences was the primary mechanism for evolutionary change.

In apparent support of the mutationist position were the experimental observations of Johannsen that selection was ineffective in quantitatively changing the size of beans descended from homozygous **pure lines**. Furthermore, even when selection was practiced on beans descended from crosses between different pure lines, size differences among their descendants seemed to show relatively little change from the range of values initially observed in the F_2 generation of the cross. It seemed therefore that marked changes in size of beans could only come from mutations with large effect (**macromutations**), rather than from selection among the small differences observed in Johannsen's experiments. Various biologists therefore extended these views to propose that new species can arise in only one or a few mutational steps driven perhaps by mutation pressure in a particular and even nonadaptive direction, a view known as **saltation**. According to the saltationists, the slow, plodding process of Darwinian selection was no longer necessary to explain evolution.

Castle and Phillips, on the other hand, demonstrated that entirely new coat color patterns could appear by selection in hooded rats: some selected lines had "less pigment than any known type other than albino," whereas others were "so extensively pigmented that they would readily pass for the 'Irish type' which has white on the belly only." Although no specific genes with quantitative effect were identified in these studies, similar results with other organisms (see Fig. 10-28) did point to the likelihood that selection could act on small, continuous characters to produce marked changes in phenotype.

The rift between **mutationists** and **selectionists** in explaining the basic mechanisms of evolution remained until the 1920's and 1930's but the gap had already been bridged by further experimental work on **quantitative characters**. For example, Nilsson-Ehle and East showed that a number of different gene pairs (multiple factors) may affect a single quantitative character so that a wide array of possible genotypes can occur, each with a different phenotype. That is, genes which segregate in typical mendelian patterns are responsible for many observed distributions of continuous traits (see Fig. 10-29). Thus, there was no real difficulty in providing a mendelian interpretation of Darwinian selection for quantitative traits, and dependence on the introduction of mutations with large effect no longer seemed necessary to explain most basic evolutionary changes.

THE NEO-DARWINIAN SYNTHESIS

Concurrent with the disputes between selectionists and mutationists, important new concepts were also being proposed by Yule, Castle, Hardy, and Weinberg. These workers emphasized that populations rather than individuals were an important evolutionary focus, and attention had to be paid to populational gene frequencies rather than only to whether a gene was present or absent. It became clear that the gametes contributed by a population to the next generation could be considered as a giant gene pool from which offspring withdrew their various genotypic combinations at random. In the absence of selection and other factors that could change gene frequencies, these frequencies tended to be conserved, as demonstrated by the **Hardy-Weinberg equilibrium** discussed below.

During the 1920's, the approach towards considering evolution as a change in gene frequencies was developed in considerable detail by Fisher, Wright, and Haldane in various papers that dealt with the effects of inbreeding, the evolution of dominance, and the effect of selection on gene frequencies, as well as the effects of mutation, migration, and genetic drift. In the early 1930's these studies culminated in a variety of papers and books which laid the foundations for **population genetics**. Along with these mathematical models were observational findings by Chetverikoff in the Soviet Union and others elsewhere indicating considerable genetic variation in natural populations upon which selection could act. All these studies helped establish the concept that it is the population which possesses the variability necessary to understand evolutionary genetic change through space and time, whereas an individual is extremely limited in these dimensions (Table 20-1): it is populations that evolve, not individuals.

This emphasis on the genetics of populations helped transform evolutionary thinking into its more modern form, often called the **Neo-Darwinian** (or **modern**) **synthesis**. At the base of this synthesis is the concept that mutation occurs randomly and furnishes the fuel for evolution by introducing genetic variability. Evolution can then be defined as a change in the frequencies of genes introduced by random mutation, with natural selection usually considered as the most important, although not only, cause for such changes. (Among other factors are migration and random genetic drift, discussed in the following chapter.) The accumulation of gene frequency differences, by what-

TABLE 20-1 Characteristics of Individuals Compared to Those of Populations

Characteristic	Individual	Population
Life span	One generation	Many generations
Spatial continuity	Limited	Extensive
Genetic characteristics	Genotype	Gene frequencies
Genetic variability	None	Considerable
Evolutionary characteristics	No changes	Can evolve (change in gene frequency)

ever means, leads eventually to more pronounced (racial) differences among populations in geographically different localities. When gene exchange between racial groups can no longer occur because of reproductive isolation, separate species become established. Furthermore, mechanisms such as mutation and selection that lead to the origin of races and species are generally no different from the mechanisms that lead to the origin of higher taxa such as genera, families, and orders, although the formation of higher taxa usually takes place over longer periods of time.

This Neo-Darwinian synthesis has made itself felt in every evolutionary biological field from anthropology to paleontology and systematics. The influence of population genetics, the essential component of this synthesis, extends also to numerous other fields such as demography, ecology, epidemiology, plant and animal breeding, and other areas in which gene variation and distribution affect the relationships and life patterns of organisms. Although population genetics has placed emphasis on mathematical models which cannot reflect reality in all of its myriad details, it has provided logically precise concepts that help us understand many common populational features, such as the frequently observed conservation of gene and genotype frequencies and the general effects of forces that change these frequencies.

POPULATIONS AND GENE FREQUENCIES

Among geneticists a population is usually defined as a group of sexually interbreeding or potentially interbreeding individuals. Since mendelian laws apply to the transmission of genes among these individuals, such a group has been termed by Wright a **mendelian population**. The size of the population may vary, but it is usually considered to be a local group (also called **deme**), each member of which has an equal chance of mating with any other member of the opposite sex. Most of the theoretical and experimental emphasis has been laid so far on populations of diploid organisms, and the discussions that follow deal mostly with these cases. However, whether diploid or haploid, populations can be said to have two important attributes: **gene frequencies** (occasionally called allele frequencies) and a **gene pool**.

Gene frequencies are simply the proportion of the different alleles of a gene in a population. To obtain these proportions we count the total number of organisms with various genotypes in the population and estimate the relative frequencies of the alleles involved. Of course, except for gametes and occasional mutation, the genetic complements of all cells in a multicellular organism are the same. One may therefore adopt the convention that a haploid organism has only one gene at any one locus, a diploid has two, a triploid three, and so on. For example we can presume that the difference between humans who can and cannot taste the chemical phenylthiocarbamide resides in a single gene difference between two alleles, *T* and *t*. Since the allele for tasting, *T*, is dominant over *t*, the heterozygous genotype, *Tt*, represents tasters and the nontasters are *tt*. A population of 200 individuals composed of 90 *TT*, 60 *Tt*, and 50 *tt* will therefore have a total of 400 genes (alleles) at this locus. As shown in Table 20-2(1), 240 of these are *T* (a frequency of .60), and 160 are *t* (a frequency of .40). The same gene frequencies can also be calculated from the frequencies of the three genotypes, according to the formula: frequency of a gene = frequency of homozygotes for that gene + 1/2 frequency of heterozygotes, who each contain one such gene out of two (Table 20-2(2)).

The gene pool is the sum total of genes in the reproductive gametes of a population. It can be considered as a gametic pool from which samples are drawn at random to form the zygotes of the next generation. Thus the genetic relationship between an entire generation and the subsequent generation is very similar to the genetic relationship between a parent and its offspring. Since the frequencies of genes in the new generation will hinge, to some degree at least, upon their frequencies in the old, one might say that gene frequencies rather than genes are inherited in populations. In what form can these gene-frequency relationships between generations be expressed and analyzed?

One of the first attempts at utilizing the concept of gene frequencies occurred in the dispute mentioned earlier between the biometricians and mendelians. It was argued that dominant alleles, no matter what their initial frequency, would be expected to reach a stable equilibrium frequency of three dominant individuals to one recessive, since this was the mendelian segre-

TABLE 20-2 Techniques for Obtaining Gene Frequencies for a Diploid Population of 200 Individuals*

1. Using numerical gene counts (there are 400 genes in 200 diploid individuals):

$$T = 180 \text{ (in } TT) + 60 \text{ (in } Tt) = \frac{240}{400} = .60$$

$$t = 100 \text{ (in } tt) + 60 \text{ (in } Tt) = \frac{160}{400} = .40$$

Total 1.00

2. Using genotype frequencies:

$$T = .45\ TT + \tfrac{1}{2}(.30\ Tt) = .45 + .15 = .60$$

$$t = .25\ tt + \tfrac{1}{2}(.30\ Tt) = .25 + .15 = .40$$

Total 1.00

*These individuals are of the following types:
90 *TT* + 60 *Tt* + 50 *tt* = 200 individuals
[.45 *TT* + .30 *Tt* + .25 *tt* = 1.00 (genotypes)]

gation pattern for these genes. The fact that many dominant alleles such as that for human brachydactyly (short fingers) were present in very low frequency was therefore considered evidence that mendelian dominants and recessives were not segregating properly in populations. Although widely accepted at first, this argument was disproved in 1908 by both Hardy in England and Weinberg in Germany who demonstrated that gene frequencies are not dependent upon dominance or recessiveness but remain essentially unchanged from one generation to the next under certain conditions. Such **conservation of gene frequencies** will be briefly discussed in this chapter, while Chapter 21 will deal with the forces that can change gene frequencies. Fuller treatment of the basic theoretical principles of population genetics along with more complete formula derivations and extensive examples can be found in various books including those of Crow, Crow and Kimura, Falconer, Hartl and Clark, Hedrick, Spiess, and Wallace.

CONSERVATION OF GENE FREQUENCIES

The principle discovered by Hardy and Weinberg may be simply illustrated using the tasting example previously mentioned. Let us, for instance, place upon an island a group of children in the ratio given above: .45 *TT*:.30 *Tt*:.25 *tt*, where gene frequencies are therefore .60 *T* and .40 *t*. Let us also assume that the number of individuals in this newly formed population is large and that tasting or nontasting has no effect upon survival (viability), fertility, or attraction between the sexes.

As these children mature they will choose their mates at random from those of the opposite sex regardless of their tasting abilities. Matings between any two genotypes can then be predicted solely on the basis of the genotypic frequencies in the population. As shown in Table 20-3, nine different types of matings can occur, of which three matings are reciprocals of others (e.g., $TT \times tt = tt \times TT$). In all, therefore, these six different mating combinations will produce offspring in the ratios shown.

Note that although the frequencies of genotypes have been altered by **random mating**, the gene frequencies among the offspring have not changed. For *T* the offspring gene frequency is equal to .36 + ½(.48) = .60, and the frequency of *t* is .16 + ½(.48) = .40, exactly the same as before. Under these conditions, no matter what the initial frequencies of the three genotypes, the gene frequencies of the next generation will be the same as those of the parental generation. For example, if the founding population of this island contained .40 *TT*, .40 *Tt*, and .20 *tt*, the gene frequency for *T* would be .40 + ½(.40) and .20 + ½(.40) for *t*, the same as before. However, as shown in Table 20-4, despite the different genotype frequencies, offspring are again produced in the ratio .36 *TT* :.48 *TT*:.16 *tt*, or a gene frequency of .60 *T*:.40 *t*.

Two important conclusions follow:

1. Under conditions of random mating (**panmixia**) in a large population where all genotypes are equally viable, gene frequencies of a particular generation depend upon the gene frequencies of the previous generation and not upon the genotype frequencies.
2. The frequencies of different genotypes produced through random mating depend only upon the gene frequencies.

Both these points mean that by confining our attention to genes rather than to genotypes we can predict both gene and genotype frequencies in future generations, providing outside forces are not acting to change their frequency and random mating occurs between all genotypes. To continue our previous illustration we may predict that under these conditions the initial gene frequencies in taster-nontaster populations will not change in the next or succeeding generations. Also, after the first generation the genotype frequencies will also remain stable, that is, at **equilibrium**.

This genotypic equilibrium, based on stable gene frequencies and random mating, is known as the Hardy-Weinberg principle (or law) and has served as the founding theorem of population genetics. Perhaps

TABLE 20-3 Gene Frequencies Produced by Random Mating Between Individuals in a Population Having the Frequencies Given in Table 20-2

		Males		
		TT = .45	*Tt* = .30	*tt* = .25
Females	*TT* = .45	.2025 (1)	.1350 (2)	.1125 (3)
	Tt = .30	.1350 (4)	.0900 (5)	.0750 (6)
	tt = .25	.1125 (7)	.0750 (8)	.0625 (9)

Parents		Offspring		
Mating	Frequency	*TT*	*Tt*	*tt*
TT x *TT* (1)	.2025	.2025		
TT x *Tt* (2) + (4)	.2700	.1350	.1350	
TT x *tt* (3) + (7)	.2250		.2250	
Tt x *Tt* (5)	.0900	.0225	.0450	.0225
Tt x *tt* (6) + (8)	.1500		.0750	.0750
tt x *tt* (9)	.0625			.0625
	1.0000	.3600	.4800	.1600

Gene frequencies among offspring: $T = (.36\ TT) + \frac{1}{2}(.48\ Tt) = .60$
$t = (.16\ tt) + \frac{1}{2}(.48\ Tt) = .40$

TABLE 20-4 Gene Frequencies Produced by Random Mating Between Individuals in a Population that Has Genotypic Frequencies of .40 *TT*, .40 *Tt*, and .20 *tt* (Gene Frequencies: .60 *T*, .40 *t*)

		Males		
		TT = .40	*Tt* = .40	*tt* = .20
Females	*TT* = .40	.1600 (1)	.1600 (2)	.0800 (3)
	Tt = .40	.1600 (4)	.1600 (5)	.0800 (6)
	tt = .20	.0800 (7)	.0800 (8)	.0400 (9)

Parents		Offspring		
Mating	Frequency	*TT*	*Tt*	*tt*
TT x *TT* (1)	.1600	.1600		
TT x *Tt* (2) + (4)	.3200	.1600	.1600	
TT x *tt* (3) + (7)	.1600		.1600	
Tt x *Tt* (5)	.1600	.0400	.0800	.0400
Tt x *tt* (6) + (8)	.1600		.0800	.0800
tt x *tt* (9)	.0400			.0400
	1.0000	.3600	.4800	.1600

Gene frequencies among offspring: $T = (.36\ TT) + \frac{1}{2}(.48\ Tt) = .60$
$t = (.16\ tt) + \frac{1}{2}(.48\ Tt) = .40$

its main contribution to evolutionary thought lies in demonstrating that genetic differences in a randomly breeding population will tend to remain constant unless acted upon by external forces: a point that is contrary to the premendelian concept that heredity involves a blending of traits which become dilute with each generation of interbreeding. Table 20-5 offers an outline of major assumptions and steps in the Hardy-Weinberg principle.

The general relationship between gene frequencies and genotype frequencies can be described in algebraic terms by means of the Hardy-Weinberg principle as follows: if p is the frequency of a certain gene in a panmictic population (e.g., T) and q the frequency of its allele (e.g., t), so that $p + q = 1$ (i.e., there are no other alleles), the equilibrium frequencies of the genotypes are given by the terms $p^2(TT)$, $2pq(Tt)$, and $q^2(tt)$. If the gene frequencies of T and t are $p = .6$ and $q = .4$, respectively, the equilibrium frequencies will then be

$$(.6)^2(TT) + 2(.6)(.4)(Tt) + (.4)^2(tt) = .36\ TT + .48\ Tt + .16\ tt$$

This relationship can be visualized by drawing a checkerboard in which the genotype frequencies are the result of random union between alleles that are in the frequencies of p and q (Fig. 20-1). The same results are also produced by **expansion of the binomial** $(p + q)^2 = p^2 + 2pq + q^2$. Therefore with any given p and q and random mating between genotypes, one generation is sufficient to establish an equilibrium condition for the frequencies of genes and genotypes. Once established, the equilibrium condition will per-

TABLE 20-5 Assumptions and Steps in the Hardy-Weinberg Equilibrium

Assumptions	Steps
1. Parents represent a random sample of the gene frequencies in the population	A. Provides gene frequency in parents
2. Genes segregate normally into gametes (heterozygotes for any gene pair produce their two kinds of gametes in equal frequency) 3. Parents are equally fertile (gametes are produced according to the frequency of the parents)	B. Provides gene frequency in gametes
4. The gametes are equally fertile (all have an equal chance of becoming a zygote) 5. The population is large (all the possible kinds of zygotes will be formed in frequencies determined by the gametic frequencies)	C. Provides gene frequency in the gametes that form the zygotes
6. Mating between parents is random (not determined by genotypic preference) 7. Gene frequencies are the same in both male and female parents	D. Provides genotype frequencies in the zygotes
8. All genotypes have equal viability	E. Provides genotype frequencies in adult progeny produced by zygotes
	F. Repeat of steps A, B, C, D, E, etc.

Adapted from Falconer.

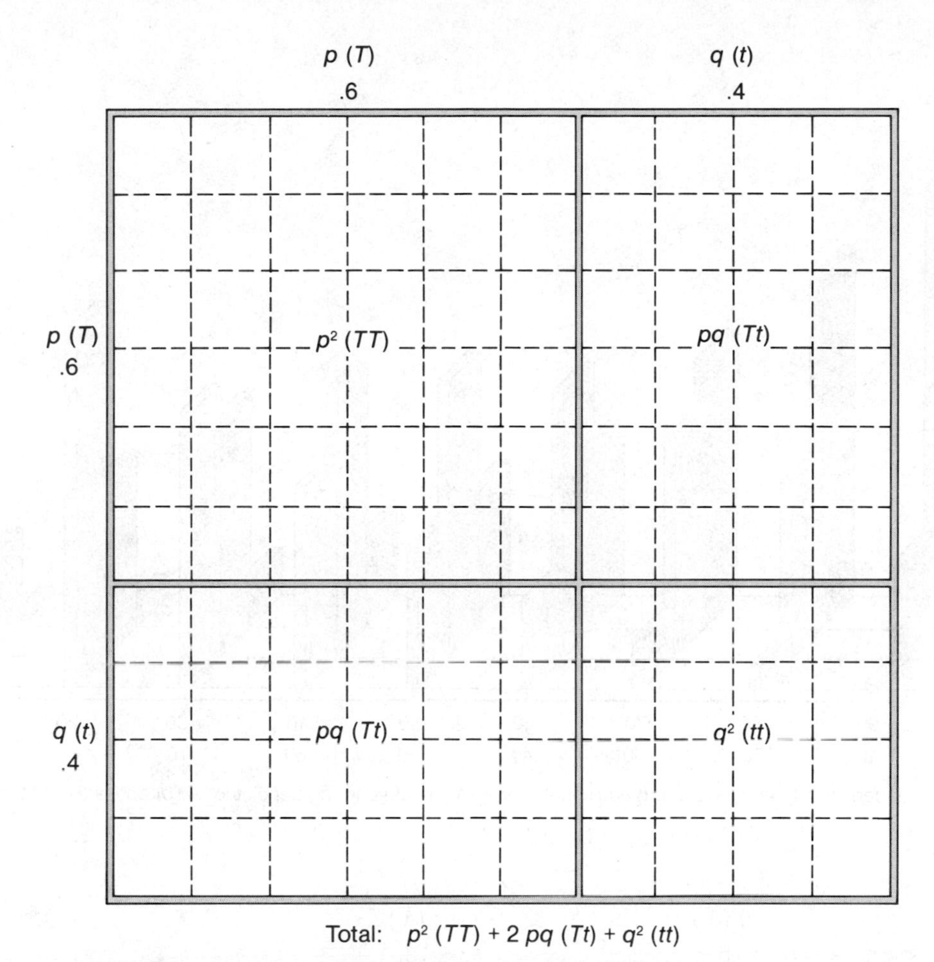

FIGURE 20-1 Genotypic frequencies generated under conditions of random mating for two alleles, *T* and *t*, at a locus when their respective frequencies are $p = .6$ and $q = .4$.

sist until the gene frequencies are changed. Figure 20-2 shows the genotypic frequencies at Hardy-Weinberg equilibrium for a two-allele locus, where the frequency of each allele ranges from zero to one. Note that the frequency of heterozygotes never exceeds .50, but is significantly higher than the frequency of homozygotes for a rare allele (e.g., when *a* is .1, *aa* is .01 but *Aa* is .18).

The presence of two alleles at a locus is only one example for which information may be desired. Instances of genes for which there are more than two alleles at a locus, a common occurrence, we must consider each allelic frequency as an element in a multinomial expansion. For example, if there are only three possible alleles at a locus, A_1, A_2, and A_3, with respective frequencies p, q, and r, so that $p + q + r = 1$, the genotypic equilibrium frequencies are determined from the **trinomial expansion** $(p + q + r)^2$. The six genotypic values are then

$$p^2 A_1A_1 + 2pq A_1A_2 + 2pr A_1A_3 + q^2 A_2A_2 + 2qr A_2A_3 + r^2 A_3A_3$$

Since each haploid gamete contains only a single allele for any one gene locus, zygotic combinations will depend only upon the frequency of each allele (Fig. 20-3), and, as when there are only two alleles, equilibrium is established in a single generation of random mating.

Attainment of Equilibrium at Two or More Loci

Establishment of equilibrium in one generation holds true as long as we consider each single gene locus separately without being concerned about what is happening at other gene loci. If, however, we consider the products of two independently assorting gene-pair differences simultaneously, for instance, *Aa* and *Bb*, the number of possible genotypes increases to 3^2 (i.e., *AABB*, *AABb*, *AaBB*, *AaBb*, etc.). As expected, more terms are now involved in the multinomial expansion, so that if we call p, q, r, and s the gene frequencies of *A*, *a*, *B*, and *b*, respectively, the equilibrium ratios of

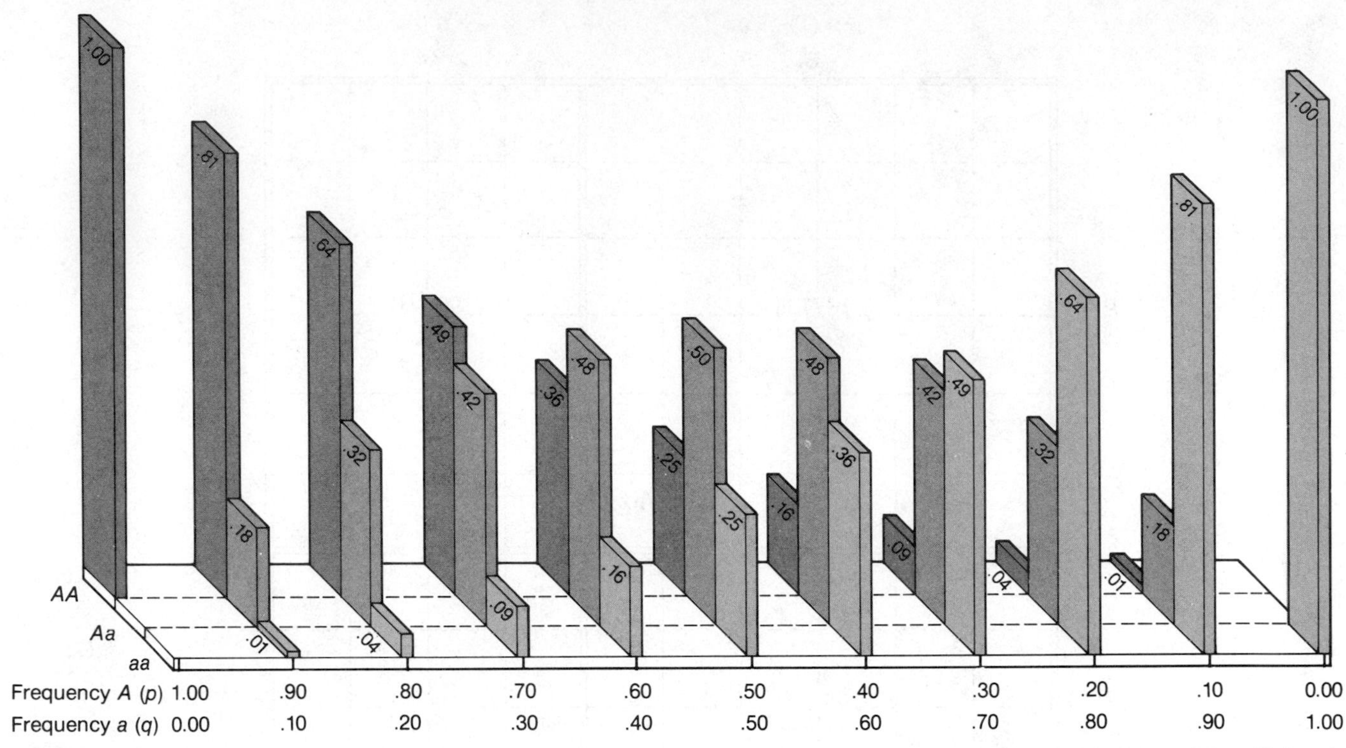

FIGURE 20-2 Genotypic frequencies at Hardy-Weinberg equilibrium for a variety of gene frequencies of A (p) and a (q). (Adapted from Wallace.)

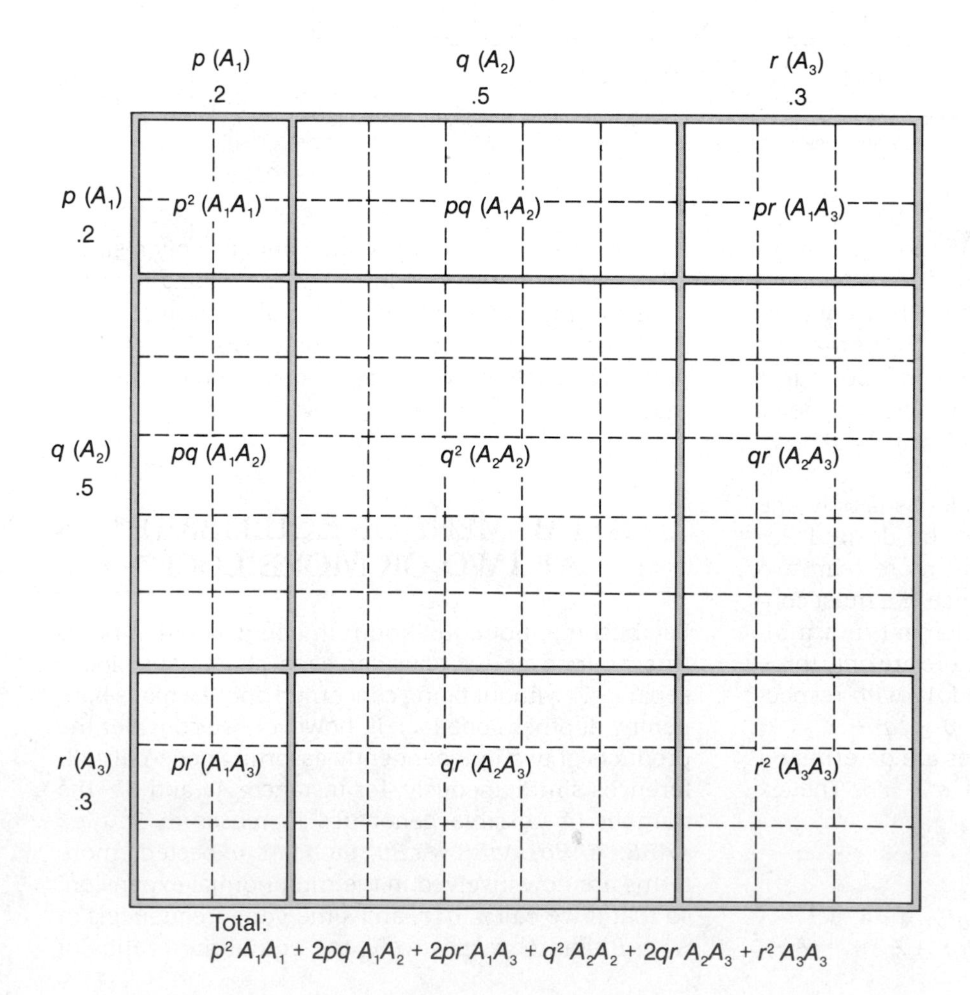

FIGURE 20-3 Genotypic frequencies generated under conditions of random mating when there are three alleles, A_1, A_2, and A_3, present at a locus. For purposes of illustration the respective gene frequencies of these alleles have been given as $p = .2$, $q = .5$, and $r = .3$. Equilibrium genotypic frequencies are therefore .04 A_1A_1, .20 A_1A_2, .12 A_1A_3, .25 A_2A_2, .30 A_2A_3, and .09 A_3A_3.

their genotypes are expressed as $(pr + ps + qr + qs)^2$, or p^2r^2 *AABB*, $2p^2rs$ *AABb*, $2p^2s^2$ *AAbb*, $2pqr^2$ *AaBB*, . . . , q^2s^2 *aabb*.

This equilibrium formula depends on the terms *pr*, *ps*, *qr*, *qs*, which are the equilibrium frequencies of the gametes *AB*, *Ab*, *aB*, and *ab*, respectively. Once the gametic frequencies have reached these equilibrium values, the equilibrium genotypic frequencies will also have been reached. The problem of attainment of equilibrium therefore resolves itself to the time that it takes for the gametic frequencies to reach these values. If we begin only with heterozygotes (*AaBb* × *AaBb*) in which the frequencies of all genes are the same (i.e., $p = q = r = s = .5$), all four types of gametes (*AB*, *Ab*, *aB*, *ab*) are immediately produced at equilibrium frequencies (.25), and genotypic equilibrium is reached within one generation. However, this is the only condition in which equilibrium is reached so rapidly. To take an extreme case, if we begin a population with the genotypes *AABB* and *aabb*, only two type of gametes are produced (*AB* and *ab*) and equilibrium for all genotypes cannot be reached in the next generation since numerous genotypes are missing (i.e., *AAbb*, *aaBB*, etc.). In general, two questions may be asked: (1) What are the expected equilibrium frequencies of gametes? (2) How rapidly are these frequencies achieved?

To deal with these questions we can divide gametes into those in **repulsion** (*Ab* and *aB*) and those in **coupling** (*AB* and *ab*). Since the frequencies of genes in gametes in repulsion are equal to the frequencies of genes in gametes in coupling, we would expect the products of the frequencies of both types of gametes to be equal at equilibrium: $(Ab) \times (aB) = (AB) \times (ab)$. For example, if the frequencies of *A* and *B* are .6 each, and the frequencies of *a* and *b* are .4 each, then at equilibrium $(.24)(.24) = (.36)(.16)$, or both products equal .0576. If there is a difference between the coupling and repulsion products in the initial population, this difference therefore represents the change in gametic frequencies that must occur for equilibrium values to be reached. If we call this difference **disequilibrium**, or *d*, and it is positive so that $(Ab)(aB) - (AB)(ab) = (+)d$, then at equilibrium this fraction will have been added to each of the coupling gametes and subtracted from each of the repulsion gametes. If *d* is negative, the reverse operation will occur. In both cases, disequilibrium will have diminished to zero.

Until the final gametic ratios are reached, half of the difference from equilibrium is reduced each generation, so that within four to five generations more than 90 percent of this difference from equilibrium frequency has been attained by all gametes, or less than 10 percent of disequilibrium value remains. Table 20-6 shows how *d* is calculated and how the changes in gametic frequencies occur until equilibrium is attained. For three gene pairs the speed of approach to equilibrium is even further diminished, and it becomes slower still as more gene pairs are involved.

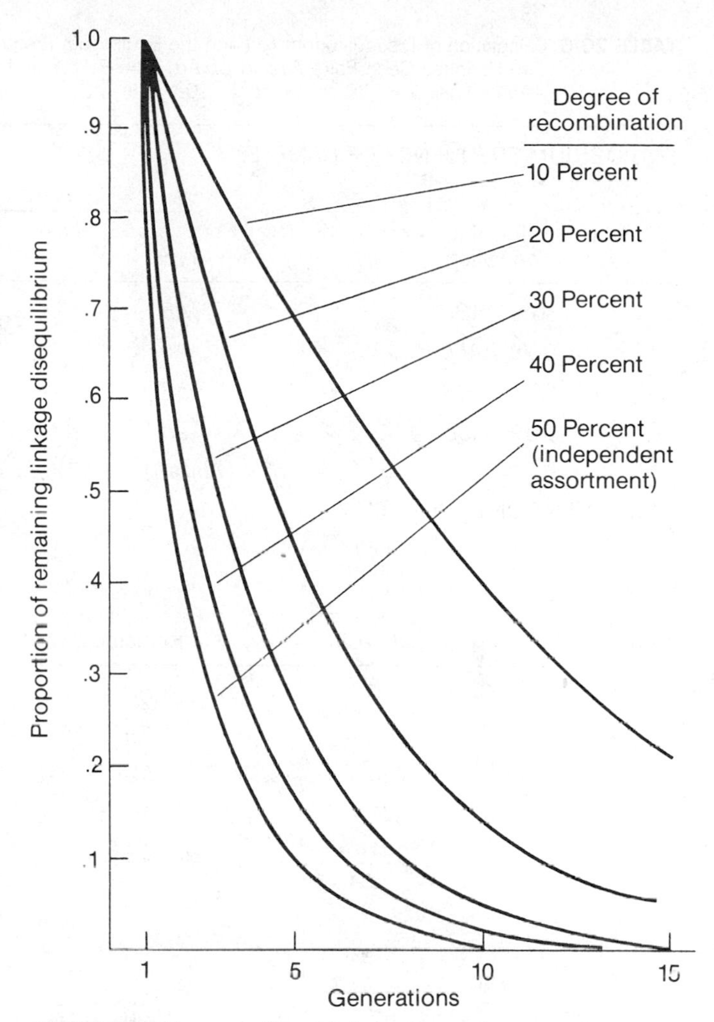

FIGURE 20-4 The proportion of linkage disequilibrium that remains in various generations (starting from an initial value of one) when different degrees of recombination occur between two loci. (Adapted from Strickberger.)

As may be expected, linkage between two loci adds complexity to the attainment of equilibrium since the chances that all the different types of dihybrid gametes will be found depend on crossover frequencies between the two loci. The closer the linkage, the longer it will take for the frequency of coupling gametes to equal the frequency of repulsion gametes. In other words, such **linkage disequilibrium** is dependent on recombination frequency, and lower recombination frequencies between linked loci delay the attainment of equilibrium accordingly (Fig. 20-4). This does not mean that the eventual equilibrium values for linked genes will be any different from those attained in the absence of linkage; *d* is dependent on gametic frequencies and not on linkage. Thus, once equilibrium is attained there is no way of distinguishing linked or unlinked genes except through tests for departures from independent assortment.

TABLE 20-6 Calculation of Disequilibrium (d) and the Equilibrium Frequencies of Gametes for a Population in Which the Frequencies of Two Unlinked Gene Pairs *Aa* and *Bb* Are $A = B = .6$ and $a = b = .4$ and the Initial Genotypic Frequencies Are $AABB = AAbb = aaBB = .30$ and $aabb = .10$

EQUILIBRIUM FREQUENCY OF GAMETES

Initial Population	Gametes: Type	Initial Frequency	Equilibrium Frequency
30% *AABB*	*AB*	.3	$.3 + d$
30% *AAbb*	*Ab*	.3	$.3 - d$
30% *aaBB*	*aB*	.3	$.3 - d$
10% *aabb*	*ab*	.1	$.1 + d$

$$d = (Ab)(aB) - (AB)(ab) = (.3)(.3) - (.3)(.1) = .06$$

ATTAINMENT OF EQUILIBRIUM

Generation	Amount Added (*AB*, *ab*) or Subtracted (*Ab*, *aB*)	Proportion of Disequilibrium Remaining	Gametic Frequencies: *AB*	*Ab*	*aB*	*ab*
1		1.0d	.3	.3	.3	.1
2	.5d	.5d	.33	.27	.27	.13
3	.75d	.25d	.345	.255	.255	.145
4	.875d	.125d	.3525	.2475	.2475	.1525
5	.9375d	.0625d	.35625	.24375	.24375	.15625
•						
•						
•						
Equilibrium	d	0.0d	.36	.24	.24	.16

From Strickberger.

In spite of these theoretical considerations, not all gametes of linked loci in natural populations reach equilibrium frequencies, and various causes have been ascribed to explain this phenomenon. For example, some linkage disequilibrium has been found for genes between which recombination is extremely rare. A number of cases have also been discovered in which linkage disequilibrium appears to be maintained because of advantages conferred upon certain linked allelic combinations. Thus, the third chromosome gene arrangements common in *Drosophila pseudoobscura* probably represent linked groups of genes that are advantageous under particular environmental conditions (p. 195). Because these genes are included within inversions that restrict recombination, their linkage can be preserved for relatively long periods of time, thereby forming **coadapted gene complexes** (Wallace).

SEX LINKAGE

For sex-linked genes the number of possible genotypes is increased because of the difference in number of sex chromosomes between the homogametic and heterogametic sexes. If females are chromosomally XX and males XY, five genotypes can occur for a sex-linked pair of alleles *A* and *a*: three in females (*AA*, *Aa*, *aa*) and two in males (*A* and *a*). If we assign the frequencies p and q to *A* and *a*, respectively, the equilibrium genotypic values in females are the same as for an autosomal gene, p^2 *AA*, $2pq$ *Aa*, and q^2 *aa*, but are expressed directly in hemizygous males as p *A* and q *a* genotypes. Thus, at equilibrium the sex-linked gene frequencies are the same in both sexes, although the genotypes differ.

Assuming all genotypes are equally viable for the sex-linked gene, a difference in gene frequencies between males and females indicates that the population is not at equilibrium. For example, in a population with the proportions .20 *A*:.80 *a* in males and .20 *AA*:.60 *Aa*:.20 *aa* in females, the frequency of *A* is .2 in males and .5 in females. Equilibrium frequencies of all five genotypes can then be calculated by considering that since there is only one X chromosome in males and two in females, the average frequency of a sex-linked gene in a breeding population with equal numbers of males and females is the sum of one third of its frequency in males plus two thirds of its frequency in females, or $p = 1/3\ (p_{\text{males}}) + 2/3\ (p_{\text{females}}) = (p_{\text{males}} + 2p_{\text{females}})/3$. In the present example this translates into an *A* (*p*) frequency of $[.2 + 2(.5)]/3 = 1.2/3 = .4$, and an *a* (*q*) frequency of .6. The equilibrium genotypic values expected are therefore .16 *AA*:.48 *Aa*:.36 *aa* in females, and .4 *A*:.6 *a* in males.

However, in contrast to single autosomal loci with two alleles, equilibrium values for these genotypes will not be reached in a single generation. Because males inherit their X chromosomes only from their mothers, the frequency of a sex-linked gene among them is the same as its maternal frequency, whereas the frequency of the gene among daughters is an average of paternal and maternal frequencies since they each inherit one paternal and one maternal X chromosome. Therefore, if the females in a founding population had a frequency of *A* equal to .5, but the males had an *A* frequency of only .2, the daughters would have an *A* frequency of $(.2 + .5)/2 = .35$, while their brothers would have the .5 frequency of their mothers. Thus, in the first generation of random mating the *A* equilibrium value of .4 will not be reached by the daughters and will be exceeded by the sons. In the second generation, the difference from equilibrium values will be diminished, but this time the sons will be below equilibrium ($A = .35$) and the daughters above [$A = (.35 + .5)/2 = .425$]. As shown in Figure 20-5, each succeeding generation will show a similar reversal but nevertheless achieve a successively closer approximation to the final equilibrium values.

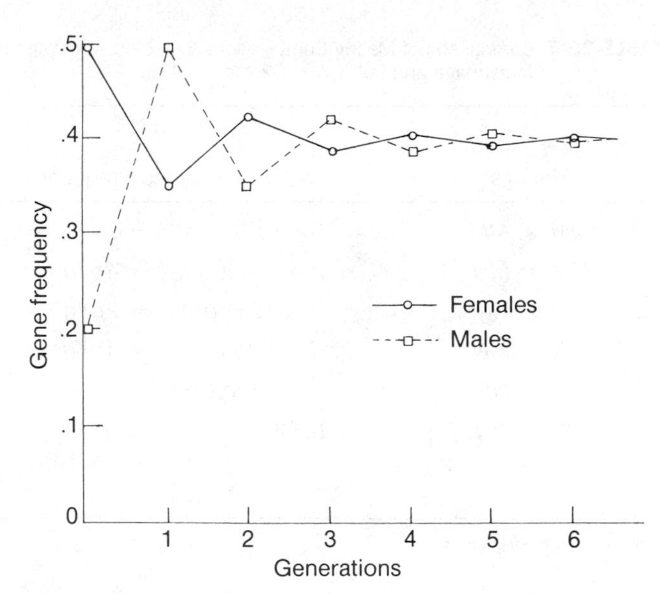

FIGURE 20-5 Frequencies of the sex-linked gene *A* in males and females in successive generations under conditions of random mating when the initial frequency of *A* is .2 in males and .5 in females.

EQUILIBRIA IN NATURAL POPULATIONS

Where all segregants of a gene at a single locus can be scored in a natural population, gene frequencies can be reliably estimated, and the observed genotype frequencies can then be easily compared to their expected equilibrium values. To take a simple example, codominance at the the *MN* blood group locus (using only two alleles, *M* and *N*) enabled Boyd to classify 104 American Ute Indians into genotype frequencies of .59 *MN*, .34 *MN*, and .07 *NN*. Since the gene frequencies are $.59 + .17 = .76$ for *M*, and $.07 + .17 = .24$ for *N*, the expected genotype frequencies are $(.76)^2 = .58$ for *MM*, $2(.76)(.24) = .36$ for *MN*, and $(.24)^2 = .06$ for *NN*. The close correlation between observed and expected genotypic values thus indicates that this population has reached Hardy-Weinberg equilibrium.

When the genotypes of mated couples in the population are known, the assumption of random mating can also be tested. Under random mating the frequencies of the different mating combinations should depend only upon the frequencies of their genotypes. An actual set of data collected by Matsunaga and Itoh provided the blood types of 741 couples (or 1,482 individuals) in a Japanese town in which were found genotypic frequencies of .274 *MM*, .502 *MN*, and .224 *NN*. (Since the gene frequencies are .525 *M* and .475 *N*, the expected equilibrium genotypic frequencies are .276 *MM*, .499 *MN*, and .225 *NN*, again indicating equilibrium for this locus.)

The demonstration for random mating among these individuals is in Table 20-7, with the number of observed matings of different combinations given in the last column. To the left of this column, the expected mating combination frequencies are calculated on the basis of the gene frequencies, *p* for *M* and *q* for *N*. For example, since the frequency of the genotype *MM* is p^2 at equilibrium, the random-mating combination $MM \times MM$ should be p^4. When the frequencies of all expected mating combinations are calculated this way,

TABLE 20-7 Comparison of Mating Combinations Expected According to Random Mating and Those Observed in 741 Couples by Matsunaga and Itoh ($p = .525$, $q = .475$)

Matings	Expected Frequency			Expected Number (Freq. × 741)	Observed Number
$MM \times MM$	$(p^2)(p^2)$	$= p^4$	$= .0760$	56.3	58
$MM \times MN$	$2 \times (p^2)(2pq)$	$= 4p^3q$	$= .2749$	203.7	202
$MM \times NN$	$2 \times (p^2)(q^2)$	$= 2p^2q^2$	$= .1244$	92.2	88
$MN \times MN$	$(2pq)(2pq)$	$= 4p^2q^2$	$= .2487$	184.3	190
$MN \times NN$	$2 \times (2pq)(q^2)$	$= 4pq^3$	$= .2251$	166.8	162
$NN \times NN$	$(q^2)(q^2)$	$= q^4$	$= .0509$	37.7	41
			1.0000	741	741

Adapted from Strickberger.

a comparison between observed and expected shows excellent agreement with the assumption of random mating.

A more common case in natural populations is when the effect of one allele at a locus is completely dominant over another so that the heterozygous genotype (e.g., *Aa*) cannot be phenotypically distinguished from the homozygous dominant (e.g., *AA*). Under such circumstances gene frequencies cannot be obtained directly, as in codominance, since two of the genotypic frequencies (*AA*, *Aa*) are unknown, but must instead rely on the recessive homozygote (*aa*) whose phenotypic frequency coincides with its genotypic frequency. That is, if we assume that Hardy-Weinberg equilibrium has been reached in such a population (p^2 *AA*, $2pq$ *Aa*, q^2 *aa*), the recessive homozygotes are present in a frequency q^2 equal to the square of the recessive gene frequency, q. If, let us say, q^2 is .49, then q is $\sqrt{.49} = .70$, and the frequency of the dominant allele p is $1 - q$, or .30. The homozygous dominants, therefore, have the frequency $p^2 = (.30)^2 = .09$, and the heterozygotes have the frequency $2pq = 2(.30)(.70) = .42$.

One of the consequences of this analysis is that when recessive phenotypes are rare, it is surprisingly common to find that the **carrier heterozygotes**, phenotypically disguised as dominants, are present in relatively high frequency. Albinism, for example, affects only about 1 in 20,000 humans in some populations, or $q^2 = 1/20{,}000 = .00005$. The gene frequency, q, of the albino gene is therefore .007, and the frequency, p, of the non-albino allele is .993. The frequency of heterozygous albino carriers is therefore $2(.993)(.007) = .014$, or approximately 1 in 70 individuals. Thus, there are $.014/.00005 = 280$ times as many heterozygotes for this trait as there are homozygotes. Similarly high proportions of carriers of other recessive traits (Table 20-8) point to the difficulty of eliminating rare deleterious recessive alleles, since they are carried mostly in the unexpressed heterozygous condition.

When more than two alleles are present at a locus, the Hardy-Weinberg equilibrium is based on a multinomial expansion such as that described on page 423. For that example of three alleles (A_1, A_2, A_3), six genotypes are expected and the gene frequency of each allele (p, q, r) can be calculated from the following equations:

$$p = \frac{2(A_1A_1) + (A_1A_2) + (A_1A_3)}{2N}$$

$$q = \frac{2(A_2A_2) + (A_1A_2) + (A_2A_3)}{2N}$$

$$r = \frac{2(A_3A_3) + (A_1A_3) + (A_2A_3)}{2N}$$

where A_1A_1, A_1A_2, A_1A_3, and so on refer to the numbers of genotypes in each category, and N refers to the total number of individuals scored.

A system of this type is found in human populations bearing different forms of the red blood cell enzyme acid phosphatase which can be scored into six different phenotypes, AA, BB, CC, AB, BC, or AC, as determined by all possible combinations of the alleles *A*, *B*, and *C* at a single locus. As shown in Table 20-9, investigations of a Brazilian population indicate that the observed phenotypic frequencies of the acid phosphatase combinations conform closely to those expected according to the Hardy-Weinberg equilibrium. Although there are exceptions (Spiess), conformities to Hardy-Weinberg equilibrium seem to be quite common for both autosomal and sex-linked genes and have been demonstrated in a variety of sexually outbreeding organisms. In general, as explained previously, these studies emphasize that populational gene and genotype frequencies do not change without cause.

TABLE 20-8 Genotype Frequencies for Human Diseases Caused by Recessive Genes

Disease	Population	Gene Frequency (q)	Frequency of Homozygotes (q^2)	Frequency of Heterozygous Carriers ($2pq$)	Ratio of Heterozygous Carriers to Homozygotes ($2pq{:}q^2 = 2p{:}q$)
Achromatopsia	Pingelap (Caroline Islands)	.22	1 in 20	1 in 2.8	7:1
Sickle-cell anemia	Africa (some areas)	.2	1 in 25	1 in 3	8:1
Albinism	Panama (San Blas Indians)	.09	1 in 132	1 in 6	21:1
Ellis-van Creveld syndrome	Old Order Amish	.07	1 in 200	1 in 8	26:1
Sickle-cell anemia	U.S. Blacks	.04	1 in 625	1 in 13	48:1
Cystic fibrosis	U.S. Whites (Conn.)	.032	1 in 1,000	1 in 16	60:1
Tay-Sachs disease	Ashkenazi Jews	.018	1 in 3,000	1 in 28	108:1
Albinism	Norway	.010	1 in 10,000	1 in 50	198:1
Phenylketonuria	United States	.0063	1 in 25,000	1 in 80	314:1
Cystonuria	Great Britain	.005	1 in 40,000	1 in 100	400:1
Galactosemia	United States	.0032	1 in 100,000	1 in 159	630:1
Alkaptonuria	Great Britain	.001	1 in 1,000,000	1 in 500	2,000:1

From Strickberger.

TABLE 20-9 Comparison of Observed Acid Phosphatase Phenotypes and Those Expected According to Hardy Weinberg Equilibrium in a Sample of 369 Brazilian Individuals

Phenotypes	AA	BB	CC	AB	AC	BC
Observed	15	220	0	111	4	19
Expected	14.4	219.9	0.4	112.2	4.4	17.7

From Lai et al.

INBREEDING

One set of conditions that interferes with the Hardy-Weinberg equilibrium is **nonrandom mating**. An important example of this occurs when related individuals of similar genotype mate preferentially with each other in a phenomenon called **inbreeding**. (An extreme form of inbreeding is when two gametes of a single individual unite to form a fertile zygote, **self-fertilization**.) Although the effect of inbreeding will not change the overall gene frequency, it will lead to an excess of homozygous genotypes. In the case of a rare recessive allele, this means that inbreeding will cause it to appear in greater homozygous frequency than under random mating, thus providing selection with an increased opportunity to act upon rare recessives.

Inbreeding is usually quantified by an **inbreeding coefficient**, F, which measures the probability that the two alleles of a gene in a diploid zygote are identical—descended from a single ancestral allele. Possession of identical alleles means, of course, homozygosity, and F can range from one (complete homozygosity) to zero (complete heterozygosity).

By this definition, if inbreeding occurs at a particular locus with only two alleles, A and a, there will be a proportion of F identical homozygotes. Of this inbred proportion, some will be AA and some aa, the frequencies of each depending upon their respective population gene frequencies p and q. Thus there will be pF AA and qF aa genotypes produced by inbreeding. In addition to these, however, the remaining individuals in this population $(1 - F)$ will bear genotypes whose frequencies are determined according to the Hardy-Weinberg equilibrium of p^2 AA, $2pq$ Aa, and q^2 aa. The three genotypes will therefore have the following frequencies:

$$AA = p^2(1 - F) + pF = p^2 - p^2F + pF = p^2 + pF(1 - p) = p^2 + pqF$$

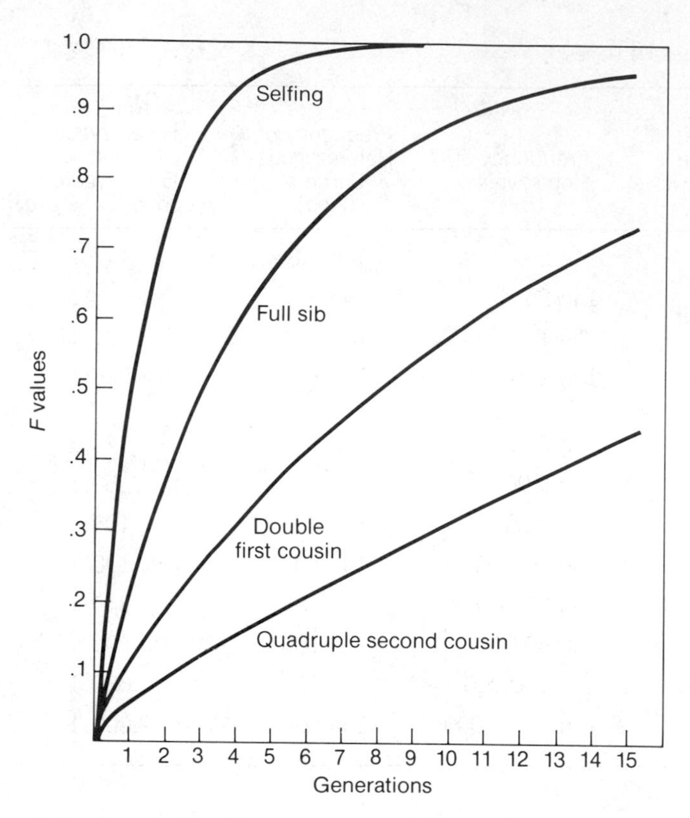

FIGURE 20-6 Inbreeding coefficients at generations 1 to 15 for four different systems of inbreeding (pedigrees given in Strickberger). Formulas for calculating F in other inbreeding systems can be obtained or derived from the basic work in this field by Wright (1921).

$$Aa = 2pq(1 - F) = 2pq - 2pqF$$

$$aa = q^2(1 - F) + qF = q^2 - q^2F + qF = q^2 + qF(1 - q) = q^2 + pqF$$

It is now easy to see that the increase in the frequency of each type of homozygote by a factor of pqF comes from an equivalent reduction in the frequency of heterozygotes ($-2pqF$). Note also that this reduction in heterozygotes affects the gene frequencies p and q equally, so that only the genotypic frequencies are changed. When inbreeding is absent, $F = 0$, and the above equations reduce to the Hardy-Weinberg frequencies p^2 AA, $2pq$ Aa, and q^2 aa. When inbreeding is complete, $F = 1$, $2pq - 2pqF = 0$, and the only remaining genotypes are pAA and qaa.

The mating system in which inbreeding is greatest is self-fertilization (e.g., hermaphrodites) where F is equal to .5 in the first generation and approaches one within four or five generations. As shown in Figure 20-6, any other mating scheme slows the rate of inbreeding.[1] Similarly, as shown in Figure 20-7, size of the population can also affect inbreeding, since the smaller the size, the greater the opportunity for related individuals to mate.

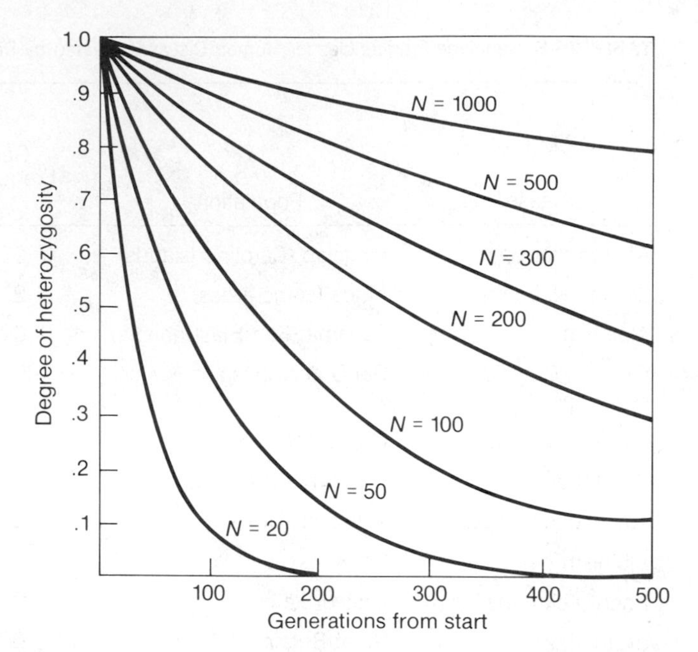

FIGURE 20-7 Degrees of heterozygosity remaining in populations of different sizes after given generations of random union between gametes. Calculations are based on 1.00 as the initial degree of heterozygosity (or when $F = 0$). (Adapted from Strickberger.)

Although some degree of inbreeding occurs in most outbreeding populations, significant amounts of inbreeding can cause **inbreeding depression** in which rare deleterious recessives may now appear with increased homozygous frequency. Thus, if a recessive disease with genotype aa occurs with frequency q^2 in a random outbred population, its frequency will be increased by pqF in an inbred population as derived above. The ratio of inbred to outbred frequency for the homozygous recessive will therefore be

$$\frac{q^2 + pqF}{q^2} = \frac{q(q + pF)}{q^2} = \frac{q + pF}{q^2}$$

[1]Systems such as brother-sister mating and first cousin mating in which individuals mate on the basis of their genetic relationship can also be called genetic assortative mating. Phenotypic similarity, however, may also be a cause for preferential mating and is certainly a practice followed in many human societies in which matings occur on the basis of characteristics such as height, color, facial form, muscular build, and intelligence. In such phenotypic assortative matings homozygosity is also increased, but only for those loci involved in the trait(s) upon which the preferred matings are based. This is in contrast to the genetic assortative mating of inbreeding which tends to increase homozygosity at all loci. A further type of mating practice is disassortative mating in which individuals of unlike genotype or phenotype form mating pairs, thereby preventing inbreeding and helping to maintain heterozygosity. There are various examples of such systems including alleles in plants which cause sterility of male gametes when they attempt to fertilize ova of the same genotype (self-sterility alleles).

Obviously, if q is large and F is small, the inbreeding increment pF will be relatively small, and the increased frequency of homozygous recessives will hardly be noticeable. However, if q is very small (rare) and p is large, then pF provides a notable increase in recessives even when F is fairly small. For example, if q is .5, first cousin mating (F = .0625) will produce an inbred-to-outbred ratio of homozygotes of

$$\frac{q + pF}{q} = \frac{.5 + (.5)(.0625)}{.5} = \frac{.53125}{.5} = 1.06$$

However, if q is .005, this ratio increases to .067/.005 = 13.4. When q = .0005, the increase of homozygotes because of first cousin mating is .0630/.0005, or 126 times that of randomly bred populations.

Should homozygous recessives have a quantitative effect on one or more traits, inbreeding would cause the measured values of these traits to tend in the direction of recessive values. Thus, in various outbred populations, such as corn, inbreeding depression can cause a reduction in height, yield, and other characters. Inbreeding depression, however, is not a universal phenomenon in all species, certainly not in many species that are normally self-fertilized and have eliminated most or all of their deleterious recessives. On the whole, there is ample evidence that most normally cross-fertilizing species will show deterioration upon consistent inbreeding, although some strains may escape because they carry relatively few deleterious recessive genes.

SUMMARY

Following the publication of the *Origin of Species*, Darwin proposed that natural selection operated on small, continuous hereditary variations, while Galton and others maintained that variations were sharp and discontinuous. The controversy was resolved when it was shown that several genes, each with small effect, can influence the expression of a single phenotypic trait that has a large effect. By the 1930's it became clear that evolution is a population phenomenon which can be represented as a change in gene frequencies in a population because of the action of various natural forces such as selection and genetic drift, and these changes can lead to differences between races, species, and higher taxa. Along with other concepts, this populational view of evolution became known as the Neo-Darwinian (modern) synthesis.

Gene frequencies and the gene pool are two major attributes of a population, which can be defined as a group of potentially interbreeding organisms. Gene frequency is the ratio of the different alleles of a gene in a population without regard to their homozygosity or heterozygosity. A gene pool consists of all alleles in the gametes of a population and therefore represents all of the genes available for the next generation.

According to the Hardy-Weinberg principle, gene frequencies are conserved in a random-mating population unless external forces act upon it, and the equilibrium genotype frequencies (e.g., $p^2 + 2pq + q^2$) derive from the gene frequencies.

If there are two or more pairs of independently assorting genes, there are many more possible genotypes, and the more gene pairs, the longer it will take to achieve overall genotypic equilibrium. In the case of linkage, the higher the frequency of recombination between linked genes, the shorter the time needed to reach equilibrium. When genes are linked on the X chromosome, gene frequencies at equilibrium will be equal in both sexes, but this may take a number of generations if there is an initial difference in frequencies between the two sexes.

In natural populations genotype frequencies can be determined quite easily if no allele is dominant, and such observations generally show that Hardy-Weinberg equilibrium has been achieved. If one of the two alleles is dominant, gene frequencies can be obtained by assuming Hardy-Weinberg equilibrium and using the frequency of homozygous recessive individuals as q^2 in the genotypic equilibrium formula. When this is done for various recessive conditions present in low frequency, it appears that the frequency of heterozygous "carriers" is surprisingly high. In the case of multiple alleles, genotype frequencies can be calculated using a multinomial expansion of the Hardy-Weinberg equation. Most studies indicate that gene pools are quite stable and generally remain at equilibrium unless selection or some other conditions interfere.

Inbreeding does not affect gene frequencies but does increase homozygosity, allowing relatively rare recessive alleles to be expressed. If these alleles are deleterious, inbreeding depression may result.

KEY TERMS

binomial expansion
biometricians
carrier heterozygotes
coadapted gene complexes
conservation of gene frequencies
continuous variation
coupling
deme
discontinuous variation
disequilibrium
equilibrium
gene frequencies
gene pool
Hardy-Weinberg equilibrium
inbreeding
inbreeding coefficient
inbreeding depression
linkage disequilibrium
macromutations
mendelian population
mendelians
modern synthesis
mutationists
Neo-Darwinian synthesis
nonrandom mating
panmixia
population genetics
pure lines
quantitative characters
random mating
repulsion
saltation
selectionists
self-fertilization
trinomial expansion

REFERENCES

Castle, W. E., and J. C. Phillips, 1914. *Piebald Rats and Selection*. Carnegie Inst. of Washington, Publ. No. 195, Washington, D.C.

Crow, J. F., 1986. *Basic Concepts in Population, Quantitative, and Evolutionary Genetics*. W. H. Freeman, New York.

Crow, J. F., and M. Kimura, 1970. *An Introduction to Population Genetics Theory*. Harper & Row, New York.

East, E. M., 1916. Studies on size inheritance in *Nicotiana*. *Genetics*, **1**, 164-176

Falconer, D. S., 1981. *Introduction to Quantitative Genetics*, 2nd ed. Longman, London.

Fisher, R. A., 1930. *The Genetical Theory of Natural Selection*. Clarendon Press, Oxford. (Second edition 1958, Dover, New York.)

Haldane, J. B. S., 1932. *The Causes of Evolution*. Harper & Row, New York. (Reprinted 1966, Cornell Univ. Press, Ithaca, N. Y.)

Hardy, G. H., 1908. Mendelian proportions in a mixed population. *Science*, **28**, 49-50.

Hartl, D. L., and A. G. Clark, 1989. *Principles of Population Genetics*, 2nd ed.. Sinauer Associates, Sunderland, Mass.

Hedrick, P. W., 1983. *Genetics of Populations*. Science Books International, Boston.

Johannsen, W., 1903. *Über Erblichkeit in Populationen und in reinen Linien*. G. Fischer, Jena.

Lai, L., S. Nevo, and A. G. Steinberg, 1964. Acid phosphatases of human red cells: predicted phenotype conforms to a genetic hypothesis. *Science*, **145**, 1187-1188.

Matsunaga, E., and S. Itoh, 1958. Blood groups and fertility in a Japanese population, with special reference to intrauterine selection due to maternal-fetal incompatibility. *Ann. Hum. Genet.*, **22**, 111-131.

Nilsson-Ehle, H., 1909. Kreuzungsuntersuchungen an Hafer und Weisen. *Lunds Univ. Aarskr. N. F. Afd. Ser. 2, Vol. 5*, No. 2, pp. 1-122.

Provine, W. B., 1971. *The Origins of Theoretical Population Genetics*. Univ. of Chicago Press, Chicago.

Spiess, E. B., 1977. *Genes in Populations*. John Wiley, New York.

Strickberger, M. W., 1985. *Genetics*, 3rd ed. Macmillan, New York.

Wallace, B., 1981. *Basic Population Genetics*. Columbia Univ. Press, New York.

Weinberg, W., 1908. Über den Nachweis der Vererbung beim Menschen. *Jahreshefte des Vereins für Vaterändlische Naturkunde in Württemburg*, **64**, 368-382.

Wright, S., 1921. Systems of mating. *Genetics*, **6**, 111-178.

———, 1931. Evolution in Mendelian populations. *Genetics*, **16**, 97-159.

Yule, G. U., 1902. Mendel's laws and their probable relations to intra-racial heredity. *New Phytologist*, **1**, 193-207, 222-238.

21

CHANGES IN GENE FREQUENCIES

n order for populations to evolve, that is, to change their gene frequencies, mutation must first introduce the nucleotide differences upon which such changes are based. The mere appearance of new genes (alleles), however, is no guarantee that they will persist or prevail over others. For example, there is no certainty that a newly mutated gene such as *a* (e.g., $A \rightarrow a$) will be transmitted, since its carrier (e.g., *Aa*) may or may not survive and may or may not mate. Even if the *Aa* mutant carrier does mate (*Aa* x *AA*), the chances of transmission of *a* to the next generation is reduced since a significant proportion (40 percent) of matings in most stable populations produce families with zero surviving offspring (*a* is lost) or only one offspring (*a* has a 50 percent chance of being lost during meiosis). Larger families may also lose the gene since, for example, even when the *Aa* x *AA* family produces two offspring the mutant gene has a 25 percent chance ($.5 \times .5$) of not being transmitted to either of them. Fisher has calculated that the chances that a newly mutated gene may be eliminated within one generation is more than 33 percent because of its possible random loss in families of such different sizes. By the time 30 generations have passed after its introduction, this probability of elimination has risen to almost 95 percent. To explain the persistence of many mutations and their increase in frequency we must therefore look elsewhere than the original mutational event.

MUTATION RATES

One factor that can be expected to affect gene frequency is the frequency of mutation. Obviously if gene *A* continually mutates to *a* and the reverse mutation never occurs, the chances improve that *a* will increase in frequency with each generation. Given a long enough period of time and a persistent mutation rate in a population of constant size, *a* can eventually replace *A*. Of course, the mutation rate does not always occur in only one direction. For example, if *u* is the mutation rate of *A* to *a*, the allele *a* may mutate back to *A* with frequency *v*. We can estimate these effects quantitatively by calling the initial frequencies of alleles *A* and *a*, p_0 and q_0, respectively, and noting that a single generation of mutation will produce a frequency of *A* equal to $p_0 + vq_0$ and a frequency of *a* equal to $q_0 + up_0$. If we now confine our attention to only one of the alleles, a, it is clear that it has gained the fraction up_0 (new a alleles) but lost the fraction vq_0 (new A alleles). In other words, the change in the frequency of a, called delta q (Δq), can be expressed as $\Delta q = up_0 - vq_0$. Thus, if p were relatively large and q small, Δq would be large and q would increase rapidly; when q became larger and p became smaller, Δq would diminish. The point at which Δq is zero—that is, the point where there is no further change and p and q are balanced in relation to their mutation frequencies—is known as the **mutational equilibrium** (frequency $a = \hat{q}$, or "q hat"): $\Delta q = 0 = up - vq$, or $up = vq$ at $\hat{q}$. However, since there are only two alleles, A and a, $p = 1 - q$, which leads to

$$up = vq$$

$$u(1 - q) = vq$$

$$u = uq + vq = q(u + v)$$

$$\hat{q} = \frac{u}{u + v}$$

The same procedure applied to the frequency of A gives $\hat{p} = v/(u + v)$, so that $\hat{p}/\hat{q} = [v/(u + v)]/[u/(u + v)] = v/u$. Thus when there is equality between the mutation rates, that is, $u = v$, the **equilibrium gene frequencies** $\hat{p}$ and $\hat{q}$ will be identical. If the mutation rates differ, the equilibrium frequencies will also differ. For example, if $u = .00005$ and $v = .00003$, the equilibrium frequency $\hat{q}$ equals $5/8 = .625$ and $\hat{p} = 3/8 = .375$. However, the rate at which this equilibrium frequency is reached by mutation is usually quite slow and can be derived by the methods of calculus from Δq as

$$(u + v)n = \log_e [(q_0 - \hat{q})/(q_n - \hat{q})]$$

where n is the number of generations required to reach a frequency q_n when starting with a frequency q_0. For the example just considered, the number of generations necessary for q to increase from a frequency of one eighth to three eighths is

$$(.00008)n = \log_e \frac{.125 - .625}{.375 - .625}$$

$$= \log_e 2.00 = .69315$$

$$n = \frac{.69315}{.00008} = 8{,}664 \text{ generations}$$

Thus the approach to equilibrium based on the usually observed mutation rates of 5×10^{-5} or less (Table 10-4) is very slow, and mutational equilibrium is probably rarely if ever reached, especially since mutation rates are probably not constant. As a rule, therefore, the attainment of mutational equilibrium does not appear to be the sole cause that accounts for existing gene frequencies. A more efficient mechanism that can help explain how gene frequencies change is to look at the effect of selection, the "scrutinizing process" proposed by Darwin.

SELECTION

The fact that genotypes can differ in viability and fertility can evidently produce important effects on their frequencies. Obviously if individuals carrying gene A are more successful in producing viable and fertile offspring than individuals carrying its allele a, then the frequency of the former will tend to increase relative to the latter. The wide variety of mechanisms that affect the reproductive success of a genotype is known collectively as **selection**, and the extent to which a genotype contributes to the offspring of the next generation is commonly known as its **fitness**, **selective value**, or **adaptive value**. That is, selection can be described as a composite of the forces that limit the reproductive success of a genotype and fitness as the comparative ability of a genotype to withstand selection. The genetic effect of selection on a particular trait in a population therefore is confined to fitness differences between the different alleles that affect that particular trait. Thus, when the selective process operates, gene frequencies will tend to change between generations, unless the population has reached a genetic equilibrium, as described later.[1]

In simplest form, fitness and selection are measured by the number of fertile offspring produced by one genotype compared to those produced by another. For example, if individuals of genotype A produce an average of 100 offspring that reach full reproductive maturity while genotype a individuals produce only 90 in the same environment, the adaptive value of a relative to A is reduced by ten offspring, or the fraction $10/100 = .1$. If we designate the adaptive value of a genotype as W and the selective force acting to reduce its adaptive value as s (the **selection coefficient**) then

[1]If we define evolution as hereditary changes over time, selection, although important, is not the only process that can cause such changes; other evolutionary mechanisms considered in this chapter (mutation, migration, and random genetic drift) can also affect gene frequencies. In other words, the relative fitness of a genotype may not be the only reason for its survival, and the statement that evolutionary theory merely proposes the "survival of the fittest" is misleading and incorrect. ("Survival of the fittest" is often claimed to be a circular, tautological, or unprovable principle because it defines those who survive as fittest, and fittest as those who survive. Some philosophers, such as Waters, therefore propose that this principle be abandoned, whereas others, such as Resnik, suggest that this principle has practical value.)

we can say that $W = 1$ and $s = 0$ for A in the above example, and $W = .9$ and $s = .1$ for a. The relationship between W and s is therefore simply $W = 1 - s$, or $s = 1 - W$.

Selection against a genotype may occur in either the haploid (gametic) or diploid (zygotic) stage or both, depending at which of these stages gene expression influences survival or fertility (Fig. 21-1). In any of these stages, selection may be obvious or subtle, its effects ranging from complete lethality or sterility ($s = 1$) to only slight reductions in adaptive value (e.g., $s = .01$). When selection occurs among haploids, there is of course no difference between dominant and recessive genes, since both kinds of alleles are phenotypically expressed in their carriers. Thus, as we might expect, the effect of selection on haploids is much more rapid and direct than on diploids because **deleterious recessive** alleles cannot be hidden from selection among heterozygotes as they are in diploids. Table 21-1 provides estimates of the number of generations necessary to change the frequency of deleterious genes in haploids under a variety of selective conditions. Note that compared to diploids (see Table 21-3) a **deleterious lethal** gene ($s = 1$) is completely eliminated in haploids in one generation and even lesser selection coefficients result in relatively rapid gene-frequency changes.

In most higher animals and plants selection takes place primarily in the diploid or zygotic stage. In diploids however, there are three possible genotypes for a single gene difference (e.g., *AA, Aa, aa*), so that the effectiveness of selection depends, among other things, upon the degree of dominance. Table 21-2 shows calculation of the change in gene frequency (Δq) of *a* for one generation when complete dominance exists and selection occurs only against the recessive *aa*.

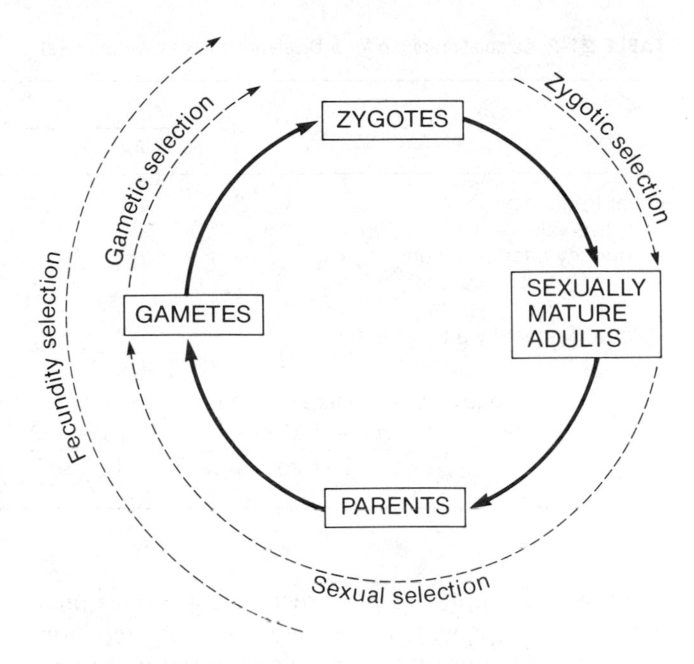

FIGURE 21-1 Types of selection acting on the life stages of an organism. (Adapted from Christiansen.)

Table 21-3 summarizes the number of generations necessary to effect changes in such deleterious recessive gene frequencies when selection is carried forth over periods of time and for various selection coefficients. Note that the initial change in gene frequency from .99 to .10 is relatively rapid for practically all selection coefficients. Further reductions in gene frequency are considerably slower: to reduce the gene frequency below .01 may take thousands of generations, even when the selection coefficient is relatively high. As indicated in the previous chapter, the reason for the relative inefficiency of selection against rare

TABLE 21-1 Number of Generations Required for Given Frequency Changes (q_0 to q_n) of a Deleterious Gene Under Different Selection Coefficients in Haploids

Change in Gene Frequency		Number of Generations for Different *s* Values					
From (q_0)	To (q_n)	$s = 1$	$s = .80$	$s = .50$	$s = .20$	$s = .10$	$s = .01$
.99	.90	1	3	5	12	24	240
.90	.75		1	2	5	11	110
.75	.50		1	2	5	11	110
.50	.25		1	2	5	11	110
.25	.10		1	2	5	11	110
.10	.01		3	5	12	24	240
.01	.001		3	5	12	24	240
.001	.0001		3	5	12	23	230

Adapted from Strickberger.

TABLE 21-2 Calculation of Δq for a Deleterious Recessive Gene (a)

	Genotypes			
	AA	Aa	aa	Total
Initial frequency	p^2	$2pq$	1	
Adaptive value	1	1	$1 - s$	
Frequency after selection	p^2	$2pq$	$q^2(1 - s)$	$p^2 + 2pq + q^2 - sq^2 = 1 - sq^2$
Relative frequency after selection	$\frac{p^2}{1 - sq^2}$	$\frac{2pq}{1 - sq^2}$	$\frac{q^2(1 - s)}{1 - sq^2}$	

Δq = relative frequency of a after selection − initial frequency of a

$$\Delta q = \frac{pq + q^2(1 - s)}{1 - sq^2} - q = \frac{q(1 - sq)}{1 - sq^2} - q = \frac{q - sq^2 - q + sq^3}{1 - sq^2} = \frac{-sq^2(1 - q)}{1 - sq^2}$$

recessives is simply that most recessive genes are present in heterozygotes where they are protected from selection: the more rarely a gene is found in a population, the more frequently does it occur in heterozygotes compared to homozygotes (see Table 20-8).

The selective situation can, of course, be reversed so that the dominant allele is selected against and the recessive is favored. When this occurs selection will obviously be more effective, since the **deleterious dominant** gene is subject to selection in all genotypes in which it occurs. For example, should a dominant allele become lethal, its frequency is reduced to zero in a single generation. However, as the selection coefficient against the dominant allele decreases, replacement by the recessive is considerably slower. For the general case, selection against a dominant allele of gene frequency p results in a change of

$$-sp(1 - p)^2/[1 - sp(2 - p]$$

as shown in Table 21-4. Note that if s is small, the denominator is close to 1, and Δp is effectively equal to $-sp(1 - p)^2$. Since $1 - p$ is q and p is $1 - q$, this means that Δp is now $-sq^2(1 - q)$, or Δp is identical to Δq for a deleterious recessive at low selection coefficients. Under these conditions, we may apply the values of Table 21-3 in reverse order. That is, for a selection coefficient of .10 in favor of a recessive allele, 90,023 generations are necessary to increase its frequency from .0001 to .001, or to reduce the frequency of the dominant allele from .9999 to .9990. Subsequent changes in frequency are more rapid as the favored recessive homozygotes become more frequent.

When dominance of the advantageous allele is incomplete, heterozygotes will show the effect of a deleterious gene since the heterozygous phenotype is at least partially deleterious. If dominance is absent completely and the heterozygote has a phenotype exactly intermediate to that of the two homozygotes, its

TABLE 21-3 Number of Generations Required for a Given Change in Frequency (q_0 to q_n) of a Deleterious Recessive Allele in Diploids Under Different Selection Coefficients

Change in Gene Frequency		Number of Generations for Different s Values					
From (q_0)	To (q_n)	$s = 1$	$s = .80$	$s = .50$	$s = .20$	$s = .10$	$s = .01$
.99	.90	}	3	5	13	25	250
.90	.75	} 1	2	3	7	13	132
.75	.50	}	2	4	9	18	177
.50	.25	2	4	6	15	31	310
.25	.10	6	9	14	35	71	710
.10	.01	90	115	185	462	924	9240
.01	.001	900	1128	1805	4512	9023	90231
.001	.0001	9000	11515	18005	45011	90023	900230

Adapted from Strickberger.

TABLE 21-4 Single Generation Changes in Gene Frequency for Diploid Genotypes Subject to Given Selection Coefficients Under Different Conditions of Dominance

Dominance Relations for the Three Given Genotypes	Adaptive Values for Genotype Frequencies Initially in Hardy-Weinberg Equilibrium: *AA* p^2	*Aa* $2pq$	*aa* q^2	Change in Gene Frequency*
Complete dominance: selection against the recessive allele	1	1	$1 - s$	$\frac{-sq^2(1 - q)}{1 - sq^2}$
Complete dominance: selection against the dominant allele	$1 - s$	$1 - s$	1	$\frac{-sp(1 - p)^2}{1 - sp(2 - p)}$
Absence of dominance: selection against the *a* allele occurs also in the heterozygote	1	$1 - s$	$1 - 2s$	$\frac{-sq(1 - q)}{1 - 2sq}$
Overdominance: selection against both homozygotes	$1 - s$	1	$1 - t$	$\frac{pq(ps - qt)}{1 - p^2s - q^2t}$

*As mentioned in Chapter 20, mathematical derivations for various formulae have been omitted for simplicity but can be found in many population genetics textbooks.

selection coefficient will be exactly half that in the deleterious homozygotes. As shown in Table 21-4, the resultant change in gene frequency in one generation $[-sq(1 - q)]/[1 - 2sq]$ is almost identical to that for gametic selection $[-sq(1 - q)]/[1 - sq]$. In other words, the absence of dominance uncovers deleterious alleles and makes all of them available for selection, allowing rapid changes in gene frequencies mostly on the order of those observed in Table 21-1. The effectiveness of selection is therefore strongly dependent upon the degree to which the deleterious gene is expressed in the heterozygote. Since most recessive genes are believed to have some heterozygous expression, selection efficiency for or against them probably falls between the extremes of slow progress for complete dominance and rapid progress for absence of dominance.

HETEROZYGOUS ADVANTAGE

The examples of selection just considered always go in one direction, toward **elimination** of the deleterious allele and establishment or **fixation** of the favored allele. As long as the selection coefficient does not change, an equilibrium between favored and unfavored alleles is impossible without new mutations. Various conditions, however, permit the establishment of an equilibrium through which both alleles may remain indefinitely within the population. One such condition, **overdominance**, occurs when the heterozygote has superior reproductive fitness to both homozygotes.[2]

In general, if the heterozygote *Aa* has an adaptive value of one while the fitnesses of the homozygotes *AA* and *aa* are reduced by the selective coefficients s and t, respectively, the change in frequency of a in a single generation is that shown in Table 21-4. When Δq is zero, equilibrium has been reached and there will be no further change in gene frequency. Note that there are three possible conditions that will cause the numerator $[pq(ps - qt)]$ to be equal to zero and therefore Δq to equal zero. Under the first two conditions, when either p or q are zero, both alleles will not be present in the population at the same time, and balance, or equilibrium, will be absent. The third condition occurs when $ps = qt$, so that the numerator of Δq is $pq(0) = 0$. When this happens, the following relationships can be derived:

$$ps = qt$$

add qs to both sides	add pt to both sides
$ps + qs = qt + qs$	$ps + pt = qt + pt$
$s(p + q) = q(s + t)$	$p(s + t) = t(p + q)$

(Now, since $p + q = 1$)

$$q = \frac{s}{s + t} \qquad p = \frac{t}{s + t}$$

[2]The superiority of the heterozygote, often called **heterosis** or **hybrid vigor**, may show itself in the improvement of fitness characters such as longevity, fecundity, and resistance to disease. An oft-cited example of heterosis is the dramatic increase in agricultural yield by hybrid corn, achieved by crossing selected inbred lines. However, it is still debated whether such hybrid vigor arises from the superiority of the heterozygote for particular gene differences (overdominance) or from other causes such as the introduction of favorable dominant alleles at particular loci which were formerly homozygous for deleterious recessives.

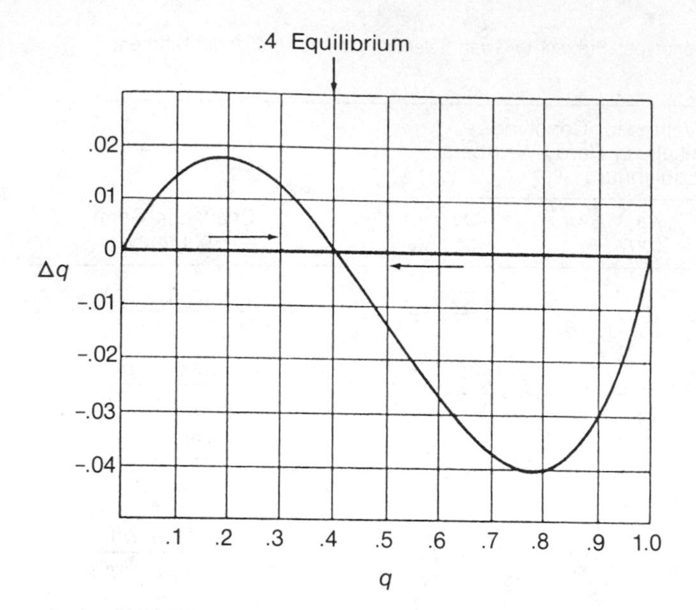

FIGURE 21-2 Change in the frequency (Δq) of allele *a* when the genotypic adaptive values are *AA* = .8, *Aa* = 1.0, *aa* = .7, and population size is infinite. These values provide a stable balanced polymorphism ($\Delta q = 0$) at $q = .4$. That is, Δq is positive (*q* increases) if *q* is less than .4 and negative (*q* decreases) if *q* is more than .4. Should one allele be accidentally eliminated, that is, $q = 0$ or 1, Δq is of course zero, but polymorphism is lost. (Adapted from Li.)

It is easy to see that if *s* and *t* are constant values, both *p* and *q* will reach a stable equilibrium: if *q* departs from the equilibrium value, selection pressure will force it back. That is, if Δq is positive, the gene frequency *q* increases, but if Δq is negative, *q* decreases, the negative or positive sign of Δq depending on whether *q* is above or below its equilibrium value. For example, when $s = .2$ and $t = .3$, the equilibrium value for *q* is $s/(s + t) = .2/(.2 + .3) = .4$. Values of *q* below .4 cause Δq to be positive, which increases *q*, whereas values of *q* above .4 cause Δq to be negative, which decreases *q*. As can be seen in Figure 21-2, the effect of such heterozygote superiority is to drive the frequencies of the two alleles in the population to a stable equilibrium at $q = .4$.[3]

SELECTION AND POLYMORPHISM

The maintenance of different genotypes through heterozygote superiority is an example of **balanced polymorphism**, a term invented by Ford to describe the preservation of genetic variability through selection. In general, a gene locus is considered to be polymorphic if at least two alleles are present, with a frequency of at least 1 percent for the second most frequent allele. Although selection coefficients are difficult to measure in natural populations, such polymorphisms are certainly ubiquitous features in practically all populations examined so far, both on the chromosomal level (see Fig. 10-32) and on the genic level (see Table 10-5). One prominent example of polymorphism caused by overdominance is that of the sickle-cell gene in humans (p. 187) where heterozygotes (Hb^A/Hb^S) survive the malarial parasite more successfully than either normal (Hb^A/Hb^A) or sickle-cell homozygotes (Hb^S/Hb^S). As shown in Figure 21-3, this gene (as well as others that appear to offer protection against malaria) is maintained in notable frequencies in geographical areas where the malarial disease is prevalent (see also Rotter and Diamond).

In laboratory populations, where genetic variability can be more easily controlled and measured, numerous experiments demonstrate the attainment of balanced polymorphism apparently by some sort of overdominance. In *Drosophila pseudoobscura*, for example, Dobzhansky and Pavlovsky have shown that the frequencies of the Standard (ST) and Chiricahua (CH) third-chromosome arrangements will come to a stable equilibrium when flies carrying these arrangements are placed together in a population cage kept continuously for a year or longer (Fig. 21-4). The superiority of the heterozygote can be seen in the relative adaptive values calculated for the various third-chromosome combinations: ST/ST = 0.90, ST/CH = 1.00, CH/CH = .41. In fact, one can calculate that even lethal recessive genes may remain in a population if they confer only a small heterozygous advantage. For example, gene *a*, lethal in *aa* homozygous condition but providing a 1 percent advantage to the *Aa* heter-

[3]Not all equilibria are permanent or stable. They are considered unstable if any disturbance of equilibrium frequencies causes one of the alleles to go to fixation. One such unstable equilibrium is possible when selection acts against the heterozygote at a gene locus with two alleles. If both homozygotes have equal adaptive value and the heterozygote is inferior, an equilibrium will be produced only when the frequency of each of the two alleles is exactly equal to .5. At this value the alleles are perfectly balanced, since equal amounts of each of the two are being removed in the heterozygote, that is, when genotype frequencies are .25 *AA*, .50 *Aa*, and .25 *aa*. However, any slight departure from these frequencies will cause the less frequent allele to have proportionally more of its genes in heterozygotes than the more frequent allele does. For example, if the gametic frequency of *A* rose accidentally to .6 and that of *a* fell to .4, then the genotypic frequencies under random mating are .36 *AA*, .48 *Aa*, and .16 *aa*, and the heterozygotes now contain a greater proportion of the *a* alleles than they do of the *A* alleles ($.24/.16 > .24/.36$). Thus, if the heterozygotes were lethal, the *A* gene frequency would become $.36/(.36 + .16) = .69$, and *a* would become $.16/(.36 + .16) = .31$. In the next generation continued lethality of the heterozygotes would lead to an increase of the *A* frequency to .83, and the *a* frequency would fall to .17. Within a relatively short time, the *A* allele would go to fixation and the *a* allele to elimination.

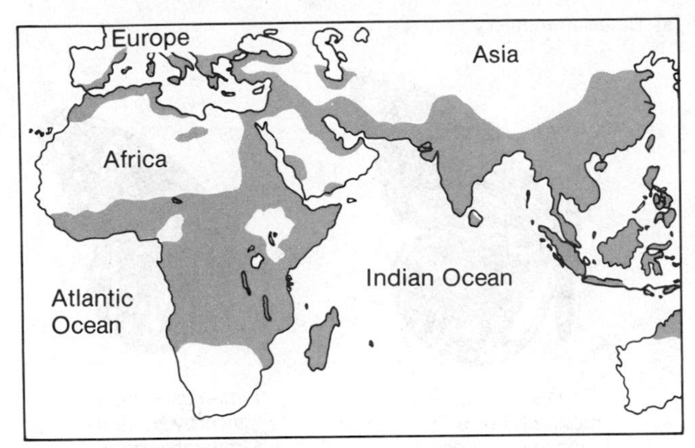

(a) Falciparum malaria

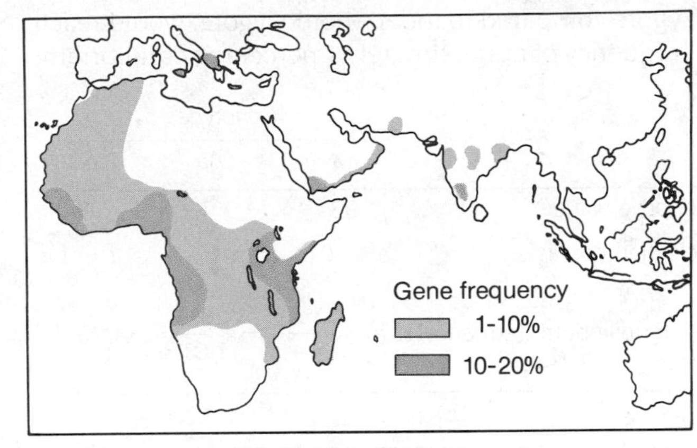

(b) Sickle cell anemia

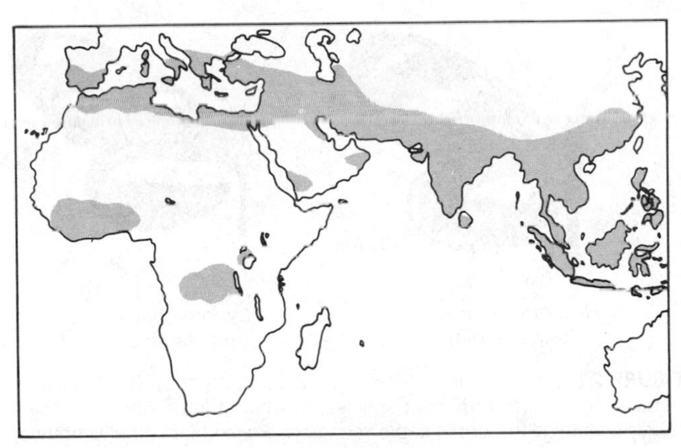

(c) Thalassemia

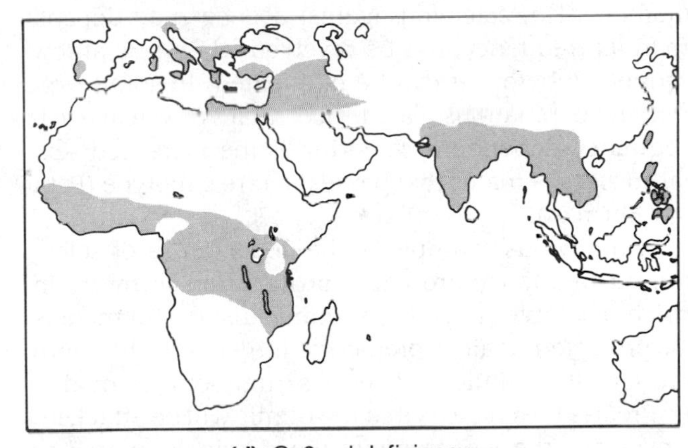

(d) G-6-pd deficiency

FIGURE 21-3 Relationship between the geographic distributions of malaria and genes that confer resistance against the disease. (a) Distribution of falciparum malaria in the Old World before 1930. (b) Distribution of the gene for sickle-cell anemia (*Hb*^{S}). (c) Distribution of the gene for β-thalassemia. (d) Distribution of the sex-linked gene for glucose-6-phosphate dehydrogenase deficiency in males in frequencies above 2 percent. (From Strickberger, adapted from Allison.)

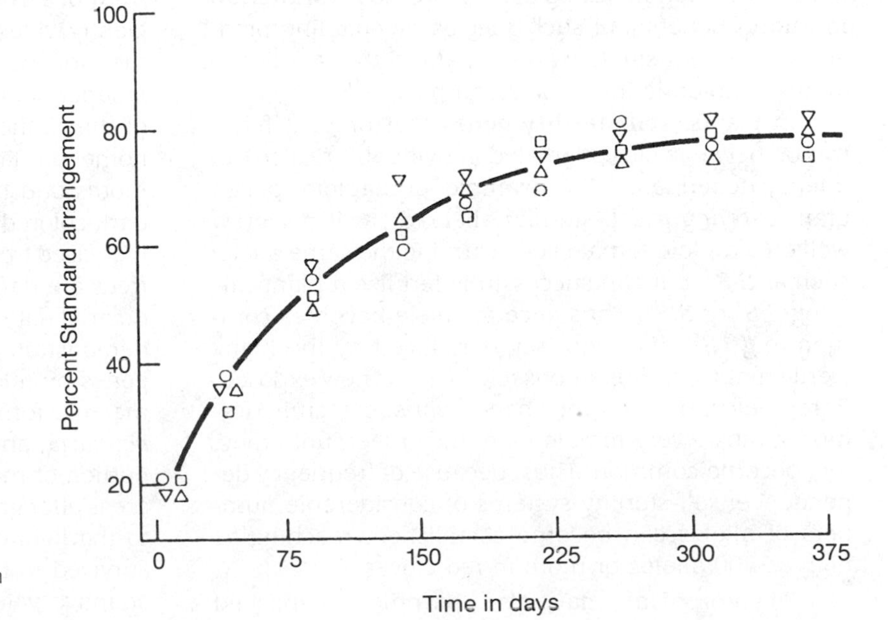

FIGURE 21-4 Results of four *Drosophila pseudoobscura* population cage experiments in which two third chromosome arrangements are competing, Standard (ST) and Chiricahua (CH). Each population is denoted as a circle, square, or triangle, and was begun with 20 percent ST and 80 percent CH, reaching equilibrium values of 80 to 85 percent ST after approximately one year. The solid curve represents the frequencies of the ST arrangement expected according to the adaptive values ST/ST = .90, ST/CH = 1.00, CH/CH = .41. (Adapted from Dobzhansky and Pavlovsky.)

ozygote compared to the *AA* homozygote, would reach a frequency of approximately 1 percent at equilibrium:

	Genotypes		
	AA	*Aa*	*aa*
Adaptive value	.99	1.00	0
Selection coefficient	$s = .01$	0	$t = 1$

$$\hat{q} \text{ (equilibrium frequency of } a) = \frac{s}{s+t} = \frac{.01}{1.01} = .0099$$

Other conditions responsible for polymorphism may include a change in selection coefficients so that genes detrimental at one time are advantageous at another. Also, selection against a gene may depend upon its frequency and be reversed when it is at low frequency, before it can be eliminated. In some *Drosophila* populations, a device that ensures such **frequency-dependent selection** is the increased sexual success of males that possess a rare genotype (Petit and Ehrman).

An obvious example of the dependence of selection on frequency are cases of **Batesian mimicry** in which palatable species that mimic distasteful models are protected against predators. In general, the more frequent the mimic and the less frequent its model, the greater the chances that the mimic will be attacked; conversely, the less frequent the mimic compared to the model, the greater the chances that the mimic will be protected. As shown in Figure 21-5, mimicry also occurs when a palatable mimic imitates a conspicuous warning (aposematic) coloration or pattern shared by two or more different unpalatable species. Mimicry between different unpalatable species (**Müllerian mimicry**) benefits all such species by enabling predators to learn a single warning pattern that applies to all these potential but distasteful prey.

In plants, **self-sterility genes** that prevent fertilization between closely related individuals are also frequency-dependent. For example, a haploid pollen grain carrying a self-sterility allele, S^1, will not grow well on a diploid female style carrying the same allele, such as S^1S^2, but can successfully fertilize a plant carrying S^2S^3 or S^3S^4. Thus once an allele becomes common (e.g., S^1), its frequency is reduced by the many sterile mating combinations to which it is now exposed. Rare alleles, on the other hand, will successfully fertilize almost every female plant they meet, until they, too, become common. Thus, because of frequency dependence, self-sterility systems of considerable numbers of alleles can become established, reaching as high as 200 alleles or more in red clover.

Polymorphism may also become established when selection coefficients are not constant but vary

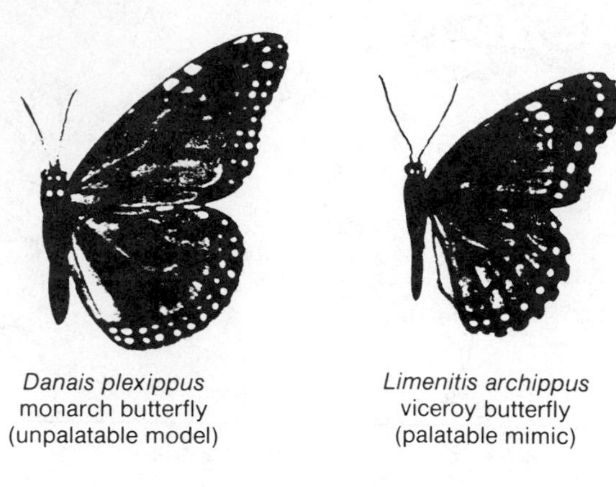

FIGURE 21-5 Mimicry in different species of butterflies. (a) Batesian mimicry by a North American species, in which the palatable viceroy butterfly mimics the unpalatable monarch. Resemblance between two South American unpalatable species in (b) provides a common warning pattern to predators and helps protect both prey species (Müllerian mimicry).

from one environment to another. A population sufficiently widespread to occupy many environments may therefore maintain a variety of genotypes each of which is superior in a particular habitat. A prominent example of this is the polymorphism associated with the phenomenon known as **industrial melanism**. Certain moths and butterflies show increased proportions of dark-colored, or melanic, forms, usually caused by the increased frequency of a dominant gene in areas where trees are darkened because of air pollution. In the industrial city of Birmingham, Kettlewell and others demonstrated the selective advantage of such melanic genes by releasing known numbers of both light and melanic forms of the British peppered moth, *Biston betularia*, and recapturing a significantly greater proportion of melanic forms. In other words, these sooty areas offer greater protection to the melanic forms than to the light-colored forms, since more of the former survived to be recaptured. As Kettlewell showed, the adaptive value of the melanic types probably lies in their ability to remain concealed on darkened tree

FIGURE 21-6 Light-colored and dark-colored tree trunks, each with a melanic and non-melanic *Biston betularia* moth. (From Kettlewell.)

trunks from bird predators (Fig. 21-6). In non-industrial areas, on the other hand, trees covered with normal gray lichens offer decided advantages to the light-colored moths. Thus there are various degrees of polymorphism in English *B. betularia* populations, ranging from high frequencies of the melanic gene in industrial areas to almost zero in rural areas.[4] Interestingly, passage of clean air legislation in Britain in 1956 has led to a reduction of industrial smoke and sulfur dioxide in many formerly polluted areas. This reduction in pollution is now correlated with "reverse evolution": there have been dramatic declines in the frequency of melanic forms of *B. betularia* and other insects (Brakefield).

Levins has pointed out that both the spatial and temporal organization of the environment may have significant effects on the extent to which a population will rely upon genetic polymorphism as an adaptive strategy. **Coarse-grained environments**, in which different individuals in a population endure different experiences, will promote greater genetic polymorphism than **fine-grained environments**, in which the environmental differences are experienced by all individuals. A discussion of other mechanisms that can maintain polymorphism is given by Hartl and Clark.

THE KINDS OF SELECTION

When selection has occurred for particular conditions over long periods of time, most populations can be considered to have achieved phenotypes that are optimally adapted to their surroundings. That is to say, many phenotypes will tend to cluster around some value at which fitness is highest. Individuals that depart from these **optimum phenotypes** may therefore be expected to possess lower fitness than those closer to the optimal values. In a classic 1899 study on sparrows that survived a storm, Bumpus showed that measurements taken on eight of nine different characteristics tended to cluster around intermediate phenotypic values, while sparrows killed by the storm showed much greater variability. In Bumpus' terms, "it is quite as dangerous to be conspicuously above a certain standard of organic excellence as it is to be conspicuously below the standard." This view has since been supported by many studies on a variety of organisms, including snails, lizards, ducks, chickens, and others (Lerner).

In humans, measurements of birth weights of newborn babies, among other characteristics, also show selection for optimum values. As can be seen in

[4]Based on dates of British amateur and museum collections, one can estimate that it took about 40 generations (one generation per year) during the nineteenth century for the frequency of non-melanic phenotypes of *B. betularia* to decrease in some industrial areas from about 98 percent to about 5 or 6 percent. Using such data we can arrive at an approximate selection coefficient for industrial melanism in this moth by noting that the non-melanic phenotypes are homozygotes (frequency q^2) for the recessive non-melanic allele (frequency q), and therefore q was reduced during this 40-generation interval from $\sqrt{.98} \approx .99$ to $\sqrt{.06} \approx .25$. From Table 21-3, it would take about 44 generations (13 + 7 + 9 + 15) to reduce q from .99 to .25 when the selection coefficient is .20. In other words, the selection coefficient against the non-melanic gene in some of these industrial areas was about .20 or somewhat greater.

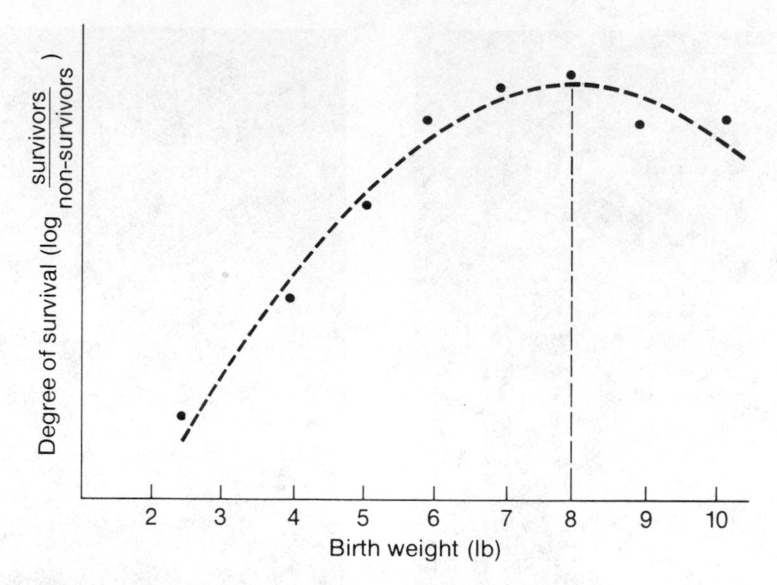

FIGURE 21-7 Relationship between birth weight and the degree of survival in female births in a London obstetric hospital. Of 6,693 births there were 6,419 survivors one month later, or a mortality rate of 274/6,693 = 4.1 percent. Since mortality in the "optimum" 8-pound class was only 1.2 percent, this means that 4.1 − 1.2 = 2.9 percent of deaths occurred among the non-optimal classes, indicating that a fairly high proportion (2.9/4.1 = 70.8 percent) of deaths between birth and one month of age is caused by selection against non-optimal phenotypes. (Adapted from Karn and Penrose.)

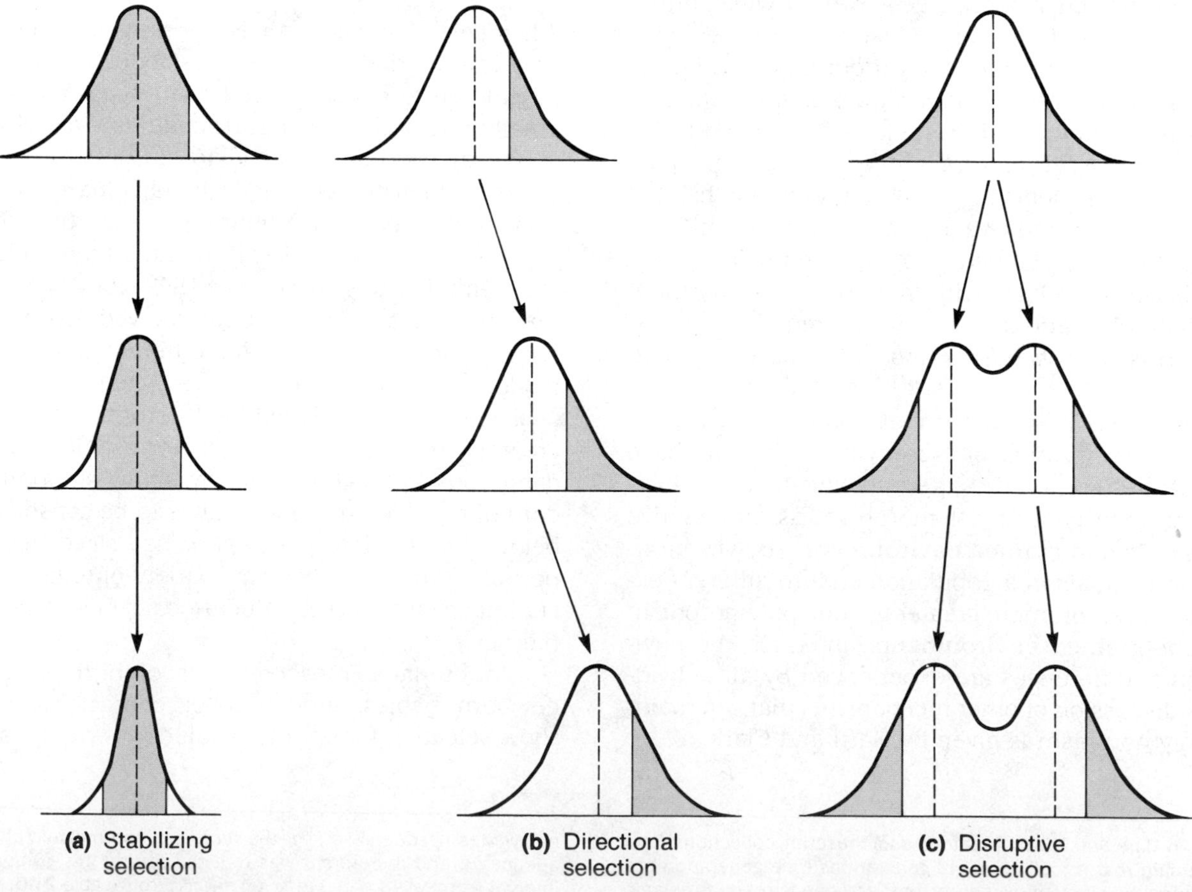

FIGURE 21-8 Three basic modes of selection and their effects on the mean (dashed lines) and variation of a normally distributed quantitative character. The horizontal axis of each bell-shaped curve represents measurements of a quantitative character (e.g., from low on the left end to high on the right end), and the vertical axis represents the number of individuals found at each measurement. Shaded areas represent the individuals chosen to be parents of the next generation.

Figure 21-7, most of the survivors cluster around a birth weight of 8 pounds, and those who depart from this value have lesser chances for survival. This reduction in frequency of extreme phenotypes has been termed **stabilizing**, or **centripetal**, **selection**, since it signifies selection for an intermediate stable value (Fig. 21-8(a)).

However, not all selection is stabilizing, since selection may well favor an extreme phenotype by proceeding in one or the other direction of the distribution of phenotypes (Fig. 21-8(b)). Such **directional selection** is commonly practiced by animal and plant breeders who select for extremes of yield, productivity, resistance to disease, and so forth (Fig. 21-9). The role of directional selection in evolution is of special importance when the environment of a population is changing and only extreme phenotypes happen to be adapted for new conditions.

Selection, whether stabilizing or directional, may act in a constant fashion if the selective environment is uniform. However, when conditions are changeable, a population may be subjected to divergent or cyclically changing (oscillating) environments to which different genotypes among its members are most suited (Gibbs and Grant). Such selection is therefore **disruptive**, or **centrifugal**, because it leads to the establishment of genetic differences in a population (Fig. 21-8(c)).

In order to ultimately affect evolution, selection, however it occurs, must change the frequencies of genes involved in fitness. This means that genetic variability must be present, and **pure lines** that are homozygously uniform for such fitness genes offer no opportunity for selection to produce any noticeable evolutionary change. This principle was formulated mathematically by Fisher as a fundamental theorem which essentially states that *the greater the genetic variability upon which selection for fitness may act, the greater the expected improvement in fitness*. One consequence of this theorem is that populations long subjected to selection—and this includes all populations—would be expected to have little remaining variability for genes affecting fitness, since such variability would have been diminished by selection. The continued existence of selection therefore implies that variability itself is favorably selected: there are continuous changes in the environment that affect formerly unselected genes whose presence offers new advantageous opportunities for improvement in fitness.[5] Such environmental changes obviously include changes in resources, supplies, waste products, and predator and parasite populations.

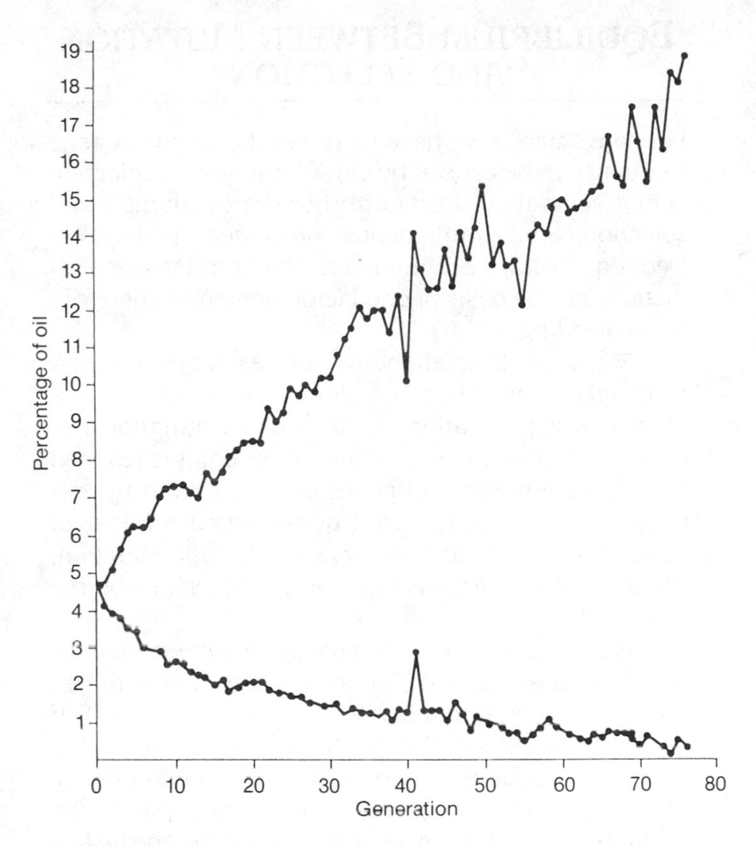

FIGURE 21-9 Selection for high and low oil content in corn kernels in an experiment begun in 1896 at the University of Illinois and continuing to the present. Selection for high oil content still continues to yield increases, whereas selection for low oil content has tapered off on reaching the 0 percent lower limit. (Adapted from Dudley.)

Van Valen has proposed that species generally compete with each other for resources so that an advantage, or improvement in fitness, for one species represents a deterioration in the environment of others. He points out that species survival is therefore very much in accord with the observation made by the Red Queen whom Alice meets in her adventures in Lewis Carroll's *Through the Looking Glass*: "Here, you see, it takes all the running you can do to keep in the same place." That is, according to the **Red Queen hypothesis**, each species continually faces new selective challenges because of changes in the fitness of populations with which it interacts, and each must continually improve its fitness in order to survive. Or as Darwin stated in *The Origin of Species*, "if some of these many species become modified and improved, others will have to be improved in a corresponding degree or they will be exterminated."

[5]Fisher's theorem is not true under all conditions, since variability for some fitness genes can persist in spite of continued selection acting upon them, such as when allelic differences are maintained through devices discussed in the previous section (e.g., frequency dependence). In general, however, populations do tend to change genetically in directions that improve fitness for their environment, and Endler lists more than 160 cases in natural populations where selection has been demonstrated. Such findings indicate that genetic variability for fitness must have provided the baseline upon which selection acts.

EQUILIBRIUM BETWEEN MUTATION AND SELECTION

For convenience we have considered changes in gene frequency to be caused by either mutation or selection acting separately. In nature, however, mutation and selection are simultaneous processes, and gene-frequency values are influenced by both factors. Predictions on the basis of one factor alone may therefore be misleading.

For example, even though a recessive gene is detrimental in homozygous condition, it may nevertheless persist in a population because of its mutation frequency. That is, a certain equilibrium point is reached at which the number of genes being removed by loss of homozygotes is replaced by the same number of genes introduced into heterozygotes through mutation. We may determine this equilibrium frequency by the following argument.

We have seen that the change in gene frequency per generation for a deleterious recessive a with frequency q is equal to a loss of $sq^2(1 - q)/(1 - sq^2)$. If s is small, the denominator can be considered 1, and the loss in frequency is then $sq^2(1 - q)$. The frequency of newly mutated a genes, however, is equal to the mutation rate (u) of $A \rightarrow a$ multiplied by the A frequency, which is $1 - q$. Thus the loss of a genes through selection is exactly balanced by the gain of newly mutated a genes when

$$sq^2(1 - q) = u(1 - q)$$

$$sq^2 = u$$

$$q^2 = \frac{u}{s}$$

$$q = \sqrt{\frac{u}{s}}$$

The equilibrium frequency of a mutant gene in a population is thus a function of both the mutation frequency and the selection coefficient. As can be seen in the hydraulic model of this relationship in Figure 21-10, when the mutation rate increases, the equilibrium gene frequency will also increase, but the equilibrium frequency decreases when the selection coefficient increases.

For a deleterious dominant allele, similar algebraic manipulations point to an equilibrium frequency of about u/s, a value almost identical to the equilibrium frequency of genes that lack dominance. Since such dominant or partially dominant genes are of considerable disadvantage to heterozygotes, Fisher proposed that their deleterious effect is probably diminished in most organisms by selection of **modifier genes** at other loci that change the degree of dominance. For example, mutant alleles at a particular locus A (e.g., A^1, A^2, A^3 ...) may act as partial dominants in the presence of the wild-type allele A^+. Since these mutant alleles are mostly deleterious, modifier genes at other loci (e.g., B^1, B^2..., or C^1, C^2..., etc.) that increase the dominance of A^+ will be selected until the effects of mutations at the A locus are relatively recessive.

Evidence for Fisher's view is seen in the successful selection for dominant and recessive modifiers demonstrated by Ford in the currant moth *Abraxas grossulariata*. In this moth a single gene, *lutea*, in homozygous condition, produces yellow instead of the normal white ground color but has an intermediate effect as a heterozygote. After four generations of selecting moths for greater and lesser expression of the *lutea* phenotype in the heterozygote, Ford was able to show that two distinct strains could be obtained: in one case *lutea* acted almost as a complete dominant and in the other case almost as a complete recessive. In each strain special modifiers had been chosen, some enhancing and some detracting from the dominance of this particular gene.

Instead of modifiers, Haldane suggested that special wild-type alleles are selected (e.g., A^{X+}, A^{Y+}, A^{Z+}) that act as dominants in the presence of a mutant allele (e.g., A^1, A^2, A^3,...), and the entire subject has received considerable discussion (e.g., Wallace). In general, it seems likely that evolution has made use of both kinds of dominance mechanisms mentioned, as demonstrated by Harland and others in two species of cotton, *Gossypium barbadense* and *G. hirsutum*. In these plants certain alleles show simple dominance when crosses are made between variants of the same species. Interspecific crosses, by contrast, show the effect of numerous modifying genes on these traits, as well as differences in the degree of dominance of particular alleles. However, in spite of these dominance-producing mechanisms, many deleterious genes are probably still not completely recessive and seem to have some effect in heterozygous condition. Thus, in natural populations the equilibrium frequencies of deleterious genes are probably higher than for dominants but lower than for pure recessives.

MIGRATION

Mutation is not the only mechanism for introducing new genes into a population. A population may receive alleles by **migration** (also called **gene flow**) from a nearby population that maintains an entirely different gene frequency. When this occurs, two factors are of importance to the recipient population: the difference

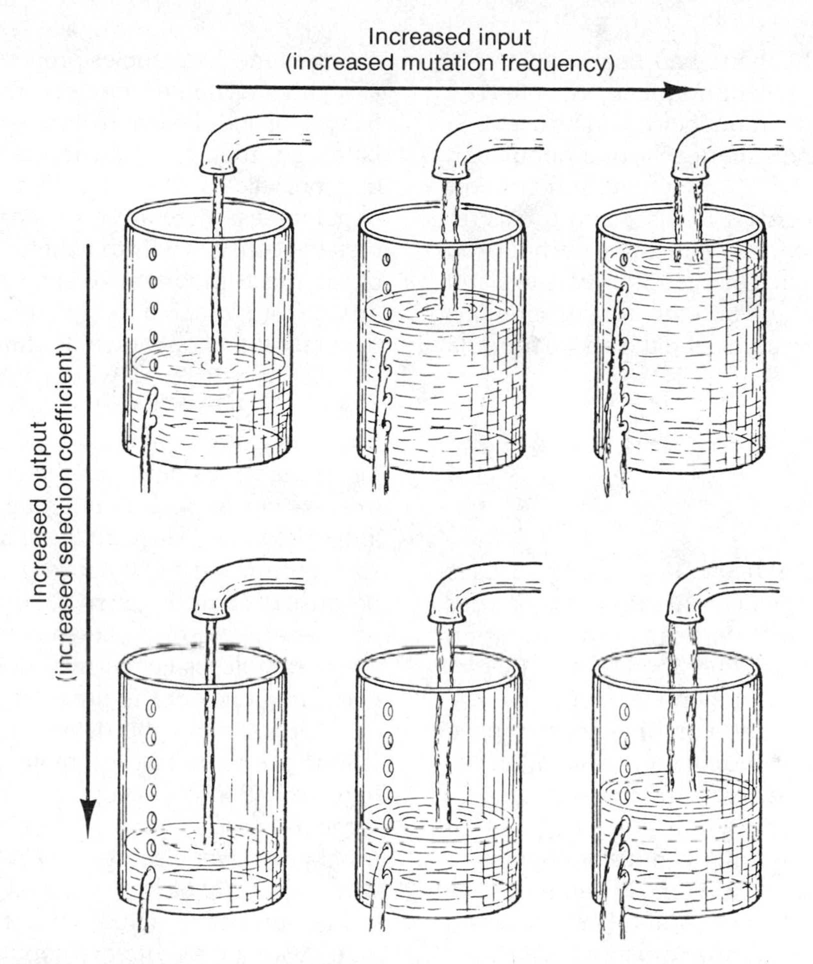

FIGURE 21-10 Hydraulic model of mutation-selection equilibrium. Each container is analogous to a population in which the water level represents the equilibrium frequency of a gene. As the water input (mutation frequency) increases, the standing water level (equilibrium gene frequency) increases. When the overflow holes are small (small selection coefficient) the water levels are higher for the same input (mutation frequency) than when the overflow holes are larger (large selection coefficient). (Adapted from Stern.)

in frequencies between the two populations and the proportion of migrant genes that are incorporated each generation. If we designate q_0 as the initial gene frequency in the recipient, or hybrid, population, Q as the frequency of the same allele in the migrant population, and m as the proportion of newly introduced genes each generation, then the gene frequency in the hybrid population will suffer a loss of q_0 equal to mq_0 and a gain of Q equal to mQ. Over n generations of migration, when the gene frequency of the hybrid population becomes q_n, one can calculate that the relationship between these factors will reach

$$q_n - Q = (1 - m)^n(q_0 - Q)$$

$$\text{or } (1 - m)^n = \frac{q_n - Q}{q_0 - Q}$$

For populations where this equation can be applied, four of these factors must be known in order to allow the fifth to be calculated. One such example can be found in human populations where blood group gene frequencies are known for both American blacks and American whites, two populations between which gene exchange has occurred. In general, although some black genes undoubtedly enter the white population, the white population is so large that this introduction probably makes little difference in white gene frequencies. On the other hand, the black population is much smaller and has remained isolated from its African origin for two or more centuries. On this basis the white population can be considered as the gene donor or migrant population (Q) and the present black population as the hybrid (q_n).

To obtain the original gene frequency of one of the Rh blood group alleles, R^0, in the black population (q_0), data of present East African blacks were used on the assumption that these data may reflect the original gene frequencies of 200 to 300 years ago. Among the East Africans, R^0 showed a frequency of .630, indicating that the frequency of this gene had been reduced in American blacks to its present frequency of .446. The cause for this reduction in frequency could be ascribed to interbreeding with the American white population where the frequency of R^0 is about .028, much lower than among blacks. According to Glass and Li this reduction had begun at the time of the inital introduction of blacks into the American colonies 300 years ago and probably continued throughout the ten generations since. Substituting these values into the above formula we obtain

$$(1 - m)^{10} = \frac{q_{10} - Q}{q_0 - Q} = \frac{.446 - .028}{.630 - .028} = .694$$

$$1 - m = \sqrt[10]{.694} = .964; \text{ then } m = .036$$

This value of m means that, excluding all other causes such as mutation, 36 genes per 1,000, or 3.6 percent of genes in the black population, were introduced from the white population each generation. Since $1 - m$ represents the proportion of non-introduced genes, $(1 - m)^{10} = .694$ is the proportion of genes that have remained of African origin over the ten-generation period. Supported by somewhat similar estimates in more recent studies, blood group gene frequencies generally indicate that the American black population is genetically about 70 to 80 percent African and 20 to 30 percent white, with some differences between Southern and Northern blacks (Adams and Ward).

Where exact information on gene frequency exchanges between populations is unavailable, and this includes most populations, there has been considerable discussion and dispute about the importance of migration. According to Mayr and to Stanley, migration can hinder local evolutionary changes by infusing genes from populations which are not adapted to local conditions. For example, some populations of mammals who live on dark, formerly volcanic lava flows have dark fur when they are isolated from neighboring populations who live on lighter colored backgrounds but do not have dark fur when they receive immigrants from the lighter-colored surroundings. On the other hand, Ehrlich and co-workers describe populations of the butterfly *Euphydras editha* which show no phenotypic changes whether or not they are subject to migration from phenotypically different populations. In the absence of genetic information, these issues are, so far, difficult to resolve (Slatkin).

RANDOM GENETIC DRIFT

The three forces considered up to now, mutation, selection, and migration, have one important quality in common; they usually act in a directional fashion to change gene frequencies progressively from one value to another. When unopposed, these forces can lead to fixation of one allele and elimination of all others; when balanced, they can lead to equilibrium between two or more alleles. However, in addition to these directional forces, there are also changes that have no predictable constancy from generation to generation. One of the most important of such **nondirectional forces** arises from variable sampling of the gene pool each generation and is known as **random genetic drift**.

This is apparent if we consider that, in the absence of directional forces to change gene frequencies, there is always a strong likelihood of obtaining a good sample of the genes of the previous generation as long as the number of parents in a population is consistently large. However, since real populations are limited in size, genetic drift will cause gene-frequency changes because of **sampling errors**. For example, if only a few parents are chosen to begin a new generation, such a small sample of genes may deviate widely from the gene frequency of the previous generation.

The extent of the deviation for all sizes of populations can be measured mathematically by the standard deviation of a proportion $\sigma = \sqrt{pq/N}$, where p is the frequency of one allele, q of the other, and N the number of genes sampled. For diploid parents, each carrying two alleles, σ is equal to $\sqrt{pq/2N}$, where N is the number of actual parents. For example, if we begin with a large diploid population, where $p = q = .5$, and continue this population each generation by using 5,000 parents, then $\sigma = \sqrt{(.5)(.5)/10{,}000} = \sqrt{.000025} = .005$. The values of such populations will therefore fluctuate mostly around $.5 \pm .005$, or between .495 and .505. On the other hand, a choice of only two parents as founders will produce a standard deviation of $\sqrt{(.5)(.5)/4} = \sqrt{.0625} = .25$, or values of $.50 \pm .25$ (from .25 to .75).

In other words, sampling accidents because of smaller population size can easily yield gene frequencies that depart considerably from the initial .5 values in a single generation. Were the population to remain small and the next generation to begin with either of these extremes, that is, a gene frequency of .25 or .75 for a particular allele, the following generation may have the frequency of that allele reduced to almost zero ($.25 \pm \sqrt{(.25)(.75)/4} = .25 \pm .22$, a range of .03 to .47) or increased to almost one ($.75 \pm \sqrt{(.75)(.25)} = .75 \pm .22$, a range of .53 to .97). Should such small populations continue each generation, the likelihood increases that one or more will eventually reach fixation for one of the alleles. The proportion of such populations that attain fixation, that is, the **rate of fixation**, will eventually reach $1/2N$. Obviously, if N is large, fixation proceeds slowly, but even large populations can show some degree of drift, as diagrammed in Figure 21-11.

This reliance of drift upon population number emphasizes the importance of what is called **effective population size** (N_e). It differs from the observed population size because not all members of a population are necessarily parents and because parentage can also be limited by a reduced number of one of the sexes. For example, if out of a total population of 1,000, the next generation is produced by 3 males mated to 300 females, the effective population size is more than 6 but still less than 303. The relationship has been expressed by Wright as $N_e = 4N_f N_m/(N_f + N_m)$, where N_f is the number of parental females and N_m the number of parental males. In the above case N_e would be $4(300)(3)/303 = 11$. Inequalities in numbers of offspring among different parents will also reduce the effective population size.

As a result of these considerations, Wright has proposed that genetic drift may be of considerable importance in producing gene-frequency changes among populations when their effective sizes are small. Among the observations that illustrate this concept is that of Buri who set up 107 separate lines of *D. melanogaster*, each line carrying two alleles at the *brown* locus (*bw* and bw^{75}) at initially equal frequencies of 50 percent. The lines were then continued for 19 generations by randomly selecting 8 males and 8 females as parents from each preceding generation (N = 16 = 32 *brown* alleles) and scoring the frequency of the two different *brown* alleles. As shown in Figure 21-12, by the first generation a number of populations already showed departures from the original 50 percent bw^{75} frequency, and genetic drift continued to increase successively so that by generation 19 more than half of the 107 populations reached fixation for either the *bw* or bw^{75} alleles.

Although the persistence of small population size over many generations is an obvious cause for genetic drift, occasional reductions in size for only one or a few generations may also have pronounced effects on gene frequencies and future evolution. At the extreme of such reductions, termed the **founder principle** by Mayr, a population may occasionally send forth only a few founders to begin a new population. Whatever gene or chromosome arrangements these founders take with them, detrimental or beneficial, all stand a good chance of becoming established in the new population because of this sudden sampling accident. Thus Carson, by careful analysis of salivary chromosome banding patterns, has shown that the more than 100 native Hawaiian "picture-winged" *Drosophila* species can be derived from founder events in which each island was settled by relatively few individuals whose descendants evolved into different species. For example, the 40 species unique to the Maui island complex (Fig. 21-13) derive from only 12 founders, 10 from Oahu and 2 from Kauai, with each single founder pro-

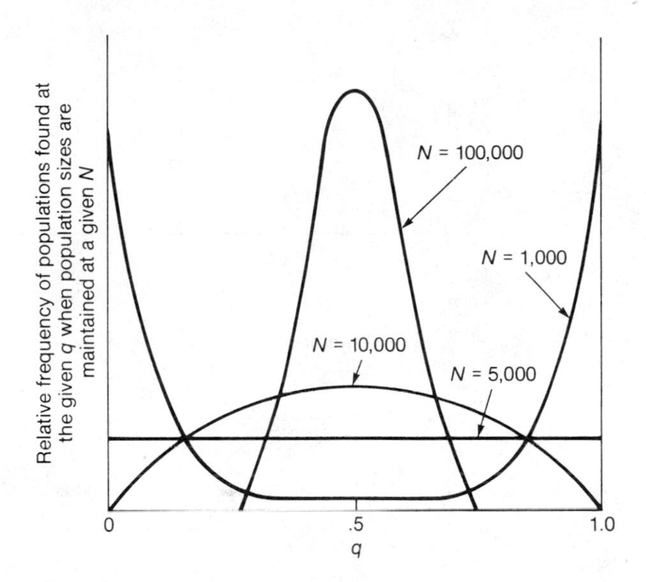

FIGURE 21-11 Distribution of equilibrium gene frequencies for populations of different sizes when selection is zero and a small amount of migration occurs into each population ($m = .0001$) from a population whose gene frequency is $q = .5$. In spite of this migration, populations of sizes $N = 1{,}000$ and $N = 5{,}000$ show a considerable amount of random genetic drift, many reaching elimination ($q = 0$) or fixation ($q = 1$). Only populations of relatively very large sizes ($N = 10{,}000$, $N = 100{,}000$) maintain the initial gene frequency $q = .5$ in appreciable proportions. (Adapted from Wright.)

viding unique chromosome arrangements that can be traced in the descendant species.

There are by now numerous examples in many organisms, including humans, of unique gene frequencies that seem best explained by such founder events, or **bottleneck effects**. It therefore seems likely that at least some populations were begun with only a few "Adams" and "Eves" carrying genotypes that may have differed greatly in frequency from their parental populations. Certainly the relatively high incidences of some genes listed in Table 20-8, such as achromatopsia among the Pingelapese and Ellis–van Creveld syndrome (polydactylous dwarfism) among the Lancaster County Amish, are difficult to explain except as founding accidents, since they appear to confer no advantage on either their homozygous or heterozygous carriers. The same conclusion is true for chromosomal translocations which are usually selected against because they can cause sterility in heterozygotes (p. 182), yet are nevertheless common features in the evolution of many mammalian lines.

Bottleneck effects may therefore counter the effects of previous selection for a short period of time—an interval during which previously favorable mutations may be lost and deleterious mutations may be fixed. However, it is difficult to imagine that any genetic trait that affects the armament with which organisms face their environment can long continue to escape selective environmental pressures. Through non-selec-

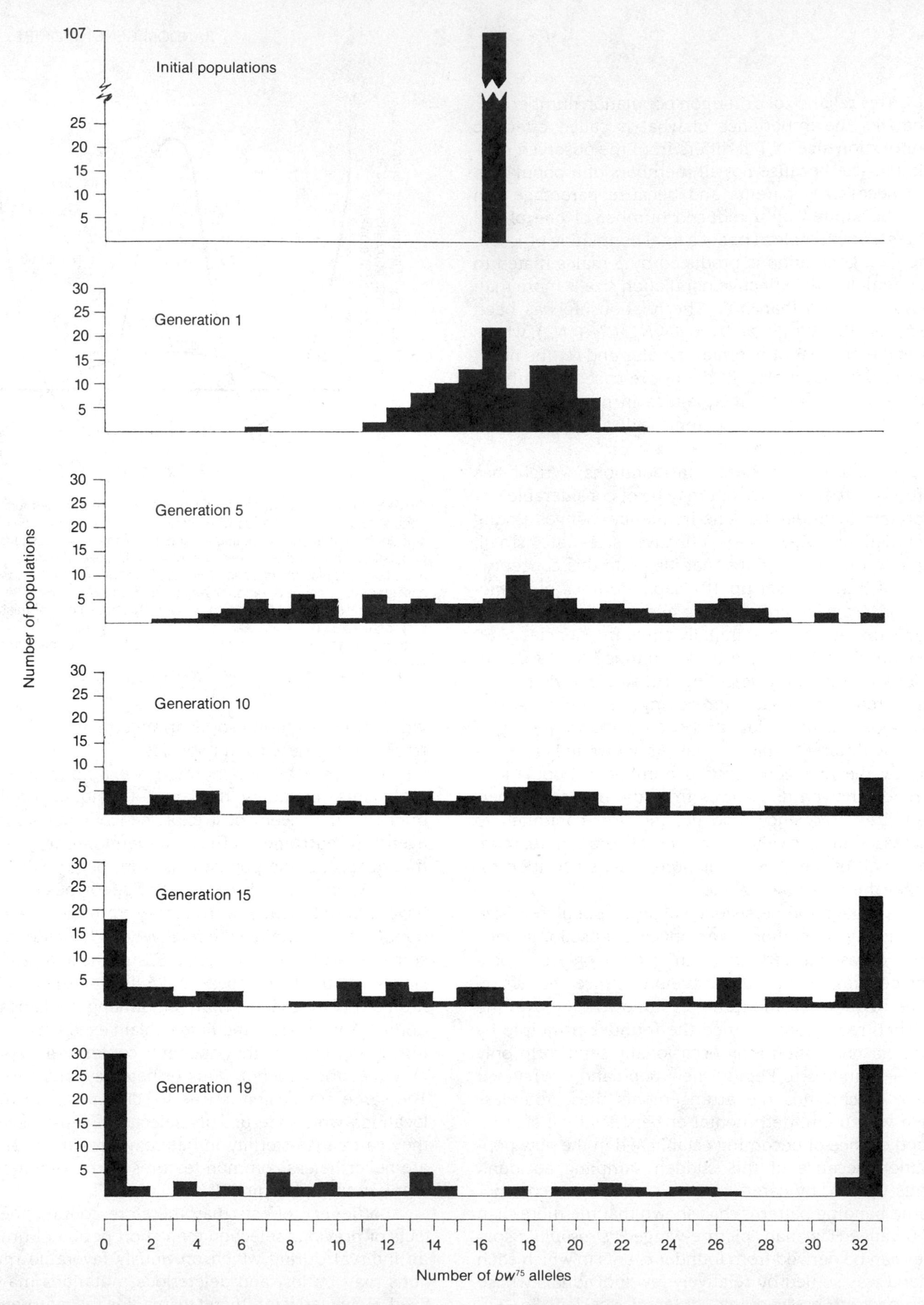

FIGURE 21-12 Distributions of the numbers of bw^{75} alleles in 107 lines of *D. melanogaster*, each with an initial frequency of .5 bw^{75}. The lines were continued for 19 generations, using 16 parents to start each generation (32 alleles at the *brown* locus), and the number of bw^{75} alleles found are given for the various lines. Note that by generation 19, the bw^{75} allele had been eliminated from 30 of these lines (0 alleles) and had reached fixation in 28 of these lines (32 alleles). (Data from Buri's series I cultures.)

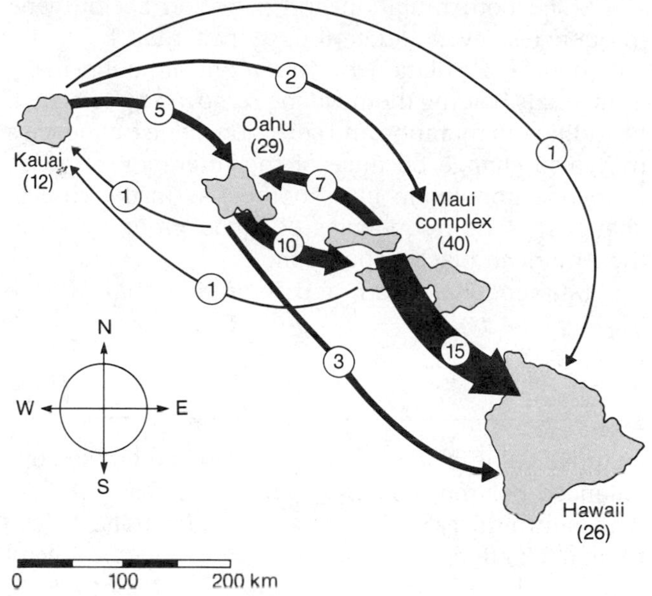

FIGURE 21-13 Colonization pattern showing the founder events proposed by Carson to explain the origin of native "picture-winged" *Drosophila* species found on the Hawaiian islands. The width of each arrow is proportional to the number of founders (circled), and the number of *Drosophila* species that are now present on each island is given in parentheses. Each successful founder is presumed to have been a fertilized female, usually from a geologically older island. The oldest of the Hawaiian islands is Kauai (about 5.6 million years old), and 10 of the 12 species that it possesses are believed to be the most ancient *Drosophila* elements in the islands. The youngest island, Hawaii, has been colonized entirely by founders from the older islands. (Adapted from Carson.)

tive genetic changes, a bottleneck may cast the evolution of a population in a new direction, but, by all we understand of evolution, this direction could not remain non-adaptive without extinction.

In general, therefore, the consequences of founding events on gene frequencies are largely unpredictable compared to variability estimates (i.e., σ) that can be arrived at when population size is constant. Other factors that cause the variability of gene frequencies to be unpredictable are unique historical events such as a change in the direction or intensity of selection because of a radical change in environment, an unusually favorable mutation, a rare hybridization event with another variety, or an unusual swamping of a population because of mass immigration. The effects of these factors on evolutionary changes may be of considerable importance, although genetic data for such events are still difficult to obtain.

SUMMARY

New alleles that appear because of mutation must persist in populations in order to effect evolutionary change. An allele a will increase in frequency if the mutation rate (A to a) exceeds the frequency of the reverse mutation (a to A) until equilibrium is reached between the alleles. Since mutational equilibrium is rarely attained, other mechanisms must influence the frequency of alleles that were introduced by mutation.

One such factor is selection, since if certain alleles improve the reproductive success of the carrier (that is, its fitness), they will tend to increase in frequency. Selection acts on the fitness differences among different alleles affecting a given trait, so gene frequencies will tend to change from generation to generation.

Alleles whose effects are deleterious will decline in frequency, the rate of decline depending on the allelic frequency and on the gene's recessiveness or dominance. For example, deleterious recessive alleles decline rapidly from high frequencies but only very slowly when at low frequencies. In any case, selection may operate to eventually remove the deleterious allele from the population, and persistence of the allele (equilibrium) depends upon factors such as the occurrence of new mutations.

Genetic variability (polymorphism) can be preserved through selection if the heterozygous genotype has greater fitness thatn either homozygous one, as in the case of sickle-cell anemia or when the frequency of an allele affects its fitness. Additionally, fitness of alleles may vary in different environments, as in the case of the peppered moth, where lighter forms are favored in unpolluted areas and melanic forms in industrial regions.

There are several ways in which selection can operate on a population: (1) if phenotypes far from the norm are less fit (stabilizing selection); (2) if an extreme phenotype has adaptive value (directional selection); and (3) if different phenotypes are favored in different enviromental circumstances (disruptive selection). For selection to occur, genetic variability must be present: the greater the genetic variability, the greater the opportunity for improvement in fitness.

Since both mutation and selection act on gene frequencies, even deleterious genes may be maintained in a population by mutation, although many individuals bearing them will be removed by selection. In addition to mutation and selection, gene frequencies may also change because of the migration of genes from one population into another, as in the case of changes in the frequencies of blood group genes in the American black population.

Mutation, selection, and migration change gene frequencies in a directional way and can lead to fixation of certain genes and loss of others or to equilibrium. However, non-directional forces such as genetic drift may be important in causing gene frequency changes. This is especially true if a new population is begun with a small sample ("founders") in which the gene frequency varies from that of the original population. Drift and other random factors may be important in establishing unpredictable variability within population.

KEY TERMS

adaptive value
balanced polymorphism
Batesian mimicry
bottleneck effects
centrifugal selection
centripetal selection
coarse-grained environments
deleterious dominant
deleterious lethal
deleterious recessive
directional selection
disruptive selection
effective population size
elimination
equilibrium gene frequencies
fine-grained environments
fitness
fixation
founder principle
frequency-dependent selection
gene flow
heterosis
heterozygous advantage
hybrid vigor
industrial melanism
migration
modifier genes
Müllerian mimicry
mutation rates
mutational equilibrium
mutation-selection equilibrium
nondirectional forces
optimum phenotypes
overdominance
polymorphism
pure lines
random genetic drift
rate of fixation
Red Queen hypothesis
sampling errors
selection
selection coefficient
selective value
self-sterility genes
stabilizing selection

REFERENCES

Adams, J., and R. H. Ward, 1973. Admixture studies and the detection of selection. *Science*, **180**, 1137-1143.

Allison, A. C., 1961. Abnormal hemoglobin and erythrocyte enzyme-deficiency traits. In *Genetical Variation in Human Populations*, G. A. Harrison (ed.). Pergamon, New York, pp. 16-40.

Brakefield, P. M., 1987. Industrial melanism: do we have the answers? *Trends in Ecol. and Evol.*, **2**, 117-122.

Brower, L. P. (ed.), 1988. *Mimicry and the Evolutionary Process*. Univ. of Chicago Press, Chicago.

Bumpus, H. C., 1899. The elimination of the unfit as illustrated by the introduced sparrow. *Biol. Lect. Woods Hole*, pp. 209-226.

Buri, P., 1956. Gene frequency in small populations of mutant *Drosophila*. *Evolution*, **10**, 367-402.

Carson, H. L., 1983. Chromosomal sequences and inter-island colonizations in Hawaiian *Drosophila*. *Genetics*, **103**, 465-482.

Christiansen, F. B., 1984. The definition and measurement of fitness. In *Evolutionary Ecology*, B. Shorrocks (ed.). Blackwell, Oxford, pp. 65-79.

Dobzhansky, Th., and O. Pavlovsky, 1953. Indeterminate outcome of certain experiments on *Drosophila* populations. *Evolution*, **7**, 198-210.

Dudley, J. W., 1977. Seventy-six generations of selection for oil and protein percentages in maize. In *Proceedings of the International Conference on Quantitative Genetics*, E. Pollak, O. Kempthorne, and T. B. Bailey, Jr. (eds.). Iowa State Univ. Press, Ames, pp. 459-473.

Ehrlich, P. R., R. White, M. C. Singer, W. W. McKechnie, and L. E. Gilbert, 1975. Checkerspot butterflies; a historical perspective. *Science*, **188**, 221-228.

Endler, J. A., 1986. *Natural Selection in the Wild*. Princeton Univ. Press, Princeton, N.J.

Fisher, R. A., 1930. *The Genetical Theory of Natural Selection*. Clarendon Press, Oxford.

Ford, E. B., 1940. Genetic research in the *Lepidoptera*. *Ann. Eugenics*, **10,** 227-252.

Gibbs, H. L., and P. R. Grant, 1987. Oscillating selection on Darwin's finches. *Nature*, **327**, 511-513.

Glass, H. B., and C. C. Li, 1953. The dynamics of racial admixture: an analysis based on the American Negro. *Amer. Jour. Hum. Genet.*, **5**, 1-20.

Harland, S. C., 1936. The genetic conception of species. *Biol. Rev.*, **11**, 83-112.

Hartl, D. L., and A. G. Clark, 1989. *Principles of Population Genetics*, 2nd ed. Sinauer Associates, Sunderland, Mass.

Karn, M. N., and L. S. Penrose, 1951. Birth weight and gestation time in relation to maternal age, parity, and infant survival. *Ann. Eugenics*, **161**, 147-164.

Kettlewell, H. B. D., 1973. *The Evolution of Melanism*. Clarendon Press, Oxford.

Lerner, I. M., 1954. *Genetic Homeostasis*. John Wiley, New York.

Levins, R., 1968. *Evolution in Changing Environments*. Princeton Univ. Press, Princeton, N.J.

Li, C. C., 1955. The stability of an equilibrium and the average fitness of a population. *Amer. Naturalist*, **89**, 281-295.

Mayr, E., 1942. *Systematics and the Origin of Species*. Columbia Univ. Press, New York.

Petit, C., and L. Ehrman, 1969. Sexual selection in *Drosophila*. *Evol. Biol.*, **3**, 177-223.

Resnik, D. B., 1988. Survival of the fittest: law of evolution or law of probability? *Biol. and Philosophy*, **3**, 349-362.

Rotter, J. I., and J. M. Diamond, 1987. What maintains the frequencies of human genetic diseases? *Nature*, **329**, 289-290.

Slatkin, M., 1985. Gene flow in natural populations. *Ann. Rev. Ecol. and Syst.*, **16**, 393-430.

Stanley, S. M., 1979. *Macroevolution: Process and Product*. W. H. Freeman, San Francisco.

Stern, C., 1973. *Principles of Human Genetics*, 3rd ed. W. H. Freeman, San Francisco.

Strickberger, M. W., 1985. *Genetics*, 3rd ed. Macmillan, New York.

Van Valen, L., 1973. A new evolutionary law. *Evol. Theory*, **1**, 1-30.

Wallace, B., 1981. *Basic Population Genetics*. Columbia Univ. Press, New York.

Waters, C. K., 1986. Natural selection without survival of the fittest. *Biol. and Philosophy*, **1**, 207-225.

Wickler, W., 1968. *Mimicry in Plants and Animals*. Weidenfeld & Nicolson, London.

Wright, S., 1951. The genetic structure of populations. *Ann. Eugenics*, **15**, 323-354.

22

STRUCTURE AND INTERACTIONS OF POPULATIONS

he structure and relationships of natural populations obviously depart from many of the ideal conditions that would make their evolutionary behavior simple to understand; that is, populations are not of constant size, nor uniformly distributed in space, nor always of the same mating pattern, nor uniformly subject to constant conditions of mutation, migration, and selection. It is also obvious that because of multiple alleles, there are usually more than three diploid genotypes for any one locus and that, through developmental interactions, the fitness conferred by these genotypes must depend upon genes at other loci.

Moreover, the environmental contexts in which populations evolve are usually changing. These include elements in their physical environment such as moisture, temperature, pressure, and sunlight-shade, as well as elements in their biological environment such as prey, predators, parasites, hosts, and competitors. The relationship of a population to various ecological factors is also more than that of a passive recipient, since a population often modifies its physical and biological environment in ways that can diminish or enhance both its own resources and those of other populations. Because of all these complexities it is little wonder that populations "must continually keep running in order to stay in the same place" (the "Red Queen" hypothesis, p. 443). It is also not surprising that the structure of populations is difficult if not impossible to predict mathematically in detail, even by the most elaborate techniques.

Nevertheless, mathematical models, like other kinds of generalities, allow information derived from populations under one set of conditions to be applied to others under similar conditions. Certainly, there are common features shared by various populations in their response to factors such as random breeding, selection, mutation, migration, and genetic drift. A popular approach has therefore been to make measurements of various characteristics present in natural and experimental populations and then, in combination with mathematical analysis, use these observations to derive some broad evolutionary concepts about the structure of populations. This chapter offers a brief survey of some such attempts at the ecological and genetic levels.

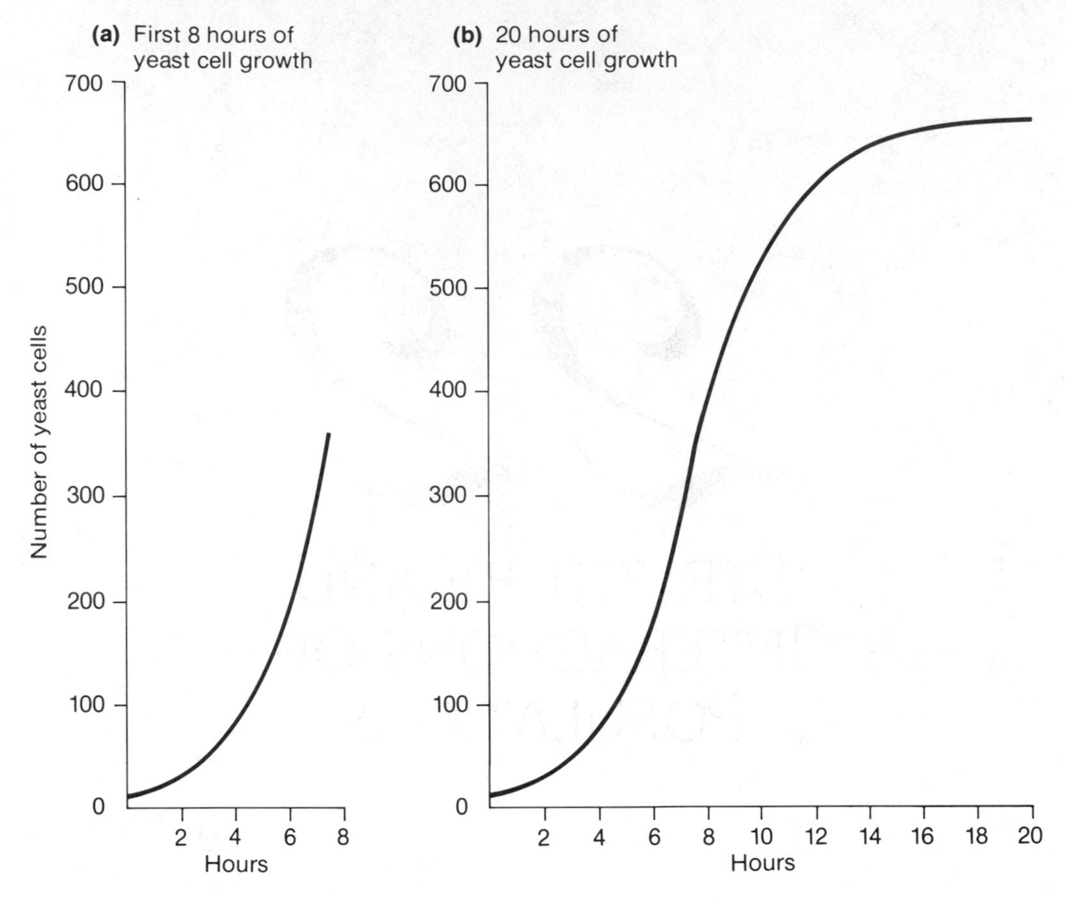

FIGURE 22-1 Numbers of yeast cells (*Saccharomyces cerevisiae*) in a defined volume of culture medium for two growth periods, beginning with approximately 10 cells per volume. (a) Exponential growth during the first 8 hours ($\Delta N = rN$, where $r = .5535$). (b) Sigmoidal growth curve approximating the logistic relationship [$\Delta N = rN \times (K - N)/K$, where $r = .5535$ and $K = 665$] for the 20-hour growth period. (Data from T. Carlson, 1913. *Biochem. Z.*, **57**, 313-334.)

SOME ECOLOGICAL ASPECTS OF POPULATION GROWTH

One area of mathematical modelling that deals with the evolutionary potential of populations applies to their reproductive powers. It is, after all, the capacity for reproduction which is counterposed against selection in the evolutionary process. In its simplest form, as in rapidly growing, asexual, unicellular species, the early stages of population growth in an environment well supplied with resources can occur exponentially so that a single individual produces 2 offspring, who then produce 4 in the next generation, 8 in the following, then 16, 32, . . . and so on, until there are 2^t individuals at t generations. Assuming the persistence of uniform reproductive properties for each individual in each generation, this provides an exponential growth curve of the type shown in Figure 22-1(a), which would be limitless if space and resources were limitless.

Quantitatively, we can describe the rate of numerical change (ΔN) in populations by noting that this change is equal to the difference between birth (b) and death (d) rates multiplied by the number of individuals (N): $\Delta N = (b - d)N$. Thus, it follows that when the birth rate is higher than the death rate ΔN is positive and population size increases; equal birth and death rates yield $\Delta N = 0$ and an unchanged N, while a death rate higher than the birth rate yields negative ΔN, causing a decrease in population size. Ignoring other causes in this simple illustration, $b - d$ is thus a primary factor determining population size and can therefore be described as the **rate of increase**, r, so that $\Delta N = rN$. In environments in which a population is free of those factors that limit its growth, it attains what is called an **intrinsic rate of natural increase**, r_m.

Environmental resources and space are not limitless, of course, nor are individuals in a dense population unaffected by the waste products and toxins produced by neighboring individuals. As a consequence, population growth is not limitless, but population size will eventually stabilize at some constant value or may even suddenly "crash" to some very low number. (As pointed out in Chapter 2, it was Malthus' popularization of the idea that the exponential growth

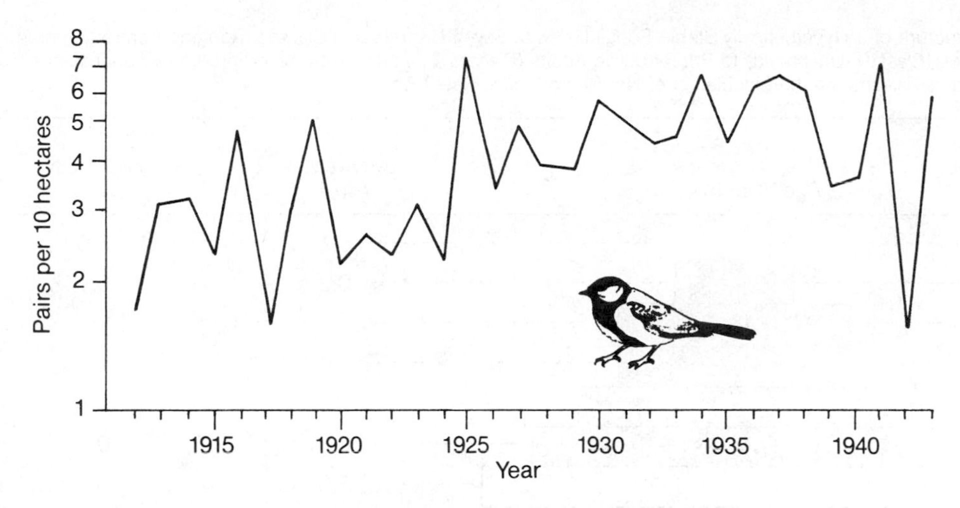

FIGURE 22-2 Fluctuations in the population size of the great tit (*Parus major*) in Holland between 1912 and 1943. (Adapted from Begon and Mortimer.)

of human populations was held in check by war, famine, and disease which led both Darwin and Wallace to the concepts of the struggle for existence and natural selection.)

In accordance with such limitations, Figure 22-1(b) shows how population size of a strain of yeast grown under a particular set of conditions levels off from its early exponential direction to a plateau of about 665 individuals. The smooth S-shaped curve that results can be described mathematically as the modification of ΔN by a factor $(K - N)/K$ in the formula $\Delta N = rN \times (K - N)/K$, so that when N is very small this factor is essentially 1, and population growth is then almost exponential, or $\Delta N = rN$ as before. However, as N increases in value closer to K, the $(K - N)/K$ factor becomes a fractional quantity closer to zero. Eventually N is sufficiently large to equal K so that $\Delta N = rN \times 0 = 0$, and population size no longer changes theoretically but becomes stabilized at the value K. The growth model that provides the relationship $\Delta N = rN(K - N)/K$ is known as the **logistic** and, for a particular environment, K is commonly called the **carrying capacity** of the population. The original equation for such logistic growth was discovered by P. F. Verhulst, and discussions of its derivation along with its possible applications can be found in population ecology textbooks such as those of Emlen and Pianka and in the historical account by Hutchinson.

In reality, populations rarely follow such smooth growth curves, and considerable fluctuations in numbers may occur (Fig. 22-2) that can often be ascribed to the impact of environmental agents. Some of these agents, such as the effects of climate, may often be independent of population size and crowding and are therefore called **density independent,** whereas others, such as the effects of metabolically produced toxins, depletion of resources, and intrapopulational aggression, are more dependent upon crowding, and are therefore called **density dependent**. Although there seem to be obvious examples where one or both of these density factors can influence population size, their quantitative effects are often difficult to measure, and their relative importance has therefore been considerably disputed among ecologists (e.g., Strong).

Aside from the effect of environmental agents, a most important aspect of a population's reproductive power is its age structure. Obviously, if reproduction is associated with a particular age of individuals, the chances to survive to that age and the number of offspring that such individuals produce are essential attributes. The first of these, **survivorship**, is measured by a factor l_x which represents that proportion of individuals who survive from age 0 to age x out of a group (cohort) who were all born during the same period. The second, **fecundity**, is represented by m_x, the average number of offspring produced by an individual of age x. The **net reproductive rate** of the entire population (R_0) is therefore the sum of all of its individuals multiplied by their fecundity at each age x:

$$R_0 = \Sigma\, l_x m_x$$

This relationship between survival and fecundity can be further elaborated in various ways (e.g., Pianka) but, in general, a population that is stable in size will have a net reproductive rate (or replacement rate) of 1, as illustrated in Table 22-1. Populations with R_0 less than 1 (e.g., decreased survival of age class 3 in Table 22-1 and the absence of compensating increases in the size and fecundity of other reproductive age classes) will decrease in size, and populations with R_0 greater than 1 will increase in size.

Older age classes such as class 6 in Table 22-1 are generally reproductively barren, with the lowest

TABLE 22-1 Age Structure of a Hypothetically Stable Population with Seven Discrete Age Classes, Ranging from Nonreproductive Juveniles (Class 0, Unshaded) to Reproductive Adults (Classes 1-5, Shaded) to Nonreproductive Senescents (Class 6, Unshaded), Illustrating the Calculation of Net Reproductive Rate (R_0)

Age Class (*x*)	Age Structure*	Survivorship (l_x)	Fecundity (m_x)	Realized Fecundity ($l_x m_x$)
6		0.1	0.0	0.00
5		0.2	0.1	0.02
4		0.4	0.6	0.24
3		0.4	1.0	0.40
2		0.6	0.3	0.18
1		0.8	0.2	0.16
0		1.0	0.0	0.00
	15 10 5 0 5 10 15 Percentage of population			R_0 = 1.00

Adapted from Pianka.
*Frequency of each age class in a stable population is related to l_x (Hutchinson) and can be noted here as the percentage graphed symmetrically on either side of the zero midline: for example, class 0 = 14.5 + 14.5 = 29.0 percent.

survival rate of all classes. In the past, their aging (**senescence**) and death were considered to have been selected as traits that benefit a population by removing individuals who might compete for its resources but who no longer contribute to its reproductive success. That is, selection occurred between populations (group selection, discussed later in this chapter) by conferring added fitness upon populations in which nonreproductives died out because of the aging process. Although population benefits of this kind probably exist, such evolutionary explanations are presently in disfavor. It has instead been suggested that senescence evolves by mechanisms that allow genes with deleterious effects on older, nonreproductive individuals to spread through a population because they are either neutral or advantageous to younger, reproductive individuals. The particular genes involved are not known, but experiments have shown that at least some populations do maintain genetic variability for aging: for example, longevity can be reduced in *Drosophila* populations whose individuals were selected for reproduction early in the life cycle, or longevity can be increased by selecting individuals who reproduce later in the life cycle (Rose). In one recent experiment, the decline in female fecundity, a measure of senescence, was apparently caused by increased homozygosity for deleterious recessive alleles (Mueller [1987]) in populations whose individuals were selected for rapid growth rates (*r*-selection, see below).

Whatever the causes for differences in age distribution, it is also apparent that the strategies used in reproduction may also differ considerably among organisms. Many annual plants and insects, for example, breed only once during their lifetimes (**semelparous**), whereas many perennial plants and vertebrates breed repeatedly (**iteroparous**). The number of offspring produced by a reproductive female at any one time also varies significantly, ranging, for example, from a single offspring in many larger mammals to the millions of eggs laid by a female codfish. Furthermore, since there is always the possibility of death before reproductive ages are reached, the sooner reproduction begins the greater the chances of producing offspring. Thus, in some organisms, reproductive stages follow soon after hatching, whereas other organisms, more dependent for survival on reaching larger size, obtaining greater experience, or needing more parental care, delay reproduction until a fairly late stage in the life cycle.

The alternatives of producing many offspring with little parental care or few offspring with greater parental care are among properties often ascribed to differences between what are called ***r* and *K* selection** strategies. In organisms such as bacteria and plant weeds that tend to exhibit rapid populational growth in facing wildly fluctuating environmental challenges and opportunities, reproductive emphasis is placed on a rapid rate of increase (*r*), and individuals in such populations are therefore considered to be *r*-selected. As stated by Darwin, "a large number of eggs is of some importance to those species which depend upon a fluctuating amount of food, for it allows them rapidly to increase in numbers." Other organisms, such as large vertebrates, face more uniform or predictable

TABLE 22-2 Characteristics Often Associated with *r*- and *K*-selection

Characteristic	*r*-selection	*K*-selection
Climate	Variable or unpredictable	Fairly constant or predictable
Diversity of resources and habitats	Usually broad	Relatively narrow
Causes for mortality	Often catastrophic and density independent	Mostly density dependent
Survivorship	Very high mortality at younger stages, with high survivorship at later stages	Either constant rate of mortality at most stages, or little mortality until a certain stage is reached
Competitive interactions	Variable, mostly weak	Usually strong
Length of life	Relatively short, usually less than 1 year	Longer, usually more than 1 year
Selection pressure for:	1. Rapid development 2. Rapid increase in numbers 3. Early reproduction 4. Small body size 5. Semelparity 6. Many small offspring 7. Increased productivity (quantity)	1. Slower development 2. Greater competitive ability 3. Delayed reproduction 4. Larger body size 5. Iteroparity 6. Fewer large offspring 7. Increase efficiency (quality)

Adapted from Pianka.

environments with population sizes that are close to the environmental carrying capacity (*K*). In these *K*-selected organisms there is density-dependent competition for food, nesting space, and other resources, thus providing selective advantages that lead to increased efficiency in the utilization of resources as well as ensuring that offspring are raised to the stage when they can compete themselves.

A comparison between some of the characteristics often deemed to be associated with *r*- and *K*-selection is given in Table 22-2. Support for the evolution of different competitive abilities under these different selective models has been demonstrated by Mueller (1988) in experimental *Drosophila* populations. It should be noted, however, that these two types of selection are not strict alternatives but that some populations may compromise between the two strategies. In Welsh populations of the periwinkle mollusc, *Littorina*, for example, larger adult size may be accompanied by increased numbers of smaller offspring in some environments, while other environments are populated by smaller adults who produce relatively fewer but larger offspring (Hart and Begon). Thus, although *r*- and *K*-selection schemes offer a pattern for understanding life histories in various populations, not all groups fit conveniently into such conventions. A fairly detailed discussion of some pros and cons of *r*- and *K*-selection theory is given by Boyce.

In general, the ecological approach briefly reviewed in this section and a later one (Group Interaction) emphasizes how populations respond to their environment in terms of their numbers and distribution. Unfortunately, in the overwhelming majority of these studies, it has been extremely difficult to discover and incorporate genetic information (e.g., gene-frequency estimates) which would provide an understanding of the correlation between such changes and evolutionary mechanisms (e.g., mutation, selection, genetic drift) discussed in previous chapters. Nevertheless, attempts at relating population ecology to population genetics are proceeding and beginning to receive considerable attention, but so far these are mostly on a theoretical plane (e.g., Roughgarden).

GENETIC LOAD AND GENETIC DEATH

In contrast to the ecological approach, with its emphasis on population distributions, numbers, and growth rates, geneticists have placed more emphasis on the amounts and kinds of genetic variability present in populations and on uniting observations on natural populations with models of their genetic structure. These studies received special impetus from the 1930's onward through the work of Chetverikoff, Dobzhansky, Fisher, Ford, Haldane, Wright, and many others. What was most interesting in these studies was the demonstration that large amounts of genetic variability are present in practically all natural populations examined. As pointed out previously (e.g., see Fig. 10-32 and

Table 10-5), considerable polymorphism has been found on both the chromosomal and genic levels, and it is this variability that obviously allows genetic evolutionary changes to proceed.

However, in spite of the many advantages of genetic variability, many of the genes maintained by natural populations may be disadvantageous to their carriers either in certain combinations or in homozygous condition. As we have seen previously in Figure 21-7, selection can account for a significant loss of nonoptimal individuals even in long-standing populations. Thus, if we consider genetic perfection as the elimination of all deleteriously inferior gene combinations, there is little doubt that most, if not all, populations are genetically imperfect.

The extent to which a population departs from a perfect genetic constitution is called its **genetic load** and is accompanied by the loss of a portion of its individuals through their **genetic death**. Genetic death is not necessarily an actual death before reproductive age but can be expressed through sterility, inability to find a mate, or any means that reduces reproductive ability relative to the optimum genotype(s). Estimates of these values are therefore in terms of the proportion of individuals eliminated because of selection against them. For example, if a gene is deleterious in homozygous condition, the frequency of homozygotes before and after selection will be:

	Genotypes		
	AA	*Aa*	*aa*
Frequency at fertilization	p^2	$2pq$	q^2
Relative adaptive value	1	1	$1 - s$
Frequency after selection	p^2	$2pq$	$q^2 - sq^2$

Thus, the loss in frequency of individuals, or incurred genetic load, is equal to sq^2. Therefore, if there were N individuals in the population before selection, sq^2N are now eliminated because of genetic load.

This value of sq^2, however, also equals the mutation rate (u) at equilibrium for $A \rightarrow a$ (p. 444), which means that the genetic load caused by a deleterious homozygous recessive is equal to its mutation rate. An important feature of this relationship pointed out by Haldane is that if the mutation rate is constant, it will make little difference to the genetic load whether s is small or large. As can be seen in each single column (constant mutation rate) of the hydraulic model in Figure 21-10, if s is small, q will be large at equilibrium, and if s is large, q will be small. High selection coefficients will therefore cause the gene to be eliminated more rapidly (low q), and low selection coefficients will permit the gene to remain longer in the population (high q). In either case, however, the genetic load is still $sq^2 = u$ and the total number of genetic deaths remains at sq^2N. Insofar as deleterious recessives are produced by mutation, therefore, any increase in mutation rate will cause a corresponding increase in genetic load and consequent increase in genetic death.

According to Crow (1962), various genetic components contribute to the genetic load, and techniques have been devised by him and others to evaluate their relative importance. The **mutational load**, discussed above, is only one of the essential factors considered responsible for genetic load. Another is the **segregational**, or **balanced, load**, restricted to those instances in which a heterozygous genotype is superior to both types of homozygotes. For a gene with two alleles, the segregational load amounts to p^2s for the *AA* homozygotes plus q^2t for the *aa* homozygotes, since s and t are the selection coefficients against these homozygotes, respectively, when the heterozygote is overdominant (see Table 21-4). If we substitute the equilibrium frequencies of p and q under those circumstances into $p^2s + q^2t$ we obtain

$$\left(\frac{t}{s+t}\right)^2 s + \left(\frac{s}{s+t}\right)^2 t = \frac{st^2 + ts^2}{(s+t)^2}$$

$$= e\frac{st(s+t)}{(s+t)^2}$$

$$= \frac{st}{s+t}$$

Thus, if s and t are both about .1, the segregational load will be .01/.2, or .05. This value is considerably higher than most mutation rates and demonstrates the increased genetic load that may be expected to be caused in randomly breeding populations by segregation as compared with the load caused by mutation ($sq^2 = u$).

Although there is little question that most or all populations carry genetic loads of one kind and another, the relative values of each type of load has not yet been determined. From the viewpoint of evolution, "genetic perfection" with its absence of genetic death is probably never reached in any species, since environments usually change with time and the advantages of different genotypes change accordingly. It is, in fact, conceivable that a population so perfectly adjusted to its environment—that is, with little or no genetic load (i.e., no variability)—may become extinct within a short period because of rapid environmental change. On the other hand, a population with a relatively large genetic load may be subjected to a new environment in which formerly deleterious genes provide it with adaptations that enable it to survive. The absence of genetic load may therefore be more detrimental to a population than its presence (Li, Brues). Therefore, although the genetic load can be measured in terms of departure from the optimum genotype, the evolutionary value of a particular optimum genotype may be a very limited one; the optimum genotype may change in time, or from place to place, or may even differ in the same place, if, for example, a division of labor occurs between the individuals of a population (e.g., male and female).

THE COST OF EVOLUTION AND THE NEUTRALIST ARGUMENT

Whatever type of load a population bears, natural selection will cause new additions to the load by favoring some genes and discarding others. Since this process of gene replacement is ongoing and pervasive, it is reasonable to ask how many individuals a population must lose, in terms of genetic death, to replace a single gene by the action of selection alone. If a dominant mutation arises that is advantageous in comparison to the more frequent recessive, we have seen that the number of genetic deaths will be sq^2N for one generation of selection, where N is the number of individuals in the population. For complete replacement of a deleterious allele, Haldane has calculated that the total number of genetic deaths will be determined by a factor D that is based primarily on the initial frequency of the favored allele, multiplied by the population number N. For a newly favored but rare dominant allele, D is about ten, so the cost of evolution (DN) for eliminating the deleterious recessive allele is about ten times the average number of individuals in a single generation. This cost will, of course, be spread over many generations, the rapidity of gene replacement depending on the selection coefficient.[1] If the rare newly favored allele is recessive, DN is increased to about $100N$ since the favored homozygotes are extremely rare. According to Haldane, many favored alleles are probably intermediate in dominance, and he therefore proposed an average death value of $30N$ for the replacement of a single allele. He suggested that a population is probably capable of sacrificing about one tenth of its reproductive powers for such selective purposes; that is, a gene substitution can occur at a rate $1/10 \times 30N$, or every 300 generations.

However, not all gene substitutions may be caused by selection, and Kimura has proposed that the rate of evolution at the molecular level is actually far more rapid than Haldane suggested. In vertebrates, for example, one can calculate that many hemoglobin protein amino acids are replaced at a rate of approximately one amino acid change per 10^7 years, yet the total amount of DNA is certainly more than necessary to code for 10^7 amino acids. This means that each vertebrate species would be expected to undergo at least one complete amino acid substitution per year if we assume that overall amino acid substitution rates are about the same for all proteins. If we accept Haldane's value of about $30N$ for the cost of a single gene substitution by selection and estimate an average of about three years per vertebrate generation, then each vertebrate population must continually expend an enormous number of genetic deaths in order to maintain its size and escape extinction. In the present example, the population would have to devote 90 times its number in each generation ($3 \times 30N$) were selection the primary cause for gene-frequency changes.

Because of this presumably high cost of selection, Kimura and co-workers, as well as King and Jukes, have instead proposed that most amino acid changes are neutral in their effect. Since selection among **neutral mutations** is absent, the fixation of such alleles obviously incurs no genetic load and would depend only on their mutation rates and on random genetic drift.[2] This neutralist hypothesis (also called by some **non-Darwinian evolution**) seems to be supported by the widely observed and extensive degrees of enzyme and protein polymorphisms, indicating that allelic differences are being maintained at many thousands of loci in many species. One could of course claim that allelic differences can be maintained without fixation and,

[1] When s is large, gene replacement is more rapid than when s is small, but the population now suffers the danger of extinction because numbers may be insufficient to ensure mating partners or survival in the face of accident. In the extreme case, when $s = 1$, replacement may occur in a single generation if the unfavored allele is dominant and only favored recessive homozygotes survive. Under such circumstances, however, extinction is fairly sure to occur if the favored gene is present in very low frequency, since there will be very few favorable homozygous recessives available for survival.

[2] If the chance for any gene to mutate is u, a population of N diploid individuals bearing $2N$ genes will have $2Nu$ newly arisen mutations. Should all or most of these mutant genes be equally neutral in their phenotypic effect, each will be present in about $1/2N$ frequency and will probably persist in this relative frequency because any one of them is no better than the other. Thus, there is a total of $2Nu$ neutral mutations, each with a $1/2N$ chance of fixation, or the chance of fixation for a particular neutral allele is $2Nu \times 1/2N = u$. On an individual basis the time necessary to attain fixation for any particular neutral allele will depend on population size, since the random drift process in small populations greatly speeds up the fixation of these genes, although it does not change their probability of fixation. Kimura and Ohta calculate that the average number of generations necessary for the fixation of a new neutral mutation is approximately four times the number of parents in each generation.

since they will never be fixed, we can exclude the high cost of evolution that Haldane has shown to be necessary for gene substitution. However, the maintenance of polymorphism by selection may itself entail an enormous and intolerable genetic load.

For example, if polymorphism was maintained by the superiority of heterozygotes, such segregational load would be $st/(s + t)$, as explained previously. If this segregational load is caused by lethality of the two homozygotes (**balanced lethals**), the selection coefficients s and t are both 1, the load is then 1/2, and the remaining fitness of the population is $1 - 1/2 = 1/2$. For two pairs of balanced lethal genes acting independently of each other (e.g., *AA* and *aa* are lethal, as are *BB* and *bb*, and only the *AaBb* heterozygotes survive), the fitness of the population is reduced to $(1 - 1/2)^2 = 1/4$. In general, no matter what the value of the selection coefficients, the average fitness of a population bearing a balanced or segregational load is therefore $[1 - st/(s + t)]^n$ where n designates the number of gene pairs at which heterozygote superiority is being maintained. Kimura and Crow have calculated that the fitness of such a population is approximately equal to $e^{-\Sigma L}$ where e (2.718) is the base of natural logarithms and ΣL designates the sum of individual loads for each gene pair involved.

Not surprisingly this load can be quite large, even if the selection coefficients acting against homozygotes are small, as long as there are many gene pairs involved in maintaining superior heterozygotes. For example, if superior heterozygotes are being maintained at 100 loci, each bearing a genetic load of .01 (e.g., $s = .02$, $t = .02$: $st/(s + t) = .0004/.04 = .01$), the average fitness of the population is reduced to about $e^{-100(.01)} = e^{-1} = .37$. For 500 gene pairs acting similarly, the fitness is approximately $e^{-5} = .007$, and it is $e^{-10} = .00005$ for 1,000 such gene pairs. This means that to maintain polymorphism at 1,000 loci, even with only a very small selective advantage for the heterozygote at each gene, only one out of 20,000 offspring would survive. Thus, for every female to produce two surviving offspring (and thereby allow the population to survive), each such female would have to produce about 40,000 young for this selective purpose alone! Most polymorphic systems must therefore consist of neutral mutations, according to Kimura and his followers, who add support to this argument with observations of high frequencies of selectively neutral mutations in a rapidly mutating strain of *Escherichia coli* (Gibson et al.).

Among further evidence used by neutralists is the presumed constancy of amino acid substitution rates in particular proteins (the "evolutionary clock"; Chapter 12). Kimura and Ohta, for example, point to the fact that the same number of changes in the α-hemoglobin chain has occurred, relative to amino acids in β-hemoglobin, whether the α-chain comes from the same species or different species. Compared to the human β-chain, the human α-chain shows 75 differences, the horse α-chain shows 77 differences, and the carp α-chain shows 77 differences. Kimura and Ohta ask, Why should the α-hemoglobin chains of humans, horses, and fish, each with a different selective history, have diverged from the human β-chain at exactly the same rate? According to the neutralists, this uniformity can most easily be explained by a common rate of neutral mutation and drift rather than common selective conditions. As striking as this evidence is, the regularity of the molecular clock is still disputed. There are certainly exceptions to constant evolutionary rates in some proteins, and different rates of nucleotide substitution are noted in comparisons among different proteins (p. 238), although Takahata has argued that even such variations in mutation rate may be explained by the neutral hypothesis.

THE SELECTIONIST ARGUMENT

To counter the neutralist view, various selection schemes have been proposed by **selectionists** which would explain the persistence of numerous polymorphisms but confer only minimal genetic loads. One such mechanism is frequency-dependent selection (p. 440) which incurs genetic loads only when a relatively rare selected allele is undergoing frequency changes but produces no genetic load when the allele has reached equilibrium (Kojima and Yarbrough). However, since polymorphism seems to exist at thousands of loci, it may be questioned whether there are also thousands of individual frequency-dependent mechanisms in the environment.

Sved, King, and others have suggested that selection in a natural population probably lumps the effects of many individual genotypes together into two main groups: the fit and the unfit. That is, there are threshold numbers of polymorphic loci in a population which serve to distinguish these two groups of genotypes but do not usually distinguish individual genotypes within each group. In such comparisons between genotypes, heterozygotes for more gene loci than the threshold number show no increased heterotic effect, and heterozygotes for fewer genes than the threshold number are presumably all equally deleterious. The presence of a threshold thus acts as a form of **truncation selection** to eliminate or truncate an entire class of phenotypes whatever their genotypes may be (Wills).

Although a threshold of this kind may shift, depending as it does on the environmental stresses placed on the population (Milkman [1967], Wallace), the genetic loads incurred by such populations may be relatively small since, because of lumping, differences between each of the many genotypes above the threshold or below the threshold do not add to the load. The likelihood that genes are not individually replaced in evolution but can be selected in linkage blocks (Franklin and Lewontin), or perhaps as functional groups, supports the view that there may be considerable interaction between genes to produce threshold effects. The possibility of such interactions also indicate that theoretical genetic loads may be erroneously calculated if it is assumed that each locus acts independently of all others. However, although the concept is attractive to some population geneticists, there is little direct evidence as yet that thresholds exist.

At present, considerable emphasis is being placed by selectionists on the following findings.

Association Between Enzyme Polymorphisms and Ecological Conditions

The argument has been presented that a strong correlation between particular enzymatic alleles and particular environmental conditions might indicate that **allozyme frequencies** are being maintained through selection. One well-known example of such correlation is the relationship between the gene for sickle cell anemia and malaria (see Fig. 21-3), where the heterozygote is superior in fitness to both homozygotes. A further documented example of overdominance is that for an alcohol dehydrogenase gene in yeast (Hall and Wills). In such cases, however, the maintenance of polymorphisms by selection obviously incurs a significant expense in the loss of homozygotes.

A system that may have milder selective effects is the polymorphism discovered by Koehn in the freshwater fish *Catostomus clarkii*. The distribution of two alleles of an esterase enzyme in this fish seems to follow a temperature cline along the Colorado River basin. The homozygote for the allele that is most frequent in the more southern (and warmer) latitudes produces an esterase enzyme that becomes more active as temperature increases, whereas the allele that is more frequent in the northern (and colder) latitudes forms an enzyme that is more active as temperature decreases. Not unexpectedly the heterozygote for the two alleles forms an enzyme that is most active at intermediate temperatures.

In *Drosophila melanogaster*, allozymes produced by the *alcohol dehydrogenase* locus (*Adh*) also show correlations between their frequencies and environmental temperatures. According to experiments of Sampsell and Sims, there is a relationship between the stability of different *Adh* allozymes under high temperatures and the fitnesses they confer upon flies by enabling them to survive high alcohol concentrations in the food medium. It seems likely that selection acting on such allozyme differences can help explain some of the gene-frequency changes observed along various north-south geographical gradients (e.g., see Oakeshott et al.).

Although other such correlations between alleles and particular environments have been shown in plants (Allard and Kahler) as well as in animals (Koehn et al., Somero), there are numerous instances where no obvious correlation exists. It is therefore difficult to prove that such correlations are always necessarily causal and may not be accidental. Neutralists have also argued that some enzyme loci being scored for polymorphism may be strongly linked to a gene locus at which selection is operating, and the protein polymorphism observed would only be the effect of such linkage disequilibrium, or **hitchhiking**.

Nonrandom Allelic Frequencies in Enzyme Polymorphisms

It has been pointed out that there are a number of enzyme polymorphisms whose advantages are unknown but whose frequencies are difficult to explain on a purely random basis. An early example of this kind was the observation by Prakash et al. that populations of *D. pseudoobscura* ranging from California to Texas show remarkably similar allozyme frequencies for a number of proteins. Although it has been suggested that such similarities arise because of migration between different populations rather than through common selective factors, migration could hardly explain similarities in gene frequencies between different species. As Ayala and Tracey pointed out, genetically isolated species of the *D. willistoni* group share common gene frequencies for the alleles of many different enzymes. Furthermore, the pattern of similarity between them is not constant, and the data seem to show that different species share common selective factors for some enzymes but not for others. On a broad scale, coordinating both genetics and ecology, Nevo has made a survey of 35 Israeli species including insects, molluscs, vertebrates, and plants, and comes to the conclusion that the amount of allozyme polymorphism in a species is correlated to factors such as life habit and climate.

Perhaps even more striking is the finding by Milkman (1973) that clones of *E. coli* isolated from the intestinal tracts of animals as diverse as lizards and humans and from localities as widespread as New Guinea and Iowa appear to share common allozyme frequencies. For each of five different enzymes, Milkman found that one particular electrophoretic band was frequent in almost all samples. Since other allozymes of these proteins exist, the finding of such a narrow distribution of allozymes would seem difficult to explain on any basis other than selection.[3]

Association Between Enzyme Function and Degree of Polymorphism

The suggestion that the function of an enzyme influences the degree of polymorphism at its locus was first made by Gillespie and Kojima. They grouped enzymes into two classes, namely those involved in restricted pathways of energy metabolism such as glycolysis (e.g., aldolase) and those that can use a variety of substrates (e.g., esterases, acid phosphatases, etc.). Their findings and others have since shown that enzymes with more restricted uses show significantly less polymorphism than enzymes whose substrates are more variable. Johnson, extending this notion further, proposed that enzymes involved in regulating metabolic pathways (e.g., glucose-6-phosphate dehydrogenase, phosphoglucomutase, etc.) are generally more polymorphic than enzymes whose functions are not primarily regulatory (e.g., malate dehydrogenase, fumarase, etc.). Although the biochemical causes that sustain these differences in polymorphism are not yet known, it seems clear that they are not random, and many of the polymorphic alleles are consequently not neutral.

Decrease in Polymorphisms for DNA Coding Sequences

In general, nucleotide sequence analyses show that polymorphisms are significantly greater in those DNA sequences that do not determine amino acid sequences compared to those DNA sequences that are transcribed and translated into amino acids. This argument suggests that selection reduces the variability in amino acid coding regions because such sequences have a greater effect on the phenotype than do non-coding regions. Or, put differently, the variability that remains in amino acid coding regions is not determined by those random "neutral" mutations mostly responsible for variability in non-coding regions. Thus, selection would be expected to operate more strongly on the first two amino acid codon positions, since these are more involved in amino acid determination than the "wobbly" third codon position, which is essentially responsible for degeneracy of the genetic code (Chapter 8). Similarly, we would expect more polymorphism in introns (intervening sequences, Chapter 9) which do not code for amino acids than in exons (expressed sequences) which do. According to Kreitman, both these expectations are fulfilled for the *D. melanogaster alcohol dehydrogenase* gene, in which he found 6 and 7 percent polymorphism among introns and third codon positions but polymorphism practically absent in exons and in the first two codon positions. Since random events can hardly explain such pronounced differences in polymorphism, these findings strongly suggest that selection must be the discriminating agent that determines which nucleotide base substitutions will become established in functionally different DNA sequences.

In summary, selective forces unquestionably affect polymorphism and help to maintain it under some circumstances. However, the number of genes on which selection operates at any one time, the linkage relationship between these genes, the kinds of selection that operate, and the size of the selection coefficients are all unknown. Because of this absence of information, neutral mutation and random genetic drift cannot be excluded as important causes for some, or even many, polymorphisms. Certainly some genetic variants may be neutral at certain times or under certain conditions but have selective value when circumstances change (Hartl and Dykhuizen).

SOME GENETIC ATTRIBUTES OF POPULATIONS

As discussed previously (see Table 20-1), populations have a unique evolutionary distinction, and many genetic factors that affect the evolution of populations may act differently from our expectations if individuals alone were considered. Among these factors are the following.

Sex

Sex may be of little value to an individual (or even a disadvantage) but can be a distinct advantage to a population. For example, it has often been pointed out that females who reproduce parthenogenetically accomplish the work of the usual two sexes more efficiently since such females need expend no resources in producing a separate male sex. The cause for sexual reproduction is therefore related, in some way, to pop-

[3] Selander and Levin nevertheless contest this view by suggesting that mutational and recombinational events in *E. coli* may have been sufficiently rare so that the genotypes of the initial founding populations were preserved over long periods of time.

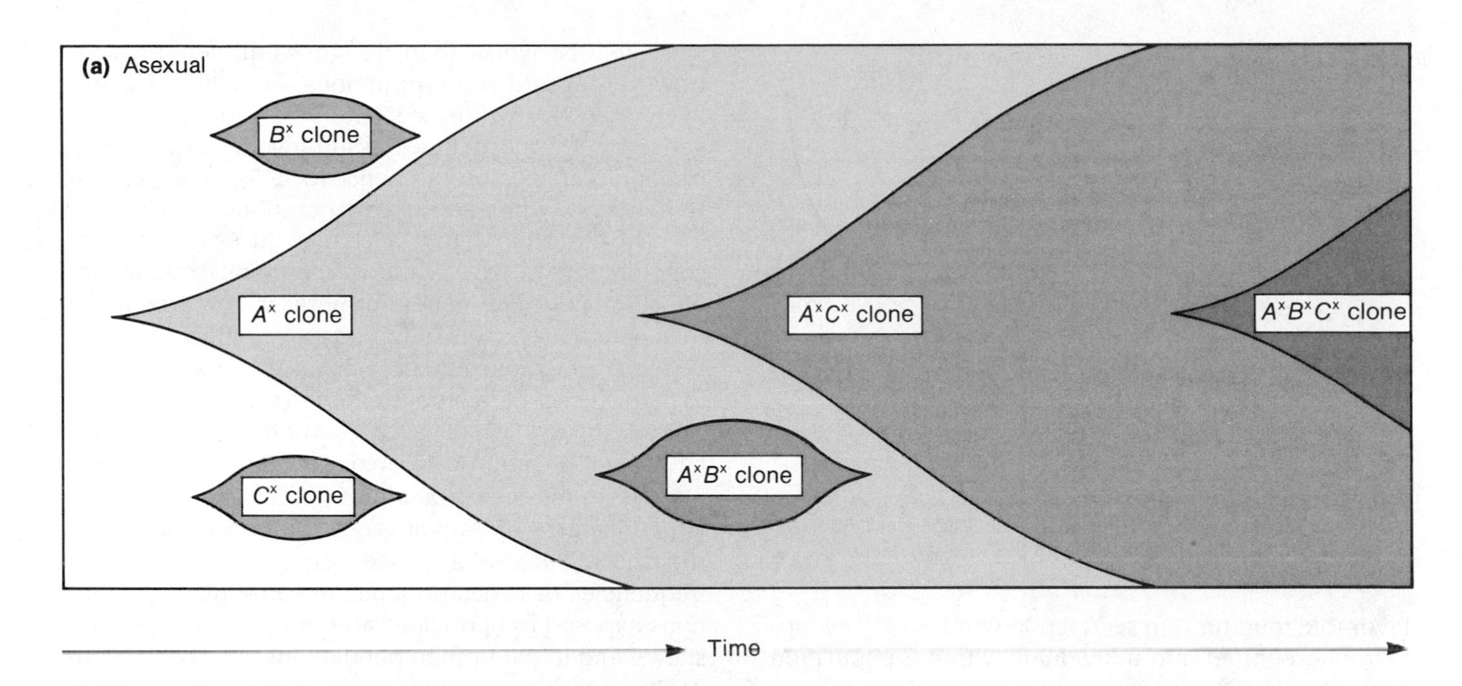

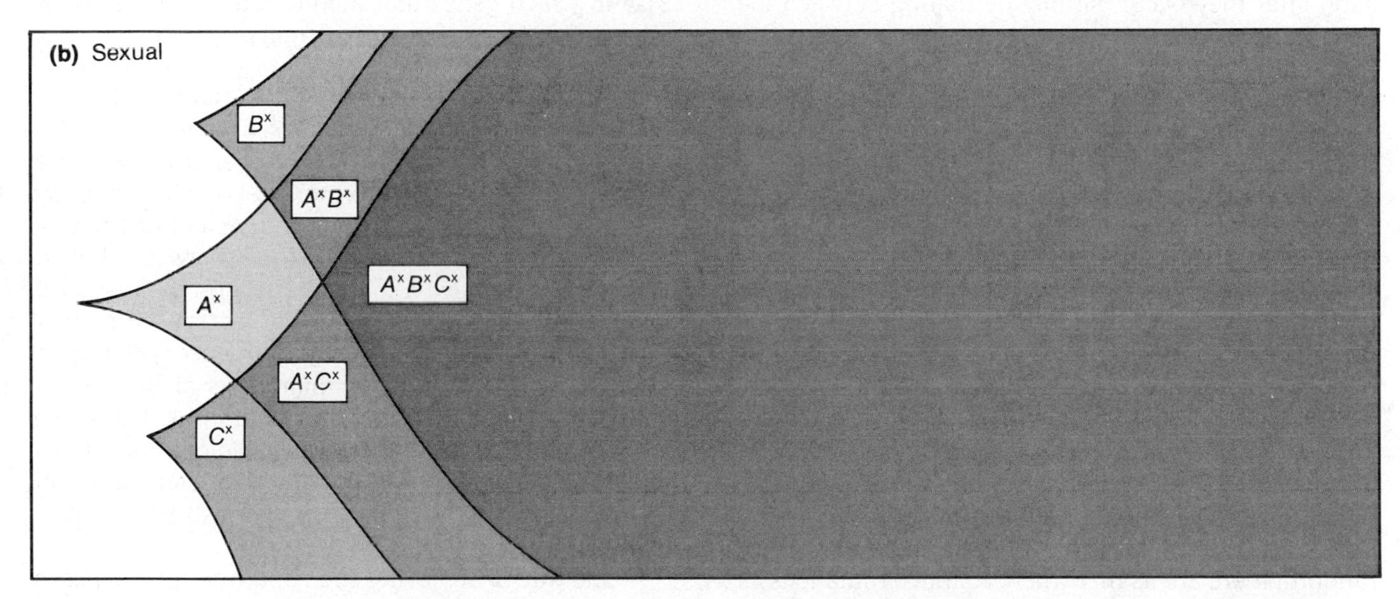

FIGURE 22-3 Muller's model of the difference between asexual and sexual populations in the speed at which they incorporate combinations of advantageous mutant alleles. In this illustration, both kinds of populations are assumed to begin with an initial state where three loci are fixed for the *A*, *B*, and *C* alleles. The advantageous alleles that then arise at these loci are designated A^x, B^x, and C^x, with individuals carrying A^x more fit than those carrying B^x or C^x. Among the other combinations, A^xC^x is presumed to be more fit than A^xB^x or B^xC^x, while all individuals carrying all three mutant alleles, $A^xB^xC^x$, are the most fit of all. (a) In asexual clonal populations, each clone is independent of the others, so attainment of the most fit genotype must await successive mutational events in a single clone ($A^x \rightarrow A^xC^x \rightarrow A^xB^xC^x$), while the least fit clones become extinct. (b) In sexual populations, the $A^xB^xC^x$ genotype is achieved relatively rapidly through recombination between individuals who carry the various advantageous alleles, and no further mutational events are necessary. (Adapted from Muller.)

ulations rather than to individual females, and a number of proposals for the advantages of sex have been offered.

Among the most popular of these proposals is that sexual crossing allows different beneficial mutations to be easily incorporated into single individuals within a population through mating and recombination, whereas in the absence of sex, combinations of such beneficial mutations are more difficult. As shown in Figure 22-3, favorable mutations such as A^x, B^x, and C^x may arise in an asexual population but will remain in separate individual lineages. Only if the asexual population becomes extremely large, which may take considerable time, is there a reasonable chance that a second favorable mutation will occur in a lineage that contains a previous favorable mutation. By contrast,

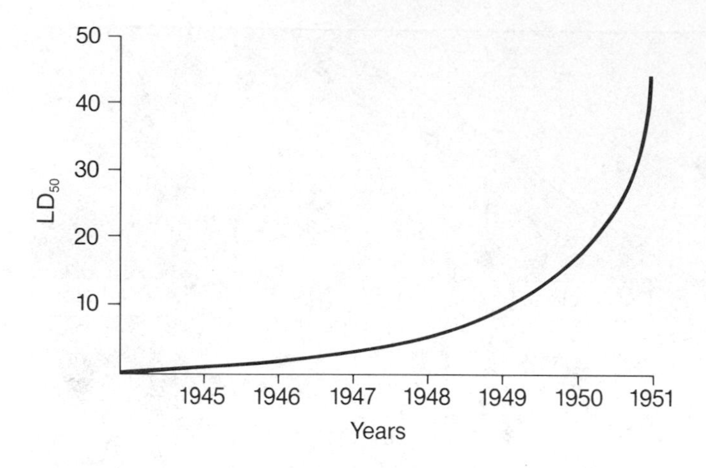

FIGURE 22-4 Resistance to DDT in houseflies collected from Illinois farms measured in terms of the lethal dose necessary to kill 50 percent of the flies (LD_{50}). (Adapted from Strickberger, data from Decker and Bruce.)

favorable mutations in sexual populations may be rapidly incorporated into individuals within a population soon after they occur, simply by mating between the separate lineages that carry the favorable mutations.

Furthermore, sexual populations may adapt more rapidly to changing environmental conditions by producing new favorable combinations of already existing gene differences through recombination, whereas asexual populations cannot rely on recombination to face such challenges. That is, sexual populations can increase their genetic variability relatively easily through recombination, but the opportunity for such increases are absent in asexual populations, which must instead depend mostly upon the variability they already have (Bell, Crow [1988], Maynard Smith).[4]

Mutation

A further important attribute of populations is mutation. One might expect that since mutation is mostly random, there are as many new adaptive mutations as there are deleterious ones. It would therefore appear as though evolution merely awaits the occurrence of superior adaptive individuals before it proceeds. In practically all populations, however, the role of new mutations is not of immediate significance. A population that has long been established in a particular environment will have many of its genes adapted for these conditions. New mutations that arise, if not neutral in their effect, will therefore rarely be better and will likely be worse than the genes already present. However, even if some mutations are better, it is unlikely that they will exceed the fitness of established genes by a large degree. Assuming these beneficial mutations are not lost by chance (p. 443), their increase in frequency will take many generations (p. 436).

On the other hand, a change in environmental conditions may have a more important evolutionary effect, since many genes formerly in low frequency may suddenly possess high adaptive value. This can be seen in the rapid genetic changes that have occurred in many insect populations exposed to pesticides such as Dieldrin and DDT (Fig. 22-4), where resistant alleles have been located on all major chromosomes (Crow [1957]); in the *Biston betularia* populations that show large increases of melanic gene frequencies in industrialized regions (p. 441, see also Lees); in increased frequencies of resistant genes in some plant populations exposed to herbicides and metallic toxins (Bradshaw); and in the human populations shown in Figure 21-3 in which genes that modify red blood cell physiology offer protection against malaria.

Linkage

Selection among individual mutant genes on separate chromosomes, however, is not the only method by which genetic progress is achieved. As pointed out by Mather and others, recombination between linked genes may also have a marked effect on the response to selection. To illustrate this point let us assume that each of four loci, *Aa*, *Bb*, *Cc*, *Dd*, influence a character quantitatively, and that all capital-letter alleles have a plus effect on the character and all small-letter alleles have a minus effect. If the phenotypic optimum is an intermediate one, as it is for many characters, it would be advantageous that the genotype also be intermediate, for example, *AaBbCcDd*.

One way of achieving such optimum genotypes is for tight linkage to be present between these four loci in the fashion of *AbCd* on one homologue and *aBcD* on the other. Thus any combination of chromosomes will always have four plus genes and four minus genes, yet the population retains the variability of all the different alleles. Note, however, that such a linkage group must have three crossovers in order for chromosomes to be formed containing all plus or all minus genes. Thus if selection changes from an intermediate phe-

[4] The relationship between the evolution of sex and the evolution of recombination is therefore a matter of considerable interest. For example, one hypothesis suggests that sex (meiosis) originated as a way of overcoming DNA damage by utilizing recombinational DNA repair mechanisms and the genetic variation that resulted was only an accidental by-product (Bernstein et al.). However, according to Maynard Smith (1988), even if recombination had a DNA repair origin, one can still argue that it was the genetic variability "by-product" which became the primary sexual mainstay. Other aspects of this topic are discussed by various contributors in the collection of Michod and Levin.

notype to an extreme phenotype (either all plus or all minus), some time may elapse before the appropriate crossovers can furnish the most adaptive combinations of genes. In other words, recombination as well as mutation can dictate the progress of selection.

ADAPTIVE PEAKS

In general, we can see that since more than one locus affects fitness in a population, there are many possible ways in which increased fitness can evolve. For simplicity, we can consider a population containing only homozygous genotypes, where the same four loci mentioned above affect a character, so that the optimum phenotype is determined by equal numbers of capital-letter and small-letter alleles. A variety of six optimum genotypes is then possible, for example, *AABBccdd*, *AAbbCCdd*, *aaBBccDD*, and so forth. According to Wright (1963, and earlier publications), each of these genotypes may be considered to occupy an **adaptive peak**, which means simply a position of high fitness associated with a specific environment. As long as no other factors change the fitness of these genotypes, each of these six peaks is of equal height, and a population consisting entirely of any one of these genotypes would therefore achieve maximum fitness for this phenotype.

When more than four loci are involved in fitness with more than two alleles at any locus, the number of possible adaptive peaks increases astronomically. For a locus with only four alleles, there are ten possible diploid gene combinations, and for 100 loci with four alleles each there are 10^{100} possible gene combinations. Even limited to this relatively small number of loci, the number of possible combinations far exceeds the number of individuals in any species and even the estimated number of protons and neutrons in the universe (2.4×10^{70}). Thus, even if only a small portion of these gene combinations are adaptive, there are undoubtedly more possible adaptive peaks than can be occupied by a species at any one time.

Again, however, we must clearly differentiate between populations and individuals; that is, a potentially high adaptive peak for a population need not coincide with a high selective peak for a genotype within the population. This discrepancy arises because the selective values of genotypes are based on competition with other genotypes but may not indicate their effect on the population. For example, it has been pointed out by Haldane, Wright, and others that **altruists** who sacrifice themselves for the benefit of shared genotypes may have low selective value as individuals, although a population bearing such altruistic genotypes may have higher reproductive values than one without them (pp. 312 and 410). Conversely, **social parasites** that increase their frequency at the expense of other genotypes in a population may have high individual selective value, although they depress the reproductive fitness of the population. Illustrations of the latter type are alleles that modify segregation ratios in their favor (**segregation distorters**) so that the gametes produced by heterozygotes carrying these alleles consist mostly of such distorters rather than possessing an equal proportion of normal non-distorter alleles. The frequency of such segregation distorters thus tends to increase in a population even though some are associated with deleterious or even lethal phenotypic effects (e.g., *tailless* alleles in mice).

For simplicity, however, let us assume that such complications are not involved in the present example of four gene pairs. Even so, the concept of many adaptive peaks with uniform height undoubtedly departs from real conditions. It is likely that the effects of each of these four pairs of genes may differ considerably, and we can, for example, assign adaptive values to the effects of the *A* and *B* genes on the basis that the greater the number of capital-letter genes in these two pairs, the greater the fitness. When combined with the previous adaptive values, a **landscape** of peaks can now be constructed, as in Figure 22-5, showing the adaptive heights of different possible genotypes. Note that there is now one peak superior to all (*AABBccdd*) and also a number of possible intermediate peaks, each surrounded by genotypes that are relatively inferior. A population may therefore increase in fitness during evolution but nevertheless reach an intermediate peak which is not necessarily the most adaptive. To move from peak to peak until a population finds the highest one demands that it travel through inferior genotypes that occupy the less **adaptive valleys** of this landscape. Such reductions in fitness are indicated by the arrows in Figure 22-5, which show the general course of travel that a population located at *aabbCCDD* might take to reach the highest peak at *AABBccdd*. There are at least two non-adaptive stages in this illustration at which the population will suffer.

Once such an adaptive landscape has evolved, further evolution will depend upon the origin of a new selective environment and the creation of new adaptive peaks. However, if conditions are not changing rapidly, the same set of adaptive peaks may remain for long periods of time. A population on one adaptive peak can then no longer reach a higher peak without going through a non-adaptive valley. Since selection can hardly occur for non-adaptation, it seems reasonable to ask, how can a population located on a relatively low adaptive peak evolve so that it occupies the highest or near-highest peaks on the adaptive landscape?

In answer to this problem Wright proposed that many populations are broken into small groups of subpopulations. These local populations, or **demes**, are

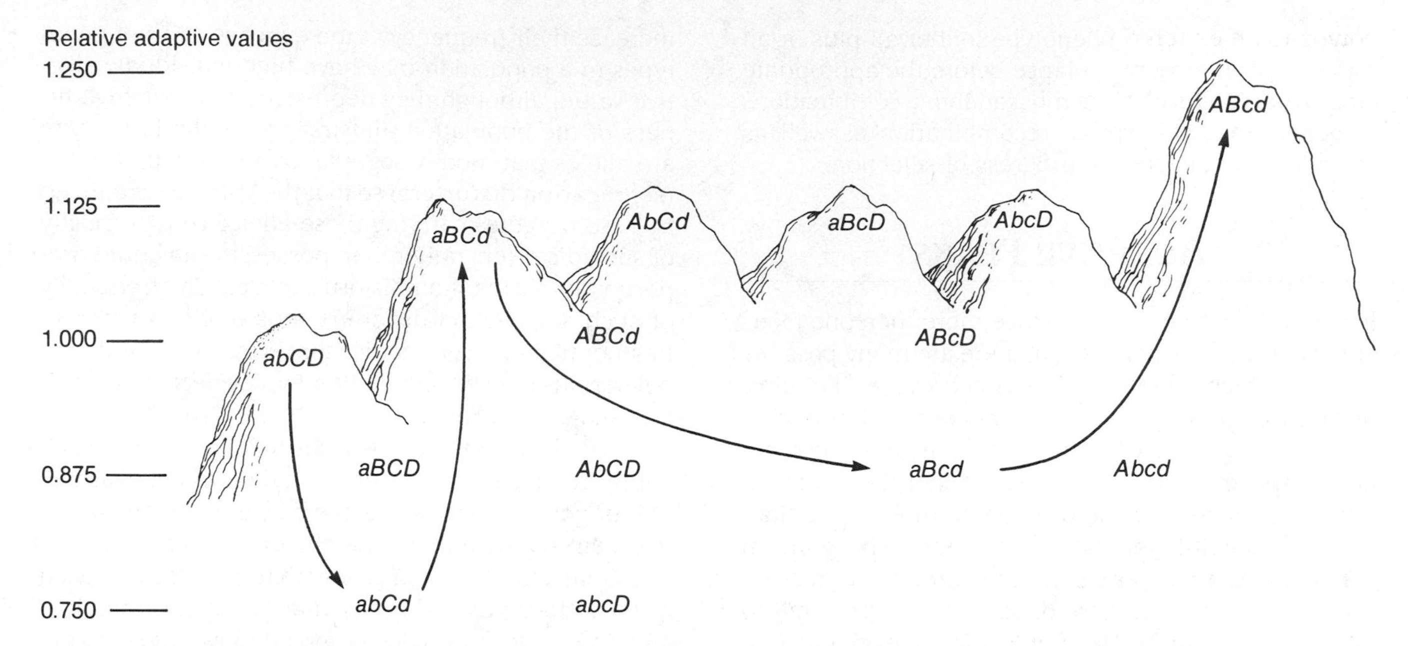

FIGURE 22-5 Adaptive landscape based on assigned adaptive values for various genotypic combinations, with arrows showing one possible path in the progress of a population from a lower adaptive peak at *aabbCCDD* to the highest adaptive peak at *AABBccdd*. (Adapted from Wright [1963].)

small enough to enable genetic differences to occur between them through the non-selective process of random genetic drift but are not so widely separated as to completely prevent gene exchange and the introduction of new genetic variability. The adaptive landscape is therefore occupied by a network of demes, some at higher peaks than others. Thus selection takes place not only between genotypes competing within demes but also between demes competing within a general environment.

The kaleidoscopic pattern of evolutionary forces acting upon these demes was called by Wright the **shifting balance process**, and can be simplified as follows:

1. Random genetic drift, acting upon polymorphism and heterozygosity at various loci, allows a number of demes to change their gene frequencies and move across non-adaptive valleys to different parts of the adaptive landscape.
2. Selection pushes some of these demes up the nearest available adaptive peak by changing gene frequencies even further, that is, by making some loci homozygous or nearly so.
3. Polymorphism retained at other loci, or variability introduced through migration and mutation, provides further opportunity for genetic drift to trigger movement across the adaptive landscape, eventually enabling still higher adaptive peaks to be occupied.
4. A deme that has attained a high adaptive peak will tend to displace other demes at lower peaks by expanding in size or to change the genetic structure of other demes through migration.
5. The fact that environmental conditions rarely remain constant continually produces new adaptive peaks and therefore new adaptive landscapes, thus encouraging populations to continually shift their genetic structures.[5]

Fisher, by contrast, suggested that most populations are large and fairly homogeneous, so that each new allele is tested in competition with all other alleles in the population. This large population thus primarily improves its fitness by small, incremental, selective steps rather than major random genetic drift.

According to Wright's scheme (Table 22-3), subdivided populations have many evolutionary advantages over a single, large, homogeneous population, and the scheme helps explain how evolution really occurs in most sexually interbreeding species. However, the dispute between the views of Fisher and

[5]As pointed out by Grant and Grant in a study of selective changes in Darwin's finches on the Galapagos Islands,
The population tracks a moving peak in an adaptive landscape under environmental fluctuations, and there is more than one individual fitness optimum within the range of phenotypes in the population.

TABLE 22-3 Comparison of Evolutionary Processes in a Single Homogeneous Population and in a Subdivided Population

	Homogeneous Population	Population Subdivided into Demes
What is selected	A gene, differing from its alleles in net selective value	Different gene frequencies
Source of variation	Gene mutation	Random drift between demes and selection toward new adaptive peaks
Process of selection	Selection among individuals	Selection among demes
Evolution under static conditions	Progress restricted to a single peak	Continued shifts as new adaptive peaks are encountered
Evolution under changing conditions	Progress up the nearest adaptive peak	Selection between different demes for occupancy of all available peaks

From Wright (1963).

Wright is not yet resolved, although it seems likely that there are populations of both types, or that some populations may be large and homogeneous during one period and subdivided during another. Most probably, as Wright (1988) stated, different aspects of populations demand different theoretical approaches:

> It is to be noted that the mathematical theories developed by Kimura, Fisher, Haldane, and myself dealt with four very different situations. Kimura's 'neutral' theory dealt with the exceedingly slow accumulations of neutral biochemical changes from accidents of sampling in the species as a whole. Fisher's 'fundamental theorem of natural selection' was concerned with the total combined effects of alleles at multiple loci under the assumption of panmixia in the species as a whole. He recognized that it was an exceedingly slow process. Haldane gave the most exhaustive mathematical treatment of the case in which the effects of a pair of alleles are independent of the rest of the genome. He included the important case of 'altruistic' genes, ones contributing to the fitness of the group at the expense of the individual. I attempted to account for occasional exceedingly rapid evolution on the basis of intergroup selection (differential diffusion) among small local populations that have differentiated at random, mainly by accidents of sampling (i.e., by local inbreeding), exceptions to the panmixia postulated by Fisher.
>
> All four are valid.

GROUP SELECTION

One important consequence of Wright's shifting balance theory has been to place emphasis on differences in survival or extinction among populations rather than only among individuals. Selection among individuals in a population is a conservative force that leads to movement of the population up a single adaptive peak, whereas selection among populations (accompanied by random genetic drift) leads to occupation of higher adaptive peaks and replacement, extinction, or colonization of populations at lower adaptive peaks.

To some extent, we have already considered selection among groups in our discussion of the advantages of sexual reproduction. Because sex may involve hazards as well as significant expenditures of resources, there are often considerable disadvantages to the individuals involved. Benefits in the evolution of sex therefore seem most likely related to the variability conferred upon the sexual population as a whole (Nunney).

A further example where selection appears to operate on a level different from individuals alone is the **kin selection** model used to explain social behavior in some groups of hymenopteran insects—ants, bees, and wasps. As previously noted (Chapter 15), these groups have haploid males and diploid females, so that all females derived from a single pair of parents are more related to each other than are mothers to their own daughters. It therefore benefits the genotype of their group, or kin, for female workers to sacrifice their own reproductive ability and rely instead on the reproductive ability of their mother by helping to raise sisters rather than producing daughters.

A similar type of selection probably also occurs in distasteful prey species where mutant individuals arise who possess an aposematic warning pattern (p. 440), but through their mortality help protect related genotypes that carry the same pattern. In the flour beetle, *Tribolium confusum*, Wade has shown experimentally that egg-eating cannibalism by larvae declines in groups in which larvae feed upon genetically related eggs. This altruistic behavior of refraining from cannibalism is apparently selected because it enhances the survival of related individuals who would be considered prey in the absence of altruism.

The extent to which group selection occurs, or

TABLE 22-4 Interaction Effects that Can Occur Between Two Populations

Type	Effect on Species A	Effect on Species B	Nature of Interaction
Neutralism	0	0	Neither population affects the other
Commensalism	0	+	Species *A* (e.g., the host) is not affected but species *B* (the commensal) benefits from the relationship
Amensalism	0	−	Species *A* is not affected but species *B* is inhibited
Mutualism	+	+	Both species benefit (e.g., Müllerian mimicry)
Predation or Parasitism	+	−	Species *A* (predator or parasite) benefits at expense of species *B* (prey or host)
Competition	−	−	Each species inhibits the other

Adapted from Pianka.

whether it occurs at all, has long been a subject of debate (Wilson). However, it is now clear from the above examples and others that group selection is receiving much more attention and approval than it has in the past.

GROUP INTERACTION

The natural communities in which organisms are found are assemblages of species or groups, each group interacting with others in various ways. In terms of survival, growth, or fecundity, a slight increase in the numbers of one particular group may cause an increase (+) in numbers of another group or a decrease (−) or have no discernible effect (0). The kinds of interactions that occur between two groups are generally classified according to the terminology of Table 22-4, with categories ranging from neutral interaction (0, 0) to mutual gain (+, +), predation (+, −), and competitive inhibition (−, −). Examples and variations of practically all such interactions have been observed by ecologists, and there is an extensive variety of models proposed to explain their mechanisms and effects, usually under the headings of population and community ecology (e.g., the introductory texts by Begon and Mortimer and by Putman and Wratten). However, although these ecological interactions have significant implications for any species, unraveling their exact evolutionary effects is still difficult. We can briefly review here a few aspects of two of these interactions, **competition** and **predation**. In their various forms, competition and predation are interactions that often decide which species will be members of a localized community of organisms (Roughgarden and Diamond).

Competition arises when two groups are dependent on the same limited environmental resource(s) so that each group causes a demonstrable reduction in numbers of the other. Such reduced availability of common resources often has important ecological or behavioral consequences. For example, in some groups natural selection will favor the evolution of **protective territorial mechanisms** that inhibit the utilization of such resources by competitors. These devices may include growth-inhibiting chemicals (e.g., toxins such as the creosote produced by some plants) or aggressive encounters (e.g., fighting in many vertebrates) often responsible for species dispersal.

Competition also often leads to ecological diversity: it can be to the advantage of competing groups that they minimize the deleterious effects of direct competition by utilizing different aspects of their common environmental resources. Among the many examples of such **resource partitioning** is that illustrated in Figure 22-6 for five species of warblers, each utilizing different parts of their spruce tree habitat. A further possible evolutionary response to competition is **character displacement**, where measurable phenotypic differences accompany the partitioning of resources among coexisting groups. One prominent example of this occurs among the Galapagos finches (see Fig. 2-4) where species coexisting on the same island show greater differences in bill sizes between them than when each species exists separately from the others on different islands. For example, one beak dimension of both *Geospiza fuliginosa* and *G. fortis* is about 10 mm on islands where each exists separately, whereas this measurement is about 8 mm for *G. fuliginosa* and 12 mm for *G. fortis* on islands where both species exist together, with neither group showing the 10 mm size.

When competition is not checked by partitioning or fluctuation of resources and two competing species utilize exactly the same resources in the same environment—that is, they both occupy the same niche[6]—laboratory experiments by Gause and others have shown that one species commonly dies out (Fig. 22-7). This finding has given rise to the **principle of com-**

[6]There is a multitude of niche definitions. Most definitions essentially propose that the niche of a species includes all the various environmental resources used by the species as well as the strategies it applies in making use of these resources.

FIGURE 22-6 Most common feeding zones (shaded) in spruce trees for five species of northeastern United States warblers of the genus *Dendroica*, based on the number of birds observed. The Cape May warbler may be quite rare unless there is a large outbreak of insects. The myrtle warbler is also rare and less specialized than the other species. The generally more common warblers, Blackburnian, bay-breasted, and black-throated green, are sufficiently different in feeding zone preferences to explain their coexistence. (Adapted from Krebs, based on MacArthur.)

petitive exclusion, which states that two species cannot continue to coexist in the same environment if they utilize it in the same way. This principle, also called Gause's Axiom or Law, had been foreshadowed by Darwin's statement in *The Origin of Species:* "Owing to the high geometrical rate of increase of all organic beings, each area is already fully stocked with inhabitants; and it follows from this, that as the favored forms increase in number, so generally will the less favored decrease and become rare."

It has, however, been seriously questioned whether competitive exclusion alone accounts for the

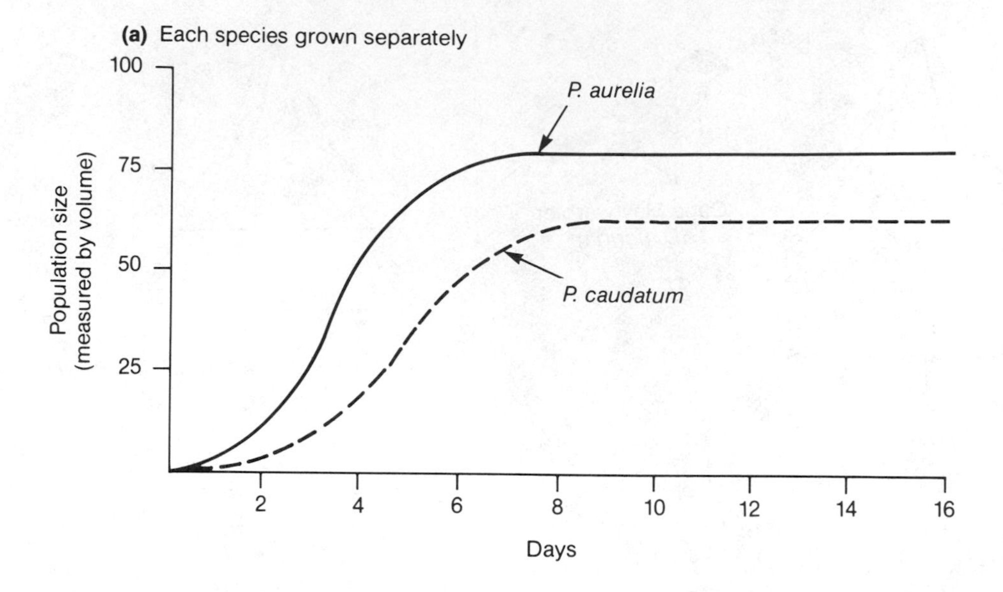

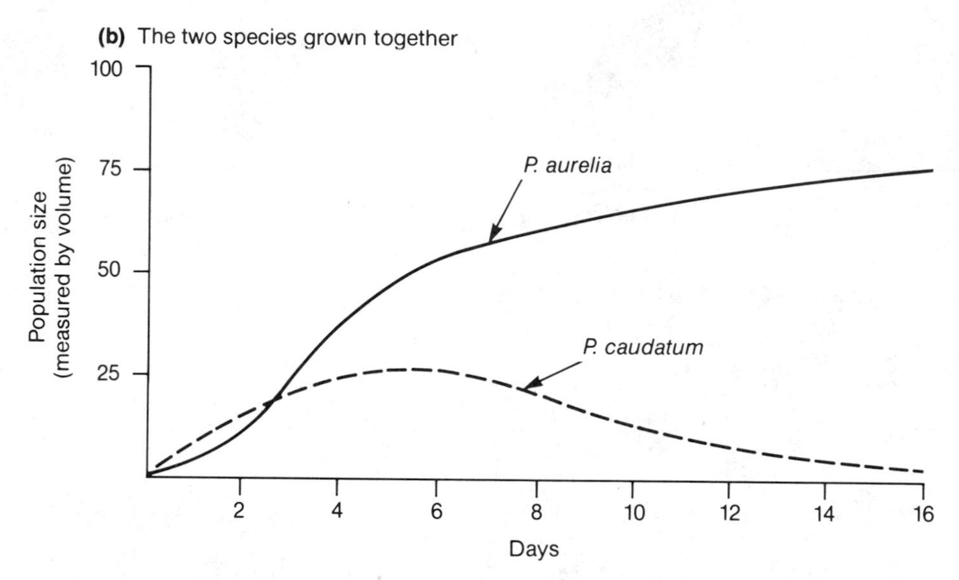

FIGURE 22-7 Growth of two species of *Paramecium* (a) in separate cultures, and (b) in mixed cultures. Although *P. aurelia* generally replaces *P. caudatum* as shown, there are mixed culture conditions in which *P. caudatum* multiplies faster than *P. aurelia.* Apparently, *P. caudatum* is more sensitive to metabolic pollutants than is *P. aurelia*, so that the removal of such pollutants encourages *P. caudatum* population growth. The results of competitive interactions may therefore be quite sensitive to external environmental factors. (Adapted from Gause.)

differences observed between coexisting species (e.g., character displacement) or for the fact that closely related and potentially competitive species often occupy different habitats. Morphological differences between related coexisting species may have evolved in the past when these species were not competitors, and related species that occupy different habitats may not have diverged because of competition but because of different food preferences, nesting sites, and so on (Den Boer). In fact, one can argue that much more coexistence is found among related species (e.g., species found in oceanic plankton) than would be expected if species were randomly distributed. Certainly, one important factor that probably diminishes exclusion between competing coexisting species is predation: predators can reduce the ability of any single dominant species to reach its full potential carrying capacity, thus making room for other competitors.[7]

In predation, the predator entirely or partially consumes its prey, thus affecting the numbers of those it

[7]This concept is similar to that described as "cropping" in Chapter 14.

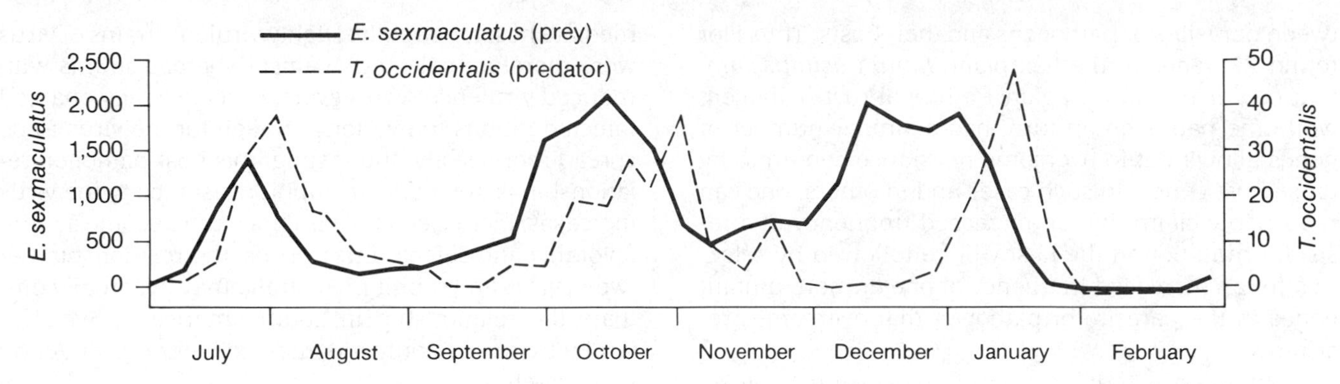

FIGURE 22-8 Three cycles of oscillating population numbers for two species of mites in a defined, controlled environment, one species (*Typhlodromus occidentalis*) being the predator and the herbivorous species (*Eotetranychus sexmaculatus*) being the prey. For each cycle, as numbers of prey increase, predator numbers follow, causing a "crash" in the prey population, which is then followed by a crash in the predator population. (Adapted from Pianka, from Huffaker.)

feeds upon. Predation may be exercised in various ways on both plants and animals, including overt attack and consumption of prey or **parasitism** (infestation and impairment of host tissues). The intimate dependence of predators on their prey often leads to a coupling of their relative abundances: an increase in numbers of prey allows an increase in numbers of predators which, in turn, can lead to a reduction in numbers of prey that can then reduce the number of predators. As shown in Figure 22-8, cyclic oscillations in population numbers may become a pattern, especially in the relationships of predators confined to a single species of prey.

Such simplicities are not, of course, the rule, since some predators need not necessarily cause a crash in abundance of prey but, like the predation of wolves on caribou herds, consume mostly those prey weakened by age or disease who have little reproductive value ("prudent predation"). Moreover, the regulation of predator numbers may be buffered by predation on more than one species of prey, thereby reducing large oscillations in predator population size and spreading the effects of predation so that prey population sizes also remain fairly stable.

From all this we can see that although various mathematical models have attempted to generalize the intricacies of predator-prey relationships, there are no simple universal solutions. As pointed out by Begon and Mortimer,

> Predators and prey do not normally exist as simple, two-species systems. To understand the abundance patterns exhibited by two interacting species, these must be viewed in realistic multi-species context. . . . Before multi-species systems are even considered, we must abandon our expectation of universal prey-predator oscillations, and look instead, much more closely, at the ways in which predators and their prey interact in practice.

Such interactions, whether of competition or predation, can be unique—dependent, for example, on the past evolutionary history of the populations involved, spatial limitations, climatic conditions, soil nutrients, and the effects of other species in the community. Each interaction must therefore be disentangled from others and explored separately—an extremely difficult task, but one which has engendered considerable interest and effort among population biologists (Diamond and Case).

COEVOLUTION

In spite of the lack of widely applicable theoretical models, it is clear that there are evolutionary interactions between species such that an evolutionary change in one species can prompt an evolutionary change in a species with which it interacts ecologically. Paleontological evidence already points strongly to an evolutionary "arms race" between herbivorous and carnivorous mammals during which size, speed, and intelligence seemed to increase sequentially in various members of both groups (Chapter 18).[8] Similar evolutionary progressions in respect to size, speed, and protective devices must also have occurred among the reptilian dinosaurs (Chapter 17) and have probably been a consistent trend in many prey-predator groups.

There are hardly any better present examples of such coevolutionary interactions than those found be-

[8]To illustrate the selective mechanism that operates on individuals under predatory pressure, Greenwood recounts the following story of two men in Greenland trapped in a tent by a polar bear. "One, a native Greenlander, began to remove his boots. 'What are you doing?' whispered his Danish companion. 'You know that even without your boots you can't run faster than a polar bear.' 'Yes,' replied the Greenlander, 'But without my boots I can run faster than *you*.' "

tween parasites or pathogens and their hosts. Thus Flor found 27 genes in the flax plant, *Linum usitatissium*, that confer resistance against a fungal rust pathogen, while the pathogen, in turn, had a similar number of genes allowing it to overcome resistance conferred by these host genes. In such cases and in others, one can reasonably claim that an increased frequency of a resistant mutation in the host will be followed by selection for an increased frequency of one or more mutant genes in the parasite or pathogen that overcomes resistance.

Not unexpectedly, coevolutionary events between host and parasite can be quite complex, and may, in fact, lead to a reduction in virulence of the parasite. One prominent example of this concerns a myxoma virus imported from South America to Australia to control the phenomenal populational growth of the European rabbit, *Oryctolagus cuniculus*. Although the virus caused only a mild disease in native South American cottontail rabbits (*Sylvilagus brasilensis*), it acted as a highly lethal pathogen among the Australian rabbits, being transmitted primarily by mosquitoes. Viral-caused lethality in 1950-51 was as high as 99 percent among infected rabbits, and it seemed as though the Australian rabbit population would eventually be eradicated by the virus or persist only at very low numbers. This expectation was not fulfilled, because, most surprisingly, although some Australian rabbits became increasingly resistant to the virus, the virus itself became less virulent. Apparently, since mosquitoes feed only on live rabbits, the rate at which the virus can be transmitted will be reduced if the virus kills its intermediate host too rapidly. Highly virulent strains of virus were therefore selected against, whereas strains with reduced virulence were favored because they enabled infected rabbits to live long enough for the virus to be spread more easily. Interesting, this host-pathogen relationship is not static: as rabbit resistance to the virus increases, increased viral virulence can become a more favorable and selected trait. Thus, the relationship between virus and rabbit in Australia may eventually emulate the relationship in South America, in which a virulent virus has only a partially deleterious effect on its native host.

Depending on how one defines the concept, numerous other examples of coevolution abound. These include mimics who evolve in step with the evolution of their models; competing species who evolve changes between them to reduce competition (e.g., character displacement); mutualistic interactions in which different species evolve so they can continue to derive mutual benefits from each other; plants that evolve mechanisms to attract specific pollinators who are simultaneously evolving mechanisms to recognize specific plants; and many others (Futuyma and Slatkin). Coevolution must therefore be a common phenomenon since the intimate ecological relationships among many species probably derive from coevolutionary events in which adaptive changes in one species are followed by adaptive changes in others. There obviously has been, and still is, a great deal of interdependence among many different life forms, and these cannot be fully understood except in an evolutionary framework (Loehle and Pechmann).

SUMMARY

The structure of populations is so complex that quantitative evaluation of its evolutionary potential is extremely difficult. Growth, however, is one characteristic that is amenable to measurement. Unlimited population growth is exponential, but as the environment imposes restrictions, the population will tend to stabilize at a size called the carrying capacity, or K. Stategies such as r selection (increasing the numbers of offspring) or K selection (increasing the selective advantage of offspring) enable populations to approach the carrying capacity.

While some variabililty is advantageous, the immediate effect of many alleles or genetic combinations may not be, and the extent to which these detrimental genotypes affect a population is called the genetic load. Both mutation and heterozygote superiority may contribute to genetic load by perpetuating deleterious recessive alleles. Although this polymorphism reduces the frequency of optimum genotypes, it may increase the adaptiveness of the lineage at some future time.

Because of the presence of extensive genetic polymorphism, not all of which can be explained as resulting from heterozygote superiority, it has been proposed that most alleles are neutral in effect and incur no genetic load. Under these circumstances they will remain in the gene pool, not because of selection but because of mutation and random genetic drift.

On the other hand, selectionists argue that relationships between ecological conditions and polymorphism, similarities in allozyme frequencies in different species, the association between polymorphism and enzyme function, and the higher frequency of polymorphism in non-coding DNA sequences than in coding sequences cannot be entirely explained by random forces but must be due to selection pressures.

Sexual reproduction, recombination of linked

genes, and mutation can all produce genetic combinations that increase fitness. When such advantageous genotypes occur, they are said to occupy adaptive peaks of varying values according to their degree of fitness. However, in order to reach higher adaptive peaks, a population must often pass through less adaptive valleys, a process which, among other factors, involves random genetic drift, according to Wright.

Although the issue is still unresolved, there are instances (sexual reproduction and altruistic social behaviors) where selection apparently occurs for the benefit of the group even though it may be detrimental to the individual.

Population interactions such as competition and predation may favor the selection of particular traits. Competition favors niche and character distinctions between groups, in some cases eliminating one group entirely, while predation has complex effects on the size of populations of both prey and predator. However intricate the interrelationships, it is clear that one species can influence the evolution of another with which it interacts.

KEY TERMS

adaptive landscape
adaptive peak
adaptive valleys
allozyme frequencies
altruists
balanced lethals
balanced load
carrying capacity
character displacement
coevolution
competition
demes
density dependence
density independence
enzyme polymorphisms
fecundity
genetic death
genetic load
group interaction
group selection
hitchhiking
intrinsic rate of increase
iteroparous
kin selection
logistic growth model
mutational load
net reproductive rate
neutral mutations
non-Darwinian evolution
parasitism
predation
principle of competitive exclusion
protective territorial mechanisms
r and *K* selection
rate of increase
resource partitioning
segregation distorters
segregational load
selectionists
semelparous
senescence
shifting balance process
social parasites
survivorship
truncation selection

REFERENCES

Allard, R. W., and A. L. Kahler, 1972. Patterns of molecular variation in plant populations. In *Proceedings of the Sixth Berkeley Symposium on Mathematical Statistics and Probability*, Vol. 5. Univ. of California Press, Berkeley, pp. 237-254.

Allison, A. C., 1961. Abnormal hemoglobin and erythrocyte enzyme-deficiency traits. In *Genetical Variation in Human Populations*, G. A. Harrison (ed.). Pergamon Press, New York, pp. 16-40.

Ayala, F. J., and M. L. Tracey, 1974. Genetic differentiation within and between species of the *Drosophila willistoni* group. *Proc. Nat. Acad. Sci.*, **71**, 999-1003.

Begon, M., and M. Mortimer, 1986. *Population Ecology*, 2nd ed. Blackwell, Oxford.

Bell, G., 1982. *The Masterpiece of Nature: The Evolution and Genetics of Sexuality*. Univ. of California Press, Berkeley.

Bernstein, H., F. A Hopf, and R. E. Michod, 1988. Is meiotic recombination an adaptation for repairing DNA, producing genetic variation, or both? In *The*

Evolution of Sex, R. E. Michod and B. R. Levin (eds.). Sinauer Associates, Sunderland, Mass., pp. 139-160.

Boyce, M. S., 1984. Restitution of *r*- and *K*-selection as a model of density-dependent natural selection. *Ann. Rev. Ecol. and Syst.*, **15**, 427-447.

Bradshaw, A. D., 1984. The importance of evolutionary ideas in ecology—and vice versa. In *Evolutionary Ecology*, B. Shorrocks (ed.). Blackwell, Oxford, pp. 1-25.

Brues, A. M., 1964. The cost of evolution vs. the cost of not evolving. *Evolution*, **18**, 379-383.

Chetverikoff, S. S., 1926. On certain aspects of the evolutionary process from the standpoint of modern genetics. (English translation: 1961, *Proc. Amer. Phil. Soc.*, **105**, 167-195.)

Crow, J. F., 1957. Genetics of insect resistance to chemicals. *Ann. Rev. Entomol.*, **2**, 227-246.

——, 1962. Population genetics: selection. In *Methodology in Human Genetics*, W. J. Burdette (ed.). Holden-Day, San Francisco, pp. 53-75.

——, 1988. The importance of recombination. In *The Evolution of Sex,* R. E. Michod and B. R. Levin (eds.). Sinauer Associates, Sunderland, Mass., pp. 56-73.

Decker, G. E., and W. N. Bruce, 1952. House fly resistance to chemicals. *Amer. Jour. Trop. Med. Hygiene*, **1**, 395-403.

Den Boer, P. J., 1986. The present status of the competitive exclusion principle. *Trends in Ecol. and Evol.*, **1**, 25-28.

Diamond, J., and T. J. Case, 1986. *Community Ecology*. Harper & Row, New York.

Dobzhansky, Th., 1970. *Genetics of the Evolutionary Process*. Columbia Univ. Press, New York.

Emlen, J. M., 1973. *Ecology: An Evolutionary Approach*. Addison-Wesley, Reading, Mass.

Flor, H. H., 1956. The complementary genic systems in flax and flax rust. *Adv. in Genetics*, **8**, 29-54.

Franklin, I., and R. C. Lewontin, 1970. Is the gene the unit of selection? *Genetics*, **65**, 707-734.

Futuyma, D. J., and M. Slatkin (eds.), 1983. *Coevolution*. Sinauer Associates, Sunderland, Mass.

Gause, G. F., 1934. *The Struggle for Existence*. Williams & Wilkins, Baltimore.

Gibson, T. C., M. L. Schleppe, and E. C. Cox, 1970. On fitness of an *E. coli* mutation gene. *Science*, **169**, 686-690.

Gillespie, J. H., and K. Kojima, 1968. The degree of polymorphisms in enzymes involved in energy production compared to that in nonspecific enzymes in two *Drosophila ananassae* populations. *Proc. Nat. Acad. Sci.*, **61**, 582-585.

Grant, B. R., and P. R. Grant, 1989. Natural selection in a population of Darwin's finches. *Amer. Naturalist,* **133,** 377-393.

Greenwood, J. J. D., 1984. The evolutionary ecology of predation. In *Evolutionary Ecology*, B. Shorrocks (ed.). Blackwell, Oxford, pp. 233-273.

Haldane, J. B. S., 1960. More precise expressions for the cost of natural selection. *Jour. Genet.*, **57**, 351-360.

Hall, J. G., and C. Wills, 1987. Conditional overdominance at an alcohol dehydrogenase locus in yeast. *Genetics*, **117**, 421-427.

Hart, A., and M. Begon, 1982. The status of general life-history strategy theories, illustrated in winkles. *Oecologia*, **52**, 37-42.

Hartl, D. L., and A. G. Clark, 1989. *Principles of Population Genetics*, 2nd ed. Sinauer Associates, Sunderland, Mass.

Hartl, D. L., and D. E. Dykhuizen, 1981. Potential for selection among nearly neutral allozymes of 6-phosphogluconate dehydrogenase in *Escherichia coli*. *Proc. Nat. Acad. Sci.*, **78**, 6344-6348.

Huffaker, C. B., 1958. Experimental studies on predation: dispersion factors and predator-prey oscillations. *Hilgardia*, **27**, 343-383.

Hutchinson, G. E., 1978. *An Introduction to Population Ecology*. Yale Univ. Press, New Haven, Conn.

Johnson, G. B., 1974. Enzyme polymorphism and metabolism. *Science*, **184**, 28-37.

Kimura, M., 1983. *The Neutral Theory of Molecular Evolution*. Cambridge Univ. Press, Cambridge.

Kimura, M., and J. F. Crow, 1964. The number of alleles that can be maintained in a finite population. *Genetics*, **49**, 725-738.

Kimura, M., and T. Ohta, 1972. Population genetics, molecular biometry, and evolution. In *Proceedings of the Sixth Berkeley Symposium on Mathematical Statistics and Probability*, Vol. 5. Univ. of California Press, Berkeley, pp. 43-68.

King, J. L., 1967. Continuously distributed factors affecting fitness. *Genetics*, **55**, 483-492.

King, J. L., and T. H. Jukes, 1969. Non-Darwinian evolution: random fixation of selective neutral mutations. *Science*, **164**, 788-798.

Koehn, R. K., 1969. Esterase heterogeneity: dynamics of a polymorphism. *Science*, **163**, 943-944.

Koehn, R. K., A. J. Zera, and J. G. Hall, 1983. Enzyme polymorphism and natural selection. In *Evolution of Genes and Proteins*, M. Nei and R. Koehn (eds.).

Sinauer Associates, Sunderland, Mass., pp. 115-136.

Kojima, K., and K. M. Yarbrough, 1967. Frequency-dependent selection at the esterase 6 locus in a population of *Drosophila melanogaster*. *Proc. Nat. Acad. Sci.*, **57**, 645-649.

Krebs, C. J., 1985. *Ecology*, 3rd ed. Harper & Row, New York.

Kreitman, M., 1983. Nucleotide polymorphism at the *alcohol dehydrogenase* locus of *Drosophila melanogaster*. *Nature*, **304**, 412-417.

Lees, D. R., 1981. Industrial melanism: genetic adaptation of animals to air pollution. In *Genetic Consequences of Man Made Change*, J. A. Bishop and L. M. Cook (eds.). Academic Press, London, pp. 129-176.

Li, C. C., 1963. The way the load ratio works. *Amer. Jour. Hum. Genet.*, **15**, 316-321.

Loehle, C., and J. H. K. Pechmann, 1988. Evolution: the missing ingredient in systems ecology. *Amer. Naturalist*, **132**, 884-899.

MacArthur, R. H., 1958. Population ecology of some warblers of northeastern coniferous forests. *Ecology*, **39**, 599-619.

Mather, K., 1953. The genetical structure of populations. *Symp. Soc. Exp. Biol.*, **7**, 66-95.

Maynard Smith, J., 1978. *The Evolution of Sex*. Cambridge Univ. Press, Cambridge.

——, 1988. The evolution of recombination. In *The Evolution of Sex*, R. E. Michod and B. R. Levin (eds.). Sinauer Associates, Sunderland, Mass., pp. 106-125.

Michod, R. E., and B. R. Levin (eds.), 1988. *The Evolution of Sex*. Sinauer Associates, Sunderland, Mass.

Milkman, R. D., 1967. Heterosis as a major cause of heterozygosity in nature. *Genetics*, **55**, 493-495.

——, 1973. Electrophoretic variation in *Escherichia coli* from natural sources. *Science*, **182**, 1024-1026.

Mueller, L. D., 1987. Evolution of accelerated senescence in laboratory populations of *Drosophila*. *Proc. Nat. Acad. Sci.*, **84**, 1974-1977.

——, 1988. Evolution of competitive ability in *Drosophila* by density-dependent natural selection. *Proc. Nat. Acad. Sci.*, **85**, 4383-4386.

Muller, H. J., 1932. Some genetic aspects of sex. *Amer. Nat.*, **66**, 118-138.

Nevo, E., 1983. Population genetics and ecology. In *Evolution From Molecules to Men*, D. S. Bendall (ed.). Cambridge Univ. Press, Cambridge, pp. 287-321.

Nunney, L., 1989. The maintenance of sex by group selection. *Evolution*, **43**, 245-257.

Oakeshott, J. G., J. B. Gibson, P. R. Anderson, W. R. Knibb, D. G. Anderson, and G. K. Chambers, 1982. Alcohol dehydrogenase and glycerol-3-phosphate dehydrogenase clines in *Drosophila melanogaster* on different continents. *Evolution*, **36**, 86-96.

Pianka, E. R., 1988. *Evolutionary Ecology*, 4th ed. Harper & Row, New York.

Prakash, S., R. C. Lewontin, and J. L. Hubby, 1969. A molecular approach to the study of genic heterozygosity in natural populations. IV. Patterns of genic variation in central, marginal and isolated populations of *Drosophila pseudoobscura*. *Genetics*, **61**, 841-858.

Putman, R. J., and S. D. Wratten, 1984. *Principles of Ecology*. Croom Helm, London.

Rose, M. R., 1985. The evolution of senescence. In *Evolution: Essays in Honor of John Maynard Smith*, P. J. Greenwood, P. H. Harvey, and M. Slatkin (eds.). Cambridge Univ. Press, Cambridge, pp. 117-128.

Roughgarden, J., 1979. *Theory of Population Genetics and Evolutionary Ecology: An Introduction*. Macmillan, New York.

Roughgarden, J., and J. Diamond, 1986. Overview: the role of species interactions in community ecology. In *Community Ecology*, J. D. Diamond and T. J. Case (eds.). Harper & Row, New York, pp. 333-343.

Sampsell, B., and S. Sims, 1982. Effect of *adh* genotype and heat stress on alcohol tolerance in *Drosophila melanogaster*. *Nature*, **296**, 853-855.

Selander, R. K., and B. R. Levin, 1980. Genetic diversity and structure in *Escherichia coli* populations. *Science*, **210**, 545-547.

Somero, G. N., 1986. Protein adaptation and biogeography: threshold effects on molecular evolution. *Trends in Ecol. and Evol.*, **1**, 124-127.

Strong, D. R., 1986. Density vagueness: abiding the variance in demography of real populations. In *Community Ecology*, J. Diamond and T. J. Case (eds.). Harper & Row, New York, pp. 257-268.

Sved, J. A., T. E. Reed, and W. F. Bodmer, 1967. The number of balanced polymorphisms that can be maintained in a natural population. *Genetics*, **55**, 469-481.

Takahata, N., 1987. On the overdispersed molecular clock. *Genetics*, **116**, 169-179.

Wade, M. J., 1980. An experimental study of kin selection. *Evolution*, **34**, 844-855.

Wallace, B., 1970. *Genetic Load: Its Biological and Con-*

ceptual Aspects. Prentice-Hall, Englewood Cliffs, N.J.

Wills, C., 1981. *Genetic Variability*. Clarendon Press, Oxford.

Wilson, D. S., 1983. The group selection controversy: History and current status. *Ann. Rev. Ecol. and Syst.*, **14**, 159-187.

Wright, S., 1963. Genic interaction. In *Methodology in Mammalian Genetics*, W. J. Burdette (ed.). Holden-Day, San Francisco, pp. 159-192.

——, 1978. *Evolution and the Genetics of Populations. Vol. 4: Variability Within and Among Natural Populations.* Univ. of Chicago Press, Chicago.

——, 1988. Surfaces of selective value revisited. *Amer. Nat.*, **131**, 115-123.

23

From Races to Species

We have learned that the interbreeding nature of a sexual species serves as an important cohesive force that holds it together and enables it to share a common gene pool. At the same time, we understand that such a species may consist of numerous individual populations with various degrees of interbreeding between them. Widely separated populations, for example, will have less opportunity to share their gene pools than those closer at hand. The structure of a species is therefore broken into various geographical subunits. Because the forces acting upon these subunits may vary in different localities, it will come as no surprise to find observable differences among populations.

In the yarrow plant *Achillea* a transect across central California shows populations differing significantly in factors such as height and growing season (Fig. 23-1). The adaptive nature of most of these differences is seen in the different responses of these populations when grown in different localities. Thus the coastland plants are weak when grown at higher altitudes, and the high-altitude forms grow poorly at much lower altitudes (Fig. 23-2). The adaptive features of many such plant populations was pointed out long ago by Turesson as the genetic response of a population to a particular ecological habitat.

Where gene frequencies can be scored, changes between localities and during various time intervals have been well documented in numerous instances. As shown in Figure 10-32 the frequencies of third-chromosome arrangements in *Drosophila pseudoobscura* differ notably in a range of environments across the American Southwest and also undergo significant seasonal changes. Further genetic changes in this species also extend over longer periods of time, such as the significant increase in the frequency of one arrangement (Pikes Peak) in many California populations from almost zero to as high as 10 percent over a 17-year period. Populations of the British peppered moth, *Biston betularia*, also show significant changes in the frequency of melanism for even longer periods, all associated with specific localities and environments (p. 441).

Races

In general, populations of the same species that differ markedly from each other have been characterized as **races**. Races share the possibility of participating in

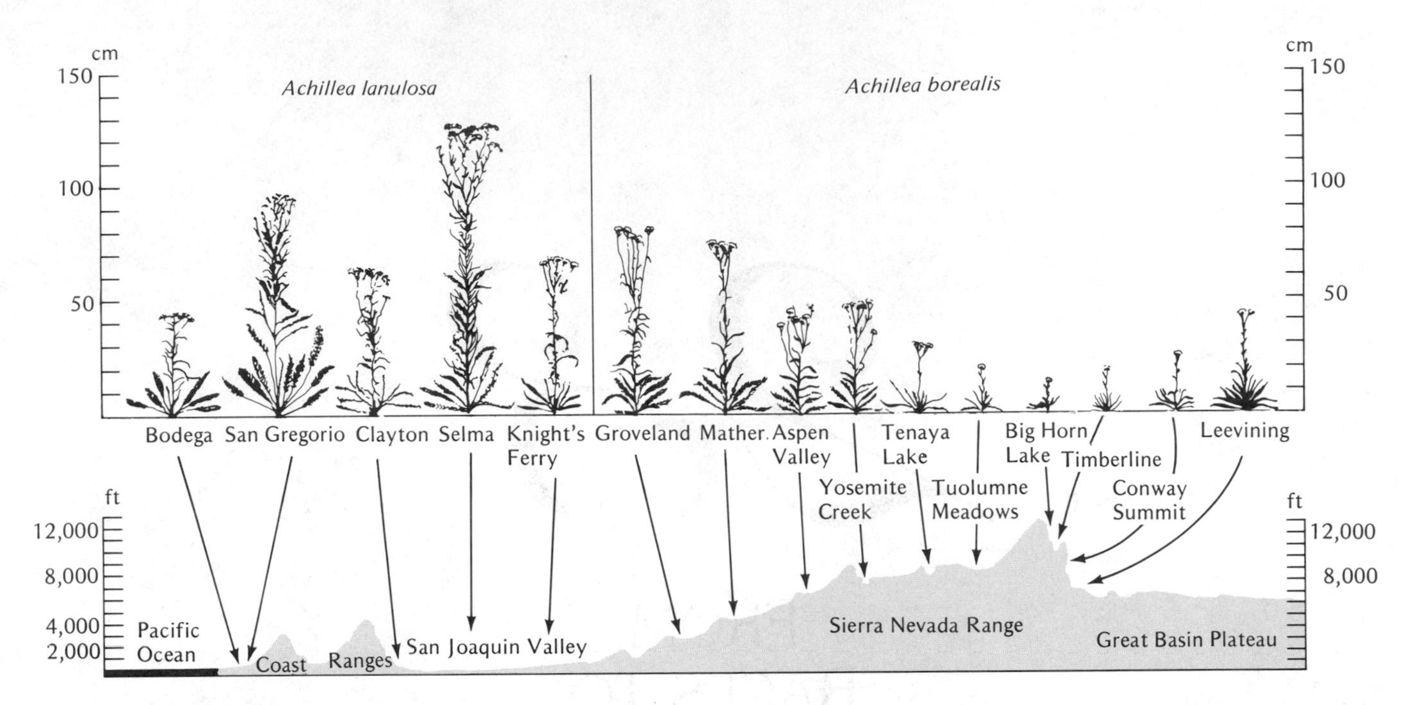

FIGURE 23-1 Representative plants from different populations of *Achillea* gathered from designated localities along a transect across central California and grown in a garden at Stanford, California. The fact that these populations, grown in a uniform environment, differ in terms of plant size, leaf shape, and other characteristics indicates that genetic differences have evolved between them. (Adapted from Clausen et al.)

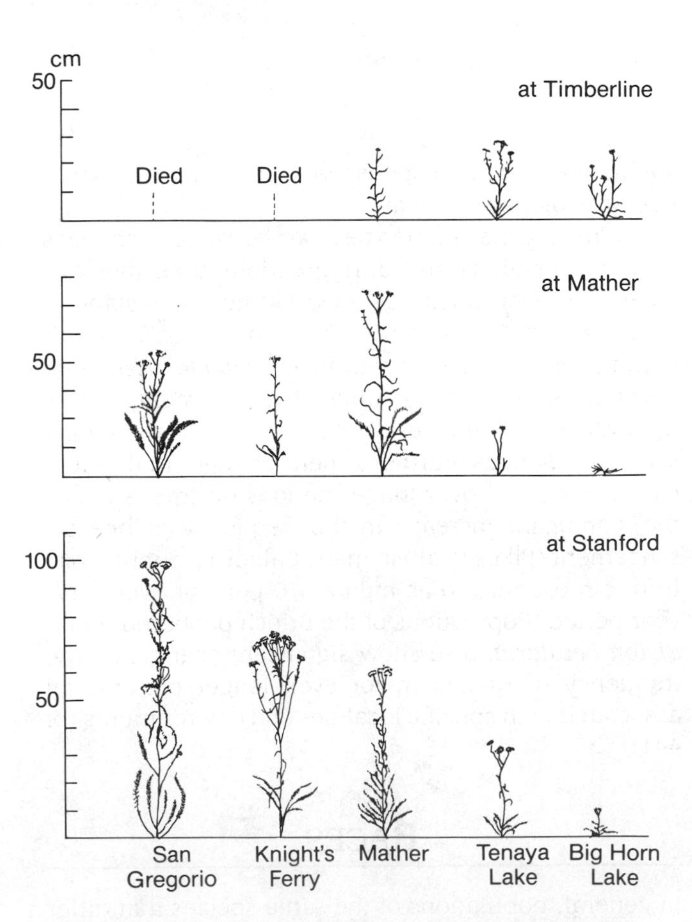

FIGURE 23-2 Responses of clones from representative *Achillea* plants originating from five localities in California and grown at three different altitudes: sea level (Stanford), 4,600 feet (Mather), and 10,000 feet (Timberline). (Adapted from Clausen et al.)

the gene pool of the entire species, although they are sufficiently separated to exhibit individually unique gene frequencies. The distinction between races is, therefore, not absolute: races may differ in the relative frequency of a particular gene, but these differences do not prohibit gene exchange.

For genes whose frequencies can be detected and scored, racial distinctions are therefore not simply discerned by the presence or absence of particular genes but are, in many instances, a matter of gene frequencies. Table 23-1, for example, shows a comparison of frequencies for a variety of gene systems in three major human racial groups. In practically all these gene systems knowledge of a particular genotype alone is not by itself sufficient to indicate to which race an individual belongs. An individual of O blood type who is also Rh positive may, for example, belong to any of the races listed if our attention is confined only to these genes.

It is interesting to note that differences among human populations have not reached the point where one population is fixed for one allele at a particular locus and another population is fixed for a different allele: when a population shows fixation for one allele, other populations are always polymorphic for it. In order to calculate genetic divergences under such circumstances, a procedure proposed by Nei is commonly used, as shown in Table 23-2. When Nei's **index of genetic distance** (*D*) is applied to human populations (Fig. 23-3), six major racial groups can be discerned:

TABLE 23-1 Frequencies of Alleles in Samples of Individuals Taken from Three Broadly Defined Human Races

Gene Locus	Allele	Caucasians (Whites)	Africans (Blacks)	Asians (Mongoloids)
PROTEINS				
Acid protein	Pa^1	.21	.14	.42
	Pa^0	.79	.86	.58
Adenylate kinase	AK^1	.96	.99	1.00
	AK^2	.04	.01	—
Esterase D	ESD^1	.89	.97	.66
	ESD^2	.11	.03	.34
Glyoxalase I	GLO^1	.44	.26	.09
	GLO^2	.56	.74	.91
BLOOD GROUPS				
ABO	A	.24	.19	.27
	B	.06	.16	.17
	O	.70	.65	.56
Duffy	Fy^a	.41	.06	.90
	Fy^b	.59	.94	.10
MN	M	.54	.58	.53
	N	.46	.42	.47
Rh (simplified to two alleles)	Rh^+	.62	.70	.95
	rh^-	.38	.30	.05

Adapted from Strickberger, data from Nei and Roychoudhury.

TABLE 23-2 Example of the Nei Procedure in Calculating Indices of Genetic Identity (I) and Genetic Distance (D) Between Caucasians and Africans for Two of the Loci Given in Table 23-1*

Locus	Allele	Caucasians	Africans
Acid protein	Pa^1	$p_1 = .21$	$p_2 = .14$
	Pa^0	$q_1 = .79$	$q_2 = .86$
ABO blood group	A	$p_1 = .24$	$p_2 = .19$
	B	$q_1 = .06$	$q_2 = .16$
	O	$r_1 = .70$	$r_2 = .65$

$$I = \frac{\text{arithmetic mean of the products of allele frequencies}}{\text{geometric mean of the homozygote frequencies}}$$

$$I = \frac{[(p_1 \times p_2) + (q_1 \times q_2) + (r_1 \times r_2)]/\text{no. of loci}}{\sqrt{[(p_1)^2 + (q_1)^2 + (r_1)^2] \times [(p_2)^2 + (q_2)^2 + (r_2)^2]/(\text{no. of loci})^2}}$$

$$I = \frac{[(.21 \times .14) + (.79 \times .86) + (.24 \times .19) + (.06 \times .16) + (.70 \times .65)]/2}{\sqrt{[.21^2 + .79^2 + .24^2 + .06^2 + .70^2] \times [.14^2 + .86^2 + .19^2 + .16^2 + .65^2]/(2)^2}}$$

$$I = \frac{1.2190/2}{\sqrt{(1.2194)(1.2434)/4}} = \frac{.6095}{\sqrt{1.5162/4}} = \frac{.6095}{.6157} = .9899$$

$$D \text{ (genetic distance)} = -\ln I = -\ln .9899 = .0101$$

From Strickberger.

* I can range from zero (no similar alleles between the two populations) to one (complete similarity of alleles and frequencies). When $I = 1$, genetic dissimilarity (D) is, of course, zero. When I approaches zero, D can increase to very large values, indicating that the alleles at some or many loci have been replaced one or more times.

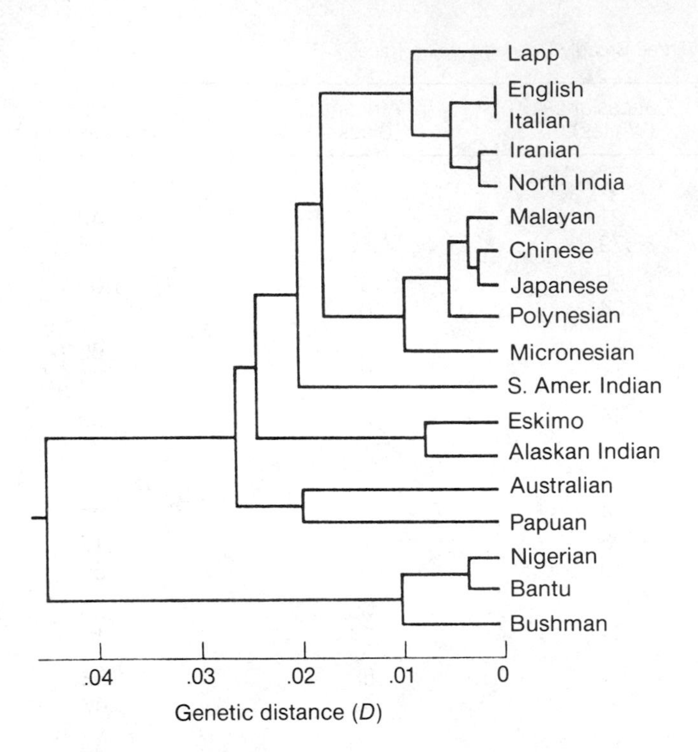

FIGURE 23-3 Phylogenetic relationships among 18 human populations based on data obtained through use of Nei's *D* measurement (Table 23-2) on a large number of loci. (Adapted from Nei and Roychoudhury.)

1. Caucasian: includes a variety of white European populations ranging from the Lapps of Scandinavia to the Mediterranean peoples of Southern Europe and North Africa.
2. African: includes various black tribes and groups indigenous to Africa.
3. Asian: includes Mongoloid peoples as well as Polynesians and Micronesians.
4. North American Eskimos and Indians.
5. South American Indians.
6. Australoid groups native to Australia and Papua.

Nei and Roychoudhury estimate that, when averaged over many loci, genetic distance in humans accumulates at some constant rate that may be roughly 3.75 million years per unit of *D*. Thus for the 85 loci that allow comparisons among the three major races, the estimated times when these races initially diverged from each other are: Caucasian from African (D = .032)—113,000 ± years ago; Caucasian from Asian (D = .019)—41,000 ± years ago; and African from Asian (D = .047)—116,000 ± years ago. The oldest divergences are apparently between the Africans and other races, a view that seems supported by the mitochondrial DNA findings previously diagrammed in Figure 19-14.

Also striking in these data is the consistently high levels of variability for the many genes that were examined. According to Nei and Roychoudhury, the proportion of loci that were polymorphic in the three races ranged from 45 to 52 percent for proteins and from 34 to 56 percent for blood groups. The average heterozygosity per locus ranged from 13 to 16 percent for proteins and from 11 to 20 percent for blood groups.

The presence of so much genetic variability in human races indicates the fictional nature of concepts such as "pure" races. Members of a race are not genetically pure in the sense of sharing a uniform genetic identity, nor does genetic uniformity even apply to members of the same family. Lewontin, in fact, points out that more than 94 percent of the genetic variability among humans comes from differences among individuals and groups of the same race and only 6 percent comes from differences between races. From a genetic point of view "purity" can only be ascribed to asexual clones derived from a single individual. In clonal reproduction, however, it is debatable whether the terms race and species are appropriate, and other terms have been devised to describe populations among microorganisms (Sonneborn).

Our criterion for evaluating the differences among populations of a species is, therefore, based essentially upon gene-frequency differences. When these differences are numerous and it is advantageous to consider populations as separate entities, we may categorize them broadly as races.[1] At times, racial differences are accompanied by observable morphological differences such as those between some human populations. At other times, observed racial differences extend only to gene or chromosomal differences such as those between Texas and California populations of *D. pseudoobscura* (Fig. 10-32).

ADAPTATIONAL PATTERNS

As we have seen, the forces producing racial differences are often adaptive; that is, at least some gene-frequency changes are the response of a population to the selective forces operating within a particular en-

[1]Taxonomists often use "subspecies," "variety," or even "subvariety" rather than "race" to designate taxonomically distinct groups within species. Since it is quite difficult to distinguish biologically and evolutionarily between these various terms (as well as between others such as "geographical race," "ecotype," and "ecological race"), the commonly accepted term "race" is used here to designate any group that can be differentiated on the basis of its unique gene frequencies.

vironment.[2] Climate, terrain, prey, and predators can obviously evince specific adaptations to differentiate a population, such as the many remarkable examples of camouflage and mimicry. There are, in fact, a number of "rules" which generalize the adaptive response of populations to certain ecological and geographical conditions. As with many other generalizations, exceptions to these **ecogeographical rules** exist, but the rules nevertheless have the value of pointing to the importance of environmental selection in exacting parallel evolutionary changes in different species.

Among the best known of these climatic rules is **Bergmann's rule** that relates body size in warm-blooded (endothermic) vertebrates to average environmental temperature. It states that races of a species in cooler climates tend to be larger than those in warmer climates. This relationship derives primarily from the fact that bodies with larger volumes have proportionately less exposed surface areas than bodies with smaller volumes. Since heat loss is related to surface area, larger bodies can retain heat more efficiently in cooler climates, whereas smaller bodies can rid themselves of heat more efficiently in warmer climates. The rule is corroborated in comparisons made between many North-South races of both terrestrial and marine birds and mammals. In some human races, the rule can be applied by noting the ratio of body weight to body surface, comparing, for example, thick-chested and short-limbed Eskimo groups to slender and long-limbed Nilotic African tribes (Fig. 23-4).

An extension of Bergmann's rule is **Allen's rule** which states that protruding body parts (e.g., tail, ears) are generally shorter in cooler climates than they are in warmer climates. Other adaptive ecogeographical regularities are given by **Gloger's rule** (races are more heavily pigmented in warm humid areas than in cool dry areas; Fig. 23-5), as well as rules that apply primarily to insects, reptiles, and amphibians (Mayr [1963]).

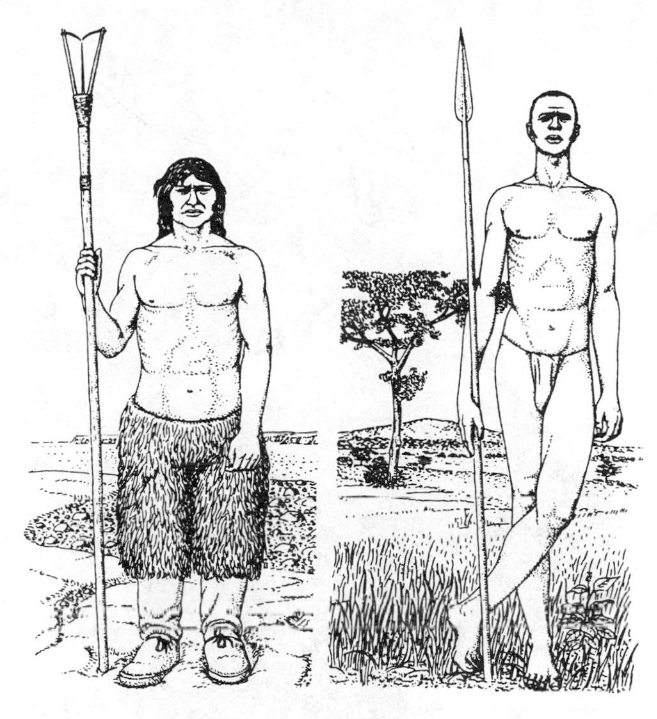

FIGURE 23-4 Difference in body proportions between an individual from a group of arctic Eskimos (left) and an individual from a black tribe in the Sudanese Nile (right), indicating differences in their adaptation to their prevailing climates. The proportionately greater bulk and smaller body surface of the Eskimo (approximately 39 kilograms per square meter) helps to conserve heat, and the proportionately greater body surface of the Sudanese (approximately 34 kilograms per square meter) helps to dissipate heat (Schreider). (From W. W. Howells, "The distribution of man." Copyright 1960, Scientific American, Inc.)

BEHAVIORAL ADAPTATIONS AND STRATEGIES

In addition to morphological adaptations, organisms relate to their surroundings by assuming various motions and positions in escaping from predators, pursuing prey, interacting with conspecifics, and so on—all of which usually come under the name "behavior." Behaviors may be **innate**, needing no prior learning

[2]Gould and Lewontin raised the question whether some or many organismal features are really the result of non-adaptive processes rather than of adaptation. For example, they suggest that the tiny forelegs of the dinosaur *Tyrannosaurus* can be explained as a "developmental correlate" that accompanies an increased size of the head and hind limbs, and they characterize adaptive explanations for this and other phenomena as "just-so stories." However, although the testing of adaptive explanations is often difficult (as is also true for non-adaptive explanations), the fact that every organismal lineage has been subjected to selection makes it extremely likely that most or all organismal characters must have been affected by selection and therefore have or had some adaptive value, to at least some degree. Even if the transmission of a character between generations can escape selection because non-selective forces are operating on its gene frequency (e.g., genetic drift, mutation, migration), it would seem that the character, if it is more than just transitory, is sooner or later bound to be subject to selection because of some effect that it has, either by itself or in conjunction with other characters, on the relationship between the organism and its environment. How many persistent phenotypic characters are there which have never had an effect, however subtle, on organismic-environmental interactions? It is certainly reasonable to claim that non-selective causes that affect the gene frequencies involved in producing such a character do not continue to operate *ad infinitum*, without the intervention of selection at some point in its history. Thus, although we often may not know the adaptive explanation for a particular character, the search for such explanations is a reasonable enterprise in evolutionary biology. To extend Gould and Lewontin's dinosaur example, why did *Tyrannosaurus* have tiny forelegs but the 40-foot-long carnosaur, *Spinosaurus*, also of the late Cretaceous, have relatively large forelegs? Further discussions of Gould and Lewontin's criticism of adaptationist explanations can be found in Mayr (1983) and in the book edited by Dupré.

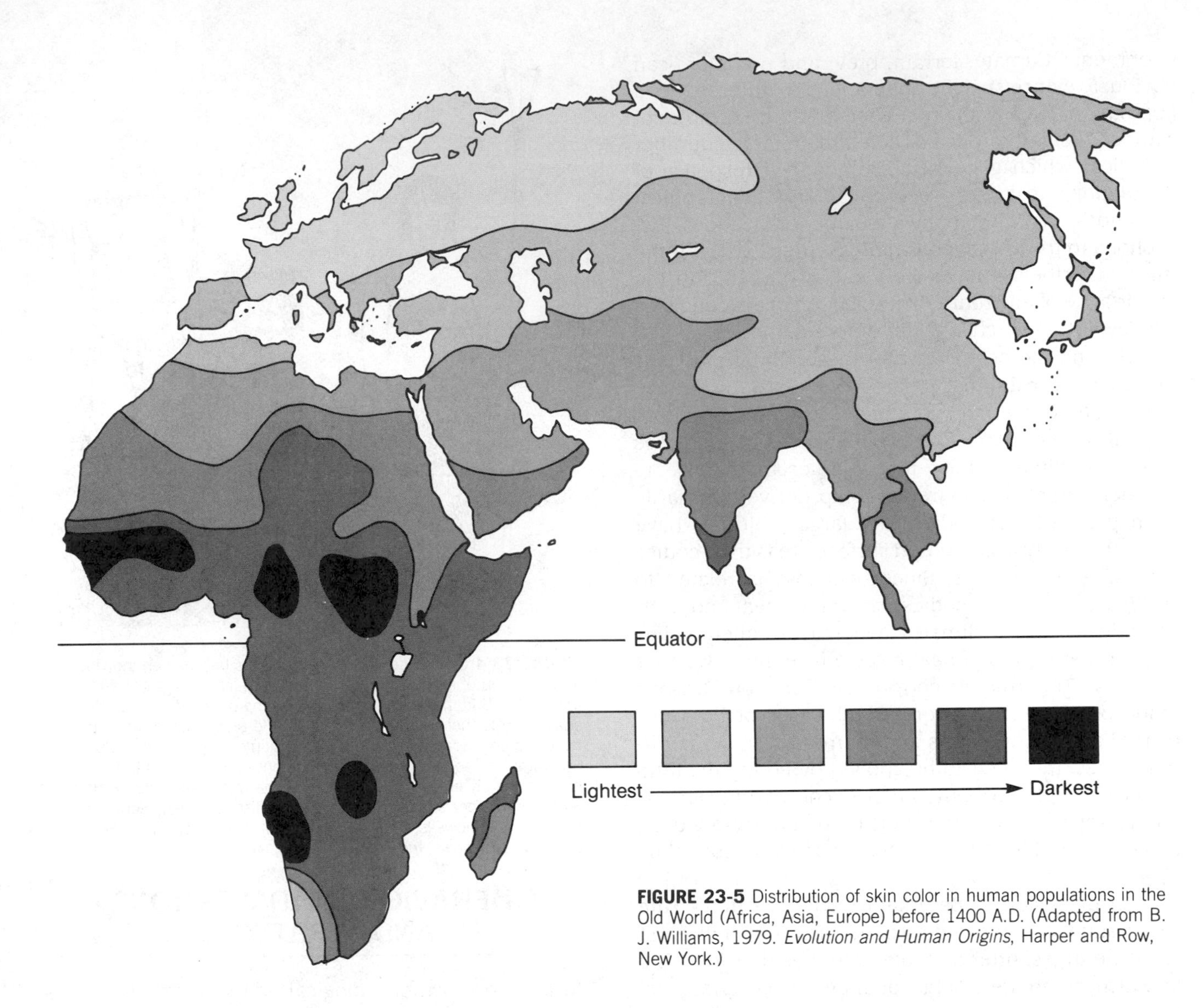

FIGURE 23-5 Distribution of skin color in human populations in the Old World (Africa, Asia, Europe) before 1400 A.D. (Adapted from B. J. Williams, 1979. *Evolution and Human Origins*, Harper and Row, New York.)

experience, or **learned**, improving over time through trial and error. Innate behaviors include **tropisms**, the directionally oriented growth patterns found in plants, fungi, and sessile animals, and **taxes**, directionally oriented locomotion among animals. Examples of such relatively simple innate behaviors are movements towards or away from light, gravity, and environmental nutrients and chemicals. Some behaviors can also be prompted by chemicals produced within organisms themselves, such as **pheromones**, which are molecular substances used by many animals to attract mates, lay down trails, and warn off competitors.

Relatively complex innate behaviors, often called **instincts**, may involve many behavioral components, such as courtship patterns in most animals and "dancing" patterns used by honeybees to communicate the direction and distance of food sources. Whether simple or complex, innate behaviors are often quite uniform within a species, and therefore seem, like other species-specific traits, to be entirely or almost entirely genetically influenced. For example, various genes and associated neuro-anatomical locations are known to be directly involved in *Drosophila* courtship behavior (Fig. 23-6).

Learned behavior is more flexible and often more complex than innate behavior since it can be modified by an individual to suit different environmental or social circumstances. Also, because these qualities are based on accommodating to a transitional and often unpredictable environment, learned behavior is apparently difficult or impossible to program entirely genetically. Instead, learned behavior derives largely from practice (e.g., play and observation) which enables individuals to modify their behavior on the basis of their own or others' past experience. Thus, for example, naive mammalian juveniles without sufficient past experience must often be protected until they can employ learned behaviors to deal with environmental and social problems or hazards.

However, even learned behavior must have ge-

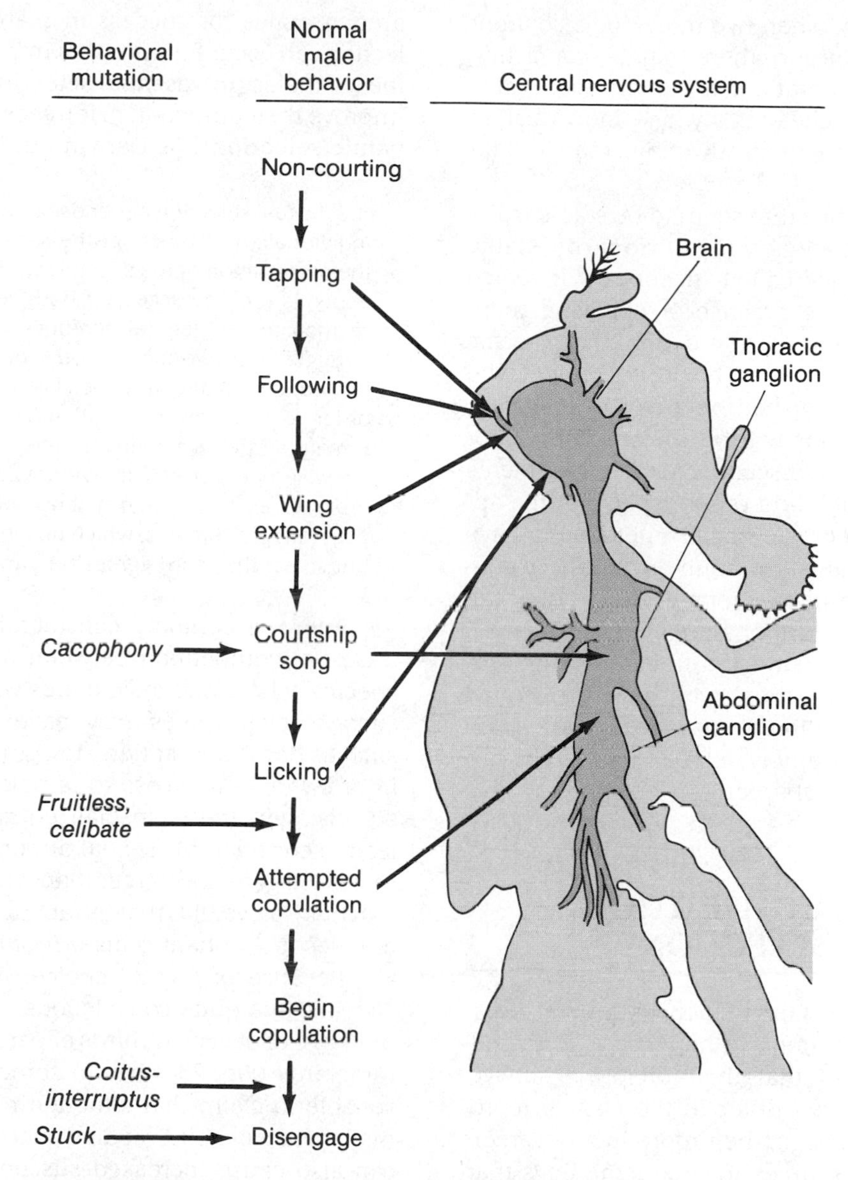

FIGURE 23-6 Sequence of the normal male courtship pattern in *D. melanogaster* and localization of some of these stages to particular sections of the central nervous system, according to the findings of Hall and others. Some of the mutations that have a primary effect on certain stages in the courtship pattern are shown in the left column. (From Strickberger.)

netic components, since many neurological, muscular, and sensory structures involved in the ability to learn and practice must be programmed into the individual. Such genetic influences have been widely investigated (Ehrman and Parsons, Fuller and Thompson). For example, experiments by Scott and Fuller have shown that genes in different races of dogs influence tameness, playfulness, and aggressiveness. In humans there is considerable evidence for genes involved in various behavioral disorders such as schizophrenia, manic depression, and mental retardation. The presence of such genetic influence and variation indicates that the capacities for most behavioral traits, like so many other adaptations, are probably the consequence of evolutionary selective forces.

An area of behavioral evolution that has received considerable attention is that of social relationships in which members of a group or species interact for purposes of breeding, feeding, and defense. Some of the behaviors involved in such social interactions (**sociobiology**) have been discussed in Chapter 19 as they relate to primates, and many, such as cooperation and dominance relations, also apply to other social groups (Wilson). In evaluating these interactions, various models have been proposed that deal with the advantages of cooperation opposed to the advantages of cheating. For example, when two individuals confront each other in a social group, there may be conflicting interests between cooperation, in which each gains some advantage, and cheating by one individual so that it gains greater immediate advantage than it could by cooperation.

Among the strategies that seem to be stable in the face of competing strategies (**evolutionarily stable strategies** or ESS) is one called "tit for tat," in which an individual follows the practice of behaving cooperatively on the first move or interaction, then repeats its opponent's previous move. Thus an opponent who acts selfishly will be punished by a selfish response, and cooperative behavior will be rewarded by a cooperative response. May discusses some experiments that support this model, and other aspects of the application of game theory to social interactions have been elaborated by Maynard Smith. When these concepts are extended to interactions among more than one pair of individuals, further strategies may develop such as **reciprocal altruism** (p. 410) in which individuals cooperate only with those who have cooperated with others. The application of sociobiology to human culture has been considerably debated and will be discussed in the following chapter.

Sexual Competition and Selection

Among the most important influences on social behavior is the mating system. Although it is to the advantage of both sexes that their offspring survive, males and females often differ in the cost of reproduction since females begin their reproductive careers by investing more resources in producing eggs than males do in producing sperm. Thus, it is to the advantage of genes carried by a female that she be discriminating in her choice of mates (**female choice**) so as not to endanger her relatively expensive gametic output. For this same reason, there is an advantage for females to increase male **parental investment** in their offspring. On the other hand, males can be more extravagant in disposing of their much more inexpensive and plentiful gametes, so it is to the advantage of genes carried by a male that he fertilize as many females as possible, often with relatively little discrimination. This sociobiological conflict of interest between the sexes leads to a variety of mating patterns that depend upon various factors, including the degree of parental care necessary for egg or infant survival and which sex (or both) provides such care.

In groups where females are primarily responsible for parental care—and this includes most vertebrate species—there is likely to be direct competition among males for success in mating. As a result, selection can occur for traits that improve combative abilities of males (**intrasexual selection**) and/or traits that improve their attraction to females (**intersexual** or **epigamic selection**). As Darwin put it,

> Sexual selection depends on the success of certain individuals over others of the same sex, in relation to the propagation of the species; whilst natural selection depends on the success of both sexes, at all ages, in relation to the general conditions of life. The sexual struggle is of two kinds; in the one it is between the individuals of the same sex, generally the males, in order to drive away or kill their rivals, the females remaining passive; whilst in the other, the struggle is likewise between the individuals of the same sex, in order to excite or charm those of the opposite sex, generally the females, which no longer remain passive, but select the more agreeable partners.

There is certainly considerable evidence for intrasexual competition between males in **polygynous species**, where one male mates with many females. In such species, males may have special armaments, such as horns and antlers, to compete with each other in order to gain access to females (Fig. 23-7(a)). Not surprisingly, these specialized masculine traits can lead to considerable sexual dimorphism. For example, males are generally larger than females in such groups, reaching a weight that is about eight times that of females in elephant seals (*Mirounga leonina*).

Because of female choice, male ornaments can also become quite conspicuous, such as the dramatic plumage of peacocks, birds of paradise, and even hummingbirds (Fig. 23-7(b)). In some cases it seems reasonable to claim that although male decorative traits may enhance their breeding success, traits of this kind can also cause increased susceptibility to predation. In fact, it is possible to show mathematically that non-adaptive male decorative traits can become so exaggerated that they are detrimental to the population as a whole, yet such traits can become even more exaggerated because they are selected by females who are the daughters of such exaggerated males (Fisher's **runaway selection**).

Nevertheless, female choice of non-adaptive traits in males is probably uncommon, and there are now reports which demonstrate that *Drosophila* fruit fly and *Colias* butterfly females choose sexual partners who confer greater fitness upon their offspring (Taylor et al., Watt et al.). Also, among passerine birds, there appears to be an association between species with bright-colored males and resistance to parasite infection (Read). As is true for other adaptations, there are sufficient variations in ecological, genetic, and evolutionary factors to prevent strict adherence to any widely applicable rules of sexual behavior (see Clutton-Brock).

(a)

(b)

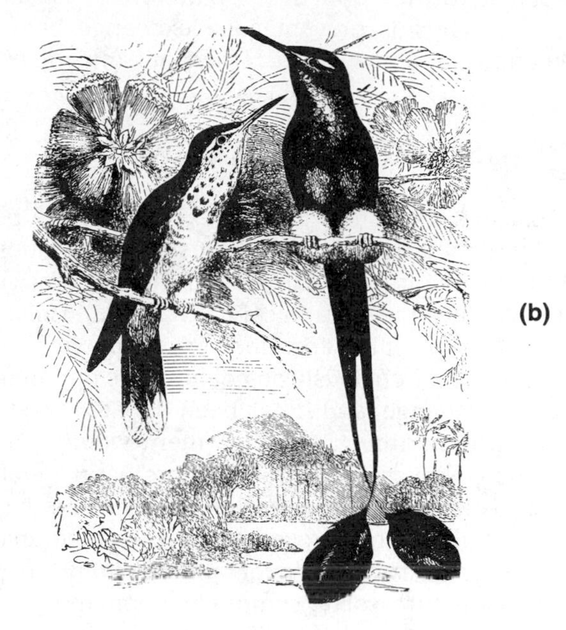

FIGURE 23-7 Two examples of sexual dimorphism in species of mammals and birds, with females on the left and males on the right. (a) Mature males of the extinct giant deer of Europe, *Megaloceros*, (also called the giant Irish elk) had antlers more than 11 feet wide, weighing about 100 pounds. (From R. J. G. Savage and M. R. Long, 1986. *Mammal Evolution.* British Museum, London.) (b) A South American hummingbird species, *Spathura underwoodi.* (From Darwin.)

Thus, for example, male features may be selected on both intra- and intersexual levels when females actively choose males that are more successful in intrasexual combat. On the whole, differences in sexual behavior between groups may therefore evolve quite rapidly and lead easily to barriers in gene exchange, a matter that will be briefly discussed below.

FROM RACES TO SPECIES BARRIERS

However racial adaptations occur, whether through morphological or behavioral changes or both, race formation is a potentially reversible process since different races may interbreed and combine again into a single populational unit. Thus, when the extent of migratory activity (gene flow) among individuals of a species is great, race formation may be impeded. Rensch, for example, has calculated that migratory species of birds have an average of less than half the number of races of non-migratory species: the greater the gene flow, the fewer the differences.

As a rule, therefore, race formation is accelerated by barriers that reduce gene exchange between populations. Initially such barriers are primarily geographical and occur when populations bud off from one another and occupy different areas or environmental habitats. The potential for gene exchange, however, enables all these different populations to be considered as members of a single species. It is only when populations have achieved sufficient differences to inhibit any gene exchange at all that they may be considered to have diverged sufficiently to have reached the level of separate species.

To biologists, the concept of a species as an interbreeding group distinct from other such groups arises in sexually reproducing organisms from the fact that such groups exist in nature and are mutually separated in many instances by "bridgeless gaps" across which interbreeding does not occur (Chapter 11). The species concept is also supported by the fact that species are recognized as distinct groups both by humans and other forms of life to whom such discrimination is essential. Predators of all kinds, for example, learn early to discriminate among varieties of prey and to select those that are palatable and can be used for food. In groups in which the recognition of species can be verbalized, including primitive human societies, the distinctions made are, in many cases, strikingly similar to the species classifications based on more sophisticated biological criteria. Thus a tribe of New Guinea islanders uses distinct names for 137 species of birds found in this region, almost equal to the exact number of 138 species recognized by ornithologists. The transition of racial differences to species differences is, therefore, usually marked by a qualitative change accompanied by reproductive separation or isolation. What mechanisms prevent gene exchange between populations, and how do such mechanisms originate?

Isolating Mechanisms

Mechanisms that prevent gene exchange have been broadly termed **isolating mechanisms**. Some authors include in this category all factors that prevent gene exchange, even geographical and spatial isolation. Such geographically separated, or **allopatric**, populations (p. 206),[3] obviously do not have the opportunity for gene exchange, and it has been debated whether, given the opportunity, many of them would remain reproductively isolated. Other authors have therefore proposed that the term isolating mechanisms be restricted to those that prevent gene exchange among populations in the same geographic locality, that is, mechanisms that isolate **sympatric** populations.

Mayr (1963) has classified sympatric isolating mechanisms into two broad categories: those that operate before fertilization can occur (**premating**), and those that operate afterward (**postmating**). Among the premating isolating mechanisms are:

1. **Seasonal** or **habitat isolation**. Potential mates do not meet because they flourish in different seasons or in different habitats. Some plant species, such as the spiderworts *Tradescantia canaliculata* and *T. subaspera*, for example, are sympatric throughout their geographical distribution, yet remain isolated because their flowers bloom at different seasons. Also, one species grows in sunlight and the other in deep shade.
2. **Behavioral** or **sexual isolation**. The sexes of two species of animals may be found together in the same locality, but their courtship patterns are sufficiently different to prevent mating. The distinctive songs of many birds, the special mating calls of certain frogs, and the sexual displays of most animals are generally attractive only to mates of the same species. Numerous plants have floral displays that attract only certain insect pollinators. Even where the morphological differences between two species is minimal, behavioral differences may suffice to prevent cross-fertilization. Thus *D. melanogaster* and *D. simulans*, designated as sibling species because of their morphological similarity, will normally not mate with each other even when kept together in a single population cage.
3. **Mechanical isolation**. Mating is attempted, but fertilization cannot be achieved because of difficulty in fitting together male-female genitalia. This type of incompatibility, long thought to be a primary isolating mechanism in animals, is no longer considered important. There is little evidence that matings in which the genitalia are markedly different are ever seriously attempted, although some exceptions exist among damselfly species and some other groups (Paulson).

Among the postmating mechanisms that prevent the success of an interpopulational cross even though mating has taken place are:

1. **Gametic mortality**. In this mechanism, either sperm or egg is destroyed because of the interspecific cross. Pollen grains in plants, for example, may be unable to grow pollen tubes in the styles of foreign species. In some *Drosophila* crosses Patterson and Stone and others have shown that an insemination reaction takes place in the vagina of the female that causes swelling of this organ and prevents successful fertilization of the egg.
2. **Zygotic mortality** and **hybrid inviability**. The egg is fertilized, but the zygote either does not develop, or it develops into an organism with reduced viability. Numerous instances of this type of incompatibility have been observed in both plants and animals. For example, Moore made crosses between 12 frog species of the genus *Rana* and found a wide range of inviability. In some crosses, no egg cleavage could be observed; in others, the cleavage and blastula stages were normal but gastrulation failed; and in still others, early development was normal but later stages failed to develop.
3. **Hybrid sterility**. The hybrid has normal viability but is reproductively deficient or sterile. This is exemplified in the mule (progeny of a male donkey and female horse) and many other hybrids. Sterility in such cases may be caused by interaction between genes from the two different sources or because of interaction between the cytoplasm from one source and the chromosomes from the other.

[3] Geographically distinct populations have been described in various ways, depending on the kind and degree of separation. The two most common terms used are allopatric for populations that are never found together and **parapatric** for populations that coexist only in one or more overlapping regions at the peripheries of their geographically separate distributions.

In general, the barriers separating species are not confined to a single mechanism. The *Drosophila* sibling species *D. pseudoobscura* and *D. persimilis* are isolated from each other by habitat (*persimilis* usually lives in cooler regions and at higher elevations), courtship period (*persimilis* is usually more active in the morning, *pseudoobscura* in the evening), and mating behavior (the females prefer males of their own species). Thus, although the distribution ranges of these two species overlap throughout large areas of western United States, these isolating mechanisms are sufficient to keep the two species apart. To date, only a few cross-fertilized females have been found in nature among many thousands of flies that have been examined. However, even when cross-fertilization occurs between these two species, gene exchange is still impeded, since the F_1 hybrid male is completely sterile and the progenies of fertile F_1 females backcrossed to males of either species show markedly lower viabilities than the parental stocks (**hybrid breakdown**).

MODES OF SPECIATION

In 1889 A. R. Wallace proposed that natural selection might favor the establishment of mating barriers between populations if the hybrids were adaptively inferior. That is, genotypes that did not mate to produce inferior hybrids would be selected over genotypes that did. According to this hypothesis, supported by Dobzhansky and others, selection for sexual isolation arises because most races and species are strongly adapted to specific environments. Hybrids between two such highly adapted populations therefore represent a genetic dilution of their parental gene complexes that can be of considerable disadvantage in the original environments. Thus genotypes that have incorporated premating isolating mechanisms would have the advantage of not wasting their gametes in producing deleterious offspring.

Full utilization of this mode of speciation demands, of course, that the different populations producing deleterious hybrids be exposed to each other in the same locality; only then could the more sexually isolated genotypes be specifically selected. Speciation should, therefore, occur in the following sequence:

1. Genetic differentiation between allopatric populations.
2. Overlap of these differentiated populations in a sympatric area.
3. Subsequent selection for intensified sexual isolating mechanisms (Fig. 23-8, left column).

Demonstration of this sequence among natural populations has been attempted by comparing the degree

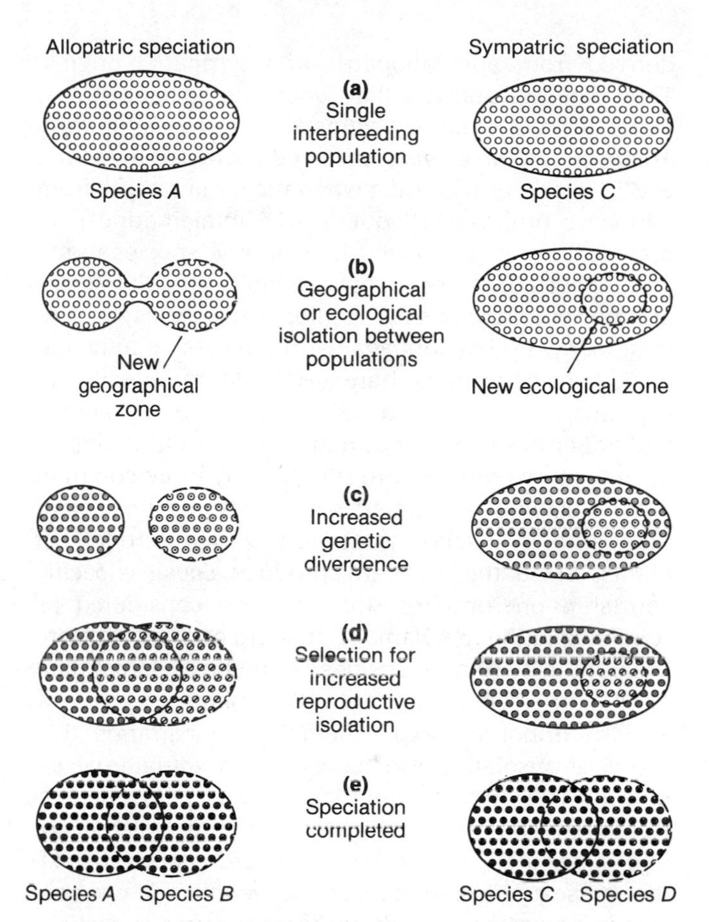

FIGURE 23-8 Two models of speciation. In allopatric speciation (left column) the population (a) splits, or buds, into one or more new geographical zones (b) which allow genetic differentiation to occur between different geographical groups by means of both random genetic drift and selection (c). When differentiation has proceeded sufficiently for each group to attain a uniquely adapted genetic identity, geographical mixture of the groups (d) can result in selection for improved reproductive isolation mechanisms between them. At the stage when gene flow between the groups can no longer occur, even when they occupy the same locality, speciation is complete (e). In sympatric speciation (right column) a population (a) splits into one or more groups that occupy different ecological zones, such as special habitats or food sources, within a single geographic locality (b). Increased genetic differentiation between the groups (c) permits selection for reproductive isolation mechanisms (d) that eventually lead to complete speciation (e). The difference between these models is the extent of physical separation involved in the initial genetic divergence between the groups. Numerous examples and variations of these models are discussed by White, and a variety of speciation models are compared in Table 1 of Barton and Charlesworth. (From Strickberger.)

of sexual isolation between different sympatric and allopatric populations; sexual isolation should be strongest among sympatric populations of different related species, since they are sufficiently close to produce deleterious hybrids, and weakest among allopatric populations of species that are too distantly separated to produce such hybrids.

In one such experiment, Wasserman and Koepfer tested the degree of sexual isolation between the sibling species *D. arizonensis* and *D. mojavensis* by attempting crosses in which the species strains were derived from both allopatric and sympatric origins. Their findings showed that when the species strains came from sympatric origins the interspecific cross *arizonensis* x *mojavensis* occurred more rarely (14 out of 377 total matings) than when the strains came from allopatric origins (119 out of 473 total matings). In plants, Grant has reported that of nine species in the annual herb *Gilia*, those that are most difficult to cross are the sympatric ones. The allopatric species, by contrast, show no barriers against intercrossing although all F_1 hybrids produced are sterile. More recently, an examination of hundreds of moth species by Phelan and Baker has shown that male scent-emitting organs used to attract females are significantly more common among species associated with the same host plant than among species associated with different host plants. Since these organs produce species-specific courtship pheromones, they can be considered as sexual-isolating mechanisms that are apparently more frequent in sympatric species (same host plants) than in allopatric species (different host plants).

An important experimental demonstration that premating isolating mechanisms can actually be increased in sympatric populations was performed by Koopman who made use of the normally isolated sibling species *D. pseudoobscura* and *D. persimilis*. Although sexual isolation exists between these two species in nature and at normal temperatures in the laboratory, cold temperatures can apparently cause a significant increase in interspecific mating. By marking each of the two species with different homozygous recessive genes, Koopman was able to recognize hybrids formed under these low-temperature conditions and remove them from interspecific population cages. He performed this operation each generation and found that, as time went on, fewer and fewer hybrids were produced. For example, after five generations the frequency of hybrids produced in the mixed populations had generally fallen to 5 percent from values that were initially as high as 50 percent. This was striking evidence that selection against hybrids had caused rapid selection for sexual isolation that led to a reduction in hybrid formation. In plants, a somewhat similar experiment was performed by Paterniani who planted a mixture of yellow sweet and white flint strains of corn: disposal of plants that produced the greatest proportion of heterozygotes led in five generations to a reduction in intercrossing from about 40 percent to less than 5 percent.

One may, of course, argue that, given sufficient time, even allopatric populations will accumulate a sufficient number of genetic differences to show sexual isolation when they are brought together in the same locality. In the *virilis* group of *Drosophila* species, Patterson and Stone have observed that the European *D. littoralis* is much more isolated from the American populations of *americana*, *texana*, and *novamexicana* than are American species in the same group. This view is also supported in some experiments that separate a single population into two or more groups for a considerable period and then test these groups for the presence of reproductive isolation. In one example, two replicate populations of *D. melanogaster*, raised in the laboratory under different conditions of temperature and humidity for six years, developed both sexual isolation and hybrid sterility (Kilias and Alahiotis).

Whatever the process through which speciation occurs between geographically separated populations (**allopatric speciation**), a number of authors suggest that it may proceed quite rapidly under some circumstances (Mayr [1954], Carson and Templeton). They put emphasis on **founding accidents**, or **bottlenecks** (Chapter 21), in which a small, isolated population is subject to forces such as random genetic drift, increased homozygosity caused by inbreeding, and changes in the adaptive landscape (Chapter 22), followed by radical changes in selection pressure. The combined effect of such forces may therefore produce novel, **coadapted gene combinations** affecting behavioral, morphological, and physiological traits that lead to reproductive isolation from neighboring and ancestral populations. On the island of Hawaii, for example, Carson uses such concepts to help explain the origin of its 25 species of "picture-winged" Drosophilidae in what may have been less than half a million years. Some of the founding events that occurred in these Hawaiian Drosophilidae, according to Kaneshiro, caused radical changes in male courtship behavior so that females in these small, isolated populations were selected to respond to such changes, whereas females in ancestral or neighboring populations were not attracted to such modified males.

In opposition to these views on bottlenecks, Barton and Charlesworth suggest that the concept of speciation caused by single founder events has little theoretical support, since such events will usually not produce an immediately significant change in an isolated population. For example, it may take many generations for random genetic drift to effectively modify gene frequencies. These authors state that "it is impossible to separate the effect of isolation, environmental differences, and continuous change by genetic drift [in moderately sized populations] from the impact of population bottlenecks [in small founder populations]." Even the impact of bottlenecks by themselves is in dispute, since a recent large-scale experiment indicates that phenotypic variation may be increased rather than decreased in populations that pass through bottlenecks (Bryant et al.).

Unfortunately, since little is known about the number and types of genes involved in speciation, the relative importance of each suggested mode of speciation remains unclear. Perhaps all one can say at present is that speciation events can occur in various ways. In some groups such as the Hawaiian Drosophilidae, speciation has been dramatically rapid and may well have involved fewer genes with greater phenotypic effects than in the slower speciation events in some other *Drosophila* groups. In other groups allopatric speciation in the absence of bottlenecks may have been more common. Such allopatric speciation modes, however, do not exclude selection for sexual isolation between sympatric populations because of hybrid sterility or inviability, although that, too, has been disputed (Butlin).

Where species barriers break down to produce viable and fertile hybrids—and there are such instances, especially in plants—**zones of hybridization** or **hybrid swarms** may occur whose genotypes and phenotypes are different from both parental species. If a unique and discrete habitat exists to which the hybrids are better adapted than the parents, it is conceivable that the new population may eventually become isolated from its parental populations. A few examples that may fit this mode of speciation are proposed by Grant.

In some cases, again especially in plants, fertile hybrids can introduce genes from one species into the other, enhancing a species' ecological range and evolutionary flexibility: a phenomenon that Anderson has termed **introgressive hybridization** (see also Levin). Furthermore, even if the plant hybrid is sterile and must propagate asexually, polyploidy may arise, enabling the hybrid to produce fertile gametes (allopolyploids; see Fig. 10-12). Since these gametes are diploid relative to the haploid gametes of the parental species, a new species is born at one stroke, fertile with itself or other such polyploid hybrids but sterile in crosses with either parental species. It is estimated that about half of all angiosperms and almost all pteridophytes (ferns) are polyploids of one type or another, with allopolyploidy the most common form.

CAN SPECIES DIFFERENCES ORIGINATE SYMPATRICALLY?

The sequence of evolutionary events in speciation seems, therefore, to begin with race formation and end with reproductive isolation. In this sequence a further disputed point among evolutionary geneticists is the degree to which geographical separation between populations is necessary to accumulate the initial genetic differences that lead to speciation. Many workers in this field believe that populations can only accumulate genetic differences when they are sufficiently spatially separated to prevent the gene exchange that might

TABLE 23-3 Results of Tests for Mating Preferences Among *D. melanogaster* Flies Selected for High Bristle Number (H) and Low Bristle Number (L) and in Which Males and Females Are Given a Free Choice of Mates

Generation of Selection	Number of Matings			
	H × H	H × L	L × H	L × L
7	12	3	4	12
8	14	2	6	10
9	10	4	6	7
10	8	4	3	13
19	27	2	8	20
	71	15	27	62

From Thoday.

eradicate these differences. They propose that the speciation process takes hold only after this important early period of geographical separation, either by the accidental origin of isolating mechanisms or by subsequent selection of isolating mechanisms because of defective hybrids.

Other workers, especially Mather and Thoday, have proposed that a population in a single locality selected for adaptation to different habitats within that locality could produce an increase in genetic variability (see disruptive selection, p. 443) that would lead to polymorphism. One such example is the polymorphism now found in the British peppered moth *Biston betularia*, and a further important example is the polymorphism of mimicry in the butterfly *Papilio dardanus* (Sheppard).

It has also been proposed that under some circumstances, especially if the selected forms can exist independently of each other in the same locality, isolation between the selected groups might result. Evidence for this view was first presented by Thoday and Gibson in selection experiments on bristle number in *D. melanogaster*. They selected flies each generation for high (H) and low (L) bristle number and found that, although random mating was permitted, mating preferences of these flies went rapidly in the direction of positive assortative mating, H **x** H and L **x** L, with relatively few H **x** L and L **x** H matings, as can be seen in Table 23-3. Increased isolation resulting from disruptive selection between populations in the same locality has since been achieved in other experiments (Coyne and Grant, Soans et al.).

However, in spite of many attempts, some results of disruptive selection have not been replicated (Scharloo), and it has been questioned whether any single locality in nature could consistently maintain divergent

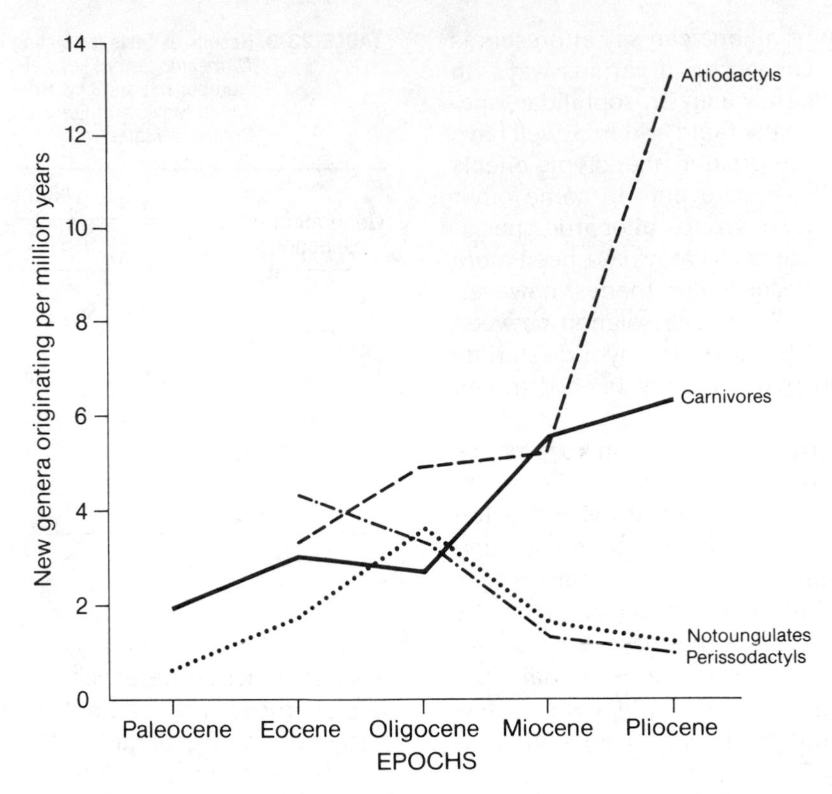

FIGURE 23-9 Evolutionary rates measured in terms of new genera originating per million years in four orders of mammals during the Tertiary period. (Adapted from Simpson.)

selective conditions for a long enough period to produce speciation (Mayr [1963]). Nevertheless, the occurrence of such **sympatric speciation** (Fig. 23-8, right column) has been used by Bush and others to explain the likelihood that various groups of insects speciated within a single geographical range by adapting to different kinds of host plants as food sources (e.g., Feder et al.). More recently, Rice and Salt have shown that a population of *D. melanogaster*, selected for radically different experimental habitats, can develop almost complete reproductive isolation between the selected groups. It seems likely that some or even many sympatric speciation events may have occurred (Barton et al.), and Seger and others have offered theoretical models to support these contentions.

Evolutionary Rates and Punctuated Equilibria

Although the underlying mechanisms of speciation are not yet easily discernible, the presence of so many fossil and existing species makes it possible to estimate **evolutionary rates** by geological and paleontological data (Chapters 6 and 13 through 19) or by biochemical changes and "molecular clocks" (Chapter 12). However, rate determinations in both cases are beset with problems. Among the questions that arise are whether rates should be measured in terms of geological periods (chronological time) or numbers of generations (biological time). Furthermore, what morphological or molecular standards should be used in comparing rates in one group with those of another? Also, how does one measure the numbers and kinds of genes involved? There is no common agreement on solutions to these problems.

On the paleontological level, taxonomic difficulties also intrude since "lumpers" and "splitters" may, respectively, combine or split groups of organisms into different taxonomic categories (Chapter 11). This might lead, for example, to different numbers of genera for the same lineages and thus to different generic evolutionary rates. Incompleteness of the fossil record and the frequent absence of evolutionary intermediate groups also clouds many paleontological rate determinations. Nevertheless, as shown in Figure 23-9, there seem to be significant differences in evolutionary rates among known fossil taxa of various mammalian groups, and this seems also true of many other lineages for which fossil information is available.

Since rates often seem to vary over time within any particular group, paleontologists usually classify rates into those most commonly found (**horotely**) flanked by slower rates (**bradytely**) and faster rates (**tachytely**). "Living fossils" (e.g., *Latimeria* and *Neopilina*, Chapter 3) are obviously at the bradytelic end of evolutionary rates in their respective groups, and artiodactyls can be considered to be at the tachytelic end of mammalian rates. Simpson has given the name **quantum evolution** to instances of very rapid evolutionary changes that are often marked by expansion into new adaptive zones and the origin of new taxa. As pointed out in Chapter 12, other biologists have proposed the term **macroevolution** to distinguish major evolutionary changes of this kind from the presumed less radical changes that occur within a species, called **microevolution**.

The question of what accounts for evolutionary rate differences, whether on the micro- or macroevolutionary levels, is not yet satisfactorily answered, but many biologists agree that such differences are probably connected to the structure of populations and to changes in the direction and intensity of selection at different times in different groups. Supporting the importance of selection is the finding that rate differences do not seem to be correlated with length of generations or with available genetic variability measured as electrophoretic protein differences (Chapter 10). For example, mammals with short generation times, such as opossums, have evolved much more slowly than those with much longer generation times, such as elephants, and the slowly evolving horseshoe crabs (*Limulus*, Fig. 11-3) show as much electrophoretic genetic variability as more rapidly evolving invertebrates.

An important point now in dispute is whether speciation and the origin of higher taxa involve macroevolutionary mechanisms uniquely different from the microevolutionary mechanisms that cause less noticeable changes within lineages. Most population geneticists propose that the speciation process may add new directions to evolution, but its mechanisms are similar to those used in non-speciation changes. Proponents of **punctuated equilibria**, such as Gould and Stanley, propose that the mode of origin of new taxa is qualitatively unique, as evidenced by the rapidity of macroevolutionary events in the fossil record: that is, punctualists see the fossil record as long intervals of microevolutionary stasis, or equilibrium, during which there is relatively little change, punctuated by rapid macroevolutionary periods during which new taxa arise.

In the 1930's and 1940's, Goldschmidt proposed that the cause for each macroevolutionary change derived from a single **macromutational** event. This view had much in common with **saltationist** doctrines, espoused by various paleontologists, that species can arise suddenly because of unknown types of events (p. 418). According to Goldschmidt, mutations with large developmental effects presumably produced some "hopeful monsters" that could then enter into new adaptive zones. Because their effects must be integrated with many other genetic changes (the genetic background), it now seems clear that single mutations large enough to cause instantaneous species differences would be inviable, and modern proponents of punctuated equilibria have generally abandoned Goldschmidt's hopeful monsters. However, as discussed in Chapter 12, punctualists have taken comfort in the finding that regulatory mutations can have significant developmental effects and probably account for important differences between various groups. Similarly, the occurrence of founding accidents, or bottlenecks, has been considered a possible cause for or accompaniment of macroevolutionary events.

From a punctualist as well as gradualist point of view, the rate of evolution in a new populational offshoot may certainly be rapid compared to changes in its parental species, which remains tied to its more traditional ecological niche. It is, however, not clear how the mechanisms used to explain macroevolution are significantly different from those used to explain microevolution. Regulatory mutations, bottlenecks, new directions and intensities of selection, as well as other presumed macroevolutionary mechanisms, are also used to help explain microevolutionary events. One can also dispute whether the fossil record offers enough information to show how rapidly macroevolutionary events have occurred. A very slow rate of change during each generation in a population may, in fact, lead to speciation in a period of geological time so short as to be undetectable by paleontologists. For example, a duration of 100 years may produce a few thousand or more generations of flies in a *Drosophila* lineage, as could 1,000 years in a rodent lineage; yet both these periods could easily produce significant morphological changes through gradual microevolutionary mechanisms, neither necessarily leaving a fossil record. Certainly the speed with which sexual isolation can develop between experimental populations shows how rapidly speciation can be achieved by microevolutionary methods while appearing to be macroevolutionary on the paleontological level.

It is therefore important to note that relatively long periods within lineages, such as 100,000 years, often go undetected in paleontology. Dawkins quotes the hypothetical example of a mouse lineage that gradually reaches the size of an elephant in 60,000 years and points out that "evolutionary change too *slow* to be detected by microevolutionists [population geneticists] can nevertheless be too *fast* to be detected by macroevolutionists [paleontologists]." Because of all these and other arguments (e.g., Ayala, Kellog, Levinton) it seems difficult to accept as yet that new evolutionary rules or mechanisms must be invoked to explain paleontological observations of punctuated equilibria. In fact, some proponents of punctuated equilibria seem to be moderating some of their former views (e.g., Eldredge).

SUMMARY

All members of a species share a common gene pool, although populations within it may vary genetically from each other. If the gene frequencies of these populations are sufficiently distinct, they are known as races. In humans, at least, there is so much polymorphism that races cannot be distinguished by the presence or absence of certain alleles but only by variations in a panoply of gene frequencies.

At least some, if not many, racial differences, both morphological and behavioral, are adaptations to dissimilar environments. Both learned and innate behaviors have a genetic component, although it is more apparent in tropisms and instincts than in learned behaviors.

New species form when genetic exchange among races is impeded. Reproductive isolating mechanisms, which provide the barriers for genetic exchange, may be of various kinds. Behavioral, seasonal, and mechanical premating mechanisms obstruct zygote formation, while with postmating mechanisms offspring will be inviable or sterile. According to some proposals, more of these mechanisms should develop in sympatric populations than in allopatric ones, which are geographically isolated from each other.

New species may form in allopatric groups by the slow accumulation of genetic differences, or if they originated from only a few individuals (founder effect), that is, a group which has been substantially diminished in size (bottleneck). Speciation is presumed to occur in sympatric populations that are under diverse selection pressures, and these groups will become distinct because of persistent preferential mating.

The rate at which new taxa form is difficult to determine, as evolutionary rates differ even within phylogenetic groups. These inconsistencies probably relate to variations in selection pressures on the population at different times. Whether or not microevolutionary forces inducing change within species are identical to marcroevolutionary forces generating new species is a matter of contention. Advocates of punctuated equilibrium believe that speciation is rapid and produced by unique forces; others feel that macroevolution is subject to the same forces (mainly natural selection) as is microevolution.

KEY TERMS

Allen's rule
allopatric populations
allopatric speciation
behavioral isolation
Bergmann's rule
bottlenecks
bradytely
coadapted gene combinations
ecogeographical rules
epigamic selection
evolutionarily stable strategies
evolutionary rates
female choice
founding accidents
gametic mortality
Gloger's rule
habitat isolation
horotely
hybrid breakdown
hybrid inviability
hybrid sterility
hybrid swarms
index of genetic distance
innate behaviors
instincts
intersexual selection
intrasexual selection
introgressive hybridization
isolating mechanisms
learned behaviors
macroevolution
macromutation
mechanical isolation
microevolution
parapatric populations
parental investment
pheromones
polygynous species
postmating isolating mechanisms
premating isolating mechanisms
punctuated equilibria
quantum evolution
races
reciprocal altruism
runaway selection
saltation
seasonal isolation
sexual isolation
sociobiology
sympatric populations
sympatric speciation
tachytely
taxes (behavior)
tropisms
zones of hybridization
zygotic mortality

REFERENCES

Anderson, E., 1949. *Introgressive Hybridization.* John Wiley & Sons, New York.

Ayala, F. J., 1983. Microevolution and macroevolution. In *Evolution From Molecules to Men*, D. S. Bendall (ed.). Cambridge Univ. Press, Cambridge, pp. 387-402.

Barton, N. H., and B. Charlesworth, 1984. Genetic revolutions, founder effects, and speciation. *Ann. Rev. Ecol. and Syst.*, **15**, 133-164.

Barton, N. H., J. S. Jones, and J. Mallet, 1988. No barriers to speciation. *Nature*, **336**, 13-14.

Bryant, E. H., S. A. McCommas, and L. M. Combs, 1986. The effect of an experimental bottleneck upon quantitative genetic variation in the housefly. *Genetics*, **114**, 1191-1211.

Bush, G. L., 1975. Sympatric speciation in phytophagous parasitic insects. In *Evolutionary Strategies of Parasitic Insects*, P. W. Price (ed.). Plenum Press, London, pp. 187-206.

Butlin, R., 1987. Speciation by reinforcement. *Trends in Ecol. and Evol.*, **2**, 8-13.

Carson, H. L., 1986. Sexual selection and speciation. In *Evolutionary Processes and Theory*, S. Karlin and E. Nevo (eds.). Academic Press, Orlando, Fla., pp. 391-409.

Carson, H. L., and A. R. Templeton, 1984. Genetic revolutions in relation to speciation phenomena: the founding of new populations. *Ann. Rev. Ecol. and Syst.*, **15**, 97-131.

Clausen, J., D. D. Keck, and W. M. Hiesey, 1948. Experimental studies on the nature of species. III. Environmental responses of climatic races of *Achillea*. *Carnegie Inst. Wash. Publ. No. 581*, 1-129.

Clutton-Brock, T. H., 1983. Selection in relation to sex. In *Evolution From Molecules to Men*, D. S. Bendall (ed.). Cambridge Univ. Press, Cambridge, pp. 457-481.

Coyne, J. A., and B. Grant, 1972. Disruptive selection on I-maze activity in *Drosophila melanogaster*. *Genetics*, **71**, 185-188.

Darwin, C., 1871. *The Descent of Man and Selection in Relation to Sex.* John Murray, London.

Dawkins, R., 1983. Universal Darwinism. In *Evolution From Molecules to Men*, D. S. Bendall (ed.). Cambridge Univ. Press, Cambridge, pp. 403-425

Dobzhansky, Th., 1970. *Genetics of the Evolutionary Process.* Columbia Univ. Press, New York.

Dupré, J. (ed.), 1987. *The Latest on the Best: Essays on Evolution and Optimality.* MIT Press, Cambridge, Mass.

Ehrman, L., and P. A. Parsons, 1981. *Behavior Genetics and Evolution.* McGraw-Hill, New York.

Eldredge, N., 1989. *Macroevolutionary Dynamics: Species, Niches, and Adaptive Peaks.* McGraw-Hill, New York.

Feder, J. L., C. A. Chilcote, and G. L. Bush, 1988. Genetic differentiation between sympatric host races of the apple maggot fly *Rhagoletis pomonella*. *Nature*, **336**, 61-64.

Fisher, R. A., 1930. *The Genetical Theory of Natural Selection.* Oxford Univ. Press, Oxford.

Fuller, J. L., and W. R. Thompson, 1978. *Foundations of Behavior Genetics.* C. V. Mosby, St. Louis.

Goldschmidt, R. B., 1940. *The Material Basis of Evolution.* Yale Univ. Press, New Haven, Conn.

Gould, S. J., 1980. Is a new and general theory of evolution emerging? *Paleobiology*, **6**, 119-130.

Gould, S. J., and R. C. Lewontin, 1979. The spandrels of San Marco and the panglossian paradigm: a critique of the adaptationist program. *Proc. Roy. Soc. London*, **205**, 581-598.

Grant, V., 1985. *The Evolutionary Process.* Columbia Univ. Press, New York.

Hall, J. C., 1979. Control of male reproductive behavior by the central nervous system of *Drosophila*: dissection of a courtship pathway by genetic mosaics. *Genetics*, **92**, 437-457.

Kaneshiro, K. Y., 1983. Sexual selection and direction of evolution in the biosystematics of Hawaiian Drosophilidae. *Ann. Rev. Entomol.*, **28**, 161-178.

Kellog, D. E., 1988. "And then a miracle occurs"—weak links in the chain of argument from punctuation to hierarchy. *Biol. and Phil.*, **3**, 3-28.

Kilias, G., and S. N. Alahiotis, 1982. Genetic studies on sexual isolation and hybrid sterility in long-term cage populations of *Drosophila melanogaster*. *Evolution*, **36**, 121-131.

Koopman, K. F., 1950. Natural selection for reproductive isolation between *Drosophila pseudoobscura* and *D. persimilis*. *Evolution*, **4**, 135-148.

Levin, D. A. (ed.), 1979. *Hybridization: An Evolutionary Perspective.* Dowden, Hutchinson and Ross, Stroudsburg, Pa.

Levinton, J., 1988. *Genetics, Paleontology, and Macroevolution.* Cambridge Univ. Press, Cambridge.

Lewontin, R. C., 1972. The apportionment of human diversity. *Evol. Biol.*, **6**, 381-398.

Mather, K., 1955. Polymorphism as an outcome of disruptive selection. *Evolution*, **9**, 52-61.

——, 1973. *Genetical Structure of Populations*. Chapman & Hall, London.

May, R. M., 1987. More evolution of cooperation. *Nature*, **327**, 15-17.

Maynard Smith, J., 1982. *Evolution and the Theory of Games*. Cambridge Univ. Press, Cambridge.

Mayr, E., 1954. Change of genetic environment and evolution. In *Evolution as a Process*, J. S. Huxley, A. C. Hardy, and E. B. Ford (eds.). Allen & Unwin, London, pp. 156-180.

——, 1963. *Animal Species and Evolution*. Harvard Univ. Press, Cambridge, Mass.

——, 1983. How to carry out the adaptationist program. *Amer. Nat.*, **121**, 324-334.

Moore, J. A., 1949. Patterns of evolution in the genus *Rana*. In *Genetics, Paleontology, and Evolution*, G. L. Jepsen, E. Mayr, and G. G. Simpson (eds.). Princeton Univ. Press, Princeton, N.J., pp. 315-355.

Nei, M., and A. K. Roychoudhury, 1982. Genetic relationship and evolution of human races. *Evol. Biol.*, **14**, 1-59.

Paterniani, E., 1969. Selection for reproductive isolation between two populations of maize, *Zea mays* L. *Evolution*, **23**, 534-547.

Patterson, J. T., and W. S. Stone, 1952. *Evolution in the Genus Drosophila*. Macmillan, New York.

Paulson, D. R., 1974. Reproductive isolation in damselflies. *Syst. Zool.*, **23**, 40-49.

Phelan, P. L., and T. C. Baker, 1977. Evolution of male pheromones in moths: reproductive isolation through sexual selection? *Science*, **235**, 205-207.

Read, A. F., 1987. Comparative evidence supports the Hamilton and Zuk hypothesis on parasites and sexual selection. *Nature*, **328**, 68-70.

Rensch, B., 1960. *Evolution Above the Species Level*. Columbia Univ. Press, New York.

Rice, W. R., and G. W. Salt, 1988. Speciation via disruptive selection on habitat preference: experimental evidence. *Amer. Nat.*, **131**, 911-917.

Scharloo, W., 1971. Reproductive isolation by disruptive selection: did it occur? *Amer. Nat.*, **105**, 83-86.

Schreider, E., 1964. Ecological rules, body-heat regulation and human evolution. *Evolution*, **18**, 1-9.

Scott, J. P., and J. Fuller, 1965. *Dog Behavior: The Genetic Basis*. Univ. of Chicago Press, Chicago.

Seger, J., 1985. Intraspecific resource competition as a cause of sympatric speciation. In *Evolution: Essays in Honour of John Maynard Smith*, P. J. Greenwood, P. H. Harvey, and M. Slatkin (eds.). Cambridge Univ. Press, Cambridge, pp. 43-53.

Sheppard, P. M., 1961. Some contributions to population genetics resulting from the study of the *Lepidoptera*. *Adv. in Genet.*, **10**, 165-216.

Simpson, G. G., 1949. *The Meaning of Evolution*. Yale Univ. Press, New Haven, Conn.

Soans, A. B., D. Pimentel, and J. S. Soans, 1974. Evolution of reproductive isolation in allopatric and sympatric populations. *Amer. Nat.*, **108**, 117-124.

Sonneborn, T. M., 1957. Breeding systems, reproductive methods, and species problems in Protozoa. In *The Species Problem*, E. Mayr (ed.). Amer. Assoc. Adv. Sci., Washington, D.C., pp. 155-324.

Stanley, S. M., 1979. *Macroevolution: Process and Product*. W. H. Freeman, San Francisco.

Taylor, C. E., A. D. Pereda, and J. A. Ferrari, 1987. On the correlation between mating success and offspring quality in *Drosophila melanogaster*. *Amer. Naturalist*, **129**, 721-729.

Thoday, J. M., 1972. Disruptive selection. *Proc. Roy. Soc. Lond. (B)*, **182**, 109-143.

Thoday, J. M., and J. B. Gibson, 1962. Isolation by disruptive selection. *Nature*, **193**, 1164-1166.

Turesson, G., 1922. The genotypical response of the plant species to the habitat. *Hereditas*, **3**, 211-350.

Wasserman, M., and H. R. Koepfer, 1977. Character displacement for sexual isolation between *Drosophila mojavensis* and *Drosophila arizonensis*. *Evolution*, **31**, 812-823.

Watt, W. B., P. A. Carter, and K. Donohue, 1986. Females' choice of "good genotypes" as mates is promoted by an insect mating system. *Science*, **233**, 1187-1190.

White, M. J. D., 1978. *Modes of Speciation*. W. H. Freeman, San Francisco.

Wilson, E. O., 1975. *Sociobiology: The New Synthesis*. Harvard Univ. Press, Cambridge, Mass.

24

Culture and the Control of Human Evolution

t the apex of our interest in evolution stands an interest in the state and future of our own species. How close are the ties between our culture and our biology? In which direction are humans evolving? Are human biological endowments satisfactory for human needs? What are the prospects for the control of human evolution? Our knowledge so far obviously offers us the opportunity to answer some aspects of these questions. However, before making this attempt, let us first consider some unique features of *Homo sapiens*.

Learning, Society, and Culture

The most distinctive feature of our species is probably our intelligence. However it is measured, it is this intelligence that provides us with flexible adaptive behaviors which are far more complex than those attained by any other species. That is, humans can consistently **learn** from their environmental experiences by incorporating such experiences into their behavior and can create new environments over which they have considerable control.

Much of human learning follows a Lamarckian pattern, in the conscious acquisition and transmission of those behavioral responses that answer the needs of specific situations. Although some learning occurs in other organisms (Chapter 23), they are primarily forced to rely for survival upon rigid and automatic responses built into their nervous systems through genetic means. Humans, by contrast, can be raised in different environments and learn to obtain food, defend themselves, provide shelter, and perform numerous tasks in many specialized ways without being dependent on specialized genotypes. Most important, humans can acquire and transmit such practices and behaviors, or **culture**, through social exchanges involving language, teaching, and imitation, both between individuals and between generations. Cultural transmission of learned behavior thus has the advantage that it eliminates the hazards encountered by individuals who must learn independently, by trial and error, to cope with environmental variables. Instead, cultural transmission allows more successful imitative learning of those adaptive practices that have been incorporated, often over more than a single lifetime, into the social and cultural heritage.

Because of such socially mediated transmission, cultural changes, unlike biological genetic changes, are not restricted to passage from one distinct generation to another, but may be proposed, accepted, and utilized during most stages in the human life cycle in interactions between both consanguineous and nonconsanguineous individuals. That is, the cultural "parents" of individuals need not be their biological parents, nor need cultural parents derive from the same geographical area as their cultural offspring. Thus, the kinds of isolation barriers that inhibit genetic exchange between biological species do not exist between human cultural groups: biological traits are transmitted **vertically** within lineages, whereas cultural traits can be transmitted both vertically and **horizontally** within and between lineages.

In short, humans have two unique hereditary systems. One is the **genetic system** that transfers biological information from biological parent to offspring in the form of genes and chromosomes. The other is the **extragenetic system** that transfers cultural information from speaker to listener, from writer to reader, from viewer to spectator, and forms our cultural heritage. Both systems are informational in that they produce their effects by instruction: the biological system through the information embodied in DNA via the coding properties of these cellular macromolecules, the cultural system through social interactions coded in language and custom and embodied in records and traditions.

RELATIVE RATES OF CULTURAL AND BIOLOGICAL EVOLUTION

The changes provided by cultural heredity over the last 10,000 years have been most impressive. We have moved from bands of hunters and fishermen mainly concerned with obtaining food to complex urban societies in which such concerns occupy relatively little time for most of us. Instead of hunting and primitive food gathering, an increased proportion of our efforts now concern cultural and technological tasks that could not have been foreseen 10,000 years ago or even a few generations ago. In fact, in some fields, one can hardly predict from one year to the next what kinds of changes will appear.

According to one estimate reported by Holzmüller, the rate at which we are gathering new experience is now doubling at least every 15 years. Our present lifetime experience is therefore equivalent to a life span of about 300 years just a few generations ago when the rate of gathering new experience was perhaps one quarter or one eighth what it is now. This remarkable rapidity in cultural and technological change, at least in the fields of science, shows even further promise of increase if we consider the increased proportion of scientists who now exist and are being trained. Price has provided the widely quoted estimate that of all scientists who have ever lived, more than 90 percent are alive today!

In contrast to the rapid changes associated with cultural and technological heredity, changes in human biological heredity during this 10,000-year period seem relatively small if at all detectable. The most distinguished possession of *Homo sapiens*, the human brain, shows no change in size over the last 100,000 years, nor is there any clear indication that there has been any qualitative change during this period. That is, our ancestors of many years ago, given our training, may well have shown the same range and distribution of mentality that we have today. Why this difference in speed between cultural and biological evolution?

By way of oversimplifying, although not too seriously, this contrast can be ascribed to differences between two distinct types of evolution: the mode of inheritance of acquired characters utilized by cultural evolution and the mode of inheritance through natural selection utilized by biological evolution. The **Lamarckian mode of cultural evolution** is an extension of the method by which humans learn. It depends upon conscious agents, that is, humans with brains, who are able to modify inherited cultural information in a direction that offers them greater adaptiveness or utility. Transmission occurs from mind to mind rather than through DNA. The information that humans receive from ancestors and contemporaries can be purposely changed to provide improved utility for themselves, their offspring, and others. Humans have thus introduced teleology (see p. 5) into evolution—changes for the sake of what they consider their own benefit. The speed with which such purposeful modification takes place and the consequent speed of cultural change are limited by numerous factors, but—theoretically, at least, and in the long run—primarily by human inventiveness. Furthermore, the generation time for cultural evolution may be as rapid as communication methods can make it. A cultural or technological improvement can now be proposed in one part of the world and implemented in another part with electronic speed.[1]

[1]Because cultural evolution relies so heavily on communication between individuals and group interaction, the whole easily becomes more than the sum of its parts, in the sense that creations by a socially coordinated group of individuals—a city, a daily newspaper, an automobile factory, a cathedral, or a movie film—are quantitatively and qualitatively more than can be created by such individuals acting alone.

In striking contrast to the speed of cultural evolution is the slow progress of natural selection. The reasons for this are apparent from previous discussions. As far as we know, there are no cellular particles that are sufficiently intelligent to detect or determine the direction of biological evolution and then change themselves accordingly. Organic evolution, as we have seen, can occur through a process of selection (among other forces) acting upon random genetic changes. According to this view, there are chance differences that arise in genes or combinations of genes (mutation and recombination) which produce a variety of effects on their carriers. These genetic differences furnish an array of genotypes among which the environment chooses for survival those that are reproductively most successful. Genetic evolution is slow since it must await fortuitous accidental genetic changes in DNA before it can proceed, and each change may take a considerable number of generations before it can be incorporated into the population.

This disparity in speed between cultural and biological evolution indicates that they evolve on separate methodological tracks, yet the biological equipment needed to transmit and utilize cultural information (memory, perception, language ability, etc.) still connects them both.

SOCIAL DARWINISM

The fact that human culture has at its source a biological foundation and that both culture and biology arise from informational systems that evolve over time has prompted various writers to suggest that there are general laws covering both society and nature, which share similar evolutionary mechanisms, especially that of natural selection.

During the nineteenth century, these ideas became embodied into concepts, later called **Social Darwinism**, that can briefly be described as follows:

1. Differences between human groups arose through natural selection.
2. Natural selection was the mechanism which led to social class structures and to national differences in respect to economic, military, and social power.

Slogans such as "struggle for existence" and "survival of the fittest," when extended to social traits, enabled various English Social Darwinists, especially Herbert Spencer, to suggest that social evolution had progressed inevitably towards increased social and moral perfection and was approaching its culmination in Victorian society.[2] The religion and social customs of Western Europe, especially England, could therefore be considered higher on the evolutionary scale than their counterparts elsewhere.

In its harsher forms, the Spencerian approach also became popular in various circles in the United States, especially through the teachings of William Graham Sumner, the best known of American Social Darwinists (Haller, Hofstadter). Sumner concluded that

> We cannot go outside of this alternative: liberty, inequality, survival of the fittest; not-liberty, equality, survival of the unfittest. The former carries society forward and favors all its best members; the latter carries society downwards and favors all its worst members.

As could be expected, many wealthy capitalists found such views to their liking. John D. Rockefeller, for example, who forged the gigantic Standard Oil trust through destruction of many smaller enterprises, justified his behavior by this argument:

> The growth of a large business is merely a survival of the fittest.... The American Beauty rose can be produced in the splendor and fragrance which bring cheer to its beholder only by sacrificing the early buds which grow up around it. This is not an evil tendency in business. It is merely the working-out of a law of nature and a law of God.

Sociologists such as Lester Ward reacted strongly against such blatant transposition of biological conduct into social conduct by pointing out that

> If we call biologic processes natural, we must call social processes artificial. The fundamental principle of biology is natural selection, that of sociology is artificial selection. The survival of the fittest is simply the survival of the strong, which implies and would better be called the destruction of the weak. If nature progresses through the destruction of the weak, man progresses through the protection of the weak.

Along with criticisms made by others such as T. H. Huxley in his *Evolution and Ethics*, we can see that the difficulties in accepting Social Darwinism stem from its assumption that society (economics, politics, etc.) operates through the same laws as biology and for the same goals. As has been pointed out repeatedly,

---

[2]Interestingly, although it was Spencer who invented the term "survival of the fittest," implying the action of selection, he remained a Lamarckian in respect to biological evolution until relatively late in life. Whatever its mechanisms, he conceived evolution as a powerful mystical force that governed all spheres of existence and therefore justified social and economic policies that supported those who were most "morally fit." For many Protestant intellectuals, it was Spencer's belief in such an evolutionary "cosmic" power that helped reconcile science to their religion and made his writings extremely popular.

this assumption is fallacious because there is no evidence that what *is* in biology, *is* or *ought to be* in society. For example, it seems obvious that the laws of inheritance of wealth and power in society are legal and man-made, whereas the laws of biological inheritance are not the result of human decision. It is also clear that, since they can be consciously selected, social goals can be directed towards almost any objective that humans choose for themselves, such as wealth, poverty, chastity, obedience, revolution, and so on. Biological goals, on the other hand, are restricted to those of organismic evolution and follow opportunistic paths without conscious or moral direction. That is, there are no moral or ethical qualities that determine survival of the biologically fittest: the "bad" and "ugly" parasite can be even more biologically fit than its host, a "good" and "beautiful" human.

Distinctions between society and biology are also reflected in the fact that the social rewards bestowed on individuals or groups may be unrelated to biological merit or even to presumed social merit. One can be mentally or physically incapacitated, or dissolute, immoral, and criminal, yet exercise considerable social and economic power. There are certainly no genes that guarantee social rewards—even dogs and cats have inherited wealth in our society. The negative correlation between fertility and those classes with greater economic resources has also often been noted. It is a common observation that professional success and improved social status in the striving middle class is often achieved—through delayed marriage—at the expense of fertility. As Jones points out, wealth reproduces itself rather than people.

In general, we can see that Social Darwinism has no valid scientific basis—the socially "fit" are not necessarily the biologically "fit." Sumner's statement that "the law of the survival of the fittest was not made by man and cannot be abrogated by man" is simply untrue on the social-cultural level. Nevertheless, in spite of these contradictions, we can also see that by presuming a "scientific" basis for social stratification, Social Darwinism has been an attractive ideology to many individuals and groups who occupy or would like to occupy "superior" social positions. It has, in fact, often been used to justify or reinforce racism, genocide, and social and national oppression, and was incorporated into the views of many writers and educators who proposed socially or racially biased directions for the genetic improvement of humans. An extreme example is that of Germany during the 1930's and 1940's where the "racial health" movement was an important ideological element in the purposeful destruction of millions of people because they were members of "inferior" racial groups. Even in the United States, with its more democratic social heritage, laws were passed during the 1920's restricting immigration from Eastern and Southern Europe because of their "inferior" or "undesirable" races. (Immigration of Asians had been halted many years earlier.)

SOCIOBIOLOGY

However, in spite of the failings of Social Darwinism, underlying biological influences on society can hardly be ignored. There are at least some, if not many, biologically induced motivations involved in friendship, sex, incest barriers, raising children, and attitudes towards strangers—behaviors that are common in all human groups whatever cultural forms they take. These and other observations have prompted modern sociobiologists such as Wilson and Alexander to argue quite logically that the capacity for culture among humans evolved from a non-cultural state by means similar to the evolution of other biological traits. Carried further, they suggest that natural selection, by favoring genetic predispositions for cultural behaviors of the kinds mentioned above, must have been a major force acting to increase the frequency of such traits. In the case of culture, the sociobiologists point out that selection is not restricted to the fitness of an individual carrier of a favorable cultural "gene," but, like altruism (pp. 312 and 410), fitness is also evaluated by its effect on the genetic relatives of individuals carrying such favorable genes ("kin selection" or "inclusive fitness").

Sociobiologists presume that their approach provides the rationale to gauge how (and perhaps to what extent) biological causes account for social behaviors. Wilson, whose book *Sociobiology* furnished a major stimulus for modern interest in this field, defines sociobiology as "the systematic study of the biological basis of social behavior" and states that, because of interactions between genes and environment, "there is no reason to regard most forms of human social behavior as qualitatively different from physiological and non-social psychological traits." It has therefore seemed to some sociobiologists and their critics that sociobiology leads to the concept that most observed human social behaviors are biologically caused—a concept very close to the views held by the Social Darwinists. Sahlins, for example, a critic of sociobiology, points out that

> Darwinism, at first appropriated to society as 'social Darwinism,' has returned to biology as a genetic capitalism.... Natural selection is ultimately transformed from the appropriation of natural resources to the expropriation of others' resources... [or] social exploitation.

The justice of such criticisms has been seriously questioned (e.g., Ruse) since many sociobiologists are certainly opposed to the claim that biological exploitation justifies social exploitation. Nevertheless, the confidence with which sociobiologists reduce humans and their social behavior to the genetic level (to Wilson there is a "morality of the gene"—"the organism [social or non-social] is only DNA's way of making more DNA"), has made it easy for some partisans to "scientifically" vindicate those forms of social domination that suit their political views.

From a historical point of view, it should be apparent that equating biological and social deeds is erroneous if we take into account the speed with which cultural changes occur and the resulting distance that has developed between human biology and culture. It seems obvious that many profound social and cultural changes, such as those involved in the transition from slavery to feudalism, or from feudalism to capitalism, or from "low" technology to "high" technology, are far too rapid to be caused by genetic changes in their human participants. Were there major differences in behavioral genes between Tudor Englishmen, Victorian Englishmen, and modern Englishmen?

Even those behaviors that seem to have a sustained adaptive evolutionary basis can be socially confounded or manipulated towards ends that seem far beyond their original reproductive goals. For example, the altruism of preserving one's genes through kin selection can be socially transformed into the altruism of being a "team player" in a corporate edifice in which none of the workers are biologically related to each other. Also, in some social circumstances, anger and rage can lead to behavior that is both self-destructive and offers no discernible biological benefit to related individuals.

Distinctions between social and biologically influenced behaviors can be supported by further observations. Assault and murder, for example, are often subject to social punishment, whatever their biological-behavioral motivations. Traffic lights are to be legally obeyed whether drivers feel impatient, impetuous, aggressive, or weary. Thus, motorists need not act altruistically in obeying traffic lights because of their altruistic genes, stockbrokers need not act selfishly in obeying specific stock exchange procedures because of their selfish genes, military personnel need not act aggressively in obeying military orders because of their aggressive genes. The importance of sociobiology to humans has therefore been subjected to considerable dispute as exemplified in collections of articles edited by Barlow and Silverberg, Caplan, Fetzer, Gregory et al., and Montagu.

It would seem that the most important lesson about humans to be learned from such disputes is the existence of a dual aspect to many forms of social behavior: behavioral patterns that were inherited biologically can be contained or modified within a cultural framework which must, at the same time, be accommodated to varying degrees within a biological framework. Both biology and culture are therefore tied to each other by many strands—some apparently quite strong, and others weak or imperceptible.

For example, the distorted behaviors caused by genetic factors involved in Down syndrome, Lesch-Nyhan syndrome, schizophrenia, manic-depressive psychosis, and various types of mental retardation indicate the importance of biological components in "normal" social behavior. Sexual behaviors in any society must also be influenced by biological factors that enable arousal and copulation. On the other hand, the biology of political affiliations such as Democrat, Republican, or Socialist is indiscernible, although the emotions that may be associated with such affiliations (loyalty, fear, aggression, altruism) have an obvious biological basis. When carried forward to activities that seem purely cultural, such as evaluating ideas in philosophy and science, the connecting strands between such evaluations and biology may be so thin and twisted as to be impossible to follow.[3]

The transformation from the evolution of biology to the evolution of culture therefore generally marks a qualitative change from one level to another. New rules and laws come into play in understanding culture that are not apparent in biology. This is obvious from previous illustrations that social laws may restrict or limit biologically motivated behavior, as well as the observation that society and culture can make use of human biology for purposes other than biological reproductive success. It is also clear that cultural differences can easily arise from the quirks of social history rather than from biological differences. For example, the particular language learned by an individual depends not upon his biology but upon the history of the society in which he lives. Similar importance of social and cultural history applies to individuals and groups in respect to their adoption of specific technologies, architectures, forms of artistic expression, and perhaps to even more biologically intimate matters such as modes of infant care and toilet training.

In general, this means that an understanding of social change is a quite different task from understanding biological change: societies and cultures cannot be simply explained as biological behaviors any more

[3]Wilson's oft-quoted statement (1978) that "genes hold culture on a leash" is therefore a gross oversimplification that distorts both biology and culture. As Kaye incisively demonstrates, such sociobiological views attribute to biology mystical goal-seeking properties, yet sociobiologists are forced to admit that many forms of culture, such as those leading to self-destructive wars, have escaped their "leash" and become mysteriously non-adaptive and abiological.

than biological behaviors can be understood as atomic interactions. The changes that have taken place in biological evolution do not provide a sufficient understanding of mechanisms, sequence, or ethics in cultural changes.[4]

BIOLOGICAL LIMITATIONS

The fact that cultural considerations can transcend biological considerations becomes apparent in dealing with the topic of human control over evolution. In which direction is evolution to be guided? What goals are to be set? These questions do not arise from unconscious biological laws but from the conscious cultural realization that we would like to improve ourselves and the world we live in and from the social technology that allows us to achieve such goals. This quest for human improvement comes mostly from the disparity between our cultural needs and our biological limitations. Let us consider a few of these.

One area of biological inadequacy stems from our advanced technology: we are becoming to a large degree sedentary in occupation, but our intestines and appetites do not adapt accordingly. Many who live in surplus societies, such as the United States, tend to put on weight and suffer from the accompanying ills. The pains and problems of childbirth are probably a consequence of our erect posture and can be aggravated by the lack of physical conditioning. The stress of many aspects of social living, ambition, and competition finds much of the human species biologically unprepared, and we suffer from anxiety, ulcers, heart disease, and other socially aggravated illnesses (called "the ulcer belt syndrome" by Comfort). Pollutions of various kinds caused by sewage, tobacco, automobiles, and industry lead to a variety of modern diseases ranging from induced cancer to emphysema and silicosis.

Perhaps one of the most important contrasts between what we would like to be and what we are lies in the difference between biological and cultural maturity. Biologically, our efficiency begins to fall soon after we reach the reproductive ages of 20 to 30 years. Our cultural efficiency, however, in the contributions we can make in various professions, often begins to increase during that period or somewhat later. Our cultural development is thus limited by our biological decline; that is, our biological heritage stresses reproductive success and hardens the arteries afterward, while our cultural development asks for continued plasticity and **longevity**.

Post-reproductive longevity, unfortunately, is a trait that tends to remain low in many organisms that reach reproductive maturity relatively early in their potential life span. Before civilization, only about half the human population passed the age of 20 and probably not more than 1 out of 10 lived beyond 40 years. These low longevity values extended into the period of the early Greeks and even into modern periods among primitive people. Life expectancy remained between 20 to 30 years until the Middle Ages, then rose somewhat and has risen sharply among Europeans and Americans in the last century, from about 40 years in 1850 to the present 75 years.

These statistics are important because they indicate that we now have among us an age group, those 40 to 50 years and older, upon whose biological attributes natural selection has never directly operated. In other words, the adaptive traits of such older individuals are those they possessed in the years prior to their reproductive periods, and their post-reproductive fitness is no longer reflected in their relative reproductive success.

For example, an individual who has produced three children and at the age of 50 develops cancer or other diseases with genetic components is by this fact no less reproductively successful than an individual of the same age who has produced three children but does not suffer from such diseases. One may, of course, argue that children with healthy grandparents are more fit than children with ill or absent grandparents since they get more attention and care. In social situations, however, such caretaking functions can be easily assigned to other individuals, and it is unlikely that grandparental attention adds to reproductive fitness.[5]

Healthy old age is therefore a trait that only relatively few genetic variants might be expected to attain. At present about 2 percent of the population reaches 90 years, and only about 1 per 1,000 individuals reaches 100. There is, however, little promise that, even with considerable medical progress, average life expectancy can be raised beyond 80 to 85 years.

A further contrast between past and present bio-

---

[4]In disparaging attempts to reduce all of biology to molecular terms, Wilson himself (1984) has pointed out that "molecular biology on its own is a helpless giant. It cannot specify the parameters of space, time, and history that are crucial to and define the higher levels of organization." This same argument can be applied to attempts to reduce an understanding of human culture to its biological components and shows the insufficiency of sociobiology when applied uncritically to humans.

[5]Muller and others have pointed out that senescence and death may have evolutionary value to populations, since such means help to ensure the turnover and replacement of older genotypes by new genotypes that may be better adapted.

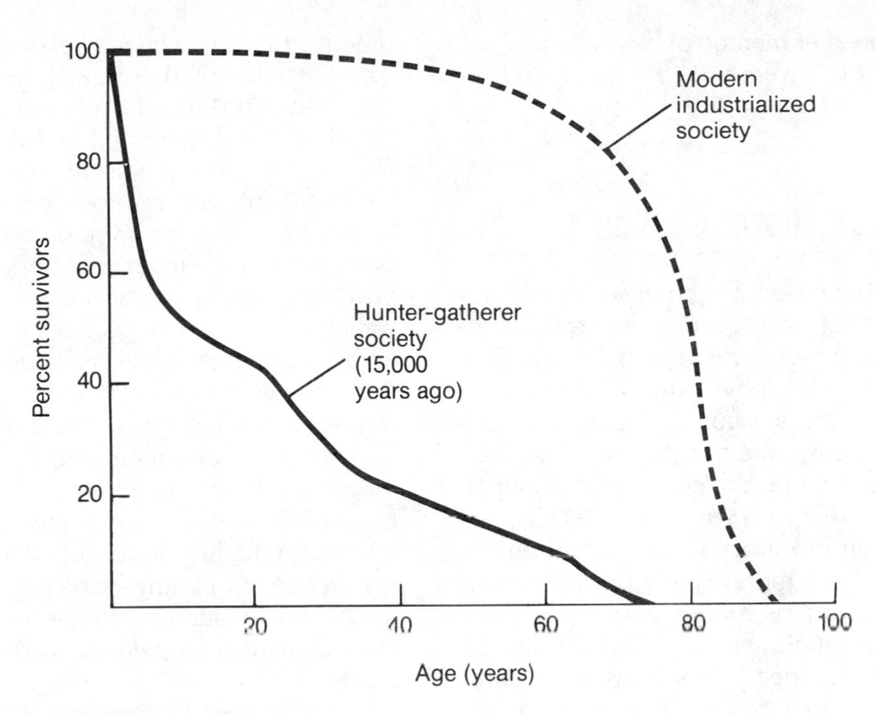

FIGURE 24-1 Survival curves for a population of hunter-gatherers who lived 15,000 years ago on the Mediterranean coast (based on skeletal remains) compared to a present day population living in an industrialized society. (Adapted from May.)

logical requirements is in fertility. **Clutch size**, a term used to describe the number of offspring born to a nesting pair of birds, is certainly as adaptive a factor in humans as in birds. That is, the ability of a mature human female under primitive conditions to produce eight or nine offspring during her reproductive period, of which an average of two survive, is of selective value in ensuring that her lineage will persist in the face of high infant mortality. However, in many modern countries infant mortality rates have markedly decreased (Fig. 24-1). In countries such as Chile, the rate of infant mortality is now about 10 percent of all births, or half of what it was 45 years ago; in highly industrialized countries such as the United States, infant mortality is about one quarter of what it is in Chile. This overall imbalance between human fecundity and survival rate has led to an exponential growth of the human population, which is now doubling at the rate of about once every 40 years, almost 7,000 times as fast as in primitive paleolithic societies!

This has led to a **population explosion**, which, if unchecked, will produce a world population of 50 billion people in only five generations. When we consider that agricultural production barely keeps up with its present requirements, the problem is indeed serious, especially if we add the widespread demand among poor people for an increased standard of living. Statistics by the United Nations show no per capita increase in food production since World War II. Many perennial famine areas still suffer from food shortages, and in some third world countries yield per acre is still low, and small agricultural increases have only come from putting more land into cultivation. The increase in agricultural acreage is, however, a limited solution because of the limited amount of usable land. In the United States today there are about 6 acres of agriculturally usable land per person to supply needs and to provide surpluses to be used by some of the agriculturally deficient countries. In England usable acreage falls to 1/10 this value and in Japan to 1/30. Considerable amounts of usable land are still available in some newly developing countries, such as in Africa, and some agricultural gains can be made by using new high-yield varieties and improved fertilizers, but even these increases will be inadequate if population growth continues. (Valuable rain forests in both Eastern and Western Hemispheres are, unfortunately, being destroyed to provide arable land for short-term purposes.)

Fortunately, a reduction in human fecundity can be instituted through presently known birth-control methods without the need for evolving such a reduction biologically. The most serious problem that lies ahead in this respect is that of educating people and institutions to utilize and promote such methods. The same approach toward other biological inadequacies, however, such as insufficient longevity and the "ulcer belt syndrome," cannot be undertaken as easily. For these traits, there is a need for some sort of controlled genetic change that would help put humans in biological harmony with their present or future cultural surroundings. As stated earlier, the question that arises is whether we can impose a direction upon our biological evolution using conscious cultural means to approximate the speedier method of inheritance of acquired characters. If this were possible, what kinds of genetic change would be desirable?

DELETERIOUS GENES

Let us briefly consider possible changes in the frequencies of genes that have obvious **deleterious effects**. The number of children classified as born markedly defective, either physically or mentally, is conservatively estimated at about 20 to 25 in 1,000 births, and the mortality rate ascribed to such congenital malformations in the United States is about 15 percent of all infant deaths. Many other defects, not immediately noted at birth, become apparent during childhood years and are more widely prevalent than may be imagined. In various studies, about 30 percent or more of hospital admissions for children and 50 percent of all childhood deaths are ascribed to birth defects or to complications that may have been caused by such defects. Although not all birth defects are genetically produced, the proportion of genetically handicapped among them is undoubtedly high. Table 24-1 lists estimates of the frequencies of some genetic disorders, which occur in approximately 3 percent of all births. If we include other genetic defects that appear later in life, such as muscular dystrophy and diabetes, the frequency is probably doubled. If in addition we include less obvious defects which nevertheless have strong genetic components, such as impaired resistance to stress and infection and other physical and psychological weaknesses, the effects of deleterious genes probably touch a majority of our population.

The ubiquity of deleterious genes with lethal effect has been dramatically demonstrated in studies made of the offspring of cousin marriages by Morton et al. and others. These studies have utilized techniques of detecting and partitioning the genetic load (p. 458) caused by inbreeding and have shown that outwardly normal individuals in our society carry a genetic load equivalent to that of approximately one to eight deleterious lethal genes **(lethal equivalents)** that, if homozygous, would cause early death.

Two important questions we can ask are, (1) what accounts for the prevalence of these deleterious genes, and (2) what, if anything, can we do to get rid of them?

The reasons for their high frequency are not yet fully agreed upon, although there is little question that they all arise originally through mutation. One opinion, held by the late Theodosius Dobzhansky and others, is that such genes, although deleterious in homozygous condition, may offer considerable advantage to their heterozygous carriers by producing some sort of hybrid vigor. According to this theory, a gene will be maintained in the population although the homozygote produced by this gene is relatively inferior in fitness (pp. 438-440). Another school, formerly headed by the late Hermann Muller, believes that such genes produce no advantage of any kind and that their frequency is now high because the usual effect of natural selection has been artificially reduced. According to Muller, genotypes that were formerly defective and would have been eliminated under more primitive conditions are now kept alive by medical techniques and enabled to pass on their defective genes to their offspring. As we know, a decrease in the selection coefficient against a particular gene causes an increase in the equilibrium frequency of the gene ($q = \sqrt{u/s}$ for a recessive gene, $p = u/s$ for a dominant gene; p. 444). Thus if deleterious genes are not eliminated by selection they will gradually increase in frequency in accord with their mutation rate. Since the mutation rate is usually low, the frequency of any particular gene will increase rather slowly, but since there are many possible deleterious genes, the genetic load will increase significantly.

It is obvious that according to Muller's theory, we will be unable to reach any biological harmony until most such genes are removed. If they are not removed and continue to increase in frequency, Muller held out the prospect that the human species would end up with two types of individuals: one kind would be so genetically crippled that they could hardly move, and the other kind would be less crippled but spend all their time taking care of the first kind. This specter is made even gloomier if we attempt to consider means by which such genes can be eliminated. Since all of us are probably carriers of at least a few deleterious recessive genes, most of which we do not know about, there is little prospect in eliminating them short of mass sterilization.

Serious as this argument may be for deleterious genes that produce severe handicaps, it is undoubtedly exaggerated for genes whose deleterious effects can be treated relatively inexpensively. Nearsightedness, for example, is a trait whose frequency has most likely increased in recent periods, but can be corrected quite simply by an optometrist. Furthermore, the fact that natural selection no longer operates to eliminate many genotypes is not necessarily an undesirable feature of modern life. Few individuals would argue today that fire and clothes should be abolished because they are artificial devices that circumvent natural selection by permitting non-furry genotypes to survive in cold climates. It would also be difficult for us to return to the "good old" pre-vaccination, pre-sanitation days of smallpox, diphtheria, typhus, cholera, and plague.

However, in spite of medical and cultural progress, the effect of numerous deleterious genes cannot be easily treated, and although Muller may have been overly alarmed about their increase in frequency, we have become more aware of their widespread exis-

TABLE 24-1 Estimated Frequencies of Some Human Genetic Disorders

Disorder	Percent
SINGLE GENE DISORDERS	
Autosomal recessive	
Mental retardation, severe	0.08
Cystic fibrosis	0.05
Deafness, severe	0.05
Blindness, severe	0.02
Adrenogenital syndrome	0.01
Albinism	0.01
Phenylketonuria	0.01
Other aminoacidurias	0.01
Mucopolysaccharidoses	0.005
Tay-Sachs disease	0.001
Galactosemia	0.0005
	0.25
X-linked	
Duchenne muscular dystrophy	0.02
Hemophilias A and B	0.01
Others	0.02
	0.05
Autosomal dominant	
Blindness	0.01
Deafness	0.01
Marfan syndrome	0.005
Achondroplasia	0.005
Neurofibromatosis	0.005
Myotonic dystrophy	0.005
Tuberous sclerosis	0.005
All others	0.015
	0.06
Total single gene disorders	0.36
CHROMOSOMAL DISORDERS	
Autosomal	
Trisomy 21	0.13
Trisomy 18	0.03
Trisomy 13	0.02
Other	0.02
	0.20
Sex chromosome	
XO and X deletions	0.02
Other severe defects	0.01
XXY	0.1
XXX	0.1
XYY	0.1
Others	0.015
	0.35
Total chromosomal disorders	0.55
MULTIFACTORIAL DISORDERS	
Congenital malformations	
Spina bifida and anencephaly	0.45
Congenital heart defects	0.4
Pyloric stenosis	0.3
Clubfoot	0.3
Cleft lip and palate	0.1
Dislocated hips	0.1
	1.65
Total multifactorial disorders	>1.65
TOTAL FREQUENCY OF LISTED GENETIC DISORDERS	>2.56

From Epstein and Golbus.

tence in recent years. Many geneticists have, therefore, turned to exploring the possibility of controlling deleterious effects by artificially changing gene frequencies.

EUGENICS

In its modern form, suggestions for improving human genetic material have come under the name **eugenics**, a term proposed by Francis Galton before the turn of the century.[6] Galton, concerned with the heredity of quantitative characters such as intelligence, became aware, after reading Darwin, that the evolution of human traits through natural selection could be substituted by their evolution through social selection. ("What nature does blindly, slowly, and ruthlessly, man may do providently, quickly, and kindly.") However, like the Social Darwinists, many early eugenicists reflected their own personal and racial biases as to which characteristics were desirable and which undesirable. C. B. Davenport, for example, a leader of the eugenics movement in the United States, exemplified this approach by using New Englanders as the standard of comparison for all American nationalities. According to Davenport and other members of his Eugenics Rec-

[6]One of the first proposals suggesting that humans could be improved through selective breeding was made by Plato in his dialogue *The Republic*. In his ideal philosopher-state only the most physically and mentally fit individuals were to be mated and their offspring raised by the state. Inferior types were to be prevented from mating or their offspring destroyed. Since family relations were absent in *The Republic*, determination of superior and inferior types could be accomplished impartially, and the governing class was selected only "from the most superior." However, Plato's notion of superiority and inferiority was quite different from that of his contemporaries, who considered a conquering people superior and a subjugated people inferior. In ancient Sparta, for example, some measure of selective breeding seems to have been practiced with the purpose of raising a "superior" military ruling class to hold in subjugation the "inferior" servant classes. The fact that cultural disparities usually existed between conquerors and subjects reinforced such notions but did not seriously reflect whether any essential biological differences were responsible for these cultural differences. Were the Spartans biologically "superior" to the Helots, Corinthians, and Athenians?

ord Office, most social characteristics had identifiable genetic components typical of particular groups (e.g., Italian violence, Jewish mercantilism, Irish pauperism, etc.). These attitudes can be said to have reached their culmination in the "racial health" movement in Germany during the 1930's when eugenic laws were promulgated establishing German Aryans as the "master race" and forbidding intermarriages with presumed racially inferior non-Aryans. Although the particular social and political causes that fostered these attitudes and their horrible consequences are not the subject of this book (see Müller-Hill), it is important to recognize that myths about the degeneration of race and intelligence because of mixture with "inferior" types are continually being perpetuated.

In the United States, one of the true racial melting pots of the world, there is no evidence for the biological superiority of any particular race in respect to intelligence. Blacks, long considered to be at the bottom of the racial pecking order, show as wide a range of intelligence as do whites. According to Pettigrew and others, racial differences in intelligence (IQ) examinations have been demonstrated to be remarkably plastic, influenced by such factors as prenatal diet, early cultural surroundings, and even the color of the interviewer in the IQ examination. Predictions that intelligence will steadily decline because of the higher reproductive rate of the lower, "unintelligent" social classes and their consequent increase in frequency are also contradicted by facts. In a Scottish survey that covered almost 90 percent of all 11-year-old children in 1932 and again in 1947, no decrease in IQ was found. On the contrary, these studies showed a significant increase in average intelligence during this interval.

The fears expressed by various writers that the abolition of privileged classes in society will lead to "hybridization" and thus to the loss or dilution of superior genotypes are therefore hardly scientifically based. The evidence at present is that high intelligence is not the exclusive genetic property of a particular social class, but rather that its expression can easily be masked in any group by deficiencies in diet, lack of cultural stimulation, and absence of opportunity. As environmental conditions improve, average intelligence scores may also be expected to improve, although genetic differences between individuals will remain. We may, in fact, predict that equality of economic and educational opportunity for all classes will enable each individual to more nearly achieve his or her true potential. Society will be the benefactor in producing more creative and inspired individuals such as Leonardo da Vinci, Voltaire, and Newton, who may otherwise die anonymously among the dispossessed sections of our society. As Dobzhansky has stated, there is little to lament in "the passing of social organizations that used the many as a manured soil in which to grow a few graceful flowers of refined culture."

The Kinds of Eugenics

Stripped of racism and provincial prejudice, eugenics may be considered a serious attempt to diminish human suffering and improve the human gene pool. It has been subdivided into two aspects: (1) **negative eugenics**, the attempt to decrease the frequency of harmful genes; and (2) **positive eugenics**, the attempt to increase the frequency of beneficial genes. Negative eugenics involves social discouragement of reproduction of genotypes that are most obviously deleterious. For example, it would be foolish and self-destructive to encourage hemophiliacs, who are being preserved by blood transfusions, to reproduce. Similarly, where known, female carriers of the hemophilia gene should be made aware of their genetic problem and encouraged not to pass it on. These eugenic programs will suffice to control suffering from a number of deleterious genes, although they will not eliminate them, and many of these educational measures are already in practice today. Many deleterious genes that are present in high frequency, however, such as diabetes, and others where the carriers are not easily known, such as cystic fibrosis, cannot be controlled in this way because of the very inefficient elimination of recessives under selection (pp. 428 and 435-436).

It might, therefore, seem somewhat more encouraging to place emphasis on positive eugenics—increasing the frequency of beneficial traits rather than merely decreasing the frequency of deleterious genes. Unfortunately many characteristics we consider desirable, such as high intelligence, esthetic sensitivity, good physical health, and longevity, are not caused by single genes that are easily identified but by complexes of many genes acting together in appropriate environments.

In other organisms in which the development of beneficial gene complexes has been attempted, the methods involve complicated schemes based on selection of parents and families along with testing of progeny under controlled environmental conditions. As discussed by Lerner and others, the results of these experiments have, in general, improved certain complex characters by some degree but have usually caused the deterioration of others. One characteristic that usually suffers most in such experiments is that overall quality called "fitness"; many highly selected lines end up physically debilitated and sterile. Muller, Crow, and others, however, have pointed to the likelihood that traits such as high intelligence and esthetic sensitivity have not been stringently selected for in the past, and considerable genetic variability for these traits probably exists. Thus, were selection to be instituted for these traits, the population might well respond rapidly without an accompanying fall in fitness.

The means of selection themselves, however, assume paramount importance in humans. Eugenic measures dictating who is to mate with whom would be intolerable, even presuming that controls on human activity can be implemented to evaluate selection progress.

As a first approach toward a more acceptable method of positive eugenics than selective mating, Muller and others have proposed the utilization of sperm banks containing the preserved frozen sperm of outstanding creative individuals. According to this method, called **germinal choice**, or **eutelegenesis**, women volunteers would choose to be artificially inseminated by males that were long dead but had highly desirable characteristics. Possible acceptance of this method has some precedence in the fact that between 5,000 and 10,000 babies fathered by sperm donors are born annually in the United States. The cause for most of these donor fertilizations lies in the sterility of the husband, although in some cases genetic incompatibilities between husband and wife (e.g., Rh factor) or genetic defects in the husband (e.g., hemophilia) are responsible. Muller proposed to extend these donor fertilizations by educating couples to desire a highly superior genetic endowment for their children and demonstrating the increased proportion of genetically gifted children that will presumably be produced by this method.

Other proposed eugenic methods involve direct manipulation of human DNA by **genetic surgery** or **genetic manipulation**, techniques that are now being used to modify the DNA of bacteria, viruses, and higher organisms. The application of these techniques to humans holds considerable future promise for directly changing human genetic material. For example, various experiments now demonstrate that human gene sequences can be inserted into retroviruses, which can then transfer these genes to mammalian cells (Friedmann). This technology may eventually advance to the point where a large repertoire of genes isolated or produced in the laboratory can be incorporated into reproductive tissues and thus change the genetic constitution of entire lineages.

Another possible eugenic method is **parthenogenesis**—to induce females bearing desirable genetic constitutions to lay diploid eggs which do not have to be fertilized. Such eggs would more truly reflect the constitutions of their mothers than fertilized eggs and thereby permit the replication of desirable maternal genotypes. Other such proposals include the transplantation of diploid nuclei from desirable genotypes into unfertilized eggs and the conversion of somatic tissues into embryonic tissues. By these methods **clones** of individuals could be created, all bearing the same genotype. However, even if such genetic uniformity were desirable, the techniques needed to implement these proposals remain to be perfected.

THE FUTURE

All one can say at present is that perhaps some form of eutelegenesis or genetic manipulation can be developed which would be acceptable and productive. It would seem that the need for improvement of the human gene pool will become more desirable the more we become aware of our biological limitations and the more our technology allows us to successfully perform genetic changes. Like other creatures, humans evolve, but unlike other creatures, humans know they evolve. The control of biological evolution lies, therefore, in changing from reproductive success caused by natural selection and other forces to reproductive success caused by human choice. Daring as it sounds, it is no more daring than the method by which many cultural advances have been and will be made.

Perhaps the most pertinent question we can ask of eugenics is its goal. Even if we assign to eugenics the most moral of motives—the good of humankind—it still remains to be determined whether this "good" is known. Can we choose the direction of human evolution with the certainty that this direction leads to what is best for our descendants? Shall we populate the world with the weak or the strong? With the sensitive or the insensitive? Fortunately, the answer to such questions will probably not be limited to an unequivocal choice of one type or the other. There will probably be the opportunity, then as now, to choose many different genotypes, among which factors such as intelligence and longevity will undoubtedly rank highly in value.

If there is a serious obstacle to future human genetic improvement, it probably lies in our **provincialisms**: our parochial social, racial, religious, economic, and political prejudices that help make us into the most dangerous and destructive beasts that terrestrial evolution has ever known. Since provincialism seems to be a common feature of every human group, its basis seems built into our biology, probably derived from the millions of years during which small bands of hominids fought and protected themselves against others. In the complex but fragile and threatening political world we now live in, these aggressive and destructive behaviors are anachronistic and no longer appropriate. That is, the sentiments of group patriotism, chauvinism, and superiority, which were so useful in the past in providing motivation for social defense and aggressive acquisition in interactions between small local groups, have now been technologically enhanced to permit military destruction on a national and international scale. The dividing line between group homicide and suicide is being rapidly obliterated.

Certainly one need for continued human survival is to become consciously aware of these underlying behaviors and purposely control them by social and cultural means (we do have traffic lights!) so they can be replaced with behaviors necessary and appropriate for the kind of society in which we would like to live—for example, to consciously devise mechanisms that increase respect for other humans as well as for our environment on an international scale. Our intellects and inventiveness, unique among all biological creatures, can certainly provide us with the means to accomplish such goals.

From this point of view, the most important present human need is to accept differences between individuals and groups without indulging in provincial, exploitative, and chauvinistic practices. If we can fulfill this cultural need, it is likely that we can progress to one great prosperous, humane, and rational world. If we cannot fulfill this need, the prospect now appears to be one of mutual destruction and a fall into unimaginable suffering and barbarism.

In short, because of our intellects and technologies, humans are graduating from being merely subjects of evolution to being co-authors of evolution. An understanding of the materials and mechanisms of change on the biological and cultural levels can provide us with the freedom to control evolution on these levels. We thus have in hand our own destiny as well as that of many other creatures on earth.

SUMMARY

Unlike most other animals, humans transfer information in the form of culture as well as hereditary information from one generation to another. The tempo of cultural evolution has been so much more rapid than human biological evolution that each person gathers new experience at a rate many times that of his or her ancestors. While cultural evolution is Lamarckian in that what is acquired can be transmitted and directed, biological evolution is dependent upon randomly occurring genetic variation.

In the nineteenth century proponents of Social Darwinism believed that cultural differences evolved primarily by natural selection, as embodied in the concept of "survival of the fittest." This belief justified many social inequities and was based on the erroneous supposition that society, which often incorporates nonbiological goals and value systems, is governed by the same laws as biological evolution.

A more sophisticated approach is that of sociobiology, the advocates of which believe that there is a biological basis for much of human culture and that natural selection favors genotypes that are predisposed toward cultural development. However, although there is a biological component in many social patterns, they are often modified by their cultural context. Cultural change may occur without biological input and cannot be explained by biological laws.

Humans desire to control all aspects of their evolution but are restricted by the incompatibility between cultural and biological fitness. They wish to lengthen life span, but natural selection cannot act on postreproductive individuals. They would like to lower fertility, but biologically high fertility has had selective value, although socially it has led to population and ecological crises. The question remains whether or not we can direct biological evolution by cultural means and, if so, what goals would be desirable.

A large proportion of people are or will be affected by deleterious genes that persist because of forces that maintain genetic variablility or because of the perpetuation by medicine of the lives of individuals suffering from genetic disorders. Eugenics represents an effort to improve the human gene pool either by reducing the frequency of such genes or by augmenting the frequency of favorable ones. Because the carriers of harmful alleles often cannot be detected, they cannot be eliminated by negative eugenics. Selecting for beneficial traits is also difficult since they are frequently multigenic, and fitness may decline in selected lines. A variety of methods, such as parthenogenesis, genetic surgery and manipulation, or artificial insemination with sperm from extaordinary individuals might be attempted in the future to improve the gene pool.

In any case, improvement of the human gene pool, intelligent direction of human evolution, and survival of the human lineage depends upon elimination of anachronistic provincial, exploitative, and chauvinistic barriers and practices.

Key Terms

biological limitations
cloning
clutch size
culture
deleterious effects
eugenics
eutelegenesis
extragenetic system
genetic manipulation
genetic surgery
genetic system
germinal choice
horizontal transmission
Lamarckian mode of cultural evolution
learning
lethal equivalents
longevity
negative eugenics
parthenogenesis
population explosion
positive eugenics
provincialism
Social Darwinism
sociobiology
vertical transmission

References

Alexander, R. D., 1979. *Darwinism and Human Affairs*. Univ. of Washington Press, Seattle.

____, 1987. *The Biology of Moral Systems*. Aldine De Gruyter, New York.

Bajema, C. J. (ed.), 1976. *Eugenics; Then and Now*. Dowden, Hutchinson & Ross, Stroudsburg, Pa.

Barlow, G. W., and J. Silverberg (eds.), 1980. *Sociobiology: Beyond Nature/Nurture?* Westview Press, Boulder.

Boyd, R., and P. J. Richerson, 1985. *Culture and the Evolutionary Process*. Univ. of Chicago Press, Chicago.

Caplan, A. L. (ed.), 1978. *The Sociobiology Debate*. Harper & Row, New York.

Comfort, A., 1963. Longevity of man and his tissues. In *Man and His Future*, G. Wolstenholme (ed.). Little Brown, Boston, pp. 217-229.

Crow, J. F., 1961. Mechanisms and trends in human evolution. *Daedalus*, **90**, 416-431.

Davenport, C. B., 1911. *Heredity in Relation to Eugenics*. Henry Holt & Co., New York.

Dobzhansky, Th., 1962. *Mankind Evolving*. Yale Univ. Press, New Haven, Conn.

Epstein, C. J., and M. S. Golbus, 1977. Prenatal diagnosis of genetic diseases. *Amer. Sci.*, **65**, 703-711.

Fetzer, J. H. (ed.), 1985. *Sociobiology and Epistemology*. Reidel, Dordrecht, Netherlands.

Friedmann, T., 1989. Progress toward human gene therapy. *Science,* **244,** 1275-1281.

Glover, J., 1984. *What Sort of People Should There Be?* Penguin Books, Harmondsworth, Middlesex, Great Britain.

Gregory, M. S., A. Silvers, and D. Sutch (eds.), 1978. *Sociobiology and Human Nature*. Jossey-Bass, San Francisco.

Haller, M. H., 1963. *Eugenics: Hereditarian Attitudes in American Thought*. Rutgers Univ. Press, New Brunswick, N. J.

Hofstadter, R., 1955. *Social Darwinism in American Thought*. Beacon Press, Boston.

Holzmüller, W., 1984. *Information in Biological Systems: The Role of Macromolecules*. Cambridge Univ. Press, Cambridge.

Jones, G., 1980. *Social Darwinism and English Thought*. Harvester Press, Brighton, Sussex, Great Britain.

Kaye, H. L., 1986. *The Social Meaning of Modern Biology*. Yale Univ. Press, New Haven, Conn.

Kevles, D. J., 1985. *In the Name of Eugenics: Genetics and the Uses of Human Heredity*. A. A. Knopf, New York.

Kitcher, P., 1985. *Vaulting Ambition: Sociobiology and the Quest for Human Nature*. M.I.T. Press, Cambridge, Mass.

Lerner, I. M., 1958. *The Genetic Basis of Selection*. John Wiley, New York.

May, R. M., 1983. Parasitic infections as regulators of animal populations. *Amer. Sci.*, **71**, 36-45.

Montagu, A. (ed.), 1980. *Sociobiology Examined*. Oxford Univ. Press, Oxford.

Morton, N. E., J. F. Crow, and H. J. Muller, 1956. An estimate of the mutational damage in man from data on consanguineous marriages. *Proc. Nat. Acad. Sci.*, **42**, 855-863.

Muller, H. J., 1963. Genetic progress by voluntarily conducted germinal choice. In *Man and His Future*, G. Wolstenholme (ed.). Little Brown, Boston, pp. 247-262.

Müller-Hill, B., 1988. *Murderous Science: Elimination by Scientific Selection of Jews, Gypsies, and others, Germany 1933-1945*. Oxford Univ. Press, Oxford.

Pettigrew, T. F., 1971. Race, mental illness and intelligence: a social psychological view. In *The Biological and Social Meaning of Race*, R. H. Osborne (ed.). W. H. Freeman, San Francisco, pp. 87-124.

Price, D. J. da S., 1963. *Little Science, Big Science*. Columbia Univ. Press, New York.

Ruse, M., 1979. *Sociobiology: Sense or Nonsense*? Reidel, Dordrecht, Netherlands.

Sahlins, M., 1977. *The Use and Abuse of Biology*. Univ. of Michigan Press, Ann Arbor.

Waddington, C. H., 1978. *The Man-Made Future*. Croom Helm, London.

Ward, L., 1893. *The Psychic Factors of Civilization*. Ginn & Co., Boston.

Wilson, E. O., 1975. *Sociobiology: The New Synthesis*. Harvard Univ. Press, Cambridge, Mass.

——, 1977. Biology and the social sciences. *Daedalus*, **106**, (4) 127-140.

——, 1978. *On Human Nature*. Harvard Univ. Press, Cambridge, Mass.

——, 1984. *Biophilia*. Harvard Univ. Press, Cambridge, Mass.

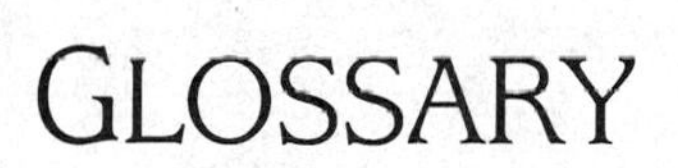
GLOSSARY

GLOSSARY

Abduction Movement of an appendage or body structure in a direction away from the midline (median sagittal) plane (e.g., extending an arm laterally).

Abiotic Substances that are of non-biological origin or environments characterized by the absence of biological organisms.

Acidic A compound that produces an excess of hydrogen ions (H^+) when dissolved in water. Using quantitative hydrogen ion measurements, such solutions have a pH value less than 7.0. (*See* **pH scale**.)

Acoelomate An animal that lacks a coelom (internal body cavity).

Acrocentric Chromosomes whose centromeres are near one end (between metacentric and telocentric locations).

Active sites Specific regions of an enzyme that bind substrates upon which the enzyme acts.

Active transport biochemical transport that requires the input of energy (e.g., hydrolysis of ATP).

Adaptation A character that has been modified and is or was maintained as a result of selection for increased fitness. A term also used for the process that produces adaptations.

Adaptive landscape A model originally devised by Sewall Wright that describes a topography in which high fitnesses correspond to peaks and low fitnesses to valleys; each position occupied by a population bearing a unique genotype.

Adaptive radiation The diversification of a single species or group of related species into new ecological or geographical zones to produce a large variety of species and groups.

Adaptive value The relative reproductive success (relative fitness) of an allele or genotype as compared to other alleles or genotypes. (*See also* **Fitness**.)

Adduction Movement of an appendage or body part towards the midline (median sagittal) plane, e.g., bringing a laterally extended arm to the side of the body.

Adenosine triphosphate (ATP) An organic compound, commonly involved in the transfer of phosphate bond energy, composed of adenosine (an adenine base + a D-ribose sugar) and three phosphate groups.

Aerobic The utilization of molecular oxygen for reactions that provide growth energy from the oxidative breakdown of food molecules.

Aerobic respiration An electron transport system in which oxygen serves as the terminal electron acceptor.

Algae Photosynthetic members of the eukaryotic kingdom of Protista.

Allele One of the alternative forms of a single gene (i.e., a particular nucleotide sequence occurring at a given locus on a chromosome).

Allen's rule The generalization that warm-blooded animals (mammals) tend to have shorter extremities (e.g., ears and tail) in colder climates than they have in warmer climates.

Allometry Differential growth rates of different body parts; during growth (increase in size) one feature may change at a rate different from that of another feature, resulting in a change of shape.

Allopatric Species or populations whose geographical distributions do not contact each other.

Allopatric speciation Speciation between populations that are geographically separated.

Allopolyploid An organism or species that has more than two sets (2n) of chromosomes (i.e., 3n, 4n, etc.) which derive from two or more different ancestral groups.

Allozyme The particular form (amino acid sequence) of an enzyme produced by a particular allele at a gene locus when there are different possible forms of the enzyme (different possible amino acid sequences), each produced by a different allele.

Alternation of generations Life cycles in which a multicellular haploid stage (1n) alternates with a multicellular diploid stage (2n).

Altruism Behavior which benefits other individuals at some cost to the altruist.

Amino acids Organic molecules of the general formula $R—CH(NH_2)COOH$, possessing both basic (NH_2) and acidic (COOH) groups, as well as a side group (R) specific for each type of amino acid. There are normally 20 different types of amino acids used in cellularly synthesized proteins.

Amino group An $-NH_2$ group.

Amniotic egg The type of egg produced by reptiles, birds, and mammals (Amniota), in which the embryo is enveloped in a series of membranes (amnion, allantois, chorion) that help sustain its development.

Anaerobic Growth (energy obtained from the oxidative breakdown of food molecules) in the absence of molecular oxygen.

Anaerobic respiration An electron transport system in which substances other than oxygen serve as the terminal electron acceptor (e.g., sulfates, nitrates, methane).

Anagenesis The evolution of new species that takes place progressively over time within a single lineage (as opposed to cladogenesis). (*See also* **Phyletic evolution**.)

Analogy The possession of a similar character by two or more quite different species or groups which arises from a developmental pathway unique to each group, i.e., the similarity is caused by factors other than their distant common genetic ancestry. (*See also* **Convergence**.)

Aneuploidy The gain or loss of chromosomes leading to a number that is not an exact multiple of the basic haploid chromosome set (n) (e.g., n + 1, 2n + 1, 2n - 1, 2n - 2, 2n + 3, etc).

Angiosperms The flowering plants, an advanced group of vascular plants with floral reproductive structures and encapsulated seeds.

Antibody A protein produced by the immune system that binds to a substance (antigen) typically foreign to the organism.

Anticodon A sequence of three nucleotides (a triplet) on transfer RNA that is complementary to the codon on messenger RNA that specifies placement of a particular amino acid in a polypeptide during translation.

Antigen A substance, typically foreign to an organism, that initiates antibody formation and is bound by the activated antibody.

Apomixis (Apomictic) Reproduction without fertilization; for example, parthenogenesis (offspring produced from unfertilized eggs in which meiosis has been partially or completely suppressed).

Apomorphy A character that has been derived from, yet differs from, the ancestral condition. (*See also* **Synapomorphy**.)

Aposematic Conspicuous warning coloration in potential prey species that advertises their toxicity or distastefulness to predators. Aposematic patterns usually contain bright colors or shades such as those found among bees, wasps, monarch butterflies, coral snakes, skunks, and poisonous salamanders.

Arboreal Living predominantly in trees.

Archaebacteria Prokaryotes that, unlike eubacteria, do not incorporate muramic acid into their cell walls and possess other distinguishing characteristics. They are considered to represent one of the early cell forms.

Archetype The concept of an ideal primitive plan ("bauplan") upon which organisms, such as vertebrates, are presumably based. Called by Richard Owen the "primal pattern" and "divine idea."

Artificial selection Selection process in which humans are the selective agents. (*See* **Selection**.)

Asexual reproduction Offspring produced by a single parent in the absence of sexual fertilization or in the absence of gamete formation.

Assortative mating Mating among individuals on the basis of their phenotypic or genotypic similarities (*positive assortative*) or differences (*negative assortative*) rather than mating among all individuals on a random basis.

Autocatalytic reaction A reaction in which the agent that promotes (catalyzes) it is formed as a product of the reaction.

Autopolyploid A species or organism that has more than two sets of chromosomes (polyploid) derived from one or more duplications in a single ancestral source.

Autosome A chromosome whose presence or absence is ordinarily not associated with determining the difference in sex (i.e., a chromosome other than a sex chromosome).

Autotroph An organism capable of synthesizing complex organic compounds needed for growth from simple inorganic environmental substrates: *photoautotroph*, an organism that can utilize light as an energy source and carbon dioxide as a carbon source; *chemoautotroph* (*chemolithotroph*), an organism that obtains energy for growth by oxidizing inorganic compounds such as hydrogen sulfide.

Bacteriophage A virus (phage) that parasitizes bacteria.

Balanced genetic load The decrease in overall fitness of a population caused by defective genotypes (e.g., homozygotes for deleterious recessives) whose alleles persist in the population because they confer selective advantages in other genotypic combinations (e.g., heterozygote advantage).

Balanced polymorphism The persistence of two or more different genetic forms through selection (e.g., heterozygote advantage) rather than because of mutation or other evolutionary forces.

Banded iron formation An iron-containing laminated sedimentary rock, often composed of layers of tiny quartz crystals (chert).

Basalt A fine-grained igneous rock found in oceanic crust and produced in lava flows.

Base (nucleotide) The nitrogenous component of the nucleotide unit in nucleic acids, consisting of either a purine (adenine [A] or guanine [G]) or a pyrimidine (thymine [T] or cytosine [C] in DNA, uracil [U] or cytosine [C] in RNA). (*See also* **Purine**, **Pyrimidine**.)

Base pairs *See* **Complementary base pairs**.

Basic (alkaline) A compound that produces an excess of hydroxyl (OH^-) ions when dissolved in water. Using quantitative hydrogen ion measurements, such solutions have a pH value greater than 7.0. (*See* **pH scale**.)

Batesian mimicry The similarity in appearance of a harmless species (the mimic) to a species which is harmful or distasteful to predators (the model).

Benthic Refers to the floor of a body of water (e.g., ocean bottom, river bed, lake bottom) and to organisms that live in it, upon it, or near it.

Bergmann's rule The generalization that animals living in colder climates tend to be larger than those of the same group living in warmer climates.

Big Bang theory The concept that the universe was born in a gigantic explosion about 15 to 20 billion years ago.

Bilateral symmetry Instances in which the left and right sides of a longitudinal (sagittal) plane that runs through an organism's midline are approximately mirror images of each other.

Binary fission Replication of an organism by its division into two mostly equal parts; the common form of asexual reproduction in prokaryotes and protistan eukaryotes.

Binomial expansion Expansion of the binomial ($a + b$) to a value n [$(a + b)^n$] where a and b represent alternative states whose sum equals the probability of 1.

Binomial nomenclature The Linnaean principle of

designating a species by two names: the name of a genus followed by the name of a species.

Biogenetic law (Haeckel) The concept that stages in the development of an individual (ontogeny) recapitulate the evolutionary history (phylogeny) of the species.

Biogeography The study of the geographical distributions of organisms. A biogeographical realm is a region characterized by a distinctive biota.

Biological species concept The view that the primary criterion for separating one species from another is their reproductive isolation.

Biota All animals (fauna) and plants (flora) of a given region or time period.

Biotic Relating to or produced by biological organisms.

Bipedal A term used mostly to describe terrestrial tetrapod locomotion that is restricted to the hind limbs when these two limbs move alternately (e.g., human walking) rather than together (e.g., kangaroo jumping).

Blastocoel The cavity of a blastula.

Blastopore The opening formed by the invagination of cells in the embryonic gastrula, connecting its cavity (archenteron) to the outside. In protostome phyla the blastopore is the site of the future mouth, whereas in deuterostomes the blastopore becomes the anus and the mouth is formed elsewhere.

Blastula A hollow sphere enclosed within a single layer of cells, occurring at an early stage of development in various multicellular animals.

Blending inheritance The abandoned concept that offspring inherit a dilution, or blend of parental traits, rather than the particles (genes) which determine those traits.

Bottleneck effect A form of genetic drift that occurs when a population is reduced in size (population crash) and later expands in numbers (population flush). The enlarged population that results may have gene frequencies that are distinctly different from those before the bottleneck. (*See also* **Founder effect**.)

Brachiation Locomotion through trees by hanging from branches and swinging alternate arms (left, right, left,...) from branch to branch, accompanied by a rotation of the body during each swing.

Brackish Water whose salt content (salinity) is intermediate between freshwater and sea water; usually at the mouths of rivers that empty into the ocean (estuaries).

Bradytelic A relatively slow evolutionary rate.

Bryophyte Mosses and liverworts, small "primitive" land plants.

Buccal Pertaining to the inside of the mouth; side of a tooth closest to the cheek.

Burrowing animal In aquatic forms, a bottom-dweller that moves through soft benthic sediments.

Calorie The amount of heat necessary to raise the temperature of 1 gm of water by 1° Centigrade at a pressure of 1 atmosphere.

Calvin cycle A cyclic series of light-independent reactions that accompany photosynthesis in which carbon dioxide is reduced to carbohydrate.

Cambrian period The interval between about 590 and 505 million years before the present, marking the plentiful appearance of fossilized organisms with hardened skeletons. It is considered the beginning of the Phanerozoic time scale (eon) and is the first period in the Paleozoic era.

Carbohydrate A compound in which the hydrogen and oxygen atoms bonded to carbons are commonly in a ratio of 2:1 (e.g., glucose ($C_6H_{12}O_6$), starch $(C_6H_{12}O_6)_n$, cellulose, $(C_6H_{10}O_5)_n$).

Carbonaceous Possessing organic (carbon) compounds. (*Carbonaceous chondrite*: a meteorite containing carbon compounds.)

Carboxylic acid An organic compound which has an acidic group consisting of a carbon with a double-bond attachment to an oxygen atom and a single-bond attachment to a hydroxyl group (O = C — OH).

Carnivores Flesh eaters; organisms (almost entirely animal, rarely plant) that feed on animals.

Carrying capacity The theoretical maximum number of organisms in a population, usually designated by *K*, that can be sustained in a given environment.

Catalyst A substance that lowers the energy necessary to activate a reaction but is not itself consumed or altered in the reaction.

Catastrophism The eighteenth and nineteenth century concept that fossilized organisms and changes in geological strata were produced by periodic, violent, and widespread catastrophic events (presumably caused by capricious supernatural forces) rather than by naturally explainable events based on laws that act uniformly through time. (*See also* **Uniformitarianism**.)

Cell wall The rigid or semi-rigid extracellular envelope (outside the plasma membrane) that gives shape to plant, algal, fungal, and bacterial cells.

Cenozoic era The period from 65 million years ago to the present, marked by the absence of dinosaurs and the radiation of mammals. This is the third and most recent era of the Phanerozoic eon and is divided into two major periods, the Tertiary and Quaternary.

Centigrade scale (°C) A scale of temperature in which the melting point of ice is taken as 0° and the boiling point of water as 100°, measured at 1 atmosphere of pressure.

Centrifugal selection *See* **Disruptive selection**.

Centripetal selection *See* **Stabilizing selection**.

Centromere The chromosome region in eukaryotes to which spindle fibers attach during cell division.

Character A feature or trait of an organism.

Character displacement Divergence in the appearance or measurement of a character between two species when their distributions overlap in the same geographical zone, compared to the similarity of the character in the two species when they are geographically separated. When common resources are limited, it is presumed that competition between overlapping species leads to divergent specializations and therefore to divergence in characters that were formerly similar.

Cheek teeth Mammalian premolar and molar teeth.

Chemoautotroph (chemolithotroph) *See* **Autotroph**.

Chert A sedimentary rock composed largely of tiny quartz crystals (SiO_2) precipitated from aqueous solutions.

Chloroplast A chlorophyll-containing, membrane-bound organelle which is the site of photosynthesis in the cells of plants and some protistans.

Chromatid One of the two sister products of a eukaryotic chromosome replication, marked by an attachment between the sister chromatids at the centromere region. When this attachment is broken during the mitotic anaphase stage, each of the sister chromatids becomes an independent chromosome.

Chromosome A length of nucleic acid comprising a linear sequence of genes that is unconnected to other chromosomes. In eukaryotes, histone proteins are bound to nuclear chromosomes, and this protein-nucleic acid complex can be made microscopically visible as deeply staining filaments.

Chromosome aberration A change in the gene sequence of a chromosome caused by deletion, duplication, inversion, or translocation.

Citric acid cycle *See* **Krebs cycle**.

Clade A cluster of taxa derived from a single common ancestor.

Cladistics A mode of classification based principally on grouping taxa by their shared possession of similar ("derived") characters that differ from the ancestral condition.

Cladogenesis "Branching" evolution involving the splitting and divergence of a lineage into two or more lineages.

Cladogram A tree diagram representing phylogenetic relationships among taxa.

Class A taxonomic rank that stands between phylum and order; a phylum may include one or more classes, and a class may include one or more orders.

Classification The grouping of organisms into a hierarchy of categories commonly ranging from species to genera, families, orders, classes, phyla, and kingdoms. In practice, the decision as to the species in which to place an organism, or the genus in which to place a species, and so forth, is most often based on phenotypic similarity to other members of the group: organisms in a species are more similar to each other than they are to organisms in other species of the same genus, species in a genus are more similar to each other than they are to species in other genera of the same family, and so forth.

Cline A gradient of phenotypic or genotypic change in a population or species correlated with the direction or orientation of some environmental feature, such as a river, mountain range, north-south transect, altitude, and so forth.

Clone A group of organisms all derived by asexual reproduction from a single ancestral individual.

Cloning (gene) Techniques for producing identical copies of a gene by inserting a DNA sequence into a cell, such as a bacterium, where it can be replicated.

Coacervate An aggregation of colloidal particles in liquid phase that persists for a period of time as suspended membranous droplets.

Coadaptation The action of selection in producing adaptive combinations of alleles at two or more different gene loci.

Coarse-grained environment A heterogeneous environment in which individuals in a population are exposed to conditions different from other individuals.

Codominance The phenotypic expression of the individual effects of two different alleles in a heterozygote (e.g., genotypes carrying both *M* and *N* alleles of the *MN* blood group show the MN blood type).

Codon The triplet of adjacent nucleotides in messenger RNA that codes for a specific amino acid carried by a specific transfer RNA or that codes for termination of translation (STOP codons). Placement of the amino acid is based on complementary pairing beween the anticodon on transfer RNA and the codon on messenger RNA.

Coelom An internal body cavity lined in eucoelomates (true coelomates) with mesodermal tissue that may contain organs such as testes and ovaries.

Coenzymes Non-protein enzyme-associated organic molecules (e.g., NAD, FAD, coenzyme A), that participate in enzymatic reactions by acting as intermediate carriers of electrons, atoms, or groups of atoms.

Coevolution Evolutionary changes in one or more species in response to changes in other species in the same community.

Cofactor A small molecule, which may be organic (i.e., coenzyme) or inorganic (i.e., metal ion), required by an enzyme in order to function.

Cohort Individuals of a population that are all the same age.

Commensalism An association between organisms of different species in which one species is benefited by the relationship but the other species is not significantly affected.

Competition Relationship between species occupying the same habitat in which each species inhibits to varying degrees the population growth of the other.

Complementary base pairs Nucleotides on one strand of a nucleic acid that hydrogen bond with nucleotides on another strand according to the rule that pairing between purine and pyrimidine bases is restricted to certain combinations: A pairs with T in DNA, A pairs with U in RNA, and G pairs with C in both DNA and RNA.

Concerted evolution The process by which a series of nucleotide sequences or different members of a gene family remain similar or identical through time.

Condensation (by dehydration) The formation of a covalent bond between two molecules by removal of H_2O.

Condylarths A mammalian order that became extinct during the Miocene period of the Cenozoic era but whose first occurrences are in the late Cretaceous period of the Mesozoic. It includes the earliest herbivorous placental mammals and the ancestors of all later herbivores.

Continental drift The movement, over time, of large land masses on the earth's surface relative to each other. (*See also* **Paleomagnetism**, **Tectonic plates**.)

Continuous variation Character variations, such as height in humans, whose distribution follows a series of small non-discrete quantitative steps from one extreme to the other. (*See also* **Quantitative character**.)

Convergence (also called *Homoplasy*) The evolution of similar characters in genetically unrelated species, mostly because they have been subjected to similar environmental selective pressures. (*See also* **Analogy**.)

Cope's rule The generalization that body size tends to increase in an animal lineage during its evolution.

Correlation The degree to which two measured characters tend to vary in the same quantitative direction (positive correlation) or in opposite directions (negative correlation).

Cosmology Study of the structure and evolution of the universe.

Covalent bond A strong chemical bond that results from the sharing of electrons between two atoms.

Creationism The belief that each different kind of organism was individually created by one or more supernatural beings whose activities are not controlled by known physical, chemical, or biological laws.

Creodonts An extinct order of early Cenozoic placental mammals that were the dominant carnivores until replaced by the modern order Carnivora during the Oligocene period.

Crepuscular A life-style characterized by activity mostly during the hours around dawn and dusk.

Crossovers (chromosome) Results of a process in which genetic material is exchanged between the chromatids of two homologous chromosomes. (*See also* **Recombination**.)

Cryptic A feature that is normally not visible.

Culture (social) The learned behaviors and practices common to a social group.

Cursorial Adapted for running on land.

Cusps (teeth) Elevations on the crowns of premolars and molars. The number, shapes, and positions of cusps are inherited characters that can provide useful phylogenetic information.

Cyanobacteria Photosynthetic prokaryotes possessing chlorophyll *a* but not chlorophyll *b*. Many are photosynthetic aerobes (oxygen producing) and some are anaerobic (non-oxygen producing). Formerly called blue-green algae.

Cytochromes Proteins containing iron-porphyrin (heme) complexes that function as hydrogen or electron carriers in respiration and photosynthesis.

Cytology The study of cells—their structures, functions, components, and life histories.

Cytoplasm All cellular material within the plasma membrane, excluding the nucleus.

Darwinism The concept, proposed by Charles Darwin, that biological evolution has led to the many different highly adapted species through natural selection acting on hereditary variations in populations.

Deciduous (teeth) Teeth that are replaced during development by permanent teeth.

Deficiency *See* **Deletion**.

Degenerate code The type of genetic code used by existing terrestrial organisms, for which there is more than one triplet codon for a particular amino acid but a specific codon cannot code for more than one amino acid. Thus, the 20 different amino acids translated in protein synthesis are coded by more than 20 codons (61) of the 64 possible different triplet codons, some by as many as six different "synonymous" codons. (*See also* **Genetic code**.)

Dehydrogenase An enzyme that catalyzes the removal of hydrogen from a molecule (oxidation).

Deleterious allele An allele whose effect reduces the adaptive value of its carrier when present in homozygous condition (recessive allele) or in heterozygous condition (dominant or partially dominant allele).

Deletion An aberration in which a section of DNA or chromosome has been lost.

Deme A local population of a species (in sexual forms, a local interbreeding group).

Density dependent The dependence of population growth and size on factors directly related to the numbers of individuals in a particular locality (e.g., competition for food, accumulation of waste products, etc.).

Density independent The dependence of population growth on factors (climatic changes, meteorite impacts, etc.) unrelated to the numbers of individuals in a particular locality.

Derived character A character whose structure or form differs (apomorphic) from that of the ancestral stock.

Deuterium An isotope of hydrogen containing one proton and one neutron, giving it twice the mass of an ordinary hydrogen atom.

Deuterostomes Coelomate phyla in which the embryonic blastopore becomes the anus.

Development The sequence of progressive changes in an organism from initial cell to maturity.

Dicotyledons Flowering plants (angiosperms) in which the embryo bears two seed leaves (cotyledons).

Differentiation Changes that occur in the structure and function of cells and tissues as the development of the organism proceeds.

Dimorphism Presence in a population or species of two morphologically distinctive types of individuals (e.g., differences between males and females, pigmented and non-pigmented forms, etc.).

Diploblastic An animal that produces only two major types of cell layers during development, ectoderm and endoderm (e.g., Cnidaria).

Diploid An organism whose somatic cell nuclei possess two sets of chromosomes (2n), thus providing two different (heterozygous) or similar (homozygous) alleles for each gene.

Directional selection Selection that causes the phenotype of a character to shift towards one of its phenotypic extremes.

Discontinuous variation Character variations that

are sufficiently different from each other that they fall into non-overlapping classes.

Disruptive selection Selection that tends to favor the survival of organisms in a population that are at opposite phenotype extremes for a particular character and eliminates individuals with intermediate values (centrifugal selection).

Diurnal A life-style characterized by activity during the day rather than at night (nocturnal).

DNA (deoxyribonucleic acid) A nucleic acid, characterized by the presence of a deoxyribose sugar in each nucleotide, that serves as the genetic material of all cells and many viruses.

Dominance (allele) Instances in which the phenotypic effect of a particular allele (e.g., *A*, the dominant) is expressed in heterozygotes (e.g., *Aa*), but the phenotypic effect of the other allele (e.g., *a*, the recessive) is not expressed.

Dominance (social) Relations within a group in which one or more individuals, sustained by aggression or other behaviors, rank higher than others in controlling the conduct of group members.

Doppler effect The shift in wavelength of light or sound that is perceived as the emitting body moves towards us (shorter wavelengths, e.g., blue-shifted) or away from us (longer wavelengths, e.g., red-shifted).

Dorsal The back side or upper surface of an animal; opposite of ventral. (In vertebrates, the surface closest to the spinal column.)

Double fertilization A distinctive feature of angiosperm plants in which two nuclei from a male pollen tube fertilize the female gametophyte, one producing a diploid embryo and the other producing polyploid (usually triploid) nutritional endosperm.

Duplication Instances in which a particular section of DNA or visible chromosome segment occurs more than once.

Ecogeographical rules Generalizations which correlate adaptational tendencies of species with environmental factors such as climate. (*See* **Allen's rule**, **Bergmann's rule**, **Gloger's rule**.)

Ecological niche The environmental habitat of a population or species, including the resources it utilizes and its association with other organisms.

Ecology The study of the relations between organisms and their environment, in terms of their numbers, distributions, and life cycles.

Ecotype A phenotypic and genotypic variant of a species associated with a particular environmental habitat. (*See also* **Race**.)

Ectoderm The outermost layer of cells that covers the early animal embryo, from which nerve tissues and outermost epidermal tissues are derived.

Ectothermic "Cold-blooded": a body temperature primarily determined by the ambient (environmental) temperature.

Ediacarian strata Geological formations containing soft-bodied invertebrate fossils found in South Australia and other places, dating to a Precambrian period lasting about 60 or more million years.

Electron carrier In oxidation-reduction reactions, a molecule that acts alternatively as an electron donor (becomes oxidized) and as an electron acceptor (becomes reduced).

Electrophoresis A technique that separates dissolved particles subjected to an electrical field according to their mobility. Electrophoretic mobility depends on the size of the particle, its geometry, and electrical charge.

Endemic A species or population that is specific (indigenous) to a particular geographic region.

Endocytosis Cellular engulfment of outside material, followed by its transfer into the cellular interior encapsulated in a membrane.

Endoderm The layer of cells that lines the primitive gut (archenteron) during the early stages of development in animals and later forms the epithelial lining of the intestinal tract and internal organs such as the liver, lung, and urinary bladder.

Endosymbiosis A relationship between two different organisms in which one (the endosymbiont) lives within the tissues or cell of the other, benefitting one or both. It is now generally thought that some eukaryotic organelles, such as mitochondria and chloroplasts, had an endosymbiotic prokaryotic origin.

Endothermic "Warm-blooded": a body temperature maintained by internal physiological mechanisms at a level independent of the ambient (environmental) temperature.

Entropy The measure of disorder of a physical system. In a closed system, to which energy is not added, the second law of thermodynamics essentially states that entropy, or energy unavailable for work, will remain constant or increase but never decrease. Living systems, however, are open systems, to which energy is added from sunlight and other sources, and order can therefore arise from disorder in such systems, i.e., energy available for work can increase and entropy can decrease.

Environment The complex of external conditions, abiotic and biotic, that affects organisms or populations.

Enzyme A protein that catalyzes chemical reactions.

Eon A major division of the geological time scale, often divided into two eons beginning from the origin of the earth 4.5 billion years ago: the Precambrian or Cryptozoic (rarity of life forms) and the Phanerozoic (abundance of life forms).

Epigamic selection Selection for mating success based on appearance or behavior during courtship.

Epigenesis The concept that tissues and organs are formed by interaction between cells and substances that appear during development, rather than being initially present in the zygote (preformed). (*See also* **Preformationism**.)

Epistasis Interactions between two or more gene loci which produce phenotypes different from those expected if each locus were considered individually.

Epoch One of the categories into which geological time is divided; a subdivision of a geological period.

Equilibrium (genetic) The persistence of the same allelic frequencies over a series of generations. Equilibria may be stable or unstable. In a stable equilibrium (e.g., when the heterozygote is superior in fitness to the homozygotes) the population returns to a particular equilibrium value when the allelic frequencies have been disturbed. In an unstable equilibrium (e.g., when the heterozygote is inferior in fitness to the homozygotes) such disturbances are not followed by a return to equilibrium frequencies.

Era A division of geological time that stands between the eon and the period: the Phanerozoic eon is divided into Paleozoic, Mesozoic, and Cenozoic eras; and each era is divided into two or more periods.

Estrus The period during which female mammals exhibit maximum sexual receptivity, usually coinciding with the release of eggs from the ovary.

Eubacteria Prokaryotes, other than archaebacteria, marked by sensitivity to particular antibiotics and by the incorporation of muramic acid into their cell walls.

Eucoelomates *See* **Coelom**.

Eugenics The concept that humanity can be improved by altering human genotypes or their frequencies.

Eukaryotes Organisms whose cells contain nuclear membranes, mitochondrial organelles, and other characteristics that distinguish them from prokaryotes. Eukaryotes may be unicellular or multicellular and include protistans, fungi, plants, and animals.

Euploidy Variations which involve changes in the number of entire chromosome sets (n) (e.g., 3n, 4n, 5n, etc.).

Eutelegenesis The use of artificial insemination to improve genetic endowment.

Eutely Constancy in the numbers of cells or nuclei from the larval stage to the adult stage.

Eutheria *See* **Placentals**.

Evolution Genetic changes in populations of organisms through time that lead to differences among them.

Evolutionary (molecular) clock The concept that the rate at which mutational changes accumulate is constant over time. To which genes or genomes this clock may apply, and whether it is really constant, are disputed.

Exon A nucleotide sequence in a gene that is transcribed into messenger RNA and spliced together with the transcribed sequences of other exons from the same gene. The continuous RNA molecule formed is then transferred to the ribosome and forms the template used in polypeptide synthesis. Exons (expressed sequences) are separated from other exons in the same gene by intervening non-translated sequences (*see* **Introns**) that are removed from the messenger RNA. Such intron-exon split genes are commonly found in eukaryotes but are almost entirely absent in prokaryotes.

Extant Currently in existence.

Extension Movement of an appendage so that the angle of the joint increases.

Extinction The disappearance of a species or higher taxon.

F (inbreeding coefficient) *See* **Inbreeding coefficient**.

Family A taxonomic category that stands between order and genus; an order may comprise a number of families, each of which contains a number of genera.

Fauna All animals of a particular region or time period.

Fecundity A measure of potential fertility, often calculated in terms of the quantity of gametes (generally eggs) produced while sexually mature.

Fermentation The anaerobic degradation of glucose (glycolysis) or related molecules, yielding energy and organic end products.

Fertility A trait measured by the number of viable offspring produced.

Filter feeder An animal that obtains its food by filtering suspended food particles from water.

Fine-grained environment A heterogeneous environment whose varied conditions can normally be experienced by a single individual during its lifetime.

Fitness Relative reproductive success: the ability of an organism (genotype) to transmit its genes to the next reproductively fertile generation, relative to this ability in other genotypes.

Fixation Achievement of a frequency of 100 percent (i.e., monomorphism) by an allele or genotype which begins in a population at a lesser frequency (i.e., polymorphism).

Fixity of species A concept held by Linnaeus and others that members of a species could only produce progeny like themselves, and therefore each species was fixed in its particular form(s) at the time of its creation.

Flexion Movement of an appendage so that the angle of the joint decreases.

Flora All plants of a particular region or time period.

Fossils The geological remains, impressions, or traces of organisms that existed in the past.

Founder effect The effect caused by a sampling accident in which only a few "founders" derived from a large population begin a new colony. Since these founders carry only a small fraction of the parental population's genetic variability, radically different gene frequencies can become established in the new colony. (*See also* **Bottleneck effect**.)

Frequency-dependent selection Instances where the effect of selection on a phenotype or genotype depends on its frequency (e.g., a genotype that is rare may have a higher adaptive value than it has when it is common).

Frozen accident The concept that an accidental event in the distant past was responsible for the presence of a universal feature in living organisms. Such events may include an accident in which the present genetic code was used by a group of early organisms that managed to survive some populational bottleneck, thereby conferring this particular code on later organisms.

Fundamentalism (religious) The belief that creation stories and the many events and rules given in religious documents (e.g., the Judaeo-Christian Bible, the Moslem Koran) are to be taken literally.

Galaxies A system of numerous stars such as the Milky Way (150 billion stars, 100,000 light years across) held together by mutual gravitational effects. Galaxies, in turn, are grouped into clusters and superclusters. Our own supercluster, centered on Virgo, contains millions of galaxies and is more than 100 million light years across.

Gamete A germ cell (usually haploid) that fuses with a germ cell of the opposite sex to form a zygote (usually diploid) in a process called fertilization.

Gametophyte The haploid gamete-producing stage of plants that have alternating generations (haploid gametophyte and diploid sporophyte). The gametophyte is produced by meiosis in the sporophyte, and its gametes are produced by mitosis.

Gamma ray A high-frequency, highly penetrating radiation emitted in nuclear reactions.

Gastrula A cuplike embryonic stage in multicellular animals that follows the blastula stage. Its hollow cavity (archenteron) is lined with endoderm and opens to the outside through a blastopore. (*See also* **Haeckel's gastrula hypothesis**.)

Gene A unit of genetic material, composed of a sequence of nucleotides, that provides a specific function to an organism: either coding for a polypeptide chain (cistron), or for a sequence of ribonucleotides used as ribosomal RNA or transfer RNA, or regulating processes associated with other genes (e.g., transcription, replication, etc.). The position a gene occupies on a chromosome is called a *locus*, and each different nucleotide sequence of a gene is called an *allele*.

Gene family Two or more gene loci in an organism whose similarities in nucleotide sequences indicate they have been derived by duplication from a common ancestral gene (e.g., the β-globin gene family, which includes β, γ, δ, and ϵ genes).

Gene flow The migration of genes into a population from other populations by interbreeding.

Gene frequency The proportion of a particular allele among all alleles at a gene locus. (Also called *allele* or *allelic frequency*.)

Gene locus The chromosomal position (nucleotide sequence) occupied by a particular gene.

Gene pair The two alleles present in a diploid organism at a specific gene locus on two homologous chromosomes.

Gene pool All of the genes present in a population during a given generation or period.

Genetic code The sequences of nucleotide triplets (codons) on messenger RNA which specify each of the different kinds of amino acids positioned on polypeptides during the translation process. With few exceptions, the genetic code used by all organisms is identical: the 20 amino acids are each specified by the same codons (total: 61 codons), and the same three triplet codons are used to terminate polypeptide synthesis. (*See also* **Degenerate code**, **Universal code**.)

Genetic crossing over *See* **Crossovers**, **Recombination**.

Genetic death The inability of a genotype to reproduce itself because of selection.

Genetic distance A measure of the divergence between populations based on their differences in frequencies of given alleles.

Genetic drift *See* **Random genetic drift**.

Genetic load The loss in average fitness of individuals in a population because the population carries deleterious alleles or genotypes.

Genetic polymorphism The presence of two or more alleles at a gene locus over a succession of generations. (Called *balanced polymorphism* when the persistence of the different alleles cannot be accounted for by mutation alone.)

Genome The complete genetic constitution of a cell or an individual.

Genotype The genetic constitution of cells or individuals, often referring to alleles of one or more specified genes.

Genus A taxonomic category that stands between family and species: a family may comprise a number of genera, each of which contains a number of species that are presumably related to each other by descent from a common ancestor. In taxonomic binomial nomenclature, the genus is used as the first of two words in naming a species: *Homo* (genus) *sapiens* (species).

Geographic isolation The separation between populations caused by geographic distance or geographic barriers.

Geographic speciation *See* **Allopatric speciation**.

Geological strata A series of layers of sedimentary rock.

Geological time scale The correlation between rocks (or the fossils contained in them) and time periods of the past.

Germ plasm Cells or tissues in a multicellular organism that are exclusively devoted to transmitting hereditary information to offspring, either asexually or by means of gametes (sex cells).

Germinal choice *See* **Eutelegenesis**.

Gloger's rule The generalization that warm-blooded (endothermic) animals tend to have more pigmentation in warm, humid areas than in cool, dry areas.

Glycolysis The energy-producing conversion of glucose to pyruvate under anaerobic conditions (fermentation). Subsequent steps may yield lactic acid or ethanol.

Gondwanaland The supercontinent in the Southern Hemisphere formed from the breakup of the larger Pangaea land mass about 180 million years ago. Gondwanaland was composed of what is now South America, Africa, Antarctica, Australia, and India.

Grade A level of phenotypic organization or adaptation reached by one or more species. Distantly related or unrelated species that reach the same grade are considered to have undergone parallel or convergent evolution.

Granite A coarse-grained igneous rock commonly intruded into continental crust.

Great Chain of Being The eighteenth century concept that instead of a static universe, there is a continuous progression of stages leading to a superior supernatural being; the transformation of the "Ladder of Nature" into a succession of moving platforms.

Grooming Body surface cleaning by use of mouth, fingers, or claws.

Group selection Selection acting on the attributes of a group of related individuals in competition with other groups rather than only on the attributes of an individual in competition with other individuals (e.g., altruism may not be beneficial to the individual altruist but can be quite beneficial to a group containing altruists).

Gymnosperms A group of vascular plants with seeds unenclosed in an ovary (naked); mainly cone-bearing trees.

Habitat The place and conditions in which an organism normally lives. (*See also* **Environment**.)

Haeckel's gastrula hypothesis The concept that metazoans developed from swimming hollow-balled colonies of flagellated protozoans which evolved an anterior-posterior orientation in searching for food. The anterior cells, specialized for digestion, invaginated through a circular blastopore to form a digestive archenteron, and this bilayered cup, called a gastrula or gastraea, was, according to Haeckel, the progenitor of the gastrula developmental stage found in some present-day metazoans.

Half-life (radioactivity) The time required for the decay of one half the original amount of a radioactive isotope: a period of one half-life reduces the isotope amount by one half so that a length of two half-lives leaves a remainder of one quarter, three half-lives, a remainder of one eighth, and so on. Each radioactive isotope has a distinctive half-life period which remains constant over time.

Haplodiploidy A reproductive system found in some animals, such as bees and wasps, in which males develop from unfertilized eggs and are therefore haploid, while females develop from fertilized eggs and are therefore diploid.

Haploid Cells or organisms that have only one set (1n) of chromosomes, meaning the presence of only a single allele for each gene.

Hardy-Weinberg principle The conservation of gene (allelic) and genotype frequencies in large populations under conditions of random mating and in the absence of evolutionary forces, such as selection, migration, and genetic drift, which act to change gene frequencies,.

Hemizygous Genes, such as those on the X chromosome in a male mammal, which are unpaired in a diploid cell.

Herbivores Animals that feed on plants.

Heritability In a general sense, the degree to which variations in the phenotype of a character are caused by genetic differences; traits with high heritabilities can be more easily modified by selection than traits with low heritabilities. (A measure of the heritability of a trait is the ratio of its genetic variance to its phenotypic variance.)

Hermaphrodite An individual possessing both male and female sexual reproductive systems.

Heterocercal Fish tail in which the vertebral axis is curved (usually upwards).

Heterodont An organism with structural and functional differences among its teeth.

Heterogametic The sex that produces two kinds of gametes for sex determination in offspring, one kind for males and the other for females. The heterogametic sex is the male in mammals and the female in birds. (*See also* **Sex chromosomes**.)

Heterosis (hybrid vigor) The increase in vigor and performance that can result when two different, often inbred strains are crossed. Since each inbred parental strain may be homozygous for different deleterious recessive alleles (e.g., $A^1A^1 \times A^2A^2$), the cause for heterosis has been ascribed by some authors to the superiority of heterozygotes (e.g., A^1A^2). (*See* **Heterozygote advantage**).

Heterotroph An organism that cannot use inorganic materials to synthesize the organic compounds needed for growth but obtains them by feeding on other organisms or their products, such as a carnivore, herbivore, parasite, scavenger, or saprophyte.

Heterozygote A genotype or individual that possesses different alleles at a particular gene locus on homologous chromosomes (e.g., *Aa* in a diploid).

Heterozygote advantage (superiority) The superior fitness of some heterozygotes (e.g., A^1A^2) relative to homozygotes (e.g., A^1A^1, A^2A^2). (*See also* **Heterosis**, **Overdominance**.)

Histones A family of small acid-soluble (basic) proteins that are tightly bound to eukaryotic nuclear DNA molecules and help fold DNA into thick chromosome filaments.

Homeotic mutations Regulatory mutations that cause the development of tissue in an inappropriate position: for example, the *bithorax* mutations in *Drosophila* that produce an extra set of wings (see Fig. 10-25).

Hominid A member of the family Hominidae, which includes humans, whose earliest fossils can now be dated to about 4 million years ago (genus *Australopithecus*). Only a single hominid species (*Homo sapiens*) presently exists.

Hominoids A group (superfamily Hominoidea) that

includes hominids (Hominidae), gibbons (Hylobatidae), and apes (Pongidae).

Homogametic The sex that produces only one kind of gamete for sex determination in offspring, thus causing sex differences among offspring to depend on the kind of gamete contributed by the heterogametic sex. The homogametic sex is the female in mammals and the male in birds. (*See also* **Sex chromosomes**.)

Homologous chromosomes Chromosomes that pair during meiosis, each pair usually possessing a similar sequence of genes.

Homology The presence of a similar phenotypic or genotypic feature in different species or groups because of their descent from a common ancestor.

Homozygote A genotype or individual that possesses the same alleles at a particular gene locus on homologous chromosomes (e.g., *AA* or *aa* in a diploid).

Horotelic Evolving at a comparatively average rate.

Hybrids (hybridization) Offspring of a cross between genetically different parents or groups.

Hybrid breakdown, inviability, sterility Hybrids that suffer from loss of fitness and reproductive failure.

Hybrid vigor *See* **Heterosis**.

Hydrogen bond A weak, non-covalent bond between a hydrogen atom and an electronegative atom such as oxygen.

Hydrogen ion A proton (H^+) which in aqueous solution exists only in hydrated form (H_3O^+, hydronium ion).

Hydrolysis Splitting of a molecule by the addition of the three atoms from a water molecule (H_2O).

Hydrophilic A compound (e.g., charged molecule) or part of a compound (e.g., polar group) that has an affinity for water molecules.

Hydrophobic Compounds such as lipids that do not readily interact with water but tend to dissolve in organic solvents.

Hydrostatic pressure The pressure exerted by a liquid. When the liquid is in an elastic, muscularly controlled container (e.g., the coelom of a worm), changes in shape of the container can be effected by muscularly generated hydrostatic pressure.

Idealism The philosophy that the universe is constituted of non-material ideas.

Igneous rock A rock such as basalt (fine grained) and granite (coarse grained), formed by the cooling of molten material from the earth's interior.

Implantation (mammals) The attachment of the embryo to the uterine wall.

Inbreeding Mating between genetically related individuals, often resulting in increased homozygosity in their offspring.

Inbreeding coefficient (F) The probability that the two alleles of a gene in a diploid organism are identical because they originated from a single allele in a common ancestor.

Inbreeding depression Decrease in the average value of a character, or in growth, vigor, fertility, and survival, as a result of inbreeding.

Inclusive fitness The fitness of an allele or genotype measured not only by its effect on an individual but also by its effect on related individuals that also possess it (kin selection).

Independent assortment A basic principle of mendelian genetics--that a gamete will contain a random assortment of alleles from different chromosomes because chromosome pairs orient randomly towards opposite poles during meiosis.

Industrial melanism The effect of soot and pollution in industrial areas in increasing the frequency of darkly pigmented (melanic) forms because of selection by predators against non-pigmented or lightly pigmented forms.

Inheritance of acquired characters The concept used by Lamarck to explain evolutionary adaptations—that phenotypic characters acquired by interaction with the environment during the lifetime of an individual are transmitted to its offspring.

Insectivore An animal that feeds primarily on insects.

Instinct An inherited (innate), relatively inflexible behavior pattern that is often activated by one or several environmental factors (releasers).

Intrinsic rate of natural increase The potential rate at which a population can increase in an environment free of limiting factors.

Introgressive hybridization The incorporation of genes from one species into the gene pool of another because some fertile hybrids are produced from crosses between the two species.

Intron A nucleotide sequence in a split gene that intervenes between two exons. The intron sequences are removed from messenger RNA, and only the exon sequences are translated into polypeptides.

Intrusion Igneous rock that is inserted within or between geological strata rather than on the earth's surface.

Inversion An aberration in which a section of DNA or chromosome has been inverted 180° so that the sequence of nucleotides or genes within the inversion is now reversed with respect to its original order in the DNA or chromosome.

Ion An atom or molecule carrying a positive or negative electrostatic charge.

Isolating mechanisms Biological mechanisms that act as barriers to gene exchange between populations. These are generally divided into two groups: premating isolating mechanisms which inhibit cross-fertilization (e.g., behavioral differences in courtship) and postmating isolating mechanisms which interfere with the success of the gamete or zygote even when cross-fertilization has occurred (e.g., hybrid inviability).

Isomerase An enzyme that catalyzes the rearrangement of atoms within a molecule.

Isotope One of several forms of an element, with a distinctive mass based on the number of neutrons in the atomic nucleus. (The number of protons and electrons is the same in different isotopes of an element.) Radioactive isotopes decay at a rate which is constant for each isotope and release ionizing radiation as they decay. (*See* **Half-life**, **Radioactive dating**).

***K* value (carrying capacity)** *See* **Carrying capacity**.

Karyotype The characteristic chromosome complement of a cell, individual, or species.

Kelvin scale (°K) A scale of temperature in which absolute zero (the point at which molecules oscillate at their lowest possible frequency, -273°C) is designated as 0°K, and the boiling point of water as 373°K.

Kilocalories (kcal) Units of 1,000 calories. (*See* **Calories**.)

Kin selection Selection effects (e.g., altruism) that influence the survival and reproductive success of genetically related individuals (kin). This contrasts with selection confined solely to an individual and its own offspring. (*See also* **Inclusive fitness**.)

Kingdom The highest inclusive category of taxonomic classification, each kingdom including phyla or sub-kingdoms. The most common presently used classification system proposes five kingdoms: Monera (prokaryotes), Protista, Fungi, Animalia, and Plantae.

Knuckle-walking Quadrupedal gait of chimpanzees and gorillas performed by curling the fingers towards the palm of the hand and using the backs (dorsal surfaces) of the knuckles to support the weight of the front part of the body.

Krebs cycle The cyclic series of reactions in the mitochondrion in which pyruvate is degraded to carbon dioxide and hydrogen protons and electrons. The latter are then passed into the oxidative phosphorylation pathway to generate ATP.

***K*-selection** Selection based on a population being maintained at or near the limit of its carrying capacity; selection is therefore theoretically for improved competitive ability rather than for rapid numerical increase.

Lactation Formation and secretion of milk in maternal mammary glands for nursing offspring, a distinctive characteristic of mammals.

Ladder of Nature A concept based on Aristotle's view (The Scale of Nature) that nature can be represented as a succession of stages or ranks that leads from inanimate matter through plants, lower animals, higher animals, and finally to the level of humans. (*See also* **Great Chain of Being**.)

Lamarckian inheritance The concept that characters acquired or lost during the life experience of an organism, as well as characters that organisms attempt to acquire in order to meet environmental needs, can be transmitted to offspring. Lamarck proposed that it is through such means that changes in organisms (evolution) takes place. (*See also* **Inheritance of acquired characters**, **Use and disuse**.)

Language A structured system of communication between individuals using vocal, visual, or tactile symbols to describe thoughts, feelings, concepts, and observations.

Larva A sexually immature stage in various animal groups, often with a form and diet distinct from those of the adult.

Laurasia The supercontinent in the Northern Hemisphere (comprising what is now North America, Greenland, Europe, and parts of Asia) formed from the breakup of Pangaea about 180 million years ago.

Learning Acquisition of a behavior through experience.

Lethal allele An allele whose effect prevents its carrier from reaching sexual maturity when present in homozygous condition for a recessive lethal or in either heterozygous or homozygous condition for a dominant lethal.

Life The capability of performing various organismic functions such as metabolism, growth, and reproduction of genetic material.

Life cycle The series of stages that take place between the formation of zygotes in one generation of a species to the formation of zygotes in the next generation. (Also *life history*: the series of stages experienced by an individual of a species, from birth to death.)

Light year The distance traveled by light, moving at 186,000 miles a second, in a solar year; approximately 6×10^{12} miles or 9.5×10^{12} kilometers.

Lineage An evolutionary sequence, arranged in linear order from an ancestral species to a descendant species (or *vice versa*).

Lingual Side of a tooth closest to the tongue.

Linkage (gene) The occurrence of two or more gene loci on the same chromosome.

Linkage equilibrium The attainment of those genotypic frequencies in a population which indicate that recombination between two or more gene loci has reached the point at which their alleles are now found in random genotypic combinations. (*Linkage disequilibrium*: the absence of linkage equilibrium [i.e., the presence of non-random associations between alleles at different loci], caused possibly by a selective advantage for those genotypes possessing specific combinations of alleles at two or more linked loci.)

Linkage map The linear sequence of known genes on a chromosome.

Lipids Organic compounds such as fats, waxes, and steroids that tend to be more soluble in organic solvents of low polarity (e.g., ether, chloroform) than in more polar solvents (e.g., water).

Living fossil An existing species whose similarity to ancient ancestral species indicates that very few morphological changes have occurred over a long period of geological time.

Locus (plural **Loci)** The site (nucleotide sequence) on a chromosome occupied by a specific gene.

Logistic growth curve Population growth that follows a sigmoid (S-shaped) curve in which numbers increase slowly at first, then rapidly, and finally level off as the population reaches its maximum size or carrying capacity for a particular environment.

Longevity The average life span of individuals in a population.

Macroevolution Evolution of taxa higher than the species level (e.g., genera, families, orders, classes), commonly entailing major morphological changes. This concept is often associated with the school of thought proposing that evolutionary events different from those responsible for changes in populations or the origin of species have caused the origin of higher taxa. (*See* **Punctuated equilibrium**.)

Macromolecules Very large polymeric molecules such as proteins, nucleic acids, and polysaccharides.

Macromutation The concept that there are single mutations whose effects are large enough to produce an instantaneous new species or perhaps even to produce organisms that signify the beginning of a higher taxonomic category.

Malthusian parameter *See* **Intrinsic rate of natural increase**.

Mammary glands One or more pairs of ventrally placed glands used by mammalian females for nursing offspring. (*See also* **Lactation**).

Marsupials Mammals of the Infraclass Metatheria possessing, among other characters, a reproductive process in which tiny live young are born,and then nursed in a female pouch (marsupium).

Meiosis The eukaryotic cell division process used in producing haploid gametes (animals) or spores (plants) from a diploid cell. Meiosis is characterized by a reduction division which ensures that each gamete or spore contains one representative of each pair of homologous chromosomes in the parental cell.

Meiotic drive *See* **Segregation distortion**.

Mendel's laws *See* **Segregation**, **Independent assortment**.

Mesoderm The embryonic tissue layer between ectoderm and endoderm in triploblastic animals that gives rise to muscle tissue, kidneys, blood, internal cavity linings, and so on.

Mesozoic era The middle era of the Phanerozoic eon, covering the approximately 185-million-year interval between the Paleozoic (ending about 250 million years ago) and Cenozoic (beginning about 65 million years ago). It is marked by the origin of mammals in the earliest period of the era (Triassic), the dominance of dinosaurs throughout the last two periods of the era (Jurassic and Cretaceous), and the origin of angiosperms.

Messenger RNA (mRNA) An RNA molecule produced by transcription from a DNA template, bearing a sequence of triplet codons used to specify the sequence of amino acids in a polypeptide.

Metabolic pathway A sequence of enzyme-catalyzed reactions which convert a precursor substance to an end product.

Metabolism Pertaining to enzyme-catalyzed reactions that occur in living organisms.

Metacentric A chromosome whose centromere is at or near the center.

Metamerism Division of the body, or a major portion of the body, into a series of similar segments along the anterior-posterior axis.

Metamorphic rock Rock which has been subjected to high but non-melting temperatures and pressures, causing chemical and physical changes.

Metamorphosis The transition from one form into another during development (e.g., a larva into a different adult form).

Metatheria *See* **Marsupials**.

Metazoa Multicellular animals.

Microevolution Evolutionary changes of the kinds usually responsible for causing differences between populations of a species (e.g., gene frequency changes and chromosomal variations). Many evolutionists suggest that accumulations of such changes over time are sufficient to explain the origin of most or all taxa.

Microsphere Microscopic membrane-bound sphere formed when proteinoids are boiled in water and allowed to cool. Some cell-like properties, such as osmosis, growth in size, and selective absorption of chemicals have been ascribed to them.

Migration The transfer of genes from one population into another by interbreeding (gene flow). (Also used to indicate movement of a population to a different geographical area or its periodic passage from one region to another.)

Mimicry Resemblance of individuals in one species (mimics) to individuals in another (models). (*See also* **Batesian mimicry**, **Müllerian mimicry**.)

Mitochondrion An organelle in eukaryotic cells that uses an oxygen-requiring electron transport system to transfer chemical energy derived from the breakdown of food molecules to ATP. Mitochondria possess their own genetic material (DNA without histones) and generate some mitochondrial proteins by utilizing their own protein-synthesizing apparatus. (Most of the mitochondrial proteins are coded by nuclear DNA and produced on cytoplasmic ribosomes.)

Mitosis The mode of eukaryotic cell division which produces two daughter cells possessing the same chromosome complement as the parent cell.

Modern synthesis (evolution theory) *See* **Neo-Darwinism**.

Modifier (gene) A gene whose effect alters the phenotypic expression of one or more genes at loci other than its own.

Molecular clock *See* **Evolutionary clock**.

Monocotyledons Flowering plants (angiosperms) in which the embryo bears one seed leaf (cotyledon).

Monomers The subunits linked together to form a polymer (e.g., nucleotides in nucleic acids, amino acids in proteins, sugars in polysaccharides).

Monomorphic A population or species which shows no genetic or phenotypic variation for a particular gene or character.

Monophyletic Derivation of a taxonomic group from a single ancestral lineage.

Monotremes Egg-laying mammals, presently restricted to Australasia; the platypus (*Ornithorhyncus*) and echidna spiny anteater (*Tachyglossus*, *Zaglossus*).

Morphology Study of the anatomical form and structure of organisms.

Müllerian mimicry Sharing of a common warning coloration or pattern among a number of species that are all dangerous or toxic to predators.

Multigene family *See* **Gene family**.

Mutation A change in the nucleotide sequence of genetic material whether by substitution, duplication, insertion, deletion, or inversion.

Mutational load That portion of the genetic load caused by production of deleterious genes through recurrent mutation.

Mutualism A relationship between different species in which the participants benefit.

Natural selection Differential reproduction or survival of replicating organisms caused by non-human environmental agencies. Since these differential selective effects often act on hereditary (genetic) variations, natural selection is one of the causes for a change in the gene frequencies of a population that lead to a new distinctive genetic constitution (evolution).

Negative eugenics Proposals to eliminate deleterious genes from the human gene pool by identifying their carriers and restraining or discouraging their reproduction.

Neo-Darwinism The theory (also called the Modern Synthesis) that regards evolution as a change in the frequencies of genes introduced by mutation, with natural selection considered as the most important, although not the only, cause for such changes.

Neoteny The retention of juvenile morphological traits in the sexually mature adult.

Neutral mutation A mutation that does not affect the fitness of an organism in a particular environment.

Neutral theory of molecular evolution The concept that most mutations that contribute to genetic variability (genetic polymorphism on the molecular level) consist of alleles that are neutral in respect to the fitness of the organism and that their frequencies can be explained in terms of mutation and random genetic drift.

Niche *See* **Ecological Niche**.

Nocturnal A life-style characterized by nighttime activity.

Non-Darwinian evolution *See* **Neutral theory of molecular evolution**.

Nondisjunction The failure of homologous chromosomes (or sister chromatids) to separate ("disjoin") from each other during one of the two meiotic anaphase stages and go to opposite poles. Because of nondisjunction, one daughter cell will receive both homologues (or sister chromatids) and the other daughter cell will receive none, leading to an increase or decrease, respectively, in chromosome number.

Non-random mating *See* **Assortative mating**.

Nonsense mutation A mutation that produces a codon which terminates the translation of a polypeptide prematurely. Such codons were previously called "nonsense" codons but are now generally called *stop codons* or *chain-termination* codons.

Nucleic acid An organic acid polymer, such as DNA or RNA, composed of a sequence of nucleotides.

Nucleotide A molecular unit consisting of a purine or pyrimidine base, a ribose (RNA) or deoxyribose (DNA) sugar, and one or more phosphate groups.

Nucleus A membrane-enclosed eukaryotic organelle that contains all of the histone-bound DNA in the cell (i.e., practically all of the cellular genetic material).

Numerical (phenetic) taxonomy A statistical method for classifying organisms by comparing them on the basis of measurable phenotypic characters and giving each character equal weight. The degree of overall similarity between individuals or groups is then calculated, and a decision is made as to their classification.

Olfactory Referring to the sense of smell.

Omnivores Animals that feed on both plants and animals.

Ontogeny The development of an individual from zygote to maturity.

Operon A cluster of coordinately regulated structural genes.

Order A taxonomic category between class and family: a class may contain a number of orders, each of which contains a number of families.

Organelles Functional intracellular membrane-enclosed bodies, such as nuclei, mitochondria, and chloroplasts.

Organic Carbon-containing compounds. Also refers to features or products characteristic of biological organisms.

Organism A living entity. (*See* **Life**.)

Orthogenesis The concept that evolution of a group of related species proceeds in a particular direction (e.g., an increase in size) because of forces other than selection.

Orthologous genes Gene loci in different species that are sufficiently similar in their nucleotide sequences (or amino acid sequences of their protein products) to suggest they originated from a common ancestral gene.

Overdominance Instances when the phenotypic expression of a heterozygote (e.g., A^1A^2) is more extreme than that of either homozygote (e.g., A^1A^1 or A^2A^2). Overdominance has been considered a cause for hybrid vigor. (*See* **Heterosis**).

Oviparous Females that lay eggs which develop outside the body.

Oxidation-reduction Reactions in which electrons are transferred from one atom or molecule (the reducing agent that is oxidized by the loss of electrons) to another (the oxidizing agent that is reduced by the gain of electrons).

Oxidative phosphorylation A process that produces ATP by transferring electrons to oxygen.

Paedomorphosis The incorporation of adult sexual features into immature developmental stages.

Paleomagnetism The magnetic fields of ferrous (iron-containing) materials in ancient rocks. Among other applications, paleomagnetism provides information on the position of land masses and continents relative to the earth's magnetic poles at the time that the rocks were formed and thus can be used to describe the historical movement of continents relative to each other (continental drift).

Paleontology The study of extinct fossil organisms.

Paleozoic The first era of the Phanerozoic eon, extending from 590 to about 250 million years ago.

Pangaea A very large supercontinent formed about 250 million years ago comprising most or all of the present continental land masses. (*See* **Gondwanaland**, **Laurasia**.)

Pangenesis The concept of heredity, held by Darwin and others, that small, particulate "gemmules," or "pangenes," are produced by each of the various tissues of an organism and sent to the gonads where they are incorporated into gametes. The increase or decrease of specific gemmules during the use or disuse of organs was proposed in order to explain the Lamarckian concept of inheritance of acquired characters.

Panmixis (Panmictic) *See* **Random mating**.

Panspermia The concept that life was introduced on earth from elsewhere in the universe.

Parallel evolution The evolution of similar characters in related lineages whose common ancestor was phenotypically different. (*See also* **Convergence**.)

Paralogous genes Two or more different gene loci in the same organism that are sufficiently similar in their nucleotide sequences (or in the amino acid sequences of their protein products) to indicate they originated from one or more duplications of a common ancestral gene. (*See also* **Gene family**.)

Parapatric Geographically adjacent species or populations whose distributions do not overlap but are in contact at one or more of their mutual boundaries.

Paraphyletic A taxonomic grouping in which some but not all of its members are descendants of a single common ancestor.

Parasitism An association between species in which individuals of one species (the parasite) obtain their nutrients by living on or in the tissues of another species (the host), often with harmful effects to the host.

Parental investment Parental provision of resources to offspring that increases the offspring's reproductive success at a cost of further reproductive success of the parents.

Parsimony method Choice of a phylogenetic tree that minimizes the number of evolutionary changes necessary to explain species divergence.

Parthenogenesis Development of an individual from an egg which has not been fertilized by a male gamete.

Partial (incomplete) dominance Instances where two different alleles of a gene in a heterozygote (e.g., A^1A^2) produce a phenotypic effect intermediate between the effects produced by the two homozygotes (e.g., A^1A^1, A^2A^2).

Pelagic Refers to an entire body of water and the organisms within it, excluding the bottom (benthic) zone.

Peptide (polypeptide) An organic molecule composed of a sequence of amino acids covalently linked by peptide bonds (a bond formed between the amino group of one amino acid and the carboxyl group of another through the elimination of a water molecule).

pH scale (The negative logarithm of the hydrogen ion (H^+) concentration in an aqueous solution.) A scale used for measuring acidity (pH less than 7) and alkalinity (pH greater than 7), given that pure water has a neutral pH of 7.

Phanerozoic eon A major division of the geological time scale marked by the relatively abundant appearance of fossilized skeletons of multicellular organisms, dating from about 590 million years ago to the present.

Phenetic Referring to phenotypic characters that can be described or measured. Also, a system of classification which groups taxa by their degree of similarity for measured or numerically evaluated characters. (*See also* **Numerical taxonomy**).

Phenotype The characters that constitute the structural and functional properties of an organism, whether genetically or environmentally determined.

Phosphorylation The addition of one or more phosphate groups (HPO_4^-) to a compound (e.g., the phosphorylation of ADP to ATP).

Photoautotroph *See* **Autotroph**.

Photosynthesis The synthesis of organic compounds from carbon dioxide and water through a process that begins with the capture of light energy by chlorophyll.

Phyletic evolution Evolutionary changes within a single non-branching lineage. Although new species are produced by this lineage over time (chronospecies) there is no increase in the number of species existing at any one time. (*See also* **Anagenesis**.)

Phylogenetic evolution Evolutionary changes that produce two or more lineages which diverge from a single ancestral lineage. (Also called "branching evolution," *see* **Cladogenesis**.)

Phylogeny The evolutionary history of a species or group of species in terms of their derivations and relationships. A phylogenetic tree is a schematic diagram that represents that evolution.

Phylum (plural **Phyla**) The major taxonomic category below the level of kingdom, used to include classes of organisms that may be phenotypically quite different but share some general features or body plan.

Placenta A mammalian organ formed by union between the female uterine lining and embryonic membranes that provides nutrition to the embryo, allows exchange of gases, and aids elimination of embryonic waste products.

Placentals Mammals of the infraclass Eutheria, possessing, among other features, a reproductive process that utilizes a placenta to nourish their young until a relatively advanced stage of development compared to other mammalian groups (i.e., monotremes and marsupials).

Planula hypothesis The concept that metazoans evolved from small primitive organisms that consisted of solid balls of cells (planulae) similar to embryonic stages of sponges and Cnidaria.

Plasma membrane The boundary membrane consisting of phospholipids and proteins surrounding the cytoplasm of a cell.

Plate tectonics The concept that the earth's crust is divided into a number of fairly rigid plates, whose movements (tectonics) relative to each other are responsible for continental drift and many crustal features. (*See also* **Tectonic plates**.)

Pleiotropy Instances when a single gene produces phenotypic effects on more than one character.

Plesiomorphy Instances when a species character is similar to that character in an ancestral species.

Polygene A gene which interacts with other polygenes to produce a quantitative phenotypic effect on a character.

Polymer A molecule composed of many repeating subunits (monomers) linked together by covalent bonds.

Polymerase An enzyme that catalyzes the synthesis of a polymer by linking together its component monomers.

Polymorphism The presence of two or more genetic or phenotypic variants in a population. Usually refers to genetic variations where the frequency of the rarest type is not maintained by mutation alone. (*See also* **Balanced polymorphism**.)

Polyphyletic The presumed derivation of a single taxonomic group from two or more different ancestral lineages through convergent or parallel evolution.

Polyploidy Variations in which the number of chromosome sets (n) is greater than the diploid number (2n) (e.g., triploidy (3n), tetraploidy (4n), etc.).

Population A geographically localized group of individuals in a species that, in sexual forms, share a common gene pool. (*See also* **Deme**.)

Preadaptation A character that was adaptive under a prior set of conditions and later provides the initial stage for evolution of a new adaptation under a different set of conditions.

Precambrian eon A major division of the geological time scale that includes all eras from the origin of the earth about 4.5 billion years ago to the beginning of the Phanerozoic eon, about 590 million years ago. The Precambrian (also known as the Cryptozoic) is marked biologically by the appearance of prokaryotes about 3.5 billion years ago and small, non-skeletonized multicellular organisms in the Ediacarian period about 100 million years before the Phanerozoic.

Preformationism The concept that an organism is preformed at conception in the form of a miniature adult and development consists of enlargement of the already preformed structures.

Progenote A hypothetical ancestral cellular form that

gave rise to prokaryotes (archaebacteria, eubacteria) and eukaryotes.

Prokaryotes Organisms such as bacteria and cyanobacteria that lack histone-bound DNA, endoplasmic reticulum, a membrane-enclosed nucleus, and other cellular organelles found in eukaryotes.

Protein A macromolecule composed of one or more polypeptide chains of amino acids, coiled and folded into specific shapes based on its amino acid sequences.

Proteinoids Synthetic polymers produced by heating a mixture of amino acids. Some show proteinlike properties in respect to enzyme activity, color-test reactions, hormonal activity, and so on.

Protista One of the four eukaryotic kingdoms; includes protozoa, algae, slime molds, and some other groups. (Called Protoctista by some authors.)

Protoplasm Cellular material within the plasma membrane.

Protostomes Coelomate phyla in which the embryonic blastopore becomes the mouth.

Prototheria *See* **Monotremes**.

Pseudocoelomates Organisms that possess a coelom derived from a persistent embryonic blastocoel, largely unlined with mesodermal tissue.

Punctuated equilibrium The view that evolution of a lineage follows a pattern of long intervals in which there is relatively little change (stasis, or equilibrium), punctuated by short bursts of speciation and macroevolutionary events during which new taxa arise.

Purine A nitrogenous base composed of two joined ring structures, one five membered and one six membered, commonly present in nucleotides as adenine (A) or guanine (G).

Pyrimidine A nitrogenous base composed of a single six-membered ring, commonly present in nucleotides as thymine (T), cytosine (C), or uracil (U).

Quadrupedal Tetrapod locomotion using all four limbs.

Quantitative character A character whose phenotype can be numerically measured or evaluated; a character displaying continuous variation.

Quantum evolution A rapid increase in the rate of evolution over a relatively short period of time.

Race A population or group of populations in a species that share a geographically and/or ecologically identifiable origin and possess unique gene frequencies and phenotypic characters which distinguish them from other races. Because of the large amount of genetic and phenotypic variability in most species, the number of racial distinctions that can be made is often arbitrary. (*See also* **Subspecies**.)

Racemic mixture A mixture of two kinds of molecules whose structures are similar but differ in that they are mirror images of each other (one kind cannot be superimposed on the other). Each of the two molecular forms rotates the plane of polarized light in a particular direction, but the racemic mixture is optically inactive.

Radioactive dating The dating of rocks by measuring the proportions of a radioactive element in an igneous intrusion and the isotopes produced by its radioactive decay. Since the rate at which a particular radioactive element decays is constant, these proportions provide an estimate of the age of the rock, which can often be confirmed by dating with other radioactive elements.

Radioactivity Emission of radiation by certain elements as their atomic nuclei undergo changes.

Random genetic drift The random change in frequency of alleles in a population. These can be caused by sampling errors which are of greater magnitude in small populations or by bottlenecks (founder effects) in which population size is suddenly reduced to a few individuals.

Random mating Mating within a population regardless of the phenotype or genotype of the sexual partner (panmixis).

Range The geographical limits of the region habitually traversed by an individual or occupied by a population or species.

Recessive allele An allele (e.g., *a*), which has no obvious phenotypic effect in a heterozygote (e.g., *Aa*), producing its phenotypic effect only when homozygous (e.g., *aa*).

Recessive lethal An allele whose presence in homozygous condition causes lethality. (*See also* **Lethal allele**.)

Reciprocal altruism A mutually beneficial exchange of altruistic behavioral acts between individuals.

Recombination A chromosomal exchange process (*see* **Crossovers**) which produces offspring that possess gene combinations different from those of their parents. (Also used by some authors to describe the results of independent assortment.)

Red Queen hypothesis The view that adaptive evolution in one species of a community causes a deterioration of the environment of other species. As a consequence, each species must evolve as fast as it can "in order to stay in the same place" (to survive).

Regulator gene A gene which controls the rate at which other genes, adjacent or distant, will synthesize their products.

Repetitive DNA DNA nucleotide sequences that are repeated many times in the genome.

Repressor protein A regulator gene product that binds to a particular nucleotide sequence and prevents transcription.

Reproductive isolation The absence of gene exchange between populations. (*See* **Isolating mechanisms**.)

Reproductive success The proportion of reproductively fertile offspring produced by a genotype relative to other genotypes. (*See also* **Fitness**.)

Restriction enzymes Enzymes that recognize particular nucleotide sequences and cut DNA molecules at or near those sequences.

Ribosomal RNA (rRNA) RNA sequences that are incorporated into the structure of ribosomes.

Ribosomes Intracellular particles composed of ribosomal RNA and proteins that furnish the site at which messenger RNA molecules are translated into polypeptides.

RNA (ribonucleic acid) A nucleic acid, characterized by the presence of a ribose sugar in each nucleotide, whose sequences serve as messenger RNA, ribosomal RNA, or transfer RNA in cells or as genetic material in some viruses.

***r*-selection** Selection in populations subject to rapidly changing environments with highly fluctuating food resources. Theoretically, selection in such populations emphasizes adaptations for rapid population growth rather than for the competitive ability experienced in *K*-selected populations.

Saltation The concept that new species or higher taxa originate abruptly because of macromutations, or because of sudden unknown causes.

Sampling error (gene frequencies) Variability in gene frequencies caused by the fact that not all samples taken from a population have exactly the same gene frequency as the population itself.

Saprophyte An organism that feeds on decomposing organic material.

Scavenger An organism that habitually feeds on animals who died naturally or accidentally or were killed by another carnivore.

Seafloor spreading Expansion of oceanic crust through the deposition of mantle material along oceanic ridges. (*See also* **Tectonic plates**.)

Sedimentary rock Rock formed by the hardening of accumulated particles (sediments) that had been transported by agents such as wind and water. The prime source of fossils. (*See also* **Geological strata**.)

Segregation The mendelian principle that the two different alleles of a gene pair in a heterozygote segregate from each other during meiosis to produce two kinds of gametes in equal ratios, each bearing a different allele.

Segregational genetic load *See* **Balanced genetic load**.

Segregation distortion (meiotic drive) Aberrant segregation ratios among the gametes produced by heterozygotes because of the presence of certain alleles (segregation distorters).

Selection A composite of all the forces that cause differential survival and differential reproduction among genetic variants. When the selective agencies are those of human choice, the process is called *artificial selection*; when the selective agencies are not those of human choice, it is called *natural selection*. (*See also* **Adaptive value, Fitness**.)

Selection coefficient (symbol *s*) A relative measure of the effect of selection, usually in terms of the loss of fitness endured by a genotype, given that the genotype with greatest fitness has a value of 1.

Self-assembly The spontaneous aggregation of macromolecules into biological configurations that can have functional value.

Selfish DNA The concept that the persistence of DNA sequences with no discernible cellular function (e.g., various repetitive DNA sequences) arises from the likelihood that, once present in the genome, they are impossible to remove without the death of the organism—that is, they act as "selfish," or "junk" DNA, which the cell has no choice but to replicate along with functional DNA.

Semipermeable membrane A membrane that selectively permits transmission of certain molecules but not others.

Senescence The process of aging.

Serial homology Similarities between parts of the *same* organism, such as the vertebrae of a vertebrate or the different kinds of hemoglobin molecules produced by a mammal. The genetic basis for such homology can often be ascribed to gene duplications which have diverged over time but still produce somewhat similar effects. (*See* **Gene family**, **Paralogous genes**).

Sex chromosomes Chromosomes associated with determining the difference in sex. These chromosomes are alike in the homogametic sex (e.g., XX) but differ in the heterogametic sex (e.g., XY).

Sex linkage Genes linked on a sex chromosome. The results of such linkage may be differences between the sexes in the appearances of certain traits. For example, a recessive allele on the X chromosome will not be expressed in the XX homogametic sex if that individual possesses a dominant allele on its second X chromosome but can be expressed in the XY heterogametic sex which has only one X chromosome.

Sex ratio The relative proportions of males and females in a population.

Sexual reproduction Zygotes produced by the union of genetic material from different sexes through gametic fertilization.

Sexual selection Selection that acts directly on mating success through direct competition between members of one sex for mates (intrasexual selection), or through choices made between them by the opposite sex (intersexual selection), or through a combination of both selective modes. In any of these cases, sexual selection may cause exaggerated phenotypes to appear in the sex upon which it is acting (large antlers, striking colors, etc.).

Sibling species Species so similar to each other morphologically that they are difficult to distinguish but which are nevertheless reproductively isolated. (Sometimes called "cryptic species.")

Social Darwinism The concept that social and cultural differences in human societies (political, economic, military, religious, etc.) arise through processes of natural selection, similar to those that account for biological differences among populations and species.

Sociobiology The study of the biological basis of social behavior.

Speciation The splitting of one species into two or more new species (*see* **Cladogenesis**, **Phylogenetic evolution**) or the transformation of one species into a new species over time. (*See* **Anagenesis, Phyletic evolution**.)

Species A basic taxonomic category for which there are various definitions (Chapter 11). Among these are an interbreeding or potentially interbreeding group of populations reproductively isolated from other groups (the biological species concept) and a lineage evolving separately from others with its own unitary evolutionary role and tendencies (Simpson's evolutionary species concept).

Split gene A gene whose nucleotide sequence is divided into exons and introns.

Spontaneous generation An early concept that complex organisms can appear spontaneously from inert materials without biological parentage.

Sporophyte The diploid spore-producing stage of plants that have alternating generations (haploid gametophyte and diploid sporophyte). The sporophyte arises from the union of gametophyte gametes and produces its haploid spores (which become gametophytes) by meiosis.

Stabilizing selection Selection that favors the survival of organisms in a population that are at an intermediate phenotypic value for a particular character and eliminates the extreme phenotypes. (Also called centripetal, or normalizing, selection.)

Stasis A period of equilibrium during which change appears to be absent, e.g., in the concept of punctuated equilibrium.

Stop codon One of the three messenger RNA codons (UAA, UAG, UGA) that terminates the translation of a polypeptide. (Also called chain-termination codon or nonsense codon.)

Strata *See* **Geological strata**.

Stromatolites Laminated rocks produced by layered accretions of benthic microorganisms (mainly filamentous cyanobacteria) that trap or precipitate sediments.

Structural gene A DNA nucleotide sequence that codes for RNA or protein.

Subspecies A taxonomic subdivision of a species often distinguished by special phenotypic characters and by its origin or localization in a given geographical region. Like other species subdivisions (*see* **Race**), a subspecies can still interbreed successfully with the remainder of the species. However, in some cases, interbreeding capabilities are unknown, and subspecies designations (e.g., *Homo sapiens neanderthalensis* and *Homo sapiens sapiens*) are based entirely on phenotype.

Survivorship The proportion of individuals born at a given time (cohort) who survive to a given age.

Symbiont A participant in the interactive association (symbiosis) between two individuals or two species. This term is often restricted to mutually beneficial associations (mutualism).

Sympatric Species or populations whose geographical distributions coincide or overlap.

Sympatric speciation Speciation that occurs between populations occupying the same geographic range.

Synapomorphy The possession by two or more related lineages of the same phenotypic character derived from a different but homologous character in the ancestral lineage.

Systematics Although defined by Simpson as the study of the diversity of organisms and all their comparative and evolutionary relationships, it is often used interchangeably with the terms *classification* and *taxonomy*.

Tachytelic A relatively rapid evolutionary rate.

Taxon (plural **taxa**) A named taxonomic unit consisting of a distinctive group of organisms placed in a taxonomic category, whether the unit is that of a species, genus, family, order, and so on.

Taxonomy The principles and procedures used in classifying organisms.

Tectonic plates The fairly rigid plates composing the earth's crust whose boundaries are marked by earthquake belts and volcanic chains. In oceanic regions, accretions to these plates occur at midoceanic ridges (seafloor spreading), and they are subducted under other plates at the deep oceanic trenches. Continental masses ride on some of these plates, accounting for continental drift and such processes as the mountain building that occurs when these plates collide.

Teleology The concept that natural processes such as development or evolution are being intentionally guided by their final stage (*telos*) or for some particular purpose.

Telocentric Chromosomes whose centromeres are located at one end.

Terminal electron acceptor The molecule that is the final acceptor of electrons in a metabolic pathway (e.g., in aerobic respiration, oxygen is the terminal electron acceptor).

Terrestrial On the ground; also on or of the planet earth.

Tetrapod Literal meaning: four-footed. Commonly used to specify a member of the land-evolved vertebrate classes: amphibia, reptiles, birds, and mammals.

Therapsids An order of synapsid mammal-like reptiles, composed mainly of fairly large herbivorous and carnivorous forms, which were dominant reptilian stocks during the Permian and Triassic periods. During the Triassic, some or many of these stocks were probably already endothermal, and one or more groups (cynodonts—dog teeth) made the transition to smaller mammalian forms such as the morganucodontids.

Theria The viviparous mammalian subclass, consisting of marsupials and placentals.

Tissue A group of cells, all performing a similar function in a multicellular organism.

Transcription The process by which the synthesis of an RNA molecule (e.g., messenger RNA) is initiated and completed on a DNA template by RNA polymerase enzyme.

Transfer RNA (tRNA) Relatively small RNA molecules (about 80 nucleotides long) that bring specific amino acids to the ribosome for polypeptide synthesis. Each kind of tRNA has a unique anticodon complementary to messenger RNA codons which specify the placement of particular amino acids in the polypeptide chain.

Translation The protein-synthesizing process that takes place on the ribosome, linking together a particular sequence of amino acids (polypeptide) on the basis of information received from a particular sequence of codons on messenger RNA.

Translocation An aberration in which a sequence of nucleotides is moved to a different position in the genome.

Transposons (transposable elements) Nucleotide sequences that produce enzymes to promote their own movement from one chromosomal site to another and may carry additional genes such as those for antibiotic resistance.

Triploblastic An animal that produces all three major types of cell layers during development—ectoderm, endoderm, and mesoderm.

Tritium An isotope of hydrogen that has three times the mass of an ordinary hydrogen atom.

Typology The study of organic diversity based on the principle that all members of a taxonomic group conform to a basic plan and variation among them is of little or no significance. (*See also* **Archetype**.)

Ultraviolet radiation Electromagnetic radiation at wavelengths between about 4 and 400 nanometers, shorter than visible light but longer than x-rays. It is absorbed by purine and pyrimidine ring structures and is therefore quite damaging to nucleic acid genetic material.

Unequal crossing over The result of improper pairing between chromatids, causing their crossover products to differ from each other in the amounts of genetic material.

Ungulate A hoofed mammal.

Uniformitarianism A concept, popularized by Lyell in geology, that none of the forces active in past Earth history were different from those active today.

Universal genetic code The utilization of the same genetic code in all living organisms. (A few codons differ from the universal code in mitochondria, mycoplasmas, and some ciliated protozoa.)

Use and disuse A concept used by Lamarck to explain evolution as resulting from the transmission of characters that became enhanced or diminished because of their use or disuse, respectively, during the life experience of individuals. (*See also* **Lamarckian inheritance**.)

Vascular plants Plants that have special water- and food-conducting vessels and tissues (xylem and phloem).

Ventral The belly or downward surface of an animal. (In vertebrates, the surface opposite the spinal column.)

Vestigial organs Organs or structures which appear to be small and functionless but can be shown to be homologous with ancestral organs and structures that were larger and functional.

Virus A small intracellular parasite, often composed of little more than nucleic acid and a few proteins, that depends upon the host cell to replicate its genetic material and to synthesize its proteins.

Vitalism The concept that the activities of living organisms cannot be explained by any underlying physical or chemical principles but arise from mystical or supernatural causes.

Viviparous Mode of reproduction in which eggs develop into live young before exiting the maternal body.

Wild type The most commonly observed phenotype or genotype for a particular character. Variations from wild type are considered as mutants.

X chromosome The name given to a sex chromosome present twice in the homogametic sex (XX) and only once in the heterogametic sex (XY or XO).

X-linked genes Genes present on the X chromosome. (*See* **Sex linkage**.)

Y chromosome A sex chromosome present only in the heterogametic sex (XY).

Zygote The cell formed by the union of male and female gametes.

INDEXES

AUTHOR INDEX

SUBJECT INDEX

A page number in **boldface** indicates mention of the subject in a figure or figure legend.